STEM CELLS HANDBOOK

EDITED BY

STEWART SELL, MD

WADSWORTH CENTER AND ORDWAY RESEARCH INSTITUTE
EMPIRE STATE PLAZA, ALBANY, NY

999 Riverview Drive, Suite 208
Totowa, New Jersey 07512
humanapress.com

For additional copies, pricing for bulk purchases, and/or information about other Humana titles, contact Humana at the above address or at any of the following numbers: Tel.: 973-256-1699; Fax: 973-256-8341; E-mail:humana@humanapr.com; Website: humanapress.com

Cover illustrations: Fig. 3A from Chapter 37.

Cover design by Patricia F. Cleary.

This publication is printed on acid-free paper. ∞
ANSI Z39.48-1984 (American National Standards Institute) Permanence of Paper for Printed Library Materials.

E-ISBN 1-59259-411-5

Printed in the United States of America. 10 9 8 7 6 5 4 3 2 1

Library of Congress Cataloging-in-Publication Data

Stem cells handbook / edited by Stewart Sell.
p. cm.
Includes bibliographical references and index.
ISBN 1-58829-113-8 (alk. paper)
1. Stem cells. I. Sell, Stewart, 1935-
QH587.S728 2003
616'.02775–dc21 2002191331

STEM CELLS HANDBOOK

Dedication

Professor Janis Klavins at the 7th Annual Meeting of IATMO in Kiev, Ukraine, 1990

Janis V. Klavins, MD, Professor Emeritus of Albert Einstein Medical College and former Chairman of Pathology at Catholic Medical Center, Queens, New York. He has won many awards, including the Schubert medal and the Three Star Order from the President of Latvia.

Professor Klavins was born in Riga, Latvia in 1921. His medical education was interrupted by both Nazi and Russia invasions during WWII and he had to flee Latvia to avoid deportation to Russia. He completed both medical and musical degrees in Germany. After WWII he and his young wife Ilga moved to Parkersburg, WV, where he worked as a physician's assistant. He completed pathology training and joined the pathology department of Tom Kinney at Western Reserve in Cleveland, OH. He moved with Dr. Kinney to Duke University and then to New York, first at Brooklyn-Cumberland Medical Center and then as Chairman of Pathology at Catholic Medical Center. He is known internationally for his vocal interpretations of Schubert's lieder.

Professor Klavins not only has been a long time champion of the stem cell concept, but also, along with Georg D. Birkmayer of Vienna, Austria, is the founder of the International Academy of Tumor Marker Oncology, an organization that encourages the application of the products of stem cells and their cancerous progeny to the clinical diagnosis and prognosis of human cancer.

Preface

The power of stem cells for tissue development, regeneration, and renewal has been well known by embryologists and developmental biologists for many years. Those presently active in research in the stem cell field owe much to previous work by embryologists and cancer researchers for their insights into what stem cells can do. In the last 4–5 years, the rapid expansion of the concept of adult tissue stem cells as pluripotent progenitors for various tissues has led to an even greater appreciation of the power of stem cells. The demonstration that both embryonic and adult tissue stem cells have the ability to produce progenitor cells for tissue renewal has opened vast possibilities for treatment of congenital deficiency diseases as well as for regeneration of damaged tissues. Older concepts of determination leading to loss of potential during differentiation of adult tissues are being replaced by newer ideas that cells with multiple potential exist in different forms in various adult organs and that cells thought to be restricted to differentiation to one cell type may be able to "transdifferentiate" into other tissue cell types. Thus, the concept of "embryonic rests" in adult tissues, hypothesized to be the cellular origin of cancer by Durante and Conheim in the 1870s, now can be expanded to include survival of pluripotential embryonic-like stem cells in adult tissues.

The goal of *Stem Cells Handbook* is to present in one resource both the background and the current understanding of what stem cells are and what they can do. The authors of the various chapters were selected for their significant contributions to and expertise in various aspects of stem cell biology. First, the function of embryonic stem cells in early development and organogenesis, and germinal stem cells in reproduction are presented, followed by how embryonic stem cells may be cloned and how they are programmed. The role of stem cells in amphibian regeneration and mammalian wound healing shows the potential of these cells for tissue renewal. The participation of stem cells in normal tissue renewal of various organ systems, including blood, nervous tissue, retina, blood vessels, heart, kidney, skin, glandular organs, gastrointestinal tract, liver, pancreas, mammary gland, prostate, and lung are then specifically adumbrated, including not only the role of stem cells in tissue renewal and carcinogenesis, but also the isolation and characterization of various stem cell types, the potential for their manipulation, and the possibilities for future therapeutic uses in experimental models and in human diseases. The remarkable properties of hematopoietic stem cells and the clinical results achieved by transplantation of bone marrow stem cells are documented in several chapters. The potential future promise for clinical applications for regeneration of the cardiovascular and nervous system as described in preclinical models is also emphasized. Of particular interest to the editor is the potential for stem cell therapy for liver, not only because the liver has special problems and importance as the major metabolic organ of the body, but also because of its potential as an objective for transplantation and gene therapy.

Finally, a codicil for a book such as this that tries to cover an active field of research is that by the time it is published there will almost certainly be advances in understanding that have already made some of the material out of date. For example, in the last few months, there have been a number of additional papers on the plasticity of adult tissue stem cells as well as the observation that some effects believed to result from stem cell plasticity may be explained by cell fusion. Only ongoing studies will resolve these questions and provide the approaches required for potential breakthroughs in application to human diseases. In the meantime, we hope that the expert chapters in *Stem Cells Handbook* will provide useful and authoritative information to aid those who seek the answers to the unanswered questions.

The editor is indebted to G. Barry Pierce for his encouragement and insights into teratocarcinoma as a stem cell tumor, to Gerri Abelev for discovering alphafetoprotein, to the late Hidematsu Hirai for his enthusiastic support of international research in oncodevelopmental biology, to Fred Becker and Emmanuel Farber for their models and concepts of chemical hepatocarcinogenesis, to Benito Lombardi and Hishasi Shinozuka for their early work on models of oval cell proliferation, to Hyam Leffert for his encyclopedic knowledge of liver cell culture, to the many postdoctoral fellows, graduate students, and technicians who did all of the real work in my laboratory, to Thomas Lanigan and Humana Press for their encouragement and patience, and especially to the distinguished authors who contributed chapters to *Stem Cells Handbook*.

***Stewart Sell,* MD**

Contents

Contributors

W. French Anderson, MD, *Gene Therapy Laboratories, USC Keck School of Medicine, Norris Cancer Center, Los Angeles, CA*

Takayuki Asahara, MD, PhD, *Cardiovascular Research and Medicine, St. Elizabeth's Medical Center, Tufts University School of Medicine, Medford, MA*

Kurtis I. Auguste, MD, *Department of Neurological Surgery, University of California, San Francisco, San Francisco, CA*

Stephen M. Baird, MD, *Professor of Pathology, University of California San Diego, and Chief Laboratory Services, VA Medical Center, San Diego, CA*

Emma M. A. Ball, PhD, *Centre for Urological Research, Monash Institute of Reproduction and Development, Monash University, Melbourne, Victoria, Australia*

Helen M. Blau, PhD, *Donald E. and Delia B. Baxter Professor, Director, Baxter Laboratory in Genetic Pharmacology, Stanford University School of Medicine, Stanford, CA*

Helmut Bonkhoff, PhD, *Institute of Pathology, University of Saarland, Saar, Germany*

Mario A. Bourdon, PhD, *La Jolla Institute of Molecular Medicine, San Diego, CA*

Luc Bouwens, PhD, *Department of Experimental Pathology, Free University of Brussels (VUB) Brussels, Belgium*

Mairi Brittan, BS, *Histopathology Unit, Cancer Research UK, Lincoln's Inn Fields, London, UK*

Arnold I. Caplan, PhD, *Department of Biology, Skeletal Research Center, Case Western Reserve University, Cleveland, OH*

Pierre Charbord, MD, *DR Inserm, Laboratoire d'Hématopoïèse, Faculté de Médecine, Batiment Bretonneau, Tours, France*

Jae-Jin Cho, DVM, PhD, *Assistant Professor, Division of Craniomaxillofacial Reconstruction, School of Dentistry, Seoul National University, Seoul, South Korea*

Kyunghee Choi, PhD, *Department of Pathology and Immunology, Washington University School of Medicine, St. Louis, MO*

Heather A. Crosby, PhD, *School of Biosciences, University of Birmingham, and Liver Research Laboratories, University Hospital Birmingham, Birmingham, UK*

Marcel M. Daadi, PhD, *Layton Bioscience, Sunnyvale, CA*

James E. Dennis, PhD, *Department of Biology, Skeletal Research Center, Case Western Reserve University, Cleveland, OH*

Regis Doyonnas, PhD, *Baxter Laboratory in Genetic Pharmacology, Stanford University School of Medicine, Stanford, CA*

Curt R. Freed, MD, *Division of Clinical Pharmacology and Toxicology, University of Colorado School of Medicine, Denver, CO*

Juri Gelovani (aka Tjuvajev), MD, PhD, *Associate Professor, Departments of Neurology and Radiology, Memorial Sloan Kettering Cancer Center and Sloan Kettering Institute, New York, NY*

Margaret A. Goodell, PhD, *Center for Cell and Gene Therapy and Department of Pediatrics, Baylor College of Medicine, Houston, TX*

Lynn M. Gruman, HT, *Department of Anatomy and Cell Biology, The University of Iowa College of Medicine, Iowa City, IA*

Peter Gruss, PhD, *Max-Planck Institute of Biophysical Chemistry, Department of Molecular Cell Biology, Göttingen, Germany*

Sanjeev Gupta, MBBS, MD, *Professor of Medicine and Pathology, Marion Bessin Liver Research Center, Department of Medicine, and Department of Pathology, Albert Einstein College of Medicine, Bronx, NY*

Anna-Katerina Hadjantonakis, PhD, *Department of Genetics and Development, College of Physicians and Surgeons of Columbia University, New York, NY*

Simon W. Hayward, PhD, *Department of Urologic Surgery; Vanderbilt Prostate Cancer Center; Department of Cancer Biology; and Vanderbilt-Ingram Comprehensive Cancer Center, Vanderbilt University Medical Center, Nashville, TN*

Mary J. C. Hendrix, PhD, *Department of Anatomy and Cell Biology, The University of Iowa College of Medicine, Iowa City, IA*

Jürgen Hescheler, PhD, *Department of Neurophysiology, Center of Physiology and Pathophysiology, University of Cologne, Cologne, Germany*

Angela R. Hess, PhD, *Department of Anatomy and Cell Biology, The University of Iowa College of Medicine, Iowa City, IA*

DOUGLAS C. HIXSON, PhD, *Carcinogenesis Laboratory, Department of Medicine, Rhode Island Hospital, Providence, RI*
SUI HUANG, MD, PhD, *Departments of Surgery and Pathology, Children's Hospital and Harvard Medical School, Boston, MA*
DONALD E. INGBER, MD, PhD, *Departments of Surgery and Pathology, Children's Hospital and Harvard Medical School, Boston, MA*
JEFFREY M. ISNER, MD, PhD (DECEASED), *Cardiovascuolar Research and Medicine, St. Elizabeth's Medical Center, Tufts University School of Medicine, Medford, MA*
SILVIU ITESCU, MD, *Director, Transplantation Immunology, Departments of Medicine and Surgery, Columbia University, New York, NY*
KATHYJO A. JACKSON, PhD, *Center for Cell and Gene Therapy and Department of Pediatrics, Baylor College of Medicine, Houston, TX*
SHERIF M. KARAM, MD, PhD, *Department of Anatomy, Faculty of Medicine and Health Sciences, Al-Ain, UAE University, United Arab Emirates*
KATHERINE S. KOCH, PhD, *Department of Pharmacology, School of Medicine, University of California, San Diego, La Jolla, CA*
MAHESH LACHYANKAR, PhD, *Department of Neurology, Harvard Medical School Harvard Institutes of Medicine, Beth-Israel Deaconess Medical Center, Boston, MA*
ROBERT LANGER, ScD, *Department of Materials Science and Engineering, and Department of Chemical Engineering, Massachusetts Institute of Technology, Boston, MA*
ERIN B. LAVIK, PhD, *Department of Materials Science and Engineering, and Department of Chemical Engineering, Massachusetts Institute of Technology, Boston, MA*
HYAM L. LEFFERT, MD, *Department of Pharmacology and Center for Molecular Genetics, School of Medicine, University of California, San Diego, La Jolla, CA*
HAIFAN LIN, PhD, *Department of Cell Biology, Duke University Medical Center, Durham, NC*
WILLIAM J. LINDBLAD, PhD, *Department of Pharmaceutical Sciences, Eugene Applebaum College of Pharmacy and Health Sciences, Wayne State University, Detroit, MI*
JEFFREY R. MANN, PhD, *Division of Biology, Beckman Research Institute of the City of Hope, Duarte, CA*
FIONA C. MANSERGH, PhD, *Cardiff School of Biosciences, Cardiff University, Cardiff, Wales, UK*
ALEKSANDRA E. MARCINIAK, BS, *Department of Neurology, Harvard Medical School Harvard Institutes of Medicine, Beth-Israel Deaconess Medical Center, Boston, MA*
MICHAEL A. MARCONI, BS, *Department of Neurology, Harvard Medical School Harvard Institutes of Medicine, Beth-Israel Deaconess Medical Center, Boston, MA*
TILL MARQUARDT, PhD, *The Salk Institute for Biological Studies, La Jolla, CA*
ERNEST A. MCCULLOCH, MD, FRS, *Cell and Molecular Biology, The Ontario Cancer Institute, Toronto, Ontario, Canada*
PAUL S. MELTZER, MD, PhD, *Human Genome Research Institute, National Institutes of Health, Bethesda, MD*
LUCIO MIELE, MD, PhD, *Department of Pathology, Loyola University Medical Center, Chicago, IL*
BRIAN J. NICKOLOFF, MD, PhD, *Department of Pathology, Loyola University Medical Center, Chicago, IL*
SARBJIT S. NIJJAR, PhD, *School of Biosciences, University of Birmingham, and Liver Research Laboratories, University Hospital Birmingham, Birmingham, UK*
JITKA OUREDNIK, PhD, *Department of Biological Sciences, College of Veterinary Medicine, Iowa State University, Ames, IA*
VACLAV OUREDNIK, PhD, *Department of Biological Sciences, College of Veterinary Medicine, Iowa State University, Ames, IA*
VIRGINIA E. PAPAIOANNOU, PhD, *Department of Genetics and Development, College of Physicians and Surgeons of Columbia University, New York, NY*
KOOK I. PARK, MD, DMSc, *Department of Neurology, Harvard Medical School Harvard Institutes of Medicine, Beth-Israel Deaconess Medical Center, Boston, MA, and Department of Pediatrics, Yonsei University College of Medicine, Seoul, Korea*
LESLEY ANN PATERSON, PhD, *Roslin Institute, Roslin, Midlothian, UK*
AMMON B. PECK, PhD, *Department of Pathology, Immunology & Laboratory Medicine, University opf Florida College of Medicine, Gainesville, FL, and Ixion Biotechnology, Inc., Alachua, FL*
ALAN O. PERANTONI, PhD, *Laboratory of Comparative Carcinogenesis, National Cancer Institute, Frederick, MD*
IAN PONTING, PhD, *Gene Therapy Laboratories, USC Keck School of Medicine, Norris Cancer Center, Los Angeles, CA*
VIJAYAKUMAR K. RAMIYA, PhD, *Ixion Biotechnology, Inc., Alachua, FL*
DERRICK E. RANCOURT, PhD, *Departments of Oncology, Biochemistry, and Molecular Biology, The University of Calgary, Calgary, Alberta, Canada*
SCOTT H. RANDELL, PhD, *Cystic Fibrosis/Pulmonary Research and Treatment Center, The University of North Carolina at Chapel Hill, Chapel Hill, NC*
JOY RATHJEN, PhD, *Department of Molecular Biosciences, The University of Adelaide, Adelaide, Australia*
PETER DAVID RATHJEN, PhD, *Department of Molecular Biosciences, and ARC SRC for Molecular Genetics of Development, The University of Adelaide, Adelaide, Australia*
D. EUGENE REDMOND, MD, *Department of Neuropsychopharmacology, Yale University, New Haven, CT, and Neural Transplantation and Regeneration Program, Axion Research Foundation, Hamden, CT*
GAIL P. RISBRIDGER, PhD, *Centre for Urological Research, Monash Institute of Reproduction and Development, Monash University, Melbourne, Victoria, Australia*
LORRAINE ROBB, MBBS, PhD, *The Walter and Eliza Hall Institute of Medical Research, Victoria, Australia*
HEATHER L. ROSE, MA, *Department of Neurology, Harvard Medical School Harvard Institutes of Medicine, Beth-Israel Deaconess Medical Center, Boston, MA*

AGAPIOS SACHINIDIS, PhD, *Department of Neurophysiology, Center of Physiology and Pathophysiology, University of Cologne, Cologne, Germany*

HEINRICH SAUER, PhD, *Department of Neurophysiology, Center of Physiology and Pathophysiology, University of Cologne, Cologne, Germany*

GINA C. SCHATTEMAN, PhD, *Department of Anatomy and Cell Biology, The University of Iowa College of Medicine, Iowa City, IA*

ELISABETH A. SEFTOR, BS, *Department of Anatomy and Cell Biology, The University of Iowa College of Medicine, Iowa City, IA*

RICHARD E. B. SEFTOR, PhD, *Department of Anatomy and Cell Biology, The University of Iowa College of Medicine, Iowa City, IA*

STEWART SELL, MD, *Wadsworth Center and Ordway Research Institute, Empire State Plaza, Albany, NY*

DON D. SHERIFF, PhD, *Department of Anatomy and Cell Biology, The University of Iowa College of Medicine, Iowa City, IA*

RICHARD L. SIDMAN, MD, *Department of Neurology, Harvard Medical School Harvard Institutes of Medicine, Beth-Israel Deaconess Medical Center, Boston, MA*

GILBERT H. SMITH, PhD, *Mammary Biology and Tumirogenesis Laboratory, Center for Cancer Research, National Cancer Institute, National Institutes of Health, Bethesda, MD*

EVAN Y. SNYDER, MD, PhD, *Department of Neurology, Harvard Medical School Harvard Institutes of Medicine, Beth-Israel Deaconess Medical Center, Boston, MA; and Stem Cells and Regeneration, The Burnham Institute, La Jolla, CA*

DAVID L. STOCUM, PhD, *Department of Biology and Center for Regenerative Biology and Medicine, Indiana University–Purdue University Indianapolis, Indianapolis, IN*

ALASTAIR J. STRAIN, PhD, *School of Biosciences, University of Birmingham, and Liver Research Laboratories, University Hospital Birmingham, Birmingham, UK*

HÉLÈNE STRICK-MARCHAND, PhD, *Unité de Génétique de la Différenciation, URA 2578 du CNRS, Institut Pasteur, Paris, France*

BARRY R. STRIPP, PhD, *Department of Environmental and Occupational Health, University of Pittsburgh, Pittsburgh, PA*

PIROSKA E. SZABÓ, PhD, *Division of Biology, Beckman Research Institute of the City of Hope, Duarte, CA*

ROSANNE M. TAYLOR, DVM, PhD, *Department of Animal Science, Faculty of Veterinary Science, University of Sydney, New South Wales, Australia*

YANG D. TENG, MD, PhD, *Departments of Neurology, Pediatrics, and Neurosurgery, the Children's Hospital, and Brigham & Women's Hospital, Harvard Medical School, Boston, MA*

ANK A. W. TEN HAVE-OPBROEK, MD, PhD, *Department of Pulmonology, Leiden University Medical Center, Leiden, The Netherlands*

MARIA WARTENBERG, PhD, *Department of Neurophysiology, Center of Physiology and Pathophysiology, University of Cologne, Cologne, Germany*

WENDY C. WEINBERG, PhD, *Laboratory of Immunobiology, Center for Biologics Evaluation and Research, Food and Drug Administration, Bethesda, MD*

MARY C. WEISS, PhD, *Unité de Génétique de la Différenciation, URA 2578 du CNRS, Institut Pasteur, Paris, France*

KARIN WILLIAMS, PhD, *Department of Urologic Surgery and Vanderbilt Prostate Cancer Center, Vanderbilt University Medical Center, Nashville, TN*

IAN WILMUT, *Roslin Institute, Roslin, Midlothian, UK*

MICHAEL A. WRIDE, PhD, *Department of Optometry and Vision Sciences, Cardiff University, Cardiff, Wales, UK*

NICHOLAS A. WRIGHT, MD, PhD, *Department of Histopathology, Barts and the London, Queen Mary's School of Medicine and Dentistry, London, UK*

HIDEYUKI YOSHITOMI, MD, PhD, *Cell and Developmental Biology Program, Fox Chase Cancer Center, Philadelphia, PA*

STUART H. YUSPA, MD, *Laboratory of Cellular Carcinogenesis and Tumor Promotion, Center for Cancer Research, National Cancer Institute, Bethesda, MD*

KENNETH S. ZARET, PhD, *Cell and Developmental Biology Program, Fox Chase Cancer Center, Philadelphia, PA*

YI ZHAO, MD, *Gene Therapy Laboratories, USC Keck School of Medicine, Norris Cancer Center, Los Angeles, CA*

1 Stem Cells

What Are They? Where Do They Come From? Why Are They Here? When Do They Go Wrong? Where Are They Going?

STEWART SELL, MD

The stem cell is the origin of life. At stated first by the great pathologist Rudolph Virchow, "All cells come from cells." The ultimate stem cell, the fertilized egg, is formed from fusion of the haploid progeny of germinal stem cells. The fertilized egg is totipotent; from it forms all the tissues of the developing embryo. During development of the embryo, germinal stem cells are formed, which persist in the adult to allow the cycle of life to continue. In the adult, tissue is renewed by proliferation of specialized stem cells, which divide to form one cell that remains a stem cell and another cell that begins the process of differentiation to the specialized function of a mature cell type. Normal tissue renewal is accomplished by the differentiating progeny of the stem cells, the so-called transit-amplifying cells. For example, blood cells are mature cells derived from hematopoietic stem cells in the bone marrow; the lining cells of the gastrointestinal tract are formed from transit-amplifying cells, progeny of stem cells in the base of intestinal glands. Nineteenth-century pathologists first hypothesized the presence of stem cells in the adult as "embryonal rests" to explain the cellular origin of cancer and more recent studies indicate that most cancers arise from stem cells or their immediate progeny, the transit-amplifying cells. Cancer results from an imbalance between the rate at which cells are produced and the rate at which they terminally differentiate or die. Understanding how to control the proliferation and differentiation of stem cells and their progeny is not only the key to controlling and treating cancer, but also to cell replacement and gene therapy for many metabolic, degenerative, and immunological diseases.

1.1. WHAT ARE THEY?

In the beginning there is the stem cell; it the origin of an organism's life. It is a single cell that can give rise to progeny that differentiate into any of the specialized cells of embryonic or adult tissues; that is, it is totipotent. The ultimate stem cell, the fertilized egg, divides five or six times to give rise to branches (lines) of cells that form various differentiated organs (Fig. 1). During these early divisions, each daughter cell retains totipotency. Then, through a series of divisions and differentiations, the embryonic stem cells (ESCs) lose potential and gain differentiated function (a process known as determination; *see* below). During normal tissue renewal in adult organs, tissue stem cells give rise to progeny that differentiate into mature functioning cells of that tissue. Stem cells with less than totipotentiality are called "progenitor cells". Except for germinal cells, which retain totipotency, most stem cells in adult tissues have reduced potential to produce cells of different types (i.e., are determined). However, there is increasing evidence for retention of some toti/multi-potent cells in the tissues of adults, especially in the bone marrow.

From: *Stem Cells Handbook*
Edited by: S. Sell © Humana Press Inc., Totowa, NJ

1.2. WHERE DO THEY COME FROM?

According to Leslie Brainerd Arey, the father of modern embryology (Arey, 1974), the first recorded attempt to understand the origin of life and the early development of the human was most likely made by Aristotle (384–322 BC). He recognized the early stages of development in the uterus and apparently was the first to contemplate the basic conflict of whether or not a new individual was formed *de novo* or was pre-formed in the mother and only enlarged during development (Arey, 1974). Aristotle deduced that the embryo was derived from the mother's menstrual blood, a conclusion that was based on the concept that living animals arose from slime or decaying matter (a hypothesis known in the middle ages as "spontaneous generation"). This concept was generally accepted for more than 2000 yr, until its validity became the major biological controversy of the 19th century. The hypothesis that life did not arise spontaneously, but rather only from preexisting life (*omne vivum ex vivo*) was pronounced by Leydig in 1855. Virchow (1855) then extended this to postulate that all cells in an organism are derived from preexisting cells (*omnis cellula e cellula*) (*see also* Oberling, 1944); all the cells of the human body arise from a preexisting stem cell, the fertilized egg. The counterhypothesis of spontaneous generation was not formally disproved until 1864,

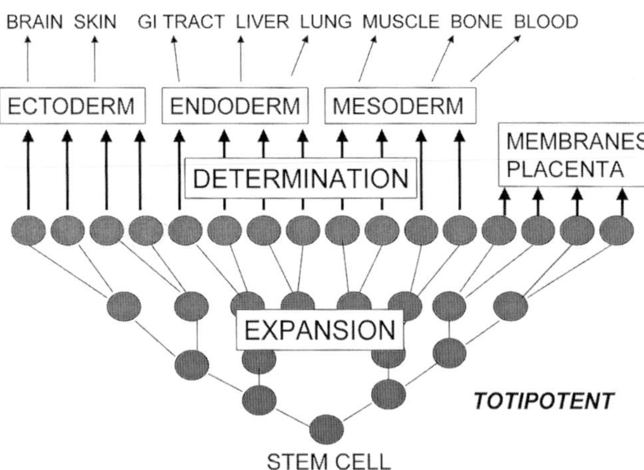

Fig. 1. Embryonic stem (ES) cell and progeny. During embryonic development, the ultimate stem cell, the fertilized egg, gives rise to progeny that retain totipotentiality as the population expands. Then determination occurs, and the cells begin to lose potential and to gain the specialized functions required to form mature organs. During blastulation the cells in the outer cells of the blastocyst (blasto- meaning embryo or germ; cyst - cavity) become determined for extraembryonic structures (placenta and membranes), whereas the ICM retains totipotentiality until the next stage of development, gastrulation (invagination of the blastula, *see* Fig. 4. During gastrulation the cells become determined to form the primary germinal layers: ectoderm, endoderm, and mesoderm. Ectoderm further differentiates to skin and brain; endoderm to gastrointestinal (GI) tract and internal organs; and mesoderm to connective tissue, bone, blood vessels, and blood-forming tissue. The relationship of cancers to the developing embryo is reflected in the use of the terms *carcinoma* for cancer arising from ectoderm-derived cells; *sarcoma* for cancer arising from mesoderm-derived cells.

Table 1.1
Terminology of Potential (Plasticity)

Prefix	*Meaning*	*Example*
Toti	All	Embryonal
Multi	Many/much	Hematopoietic
Pluri	Several/many	Hematopoietic
Oligo	Few/little	GI stem cell
Quadri	Four	GI stem cell
Tri	Three	Bronchial lining
Bi	Two	Bile duct
Uni	One	Prostate

Source: The Random House Dictionary of the English Language. Recent evidence indicates that some cells in the tissues used as examples of determined potential may have more potential than previously appreciated. For example, skin stem cells, believed to be uni- or bipotent, may contain multipotent progenitor cells (Liang and Bickenbach, 2002).

when Louis Pasteur performed carefully controlled experiments that demonstrated the failure of microorganisms to grow (corruption) in sterilized broth in vessels having long necks that prevented ambient organisms from entering (Debre, 1998). At present, the question is posed in the context of the conflict over abortion: "When does life begin?" According to the principles derived from Leydig, Virchow, and Pasteur, life as we know it neither ends nor begins but is continuous (Fig. 2). The adult human, for example, is only one stage in the cycle of human life.

Until the 1800s, the dominant hypothesis was that pre-formed individuals resided in the egg or the sperm. This pre-formed individual was called a homunculus. The homunculus in the egg was activated to develop after stimulation with sperm, or, conversely, the homunculus in the male sperm was activated to develop when provided an appropriate environment in the uterus. By the early 1900s, this concept had been proven to be incorrect, the embryo was shown to be formed by the fertilization of an egg, which developed in the ovary of the female, by fusion with a sperm provided by the male (*see* Needham, 1959). The product of the union of a sperm with an egg is the primordial totipotent stem cell (*see* the terminology in Table 1). How the cycle of life originally began is a subject of controversy. The two major alternate hypotheses are that human life was either created by Divine intervention in 7 d or else it evolved from primordial chemical biosynthesis, followed by natural selection from preceding life forms, over millions of years.

1.3. WHY ARE THEY HERE?

Until recently, most of what we understood about how the adult develops from the primordial stem cell was derived from classic studies in developmental anatomy (Arey, 1974). Following fertilization, the egg undergoes a process of cell divisions and cell migrations known as cleavage. In this early process, each daughter cell receives the full chromosome complement of the original cell, and each daughter cell appears to be the same. This is known as symmetric division, in contrast to the properties of somatic stem cells, which exhibit asymmetric division (Thrasher, 1966; Merok and Sherley, 2001) (Fig. 3).

The daughter cells, called blastomeres, stick together to form a cluster of cells known as a morula (from Morus, mulberry). At each division the blastomeres are reduced in size, but transplantation studies indicate that each embryonic blastomere is able to produce all differentiated cell types; that is, it is totipotent. Eventually, as the number of blastomeres approaches 32 or 64 cells, a cell-free center appears in the expanding cluster of blastomeres, and a hollow sphere of cells is formed (blastocyst). In mammals, the outer cells form the embryonic membranes and the placenta, whereas the mass of cells within the blastocyst, the inner cell mass (ICM), forms the embryo. At this stage not all the cells are still totipotent, as some of the outer cells become committed to membranes or the placenta. As ICM develops, the daughter cells begin to acquire properties different from one another, so that specific regions are formed that are destined to become different components of the developing embryo, a process known as gastrulation (Fig. 4). During gastrulation, the totipotency of the cells of the ICM is lost, and the blastula is rearranged by invagination of cells from the outer blastocyst to form layered "germ" zones known as ectoderm (outer skin), mesoderm (middle skin), and endoderm (inner skin), which are destined to form the adult organs. Skin, dermal appendages (including breast), and brain and neural tissue are derived from ectoderm; connective tissue, muscle, bone, and blood vessels from mesoderm; and the GI tract and internal glandular organs from endoderm (Arey, 1974).

The process of the loss of potential and the gain of specialized function is known as *determination*. In this process, the *totipotent* stem cells of the blastomere give rise to multi/pluripotent cells of the germ layers. These, in turn, give rise to progenitor cells of the developing organs. Tissue determination is accom-

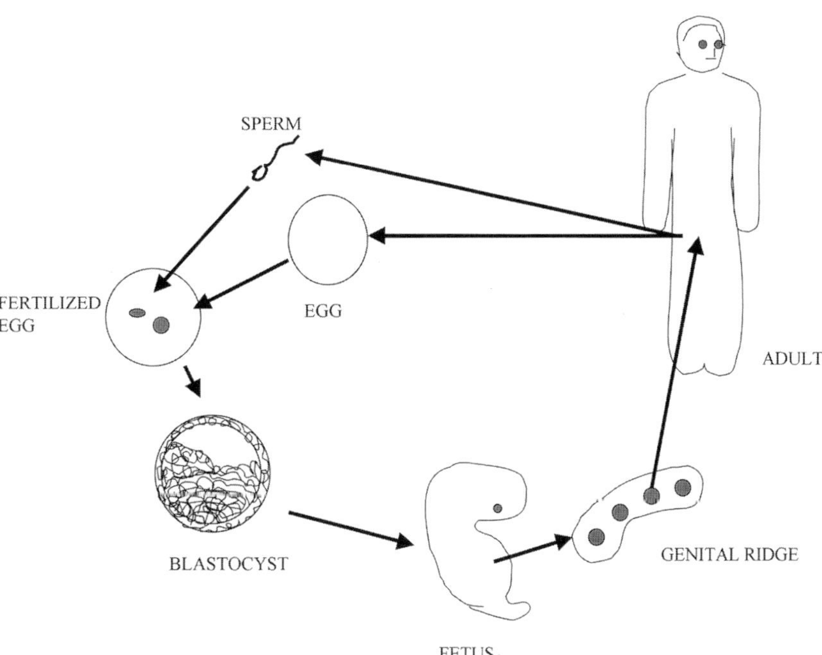

Fig. 2. Cycle of life. Human life is continuous. The life of an individual begins with fertilization of the egg and formation of a fetus. Totipotent cells in the developing fetus migrate to the genital ridge and in adults produce germinal stem cells in the gonads. Germinal cells give rise to gametes (egg and sperm) by reduction division (meiosis), resulting in cells containing half the chromosomes of an adult. Genetic reconstitution occurs when the sperm fertilizes the egg. In this process life is continuous; it is neither created nor destroyed.

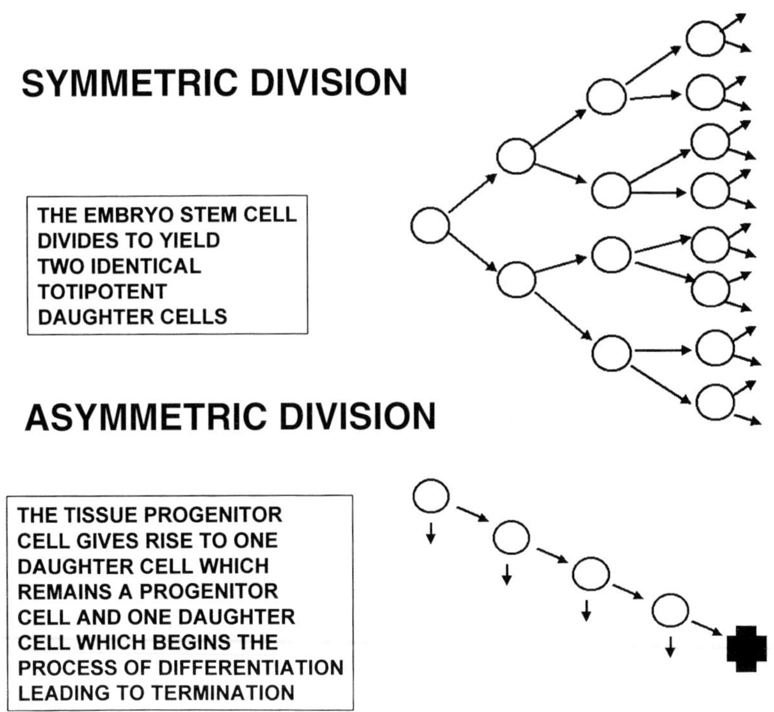

Fig. 3. Symmetric and asymmetric division. During early embryonic development, each cell divides and gives rise to two daughter cells with the same potential: symmetric division. During normal tissue renewal in the adult, each progenitor cell gives rise to one daughter cell that remains a progenitor cell, and one daughter cell that begins the process of determination to a terminally differentiated cell—asymmetric division. The number of cells increases exponentially during early embryogenesis, but the cell number remains constant during normal tissue renewal, as the number of new progenitor cells equals the number of cells destined to die.

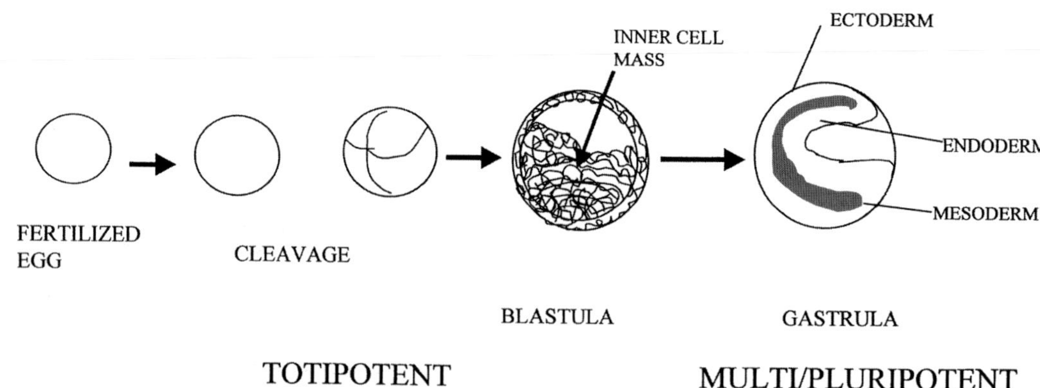

Fig. 4. Early development of embryo. Division of the fertilized egg results in the formation of a ball-like structure with a cavity on one end (the blastula). Until this stage, the cells divide by symmetric division, and all cells produced are totipotent. Invagination of one pole of the blastula leads to formation of the gastrula and establishment of the primitive germinal cell layers. During gastrulation and later formation of the fetus, the daughter cells lose potential as they gain specialized function.

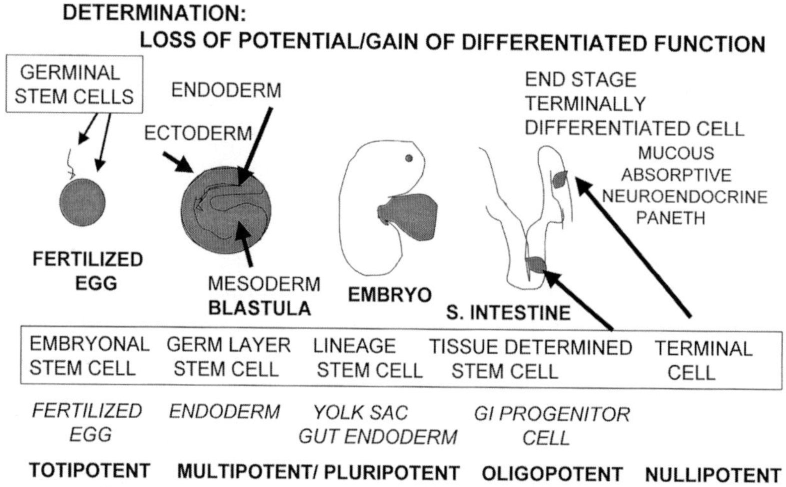

Fig. 5. Levels of progenitor cells during differentiation of small intestine. The small intestinal epithelium develops from a series of determinations from the embryonal stem cell (ESC), blastula, and endoderm. The intestinal progenitor is quadripotent; it can give rise to progeny that may become mucous, absorptive, neuroendocrine, or Paneth cells. Each of these becomes terminally differentiated in its fully mature form.

plished in progenitor cells through interaction with other cell types. For example, determination of primitive gut cells to liver (Matsumoto et al., 2001) or pancreas (Lammert et al., 2001) takes place in association with developing endothelium of blood vessels (Weinstein, 2002). As the cells become specialized into tissue, their potential becomes more limited, and they finally differentiate to terminally differentiated cells.

The terminology used regarding potential is listed in Table 1. The term *totipotent* should be reserved for those stem cells that can give rise to all of the differentiated tissues of the body as well as the placenta and membranes (ES cells, germinal stem cells). The terms *multipotential* and *pluripotential* are essentially synonymous: both refer to the ability of a given stem cell to form many different cell types. In an official National Institutes of Health primer released in 2000, pluripotent is defined as "capable of giving rise to most tissues of an organism," and totipotent as "having unlimited capacity" (www.nih.gov/news/stemcell/primer.htm). As depicted in Fig. 5, the potential decreases with the progressive development of the embryo.

The specialized cells in the adult that give rise to egg and sperm are called germinal cells. Germinal cells retain totipotentiality. Cancers derived from germinal cells may contain placental as well as adult tissues derived from more than one germ layer (teratocarcinomas; *see* below). Egg and sperm are derived from germinal cells by a special form of cell division called meiosis. As a result of meiosis, the egg and sperm each contain half of the chromosomes of their respective parent cells. With the joining of a sperm and an egg, a newly formed animal inherits its characteristics from the genetic material provided by both parents.

In classic embryology, determination was understood as a one-way process: once a cell is fully differentiated in the adult, it is strikingly stable (Surani, 2001) and not able to "de-differentiate". On the other hand, under some circumstances, this differentiated state does appear to be reversible. Transfer of nuclei from differentiated adult organs into an oocyte can result in restoration of the totipotency of the nucleus in the oocyte (Wilmut et al., 1997; Surani, 2001). This appears to be true even for mature T- and B-cells, which have rearranged T-cell receptor and immunoglobulin genes, respectively (Hochedlinger and Jaenisch, 2002); every tissue in mice cloned from a B-cell has the same immunoglobulin sequence as the original donor nucleus. It has even been reported that fibroblasts may be *re-programmed* to express T-cell or neuronal markers using cell extracts (Hakelien et al., 2002). However, others have found more stringent restrictions for re-programming.

Nuclear transplantation was first carried out in *Amoeba* (Comandon et al., 1930), and was extended to the frog *Rana pipiens* in 1952 (Briggs and King, 1952). In 1962, it was demonstrated that the nucleus from the intestine of a feeding tadpole could provide all of the information required for an ovum to develop into an adult frog (Gurdon, 1962). Later, nuclei from a number of organs were used to reproduce this result, but the technique was successful only if the donor cells were cultured for a few days in vitro before the nuclei were obtained (Laskey and Gurdon, 1970). Thus, although the proportion of successful transplants is small, <1–2%, at least some of the cells of adult organs of vertebrates contain nuclei that carry an entire set of the normal genes also found in the normal fertilized ovum. The question then becomes: Which cells in adult organs can supply nuclei that provide all of the information needed to produce a complete individual when used for nuclear transplantation into an anucleate ovum?

It appears that this may be the capability of tissue stem cells in the adult. In experiments in which nuclei from the tadpole are used, totipotentiality was progressively lost during gastrulation (King and Briggs, 1955). It has been stated, "The first generalization is that nuclei from more advanced developmental stages and from more differentiated cells always promote less normal nuclear-transplant embryo development (quantitatively and qualitatively) than nuclei from early developmental stages or undifferentiated cells" (Gurdon, 1974, p. 28). In the case of the cloning of sheep, nuclei from a d-9 embryo, a 26-d fetus, and the mammary gland of a 6-yr-old ewe in the last trimester of pregnancy were used (Wilmut et al., 1997), the latter implies that nuclei from adult mammals retain all of the information of the germ cells. However, the exact nature of the cells in the mixture used for cloning that contributed the nuclei in the transplant from the pregnant ewe is not defined. It is difficult to imagine that the nucleus of a secretory glandular cell could be re-programmed in such a manner. On the other hand, the mammary epithelium during the third semester is actively proliferating, and the proliferating gland contains active progenitor cells. Thus, it seems likely that the nuclei from an adult tissue that are capable of re-programming in an oocyte could actually be the tissue progenitor cells, or even the putative circulating totipotent stem cells of adults (Van der Kooy and Weiss, 2000). In the latter case, re-programming may not even be necessary, as the nucleus of this cell would already be totipotent.

1.3.1. EMBRYONIC STEM CELLS Under optimal conditions, cells from the ICM of the preimplantation blastocyst are able to proliferate indefinitely (Evans and Kaufman, 1981). As well, under inducing conditions, they can undergo determination and differentiate into other tissue types. In contrast, after formation of germ layers, most somatic progenitor cells have limited life-spans, and they exhibit decreasing differentiation potential as mature organs are formed (Merok and Sherley, 2001). Cole and Edwards (1967) were able to isolate ESCs from pre-implantation blastocysts of rabbits using feeder layers, and outgrowths of these cells differentiated into blood islands, muscle, connective tissue, neurons, and macrophages. Gardner (1968) demonstrated that, after injection of ESCs into a normal blastocyst, the cells could cocolonize in the developing embryo and form a chimeric individual. Essentially, Gardner demonstrated, that prior to implantation, the cells of the ICM were pluripotent. Edwards and his colleagues were able to obtain human oocytes after gonadotropin stimulation, fertilize the eggs in vitro, and grow the fertilized eggs in vitro to the blastocyst stage (Edwards et al., 1980, Steptoe et al., 1980). Subsequent transfer of the in vitro–fertilized embryos to the uteri of infertile patients supplemented with luteal support, eventually led to the successful clinical application of in vitro fertilization.

The potential of the in vivo use of ESCs for therapy was demonstrated when it was shown that injection of ESCs into lethally irradiated mice could restore the lost bone marrow stem cells (Hollands, 1987). In 1998, the in vitro culture of human ESCs that could differentiate into gut epithelium, cartilage, bone, muscle, neurons, and other cell types was reported (Thompson et al., 1998, Shamblott et al., 1998). The potential for human ESCs to be cultured in vitro, the possibility of producing human embryonic cell lines (Schuldiner et al., 2001), and the likelihood that these cells can be directed to differentiate into different cell types (Pittenger et al., 1999), have sparked the tremendous contemporary interest in ESC research for replacement of lost or damaged tissue cells (Donovan and Gearhart, 2001). Because ESCs give rise to germ cells, germ cells to egg and sperm, egg and sperm to a fertilized egg, and a fertilized egg to embryonic cells (Fig. 2), it is expected that any of the diploid cells in this cycle can give rise to any of the cells of the adult individual.

A problem in using ESCs for replacement of adult tissues concerns the low efficiency and long time required for ESCs to differentiate into functional adult cells. These issues may be addressed either by using adult precursor cells (*see* below) or by directing ESCs to a specialized tissue pathway. ESCs require a series of signals in order to produce progeny of a more highly differentiated type. For example, specialized culture conditions, including exposure to and withdrawal from fibroblast growth factor (FGF) (Zhang et al., 2001a), or culture with FGF and other growth factors (Reubinoff et al., 2001) allow generation of neural precursor cells, which may subsequently be shown to incorporate into the developing brain, at least in the mouse. This requires identification of in vitro conditions for different potential uses of ESCs.

1.3.2. GERMINAL STEM CELLS Early in embryogenesis, a few cells are designated to become germinal cells (Meachem et al., 2001). These cells migrate into the primitive gonad (genital ridge) and differentiate into female or male germ cell precursors, depending on the presence of two X chromosomes (female) or one X and one Y chromosome (male). They can be recognized by expression of the transcription factor Oct4 and of alkaline phosphatase (Peace and Scholer, 2000). In *Drosophila*, germinal cells, both in the ovary (Sprading et al., 2001; Xie and Sprading, 2001) and in the testes (Kegler et al., 2001; Kiger and Fuller, 2001; Tulina and Matunis, 2001), reside in specialized tissue ("niches") that controls the division of the stem cells; in this way, one daughter cell remains a stem cell and the other becomes a tissue-amplifying cell that enters into the process of producing sperm or egg. In the testes, the germinal cells surround a central core of supporting somatic cells known as the "hub"; in the ovary, the germinal cells are wrapped inside somatic cells in a structure known as the "ovariole". Maintenance of the stem cell requires activation of the JAK-STAT signaling pathway (Kiger et al., 2001; Tulina and Matunis, 2001); loss of STAT activity is associated with loss of stem cell properties. In the ovary, overexpression of the *decapentaplegic* (*dpp*) gene, a homolog of the human bone morphogenic protein genes 2 and 4, prevents germ cell differentiation and can lead to formation of stem cell tumors, whereas reduction of *dpp* causes germ cells to differentiate and exit the special germarium niche (Xie and Sprading, 2001). In the adult, germinal stem cells maintain their totipotentiality, and loss of normal differentiation signals may lead to uncontrolled proliferation and tumor formation.

Studies beginning in the 1970s using transplantation of germinal cells clearly demonstrated the totipotentialty and tumorigenicity of germinal cells. Stevens (1970) transplanted the germinal cells from the genital ridges of 21-d old fetal mice into the testes of adult syngeneic mice. The cells from these transplants produced teratocarcinomas. Teratocarcinomas are cancers of mixed cell types and may contain essentially all cellular elements of adult tissues as well as placenta and yolk sac. The growth of the tumor appears to be related to loss of the growth restrictions (Pierce et al., 1978) that are placed on the developing germinal cells by their normal environment in the blastocyst (Sprading et al., 2001). The tumor stem cells of these transplantable teratocarcinomas are able to differentiate into normal mature cells (Pierce and Wallace, 1971), including brain, bone, teeth, bone marrow, eyes, secretory glands, muscle, skin, and intestine, arising from all three germ layers (Kleinsmith and Pierce, 1964). During normal development, the environment of the blastocyst is able to control completely the differentiation of these tumor cells. Introduction of the transplantable malignant tumor cells into normal blastocysts can lead to the development of mosaic mice, with normally functioning tissues derived from the tumor cells (Mintz and Illmensee, 1975). Although controversial, their studies even reported that a mosaic male mouse was able to father normal offspring (Fig. 6). These studies demonstrate not only the totipotentiality of germinal cells and the teratocarcinomas derived from them, but they also illustrate the principle that the differentiation potential of totipotential cells—and in fact their ability to produce cancers—is controlled by environmental signals provided by the appropriate normal tissues, as predicted in 1911 by Rippert. Thus, cancer can variously be viewed as a problem of developmental biology (Pierce et al., 1978), a problem of the control of the progenitor cell by its niche (Sprading et al., 2001), or by chemical morphogens (Grudon and Bourillot, 2001).

1.3.3. SOMATIC PROGENITOR CELLS AND NORMAL TISSUE RENEWAL The cells of normal adult organs are continually being replaced. This replacement is accomplished by proliferation of *progenitor cells* or transit-amplifying cells. Transit-amplifying cells are progeny of the tissue stem cell. They provide an expanded population of mitotically competent tissue determined progenitor cells and produce progeny that differentiate into more mature cells that can no longer proliferate and eventually die (Sell and Pierce, 1994). In classic embryology, determination is considered a unidirectional pathway. For example, when primitive endoderm becomes committed to forming liver, it is not able to dedifferentiate or to transdifferentiate into another tissue type. However, as will be explained in by many of the chapters in this book, there is increasing evidence of plasticity of progenitor cells (*see* Table 2). The question of progenitor-cell plasticity is particularly relevant in a consideration of the potential of progenitor cells in adult organs. In addition, there is increasing evidence that adult stem cells in the bone marrow may be able to transdifferentiate into other tissue types, or that the bone marrow contains small numbers of multipotent undifferentiated stem cells (*see* Wulf et al., 2001).

In the process of normal tissue renewal, in contrast to the synchronous division that occurs during embryological development, division is asynchronous, with preservation of the progenitor cell (Fig. 3). Cohorts of transit-amplifying cells or progenitor cells, that are the daughter cells of the more primitive stem cell, accomplish normal renewal. This true tissue stem cell normally does not divide but it can be stimulated to proliferate if a loss of transit-amplifying cells occurs (*see* Subheading 1.3.8.). When a tissue progenitor cell divides, one daughter cell remains a progenitor cell, whereas the other begins the process of determination or differentiation, eventually producing a terminally differentiated cell, which in time dies. To maintain the total number of cells in an adult organ in equilibrium, the number of progenitor cells that divide essentially equals the number of cells that differentiate and die. A comparison of the rates of normal tissue turnover in various organs is given in Table 2.

1.3.4. HEMATOPOIETIC STEM CELLS In 1917, Pappenhein postulated the existence of an undifferentiated stem cell for blood cells (*gemeinsame Stammzelle*). Blood is one of the most rapidly replaced tissues in the body (Reya et al., 2001). The hematopoietic or blood-forming cells are located in the bone marrow. The lineage of blood cells extends from a resting stem cell, to transit-amplifying precursor cells, to mature circulating blood cells (Fig. 7). Circulating blood cells vary from polymorphonclear cells, which only live for a day or two, to erythrocytes, which may survive for several months (*see* review by Graf, 2002). The majority of circulating blood cells cannot proliferate. For example, erythrocytes have excluded their nuclei and polymophonulclear cells have inactivated their nuclei and condensed them into clumps. Because of the large number of polymorphonuclear cells and their rapid turn over, an enormous number of precursor cells is required. These are present in the bone marrow as transit-amplifying blast cells, as well as their progeny existing at various stages of differentiation into mature polymorphonuclear cells. The ability of the bone marrow progenitor cells to form colonies of various types of blood cells was first recognized by Till and McCulloch (1961) as colony-forming units (Siminovich et al., 1963). By the use of such techniques, it was possible to show that an infinitesimally small number of progenitor cells can both reproduce themselves and give rise to the complete spectrum of phenotypic colonies of blood cells (Carnie et al., 1976). The number of proliferating hematopoietic progenitor cells has been estimated to be 0.05% of the number of bone marrow cells (Goodell et al., 1996).

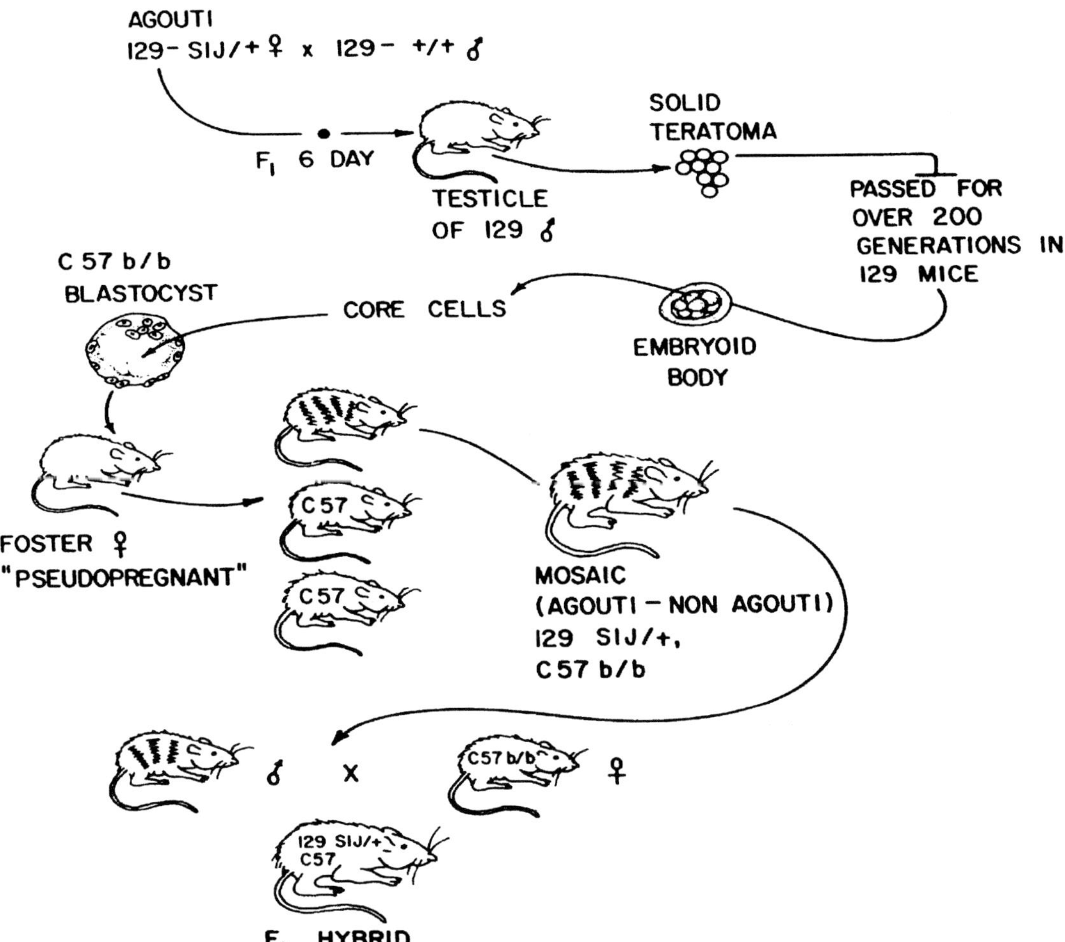

Fig. 6. Development of totipotent teratocarcinoma cells from normal genital ridge cells and control of differentiation by normal blastocysts. Transplantable carcinomas derived from the genital ridge of normal fetal tissue (Stevens, 1970) have been maintained for more than 200 generations by serial transplantation or cultivation in vitro. The core cells of the embryoid bodies formed by the cultured tumor cells are the tumor stem cells. When these are introduced into the ICM of a normal blastocyst, mosaic mice containing a mixture of blastocyst and tumor origins may be produced. As determined by histocompatibility antigen type, red blood cell type, immunoglobulin allotypes, and isoenzymes, the cells derived from the tumor cells can carry on their normal differentiated functions. It has been reported that mating of at least one male mosaic mouse was able to serve as the father of normal offspring with the genotype of the tumor cells (Mintz and Illmensee, 1975). Thus, the transplantable tumor may be considered to be the grandparent of these F1 mice. (From Sell, 1978.)

Until recently, the most primitive bone marrow progenitor cell was believed to be pluripotent, giving rise to stromal cells and lymphocytic cells, as well as RBCs, white blood cells (WBCs), and megakaryocytes (platelets) (Weatherall, 1997). In addition to the hematopoietic precursor, bone marrow also contains a mesenchymal progenitor cell that can give rise to many other cell types, such as osteocytes, adipocytes, muscle cells, astrocytes, and neurons, as well as stromal cells that support hematopoiesis (Prockop, 1997; Minguell et al., 2001). However, the accumulating evidence is that, not only does the bone marrow contain a pluri/multipotent blood-forming stem cell, but it also contains a cell that has the capacity to circulate to other organs and replace different nonhematopoietic tissues (Wuff et al., 2001). After bone marrow transplantation, the donor bone marrow cells contain primitive toti/multipotent stem cells, which can give rise to circulating cells that have the potential to differentiate into endothelium (Choi, 1998; Lin et al., 2000; Rifai, 2000), muscle (Ferrari et al., 1998),

Table 2
Postulated Turnover Rates and Potentials of Progenitor Cells in Some Adult Organs

Organ	*Replacement time*	*Potential*
Bone marrow (polys)	1–2 days	Multi/pluripotential (?toti)
GI lining	2–4 d	Quadripotential (four types)
Skin	2 wk	Bipotential
Liver	1–2 yr	Bipotential
Brain (neurons)	Lifetime	Unipotential[a]

[a]Mature neurons cannot divide, but neuronal progenitor cells are present in some parts of the adult brain. A generalization that can be made from these data is that, the faster the cell turnover in a given organ, the more likely is the presence of a stem or progenitor cell with potential to form other cell types.

TOTIPOTENT STEM CELL

PLURIPOTENT PROGENITORS

OLIGOPOTENT MYELOGENOUS PROGENITORS

LINEAGE RESTRICTED TRANSIT-AMPLIFYING CELLS

TERMINALLY DIFFERENTIATED

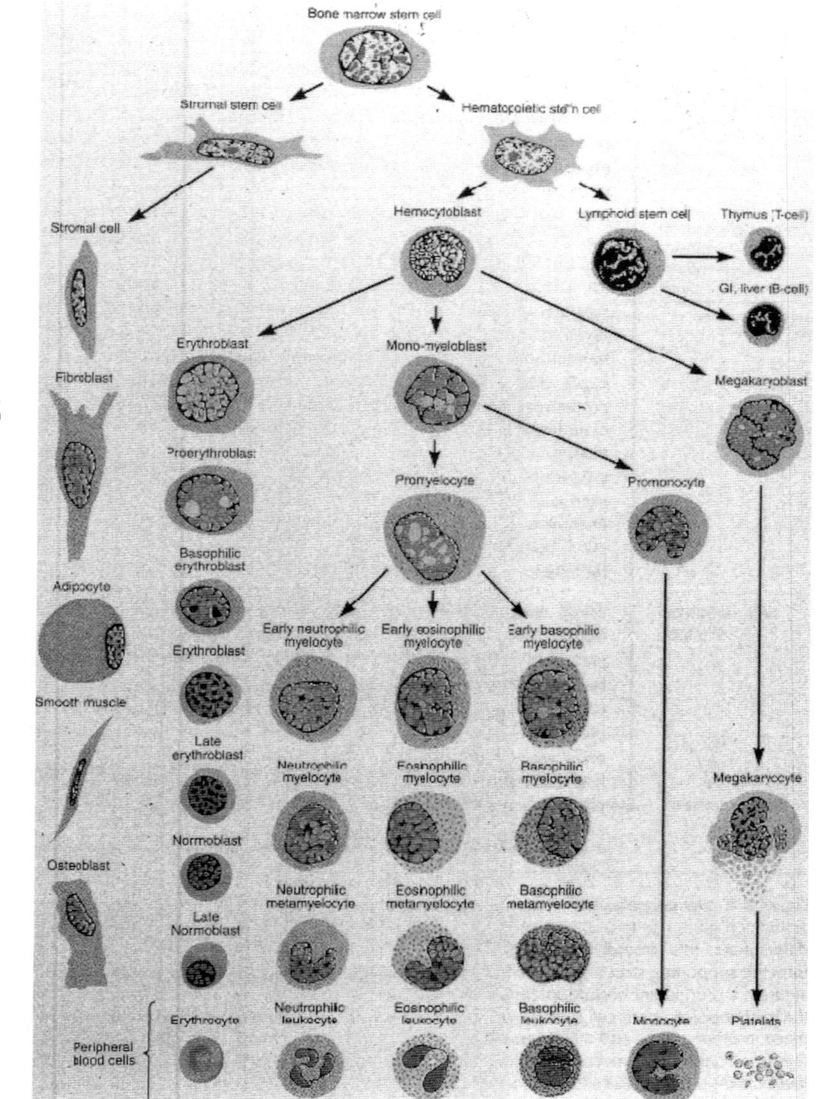

Fig. 7. Diagram of hematopoietic cell lineage. The bone marrow contains a totipotent stem cell that gives rise to pluripotent mesenchymal and hematopoietic lineages. The hematopoietic progenitor cell is believed to produce progeny that enter the blood cell pathway (hemocytoblast) or the lymphocytic pathway (lymphoid stem cell). The hemocytoblast is the common progenitor cell for erythrocytes, while blood cells, and megakaryocytes (platelets). The WBC precursor (monomyeloblast) gives rise to transit-amplifying cells for monocytes (promonocytes) and polymorphonuclear cells (promyelocytes). The promyelocyte produces progenitors of neutrophils, eosinophils, and basophils. During this process, an exponential increase in cells occurs at each stage. Thus, the totipotent stem cell is extremely rare, whereas there are enormous numbers of erythrocytes and polymorphonuclear neutrophils. (from Sell, 2001a).

liver (Petersen et al., 1999; Legasse et al., 2000; Theise et al., 2000), pancreatic islet beta cells (Ianus et al., 2003; Lee and Stoffel, 2003), heart (Jackson et al., 2001; Quaini et al., 2002), brain (Eglitis and Mezey, 1999; Priller et al., 2001), type I (Kotton et al., 2001) and type II pneumocytes (Krause et al., 2001), kidney (Poulsom et al., 2001), ocular retina of the eye (Grant et al., 2002), glomerular mesangial cells (Masuya et al., 2003), and other organs (Krause et al, 2001). Although these bone-marrow-derived cells have markers of the hematopoietic stem cell (HSC) (Legasse et al., 2000; Krause et al., 2001), it has not been ruled out that this multipotent cell may be of stromal origin (Donovan and Gearhart, 2001). Serial transplantation indicates that a single bone marrow cell may give rise to many different tissue types (Krause et al., 2001) and suggests that a common precursor must exist, not only for the stromal and hematopoietic lineages, but also for other germ layer–derived cell types. It is this putative totipotent bone marrow cell that has stimulated the great revival of interest in adult stem cells in the last few years. The putative totipotent stem cells arising from the bone marrow of adults may have just as much promise as ESCs for stem cell therapy (Gunsilius et al., 2001; Labat et al., 2001; Jang et al., 2002).

There are some caveats to the generally accepted assumption of pluripotentiality of tissue stem cells, including HSCs. For example, although it was reported that transplanted bone-marrow stem cells provided new endothelium in transplant models of atherosclerosis, it appears, after careful analysis, that that the new endothelial cells are actually derived from preexisting endothelium (Hillebrands et al., 2002). In a study of muscle-derived cells contributing to bone marrow, the muscle-derived bone marrow stem cells were found to be derived from HSCs present in the muscle (McKinney-Freeman et al., 2002). Whereas Bjornson and colleagues (1999) generated substantial numbers of hematopoietic precursor cells from neural stem cells, Moorshead and colleagues (2002) could not reproduce this result (*see Nature Med.* 8:535–537, 2002). In addition, recent studies have shown that in vitro culture of adult mouse brain or bone marrow stem cells with embryonic cells generates hybrid cells containing chromosomes of both adult and embryonic cells (Terada et al., 2002; Ying et al., 2002), thus raising the question of whether adult stem cells may appear to make other tissues by fusing with existing tissue cells. Both of these studies employed co-culture of adult cells with embryonic cells, which is not the case in bone marrow transplants in adults. However, this finding has stimulated in vivo studies on the possibility that transplanted donor bone marrow cells fuse with recipient cells in different organs. The results are conflicting. In a chronic model of restoration of damaged liver fusion of bone marrow cells and liver cells is described (Vassilopoulos et al., 2003; Wang et al., 2003), whereas fusion is not seen in restored pancreatic islet cells (Ianus et al., 2003) or in male cells from bone marrow donors in the buccal mucosa of female recipients (Tran et al., 2003). Thus, the possibility of fusion in vivo remains to be resolved.

Assuming that toti/multipotent bone marrow stem cells actually exist, a number of basic questions remain unanswered. Is there a single cell type in the bone marrow that is totipotent, or are there different progenitor cells that can give rise to different cell types in other organs? The first data relevant to this question indicate that there is a single cell that can do it all (Krause et al., 2001). Is the ability of a bone marrow stem cell to contribute to an epithelial organ a very rare or a more common event? Krause and colleagues (2001) find this to be fairly common, whereas Wagers and colleagues (2002) claim that this is an extremely rare event. The ability of bone-marrow-derived cells to contribute to epithelial tissue may depend on the age of the donor and the method of purification of the stem cells. Does the progenitor cell normally circulate, or is it liberated from the bone marrow by humoral signals after injury? The evidence is that this cell circulates normally but may be increased after injury. Are there normally progenitor cells in mature organs, such as liver or kidney, that can recirculate back to the bone marrow? Are these cells derived from the bone marrow or can totipotent progenitor cells be generated in differentiated organs (Blau et al., 2001)? The ability of brain cells to become blood cells (Bartlett, 1982; Bjornson et al., 1999) suggests that the answer to this is "yes," but recirculation of blood-derived cells to the brain, and back to the bone marrow, cannot be ruled out. Can progenitor cell lines that could be used for transplantation, for organ replacement, or for delivery of gene therapy be derived from bone marrow or other organs? Extensive culture of neural stem cells may be required before such cells can differentiate into blood cells, suggesting that rare transdifferentiation events may occur upon culturing that may account for neural to blood differentiation (Morshead et al., 2002). This finding suggests that normal brain precursor cells cannot contribute to blood-cell lineages in vivo. Answers to these questions, and clinical application of the use of tissue stem cells, will undoubtedly be forthcoming in the next few years.

Table 3
History of Stem Cell Origin of Cancer

1862	(Virchow)—Teratocarcinoma is made up of embryonic cells.
1874	(Durante and 1875 Cohnheim)—Cancer in adults arises from embryonal rests.
1899	(Wilms)—Wilms' tumor (nephroblastoma) arises from fetal primitive mesoderm.
1904	(Beard)—Cancer arises from displaced trophoblast of activated germinal cells.
1910	(Wright)—Neuroblastoma is a tumor of fetal neural crest.
1911	(Rippert)—Altered environment allows embryonic cells to escape growth control.
1921	(Rotter)—Primitive germ cells lodge in other organs during development.
1937	(Furth)—Tumor stem cells are cloned from leukemias.
1955	(Makino and Kano)—Leukemia arises from a stem cell.

1.4. WHEN DO THEY GO WRONG?

The idea that an adult human might harbor primitive stem cells was first conceived from studies on the origin of cancer. Historically, theories of the origin of cancer have ranged from parasitic infections, irritants, or viruses, to disturbance of normal tissue equilibrium (MacCallum, 1924; Oberling, 1944). However, these theories generally addressed etiology rather than the cells of origin of cancer. The idea that adult differentiated tissue could contain cells with embryonic potential most likely arose from the first studies of the pathology of cancer (Table 3). With the aid of the microscope, pathologists of the mid-19th century recognized that tumors in adults looked like embryonic tissue, and Recamier (1829) hypothesized that tumors might arise from embryonic cells present in the adult. The father of pathology, Rudolf Virchow, observed that teratocarcinomas were made up of tissues also found in developing embryos (Virchow, 1863). The term *de-differentiation* was coined by pathologists to describe this resemblance, and pathologists today still refer to the "degree of de-differentiation", to grade tumors as fast growing (poorly differentiated) or more slow growing (well differentiated). However, it is unfortunate that this term implies that tumors arise from actual de-differentiation of mature tissues, which is unlikely.

1.4.1. TUMOR STEM CELLS In theory, a malignant tumor may arise from any cell in the body that can divide (*see* Subheading 1.4.4.). Cancers at all ages may arise from "maturation arrest' of stem cells (Sell and Pierce, 1994); the degree of differentiation depends on the stage of differentiation at which the majority of the cancer cells become arrested. This idea is a variation on the theme of the "embryonal rest" theory of cancer. An Italian, Durante (1874), first postulated that cancer in adults developed from embryonal tissues, but the credit for the concept that cancers arise from putative embryonal cells present in mature tissues has been given to Cohnheim (1875), and the embryonal rest theory is attributed to him.

Tumors arising in children provide clues to the relationship to residual embryonic cells and cancer. In the late 1800s, a tumor of the kidney of children that contained a mixture of embryonal cell types was identified. In 1899, Wilms wrote a definitive monograph on this composite tumor and concluded that it arose from a fragment of the primitive undifferentiated mesodermal tissue, which normally is the anlagen of the myotome (striated muscle), nephrotome (Wolffian body), and smooth muscle (Wilms, 1899). James Homer Wright, the founder of the pathology laboratories at Massachusetts General Hospital, noted that the cells of Wilms' tumors of the kidney resembled embryonal cells and contained rosettes of cells similar to embryonal glomeruli (Jame Homer Wright rosettes) as well as sarcomatous tumor and nonstriated muscle. Wilms' tumor is also called mixed tumor, adenosarcoma or nephroblastoma (Fraser, 1920; Beckwith and Palmer, 1978). It only occurs in children under 7 yr of age; microscopically, it is composed of a mixture of undifferentiated spindle cells and immature epithelial tubules. Wilms' tumors appear to arise from embryonal cells in infants; these embryonic rests differentiate with age and essentially disappear, so that they are no longer a source of cancers after 7 yr of age (Beckwith and Palmer, 1978; Kretschmer, 1938).

A similar embryonic tumor of sympathetic ganglia, called neuroblastoma, was also recognized by James Homer Wright (1910). Neuroblastoma appears to arise in fetal neural crest cells of the sympathetic nervous system. These cancers have a peak age of presentation of 2 yr and are composed of embryonal-type neural cells that become more differentiated with age. In young infants, they are anaplastic and aggressive, whereas in older children they appear as more differentiated tumors or even as benign ganglioneuromas (Shimada et al., 1984). It appears that the undifferentiated embryonic cells giving rise to these tumors can be differentiated into benign mature cells by the environment of the growing infant.

In 1911, Beard hypothesized that tumors arise from displaced placental tissue or activated germinal cells in adult tissues (*see* Oberling, 1944). Rippert (1911) suggested that the critical factor is the isolation of these potential cancer cells from their normal environment. He pointed out that it was incorrect to consider the growing margin of an epithelial cancer as a transitional zone between normal and tumor tissue, in which the normal tissue is being converted into tumor tissue—an idea with a considerable following at the time. Rotter (1921) raised the possibility that primitive sex cells (germinal cells) wander through the tissue of the developing embryo and can accidentally lodge anywhere outside the ultimate sex glands. Such cells could serve as the origin of tumors. According to these hypotheses, all cancers may be considered to arise from embryonal-like progenitor cells or partially determined progeny cells present in the adult. Stem cells have the property to invade, migrate, and grow in tissue sites distant from which they arose. These are the same properties as metastatic tumor cells (Fidler, 1986). With the growing recognition that cancers in adults could be caused by chemicals or viruses, the embryonal rest origin of cancer was generally replaced by the process of de-differentiation from adult tissue as the cellular source of cancers. Thus, the concept that cancer arose from some change in the character of the cells was popularized. In fact, the term "anaplasia" was first coined to describe the change in cells that would allow cancerous growth by de-differentiation. However, as now used anaplasia (*ana*, backward; *phasia*, to form) is not considered a process that leads to tumor growth, but is used rather as its morphological correlate, describing the loss of differentiated features (without form). In modern cancer biology, genetic changes leading to alterations of growth regulatory functions are widely accepted as causing the altered growth seen in cancers (Boveri, 1929). The question remains as to the identity of the cells in which these genetic changes arise. Frank and Nowak (2003) argue, from a mathematical analysis of the total number of mutation events during the history of a cell, that mutations are likely to accumulate rapidly in cells during their symmetric linear growth phase (i.e., during early development). This results in seeding of a young individual with a small fraction of mutated stem cells.

Clearly, any alteration in a cell leading to cancer must arise in a cell that has the potential to divide and not to be lost during normal tissue turnover. For example, the cells above the basal layer of the skin or in the upper mucosa of the gastrointestinal tract that undergo a malignant mutation may be sloughed off before they can give rise to a tumor mass. In order to produce a cancer there must be a way for the transformed cells to remain in the body. Perhaps the best example of arrest of differentiation in cancer is leukemia.

1.4.2. LEUKEMIAS AS MODELS OF MATURATION ARREST The demonstration that tumor growth depends on a population of proliferating stem cells was first described in transmissible leukemias of mice. Jacob Furth (Furth and Kahn, 1937) was able to transplant leukemia to other mice using a single undifferentiated cell. Makino and Kano (1955) were also able to obtain clones of tumor cells from single cells, and they postulated the existence of tumor stem cells. Human leukemias feature gene rearrangements that are present in all the cells of the tumor (Rowley, 1999), suggesting the tumor's origin in a single cell that had undergone the rearrangement. The effect of a genetic change in the precursor cell of the population is exemplified by the malignant increase of multiple cell types in chronic mono/myelogenous leukemias, including various types of polymorphonuclear cells (neutrophils, eosinophils, and basophils), as well as monoctyes, erythrocytes, and platelets (megakaryocytes), all arising from a malignant precursor containing the same genetic lesion or lesions. Thus, the malignant cell is the progenitor cell with the capacity to differentiate into different blood cell types.

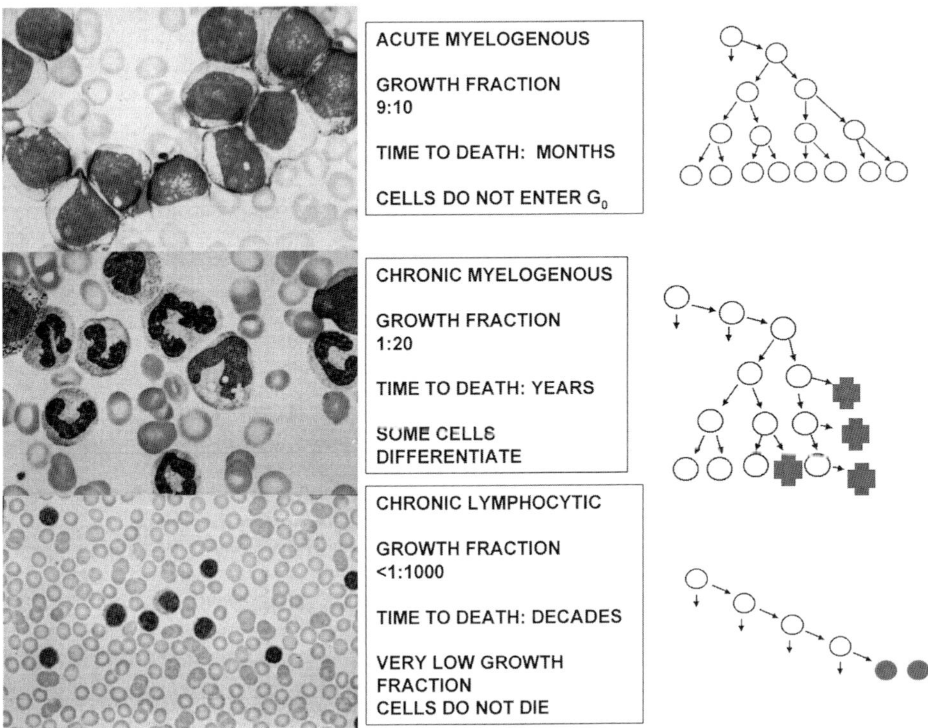

Fig. 8. Growth fractions, morphology, and clinical course of three selected leukemias. In acute myelogenous leukemia, the tumor cells are arrested in an active growth phase. Cells that divide do not enter G_0, but pass directly into the next cell cycle. Few if any differentiated cells are seen. The growth fraction is very high; the expansion of cells is essentially exponential, and the time to death, if the disease is not treated, is a few months. In chronic myelogenous leukemia, the arrest is at the level of the transit-amplifying cells. The number of cycling cells is much smaller, and a much higher proportion of the tumor cells undergoes differentiation. The growth fraction is small, and the time to death is years. In CLL, maturation arrest occurs in nondividing cells. The functional change is a lack of cell death. The growth fraction is vanishing small, and the time to death is decades.

The response to chemotherapy of acute leukemias also implies the involvement of a primitive stem cell, as repeated cycles of antiproliferative treatment are required to affect cells that are one specific time of therapy. Because the therapy acts on actively proliferating cells, it would be predicted that one cycle of drug treatment will be sufficient to eliminate actively proliferating cells present at the time of treatment. However, the malignant genetic change is also present in the "resting" or G_0 stage stem cells that are not proliferating when the drug is administered. When one cycle of treatment is completed, these resting cells will be stimulated to proliferate. When in an active state, they will now be susceptible to treatment. Successful treatment may require four or more cycles of antiproliferative therapy to catch all of the leukemic cells. Even so, up to 30% of patients will require total bone marrow ablation and stem cell transplantation from a normal donor. This is likely because the genetic lesion is present in the primitive resting stem cell in the bone marrow that is not susceptible to antiproliferative treatment. This lesion must eventually be destroyed by irradiation or the leukemia will reappear.

The characteristics of various leukemias illustrate how the stage of maturation arrest at which manifestation of a genetic change occurs determines the natural history (growth rate) of the tumor. A few of the known gene rearrangements in leukemias and lymphomas are listed in Table 4. Leukemias feature genetic lesions that affect either cell proliferation (transactivation) or cell death (apoptosis). With respect to determining the stage of differentiation at which the genetic rearrangement is found, it is generally identifiable in the early myelogenous progenitor or even in the more primitive bone marrow stem cells. Thus, the genetic lesion causing leukemias occurs in the stem cells, but the growth characteristics of the tumor are determined by the stage of maturation of the affected cell lineage.

The rate of cell proliferation (growth fraction) of the leukemia is reflected in the morphology of the cells (Fig. 8). A high percentage of large blast cells is a characteristic of fast-growing acute

Table 4
Some Gene Rearrangements in Leukemias and Lymphomas[a]

Leukemia	*Translocation*		*Effect of translocation*
CML (Philadelphia)	t9:22	bcr/abl	Tyrosine kinase activation
AML	t8:21	IL-3R	Tyrosine kinase activation
ALL	t12:22	TEL/MN1	Transcription activation
	t9:12	TEL/abl	Tyrosine kinase activation
Burkett's (B-cell)	t8:14	IgG/myc	Transcription activation
B-cell lymphoma	t14/	IgG/bcl2	Blocks apoptosis
CLL	?	?	Blocks apoptosis

[a]CML, chronic myelogenous leukemia; AML, acute myeloblastic leukemia; ALL, acute lymphoblastic leukemia; CLL, chronic lymphocytic leukemia.

TWO STAGE MECHANISM OF CARCINOGENESIS:

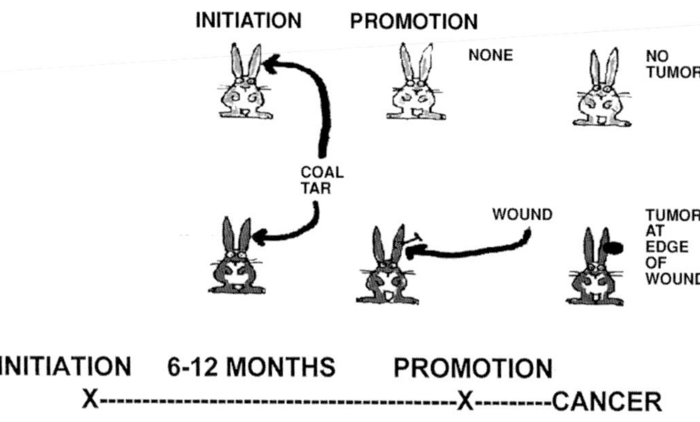

SKIN CELLS REPLACED EVERY 14 DAYS
6 MONTHS AFTER INITIATION THE ONLY CELLS REMAINING BEARING EFFECTS OF INITIATION WOULD HAVE TO BE THE STEM CELLS

Fig. 9. Initiation and promotion: two-step model of carcinogenesis. In the original experiments of Peyton Rous (Rous and Kidd, 1942; Friedwald and Rous, 1944), initiation was accomplished by painting the ear of a rabbit with coal tar. This produces genetic lesions in the epidermal cells by causing DNA/carcinogen adducts that cannot be repaired. If no further insult occurs, tumors do not develop. However, if the site is wounded by scraping with a cork borer, epithelial cancer appears at the edge of the wound. Later, it was shown that the initiation event could be separated from the promotion step by up to a year (Berenblum and Shubik, 1947; Berenblum, 1954; Foutwell, 1964). Because the epidermal cells of the skin are completely replaced every 2 wk (Potten and Morris, 1988; Watt, 1989), only the self-renewing progenitor cell population would survive during the interval between initiation and promotion. Thus, initiation must take place in the self-renewing stem cell population of the skin.

leukemias; a high percentage of more mature cells is a characteristic of chronic leukemia. Blast cells represent cells that are in the cell cycle; mature cells represent cells that are not in the cycle. Thus, acute leukemias are composed of blast cells, while chronic leukemias are composed of more highly differentiated cells. If a leukemia is composed of blast cells, it means that it has a high growth fraction (a high percentage of cells in cycle at any given time). If a leukemia is composed of mature cells, as in chronic lymphocytic leukemia (CLL), it has a low growth fraction. A mixture of blast and more mature cells implies an intermediate growth fraction. Extrapolation of this principle predicts that the degree of differentiation and the clinical behavior of a cancers is directly related to the stage of maturation arrest of the cells in the tumor lineage (Fig. 6).

1.4.3. EPIDERMAL CARCINOGENESIS AND STEM CELLS

Similar to leukemias, the biologic behavior of a skin cancer depends on the stage of differentiation at which the cell undergoes malignant change and maturation arrest. The lineage of the skin includes the precursor cells in the hair follicule, more determined basal cells in the epidermis, suprabasal proliferating cells (transit-amplifying cells) and non-proliferating cells above the transit amplifying cells. Although an oversimplification, transformation of the primitive skin progenitor cells in the bulge of the hair follicle gives rise to invasive hair follicule or basal cell carcinomas (Brown et al, 1998). Activation of the more determined basal cells of the skin by overexpression of Ras produces squamous cell carcinoma (Arnold and Watt, 2001; Waikel et al.), and expression of the c-myc gene in normally non-proliferating suprabasal cells reactivates the cell cycle and leads to hyperplasia (papillomas), but these do not progress to invasive tumors (Pelengaris et al., 1999). In squamous cell carcinoma of the uterine cervix associated with human papilloma virus infection, the initially infected and transformed cell is the basal stem cell, but productive infection requires keratinocyte differentiation (Garland, 2002; Munger and Howley, 2002).

Perhaps the best evidence that tumors arise from stem or progenitor cells is provided by the two-step model of skin cancer first described by Peyton Rous in 1941 (Fig. 9). The two steps are initiation and promotion (Friedwald and Rous, 1944; Berenblum and Shubik, 1947; Berenblum, 1954). In the classic model, benz(o)pyrine, the initiator, is painted onto the skin. This chemical binds to DNA in the skin cells, causing a permanent genetic alteration (initiation). However, cancers will not arise unless a proliferative stimulus is also given (promotion). This is provided by treating the skin with phorbol ester. Thus, the initiation event induces genetic damage, and the promoter then stimulates the damaged cells to proliferate, leading to cancer. Initiation must occur before promotion. If promotion is performed prior to initiation, cancers will not develop.

The time between initiation and promotion is the critical factor in implicating the stem cell as the initiated cell. This interval can be days or even months or years. For tumors to grow out in this model, the initiated cells must survive from the time of initiation to the time of promotion. Given the well-established fact that all cells in the skin, except the self-renewing progenitor cells, turn over completely every 2–3 wk (Potten and Morris, 1988; Watt, 1989), it is clear that the only way in which the initiated cells could still be present if months or years had passed since initiation, would be for initiation to have occurred in the self-renewing progenitor cell population. Thus, in the initiation-promotion model for skin carcinogenesis, the initiated cell must be a self-renewing progenitor cell.

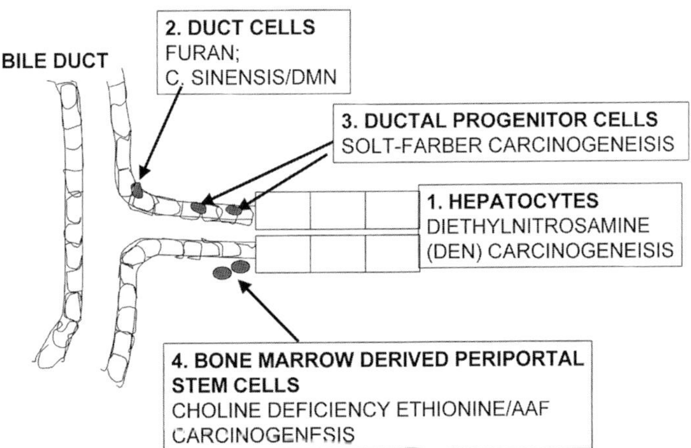

Fig. 10. Different levels of progenitor cells in adult liver. Following various models of liver injury or chemical hepatocarcinogenesis different cell types respond: (1) hepatocytes, (2) bile duct cells, (3) ductal progenitor cells, or (4) periductular oval cells, which may arise from circulating bone marrow precursor cells. (Modified from Sell, 2001b.)

If this is true, then why is it that the initiated cells do not grow out and produce cancer during normal tissue renewal? Why is promotion required for initiated cells to grow into cancers? The answers to these questions are not known. Presumably, normal tissue renewal does not involve proliferation of the long-term self-renewing cells. As for the hematopoietic cell lineage, most normal cell renewal involves not the more primitive skin determined stem cells, but rather the transit-amplifying cells. Thus, it is likely that the initiated transit-amplifying cells are not self-renewing. The initiated cell that gives rise to cancer must be a resting stem cell that is only called on to proliferate under the stress of promotion.

The concept of the stem cell origin of epithelial cancer is supported by the phenomenon of *field cancerization* (Slaughter et al., 1953). Field cancerization was first recognized by the presence of biologically abnormal tissue surrounding oral squamous cell carcinoma, and the development of multiple primary tumors and locally recurrent cancer in these areas. Field cancerization is associated with acquisition of a genetic change in a stem cell with formation of an altered patch of epithelium, which can be recognized histologically (Carlson et al., 2002). Patches of field cancerization may be identified by a zone of cells surrounding various epithelial cancers containing mutations of TP53 or loss of heterozygosity (Braakhuis et al., 2003). Field canceration appears to be a manifestation of an initiation event. If followed by expansion of the initiated population (promotion), a population of cells at high at risk for developing further genetic changes leading to invasive cancer is produced. When recognized clinically, it is advisable not only to remove any obvious tumors, but also to remove the surrounding cancerized field, if possible (Braakhuis et al., 2003).

1.4.4. HEPATIC CARCINOGENESIS AND PROGENITOR CELLS The cellular response to various regimens for chemical injury and hepatocarcinogenesis indicates that different cells in the hepatic lineage may restore lost hepatocytes or give rise to hepatocellular carcinomas (Dunsford et al., 1989; Sell, 2001b) (Fig. 10). A role for stem cells in the liver has only been appreciated for the last 25 yr. The finding that after surgical removal of up to two-thirds of the liver was compensated in a few days by proliferation of hepatocytes with no evidence of involvement of progenitor cells led to the conclusion that progenitor cells do not take part in liver renewal (Higgins and Anderson, 1931; Grisham, 1962; Bucher and Malt, 1971). On the other hand, carcinogenic studies in the 1930s and 1940s demonstrated that hyperplasia of bile duct-like small round cells in the portal zone of the liver preceded hepatocellular carcinoma (HCC) induced by butter yellow (Yoshida, 1935; Kinosita, 1937; Orr, 1940; Opie, 1944). These cells became known as oval cells (Farber et al, 1956). A relationship between the hyperplastic duct-like oval cells and hepatocellular carcinoma (HCC) was considered, but it was generally concluded that HCC arose from transformation of hyperplastic liver cells, not the oval cells (Farber, 1956; Farber, 1984). However, Gillman et al. (1954) concluded that the duct epithelium might undergo malignant transformation into HCC and might even differentiate into normal parenchymal liver cells. The concept that bile ducts may contain progenitor cells that could give rise to liver cells was extended by Wilson and Leduc (1958), who reported the proliferation of cholangioles, and the differentiation of terminal cholangioles into hepatocytes in mice fed a methionine-rich diet containing bentonite. However, Grisham and Porta (1964), in an autoradiographic study of bile-duct proliferation induced by various stimuli in rats concluded that hepatocytes were clearly distinct from biliary cells and that "cells possessing characteristics intermediate between the two were not seen in this study. ... It is concluded that in the adult rat under the conditions studied each type of cell is specific and reproduces only its own kind." For many years, models of chemical hepatocarcinogenesis were interpreted to support dedifferentiation of hepatocytes giving rise to HCC under the influence of an initiation/promotion type of chemical carcinogenesis regimen (Farber, 1984; Alterman, 1992). Because it is well known that mature hepatocytes can proliferate in response to injury, it is plausible that that hepatocytes can give rise to cancers. In carcinogenic models, foci of altered hepatocytes were the first observed change (Farber, 1956; Teebor and Becker, 1971). These foci were thought to give rise to so-called "preneoplastic nodules," these nodules in turn give rise to HCC. Thus, HCC developed from dedifferentiation of hepatocytes.

The kinetics of the early elevation of the fetal protein, α-fetoprotein (AFP) during chemical hepatocarcinogenesis gave the first clue that various cell populations are involved (Sell and Becker, 1978; Dunsford et al., 1989). During hepatocarcinogenesis induced by diethylnitrosamine, AFP first appears in atypical hepatocytes and then in microcarcinomas, and finally in frank hepatocellular carcinomas, suggesting that HCC arises from hepatocytes in this model. However, when hepatocyte proliferation is inhibited by administration of the chemical *N*-2-acteylaminofluorene (AAF) and the liver is exposed to diethylnitrosamine (initiator) followed by promotion by partial hepatectomy (the Solt–Farber model; Solt et al., 1977), there is proliferation of small, so-called "oval cells" (Farber, 1956) in ductular structures. This proliferation precedes HCC, suggesting that HCC arises from ductular progenitor cells. Finally, when rats are fed a choline-deficient diet and exposed to AAF or ethionine, or are fed cycles of a diet containing AAF, there is proliferation of periductular oval cells, which appear to be the precursor cells for the HCCs. Thus, HCCs may arise from hepatocytes, ductular progenitor cells, or periductular stem cells (Fig. 10). Given the finding that hepatocytes can arise from bone marrow precursors (Petersen et al., 1999; Theise et al., 2000; Krause et al., 2001), it seems likely that circulating bone marrow–derived stem cells that enter the liver will do so in the periductular region from the hepatic artery. It is therefore possible that these circulating cells give rise to HCC under the conditions of the third group of regimens described (Sell, 2001b). Given the results suggesting fusion of bone marrow stem cells with hepatocytes described above, is it possible that some HCC arise from fusion between bone marrow derived cells and hepatocytes? This could explain the marked variations in ploidy seen in HCC.

1.4.5. STEM CELLS AND ADENOCARCINOMAS The observations on the stem cell origin of teratocarcinomas and HCCs may be extrapolated to other organs (Sell and Pierce, 1994). Exocrine glands consist of acini of glandular cells drained by a collecting duct. The key progenitor cells are located between the gland and the duct. For example, it is known from both histological examination of human lesions (Wellings et al., 1959) and from examination of "preneoplastic" lesions of experimental mouse models (Haslan and Bern, 1977; Medina, 1988), that breast cancers arise from basal progenitor cells located in the terminal ductular lobular unit. These are the same cells that proliferate during pregnancy and produce progeny that eventually differentiate to form the lactating mammary gland. Prostate cancer also arises from basal stem cells (Bonkhoff and Remberger, 1996; Kadkol et al., 1998; Dubey et al., 2001). Intestinal carcinogenesis is directed toward functionally anchored stem cells in the crypts (Kim et al., 1993), and mutated stem cells appear to be able to spread through the mucosa, by fusion of crypts, before becoming invasive (Garcia et al., 1999). Mutation in the adenomatous polyposis coli gene is associated with increased expression of survivin, which allows expansion of crypt stem cells and development of colon cancer (Zhang et al., 2001b). Finally, mixed tumors of epithelial and mesenchymal components (carcinosarcomas) appear to arise from a common monoclonal origin (Thompson et al., 1996). In each instance, carcinomas arise from tissue progenitor cells. Is it possible that these cells can arise from circulating stem cells that have the ability to cross basement membranes and localize in different glandular tissues?

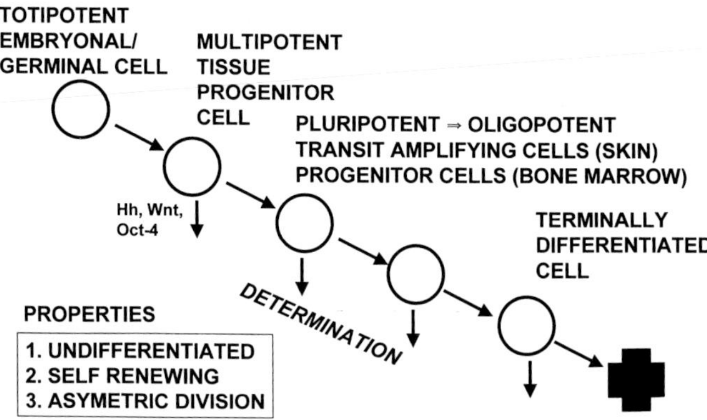

Fig. 11. Stem cell model of tissue renewal. The fertilized egg and adult germinal cells are both totipotent stem cells. ESCs derived from the fertilized egg are also totipotent and give rise to progeny that begin the process of determination to become specialized, differentiated tissue cells. In normal tissue replacement, there is a multipotent, possibly totipotent tissue progenitor cell, which produces a population of transit-amplifying cells. The transit-amplifying cells are responsible for replacing terminally differentiated tissue cells. Thus, in the adult, there are various levels of stem or progenitor cells having various potentials for proliferation and differentiation. In some cases, totipotent embryonic cells may be the best choice to mediate stem cell/progenitor cell therapies; in other cases, stem cell/progenitor cell therapy may best be mediated by tissue-determined progenitor cells. (Modified from Pierce and Wallace, 1971.)

1.5. WHERE ARE THEY GOING?

The chapters in this book cover many different aspects of stem cell biology and illustrate the various levels of determination, not only during development from ESCs, but also during normal tissue renewal; this is illustrated in Fig. 11. Most public attention has focused on the use of ESCs for possible replacement of damaged tissue or for gene therapy. This is because ESCs cells are capable of forming any other cell type. However, it is not clear that these cells will respond to signals derived from the microenvironment of adult organs to differentiate into the mature cell types desired; and if indeed they do respond, how many cells will differentiate and how long will the process take? In addition, there is an ongoing ethical controversy concerning the use of human embryos or cell lines derived from human embryos, for such purposes. On the other hand, there is increasing evidence that some form of tissue stem cell from adult individuals may be just as good, or even better, for some purposes. As will be illustrated in many of the chapters in this book, there are potential stem cells in the bone marrow, brain, circulating blood, and other organs that may be toti/multipotential. In addition, there are tissue-determined progenitor cells in yet other organs, such as muscle, skin, GI tract, and liver that may actually be better for more specific cell replacement or gene therapy in that they would be more efficient in replacing a given tissue.

Such progenitor cells could be obtained from injured organs, whose cells may have more potential than do the cells from normal organs. For example, it is possible to culture and then clone progenitor cells from animals with acute liver injury; these cells may have the potential to differentiate into not only liver cells, but also other related cell types. It is not yet clear whether these injury-induced progenitor cells are derived from the very few progenitor cells in the mature liver, or rather from circulating blood stem cells, which enter the liver in larger numbers is response to the injury. In any case, there is every reason to believe that stem cell therapy is not limited to ESCs, and may in fact be more feasible using adult stem cells.

The potential of stem cells to replace tissue cells damaged by either genetic lesions or epigenetic injury is controlled both by the innate potential of the stem cells and by the environment in which they are placed. This is similar to the "seed and soil" hypothesis of Paget (1889) for metastatic cancer cells. The ability of a stem cell to function in a given location is dependent upon its potential (seed), as well as upon microenvironmental signals, including: cell–matrix interactions, cell–cell interactions, growth factors, hormones, and circulation gradients (soil) (Wulf et al., 2001). Stem cells become restricted in the variety of differentiated progeny that they can produce during the determination process. For toti/multipotent stem cells to exist in the adult, preservation of the ESC potential must occur in certain cells, the multipotent tissue progenitor cells (Krause et al., 2001). On the other hand, a loss of potential has been clearly shown for neuronal cell lineages, as cells from the developing brain remember whence they originated (Temple, 2001). Yet, ESCs are able to re-populate damaged neural tissue and to provide cells that will remyelinate damaged spinal cord, in experimental models (Brustle et al., 1999; Horner and Gage, 2000; Mansergh et al., 2000); bone marrow stem cells may also repair brain lesions (Li et al., 2001; Zhao et al., 2002).

There appears to be no clear-cut answer to the question of what stem cells would be best for which types of therapeutic approaches. However, if one thinks about cells to replace injured liver, or to grow up cell lines for liver bioreactors, logic suggtests that cells already some distance down the liver lineage pathway might be more suitble than totipotent ESCs; likewise, cells committed to neuronal lineages might be better for replacement of damaged neuronal cells. Thus, it appears that the time has come to develop and test "designer stem cells" for specific therapeutic challenges (*see* review by Dove, 2002).

REFERENCES

Alterman, K. (1992) The stem cells of the liver—a selective review. *J. Cancer Res. Clin. Oncol.* 118:87–115.

Arey, L. B. (1974) *Developmental Anatomy: A Textbook and Laboratory Manual of Embryology*, 7th ed. W.B. Saunders, Philadelphia, PA.

Arnold, I. And Watt, F. M. (2001) c-Myc activation in transgenic mouse epidermis results in mobilization of stem cells and differentiation of their progeny. *Curr. Biol.* 11:558–568.

Bartlett, P. F. (1982) Pluripotential hemopoietic stem cells in the adult mouse brain. *PNAS* 79:2722–2725.

Beckwith, J. B. and Palmer, N. F. (1978) Histopathology and prognosis of Wilm's tumor: results from the first national Wilm's tumor study. *Cancer* 41:1937–1948.

Berenblum, I. (1954) Carcinogenesis and tumor pathogenesis. *Adv. Cancer Res.* 2:129–175.

Berenblum, I. and Shubik, P. (1947) A new quantitative approach to the study of the stages of chemical carcinogenesis in the mouse's skin. *Br. J. Cancer* 1:384–391.

Bjornson, C. R. R., Rietz, R. L., Reynolds, B. A., Magli, M. C., and Vesonvi, A. L. (1999) Turning brain into blood: a hematopoietic fate adopted by adult nerual stem cells in vivo. *Science* 183:534–537.

Blau, H. M., Brazelton, T. R., and Weimann, J. M. (2001) The evolving concept of a stem cell: entity or function? *Cell* 105:829–841.

Bonkhoff, H. and Remberger, K. (1996) Differentiation pathways and histogenetic aspects of normal and abnormal prostatic growth: a stem cell model. *Prostate* 28:98–106.

Boutwell, R. K. (1964) Some biological aspects of skin carcinogenesis. *Prog. Exp. Tumor Res.* 4:207–250.

Boveri, T. (1929) *The Origin of Malignant Tumors*, (translation by Marcella Boveri), Williams and Wilkins, Baltimore, MD.

Braakhuis, B. J. M., Tabot M. P., Kummer, J. A., Leemans, C. R., Brakenhoff, R. H. (2003) A genetic explanation of Slaughter's concept of field cancerizaiton: evidence and clinical implications. *Cancer Res.* 63:1727–1730.

Briggs, R. and King, T. J. (1952) Transplantation of living nuclei form blastula cells into enucleated frogs' eggs. *Proc. Natl. Acad. Sci. USA* 38:455–463.

Brown, K., Strathdee, D., Bryson, S., lambie, W., Balmain, A. (1998) The malignant capacity of skin tumours induced by expression of a mutant H-ras transgene depends on the cell type targeted. *Curr. Biol.* 8:516–514.

Brustle, O., Jones, K. N., Learish, R. D., et al. (1999) Embryonic stem cell–derived glial precursors: a source of myelinating transplants. *Science* 285:754–756.

Bucher, N. L. R. and Malt, R. A. (1971) Regeneration of the Liver and Kidney. Little, Brown and Co, Boston, MA.

Carlson, J. A., Scott D., Wharton, J., and Sell, S. (2001) Incidental h8istopathologic patterns: possible evidence of "Field Cancerization" surround skin tumors. *Am. J. Dermatopath.* 23:494–497.

Carnie, A. M., Lala, P. K., and Osmond, D. G., eds. (1976) *Stem Cells of Renewing Cell Populations*, Academic, New York, NY.

Choi, K. (1998) Hemangioblast development and regulation. *Biochem. Cell Biol.* 76:947–956.

Cohnheim, J. (1875) Congenitales, quergestreiftes muskelsarkon der nireren. *Virchows Arch.* 65:64.

Cole, R, Edwards, R. G., and Paul, J. (1966) Cytodifferentiation and embryogenesis in cell colonies and tissue cultures derived from ova and blastocysts of the rabbit. *Dev. Biol.* 13:385–407.

Comandon, J. and deFonbrune, P. (1930) Greffe nucleaire totale, simple ou multiple, chez une Amibe. *C. R. Soc. Biol. Paris* 130:744–748.

Debre, P. (1998) Louis Pasteur (translated by Elborg Forster) Johns Hopkins University Press, Baltimore, pp. 148–176.

DeOme, K. B., Faulkin, L. J., Jr, Bern, H. A., and Blair, P. E. (1959) Development of mammary tumors from hyperplastic alveolar nodules transplanted into gland-free mammary at pads of female C3H mice. *Cancer Res.* 19:515–520.

Donovan, P. J. and Gearhart, J. (2001) The end of the beginning for pluripotent stem cells. *Nature* 414:92–97.

Dubey, P., Wu, H., Reiter, R. E., and Witte, O. N. (2001) Alternative pathways to prostate carcinoma activate prostate stem cell antigen expression. *Cancer Res.* 61:3256–3261.

Dunsford, H. A., Karnasuta, C., Hunt, J. M., and Sell, S. (1989) Different lineages of chemically induced hepatocellular carcinoma in rats defined by monoclonal antibodies. *Cancer Res.* 49:4898–4900.

Durante, F. (1874) Nesso fisio-pathologico tra la struttura dei nei materni e la genesi di alcuni tumori maligni. *Arch. Memori Osservazioni Chirugia Practica* 11:217–226.

Edwards, R. G., Steptoe, P. C., and Purdy, I. M. (1980) Establishing full term human pregnancies using cleaving embryos gorwn in vitro. *Br. J. Obstet. Gynecol.* 87:737–756.

Eglitis, M. A. and Mezey, E. (1999) Hematopoietic cells differentiate into both microglia and macroglia in the brains of adult mice. *Proc. Natl. Acad. Sci. USA* 94:4080–4085.

Evans, M. J. and Kaufman, M. H. (1981) Establishment in culture of pluripotential cells from mouse embryos. *Nature* 292:154–156.

Farber E. (1956) Similarities in the sequence of early histologic changes induced in the liver by ethionine, 2-acetylaminofluorene and 3'-methyl-4-dimethylaminoazobenzene. *Cancer Res.* 16:142–148.

Farber E. (1984) Precancerous steps in carcinogenesis: their physiological adaptive nature. *Biochim. Biophys. Acta* 738:171–180.

Ferrari, G., Cusella-DeAngelis, G., Coletta, M., et al. (1998) Muscle regeneration by bone marrow-derived myogenic progenitors. *Science* 279:1528–1530.

Fidler, I. F. (1978) Tumor heterogeneity and the biology of cancer invasion and growth. *Cancer Res.* 38:2651–2660.

Frank, S. A. and Nowak, M. A. (2003)Developmental predisposition to cancer. *Nature* 422:494.

Fraser J. (1920) Adeno-sarcomatous tumors of the kidney: a clinico-pathological study. *Edinburgh Med. J.* 24:372–391.

Friedwald, W. F. and Rous, P. (1944) The initiating and promoting elements in tumor production: an analysis of the effects of tar, benzpyrene and methylcholanthrene on rabbit skin. *J. Exp. Med.* 80: 101–126.

Furth, J. and Kahn, M. C. (1937) The transmission of leukemia of mice with a single cell. *Am. J. Cancer* 31:276–282.

Garcia, S. B., Park, H. Y., Novelli, M., and Wright, N. A. (1999) Field cancerization, clonality, and epithelial stem cells: the spread of mutated clones in epithelial sheets. *J. Pathol.* 187:61–81.

Gardner, R. L. (1968) Mouse chimeras obtained by injection of cells into the blastocyst. *Nature* 220:596–597.

Garland, A. M. (2002) Human papillomavirus update with a particular focus on cervical disease. *Pathology* 34:213–224.

Gillman, J., Gilbert, C., and Spence, I. (1954) Some factors regulating the structural integrity of the intrahepatic bile ducts with special reference to primary carcinoma of the liver and vitamin A. *Cancer* 7:1109–1154.

Goodell, M. A., Brose, K., Paradis, G., Conner, A. S., Mulligan, R. C. (1996) Isolation and function properties of murine hematopoietic stem cells that are replicating in vivo. *J. Exp. Med.* 183:1797–1806.

Grisham, J. W. (1962) A morphologic study of deoxyribonucleic acid synthesis and cell proliferation in regenerating rat liver: autoradiography with thymidine-H3. *Cancer Res.* 22:842–849.

Grisham, J. W. and Porta, E. A. (1964) Origin and fate of proliferated hepatic ductal cells in the rat: electron microscopic and autoradiographic studies. *Exp. Mole. Pathol.* 3:242–261.

Gunsilius, E., Gastl, G., and Petzer, A. L. (2001) Hematopoietic stem cells. *Biomed. Pharmocother.* 55:186–194.

Gurdon, J. B. (1962) The developmental capacity of nuclei taken from intestinal epithelium cells of feeding tadpoles. *J. Embryol. Exp. Morphol.* 10:622–640.

Gurdon, J. B. (1974) The control of gene expression in animal development. Harvard University Press, Cambridge, MA, pp. 3–36.

Gurdon, J. B. and Bourillot, P.-Y. (2001) Morphogen gradient interpretation. *Nature* 413:797–803.

Hakelein, A.-M., Landsverk, H. B., Robi, M. M., Skalhegg, B. S., and Collas, P. (2002) Reprogramming fibroblasts to express T-cell functions using cell extracts. *Nature Biotech.* 20:460–466.

Haslan, S. Z. and Bern, H. A. (1977) Histopathogenesis of 7,12-dimethylbenz[a]anathracene-induced rat mammary tumors. *Proc. Natl. Acad. Sci. USA* 74:4020–4024.

Hayflick, L. (1965) The limited in vitro lifetime of human diploid cell strains. *Exp. Cell Res.* 37:614–636.

Higgins, G. M. and Anderson, R. M. (1931) Experimental pathology of the liver: restoration of the liver of the white rat following partial surgical removal. *AMA Arch. Pathol.* 12:186–202.

Hochedlinger, K. and Jaenisch, R. (2002) Monoclonal mice generated by nuclear transfer from mature B and T donor cells. *Nature* 415: 1035–2038.

Hollands, P. (1987) Differentiation of embryonic haematopoietic stem cells from mouse blastocysts growing in vitro. *Development* 99:69–76.

Horner, P. J. and Gage, F. H. (2000) Regenerating the damaged central nervous system. *Nature* 407:963–970.

Ianus, A., Holz, G. G., Theise, N. D., and Hussain, M. A. (2003) In vivo derivation of glucose-competent pancreatic endocrine cells from bone marrow without evidence of cell fusion. *J. Clin. Invest.* 111:843–850.

Jackson, K. A., Majka, S. M., Wang, H., et al. (2001) Regeneration of ischemic cardiac muscle and vascular endothelium by adult stem cells. *J. Clin. Invest.* 107:1395–1402.

Jiang, Y., Jahagirdar, B. N., Reinhardt, R. L., et al. (2002) Pluripotency of mesenchymal stem cells derived from adult marrow. *Nature*. Advance online publication, June 20, 2002 (doi:10.1038/nature 008701).

Kadkol, S. S., Brody, J. R., Epstein, J. I., Kuhajda, F. P., and Pasternack, G. R. (1998) Novel nuclear phosphoprotein pp32 is highly expressed in intermediate- and high-grade prostate cancer. *Prostate* 34:231–237.

Kiger, A. A. and Fuller, M. T. (2001) Male germ-line stem cells. In: *Stem Cell Biology* (Marshak, D. R., Gardner, R. L., and Gottleib, D., eds.), Cold Spring Harbor Laboratory Press, Cold Spring Harbor, NY, pp. 149–187.

Kiger, A. A., Jones, D.L., Schultz, C., Rogers, M. B., and Fuller, M. T. (2001) Stem cell self-renewal specified by JAK-Stat activation in response to a support cell cue. *Science* 294:2542–2545.

Kim, X. H., Roth, K. A., Moser, A. R., and Gordon, J. I. (1993) Transgenic mouse models that explore the multistep hypothesis on intestinal neoplasia. *J. Cell Biol.* 123:877–893.

King, T. J. and Briggs, R. (1955) Changes in the nuclei of differentiating gastrula cells as demonstrated by nuclear transplantation. *Proc. Natl. Acad. Sci. USA* 41:321–325.

Kinosita, R. (1937) Special report: studies on the cancerogenic chemical substances. *Trans. Japan Pathol. Soc.* 27:665–725.

Kleinsmith, L. J. and Pierce, G. B. (1964) Multipotentiality of single embryonal carcinoma cells. *Cancer Res.* 24:1544–1551.

Krause, D. S., Theise, N. D., Collector, M. I., et al. (2001) Multi-organ, multi-lineage engraftment by a single bone marrow–derived stem cell. *Cell* 105:369–377.

Kretschmer, H. L. (1938) Malignant tumors of the kidney in children. *J. Urology* 39:250–275.

Labat, M. L. (2001) Stem cell and the promise of eternal youth: embryonic versus adult stem cells. *Biomed. Pharmacother.* 55:179–185.

Ladd, W. E. and White, R. R. (1941) Embryoma of the kidney (Wilms' tumor). *JAMA* 117:1858–1863.

Lammert, E., Cleave,r O., and Melton, D. (2001) Induction of pancreatic differentiation by signals from blood vessels. *Science* 294:564–567.

Laskey, R. A. and Gurdon, J. B. (1970) Genetic content of adult somatic cells tested by nuclear transplantation from cultured cells. *Nature* 228: 1332–1334.

Lee, V., and Stoffel, M. (2003) Bone marrow: an extra-pancreatic hideout for the elusive pancreatic stem cell? *J. Clin. Invest.* 111:799–801.

Li, Y., Chen, J., Wang, L., Lu, M., and Chopp, M. (2001) Treatment of stroke in rat with intracarotid administration of marrow stromal cells. *Neurology* 56:1666–1672.

Liang, L. and Bickenbach, J. R. (2002) Somatic epidermal stem cells can produce multiple cell lineages during development. *Stem Cells* 20: 21–31

Lin, Y., Weisdorf, D. J., Solovey, A., and Hebbel, R. P. (2000) Origins of circulating endothelial cells and endothelial outgrowth from blood. *J. Clin. Invest.* 105:71–77.

MacCallum, W. G. (1924) *A Text-Book of Pathology*, W.B. Saunders, Philadelphia, PA, pp. 1115–1122.

Makino, S. and Kano, K. (1955) Cytological studies of tumors. XIV. Isolation of single-cell clones from a mixed-cell tumor of the rat. *JNCI* 15:1165–1181.

Mansergh, F. C., Wride, M. A., and Rancourt, D. E. (2000) Neurons from stem cells: implications for understanding nervous system development and repair. *Biochem. Cell Biol.* 78:613–628.

Masuya, M., Drake, C. J., Fleming, P. A., et al. (2003) Hematopoietic origin of glomerular mesangila cells. *Blood* 101:2215–2218.

Matsumoto, K., Yoshitomi, H., Rossant, J., and Zaret, K. S. (2001) Liver organogenesis promoted by endothelial cells prior to vascular function. *Science* 294:559–563.

Meachem, S., von Schonfeldt, V., and Schlatt, S. (2001) Spermatogonia: stem cells with a great perspective. *Reproduction* 121:825–834.

Medina, D. (1988) The preneoplastic state in mouse mammary tumorigenesis. *Carcinogenesis* 9:113–119.

Merok, J. R. and Sherley, J. L. (2001) Breaching the kinetic barrier to in vitro somatic stem cell propagation. *J. Biomed. Biotech.* 1:25–27.

Minguell, J. J., Erices, A., and Conget, P. (2001) Mesenchymal stem cells. *Exp. Biol. Med.* 226:506–520.

Mintz, B. and Illmensee, K. (1975) Normal genetically mosaic mice produced from malignant teratocarcinoma cells. *Proc. Natl. Acad. Sci. USA* 72:3583–3589.

Morshead, C. M., Benveniste, P., Iscove, N. N., and van der Kooy, D. (2002) Hematopoietic competence is a rare property of neural stem cells that may depend on genetic and epigenétic alterations. *Nat. Med.* 2:268–273.

Munger, K. and Howley, P. M. (2002) Human papillomavirus immortalization and transformation functions. *Virus Res.* 89:213–238.

Needham, A. (1959) *History of Embryology*, Cambridge University Press, New York, NY.

Oberling, C. (1944) *The Riddle of Cancer*, Yale University Press, New Haven, CT.

Opie, E. L. (1944) The pathogenesis of tumors of the liver produced by butter yellow. *J. Exp. Med.* 80:231–246.

Orr, E. L. (1940) The histology of the rat's liver during the course of carcinogenesis by butter-yellow (p-dimethylaminoazobenzene). *J. Pathol. Bacteriol.* 50:393–408.

Paget, S. (1889) The distribution of secondary growth in cancer of the breast. *Lancet* 1:571–573.

Pappenheim, A. (1917) Prinzipien der neueren morphologischen Haematozytologie nach zytogenetischer Grundlage. *Foloa Haematologica* 21:91–101.

Peace, M. and Scholer, H. R. (2000) Control of totipotency and germline determination. *Mol. Reprod. Dev.* 55:452–457.

Petersen, B. E., Bowen, W. C., Patrene, K. D., et al. (1999) Bone marrow as a potential source of hepatic oval cells. *Science* 284:1168–1170.

Pierce, G. B., Shikes, R., and Fink, L. M. (1978) *Cancer: a problem of developmental biology*. Prentice Hall, Engelwood Cliffs, NJ, pp. 1–242.

Pierce, G. B. and Wallace, C. (1971) Differentiation of malignant to benign cells. *Cancer Res.* 31:127–134.

Pittenger, M. F., McKay, A. M., Beck, S. C., et al. (1999) Multilineage potential of adult and human mesenchymal stem cells. Multilineage potential of adult human mesechymal stem cells. *Science* 284:143–147.

Potten, C. S. and Morris, R. J. (1988) Epithelial stem cells in vivo. *J. Cell Sci.* (Suppl.) 10:45–62.

Poulsom, R., Forbes, S. J., Hodivala-Dilke, K., et al. (2001) Bone marrow contributes to renal parenchymal turnover and regeneration. *J. Pathol.* 195:229–235.

Priller, J., Persons, D. A., Klett, F. F., Kempermann, G., Kreutzberg, G. W., and Dirnagl, U. (2001) Neogenesis of cerebellar Purkinje neurons form gene-marked bone marrow cells in vivo. *J. Cell Biol.* 155:733–738.

Prockop, D. J. (1997) Marrow stromal cells as stem cells for non-hematopoietic tissues. *Science* 276:71–74.

Quaini, F., Urbanek, K., Beltrami, A. P., et al. (2002) Chimerisms of the transplanted heart. *N. Engl. J. Med.* 346:5–15.

Recamier, J. C. A. (1829) *Recherches sur le Traitement du Cancer, etc.* Paris, France.

Reubinoff, B. E., Itsykson, P., Turetsky, T., et al. (2001) Neural progenitors from human embryonic cells. *Nat. Biotech.* 19:1134–1140.

Reya, T., Morrison, S. J., Clake, M. F., and Weissman, I. L. (2001) Stem cells, cancer and cancer stem cells. *Nature* 414:105–111.

Rifai, S. (2000) Circulating endothelial precursors: mystery, reality and promise. *JCI* 105:17–19.

Rippert, V. (1911) Ueber ein myosarcoma striocellulare des nierenbeckens und des ureters: Geschwulstlehre, Bonn, 1904; Das carcinom des menschen, Bonn.

Rotter, W. (1921) Histogenese der malignen Geschwulste. *Ztschr. Krebsforschung*. 18:171–208.

Rous, P. and Kidd, J. G. (1942) Conditional neoplasms and subthreshold neoplastic states. *J. Exp. Med.* 73:365–372.

Rowley, J. D. (1999) The role of chromosomal translocations in leukemogenesis. *Semin. Hematol.* 36:59–72.

Schuldiner, M., Yanuka, O., Itskovitz-Eldor, J., Melton, D. A., and Benvenisty, N. (2001) Effects of eight growth factors on the differentiation of cells derived from human embryonic stem cells. *Proc. Natl. Acad. Sci. USA* 97:11,307–11,312.

Sell, S. (1978) The biologic and diagnostic significance of oncodevelopmental gene products. In: *The Handbook of Cancer Immunology*, vol. 3 (Waters, H., ed.) Garland SIPM Press, New York, NY, pp. 1–69.

Sell, S. (2001a) Heterogeneity and plasticity of hepatocyte lineage cells. *Hepatology* 33:738–750.

Sell, S. (2001b) *Immunology, Immunopathology and Immunity*, 6th ed., ASM, Washington, DC.

Sell, S. and Becker, F. F. (1978) Alphafetoprotein. *J. Natl. Cancer Inst.* 60:19–26.

Sell, S. and Pierce, G. B. (1994) Maturation arrest of stem cell differentiation is a common pathway for the cellular origin of teratocarcinomas and epithelial cancers. *Lab. Invest.* 70:6–22.

Shamblott, M. J., Axelman, J., Littlefield, J. W., et al. (1998) Derivation of pluripotent stem cells from cultured human primordial germ cells. *Proc. Natl. Acad. Sci. USA* 95:13,726–13,731.

Shimada, H., Chatten, J., Newton, W. A., et al. (1984) Histopathologic prognostic factors in neuroblastic tumors: definition of subtypes of ganglioneuroblastoma and an age-linked classification of neuroblastomas. *JNCI* 73:405–416.

Siminovich, L., McCulloch, E. A., and Till, J. E. (1963) The distribution of colony-forming cells among spleen colonies. *J. Cell Comp. Physiol.* 62:327–336.

Slaughter, D. P., Southwick, H. W., and Smejdal, W. (1953) "Field cancerization" in oral stratified squamous epithelium. *Cancer* 6:963–968.

Solt, D. B., Medline, A., and Farber, E. (1977) Rapid emergence of carcinogen-induced hyperplastic lesions in a new model for the sequential analysis of liver carcinogenesis. *Am. J. Pathol.* 88: 595–609.

Sprading, A., Drummond-Barbosa, D., and Kai, T. (2001) Stem cells find their niche. *Nature* 414:98–104.

Steptoe, P. C., Edwards, R. G., and Purdy, I. M. (1980) Clinical aspects of pregnancies established with cleaving embryos grown in vitro. *Br. J. Obstet. Gynecol.* 87:757–768.

Stevens, L. C. (1970) The development of transplantable teratocarcinomas from intratesticular grafts of pre- and post-implantation of mouse embryos. *Dev. Biol.* 21:364–382.

Surani, M. A. (2001) Reprogramming of genome function through epigenetic inheritance. *Nature* 414:122–128.

Teebor, G. E. and Becker, F. F. (1971) Regression and persistence of hyperplastic hepatic nodules induced by N-2-fluorenylacetamide and their relationship to hepatocarcinogenesis. *Cancer Res.* 31:1–2.

Temple, S. (2001) The development of neural sem cells. *Nature* 414: 112–117.

Terada, N., Hamazaki, T., Oka, M,. et al. (2002) Bone marrow cells adopt the phenotype of other cells by spontaneous cell fusion. *Nature*. Advance online publication, March13, 2002. (DOI 10.1038/nature730).

Theise, N. D., Badve, S., Saxena, R., et al. (2000) Derivation of hepatocytes from bone marrow cells in mice after radiation-induced myeloablation. *Hepatology* 31:235–240.

Thompson, J. A., Itskovitz-Eldor, J., Shapiro, S. S., Waknitz, M. A., Swinergiel Marshall, V. S., and Jones, J. M. (1998) Embryonic stem cell lines derived from human blastocysts. *Science* 282:1145–1147.

Thompson, L., Chang, B., and Barsky, S. H. (1996) Monoclonal origins of malignant mixed tumors (Carcinosarcomas). *Am. J. Surg. Pathol.* 20:277–285.

Thrasher, J D. (1966) Analysis of renewing epithelial cell populations. In: *Methods in Cell Physiology*, vol. 2 (Prescott, D. M., ed.), Academic, New York, pp. 323–357.

Till, J. E. and McCulloch, E. A. (1961) A direct measurement of the radiation sensitivity of normal mouse bone marrow cells. *Radiat. Res.* 14:213–222.

Tran, S. D., Pillemer, S. R., Dutra, A., et al. (2003) Differentiation of human bone marrow-derived cells into buccal epithelial cell in vivo; amolecular analytical study. *The Lancer* 361:1084–1088.

Tulina, N. and Matunis, E. (2001) Control of stem cell self-renewal in *Drosophila spermatogenesis* by Jak-STAT signaling. *Science* 294: 2546–2549.

Van der Kooy, D. and Weiss, S. (2000) Why stem cells? *Science* 287: 1439–1441.

Van Duuren, B. L., Sivak, A., Katz, C., Seidman, I., and Melchionne, S. (1975) The effect of ageing and interval between primary and secondary treatment in two-stage carcinogenesis on mouse skin. *Cancer Res.* 35:502–505.

Vassilopoulos, G., Wang, P.-R., and Russell D. W. (2003) Transplanted bone marrow regenerates liver by cell fusion. *Nature*. 422:901–904.

Virchow, R. (1855) Editoral, *Archiv fuer pathologische Anatomie und Physiologie und fuer klinische Medizin.* 8:23.

Virchow, R. (1863) *Die krankhafter Geschwulste*, Bd 1. Hirschwald, Berlin.

Wagers, A. J., Sherwood, R. I., Christensen, J. L., and Weissman, I. L. (2002) Little evidence for developmental plasticity of adult hematopoietic stem cells. *Science* 297:2256–2262.

Waikel, R. L., Kawachi, Y., Waikel, P. A., Wang, X. J., and Roop, D. R. (2001) Deregulated expression of c-Myc depletes epidermal stem cells. *Nat. Genet.* 28:165–168.

Wang, X., Willenbring, H., Akkairi, Y., et al. (2003) Cell fusion is the principal source of bone-marrow-derived hepatocytes. *Nature*. 422:897–901.

Watt, F. M. (1989) Terminal differentiation of epidermal keratinocytes. *Curr. Opin. Cell. Biol.* 1:1107–1115.

Weatherall, D. J. (1997) ABC of clinical haematology: the hereditary anaemias. *Br. Med. J.* 314:492–498.

Wellings, S. R., Jensen, H. M., and Marcum, R. G. (1975) An atlas of subgross pathology of the human breast with special reference to possible precancerous lesions. *J. Natl. Cancer Inst.* 55:231–273.

Wilms, M. (1899) *Die Mischgeschwuelste*, Arthur Georgi, Leipzig.

Wilmut, I., Schnieke, A. E., McWhir, J., Kind, A. J., and Campbell, K. H. S. (1997) Viable offspring derived from fetal and adult mammalian cells. *Nature* 385:810–813.

Wilson, J. W. and Leduc, E. H. (1958) Role of cholangioles in restoration of the liver of the mouse after dietary injury. *J. Pathol. Bacteriol.* 76: 441–449.

Wright, J. H. (1910) Neurocytoma or neuroblastoma, a kind of tumor not generally recognized. *J. Exp. Med.* 12:556–563.

Wuff, G. G., Jackson, K. A., and Goodell, M. A. (2001) Somatic stem cell plasticity: current evidence and emerging concepts. *Exp. Hematol.* 29: 1361–1370.

Xie, T. and Sprading, A. (2001) The Drosophila Ovary: An In Vivo Stem Cell System. In: *Stem Cell Biology* (Marshak, D. R., Gardner, R. L., and Gottleib, D., eds.), Cold Spring Harbor Laboratory Press, Cold Spring Harbor, NY, pp. 129–148.

Yin, L., Sun, M., Illic, Z., Leffert, H. L., and Sell, S. (2002) Derivation, characterization and phenotypic variation of hepatic progenitor cell lines isolated from adult rats. *Hepatology* 35:315–324.

Ying, Q.-L., Nichols, J., Evans, E. P., and Smith, A. G. (2002) Changing potency by spontaneous fusion. *Nature*. Advance online publication, March 13, 2002 (DOI 10.1038/nature729).

Yoshida, T. (1935) o-Amidoazotoluol; uber die expermentelle annulare Lebercirrhose des Kaninchens. *Japan Pathol. Soc.* 25,409–411.

Zhang, S.-C., Wering, M., Duncan, I. D., Brustle, O., and Thompson, J. A. (2001) In vitro differentiation of transplantable neural precursors from human embryonic stem cells. *Nat. Biotech.* 19:1129–1133.

Zhang, T., Otevrel, T., Gao, Z., et al. (2001) Evidence that APC regulates survivin expression. A possible mechanism contribution to the stem cell origin of colon cancer. *Cancer Res.* 62:8664–8667.

Zhao, L. R., Duan, W. M., Reyes, M., Keene, C. D., Verfaillie, C. M., and Low, W. C. (2002) Human bone marrow stem cells exhibit neural phenotypes and ameliorate neurological deficits after grafting into the ischemic brain of rats. *Exp. Neurol.* 174:11–20.

2 Stem Cells from Early Mammalian Embryos

Common Themes and Significant Differences

VIRGINIA E. PAPAIOANNOU, PhD
AND ANNA-KATERINA HADJANTONAKIS, PhD

A description of the potential of cells during embryonic development indicates that the first restriction of the totipotentiality in the zygote occurs during formation of the blastocyst when the outer layer of cells forms the trophectoderm, which will give rise to the placenta, and the inner cells form the inner cell mass (ICM), which will develop into the fetus and fetal membranes. At this point, the ICM cells lose the ability to form trophectoderm and are considered pluripotent in that they can form all the cells of the body of the embryo, but not the placenta. A population of pluripotent cells persists for several days, but later in development, pluripotency is limited to the primordial germ cells (PGCs), which will eventually give rise to the gametes. A variety of methods have been devised to harness the stem cell capacity of early embryo-derived cells in vitro. Historically, mouse teratocarcinoma stem cells and embryonal carcinoma stem cells were the first to be derived. They were isolated from spontaneously occurring germ cell tumors or from tumors derived from embryo ectopic explants, respectively. They can be propagated indefinitely in vitro and have the capacity to differentiate into many cell types. More recently, three stem cell types have been isolated directly from the early embryo, without the intervening tumor growth: embryonic stem (ES) cells from the ICM of the blastocyst, trophectoderm stem (TS) cells from the trophectoderm of the blastocyst or the extraembryonic ectoderm, and embryonic germ (EG) cells from the PGCs. ES cells and EG cells are similar in their pluripotency and capacity for indefinite self-renewal, whereas TS cells have a more restricted developmental potential. No truly totipotent stem cell line has yet been derived, but the phenotype of ES lines can be changed to TS under appropriate culture conditions, indicating that transdifferentiation is possible. Recently, ES and EG cell lines, which share many but not all of the characteristics of mouse stem cell lines, have been derived from human embryos.

2.1. MULTIPOTENTIAL CELLS IN EARLY MAMMALIAN DEVELOPMENT

Throughout mammalian development, there exist only a few proven totipotent cells. Following fertilization, the zygote has the potential to differentiate into all the cell types of the fetal membranes and placenta, as well as all the cell types of the body, a capacity that defines totipotency. As the cleavage divisions take place, the earliest blastomeres also share this capacity to give rise to all cell types. However, by the time the blastocyst forms, cells have become limited and fixed in their potential such that only pluripotent cells, and not totipotent cells, are present. In the course of development, the progeny cells of the zygote will continue to divide and multiply and differentiate into cell types with different characteristics. As a part of this progression, specific populations of cells will be imbued with the dual capacity of producing more cells exactly like themselves and of producing cells different from themselves through cellular differentiation. These self-renewing cells with the potential to differentiate are termed *stem cells* and are an essential part of normal growth and development. Their existence and persistence for different periods of time during embryogenesis have been the object of curiosity and the subject of intense investigation since the earliest days of embryology. Questions regarding the potency of stem cells, their determination, and the control of their differentiation have been posed in order to better understand the great mystery of embryogenesis. Recently, interest in pluripotent stem cells has taken a more practical turn as potential clinical applications in the repair or replacement of failing or damaged organs enters the realm of possibility. Whether stem cells from the embryo can be harnessed and utilized in this way has yet to be demonstrated, but the ease of their isolation, purification, and propagation, as well as the possibility of control over their ultimate differentiation, will determine the eventual usefulness of embryo-derived stem cells for these purposes.

2.2. ROLE OF EMBRYONIC STEM CELLS IN DEVELOPMENT

2.2.1. LINEAGE SPECIFICATION AND PLURIPOTENT CELLS WITHIN THE EARLY EMBRYO The totipotent cells of the cleaving mammalian embryo compact to form a morula and become radially polarized. This ball of cells gives rise to the cystic

From: *Stem Cells Handbook*
Edited by: S. Sell © Humana Press Inc., Totowa, NJ

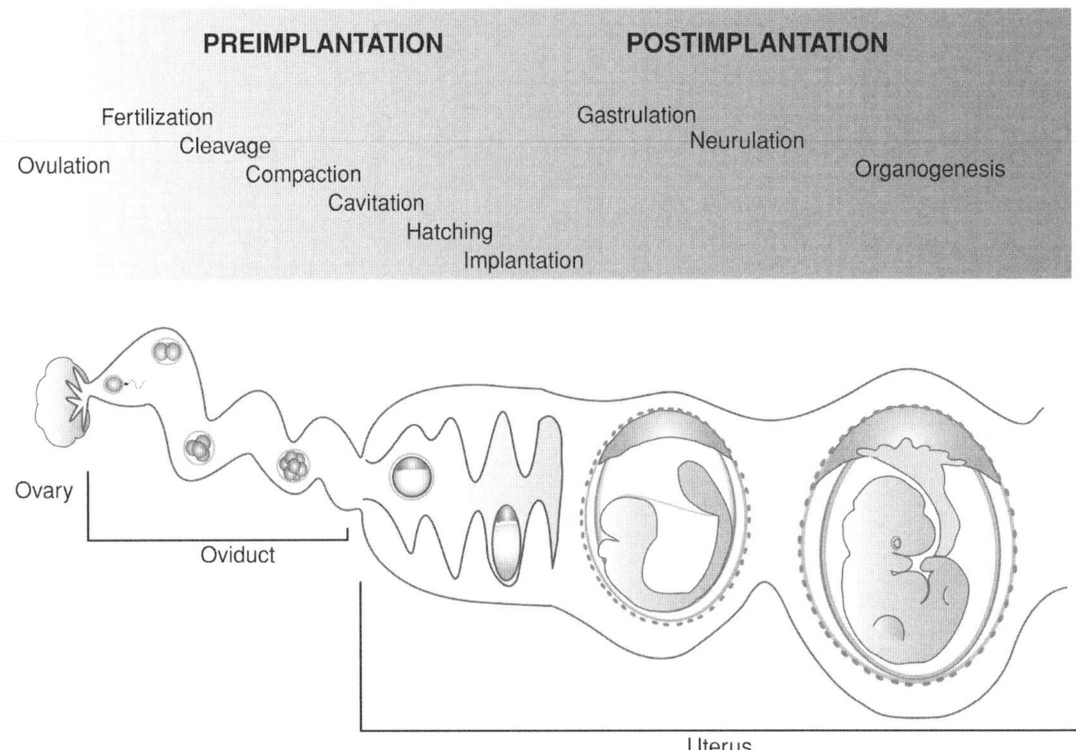

Fig. 1. Overview of mouse embryonic development: from fertilization to midgestation. After fertilization has taken place within the oviduct, the preimplantation stage embryo makes its way to the uterus. The first cell division, or cleavage, occurs approx 1 d after fertilization. Thereafter, cleavages take place approximately every 12 h. Transcription from the zygotic genome begins at the two-cell stage. Totipotency is lost by the time the embryos have entered the uterus. After hatching out of the zona pellucida, the mature blastocyst induces a decidual response and implants into the receptive uterus. The latter half of embryonic development takes place within the uterus, with a maternal–fetal interface established between the trophoblast of the embryo and the maternal deciduum, eventually forming the mature placenta.

structure called the blastocyst by a process of cavitation at about 3.5 d postcoitum (Fig. 1). The outermost cells differentiate into an epithelial layer known as the trophectoderm, whereas the inner cells, known as the inner cell mass (ICM), remain relatively undifferentiated. During this differentiation between trophectoderm and ICM, both cell types become developmentally restricted: further differentiation of trophectoderm cells is limited to a number of placental cell types, whereas the ICM cells are no longer capable of forming trophectoderm, but will make every other cell type in the normal course of development (Fig. 2). ICM cells are thus considered pluripotent, rather than totipotent, because they are restricted from the trophectoderm differentiation pathway. Experimental work elucidating these restrictions has taken into account not only what the cells do in their normal, undisturbed development (their normal fate), but also what the cells are capable of doing under experimental conditions (their potential). To date, in the mouse, at any rate, no experimental circumstances have been found that allow normal trophectoderm or ICM to deviate from their mutually exclusive developmental potential repertoires, with the sole exception that ICM cells lacking *Oct4* appear to differentiate as trophectoderm (Nichols et al., 1998).

ICM cells eventually differentiate into all cell types of the fetus and fetal membranes, apart from the trophoblast. Shortly after blastocyst formation, the primitive endoderm differentiates on the blastocoelic surface of the ICM. This cell layer is highly restricted in its fate and potential and forms only the visceral and parietal endoderm of the fetal membranes (Fig. 2). The remainder of the ICM, which is known as the epiblast, remains pluripotent, having the potential to form the entire fetal body as well as some of the fetal membranes. This potential is realized through the process of gastrulation, whereby cells rearrange by morphogenetic movements through the primitive streak to form the three germ layers: ectoderm, mesoderm, and endoderm. Pluripotency is briefly retained only in the undifferentiated ICM of the blastocyst and its later derivative, the epiblast. As tissues are progressively formed in development, embryonic cell potential gradually becomes narrowed down until almost all of the cells of the embryo have acquired a highly restricted developmental potential. The property of pluripotency then resides only in a relatively small population of cells, the primordial germ cells (PGCs), which are the progenitors of the male and female germ cells and, eventually, the gametes (Fig. 3).

2.2.2. HARNESSING THE POTENTIAL OF EMBRYONIC CELLS Embryos increase in size at a phenomenal rate while also differentiating into a variety of cell types. One means to accomplish this feat is the rapid expansion of multipotential cell populations in specific lineages at different stages of development, prior to their further differentiation. This results in what can be thought of as transitory populations of stem cells with the capacity to reproduce similar cells, with similar developmental potential (Papaioannou et al., 1978). These stem cell populations exist only briefly in the rapidly changing embryo, but their stem cell

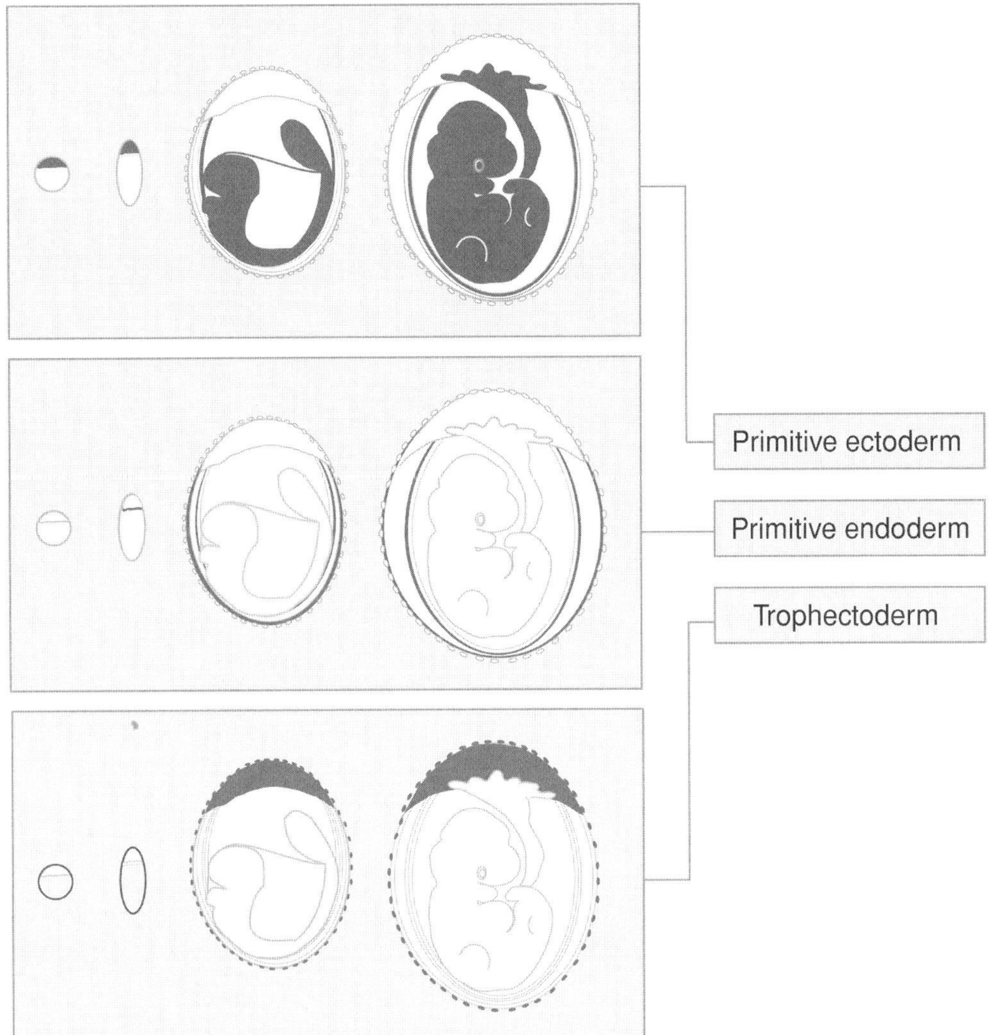

Fig. 2. Derivatives of the first three cell types to differentiate in the mouse embryo. The fetus and the extraembryonic mesoderm develop exclusively from the ICM of the blastocyst and the primitive ectoderm or epiblast of the egg cylinder–stage embryo. The primitive endoderm, which differentiates from the ICM at the late blastocyst stage, gives rise to the extraembryonic endoderm. The trophoblast of the blastocyst gives rise to extraembryonic membranes and the placenta.

capacity can be revealed and exploited by experimental intervention, which essentially "captures" the cells in a proliferative, stem cell state. Study of these cells constitutes the field of embryonic stem (ES) cell research.

ES cell research began with the study of spontaneous gonadal tumors of germ cell origin, the teratocarcinomas. In the 1970s, it was discovered that teratocarcinomas could also be derived following the grafting of early embryos into ectopic sites in adults. These tumors consist of many differentiated embryonic cell types, as well as undifferentiated stem cells that can be propagated and/or differentiated in culture, and can take part in embryonic development in chimeras in vivo (Solter et al., 1970, Stevens; 1970, 1983; Papaioannou and Rossant, 1983b). The stem cells of teratocarcinomas were termed embryonal carcinoma (EC) cells and were thought to be the counterpart of the pluripotent cells within the early embryo. Work with EC cells spurred the search for a means of deriving stem cells directly from embryos without the intervening step of tumor formation. In the following decades, stem cell lines were successfully isolated from the pluripotent cells of preimplantation embryos and also from PGCs. The resulting cell lines are called ES and embryonic germ (EG) cells, respectively (Hadjantonakis and Papaioannou, 2001). In addition, stem cells with a more restricted potential were isolated from the trophectoderm cell lineage. These cells are called trophoblast stem (TS) cells (Tanaka et al., 1998). Each of these embryo-derived stem cell types are considered in detail in the following sections.

2.3. STEM CELLS DERIVED FROM EARLY MOUSE EMBRYOS

2.3.1. EC CELLS EC cells were first derived by culturing cells from spontaneous teratocarcinomas that occur in the testes or ovaries of certain strains of mice. These spontaneous tumors arise from the germ cells and are capable of differentiation into many cell types such that the tumors resemble embryonic development gone awry. A tumor might contain several types of epithelia, areas of bone and cartilage, muscle, fat, and hair, all mixed together in

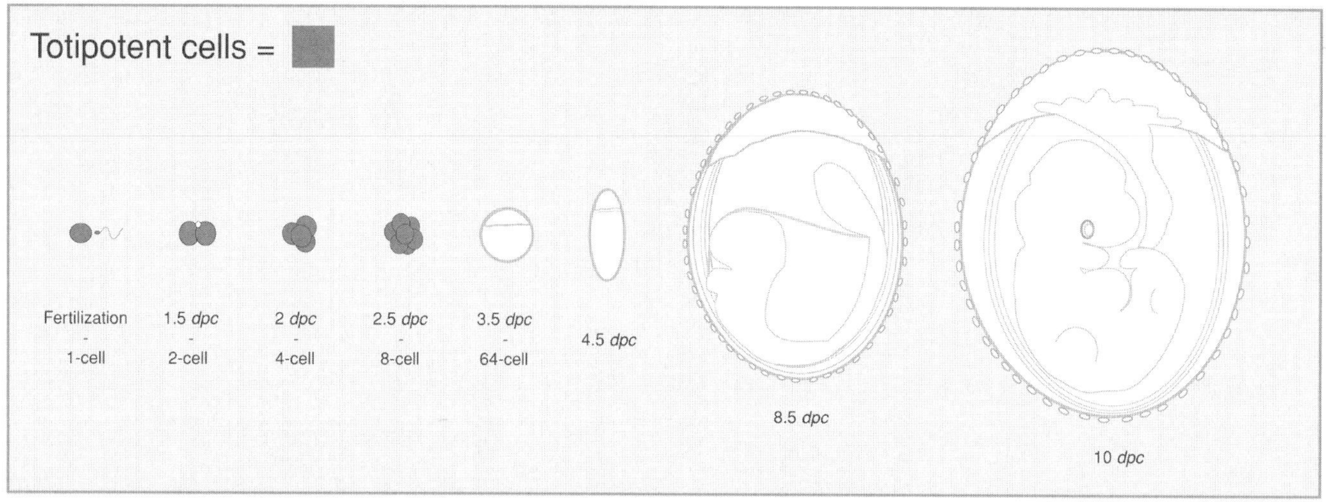

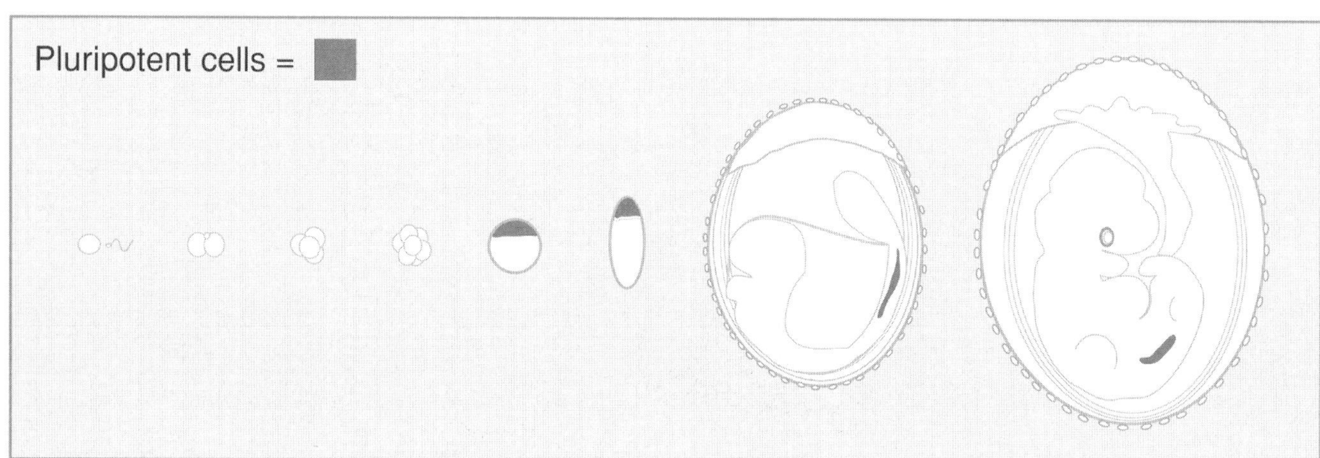

Fig. 3. Totipotent and pluripotent cells in mouse embryo. Only the zygote and the blastomeres of cleavage-stage embryos are truly totipotent. After the differentiation of the first lineage, the trophectoderm, totopotency is lost. The trophectoderm makes up the outer layer of cells in the cavitated embryo and is limited to making placental cell types. The inner cells, which are clustered at one end of the blastocyst, comprise the ICM. These cells are pluripotent; they can give rise to the derivatives of the ICM but not to trophoblast. At later stages of gestation, pluripotency resides only in the PGCs of the gonads

a chaotic array, in addition to areas of rapidly dividing, undifferentiated cells that account for the bulk of the tumor's growth and are the cells that can be cultured in vitro as stem cells (Stevens, 1970, 1983). The tumors are not metastatic but are transplantable from one animal to another, and the EC cells derived from them can be propagated indefinitely in vitro.

In the 1970s, it was discovered that normal embryos from the blastocyst through early postimplantation stages, as well as isolated genital ridges, the primordia of the gonads, give rise to identical tumors when transplanted to ectopic sites in histocompatible hosts. In other words, teratocarcinomas can be derived from all stages of embryogenesis in which a high proportion of pluripotent cells can be identified (Fig. 3). These embryo-derived tumor stem cells can be isolated and maintained in vitro (Fig. 4) and are identical to the EC cells derived from spontaneous tumors (Solter et al., 1970; Stevens, 1970, 1983). EC cells can proliferate indefinitely in the undifferentiated state and still retain the ability to differentiate under specific conditions, although there is considerable variability among different cell lines in the range of developmental potential, with some lines being very limited. One explanation for this variable restriction of potential seen in different EC cell lines is aneuploidy, a condition that might arise during the transition from embryonic growth to tumor growth. Aneuploid cells, or cells that have lost differentiation and growth checkpoints, may be at a proliferative advantage during the formation of tumors.

EC cells have close parallels to embryonic cells. The protein synthesis profile of several EC cell lines closely resembles that of the epiblast of the egg cylinder–stage embryo. When placed into the embryonic environment by injection into a blastocyst, a stringent test for normal developmental potential, EC cells resemble normal embryonic cells in their ability to participate in development to form chimeras (Brinster, 1974, Mintz and Illmensee, 1975; Papaioannou et al., 1975). However, the contribution of EC cells to chimeras is less extensive and less uniform than that of embryonic cells in a similar chimeric situation, and EC cells seldom, if ever, differentiate into germ cells (Papaioannou and Rossant, 1983a). Perhaps the most dramatic difference between EC cells and embryonic cells in chimeras is that EC cells frequently continue to proliferate in an undifferentiated state even after they have completed embryogenesis, with the result that multiple tumors

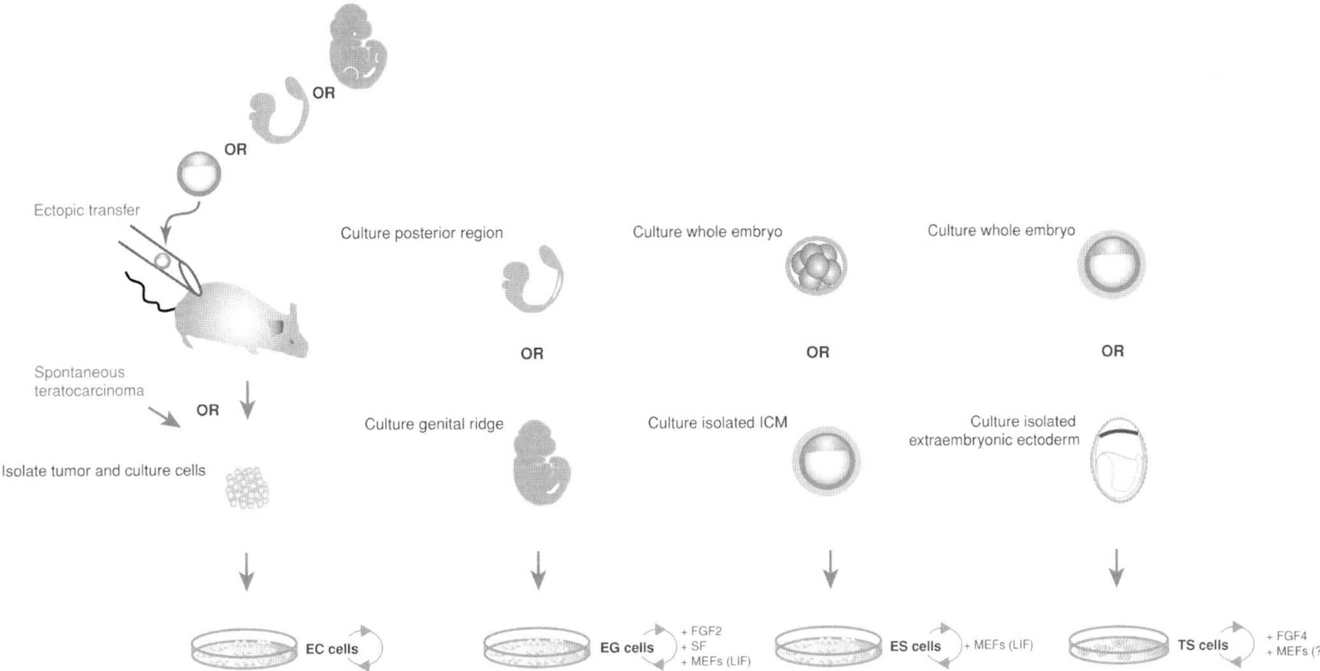

Fig. 4. Derivation of stem cell lines from early mouse embryos. EC, ES, EG, and TS cells having developmental potential mirroring that of early embryonic lineages can be isolated from embryos by a combination of dissection and specialized culture conditions (*see* text for details).

form postnatally in the chimeras. This incomplete regulation of proliferative potential by the normal embryonic environment may reflect genetic changes in the EC cells that occurred during the initial formation of teratocarcinoma or during the derivation or culture of the EC cells (Papaioannou and Rossant, 1983b). Nonetheless, the existence of stem cells in these embryo-derived tumors and the tantalizing similarity between EC cells and the putative ES cells fueled hopes of isolating stem cells directly from embryos, bypassing the tumor formation step.

2.3.2. ES CELLS In the early 1980s, three groups independently succeeded in deriving stem cell lines directly from early mouse embryos using different blastocyst culture conditions (Evans and Kaufman, 1981; Martin, 1981; Axelrod, 1984). These primary cell lines, called ES cell lines, corresponded closely to the stem cells of the ICM and epiblast predicted by EC cell work. ES cell lines can be routinely derived from embryos by in vitro outgrowth of blastocysts, followed by disruption of the ICM and culture of the disaggregated cells in the presence of the cytokine leukemia inhibitory factor (LIF) (Smith et al., 1988), or by growth on murine embryonic fibroblasts (MEFs), which provide a source of LIF (Fig. 4).

ES cells can be maintained as permanent, undifferentiated cell lines when propagated in the presence of LIF. They retain a normal, euploid karyotype, on the whole, as well as the capacity to differentiate into multiple cell types in vitro, in teratomas following transplantation to ectopic sites in host mice, and in chimeras following blastocyst injection. There is considerable strain variability in the ease of establishment of ES cell lines, with 129/Sv being the most successfully used strain (Smith, 2001). The majority of established ES cell lines are male (XY), since the XX karyotype appears to be less stable and one X chromosome is frequently lost. Gene and protein expression studies further define the identity of ES cells. For example, *Oct4* (also known as *Pou5f1*), a marker of pluripotency in the ICM of normal embryos, is expressed in ES cells and is downregulated upon differentiation. ES cells can contribute to all cell types in chimeras, with the exception of the trophectoderm and extraembryonic endoderm lineages, including the germ cells (Beddington and Robertson, 1987), a potential similar but not identical to that of ICM cells (Fig. 2). ES cells thus appear to represent the in vitro counterpart of the transient stem cell populations of the ICM and/or epiblast of the embryo.

When LIF is withdrawn during culture, ES cells differentiate and pluripotency is lost. LIF acts by binding a heterodimeric receptor complex comprising the LIF receptor (LIFR) and glycoprotein 130 (gp130). Receptor binding results in activation of gp130 signaling through the JAK/STAT pathway, which is essential for maintenance of pluripotency in vitro (Smith, 2001). Several cytokines related to LIF, including ciliary neurotrophic factor, cardiotropin 1, and oncostatin M, bind the LIFR/gp130 heterodimer and can substitute for LIF in vitro. Additionally, a combination of interleukin-6 and a soluble version of its receptor can substitute for LIFR action. This combination of ligand and receptor can activate gp130 homodimers and can be used to derive and maintain ES cells. In an interesting twist, the parallel between ES cells and the pluripotent cells of the ICM/epiblast of the embryo was challenged by the finding that gp130 signaling is not essential for early embryonic development, as mice carrying mutations in *LIF*, *LIFR*, or *gp130* develop beyond periimplantation stages (*see* Nichols et al., 2001 for a review). However, additional studies suggest that gp130 signaling is essential for maintaining pluripotency of the epiblast during implantational delay or diapause (Nichols et al., 2001), an explanation that would restore the parallel between ES and embryonic cells.

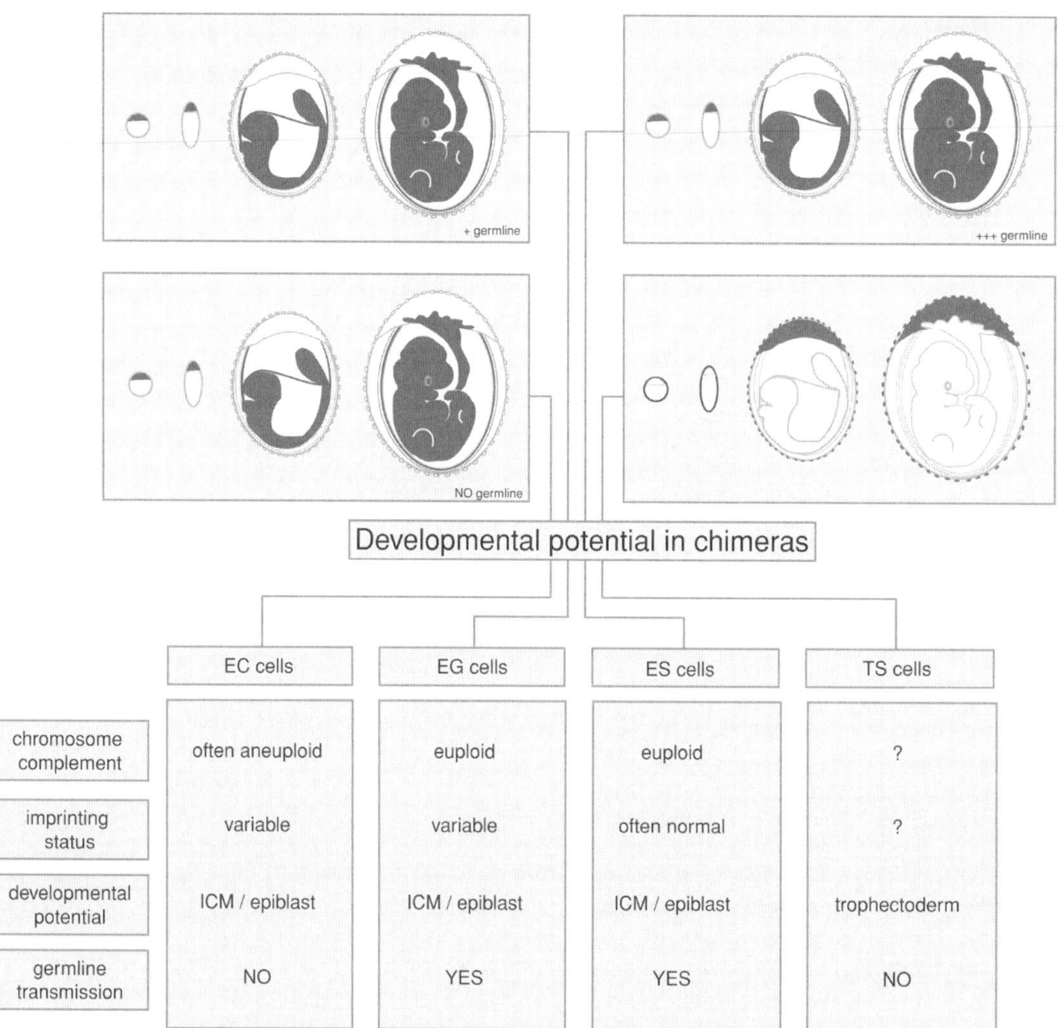

Fig. 5. Developmental potential and comparison of properties of mouse embryo–derived stem cells.

2.3.3. EG CELLS Spontaneous teratocarcinomas arising within gonads and teratocarcinomas developing from transplanted genital ridges were indications that stem cells could be derived directly from primordial germ cells prior to their differentiation into the highly specialized gametes. The discovery that Steel factor (SF), the c-*kit* ligand, is essential for the survival and proliferation of migrating PGCs in the embryo (reviewed by Morrison-Graham and Takahashi, 1993) provided a clue to the necessary in vitro culture conditions for deriving stem cells of primordial germ cell origin. In 1992, two groups were successful in exploiting the stem cell properties of primordial germ cells to derive cell lines, which became known as EG cells. As with the derivation of ES cells, LIF was an important ingredient in establishing permanently growing, pluripotent cell lines, but in addition, basic fibroblast growth factor (FGF) (also known as FGF2) and SF were required (Fig. 4). Both male and female EG cell lines can be isolated directly from PGCs prior to their migration in gastrulating embryos or following their arrival in the genital ridges (Matsui et al., 1992; Resnick et al., 1992). These cells have many characteristics of ES cells with respect to their differentiation potential in vitro (Rohwedel et al., 1996; Ohtaka et al., 1999) and their contribution in chimeras, and, like ES cells, are capable of contributing to the germ line of chimeric mice (Labosky et al., 1994; Stewart et al., 1994)(Fig. 5).

The status of imprinted genes is one important respect in which EG cells differ from ES cells. The expression of imprinted genes is dependent on their parental origin as reflected in the heritable, differential methylation of maternally or paternally derived alleles (Surani, 1998; Howell et al., 2001). The imprint is, however, reversible, as it is erased and established anew in the germ cells at each generation. When and how this occurs is still a subject of intense investigation, but it appears to have profound consequences for the expression of genes in EG cell lines. Some chimeras developing with EG cell contributions are normal and transmit the EG cell–derived genotype to the next generation. Others, however, show fetal overgrowth and skeletal abnormalities, features characteristic of, although less severe than, chimeras made with androgenetically derived ES cells, which have a paternal imprint (Labosky et al., 1994; Tada et al., 1998). This observation indicates variability in the expression of imprinted genes in independently derived EG cell lines. EG cell lines derived from early PGCs and those derived from later genital ridges show differences in the methylation status of at least one gene, *Igf2r*, leading to the conclusion that the imprint is erased between these two developmental time points, and the later-derived lines have either no imprint or a paternal imprint (Labosky et al., 1994). A survey of a larger number of imprinted genes found a similar paternal imprint pattern in male and female EG lines derived from later PGCs, although there were several exceptions in which genes showed a pattern similar to normal somatic tissue (Tada et al., 1998). With respect to experimental or therapeutic uses of EG cells, this important variable in the state of imprinted genes is essential to consider.

2.3.4. TS CELLS At first glance, the trophectoderm would appear to be a highly differentiatied tissue. Certainly, within the trophectoderm, many cells rapidly become terminally differentiated, postmitotic giant cells that undergo endoreduplication of DNA. However, cells directly in contact with the ICM, which are known as the polar trophectoderm, normally remain diploid and continue dividing in a relatively undifferentiated state, eventually forming the ectoplacental cone and extraembryonic ectoderm, which in turn form components of the placenta, including the secondary trophoblast giant cells (Fig. 2). The polar trophectoderm constitutes a limited-potential stem cell population, balancing proliferation (self-renewal) with differentiation into highly specialized, physiologically active, postmitotic cells.

It has long been known that the maintenance of proliferation in the polar trophectoderm is dependent on signals from the ICM and its later derivatives, and that without these signals, the cells differentiate into giant cells or other terminally differentiated cells of placental lineages (Gardner et al., 1973; Rossant and Ofer, 1977; Rossant and Tamura-Lis, 1981). Signaling through the FGF pathway has been strongly implicated in mediating the interaction between ICM and trophectoderm that maintains a proliferative cell population. This evidence comes both from the expression patterns of the FGF ligands and receptors (FGFRs), and from mutant and transgenic mice. In the case of the *Fgf4* and *FGFR2* null embryos, no diploid trophoblast cells are detected either in vivo or in embryo outgrowths in vitro, as all the cells become trophoblast giant cells (Feldman et al., 1995; Arman et al., 1998,). A presumed hypomorphic allele of *FGFR2* allows survival past the peri-implantation period, but embryos die later in gestation with multiple defects, including a deficiency of trophoblast cells and a complete lack of the labyrinthine component of the placenta (Xu et al., 1998). Expression of a transgenic, dominant-negative FGF receptor in polar trophectoderm causes the cells to cease division and differentiate into trophoblastic giant cells (Chai et al., 1998). Another line of evidence comes from studies on a targeted mutation in *Oct4*, a gene coding for a transcription factor that synergizes with Sox2 to cooperatively bind the *Fgf4* promoter and regulate gene expression. Oct4-deficient embryos develop to the blastocyst stage, but the inner cells differentiate along the trophoblast lineage and trophoblast proliferation is not maintained. When FGF4 is added to cultures of the inner cells of *Oct4* mutant embryos, ICM pluripotency is not restored, but, instead, dividing, undifferentiated cells that appear to be diploid trophoblast cells emerge from the differentiated trophoblast layer (Nichols et al., 1998).

These observations pointed the way to the in vitro "capture" of a stem cell for the trophectoderm lineages by implicating FGF4 in the maintenance of the diploid trophectoderm precursors. FGF4 added to the culture medium of isolated extraembryonic ectoderm suppresses differentiation into giant cells and maintains the population of undifferentiated trophoblast precursors (Nichols et al., 1998). Furthermore, culture of extraembryonic ectoderm on MEFs with the addition of FGF4 facilitates the isolation of diploid epithelial cell lines that are capable of indefinite growth in vitro (Fig. 4). These cell lines differentiate into trophoblast giant cells upon the removal of either FGF4 or MEFs (Tanaka et al., 1998). When these cells, called TS cells, are tested for their developmental potential by injection into blastocysts, they contribute exclusively to trophoblast subtypes in chimeras (Tanaka et al., 1998). Gene expression studies confirm the molecular identity of TS cells and indicate that as they differentiate in vitro, they closely recapitulate the gene expression profile of the trophoblast lineage in vivo (Kunath et al., 2001), lending further support to the notion that these cells are indeed the in vitro counterparts of the proliferating polar trophectoderm.

TS cells can be derived from blastocysts, from the extraembryonic ectoderm at 6 to 7 d postcoitum, and from the chorionic ectoderm of 7.5 d postcoitum embryos at high efficiency regardless of the strain of mouse or sex of the embryo (Tanaka et al., 1998; and our unpublished observations) (Fig. 4). FGF1 or FGF2 can substitute for FGF4 in vitro, but the factor(s) supplied by the MEFs has not yet been identified. However, that MEF-conditioned medium can substitute for MEFs, suggests that a soluble factor is involved (Tanaka et al., 1998).

2.3.5. COMPARISON OF EMBRYO-DERIVED STEM CELLS Several different stem cell types representing either the developmental potential of the trophectoderm or the ICM/epiblast have been isolated from early mammalian embryos and "captured" in vitro (Figs. 4 and 5). The culture conditions that lead to the successful culture of embryo-derived stem cells are different for each stem cell type, and the properties of the in vitro cells reflect their origin from embryonic cells with different potential. Nonetheless, stem cell potential may be different in vivo than it is in vitro. For example, ES and EG cells that mirror the developmental potential of the ICM/epiblast of the embryo are capable of differentiating into extraembryonic endoderm in vitro. In vivo in chimeras, however, neither of these stem cells contributes to the extraembryonic endoderm lineages. Thus, both types of stem cell lines recapitulate in vitro the developmental potential of the ICM, which still has the potential for extraembryonic endoderm differentiation, but in vivo they recapitulate the developmental potential of the later epiblast, which does not differentiate into extraembryonic endoderm.

No truly totipotent cell type representing the developmental potential of the zygote or early blastomeres of the cleavage-stage embryo has yet been isolated and captured in vitro. Similarly, no cell line representing a stem cell population for the primitive endoderm, the third primitive lineage present in the blastocyst (Fig. 2), has been established to date. However, given the substantial progress made in recent years, it may be just a matter of time until conditions are found that will favor the derivation of endodermal stem cells or even totipotent ES cells.

2.3.6. STEM CELLS FROM CLONED EMBRYOS—NUCLEAR TRANSFER ES CELLS Cloning is a mode of asexual reproduction resulting in offspring bearing the identical nuclear genome to their parent. In recent years, mammalian cloning has been achieved by the introduction of somatic cell nuclei into enucleated oocytes. Wakayama et al. (2001) were the first to report the derivation of ES cell lines from cloned blastocysts (Fig. 6). These cells are referred to as nuclear transfer (NT) ES cells. NT ES cells are identical in their developmental potential to normal ES

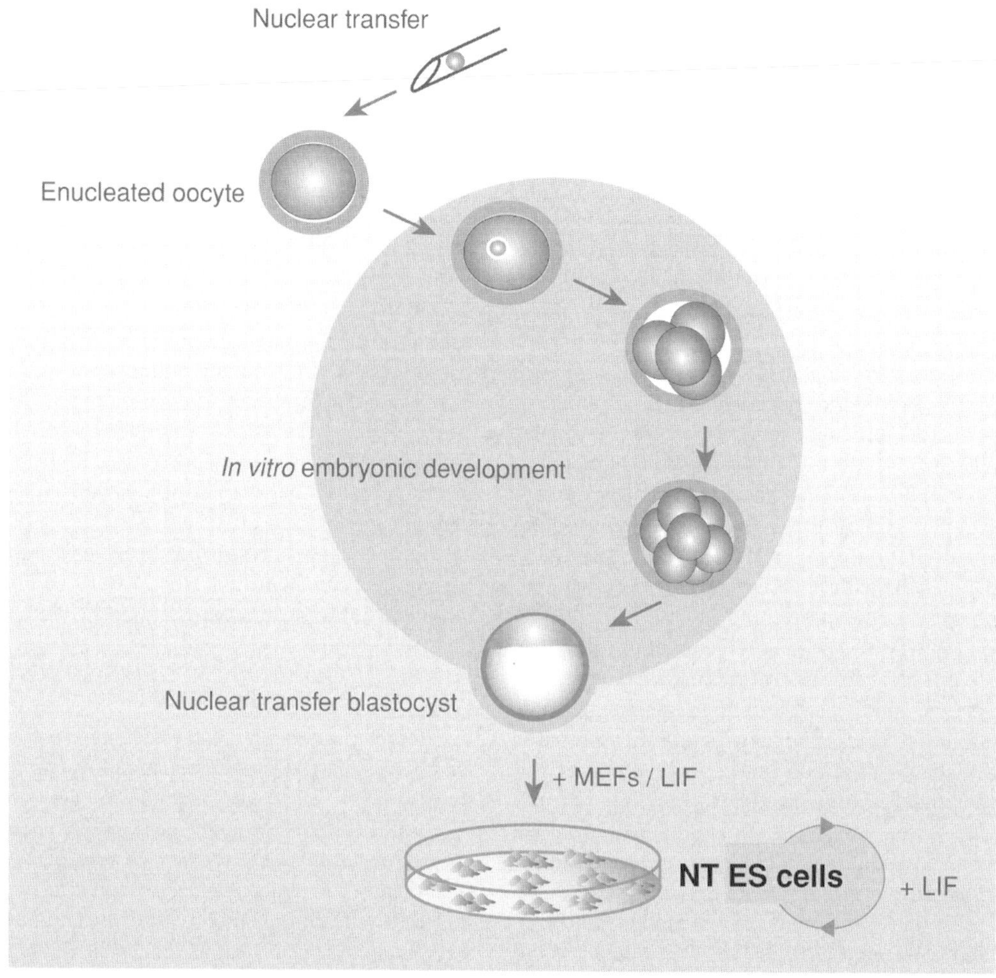

Fig. 6. NT ES cells: ES cells from cloned embryos. A somatic cell nucleus is transferred into an enucleated oocyte and allowed to develop in vitro to the blastocyst stage. The cloned embryo is then cultured on MEFs in the presence of LIF for the derivation of ES cells.

cells in that they can be maintained as pluripotent stem cells when propagated in the presence of LIF, and they are able to contribute widely in chimeras. However, little is known as yet about their imprinting status.

The ability to derive ES cells from cloned embryos allows the partnering of the technology of cloning with that of genetic engineering in ES cells and thus offers unprecedented opportunities for exploitation. In light of the ethical and scientific controversies surrounding cloning, it is worthwhile to make a distinction between cloning with the intent of generating individuals (reproductive cloning), and cloning with the intent of deriving genetically matched pluripotent ES cells (therapeutic cloning) (Hadjantonakis and Papaioannou, 2002). NT ES cells can be used in therapeutic regimes, in which ES cells generated from sick adult individuals could provide the basis for their own therapy by providing genetically matched stem cells. Genome manipulation could be undertaken to correct defects in the stem cells and the resulting cells could be used for the treatment of the sick individual through cell-based therapy (Fig. 7). In fact, a recent study describes the first successful application of therapeutic cloning, in which immunodeficient mice were treated with their own genetically repaired NT ES cells (Rideout et al., 2002).

2.3.7. EXPLOITING MOUSE ES CELLS Mouse ES cells have been invaluable tools in embryological studies of cell fate and cell lineage, and they have also provided a versatile tool for gene manipulation. Pluripotent ES cells are capable of differentiation into the germ cells of a chimera, even after extensive in vitro culture, electroporation of genetic material, and drug selection. This capacity of ES cells has allowed for specific genetic changes to be engineered into ES cells, selected in vitro, and subsequently introduced into mice through germline transmission from ES cell chimeras (Robertson, 1991). The development of this "gene-targeting" technology and its exploitation during the past 15 years has been the single most important use for ES cells to date. Both directed genome alterations, such as "knock-outs," "knock-ins," single base changes, and gene replacements, as well as random alterations such as "gene traps," insertional mutagenesis, and chemically induced mutagenesis, have become commonplace and thousands of mutations have been produced (*see* Website: http://tbase.jax.org).

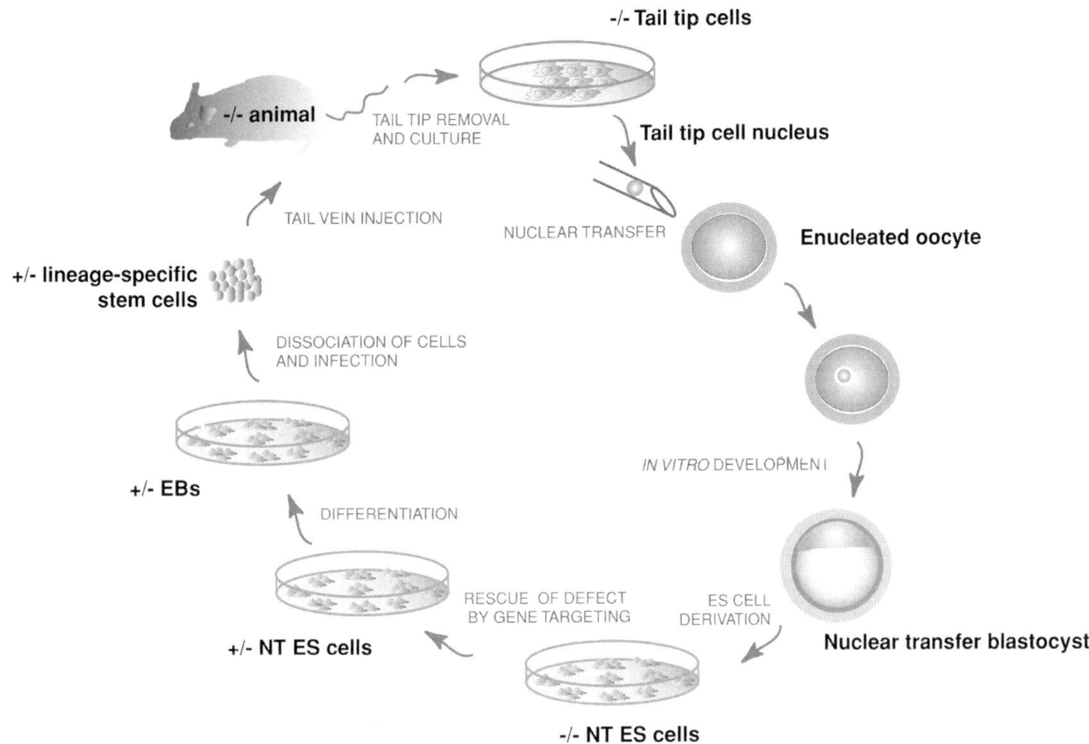

Fig. 7. Scheme for treatment of genetic disorders, combining therapeutic cloning with gene therapy. Somatic cells from a mutant (–/–) animal are cultured and used as nuclear donors for injection into enucleated oocytes. Cloned embryos are used to derive ES cells. The genetic defect is then corrected (+/–) by gene targeting in the ES cells, which are then differentiated and returned to the mutant mouse.

Once mutations are in the germline, ES cells can again be put to use in the study of mutant effects on development of embryos and/or organs. Combinations of mutant ES cells and normal cells in chimeras address questions about the cell autonomy of a mutant effect and the phenotypic consequences of the mutant effect on cell–cell interactions and induction during development. Thus, in addition to being the means through which specific mutations are introduced into the mouse germline through targeted mutagenesis, ES cells are also a tool in the analysis of those mutations.

2.4. TRANSDIFFERENTIATION OF EMBRYO-DERIVED STEM CELLS: ALTERING DEVELOPMENTAL POTENTIAL

2.4.1. ES CELLS TURN INTO TS CELLS The coordinate regulation of specific signaling pathways is pivotal for the first differentiation event involving the specification of trophectoderm vs ICM lineages. Similarly, the isolation and maintenance of stem cells representing those lineages is promoted by the regulation of intracellular signaling cascades though culturing under appropriate conditions (Fig. 4). This raises the possibility that transdifferentiation between ES and TS cells could be encouraged by controlling the on/off status of key signaling pathways and transcriptional regulators. For example, FGFR/ mitogen-activated protein kinase (MAPK) signaling OFF; gp130/JAK/STAT signaling ON; OCT4 ON characterizes the ICM lineage, whereas diploid trophoblast lineage would be specified by FGFR/MAPK signaling ON; gp130/JAK/STAT signaling OFF; OCT4 OFF. Although culture conditions that orchestrate a switch from ES to TS cells or vice versa have yet to be formulated, a genetic switch has been elucidated. Thus, when the in vitro culture conditions of ES cells are changed to those of TS cells (i.e., withdrawal of LIF and addition of FGF4 and MEF-conditioned media), the ES cells transdifferentiate. The central premise is that *Oct4* expression is extrinsically manipulated. In ES cell lines, when *Oct4* is repressed, the ability of the cells to self-renew is lost and they differentiate into what resemble trophoblast giant cells, even in the presence of LIF (Niwa et al., 2000). Moreover, if *Oct4* expression is repressed and the culture conditions promoting TS cell maintenance are substituted for those promoting ES cells, the cells continue to self-renew but now resemble TS cells in morphology and adopt a TS-cell-specific gene expression profile (Fig. 8). This suggests that ES cells have the potential to adopt a TS cell phenotype, and that *Oct4* expression is pivotal in dictating their status and developmental potential in this transdifferentiation event.

However, many questions still remain. For example, it is not yet known whether TS cells can turn into ES cells on upregulation of *Oct4* coupled with addition of LIF and withdrawal of FGF4. These studies demonstrate that culture conditions can only influence extracellular signaling, and altering extrinsic signals may indeed be necessary but not sufficient to enable a change in developmental potential such as that associated with transdifferentiation. Only by altering the expression of a pivotal intrinsic transcriptional regulator in concert with a change in culture conditions could the change of the cells' developmental potential be achieved. Thus, lineage-specific stem cells may engage a gene expression profile that results in the shutting down of pluripotency-

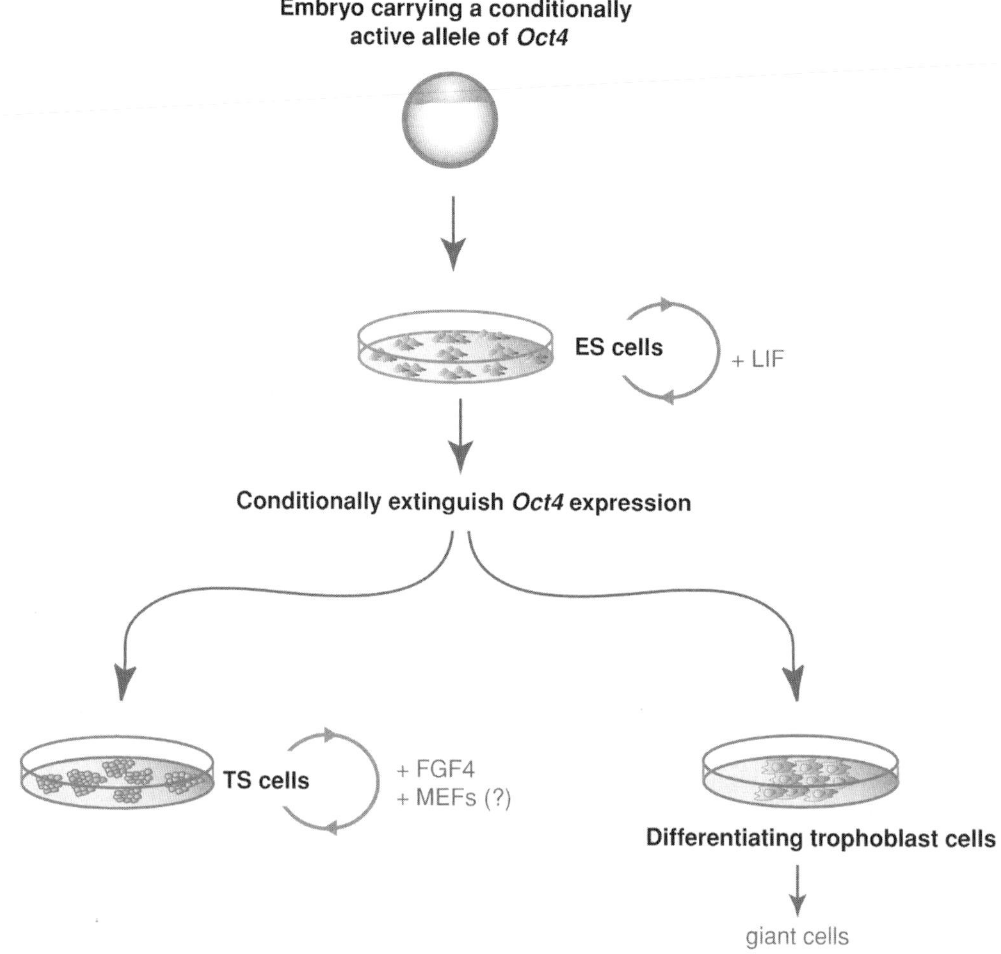

Fig. 8. Interconversion of stem cell identity. The expression of *Oct4* is pivotal for maintenance of ES cell identity. Loss of Oct4 and stimulation of the MAPK pathway by exposure to FGF4 can change the developmental potential and identity of stem cells from the primitive ectoderm to that of the trophectoderm.

associated genetic pathways, and only by genetically manipulating the expression of key intrinsic regulators can a change in developmental potential be achieved and hidden pluripotency uncovered.

2.4.2. TRANSDIFFERENTIATION OF LINEAGE-SPECIFIC STEM CELLS AND EXPANSION OF POTENCY: FACT OR ARTIFACT? It has been generally accepted that only embryo-derived stem cells possess the ability to generate the full spectrum of cell types present in the adult, and that adult-derived, lineage-specific stem cells, such as bone marrow stem cells, are more restricted in the cell types to which they can give rise. However, there have been several recent reports demonstrating an exceptional plasticity in lineage-specific stem cells when cultured under appropriate conditions (Brazelton et al., 2000; Tropepe et al., 2000). This plasticity has been attributed to a proposed transdifferentiation of adult stem cells, which broadens their developmental potential. It has therefore been suggested that many lineage-specific stem cells may indeed possess the developmental potential of embryo-derived stem cells, provided conditions are found that will trigger this transdifferentiative event, thereby unlocking a hidden pluripotency.

Pluripotency of cells, regardless of origin, would have profound implications for stem cell research and application in clinical regimes. The notion that adult-derived, lineage-restricted stem cells might be coaxed into mimicking embryo-derived stem cells has fueled arguments about stem cell research, encouraging emphasis on adult-derived, lineage-specific stem cells at the expense of work with embryonic material. However, an alternative explanation has been put forward that casts doubt on the validity of studies citing transdifferentiation as the reason behind the apparent increased plasticity observed after in vitro culture of several lineage-specific adult stem cell types. Recent studies show evidence of cell fusion and evoke this as a mechanism of achieving broadened developmental potential. We have discussed how coordinate regulation of specific signaling pathways, as dictated by the environment in vitro or in vivo, is instrumental in shaping the developmental potential of stem cells. Two groups of investigators set out to investigate this issue and to establish whether coculture with ES cells could bestow a broadened developmental potential on lineage-specific stem cells (namely, neural and bone marrow stem cells) (Terada et al., 2002; Ying et al., 2002). The ES cells did not provide a conditioned environment that provided extrinsic cues leading to a change in the developmental potential of the lineage-specific stem cells. Instead, the adult stem cells spontaneously fused with the ES cells. Such an event could certainly lead to an increased developmental potential, which may have been incorrectly attributed to transdiffer-entiation in the previous studies. The critical factor in recognizing cell fusion was the use of genetically tagged cells, thereby allowing tracking of both compartments at all stages of the experiment.

After several weeks of coculture, subclones that had markers for both the ES cells and the lineage-specific stem cells could be recovered. These subclones also appeared to have an increased DNA content and/or harbor four sex chromosomes, leading to the conclusion that the two cell types had fused during the course of in vitro coculture. Analysis of marker gene expression or introduction of these cells into chimeras demonstrated that they had in fact adopted a broader developmental potential than the parent, adult-derived stem cell, an observation that would previously have been attributed to transdifferentiation.

These findings have important and widespread ramifications, and, ultimately, much of the previous work concerning the acquired pluripotency of adult-derived stem cells will need to be reevaluated. Perhaps lineage-specific stem cells fuse with cells resident in an organism, thereby generating polyploid cells with an altered developmental potential that can contribute to a broader set of tissues. However, transdifferentiation has been noted to occur at frequencies of up to 50%, whereas cell fusion was detected at frequencies of 10^{-5}–10^{-4}. Therefore, the rarity of cell fusion may be unable to account for transdifferentiation observed in lineage-specific stem cells. We currently do not know whether cell fusion or some other such mechanism of cell conversion could operate in vivo and account for the observed increased developmental potential of lineage-specific stem cells (*see* Holden, 2003; Medvinsky and Smith, 2003). Ultimately, the reevaluation of the in vivo data, such as by isolating green fluorescent protein–positive stem cells from chimeras and karyotyping them, should shed some light on the issue.

As a final note, it is worth highlighting an interesting feature that is apparent from these studies. The fused cells appear to more closely resemble ES cells in their phenotype and developmental potential than the lineage-specific stem cells. Does this mean that the embryonic genetic program is dominant, or that given the choice, pluripotency and a broadened developmental potential is chosen over a restricted one? These observations underscore the need for continued research using stem cells derived from early embryos, as they are currently the only cells that unequivocally exhibit pluripotency.

2.5. HUMAN EMBRYONIC STEM CELLS

In recent years, many species, including humans, have been included in stem cell research. Stem cell lines have been derived from human blastocysts (Thomson et al., 1998) and PGCs (Shamblott et al., 1998) with techniques very similar to those developed for mouse embryos. These lines have been designated with the same terminology of ES and EG cells, respectively. Like their mouse counterparts, these stem cells maintain a stable karyotype and perhaps can proliferate indefinitely in vitro. They can be clonally derived and still maintain pluripotency when differentiated in vitro or as teratomas in vivo in nude mice. Their differentiation potential cannot, however, be tested in vivo in chimeras for obvious ethical reasons. Although they are basically similar, there are some differences between mouse and human ES cells in their culture requirements, the morphology of the undifferentiated cells in vitro, and the expression of surface antigens (Brinvalou et al., 2003). Furthermore, human ES cells have the potential to form differentiated trophoblast in vitro, a feature not seen for mouse ES cells (Thomson and Odorico, 2000; Odorico et al., 2001).

The status of imprinted genes has not yet been investigated in human ES or EG cells, but the differences documented for mouse stem cells and embryonic cells indicate that this will be an important consideration for future uses of stem cells. Although there are serious ethical considerations in the derivation and use of human stem cells from embryos, their potential value for therapeutic uses as a readily available, renewable source of stem cells for a variety of organs and tissues has sparked an intensive research effort as well as an active public debate. The possibility of repairing or replacing failing organs with stem cells holds enormous appeal, especially if the stem cells can be engineered in vitro to avoid the host's immune response.

2.6. PROSPECTS FOR GENE THERAPY

The extensive background of knowledge on mouse embryo–derived stem cells provides an experimental model for human ES cell research and a means of testing ideas on the biological basis of therapeutic interventions involving stem cells. Given the extensive database of genomic and developmental information and the long history of mutagenesis and ES cell experimentation, the mouse is the ideal model organism for this purpose. The differentiation pathways leading to many specific cell types have been elucidated for stem cells in vitro and are supported by studies on developmental potential as assayed through contribution in vivo (reviewed by Odorico et al., 2001). Even highly organized structures such as insulin-expressing cells with the three-dimensional structure of pancreatic islets have been differentiated from ES cells, demonstrating their potential for the assembly of functional organs (Lumelsky et al., 2001). The feasibility of using ES cells to effect cures through tissue transplantation has been tested and validated (Soria et al., 2000). Through the manipulation of histocompatibility genes, mouse stem cells can be engineered to escape the immune response (Grusby et al., 1993).

With the advent of cloning mammals by nuclear transplantation into enucleated egg cytoplasm, another technical innovation to avoid immune rejection is being tested in mice. Embryos cloned by nuclear transfer of somatic cell nuclei have been used as a source for the derivation of new ES cell lines that retain full developmental potential (Munsie et al., 2000; Wakayama et al., 2001). Using this method, unique ES cell lines could be "tailor made" through cloning from a donor's somatic nuclei for eventual stem cell therapy of that same donor. Because the nuclear material originated with the donor, there would be no danger of immune rejection (Rideout et al., 2002) (Fig. 7). Studies like these demonstrate the potential for therapeutic uses of stem cells, and it is through the use of mouse stem cells that the strategies and therapies can be developed with a view to possible application to humans.

ACKNOWLEDGMENTS

We wish to thank members of the laboratory and Sarah Goldin in particular for helpful discussion. This work was supported by grants from the National Institutes of Health (GM60561), the National Science Foundation (IBN-9985953), and the Muscular Dystrophy Association. A.-K.H. is an American Heart Association Fellow.

REFERENCES

Arman, E., Haffner-Krausz, R., Chen, Y., Heath, J. K., and Lonai, P. (1998) Targeted disruption of fibroblast growth factor (FGF) receptor 2 suggests a role for FGF signaling in pregastrulation mammalian development. *Proc. Natl. Acad. Sci. USA* 95:5082–5087.

Axelrod, H. R. (1984) Embryonic stem cell lines derived from blastocysts by a simplified technique. *Dev. Biol.* 101:225–228.

Beddington, R. S. and Robertson, E. J. (1989) An assessment of the developmental potential of embryonic stem cells in the midgestation mouse embryo. *Development* 105:733–737.

Brazelton, T. R., Rossi, F. M. V., Keshet, G. I., and Blau, H. M. (2000) From marrow to brain: Expression of neuronal phenotypes in adult mice. *Science* 290:1775–1779.

Brinster, R. L. (1974) The effect of cells transferred into the mouse blastocyst on subsequent development. *J. Exp. Med.* 140:1049–1056.

Brivanlou, A. H., Gage, F. H., Jaenisch, R., Jessell, T., Melton, D., Rossant, J. (2003) Setting standards for human embryonic stem cells. *Science* 300:913–916.

Chai, N., Patel, Y., Jacobson, K., McMahon, J., McMahon, A., and Rappolee, D. (1998) FGF is an essential regulator of the fifth cell division in preimplantation mouse embryos. *Dev. Biol.* 198:105–115.

Evans, M. J. and Kaufman, M. H. (1981) Establishment in culture of pluripotential cells from mouse embryos. *Nature* 292:154–156.

Feldman, B., Poueymirou, W., Papaioannou, V. E., DeChiara, T., and Goldfarb, M. (1995) Requirement of FGF-4 for postimplantation mouse development. *Science* 267:246–249.

Gardner, R. L., Papaioannou, V. E., and Barton, S. C. (1973) Origin of the ectoplacental cone and secondary giant cells in mouse blastocysts reconstituted from isolated trophoblast and inner cell mass. *J. Embryol. Exp. Morphol.* 30:561–572.

Grusby, M. J., Auchincloss, H., Jr., Lee, R., et al. (1993) Mice lacking major histocompatibility complex class I and class II molecules. *Proc. Natl. Acad. Sci. USA* 90:3913–3917.

Hadjantonakis, A.-K. and Papaioannou, V. E. (2001) The stem cells of early mouse embryos. *Differentiation* 68:159–166.

Hadjantonakis, A.-K. and Papaioannou, V. E. (2002) Can mammalian cloning combined with embryonic stem cell technologies be used to treat human diseases? *Genome Biol.* 3:1023.1–1023.6.

Holden, C. (2003) Cells find destiny though merger. *Science* 300:35.

Howell, C. Y., Bestor, T. H., Ding, F., et al. (2001) Genomic imprinting disrupted by a maternal effect mutation in the Dnmt1 gene. *Cell* 104: 829–838.

Kunath, T., Strumpf, D., Rossant, J., and Tanaka, S. (2001) Trophoblast stem cells. In: *Stem Cell Biology* (Marshak, D. R., et al. eds.), Cold Spring Harbor Laboratory Press, Cold Spring Harbor, NY, pp. 267–287.

Labosky, P. A., Barlow, D. P., and Hogan, B. L. (1994) Mouse embryonic germ (EG) cell lines: transmission through the germline and differences in the methylation imprint of insulin-like growth factor 2 receptor (*Igf2r*) gene compared with embryonic stem (ES) cell lines. *Development* 120:3197–3204.

Lumelsky, N., Blondel, O., Laeng, P., Velasco, I., Ravin, R., and McKay, R. (2001) Differentiation of embryonic stem cells to insulin-secreting structures similar to pancreatic islets. *Science* 292:1389–1394.

Martin, G. (1981) Isolation of a pluripotent cell line from early mouse embryos cultured in medium conditioned by teratocarcinoma stem cells. *Proc. Natl. Acad. Sci. USA* 78:7634–7638.

Matsui, Y., Zsebo, K., and Hogan, B. L. (1992) Derivation of pluripotential embryonic stem cells from murine primorial germ cells in culture. *Cell* 70:841–847.

Medvinsky, A. and Smith, A. (2003) Stem cells: Fusion brings down barriers. *Nature* 422:823–825.

Mintz, B. and Illmensee, K. (1975) Normal genetically mosaic mice produced from malignant teratocarcinoma cells. *Proc. Natl. Acad. Sci. USA* 72:3585–3589.

Morrison-Graham, K. and Takahashi, Y. (1993) Steel factor and c-Kit receptor: from mutants to a growth factor system. *BioEssays* 15: 77–83.

Munsie, M. J., Michalska, A. E., O'Brien, C. M., Trounson, A. O., Pera, M. F., and Mountford, P. S. (2000) Isolation of pluripotent embryonic stem cells from reprogrammed adult mouse somatic cell nuclei. *Curr. Biol.* 10:989–992.

Nichols, J., Chambers, I., Taga, T., and Smith, A. (2001) Physiological rationale for responsiveness of mouse embryonic stem cells to gp130 cytokines. *Development* 128:2333–2339.

Nichols, J., Zevnik, B., Anastassiadis, K., et al. (1998) Formation of pluripotent stem cells in the mammalian embryo depends on the POU transcription factor Oct4. *Cell* 95:379–391.

Niwa, H., Miyazaki, J., and Smith, A. G. (2000) Quantitative expression of Oct-3/4 defines differentiation, dedifferentiation or self-renewal of ES cells. *Nat. Genet.* 24:372–376.

Odorico, J. S., Kaufman, D. S., and Thomson, J. A. (2001) Multilineage differentiation from human embryonic stem cell lines. *Stem Cells* 19: 193–204.

Ohtaka, T., Matsui, Y., and Obinata, M. (1999) Hematopoietic develoment of primordial germ cell-derived mouse embryonic germ cells in culture. *Biochem. Biophys. Res. Commun.* 260:475–482.

Papaioannou, V. E. and Rossant, J. (1983a) Appendix Table 3, EC-embryo chimeras. In: *Teratocarcinoma Stem Cells* (Silver, L. M., et al. eds.), vol. 10, Cold Spring Harbor Laboratory Press, Cold Spring Harbor, NY, pp. 734–735.

Papaioannou, V. E. and Rossant, J. (1983b) Effects of the embryonic environment on proliferation and differentiation of embryonal carcinoma cells. *Cancer Surv.* 2:165–183.

Papaioannou, V. E., McBurney, M. W., Gardner, R. L., and Evans, M. J. (1975) Fate of teratocarcinoma cells injected into early mouse embryos. *Nature* 258:70–73.

Papaioannou, V. E., Rossant, J., and Gardner, R. L. (1978) Stem cells in early mammalian development. In: *Stem Cells and Tissue Homeostasis. British Society for Cell Biology Symposium 2* (Lord, B. I., et al., eds.), Cambridge University Press, Cambridge, MA, pp. 49–69.

Resnick, J. L., Bixler, L. S., Cheng, L., and Donovan, P. J. (1992) Long-term proliferation of mouse primordial germ cells in culture. *Nature* 359:550–551.

Rideout, W. M. III, Hochedlinger, K., Kyba, M., Daley, G. Q., and Jaenisch, R. (2002) Correction of a genetic defect by nuclear transplantation and combined cell and gene therapy. *Cell* 109:17–27.

Robertson, E. J. (1991) Using embryonic stem cells to introduce mutations into the mouse germ line. *Biol. Reprod.* 44:238–245.

Rohwedel, J., Sehlmeyer, U., Shan, J., Meister, A., and Wobus, A. M. (1996) Primordial germ cell-derived mouse embryonic germ EG cells *in vitro* resemble undifferentiated stem cells with respect to differenciation capacity and cell cycle distribution. *Cell Biol. Int.* 20: 579–587.

Rossant, J. and Ofer, L. (1977) Properties of extra-embryonic ectoderm isolated from postimplantation mouse embryos. *J. Embryol. Exp. Morphol.* 39:183–194.

Rossant, J. and Tamura-Lis, W. (1981) Effect of culture conditions on diploid to giant-cell transformation in postimplantation mouse trophoblast. *J. Embryol. Exp. Morphol.* 62:217–227.

Shamblott, M. J., Axelman, J., Wang, S., et al. (1998) Derivation of pluripotent stem cells from cultured human primordial germ cells. *Proc. Natl Acad. Sci. USA* 95:13,726–13,731.

Smith, A. (2001) Embryonic stem cells. In: *Stem cell Biology* (Marshak, D. R., et al., eds.), Cold Spring Harbor Laboratory Press, Cold Spring Harbor, NY, pp. 205–230.

Smith, A. G., Heath, J. K., Donaldson, D. D., Wong, G. G., Moreau, J., Stahl, M., and Rogers, D. (1988) Inhibition of pluripotential embryonic stem cell differentiation by purified polypeptides. *Nature* 336:688–690.

Solter, D., Skreb, N., and Damjanov, I. (1970) Extrauterine growth of mouse egg-cylinders results in malignant teratoma. *Nat. Lond.* 227: 503, 504.

Soria, B., Roche, E., Berna, G., Leon-Quinto, T., Reig, J. A., and Martin, F. (2000) Insulin-secreting cells derived from embryonic stem cells normalize glycemia in streptozotocin-induced diabetic mice. *Diabetes* 49:1–6.

Stevens, L. C. (1970) The development of transplantable teratocarcinomas from intratesticular grafts of pre- and postimplantation mouse embryos. *Dev. Biol.* 21:364–382.

Stevens, L. C. (1983) The origin and development of testicular, ovarian and embryo-derived teratomas. In: *Teratocarcinoma Stem Cells*, vol. 10, (Silver, L. M., et al., eds.), Cold Spring Harbor Laboratory Press, Cold Spring Harbor, NY, pp. 23–36.

Stewart, C. L., Gadi, I., and Bhatt, H. (1994) Stem cells from primordial germ cells can reenter the germ line. *Dev. Biol.* 161:626–628.

Surani, M. A. (1998) Imprinting and the initiation of gene silencing in the germ line. *Cell* 93:309–312.

Tada, T., Tada, M., Hilton, K., et al. (1998) Epigenotype switching of imprintable loci in embryonic germ cells. *Dev. Genes Evol.* 207: 551–561.

Tanaka, S., Kunath, T., Hadjantonakis, A.-K., Nagy, A., and Rossant, J. (1998) Promotion of trophoblast stem cell proliferation by FGF4. *Science* 282:2072–2075.

Terada, N., Hamazaki, T., Oka, M., et al. (2002) Bone marrow cells adopt the phenotype of other cells by spontaneous cell fusion. *Nature* 416: 542–545.

Thomson, J. A. and Odorico, J. S. (2000) Human embryonic stem cell and embryonic germ cell lines. *TIBTECH* 18:53–57.

Thomson, J. A., Itskovitz-Eldor, J., Shapiro, S. S., et al. (1998) Embryonic stem cell lines derived from human blastocysts. *Science* 282: 1145–1147.

Tropepe, V., Coles, B. L. K., Chiasson, B. J., et al. (2000) Retinal stem cells in the adult mammalian eye. *Science* 287:2032–2036.

Wakayama, T., Tabar, V., Rodriguez, I., Perry, A. C. F., Studer, L., and Mombaerts, P. (2001) Differentiation of embryonic stem cell lines generated from adult somatic cells by nuclear transfer. *Science* 292: 740–743.

Xu, X., Weinstein, M., Li, C., et al. (1998) Fibroblast growth factor receptor 2 (FGFR2)–mediated reciprocal regulation loop between FGF8 and FGF10 is essential for limb induction. *Development* 125: 753–765.

Ying, Q., Nichols, J., Evans, E. P., and Smith, A. G. (2002) Changing potency by spontaneous fusion. *Nature* 416:545–548.

3 Embryonic Stem Cells

Isolation and Application of Pluripotent Cells from Pregastrulation Mammalian Embryo

JOY RATHJEN, PhD AND PETER DAVID RATHJEN, PhD

Pluripotent embryonic stem (ES) cells have been isolated from both mouse and human blastocysts. In vitro, these cells can be differentiated to a diverse range of functional progenitor and terminally differentiated cells. Characterization of this capability has led to recognition of the roles of growth factors and cell-cell and cell-extracellular matrix interactions in the determination of cell fate. The development of increasingly more sophisticated differentiation protocols for the formation of mature cell types will extend the utility of ES cells to allow the production of cell populations, such as those enriched in hematopoietic cells, cardiomyocytes, and neuronal cells, for use in cell replacement therapies for human disease.

3.1. HISTORY AND BIOLOGY OF EMBRYONIC STEM CELLS

3.1.1. ISOLATION AND MAINTENANCE OF EMBRYONIC STEM CELLS FROM THE MOUSE Isolation and maintenance in vitro of pluripotent embryonic stem (ES) cells from mammalian embryos was reported independently by Evans and Kaufman (1981) and Martin (1981). Essentially similar, the two approaches utilized coculture with a fibroblast feeder layer to maintain pluripotent cells from the preimplantation mouse blastocyst in an undifferentiated state. Unlike somatic cells, karyotypically stable cell lines (Pera et al., 2000) could be established and expanded as essentially homogeneous populations of cells indefinitely in culture, without a requirement for immortalization. The pluripotential nature of the cells was identified initially by their morphological similarity to embryonal carcinoma (EC) cell lines and their differentiation capability, in vitro as embryoid bodies (EBs) (Evans and Kaufman, 1981) and in vivo as teratocarcinomas (Martin, 1981). This identification was confirmed subsequently by demonstration that ES cells express a number of pluripotent cell markers (Rosner et al., 1990; Rogers et al., 1991; Pelton et al., 2002) and can contribute to all tissues of the embryo and adult after reintroduction into the early mouse embryo (Robertson et. al., 1986; Thomas and Capecchi, 1987).

The development of feeder-independent growth conditions for ES cells, using conditioned medium from buffalo rat liver cells (Smith and Hooper, 1987), led to identification of a diffusible differentiation inhibitory activity, the cytokine leukemia inhibitory factor (LIF) (Smith et al., 1988; Williams et al., 1988), sufficient for the isolation and maintenance of pluripotent cells from the mouse blastocyst (Nichols et al., 1990; Pease et al., 1990). LIF binds the LIF^R/gp130 receptor complex, resulting in activation of STAT3, SHP-2, Hck, ERK1 and ERK2 within the pluripotent cells (Burdon et al., 1999; Ernst et al., 1999). Activation of STAT3 appears to be of particular significance because expression of a dominant interfering mutant in ES cells results in loss of pluripotence (Niwa et al., 1998), while expression of a conditionally active STAT3 is sufficient for maintenance of pluripotence (Matsuda et al., 1999). Other members of the interleukin-6 (IL-6) cytokine family, which signal through the common gp130 receptor, can functionally replace LIF in pluripotent cell maintenance (Gearing and Bruce, 1992; Conovar et al., 1993; Nichols et al., 1994; Rose et al., 1994; Yoshida et al., 1994; Pennica et al., 1995). A comprehensive review of the role of LIF signaling in ES cells is beyond the scope of this chapter but can be found in Niwa (2001).

3.1.2. ES CELL ISOLATION FROM NONMOUSE MAMMALS Although ES cell isolation from mouse was reported more than 30 yr ago, the anticipated application of this technology to other mammalian species, or even to a broad range of laboratory mouse strains, has not been realized. The isolation of ES-like cells from a diverse range of mammalian species has been reported; however, these cells were identified on the basis of morphology and a limited repertoire of markers and have not been ratified by the production of chimeras and/or contribution of ES cell progeny to the germ lineage (reviewed by Prelle et al., 1999). Proven pluripotent cell lines—i.e., those that are capable of germline colonization—have been isolated readily from the blastocysts of the permissive 129 laboratory mouse strain but only rarely from blastocysts of other laboratory mouse strains (Brook and Gardner, 1997). Improvements in isolation efficiency, which can be

From: *Stem Cells Handbook*
Edited by: S. Sell © Humana Press Inc., Totowa, NJ

achieved by removal of contaminating differentiated cell populations by surgical manipulation (Brook and Gardner, 1997) or transgenesis followed by selection (McWhir et al., 1996; Mountford et al., 1998), have allowed the establishment of ES cell lines from a number of previously nonpermissive strains of mice.

3.1.2.1. Isolation of ES Cells from Humans The isolation of pluripotent cells from human blastocysts (Thomson et al., 1998; Reubinoff et al., 2000) potentially provides a system for characterizing human embryonic development as well as a source of cells for replacement therapies. Although ethical constraints prevent formal ratification of these lines by chimera formation, human ES cells share many features with their mouse counterparts. They can be proliferated for extended periods of time without immortalization and maintain karyotypic and phenotypic stability during long-term culture. The cells express genetic and functional markers characteristic of pluripotent cells and exhibit a broad differentiation potential when differentiated as teratocarcinomas in severe combined immunodeficiency (SCID) mice or in vitro (Thomson et al., 1998; Itskovitz-Eldor et al., 2000; Reubinoff et al., 2000).

Human ES cells appear to differ from mouse ES cells in several respects. They form flat colonies with distinct cell borders compared to the characteristic tightly packed, domed colony morphology of mouse ES cells. Furthermore, while mouse ES cells can be maintained in effective homogeneity, human ES cells have a reduced phenotypic stability and the cells differentiate readily in culture, leading to heterogeneous colonies comprising pluripotent and differentiated cells (Pera et al., 2000). Finally, the cytokine requirements of the two cells differ. Establishment of human ES lines requires coculture with mouse embryonic fibroblasts (MEFs) but unlike mouse, these cells cannot be functionally replaced with LIF (Thomson et al., 1998; Reubinoff et al., 2000). Feeder-independent conditions for the maintenance of human ES cells rely on a laminin-containing matrix and conditioned medium from MEFs, suggesting that alternative signaling pathways operate in the maintenance of human ES cells (Xu et al., 2001). A possible requirement for other growth factors such as human basic fibroblast growth factor (FGF) has been suggested, particularly during growth in serum-free medium (Amit et al., 2000), but a lack of consistency among individual human ES cell isolates makes assessment of individual growth requirements difficult. Furthermore, individual human ES cell lines appear to differ in their suitability for clonal propagation or formation of EB structures during differentiation (Reubinoff et al., 2000).

3.1.3. ROLE OF LIF IN THE MAINTENANCE OF PLURIPOTENCE Responsiveness to gp130-mediated signal transduction appears to be a property peculiar to ES cell lines isolated from mice. In this regard, it is interesting that although expression of *Lif*, *LIFR*, and *gp130* has been demonstrated in the pregastrulation mouse embryo (Rathjen et al., 1990; Nichols et al., 1996), knockout experiments have failed to show a requirement for members of the IL-6 cytokine family, or the gp130 signaling pathway, in the establishment and/or maintenance of pluripotent cells during embryogenesis (Li et al., 1995; Ware et al., 1995; Yoshida et al., 1996). This suggests the existence of alternate signaling pathways for pluripotent cell elaboration within the embryo (Dani et al., 1998). Interestingly, a role for LIF in embryonic diapause has been described that may rationalize the bioactivity of this molecule on pluripotent cells in vitro (Nichols et al., 2001).

3.1.4. EMBRYONIC ORIGIN OF ES CELLS The pluripotent lineage of the mouse arises from the central blastomeres of the cleavage stage embryo and gives rise to developmentally restricted populations comprising the ectoderm, mesoderm, endoderm, and germ cells at gastrulation. During this time, the cells undergo a number of differentiation events and several distinguishable populations of pluripotent cells exist (Pelton et al., 2002; Rodda et al., 2002), any or all of which may serve as the founder population of ES cells. The window of opportunity for ES cell isolation with current technology appears to be restricted to the inner cell mass (ICM) prior to formation of the proamniotic cavity and does not encompass the pluripotent cells of the primitive ectoderm (Rossant, 1993; Brook and Gardner, 1997; J. Rathjen, unpublished). Furthermore, the analysis of extraembryonic tissues in ES cell chimeras demonstrated that ES cells contributed to both trophectoderm and extraembryonic endoderm, a developmental potential consistent with ICM but not later primitive ectoderm (Beddington and Robertson, 1989).

The blastocyst comprises three distinct cell populations, the extraembryonic endoderm and trophectoderm lineages, and the pluripotent cell lineage. Early approaches to ES cell isolation (Evans and Kaufman, 1981; Martin, 1981; Handyside et al., 1989) were unable to exclude the possibility that ES cells were derived from the extraembryonic lineages. However, explants from isolated cell populations, prepared by microsurgical dissection, demonstrated conclusively that ES cell lines arise from the pluripotent cells of the ICM (Brook and Gardner, 1997) and not from trophectoderm or endoderm.

3.2. GENE EXPRESSION MARKERS USED TO RECOGNIZE ES CELLS

3.2.1. MARKERS FOR MOUSE ES CELLS The POU-domain transcription factor Oct4 was identified as binding to a DNA motif selectively activated in undifferentiated EC cells (Rosner et al., 1990; Schöler et al., 1990). *Oct4* transcripts are restricted to the pluripotent cells of the embryo and pluripotent cells in culture and are rapidly downregulated on differentiation, although low-level *Oct4* expression can be detected in the ectoderm/neuroectodermal lineage both in vitro and in vivo (Rosner et al., 1990; Schöler et al., 1990; Rathjen et al., 2002). Gene knockout experiments suggest that *Oct4* plays an obligate role in establishment of pluripotent cell identity, with an absence of pluripotent cells within *Oct4*$^{-/-}$ blastocysts (Nichols et al., 1998). Furthermore, experimental manipulation of *Oct4* in ES cells suggests that expression of *Oct4* within certain limits is important for the maintenance of pluripotence, and that perturbation of Oct4 expression can influence the choice of cell fate following differentiation (Niwa et al., 2000). *Oct4* has been shown to regulate the expression of genes expressed within the pluripotent cell lineage such as *Rex1* and *Fgf4*; however, the functional significance, if any, of these genes in maintenance of the pluripotent state is not clear (Ben-Shushan et al., 1998; Nichols et al., 1998; Niwa, 2001). Homologs of *Oct4* are expressed in the pluripotent cells of other mammalian species including humans, suggesting a conservation of expression and possibly function within mammals (Abdel-Rahman et al., 1995; van Eijk et al., 1999; Reubinoff et al., 2000). The restricted expression pattern and functional importance in vitro and in vivo (Nichols et al., 1998; Niwa et al., 2000) make *Oct4* a powerful marker for pluripotent cells.

Other markers that are expressed broadly throughout mouse pluripotent cell populations in vitro and in vivo include the stage-specific mouse embryonic antigen-1 (SSEA-1) (Solter and Knowles,

1978), telomerase (Armstrong et al., 2000), and alkaline phosphatase enzyme activity (Hahnel et al., 1990; Pease et al., 1990).

During embryogenesis, cells of the ICM differentiate to a second pluripotent cell population, the primitive ectoderm. Several gene expression markers, such as *Rex1*, *CRTR-1*, and *Psc1*, have been identified that distinguish these two populations, provide a source of specific ES cell/ICM markers, and allow correlation of the gene expression of ES cells and their founder cells within the ICM (Pelton et al., 2002). All three markers are expressed by the pluripotent cells of the ICM and ES cells but are downregulated prior to primitive ectoderm formation in vivo or with formation of early primitive ectoderm-like (EPL) cells in vitro (Rosner et al., 1990; Pelton et al., 2002). Discrimination of ES and EPL cells from the pluripotent cells of the primordial germ lineage can be achieved with use of the marker E-cadherin/uvomorulin, which is downregulated in pluripotent cells on establishment of the germ lineage (Sefton et al., 1992).

3.2.2. MARKERS FOR HUMAN ES CELLS With the isolation of ES cells from human blastocysts came the identification of human markers expressed in pluripotent ES cells. In addition to expression of *Oct4* (Reubinoff et al., 2000), a bank of cell-surface antigens with restricted expression—SSEA-4, TRA 1-60, GCTM-2, TRA1-81, and SSEA-3—has been used in the identification of human ES cells (Thomson et al., 1998; Reubinoff et al., 2000). These antigens were initially identified on the basis of expression in pluripotent human EC cell lines or in nonhuman primate ES cell lines. Interestingly, SSEA-1, which is used in the identification of mouse ES cells, is not reactive with human ES cells (Thomson et al., 1998; Reubinoff et al., 2000; Kaufman et al., 2001). Additionally, CD90, CD133, and CD117 are expressed by human ES cells, as is *HTERT*, the human telomerase gene (Xu et al., 2001). It is frustrating that, with the exception of *Oct4* expression and some preliminary evidence demonstrating restricted activity of the mouse *Rex1* promoter in human ES cells (Eiges et al., 2001), there is so little overlap in the gene identification markers used to identify mouse and human ES cells. This precludes meaningful discussion of the comparative similarity between mouse and human ES cell populations, and between human ES cells and the pluripotent cells of the embryo.

3.3. DIFFERENTIATION OF MOUSE ES CELLS IN CULTURE

Mouse ES cell differentiation can be achieved in vitro via several techniques. The withdrawal of gp130 agonists from ES cells in adherent culture results in loss of pluripotence, as indicated by downregulation of *Oct4* expression, and formation of heterogeneous populations of cells morphologically distinct from ES cells (Smith, 1991). Although these cell populations have not been extensively characterized, they appear to comprise a limited number of cell types, predominantly a fibroblastic-like cell and a cell with characteristics of extraembryonic endoderm. Differentiation in this system is incomplete and after continued culture "nests" of pluripotent cells appear, thought to be maintained by "feedback" LIF production from the differentiated cells within the culture (Rathjen et al., 1990). Induction of ES cell differentiation in adherent culture can also be achieved by the addition of chemical agents, such as retinoic acid (RA) and dimethylsulfoxide, but, again, the repertoire of cell types produced is limited and largely uncharacterized. The projected utility of these differentiation regimes is limited by the restricted range of cell types produced and the dubious relevance to lineage establishment in vivo.

Withdrawal of gp130 agonists and cellular aggregation of ES cells in suspension culture—EB formation—results in a program of differentiation that recapitulates many aspects of early mammalian embryogenesis (Doetschman et al., 1985). Initially, outer cells of the aggregates differentiate to primitive endoderm, a differentiation event analogous to the formation of primitive endoderm by cells of the ICM exposed to the blastocelic cavity. Primitive endoderm in both EBs and the embryo differentiates to form parietal and visceral endoderm. Like pluripotent cells of the ICM, pluripotent inner cells of EBs differentiate via formation of a primitive ectoderm intermediate followed by loss of pluripotence and differentiation to the three primary germ lineages: ectoderm, mesoderm, and endoderm. With continued culture of EBs, terminally differentiated cell populations representative of the three germ lineages can be detected (Doetschman et al., 1985). The processes of lineage establishment in EBs appears to follow the events of embryogenesis and, consistent with the differentiation of ES cells in chimeric animals, suggests that ES cells can recognize and respond to the signals regulating embryonic development.

EBs differ from the embryo in the lack of positional cues, such that the cell populations form as a mass of tissue without dorsal/ventral or anterior/posterior organization, not unlike a teratocarcinoma. The complex differentiation environment, lack of organization, and inherent heterogeneity of the system have limited the use of EB technology for investigation of lineage establishment. Furthermore, although EBs form relatively large structures, the abundance of particular cell types can be low, limiting the applicability of EB differentiation to the production of cell populations for either experimental investigation or predicted uses in cell therapy (Smith, 1998). Culture of EBs in the presence of differentiation inducers, in medium supplemented with biologically derived signals such as conditioned medium or purified growth factors, or in selective medium has been used to enrich for specific cell populations or cellular lineages during ES cell differentiation. For example, RA induces the formation of neural lineages during ES cell differentiation and has been used successfully in protocols for the efficient generation of neural precursors and neurons (Fraichard et al., 1995; Strübing et al., 1995; Bain et al., 1996)

In summary, the unique properties of mouse ES cells provide a system of considerable versatility compared with other stem or progenitor cell populations:

1. *Differentiation potential*: As a population equivalent to the founder cells of the mammalian embryo, ES cells have an unrestricted capacity to form embryonic and adult cell types and provide a source of essentially normal cells for experimental analysis and transplantation.
2. *Biological relevance*: The demonstrated ability of ES cells to respond normally to all signals that regulate embryogenesis allows the development of differentiation systems that recapitulate embryonic events, and an opportunity to identify and characterize the molecular requirements for cell differentiation, cell fate specification, and cell proliferation. Developmental intermediates and developmentally distinct cellular subpopulations that accurately reflect in vivo counterparts, many of which occur transiently during embryonic development and/or in small numbers, can be identified and characterized.

3. *Experimental utility*: The availability of a system in which both the differentiation environment and genotype of a homogeneous precursor population can be manipulated with precision provides unrivaled opportunities for investigation of the molecular, genetic, and cellular basis for cell fate specification during mammalian development.

We describe next examples that illustrate how manipulation of mouse ES cell differentiation in vitro has been used to investigate questions of developmental importance, notably lineage establishment, and to generate cell populations for experimental analysis or exploration of potential cell replacement therapies.

3.4. SYSTEMS FOR MOUSE ES CELL DIFFERENTIATION: EXPLOITING PLURIPOTENCE IN CULTURE FOR THE STUDY OF EMBRYOGENESIS

3.4.1. FORMATION OF THE EARLY EXTRAEMBRYONIC LINEAGES The first differentiation event in EBs is formation of extraembryonic endoderm lineages from outer pluripotent cells. Disruption of FGF signaling, achieved by expression of a dominant-negative FGF receptor in ES cells, results in failure to form extraembryonic endoderm during EB differentiation (Chen et al., 2000), implicating FGF-mediated signaling in establishment of this lineage. Although the source and molecular identity of the FGF signal were not elucidated in this study, pluripotent cells have been shown to express FGF4 both in vitro and in vivo (Heath et al., 1989; Niswander and Martin, 1992).

Interruption of bone morphogenetic protein (BMP) signaling within EBs, by expression of a dominant negative BMP receptor, did not prevent formation of the extraembryonic endodermal lineage but interfered with the differentiation of visceral endoderm from primitive endoderm (Coucouvanis and Martin, 1999), suggesting sequential action of the FGF and BMP signaling pathways in establishment of this lineage. It has been hypothesized that the formation of visceral or parietal endoderm from primitive endoderm is determined by proximity to the pluripotent cells (Lake et al., 2000) or trophectodermal cells of the embryo, respectively (Hogan and Tilly, 1981). Consistent with this, BMP-4 is expressed by the pluripotent cells of the embryo and EBs at the time of visceral endoderm formation (Coucouvanis and Martin, 1999).

3.4.2. PLURIPOTENT CELL DEVELOPMENT After establishment of the extraembryonic lineages, pluripotent cells of the ICM differentiate to form a second pluripotent cell population, the primitive ectoderm. Primitive ectoderm formation can be recapitulated in vitro by the formation of EPL cells from ES cells (Rathjen et al., 1999), or during EB differentiation (Hébert et al., 1991; Shen and Leder, 1992). The formation of EPL cells occurs in response to medium conditioned by the human hepatocellular carcinoma cell line, HepG2 (MEDII), and results in the formation of a homogeneous population of cells with morphology, gene expression profile, cytokine responsiveness, and differentiation potential distinct from ES cells and characteristic of the embryonic primitive ectoderm (Rathjen et al., 1999; Lake et al., 2000).

Several lines of evidence from the embryo suggest that signaling from the adjacent extraembryonic endoderm is required for differentiation of ICM to primitive ectoderm (reviewed in Pelton et al., 1998). In vitro, primitive ectoderm formation in EBs lacking the primitive or visceral endoderm lineages is disrupted, evidenced by continued expression of the ICM-specific marker *Rex1* (Faloon et al., 2000) or a failure to establish the stratified epithelial layer of pluripotent cells characteristic of primitive ectoderm (Coucouvanis and Martin, 1999). Furthermore, despite their diverse embryological origin, similarities of gene expression and function between liver and visceral endoderm have been acknowledged (Meehan et al., 1984; Rossant, 1995; Barbacci et al., 1999), suggesting that EPL cell–inducing factors expressed by the HepG2 cell line substitute for inductive signals from visceral endoderm in vivo. Consistent with this the active components of MEDII, a small diffusible peptide and an extracellular matrix (ECM) component (Rathjen et al., 1999), are respectively expressed by a visceral endoderm-like cell line in vitro and localized to the basement membrane supporting the pluripotent cells in vivo and in vitro (Bettess, 2001).

Primitive ectoderm formation is accompanied by formation of the proamniotic cavity in both the embryo and EBs. Cavitation in EBs is initiated with the formation of multiple foci of cell death throughout the pluripotent cell mass that merge to form a single cavity surrounded by a monolayer of primitive ectoderm, an orchestrated process involving both cell death and survival (Coucouvanis and Martin, 1995). EBs formed from *LAMC1*$^{-/-}$ ES cells fail to cavitate or establish a basement membrane between the pluripotent and endoderm cell populations. Cavity formation and morphological rearrangement of the pluripotent cells to an epithelial layer in these EBs was restored with addition of ECM components to the differentiation environment (Murray and Edgar, 2000), suggesting a role for the basement membrane in the process of cavitation and primitive ectoderm formation. Similarly, addition of ECM components to endoderm-deficient EBs, formed from ES cells expressing the dominant-negative FGF receptor, restored cavitation (Li et al., 2001; *see also* Rathjen et al., 2002). These data suggest that the signals required for both pluripotent cell death and survival during cavitation are inherent in the basement membrane separating the extraembryonic endoderm and pluripotent cells (Rodda et al., 2002) and implicate both visceral endoderm and the basement membrane in primitive ectoderm formation.

3.4.3. LOSS OF PLURIPOTENCE: FORMATION OF THE EMBRYONIC GERM LAYERS Although germ layer formation occurs during EB development, the lack of organization and complexity of the EB environment has impeded investigation of cell specification at gastrulation in vitro. However, EPL cells provide a source of primitive ectoderm that can be manipulated to recapitulate formation of individual germ layers in vitro in the absence of alternate germ lineages and extraembryonic endoderm and has allowed evaluation of environmental signals in these processes. With variation of the differentiation environment, EPL cells can be induced to differentiate to the mesodermal or ectodermal lineages (Rathjen et al., 1999, 2001, 2002; Lake et al., 2000). Reduction of EPL cells to a single cell suspension and formation of EPL cell EBs result in the formation of multipotent mesodermal progenitors and differentiated mesodermal cell types but not cell populations representative of either visceral endoderm or ectoderm (Lake et al., 2000). Disruption of cell:cell contacts with the removal of cell:ECM association and MEDII/visceral endoderm signaling recapitulates events at the primitive streak during mesoderm formation, where the pluripotent cells delaminate, undergo an epithelial-to-mesenchymal transition, and migrate away from the endoderm.

By contrast, suspension culture of ES cells in MEDII (EBM) induces formation of EPL cells followed by differentiation of the pluripotent cells into aggregates comprising a homogeneous population of neurectoderm (Rathjen et al., 2002). In this environment, the pluripotent cells maintain cell:cell and cell:ECM associations and differentiate in the presence of a source of visceral endoderm-like signaling, recapitulating the environment of cells fated to neuroectoderm in the embryo (Quinlan et al., 1995). In chimeric embryos, *FgfR1*$^{-/-}$ pluripotent cells, which are unable to migrate through the primitive streak and thereby maintain cell:cell and cell:ECM contact, accumulate on the border of the streak, and give rise to a second site of neurectoderm formation (Ciruna et al., 1997). Evidence in vitro and in vivo therefore points to a key role for these environmental cues in pluripotent cell differentiation and determination of embryonic germ layers.

Unlike the situation in *Xenopus laevis* in which addition of exogenous factors to undifferentiated cells can result in specification of cell fate, equivalent factors have not been shown to effect differentiation of ES cells in a biologically meaningful manner. Although the reasons for this are unclear, it has been suggested that the complex nature of the culture medium, which characteristically contains high levels of serum, interferes with signaling from exogenous molecules (Johansson and Wiles, 1995). Alternatively, ES cells, with their demonstrated similarity to the ICM but not primitive ectoderm, potentially represent an inappropriate starting cell population for differentiation in response to factors involved in gastrulation. ES cell differentiation in chemically defined medium has suggested potential roles for members of the transforming growth factor-β (TGF-β) family in mesoderm induction (Johansson and Wiles, 1995; Wiles and Johansson, 1999). In this system, differentiation in unsupplemented medium results in a rapid loss of pluripotent cell markers and upregulation of the neural marker *Pax6*. Supplementation of the differentiation environment with BMP-4 and activin A resulted in upregulation of mesodermal markers and, in the case of BMP-4, nonexpression of *Pax6*. The timing of developmental arrest and lack of mesoderm formation in embryos with targeted disruptions in the BMP signaling pathway (Mishina et al., 1995; Winnier et al., 1995; Beppu et al., 2000) are consistent with this proposed role of BMP in pluripotent cell differentiation.

3.4.4. FORMATION OF SOMATIC LINEAGES The establishment of somatic lineages during EB differentiation has received enormous attention and a comprehensive description of the field is beyond the scope of this chapter. The power of ES cell–based somatic cell differentiation systems is illustrated by reference to selected examples including the hemopoietic and neural lineages.

3.4.4.1. Haemopoietic Lineages The early events of hemopoietic differentiation within EBs have been shown to proceed via consecutive, identifiable intermediate cell populations with progressively more restricted differentiation potentials. Following loss of pluripotence, establishment of the hemopoietic lineage occurs with formation of the transitional colony, a heterogeneous and developmentally plastic progenitor cell population expressing both *Flk1* and *brachyury*, with the potential to differentiate into hemopoietic, endothelial, and other mesodermal lineages (Robertson et al., 2000). The transitional colony gives rise to a second hemopoietic cell population, an *Flk1*$^{+}$ blast cell or hemangioblast, with a developmental potential restricted to the hemopoietic and endothelial lineages (Kennedy et al., 1997; Choi et al., 1998; Nishikawa et al., 1998; Robertson et al., 2000). Differentiation of ES cell–derived transitional colonies to blast cell colonies requires the Flk1 ligand vascular endothelial growth factor (VEGF) (Kennedy et al., 1997; Nakayama et al., 2000; Robertson et al., 2000), and the transcription factor *SCL/tal1* (Robertson et al., 2000). Activation of the FGF signaling pathway is not required for induction of the blast cell colony but has been shown to stimulate proliferation of the *Flk1*$^{+}$ blast cell (Faloon et al., 2001).

Initial induction of the hemopoietic lineage occurs spontaneously in fetal calf serum; however, the use of serum-free differentiation conditions has defined a role for BMP signaling in the formation of hemopoietic competent mesoderm (Johansson and Wiles, 1995; Nakayama et al., 2000; Adelman et al., 2002). Characterization of the mesodermal subpopulations formed from ES cells in response to two members of the TGF-β family suggested a role for BMP-4 in induction of cells fated to form ventral mesoderm, the precursor for hemopoietic lineages, while activin A induced dorsoanterior-like mesoderm (Johansson and Wiles, 1995). Formation of hemo-poietic cell populations in EBs can be enriched by the sequential action of BMP-4 and VEGF, which increase formation of a multipotent mesoderm progenitor cell population and further differentiation to the hemopoietic lineage, respectively (Nakayama et al., 2000; Adelman et al., 2002). This is consistent with the expression pattern of VEGF during mouse development (Miquerol et al., 1999) and compromised hemopoietic and vascular development in *VEGF-A*$^{+/-}$ mice (Ferrara et al., 1996).

The ability to form, purify, and analyze defined intermediary populations in EBs has allowed analysis of differentiation potential and the opportunity to construct cellular phylogenies. Contrary to the conclusions from conventional embryology, ES cell–derived blast cells have been shown to act as precursors for both the endothelial and mural cell populations of the vasculature, which can be induced by VEGF and platelet-derived growth factor-BB, respectively (Yamashita et al., 2000). This suggests that both lineages that comprise the vasculature can be formed from the hemopoietic progenitor cells.

3.4.4.2. Neural Lineages A comprehensive review of neural induction from ES cells can be found in Rathjen and Rathjen (2002). The most common methodology for formation of neural populations from ES cells is culture of ES cell aggregates in RA at concentrations between 10^{-7} and 10^{-6} *M*, which results in formation of populations enriched (approx 70%) in cells exhibiting properties of neural and glial lineages (Fraichard et al., 1995; Strübing et al., 1995; Bain et al., 1996; Li et al., 1998). The use of this system for investigation of neural specification is compromised by a lack of evidence for involvement of RA signaling in establishment of this lineage in vivo, and the extensive cell death that accompanies RA treatment of pluripotent cells (Li et al., 1998), which prevents definitive identification of the RA-responsive population and analysis of inductive events.

Two independent sources of biologically derived neural inducing activities have been identified (Kawasaki et al., 2000; Rathjen et al., 2002). These suggest that in addition to the requirements for maintenance of cell:cell and cell:matrix association during pluripotent cell differentiation (Rathjen et al., 2002), neural lineage formation requires additional inductive signals. Adherent coculture of ES cells with the stromal cell line PA6 results in efficient neural differentiation, with 92% of the resultant colonies expressing neural markers, and <2% of colonies containing cells expressing mesodermal markers (Kawasaki et al., 2000). Differentiation appeared to recapitulate embryogenesis with formation of an intermediate population that formed both neural ectoderm and surface ectoderm, the latter in response to exogenous BMP-4. Neural induction was shown to require a PA6-associated activity named stromal cell–derived neural inducing activity, which was in part associated with the surface of the PA6 cells, with potential involvement of a soluble component.

Differentiation of ES cells as EBM resulted in a homogeneous population of aggregates comprising a convoluted, columnar, stratified epithelium reminiscent of the embryonic neural plate/neural tube (Rathjen et al., 2002). Analysis at a cellular and molecular level showed expression of markers of neural progenitors N-Cam (97% of cells), nestin, *Sox1*, and *Sox2* within the population and a differentiation potential that encompassed neurons, glia, and neural crest, characteristics consistent with a population of neural progenitors. The formation of populations of glial or neural crest cells could be directed in homogeneous fashion in response to biologically relevant signaling molecules (Rathjen et al., 2002). In addition to the near homogeneous expression of neural markers, gene expression representative of progenitor/differentiated mesoderm or extraembryonic endoderm could not be detected at any stage during EBM development (Rathjen et al., 2002). Differentiation in response to MEDII followed closely establishment of the neural lineage in the embryo, with the sequential formation of primitive ectoderm, neural plate and neural tube equivalent cell populations, demonstrated by the synchronous temporal and sequential regulation of genetic markers for each of these populations. The synchrony inherent in this differentiation regime allows characterization of intermediate cell populations such as definitive ectoderm.

3.4.4.3. Other Lineages Identification of differentiation-inducing activities by addition of purified factors to differentiating ES cells has been limited by the repertoire of known molecules. Characterization of novel factors and factor combinations has been achieved by identification of tissue types, cell lines, and conditioned medium capable of directing the differentiation of ES cells to specific cellular lineages. For example, keratinocyte formation in EBs can be stimulated by conditioned medium from human dermal fibroblasts (Bagutti et al., 2001), an enhancement that has been attributed to the presence and activity of keratinocyte growth factor and FGF10 in the medium. Conditioned medium and cell lines have been identified that direct ES cell differentiation to specific cellular lineages, such as the hemopoietic (OP9 cells, Nakano et al., 1994) and neural lineages (PA6 cells, Kawasaki et al., 2000; MEDII, Rathjen et al., 2002), providing opportunities to identify and characterize the factors involved as a prelude to establishment of developmental significance in vivo.

Recapitulation in vitro of lineage establishment in the mammalian embryo by exploitation of ES and EPL cell differentiation has led to a more advanced understanding of these processes, as well as the genetic and environmental signals that regulate them. The generally strong correlation between regulation of developmental events in vitro and in vivo provides confidence that the experimental versatility of the ES cell system will facilitate analysis at an increasingly sophisticated level. The ability to modify the environment of transient differentiation intermediates by addition of biological factors or inductive tissues/cells can be expected to lead to identification of the cellular origin and molecular identity of inductive signals that regulate embryogenesis. ES cells and EPL cells with specific genetic modifications can be used to investigate gene function in a biologically faithful environment free from the constraints of normal embryogenesis. Additional application of ES cell differentiation systems is likely to include gene discovery in elusive cell types by genomic and proteomic approaches.

3.5. SYSTEMS FOR CELL PURIFICATION FROM COMPLEX MIXTURES: MOUSE ES CELLS AS A PROGENITOR POPULATION

As a pluripotent population capable of indefinite renewal, ES cells can also be used as a starting point for the production of experimentally relevant cell populations for analysis of downstream differentiation events, cellular characterization, and transplantation. In cases in which the route of cell generation need not follow the embryonic strategy, the level of partially or fully differentiated ES cell descendants can be enriched by a variety of selection and/or purification strategies.

Several strategies have been employed to achieve populations highly enriched in cells of predetermined identity. Enhancement of defined cell populations within EBs has been achieved by the application of selective growth conditions (Okabe et al., 1996; Tropepe et al., 2001), use of chemical inducers of differentiation (Bain et al., 1996), modification of the ES cell genome to allow expression of selectable markers under the control of cell- or lineage-restricted promoters (Klug et al., 1996; Muller et al., 2000), addition of cytokines to differentiating mixtures of cells (Choi et al., 1998; Fairchild et al., 2000), or combinations of these approaches (Li et al., 1998; Brüstle et al., 1999; Lee et al., 2000). Furthermore, coculture of ES cells with feeder layers has been used to induce differentiation and result in enriched populations of some cell lineages (Nakano et al., 1994; Kawasaki et al., 2000; Rathjen et al., 2002). Populations enriched in neural progenitors and neurons (Bain et al., 1996; Okabe et al., 1996; Li et al., 1998), cardiomyocytes (Klug et al., 1996; Muller et al., 2000), and haemapoietic lineages (Nakano et al., 1994; Choi et al., 1998; Fairchild et al., 2000) have been reported. Although these approaches result in near homogeneous populations, in general, establishment of the cellular lineages relies on initial stochastic differentiation within EBs and does not exclude the formation of other cell populations, particularly visceral endoderm. The resultant competing and potentially inappropriate differentiation-inducing signals within EBs may complicate the production of biologically representative cell populations (Rathjen and Rathjen, 2002).

Analysis of the cell populations produced by selective procedures, both in vitro and in vivo, suggests a high degree of functional relevance. Cardiomyocytes enriched from EBs were found to be similar to fetal cardiomyocytes in the characteristics and distribution of ion currents (Doevendans et al., 2000), and biophysical and pharmacological properties of the Ca^{2+} current I_{Ca} were similar to I_{Ca} in terminally differentiated cardiomyocytes (Kolossov et al., 1998). Implantation of ES cell–derived cardiomyocytes to the hearts of adult dystrophic mice showed that these cells formed intracardiac grafts stable for at least 7 wk in vivo (Klug et al., 1996). Similarly, analysis of neurons formed in vitro demonstrated these cells to be characterized by complex electrophysiology, immunocytochemistry, and expression of neuronal signaling molecules such as dopamine, and functionally comparable to neurons in vivo (reviewed in Guan et al., 2001). Implantation of these cells into the brains or spinal cords of healthy or diseased animals has shown survival; integration; and, in some cases, function in vivo (Brüstle et al., 1999; McDonald et al., 1999; Kawasaki et al., 2000; Björklund et al., 2002).

3.6. DIFFERENTIATION OF HUMAN ES CELLS IN VITRO

The developmental potential of human ES cells was demonstrated initially by teratocarcinoma formation following injection into SCID mice (Thomson et al., 1998; Reubinoff et al., 2000). The teratocarcinomas were similar to those produced from mouse ES cells, with the formation of complex tissue morphologies comprising cell populations representative of all three germ layers. Controlled differentiation of human ES cells in vitro has proved more difficult than with mouse ES cells, compromised by the heterogeneous nature of the starting population and difficul ties associated with reduction of the cells to a single cell suspension and in some lines of human ES cells by an inability to survive as EBs (Reubinoff et al., 2001). Early work suggests that regulation of human ES and mouse ES cell differentiation in vitro is similar. Exposure of human ES cells to RA results in the formation of cell populations enriched in the neural lineage (Carpenter et al., 2001, Schuldiner et al., 2001). Alternatively, neural progenitor cells can be enriched by culture in serum-free medium supplemented with FGF2, conditions determined by use of mouse ES cells to favor neural differentiation and survival of neural progenitors in culture. Neural progenitors differentiated in vitro to both neuronal and glial cells, as determined by immunocytochemistry, and survived and differentiated after implantation to the brains of mice (Reubinoff et al., 2001; Zhang et al., 2001).

The formation of mesodermal cell lineages from human ES cells in culture has also been demonstrated. Foci of beating cardiomyocytes can be detected in human ES cell–derived EBs although the frequency and timing of formation differ from mouse ES cells, with only 8% of human ES cell–derived EBs containing cardiomyocytes on d 20 of differentiation compared with approaching 40% in mouse ES cell–derived EBs by d 12. (Lake et al., 2000; Kehat et al., 2001). Despite these differences, cardio-myocytes produced from human ES cells are structurally and functionally comparable to cells formed during embryogenesis (Kehat et al., 2001). Hemopoietic colony-forming cells can be formed from human ES cells when cocultured with the murine bone marrow cell line S17 or the yolk sac endothelial cell line C166 (Kaufman et al., 2001).

Alteration of human ES cell differentiation in response to exogenous signaling molecules was demonstrated by reverse transcriptase-polymerase chain reaction analysis of cell populations resulting from differentiation of human ES cells in environments supplemented with a range of purified growth factors (Schuldiner et al., 2000). Although the outcomes of these experiments were highly heterogeneous, it is reassuring that the addition of BMP-4 resulted in a population expressing markers of skin, hemopoiesis, and extraembryonic endoderm, consistent with expectations based on differentiation of mouse ES cells (Coucouvanis and Martin, 1999; Kawasaki et al., 2000; Nakayama et al., 2000).

3.7. THERAPEUTIC APPLICATION OF ES CELL TECHNOLOGY BY CELL TRANSPLANTATION

The treatment of a number of human disease states resulting from cell loss or cell dysfunction is limited by the lack of suitable cell populations for therapeutic transplantation. ES cell differentiation is one potential source of transplantable cells for use as cell therapeutics. In combination the diverse differentiation potential, extensive proliferative capacity, and tractability for sophisticated genetic modification suggest that an ES cell population can be viewed as an unlimited number of any cells with any genetic constitution. Moreover, analysis of many of the cell populations produced from ES cells by differentiation has shown them to be not only representative of embryonic or adult populations by morphology and gene transcription, but also to share many functional characteristics and, in the case of progenitor and intermediate cells, manifest the appropriate developmental potential. The ability to form biologically functional cell populations in response to normal signaling pathways and via developmentally relevant intermediate populations suggests that ES cell differentiation is a superior route to the production of cells suited to integration, response and function within developing, developed, or diseased tissue.

Transplantation of ES cell–derived cell populations into normal and disease model animals has demonstrated the potential for these cells in disease control. ES-derived neurons and neural precursors (Brüstle et al., 1999; McDonald et al., 1999; Arnhold et al., 2000; Kawasaki et al., 2000; Liu et al., 2000; Björkland et al., 2002), cardiomyocytes (Klug et al., 1996), mast cells (Tsai et al., 2000), and insulin-secreting cells (Soria et al., 2000) have been transplanted successfully into appropriate recipient sites and shown to survive; integrate; and, to some measurable extent, function within host tissue (Brüstle et al., 1999; McDonald et al., 1999; Liu et al., 2000; Tsai et al., 2000; Björkland et al., 2002). For example, differentiating mouse ES cells and ES cell–derived neural precursors have been implanted to the brains of rats, into both ventricles and sites of solid tissue, and found to survive and incorporate into the recipient brain; migrate away from the site of injection; and differentiate to neurons, astrocytes, and oligodendrocytes (Brüstle et al., 1997; Arnold et al., 2000). ES cell–derived glial precursors implanted to spinal cords of myelin-deficient rats differentiated to myelinating oligodendrocytes and astrocytes both at the site of injection and over a distance of several millimeters from the site of injection (Brüstle et al., 1999; Liu et al., 2000). Similarly, integration and differentiation of human ES cell–derived neural precursors in the mouse brain has been demonstrated (Reubinoff et al., 2001; Zhang et al., 2001). By contrast, ES-derived hemopoietic precursors have not been reported to integrate into the hemopoietic tissue of the host or rescue and repopulate lethally irradiated mice. This presumably reflects a failure of the differentiation environment in vitro to generate functional precursors for reconstituting adult hemapoiesis.

Functional recovery of disease model animals has been reported after implantation of ES cell–derived cell populations, providing important proof of concept for the projected applications of cell therapy. Differentiating ES cells have been implanted into the sites of cellular damage in rats with 6-ODHA lesions in the striatum, a model for Parkinson disease, where they integrated, differentiated appropriately to dopaminergic neurons; and, most important, reduced the phenotypic consequences of the lesion (Björklund et al., 2002). Behavioral recovery was accompanied by alterations in brain chemistry consistent with formation of functional dopaminergic neurons. Similarly, implantation of RA–induced ES cell–derived neural progenitors to sites of lesion in mechanically damaged spinal cords effected an improvement in locomotor function, consistent with differentiation of implanted cells to neurons, astrocytes, and oligodendrocytes (McDonald et al., 1999; Liu et al., 2000).

3.8. PROSPECTS FOR THE FUTURE

Early conclusions from ES cell differentiation in vitro complement conclusions drawn from the molecular characterization of mouse development and provide confidence that the system is relevant to the study of mammalian embryogenesis. In combination with the ability to modify precisely both the differentiation environment and the genome, ES cell differentiation therefore provides a system of unrivaled potential for the characterization of lineage establishment in vivo, and for the production of a diverse range of biologically normal differentiated cell populations for experimental analysis and therapeutic application.

Directed differentiation of the cells, although in its infancy, can be used to analyze establishment of the primary germ lineages and their derivatives in simplified culture environments essentially free of the complications of endogenous signaling from contaminating cell populations. Coupled with genetic modification, these approaches provide a powerful tool for identification of inductive molecules, signaling pathways, "master genes," and transient intermediate populations in lineage formation. Combined with the more conventional approaches of mouse developmental biology, this promises a more sophisticated experimental approach to understanding mammalian embryology.

Formation of somatic lineages from ES cells, representative of differentiation intermediates or fully differentiated cell populations, provides an unlimited resource for cell characterization and function studies. Particular experimental advantages are predicted for the analysis of rare or transient cell populations in the embryo, and the analysis of gene function in circumstances in which it is obscured by embryological constraints in vivo. Establishment of the differentiation potential of intermediate populations is predicted to underpin the development of sophisticated cellular phylogenies and will allow precise tracing of the contribution of individual lineages to complex cellular structures. Furthermore, these cell populations can be implanted into animal disease models to establish proof of concept for cell therapies and assess the suitability of individual cell populations for disease control.

Isolation of pluripotent cells from the human opens the way to study the processes of human embryology without resort to human embryos. Comparative analysis of human and mouse ES cell differentiation will allow the generality of these concepts to mammalian development to be explored. The availability of an unlimited source of untransformed human cells with genomic alterations of predicted therapeutic value presents opportunities for greater understanding of human physiology and disease and is likely to underpin the development of novel cell and genetic therapies for correction of currently intractable disease states.

ACKNOWLEDGMENTS

The field of ES cell differentiation covers many areas of biology and has resulted in an enormous literature. As a consequence, we were not able to include the work of many researchers who have made valuable contributions to this field. We would, however, like to thank past and present members of the laboratory for discussions on the finer points of ES cell differentiation that have helped shape our views of the field. Research in Peter Rathjen's laboratory is supported by the Australian Research Council and the National Health and Medical Research Council. We especially acknowledge the contribution of Ray Ryce to Parkinson-related research.

REFERENCES

Abdel-Rahman, B., Fiddler, M., Rappolee, D., and Pergament, E. (1995) Expression of transcription regulating genes in human preimplantation embryos. *Hum. Reprod.* 10:2787–2792.

Adelman, C. A., Chattopadhyay, S., and Bieker, J. (2002) The BMP/BMPR/Smad pathway directs expression of the erythroid-specific EKLF and GATA1 transcription factors during embryoid body differentiation in serum-free medium. *Development* 129:539–549.

Amit, M., Carpenter, M. K., Inokuma, M. S., et al. (2000) Clonally derived human embryonic stem cell lines maintain pluripotency and proliferative potential for prolonged periods of culture. *Dev. Biol.* 227:271–278.

Armstrong, L., Lako, M., Lincoln, J., Cairns, P. M., and Hole, N. (2000) *mTert* expression correlates with telomerase activity during the differentiation of murine embryonic stem cells. *Mech. Dev.* 97: 109–116.

Arnhold, S., Lenartz, D., Kruttwig, K., et al. (2000) Differentiation of green fluorescent protein-labelled embryonic stem cell-derived neural precursor cells into Thy-1-positive neurons and glia after transplantation into adult rat striatum. *J. Neurosurg.* 93:1026–1032.

Bagutti, C., Hutter, C., Chiquet-Ehrismann, R., Fassler, R., and Watt, F. M. (2001) Dermal fibroblast-derived growth factors restore the ability of β_1 integrin-deficient embryonal stem cells to differentiate into keratinocytes. *Dev. Biol.* 231:321–333.

Bain, G., Kitchens, D., Yao, M., Huettner, J. E., and Gottlieb, D. I. (1996) Embryonic stem cells express neuronal properties in vitro. *Dev. Biol.* 168:342–357.

Barbacci, E., Reber, M., Ott, M., Breillat, C., Huetz, F., and Cereghini, S. (1999) Variant hepatocyte nuclear factor 1 is required for visceral endoderm specification. *Development* 126:4795–4805.

Beddington, R. S. P. and Robertson, E. J. (1989) An assessment of the developmental potential of embryonic stem cells in the midgestation mouse embryo. *Development* 105:733–737.

Ben-Shushan, E., Thompson, J. R., Gudas, L. J., and Bergman, Y. (1998) *Rex-1*, a gene encoding a transcription factor expressed in the early embryo, is regulated via Oct-3/4 and Oct-6 binding to an octamer site and a novel protein, Rox-1, binding to an adjacent site. *Mol. Cell. Biol.* 18:1866–1878.

Beppu, H., Kawabata, M., Hamamoto, T., et al. (2000) BMP type II receptor is required for gastrulation and early development of mouse embryos. *Dev. Biol.* 221:249–258.

Bettess, M. D. (2001) Purification, identification and characterisation of signals directing embryonic stem (ES) cell differentiation. In: *Department of Molecular Biosciences* Adelaide University, Adelaide, South Australia.

Björklund, L. M., Sanchez-Pernaute, R., Chung, S., et al. (2002) Embryonic stem cells develop into functional dopaminergic neurons after transplantation in a Parkinson rat model. *Proc. Natl. Acad. Sci. USA* 99:2344–2349.

Brook, F. A. and Gardner, R. L. (1997) The origin and efficient derivation of embryonic stem cells in the mouse. *Proc. Natl. Acad. Sci. USA* 94: 5709–5712.

Brüstle, O., Spiro, A. C., Karram, K., Choudray, K., Okabe, S. and McKay, R. D. G. (1997) In vitro–generated neural precursors participate in mammalian brain development. *Proc. Natl. Acad. Sci. USA* 94: 14,809–14,814.

Brüstle, O., Jones, K. N., Learish, R. D., et al. (1999) Embryonic stem cell–derived glial precursors: a source of myelinating transplants. *Science* 285:754–756.

Burdon, T., Stracey, C., Chambers, I., Nichols, J., and Smith, A. (1999) Suppression of SHP-2 and ERK signalling promotes self-renewal of mouse embryonic stem cells. *Dev. Biol.* 210:30–43.

Carpenter, M. K, Inokuma, M. S., Denham, J., Mujtaba, T., Chiu, C. P., and Rao, M. S. (2001) Enrichment of neurons and neural precursors from human embryonic stem cells. *Exp. Neurol.* 172:383–397.

Chen, Y., Li, X., Swarakumar, V. P., Seger, R., and Lonai, P. (2000) Fibroblast growth factor (FGF) signaling through PI 3-kinase and

Akt/PKB is required for embryoid body differentiation. *Oncogene* 19:3750–3756.

Choi, K., Kennedy, M., Kazarov, A., Papadimitriou, J. C., and Keller, G. (1998) A common precursor for hematopoietic and endothelial cells. *Development* 125:725–732

Ciruna, B. G., Schwartz, L., Harpal K., Yamaguchi, T. P., and Rossant, J. (1997) Chimeric analysis of fibroblast growth factor receptor-1 (Fgfr1) function: a role for FGFR1 in morphogenetic movement through the primitive streak. *Development* 124:2829–2841.

Conovar, J. C., Ip, N. Y., Poueymirou, W. T., et al. (1993) Ciliary neurotrophic factor maintains the pluripotentiality of embryonic stem cells. *Development* 119:559–565.

Coucouvanis, E. and Martin, G. R. (1995) Signals for death and survival: a two-step mechanism for cavitation in the vertebrate embryo. *Cell* 83:279–287.

Coucouvanis, E. and Martin, G. R. (1999) BMP signaling plays a role in visceral endoderm differentiation and cavitation in the early mouse embryo. *Development* 126:535–546.

Dani, C., Chambers, I., Johnstone, S., et al. (1998) Paracrine induction of stem cell renewal by LIF-deficient cells: a new ES cell regulatory pathway. *Dev. Biol.* 203:149–162.

Doetschman, T. C., Eistetter, H., Katz, M., Schmidt, W., and Kemler, R. (1985) The in vitro development of blastocyst-derived embryonic stem cell lines: formation of visceral yolk sac, blood islands and myocardium. *J. Embryol. Exp. Morphol.* 87:27–45.

Doevendans, P. A., Kubalak, S. W., An, R. H., Becker, D. K., Chien, K. R., and Kass, R. S. (2000) Differentiation of cardiomyocytes in floating embryoid bodies is comparable to fetal cardiomyocytes. *J. Mol. Cell. Cardiol.* 32:839–851.

Eiges, R., Schuldiner, M., Drukker, M., Yanuka, O., Itskovitz-Eldor, J., and Benvenisty, N. (2001) Establishment of human embryonic stem cell-transfected clones carrying a marker for undifferentiated cells. *Curr. Biol.* 11:514–518.

Ernst, M., Novak, U., Nicholson, S. E., Layton, J. E., and Dunn, A. R. (1999) The carboxyl-terminal domains of gp130-related cytokine receptors are necessary for suppressing embryonic stem cell differentiation. *J. Biol. Chem.* 274:9729–9737.

Evans, M. J. and Kaufman, M. H. (1981) Establishment in culture of pluripotential cells from mouse embryos. *Nature* 292:154–156.

Fairchild, P. J., Brook, F. A., Gardner, R. L., et al. (2000) Directed differentiation of dendritic cells from mouse embryonic stem cells. *Curr. Biol.* 10:1515–1518.

Faloon, P., Arentson, E., Kazarov, A., et al. (2000) Basic fibroblast growth factor positively regulates hematopoietic development. *Development* 127:1931–1941.

Ferrara, N., Carver-Moore, K., Chen, H., et al. (1996) Heterozygous embryonic lethality induced by targeted inactivation of the VEGF gene. *Nature* 380:439–442.

Fraichard, A., Chassande, O., Bilbaut, G., Dehay, C., Savatier, P., and Samarut, J. (1995) In vitro differentiation of embryonic stem cells into glial cells and functional neurons. *J. Cell Sci.* 108:3181–3188.

Gearing, D. P. and Bruce, G. (1992) Oncastatin M binds the high-affinity leukemia inhibitory factor receptor. *New Biol.* 4:61–65.

Guan, K., Chang, H., Rolletschek, A., and Wobus, A. M. (2001) Embryonic stem cell-derived neurogenesis. Retinoic acid induction and lineage selection of neuronal cells. *Cell Tissue Res.* 305:71–76.

Hahnel, A. C., Rappolee, D. A., Millan, J. L., et al. (1990) Two alkaline phosphatase genes are expressed during early development in the mouse embryo. *Development* 10:555–564.

Handyside, A. H., O'Neill, G. T., Jones, M., and Hooper, M. L. (1989) Use of BRL-conditioned medium in combination with feeder layers to isolate a diploid embryonal stem cell line. *Roux's Arch. Dev. Biol.* 198:48–55.

Heath, J. K., Paterno, G. D., Lindon, A. C., and Edwards, D. R. (1989) Expression of multiple heparin-binding growth factor species by murine embryonal carcinoma and embryonic stem cells. *Development* 107:113–122.

Hébert, J. M., Boyle, M., and Martin, G. M. (1991) mRNA localization studies suggest that murine FGF-5 plays a role in gastrulation. *Development* 112:407–415.

Hogan, B. L. and Tilly, R. (1981) Cell interactions and endoderm differentiation in cultured mouse embryos. *J. Embryol. Exp. Morphol.* 62:79–394.

Itskovitz-Eldor, J., Schuldiner, M., Karsenti, D., et al. (2000) Differentiation of human embryonic stem cells into embryoid bodies compromising the three embryonic germ layers. *Mol. Med.* 6:88–95.

Johansson, B. M and Wiles, M. V. (1995) Evidence for involvement of activin a and bone morphogenetic protein 4 in mammalian mesoderm and hematopoietic development. *Mol. Cell Biol.* 15:141–151.

Kaufman, D. S., Hanson, E. T., Lewis, R. L., Auerbach, R., and Thomson, J. A. (2001) Hematopoietic colony-forming cells derived from human embryonic stem cells. *Proc. Natl. Acad. Sci. USA* 98:10,716–10,721.

Kawasaki, H., Mizuseki, K., Nishikawa, S., et al. (2000) Induction of midbrain dopaminergic neurons from ES cells by stromal cell–derived inducing activity. *Neuron* 28:31–40.

Kehat, I., Kenyagin Karsenti, D., Snir, M., et al. (2001) Human embryonic stem cells can differentiate into myocytes with structural and functional properties of cardiomyocytes. *J. Clin. Invest.* 108:407–414.

Kennedy, M., Firpo, M., Choi, K., et al. (1997) A common precursor for primitive erythropoiesis and definitive haematopoiesis. *Nature* 386: 488–493.

Klug, M. G., Soonpaa, M. H., Koh, G. Y., and Field, L. J. (1996) Genetically selected cardiomyocytes from differentiating embryonic stem cells form stable intracardiac grafts. *J. Clin. Invest.* 98:216–224.

Kolossov, E., Fleischmann, B. K., Liu, Q., et al. (1998) Functional characteristics of ES cell–derived cardiac precursor cells identified by tissue-specific expression of the green fluorescent protein. *J. Cell. Biol.* 143:2045–2056.

Lake, J., Rathjen, J., Remiszewski, J., and Rathjen P. D. (2000) Reversible programming of pluripotent cell differentiation. *J. Cell Sci.* 113:555–566.

Lee, S. H., Lumelsky, N., Studer, L., Auerbach, J. M., and McKay, R. D. (2000) Efficient generation of midbrain and hindbrain neurons from mouse embryonic stem cells. *Nat. Biotechnol.* 18:675–679.

Li, M., Sendtner, M., and Smith, A. (1995) Essential function of LIF receptor in motor neurons. *Nature* 378:724–727.

Li, M., Pevny, L., Lovell-Badge, R., and Smith, A. (1998) Generation of purified neural precursors from embryonic stem cells by lineage selection. *Curr. Biol.* 8:971–974.

Li, X., Chen, Y., Scheele, S., et al. (2001) Fibroblast growth factor signaling and basement membrane assembly are connected during epithelial morphogenesis of the embryoid body. *J. Cell Biol.* 153:811–822.

Liu, S., Qu, Y., Stewart, T. J., et al. (2000) Embryonic stem cells differentiate into oligodendrocytes and myelinate in culture and after spinal cord transplantation. *Proc. Natl. Acad. Sci. USA* 97:6126–6131.

Martin, G. R. (1981) Isolation of a pluripotent cell line from early mouse embryos cultured in medium conditioned by teratocarcinoma stem cells. *Proc. Natl. Acad. Sci. USA* 78:7634–7638.

Matsuda, T., Nakamura, T., Nakao, K., et al. (1999) STAT3 activation is sufficient to maintain an undifferentiated state of mouse embryonic stem cells. *EMBO J.* 18:4261–4269.

McDonald, J. W., Liu, X.-Z., Qu, Y., et al. (1999) Transplanted embryonic stem cells survive, differentiate and promote recovery in injured rat spinal cord. *Nat. Med.* 5:1410–1412.

McWhir, J., Schnieke, A. E., Ansell, R., et al. (1996) Selective ablation of differentiated cells permits isolation of embryonic stem cell lines from murine embryos with a non-permissive genetic background. *Nat. Genet.* 14:223–226.

Meehan, R. R., Barlow, D. P., Hill, R. E., Hogan, B. L. M., and Hastie, R. E. (1984) Pattern of serum protein gene expression in mouse visceral yolk sac and foetal liver. *EMBO J.* 3:1881–1885.

Miquerol, L., Gertsenstein, M., Harpal, K., Rossant, J., and Nagy, A. (1999) Multiple developmental roles of VEGF suggested by a LacZ-tagged allele. *Dev. Biol.* 212:307–322.

Mishina, Y., Suzuki, A., Ueno, N., and Behringer, R. R. (1995) *Bmpr* encodes a type I bone morphogenetic protein receptor that is essential for gastrulation during mouse embryogenesis. *Genes Dev.* 9: 3027–3037

Mountford, P., Nichols, J., Zevnik, B., O' Brien, C., and Smith, A. (1998) Maintenance of pluripotential embryonic stem cells by stem cell selection. *Reprod. Fertil. Dev.* 10:527–533.

Muller, M., Fleischmann, B. K., Selbert, S., et al. (2000) Selection of ventricular-like cardiomyocytes from ES cells *in vitro*. *FASEB J.* 14: 2540–2548.

Murray, P. and Edgar, D. (2000) Regulation of programmed cell death by basement membranes in embryonic development. *J. Cell Biol.* 150: 1215–1221.

Nakano, T., Kodama, H., and Honjo, T. (1994) Generation of lymphohematopoietic cells from embryonic stem cells in culture. *Science* 265: 1098–1101.

Nakayama, N., Lee, J., and Chiu, L. (2000) Vascular endothelial growth factor synergistically enhances bone morphogenetic protein-4-dependent lymphohematopoietic cell generation from embryonic stem cells in vitro. *Blood* 95:2275–2283.

Nichols, J., Evans, E. P., and Smith, A. G. (1990) Establishment of germ-line-competent embryonic stem (ES) cells using differentiation inhibiting activity. *Development* 110:1341–1348.

Nichols, J., Chambers, I., and Smith, A. (1994) Derivation of germline competent embryonic stem cells with a combination of interleukin-6 and soluble interleukin-6 receptor. *Exp. Cell Res.* 215:237–239.

Nichols, J., Davidson, D., Taga, T., Yoshida, K., Chambers, I., and Smith, A. (1996) Complementary tissue-specific expression of LIF and LIF-receptor mRNAs in early mouse embryogenesis. *Mech. Dev.* 57:123–131.

Nichols, J., Zevnik, B., Anastassiadis, K., et al. (1998) Formation of pluripotent stem cells in the mammalian embryo depends on the POU transcription factor Oct4. *Cell* 95:379–391

Nichols. J., Chambers, I., Taga, T., and Smith, A. (2001) Physiological rationale for responsiveness of mouse embryonic stem cells to gp130 cytokines. *Development* 128:2333–2339.

Nishikawa, S. I., Nishikawa, S., Hirashima, M., Matsuyoshi, N., and Kodama, H. (1998) Progressive lineage analysis by cell sorting and culture identifies FLK1$^+$ VE-cadherin$^+$ cells at a diverging point of endothelial and hemopoietic lineages. *Development* 125:1747–1757.

Niswander, L. and Martin, G. R. (1992) Fgf-4 expression during gastrulation, myogenesis, limb and tooth development in the mouse. *Development* 114:755–768.

Niwa, H. (2001) Molecular mechanisms to maintain stem cell renewal of ES cells. *Cell Struct. Funct.* 26:137–148.

Niwa, H., Burdon, T., Chambers, I., and Smith, A. (1998) Self-renewal of pluripotent embryonic stem cells is mediated via activation of STAT3. *Genes Dev.* 12:2048–2060.

Niwa, H., Miyazaki, J., and Smith, A. G. (2000) Quantitative expression of Oct-3/4 defines differentiation, dedifferentiation or self-renewal of ES cells. *Nat. Genet.* 24:372–376.

Okabe, S., Forsberg-Nilsson, K., Spiro, A. C., Segal, M., and McKay, R. D. (1996) Development of neuronal precursor cells and functional postmitotic neurons from embryonic stem cells in vitro. *Mech. Dev.* 59:89–102.

Pease, S., Braghetta, P., Gearing, D., Grail, D., and Williams, R. L. (1990) Isolation of embryonic stem (ES) cells in media supplemented with recombinant leukemia inhibitory factor (LIF). *Dev. Biol.* 141: 344–352.

Pelton, T. A., Bettess, M. D., Lake, J., Rathjen, J., and Rathjen, P. D. (1998) Developmental complexity of early mammalian pluripotent cell populations in vivo and in vitro. *Reprod. Fertil. Dev.* 10: 535–549.

Pelton, T. A., Sharma, S., Schulz, T. C., Rathjen, J., and Rathjen, P. D. (2002) Transient pluripotent cell populations during primitive ectoderm formation: correlation of in vivo and in vitro pluripotent cell development. *J. Cell Sci.* 115:329–339.

Pennica, D., Shaw, K. J., Swanson, T. A., et al. (1995) Cardiotrophin-1: biological activities and binding to the leukemia inhibitory factor receptor/gp130 signalling complex. *J. Biol. Chem.* 270:10,915–10,922.

Pera, M. F., Reubinoff, B., and Trounson, A. (2000) Human embryonic stem cells. *J. Cell Sci.* 113:5–10.

Prelle, K., Vassiliev, I. M., Vassilieva, S. G., Wolf, E., and Wobus, A. M. (1999) Establishment of pluripotent cell lines from vertebrate species—present status and future prospects. *Cells Tissues Organs* 165:220–236.

Quinlan, G. A., Williams, E. A., Tan, S.-S., and Tam, P. L. (1995) Neurectodermal fate of epiblast cells in the distal region of the mouse egg cylinder: implications for body plan orgaization during early embryogenesis. *Development* 121:87–98.

Rathjen, J. and Rathjen, P. D. (2002) Formation of neural cell populations by differentiation of embryonic stem cells in vitro. *Sci. World* 2: 690–700.

Rathjen, J., Lake, J.-A., Bettess, M. D., Washington, J. M., Chapman, G., and Rathjen, P. D. (1999) Formation of a primitive ectoderm like cell population from ES cells in response to biologically derived factors. *J. Cell Sci.* 112:601–612.

Rathjen, J., Dunn, S., Bettess, M. D., and Rathjen, P. D. (2001) Lineage specific differentiation of pluripotent cells in vitro: a role for extraembryonic cell types. *Reprod. Fertil. Dev.* 13:15–22.

Rathjen, J., Haines, B. P., Hudson, K. M., Nesci, A., Dunn, S., and Rathjen, P. D. (2002) Directed differentiation of pluripotent cells to neural lineages: homogeneous formation and differentiation of a neurectoderm population. *Development* 129:2649–2661.

Rathjen, P. D., Nichols, J., Toth, S., Edwards, D. R., Heath, J. K., and Smith, A. G. (1990) Developmentally programmed induction of differentiating inhibiting activity and the control of stem cell populations. *Genes Dev.* 4:2308–2318.

Reubinoff, B. E., Pera, M. F., Fong, C.-Y., Trounson, A., and Bongso, A. (2000) Embryonic stem cell lines from human blastocysts: somatic differentiation in vitro. *Nat. Biotechnol.* 18:399–404.

Reubinoff, B. E., Itsykson, P., Turetsky, T., et al. (2001) Neural progenitors from human embryonic stem cells. *Nat. Biotechnol.* 19: 1134–1140.

Robertson, E., Bradley A., Kuehn, M., and Evans, M. (1986) Germ-line transmission of genes introduced into cultured pluripotential cells by retroviral vector. *Nature* 323:445–448.

Robertson, S. M., Kennedy, M., Shannon, J. M., and Keller, G. (2000) A transitional stage in the commitment of mesoderm to hematopoiesis requiring the transcription factor SCL/tal-1. *Development* 127: 2447–2459.

Rodda, S. J., Kavanagh, S. J., Rathjen, J., and Rathjen, P. D. (2002) Embryonic stem cell differentiation and the analysis of mammalian development. *Int. J. Dev. Biol.* 46:449–458.

Rogers, M. B., Hosler, B. A., and Gudas, L. (1991) Specific expression of a retinoic acid-regulated, zinc-finger gene, *Rex-1*, in preimplantation embryos, trophoblast and spermatocytes. *Development* 113: 815–824.

Rose, T. M., Weiford, D. M., Gunderson, N. L., and Bruce, A. G. (1994) Oncostatin M (OSM) inhibits the differentiation of pluripotent embryonic stem cells in vitro. *Cytokine* 6:48–54.

Rosner, M. H., Vigano, M. A., Ozato, K., et al. (1990) A POU-domain transcription factor in early stem cells and germ cells of the mammalian embryo. *Nature* 345:686–692.

Rossant, J. (1993) Immortal germ cells? *Curr. Biol.* 3:47–49.

Rossant, J. (1995) Development of the extraembryonic lineages. *Semin. Dev. Biol.* 6:237–247.

Schöler, H. R., Dressler, G. R., Balling, R., Rohdewohld, H., and Gruss, P. (1990) *Oct-4*: a germline-specific transcription factor mapping to the mouse t-complex. *EMBO J.* 9:2185–2195.

Schuldiner, M., Yanuka, O., Itskiovitz-Eldor, J., Melton, D. A., and Benvenisty, N. (2000) Effects of eight growth factors on the differentiation of cells derived from human embryonic stem cells. *Proc. Natl. Acad. Sci. USA* 97:11,307–11,312.

Schuldiner, M., Eiges, R., Eden, A., et al. (2001) Induced neuronal differentiation of human embryonic stem cells. *Brain Res.* 913:201–205.

Sefton, M., Johnson, M. H., and Clayton, L. (1992) Synthesis and phosphorylation of uvomorulin during mouse early development. *Development* 115:313–318.

Shen, M. M. and Leder, P. (1992) Leukemia inhibitory factor is expressed by the preimplantation uterus and selectively blocks primitive ectoderm formation in vitro. *Proc. Natl. Acad. Sci. USA* 89:8240–8244.

Smith, A. (1998) Cell therapy: in search of pluripotency. *Curr. Biol.* 8:R802–R804.

Smith, A. G. (1991) Culture and differentiation of embryonic stem cells. *J. Tissue Cult. Methods* 13:89–94.

Smith, A. G. and Hooper, M. L. (1987) Buffalo rat liver cells produce a diffusible activity which inhibits the differentiation of murine embryonal carcinoma and embryonic stem cells. *Dev. Biol.* 121:1–9.

Smith, A., Heath, J. K., Donaldson, D. D., et al. (1988) Inhibition of pluripotential embryonic stem cell differentiation by purified polypeptides. *Nature* 336:688–690.

Solter, D. and Knowles, B. B. (1978) Monoclonal antibody defining stage-specific mouse embryonic antigen (SSEA-1). *Proc. Natl. Acad. Sci. USA* 75:5565–5569.

Soria, B., Roche, E., Berna, G., Leon-Quinto, T., Reig, J. A., and Martin, F. (2000) Insulin-secreting cells derived from embryonic stem cells normalize glycemia in streptozotocin-induced diabetic mice. *Diabetes* 49:157–162.

Strübing, C., Ahnert-Hilger, G., Shan J., Wiedenmann, B., Hescheler, J., and Wobus, A. M. (1995) Differentiation of pluripotent embryonic stem cells into the neuronal lineage in vitro gives rise to mature inhibitory and excitatory neurons. *Mech. Dev.* 53:275–287.

Thomas, K. R. and Capecchi, M. R. (1987) Site-directed mutagenesis by gene targeting in mouse embryo-derived stem cells. *Cell* 51:503–512.

Thomson, J. A., Itskovitz-Eldor, J., Shapiro, S. S., et al. (1998) Embryonic stem cell lines derived from human blastocysts. *Science* 282: 1145–1147.

Tropepe, V., Hitoshi, S., Sirard, C., Mak, T. W., Rossant, J., and van der Kooy, D. (2001) Direct neural fate specification from embryonic stem cells: a primitive mammalian neural stem cell stage acquired through a default mechanism. *Neuron* 30:65–78.

Tsai, M., Wedemeyer, J., Ganiatsas, S., Tam, S. Y., Zon, L. I., and Galli, S. J. (2000) In vivo immunological function of mast cells derived from embryonic stem cells: an approach for the rapid analysis of even embryonic lethal mutations in adult mice in vivo. *Proc. Natl. Acad. Sci. USA* 97:9186–9190.

van Eijk, M. J., van Rooijen, M. A., Modina, S., et al. (1999) Molecular cloning, genetic mapping, and developmental expression of bovine POU5F1. *Biol. Reprod.* 60:1093–1103.

Ware, C. B., Horowitz, M. C., Renshaw, B. R., et al. (1995) Targeted disruption of the low-affinity leukemia inhibitory factor receptor gene causes placental, skeletal, neural and metabolic defects and results in perinatal death. *Development* 121:1283–1299.

Wiles, M. V. and Johansson, B. M. (1999) Embryonic stem cell development in a chemically defined medium. *Exp. Cell Res.* 247: 241–248.

Williams, R L., Hilton, D. J., Pease, S., et al. (1988) Myeloid leukaemia inhibitory factor maintains the developmental potential of embryonic stem cells. *Nature* 336:684–687.

Winnier, G., Blessing, M., Labosky, P. A., and Hogan, B. L. M. (1995) Bone morphogenetic protein-4 is required for mesoderm formation and patterning in the mouse. *Genes Dev.* 9:2105–2116.

Xu, C., Inokuma, M. S., Denham, J., et al. (2001) Feeder-free growth of undifferentiated human embryonic stem cells. *Nat. Biotechnol.* 19: 971–974.

Yamashita, J., Itoh, H., Hirashima, M., et al. (2000) Flk1-positive cells derived from embryonic stem cells serve as vascular progenitors. *Nature* 408:92–93.

Yoshida, K., Chambers, I., Nichols, J., et al. (1994) Maintenance of the pluripotent phenotype of embryonic stem cells through direct activation of gp130 signalling pathways. *Mech. Dev.* 45:163–171.

Yoshida, K., Taga, T., Saito, M., et al. (1996) Targeted disruption of gp130, a common signal transducer for the interleukin 6 family of cytokines, leads to myocardial and hematological disorders. *Proc. Natl. Acad. Sci. USA* 93:407–411.

Zhang, S.C., Wernig, M., Duncan, I. D., Brustle, O., and Thomson, J. A. (2001) In vitro differentiation of transplantable neural precursors from human embryonic stem cells. *Nat. Biotechnol.* 19: 1129–1133.

4 From Stem Cells to Functional Tissue Architecture

What Are the Signals and How Are They Processed?

SUI HUANG, MD, PhD AND DONALD E. INGBER, MD, PhD

Specific regulatory inputs from the cell's environment influence cell and tissue differentiation; soluble growth factors and insoluble extracellular matrix interactions, as well as less-specific signals, such as mechanical forces, can switch cells among different phenotypic fates. This behavior may be explained by the existence of different "attractor" states, which represent stable phenotypes, including growth, differentiation, quiescence, apoptosis, and distinct differentiation lineages. The attractor landscape, which is particular for each cell type, is determined by the underlying hard wiring of the cell's regulatory interactions. Thus, different regulatory inputs may select for proliferation, differentiation, quiescence, or apoptosis of the stem cells and the ultimate formation of mature tissues by switching among different attractor states, rather than by activating linear signaling pathways.

4.1. INTRODUCTION

Tissue development in the embryo results from the growth and differentiation of pluripotent stems cells that form increasingly specialized cell types and exhibit different "cell fates": they divide, differentiate, or die. In solid tissues with specialized form, stem cell differentiation is accompanied by progressive remodeling of preexisting extracellular matrix (ECM) scaffolds, and it is the coordination between these processes in both time and space that ensures correct pattern formation. Maintenance of tissue architecture throughout adult life also requires fine control over both tissue mass and ECM pattern integrity, which are ensured through spatially coordinated growth and differentiation of tissue-specific stem cells.

Recently, tissue-specific stem cells have been isolated and arrays of soluble cytokines and growth factors that regulate their differentiation have been described. Pluripotent stem cells that can differentiate into various cell types also have been isolated, and some can spontaneously assemble into structures that resemble mature tissues. At the same time, ECM molecules have been shown to have potent effects on cell proliferation and differentiation, and to direct assembly of three-dimensional (3D) tissues in vitro. Yet, ECM is often not considered when discussing how stem cell behavior is regulated.

The challenge for the future in the stem cell field is to understand how these cells can spontaneously generate 3D functional tissue architecture "without a template" (Pedersen, 1999). To address this challenge fully, we must answer some fundamental questions. First, what is the very nature of the signals conveyed by both soluble and insoluble factors that govern where and when cells will be switched between the different fates that are necessary for cells to organize into tissues and for tissues to organize into organs? Second, how do cells process multiple simultaneous inputs to produce a single coordinated cell fate response?

The major point of this chapter is that focusing on isolated cells and individual regulatory factors will not answer these questions. Instead, we must consider how individual cells chemically and physically interact with their local tissue environment as well as how cells use regulatory networks to process information provided by multiple environmental cues. As a first attempt to confront these issues, we present a comprehensive model of cell signaling networks that provides a new conceptual framework for understanding how many different signaling pathways integrate inside the cell and collectively give rise to meaningful cell behavior. We also describe how local cues, in particular ECM and mechanical forces that produce changes in cell shape, provide a structural basis for the self-organizing process of tissue patterning.

4.2. CONTROL OF PATTERN FORMATION

4.2.1. GLOBAL VS LOCAL REGULATORY SIGNALS The precise coordination of cell growth and differentiation that is necessary to create functional tissue architecture requires that each cell correctly interpret multiple temporal and spatial cues from its microenvironment. But how does a multipotent precursor cell know when and where to stop dividing and to start to differentiate? In other words, what are the environmental signals? In past studies on morphogenesis, this question has generally been addressed by focusing on global signal sources and, thus, it has been framed as a problem of "positional information." For example, great focus has been placed on the identification of signals that determine

From: *Stem Cells Handbook*
Edited by: S. Sell © Humana Press Inc., Totowa, NJ

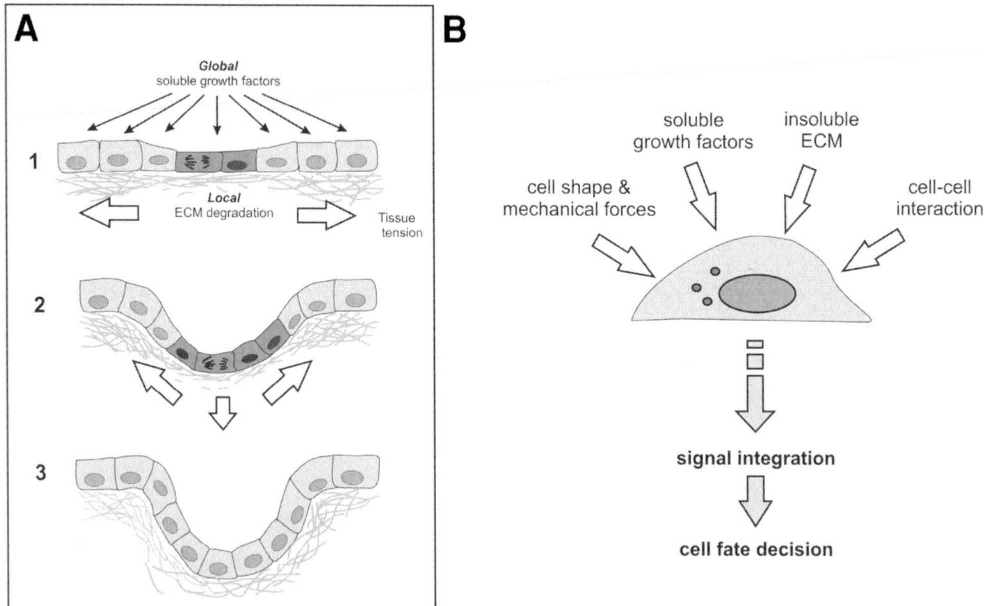

Fig .1. Integration of local signals during tissue pattern formation **(A)** Model for formation of an epithelial bud. Local, mesenchyme-directed increase of ECM turnover and thinning results in a region of stretch because of tension in the tissue (Ingber and Jamieson, 1985) (1). Stretched cells become responsive to soluble growth factors and proliferate (2), thereby providing the increase of epithelial cell mass needed to accommodate the out branching (3). Note that neighboring, not distorted, cells remain quiescent despite adherence to the ECM and exposure to the same soluble growth factor. **(B)** Multiple environmental clues are integrated by the cell to produce a distinct cell fate decision.

long-range patterning, such as the anterior–posterior axes of the embryonic body plan or the proximal–distal axes in limbs (Wolpert, 1996). At this level, concentration gradients of soluble regulator molecules ("morphogens") are thought to convey the positional information that regulates each cell's fate (Gurdon and Bourillot, 2001). The cellular response to these signals would, in turn, control overall organ position, size, and shape.

By contrast, the intricate functional design of most living tissues, such as glandular epithelium or branching networks of capillary blood vessels, has structures with characteristic lengths at the same size scale of individual cells (Huang and Ingber, 1999). To generate structures that exhibit localized bending, budding, or branching, one cell or a small group of cells has to divide to expand tissue mass, while its immediate neighbors must remain quiescent and differentiated (Fig. 1A). This spatial heterogeneity of cell fate control is the basis for creating and preserving micro-anatomical patterns that are characteristic of all tissues. Thus, individual stem cells that are responsible for replenishing lost cells and maintaining stable tissue form have to continuously monitor their environment and make their cell fate decisions based on *local cues* in addition to global information.

Moreover, many living tissues exhibit fractal patterns; this finding suggests that pattern formation results from iterative modular processes that occur at multiple sites and at multiple size scales within the same microenvironment. Fractal patterns cannot be created solely by gradients of soluble chemicals over an entire body part. Instead, computer modeling suggests that these patterns emerge from interactions among cells whose behaviors are governed by simple local *rules* (Goldberger and West, 1987). Such rules may depend on local regulatory inputs that result from the need for an increase in cell mass of a specific cell type at a given site. One example for the complementary role of global and local signals is liver regeneration: following partial hepatectomy, liver hepatocytes start to proliferate. Cell division is stimulated by the release of diffusible growth factors, which, via feedback control loops, also ensure that the regenerative process stops when physiological tissue mass is achieved. However, at the microscopic level, cell proliferation is governed largely by local neighborhood rules such that parenchymal and stromal cells, vessels, and their interconnecting ECM are arranged appropriately in order to rebuild a rudimentary but functional tissue architecture, even if the precise preexisting lobular structure is not fully restored.

Advances in cell and molecular biology as well as tissue engineering have shed light on the nature of these local cues. A central notion is that "solid-state" signals play a regulatory role in the cell fate decision, in addition to soluble growth factors and cytokines. These structural signals include adhesive cues from cell binding to ECM and to other cells as well as mechanical cues that come from physical distortion of the cells or their ECM. In other words, both molecularly encoded information (e.g., the specific 3D structure of receptor ligands) and "nonspecific" physical factors devoid of genetic information impinge on the cell's internal signaling network to govern when and where it is appropriate to divide, differentiate, or die.

4.2.2. CELL FATE SWITCHING: INSTRUCTIVE VERSUS SELECTIVE CONTROL If the cell fate decision is not simply instructed by binding of regulatory molecules to their cell-surface receptors, then the challenge is to understand how the cell integrates the multitude of molecular and mechanical signals to generate a coherent cell behavior, like the decision of a stem cell either to divide or to differentiate along a specific lineage. Take the best-studied example of cell fate regulation, in which an external regu-

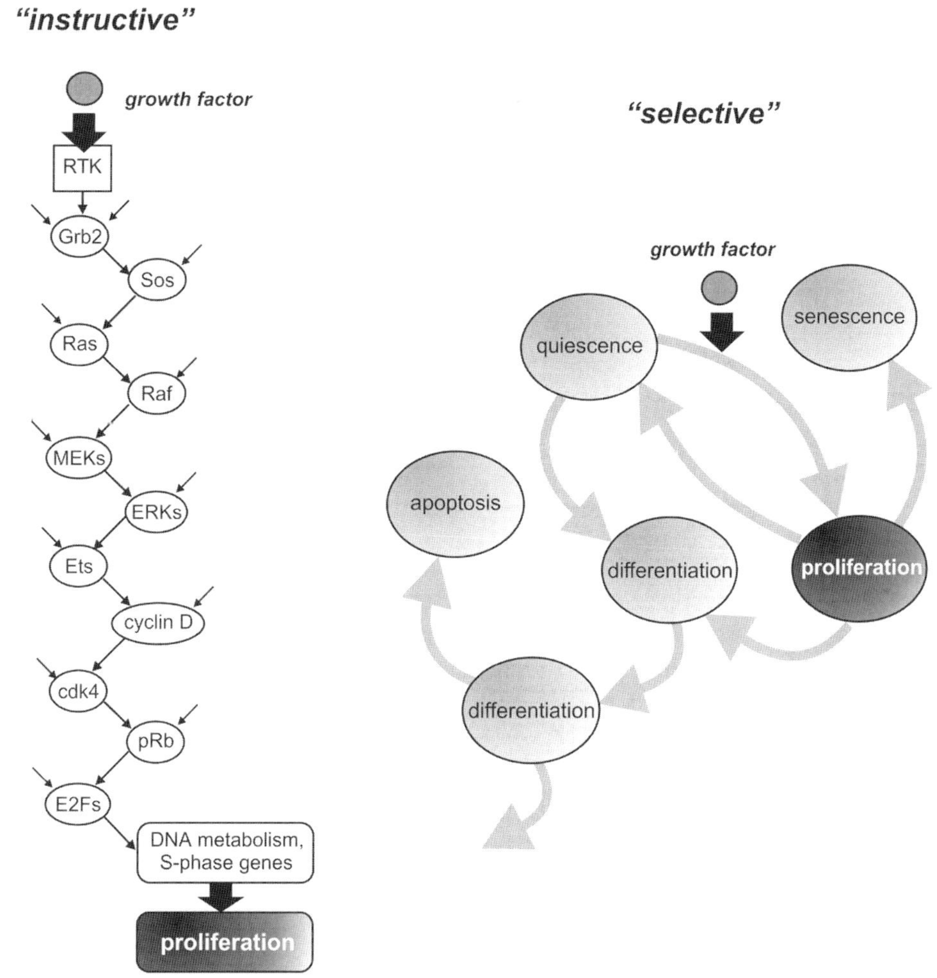

Fig. 2. Concepts of instructive and selective regulation. **(A)** In instructive regulation, a linear cascade of interaction events mediated by high-affinity molecular recognition conveys the message of the growth factor down to the nucleus, leading to the activation of the relevant genes to implement the "instructed" behavior. **(B)** In selective regulation, global cell behavior is determined by a finite repertoire of distinct cell states ("fates") from which the cell has to select in response to external signals. Note that internal rules determine which transitions (thick gray arrows) between the states are possible.

latory signal is a growth factor (e.g., fibroblast growth factor [FGF]) that stimulates the cell to exit its quiescent state and enter the cell cycle. Much of the work has been devoted to elucidating the molecular targets and establishing the biochemical link between input signal and cellular response, based on the concept of *instructive* regulation. Accordingly, the signaling molecule carries information in its 3D structure and transfers it to the cell via specific molecular recognition between ligand and cell-surface receptor, followed by a chain of similar, high-affinity, molecular binding events (Brivanlou and Darnell, 2002).

In the case of mitogenic stimulation, the molecular basis of these events is well characterized. For example, many ligand-activated growth factor receptors activate the Ras protein via adapter proteins (Fig. 2A). Active Ras then triggers a cascade of phosphorylation events: Raf →Mek1 → Erk1/2 (Chang and Karin, 2001). The phosphorylated mitogen-activated protein kinase Erk1/2 translocates into the nucleus and activates the Ets family transcription factors, which are involved in transcriptional activation of *cyclin D1* gene (Albanese et al., 1995; Lavoie et al., 1996). Cyclin D1 protein promotes progression of the cell division cycle by activating cyclin-dependent kinases (cdk4/6) that initiate progressive phosphorylation of the "gate keeper" of the cell division cycle, the retinoblastoma protein, pRb. Once hyperphosphorylated, pRb releases an E2F transcription factor that acts as a transactivator of genes required for S-phase initiation, such as DNA synthesis enzymes, and S-phase cyclins (Sherr, 1996).

The concept that the ligand-mediated signal trickles down a linear cascade of precise protein–protein interactions to the "destination" (gene activation) is reminiscent of a Rube-Goldberg machine in which every element of a chain of events that is apparently unnecessarily long has a dedicated function. Yet, this molecular description of the cell-signaling machinery is the current paradigm of signal transduction by an "instructive" cue because it impressively establishes a link between a specific extracellular signal carried by a growth factor and the corresponding cellular response—cell-cycle entry. However, it has not been possible to establish such a crisp causal chain for differentiation events or for many other aspects of cell behavior (e.g., many apoptosis events). Moreover, even in the case of cyclin D induction during growth promotion, the picture now appears to be more complicated, since other temporally shifted pathways that also impinge on cyclin D1 are equally important for successful switching into a proliferation mode. They include phosphatidyl inositol 3 kinase and small guanosine 5'-triphosphatase–dependent pathways (Marshall, 1999; Assoian and Schwartz, 2001), as well as signals that depend on appropriate mechanical signals from the cytoSkeleton (CSK) (Huang and Ingber, 2002). In other words, other "nonspecific" cues, such as mechanical distortion of cells, can harness the same molecular machinery responsible for cell-cycle regulation as molecular growth factors (*see* below). The linear view of cell regulation also has come under attack; the Genome Project has revealed that information must be encoded in some complex combinatorial way that would not permit one-to-one encoding for each protein in its control of complex cell functions (Claverie, 2001; Venter et al., 2001).

The concept of linear pathways of instructive signals fails to capture essential features of cell regulation dynamics for several reasons. First, from an evolutionary point of view, a "hard-coded" information-processing chain devoted to the transmission of a single message is difficult to explain because of the rigidity and precision required. Instead, it has been suggested that evolvability requires that regulatory pathways use "weak linkages" in which signals do not need to be instructive (Kirschner and Gerhart, 1998). Second, accumulating evidence of cross talk between pathways, as mentioned earlier for cyclin D1 induction, has led to the picture of a *network* of interactions rather than point-to-point communication lines between receptors and genes (Marcotte, 2001). In fact, many regulatory proteins defy our habit of assigning an instructive role that tells the cell what to do (e.g. FGF); instead, they are pleiotropic and exert multiple distinct functions. For instance, Ras not only induces proliferation, as originally described, but also apoptosis and senescence—cell fates quite distinct from proliferation (Huang and Ingber, 2000). Third, as already described, the environmental cues that determine cell fate are not only limited to soluble, high-affinity factors, but also include a whole array of nonspecific signals, such as mechanical forces and topological constraints. These various signals, which simultaneously impinge on every cell, have to be integrated into a single cell fate decision and, hence, activate a common, coherent biochemical response. Finally, cellular responses are subjected to natural dynamic constraints and thus obey certain *rules* that obviate the need for strict instruction. This so-called "rule-like behavior" means that external signals have no absolute "freedom of action," but can only steer cell behavior in some predetermined ways.

Rules of cell behavior are well known in developmental biology, which takes a supracellular, rather than subcellular, view of cell fate decisions. From this perspective, regulation of cell behavior is a *selective* rather than an instructive process (Wolpert, 1994). In a given tissue context, cells have only a limited number of behavioral programs or cell fates available to "choose from." The regulatory cues simply serve to activate endogenous mechanisms that select among these different fates within the cell's behavioral repertoire. For instance, a stem cell in a given tissue can choose among a finite set of distinct alternatives, such as to proliferate (i.e., undergo division for self-renewal), terminally differentiate, undergo cell death—depending on the tissue context (Fig. 2B). These cell fates are functionally distinct behaviors that are somehow "preprogrammed" in the cell's internal regulatory network. No explicit instruction as to how to implement a cell fate is conveyed by the signal. Cell biology is full of evidence for selective cell fate regulation that might have been underappreciated owing to past overemphasis on the identification of specific molecular pathways. We discuss some of the most salient examples in the next section.

4.3. SELECTIVE REGULATION AND ROBUSTNESS OF CELL BEHAVIORAL STATES

4.3.1. MUTUAL EXCLUSION OF PROLIFERATION AND DIFFERENTIATION The often used metaphor of "switch" to describe how cells change among different stable behavioral states or fates reflects the implicit perception of cell fates as distinct, mutually exclusive states, separated by all-or-none transitions. One important cell fate switch is the transition from a growth state to a differentiated state: cells undergoing "terminal differentiation" commonly lose the capacity to proliferate (Goss, 1967). This process is not the result to a gradual change from one state A to another state B, but a manifestation of the existence of two mutually exclusive fates that the cell has to select. This is illustrated by the frequent finding that a variety of treatments that arrest the cell cycle also induce differentiation in cell culture. Thus, exit from the proliferative state is often directly coupled to entry into the differentiation program; the cell does not need a specific instruction for how to differentiate. For instance, removal of serum triggers differentiation of precursor cells into different types of mature "postmitotic" cells, such as myocytes, adipocytes or neurons (Harrison et al., 1985; Olson, 1992; Brüstle et al., 1999). Even inducing cell-cycle arrest by overexpressing the cell-cycle inhibitor, p21Cip, which blocks G1 cdk activity can trigger differentiation in some cell lines (Steinman et al., 1994; Parker et al., 1995).

4.3.2. ROBUSTNESS OF DIFFERENTIATION Many in vitro differentiation models also support the idea that cell fate switching is a selective process. As discussed in Chapter 1, embryonic stem (ES) cells derived from the blastocyst inner cell mass and, cultivated in vitro, spontaneously differentiate into various lineages, giving rise to different specialized cell types, such as glandular epithelial cells, myocytes, chondrocytes, and endothelial cells. This process indicates that the proliferative state of these stem cells is inherently unstable, and that cells randomly "fall" into the preexisting differentiation programs that are inherently robust.

Adult cell differentiation models similarly reveal the intrinsic robustness of cell fates, underscoring the model of selective regulation. A frequent observation is that differentiation can be induced by a multitude of reagents that do not necessarily need to be specific "differentiation factors." For instance, various nonspecific chemicals that do not act instructively by binding high-affinity receptors, such as dimethylsulfoxide, ethanol, methanol, sodium butyrate, or ions, can induce the same terminal differentiation as do physiological signals. This has been described for the differentiation of various cells, including myeloid, neuronal, and cartilage precursor cells (Boyd and Metcalf, 1984; Spremulli and Dexter, 1984; Langdon and Hickman, 1987; Messing, 1993; Newmark et al., 1994; Yu and Quinn, 1994; Waclavicek et al., 2001). A well-studied system is keratinocyte differentiation triggered by calcium, in which it was shown that multiple generic signaling pathways, such as protein kinase C pathways, are involved (Bikle et al., 2001).

Conversely, in cell fate control by physiological chemicals that act on specific pathways, such as retinoic acid (RA) or cyclic adenosine monophosphate (camp), the molecules do not actually specify the destination of differentiation but, instead, can cause differentiation in many different tissues. These differentiating agents carry no specific instructions, as the destination cell fate is "programed" in the state of the precursor cell. Thus, the product of the switch depends on the existing cell state and on environmental factors. For instance, RA, which activates a nuclear receptor, is a potent differentiation agent in many cells, including hematopoietic and neuronal cells (Breitman et al., 1980; Linney et al., 1992; McCaffery and Drager, 2000). Similarly, cAMP, which activates protein kinase A–dependent pathways, induces differentiation in a large number of cell lines, including neurons, melanocytes, hematopoietic cells, and epidermal cells (Voorhees et al; 1973; Deshpande and Siddiqui, 1976; Waymire et al., 1978; Giuffre et al., 1988; Jiang et al., 1997; Hansen et al., 2000). Although many of the experimental systems use tumor-derived cell lines as a source for the undifferentiated precursor cells, these findings are clearly indicative of qualitative (all-or-none) switches from a proliferative to a differentiated state in which the information for the destination cell fate preexists within the precursor cell; the inducing chemical acts only as a nonspecific *trigger* without instructive information.

4.3.3. WADDINGTON'S EPIGENETIC LANDSCAPE AND CELL FATE SELECTION The nature of the switch among different cellular phenotypes and its dynamics support the concept that cell fate regulation is selective and not instructive. This idea of selection of preexisting, intrinsically robust fates was actually captured in Waddington's "epigenetic landscape," a model for "channeling" in development that was first proposed more than 60 yr ago (Fig. 3) (Waddington, 1956; Stern, 2000). In this view of embryonic regulation, the dynamics of development occurs on a landscape with hills and valleys. The phenotypic state of a cell at any time is indicated by the position of a marble on that landscape. The marble will spontaneously roll down the valleys, which represent stable developmental paths leading to a distinct phenotype. In this developmental landscape, the "pits" (or the local minima in the "potential landscape") would correspond to the distinct, stable phenotypes within a given repertoire of fates that the cell may experience.

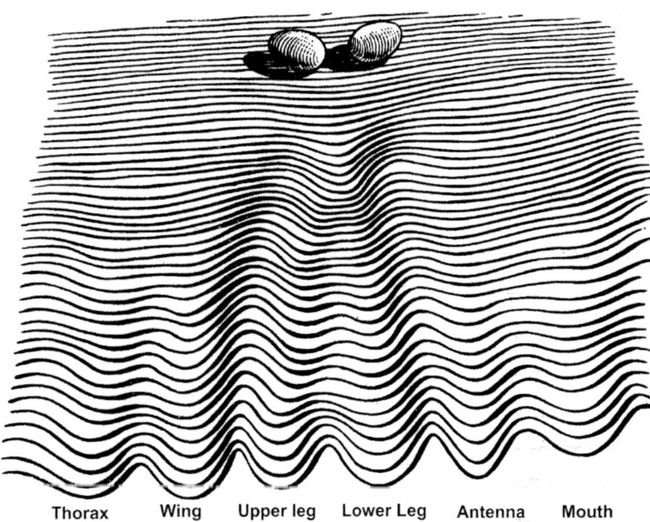

Fig. 3.Waddington's epigenetic landscape. (From C.H. Waddington, 1956. Reprinted with permission from Harper Collins, London.)

4.4. CELL FATE REGULATION BY SOLID-STATE FACTORS

4.4.1. MECHANICAL FORCES AND CELL DISTORTION The fact that chemical triggers of stem cell differentiation are not necessarily physiological, designated inducers of differentiation but, instead, can be nonphysiological agents that tip the cell into a robust and preprogrammed fate that is essentially "waiting in the wing," paves the way to understanding why physical cues devoid of molecularly encoded information can similarly control specific behavioral programs. The robustness of developmental processes in generating particular macroscopic 3D structures suggests that during the formation of such complex structures by coordinated cell fate switches, the cells have to continuously reinterpret signals from the physical world, such as mechanical forces to which the tissue is exposed. This feedback control is essential in allowing the organism to form the higher-level structures that are subjected to the laws of macroscopic mechanics (Chicurel et al., 1998; Huang and Ingber, 2000). In fact, mechanical tension plays a central role in shaping tissue patterns in all organs; examples include the development of bronchial trees (Moore et al., 2002) and cortex gyration in the brain (Van Essen, 1997). But how are physical cues translated into precise cell fate decisions that comply with the physical constraints imposed on the cells by the tissue microenvironment?

4.4.2. ECM, CELL SHAPE AND CONTROL OF PROLIFERATION The common view of mitogenic stimulation (i.e., switch from quiescence to growth) holds that cooperative stimulation of cell-surface growth factor receptors by soluble mitogens and of transmembrane adhesion receptors, known as integrins, by insoluble ECM proteins is required to promote cell-cycle progression (Huang and Ingber, 1999; Assoian and Schwartz, 2001). This dual requirement is the basis for the concept of "anchorage dependence" for growth of nontransformed cells. However, it has long been suggested that cell shape or mechanical distortion also controls the switch between quiescence and growth (Folkman and Moscona, 1978; Ingber et al., 1981; Ingber, 1990). Thus, the cell–ECM interaction has two components: a chemical component, which is mediated by the ligand-induced activation of integrins and associated biochemical signaling cascades, and a mechanical component, which is mediated by integrin's role in physically resisting cell-generated tractional forces (Ingber and Jamieson, 1985; Ingber and Folkman, 1989). This latter structural effect of ECM binding to integrins is responsible for the rearrangement of the actin CSK and ensuing cell shape changes that are observed in response to cell adhesion to the ECM. However, in living tissues, integrins also can modulate cell and CSK form based on their ability to transmit external mechanical loads from the ECM (e.g., owing to tissue or organ distortion) to the CSK (Ingber, 1991; Wang et al., 1993).

4.4.3. EXPERIMENTAL APPROACH FOR CONTROLLING CELL SHAPE AS AN INDEPENDENT VARIABLE To study the role of the mechanical component in cell fate regulation, it is necessary to experimentally separate signals elicited by binding of integrin receptors from those resulting from mechanical force application to cells and their ECM adhesions. Application of microfabrication technologies has provided a means to accomplish this goal by varying cell distortion independently of cell-surface receptor binding to soluble and insoluble regulatory molecules. For example, microcontact printing methods first

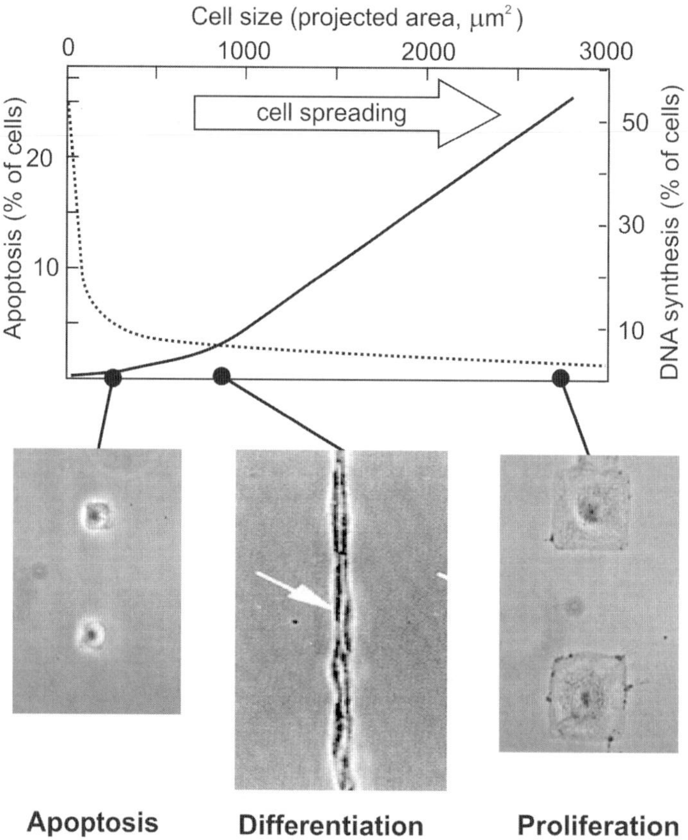

Fig. 4. Control of cell proliferation, apoptosis, and differentiation in endothelial cells by cell distortion. Variation in size and shape of ECM islands (expressed as projected cell area, horizontal axis) in the presence of growth factors controls proliferation (solid line in graph, vertical axis on right), apoptosis (dashed line, vertical axis on left), or differentiation into capillary tubes. (**Bottom**) Phase contrast microscopy photograph of microvascular endothelial cells kept on patterned substrates at a distinct shape and size as indicated by horizontal axis of projected cell area. The arrow indicates a differentiated endothelial cell forming a capillary (Chen et al., 1997; Dike et al., 1999).

developed by the laboratory of George Whitesides at Harvard University for microelectronics applications (Whitesides et al., 2001) can be used to pattern adhesive ECM islands of varying sizes and shapes on the scale of individual cells onto otherwise nonadhesive substrates; living cells adhere and take the shapes of these adhesive islands (Singhvi et al., 1994; Chen et al., 1997). By plating the cells on different sized islands, cell shape (spreading across the substrate) can be varied while keeping the concentration of the soluble (growth factors) and immobilized (ECM) mitogenic ligands constant.

Using this approach, it was shown that endothelial cells that are allowed to extend on large ECM islands switch into the growth mode, whereas cells prevented from extending their CSK were cell-cycle-arrested, and thus remained quiescent, despite stimulation of both their surface growth factor receptors and integrins by binding of these mitogenic ligands (Fig. 4). That CSK extension is a crucial factor is evidenced by the finding that the ECM needs to be either rigid or stiffened under isometric tension in order to resist the cells' inherent contractile forces (Mochitate et al., 1991). More important, cells that are arrested by modulation of cell shape exhibit the same profile of molecular changes that characterizes other types of cell-cycle arrest (e.g., following growth factor withdrawal or lack of adhesion) (Assoian and Schwartz, 2001). Thus, cells on small ECM islands fail to induce cyclin D1 and to downregulate the inhibitor p27Kip (Huang et al., 1998), demonstrating that without explicit instruction mediated by high-affinity growth factors, cell shape control impinges on the same molecular machinery as do specific growth factors via their molecular recognition pathways (Huang and Ingber, 2000).

4.4.4. CELL SHAPE: ONE VARIABLE CONTROLS VARIOUS CELL FATES Cell distortion may represent a fundamental control element for switching between cell fates. For example, when cells are forced onto even smaller size islands so that they almost round up, they undergo apoptosis even though they still remain attached to the ECM and receive growth factor stimulation, which normally would induce proliferation (Chen et al., 1997). Interestingly, cells cultured on an intermediate island size, such that they neither divide nor die, can be triggered to differentiate. Endothelial cells cultured under these conditions form 3D capillary tubes (Fig. 4) (Dike et al., 1999) and hepatocytes secrete blood proteins, such as albumin, at high levels (Singhvi et al., 1994).

Thus, one physical parameter—cell distortion—can control the switch to multiple cell fates, including division, differentiation, and death. However, somehow a continuous variation in this shape parameter is interpreted by the cell such that it is translated into discrete cell fates. This finding has two major implications for stem cell biology. First, it confirms the idea that cell fates are robust, preexisting cellular "programs" that can be "selected" by a variety of environmental inputs. Second, it provides a mechanism by which the process of pattern formation is "self-aware" of its physical progress because of a control loop in which mechanical signals generated by the growth of the developing tissue feed back to control cell fate decisions.

More important, experimental control of cell life, differentiation, and death by cell distortion may mimic what occurs in the tissue microenvironment. In developing tissues, such as branching capillary networks or budding epithelial glands (Fig. 1A), the cells that commonly exhibit the highest growth rates are those that are adherent to localized regions of the ECM (basement membrane) that undergo the most rapid turnover and are physically thin, whereas nearby cells along thicker regions of the same basement membrane remain smaller and turn on differentiation (Huang and Ingber, 1999). Because all tissues exist in a state of isometric tension (i.e., are physically tensed or prestressed), thinner regions of the ECM should extend more than the rest, as should the living cells that are adherent in these regions. Thus, although the microenvironment may be saturated with soluble mitogens and all cells may be bound to ECM, only those cells that physically distort in regions of ECM thinning may respond to these soluble growth cues by switching to the proliferation state. In this manner, local changes in tissue structure owing to ECM remodeling and cell-generated tractional forces may guide local tissue patterning (Ingber and Jamieson, 1985; Huang and Ingber, 1999). Given that regions at the bottom of the crypts of the intestinal epithelium that exhibit the highest stem cell proliferative activity also display thinning of the underlying basement membrane, a similar form of local control may influence where growth of stem cells occurs in adult tissues.

4.4.5. CELL–CELL INTERACTION AND CELL FATE CONTROL There is another solid-state factor that contributes significantly to control of growth and differentiation in the tissue microenvironment: cell-cell adhesion (Aplin et al., 1998). This environmental input again has two components: one being physical (nonspecific) and the other molecular (specific). Physical cell–cell contacts modulate CSK changes and influence cell shape by modulating transfer of physical forces across the cell surface from other cells, much like integrins do for forces transmitted from ECM. Thus, e.g., restriction of cell spreading by high levels of cell–cell contact formation may influence cell fate much like prevention of cell spreading by experimental ECM islands (Folkman and Moscona, 1978). In addition, this "contact inhibition" or, more precisely, "cell density–dependent inhibition" is also mediated by specific cell-surface receptors that engage in high-affinity interactions and signal transduction. One of the best-studied classes of cell–cell adhesion molecules is the cadherin family of homophilic receptors. E-cadherin and N-cadherin have been shown to have a role in signal transduction in addition to mediating cell–cell adhesion in epithelium and neuronal cells, respectively. In fact, activation of these cadherins by ligands has been reported to induce p27Kip-mediated cell-cycle arrest (St Croix et al., 1998; Levenberg et al., 1999), much like prevention of cell spreading on small ECM islands (Huang et al., 1998). However, this chemical component of density-dependent inhibition of growth may play a less significant role in vivo because stem cells, which are responsible for maintaining tissue mass through proliferation, are commonly surrounded by well-formed cell–cell junctions within living tissues (e.g., within crypts of the intestine or hair follicles of the skin).

In contrast to growth control, cell–cell interactions mediated by specific molecular binding events play a dominant role in the control of differentiation. Specifically, in asymmetric stem cell division (Wolpert, 1988), in which only one daughter cell switches into differentiation, distinct molecular signals on neighboring cells, such as the Notch signaling system, ensure the disparate cell fates of the two progenies. Again, it appears that these signals act more as triggers of latent programs since the same molecules are used in different tissues to produce different types of cell specialization (Lin and Schagat, 1997; Artavanis-Tsakonas et al., 1999).

Taken together, the switch between growth, differentiation, and quiescence is governed by a set of inputs from the cell's environment (Fig. 1B) that includes soluble mitogens, insoluble adhesive cues from ECM and cell–cell adhesion molecules, and mechanical forces that can produce CSK distortion and modulate cell shape. Any one of these signals is not alone sufficient to decide cell fate; behavioral control is an integrated response that simultaneously takes all of these signals into account. For example, for entry into the growth state, all three of the following factors need to be present: (1) high levels of soluble mitogens, (2) high density of ECM ligands, and (3) mechanical forces that produce distortion of the CSK. In other words, any one signal has veto power. This safeguard against uncontrolled growth requires that the cell integrate these diverse signals and process them to yield a single response that is best suited for the local tissue microenvironment.

4.5. CELLULAR SIGNAL PROCESSING THROUGH USE OF DYNAMIC NETWORKS

4.5.1. FROM PATHWAYS TO DYNAMIC NETWORKS We have described the existence of distinct, mutually exclusive, robust cell fates as the manifestation of some universal underlying rules that constrain cellular behavior and argued that the concept of instructive, linear signaling pathways fails to explain this behavior. At the same time, genome-scale characterization of cellular signaling reveals the existence of a cellular regulatory network that replaces the traditional pathway paradigm (Marcotte, 2001). Thus, the challenge for the future is to understand how this complex network of interactions gives rise to a phenotypic landscape like that described by Waddington (Fig. 3), and the associated prevalence of "rulelike" cell behavior.

Although the large size and complexity of the network of interactions might appear intractable, it turns out that analysis of generic networks from the field of complexity science reveals some interesting general features. The most important notion is that any network of interactions will exhibit ordered (rather than "chaotic") global dynamics given that the "wiring diagram" exhibits certain features, such as overall sparse connectivity (Kauffman, 1993), which in fact was recently found for yeast for which such genomewide wiring diagrams are available (Jeong et al., 2001; Wagner, 2002). To understand this, let us focus on the protein–protein interactions and assume that we knew the structure of the wiring diagram ("interactome")—i.e., the entire interaction map depicting how every protein interacts with the others, and via what modality (e.g., inhibition, activation, synergism). As postgenomic analysis of genes and proteins continues to expand, such maps become increasingly available, although most still are only partial versions at present (Hazbun and Fields, 2001; Wu et al., 2002).

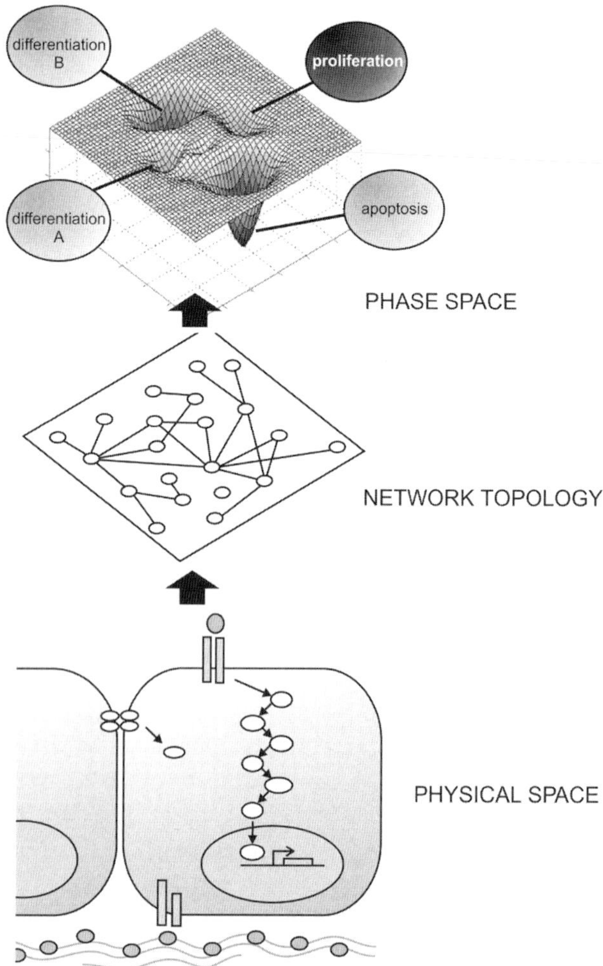

Fig. 5. Relationship among physical space, network topology, and phase space: various ways of representing how cells process signals. At the bottom, the biologists' familiar schematic depiction of the physical space with a cell and its nucleus is shown. The small arrows indicate increasing abstraction of representation. The middle graph represents a protein interaction network in which each open circle represents one protein, with the connecting lines indicating their interactions. The abstract phase space is visualized in the top graph as a 3D "potential surface" with valleys and pits. Each point on the landscape represents an entire genomewide profile of protein activities whose dynamics corresponds to cell behavior and is constrained by the structure of the landscape. Note that the pits (attractors) correspond to the stable cell fates illustrated in Fig. 2B.

With such an interaction map in hand, one would then like to translate it into a phenotypic landscape that captures the biological essence of discrete cell fates, as illustrated in Waddington's epigenetic landscape for the case of embryonic development (Fig. 3) and manifested in the selective mode of cell fate regulation (Fig. 2B). The central notion for doing so is that the interaction networks are not static maps. Rather, they are dynamic because every protein within that web of interactions exhibits a numerical value that indicates its actual expression level or activation status, and these values change over time. The profile consisting of such expression and activity values across all the proteins directly maps into the cellular phenotype. Because their activities are interconnected by the regulatory interactions that define the network (e.g., stimulatory or inhibitory interactions), the profile of protein activities constitutes a *network state* for a given time point. Now, the change of the profile (=network state) over time cannot be arbitrary but is highly constrained by the architecture of that network and the modality of interactions among the elements (e.g., inhibitory, stimulatory). In the next section, we introduce a formal tool that will help us to understand how the constraints of network dynamics are imposed by the underlying regulatory interactions.

4.5.2. THE PHASE SPACE CONCEPT Physicists who study "dynamic systems" composed of interacting parts have developed tools to describe the dynamic behavior of a system based on its network structure, and to represent its global possibilities and constraints (Kaplan and Glass, 1995). The goal in biology would be to accomplish a similar aim by determining how the constrained dynamic behavior of cells, characterized by the switching among different stable phenotypes, may be regarded as a manifestation of the underlying signal transduction networks.

In the case of the cellular regulatory network, we are clearly far from knowing all the quantitative and qualitative biochemical details of molecular interactions. Nevertheless, it is possible from the knowledge of general network features to think of biological regulatory networks as dynamic systems in order to arrive at some universal principles (Huang, 2001). The essential concept that we now borrow from the physical sciences is that of the *phase space* (or state space): an abstract space that contains all the possible *network* states. Much as the familiar network charts that biologists use to depict the wiring diagram of molecular interactions is an abstraction from the actual physical space, the phase space is an abstraction from such interaction maps and represents the dynamic behavior of the regulatory network (Fig. 5). To envision the phase space of the cellular network, imagine that every network state is defined by one possible combination of the activity status of all N proteins that form the regulatory network. Some of the combinations of protein activity states might represent a physiological cell fate (e.g., growth or differentiation). All possible combinations of activity states of the proteins, i.e., all possible network states, together form the phase space, which has N dimensions, one for each protein. With tens of thousands of cellular proteins, it is obvious that the number of possible combinations of activity states across the genome, and hence the size of the phase space, is huge. However, because the proteins are not separate entities that can change their activation state (e.g., expression level or phosphorylation state) independently, most of the combinations of activities (i.e., most network states) cannot occur. For instance, because protein p27 inhibits cdk4 activity (Sherr and Roberts, 1999), all theoretically possible network states in which we have, simultaneously, high levels of p27 and high cdk4 activity would not be possible, although such states are part of the theoretical phase space. Such "unstable" network states would be forced to transition into a state in which, e.g., p27 is high and cdk4 activity is absent (Huang and Ingber, 2000).

4.5.3. STRUCTURE IN PHASE SPACE: ATTRACTORS By extending this two-protein interaction to the entire genomewide network, it becomes plausible that the overall dynamics of the network is highly constrained by the underlying protein–protein interactions. Unstable network states are forced to move in the phase space until they end in "stable" states that represent protein activity combinations that comply with the rules of their underlying regulatory interactions. Multiple theoretically possible, but unstable, network states can converge onto the same stable state, and "gradients" pointing toward these stable states are be formed. In the technical language, these stable states to which the unstable states "drain" are the *attractor* states of the system. The set of all the nonstable states that drain into the same attractor forms the *basin of attraction* of that attractor. Thus, the phase space is not structureless but is divided into valleys and pits. Because cell behavior is a manifestation of changes in the network state, which is equivalent to moving in phase space, its structure determines the constraints of the phenotypic behavior of the cell. One has to keep in mind that despite the landscape picture that we use here to capture visually the concept of network dynamics (as shown in Fig. 5), the phase space is not in 3D but in N-dimensional space. Nevertheless, this mental picture now provides us with a molecular explanation of Waddington's epigenetic landscape that thus can be interpreted as the phase space of an underlying regulatory network.

4.6. IMPLICATIONS FOR STEM CELL BIOLOGY

4.6.1. CELL FATES AS ATTRACTORS From here it is straightforward to interpret the *attractors of the phase space as the cell fates of the phenotype space*. Then the structure of the "attractor landscape" with its ridges, peaks, valleys, and pits essentially reflects the constraints on the phenotype dynamics imposed by molecular regulatory interactions within the underlying network, which give rise to distinct cell fates. In fact, already in the 1940s, Delbrück (1949) discussed the possibility that differentiation states are attractors in a genetic circuit that exhibits multiple stable states ("multistability"). Similar ideas were later expressed by Monod and Jacob (1961) and others (for review, *see* Thomas, 1998), and more prominently by Kauffman (1969, 1993), who extended the concept to large genetic networks and postulated that distinct cell types correspond to attractors that have come about through evolution. With recent insights into the detailed dynamics of cell fate regulation in mammalian cells, a finer picture has emerged: cell fate switching, such as among growth, differentiation into distinct cellular phenotypes, and commitment to apoptosis, similarly obeys constrained dynamics that are governed by underlying interactions of the cellular regulatory network. In other words, these cell fates correspond to the attractor states that emerge from a network of molecular interactions (Huang, 1999).

4.6.2. BIOLOGICAL SIGNIFICANCE OF THE ATTRACTOR LANDSCAPE The concept of an attractor landscape that structures the phase space of a network establishes a link between the observed selective nature of cell fate regulation and the molecular network of signal transduction. Viewing cell regulation through the optics of the phase space may help to explain a series of fundamental observations in stem cell biology that has escaped explanation in the narrow perspective of instructive pathways.

4.6.2.1. Intrinsic Robustness If the attractors that emerge from the cell's molecular signaling networks represent cell fates, then it becomes clear why cell fates are intrinsically robust and can resist many random perturbations. This also provides a natural explanation for the observation that a cell fate can be achieved "from many different directions," triggered by a large set of stimuli. Boundaries of basin of attraction represent a natural decision-making criterion, akin to "watersheds." This allows the integration of multiple, often conflicting signals to produce a single, coherent self-organized cell fate. Remember that each position on the landscape represents an entire pattern of molecular activation in the cell. Thus, as the marble on the landscape rolls down a valley, the network state is forced to move along a certain trajectory (a sequence of changing network states in the phase space) that precisely determines which protein or gene is activated when. Thus, what in the traditional instructive pathway point of view appears as an amazingly well-orchestrated process is, in fact, an inevitable process reflecting the state of the network state that follows the structure of the phase space as it moves down a valley toward an attractor.

4.6.2.2. Pleiotropy in Signal Transduction If cell fates are attractors, then regulatory inputs from the cell's environment, such as soluble factors, insoluble ECM, or mechanical forces are perturbations to the network that may cause attractor switching. Thus, the target of selective regulation is to find the "right attractor" in a given situation. Because attractors are intrinsically stable, network simulations show that it typically will be necessary to change the status of multiple network elements (i.e., the activity levels of multiple genes and proteins) to overcome the boundaries between different attractors (Huang, 2002). The need to activate such "transition-causing sets" of genes and proteins might explain why even specific growth factors or differentiating agents elicit a pleiotropic biochemical response that branches out to affect multiple genes and proteins. For instance, platelet-derived growth factor or FGF receptor activation changes the expression of up to a hundred genes (Fambrough et al., 1999). By simultaneously altering multiple molecular elements within the network, one attractor may be switched to another (e.g., from quiescence to growth). Such sets of genes and proteins appear to overlap, and an individual molecule might be part of various transition-causing sets. This might explain the promiscuous function of many signaling proteins such as Ras, cAMP, and Myc, in which protein "function" depends on the cellular context (Huang and Ingber, 2000).

The existence of stable attractors and the requirements for pleiotropy also might help to explain why it is relatively simple to find extracellular stimuli that can induce a cell fate switch, but also why attempts to find downstream factors, such as transcription factors that control a switch, have proven much less fruitful. Only a few "master switches" have been found, such as myoD, PPARγ, or GATA1, that are transcription factors and are (given a certain initial cellular context) sufficient *eo ipso* to induce differentiation of a precursor cell, or even trigger transdifferentiation among different cell lineages (Zon et al., 1991; Tontonoz et al., 1994; Walsh and Perlman, 1997; Morrison and Farmer, 1999). Such master switch genes themselves have broad downstream effects, including activation of multiple proteins and other genes that are required to promote the attractor switch, again a result consistent with the attractor paradigm.

4.6.2.3. Default Fates From the landscape metaphor, it is plausible that attractors with a large basin of attraction are more stable than those with a small basin. Although we are technically not yet able to measure the effective structure of the attractor landscape of the mammalian genome, we can assume that the differentiated state occupies a very large basin that covers large regions of the phase space accessible to a given cell type, since many nonspecific agents can trigger differentiation. This picture thus provides a conceptual explanation for the widely held notion that the differentiated state is the "default cell fate," which needs no specific instruction for the cell to realize, and is often achieved spontaneously if no specific countermeasures are taken, as may be readily observed in ES cell cultures.

4.6.2.4. Differential Barriers to Cell Fate Switching Transitions among attractors require a complex, high-dimensional signal that consists of changes in the activities of multiple proteins and genes. Moreover, the higher the barrier between two attractor states, the higher the requirements for signal pleiotropy: more genes and proteins need to be turned on or off in order to achieve a transition. For instance, it is in general "easier" to switch from growth to differentiation than from a differentiated state to growth. For most cell types in higher mammals, this transition is almost unidirectional toward differentiation. In fact, when gene expression profile changes were measured (Fambrough et al., 1999; Tamayo et al., 1999), a much more dramatic genomewide change was observed after stimulation of quiescent cells to enter the cell cycle than when a precursor cell was stimulated to differentiate (Huang and Ingber, unpublished observation). In more general terms, the "ease" and directionality of transitions among the cell fates appear to be governed by rules that emerge from the underlying dynamic networks that comprise cellular signal transduction.

One rule is that not every attractor can be reached from every other attractor (Fig. 2B). Some transitions occur with ease and may require only a few changes in protein activities, others require simultaneous activation of large sets of genes and proteins, and still others require very specific signals or never occur. The feasibility of any particular cell fate switch depends on the position of the attractor states in the phase space and the position of the cell at the time of stimulation on the phenotype landscape. It is with this background in mind that the occurrence of spontaneous transdifferentiation events, such as from prostate tumor cells to neuronal cells, or from pancreas to liver, needs to be seen (Bang et al., 1994; Shen et al., 2000). It will be interesting in the future to measure the phase space structure to reveal the actual neighborhood relationships among these differentiation attractors of various tissues.

4.6.2.5. Contextual Signals to Maintain Differentiated Phenotype Although much of development and differentiation appears to be driven by robust intrinsic processes that do not depend on precise instructive signals, the opposite can also be observed. An excellent example is the loss of differentiated phenotypic traits when cells are removed from their tissue context and cultivated in vitro. This cellular response also can be explained by the concept of the attractor landscape that captures the necessity of both intrinsic and extrinsic signals for regulation of cell behavior. The constitutive activity state of proteins can affect the apparent shape of the attractor landscape in terms of the ability of the cell to attain or maintain any particular attractor state. For instance, the differentiation attractor may be stable only if a certain set of cellular proteins and genes is activated; these molecules may, in turn, be constitutively activated by environmental signals that are always present in vivo, but absent in vitro. Removing cells from their tissue context, therefore, might destroy that attractor by locally distorting the attractor landscape and thus yield cells that lose their natural differentiated phenotype when placed in culture. These cells may be stable in their new phenotype in vitro because it represents a different attractor that is similar but not identical to the one available *in situ*. This could, e.g., correspond to a trough along a valley wall in the attractor landscape, rather than the lowermost pit.

4.6.2.6. Genomewide Reprogramming Although theoretically possible, in reality the cell normally cannot travel freely anywhere in the phase space because the typical size of the transition-causing set of proteins and genes after a perturbation with a single chemical or growth factor is limited. Thus, in response to "typical" environmental signals that the cell in the tissue can be exposed to, a given cell state has only a limited number of cell fate alternatives available ("neighboring attractors") among which it can choose. However, complete genome reprogramming can occur by exposing the nucleus to a new cytoplasm, as during somatic cloning (Surani, 2001). This might represent a case in which large, nonphysiological distances in phase space are traversed to place the network state in an attractor that effectively corresponds to the totipotent zygote.

4.7. CONCLUSION

In this chapter, we introduced the concept that control of cell behavior obeys relatively simple rules that give rise to selective regulation of robust cell fates. These rules are the consequence of the natural constraints imposed by the underlying intracellular regulatory network of molecular interactions and are not explicitly encoded in linear instructive pathways. The biological importance of this form of regulation based on dynamic networks is that it allows the cell to integrate a large variety of microenvironmental signals, both carried by specific high-affinity messenger molecules and arising from nonspecific physical interactions devoid of molecular specificity. The latter type of cues is crucial for pluripotent cells to tune their cell fate decisions to adapt to their physical environment, such as to assemble complex macroscopic 3D structures that generate and resist mechanical forces.

The underlying concept of attractor landscapes presented here provides an intuitive explanation for many observations of stem cell behavior that defy conventional linear pathway models. In particular, it unites basic principles of developmental cell biology, such as intrinsic programs, extrinsic cues, and probabilistic mechanisms, under a common overarching concept. Nevertheless, that should by no means obviate the need for detailed characterization of the specific molecular features that are distinct for each individual biological process. Despite the fundamental principle of attractors and cell fates, some types of specialized cellular tasks, such as the induction of cytokine release by a cytokine, may still rely on an instructive information-processing system. Clearly, understanding the specifics at the molecular level is critical to provide handles for therapeutic intervention by drugs. However, overemphasis of the linear pathway paradigm harbors the danger of missing the forest for the trees. Thus, the general concepts discussed in this chapter may help to organize and synthesize the vast amounts of molecular data that one might be tempted to hastily accumulate in this coming age of high-throughput genomics and proteomics. A future goal, then, should be to identify the available paths in the attractor landscape that connect the stable attractor states, instead of characterizing individual molecular pathways and assigning them "instructive functions." Ultimately, such a road map that links the various cell attractors will tell us which potential destinations a given stem cell can possibly reach and how it might be induced to get there.

REFERENCES

Albanese, C., Johnson, J., Watanabe, G., et al. (1995) Transforming p21ras mutants and c-Ets-2 activate the cyclin D1 promoter through distinguishable regions. *J. Biol. Chem.* 270:23,589–23,597.

Aplin, A. E., Howe, A., Alahari, S. K., and Juliano, R. L. (1998) Signal transduction and signal modulation by cell adhesion receptors: the role of integrins, cadherins, immunoglobulin-cell adhesion molecules, and selectins. *Pharmacol. Rev.* 50:197–263.

Artavanis-Tsakonas, S., Rand, M. D., and Lake, R. J. (1999) Notch signaling: cell fate control and signal integration in development. *Science* 284:770–776.

Assoian, R. K. and Schwartz, M. A. (2001) Coordinate signaling by integrins and receptor tyrosine kinases in the regulation of G1 phase cell-cycle progression. *Curr. Opin. Genet. Dev.* 11:48–53.

Bang, Y. J., Pirnia, F., Fang, W. G., et al. (1994) Terminal neuroendocrine differentiation of human prostate carcinoma cells in response to increased intracellular cyclic AMP. *Proc. Natl. Acad. Sci. USA* 91:5330–5334.

Bikle, D. D., Ng, D., Tu, C. L., Oda, Y., and Xie, Z. (2001) Calcium- and vitamin D-regulated keratinocyte differentiation. *Mol. Cell. Endocrinol.* 177:161–171.

Boyd, A. W. and Metcalf, D. (1984) Induction of differentiation in HL60 leukaemic cells: a cell cycle dependent all-or-none event. *Leuk. Res.* 8:27–43.

Breitman, T. R., Selonick, S. E., and Collins, S. J. (1980) Induction of differentiation of the human promyelocytic leukemia cell line (HL-60) by retinoic acid. *Proc. Natl. Acad. Sci. USA* 77:2936–2940.

Brivanlou, A. H. and Darnell, J. E. Jr. (2002) Signal transduction and the control of gene expression. *Science* 295:813–818.

Brüstle, O., Jones, K. N., Learish, R. D., et al. (1999) Embryonic stem cell-derived glial precursors: a source of myelinating transplants. *Science* 285:754–756

Chang, L. and Karin, M. (2001) Mammalian MAP kinase signalling cascades. *Nature* 410:37–40.

Chen, C. S., Mrksich, M., Huang, S., Whitesides, G. M., and Ingber, D. E. (1997) Geometric control of cell life and death. *Science* 276: 1425–1428.

Chicurel, M. E., Chen, C. S., and Ingber, D. E. (1998) Cellular control lies in the balance of forces. *Curr. Opin. Cell. Biol.* 10:232–239

Claverie, J. M. (2001) Gene number: what if there are only 30,000 human genes? *Science* 291:1255–1257.

Delbrück, M. (1949) Discussion. In: *Unités Biologiques Douées de Continuité Génétique*. Colloques Internationaux du Centre National de la Recherche Scientifique (CNRS), Paris.

Deshpande, A. K. and Siddiqui, M. A. (1976) Differentiation induced by cyclic AMP in undifferentiated cells of early chick embryo in vitro. *Nature* 263:588–591.

Dike, L. E., Chen, C. S., Mrksich, M., Tien, J., Whitesides, G. M., and Ingber, D. E. (1999) Geometric control of switching between growth, apoptosis, and differentiation during angiogenesis using micropatterned substrates. *In Vitro Cell Dev. Biol. Anim.* 35: 441–448.

Fambrough, D., McClure, K., Kazlauskas, A., and Lander, E. S. (1999) Diverse signaling pathways activated by growth factor receptors induce broadly overlapping, rather than independent, sets of genes. *Cell* 97:727–741.

Folkman, J. and Moscona, A. (1978) Role of cell shape in growth control. *Nature* 273:345–349.

Giuffre, L., Schreyer, M., Mach, J. P., and Carrel, S. (1988) Cyclic AMP induces differentiation in vitro of human melanoma cells. *Cancer* 61:1132–1141.

Goldberger, A. L. and West, B. J. (1987) Applications of nonlinear dynamics to clinical cardiology. *Ann. NY Acad. Sci.* 504:195–213.

Goss, R. J. (1967) The strategy of growth. In: *Control of Cellular Growth in the Adult Organism*. (Teir, H. and Rytömaa, T., eds.), Academic Press, London, UK, pp. 3–27.

Gurdon, J. B. and Bourillot, P. Y. (2001) Morphogen gradient interpretation. *Nature* 413:797–803.

Hansen, T. O., Rehfeld, J. F., and Nielsen, F. C. (2000) Cyclic AMP-induced neuronal differentiation via activation of p38 mitogen-activated protein kinase. *J. Neurochem.* 75:1870–1877.

Harrison, J. J., Soudry, E., and Sager, R. (1985) Adipocyte conversion of CHEF cells in serum-free medium. *J. Cell. Biol.* 100:429–434.

Hazbun, T. R. and Fields, S. (2001) Networking proteins in yeast. *Proc. Natl. Acad. Sci. USA* 98:4277–4278.

Huang, S. (1999) Gene expression profiling, genetic networks and cellular states: an integrating concept for tumorigenesis and drug discovery. *J. Mol. Med.* 77:469–480.

Huang, S. (2001) Genomics, complexity and drug discovery: insights from Boolean network models of cellular regulation. *Pharmacogenomics* 2:203–222.

Huang, S. (2002) Regulation of cellular states in mammalian cells from a genome-wide view. In: *Gene Regulation and Metabolism: Post-Genomic Computational Approaches* (Collado-Vides, J. and Hofestädt, R., eds.) MIT Press, Cambridge, MA.

Hung, S., Chen, S. C., Ingber D. E. (1998) Cell-shape-dependent ccontrol of p27Kip and cell cycle progression in human capillary endothelial cells. *Mol. Biol. Cell.* 9:3179–3193.

Huang, S. and Ingber, D. E. (1999) The structural and mechanical complexity of cell-growth control. *Nat. Cell. Biol.* 1:E131–E138.

Huang, S. and Ingber, D. E. (2000) Shape-dependent control of cell growth, differentiation and apoptosis: switching between attractors in cell regulatory networks. *Exp. Cell Res.* 261:91–103.

Huang, S. and Ingber, D. E. (2002) A discrete cell cycle checkpoint in late G1 that is cytoskeleton-dependent and MAP kinase (Erk)-independent. *Exp. Cell Res.* 275:255–264.

Ingber, D. E. (1990) Fibronectin controls capillary endothelial cell growth by modulating cell shape. *Proc. Natl. Acad. Sci. USA* 87:3579–3583.

Ingber, D. E. (1991) Integrins as mechanochemical transducers. *Curr. Opin. Cell. Biol.* 3:841–848.

Ingber, D. E. and Folkman J. (1989) How does extracellular matrix control capillary morphogenesis? *Cell* 58:803–805.

Ingber, D. E. and Jamieson, J. D. (1985) Cells as tensegrity strutures: architectural regulation of histodifferentiation by physical forces transduced over basement membrane. In: *Gene Expression During Normal and Malignant Differentiation*. (Anderson, L. C., Gahmberg, C. G., and Ekblom, P., eds.), Academic, Orlando, FL, pp. 13–32.

Ingber, D. E., Madri, J. A., and Jamieson, J. D. (1981) Role of basal lamina in neoplastic disorganization of tissue architecture. *Proc. Natl. Acad. Sci. USA* 78:3901–3905.

Jeong, H., Mason, S. P., Barabasi, A. L., and Oltvai, Z. N. (2001) Lethality and centrality in protein networks. *Nature* 411:41,42.

Jiang, L., Foster, F. M., Ward, P., Tasevski, V., Luttrell, B. M., and Conigrave, A. D. (1997) Extracellular ATP triggers cyclic AMP-dependent differentiation of HL-60 cells. *Biochem. Biophys. Res. Commun.* 236:626–630.

Kaplan, D. and Glass, L. (1995) *Understanding Nonlinear Dynamics*, Springer-Verlag, New York.

Kauffman, S. A. (1969) Metabolic stability and epigenesis in randomly constructed genetic nets. *J. Theor. Biol.* 22:437–467.

Kauffman, S. A. (1993) *The Origins of Order*, Oxford University Press, New York.

Kirschner, M. and Gerhart, J. (1998) Evolvability. *Proc. Natl. Acad. Sci. USA* 95:8420–8427.

Langdon, S. P. and Hickman, J. A. (1987) Correlation between the molecular weight and potency of polar compounds which induce the differentiation of HL-60 human promyelocytic leukemia cells. *Cancer Res.* 47:140–144.

Lavoie, J. N., L'Allemain, G., Brunet, A., Müller, R., and Pouysségur, J. (1996) Cyclin D1 expression is regulated positively by the p42/p44MAPK and negatively by the p38/HOGMAPK pathway. *J. Biol. Chem.* 271:20,608–20,616.

Levenberg, S., Yarden, A., Kam, Z., and Geiger, B. (1999) p27 is involved in N-cadherin-mediated contact inhibition of cell growth and S-phase entry. *Oncogene* 18:869–876.

Lin, H. and Schagat, T. (1997) Neuroblasts: a model for the asymmetric division of stem cells. *Trends Genet.* 13:33–39.

Linney, E. (1992) Retinoic acid receptors: transcription factors modulating gene regulation, development, and differentiation. *Curr. Top. Dev. Biol.* 27:309–350.

Marcotte, E. M. (2001) The path not taken. *Nat. Biotechnol.* 19: 626–627.

Marshall, C. (1999) How do small GTPase signal transduction pathways regulate cell cycle entry? *Curr. Opin. Cell Biol.* 1:732–736.

McCaffery, P. and Drager, U. C. (2000) Regulation of retinoic acid signaling in the embryonic nervous system: a master differentiation factor. *Cytokine Growth Factor Rev.* 11:233–249.

Messing, R. O. (1993) Ethanol as an enhancer of neural differentiation. *Alcohol Suppl.* 2:289–293.

Mochitate, K., Pawelek, P., and Grinnell, F. (1991) Stress relaxation of contracted collagen gels: disruption of actin filament bundles, release of cell surface fibronectin, and down-regulation of DNA and protein synthesis. *Exp. Cell Res.* 193:198–207.

Monod, J. and Jacob, F. (1961) General conclusions: teleonomic mechanisms in cellular metabolism, growth and differentiation. *Cold Spring Harbor Symp. Quant. Biol.* 26:389–401.

Moore, K. A., Huang, S., Kong, Y. P., Sunday, M. E., and Ingber, D. E. (2002). Control of embryonic lung branching morphogenesis by the Rho activator, cytotoxic necrotizing factor-1. *J. Surg. Res.* 104:95–100.

Morrison, R. F. and Farmer, S. R. (1999) Insights into the transcriptional control of adipocyte differentiation. *J. Cell Biochem. Suppl.* 32–33: 59–67.

Newmark, H. L., Lupton, J. R., and Young, C. W. (1994) Butyrate as a differentiating agent: pharmacokinetics, analogues and current status. *Cancer Lett.* 78:1–5.

Olson, E. N. (1992) Interplay between proliferation and differentiation within the myogenic lineage. *Dev. Biol.* 154:261–272.

Parker, S. B., Eichele, G., Zhang, P., et al. (1995) p53-independent expression of p21Cip1 in muscle and other terminally differentiating cells. *Science* 267:1024–1027.

Pedersen, R. A. (1999) Embryonic stem cells for medicine. *Sci. Am.* 280: 68–73.

Shen, C.-N., Slack, J. M. W., and Tosh, D. (2000) Molecular basis of transdifferentiation of pancreas to liver. *Nat. Cell. Biol.* 2:879–887.

Sherr, C. J. (1996) Cancer Cell Cycles. *Science* 274:1672–1677.

Sherr, C. J. and Roberts, J. M. (1999) CDK inhibitors: positive and negative regulators of G1-phase progression. *Genes Dev.* 13:1501–1512.

Singhvi, R., Kumar, A., Lopez, G.P., et al. (1994) Engineering cell shape and function. *Science* 264:696–698.

Spremulli, E. N. and Dexter, D. L. (1984) Polar solvents: a novel class of antineoplastic agents. *J. Clin. Oncol.* 2:227–241.

St Croix, B., Sheehan, C., Rak, J. W., Florenes, V. A., Slingerland, J. M., and Kerbel, R. S. (1998) E-Cadherin-dependent growth suppression is mediated by the cyclin-dependent kinase inhibitor p27(KIP1). *J. Cell Biol.* 142:557–571.

Steinman, R. A., Hoffman, B., Iro, A., Guillouf, C., Liebermann, D. A., and el-Houseini, M. E. (1994) Induction of p21 (WAF-1/CIP1) during differentiation. *Oncogene* 9:3389–3396.

Stern, C. D. (2000) Conrad H. Waddington's contributions to avian and mammalian development, 1930–1940. *Int. J. Dev. Biol.* 44:15–22.

Surani, M. A. (2001) Reprogramming of genome function through epigenetic inheritance. *Nature* 414:122–128.

Tamayo, P., Slonim, D., Mesirov, J., et al. (1999) Interpreting patterns of gene expression with self-organizing maps: methods and application to hematopoietic differentiation. *Proc. Natl. Acad. Sci. USA* 96:2907–2912.

Thomas, R. (1998) Laws for the dynamics of regulatory networks. *Int. J. Dev. Biol.* 42:479–485.

Tontonoz, P., Hu, E., and Spiegelman, B. M. (1994) Stimulation of adipogenesis in fibroblasts by PPAR gamma 2, a lipid-activated transcription factor. *Cell* 79:1147–1156.

Van Essen, D. C. (1997) A tension-based theory of morphogenesis and compact wiring in the central nervous system. *Nature* 385:313–318.

Venter, J. C., Adams, M. D., Myers, E. W., et al. (2001) The Sequence of the Human Genome. *Science* 291:1304–1351.

Voorhees, J. J., Duell, E. A., Bass, L. J., and Harrell, E. R. (1973) *Natl. Cancer Inst. Monogr.* 38:47–59.

Waclavicek, M., Berer, A., Oehler, L., et al. (2001) Calcium ionophore: a single reagent for the differentiation of primary human acute myelogenous leukaemia cells towards dendritic cells. *Br. J. Haematol.* 114:466–473.

Waddington, C. H. (1956) *Principles of Embryology*, Allen & Unwin, London, UK.

Wagner, A. (2002) Estimating coarse gene network structure from large-scale gene perturbation data. *Genome Res.* 12:309–315.

Walsh, K. and Perlman, H. (1997) Cell cycle exit upon myogenic differentiation. *Curr. Opin. Genet. Dev.* 7:597–602.

Wang, N., Butler, J. P., and Ingber, D. E. (1993) Mechanotransduction across the cell surface and through the cytoskeleton. *Science* 260: 1124–1127.

Waymire, J. C., Gilmer-Waymire, K., and Haycock, J. W. (1978) Cyclic-AMP-induced differentiation in neuroblastoma is independent of cell division rate. *Nature* 276:194–195.

Whitesides, G. M., Ostuni, E., Takayama, S., Jiang, X., and Ingber, D. E. (2001) Soft lithography in biology and biochemistry. *Annu. Rev. Biomed. Eng.* 3:335–373.

Wolpert, L. (1994) Do we understand development? *Science* 266: 571–572.

Wolpert, L. (1996) One hundred years of positional information. *Trends Genet.* 12:359–364.

Wolpert, L. (1998) Stem cells: a problem in asymmetry. *J. Cell Sci. Suppl.* 10:1–9.

Wu, C. H., Huang, H., Arminski, L., et al. (2002) The Protein Information Resource: an integrated public resource of functional annotation of proteins. *Nucl. Acids Res.* 30:35–37.

Yu, Z. W. and Quinn, P.J. (1994) Dimethyl sulphoxide: a review of its applications in cell biology. *Biosci. Rep.* 14:259–281.

Zon, L. I., Youssoufian, H., Mather, C., Lodish, H. F., and Orkin, S. H. (1991) Activation of the erythropoietin receptor promoter by transcription factor GATA-1. *Proc. Natl. Acad. Sci. USA* 88:10,638–10,641.

5 Germline Stem Cells

HAIFAN LIN, PhD

Germline stem cells (GSCs) serve as the self-renewing source for the continuous production of gametes in diverse organisms. These stem cells can be classified into two types—stereotypic and populational–according to how they self-renew. Stereotypic GSCs self-renew by asymmetric division, producing a daughter stem cell and a differentiated daughter cell. An example is the *Drosophila* ovary. Each ovariole contains two to three GSCs at its apical region. These cells divide to produce one daughter cell that remains a stem cell and another that begins differentiation as it is displaced from the cap cell. Populational stem cells divide symmetrically and self-perpetuate only at the populational level. In the gonads of *Caenorhabditis elegans*, the mitotically active germline nuclei share a common cytoplasm at the distal end of the gonad and are the self-renewing population of stem cells for gametogenesis. The mitotic nuclei show a gradient of mitotic ability, with those more than 20 nuclei away from the distal tip cell eventually entering meiosis and differentiating into haploid gametes. In mammalian testes, GSCs exist as single cells among type A spermatogonia (type A_S). An A_S spermatogonium will either divide completely to form two daughter A_S spermatogonia or undergo incomplete cytokinesis to produce a pair of connected daughter cells (A_{pr}) that are both committed to differentiation. In most animals, GSCs are established during preadult gonadogenesis following the proliferation and migration of embryonic primordial germ cells. However, it is not known how the stem cell fate is established in the preadult gonad.

Current research suggests that the self-renewing division of GSCs is largely controlled by signaling from the surrounding somatic cells that form the stem cell niche. In *Drosophila*, the somatic signaling requires the *Yb* and *piwi* genes as well as the *dpp* (TGFβ) signaling pathway. In addition, the *hedgehog* signaling pathway is also involved in regulating GSC division. In *C. elegans*, the Leg-2/Glp-1 (Notch-like) pathway plays an essential role in germline maintenance. In mammalian systems, the most upstream signals are follicle-stimulating hormone (FSH) and luteinizing hormone (LH) from the hypothalamic-pituitary axis. These hormones regulate the activities of Leydig and Sertoli cells. In particular, Leydig cells produce testosterone to stimulate Sertoli cells for the production of a number of growth and differentiation factors such as basic fibroblast growth factor (bFGF), TGF-ß1, activinβ1, inhibin βB, Müllerian inhibiting substance, and Desert Hh. Germ cells express Wnt1, bFGF, nerve growth factor (NGF), bone morphogenetic protens, BMP-8A, and BMP-8B. How these factors interact is not well known. However, the Steel/c-KIT signaling does not appear to be involved in GSC division or maintenance but may be required for the survival and/or proliferation of the differentiating type A spermatogonia. In addition, cell-cell junctions among Sertoli cells, Leyding cells, and spermatogonia, mediated by Connexin 43, are important for initiating proliferation of quiescent gonocytes for spermatogenesis. Complicated feedback loops among GSCs, Sertoli cells, Leydig cells, and the pituitary have also been identified.

The somatic signaling appears to control the stem cell division via intracellular mechanisms such as the cell-cycle machinery, differential gene expression, and asymmetric cytoskeletal organization. At present, little is known about the cell-cycle machinery in stem cell division. Type A cyclins appear to be involved in meiosis but not mitosis, and PCNA is involved in proliferation of spermatogonia but not in meiosis. The generation of asymmetric stem cell division is regulated by factors that are differentially expressed and/or segregated between the two daughter cells. In *Drosophila*, genes called *pumilio*, *nanos*, and *stonewall* are involved in GSC maintenance, whereas *bag of marbles* (*bam*), and *benign gonial cell neoplasm* (*bgcn*) promote differentiation. In mammals, a candidate gene for self-renewal of spermatogonial stem cells is the human *deleted in azoospermia* (*DAZ*) gene cluster, and other genes have been identified to control differentiation and apoptosis, such as p53. The asymmetric division of GSCs in the *Drosophila* ovary also involves the membrane skeletal proteins in the form of the spectrosome that anchors one pole of the mitotic spindle to define the divisional orientation. Such orientation is not required for the maintenance of the stem cells but is important for the proper asymmetry of their division. Current findings provide a framework for the further study of stem cell renewal in general.

From: *Stem Cells Handbook*
Edited by: S. Sell © Humana Press Inc., Totowa, NJ

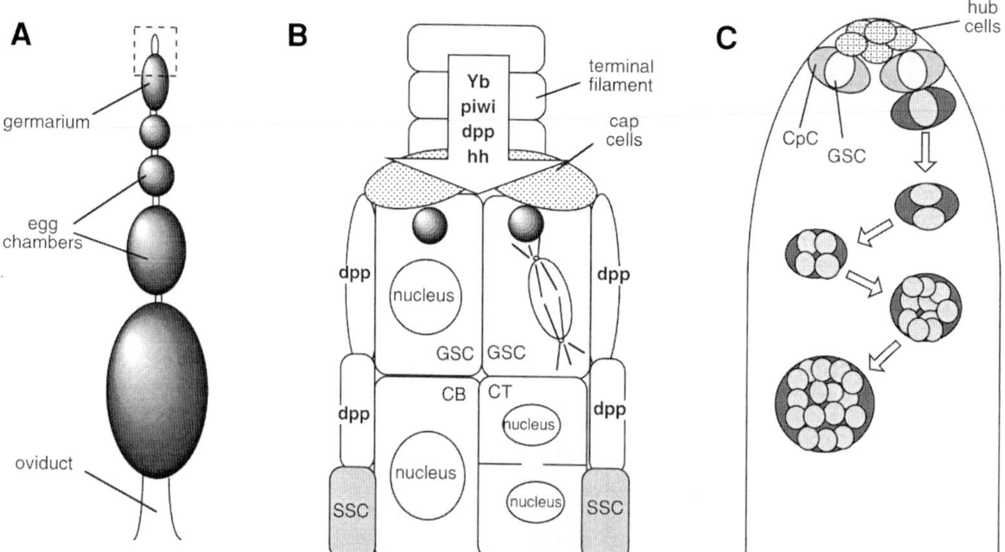

Fig. 1. Asymmetric division of GSCs in the *Drosophila* ovary and testis. **(A)** Ovariole with a string of developing egg chambers produced by germarium; **(B)** magnified view of apical region of germarium boxed area in (A). There are usually two GSCs in the germarium, in contact with apical somatic cells expressing *Yb*, *piwi*, *dpp*, and *hh*. In GSCs, spectrosomes (large, round shaded spheres), containing spectrin (Sp) and Hts proteins, reside in the apical region of the cytoplasm both at interphase and during mitosis, apposed to the signaling somatic cells. During mitosis, the spectrasome anchors one pole of the spindle so that the divisional plan is approximately perpendicular to the apicobasal axis of the germarium. As a result, the daughter stem cell remains in contact with the somatic cells while the cystoblast (CB) becomes one cell away from the somatic cells. The somatic stem cells (SSCs) are located in the middle region of the germarium, two to five cells away from the signaling cells. CT: a two-cell cyst resulted from the incomplete division of the cystoblast. Not shown are genes such as *pum*, *nos*, *bam*, and *bgcn* that are differentially expressed between the stem cell and the cystoblast. (**C**) Apical tip of *Drosophila* testis with only two of five to nine GSCs shown, one at interphase, and the other at telophase. Shown also are the four synchronous divisions of the gonialblast with incomplete cytokinesis that generate 16 interconnected spermatogonia encased by the two somatic cyst cells. CpC, somatic cyst progenitor cell.

5.1. INTRODUCTION

The hallmark of a stem cell is its ability to self-renew and to generate a large number of differentiated cells. Nowhere is the significance of stem cells more apparent than in the germline, where a small number of stem cells constantly divide in a self-renewing fashion, producing numerous gametes with the remarkable ability to create new individuals. Although the existence of stem cells in the germline was proposed by E. B. Wilson (1896) and his contemporaries more than a century ago, mechanisms that control the stem cell property remain enigmatic. Traditionally, germline stem cells have been treated as a minor and simple aspect of gametogenesis in reproductive and developmental studies. These studies provided brief, and often speculative, descriptions of germline stem cell (GSC) division, regarding it as one of many processes involved gametogenesis.

Within the past 10 yr, the central role of GSCs in gametogenesis has been increasingly appreciated, and their effectiveness as a model for stem cell research has been recognized. This recognition has led to a surge of research activities that effectively combine genetic, cell biological, and molecular approaches to study GSCs in genetic model systems such as *Drosophila*, *Caenorhabditis elegans*, and mice. These studies have in turn led to a series of exciting discoveries in the past several years that provide breakthrough insights into the mechanisms underlying the self-renewing division of GSCs. This chapter summarizes and evaluates these discoveries across major model systems in a balanced fashion, with emphasis on its underlying mechanism. For more information on various aspects of GSC division, please *see* earlier reviews (Lin, 1997, 1998; Russell and Griswold, 2000; Kiger and Fuller, 2001; Xie and Spradling, 2001)

5.2. EXISTENCE OF STEM CELLS IN THE GERMLINE: TWO TYPES, ONE PURPOSE

GSCs exist in many organisms, ranging from hydrazoan polyps, nematodes, arthropods, amphibians, birds, fishes, to reptiles and mammals (reviewed in Lin, 1997). Exceptions to this rule are some vertebrates, including mammals, that do not contain stem cells in the female germline (reviewed in Lin, 1997). GSCs can be classified into two types, stereotypic vs populational, according to how they self-renew. The main features of these two types of stem cells are discussed below.

5.2.1. STEREOTYPIC STEM CELLS The first type of stem cells, called stereotypic GSCs (Lin, 1997), self-renew strictly by asymmetric divisions that produce a daughter stem cell and a differentiated daughter cell. A well-illustrated example of stereotypic stem cells is found in the *Drosophila* ovarian germline. The *Drosophila* ovary is composed of *16–18* functional units called ovarioles, each of which contains two to three GSCs at its apical tip in a specialized structure called the germarium, in contact with a group of somatic signaling cells called the cap cells (Deng and Lin, 1997; Lin and Spradling, 1997) (Fig. 1). These stem cells undergo self-renewing asymmetric divisions during each cell cycle, producing a daughter stem cell remaining in contact with the cap cell and a differentiated daughter cell, called the cystoblast, that becomes displaced one cell away from the cap cell. Laser ablation of these stem cells ceases the production of new egg chambers, revealing that they are the exclusive source for germline renewal (Lin and Spradling, 1993).

The asymmetry of the *Drosophila* ovarian GSC division is manifested cytologically not only by the relative position of the two daughter cells with respect to the cap cell, but also by the differential segregation of an organelle in these stem cells called the spectrosome (Lin et al., 1994; Lin and Spradling, 1995; Deng and Lin, 1997). The spectrosome, a spherical structure 1.5 μm in diameter at interphase, resides in the apical region of the cytoplasm closely apposed to the cap cell (Fig. 1B). It contains membrane vesicles and membrane skeletal proteins such as α- and β-spectrin, ankyrin, and adducin that are essential for its organization (Lin et al., 1994; de Cuevas et al., 1996). The spectrosome also contains regulatory molecules such as cyclin A and the Bag-of-Marble (BAM) protein required for cystoblast differentiation (McKearin and Ohlstein, 1995; Lilly et al., 2000). In addition, it is associated with a centrosome (Lin et al., 1994). During mitosis, the spectrosome remains in its apposition to the cap cell and anchors one pole of the mitotic spindle to orient the spindle parallel to the germarial axis (Deng and Lin, 1997; de Cuevas and Spradling, 1998) (Fig. 1B). At telophase, the spectrosome grows in size and elongates toward the future cystoblast that is destined to be one cell away from cap cells. The cleavage furrow then bisects the elongated spectrosome asymmetrically so that two-thirds of it remains in the daughter stem cell and one-third of it is segregated to the cystoblast. The elongated spectrosome reflects the uncoupling of cytokinesis with the rest of the cell-cycle machinery. GSCs without spectrasome divide with completely randomized spindle orientation and produce ill-differentiated daughter cells (Deng and Lin, 1997). Thus, the spectrosome anchors one of the spindle poles to define the orientation of the stem cell division. Such defined orientation appears to be essential for the proper differentiation of the daughter cells.

Stereotypic GSCs also exist in the *Drosophila* testis, as revealed by morphological studies and confirmed by genetic clonal analysis (Harfy et al., 1979; Gonczy and DiNardo, 1996). In the *Drosophila* testis, five to nine GSCs exist in the adult testis, apposed to a group of nonmitotic somatic cells called hub cells (Fig. 1C). Each GSC is flanked by a pair of cyst progenitor cells—the somatic stem cells that are also incontact with the hub cells. Both germline and somatic stem cells divide asymmetrically in synchrony in a radial orientation with respect to the top of the hub, producing differentiated daughter cells that are displaced one cell away from the hub. This stereotypic pattern of asymmetry is very similar to the division of ovarian GSCs. The differentiated germ cell, called the gonialblast, will further divide to produce interconnected spermatogonia. The two differentiated somatic cells grow without further division, encasing the spermatogonia cluster. The daughter GSC and the gonialblast are connected by a transient cytoplasmic bridge that persists well into the next cycle, a situation somewhat resembling that of GSCs in the *Drosophila* ovary and mammalian testis (*see* below). A salient feature of gametogenesis driven by stereotypic stem cells is that stem cell division generates a defined cell lineage in which a differentiated daughter cell divides a fixed number of times before entering meiosis and postmeiotic differentiation.

5.2.2. POPULATIONAL STEM CELLS The second type of GSCs, called populational GSCs (Lin, 1997), self-perpetuate only at the populational level, whereby the collective behavior of mitotic cells in the gonad maintains a steady source of germline, even though individual cells do not divide asymmetrically like stereotypic stem cells. For example, in *C. elegans*, the mitotically active germline nuclei share a common cytoplasm at the distal end of the gonad and serve as a population of stem cells for gametogenesis by virtue of their proximity to a somatic signaling cell, the distal tip cell (DTC) (Hirsh et al., 1976; Kimble and White, 1981) (Fig. 2). Laser ablation of the 5–10 most distal germ cells in L3 gonads affects, but does not eliminate, the proliferative germ cell population . This is because mitotic germline nuclei show a gradient of mitotic ability, with those closest to DTCs being most mitotically active and those more than 20 nuclei away eventually enter meiosis to become terminally differentiated.

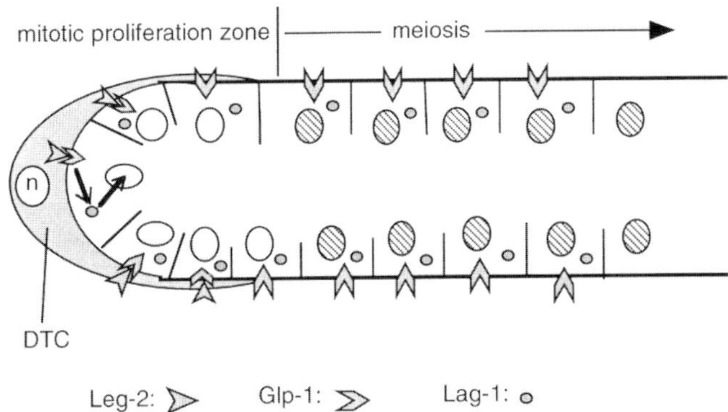

Fig. 2. Somatic signaling pathway at distal region of *C. elegans* gonad. *See* text for details. In the germline, mitotic germline nuclei that form a "stem cell" population are indicated as white nuclei, while meiotic nuclei are shaded. The mitotic proliferation zone spans a 20-nucleus distance along the gonadal arm. n, nucleus.

Mammalian testes also appear to contain populational GSCs. These spermatogenic stem cells are a subpopulation of spermatogonia that are located at the basal layer of the seminiferous tubules, resting on the basal lamina and flanked by Sertoli cells whose occluding junctions form the "blood–testis barrier" (Fig. 3). Outside the basal lamina are thin myoid cells and the steroid-secreting Leydig cells. According to the current model, the A_s model, GSCs exist as single cells among type A spermatogonia and are thus often called Type A_s (single) spermatogonia (reviewed in de Rooij, 2001). This conclusion is based not only on morphological analyses, but also on kinetic studies involving ^{3}H-dNTP autoradiographic labeling and radiation insult. A_s spermatogonia exist in low abundance. For example, in mouse, morphological studies estimate that there are 2.5 million differentiating spermatogonia and 0.3 million undifferentiated spermatogonia per testis, among which only about 35,000 are stem cells. This corresponds to about 0.03% of all germ cells in the testis . An A_s spermatogonium either divides completely to form two daughter A_s spermatogonia to renew itself or undergoes incomplete cytokinesis to produce a pair of connected daughter cells called A_{pr} (pair) spermatogonia that are both committed to differentiation (Fig. 3). A_{pr} in turn divides into chains of interconnected A_{al} (aligned) spermatogonia. A_{al} then undergo complete divisions and differentiate into A_1, A_2, A_n (n=4 in mice); type In (intermediate); and type B spermatogonia (reviewed in de Rooij, 2001. A_s, A_{pr}, A_{al}, A_1, A_2, A_3, and A_4 spermatogonia can each be identified by their chromatin, especially heterochromatin patterns in the nucleus (Chiarini-Garcia and Russel, 2001), as can type In and B spermatogonia (Martins and Silva, 2001).

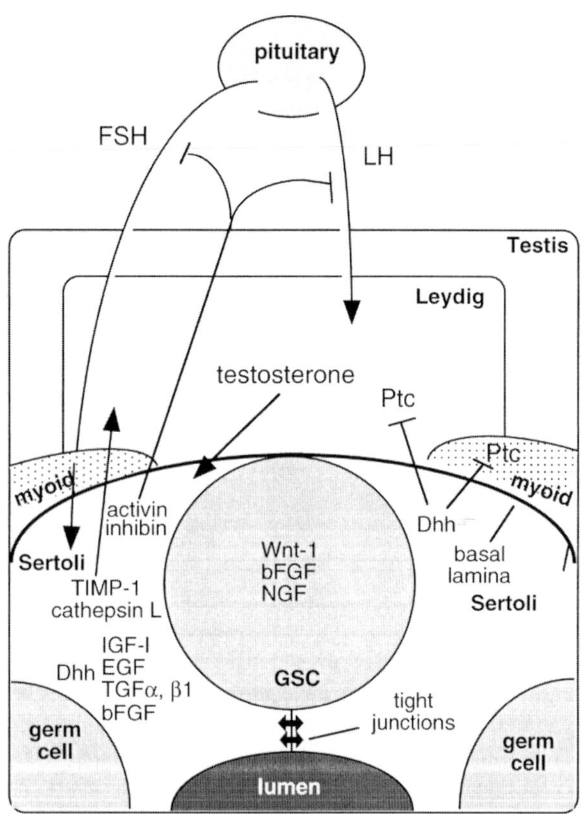

Fig. 3. Cell-cell signaling pathway during mammalian spermatogenesis. *See* text for details. GSC, spermatogonial GSC.

A main feature of the A_s model is that stem cells renew by symmetric divisions. Glial cell line–derived neurotrophic factor from Sertoli cells controls the ratio between self-renewal and differentiation of the stem cells (reviewed in de Rooij, 2001). Such regulation is not fine tuned. Thus, the density of spermatogonial stem cells varies during different stages of the seminiferous tubule cycle. As a result, the distribution of A_1 spermatogonia is very uneven along the tubule. Regulated apoptosis of A_2, A_3, or A_4 spermatogonia then occurs to eliminate the surplus cells, allowing an equal density of spermatocytes along the tubule (*see* below).

Regardless of whether the A_s model or any of the two other alternative models is correct (reviewed in Meistricha nd van Beek, 1993; Lin, 1997), spermatogonial transplantation assay (Brinster and Zimmerman, 1994) in combination with molecular marking provides a functional assay to distinguish stem cells from spermatogonia that are committed to differentiation. Spermatogonia that express α6- and β1- integrins but not c-Kit or αv-integrin possess the best ability to restore spermatogenesis in recipient testes (Shinohara et al., 1999, 2000). Interestingly, the presence of these integrins on spermatogonial stem cells suggests that these cells share elements of a common molecular machinery with stem cells in other tissues.

Populational GSCs also appear to exist in *Hydra oligactis* and *H. magnipapillata* (Littlefield, 1985; Nishimiya-Fujisawa and Sugiyama, 1993). The known mechanism for the self-renewal of populational GSCs differs from that of stereotypic GSCs (*see* below). At present, it is too early to determine whether this difference is significant.

5.3. ESTABLISHMENT OF STEM CELLS IN THE GERMLINE: ROME WAS NOT BUILT IN A DAY

Germline lineage analysis in many organisms has clearly demonstrated that primordial germ cells (PGCs) are the precursor of GSCs. The origin of PGCs and their migration that leads to gonad formation have been well documented in both invertebrate and mammalian systems (reviewed in Wylie, 1999). However, owing to the lack of functional assays for the stem cell property, little is known about when PGCs first acquire the self-renewing property to become stem cells, let alone how this transition is regulated. At present, few, if any, studies have been focused on these specific areas of question. Nevertheless, descriptive knowledge on germline development in general has provided sufficient information that sheds light on the timing of stem cell establishment in the germline. PGCs in higher metazoans, such as *Drosophila* and vertebrates, are often formed outside the gonad or even outside the embryo. They then migrate to the gonadal primordia, where they interact with somatic gonadal cells to form the embryonic gonad. The migration is often accompanied and/or followed by cell proliferation. Thus, it is reasonable to assume that the stem cell fate is established in the germline after gonadal formation but before the initiation of their self-renewing divisions that leads to gametogenesis.

5.3.1. *DROSOPHILA* The *Drosophila* embryonic gonad initially contains 11.7 ± 3.0 (female) to 14.8 ± 3.8 (male) PGCs (Poirie et al., 1995). In the male gonad, these PGCs appear to become GSCs during embryogenesis, since differentiating spermatogonia are found in the first instar larva (Kerkis, 1993). By the third instar larval stage, there appear to be 16–18 GSCs. However, the number of GSCs decreases to five to nine in the young adult (3-d-old) testis (Kerkis, 1933; Hardy et al., 1979). It is not known how this decrease occurs and what its significance is.

In the *Drosophila* female gonad, only a subset of PGCs in the gonad become stem cells. In contrast to the male gonad, PGCs in the female gonad continue to proliferate during larval stages without entering oogenesis, so that by the late third instar larval stage each gonad contains 60–110 PGCs (King, 1970; Parisi and Lin, 1999). The gonad then differentiates into the adult ovary, partitioning PGCs into approx 17 functional units called ovarioles (King, 1970). Each ovariole contains four to seven PGCs but only two to three of them will become stem cells, as indicated by germline clonal analysis and mitotic cycle studies (Wieschaus and Szabad, 1979; Deng and Lin, 1977; Lin and Spradling, 1997). The remaining PGCs are thought to differentiate directly into cystoblasts (King, 1970). However, the true fate of these cells remains unknown. Thus, GSCs may not be established until the larval–pupal transition. Consistent with this possibility, PGCs are sensitive to P-M hybrid dysgenesis (a temperature-sensitive process) from the last few hours of embryogenesis to the second instar larval stages but not at pupal stages following stem cell formation (Engels and Preston, 1079; Schaefer et al., 1979). This suggests that PGCs up to the second instar stage possess different properties from GSCs at the larval–pupal transition. Iinterestingly the resistance of these GSCs to P-M hybrid dysgenesis bears a resemblance to the resistance of mammalian "quiescent" spermatogonia to irradiation insult (*see* next section), even though it is not known whether this resemblance reflects any mechanistic similarities.

At present, it is not known how a subset of PGCs becomes GSCs. It is possible that different PGCs possess different developmental potentials on their formation, possibly owing to their unequal inheritance of germ plasm components. It is also possible that PGCs are initially equipotent; a subset of them are then induced to the stem cell fate by somatic signaling in the gonad. Consistent with the second possibility, somatic cells in embryonic male and female gonads already exhibit differential gene expression with respect to the anterior–posterior axis (Van Doren et al., 2002). Thus, it is possible that the anterior embryonic gonadal somatic cells might initiate the induction of GSC fate. This is more readily comprehensible for the male gonad and may also be true for the female gonad. In the female gonad, although the asymmetric division of *Drosophila* female GSCs does not occur until the larval–pupal transition stage, the determination of the stem cell fate may occur as early as the embryonic gonadal stage. It is possible that somatic cells at this stage already signal a subset of PGCs to become stem cells. These stem cells then proliferate via equal divisions to expand the stem cell population during larval development, before the onset of asymmetric division at the larval–pupal transition. The somatic signaling pathways mediated by *Yb*, *piwi*, and *decapentaplegic* (*dpp*, the homolog of transforming growth factor β [TGF-β]) are required for the self-renewing division of GSCs during oogenesis in *Drosophila* (Cox et al., 1998; Xie and Xpradling, 1998; King and Lin, 1999; King et al., 2001; *see* below). In addition, *piwi* and *dpp* are also required for spermatogenesis. It is possible that these signaling mechanisms are also involved in the initial establishment of the stem cell fate. Consistent with this, *escargot* (*esg*), encoding a transcription factor with zinc-finger motifs, is expressed in the male embryonic gonad, both in somatic cells and in PGCs (Streit et al., 2002). The germline *esg* expression is not required for sex determination and spermatogenesis. However, somatic *esg* expression is required for the maintenance of male GSCs, possibly owing to its embryonic expression.

A cell-autonomous gene, *pumilio* (*pum*), appears to be required for both the establishment and maintenance of GSC fate (Lin and Spradling, 1997; Forbes and Lehmann, 1998; Parisi and Lin, 1999). *pum* encodes an RNA-binding protein known to be a translational repressor during early embryogenesis (Murata and Wahrton, 1995; Wharton et al., 1998). *pum* is essential for GSC maintenanace (Lin and Spradling, 1997; Forbes and Lehmann, 1998; Parisi and Lin, 1999). It is also required for embryonic and larval PGC development and for stem cell establishment (Asaoka-taguchi et al., 1999; Parisi and Lin, 1999). The preoogenic and oogenic requirements are manifested in a continuous fashion (Parisi and Lin, 1999), thus possibly including the requirement for transition from PGCs to GSCs.

5.3.2. MAMMALS In mammals, GSCs exist only in males. Morphological studies have revealed a defined sequence of germ cell development that leads to stem cell formation and their subsequent divisions during spermatogenesis. In the mouse, the ancestral cells of PGCs are located in the epiblast, but lineage restriction does not occur until these cells have moved to the extraembryonic region. PGCs first become identifiable at 7.25–7.5 d postcoitum (dpc), when they are located in a cluster at the base of the future allantois (Chiquoine, 1954; Eddy, 1970; Jeon and Kennedy, 1973; Clark and Eddy, 1975). They then migrate to the urogenital ridges, accompanied by proliferation. By 11.5 dpc, PGCs reach the gonadal primordium in the genital ridges where they continue to proliferate. Meanwhile, they start to undergo substantial growth in cell size. Following this phase of proliferation and growth, the male PGCs, now known as gonocytes or prospermatogonia, become mitotically arrested around 13 to 14 dpc (whereas female PGCs stop mitosis between 13 to 15 dpc and immediately enter meiosis; *see* McLaren, 1984). The gonocytes move between adjacent Sertoli cells and become located in the center of the seminiferous cords. At the onset of postnatal spermatogenesis, gonocytes extend processes toward the basal lamina and relocate themselves peripherally within the cords to assume a position apposed to the basal lamina. This relocation is accompanied by the initiation of DNA synthesis, a critical event marking the entry into spermatogenesis (for rats *see* McGuinness and Orth, 1992). During this time, the Golgi complex increases in size and reorients toward the periphery of the cords; the gonocyte differentiates into a transitional cell type called the primitive spermatogonia, which then start to enter mitosis 3 d postnatally to yield type A spermatogonia (Nebel et al., 1961). Thus, gonocytes are the progenitors of the definitive spermatogonia with stem cell properties. Spermatogenesis starts at postnatal d 5 or 6 when the primitive spermatogonia begin to proliferate (Bellve et al., 1977). These progenitor cells undergo a series of mitoses to produce types A1–4, intermediate, and type B spermatogonia (Bellve, 1979). On d 9, the type B spermatogonia divide once to give rise to preleptotene spermatocytes, which then enter the first meiotic divisions by d 18 and rapidly divide to produce haploid round spermatids that enter spermiogenesis. This pattern of germline development and spermatogenesis is highly conserved among mammals and birds, even though the exact timing and the number of premeiotic spermatogonial divisions varies from species to species (e.g., Hilscher et al., 1974; reviewed in Bellve, 1979; Jones and Lin, 1993).

The above-reviewed morphological studies suggest that stem cell properties may be acquired by the gonocyte when they become primitive spermatogonia. However, little is known at the molecular or functional levels about how gonocytes are different from spermatogonia. The relocation behavior and the cellular changes of gonocytes at the onset of spermatogenesis may reflect the difference between these mobile cells and the primitive spermatogonia, the likely GSCs. If so, the gonocyte–spermatogonial transition should be a critical step for establishing the stem cell fate. However, relocating gonocytes already initiate DNA replication, suggesting that prerelocating gonocytes may be G1-arrested primitive spermatogonia (Huckins and Clermont, 1968). If so, stem cell properties would have to be established prior to or during gonocyte formation. At present, few experimental data are available to distinguish between these two possibilities. For example, markers specific to gonocyte or primitive spermatogonia have not been identified. Characterization of the expression and activity of cell-cycle molecules during gonocyte–spermatogonial transition should help in testing whether gonocytes are G1-arrested primitive spermatogonia, while transplantating gonocytes into the testis using the technique of Brinster and Zimmermann (1994) should test whether gonocytes possess stem cell properties.

Regardless of whether gonocytes are G1-arrested primitive spermatogonia, a critical event for establishing the identity of spermatogonial stem cells is the transition from proliferating PGCs to mitotically quiescent gonocytes. This drastic transition is marked not only by the abrupt entry into mitotic dormancy but also by the subsequent loss of most PGC markers starting at 14.5 dpc (Donovan et al., 1986), the loss of PGCs' adhesiveness to fibronectin and laminin substrates (De Felici and Dolci, 1989), and substantial cell growth. For example, rat PGC increases in size from 530 to 2380 μm^3 of postnatal gonocytes (Beaumont and Mandl, 1963). There also appears to be a decrease in the level of the *c-kit* receptor tyrosine kinase RNA that roughly correlates with the entry of PGC into nonmitotic gonocytes (Orr-Urtreger et al., 1990; Manova and Bachvarova, 1991), even though the signal is present during the transition and persists up to neonatal gonocytes (Orth et al., 1996). In addition, a new carbohydrate epitope recognized by monoclonal antibody 4B6.3E10 appears on the surface of gonocytes from 17 dpc and persists until the day of birth (van Dissel-Emiliani et al., 1993). In human PGCs, the glycogen content is depleted on transition to gonocytes (Falin, 1969; Fujimoto et al., 1977). All these changes could be indicative of a major switch in cell fate or developmental properties. It would be very interesting to study what causes these switches from PGCs to gonocytes and how these switches are related to the establishment of the stem cell fate.

Finally, it is possible that the migrating PGCs already possess certain stem cell properties. This possibility is supported by the observation that male and female mouse PGCs at the gestation stage share indistinguishable stem cell properties, as assayed by key regulators of PGC survival and proliferation such as Steel factor (SLF, the c-Kit ligand), leukemia inhibitory factor (LIF), and basic fibroblast growth factor (bFGF) (reviewed in De Felici and Pesce, 1994; Donovan, 1994). These factors may play a rather direct role in stem cell fate determination. For example, mouse PGCs express low levels of transcripts for c-Kit from their first appearance in the 7.5-d embryo. The expression in PGCs continues through early proliferation and migration stage, increases as they reach the gonadal primordium, and then decreases as they become nonproliferative (Orr-Urtrreger et al., 1990; Manova and Bachvarova, 1991). Consistent with this expression pattern, in embryos homozygous for severe mutations in the *white-spotting* (*W*) locus that encodes c-Kit, the number of germ cells does not increase after 8 days of development, suggesting that its ligand, the SLF, is a proliferative signal for PGCs shortly after their segregation from somatic cells. This signal may be required throughout the proliferative phase of early germ cells and also in postnatal stages of germ cell development, because in the testis c-Kit is expressed in spermatogonia from 6 d of age onward, where it appears to be present from at least as early as type A2 spermatogonia through type B spermatogonia and into pachytene spermatocytes (Manova et al., 1990; Sorrentino et al., 1991). However, these results do not suggest a role of c-Kit in gonocyte development. Techniques such as gonocyte and spermatogonia isolation (van Dissel-Emiliani et al., 1989; Bellve, 1993) combined with germ cell transplantation assay (Brinster and Zimmermann, 1994) should effectively determine the potential stem cell properties of gonocytes and primitive spermatogonia.

In addition to signaling molecules such as SLF, other mechanisms required for germline development may play a role in GSC formation. For example, the Oct-4 POU transcription factor is expressed specifically throughout the germline and appears to be a master switch involved in the specification and development of the germline (reviewed in Pesce et al., 1998). Naturally, it is reasonable to assume that Oct-4 is directly or indirectly involved in specifying the stem cell fate in the germline. Murine P1 protein, a homolog of the yeast MCM3 protein required for initiating DNA replication, is expressed in spermatogonia with transient accumulation in the heterochromatic region of all chromosomes at late G1 (Starborg and Hoog, 1995). It is quite possible that the specification of the stem cell fate is a gradual and reversible process involving the DNA replication, cell cycle, and differentiation mechanisms as induced by the extracellular signals. The recent development of cellular and transgenic technologies should allow these areas of question to be addressed effectively. For example, the testicular transplantaion method of Brinster (Brinster and Zimmerman, 1994) can be used to assay the stem cell property of various stages of PGCs and gonocytes of various genetic background. Likewise, cell type- and stage-specific conditional gene knockout or transgene expression afforded by the inducible Cre-lox technology (reviewed in Nagy, 2000, *see also* Vallier et al., 2001) provides another effective approach to assay the stem cell property of the germ cells at various stages of development and to investigate its underlying molecular mechanisms.

5.3.3. *C. ELEGANS* In the *C. elegans* germline, the syncytial "stem cells" are derived from two precursor germ cells, Z2 and Z3, in the gonadal primordia specified in the embryo (Kimble and Hirsh, 1979; Sulston et al., 1983). The segregation of the germline starts at the first cleavage division when the posterior P1 cell inherits the P granule and other germline determinants. This process continues during subsequent cleavage divisions so that at the 28-cell stage only the P4 cell contains P granules. P4 then divides once to produce Z2 and Z3 cells, both of which become the germline precursors for the postembryonic gonad. In the embryo, Z2 and Z3 associate with Z1 and Z4, the two somatic gonadal precursor cells, to form the gonadal primordium. The two germ cells divide variably throughout larval and adult stages, producing more than 1000 germ cells.

Z2 and Z3 appear to be developmentally distinct from both their precursor cells and their progeny. The difference between Z2/Z3 and their precursor cells is that only their precursor cells, not Z2 or Z3, are transcriptionally silent (Seydoux et al., 1996; Seydoux and Dunn, 1997). Moreover, Z2 and Z3 and possibly their intermediate daughters also are different from their mitotically active progeny. Z2 and Z3 do not require *glp-1* for their divisions, yet *glp-1* is continuously required for the germline proliferation after the L1 stage (Austin and Kimble, 1987). This differential requirement of *glp-1* function as determined by the temperature-shift experiment is supported further by the observations that the null *glp-1* mutation reduces the production of germ cells to only four to eight but does not eliminate the germline at the L1 stage (Austin and Kimble, 1987). In addition, ablation of DTCs caused *glp-1* mutant-like defects (Kimble and White, 1981). Thus, Z2, Z3, and possibly their immediate progeny at L1 appear to represent a special stage of germline development that leads to the establishment of the stem cell population in the germline. If so, the formation of these stem cell precursors would be dependent on the removal of transcriptional suppression but not on activation of the GLP-1-mediated proliferation mechanism. Furthermore, the Z2/Z3-derived stem cell population would have to be expanded and modulated during the subsequent larval development, since ablating the 5–10 most distal germ cells in early L3 gonads does not eliminate proliferation of the germ cell population (McCarter et al., 1997). Hence, the stem cell property in the *C. elegans* germline appears to be established through a two-step process involving Z2 and Z3 as critical intermediate precursors.

5.4. EXTRINSIC MECHANISMS: THE STEM CELL NICHE

It has long been implicated that the self-renewing division of GSCs is regulated by extrinsic signaling (reviewed in Lin, 1997).

In mammalian systems, extrinsic signaling is mediated by hormones, growth factors, cytokines, and short-range cell–cell signaling pathways such as those involving Steel/c-kit. In *C. elegans*, a laser ablation experiment has elegantly demonstrated that somatic signaling is required for germline maintenance (Kimble and White, 1981). Somatic signaling also plays an essential role in maintaining GSCs in *Drosophila* (Cox et al., 1998; King and Lin, 1999; Xie and Spradling, 2000). Recent studies in these model systems suggest that the somatic signaling cells constitute a microenvironment, the so-called stem cell niche, that regulates the stem cell fate in the germline. The following section review the current knowledge on the extrinsic mechanisms, starting from the most upstream regulatory signals to molecular mechanisms underlying the stem cell niche.

5.4.1. SIGNALS FROM BEYOND THE NICHE: HORMONES AND CYTOKINES Hormones are known to regulate gametogenesis from *Drosophila* to mammals. Their specific role in regulating GSC division has also been implicated. For example, in *Drosophila*, it has been shown that ovarian metamorphosis, coinciding with the initiation of oogenesis, is regulated by ecdysone and juvenile hormone (reviewed in King, 1970). Thus, these hormones should play a critical role in stem cell development. Recently, it has been shown that dietary condition regulates the division rate of both germline and somatic stem cells in the ovary (Drummond-Barbosa and Spradling, 2001). Such nutritional regulation requires an intact insulin pathway.

In mammals, steroid and gonodotropic hormones are involved in regulating spermatogenesis, with functions largely known to regulate postspermatogonial development. During spermatogenesis, luteinizing hormone (LH) and follicle-stimulating hormone (FSH) produced by the gonodotrophs of the anterior pituitary gland target their receptors on Leydig and Sertoli cells, respectively (reviewed in Griswold, 1995). On LH stimulation, Leydig cells secrete testosterone into the interstitial compartment, which then diffuses to reach Sertoli cells. The FSH and testosterone stimulation lead Sertoli cells to secrete molecules required for germ cell differentiation. However, these hormones apparently do not affect the proliferation or differentiation of type A spermatogonia (Meistrich and van Beck, 1993).

In addition to FSH and LH, a number of other growth factors such as the seminiferous growth factor, and acidic and basic fibroblast growth factors (aFGF and bFGF) have been shown to stimulate the proliferation of cultured cells of testicular origin (Braunhut et al., Zheng et al., 1990). Hence, these factors may serve as autocrine or paracrine modulators of testicular functions, including the regulation of GSC divisions. Alternatively, LIF and ciliary neurotropic factor (CNTF) significantly enhance survival, but not proliferation, of Sertoli cells and proliferating gonocytes in a Sertoli cell–gonocyte coculture system (De Miguel et al., 1996). Moreover, LIF and FGF-2 stimulate the proliferation of quiescent gonocytes (van Dissel-Emiliani et al., 1996). These results indicate that LIF or CNTF may play a role in initiating GSC division. Finally, retinoids such as vitamin A are required for the maintenance of spermatogenesis (Meistrich anf van Beek, 1993). In the vitamin A–deficient rat testis, many of the proliferating A spermatogonia degenerate (van Pelt et al., 1995), indicating its requirement for stem cell maintenance. In Japanese eel (*Anguilla japonica*), estrogen estradiol-17β, a natural estrogen in vertebrates, is found to be present in the serum, and its receptor was expressed in the testis during spermatogenesis (Miura et al., 1999). This estrogen appears to promote spermatogonial stem cell renewal. The exact effect of these signaling molecules on stem cells awaits further investigation.

5.4.2. THE STEM CELL NICHE: A FLY OVARIAN VERSION In the *Drosophila* ovary, the role of somatic cells in regulating GSC division was indicated first by the laser ablation of the apical portion of the terminal filament, which increased the rate of GSC division (Lin and Spradling, 1993). The essential role of somatic signaling has been revealed by the genetic analysis of *Yb*, *piwi*, *dpp*, and *hh* genes (Cox et al., 1998; Xie and Spradling, 1998; King and Lin, 1999; King et al., 2001). Loss-of-function mutations in *piwi* or *Yb* cause the failure of GSC maintenance. GSCs in the mutants often differentiate into germline cysts without divisions. Sometimes, they undergo a limited round of aberrant divisions. Either way, functional GSCs are depleted. Interestingly, *Yb* appears to be specifically expressed in the terminal filament and cap cells (King and Lin, 1999); removing *Yb* function in the germline does not affect GSC division. *piwi* is expressed in both somatic and germline cells; removing *piwi* from the germline does not affect the maintenance of GSCs, even though their division rate is slowed down by approximately fourfold (Cox et al., 1998, 2000). *dpp* appears to be expressed in cap cells and inner sheath cells that are posterior to cap cells (Xie and Spradling, 2000). *hh* is specifically expressed in terminal filament and cap cell (Forbes et al., 1996) and plays a somewhat redundant role in regulating GSC division (King et al., 2001). Thus, the study of *Yb*, *piwi*, *dpp*, and *hh* reveals the essential role of somatic signaling mediated by the terminal filament and cap cells in regulating GC division.

In particular, cap cells appear to play a central role in this process. This is suggested by four lines of evidence. First, cap cells are in direct contact with GSCs (Lin and Spradling 1997); failure in anchoring one pole of the stem cell spindle to near the cap cells affects the symmetry of the stem cell division (Ding and Lin, 1997). Second, the *zero population growth* (*zpg*) gene encodes a germline-specific gap junction protein, Innexin 4, that is required for survival of differentiating early germ cells during oogenesis and spermatogenesis (Tazuke et al., 2002). ZPG protein is found on germ cell surfaces, both those adjacent to surrounding somatic cells, including cap cells, and those adjacent to other germ cells. The *zpg*-null mutant contains small numbers of early germ cells, resembling stem cells or their early products, but lack later stages of germ cell differentiation. This suggests that gap junctions between germ cells and their contacting somatic cells are important for proper GSC maintenance. Third, *Yb* mutation completely eliminates *hh* expression in cap cells but only partially affects *hh* expression in terminal filament cells (King et al., 2001). Fourth, ablation of most of the terminal filament cells increases the rate of GSC division instead of causing failure in GSC maintenance (Lin and Spradling, 1993). This suggests that terminal filament and cap cells have different roles, with cap cells required for GSC maintenance. Thus, cap cells likely define a niche essential for stem cell maintenance in the germline.

The size of this stem cell niche appears to correspond to the number of stem cells in the germline (Xie and Spradling, 2000). Germaria with one, two, and three GSCs contain an average of 4.2, 5.3, and 6.6 cap cells, respectively. GSCs lost by normal or induced differentiation are efficiently replaced by symmetric division of another stem cell along the cap cells, repopulating the lost stem cells.

The molecular nature of somatic signaling from the niche is being effectively revealed. DPP is the *Drosophila* of homolog of BMP2/4. Loss of *dpp* function causes *Yb/piwi*-like germline maintenance defects, while over expression of *dpp* produces GSC tumors (Xie and Spradling, 1998). Genetic clonal analyses on the receptors and downstream transducers of the *dpp* signal, such as *punt* (encoding a type II receptor) and *mad/Med* (encoding a dimeric transcription activator), show that the *dpp* signal is directly received by GSCs. Somewhat similar to *dpp*, overexpression of *Yb* or *piwi* in somatic cells also increases the number of germline stemlike cells by 2.5-fold (Cox et al., King et al., 2001). Consistent with its somewhat redundant role in regulating GSC division, overexpression of *hh* in somatic cells causes a slight (56%) increase in GSC numbers. Thus, the somatic signaling from the stem cell niche modulates the GSC activity in a dose-dependent fashion.

Although the biochemical nature of YB and PIWI proteins is not known, they do not show sequence features of cell-surface or secreted molecules. In fact, PIWI is localized in the nucleoplasm in both somatic and germline cells in the ovary (Cox et al., 2000). PIWI is the founding member of the evolutionarily conserved PIWI (a.k.a. Argonante) family proteins that have been shown to play crucial roles in stem cell division, gametogenesis, and RNA silencing in diverse organisms in both animal and plant kingdoms (Cox et al., 1998; Benfey, 1999; Tabara et al., 1999; Cerutti et al., 2000). A murine homolog of PIWI, called MIWI, appears to bind to mRNA of its target genes to regulate spermiogenesis (Deng and Lin, 2002). Therefore, it is possible that PIWI plays different roles in signaling cells and stem cells by regulating different sets of mRNAs. In the somatic signaling cells, YB is required for PIWI expression (King et al., 2001). Since PIWI is not required for HH and DPP expression (King et al., 2001; Szakmary and Lin, unpublished data), it must lead to the production of yet another signaling molecule(s) essential for GSC maintenance. From these studies, a salient feature of the ovarian niche is emerging: it involves multiple signaling pathways, with at least two of them (DPP- and YB/PIWI-mediated pathways) essential for the stem cell maintenance in the germline.

Interestingly, the cap cells also control the division of somatic stem cells in the germarium that generate follicle cells to envelop the GSC products into egg chambers (King et al., 2001). *Yb* is required for both germline and somatic stem cell divisions; loss of *Yb* function eliminates GSCs and drastically reduces somatic stem cell division, while *Yb* overexpression increases GSC number and causes somatic stem cell overproliferation. *Yb* achieves this dual control role by regulating the expression of *piwi* and *hh* in the somatic signaling cells (King et al., 2001). The *piwi*- and *hh*-mediated bifurcating signaling pathways then control GSC and somatic stem cell division, respectively, with *hh* signaling also having a minor effect in GSC division.

Yb-mediated signaling is female-specific because *Yb*-null mutations only affect females and do not display a male phenotype (King and Lin, 1999). While *piwi*-mediated signaling is required for both male and female GSCs. Molecular analysis of *Yb* and *piwi* and their interacting genes should lead to identification of the molecular pathways involved in the somatic signaling, including the signal(s) itself.

5.4.3. THE STEM CELL NICHE IN *DROSOPHILA* TESTIS: A GROUP EFFORT In the *Drosophila* testis, the hub cells appear to be the somatic signaling center that regulates both germline and somatic stem cell division. A key signaling molecule from the hub cells is the ligand unpaired, which activates the Janus kinase-signal transducer and activator of transcription pathway in adjacent GSCs to specify their self-renewal and maintenance (Kiger et al., 2001). Unexpectedly, the *dpp* signaling pathway appears to play a different role in the *Drosophila* testis (Matunis et al., 1997). First, the *dpp* signaling in the testis is not required for germline division or maintenance but restricts the proliferation of the differentiated daughter. Second, the receptors and downstream transducers of the *dpp* signal are not required in the germline but in their flanking somatic cyst cells. The loss of *punt* and *schnurri* (*shn*) function in the somatic cyst cells results in overproliferation of the differentiated daughter of the GSCs. *shn* encodes a zinc finger transcription factor (an MBP-1/2 homolog) that acts downstream of the *dpp* signal to regulate the expression of *dpp*-responsive genes. Similarly, *raf* is required in somatic cyst cells to restrict self-renewal and control stem cell number by ensuring asymmetric division (Tran et al., 2000). In testes that are deficient for Raf activity, there are excess stem cells and gonialblasts. Moreover, the GSC population remains active for a longer fraction of lifespan than in wild type. A third signaling molecule involved in restricting GSC overproliferation is the *Drosophila* epidermal growth factor receptor (EGFR). Loss of EGFR activity in somatic cyst cells disrupted the balance of self-renewal vs differentiation in the germline, resulting in an increased number of GSCs (Kiger et al., 2000). It appears that activation of this receptor specifies normal behavior of somatic cyst cells. The somatic cyst cells in turn provide information that either prevents overproliferation or promotes differentiation of GSCs and/or their daughter cells that the cyst cells enclose. Therefore, while hub cells are required for the self-renewal of GSCs, somatic cyst cells could be considered an antiproliferative part of the stem cell niche that ensures the proper asymmetry of the stem cell division.

5.4.4. THE STEM CELL NICHE FOR THE *C. ELEGANS* GERMLINE: A SOLO OF DTCS The stem cell niche in *C. elegans* is represented by DTCs whose process extends to a 20-nucleus distance toward the proximal direction, stopping at the mitosis/meiosis border (Fitzgerald and Greenwald, 1995). It has been long established that *lag-2/glp-1* signaling pathway plays a key role in germline maintenance in *C. elegans*, where the LAG-2 somatic signal on the surface of DTCs interacts directly with the GLP-1 receptor (reviewed in Kimble and Simpson, 1997). LAG-2 is a transmembrane protein homologous to Delta (Dl) and Serrate (Ser) in *Drosophila*, while GLP-1 is homologous to Notch, the receptor of Dl and Ser. The GLP-1-mediated signal is transduced in the germ cells through interaction with the LAG-1 protein, a homolog of the *Drosophila* DNA-binding protein Suppressor of Hairless (Su[H]) and the mammalian DNA-binding protein CBF1, which are downstream interactors of the Notch protein in *Drosophila* and mammals, respectively (reviewed in Kopan and Turner, 1996). LAG-1 maintains a mitotic population of germ nuclei by suppressing the function of *gld-1* and *gld-2*, which promotes meiosis (Kadyk and Kimble, 1998). Thus, *gld-1* and *gld-2* may be key differentiation factors in the *C. elegans* germline. The DTC as a single cell is perfectly designed as a stem cell niche—LAG-2 protein is distributed evenly on the surface of the cell, including its cellular processes. The cellular processes extend variably to multinuclear distances, forming a nice LAG-2 gradient that defines a 20-nucleus-long mitotic proliferation zone, with nuclei closer to DTCs acquiring higher mitotic activity. Although the LAG-2/

GLP-1 signaling pathway is conserved between *C. elegans* and *Drosophila*, it appears to have no obvious function in regulating *Drosophila* GSC division (Ruohola et al., 1991). This dichotomy again reflects the divergence of the signaling mechanisms employed by different GSC systems.

5.4.5. THE STEM CELL NICHE IN THE MAMMALIAN TESTIS: MULTIPLE CELL TYPES IN ACTION The stem cell niche in the testis can be elegantly demonstrated by spermatogonial transplantation (Dobrinski et al., 1999; Nagano et al., 1999). When spermatogonia are transplanted into seminiferous tubules of host testes depleted of the GSC population, the donor spermatogonia migrate to the wall of the tubule, forming colonies apparently derived from single stem cells, restoring spermatogenesis. This homing event occurs even if the site of injection is as distant as the rete testis or efferent duct (Jiang, 2001). Moreover, the stem niche is sufficiently potent to attract spermatogonial stem cells to colonize, although there is incompletion of spermatogenesis, following xenotransplantation between two different species, such as baboon and mice, with more than 100 million yr of evolutionary separation (Nagano et al., 2001). Clearly, the testicular stem cell niche plays a decisive role in the self-renewal of spermatogonial stem cells.

Then, what constitutes the potent stem cell niche in the mammalian testis? It has been well established that intratesticular cell–cell interactions play a key role in regulating GSC division in mammals (Fig. 3). In the testis, Leydig cells produce testosterone that supports spermatogenesis through Sertoli cells. Conversely, steroidogenesis in Leydig cells is regulated by pituitary gonadotropins as well as a 70-kDa protein complex containing tissue inhibitor of metalloproteinase-1 (TIMP-1) and the proenzyme form of cathepsin L secreted from Sertoli cells (Robertson et al., 1993; Boujrad et al., 1995; Bitgood et al., 1996). In addition, Sertoli cells produce a number of growth and differentiation factors, such as insulin-like growth factor-1 (IGF-1; somatomedin C), TGF-α, EGF, bFGF, TGF-β1, activinβA, inhibinβB, Müllerian inhibiting substance (MIS), and Desert Hh (Dhh) (Bardin et al., 1993). Germ cells express Wnt1, bFGF, nerve growth factor (NGF), BMP8A, and BMP8B (Shackleford and Varmus, 1987; Manova and Bachvarova, 1991; Parvinen et al., 1992; Han et al., 1993). Some of these factors interregulate each other. For example, IGF-1 is able to induce the maturation of Leydig cell function and the IGF-1 production is stimulated by growth hormone (Hansson et al., 1989; Chatelain et al., 1991). At present, little is known about how these molecules serve as paracrines or autocrines to regulate GSC division. Inactivation of inhibin α and activinβB, FSH, LH, MIS, and TGF-α genes does not cause an apparent defect in GSC division (reviewed in Jenkin et al., 1995; *see also* Luetteke et al., 1993; . However, phenotypic analysis of Sl/W loci and the inactivation of BMP-8b and Dhh has started to determine whether they are involved in germline proliferation and maintenance (Bitgood et al., 1996; Zhao et al., 1996), as discussed next.

Contrary to common belief, Steel/c-KIT signaling does not appear to be involved in GSC division or maintenance (Rossi et al., 2000). Sertoli cells produce SLF on FSH stimulation. SLF then binds the c-KIT receptor on differentiating spermatogonia to maintain spermatogenesis. Intraperitoneal and iv injections of an anti-c-KIT antibody into prepubertal and adult mice completely block the mitosis of differentiating type A spermatogonia but not the mitosis of the gonocytes and undifferentiated type A spermatogonia. These results indicate that c-KIT is required for the survival and/or proliferation of the differentiating type A spermatogonia, but not the undifferentiated type A spermatogonia (the GSCs). Thus, the known role of the Steel/c-KIT-mediated mechanism is in the pre- and poststem cell processes, in contrast to its name as the stem cell factor.

On the other hand, the *hh/ptc* signaling pathway may be involved in GSC maintenance in mammalian systems. This conservation, if it indeed exists, involves a more complicated regulatory circuitry of cell types. Among *hh* homologs in mammals, *Dhh* is expressed specifically during spermatogenesis in the Sertoli cell precursors shortly after the activation of *Sry*; its expression persists in the Sertoli cells in the testis into the adult (Bitgood and McMahon, 1995; Bitgood et al., 1996). Complete inactivation of *Dhh* leads to germ cell degeneration (Bitgood et al., 1996). The *Dhh*-null prepubertal and adult testes display abnormal peritubular tissue and severely restricted spermatogenesis (Clark et al., 2000). In particular, 92.5% of the mutant testes are feminized and lack adult-type Leydig cells. These testes contain numerous undifferentiated fibroblastic cells in the interstitium and abundant collagen. The basal lamina are focally absent between the myoid cells and Sertoli cells. In addition, extracordal gonocytes, apolar Sertoli cells, and anastomotic seminiferous tubules are evident. Consistent with these defects, *ptc*, a likely target of *Dhh* in the testis, displays male-specific transcription in the Leydig cells and myoid cells instead of spermatogonial cells (Bitgood et al., 1996; Clark et al., 2000). PTC expression is lost in *Dhh* mutants. These observations suggest that *Dhh* plays an essential role in myoid cell development, tubular morphogenesis, and the differentiation of Leydig cells. The regulatory effect of *Dhh* on mammalian spermatogenesis is indirectly achieved through interstitial cells.

The *dpp*-equivalent signaling pathway is also involved in mammalian spermatogenesis, but, again, with a different function. BMP-8a and BMP-8b are expressed initially in early germ cells at the onset of spermatogenesis in early puberty and then abundantly expressed in stage 6–8 round spermatids (Zhao and Hogan, 1996). Consistent with this biphasal expression, BMP-8b knockout (KO) mice exhibit defects both in reduced, or failure of, germ cell proliferation at the onset of spermatogenesis at puberty and the significantly increased apoptosis of spermatocytes (Zhao et al., 1996). Thus, BMP-8b is crucial for GSC proliferation and maintenance. However, BMP-8a KO mice do not show obvious germ cell defects during the initiation of spermatogenesis. However, 47% of adult homozygous BMP-8a males show germ cell degeneration, establishing a role of BMP-8a in the maintenance of spermatogenesis (Zhao et al., 1998). Phenotypic analysis suggests that BMP-8b is likely to function locally through either an autocrine effect on germ cells or a short paracrine effect on Sertoli cells. Consistent with either possibility, activin type II receptors (ActRIIA and ActRIIB) and receptors of BMP-2 and BMP-7 (Yamashita et al., 1995) are expressed within seminiferous tubules, with ActRIIB in spermatogonia and early meiotic germ cells and Sertoli cells. Since BMP-8B is within the same subfamily of BMP-7 (Griffith et al., 1996), it is possible that BMP-8B binds to ActRIIB receptor.

The effectors of the BMP signaling pathway, *madr1* and *madr2*, two mouse homologs of the *Drosophila mother against dpp* (*mad*) genes, have different patterns of expression in the testis. *madr1* is expressed mainly in the pachytene spermatocytes and round spermatids. In the adult BMP-8b mutant testis, the *madr1*-expressing pachytene spermatocytes first show increased apoptosis (Zhao and

Hogan, 1997). These data suggest that MADR1 serves as a downstream component of the BMP-8 signaling pathway during the differentiation of meiotic male germ cells. MADR-2, on the other hand, is present mainly in spermatogonia and early meiotic cells and Sertoli cells (Zhao et al., 1996) and, thus, may be involved in GSC division.

Two other related signaling molecules, glial cell line–derived neurotrophic factor (GDNF) and neurturin (NRTN), are expressed by Sertoli cells (Meng et al., 2000, 2001b) and play important roles in spermatogonial renewal and subsequent stages of spermatogenesis. Their coreceptors, GDNF family receptor $\alpha 1$ and $\alpha 2$, are expressed at different germ cells, suggesting different targets for the ligands. GDNF regulates cell fate decisions of undifferentiated spermatogonia, including the stem cells, during spermatogenesis in a dose-sensitive manner. GDNF-haploid mice show depletion of stem cell reserves, whereas mice overexpressing GDNF display accumulation of undifferentiated spermatogonia (Meng et al., 2000). These accumulated spermatogonia do not respond properly to differentiation signals and undergo apoptosis on retinoic acid treatment. They develop into seminomas—invasive testicular tumors retaining germ cell characters, in older GDNF-overexpressing mice (Meng et al., 2001a). Thus, GDNF contributes to paracrine regulation of spermatogonial self-renewal and differentiation.

Different from GDNF, NRTN-deficient mice show apoptosis of spermatogenic cells at low penetrance, while overexpression of NRTN initially causes gradual depletion of spermatogenic cells. After 5 wk of age, spermatogenic defects disappear and the NRTN-overexpressing mice were fertile. The data suggest that NRTN might regulate the survival and differentiation of spermatocytes and spermatids, but the low penetrance indicates that NRTN is not as essential as GDNF for spermatogenesis (Meng et al., 2001b).

In addition to the above signaling pathways, cell–cell junctional molecules have been implicated in spermatogonial stem cell maintenance. For example, Connexin43 (Cx43), a gap junction protein encoded by the *Gja1* gene, is expressed in several cell types of the testis (Roscoe et al., 2001). Cx43 gap junctions couple Sertoli cells with each other, Leydig cells with each other, and spermatogonia/spermatocytes with Sertoli cells. Cx43 null mutant neonates show a germ cell deficiency that arises during fetal life. Grafted mice testes lacking Cx43 display normal steroidogenesis, but the germline deficiency persists postnatally, giving rise to a "Sertoli-cell-only" phenotype (Roscoe et al., 2001). These results indicate that intercellular communication via Cx43 channels is required for postnatal expansion of the male germline, possibly by directly mediating communications between spermatogonial and other cells. In addition, Sertoli cell–gonocyte coculture experiments revealed the presence of numerous adhesion plaques between these cells, suggesting that Sertoli cells and gonocytes are able to communicate in vitro (van Dissel-Emiliani et al., 1993). This simple culture system may provide an effective model for studying the function of the cell–cell signaling pathways between Sertoli cells and gonocytes that are required for initiating the cell cycle of the quiescent gonocytes to form type A spermatogonia for the onset of spermatogenesis. Moreover, it may provide a functional assay to define a minimal stem cell niche for the mammalian testis.

5.5. INTRACELLULAR STEM CELL MECHANISM: SYMMETRY VS ASYMMETRY

The very nature of stem cells determines that they should possess a robust cell-cycle program that is tightly regulated by signaling from the niche. For stereotypic stem cells, it is also important to have a precise asymmetry-generating mechanism. The asymmetric mechanism may consist of asymmetrically expressed and/or segregated cell fate regulators and a cellular machinery responsible for their asymmetric segregation or for establishing an asymmetric exposure of the two daughter cells to extrinsic signals. At present, less is known about the intracellular mechanisms than the extrinsic signaling. Nevertheless, rapid progress in the past several years has allowed various facets of the intracellular mechanism to be elucidated.

5.5.1. CELL CYCLE MACHINERY The study of cell-cycle regulators in controlling GSC division has been largely at the descriptive level. In mice, G1 cyclins D1, D2, and D3; G2 cyclin A2 as well as CDK4 D3; and Wee-1 are expressed in spermatogonia, with some of them also expressed in later stages of spermatogenesis (Ravnik et al., 1995; Wu and Wolgemuth, 1995; Ravnik and Wolgemuth, 1996; Sweeney et al., 1996; Zhang et al., 1999). However, targeted disruption of an A-type cyclin gene suggests that it has a role in meiosis, but not in stem cell divisions (Liu et al., 2000). Cdc25A and Cdc25C phosphatases are also weakly expressed in the spermatogonia, but Cdc25C-null mice are fertile (Chen et al., 2001). The proliferative cell nuclear antigen protein, involved in DNA synthesis and repair, is detected in proliferating spermatogonia but not in meiotic spermatocytes (Chapman and Wolgemuth, 1994).

Recent mouse KO studies suggest that several cell-cycle components may have a role in stem cell division. For example, p18(Ink4c) and p19(Ink4d), two inhibitors of cyclin D–dependent kinases (Cdks) are expressed in the seminiferous tubules of postnatal wild-type mice, being largely confined to postmitotic spermatocytes undergoing meiosis. Male mice deficient in both *Ink4c* and *Ink4d* are sterile (Zindy et al., 2001). The sterility is owing to the delayed exit of spermatogonia from the mitotic cell cycle, leading to the retarded appearance of meiotic cells that do not properly differentiate and instead undergo apoptosis at an increased frequency. These data indicate that p18(Ink4c) and p19(Ink4d) collaborate in regulating spermatogenesis, helping to ensure mitotic exit and the normal meiotic maturation of spermatocytes.

Another example is afforded by p27kip1, a cyclin-dependent kinase inhibitor that regulates the G1/S transition of the cell cycle. p27kip1 is expressed when gonocytes become quiescent on d 16 dpc, suggesting that p27kip1 is an important factor for the G1/G0 arrest in gonocytes. In p27kip1 KO mice, the numbers of type A spermatogonia are increased, and abnormal preleptotene spermatocytes are present, some of which seemingly try to enter a mitotic division instead of entering the meiotic prophase (Beumer et al., 1999). These observations indicate that p27kip1 suppresses spermatogonial proliferation, or apoptosis, and the onset of the meiotic prophase in preleptotene spermatocytes. However, because p27kip1 is only expressed in Sertoli cells, the role of p27kip1 in both spermatogonia and preleptotene spermatocytes must be indirect.

In *Drosophila*, *roughex* and cyclin A regulate the normal progression of male meiosis while cyclin E controls the female nurse cell endocycle. However, their role in GSC division has not been addressed (Gonczy et al., 1994; Lilly and Spradling, 1996; Lilly et al., 2000). One gene, *shut-down* (*shu*), might link cell cycle to extrinsic signaling (Munn and Steward, 2000). The *shu* mutant displays a *piwi/Yb*-like phenotype, but, the mutation is germline dependent. Thus, *shu* appears to encode a cell-autonomous factor essential for the normal function of the GSCs and subsequent oogenesis. Consistent with this, the *shu* RNA and protein are strongly expressed in the GSCs and in 16-cell cysts. The RNA is also present in the germ cells throughout embryogenesis. The SHU protein shares similarity to the heat-shock protein-binding immunophilins. Both immunophilins and SHU contain an FK506-binding protein domain and a tetratricopeptide repeat. In plants, high-molecular-weight immunophilins have been shown to regulate cell divisions in the root meristem in response to extracellular signals. These results suggest that *shu* may regulate germ cell divisions in the germarium.

In *C. elegans*, a subset of cyclins is mostly expressed in the germline (Kreutzer et al., 1995). This expression pattern provides a basis for functional analysis of these molecules in GSC division. In addition, *cul-2*, a *C. elegans* homolog of human cullin, is expressed in proliferating germ cells and is required at two distinct points in the cell cycle: the G1-to-S-phase transition and mitosis. *cul-2* mutant germ cells undergo a G1-phase arrest that correlates with accumulation of CKI-1, a member of the CIP/KIP family of cyclin-dependent-kinase inhibitors (Feng et al., 1999). These two genetic models provide effective platforms for further analyses of the role of cell-cycle genes in stem cell division.

5.5.2. ASYMMETRICALLY EXPRESSED AND/OR SEGREGATED CELL FATE REGULATORS The generation of asymmetry during stem cell division is expected to be regulated by factors that are differentially expressed and/or segregated between the two daughter cells. Such a differential gene expression could be controlled at the DNA or genetic information level, such as gene conversion during mating-type switch in the yeasts (Klar and Bonaduce, 1993; Amon, 1996). However, this level of regulation is unlikely to occur in GSCs because these cells need to maintain their genome to produce gametes of the same genetic constitution. If such regulation happens, it would have to be reverted precisely. However, differential gene expression can be manifested at the level of asymmetric DNA replication, like what occurs in the fission yeast (Dalgaard and Klar, 2001), epigenetic modification of the chromatin, transcription, posttranscriptional RNA processing, translation, and post-translational processing of the protein products.

In *Drosophila*, although asymmetrically segregating cell fate determinants, such as NUMB and PROSPERO, have been identified in neuroblasts (reviewed in Lin and Schagat, 1997), such molecules have not been clearly demonstrated for GSCs. Nevertheless, several *Drosophila* genes have recently been shown to be required cell autonomously for either the maintenance of female GSCs or the proper differentiation of cystoblasts. Among them, *pumilio* (*pum*) is essential for GSC maintenance. *pum* mutant females often contain an abnormal number of GSCs at the onset of oogenesis, owing to the requirement of *pum* during preoogenic germline development (Parisi and Lin, 1999). Moreover, the formed GSCs fail to maintain themselves during oogenesis (Lin and Spradling, 1997; Forbes and Lehmann, 1998; Parisi and Lin, 1999). PUM is an RNA-binding protein known to mediate translational repression in the embryo (Murata and Wharton, 1995; Wharton et al., 1998). It is present at a high level in GSCs but at a low level in cystoblasts (Forbes and Lehmann, 1998; Parisi and Lin, 1999). Thus, PUM may be an asymmetrically expressed cell fate regulator that selectively suppresses the translation of certain RNAs in the stem cell to prevent it from differentiation. Several *pum* homologs have been identified in mammals and are expressed in the testis (e.g., Nagase et al., 1996; H. L., unpublished data), suggesting that a *pum*-like mechanism may exist in mammalian spermatogonial stem cells as well.

The partner of *pum* in the embryo, *nanos* (*nos*), may be involved in germline maintenance since 50% of the newly enclosed *nos* mutant adult ovaries are germlineless (Forbes and Lehmann, 1998; Bhat, 1999). This defect, however, could also be owing to an earlier requirement of *nos* for germline development prior to oogenesis. During oogenesis, *nos* has been reported to function in the differentiation of cystoblasts and germline cysts since *nos* mutations block germline cyst development in adult ovaries without the immediate loss of germline (Forbes and Lehmann, 1998). However, this could be a milder manifestation of the failure in germline maintenance, either due to the hypomorphic nature of the mutation or the slightly redundant role of *nos* in the process.

In addition to *pum* and *nos*, the putative transcription factor *stonewall* (*stwl*) is required for preoogenic development and for the maintenance of female GSCs. Some *stwl* germaria are germlineless, while others display a rapid decrease in egg chamber production, in addition to subsequent oogenic defects (Akiyama, 2002).

Contrary to the role of *pum*, *nos*, and *stwl*, the *bag of marbles* (*bam*) gene is required cell autonomously to promote stem cell differentiation in both females and males (McKearin and Spradling, 1990; mcKearin a nd Ohlstein, 1995; Gonczy et al., 1997; Ohlstein and McKearin, 1997). *bam* mutant germ cells fail to differentiate but, instead, proliferate like stem cells or ill-differentiated germline cells (McKearin and Spradling, 1990; McKearin and Ohlstein, 1995). Heat shock–induced ectopic expression of *bam* is sufficient to eliminate GSCs (Ohlstein and McKearin, 1997). Differential expression of *bam* mRNA and the cytoplasmic form of the BAM protein in the cystoblast but not the GSCs suggests that BAM is an asymmetrically expressed cell-autonomous regulator for the cystoblast fate.

Like *bam*, the *benign gonial cell neoplasm* (*bgcn*) gene appears to be another cell-autonomous factor required for the cystoblast fate. Mutations of *bgcn* block the differentiation of GSC product, causing *bam*-like phenotype (Gateff, 1982). In addition, *bgcn* mutations dominantly enhance a *bam* mutant phenotype, revealing the functional interdependence of these two genes (Ohlstein et al., 2000). *bgcn* mRNA is expressed in a small number of germline cells in the germarium, including the stem cells (Ohlstein et al., 2000). *bgcn* encodes a protein that is related to the DExH-box family of RNA-dependent helicases but lacks critical residues for adenosine triphosphatase and helicase functions. Hence, it is possible that BGCN and BAM may be involved in regulating the translation of the cystoblast differentiation genes.

In mammals, a promising candidate for an intrinsic factor required for the self-renewal of spermatogonial stem cells is the human *deleted in azoospermia* (*DAZ*) gene cluster in the *azoospermia factor* (*AZF*) region on Y chromosome (Saxena et al., 1996). *DAZ*, a candidate of *AZF*, encodes a putative RNA-binding protein with strong homology to the *Drosophila* BOULE protein required for meiosis during spermatogenesis (Eberhart et al., 1996). *DAZ* is expressed specifically in the germ cells of the adult human testis and most abundantly in spermatogonia (Menke et al., 1997). The mouse homolog of *DAZ*, *Dazh* (or *Dazla*), is also expressed in germ cells, starting at 12.5 dpc in embryonic gonads before germ cell sex differentiation (Cooke et al., 1996; Reijo et al., 1996; Seligman and Page, 1998). The *Dazh* level decreases in female embryos following the entry of oogonia into meiosis but persists in the male gonad and is present in d 1 neonatal male mice whose germ cells are gonocytes. Subsequently, *Dazh* expression increases as spermatogonial stem cells appear, reaches the peak as spermatogenic cells first enters meiosis, and persists at this level thereafter (Reijo et al., 1996). Important information on *Dazh* function comes from the analysis of *Dazh*-KO mice, whose embryonic gonads are normal up to 15 dpc, but by 19 dpc germ cells in seminiferous tubules (and in ovaries as well) are significantly reduced in both heterozygous and homozygous *Dazh*-deficient mice (Ruggiu et al., 199&). This result suggests that *Dazh* is quantitatively required for the initial maintenance of gonocytes—which are either mitotically arrested GSCs or the immediate precursors of GSCs. Recently, DAZ, and an autosomally encoded DAZ-like protein, DAZL-1, have been shown to bind ribohomopolymers in vitro (Tsui et al., 2000b). DAZL-1 is also associated with polysomes. Thus, the DAZ family proteins may regulate spermatogenesis by regulating the translation of certain spermatogenic mRNAs. Proteins interacting with DAZ and DAZL-1 have been identified (Tsui et al., 2000a; Dai et al., 2001). Further analyses of these proteins will provide new opportunities to further define the DAZ-mediated mechanism in spermatogenesis, possibly in regulating GSC division.

In addition to *DAZ*, the *Hertwig's anemia* (*an*) mutation does not affect the migration of PGCs but causes their death in the testis. Hence, the *an* gene may be required for the PGC-to-gonocyte transition (Russell, 1985). The *atrichosis* (*at*) mutation causes Sertoli-only phenotype, suggesting the role of *at* in PGC or gonocyte development (Chubb and Nolan, 1984). The *juvenile spermatogonia depletion* (*jsd*) mutation does not affect the first wave of spermatogenesis. However, the spermatogenesis fails to be maintained, leading to degeneration of germ cells after midpuberty (Beamer et al., 1998; Barton et al., 1989; Mizunuma et al., 1992). Thus, this gene plays a clearcut role in the self-renewal of GSCs. Recent germ cell transplantation experiments suggest that *jsd* is a cell-autonomous gene (Boettger-Tong et al., 2000; Ohta et al., 2001). Molecular characterization of *jsd* should shed interesting light on the cell-autonomous mechanism that controls the self-renewal of spermatogonial stem cells.

For long-term self-renewal, spermatogonial stem cells must derive a way to protect their telomere. Consistent with this notion, the telomerase activity is expressed at high levels in the type A spermatogonial stem cells, is downregulated during spermatogenesis, and is absent in the differentiated spermatozoa (Ravindranath et al., 1997).

In *C. elegans*, a DNA microarray approach has been taken to profile gene expression patterns in the germline (Reinke et al., 2000). More than 1416 germline-enriched transcripts have been identified that define three groups of germline-expressing genes—the germline intrinsic group, the sperm-enriched group, and the oocyte-enriched group. The germline intrinsic group, defined as genes expressed similarly in germlines making only sperm or only oocytes, contains a family of *piwi*-related genes that may be important for stem cell proliferation.

5.5.3. APOPTOSIS Because mammalian GSCs are populational stem cells, they utilize apoptosis as a means to achieve homeostasis between self-renewal vs differentiation. The Bcl-2 family members Bax and Bcl-x(L) are involved in this density regulation of spermatogonia along the tubule (reviewed in de Rooij, 2001). Several mechanisms are available to cope with major or minor shortages in germ cell production. After severe cell loss, stem cell renewal is preferred over differentiation and the period of proliferation of A(s), A(pr), and A(al) spermatogonia is extended. Minor shortages are dealt with, at least in part, by less apoptosis among A2–A4 spermatogonia. The regulation of apoptosis involved the tumor suppressor p53 (Beumer et al., 1998; Van Buul et al., 2001). During normal spermatogenesis in the mouse, spermatogonia do not express p53. However, following irradiation, p53 becomes expressed in spermatogonia. p53 KO mouse testes contain about 50% more A1 spermatogonia as well as increased numbers of giant-sized spermatogonial stem cells. In addition, the differentiating A2-B spermatogonia are more radioresistant compared with their wild-type controls, indicating that p53 is involved in eliminating lethally irradiated differentiating spermatogonia. These results suggest that p53 is an important factor in normal spermatogonial cell production as well as in the regulation of apoptosis after DNA damage.

Apoptosis during spermatogenesis is regulated by extrinsic signals, as shown by experiments on in vitro seminiferous tubule cultures. FSH prevents germ cells from undergoing apoptosis in vitro, while the SLF level is increased dramatically in response to FSH stimulation. SLF in turn supports germ cell survival during spermatogenesis by upregulating prosurvival Bcl-2 family proteins, Bcl-w and Bcl-xL, and downregulating proapoptosis Bcl-2 family proteins, such as Bax (Yan et al., 2000a, 200b).

5.5.4. CELLULAR MACHINERY FOR ASYMMETRIC LOCALIZATION/ORIENTATION For stereotypic stem cells, it is crucial to have a cellular mechanism that ensures the proper segregation and localization of cytoplasmic components. The role of cytoskeletal systems in mammalian and *C. elegans* GSC division still remains unexplored. In *Drosophila*, it is well established that cytoskeletal systems play a crucial role in such asymmetric segregation and localization as well as in spindle orientation during mitosis (reviewed in Amon, 1996; Guo and Kemphues, 1996; Lin and Schagat, 1997). In *Drosophila* ovarian GSCs, the cytoskeleton has been shown to play an important role in the asymmetric GSC divisions (Lin et al., 1994; Lin and Spradling, 1995; de Cuevas et al., 1996; Deng and Lin, 1997; McGrail and Hays, 1997). As reviewed in the beginning of this chapter, the spectrosome, enriched in membrane skeletal protein, ensures the divisional asymmetry by anchoring the mitotic spindle orientation and by localizing important cell fate and cell-cycle molecules such as BAM and cyclin A. At telophase, the spectrosome grows in size and elongates toward the future cystoblast and eventually becomes asymmetrically bisected by delayed cytokinesis occurring at the next cell cycle. Delayed cytokinesis may be a common feature of GSC division, since it is also seen for dividing spermatogonial stem cells in both *Drosophila* and mammals.

It is important to point out that even if the spectrosome is important for the asymmetry of GSC division, such asymmetry may not be essential for the self-renewal of GSCs in the *Drosophila* ovary. In the absence of the spectrosome and its derivative structure, the fusome, a normal number of egg chambers, though ill differentiated, are still produced (Lin and Spradling, 1997). This possibly reflects the fact that, although GSC divisions are randomly oriented without the spectrosome, the topology of the stem cell niche still allows about 50% of the daughter cells to be included in the niche. Consequently, a self-renewing population of germ cells is still maintained. Thus, without the spectrosome, the GSCs in the *Drosophila* ovary lose the stereotypic stem cell quality and become populational stem cells. The spectrosome thus has two functions: first, it ensures the precise asymmetry needed for the stem cells to undergo stereotypic fashion of self-renewal, and second, it ensures the asymmetry that is required for the proper differentiation of the cystoblast. Studies on the spectrosome have revealed the critical role of membrane skeletal proteins in establishing a divisional asymmetry that may be important in other stem cell systems.

Then, what links the spindle pole to the spectrosome? What orients the spectrosome toward the cap cell? The answer to these questions lies in cytoplasmic dynein (McGrail and Hays, 1997). In germline cysts, cytoplasmic dynein is required in linking one pole of the mitotic spindle to the fusome. Given the structural similarities between the fusome and the spectrosome, it is likely that cytoplasmic dynein plays the same role in GSCs. Studying how the spectrosome is anchored near cap cells may offer opportunities to reveal how somatic signaling induces polarity in the target stem cells.

5.6. CONCLUSION

GSCs offer excellent opportunities for studying fundamental questions in stem cell biology. Indeed, the pluripotency of stem cells in adult tissues may be a partial manifestation of the totipotency of the germline. Thus, the two systems may, after all, share common mechanisms that guarantee their multipotentials in development. By choosing to study stem cells in the germline, one places his/her research in a unique position that allows both stem cell and germline biology to be effectively explored. Owing to the effective combination of genetic, cell biological, molecular, and developmental approaches that are uniquely afforded by model organisms such as *C. elegans*, *Drosophila*, and mice, works in recent years on GSCs have clearly revealed some of the key intra- and intercellular mechanisms for the self-renewal of GSCs. The surrounding signaling cells have been shown to constitute stem cell niches that play an instructive role in determining the fate and behavior of GSCs. Some important cell-cell signaling molecules and intrinsic factors have been identified, and their regulatory relationship has begun to be explored in the context of gene expression, cell-cycle progression, cytoskeletal organization, and mitotic behavior of stem cells.

Elucidation of the mechanisms for the self-renewing division of GSCs is now occurring at an exciting rate. New genetic and microarray screens will identify more genes involved in GSC mechanisms. Meanwhile, further analysis of known genes and pathways involved in GSC division will reveal exactly how cell-cell signaling and intracellular events occur in an orchestrated manner to ensure the stem cell property. These genetic and molecular analyses will also reveal how GSC division and maintenance are related to general cell-cycle and germline mechanisms. The knowledge gained from these studies should allow researchers to effectively reveal the defining mechanisms of stem cells and the germline—two great mysteries in biology and medicine.

ACKNOWLEDGMENTS

I thank colleagues for sending preprints and reprints of their work and for sharing unpublished results. I apologize to those authors whose work is not cited owing to space limitation. I am grateful to Akos Szakmary and Zhong Wang for their critical reading of the manuscript. This work presented in this chapter as well as the work in my laboratory on GSCs was supported by National Institutes of Health grants HD33760 and HD42012.

REFERENCES

Akiyama, T. (2002) Mutations of *stonewall* disrupt the maintenance of female germline stem cells in *Drosophila melanogaster*. *Dev. Growth Differ.* 44:97–102.

Amon, A. (1996) Mother and daughter are doing fine: asymmetric cell division in yeast. *Cell* 84:651–654.

Asaoka-Taguchi, M., Yamada, M., Nakamura, A., Hanyu, K., and Kobayashi, S. (1999) Maternal Pumilio acts together with Nanos in germline development in *Drosophila* embryos. *Nat. Cell. Biol.* 1:431–437.

Austin, J. and Kimble, J. (1987) *Glp-1* is required in germ line for regulation of the decision between mitosis and meiosis in *C. elegans*. *Cell* 51:589–599.

Bardin, C. W., Gunsalu, G. L., and Cheng, C. Y. (1993) The cell biology of the Sertoli cell. In: *Cell and Molecular Biology of the Testis* (Desjardins, C. and Ewing, L. L., eds.), Oxford University Press, New York, pp. 189–219.

Darton, D. E., Yang-Feng, T. L., Mason, A. J., Seeburg, P. H., and Francke, U. (1989) Mapping of genes for inhibin subunits alpha, beta A, and beta B on human and mouse chromosomes and studies of *jsd* mice. *Genomics* 5:91–99.

Beamer, W. G., Cunliffe-Beamer, T. L., Shultz, K. L., Langley, S. H., and Roderick, T. H. (1988) *Juvenile spermatogonial depletion* (*jsd*): a genetic defect of germ cell proliferation of male mice. *Biol. Reprod.* 38:899–908.

Beaumont, H. M. and Mandl, A. M. (1963) A quantitative study of primordial germ cells in the male rat. *J. Embryol. Exp. Morph.* 11: 715–740.

Bellve, A. R. (1979) The molecular biology of mammalian spermatogenesis. In: *Oxford Review of Reproductive Biology* (Finn, C. A., ed.), Oxford University Clarendon Press, Oxford, pp. 159–261.

Bellve, A. R. (1993) Purification, culture, and fractionation of spermatogenic cells. *Methods Enzymol.* 225:84–113.

Bellve, A. R., Cavicchia, J. C., Millette, C. F., O'Brian, D. A., Bhatnagar, Y. M., and Dym, M. (1977) Spermatogenic cells of the prepubertal mouse: isolation and morphological characterization. *J. Cell. Biol.* 74: 68–85.

Benfey, P. N. (1999) Stem cells: A tale of two kingdoms. *Curr. Biol.* 9: R171–172.

Beumer, T. L., Roepers-Gajadien, H. L., Gademan, I. S., et al. (1998) The role of the tumor suppressor p53 in spermatogenesis. *Cell Death Differ.* 5:669–677.

Beumer, T. L., Kiyokawa, H., Roepers-Gajadien, H. L., et al. (1999) Regulatory role of p27kip1 in the mouse and human testis. *Endocrinology* 140:1834–1840.

Bhat, K. M. (1999) The posterior determinant gene *nanos* is required for the maintenance of the adult germline stem cells during *Drosophila* oogenesis. *Genetics* 151:1479–1492.

Bitgood, M. J. and McMahon, A. P. (1995). *Hedgehog* and *Bmp* genes are coexpressed at many diverse sites of cell-cell interaction in the mouse embryo. *Dev. Biol.* 172:126–138.

Bitgood, M. J., Shen, L., and McMahon, A. P. (1996) Sertoli cell signaling by desert hedgehog regulates the male germ line. *Curr. Biol.* 6: 298–304.

Boettger-Tong, H. L., Johnston, D. S., Russell, L. D., Griswold, M. D., and Bishop, C. E. (2000). *Juvenile spermatogonial depletion* (*jsd*) mutant seminiferous tubules are capable of supporting transplanted spermatogenesis. *Biol. Reprod.* 63:1185–1191.

Boujrad, N., Ogwuegbu, S. O., Garnier, M., Lee, C. H., Martin, B. M., and Papadopoulos, V. (1995). Identification of a stimulator of steroid hormone synthesis isolated from testis. *Science* 268:1609–1612.

Braunhut, S. J., Rufo, G. A., Ernisee, B. J., Zheng, W. X., and Bellve, A. R. (1990). The seminiferous growth factor induces proliferation of TM4 cells in serum-free medium. *Biol. Reprod.* 42:639–648.

Brinster, R. L. and Zimmermann, J. W. (1994). Spermatogenesis following male germ-cell transplantation. *Proc. Natl. Acad. Sci. USA* 91: 11,298–11,302.

Cerutti, L., Mian, N., and Bateman, A. (2000). Domains in gene silencing and cell differentiation proteins: the novel PAZ domain and redefinition of the Piwi domain. *Trends Biochem. Sci.* 25:481–482.

Chapman, D. L. and Wolgemuth, D. J. (1994). Expression of proliferating cell nuclear antigen in the mouse germ line and surrounding somatic cells suggests both proliferation-dependent and -independent modes of function. *Int. J. Dev. Biol.* 38:491–497.

Chatelain, P. G., Sanchez, P., and Saez, J. M. (1991) Growth hormone and insulin-like growth factor I treatment increase testicular luteinizing hormone receptors and steroidogenic responsiveness of growth hormone deficient dwarf mice. *Endocrinology* 128:1857–1862.

Chen, M. S., Hurov, J., White, L. S., Woodford-Thomas, T., and Piwnica-Worms, H. (2001) Absence of apparent phenotype in mice lacking Cdc25C protein phosphatase. *Mol. Cell. Biol.* 21:3853–3861.

Chiarini-Garcia, H. and Russell, L. D. (2001) High-resolution light microscopic characterization of mouse spermatogonia. *Biol. Reprod.* 65: 1170–1178.

Chiquoine, A. D. (1954) The identification, origin, and migration of the primordial germ cells in the mouse embryo. *Anat. Rec.* 118:135–146.

Chubb, C. and Nolan, C. (1984) Genetic control of steroidogenesis and spermatogenesis in inbred mice. *Ann. NY Acad. Sci.* 438:519–522.

Clark, A. M., Garland, K. K., and Russell, L. D. (2000) *Desert hedgehog* (*Dhh*) gene is required in the mouse testis for formation of adult-type leydig cells and normal development of peritubular cells and seminiferous tubules. *Biol. Reprod.* 63:1825–1838.

Clark, J. M. and Eddy, E. M. (1975) Fine structural observations on the origin and associations of primordial germ cells of the mouse. *Dev. Biol.* 47:136–155.

Cooke, H. J., Lee, M., Kerr, S., and Ruggiu, M. (1996) A murine homologue of the human *DAZ* gene is autosomal and expressed only in male and female gonads. *Hum. Mol. Genet.* 5:513–516.

Cox, D. N., Chao, A., Baker, J., Chang, L., Qiao, D., and Lin, H. (1998) A novel class of evolutionarily conserved genes defined by *piwi* are essential for stem cell self-renewal. *Genes Dev.* 12:3715–3727.

Cox, D. N., Chao, A., and Lin, H. (2000). *piwi* encodes a nucleoplasmic factor whose activity modulates the number and division rate of germline stem cells. *Development* 127:503–514.

Dai, T., Vera, Y., Salido, E. C., and Yen, P. H. (2001) Characterization of the mouse *Dazap1* gene encoding an RNA-binding protein that interacts with infertility factors DAZ and DAZL. *BMC Genomics* 2:6.

Dalgaard, J. Z. and Klar, A. J. (2001) Does *S. pombe* exploit the intrinsic asymmetry of DNA synthesis to imprint daughter cells for mating-type switching? *Trends Genet.* 17:153–157.

de Cuevas, M., Lee, J., and Spradling, A. C. (1996) a-spectrin is required for germline cell division and differentiation in the *Drosophila* ovary. *Development* 122:3959–3968.

de Cuevas, M. and Spradling, A. C. (1998) Morphogenesis of the *Drosophila* fusome and its implications for oocyte specification. *Development* 125:2781–2789.

De Felici, M. and Dolci, S. (1989) In vitro adhesion of mouse fetal germ cells to extracellular matrix components [published erratum appears in *Cell Differ. Dev.* 1989 27(2):149]. *Cell Differ. Dev.* 26:87–96.

De Felici, M. and Pesce, M. (1994) Growth factors in mouse primordial germ cell migration and proliferation. *Prog. Growth Factor Res.* 5: 135–143.

De Miguel, M. P., De Boer-Brouwer, M., Paniagua, R., van den Hurk, R., De Rooij, D., and Van Dissel-Emiliani, F. M. (1996) Leukemia inhibitory factor and ciliary neurotropic factor promote the survival of Sertoli cells and gonocytes in coculture system. *Endocrinology* 137: 1885–1893.

de Rooij, D. G. (2001) Proliferation and differentiation of spermatogonial stem cells. *Reproduction* 121:347–354.

Deng, W. and Lin, H. (1997) Spectrosomes and fusomes are essential for anchoring mitotic spindles during asymmetric germ cell divisions and for the microtubule-based RNA transport during oocyte specification in *Drosophila. Dev. Biol.* 189:79–94.

Deng, W., and Lin, H. (2002) *miwi*, a murine homolog of *piwi*, encodes a cytoplasmic protein essential for spermatogenesis. *Dev. Cell* 2: 819–830.

Dobrinski, I., Ogawa, T., Avarbock, M. R., and Brinster, R. L. (1999) Computer assisted image analysis to assess colonization of recipient seminiferous tubules by spermatogonial stem cells from transgenic donor mice. *Mol. Reprod. Dev.* 53:142–148.

Donovan, P. J. (1994) Growth factor regulation of mouse primordial germ cell development. *Curr. Top. Dev. Biol.* 29:189–225.

Donovan, P. J., Scott, D., Cairns, L. A., Heasman, J., and Whylie, C. C. (1986) Migratory and postmigratory mouse primordial germ cells behave differentely in culture. *Cell* 44:831–838.

Drummond-Barbosa, D. and Spradling, A. C. (2001) Stem cells and their progeny respond to nutritional changes during *Drosophila* oogenesis. *Dev. Biol.* 231:265–278.

Eberhart, C. G., Maines, J. Z., and Wasserman, S. A. (1996) Meiotic cell cycle requirement for a fly homologue of human *Deleted in Azoospermia. Nature* 381:783–785.

Eddy, M. (1970). Cytochemical observations on the chromatoid body of the male germ cells. *Biol. Reprod.* 2:114–128.

Engels, W. R. and Preston, C. R. (1979) Hybrid dysgenesis in *Drosophila melanogaster*: the biology of female and male sterility. *Genetics* 92: 161–174.

Falin, L. I. (1969) The development of genital glands and the origin of germ cells in human embryogenesis. *Acta Anat.* 72:195–232.

Feng, H., Zhong, W., Punkosdy, G., et al. (1999) CUL-2 is required for the G1-to-S-phase transition and mitotic chromosome condensation in *Caenorhabditis elegans. Nat. Cell. Biol.* 1:486–492.

Fitzgerald, K. and Greenwald, I. (1995). Interchangeability of *Caenorhabditis elegans* DSL proteins and intrinsic signalling activity of their extracellular domains in vivo. *Development* 121:4275–4282.

Forbes, A. and Lehmann, R. (1998) Nanos and Pumilio have critical roles in the development and function of *Drosophila* germline stem cells. *Development* 125:679–690.

Forbes, A. J., Lin, H., Ingham, P. W., and Spradling, A. C. (1996) *hedgehog* is required for the proliferation and specification of ovarian somatic cells prior to egg chamber formation in *Drosophila. Development* 122: 1125–1135.

Fujimoto, T., Miyayama, Y., and Fuyuta, M. (1977) The origin, migration and fine morphology of human primordial germ cells. *Anat. Rec.* 188: 315–330.

Gateff, E. (1982) *Gonial Cell Neoplasm of Genetic Origin Affecting Both Sexes of* Drosophila melanogaster. Alan R. Liss, New York.

Ginsburg, M., Snow, M. H. L., and McLaren, A. (1990). Primordial germ cells in the mouse embryo during gastrulation. *Development* 110: 521–528.

Gonczy, P., and DiNardo, S. (1996) The germ line regulates somatic cyst cell proliferation and fate during *Drosophila* spermatogenesis. *Development* 122:2437–2447.

Gonczy, P., Thomas, B. J., and DiNardo, S. (1994) *Roughex* is a dose-dependent regulator of the second meiotic division during *Drosophila* spermatogenesis. *Cell* 77:1015–1025.

Gonczy, P., Matunis, E., and DiNardo, S. (1997) *bag-of-marble* and *benign gonial cell neoplasm* act in the germ line to restrict proliferation during *Drosophila* spermatogenesis. *Development* 124:4361–4371.

Griffith, D. L., Keck, P. C., Sampath, T. K., Rueger, D. C., and Carlson, W. D. (1996) Three-dimensional structure of recombinant human osteogenic protein 1: structural paradigm for the transforming growth factor beta superfamily. *Proc. Natl. Acad. Sci. USA* 93:878–883.

Griswold, M. D. (1995). Interactions between germ cells and Sertoli cells in the testis. *Biol. Reprod.* 52:211–216.

Guo, S. and Kemphues, K. J. (1996). Molecular genetics of asymmetric cleavage in the early *Caenorhabditis elegans* embryo. *Curr. Opin. Genet. Dev.* 6:408–415.

Han, I. S., Sylvester, S. R., Kim, K. H., et al. (1993) Basic fibroblast growth factor is a testicular germ cell product which may regulate Sertoli cell function. *Mol. Endocrinol.* 7:889–897.

Hansson, H. A., Billig, H., and Isgaard, J. (1989) Insulin-like growth factor I in the developing and mature rat testis: immunohistochemical aspects. *Biol. Reprod.* 40:1321–1328.

Hardy, R. W., Tokuyasu, K. T., Lindsley, D. L., and Garavito, M. (1979) The germinal proliferation center in the testis of *Drosophila melanogaster. J. Ultrastruct. Res.* 69:180–190.

Hilscher, B., Hilscher, W., Bulthoff-Ohnolz, B., et al. (1974) Kinetics of gametogenesis. I. Comparative histological and autoradiographic studies of oocytes and transitional prospermatogonia during oogenesis and prespermatogenesis. *Cell Tissue Res.* 154:443–470.

Hirsh, D., Oppenheim, D., and Klass, M. (1976) Development of the reproductive system of *Caenorrhabditis elegans*. *Dev. Biol.* 49: 200–219.

Huckins, C. and Clermont, Y. (1968) Evolution of gonocytes in the rat testis during late embryonic and early post-natal life. *Arch. Anat. Histol. Embryol.* 51:341–354.

Jenkin, G., McFarlane, J., and de Kretser, D. M. (1995). Inhibin and activin in embryonic and fetal development in ruminants. *J. Reprod. Fertil. Suppl.* 49:177–186.

Jeon, K. W. and Kennedy, J. R. (1973) The primordial germ cells in early mouse embryos; light and electron microscope studies. *Dev. Biol.* 31: 275–284.

Jiang, F. X. (2001) Male germ cell transplantation: promise and problems. *Reprod. Fertil. Dev.* 13:609–614.

Jones, R. C. and Lin, M. (1993) Spermatogenesis in birds. *Oxf. Rev. Reprod. Biol.* 15.233–264.

Kadyk, L. C. and Kimble, J. (1998) Genetic regulation of entry into meiosis in *Caenorhabditis elegans*. *Development* 125:1803–1813.

Kerkis, J. (1933) Development of gonads in hybrids between *Drosophila melanogaster* and *D. simulans*. *J. Exp. Zool.* 66:477–509.

Kiger, A. A. and Fuller, M. T. (2001) Male germ-line stem cells. In: *Stem Cell Biology* (Marshak, D. R., Gardner, R. L., and Gottlieb, D., eds.), Cold Spring Harbor Laboratory Press, Cold Spring Harbor, NY, pp. 149–187.

Kiger, A. A., White-Cooper, H., and Fuller, M. T. (2000). Somatic support cells restrict germline stem cell self-renewal and promote differentiation. *Nature* 407:750–754.

Kiger, A. A., Jones, D. L., Schulz, C., Rogers, M. B., and Fuller, M. T. (2001) Stem cell self-renewal specified by JAK-STAT activation in response to a support cell cue. *Science* 294:2542–2545.

Kimble, J. and Hirsh, D. (1979) Postembryonic cell lineages of the hermaphrodite and male gonads in *C. elegans*. *Dev. Biol.* 70:396–417.

Kimble, J. and Simpson, P. (1997) The LIN-12/Notch signaling pathway and its regulation. *Annu. Rev. Cell. Dev. Biol.* 13:333–361.

Kimble, J. and White, J. (1981) On the control of germ cell development in *Caenorhabditis elegans*. *Dev. Biol.* 81:208–219.

King, F. J. and Lin, H. (1999) Somatic signaling mediated by *fs(1)Yb* is essential for germline stem cell maintenance during *Drosophila* oogenesis. *Development* 126:1833–1844.

King, F. J., Szakmary, A., Cox, D. N., and Lin, H. (2001) *Yb* modulates the divisions of both germline and somatic stem cells through *piwi*- and *hh*-mediated mechanisms in the *Drosophila* ovary. *Mol. Cell.* 7: 497–508.

King, R. C. (1970). *Ovarian Development in Drosophila Melanogaster*. Academic Press, New York.

Klar, A. J. S. and Bonaduce, M. J. (1993) The mechanism of fission yeast mating-type interconversion: evidence for two types of epigenetically inherited chromosomal imprinted events. *Cold Spring Harbor Symp. Quant. Biol.* LVIII:457–465.

Kopan, R. and Turner, D. L. (1996) The Notch pathway—democracy and aristocracy during the selection of cell fate. *Curr. Opin. Neurobiol.* 6:594–601.

Kreutzer, M. A., Richards, J. P., De Silva-Udawatta, M. N., et al. (1995) *Caenorhabditis elegans cyclin A*- and *B*-type genes: a cyclin A multigene family, an ancestral *cyclin B3* and differential germline expression. *J. Cell. Sci.* 108:2415–2424.

Lawson, K. A. and Hage, W. J. (1994) Clonal analysis of the origin of primordial germ cells in the mouse. In: *Germline Development*, CIBA Foundation Symposium 182 (March, J. and Goode, J., eds.), John Wiley & Sons, Chichester, pp. 68–92.

Lilly, M. and Spradling, A. C. (1996) The *Drosophila* endocycle is controlled by cyclin E and lacks a checkpoint ensuring S-phase completion. *Genes Dev.* 10:2514–2526.

Lilly, M. A., de Cuevas, M., and Spradling, A. C. (2000). Cyclin A associates with the fusome during germline cyst formation in the *Drosophila* ovary. *Dev. Biol.* 218:53–63.

Lin, H. (1997) The tao of stem cells in the germline. *Annu. Rev. Genet.* 31:455–491.

Lin, H. (1998) The self-renewing mechanism of stem cells in the germline. *Curr. Opin. Cell. Biol.* 10:687–693.

Lin, H. and Schagat, T. (1997) Neuroblasts: a model for asymmetric division of stem cells. *Trends Genet.* 13:33–39.

Lin, H. and Spradling, A. (1993) Germline stem cell division and egg chamber development in transplanted *Drosophila* germaria. *Dev. Biol.* 159:140–152.

Lin, H. and Spradling, A. (1995). Fusome asymmetry and oocyte determination. *Dev. Genet.* 16:6–12.

Lin, H. and Spradling, A. C. (1997) A novel group of *pumilio* mutations affects the asymmetric division of germline stem cells in the *Drosophila* ovary. *Development* 124:2463–2476.

Lin, H., Yue, L., and Spradling, A. S. (1994) The *Drosophila* fusome, a germline-specific organelle, contains membrane skeletal proteins and functions in cyst formation. *Development* 120:947–956.

Littlefield, C. L. (1985) Germ cells in *Hydra oligactis* males. I. Isolation of a subpopulation of interstitial cells that is developmentally restricted to sperm production. *Dev. Biol.* 112:185–193.

Liu, D., Liao, C., and Wolgemuth, D. J. (2000) A role for cyclin A1 in the activation of MPF and G2-M transition during meiosis of male germ cells in mice. *Dev. Biol.* 224:388–400.

Luetteke, N. C., Qiu, T. H., Peiffer, R. L., Oliver, P., Smithies, O., and Lee, D. C. (1993) TGF alpha deficiency results in hair follicle and eye abnormalities in targeted and waved-1 mice. *Cell* 73:263–279.

Manova, K. and Bachvarova, R. F. (1991) Expression of c-kit encoded at the *W* locus of mice in developing embryonic germ cells and presumptive melanoblasts. *Dev. Biol.* 146:312–324.

Manova, K., Nocka, K., Besmer, P., and Bachvarova, R. F. (1990) Gonadal expression of c-kit encoded at the *W* locus of the mouse. *Development* 110:1057–1069.

Martins, M. R. and Silva, J. R. (2001) Ultrastructure of spermatogonia and primary spermatocytes of C57BL6J mice. *Anat. Histol. Embryol.* 30:129–132.

Matunis, E., Tran, J., Gonczy, P., and DiNardo, S. (1997) *punt* and *schnurri* control progression through the germ line stem cell lineage by regulating a somatically-derived signal which restricts germ cell proliferation. *Development* 124:4361–4371.

McCarter, J., Bartlett, B., Dang, T., and Schedl, T. (1997) Soma-germ cell interactions in *Caenorhabditis elegans*: multiple events of hermaphrodite germline development require the somatic sheath and spermathecal lineages. *Dev. Biol.* 181:121–143.

McGrail, M. and Hays, T. S. (1997) The microtubule motor cytoplasmic dynein is required for spindle orientation during germline cell divisions and oocyte differentiation in *Drosophila*. *Development* 124:2409–2419.

McGuinness, M. P. and Orth, J. M. (1992) Reinitiation of gonocyte mitosis and movement of gonocytes to the basement membrane in testes of newborn rats in vivo and in vitro. *Anat. Rec.* 233:527–537.

McKearin, D. and Ohlstein, B. (1995) A role for the *Drosophila* Bag-of-marbles protein in the differentiation of cystoblasts from germline stem cells. *Development* 121:2937–2947.

McKearin, D. M. and Spradling, A. C. (1990) *Bag-of-marbles*: a *Drosophila* gene required to initiate both male and female gametogenesis. *Genes Dev.* 4:2242–2251.

McLaren, A. (1984) Meiosis and differentiation of mouse germ cells. *Symp. Soc. Exp. Biol.* 38:7–23.

Meistrich, M. L. and van Beek, M. E. A. B. (1993) Spermatogonial stem cells. In: *Cell and Molecular Biology of the Testis* (Desjardins, C. and Ewing, L. L., eds.), Oxford University Press, New York, pp. 266–295.

Meng, X., Lindahl, M., Hyvonen, M. E., et al. (2000) Regulation of cell fate decision of undifferentiated spermatogonia by GDNF. *Science* 287:1489–1493.

Meng, X., de Rooij, D. G., Westerdahl, K., Saarma, M., and Sariola, H. (2001a) Promotion of seminomatous tumors by targeted overexpression of glial cell line-derived neurotrophic factor in mouse testis. *Cancer Res.* 61:3267–3271.

Meng, X., Pata, I., Pedrono, E., Popsueva, A., de Rooij, D. G., Janne, M., Rauvala, H., and Sariola, H. (2001b) Transient disruption of spermatogenesis by deregulated expression of neurturin in testis. *Mol. Cell. Endocrinol.* 184:33–39.

Menke, D. B., Mutter, G. L., and Page, D. C. (1997) Expression of DAZ, an azoospermia factor candidate, in human spermatogonia. *Am. J. Hum. Genet.* 60:237–241.

Miura, T., Miura, C., Ohta, T., Nader, M. R., Todo, T., and Yamauchi, K. (1999) Estradiol-17beta stimulates the renewal of spermatogonial stem cells in males. *Biochem. Biophys. Res. Commun.* 264:230–234.

Mizunuma, M., Dohmae, K., Tajima, Y., Koshimizu, U., Watanabe, D., and Nishimune, Y. (1992) Loss of sperm in *juvenile spermatogonial depletion* (*jsd*) mutant mice is ascribed to a defect of intratubular environment to support germ cell differentiation. *J. Cell Physiol.* 150: 188–193.

Munn, K. and Steward, R. (2000) The *shut-down* gene of *Drosophila melanogaster* encodes a novel FK506-binding protein essential for the formation of germline cysts during oogenesis. *Genetics* 156:245–256.

Murata, Y. and Wharton, R. P. (1995) Binding of pumilio to maternal hunchback mRNA is required for posterior patterning in *Drosophila* embryos. *Cell* 80:747–756.

Nagano, M., Avarbock, M. R., and Brinster, R. L. (1999) Pattern and kinetics of mouse donor spermatogonial stem cell colonization in recipient testes. *Biol. Reprod.* 60:1429–1436.

Nagano, M., McCarrey, J. R., and Brinster, R. L. (2001) Primate spermatogonial stem cells colonize mouse testes. *Biol. Reprod.* 64: 1409–1416.

Nagase, T., Seki, N., Ishikawa, K., et al. (1996) Prediction of the coding sequences of unidentifed human genes. VI. The coding sequences of 80 new genes (KIAA0201-KIAA0280) deduced by analysis of cDNA clones from cell line KG-1 and brain. *DNA Res.* 3:321–329.

Nagy, A. (2000) Cre recombinase: the universal reagent for genome tailoring. *Genesis* 26:99–109.

Nebel, B. R., Amarose, A. P., and Hackett, E. M. (1961) Calendar of gametogenic development in the prepuberal male mouse. *Science* 134: 832–833.

Nishimiya-Fujisawa, C. and Sugiyama, T. (1993) Genetic analysis of developmental mechnisms in *Hydra*. XX. Cloning of interstitial stem cells restricted to the sperm differentiation pathway in *Hydra magnipapillata*. *Dev. Biol.* 157:1–9.

Ohlstein, B., and McKearin, D. (1997) Ectopic expression of the *Drosophila* Bam protein eliminates oogenic germline stem cells. *Development* 124:3651–3662.

Ohlstein, B., Lavoie, C. A., Vef, O., Gateff, E., and McKearin, D. M. (2000) The *Drosophila* cystoblast differentiation factor, benign gonial cell neoplasm, is related to DExH-box proteins and interacts genetically with bag-of-marbles. *Genetics* 155:1809–1819.

Ohta, H., Yomogida, K., Tadokoro, Y., Tohda, A., Dohmae, K., and Nishimune, Y. (2001) Defect in germ cells, not in supporting cells, is the cause of male infertility in the jsd mutant mouse: proliferation of spermatogonial stem cells without differentiation. *Int. J. Androl.* 24: 15–23.

Orr-Urtreger, A., Avivi, A., Zimmer, Y., Givol, D., Yarden, Y., and Lonai, P. (1990) Developmental expression of c-kit, a proto-oncogene encoded by the *W* locus. *Development* 109:911–923.

Orth, J. M., Jester, W. F., and Qiu, J. P. (1996) Gonocytes in testes of neonatal rats express the *c-kit* gene. *Mol. Reprod. Dev.* 45:123–131.

Ozdzenski, W. (1967) Observations on the origin of primordial germ cells in the mouse. *Zool. Pol.* 17:367–379.

Parisi, M. P. and Lin, H. (1999) The *Drosophila pumilio* gene encodes two functional protein isoforms that play multiple roles in germline development, gametogenesis, oogenesis and embryogenesis. *Genetics* 153: 235–250.

Parvinen, M., Pelto-Huikko, M., Soder, O., et al. (1992) Expression of beta-nerve growth factor and its receptor in rat seminiferous epithelium: specific function at the onset of meiosis. *J. Cell. Biol.* 117:629–641.

Pesce, M., Gross, M. K., and Scholer, H. R. (1998) In line with our ancestors: Oct-4 and the mammalian germ. *Bioessays* 20:722–732.

Poirie, M., Niederer, E., and Steinmann-Zwicky, M. (1995) A sex-specific number of germ cells in embryonic gonads of *Drosophila*. *Development* 121:1867–1873.

Powell-Braxton, L., Hollingshead, P., Warburton, C., et al. (1993) IGF-I is required for normal embryonic growth in mice. *Genes Dev.* 7:2609–2617.

Ravindranath, N., Dalal, R., Solomon, B., Djakiew, D., and Dym, M. (1997) Loss of telomerase activity during male germ cell differentiation. *Endocrinology* 138:4026–4029.

Ravnik, S. E. and Wolgemuth, D. J. (1996) The developmentally restricted pattern of expression in the male germ line of a murine cyclin A, cyclin A2, suggests roles in both mitotic and meiotic cell cycles. *Dev. Biol.* 173:69–78.

Ravnik, S. E., Rhee, K., and Wolgemuth, D. J. (1995) Distinct patterns of expression of the D-type cyclins during testicular development in the mouse. *Dev. Genet.* 16:171–178.

Reijo, R., Seligman, J., Dinulos, M. B., et al. (1996) Mouse autosomal homolog of *DAZ*, a candidate male sterility gene in humans, is expressed in male germ cells before and after puberty. *Genomics* 35: 346–352.

Reinke, V., Smith, H. E., Nance, J., et al. (2000) A global profile of germline gene expression in *C. elegans*. *Mol. Cell* 6:605–616.

Robertson, D. M., Risbridger, G. P., and de Krester, D. M. (1993) Inhibins and inhibin-related proteins. In: *Cell and Molecular Biology of the Testis* (Desjardins, C. and Ewing, L. L., eds) Oxford University Press, New York, pp. 220–237.

Roscoe, W. A., Barr, K. J., Mhawi, A. A., Pomerantz, D. K., and Kidder, G. M. (2001) Failure of spermatogenesis in mice lacking connexin43. *Biol. Reprod.* 65:829–838.

Rossi, P., Sette, C., Dolci, S., and Geremia, R. (2000) Role of c-kit in mammalian spermatogenesis. *J. Endocrinol. Invest.* 23:609–615.

Ruggiu, M., Speed, R., Taggart, M., et al. (1997) The mouse *Dalza* gene encodes a cytoplasmic protein essential for gametogenesis. *Nature* 389:73–77.

Ruohola, H., Bremer, K. A., Baker, D., Swedlow, J. R., Jan, L. Y., and Jan, Y. N. (1991) Role of neurogenic genes in establishment of follicle cell fate and oocyte polarity during oogenesis in *Drosophila*. *Cell* 66:433–449.

Russell, E. S. (1985) A history of mouse genetics. *Ann. Rev. Genet.* 19: 1–28.

Russell, L. D. and Griswold, M. D. (2000) Spermatogonial transplantation—an update for the millennium. *Mol. Cell. Endocrinol.* 161: 117–120.

Saxena, R., Brown, L. G., Hawkins, T., et al. (1996) The DAZ gene cluster on the human Y chromosome arose from an autosomal gene that was transposed, repeatedly amplified and pruned. *Nat. Genet.* 14:292–299.

Schaefer, R. E., Kidwell, M. G., and Fausto-Sterling, A. (1979) Hybrid dysgenesis in *Drosophila melanogaster*: morphological and cytological studies of ovarian dysgenesis. *Genetics* 92:1141–1152.

Seligman, J. and Page, D. C. (1998) The *Dazh* gene is expressed in male and female embryonic gonads before germ cell sex differentiation. *Biochem. Biophy. Res. Commun.* 245:878–882.

Seydoux, G. and Dunn, M. A. (1997) Transcriptionally repressed germ cells lack a subpopulation of phosphorylated RNA polymerase II in early embryos of *Caenorhabditis elegans* and *Drosophila melanogaster*. *Development* 124:2191–2201.

Seydoux, G., Mello, C. C., Pettitt, J., Wood, W. B., Priess, J. R., and Fire, A. (1996) Repression of gene expression in the embryonic germ lineage of *C. elegans*. *Nature* 382:713–716.

Shackleford, G. M. and Varmus, H. E. (1987) Expression of the proto-oncogene *int-1* is restricted to postmeiotic male germ cells and the neural tube of mid-gestational embryos. *Cell* 50:89–95.

Shinohara, T., Avarbock, M. R., and Brinster, R. L. (1999) beta1- and alpha6-integrin are surface markers on mouse spermatogonial stem cells. *Proc. Natl. Acad. Sci. USA* 96:5504–5509.

Shinohara, T., Orwig, K. E., Avarbock, M. R., and Brinster, R. L. (2000) Spermatogonial stem cell enrichment by multiparameter selection of mouse testis cells. *Proc. Natl. Acad. Sci. USA* 97:8346–8351.

Sorrentino, V., Giorgi, M., Geremia, R., Besmer, P., and Rossi, P. (1991) Expression of the *c-kit* proto-oncogene in the murine male germ cells. *Oncogene* 6:149–151.

Starborg, M. and Hoog, C. (1995) The murine replication protein 1 is differentially expressed during spermatogenesis. *Eur. J. Cell. Biol.* 68:206–210.

Streit, A., Bernasconi, L., Sergeev, P., Cruz, A., and Steinmann-Zwicky, M. (2002) *mgm1*, the earliest sex-specific germline marker in Drosophila, reflects expression of the gene esg in male stem cells. *Int. J. Dev. Biol.* 46:159–166.

Sulston, J. E., Schierenberg, E., White, J. G., and Thomson, J. N. (1983) The embryonic cell lineage of the nematode *Caenorhabditis elegans*. *Dev. Biol.* 100:64–119.

Sweeney, C., Murphy, M., Kubelka, M., et al. (1996) A distinct cyclin A is expressed in germ cells in the mouse. *Development* 122:53–64.

Tabara, H., Sarkissian, M., Kelly, W. G., et al. (1999) The *rde-1* gene, RNA interference, and transposon silencing in *C. elegans*. *Cell* 99: 123–132.

Tazuke, S. I., Schulz, C., Gilboa, L., et al. (2002) A germline-specific gap junction protein required for survival of differentiating early germ cells. *Development* 129:2529–2539.

Tegelenbosch, R. A. and de Rooij, D. G. (1993) A quantitative study of spermatogonial multiplication and stem cell renewal in the C3H/101 F1 hybrid mouse. *Mutat. Res.* 290:193–200.

Tran, J., Brenner, T. J., and DiNardo, S. (2000) Somatic control over the germline stem cell lineage during *Drosophila* spermatogenesis. *Nature* 407:754–757.

Tsui, S., Dai, T., Roettger, S., Schempp, W., Salido, E. C., and Yen, P. H. (2000a) Identification of two novel proteins that interact with germ-cell-specific RNA-binding proteins DAZ and DAZL1. *Genomics* 65: 266–273.

Tsui, S., Dai, T., Warren, S. T., Salido, E. C., and Yen, P. H. (2000b) Association of the mouse infertility factor DAZL1 with actively translating polyribosomes. *Biol. Reprod.* 62:1655–1660.

Vallier, L., Mancip, J., Markossian, S., et al. (2001) An efficient system for conditional gene expression in embryonic stem cells and in their in vitro and in vivo differentiated derivatives. *Proc. Natl. Acad. Sci. USA* 98:2467–2472.

Van Buul, P. P., Van Duyn-Goedhart, A., Beumer, T., and Bootsma, A. L. (2001) Role of cell cycle control in radiosensitization of mouse spermatogonial stem cells. *Int. J. Radiat. Biol.* 77:357–363.

van Dissel-Emiliani, F. M., de Boer-Brouwer, M., Spek, E. R., van der Donk, J. A., and de Rooij, D. G. (1993) Survival and proliferation of rat gonocytes in vitro. *Cell Tissue Res.* 273:141–147.

van Dissel-Emiliani, F. M., de Rooij, D. G., and Meistrich, M. L. (1989) Isolation of rat gonocytes by velocity sedimentation at unit gravity. *J. Reprod. Fertil.* 86:759–766.

van Dissel-Emiliani, F. M., De Boer-Brouwer, M., and De Rooij, D. G. (1996) Effect of fibroblast growth factor-2 on Sertoli cells and gonocytes in coculture during the perinatal period. *Endocrinology* 137: 647–654.

Van Doren, M., DeFalco, T., Verney, G., Russell, S., and Jenkins, A. (2002) The development of sexual dimorphism in the *Drosophila* gonad. Paper presented at the 43rd Annual *Drosophila* Research Conference Program and Abstracts, San Diego, CA.

van Pelt, A. M., van Dissel-Emiliani, F. M., Gaemers, I. C., van der Burg, M. J., Tanke, H. J., and de Rooij, D. G. (1995) Characteristics of a spermatogonia and preleptotene spermatocytes in the vitamin A-deficient rat testis. *Biol. Reprod.* 53:570–578.

Wharton, R. P., Sonoda, J., Lee, T., Patternson, M., and Murata, Y. (1998) The pumilio RNA-binding domain is also a translational regulator. *Mol. Cell* 1:863–872.

Wieschaus, E. and Szabad, J. (1979) The development and function of the female germline in *Drosophila melanogaster*: a cell lineage study. *Dev. Biol.* 68:29–46.

Wilson, E. B. (1896) *The Cell in Development and Heredity*, 1st ed. Macmillan, New York.

Wu, S. and Wolgemuth, D. J. (1995) The distinct and developmentally regulated patterns of expression of members of the mouse *Cdc25* gene family suggest differential functions during gametogenesis. *Dev. Biol.* 170:195–206.

Wylie, C. (1999) Germ cells. *Cell* 96:165–174.

Xie, T. and Spradling, A. C. (1998) *Decapentaplegic* is essential for the maintenance and division of germline stem cells in the *Drosophila* ovary. *Cell* 94:251–260.

Xie, T., and Spradling, A. C. (2000) A niche maintaining germ line stem cells in the *Drosophila* ovary. *Science* 290:328–330.

Xie, T. and Spradling, A. C. (2001) The *Drosophila* ovary: an *in vivo* stem cell system. In: *Stem Cell Biology* (Marshak, D. R. Gardner, R. L. and Gottlieb, D. eds.), Cold Spring Harbor Laboratory Press, Cold Spring Harbor, NY, pp. 129–148.

Yamashita, H., ten Dijke, P., Huylebroeck, D., et al. (1995) Osteogenic protein-1 binds to activin type II receptors and induces certain activin-like effects. *J. Cell. Biol.* 130:217–226.

Yan, W., Suominen, J., Samson, M., Jegou, B., and Toppari, J. (2000a) Involvement of Bcl-2 family proteins in germ cell apoptosis during testicular development in the rat and pro-survival effect of stem cell factor on germ cells in vitro. *Mol. Cell. Endocrinol.* 165:115–129.

Yan, W., Suominen, J., and Toppari, J. (2000b) Stem cell factor protects germ cells from apoptosis in vitro. *J. Cell. Sci.* 113(Pt. 1):161–168.

Zhang, Q., Wang, X., and Wolgemuth, D. J. (1999) Developmentally regulated expression of cyclin D3 and its potential in vivo interacting proteins during murine gametogenesis. *Endocrinology* 140: 2790–2800.

Zhao, G. Q. and Hogan, B. L. M. (1996) Evidence that mouse *BMP8a* (*op2*) and *BMP8b* are duplicated genes that play a role in spermatogenesis and placental development. *Mech. Dev.* 57:159–168.

Zhao, G. Q., and Hogan, B. L. M. (1997) Evidence that *mother-against-dpp-related 1* (*madr1*) plays a role in the initiation and maintenance of spermatogenesis in the mouse. *Mech. Dev.* 61, 63-73.

Zhao, G. Q., Deng, K., Labosky, P. A., Liaw, L., and Hogan, B. L. (1996) The gene encoding bone morphogenetic protein 8B is required for the initiation and maintenance of spermatogenesis in the mouse. *Genes Dev.* 10:1657–1669.

Zhao, G.-Q., Liaw, L., and Hogan, B. L. M. (1998) Bone morphogenetic protein 8A plays a role in the maintenance of spermatogenesis in the integrity of the epididymis. *Development* 125:1103–1112.

Zheng, W. X., Butwell, T. J., Heckert, L., Griswold, M. D., and Bellve, A. R. (1990) Pleiotypic actions of the seminiferous growth factor on two testicular cell lines: comparisons with acidic and basic fibroblast growth factors. *Growth Factors* 3:73–82.

Zindy, F., den Besten, W., Chen, B., et al. (2001) Control of spermatogenesis in mice by the cyclin D-dependent kinase inhibitors p18(Ink4c) and p19(Ink4d). *Mol. Cell. Biol.* 21:3244–3255.

6 Stem Cells and Cloning

Ian Wilmut and Lesley Ann Paterson, PhD

Development to term of a cloned offspring derived from an embryo created by somatic cell nuclear transfer is a remarkable demonstration of adult cellular plasticity. Reconstruction of embryos by transfer of human cells into enucleated oocytes (therapeutic cloning) is one potential source of pluripotent embryonic stem (ES) cells for cell replacement therapy. Furthermore, ES cell lines that are derived by this method should be histocompatible with the patient who supplied the donor cells. In addition, therapeutic cloning from diseased individuals will provide ES cell lines that can be used as in vitro cellular models for research purposes. In animal reproductive cloning (i.e., creation of an entire offspring), nuclei from a variety of tissues have been used for transfer into metaphase II–arrested enucleated oocytes, although the success rate is very low. Losses occur throughout the process, from the nuclear transfer procedure; from poorly developed embryos, loss of fetuses, and death at birth or newborns within 24 h; as well as from premature postnatal death. Development of viable individuals from nuclear transfer is typically between 1 and 4%; nevertheless, therapeutic cloning requires development only to the 6-day embryo stage, and thus we should expect efficiency to be greater. The results from animal reproductive cloning studies have revealed important aspects about the nuclear transfer process. The task of the oocyte in reprogramming a diploid somatic cell nucleus is a formidable one, and it is not surprising that the success rate is low. Abnormalities in DNA methylation patterns of the chromatin have been found in cloned embryos. Considering that epigenetic changes may be associated with development of cancers, care must be taken to screen for the dysprogramming of genes in ES lines derived from cloned embryos. Thus, although the therapeutic promise is very great, the safety of transplanted cloned cells must be assessed. In addition, the creation of in vitro cellular models of disease by therapeutic cloning will provide an exceptional tool for research and drug screening.

6.1. INTRODUCTION

Human stem cells offer great potential for lifelong treatment of deteriorating and debilitating diseases such as Parkinson's disease, diabetes, multiple sclerosis, and heart disease. There are primarily two stem cell sources available: those derived from embryos and those derived from adult tissues (including the umbilical vein).

Investigations carried out using stem cells in mice have revealed their powerful therapeutic capabilities. Transfer of adult neural stem cells (Teng et al., 2002) and neural differentiated embryonic cells (McDonald et al., 1999) into spinal cord injury in rats resulted in hind-limb weight bearing, stepping, and enhanced coordination. Oligodendrocytes and insulin-secreting cells, both derived from embryonic stem (ES) cells, replaced lost myelin in rat spinal cords (Liu et al., 2000) and normalized glycemia in diabetic mice (Soria et al., 2000), respectively. In addition, following transplantation of partially differentiated mouse ES cells into the rat model of Parkinson's disease, approximately half of the rats with surviving grafts contained dopamine neurons (Bjorklund et al., 2002). These demonstrations show the vast potential of cell replacement therapy in humans.

6.1.1. ADULT AND ES CELLS There are advantages and disadvantages of obtaining stem cells from adult sources or embryos as sources for therapy (*see* Table 1). ES cells are pluripotent, immortal cell lines derived from the inner cell mass (ICM) of embryos, with the potential to differentiate into cells from all three embryonic lineages, even after prolonged culture (Thomson et al., 1998). Adult stem cells reside in certain tissues of the body and exist to replenish the tissues in which they reside. These cells show limited differentiation potential and are therefore said to be multipotent. However, the plasticity of these stem cells lines is potentially much greater than previously believed, and investigations have shown that adult stem cells can "transdifferentiate" into cells of different embryonic lineages: from muscle to blood and vice versa, from blood to liver, and from brain to blood and muscle (For a review, *see* Clarke and Frisen, 2001). Nevertheless, recent publications have highlighted the need for caution in interpretation of the apparent plasticity of adult stem cells. Two independent groups cultured mouse ES cells alongside mouse brain (Ying et al., 2002) and bone marrow cells (Terada et al., 2002). Both sets of results showed that apparently reprogrammed adult cells were

From: *Stem Cells Handbook*
Edited by: S. Sell © Humana Press Inc., Totowa, NJ

Table 1
Comparison of Embryo and Adult Stem Cell Sources

	Banked ES cells	*Cloned ES cells*	*Adult stem cells*
Histocompatible		+++	+++
No. of Cells	+++	+++	?
Differentiation	++	++	?
Life Span	+++	+++	+
Normality/ Safety	?	?	?
Availability	+++	++	+

polyploid and were in fact hybrids, created by fusion with the ES cells. This does not prove that the previous demonstrations of adult stem cell plasticity are false but that further analysis, such as karyotyping, must be carried out after transdifferentiation experiments. It also highlights the need to continue all lines of research using both embryo and adult sources of cell lines, to discover the full potential, applications, and limitations of each.

In this chapter, we consider the applications and uses of ES cells, particularly regarding embryos created by nuclear transfer. The derivation and application of adult stem cell therapy are discussed in several other chapters in this book.

6.2. ES CELL DIFFERENTIATION

As with their murine counterparts (Evans and Kaufman, 1981), human ES cells are pluripotent, immortal, continually proliferating cell lines with the capability to form all three of the embryonic germ layers (Thomson et al., 1998). Differentiation into many "useful" human cell types has been demonstrated, by either spontaneous or directed differentiation, including insulin-producing cells (Assady et al., 2001), nerve cells (Reubinoff et al., 2001; Zhang et al., 2001), and hematopoietic cells (Kaufman et al., 2001).

6.3. SOURCES OF ES CELLS

ES cells can be harvested from embryos, surplus to in vitro fertilization (IVF), as was used to derive the first human ES cell line (Thomson et al., 1998); from embryos created from donated oocytes and sperm (Lanzendorf et al., 2001); and from cloned embryos. Each of these sources has advantages and disadvantages. There are thousands of IVF embryos in storage and surplus to fertility treatment requirements and would therefore provide adequate numbers for research purposes (with the appropriate fully informed consent). Furthermore, since these embryos are already destined for destruction, they may be considered, by society and governmental regulations, as a legal and suitable source of ES cells. Nevertheless, during IVF treatments a number of embryos are made, and the best-quality ones are chosen for implantation, leaving the potentially lesser-quality embryos for ES cell derivation. Furthermore, these embryos are frozen, which could further limit their developmental potential. The creation of embryos from donated sperm and oocytes in the laboratory for ICM cell harvesting may well provide higher-quality embryos and further assist in successful ES derivation by using unfrozen embryos.

Some tissues and organs such as some brain tissues and the eye are immunoprivileged so that immunologically different cells are not rejected. Therefore, banked ES cell lines derived from IVF embryos (surplus to fertility treatment or created specifically) could be used to derive the appropriate cell types. However, banked ES cell–derived tissue would be immunologically different for most patients and so would be rejected when grafted into, e.g., the liver and heart. In this case, immunosuppression treatment would be required for transplant of cells that do not illicit an immune response.

6.4. THERAPEUTIC CLONING

One strategy to avoid immunorejection is to use nuclear transfer to insert a cell nucleus, donated from the patient, into an enucleated oocyte and then activate it to begin cell division and thus derive histocompatible ES cells from the resulting blastocysts. Another very powerful application of therapeutic cloning is the derivation of disease-specific ES cell lines for research purposes, e.g., by creating nuclear transfer embryos from the cells of patients with motor neuron disease. This would provide an invaluable in vitro cellular model of the disease for a wide variety of research purposes, and as a practically limitless cell source for drug discovery.

6.5. NUCLEAR TRANSFER

6.5.1. THE PRESENT SITUATION Research into the derivation of ES cells from nuclear transfer embryos is in its infancy. Nevertheless, lessons from nuclear transfer in animal reproductive cloning have already revealed both successes and limitations of this procedure.

The actual protocol used for nuclear transfer is dependent on the species but follows a basic set of techniques (Wilmut et al., 1997). The nucleus or whole cell is transferred to a metaphase II–arrested enucleated oocyte, previously stripped of its own genetic material. The reconstructed embryo is triggered to begin embryo development by an electrical pulse or chemicals. Surviving nuclear transfer embryos are incubated in vitro, and those in which development is successful are transferred to recipients for development to term.

Seven species of animal have been cloned to date using somatic cell nuclear transfer: sheep (Wilmut et al., 1997), cow (Cibelli et al., 1998), mouse (Wakayama et al., 1998), pig (Polejaeva et al., 2000), goat (Keefer et al., 2000), cat (Shin et al., 2002), and rabbit (Chesne et al., 2002). A variety of adult cells have been used including fibroblasts, cumulus cells, granulosa cells, muscle cells, and blood cells. Although no particular adult cell type has had a significantly high success rate, cumulus and fetal fibroblasts have been successful for several species. A significantly higher proportion of mouse ES cell–derived blastocysts reach development to term, and as opposed to somatic cells (Wakayama et al., 1999), inbred cells (both somatic and ES) are significantly less successful for cloning than heterozygous cells (Eggan et al., 2001; Wakayama and Yanagimachi, 2001).

Overall, irrespective of the species or cell type, nuclear transfer remains very inefficient, and typically between 0 and 5% of reconstructed oocytes produce offspring. The losses occur throughout the nuclear transfer process (Heyman et al., 2002), from poorly developed embryos, loss of fetuses through miscarriage, and death at birth or of newborns within 24 h, which are often attributed to cardiopulmonary defects (Wilmut et al., 1997; Hill et al., 1999). Large offspring syndrome, a condition in which large size is the most recognizable feature of a variety of abnormalities, has been frequently seen in cloned animals (Sinclair et al., 2000). Even after apparently normal birth of livestock, premature postnatal deaths have been reported (Renard et al., 1999; Lanza et al., 2001) and two independent studies have

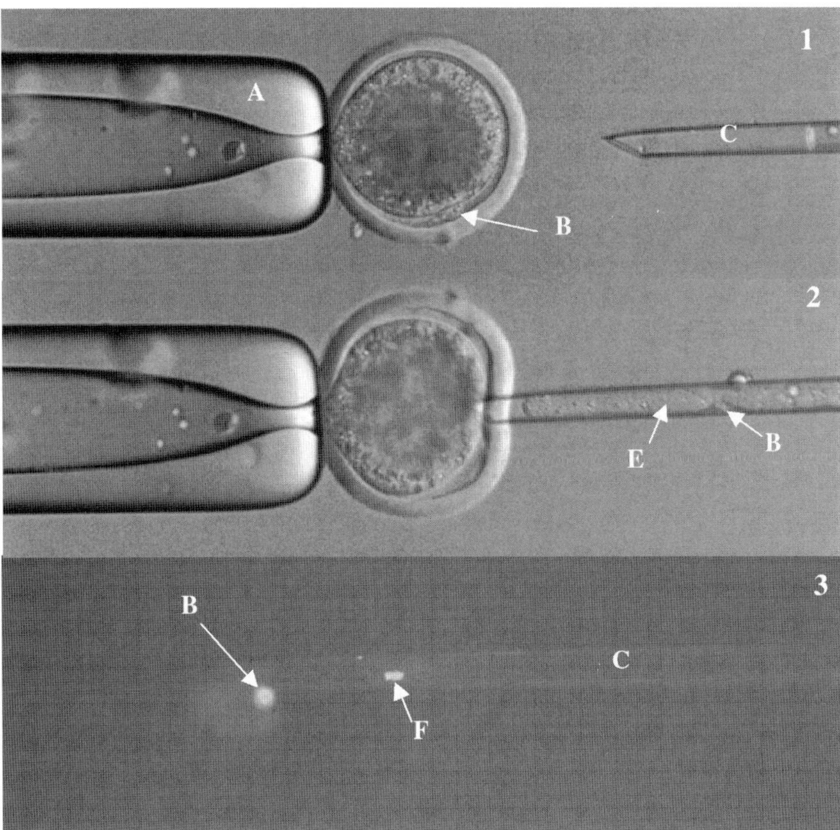

Fig. 1. Enucleation. The holding pipet (A) secures the Hoescht stained oocyte in place. The polar body (B) and metaphase chromosomes (usually located next to the polar body) are removed by the enucleation pipet (C) by piercing the zona pellucida and then sucking out the chromosomes, the polar body (B), and some cytoplasm (E). The enucleation pipet (C) is irradiated with UV light to determine that both polar body (B) and metaphase chromosomes (F) have been removed. (Photograph by W. Ritchie.)

revealed that cloned mice have a significantly shorter life-span (Ogonuki et al., 2002) and can develop obesity in adult life (Tamashiro et al., 2002). In the latter case, the phenotype is not passed onto the offspring, indicating that this abnormality was due to an epigenetic effect (*see* below for a discussion on molecular mechanisms). Nevertheless, only nuclear transfer embryos will be required for ES cell derivation, and the efficiency for successful blastocyst derivation is much higher than production of viable cloned offspring.

6.5.2. NUCLEAR TRANSFER METHOD For each cloned species, the optimal nuclear transfer technique may vary in one or more steps of the procedure. Either the entire donor cell can be fused with the enucleated oocyte (Fig. 1), which is commonly used for sheep and cattle (e.g., Wilmut et al., 1997), or the donor nucleus is extracted from the cell and inserted into the oocyte using a piezo-injection technique, which is generally used with mice (Wakayama et al., 1998). Furthermore, following construction of the nuclear transfer embryos, a delay to activation (from 15 min to hours) has been shown to be essential in mice (Wakayama et al., 1998), but not an absolute necessity in cattle.

Although single nuclear transfer directly into zygotes is an ineffective technique (Wakayama et al., 2000), serial nuclear transfer has been successfully used for pigs and mice (Polejaeva et al., 2000; Ono et al., 2001). In this case, following the first nuclear transfer into an enucleated oocyte, the "pronucleus" that forms is then transferred to a zygote (previously created by IVF and stripped of its own genetic material) for development. The beneficial effect on cloned embryo development and development to term may be reflected in exposure of the transferred somatic nucleus first to the oocyte cytoplasm for initial reprogramming and then to the naturally fertilized zygotic cytoplasm for continued development. Furthermore, rabbit was considered one of the more "difficult-to-clone" mammals and recent success with this species was attributed to extending the asynchrony between donor and recipient females and shortening the time for activation (Chesne et al., 2002). Despite repeated attempts by several laboratories, somatic cell nuclear transfer has been unsuccessful in dogs, monkeys, and rats.

6.5.3. CELL CYCLE Cell cycle of the donor cell and oocyte must be taken into account to ensure that normal ploidy is maintained (reviewed in Campbell et al., 1996). By activating an oocyte prior to nuclear transfer, a donor nucleus at any stage in the cell cycle can be used, although these cytoplasts tend to have poorer developmental rates (Wakayama and Yanagimachi, 2001). The vast majority of cloning experiments use cells in G1 or G0 with a metaphase II–arrested oocyte. Reprogramming of G2/M nuclei is also possible and has been achieved in mouse with both somatic (Ono et al., 2001) and ES cells (Wakayama et al., 1999). Because metaphase nuclei contain double the normal chromosome complement, cytochalasin B is omitted from the embryo culture medium, allowing production of a polar body and halving of the chromosome number.

Despite attempts at enhancing nuclear transfer efficiency by altering various aspects of the nuclear transfer method, using different methods, cell type cycles, and activation regimes, the overall efficiency remains low. To understand the poor development of somatic and ES cell–cloned embryos, and therefore take appropriate steps to improve it, the nuclear reprogramming mechanisms that occur following transfer must be understood.

6.5.4. MOLECULAR MECHANISMS Spermatozoa and oocytes are highly specialized haploid cells that have been equipped and prepared specifically to fuse at fertilization to form a one-celled embryo capable of development to term. An incredible amount, therefore, is asked of an enucleated oocyte to reprogram a diploid differentiated somatic cell nucleus into a functioning embryonic one. The fact that nuclear transfer is so inefficient is not surprising; the fact that nuclear transfer is ever successful is remarkable. Little is known of the early molecular events with naturally fertilized embryos let alone cloned embryos, but this research area has received more interest in recent years.

For nuclear reprogramming to be effective, a number of events must occur, beginning with breakdown of the nucleus, stripping the DNA of its somatic histones (as opposed to the protamines in which the sperm nucleus is packed), correct epigenetic reprogramming (physically marking the DNA but without altering the nucleotide sequence) such as chromatin remodeling, and DNA methylation to ensure appropriate gene transcription at the right time. One of the mechanisms highlighted for the low efficiency of somatic cell nuclear transfer and the developmental abnormalities that result is due to abnormal gene expression as a result of epigenetic errors. In the search for molecular mechanisms that are causing the problems in cloned embryos, fetuses, and offspring, DNA methylation has been investigated (reviewed in Fairburn et al., 2002), although other marking methods could also be significant (*see* Young and Fairburn, 2000).

DNA methylation patterns are very dynamic in early mammalian embryos, and the potential for errors to be introduced in nuclear transfer is substantial. Cloned bovine embryos have highly irregular patterns of methylation similar to the donor cells from which they were derived and very different from that of "normal" embryos produced by fertilization (Bourc'his et al., 2001; Dean et al., 2001; Kang et al., 2001). Overall, the studies show partial but incomplete methylation reprogramming in cloned embryos (*see* Fairburn et al., 2002), and so at least the majority of nuclear transfer embryos are unable to properly reprogram the nucleus to achieve proper embryonic methylation status. By contrast, cloned porcine embryos do not appear to show highly aberrant patterns of methylation (Kang et al., 2001). Furthermore, analyses of tissue-specific methylation patterns in cloned mice also revealed aberrant methylation patterns, compared to fertilized embryos (Ohgane et al., 2001).

Nevertheless, faulty epigenetic reprogramming can allow development to term and even then may not be apparent in the phenotype. Recent research shows that mice cloned from ES cells can survive until adulthood and appear phenotypically normal despite epigenetic aberrations in their genome (Humpherys et al., 2001). In all of the mice cloned, at least one of the six imprinted genes that were examined was not properly reprogrammed (Humpherys et al., 2001). Further research is required to determine whether this effect is also seen with adult somatic cells as the pluripotent nature of ES cells may render these cells more susceptible to improper reprogramming (*see also* Inoue et al., 2002). Not all reprogramming mechanisms are irregular in all cloned embryos. Experiments have demonstrated normal reprogramming of readjustment of telomere length and X chromosome inactivation (*see* Rideout et al., 2001, for a review).

Aside from improper expression of imprinted genes, the dysregulation of nonimprinted genes may also hamper proper development. A study showed that transcription in three regular genes, all of which have roles in either implantation and/or postimplantation development, was delayed or absent in a number of cloned bovine embryos during preimplantation development (Daniels et al., 2001). Hence, insufficient nuclear reprogramming of the genome from both imprinted and nonimprinted genes may be responsible for poor embryo and fetal development in cloned animals.

Epigenetic errors could potentially be introduced into cloned embryos at a number of levels: through intrinsic errors in the donor cells or errors induced by the culture of donor cells, through incomplete reprogramming of donor cell chromatin into an embryonic state after reconstruction (either through variations in the reprogramming capacity of the recipient oocyte or by the method of nuclear transfer), or through the subsequent method of reconstructed embryo culture. Evidence from using mouse ES cells suggests that any induced aberrations by nuclear transfer/embryo culture are similarly unlikely to be remedied and therefore will carry on through the fetus and, survival permitting, the offspring (*see* Young and Fairburn, 2000).

6.5.5. SAFETY There is no method at present to screen for all the known imprinted genes let alone the ones yet to be identified. Many epigenetic changes are associated with the development of cancer (Wajed et al., 2001), and therefore it is crucial that powerful screening methods are developed to ensure the safety of emerging cloning and stem cell technologies. There are already safety concerns from the use of ES cells (Cervantes et al., 2002), particularly their potential to become tumorigenic.

Although therapeutic cloning will not require the production of offspring, the fate of postnatal clones is still significant. The phenotype of clones reflects that nuclear reprogramming is incomplete and that, depending on the extent, aberrant epigenetic effects that alter gene expression can either be fatal or permit life but with deleterious side effects that sometimes can not be noted until later in life. Nuclear transfer embryos may be considered a further risk; incomplete reprogramming of the inserted nucleus may lead to perturbation of a gene(s) that could cause cell dysfunction. Furthermore, genes with aberrant methylation may not reveal themselves in the ES line, but the effects may come into play once differentiation into a particular cell type is induced. In addition to the ES cell lines containing aberrant functioning genes, the deleterious effects may only become apparent on differentiation to a particular cell type.

6.6. FUTURE PROSPECTS

Development to term of an offspring derived from a cloned embryo is a remarkable demonstration of plasticity, especially of even fully differentiated cells. This ability to become reprogrammed is dependent on placing the differentiated nucleus in the correct environment, i.e., oocyte cytoplasm. Investigations into the early events of nuclear reprogramming, and the oocyte factors involved, will enhance our understanding of cellular differentiation. Increasing our knowledge in this area will not only improve nuclear transfer efficiency, but also cellular plasticity may lead to methods allowing the transdifferentiaion of adult stem cells into other cell types for cell replacement therapy.

ES cells were derived relatively recently, and further research is required in derivation, culture, and differentiation of these cells. Further research needs to be directed toward creation and culture of human nuclear transfer embryos with the capability to reach the blastocyst stage, at which stage ES cells can be derived. Therapeutic cloning has the potential to produce histocompatible cells for cell replacement therapies and to be an excellent medical and research tool for an in vitro cell model for study of serious diseases.

ACKNOWLEDGMENTS

We are grateful for the financial support from BBCRC and Geron Corporation.

REFERENCES

Assady, S., Maor, G., Amit, M., Itskovitz-Eldor, J., Skorecki, K. L. and Tzukerman, M. (2001) Insulin production by human embryonic stem cells. *Diabetes* 50:1691–1697.

Bjorklund, L. M., Sanchez-Pernaute, R., Chung, S., et al. (2002) Embryonic stem cells develop into functional dopaminergic neurons after transplantation in a Parkinson rat model. *Proc. Natl. Acad. Sci. USA* 99:2344–2349.

Bourc'his, D., Le Bourhis, D., Patin, D., et al. (2001) Delayed and incomplete reprogramming of chromosome methylation patterns in bovine cloned embryos. *Curr. Biol.* 11:1542–1546.

Campbell, K. H., Loi, P., Otaegui, P. J., and Wilmut, I. (1996) Cell cycle co-ordination in embryo cloning by nuclear transfer. *Rev. Reprod.* 1:40–46.

Cervantes, R. B., Stringer, J. R., Shao, C., Tischfield, J. A., and Stambrook, P. J. (2002) Embryonic stem cells and somatic cells differ in mutation frequency and type. *Proc. Natl. Acad. Sci. USA* 99: 3586–3590.

Chesne, P., Adenot, P. G., Viglietta, C., Baratte, M., Boulanger, L., and Renard, J. P. (2002) Cloned rabbits produced by nuclear transfer from adult somatic cells. *Nat. Biotechnol.* 20:366–369.

Cibelli, J. B., Stice, S. L., Golueke, P. J., et al. (1998) Cloned transgenic calves produced from nonquiescent fetal fibroblasts. *Science* 280: 1256–1258.

Clarke, D. and Frisen, J.(2001) Differentiation potential of adult stem cells. *Curr. Opin. Genet. Dev.* 11:575–580.

Daniels, R., Hall, V. J., French, A. J., Korfiatis, N. A., and Trounson, A. O. (2001) Comparison of gene transcription in cloned bovine embryos produced by different nuclear transfer techniques. *Mol. Reprod. Dev.* 60:281–288.

Dean, W., Santos, F., Stojkovic, M., et al. (2001) Conservation of methylation reprogramming in mammalian development: aberrant reprogramming in cloned embryos. *Proc. Natl. Acad. Sci. USA* 98: 13,734–13,738.

Eggan, K., Akutsu, H., Loring, J., et al. (2001) Hybrid vigor, fetal overgrowth, and viability of mice derived by nuclear cloning and tetraploid embryo complementation. *Proc. Natl. Acad. Sci. USA* 98: 6209–6214.

Evans, M. J. and Kaufman, M. H. (1981) Establishment in culture of pluripotential cells from mouse embryos. *Nature* 292:154–156.

Fairburn, H. R., Young, L. E., and Hendrich, B. D. (2002) Epigenetic reprogramming: How now, cloned cow. *Curr. Biol.* 12:R68–R70.

Heyman, Y., Chavatte-Palmer, P., LeBourhis, D., Camous, S., and Renard, J.P. (2002) Frequency and occurrence of late-gestation losses from cattle cloned embryos. *Biol. Reprod.* 66:6–13.

Hill, J. R., Roussel, A. J., Cibelli, J. B., et al. (1999). Clinical and pathologic features of cloned transgenic calves and fetuses (13 case studies). *Theriogenology* 51:1451–1465

Humpherys, D., Eggan, K., Akutsu, H., et al. (2001). Epigenetic instability in ES cells and cloned mice. *Science* 293:95–97.

Inoue, K., Kohda, T., Lee, J., et al. (2002) Faithful expression of imprinted genes in cloned mice. *Science* 295:297.

Kang, Y. K., Koo, D. B., Park, J. S., et al. (2001) Aberrant methylation of donor genome in cloned bovine embryos. *Nat. Genet.* 28:173–177.

Kaufman, D. S., Hanson, E. T., Lewis, R. L., Auerbach, R., and Thomson, J. A. (2001) Hematopoietic colony-forming cells derived from human embryonic stem cells. *Proc. Natl. Acad. Sci. USA* 98: 10,716–10,721.

Keefer, C. L., Keyston, R., Bhatia, B., et al. (2000) Efficient production of viable goat offspring following nuclear transfer using adult somatic cells. *Biol. Reprod.* 62(Suppl.):218.

Kubota, C., Yamakuchi, H., Todoroki, J., et al. (2000) Six cloned calves produced from adult fibroblast cells after long-term culture. *Proc. Natl. Acad. Sci. USA* 97:990–995.

Lanza, R. P., Cibelli, J. B., Faber, D., et al. (2001) Cloned cattle can be healthy and normal. *Science* 294:1893–1894

Lanzendorf, S. E., Boyd, C. A., Wright, D. L., Muasher, S., Oehninger, S., and Hodgen, G. D. (2001) Use of human gametes obtained from anonymous donors for the production of human embryonic stem cell lines. *Fertil. Steril.* 76:132–137.

Liu, S., Qu, Y., Stewart, T. J., Howard, M. J., Chakrabortty, S., Holekamp, T. F., and McDonald, J. W. (2000). Embryonic stem cells differentiate into oligodendrocytes and myelinate in culture and after spinal cord transplantation. *Proc Natl Acad Sci USA* 97:6126–6131.

McDonald, J. W., Liu, X.Z., Qu, Y., et al. (1999) Transplanted embryonic stem cells survive, differentiate and promote recovery in injured rat spinal cord. *Nat. Med.* 5:1410–1412.

Ogonuki, N., Inoue, K., Yamamoto, Y., et al. (2002) Early death of mice cloned from somatic cells. *Nat. Genet.* 30:253–254.

Ohgane, J., Wakayama, T., Kogo, Y., et al. (2001) DNA methylation variation in cloned mice. *Genesis* 30:45–50.

Ono, Y., Shimozawa, N., Ito, M., and Kono, T. (2001). Cloned mice from fetal fibroblast cells arrested at metaphase by a serial nuclear transfer. *Biol. Reprod.* 64:44–50.

Polejaeva, I. A., Chen, S. H., Vaught, T. D., et al. (2000) Cloned pigs produced by nuclear transfer from adult somatic cells. *Nature* 407: 86–90.

Renard, J. P., Chastant, S., Chesne, P., et al. (1999) Lymphoid hypoplasia and somatic cloning. *Lancet* 353:1489–1491.

Reubinoff, B. E., Itsykson, P., Turetsky, T., et al. (2001) Neural progenitors from human embryonic stem cells. *Nat. Biotechnol.* 19: 1134–1140.

Rideout, W. M. 3rd, Eggan, K., Jaenisch, R. (2001) Nuclear cloning and epigenetic reprogramming of the genome. *Science* 293:1093–1098.

Shin, T., Kraemer, D., Pryor, J., et al. (2002) A cat cloned by nuclear transplantation. *Nature* 415:859.

Sinclair, K. D., Young, L. E., Wilmut, I., and McEvoy, T. G. (2000) In-utero overgrowth in ruminants following embryo culture: lessons from mice and a warning to men. *Hum. Reprod. Suppl.* 5:68–86.

Soria, B., Rosche, E., Berna, G., Leon-Quinto, T., Reig, J. A., and Martin, F. (2000) Insulin-secreting cells derived from embryonic stem cells normalize gylcemia in streptozotocin-induced diabetic mice. *Diabetes* 49:157–162.

Tamashiro, K. L., Wakayama, T., Akutsu, H., et al. (2002). Cloned mice have an obese phenotype not transmitted to their offspring. *Nat. Med.* 8:262–267.

Teng, Y. D., Lavik, E. B., Qu, X., et al. (2002) Functional recovery following traumatic spinal cord injury mediated by a unique polymer scaffold seeded with neural stem cells. *Proc. Natl. Acad. Sci. USA* 99:3024–3029.

Teng, Y. D., Lavik, E. B., Qu,X., et al.(2002) Functional recovery following traumatic spinal cord injury mediated by a unique polymer scaffold seeded with neural stem cells. *Proc. Natl. Acad. Sci. USA* 99: 3024–3029.

Terada, N., Hamazaki, T., Oka, M., et al. (2002) Bone marrow cells adopt the phenotype of other cells by spontaneous cell fusion. *Nature* 416: 542–545.

Thomson, J. A., Itskovitz-Eldor, J., Shapiro, S. S., et al. (1998) Embryonic stem cells lines derived from human blastocysts. *Science* 282: 1145–1147.

Wajed, S. A., Laird, P. W., and DeMeester, T. R. (2001) DNA methylation: an alternative pathway to cancer. *Ann. Surg.* 234:10–20.

Wakayama, T., Rodriguez, I., Perry, A. C., Yanagimachi, R., and Mombaerts, P. (1999) Mice cloned from embryonic stem cells. *Proc. Natl. Acad. Sci. USA* 96:14,984–14,989.

Wakayama, T. and Yanagimachi, R. (2001) Effect of cytokinesis inhibitors, DMSO and the timing of oocyte activation on mouse cloning using cumulus cell nuclei. *Reproduction* 122:49–60.

Wakayama, T., Perry, A. C. F., Zuccotti, M., Johnson, K. R., and Yanagimachi, R. (1998) Full-term development of mice from enucleated oocytes injected with cumulus cell nuclei. *Nature* 394: 369–374.

Wakayama, T., Tateno, H., Mombaerts, P., and Yanagimachi, R.(2000) Nuclear transfer into mouse zygotes. *Nat. Genet.* 24:108–109.

Wells, D. N., Misica, P. M., Tervit, H. R. (1999) Production of cloned calves following nuclear transfer with cultured adult mural granulose cells. *Biol. Reprod.* 60:996–1005.

Wilmut, I., Schnieke, A. E., McWhir, J., Kind, A. J., and Campbell, K. H. S. (1997) Viable offspring derived from fetal and adult mammalian cells. *Nature* 385:810–813.

Ying, Q., Nichols, J., Evans, E. P., and Smith, A. G. (2002) Changing potency by spontaneous fusion. *Nature* 416:545–548.

Young, L. E. and Fairburn, H. R. (2000) Improving the safety of embryo technologies: possible role of genomic imprinting. *Theriogenology* 53:627–648.

Zhang, S. C., Wernig, M., Duncan, I. D., Brustle, O., and Thomson, J. A. (2001) In vitro differentiation of transplantable neural precursors from human embryonic stem cells. *Nat. Biotechnol.* 19:1129–1133.

7 Genomic Imprinting in Mouse Embryonic Stem and Germ Cells

Jeffrey R. Mann, PhD and Piroska E. Szabó, PhD

Genomic imprinting is an epigenetic system determining the parent-of-origin-specific monoallelic expression of a subset of genes, with imprints being imparted in the germline according to germ cell sex. Monoallelic expression begins early in development and is maintained throughout life, regulating the expression of genes important for somatic cell growth and differentiation. An important epigenetic mechanism for controlling the monoallelic expression of imprinted genes is the differential allelic methylation state of cis-acting gene DNA regulatory elements—this differential methylation leads to differential expression *in trans*. Genetic imprinting may be altered during the derivation and propagation of embryonic stem and germ cells in vitro and, therefore, is an important consideration in the use of such stem cells, and possibly somatic-derived stem cells, for therapy.

7.1. INTRODUCTION

Genomic imprinting is an epigenetic system that determines the somatic monoallelic expression of a small subset of genes according to parent of origin—the imprinted genes. Monoallelic expression of imprinted genes begins very early in development and is maintained throughout life. It is therefore a very stable epigenetic system (Bartolomei and Tilghman, 1997; Solter, 1998; Reik and Walter, 2001). Why should a discussion of such a system be relevant to a discussion of stem cells? One reason is that stem cells are often maintained for extended periods in abnormal situations such as culture in vitro and, therefore, aside from genetic mutation, can undergo significant abnormal epigenetic change, or "epimutation" (Holliday et al., 1996). Because genomic imprinting, and other epigenetic systems, regulate the expression of a number of genes important for cell growth and differentiation, there are clearly important implications for the use of cultured stem cells in therapy. Here, we describe what is known of the stability of genomic imprinting, with particular reference to mouse embryonic stem (ES) cells and embryonic germ (EG) cells, and discuss the implications for stem cell propagation in vitro.

7.2. OVERVIEW OF GENOMIC IMPRINTING

7.2.1. IMPRINTING AS AN EPIGENETIC SYSTEM

Epigenetics can be thought of as "the study of mitotically and/or meiotically heritable changes in gene function that cannot be explained by changes in DNA sequence" (Russo et al., 1996). Epigenetics is not so much interested in what activates or represses a gene, or in gene function, but in how a gene maintains a particular activity state from one cell generation to the next. The numbers of cell divisions involved are not important—they could be very small in rapidly differentiating systems such as the early embryo, or very large in a culture of a transformed differentiated cell type. Pluripotent stem cells are unusual in that an intermediate differentiated state is stabilized almost indefinitely. Then, suddenly, through either asymmetric or symmetric division, further differentiation ensues.

What form do epigenetic mechanisms take? An excellent overview is provided by Riggs and Porter (1996), and their list includes feedback loops of *trans*-acting factors, for which there is the best evidence; DNA methylation and chromatin structure, for which there is growing evidence; and nuclear compartmentalization of chromosomes or subchromosomal regions, and replication timing, for which there is little evidence. Genomic imprinting and X chromosome inactivation are special cases in that stable differential activity states exist within the same cell. Thus, *cis*-acting mechanisms would seem to be important for maintaining these states. With respect to genomic imprinting, DNA methylation has been studied almost exclusively, although there are now some studies relating histone modifications to allele-specific expression (Pedone et al., 1999; Hu et al., 2000; El Kharroubi et al., 2001; Gregory et al., 2001; Kohda et al., 2001; Xin et al., 2001; Yoshioka et al., 2001). Modifications involving histone acetylation are likely to be specific, because induction of global core histone acetylation in cultured mouse embryonic fibroblasts does not affect monoallelic expression of imprinted genes (Fig. 1).

7.2.2. DNA METHYLATION IN GENOMIC IMPRINTING

Before we consider imprinting in stem cell systems, we first briefly describe the current knowledge regarding the role of methylation in imprinting. Imprinted genes possess differentially methylated

From: *Stem Cells Handbook*
Edited by: S. Sell © Humana Press Inc., Totowa, NJ

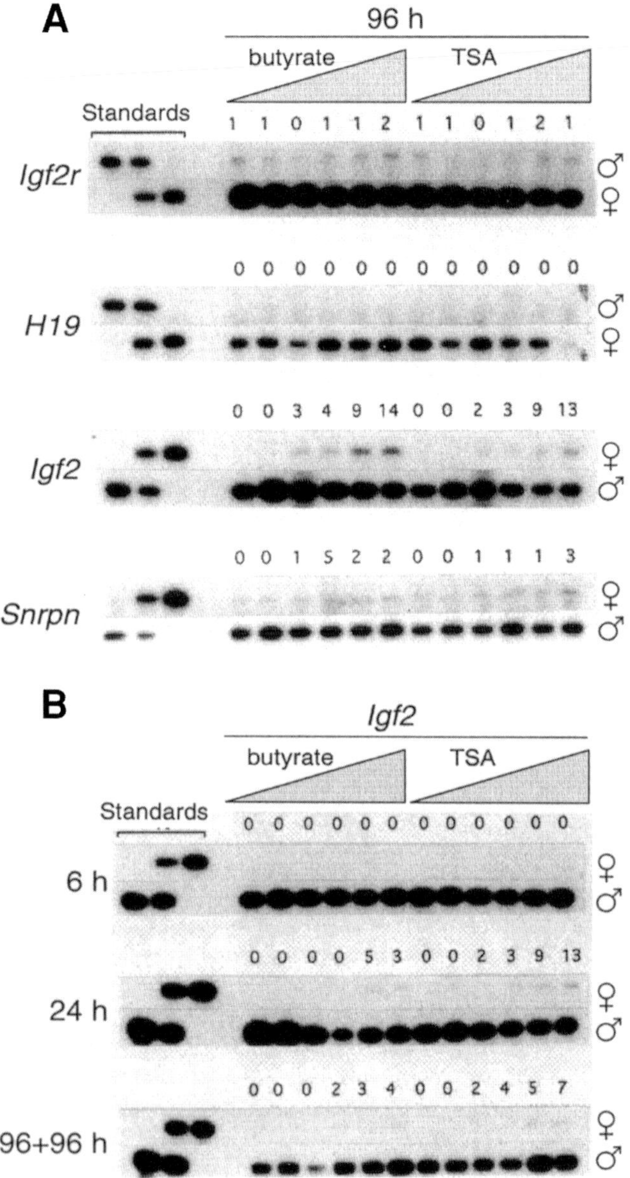

Fig. 1. Effect of histone deacetylase inhibitors on expression of imprinted genes. Exponentially growing (C57BL/6J × CAST/Ei) F_1 hybrid mouse primary embryo fibroblasts were seeded at 10^5 cells/well in a 12-well plate. After 25 h, medium was replaced and Nabutyrate (butyrate) and trichostatin A (TSA) were added at 0, 1, 2.5, 5, 10, and 20 m*M*, and 2, 10, 50, 100, 250, and 500 ng/mL, respectively. Cells were then grown for the time indicated before preparing RNA. Allele-specific analysis of expression was carried out as described, using the reverse treanscriptase polymerase chain reaction single-nucleotide primer extension assay (Szabó and Mann, 1995b). For each gene, the top and bottom row of bands represent the presumptive inactive and active allele, respectively, with the parental derivation given at the right. The value above each pair of bands is the percentage of RNA, contributed by the presumptive inactive allele, to the total RNA present for that gene. Standards are 1:0, 1:1, and 0:1 mixtures of allelic RNAs. **(A)** Effect on four imprinted genes after 96 h of exposure to drugs; **(B)** effect on *Igf2* at three different times of exposure. 96+96 h—96 h of exposure followed by 96 h of nonexposure. The only gene for which some effect was obtained was *Igf2*. However, this was achieved at a concentration of drugs in great excess of what is required to achieve global acetylation, and therefore the results were probably nonspecific.

DNA regions (DMRs); that is, for one parental allele a particular site or region is methylated, whereas for the other parental allele it is not. This observation has strongly suggested a role for DNA methylation in the regulation of monoallelic expression (Mann et al., 2000). Indeed, evidence is mounting: DNA methyltransferase (cytosine-5) (DNMT)–deficient mouse embryos lacked methylation at DMRs, and monoallelic expression of imprinted genes was disrupted (Li et al., 1993; Caspary et al., 1998). In addition, deletion studies in mice have shown that the primary DMRs of the insulin-like growth factor-2 (*Igf2*) and *H19* genes (Thorvaldsen et al., 1998; Kaffer et al., 2000; Srivastava et al., 2000; Szabó et al., 2002b), and the insulin-like growth factor-2 receptor (*Igf2r*) gene (Wutz et al., 1997, 2001), are critical for monoallelic expression. Consequently, these DMRs are now also called imprinting control regions/elements (ICRs/ICEs). While these experiments involved deletion of the ICR, a third study involved deletion of a repeat element, associated with the ICR, of the RAS protein-specific guanine nucleotide–releasing factor 1 (*Rasgrf1*) gene (Yoon et al., 2002). This deletion led to loss of both differential ICR methylation and monoallelic expression, providing evidence for a role of methylation *per se*. Putative ICRs have been assigned for other murine imprinted genes, but to date, high-resolution deletions have not been reported.

Studies of genomic imprinting can be placed into one or more of three categories:

1. Establishment of the imprint, which presumably takes place when the maternal and paternal genomes are separate—in the two germlines and in the zygote. Furthermore, it could involve a preferential retention of differential methylation during the widespread loss of paternal methylation in the zygote (Oswald et al., 2000; Santos et al., 2002).
2. Maintenance of the imprint during germ cell and somatic cell development. For sperm, imprinting information is likely to be maintained by methylation, as its nucleus is packaged by protamines; that is, there would appear to be no opportunity for the maintenance of cell memory by histone modifications (Jenuwein and Allis, 2001).
3. Development, or extension, of the imprint, when further modifications in *cis* are induced. This probably mostly occurs during the phase of intensive *de novo* methylation beginning at the periimplantation stage. Often, promoters become methylated on the inactive alleles of imprinted genes. This may aid in initiating or maintaining the inactive state, although evidence is lacking (Mann et al., 2000; Paulsen and Ferguson-Smith, 2001).

Some notable examples of studies in these categories are:

1. *Establishment:* Aside from logical necessity, the requirement for germline establishment of methylation imprints was demonstrated when the *Dnmt* cDNA was knocked-back-in to hypomethylated *Dnmt* –/– ES cells. While global methylation was restored in these ES cells, and in somatic cells derived from them, methylation at DMRs and monoallelic expression was not (Tucker et al., 1996). In a recent study, knockout of the DNA (cytosine-5)-methyltransferase 3-like gene (*Dnmt3l*) revealed that its product is required for the positive methylation of ICRs in the maternal germline, and for monoallelic somatic expression (Bourchis et al., 2001).

2. *Maintenance:* Monoallelic expression of *Igf2* is regulated by hypermethylation of the paternal *Igf2/H19* ICR—this methylation inhibits CTCF binding at the ICR and thereby prevents the region from functioning as a chromatin insulator (Bell and Felsenfeld, 2000; Hark et al., 2000; Kanduri et al., 2000a, 2000b; Szabó et al., 2000a, 2002b; Holmgren et al., 2001; Reed et al., 2001). Paternal-specific repression of the *Igf2r* gene is maintained in *cis* through expression of an antisense RNA. This RNA emanates from the ICR—contained within an intron of *Igf2r*—and extends beyond the *Igf2r* promoter (Sleutels et al., 2002). It is likely that expression of this antisense RNA is regulated by methylation (Wutz et al., 1997, 2001).
3. *Extension:* Inactivation of the paternal *H19* promoter occurs after fertilization and is dependent on the presence of the paternal ICR (Srivastava et al., 2000).

These studies are revealing that monoallelic expression is regulated by the methylation-dependent activity of *cis*-acting gene regulatory elements (ICRs). These elements, in *trans*, are switched off and on by the presence and absence of DNA methylation, respectively. The uniqueness of this system does not appear to be in the type of regulatory element involved, e.g., chromatin insulators and promoters of antisense transcripts, or in the epigenetic mechanisms regulating the activity of these elements, such as DNA methylation. What is unique is the differential acquisition and retention of these epigenetic controls with respect to the two parental alleles.

7.3. IMPRINTING DYSREGULATION AND ITS CONSEQUENCES

For imprinted genes, dysregulation often involves the activation of the normally silent allele to give two active alleles—this biallelic expression is often accompanied by overexpression—or the opposite—repression of the normally active allele to give two inactive alleles (nullallelic expression). These effects have been observed in mice deficient in DNMT activity; for example, loss of methylation in *cis* leads to activation of silent paternal *Igf2* allele and, at the same time, repression of the active maternal *Igf2r* allele (Li et al., 1993). Nullallelic activity of some imprinted genes leads to abnormal phenotype, e.g., *Igf2*—small size (DeChiara et al., 1991), achaete-scute complex homolog-like 2 (*Drosophila*) (*Ascl2*)—failed placental spongiotrophoblast development (Guillemot et al., 1995), *Igf2r*—large size and other defects (Lau et al., 1994; Wang et al., 1994). These observations show that some imprinted genes have essential functions but provide little information on the function of imprinting itself, i.e., why only one copy of the gene is active. Observing the consequences of biallelicexpression, or overexpression, might provide more information in this regard. For *Igf2* and *Igf2r*, biallelic expression results in large (Leighton et al., 1995) and small (Wutz et al., 2001) size, respectively, showing that the level of expression of these genes is critical. At least one hypothesis of evolution has been proposed for why this might be the case in the context of monoallelic expression, and it is based on the importance of these genes for growth (Moore and Haig, 1991). However, to date, biallelic expression of no other imprinted gene has been shown to have deleterious effects, and why these genes should be expressed monoallelically is more obscure.

7.4. GENOMIC IMPRINTING IN ES CELLS

7.4.1. ORIGIN OF ES CELLS The inner cell mass (ICM) gives rise to the embryo, the amnion, the yolk sac, and the chorioallantoic portion of the placenta. When explanted and dissociated in cell culture, given the right conditions, it gives rise to ES cells (Evans and Kaufman, 1981; Martin, 1981). ES cells most closely resemble early ICM cells in their ability to contribute to various cell lineages in chimeras (Beddington and Robertson, 1989). Hence, they can be thought of as ICM, or ICM-like, cells locked into continuing cycles of division in the undifferentiated state. This is achieved through activation of the receptor complex (leukemia inhibitory factor receptor [LIFR]–interleukin-6 signal transducer [IL6ST]) signaling pathway by exogenous LIF ligand, and other factors in the culture medium (Stewart et al., 1992; Yoshida et al., 1994).

7.4.2. RETENTION OF IMPRINTS From many studies, it is apparent that at least some genomic imprints are retained during during ES cell derivation. First, The relative level of expression of imprinted genes in parthenogenetic (two maternal genomes) and androgenetic (two paternal genomes) ES cells, and in differentiating embryoid bodies (EBs) derived by suspension culture of these cells, is skewed in the direction expected. For example, the amount of *Igf2* RNA in parthenogenetic and androgenetic ES cells and EBs is less and more than in their wild-type (one maternal and one paternal genome) counterparts, respectively. In addition, allele-specific expression in wild-type EBs is often skewed in the direction expected (Fig. 2). Second, viable mice can be derived, or cloned, wholly from ES cells, being achieved by aggregating them with tetraploid morula. Nevertheless, success depends on the quality of the ES cell line (Nagy et al., 1993). Third, chimeras made with androgenetic ES cells (Mann et al., 1990), compared with chimeras made with androgenetic morulae or ICMs (Barton et al., 1991; Mann and Stewart, 1991; Keverne et al., 1996), have a similar, and dramatic, phenotype. Only chimeras with a very low level of androgenetic contribution survive to term, and these possess pronounced skeletal abnormalities. The latter defects are dependent on *Igf2* expression in androgenetic cells (McLaughlin et al., 1997). Androgenetic ES cells can contribute extensively to many, if not all, lineages in chimeras (Mann et al., 1990), including the germline (Narasimha et al., 1997). Finally, chimeras made with parthenogenetic morulae fare much better in development than androgenetic chimeras, although growth retardation and a low contribution of parthenogenetic cells to certain tissues is observed (Stevens, 1978; Nagy et al., 1987; Thomson and Solter, 1988, 1989; Paldi et al., 1989; Fundele et al., 1990; Fundele and Surani, 1994; Allen et al., 1995; Keverne et al., 1996). Parthenogenetic ES cells can give extensive chimerism (Evans et al., 1985), and, paradoxically, the defects observed can be less severe (Allen et al., 1994). This could involve the loss of imprints leading to the activation of maternally repressed imprinted genes.

7.4.3. LOSS OF IMPRINTS It is clear that in some respects, ES cells differ from ICM cells. This is not surprising, seeing that ES cells undergo a relatively large number of cell divisions before they are induced to differentiate, and, these divisions take place in cell culture. By contrast, the ICM/primitive ectoderm undergoes just a few cell divisions before differentiation at gastrulation. Thus, in ES cells, there is ample opportunity for genetic mutation and epigenetic change to occur, both of which could adversely affect the differentiative program. In deriving a new cell line, from the

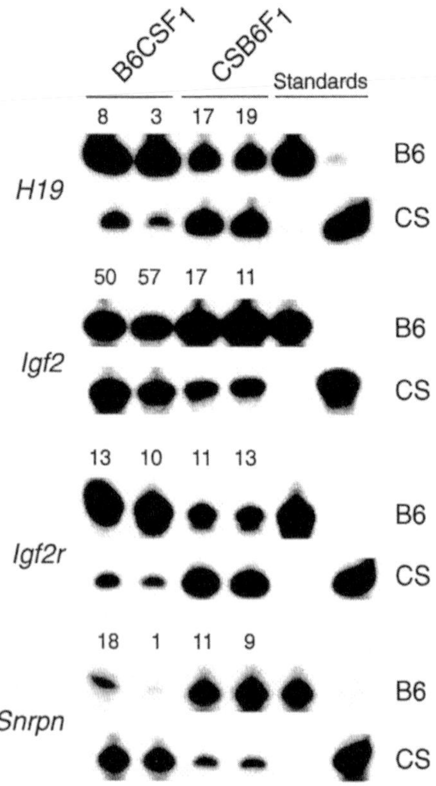

Fig. 2. Allele-specific expression of imprinted genes in EBs. Differentiating EBs were derived by culturing ES cells in suspension for 9 d, as described by Szabó and Mann (1994). Two (C57BL/6J × CAST/Ei)F_1 (B6CSF_1) ES cell lines, and two ES cell lines of the reciprocal mating (CSB6F_1), were used. Other details are as described in the legend to Fig. 1.

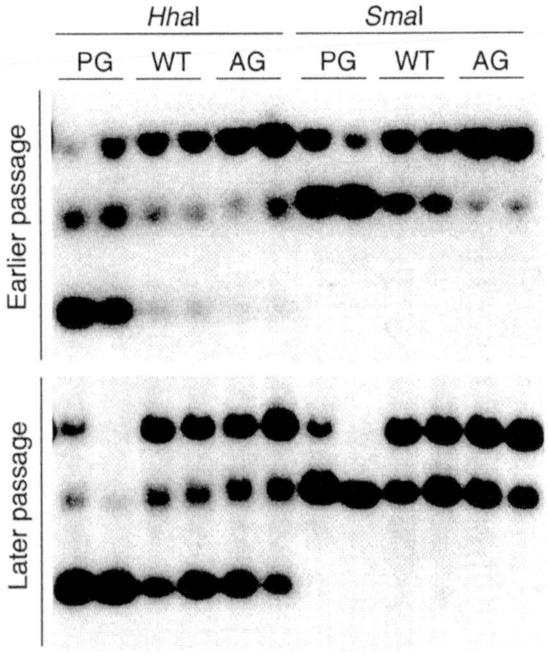

Fig. 3. Methylation of *H19* promoter in earlier- and later-passage ES cells. Southern blots showing the degree of methylation at one *HhaI* and one *SmaI* site within the *H19* promoter are shown. Two parthenogenetic (PG), two wild-type (WT), and two androgenetic (AG) cell lines were examined at earlier passage (7—12), and these same cell lines were examined at later passage (22—32). At earlier passage, note hypomethylation in the two PG cell lines (high level of digestion to small band sizes), and hypermethylation in the two AG cell lines (low level of digestion). At later passage, or approx 70 cell divisions later, all cell lines showed a greater level of digestion, indicating loss of methylation. (Data taken from Szabó and Mann, 1994.)

time that an ICM is explanted, to when a new cell line is frozen down, approx 20 cell divisions may take place over a 15-d period (Mann, 2002). In a typical gene-targeting experiment, ES cells will have undergone a further 30 or more divisions before a recombinant clone is used for chimera production.

In at least some instances, the parent- or allele-specific expression and methylation of imprinted genes in ES cells is different from that in ICM cells. First, as mentioned, the degree of difference depends on the time in culture. In later-passage ES cells, allele-specific expression and methylation deviated significantly from the situation at earlier passage (Szabó and Mann, 1994; Dean et al., 1998). For example, paternal-specific hypermethylation of the *H19* promoter—this methylation being an indicator, at least, of the closed chromatin structure of this promoter (Szabó et al., 2000)—was progressively lost during ES cell passage (Fig. 3) (Szabó and Mann, 1994). Interestingly, this region is not paternally methylated in the ICM; paternal methylation of this region occurs shortly after implantation (Sasaki et al., 1995). Second, during development, the maternal allele of the imprinted small nuclear ribonucleoprotein N (*Snrpn*) gene is always silent—no significant activity, relative to the paternal allele, was measured (Szabó and Mann, 1995a). However, in cultured EBs derived from earlier-passage ES cells, this allele exhibited measurable relative expression (Fig. 2). Furthermore, the maternal *Igf2* allele, the allele that becomes inactive, showed a higher relative level of expression than what was observed in early differentiated derivatives of the 6.5 d postcoitus (dpc) egg cylinder (Fig. 2) (Szabó and Mann, 1995a). Thus, losses of regulation in ES cells can occur quickly. Other imprinted genes also appear to be dysregulated at the levels of expression and methylation (Dean et al., 1998), although direct comparisons with ICM cells are lacking. Finally, cloned fetuses, derived by combining earlier-passage ES cells with tetraploid ova, can possess abnormal patterns of allele-specific methylation and expression of imprinted genes, again indicating rapid loss of regulation during ES cell derivation or culture (Dean et al., 1998). Similar results were obtained when cloned mice were derived from the transplantation of ES cell nuclei into oocytes (Humpherys et al., 2001). In this latter study, how much dysregulation was due to the cloning procedure is unclear, because no data exists for the performance of ICM nuclei under the same conditions.

7.5. GENOMIC IMPRINTING IN EG CELLS

7.5.1. ORIGIN OF EG CELLS EG cells are derived from primordial germ cells (PGCs). PGCs arise at approx 7.5 dpc at an extraembryonic location and, then, while dividing mitotically and increasing in number, migrate to the genital ridges, with all arriving by 11.5 dpc (Ginsburg et al., 1990). It is apparent that EG cell lines can be derived from embryos at any stage from 8.5 to 12.5 dpc, or at essentially all stages of PGC development. While LIF is sufficient for the derivation of ES cells, EG cell derivation requires LIF, or some other cytokine able to interact with the LIFR-IL-6ST receptor complex, KIT ligand, and fibroblast growth factor 2. However, once derived, LIF is sufficient for propagation (Donovan et al., 2001).

7.5.2. STATUS OF IMPRINTS A number of observations provide evidence that a significant loss of genomic imprints can occur during the derivation of EG cell lines from migratory PGCs. Imprints are largely intact in migratory PGCs, unlike the situation in PGCs of the genital ridge. In the former, the expression of *Snrpn*, *H19*, and *Igf2* was found to be monoallelic, and close to monoallelic, at 9.5 and 10.5 dpc (migratory stages), respectively (Szabó et al., 2002a). By contrast, at 11.5 dpc (nonmigratory stage), expression was essentially biallelic (Szabó et al., 2002a), confirming results obtained with manually purified PGCs (Szabó and Mann, 1995b). In accord with these data, preliminary observations indicate that the *Igf2/H19* ICR is differentially methylated in 9.5 dpc-PGCs, but biallelically hypomethylated in 11.5-dpc PGCs (M. R. Reed and J. R. Mann, unpublished data). By contrast, *Igf2r* is biallelically expressed in migrating PGCs (Szabó et al., 2002a), as in the totipotent lineage at earlier stages: during cleavage, at 2.5 dpc; in the blastocyst ICM, at 3.5 dpc; and in the primitive ectoderm, at 6.5 dpc (Szabó and Mann, 1995a). Nevertheless, the *Igf2r* ICR is differentially methylated in 9.5-dpc PGCs, but biallelically hypomethylated in 11.5-dpc PGCs (M. R. Reed and J. R. Mann, unpublished data). Thus, inherited *Igf2r* methylation imprints are maintained in migrating PGCs but are not "read" (Efstratiadis, 1994), or extended, in such a way as to achieve monoallelic expression.

Approximately half of the EG cell lines derived from 8.5-dpc PGCs, and all cell lines derived from 12.5-dpc PGCs, were hypomethylated at both copies of the *Igf2r* ICR (Labosky et al., 1994). Hypomethylation of the *Igf2r* ICR was also observed in two EG cell lines derived from 9.5-dpc PGCs. Furthermore, these two lines were hypomethylated on both copies of DMR sequences of the imprinted cyclin-dependent kinase inhibitor 1C (P57) (*Cdkn1c*) and KCNQ1 overlapping transcript 1 (*Kcnq1ot1*) genes, although biallelic hypomethylation was observed for the *Igf2/H19* ICR in only one line (Durcova-Hills et al., 2001). As might be expected, EG cell lines derived from 11.5- and 12.5-dpc PGCs were biallelically hypomethylated at DMR sequences for essentially all imprinted genes examined—*Igf2r*, *Cdckn1c*, *Kcnq1ot1*, maternally expressed transcript (*Mest*), paternally expressed gene 3 (*Peg3*), and neuronatin (*Nnat*) (Tada et al., 1998; Durcova-Hills et al., 2001). The exception was the *Igf2/H19* ICR—lines from female embryos exhibited biallelic hypomethylation (Tada et al., 1998), while lines from male embryos exhibited differential methylation (Tada et al., 1998; Durcova-Hills et al., 2001). In these latter studies, analyses were not allele specific, and, therefore, it was not possible to determine whether this sex-specific difference reflected a higher rate of erasure in female PGCs, or the establishment of new imprints in male PGCs.

The loss of methylation imprints in EG cells, derived from migrating PGCs, appears to be more pronounced than the loss in ES cells, for which methylation imprints are often retained (Labosky et al., 1994; Szabó and Mann, 1994; Tucker et al., 1996). This may reflect a fundamental difference between ICM cells and migrating PGCs. The former are undergoing extensive *de novo* methylation at the periimplantation stage (Monk et al., 1987; Santos et al., 2002), while the latter are poised to undergo demethylation at sequences of imprinted genes at the end of the migratory phase. They may also undergo extensive global demethylation at this time. On this point, although 12.5-dpc PGCs are globally hypomethylated, at imprinted and nonimprinted gene sequences (Monk et al., 1987; Brandeis et al., 1993), the global methylation status of migrating PGCs is yet to be determined.

7.6. GENOMIC IMPRINTING AND MAMMALIAN CLONING

The success of cloning sheep from adult somatic nuclei (Wilmut et al., 1997) came as quite a surprise, given the relative lack of success in previous studies—adult amphibians were unable to be cloned from adult somatic nuclei (Laskey and Gurdon, 1970; Gurdon et al., 1975; Wabl et al., 1975; Di Berardino and Hoffner, 1983), and the first attempts at cloning adult mice from blastomere nuclei were unsuccessful (McGrath and Solter, 1984). Now, totipotency can be seen to be predominantly a property of the egg cytoplasm, rather than of the gamete nuclei. Nevertheless, the frequency at which somatic nuclei can support normal development in oocyte transplantation experiments is very low (Wakayama and Yanagimachi, 2001). The reason for this low frequency is obscure but is commonly ascribed to incomplete nuclear reprogramming. It can be overlooked that high failure rates are also obtained when the nuclei of blastomeres are transplanted (McGrath and Solter, 1984; Tsunoda et al., 1987), even though these cells are totipotent (Tarkowski, 1959; Kelly, 1975). Thus, it is clear that a nucleus with a totipotent program is not sufficient to obtain normal development at high frequency after oocyte transplantation—other things are required. Nuclear stage in relation to the cell cycle is one factor (Campbell et al., 1996), but the status of other subtle interrelationships between the cytoplasm and nucleus at transplantation are probably also important, and these remain to be defined.

While imprinting is dysregulated in animals cloned using ES cells (Dean et al., 1998; Humpherys et al., 2001), to date, there is no evidence that it is dysregulated in animals cloned from somatic cell nuclei. Indeed, imprinting can be remarkably stable, given that mice can be cloned using cumulus cell nuclei through six generations (Wakayama et al., 2000)–assuming that imprinting dysregulation will lead to embryonic death. Furthermore, mice cloned from Sertoli and cumulus cell nuclei displayed a normal mRNA concentration for a number of imprinted genes, although placentae did not (Inoue et al., 2002). To date, there are no studies on the status of imprinted gene expression in dying embryos cloned using somatic nuclei; therefore, it remains possible that imprinting dysregulation does contribute to the high rate of failure. It has been observed that cow blastocysts, derived from the transplantation of fetal fibroblast nuclei into oocytes, possess aberrant methylation patterns (Kang et al., 2001). The nuclei of postmigratory 11.5-dpc PGCs probably lack the majority of imprints (Szabó and Mann, 1995b), and these nuclei may not be able support normal development on transplantation to oocytes; the failure may in part be due to the lack of imprints (Kato et al., 1999; Lee et al., 2002). It should be noted that the simple culture of ova can affect the methylation and expression of imprinted genes (Doherty et al., 2000; Khosla et al., 2001; Young et al., 2001), and this procedure is often employed in cloning procedures.

7.7. CONCLUSIONS, AND IMPLICATIONS FOR STEM CELLS FROM ADULTS

It is apparent that ES and EG cells have abnormal imprinting profiles, or changes in allele-specific methylation and expression patterns. The changes appear to be more significant in EG cells, and this is probably due to their derivation from PGCs that have undergone a significant amount of imprinting erasure, or are poised to do so. Because some imprinted genes are involved in growth and control of the cell cycle (e.g., *Igf2*, *Igf2r*, and *Cdckn1c*, and their dysregulation is implicated in cancer (Schofield et al., 2001; Plass and Soloway, 2002), caution needs to be exercised in using ES and EG cells, and their differentiated derivatives, in therapeutic strategies involving transplantation of these cells, or their differentiated derivatives. Even greater caution should be exercised in using ES cells derived from parthenogenetic blastocysts (Cibelli et al., 2002), because these already have a highly abnormal imprinting profile. Indeed, the use of such ES cells would seem to offer no advantage; whether these cells have a higher level of ethical acceptability must take into consideration that murine, and therefore probably human parthenogenomes, can develop into embryos with primitive brains, which is well beyond the legal limit for human embryo experimentation in countries even with liberal legislation. In human, this limit is 14 d, or the primitive-streak stage (McLaren, 1985), which is equivalent to approx 7.5 dpc in the mouse. Stem cells derived from adult individuals could also develop changes in their imprinting profile after extended periods in culture—to date, there is no information in this regard. Examination of the methylation status of ICRs and the expression status of imprinted genes involved in control of growth and the cell cycle would seem to be an important step in the assessment of stem cell lines to be used in therapy. When allele-specific analyses are not possible owing to the lack of polymorphisms, total methylation and expression levels can provide indications of allele-specificity—fifty percent methylation is usually taken to mean that one allele is hypermethylated, while the other is hypomethylated, and overexpression is often translated as biallelic expression.

ACKNOWLEDGMENT

We thank Minoru Yoshida for the gift of trichostatin A.

REFERENCES

Allen, N. D., Barton, S. C., Hilton, K., Norris, M. L., and Surani, M. A. (1994) A functional analysis of imprinting in parthenogenetic embryonic stem cells. *Development* 120:1473–1482.

Allen, N. D., Logan, K., Lally, G., Drage, D. J., Norris, M. L., and Keverne, E. B. (1995) Distribution of parthenogenetic cells in the mouse brain and their influence on brain development and behavior. *Proc. Natl. Acad. Sci. USA* 92:10,782–10,786.

Bartolomei, M. S. and Tilghman, S. M. (1997) Genomic imprinting in mammals. *Annu. Rev. Genet.* 31:493–525.

Barton, S. C., Ferguson-Smith, A. C., Fundele, R., and Surani, M. A. (1991) Influence of paternally imprinted genes on development. *Development* 113:679–687.

Beddington, R. S. and Robertson, E. J. (1989) An assessment of the developmental potential of embryonic stem cells in the midgestation mouse embryo. *Development* 105:733–737.

Bell, A. C. and Felsenfeld, G. (2000) Methylation of a CTCF-dependent boundary controls imprinted expression of the *Igf2* gene. *Nature* 405:482–485.

Bourchis, D., Xu, G.-L., Lin, C.-S., Bollman, B., and Bestor, T. H. (2001) *DnmtL* and the establishment of maternal genomic imprints. *Science* 294:2536–2539.

Brandeis, M., Kafri, T., Ariel, M., et al. (1993) The ontogeny of allele-specific methylation associated with imprinted genes in the mouse. *EMBO J.* 12:3669–3677.

Campbell, K. H., Loi, P., Otaegui, P. J., and Wilmut, I. (1996) Cell cycle co-ordination in embryo cloning by nuclear transfer. *Rev. Reprod.* 1: 40–46.

Caspary, T., Cleary, M. A., Baker, C. C., Guan, X. J., and Tilghman, S. M. (1998) Multiple mechanisms regulate imprinting of the mouse distal chromosome 7 gene cluster. *Mol. Cell. Biol.* 18:3466–3474.

Cibelli, J. B., Grant, K. A., Chapman, K. B., et al. (2002) Parthenogenetic stem cells in nonhuman primates. *Science* 295:819.

Dean, W., Bowden, L., Aitchison, A., et al. (1998) Altered imprinted gene methylation and expression in completely ES cell-derived mouse fetuses: association with aberrant phenotypes. *Development* 125:2273–2282.

DeChiara, T. M., Robertson, E. J., and Efstratiadis, A. (1991) Parental imprinting of the mouse insulin-like growth factor II gene. *Cell* 64: 849-859.

Di Berardino, M. A. and Hoffner, N. J. (1983) Gene reactivation in erythrocytes: nuclear transplantation in oocytes and eggs of *Rana*. *Science* 219:862–864.

Doherty, A. S., Mann, M. R., Tremblay, K. D., Bartolomei, M. S., and Schultz, R. M. (2000) Differential effects of culture on imprinted *H19* expression in the preimplantation mouse embryo. *Biol. Reprod.* 62: 1526–1535.

Donovan, P. J., De Miguel, M. P., Hirano, M. P., Parsons, M. S., and Lincoln, A. J. (2001) Germ cell biology—from generation to generation. *Int. J. Dev. Biol.* 45:523–531.

Durcova-Hills, G., Ainscough, J., and McLaren, A. (2001) Pluripotential stem cells derived from migrating primordial germ cells. *Differentiation* 68:220–226.

Efstratiadis, A. (1994) Parental imprinting of autosomal mammalian genes. *Curr. Opin. Genet. Dev.* 4;265–280.

El Kharroubi, A., Piras, G., and Stewart, C. L. (2001) DNA demethylation reactivates a subset of imprinted genes in uniparental mouse embryonic fibroblasts. *J. Biol. Chem.* 276:8674–8680.

Evans, M., Bradley, A., and Robertson, E. (1985) EK cell contribution to chimeric mice: from tissue culture sperm. In: *Genetic Manipulation of the Early Mouse Embryo*, Cold Spring Harbor Banbury Report 20 (Jaenisch, R. and Costantini, F., eds.), Cold Spring Harbor Laboratory Press, Cold Spring Harbor, NY, pp. 93–102.

Evans, M. J. and Kaufman, M. H. (1981) Establishment in culture of pluripotential cells from mouse embryos. *Nature* 292:154–156.

Fundele, R. H. and Surani, M. A. (1994) Experimental embryological analysis of genetic imprinting in mouse development. *Dev. Genet.* 15: 515–522.

Fundele, R. H., Norris, M. L., Barton, S. C., et al. (1990) Temporal and spatial selection against parthenogenetic cells during development of fetal chimeras. *Development* 108:203–211.

Ginsburg, M., Snow, M. H., and McLaren, A. (1990) Primordial germ cells in the mouse embryo during gastrulation. *Development* 110: 521–528.

Gregory, R. I., Randall, T. E., Johnson, C. A., et al. (2001) DNA methylation is linked to deacetylation of histone H3, but not H4, on the imprinted genes *Snrpn* and *U2af1-rs1*. *Mol. Cell. Biol.* 21:5426–5436.

Guillemot, F., Caspary, T., Tilghman, S. M., et al. (1995) Genomic imprinting of *Mash2*, a mouse gene required for trophoblast development. *Nat. Genet.* 9:235–241.

Gurdon, J. B., Laskey, R. A., and Reeves, O. R. (1975) The developmental capacity of nuclei transplanted from keratinized skin cells of adult frogs. *J. Embryol. Exp. Morphol.* 34:93–112.

Hark, A. T., Schoenherr, C. J., Katz, D. J., Ingram, R. S., Levorse, J. M., and Tilghman, S. M. (2000) CTCF mediates methylation-sensitive enhancer-blocking activity at the *H19/Igf2* locus. *Nature* 405: 486–489.

Holliday, R., Ho, T., and Paulin, R. (1996) Gene silencing in mammalian cells. In: *Epigenetic Mechanisms of Gene Regulation* (Russo, V. E. A., Martienssen, R. A., and Riggs, A. D., eds.), Cold Spring Harbor Laboratory Press, Cold Spring Harbor, NY, pp. 47–59.

Holmgren, C., Kanduri, C., Dell, G., Ward, A., Mukhopadhya, R., Kanduri, M., Lobanenkov, V., and Ohlsson, R. (2001) CpG methylation regulates the *Igf2/H19* insulator. *Curr. Biol.* 11:1128–1130.

Hu, J. F., Pham, J., Dey, I., Li, T., Vu, T. H., and Hoffman, A. R. (2000) Allele-specific histone acetylation accompanies genomic imprinting of the insulin-like growth factor II receptor gene. *Endocrinology* 141: 4428–4435.

Humpherys, D., Eggan, K., Akutsu, H., Hochedlinger, K., Rideout, W. M. 3rd, Biniszkiewicz, D., Yanagimachi, R., and Jaenisch, R. (2001) Epigenetic instability in ES cells and cloned mice. *Science* 293:95–97.

Inoue, K., Kohda, T., Lee, J., et al. (2002) Faithful expression of imprinted genes in cloned mice. *Science* 295:297.

Jenuwein, T. and Allis, C. D. (2001) Translating the histone code. *Science* 293:1074–1080.

Kaffer, C. R., Srivastava, M., Park, K. Y., et al. (2000) A transcriptional insulator at the imprinted *H19/Igf2* locus. *Genes Dev.* 14:1908–1919.

Kanduri, C., Holmgren, C., Pilartz, M., et al. (2000a) The 5' flank of mouse *H19* in an unusual chromatin conformation unidirectionally blocks enhancer-promoter communication. *Curr. Biol.* 10:449–457.

Kanduri, C., Pant, V., Loukinov, D., et al. (2000b) Functional association of CTCF with the insulator upstream of the *H19* gene is parent of origin-specific and methylation-sensitive. *Curr. Biol.* 10:853–856.

Kang, Y. K., Koo, D. B., Park, J. S., et al. (2001) Aberrant methylation of donor genome in cloned bovine embryos. *Nat. Genet.* 28:173–177.

Kato, Y., Rideout, W. M., Hilton, K., Barton, S. C., Tsunoda, Y., and Surani, M. A. (1999) Developmental potential of mouse primordial germ cells. *Development* 126:1823–1832.

Kelly, S. J. (1975) Studies on the potency of early cleavage blastomeres of the mouse. In: *The Early Development of Mammals* (Balls, M. and Wild, A. E., eds.), Cambridge University Press, London, pp. 97–106.

Keverne, E. B., Fundele, R., Narasimha, M., Barton, S. C., and Surani, M. A. (1996) Genomic imprinting and the differential roles of parental genomes in brain development. *Brain Res. Dev. Brain Res.* 92:91–100.

Khosla, S., Dean, W., Brown, D., Reik, W., and Feil, R. (2001) Culture of preimplantation mouse embryos affects fetal development and the expression of imprinted genes. *Biol. Reprod.* 64:918–926.

Kohda, A., Taguchi, H., and Okumura, K. (2001) Visualization of biallelic expression of the imprinted *SNRPN* gene induced by inhibitors of DNA methylation and histone deacetylation. *Biosci. Biotechnol. Biochem.* 65:1236–1239.

Labosky, P. A., Barlow, D. P., and Hogan, B. L. (1994) Mouse embryonic germ (EG) cell lines: transmission through the germline and differences in the methylation imprint of insulin-like growth factor 2 receptor (*Igf2r*) gene compared with embryonic stem (ES) cell lines. *Development* 120:3197–3204.

Laskey, R. A. and Gurdon, J. B. (1970) Genetic content of adult somatic cells tested by nuclear transplantation from cultured cells. *Nature* 228: 1332–1334.

Lau, M. M. H., Stewart, C. E. H., Liu, Z., Bhatt, H., Rotwein, P., and Stewart, C. L. (1994) Loss of the imprinted IGF2/cation-independent mannose 6-phosphate receptor results in fetal overgrowth and perinatal lethality. *Genes Dev.* 8:2953–2963.

Lee, J., Inoue, K., Ono, R., et al. (2002) Erasing genomic imprinted memory in PGC clones. *Development* 129:1807–1817.

Leighton, P. A., Ingram, R. S., Eggenschwiler, J., Efstratiadis, A., and Tilghman, S. M. (1995) Disruption of imprinting caused by deletion of the *H19* gene region in mice. *Nature* 375:34–39.

Li, E., Beard, C., and Jaenisch, R. (1993) Role for DNA methylation in genomic imprinting. *Nature* 366:362–365.

Mann, J. R. (2002) Deriving and propagating mouse embryonic stem cell lines for studying genomic imprinting. *Methods Mol. Biol.* 181: 21–39.

Mann, J. R. and Stewart, C. L. (1991) Development to term of mouse androgenetic aggregation chimeras. *Development* 113:1325–1333.

Mann, J. R., Gadi, I., Harbison, M. L., Abbondanzo, S. J., and Stewart, C. L. (1990) Androgenetic mouse embryonic stem cells are pluripotent and cause skeletal defects in chimeras: implications for genetic imprinting. *Cell* 62:251–260.

Mann, J. R., Szabó, P. E., Reed, M. R., and Singer-Sam, J. (2000) Methylated DNA sequences in genomic imprinting. *Crit. Rev. Eukaryot. Gene Expr.* 10:241–257.

Martin, G. R. (1981) Isolation of a pluripotent cell line from early mouse embryos cultured in medium conditioned by teratocarcinoma stem cells. *Proc. Natl Acad. Sci. USA* 78:7634–7638.

McGrath, J. and Solter, D. (1984) Inability of mouse blastomere nuclei transferred to enucleated zygotes to support development in vitro. *Science* 226:1317–1319.

McLaren, A. (1985) Where to draw the line? *Proc. R. Soc. Great Britain* 56:101–121.

McLaughlin, K. J., Kochanowski, H., Solter, D., Schwarzkopf, G., Szabó, P. E., and Mann, J. R. (1997) Roles of the imprinted gene *Igf2* and paternal duplication of distal chromosome 7 in the perinatal abnormalities of androgenetic mouse chimeras. *Development* 124: 4897–4904.

Monk, M., Boubelik, M., and Lehnert, S. (1987) Temporal and regional changes in DNA methylation in the embryonic, extraembryonic and germ cell lineages during mouse embryo development. *Development* 99:371–382.

Moore, T. and Haig, D. (1991) Genomic imprinting in mammalian development: a parental tug-of-war. *Trends Genet.* 7:45–49.

Nagy, A., Paldi, A., Dezso, L., Varga, L., and Magyar, A. (1987) Prenatal fate of parthenogenetic cells in mouse aggregation chimaeras. *Development* 101:67–71.

Nagy, A., Rossant, J., Nagy, R., Abramow-Newerly, W., and Roder, J. C. (1993) Derivation of completely cell culture-derived mice from early-passage embryonic stem cells. *Proc. Natl Acad. Sci. USA* 90: 8424–8428.

Narasimha, M., Barton, S. C., and Surani, M. A. (1997) The role of the paternal genome in the development of the mouse germ line. *Curr. Biol.* 7:881–884.

Oswald, J., Engemann, S., Lane, N., et al. (2000) Active demethylation of the paternal genome in the mouse zygote. *Curr. Biol.* 10:475–478.

Paldi, A., Nagy, A., Markkula, M., Barna, I., and Dezso, L. (1989) Postnatal development of parthenogenetic—fertilized mouse aggregation chimeras. *Development* 105:115–118.

Paulsen, M. and Ferguson-Smith, A. C. (2001) DNA methylation in genomic imprinting, development, and disease. *J. Pathol.* 195: 97–110.

Pedone, P. V., Pikaart, M. J., Cerrato, F., et al. (1999) Role of histone acetylation and DNA methylation in the maintenance of the imprinted expression of the *H19* and *Igf2* genes. *FEBS Lett.* 458:45–50.

Plass, C. and Soloway, P. D. (2002) DNA methylation, imprinting and cancer. *Eur. J. Hum. Genet.* 10:6–16.

Reed, M. R., Huang, C. F., Riggs, A. D., and Mann, J. R. (2001) A complex duplication created by gene targeting at the imprinted *H19*locus results in two classes of methylation and correlated *Igf2* expression phenotypes. *Genomics* 74:186–196.

Reik, W. and Walter, J. (2001) Genomic imprinting: parental influence on the genome. *Nat. Rev. Genet.* 2:21–32.

Riggs, A. D. and Porter, T. N. (1996) Overview of epigenetic mechanisms. In: *Epigenetic Mechanisms of Gene Regulation* (Russo, V. E. A., Martienssen, R. A., and Riggs, A. D., eds.), Cold Spring Harbor Laboratory Press, Cold Spring Harbor, NY, pp. 29–45.

Russo, V. E. A., Martienssen, R. A., and Riggs, A. D. (1996) Introduction. In: *Epigenetic Mechanisms of Gene Regulation* (Russo, V. E. A., Martienssen, R. A., and Riggs, A. D., eds.), Cold Spring Harbor Laboratory Press, Cold Spring Harbor, NY, pp. 47–59.

Santos, F., Hendrich, B., Reik, W., and Dean, W. (2002) Dynamic reprogramming of DNA methylation in the early mouse embryo. *Dev. Biol.* 241:172–182.

Sasaki, H., Ferguson-Smith, A. C., Shum, A. S., Barton, S. C., and Surani, M. A. (1995) Temporal and spatial regulation of H19 imprinting in normal and uniparental mouse embryos. *Development* 121: 4195–4202.

Schofield, P. N., Joyce, J. A., Lam, W. K., et al. (2001) Genomic imprinting and cancer: new paradigms in the genetics of neoplasia. *Toxicol. Lett.* 120:151–160.

Sleutels, F., Zwart, R., and Barlow, D. P. (2002) The non-coding *Air*RNA is required for silencing autosomal imprinted genes. *Nature* 415:810–813.

Solter, D. (1998) Imprinting. *Int. J. Dev. Biol.* 42:951–954.

Srivastava, M., Hsieh, S., Grinberg, A., Williams-Simons, L., Huang, S. P., and Pfeifer, K. (2000) *H19* and *Igf2* monoallelic expression is regulated in two distinct ways by a shared *cis* acting regulatory region upstream of *H19*. *Genes Dev.* 14:1186–1195.

Stevens, L. C. (1978) Totipotent cells of parthenogenetic origin in a chimaeric mouse. *Nature* 276:266–267.

Stewart, C. L., Kaspar, P., Brunet, L. J., et al. (1992) Blastocyst implantation depends on maternal expression of luekemia inhibitory factor. *Nature* 359:76–79.

Szabó, P. and Mann, J. R. (1994) Expression and methylation of imprinted genes during in vitro differentiation of mouse parthenogenetic and androgenetic embryonic stem cell lines. *Development* 120:1651–1660.

Szabó, P. E., and Mann, J. R. (1995a) Allele-specific expression and total expression levels of imprinted genes during early mouse development: implications for imprinting mechanisms. *Genes Dev.* 9:1–12.

Szabó, P. E., and Mann, J. R. (1995b) Biallelic expression of imprinted genes in the mouse germ line: implications for erasure, establishment, and mechanisms of genomic imprinting. *Genes Dev.* 9:1857–1868.

Szabó, P. E., Tang, S.-H. E., Rentsendorj, A., Pfeifer, G., and Mann, J. R. (2000) Maternal-specific footprints at putative CTCF sites in the *H19* imprinting control region give evidence for insulator function. *Curr. Biol.* 10:607–610.

Szabó, P. E., Hübner, K., Schöler, H. R., and Mann, J. R. (2002a) Allele-specific expression of imprinted genes in mouse migratory primordial germ cells. *Mech. Dev.*, in press.

Szabó, P. E., Tang, S. H., Reed, M. R., Silva, F. J., Tsark, W. M., and Mann, J. R. (2002b) The chicken b-globin insulator element conveys chromatin boundary activity but not imprinting at the mouse *Igf2/H19* domain. *Development* 129:897–904.

Tada, T., Tada, M., Hilton, K., et al. (1998) Epigenotype switching of imprintable loci in embryonic germ cells. *Dev. Genes Evol.* 207: 551–561.

Tarkowski, A. K. (1959) Experiments on the development of isolated blastomeres of mouse eggs. *Nature* 184:1286–1287.

Thomson, J. A. and Solter, D. (1988) The developmental fate of androgenetic, parthenogenetic, and gynogenetic cells in chimeric gastrulating mouse embryos. *Genes Dev.* 2:1344–1351.

Thomson, J. A. and Solter, D. (1989) Chimeras between parthenogenetic or androgenetic blastomeres and normal embryos: allocation to the inner cell mass and trophectoderm. *Dev. Biol.* 131:580–583.

Thorvaldsen, J. L., Duran, K. L., and Bartolomei, M. S. (1998) Deletion of the *H19* differentially methylated domain results in loss of imprinted expression of *H19* and *Igf2*. *Genes Dev.* 12:3693–3702.

Tsunoda, Y., Yasui, T., Shioda, Y., Nakamura, K., Uchida, T., and Sugie, T. (1987) Full-term development of mouse blastomere nuclei transplanted into enucleated two-cell embryos. *J. Exp. Zool.* 242:147–151.

Tucker, K. L., Beard, C., Dausmann, J., et al. (1996) Germ-line passage is required for establishment of methylation and expression patterns of imprinted but not of nonimprinted genes. *Genes Dev.* 10: 1008–1020.

Wabl, M. R., Brun, R. B., and Du Pasquier, L. (1975) Lymphocytes of the toad Xenopus laevis have the gene set for promoting tadpole development. *Science* 190:1310–1312.

Wakayama, T. and Yanagimachi, R. (2001) Mouse cloning with nucleus donor cells of different age and type. *Mol. Reprod. Dev.* 58:376–383.

Wakayama, T., Shinkai, Y., Tamashiro, K. L., et al. (2000) Cloning of mice to six generations. *Nature* 407:318–319.

Wang, Z.-Q., Fung, M. R., Barlow, D. P., and Wagner, E. F. (1994) Regulation of embryonic growth and lysosomal targeting by the imprinted *Igf2/Mpr* gene. *Nature* 372:464–467.

Wilmut, I., Schnieke, A. E., McWhir, J., Kind, A. J., and Campbell, K. H. (1997) Viable offspring derived from fetal and adult mammalian cells. *Nature* 385:810–813.

Wutz, A., Smrzka, O. W., Schweifer, N., Schellander, K., Wagner, E. F., and Barlow, D. P. (1997) Imprinted expression of the *Igf2r* gene depends on an intronic CpG island. *Nature* 389:745–749.

Wutz, A., Theussl, H. C., Dausman, J., Jaenisch, R., Barlow, D. P., and Wagner, E. F. (2001) Non-imprinted *Igf2r* expression decreases growth and rescues the *Tme* mutation in mice. *Development* 128: 1881–1887.

Xin, Z., Allis, C. D., and Wagstaff, J. (2001) Parent-specific complementary patterns of histone H3 lysine 9 and H3 lysine 4 methylation at the Prader-Willi syndrome imprinting center. *Am. J. Hum. Genet.* 69: 1389–1394.

Yoon, B. J., Herman, H., Sikora, A., Smith, L. T., Plass, C., and Soloway, P. D. (2002) Regulation of DNA methylation of *Rasgrf1*. *Nat. Genet.* 30:92–96.

Yoshida, K., Chambers, I., Nichols, J., et al. (1994) Maintenance of the pluripotential phenotype of embryonic stem cells through direct activation of gp130 signalling pathways. *Mech. Dev.* 45:163–171.

Yoshioka, H., Shirayoshi, Y., and Oshimura, M. (2001) A novel in vitro system for analyzing parental allele-specific histone acetylation in genomic imprinting. *J. Hum. Genet.* 46:626–632.

Young, L. E., Fernandes, K., McEvoy, T. G., et al. (2001) Epigenetic change in *IGF2R* is associated with fetal overgrowth after sheep embryo culture. *Nat. Genet.* 27:153–154.

8 Stem Cells in Amphibian Regeneration

DAVID L. STOCUM, PhD

Salamanders, newts, and frogs can regenerate limbs, spinal cord, neural retinal and lens of the eye, cardiac muscle, and so on from adult stem cells that dedifferentiate from differentiated cells at the site of injury and then differentiate in response to signals derived from serum and adjacent tissues. Regeneration of the limb requires degradation of the extracelluar matrix, which liberates osteocytes, chondrocytes, myofibers, Schwann cells, and fibroblasts to undergo dedifferentiation to form a blastema in the presence of serum-derived thrombin. Proliferation and redifferentiation of the blastema cells to form a new limb requires an overlying epidermis, which grows over the wound site, as well as regenerating nerves, which provide fibroblast growth factors. This regeneration provides a prototype for the plasticity of adult tissue cells.

8.1. INTRODUCTION

All plants and animals reproduce, either asexually or sexually, by means of stem cells. They also use stem cells to regenerate throughout life. These adult stem cells are conventionally defined as undifferentiated cells that divide asymmetrically to produce another stem cell (self-renewal) and a cell of more restricted potential that proliferates and differentiates into one or more functional phenotypes. Certain invertebrates such as coelenterates and flatworms reproduce asexually by fission. They can also regenerate whole animals from fragments obtained by transection, suggesting that they use the same or similar mechanisms for asexual reproduction and regeneration (Alvarado, 2000). In vertebrates, regeneration is restricted to a few tissues whose cells are subject to continual turnover, such as blood and lymphoid cells; epidermis; epithelial linings of the digestive and respiratory systems; olfactory neurons; feathers; hair; and scales or sporadic injury, such as liver, bone, muscle, and fingertips. Many vertebrate tissues, however, respond to injury by forming scar tissue that compromises function in proportion to the extent of scarring. Despite these restrictions, stem cells have been identified recently in several mammalian tissues that do not regenerate spontaneously, suggesting latent capacity for regeneration that has been suppressed (Stocum, 2001).

Larval and adult urodeles (salamanders and newts) and anuran (frogs and toads) tadpoles regenerate the same tissues as birds and mammals (with the exception of hair, feathers, and scales!) via conventional, undifferentiated adult stem cells. Unlike mammals, however, they can regenerate the spinal cord from ependymal stem cells (Chernoff, 1996; Benraiss, 1999). Most strikingly, amphibians can also regenerate the neural retina (NR) and lens of the eye, cardiac muscle, intestine, tail, jaws, and limbs (Stocum, 1995; Brockes, 1997; Geraudie and Ferretti, 1998). The stem cells that regenerate these tissues, however, are created in a unique way: by the dedifferentiation (loss of phenotype-specific structure) of differentiated cells at the site of injury. Depending on the tissue or structure of origin, these dedifferentiated cells can be organized as proliferating layers (NR), or as a blastema associated with (limb, tail, jaws) or without (lens, tip of heart ventricle, intestine) an epithelial covering. They may be unipotent, redifferentiating only into the cell type of origin, or able to transdifferentiate into cell types other than the one of origin. They are self-renewing, not in the conventional sense, but in the sense that they can be produced anew in successive rounds of regeneration.

Understanding the mechanisms that regulate stem cell production by dedifferentiation is an important problem with significant implications for regenerative medicine (Stocum, 2001). In this chapter, I review the mechanism of production of stem cells by dedifferentiation during the regeneration of amphibian limbs, NR, and lens.

8.2. LIMB REGENERATION

8.2.1. ORIGIN OF STEM CELLS The production of stem cells by dedifferentiation has been most extensively studied in regenerating urodele limbs. The amputated limbs of salamanders and newts regenerate via formation of a blastema of mesenchyme-like stem cells under the wound epidermis that migrates over the amputation surface (Fig. 1). The blastema is similar in form to that of the early embryonic limb bud. It grows by stem cell prolifera-

From: *Stem Cells Handbook*
Edited by: S. Sell © Humana Press Inc., Totowa, NJ

Fig. 1. Light micrograph of longitudinal section through regenerating axolotl forelimb amputated 7 d previously through the distal radius and ulna. Hematoxylin and eosin (×100). The blastema is a conical mass of stem cells derived by dedifferentiation of cartilage, muscle, dermal, and Schwann sheath cells. Note the thickened cap of apical wound epidermis (arrow) at the tip of the blastema. Signals from the wound epidermis and the nerves that have regenerated into the blastema (which cannot be seen) keep the stem cells in a dedifferentiated and proliferating state until redifferentiation/ transdifferentiation sets in. R, radius; U, ulna.

tion and differentiates into precisely the limb parts that were lost. Irradiation experiments, and experiments in which labeled tissues were grafted into the limb prior to amputation or into the early blastema, revealed that the cells of the blastema are derived from tissues local to the amputation plane. Butler and O'Brien (1942) X-irradiated whole Ambystoma larvae while keeping a small segment of the limb shielded; conversely, they irradiated the same segment while shielding the rest of the animal. The limbs were then amputated through the irradiated or nonirradiated segment. The limb regenerated only if the segment had been shielded, thus demonstrating the origin of the blastema from cells in the immediate vicinity of the amputation.

Formation of the blastema by dedifferentiation was conclusively demonstrated by tracking labeled differentiated cells in regenerating limbs. Steen (1968) grafted triploid long bone cartilage into diploid limbs and demonstrated the presence of dedifferentiated triploid cells in the blastema. Lo et al. (1993) microinjected the fluorescent lineage tracer rhodamine-conjugated lysinated dextran into cultured newt myotubes derived by the fusion of myoblasts that had proliferated in the presence of ^{3}H-thymidine, thus double labeling the myotubes (nucleus plus cytoplasm). The labeled myotubes were pelleted and implanted under the wound epidermis of regenerating hind limbs, 3–7 d after amputation through the femur. By the conical blastema stage (Fig. 1), the myotubes had cleaved and proliferated to form hundreds to thousands of undifferentiated, double-labeled mononucleate cells, confirming previous electron microscopic observations indicating that myofibers fragmented into mononucleate cells that dedifferentiated and became part of the blastema (Hay, 1959). Similar results were obtained with myotubes labeled with a retrovirus expressing human alkaline phosphatase, or with PKH26, a lipophilic tracker dye (Kumar et al., 2000). Counts of triploid cells in blastemas derived from diploid axolotl limbs in which the skin or cartilage was replaced with triploid skin or cartilage showed that dermal fibroblasts contribute 43% of the cells of the blastema, while cartilage contributes only 2% (Muneoka et al., 1986). The remainder is derived from nondermal connective tissue, myofibers, and Schwann cells, but the percentages of these contributions are unknown. The percentage of contributions from these tissues in adult newt limbs may be different, but this has not been examined. Labeled epidermal cells do not contribute to the stem cells of the blastema (Riddiford, 1960; Namenwirth, 1974).

Several specific markers characterize the dedifferentiated cells and wound epidermis of the blastema (Brockes, 1997; Geraudie and Ferretti, 1998). The 22/18 antigen is an intermediate filament of unknown type that is associated with dedifferentiated cells derived from muscle and Schwann cells. Keratins 8 and 18 are expressed in all the stem cells of the blastema. Another type II keratin, *Notopthalmus viridescens* keratin (NvK) II, is expressed in both dedifferentiated cells and wound epidermis. Two antigens that appear to be actin-binding proteins, wound epidermis (WE) 3 and 4, are expressed exclusively in the wound epidermis (Tassava et al., 1993). The function of all these proteins in regeneration is unknown.

8.2.2. MECHANISM OF STEM CELL FORMATION

8.2.2.1. Degradation of Extracellular Matrix On amputation, the extracellular matrix (ECM) of bone, cartilage, muscle, nerve sheath, and dermal tissue is degraded, liberating osteocytes, chondrocytes, myofibers, Schwann cells, and fibroblasts from their tissue organization. The liberated cells then undergo dedifferentiation to become mesenchyme-like stem cells. Degradation of the ECM is accomplished by proteases. These include acid hydrolases such as cathepsin D and acid phosphatase (Ju and Kim, 1998), β-glucuronidase and carboxylic ester hydrolases (Schmidt, 1966), as well as matrix metalloproteinases (MMPs) such as MMP 2 and 9 (gelatinases) and MMP3/10a and b (stromelysins) (Grillo et al., 1968; Dresden and Gross, 1970; Miyazaki et al., 1996; Park and Kim, 1999; Yang et al., 1999). MMPs are secreted in proenzyme form and must themselves be activated by other proteases. The most likely protease to do this in regenerating limbs (and other regenerating tissues?) is plasmin, which is itself derived from plasminogen by the action of tissue plasminogen activator, a serine protease in the blood. Plasmin can convert both procollagenases and prostromelysins to their active forms. Thus, it might be expected that high levels of tissue plasminogen activator and plasmin would be present during blastema formation, but this hypothesis has not yet been tested.

Acid hydrolases are released after amputation from injured and dying cells, and by chondroclasts and osteoclasts, which degrade cartilage and bone matrix in regenerating limbs (Stocum, 1979). The cell types that produce MMPs and perhaps other proteases have not been clearly defined, but candidates are macrophages, the wound epidermis, and the blastema stem cells themselves. The matrix of the mature limb is replaced by a more limb bud–type matrix rich in collagen I, hyaluronate, and fibronectin (Stocum, 1995). Matrix degradation ceases coincident with the start of differentiation in the growing blastema. The cessation of matrix degradation involves the downregulation of acid hydrolases and MMPs (Miyazaki et al., 1996; Ju and Kim, 1998; Park and Kim 1999, Yang et al., 1999), and is likely to involve the upregulation of tissue inhibitors of metalloproteases, though expression of these inhibitors has not yet been examined.

8.2.2.2. Mechanism of Dedifferentiation What is the mechanism of dedifferentiation, once a cell is released from its tissue organization? Altered cell shape and soluble factors in serum at the wound site, appear to play important roles. Cells of intact tissues are under tension due to their connections to ECM molecules by adhesion molecules such as integrins. Proteolytic degradation of the ECM would break these contacts, leading to changes in cell shape and reorganization of the cortical cytoskeleton. This reorganization could, in turn, activate signal transduction pathways that induce the upregulation of genes encoding enzymes ("dedifferases") that dismantle the internal structure of cells, as well as the downregulation of phenotype-specific proteins. Consistent with this idea, the ECM molecules tenascin, osteonectin, and thrombospondin promote reorganization of the actin cytoskeleton in cultured bovine arterial endothelial cells by modulating adhesive contacts (Sage and Bornstein, 1991). Tenascin is known to be upregulated in regenerating newt limbs (Onda et al., 1991). Furthermore, gelatinase A (GL-A), an MMP involved in connective-tissue remodeling and tumor invasion, is activated in cultured fibroblasts when they are treated with cytochalasin D. The activation is mediated by a membrane-type MMP that cleaves the GL-A propeptide (Tomasek et al., 1997).

8.2.2.3. Muscle Dedifferentiation The production of blastemal stem cells from myofibers is of particular interest because of its complexity compared to other cell types. Myofibers are multinucleate and must somehow produce mononucleate cells while disassembling a complex contractile apparatus, as well as reentering the cell cycle. Other cell types in the limb are already mononucleate and their structure is not as highly differentiated as that of the myofiber. The reentry of myonuclei into the cell cycle and the cellularization of myofibers have both been the subject of recent analysis.

8.2.2.3.1. Reentry of Myonuclei into Cell Cycle Serum factors appear to play an important role in the reentry of newt skeletal muscle nuclei into the cell cycle. Serum from a variety of animals contains a variety of growth factors that stimulate cultured newt and mouse myoblasts to enter S by promoting phosphorylation of the retinoblastoma (Rb) protein, leading to release of E2F transcription factors that activate genes essential for DNA synthesis. Serum-stimulated Rb phosphorylation and entry into S phase is also observed in the nuclei of cultured newt myotubes, but not in the nuclei of mouse C2C12 myotubes (Tanaka et al., 1997). However, the same growth factors that stimulate DNA synthesis in both newt and mouse myoblasts are unable to do so in newt myotubes, suggesting that serum contains another kind of factor that stimulates DNA synthesis in these multinucleate cells.

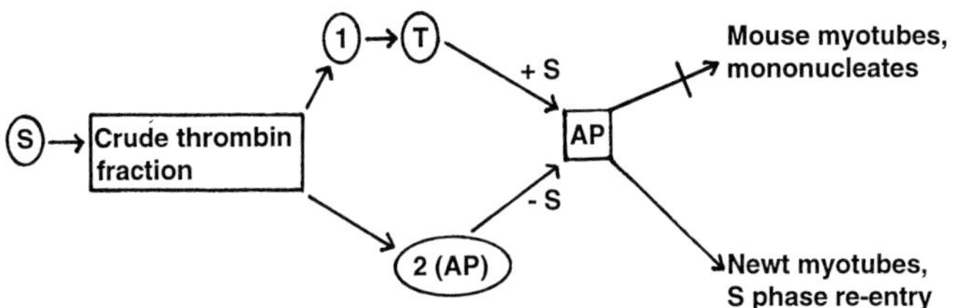

Fig. 2. Diagram illustrating effects of two serum (S) fractions (1 and 2), purified thrombin (T), and a purified protein activated by thrombin (AP in square) on mouse myotubes and myoblasts, and on newt myotubes. Purified thrombin activates the protein when incubated with serum (+S) and stimulates newt myotube nuclei (but not mouse myotube or myoblast nuclei) to synthesize DNA. Fraction 2 contains the already activated protein, which stimulates DNA synthesis in newt myotubes in the absence of serum.

Fractionation of serum and testing of the fractions on cultured myoblasts and myotubes revealed that the crude thrombin fraction contained the stimulatory activity (Fig. 2). The activity of this fraction was abolished by incubating it with thrombin inhibitors such as hirudin. The crude thrombin fraction was further separable into two peaks (Fig. 2). Peak 1 had the stimulatory activity; further fractionation of this peak yielded pure thrombin. Using a fluorogenic substrate on cryostat sections of regenerating newt limbs, thrombin activity was demonstrated to be elevated during the dedifferentiation stage and was abolished when the sections were incubated with thrombin inhibitors (Tanaka et al., 1999).

Thrombin does not act directly on newt myotubes, however, but indirectly through a serum protein activated by thrombin (Tanaka et al., 1999). This was shown by the fact that neither crude thrombin preparations nor pure thrombin could elicit DNA synthesis in cultured newt myotubes unless serum was also present. The protein was shown to reside in the second peak of the fractionated crude thrombin (Fig. 2). This peak retained 40% of the stimulatory activity of the crude thrombin fraction under serum-free conditions (Tanaka et al., 1999). The action of the protein seems to be specific to newt myotubes. Both serum and the crude thrombin fraction promote complete cell cycling in newt and mouse myoblasts, but neither the peak one nor two fractions have any effect on myoblasts.

Newt myotubes are thus clearly different from mouse myotubes in the ability to receive and transduce a signal provided by a thrombin-activated protein in serum that leads to cell-cycle entry and DNA synthesis. The protein is both necessary and sufficient to stimulate newt myotube nuclei to enter S phase, but it is not sufficient to drive them through mitosis; the nuclei arrest in the G_2 phase of the cycle. Mouse myonuclei will synthesize DNA in response to serum stimulation, however, if they are part of a heterokaryon made by fusing C2C12 and newt myoblasts (Velloso et al., 2001). These observations suggest that mouse myotubes and newt and mouse myoblasts do not respond to the serum protein because they either lack the receptor for the protein, have an inactive receptor, or lack components of the transduction system that lead to Rb phosphorylation. So far, the protein has escaped identification, but it is not any of the usual substrates associated with the blood-clotting function of thrombin (Tanaka et al., 1999).

8.2.2.3.2. Relationship of Cell-Cycle Reentry to Cellularization A major question is, Is reentry into the cell cycle of newt myofiber nuclei during regeneration mechanistically coupled to the cleavage of myofibers into mononucleate cells? To investigate this potential interdependence, cultured myotubes were either X-irradiated or transfected with the *p16* gene for cyclin-dependent kinase 4/6 inhibitor to inhibit serum-stimulated entry into S phase (Velloso et al., 2000). The inhibited myotubes were then labeled with fluorescein-dextran and implanted into limb blastemas along with control myotubes labeled with Texas Red-Dextran. Both control and experimental myotubes formed mononucleate cells, but only control cells synthesized DNA and underwent mitosis, as measured by incorporation of bromodeoxyuridine, indicating that myotube cellularization and entry into the cell cycle are independent of one another.

While serum factors are sufficient to drive the nuclei of cultured newt myotubes into S, the cell cycle is not completed unless the myotubes cleave into mononucleate cells. The mechanism of cellularization is not clear, but it would appear that mononucleate cell status is obligatory for completion of the cell cycle. Thus, it would be predicted that the thrombin-activated serum protein is sufficient to trigger complete cycling in mononucleate cells from nonmyogenic limb tissues.

2.2.3.3. Newt Mechanisms are Partly Conserved in Mouse Myotubes Although mouse myofibers do not cellularize and dedifferentiate after injury in vivo, they have the latent ability to do so, as shown by two experimental manipulations of C2C12 myotubes in vitro. The first is transfection of the myotubes with the gene *msx1*, which is expressed in the undifferentiated limb bud mesenchyme and in the regeneration bllastema (Hill, 1989; Robert et al., 1989; Crews et al., 1995; Simon et al., 1995) and inhibits myogenesis in vitro when expressed ectopically in cultured mouse myoblasts (Song et al., 1992; Woloshin et al., 1995). Forced expression of *msx1* in cultured C2C12 myotubes induces morphological and molecular dedifferentiation, reducing the expression of muscle regulatory proteins to undetectable levels in 20–50% of the myotubes (Odelberg et al., 2001). About 9% of the dedifferentiated myotubes cleave to produce either smaller multinucleated myotubes or proliferating mononucleate cells. Clonal populations of these mononucleate cells are multipotent, able to differentiate in vitro into cells that express chondrogenic, adipogenic, myogenic, and osteogenic molecular markers, like mesenchymal stem cells (MSCs) of the bone marrow stroma (Pittinger et al., 1999; Odelberg et al., 2001).

The second experimental manipulation is morphological dedifferentiation of C2C12 myotubes induced by extract derived from amputated newt limbs undergoing dedifferentiation (McGann et al., 2001). Muscle differentiation proteins were reduced to undetectable levels in 15–30% of the treated myotubes, and 18% exhibited entry of nuclei into S phase. Eleven percent of the myotubes cleaved and about 50% of these continued cleaving to produce proliferating mononucleated cells. Control myotubes exposed to extract of unamputated limbs did not cleave and dedifferentiate. These data indicate that the newt regeneration extract supplies the mouse myotubes not only with the thrombin-activated signal for entry into S phase, but also something that enables reception of the signal or activation of signal transduction pathways leading to cleavage and dedifferentiation. Regardless of the mechanism involved, it is clear that a large part of the dedifferentiation potential of amphibian myofibers is conserved in mammalian myotubes and perhaps other mammalian cell types.

8.2.3. PROLIFERATION OF BLASTEMA STEM CELLS

8.2.3.1. Wound Epidermis and Nerves are Required for Proliferation Continuous- and pulse-labeling studies with ^{3}H-thymidine indicate that >90% of blastemic cells will ultimately cycle during the regeneration of either larval or adult regenerating limbs. The actively cycling fraction of cells at any given time is lower in adult limbs, however, thus partly accounting for the fact that regeneration of adult limbs takes longer (Goldhamer and Tassava, 1987; Tomlinson and Barger, 1987). Proliferation of blastemic cells depends on factors contributed by the wound epidermis, regenerating nerves, and blastemic mesenchymal cells themselves. In newts and larval *Ambystoma*, denervation of the blastema at any time results in cessation of growth (Schotte and Butler, 1941; Singer and Craven, 1948). Preventing contact between the wound epidermis and underlying blastemic stem cells results in a decrease in proliferation. This inhibition of proliferation is always accompanied by an inability to complete the proximodistal pattern of the regenerate (Stocum and Dearlove, 1972) and premature differentiation of cartilage (Globus et al., 1980), suggesting that the epidermis provides signals essential for maintaining blastemic stem cells in a dedifferentiated state. The *Msx-2* gene may have this function, because it is strongly expressed in the wound epidermis of regenerating axolotl limbs (Carlson et al., 1998). Other genes encoding known inhibitors of differentiation that are upregulated in the wound epidermis of the regenerating newt limb are the helix-loop-helix (HLH) genes *Id2* and *HES1*. A third inhibitor of differentiation gene, *Id3*, is expressed equally in the blastemal epidermis and mesenchyme (Shimizu-Nishikawa et al., 1999).

8.2.3.2. Fibroblast Growth Factors are Important Regulators of Blastemal Stem Cell Proliferation The wound epidermis and the nerves promote the proliferation of blastema stem cells through members of the fibroblast growth factor (FGF) family. Several FGFs are able to substitute for the outgrowth-promoting effect of the apical ectodermal ridge in chick embryos (Niswander et al., 1994; Fallon et al., 1994). FGF-1, -2, and -8 are expressed in the axolotl distal limb bud ectoderm and apical epidermis of the regenerating limb (Boilly et al., 1991; Mullen et al., 1996; Han et al., 2001). FGF-1 and FGF-2 elevate the mitotic index of blastema mesenchyme in the absence of nerves or epidermis in vitro (Albert et al., 1987), or in vivo (Chew and Cameron, 1983). FGFR-1 (receptor for FGF-1, -2, -4) is expressed in the blastemal mesenchyme of regenerating newt limbs, but not the wound epidermis, whereas the keratinocyte growth factor receptor (KGFR) variant of FGFR-2 (receptor for FGF-1, -7) is expressed in the wound epidermis and the *bek* variant (receptor for FGF-2, -8) is expressed in the mesenchyme (Poulin et al., 1993; Poulin and Chiu, 1995). FGF-8 is expressed in the apical wound epidermis and FGF-10 in the mesenchyme of regenerating hind limbs in *Xenopus* tadpoles, but not in the amputated hindlimbs of tadpoles approaching metamorphosis, which fail to regenerate (Yokoyama et al., 2001). Although FGF-8 is essential for normal limb development (Kuhlman and Niswander, 1997; Meyers et al., 1998), exogenously supplied FGF-8 does not promote any substantial regeneration in regeneration-incompetent *Xenopus* limbs. Exogenous FGF-10 restores the regeneration of foot structures while simultaneously inducing expression of FGF-8 in the apical epidermal cap, but fails to restore more proximal elements (Yokoyama et al., 2001). FGFR-1 and

FGFR-2 are expressed in the wound epidermis and blastemal mesenchyme of regeneration-competent, but not regeneration-incompetent, *Xenopus* hind limbs, and antibodies to these receptors inhibit regeneration in regeneration-competent limbs (D'Jamoos et al., 1998). These observations indicate that FGF-10 and FGF-8 are key molecules in the regeneration of *Xenopus* hind limbs, and that the expression of FGF-8 is dependent on FGF-10. They also suggest that other members of the FGF family, or growth factors outside this family, are necessary for complete regeneration.

Other molecules synthesized by the apical wound epidermis that might play a role in blastema cell proliferation are an unidentified antigen called 9G1 (Onda and Tassava, 1991) and the product of the *Dlx-3* gene, a homeobox homolog of *Drosophila distal-less*, which is required for leg outgrowth in the fly (Cohen and Jurgens, 1989). Expression of 9G1 is downregulated by denervation, but this was shown to be through an effect of the nerves on the blastemic mesenchyme (Onda and Tassava, 1991). *Dlx-3* is strongly expressed in the wound epidermis of regenerating axolotl limbs and tail. Expression peaks just prior to redifferentiation and decreases to zero by late digit stages (Mullen et al., 1996).

FGF-2 is also present in the nerves reinnervating the blastema of regenerating axolotl limbs and is downregulated in response to denervation. FGF-2 can substitute for the nerve in denervated limbs when delivered in beads implanted into early blastemas, allowing regeneration to progress to digit stages (Mullen et al., 1996). The expression of FGF-2 and *Dlx-3* by the wound epidermis may be dependent on the nerves. Sensory nerves reinnervating the blastema extend into the epidermis. Denervation downregulates the expression of both FGF-2 and *Dlx-3* in the wound epithelium during nerve-dependent stages of regeneration. *Dlx-3* expression is not affected by denervation in nerve-independent stages; whether or not FGF-2 expression is affected is unknown (Mullen et al., 1996). Whether denervation affects the expression of FGF-2 and *Dlx-3* in the wound epidermis directly or through effects on the blastemic mesenchyme is also unknown. Other molecules that may be essential for the neurotrophic activity are glial growth factor (Brockes, 1984), substance P (Globus, 1988), and transferrin (Mescher, 1996). The growth-promoting activity of neural extracts on organ cultures of denervated axolotl blastemas is completely removed by antitransferrin antiserum and restored by purified axolotl transferrin (Mescher et al., 1997).

8.2.4. PLASTICITY OF BLASTEMA STEM CELLS As they continue to proliferate, the stem cells of the blastema withdraw from the cell cycle in a proximal-to-distal wave and differentiate to restore the original pattern and structure of the missing limb segments. During this process, fibroblasts and myogenic cells transdifferentiate into cartilage and cartilage transdifferentiates into fibroblasts of a variety of connective tissues, and perhaps into muscle, as shown by grafting experiments employing labeled cells.

X-irradiated axolotl limbs in which triploid skin was grafted in place of the normal diploid skin produced blastemas composed entirely of triploid stem cells after amputation (Dunis and Namenwirth, 1977). The triploid stem cells, which were derived from dermal fibroblasts, transdifferentiated into chondrocytes to provide the regenerate skeleton but did not transdifferentiate into muscle. Steen (1968) showed that blastemic stem cells derived from triploid cartilage grafted into diploid axolotl hosts differentiated only into chondrocytes. However, a large fraction of the regenerated skeleton was composed of diploid cells, indicating that stem cells from nonskeletal tissues had transdifferentiated into chondrocytes. The same result was obtained with limbs containing triploid muscle grafts, suggesting that stem cells derived from muscle transdifferentiate into chondrocytes during regeneration. To determine whether these cells were derived from mysial sheaths, myofibers, or both, Casimir et al. (1988) used as markers two hypomethylated sites on newt DNA encoding the heavy myosin chain of cardiac and skeletal muscle, called "hypo A" and "hypo B." In the unamputated limb, hypo A is more highly represented in skeletal and cardiac muscle DNA, and in the DNA of cultured myogenic cells, while hypo B is represented in muscle (myogenic plus connective tissue) DNA, but is not represented in the DNA of cultured myogenic cells. Thus, hypo A was used as a marker for myogenic cells, and hypo B as a marker for connective-tissue cells. DNA from the cartilage of limb regenerates exhibited elevated levels of both markers over that of the unamputated limb skeleton, suggesting that both connective-tissue fibroblasts and myogenic cells transdifferentiate into chondrocytes during regeneration. Other evidence indicates that blastemic stem cells derived from myofibers can give rise to nonmuscle cells. Myofibers from a nucleolar mutant of *Xenopus* (1-nu) gave rise to chondrocytes and other cell types during regeneration after implantation into wild-type *Xenopus* tadpole hind limbs, as did rhodamine-conjugated lysinated dextran and ^{3}H-T-labeled newt myotubes implanted under the wound epidermis of regenerating newt limbs (Steen, 1973; Lo et al., 1993). Consistent with these results, isolated muscle from the medusa *Podocoryne carnea* can transdifferentiate into nine different nonmuscle cell types (Schmid and Alder, 1984; Alder and Schmid, 1987), and clonal myoblasts from late embryonic rat or chick embryos differentiate as chondrocytes when cultured on demineralized bone matrix (Nathanson, 1979).

Skeletal cells can also transdifferentiate into nonskeletal cells. Triploid cartilage grafted in place of the cartilage of X-irradiated diploid limbs transdifferentiated into perichondrium, and fibroblasts of joint connective tissues, dermis, and skeletal muscle connective tissue (Namenwirth, 1974). Blastemic cells derived from cartilage implanted into X-irradiated limbs were reported to transdifferentiate into muscle (Maden and Wallace, 1975; Desselle and Gontcharoff, 1978), although it is possible that the muscle was derived from host myogenic cells that migrated to the amputation site from more proximal locations (Hinterberger and Cameron, 1990).

8.3. NR AND LENS REGENERATION The embryos of several vertebrate species, including fish, frogs, birds, and rats, as well as the tadpoles of the anuran *Xenopus laevis* and the adult newt, can regenerate the lens and NR (Reyer, 1977; Stroeva and Mitashov, 1983; Park and Hollenberg, 1993; Mitashov, 1996). Regeneration is accomplished by the dedifferentiation of pigmented retinal or corneal epithelial cells to stem cells that proliferate and then transdifferentiate into NR or lens cells. Members of the FGF family play a prominent role in regeneration of the NR and lens, as they do in limb regeneration.

8.3.1. LENS REGENERATION

8.3.1.1. Dedifferentiation and Transdifferentiation of Pigmented Epithelial Cells The newt is the only adult urodele and *X. laevis* tadpoles are the only anurans that can regenerate the lens. The newt regenerates the lens from pigmented epithelial cells of the dorsal iris after lensectomy (Fig. 3) or, if the dorsal iris is missing, from the retinal pigmented epithelium (RPE) (Reyer, 1977; Stroeva and Mitashov, 1983; Mitashov, 1996). *Xenopus*

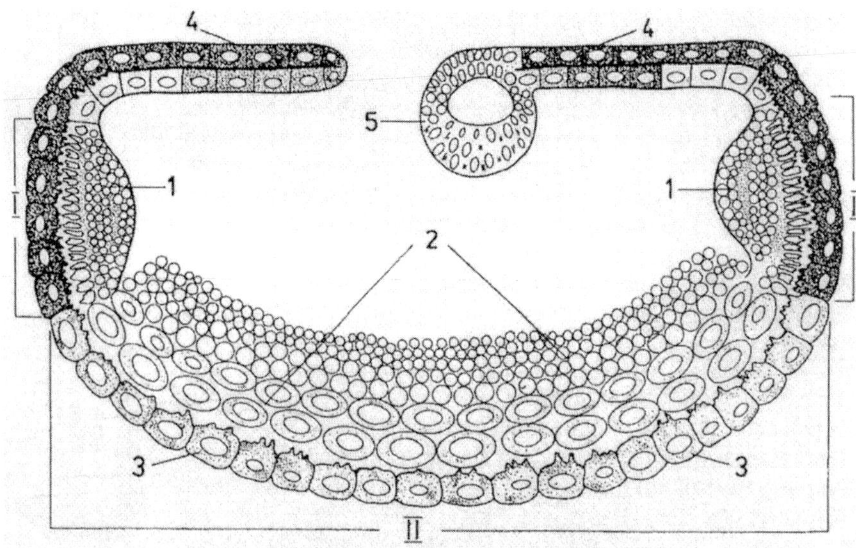

Fig. 3. Diagram of regenerating lens and NR in the adult newt eye after lentectomy or degeneration of NR. I, peripheral zone; II central zone of the retina. The peripheral part of the NR regenerates from precursor neuroepithelial cells in the ora serrata (1). The large, central part of the retina (2) regenerates from the retinal pigmented epithelium (3). The lens (5) regenerates from the dorsal iris (4). (Reproduced with permission from Mitashov, 1996.)

tadpoles regenerate the lens from the inner corneal epithelium (Freeman, 1963); this is associated with degradation of the ECM by MMPs, as in regenerating limbs (Carinato et al., 2000). The capacity for lens regeneration in the newt iris declines from dorsal to ventral. Pieces of ventral iris fail to form lens under experimental conditions in which dorsal iris readily regenerates lens (Reyer, 1977). The cells of the ventral iris, however, have not lost the capacity for lens regeneration, because they can form lens when dissociated and cultured in vitro (Eguchi, 1998).

The earliest effect of lensectomy in the newt eye appears to be an increased synthesis of RNA by the amplification of rRNA genes and their increased rate of transcription, 2 d after operation (Reese et al., 1969; Collins, 1972). Dedifferentiation and proliferation has begun by 4 d (Yamada and Roesel, 1969, 1971; Reyer, 1971). Expression of the c-*myc* proto-oncogene is enhanced, as expected, by increased proliferative activity (Agata et al., 1993). The cells take on an irregular shape and eliminate melanosomes, which are phagocytized by macrophages that invade the iris epithelium (Eguchi, 1963; Yamada and Dumont, 1972; Reyer, 1982, 1990 a, b). The numbers of mitochondria, free ribosomes, and microfilaments increase (Reyer, 1977). Proteoglycans are lost from the surface of dedifferentiating dorsal iris cells (Zalik and Scott, 1972, 1973). Eguchi (1998) identified a specific-cell surface glycoprotein, 2NI-36, that disappears from the surface of dorsal, but not ventral, iris cells. Ventral iris treated with antibody to this antigen and implanted into a lentectomized eye can regenerate a lens. These observations suggest that the disappearance or inhibition of 2NI-36 is necessary and sufficient to trigger dedifferentiation of pigmented epithelial cells. 2NI-36 is also found in many other tissues of the newt, including the RPE, and has been postulated to exert a general stabilizing effect on differentiation (Imokawa and Eguchi, 1992; Imokawa et al., 1992).

With continued proliferation, the dedifferentiated dorsal iris cells transdifferentiate into lens epithelial cells that express α and β crystallins, forming a small lens vesicle at the edge of the dorsal iris (Yamada and McDevitt, 1974). The lens epithelial cells facing the retina subsequently withdraw from the cell cycle, express γ crystallins, and differentiate into lens fibers (Takata et al., 1966; Eguchi, 1967). The new lens enlarges to normal size by the continuing division of lens epithelial cells and their differentiation into lens fibers.

Lens regeneration requires a non-species-specific interaction of the dorsal iris with the NR. Fragments of dorsal iris cultured in the dorsal fin or brain ventricle (Reyer et al., 1973) or isolated in vitro (Yamada, 1973) do not regenerate lens, although the cells sometimes undergo depigmentation. Newt dorsal iris cultured in vitro with frog NR regenerates lens, whereas newt ventral iris does not (Yamada et al., 1973). *Xenopus* corneal tissue will not form lens when isolated in vitro but will do so when cultured with NR (Bosco et al., 1993). During embryonic development, the optic vesicle exerts a late inductive influence on presumptive lens ectoderm. Both lens and nonlens embryonic ectoderm implanted into the cavity of mature lensectomized *Xenopus* eyes is stimulated to form lens cells in a fashion resembling the embryonic induction by the optic vesicle (Henry and Mittleman, 1995). Because the dorsal iris or cornea does not regenerate lens unless the lens is removed, it has been proposed that the lens produces a factor that normally inhibits iris or corneal cells from dedifferentiating. The evidence for such a factor, however, is conflicting (Reyer, 1977).

8.3.1.2. Regulation of Lens Regeneration by FGF Several lines of evidence suggest that the lens regeneration–promoting factor produced by the NR may be FGF-1. FGF-1 transcripts are expressed in the intact eye and are upregulated in regenerating lens cells of newts (Del-Rio Tsonis et al., 1997). FGF-1 stimulates cultured newt dorsal iris and *Xenopus* corneal cells to transdifferentiate into lens cells, although this transformation is not associated with the formation of a normally organized lens (Cuny et al., 1986; Bosco et al., 1997). FGF-2 does not have this effect in vivo (Del-Rio Tsonis et al., 1995), but it can induce the transdifferentiation of chick RPE cells to lens cells in vitro (Mochii et al., 1998). Exogenous FGF-1, which binds to FGFR-1 and to the KGFR variant of FGFR-2, and FGF-4, which binds to the *bek* variant of FGFR-2, cause structural abnormalities in regenerating newt lenses when applied via implanted beads, a result similar to that seen in the developing lenses of FGF transgenic mice (Del-Rio Tsonis et al., 1997).

In situ hybridization experiments have shown that transcripts for FGFR-1 are expressed in the NR and lens epithelium of the intact adult newt eye. After lensectomy, expression of the FGFR-1 gene is maintained in the retina but is upregulated during dedifferentiation of dorsal iris cells and throughout their transdifferentiation into lens cells (Del-Rio Tsonis et al., 1998). Transcripts are also expressed in the ciliary body and the ventral iris, although at a lower level than in the dorsal iris. A similar pattern of expression of FGFR-2 and FGFR-3 is observed in the intact eye and regenerating lens (Del-Rio Tsonis et al., 1997). Despite the widespread expression of FGFR-1 transcripts, immunostaining showed that the FGFR-1 protein is confined to the dedifferentiating cells of the newt dorsal iris that form the lens, indicating that expression of the gene is regulated at the level of translation (Del-Rio Tsonis et al., 1998). A further indication of the central importance of FGFR-1 (and FGF-1) is that lens regeneration is inhibited by SU5402, a 3-substituted indolin-2-one that specifically inhibits the autophosphorylation, and thus activation, of this receptor (Del-Rio Tsonis et al., 1998). McDevitt et al. (1997) examined by immunoblotting the protein expression patterns of FGFR-2 and -3, as well as FGFR-1. They reported no regional, temporal, qualitative, or quantitative differences for FGFR-1 and -2 expression. A higher level of FGFR-3, however, was observed in intact dorsal iris and in dorsal iris cells as they engage in lens regeneration.

8.3.1.3. Transcription Factors Regulating Lens Regeneration Several transcription factors are crucial to lens regeneration. *pax-6*, a master gene for eye development, is expressed in the intact retina of the adult newt, and in the cells of the dorsal iris that form new lens at all stages of lens regeneration. Expression of *pax-6* is also observed in the developing retina and lens of the embryonic and larval axolotl (Del-Rio Tsonis et al., 1995). The expression patterns of *pax-6* and *Mitf*, which encodes a basic HLH-leucine zipper protein, are complementary in chick embryo pigmented epithelial cells, both in vivo and in vitro. Overexpression of *Mitf* inhibits FGF-2-induced dedifferentiation and transdifferentiation of cultured chick pigmented epithelial cells to lens cells, and simultaneously inhibits expression of pax-6. Although not yet demonstrated in amphibians, these results suggest that the negative regulation of *pax-6* by *Mitf* is a crucial event in preventing lens regeneration, i.e., maintaining the differentiation of dorsal iris cells (Mochii et al., 1998). This idea is further strengthened by the fact that the axolotl cannot regenerate the lens at older stages of development, an inability that is correlated with a decline in the expression of *pax-6* (Del Rio-Tsonis et al., 1995).

8.3.1.4. Retinoic Acid and Lens Regeneration Eye development, including the lens, is highly dependent on retinoic acid (RA) signaling through its nuclear RA receptors (RARs). The RARs and Pax-6 activate expression of the αB crystallin gene. RA signaling is dependent on Pax-6, as shown by the fact that in mice mutated for *pax-6*, RA signaling in the eye is decreased and the developing lens cannot respond to exogenous RA (Enwright and Grainger, 2000). Inhibition of RA synthesis by disulfiram, or RAR function by the RAR antagonist 193109, resulted in the inhibition, retardation, or abnormal morphogenesis of regeneration, although in some cases ectopic lenses regenerated from the ventral iris or the cornea (Tsonis et al., 2000). Collectively, these observations suggest that RARs and Pax-6 act together in promoting the dedifferentiation and transdifferentiation of dorsal iris cells to lens cells, but that elsewhere in the eye (ventral iris and cornea) RARs function to inhibit ectopic lens regeneration. RAR-δ, which mediates proximodistal positional effects of RA in regenerating newt limbs (Pecorino et al., 1996), may also mediate the effects of RA on lens regeneration. RAR-δ transcripts are expressed at low levels in the ganglion cell layer of the retina, but not elsewhere in the eye. After lensectomy the RAR-δ gene is expressed in the dedifferentiated cells that form the new lens vesicle, and the level of expression increases to its highest level during lens fiber differentiation (Tsonis et al., 2000).

8.3.2. NR REGENERATION

8.3.2.1. Dedifferentiation and Transdifferentiation of Pigmented Epithelial Cells Cutting the optic nerve and blood vessels to the eye in adult newts and anuran tadpoles results in degeneration of the NR, followed by its regeneration by dedifferentiated RPE and ciliary body (ora serrata) cells (Fig. 3), which also act as scavengers to clean up cell debris (Klein et al., 1990; Mitashov, 1996; Raymond and Hitchcock, 1997). The dedifferentiated cells organize into layers, with those cells next to the choroid layer redifferentiating as RPE cells to maintain the pigmented retina, while those cells facing the vitreous chamber transdifferentiate into the different NR cell layers: ganglion cell layer, inner and outer nuclear layers, and photoreceptor layer (Stone, 1950; Keefe, 1973a, b, c; Levine, 1977; Reyer, 1977). The optic nerve is reformed by regenerating ganglion cell axons that project correctly to the optic tectum, allowing recovery of vision (Sperry, 1944; Gaze, 1959).

Dedifferentiation of RPE cells involves changees in their shape, internal structure, and gene activity, regulated by the ECM and soluble signals. RPE cells change shape from cuboidal to columnar within 2 d after removal or destruction of the NR, followed over the next week by cell enlargement and depigmentation by extrusion of melanosomes, which are removed by phagocytic cells. Although not well-studied, it is likely that the change in shape of the RPE cells is related to enzymatic ECM breakdown and loss of epithelial cell–cell contacts and that these shape changes lead to melanosome extrusion and altered patterns of gene activity through reorganization of the cytoskeleton, as postulated for dedifferentiating limb cells. Laminin appears to be an important regulator of NR regeneration. Dedifferentiation and transdifferentiation of *Rana* tadpole RPE cells is promoted by growing them on a laminin substrate in vitro (Reh et al., 1987). During the in vivo regeneration of NR in *Rana* tadpoles, the first new NR cells arise in association with the vitreal vascular membrane (Reh and Nagy, 1987), which contains a high concentration of laminin (Reh et al., 1987).

8.3.2.2. Regulation of NR Regeneration by FGF FGF-2, as well as nerve growth factor (NGF), insulin-like growth factor-1 (IGF-1) and IGF-2, and transforming growth factor-β (TGF-β), are expressed in the eye tissues of birds and mammals (Noji et al., 1990; Park and Hollenberg, 1993). Studies on chick embryos suggest that FGF-2 is an important signaling molecule in NR regeneration. During optic cup formation in stage 9 to 10 embryos, the prospective RPE (outer layer of the optic cup) differentiates as NR if its contact with the prospective NR (inner layer of the optic cup) is broken (Orts-Lorca and Genis-Galvez, 1960), suggesting that the prospective NR normally inhibits the prospective RPE from differentiating as NR. By stage 22–24, the RPE and NR are differentiated and the RPE will not form NR when the NR is completely removed. However, the RPE can be stimulated to regenerate the NR by implanting a piece of NR back into the vitreous chamber (Coulombre and Coulombre, 1965, 1970) or by implanting blocks of ethylene/vinyl coacetate polymer impregnated with 10 ng of FGF-2 or, at higher concentrations, FGF-1 (Park and Hollenberg, 1991). In both cases, the polarity of the regenerated NR is reversed, with the photoreceptor layer facing the implant, implying that RPE cells experiencing the highest concentration of FGF become photoreceptors. Incubation of stage 9 to 10 chick optic vesicles in medium containing FGF-2 also induced the prospective RPE to differentiate as NR, producing an eye with two NRs of normal polarity (Pittack et al., 1997). FGF-2 has also been shown to induce RPE cells to form NR cells in vitro (Pittack et al., 1991).

Collectively, these results suggest that during chick embryogenesis the RPE and NR exert a reciprocal inhibitory effect on one another. The NR prevents the RPE from becoming NR through as-yet unidentified signals, while the RPE prevents the NR from producing FGF-2. If the NR is separated from the RPE or removed and a piece put back, it produces FGF-2, which stimulates RPE cells to regenerate NR. Because in amphibians the RPE regenerates the NR after complete destruction of the original NR, either no FGF signal is required for regeneration; or enough FGF is released from the degenerating NR, or the RPE is released from an inhibitory influence of the NR and itself produces FGF, which acts in an autocrine fashion. Although FGFRs are expressed in the intact retina of the adult newt, there are no data on expression of FGFs or receptors in the regenerating amphibian NR.

8.3.2.3. Other Signaling Molecules and Transcription Factors that Regulate NR Regeneration Identification of other signaling molecules and transcription factors involved in dedifferentiation, proliferation, and transdifferentiation of RPE cells is just beginning. Overexpression of *Mitf* has the same inhibitory effects on FGF-2-induced dedifferentiation and transdifferentiation of cultured chick RPE cells into NR cells as it does on their dedifferentiation and transdifferentiation into lens cells (Mochii et al., 1998). Conversely, induction of dedifferentiation and transdifferentiation of RPE cells by FGF-2 inhibits *Mitf* expression. These observations suggest that downregulation of this gene might be essential for NR regeneration, as well as lens regeneration, in amphibians. The sonic hedgehog (*shh*) and patched (*ptc*) genes are expressed in adjacent domains in the developing mouse retina (Jensen and Wallace, 1997). Treatment of cultured mouse NR cells with the amino-terminal fragment of the Shh protein promotes their proliferation. Finally, dedifferentiated and proliferating newt RPE cells express an antigen that binds the mouse monoclonal antibody RPE-1. Expression of RPE-1 is extinguished only when the cells express NR molecular markers (Negeshi et al., 1992; Mitashov, 1996), suggesting that at least part of the expression pattern of dedifferentiated RPE cells overlaps with that of NR cells during proliferation and transdifferentiation.

8.4. CONCLUSION AND RESEARCH QUESTIONS

Amphibians create stem cells for limb, lens, and NR regeneration by a process of histolysis and dedifferentiation that involves the proteolytic degradation of ECM, loss of phenotypic structure, and renewed cell cycling. Dedifferetiation involves changes in the shape of cells liberated from their tissue matrix; loss of cell-surface proteins and proteoglycans, and, in the case of myofibers, the dismantling of a highly differentiated contractile system. The FGF family of growth factors and receptors, as well as other signaling molecules and transcription factors, plays a major role in promoting and regulating dedifferentiation and proliferation. Labeling studies have shown that myonuclei can synthesize DNA in the absence of cellularization but do not traverse the rest of the cell cycle unless cellularization has occurred. Mammalian muscle does not normally respond to a thrombin-activated serum factor that triggers DNA synthesis in newt myotubes but conserves a substantial portion of the mechanisms of cellularization, cell-cycle reentry, and dedifferentiation. Stem cells derived from pigmented epithelial cells can transdifferentiate into either lens or NR cells. Stem cells derived from skeletal elements of the limb can transdifferentiate into fibroblasts of connective tissue and perhaps muscle, while fibroblasts and myogenic cells can transdifferentiate into chondrocytes. These findings raise a number of interesting research questions, as follows:

1. *What are the mechanisms of dedifferentiation and transdifferentiation?* What is the pattern of gene activity that is associated with dedifferentiation? What is the pathway of intracellular signal transduction that leads to this pattern? What are the physical and chemical signals that are transduced? What signals are involved in changing the pattern of gene activity in a dedifferentiated cell to a pattern specifying a cell type different from the one of origin? A first step in answering these questions can be made by comparisons of arrays of genes active in tissues before and after injury. Beyond this, there must be gain- and loss-of-function studies to clarify the roles of various genes.
2. *What factors are required for dedifferentiated cell cycling?* In the regenerating limb, factors from wound epidermis and regenerating nerves provide growth and trophic factors that are necessary for cycling of dedifferentiated cells. These factors are also found in serum. What is their functional and/or interactive relationship to the thrombin-activated serum factor? Does the thrombin-activated serum factor that triggers reentry of myotube nuclei into the cell cycle do the same in other regenerating urodele tissues such as lens and NR? Are the same growth factors and trophic molecules involved in the proliferation of dedifferentiated cells derived from different structures but supplied in different ways, e.g., by the epidermis and nerves for limb, but by the dedifferentiated cells themselves for lens and retina?
3. *Does dedifferentiation in diverse amphibian tissues involve activation of a common set of genes?* It is likely that a common set of cell-cycling genes is activated in dedifferentiating cells, but what other common genes might be activated or upregulated, such as genes encoding internally acting "dedifferases," externally acting proteases that degrade ECM, genes that encode ECM molecules characteristic of a more embryonic environment, or genes encoding growth factors that maintain proliferation? Again, the comparative approach to gene activity will be valuable in answering this question.
4. Do conventional stem cells contribute to the blastema? Conventional neuroepithelial stem cells of the ciliary body

reconstitute the peripheral part of the regenerating NR (Mitashov, 1996). Do conventional stem cells contribute to the regeneration of other tissues that produce stem cells by dedifferentiation? Conventional undifferentiated stem cells reside in the bone marrow, stroma (marrow stromal or mesenchymal stem cells), and muscle (satellite cells) of amphibian limbs for use in the repair of fractures or muscle injuries. While a contribution of these cells to the limb regeneration blastema is not ruled out, there is no direct evidence for such a contribution. Testing this possibility is difficult because it requires a way to selectively label mesenchymal stem or satellite cells so that they can be followed during limb regeneration. Muscle explants might possibly be used to label satellite cells. The nuclei of satellite cells in muscle explants have been shown to be selectively labeled by ^{3}H thymidine during the first 6 d of culture (Cameron et al., 1986). The muscle explant might then be transplanted back into an amputated limb and the labeled cells subsequently followed. Selectively labeling MSCs may prove more difficult. However, if salamanders with marked bone marrow cells could be made, such as by repopulating the marrow of irradiated animals with marked marrow cells (triploid, transgenic for LacZ or green fluorescent protein, opposite sex chromosomes), or by grafting marked hematopoietic regions in early embryos, it might be possible to determine whether marrow stromal cells of long bones contribute to the regeneration blastema after amputation. A significant mechanistic difference between tissue regeneration and limb regeneration would be indicated if conventional stem cells were used only in the process of individual tissue regeneration, while whole limb regeneration depended primarily on dedifferentiation to produce stem cells. During regeneration, are some of the blastema stem cells arrested in their differentiation to become satellite cells and periosteal stem cells that also repopulate the bone marrow, or are all of these stem cells derived from preexisting stem cells of the undamaged limb tissues and migrate into the differentiating muscles and bones of the regenerate?

5. *Are limb blastema stem cells that form cartilage and bone similar to the MSCs of the bone marrow?* These cells have a common embryological history, both being derived ultimately from limb bud mesenchyme that becomes periosteum. As regeneration proceeds to maturity, some blastemic cells would redifferentiate into periosteal fibroblasts, whereas others would remain undifferentiated in the lower layers of the periosteum and become MSCs that enter the bone marrow cavity of the regenerated calcified cartilage with invading blood vessels. Are the markers expressed in the MSCs of bone similar to those expressed by blastema stem cells?
6. *How plastic are stem cells derived by dedifferentiation?* Could the dedifferentiated cells of limb, retinal pigmented epithelium, and other tissues turn into neural cells or blood cells if transplanted into a cord lesion or asked to reconstitute some other tissue?
7. *What are the missing components of the dedifferentiative mechanism in mammalian cells?* The thrombin-activated protein that triggers reentry into the cell cycle in newt myotubes is found in the sera of mammalian species (Tanaka et al., 1997). Why, then, do mammalian myotubes (and other cells?) not respond to this protein? Is it because the receptor for the protein is missing? Is the receptor present but requires the protein for activation, as well as other factors for cellularization ? Is this why the extract of regenerating newt limbs induces cellularization, morphological and molecular dedifferentiation, and proliferation of mouse myotubes in culture (McGann et al., 2001)?

The answers to these questions will greatly advance our knowledge of the mechanism by which amphibian cells dedifferentiate in response to injury and perhaps enable us to induce the production of stem cells in the same way in nonregenerating mammalian tissues.

ACKNOWLEDGMENT

I thank Holly L. D. Nye, Department of Cell and Structural Biology, University of Illinois at Urbana-Champaign, for critically reviewing the manuscript and suggesting changes.

REFERENCES

Agata, K., Kobayashi, H., Itoh, Y., Mochii, M., Sawada, K., and Eguchi, G. (1993) Genetic characterization of the multipotent dedifferentiated state of pigmented epithelial cells in vitro. *Development* 118: 1025–1030.

Albert, P., Boilly B., Courty, J., and Barrritault, D. (1987) Stimulation in cell culture of mesenchymal cells of newt blastemas by EDGF1 or II (basic or acidic FGF). *Cell Differ.* 21:63–68.

Alder, H. and Schmidt, V. (1987) Cell cycles and in vitro transdifferentiation and regeneration of isolated striated muscle of jellyfish. *Dev. Biol.* 124:358–369.

Alvarado, A. S. (2000) Regeneration in the metazoans: why does it happen? *Bioessays* 22:578–590.

Benraiss, A., Arsanto, J. P., Coulon, J., and Thouveny, Y. (1999) Neurogenesis during spinal cord regeneration in newts. *Dev. Genes Evol.* 209:363–369.

Boilly, B., Cavanaugh, K. P., Thomas, D., Hondermarck, H., Bryant, S. V., and Bradshaw, R. A. (1991) Acidic fibroblast growth factor is present in regenerating limb blastemas of axolotls and binds specifically to blastema tissues. *Dev. Biol.*145:302–310.

Bosco, L., Valle, C., and Willems, D. (1993) In vivo and in vitro experimental analysis of lens regeneration in larval *Xenopus laevis*. *Dev. Growth Differ.* 35:257–270.

Bosco, L., Venturini, G., and Willems, D. (1997) In vitro lens transdifferentiation of *Xenopus laevis* outer cornea induced by fibroblast growth factor (FGF). *Development* 124:421–428.

Brockes, J. P. (1984) Mitogenic growth factors and nerve dependence of limb regeneration. *Science* 225:1280–1287.

Brockes, J. P. (1997) Amphibian limb regeneration—rebuilding a complex structure. *Science* 276:81–87.

Butler, E. G. and O'Brien, J. P. (1942) Effects of localized X-irradiation on regeneration of the urodele limb. *Anat. Rec.* 84:407–413.

Cameron, J. A., Hilgers, A. R., and Hinterberger, T. J. (1986) Evidence that reserve cells are a source of regenerated adult newt muscle *in vitro*. *Nature* 321:607–610.

Carinato, M. Walter, B. E., and Henry, J. J. (2000) *Xenopus laevis* gelatinase B (Xmmp-9): development, regeneration, and wound healing. *Dev. Dyn.* 217:377–387.

Carlson, M. R. J., Bryant, S. V., and Gardiner, D. M. (1998) Expression of *Msx-2* during development, regeneration, and wound healing in axolotl limbs. *J. Exp. Zool.* 282:715–723.

Casimir, C. M., Gates, P. B., Patient, R. K., and Brockes, J. P. (1988) Evidence for dedifferentiation and metaplasia in amphibian limb regeneration from inheritance of DA methylation. *Development* 104:657–668.

Chernoff, E. A. G. (1996) Spinal cord regeneration: a phenomenon unique to urodeles? *Int. J. Dev. Biol.* 40:823–832.

Chew, K. E. and Cameron, J. A. (1983) Increase in mitotic activity of regenerating axolotl limbs by growth factor-impregnated implants. *J. Exp. Zool.* 226:325–329.

Cohen, S. M. and Jurgens, G. (1989) Proximal-distal pattern formation in *Drosophila*: cell autonomous requirement for distal-less gene activity in limb development. *EMBO J.* 8:2045–2055.

Collins, J. M. (1972) Amplification of ribosomal ribonucleic acid cistrons in the regenerating lens of *Triturus*. *Biochemistry* 11:1259–1263.

Coulombre, J. L. and Coulumbre, A. J. (1965) Regeneration of neural retina from the pigmented epithelium in the chick embryo. *Dev. Biol.* 12:79–92.

Coulombre, J. L. and Coulombre, A. J. (1970) Influence of mouse neural retina on regeneration of chick neural retina from chick embryonic pigmented epithelium. *Nature* 228:559–560.

Crews, L., Gates, P. B., Brown, R., et al. (1995) Expression and activity of the newt Msx-1 gene in relation to limb regeneration. *Proc. R. Soc. Lond. (Biol.)* 259:161–171.

Cuny, R., Jeanny, J. C., and Courtois, Y. (1986) Lens regeneration from cultured newt irises stimulated by retina-derived growth factors (EDGFs). Differentiation 32:221–229.

Del-Rio Tsonis, K., Washabaugh, C. H., and Tsonis, P. A. (1995) Expression of *pax-6* during urodele eye development and lens regeneration. *Proc. Natl. Acad. Sci. USA* 92:5092–5096.

Del-Rio Tsonis, K., Jung, J. C., Chiu, I.-M., and Tsonis, P. A. (1997) Conservation of fibroblast growth factor function in lens regeneration. *Proc. Natl. Acad. Sci. USA* 94:13,701–13,706.

Del-Rio Tsonis, K., Trombley, M. T., McMahon G., and Tsonis, P. A. (1998) Regulation of lens regeneration by fibroblast growth factor receptor 1. *Dev. Dyn.* 213:140–146.

Desselle, J.-C. and Gontcharoff, M. (1978) Cytophotometric detection of the participation of cartilage grafts in regeneration of X-rayed urodele limbs. *Biol. Cell.* 33:45–54.

D'Jamoos, C. A., McMahon, G., and Tsonis, P. A. (1998) Fibroblast growth factor receptors regulate the ability for hindlimb regeneration in *Xenopus laevis*. *Wound Rep. Reg.* 6:388–397.

Dresden, M. H. and Gross, J. (1970) The collagenolytic enzyme of the regenerating limb of the newt *Triturus viridescens*. *Dev. Biol.* 22: 129–137.

Dunis, D. A. and Namenwirth, M. (1977) The role of grafted skin in the regeneration of X-irradiated axolotl limbs. *Dev. Biol.* 56:97–109.

Eguchi, G. (1963) Electron microscopic studies on lens regeneration. I. Mechanism of depigmentation of the iris. *Embryologia* 8:47-62.

Eguchi, G. 1967 In vitro analyses of Wolffian lens regeneration: differentiation of the regenerating lens rudiment of the newt, *Triturus pyrrhogaster*. *Embryologia* 9:246–266.

Eguchi, G. (1998) Transdifferentiation as the basis of eye lens regeneration. In: *Cellular and Molecular Basis of Regeneration* (Ferretti, P. and Geraudie, J., eds.), John Wiley & Sons, New York, pp. 207–229.

Enwright, J. F. and Grainger, R. M. (2000) Altered retinoid signaling in the heads of small eye mouse embryos. *Dev. Biol.* 221:10–22.

Fallon, J. F., Lopez, A., Ros, M. A., Savage, M. P., Olwin, B. B., and Simandl, B. K. (1994) Apical ectodermal ridge growth signal for chick limb development. *Science* 264:104–107.

Freeman, G. (1963) Lens regeneration from cornea in *Xenopus laevis*. *J. Exp. Zool.* 154:39–65.

Gaze, R. M. (1959) Regeneration of the optic nerve in *Xenopus laevis*. *Quart. J. Exp. Physiol.* 44:290–308.

Geraudie, J. and Ferretti, P. (1998) Gene expression during amphibian limb regeneration. *Int. J. Cytol.* 180:1–50.

Globus, M. (1988) A neuromitogenic role for substance P in urodele limb regeneration. In: *Regeneration and Development* (Inoue, S., Shirai, T., Egar, M., et al., eds.), Okada Printing and Publishing, pp. 675–685.

Globus, M., Vethamany-Globus, S., and Lee, Y. C. I. (1980) Effect of apical epidermal cap on mitotic cycle and cartilage differentiation in the newt, *Notophthalmus viridescens*. *Dev. Biol.* 75:358–372.

Goldhamer, D. J. and Tassava, R. A. (1987) An analysis of proliferative activity in innervated and denervated forelimb regenerates of the newt, *Notopthalmus viridescens*. *Development* 100:619–628.

Grillo, H., Lapiere, C. M., Dresden, M. H., and Gross, J. (1968) Collagenolytic activity in regenerating forelimbs of the adult newt (*Triturus viridescens*). *Dev. Biol.* 17:571–583.

Han, M.-J., An, J.-Y., and Kim, W.-S. (2001) Expression patterns of *Fgf-8* during development and limb regeneration of the axolotl. *Dev. Dyn.* 220:40–48.

Hay, E. D. (1959) Electron microscopic observations of muscle dedifferentiation in regenerating *Amblystoma* limbs. *Dev. Biol.* 3:26–59.

Henry, J. J. and Mittleman, J. M. (1995) The matured eye of *Xenopus laevis* tadpoles produces factors that elicit a lens-forming response in embryonic ectoderm. *Dev. Biol.* 171:39–50.

Hill, R. E., Jones, P. F., Rees, A. R., et al. (1989) A new family of mouse homeobox-containing genes: molecular structure, chromosomal location, and developmental expression of Hox-7.1. *Genes Dev.* 3:26–37.

Hinterberger, T. J. and Cameron, J. A. (1990) Myoblasts and connective-tissue cells in regenerating amphibian limbs. *Ontogenez* 21(4): 341–357.

Imokawa, Y. and Eguchi, G. (1992) Expression and distribution of regeneration-responsive molecule during normal development of the newt, *Cynops pyrrhogaster*. *Int. J. Dev. Biol.* 36:407–412.

Imokawa, Y., Ono, S.-I., Takeuchi, T., and Eguchi, G. (1992) Analysis of a unique molecule responsible for regeneration and stabilization of differentiated state of tissue cells. *Int. J. Dev. Biol.* 36:399–405.

Jensen, A. M. and Wallace, V. A. (1997) Expression of Sonic hedgehog and its putative role as a precursor cell mitogen in the developing mouse retina. *Development* 124:363–371.

Ju, B.-G. and Kim, W.-S. (1994) Pattern duplication by retinoic acid treatment in the regenerating limbs of Korean salamander larvae, *Hynobius leechi*, correlates well with the extent of dedifferentiation. *Dev. Dyn.* 199:253–267.

Ju, B.-G. and Kim, W.-S. (1998) Upregulation of cathepsin D expression in the dedifferentiating salamander limb regenerate and enhancement of its expression by retinoic acid. *Wound Rep. Reg.* 6:S349–S358.

Keefe, J. R. (1973a) An analysis of urodelan retinal regeneration: I. Studies of the cellular source of retinal regeneration in *Notopthalmus viridescens* utilizing ^{3}H-thymidine and colchicines. *J. Exp. Zool.* 184:185–206.

Keefe, J. R. (1973b) An analysis of urodelan retinal regeneration: II. Ultrastructural features of retinal regeneration in *Notophthalmus viridescens*. *J. Exp. Zool.* 184:207–232.

Keefe, J. R. (1973c) An analysis of urodelan retinal regeneration: III. Degradation of extruded melanin granules in *Notopthalmus viridescens*. *J. Exp. Zool.* 184:233–238.

Klein, L. R., MacLeish, P. R., and Wiesel, T. N. (1990) Immunolabeling by a newt retinal pigment epithelium antibody during retinal development and regeneration. *J. Comp. Neurol.* 293:331–339.

Kuhlman, J. and Niswander, L. (1997) Limb deformity proteins: role in mesoderm induction of the apical ectodermal ridge. *Development* 124: 133–139.

Kumar, A., Velloso, C. P., Imokawa, Y., and Brockes, J. P. (2000) Plasticity of retrovirus-labelled myotubes in the newt limb regeneration blastema. *Dev. Biol.* 218:125–136.

Levine, R. (1977) Regeneration of the retina in the adult newt, *Triturus cristatus*, following surgical division of the eye by a post-limbal incision. *J. Exp. Zool.* 200:41–54.

Lo, D. C., Allen, F., and Brockes, J. P. (1993) Reversal of muscle differentiation during urodele limb regeneration. *Proc. Natl. Acad. Sci. USA* 90:7230–7234.

Maden, M. and Wallace, H. (1975) The origin of limb regenerates from cartilage grafts. *Acta Embryol. Exp. (Palermo)* 2:77–86.

McDevitt, D. S., Brahma, S. K., Courtois, Y., and Jeanny, J.-C. (1997) Fibroblast growth factor receptors and regeneration of the eye lens. *Dev. Dyn.* 208:220–226.

McGann, C. J., Odelberg, S. J., and Keating, M. T. (2001) Mammalian myotube dedifferentiation induced by newt regeneration extract. *Proc. Natl. Acad. Sci. USA* 98:13,699–13,703.

Mescher, A. L. (1996) The cellular basis of limb regeneration in urodeles. *Int. J. Dev. Biol.* 40:785–796.

Mescher, A. L., Connell, E., Hsu, C., Patel, C., and Overton, B. (1997) Transferrin is necessary and sufficient for the neural effect on growth in amphibian limb regeneration blastemas. *Dev. Growth Differ.* 39:677–684.

Meyers, E. N., Lewendoski, M., and Martin, G. R. (1998) Generation of an Fgf8 mutant allelic series using a single targeted mouse line carrying Cre and Fle recombinate recognition sites. *Nat. Genet.* 18:136–141.

Mitashov, V. I. (1996) Mechanisms of retina regeneration in urodeles. *Int. J. Dev. Biol.* 40:833–844.

Miyazaki, K., Uchiyawa, K., Imokawa, Y., and Yoshizato, K. (1996) Cloning and characterization of cDNAs for matrix metaloproteinases of regenerating newt limbs. *Proc. Natl. Acad. Sci. USA* 93:6819–6824.

Mochii, M., Mazaki, Y., Mizuno, N., Hayashi, H., and Eguchi, G. (1998) Role of Mitf in differentiation and transdifferentiation of chicken pigmented epithelial cell. *Dev. Biol.* 193:47–62.

Mullen, L. M., Bryant, S. V., Torok, M. A., Blumberg, B., and Gardiner, D. M. (1996) Nerve dependency of regeneration: the role of *Distal-less* and FGF signaling in amphibian limb regeneration. *Development* 122:3487–3479.

Muneoka, K., Fox, W. F., and Bryant, S. V. (1986) Cellular contribution from dermis and cartilage to the regenerating limb blastema in axolotls. *Dev. Biol.* 116:256–260.

Namenwirth, M. (1974) The inheritance of cell differentiation during limb regeneration in the axolotl. *Dev. Biol.* 41:42–56.

Nathanson, M. (1979) Skeletal muscle metaplasia: formation of cartilage by differentiated skeletal muscle. In: *Muscle Regeneration* (Mauro, A., ed.), Raven, New York, pp. 83–90.

Negeshi, K., Shinagawa, S., Ushijima, M., Kaneko, Y., and Saito, T. (1992) An immunohistochemical study of regenerating newt retinas. *Dev. Brain Res.* 68:255–264.

Niswander, L., Tickle C., Vogel, A., Booth, I., and Martin, G. R. (1993) FGF-4 replaces the apical ectodermal ridge and directes outgrowth and patterning of the limb. *Cell* 75:579–587.

Noji, S., Matsuo, T., Koyama, E., et al. (1990) Expression pattern of acidic and basic fibroblast growth factor genes in adult rat eyes. *Biochem. Biophys. Res. Commun.* 168:343–349.

Odelberg, S. J., Kollhof A., and Keating, M. T. (2001) Dedifferentiation of mammalian myotubes induced by msx-1. *Cell* 103:1099–1109.

Onda, H. and Tassava, R. A. (1991) Expression of the 9G1 antigen in the apical cap of axolotl regenerates requires nerves and mesenchyme. *J. Exp. Zool.* 257:336–349.

Onda, H., Poulin, M. L., Tassava, R. A.,and Chiu, I.-M. (1991) Characterization of a newt tenascin cDNA and localization of tenascin during newt limb regeneration by in situ hybridization. *Dev. Biol.* 148:219–232.

Orts-Lorca, F. and Genis-Galvez, J. M. (1960) Experimental production of retinal septa in the chick embryo: differentiation of pigment epithelium into neural retina. *Acta Anat.* 42:31–70.

Park, C. and Hollenberg, M. J. (1991) Induction of retinal regeneration *in vivo* by growth factors. *Dev. Biol.* 148:322–333.

Park, C. and Hollenberg, M. J. (1993) Growth factor–induced retinal regeneration *in vivo*. *Int. Rev. Cytol.* 146:49–71.

Park, I.-S. and Kim, W.-S. (1999) Modification of gelatinase activity correlates with the dedifferentiation profile of regenerating axolotl limbs. *Mol. Cells* 9:119–126.

Pecorino, L. T., Entwistle, A., and Brockes, J. P. (1996) Activation of a single retinoic acid receptor isoform mediates proximodistal respecification. *Curr. Biol.* 6:563–569.

Pittack, C., Jones, M., and Reh, T. A. (1991) Basic fibroblast growth factor induces retinal pigment epithelium to generate neural retina *in vitro*. *Development* 113:577–588.

Pittack, C., Grunwald, G. B., and Reh, T. A. (1997) Fibroblast growth factors are necessary for neural retina but not pigmented epithelium differentiation in chick embryos. *Development* 124:805–816.

Pittinger, M. F., Mackay, A.M., Beck, S., et al. (1999) Multilineage potential of adult human mesenchymal stem cells. *Science* 284:143–147.

Poulin, M. L. and Chiu, I.-M. (1995) Re-programming of expression of the KGFR and *bek* variants of fibroblast growth factor receptor 2 during limb regeneration in newts (*Notopthalmus viridescens*). *Dev. Dyn.* 202:378–387.

Poulin, M. L., Patrie, K. M., Botelho, M. J., Tassava, R. A., and Chiu, I.-M. (1993) Heterogeneity in expression of fibroblast growth factor receptors during limb regeneration in newts (*Notopthalmus viridescens*). *Dev. Dyn.* 202:378–387.

Raymond, P. A. and Hitchcock, P. F. (1997) Retinal regeneration: common principles but a diversity of mechanisms. *Adv. Neurol.* 72:171–184.

Reese, D. H., Puccia, E., and Yamada, T. (1969) Activation of ribosomal RNA synthesis in initiation of Wolffian lens regeneration. *J. Exp. Zool.* 170:259– 268.

Reh, T. A. and Nagy, T. (1987) A possible role for the vascular membrane in retinal regeneration in Rana catesviena tadpoles. *Dev. Biol.* 122: 460–482.

Reh, T. A., Nagy, T., and Gretton, H. (1987) Retinal pigment eoithelial cells induced to transdifferentiate to neurons by laminin. *Nature* 330: 68–71.

Reyer, R. W. (1971) DNA synthesis and the incorporation of labeled iris cells into the lens during lens regeneration in adult newts. *Dev. Biol.* 24:533–558.

Reyer, R. W. (1977) The amphibian eye: development and regeneration. In: *Handbook of Sensory Physiology* vol. VII/5 (Crescitelli, F., ed.), Springer-Verlag, Berlin, pp. 309–390.

Reyer, R. W. (1982) Dedifferentiation of iris epithelium during lens regeneration in newt larvae. *Am. J. Anat.* 163:1–23.

Reyer, R. W. (1990a) Macrophage invasion and phagocytic activity during lens regeneration from the iris epithelium in newts. *Am. J. Anat.* 188:329–344.

Reyer, R. W. (1990b) Macrophage mobilization and morphology during lens regeneration from the iris epithelium in newts: studies with correlated scanning and transmission electron microscopy. *Am. J. Anat.* 188:345–365.

Riddiford, L. M. (1960) Autoradiographic studies of tritiated thymidine infused into the blastema of the early regenerate in the adult newt, Triturus. *J. Exp. Zool.* 144:25–32.

Robert, B., Sassoon, D., Jacq, B., Gehring, W., and Buckingham, M. (1989) Hox-7, a mouse homeobox gene with a novel pattern of expression during embryogenesis. *EMBO J.* 8:91–100.

Sage, H. and Bornstein, P. (1991) Extracellular matrix proteins that modulate cell-matrix interactions. *J. Biol. Chem.* 266:831–834.

Schmid, V. and Alder, H. (1984) Isolated mononucleated, striated muscle can undergo pluripotent transdifferentiation and form a complex regenerate. *Cell* 38:801–809.

Schmidt, A. J. (1966) *The Molecular Basis of Regeneration: Enzymes*, Illinois Monographs Med Sci, vol. 6, no. 4, University of Illinois Press, Urbana.

Schotte, O. E. and Butler, E. G. (1941) Morphological effects of denervation and amputation of limbs in urodele larvae. *J. Exp. Zool.* 87:279–321

Shimizu-Nishikawa, K., Tazawa, I., Uchiyama, K., and Yoshizato, K. (1999) Expression of helix-loop-helix type negative regulators of differentiation during limb regeneration in urodeles and anurans. *Dev. Growth Differ.* 41:731–743.

Simon, H.-G., Nelson, C., Goff, D., Laufer, E., Morgan, B. A., and Tabin, C. (1995) Differential expression of myogenic regulatory genes and Msx-1 during dedifferentiation and redifferentiation of regenerating limbs. *Dev. Dyn.* 202:1–12.

Singer, M. and Craven, L. (1948) The growth and morphogenesis of the regenerating forelimb of adult Triturus following denervation at various stages of development. *J. Exp. Zool.* 108:279–308.

Song, K., Wang, Y., and Sassoon, D. (1992) Expression of *Hox-7.1* in myoblasts inhibits terminal differentiation and induces cell transformation. *Nature* 360:477–481.

Sperry, R. W. (1944) Optic nerve regeneration with return of vision in amphibians. *J. Neurophysiol.* 7:57–69.

Steen, T. P. (1968) Stability of chondrocyte differentiation and contribution of muscle to cartilage during limb regeneration in the axolotl (*Siredon mexicanum*). *J. Exp. Zool.* 167:49–78.

Steen, T. P. (1973) The role of muscle cells in Xenopus limb regeneration. *Am. Zool.* 13:1349–1350.

Stocum, D. L. (1979) Stages of forelimb regeneration in *Ambystoma maculatum*. *J. Exp. Zool.* 209:395–416.

Stocum, D. L. (1995) *Wound Repair, Regeneration, and Artificial Tissues*, R. G. Landes, Austin, TX.

Stocum, D. L. (2001) Rx for tissue restoration: regenerative biology and medicine. *Korean J. Biol. Sci.* 5:91–99.

Stocum, D. L. and Dearlove, G. E. (1972) Epidermal-mesodermal interaction during morphogenesis of the limb regeneration blastema in larval salamanders. *J. Exp. Zool.* 181:49–62.

Stone, L. S. (1950) The role of retina pigment cells in regenerating neural retina of adult salamander eyes. *J. Exp. Zool.* 113:9–31.

Stroeva, O. G. and Mitashov, V. I. (1983) Retinal pigment epithelium: proliferation and differentiation during development and regeneration. *Int. Rev. Cytol.* 83:221–293.

Takata, C., Albright, J. F., and Yamada, T. (1966) Lens antigens in a lens-regenerating system studied by the immunofluorescent technique. *Dev. Biol.* 9:385–397.

Tanaka, E. M., Gann, A. F., Gates, P. B., and Brockes, J. P. (1997) Newt myotubes re-enter the cell cycle by phosphorylation of the retinoblastoma protein. *J. Cell Biol.* 136:155–165.

Tanaka, E. M., Dreschel, D. N., and Brockes, J. P. (1999) Thrombin regulates S phase re-entry by cultured newt myoblasts. *Curr. Biol.* 9: 792–799.

Tassava, R. A., Castilla, M., Arsanto, J.-P., and Thouveny, Y. (1993) The wound epithelium of regenerating limbs of *Pleurodeles waltl* and *Notophthalmus viridescens*: studies with mAbs WE3 and WE4, phalloidin, and Dnase I. *J. Exp. Zool.* 267:180–187.

Tomasek, J. J., Halliday, N. L., Updike, D. L., et al. (1997) Gelatinase A activation is regulated by the organization of the polymerized actin cytoskeleton. *J. Biol. Chem.* 272:7482–7487.

Tomlinson, B. and Barger, P. M. (1987) A test of the punctuated-cycling hypothesis in Ambystoma forelimb regenerates: the roles of animal size, limb regeneration, and the aneurogenic condition. *Differentiation* 35:6–15.

Tsonis, P. A., Trombley, M. T., Rowland, T., Chandraratna, R. A. S., and Del-Rio Tsonis, K. (2000) Role of retinoic acid in lens regeneration. *Dev. Dyn.* 219:588–593.

Velloso, C. P., Kumar, A., Tanaka, E. M., and Brockes, P. (2000) Generation of mononucleate cells from post-mitotic myotubes proceeds in the absence of cell cycle progression. *Differentiation* 66: 239–246.

Velloso, C. P., Simon, A., and Brockes, J. P. (2001) Mammalian postmitotic nuclei reenter the cell cycle after serum stimulation in newt/mouse hybrid myotubes. *Curr. Biol.* 11:855–858.

Woloshin, P., Song, K., Degnin, A., Killary, D. J., Goldhamer, D.J., Sassonn, D., and Thayer, M. J. (1995) MSX1 inhibits Myo expression in fibroblast X10T1/2 cell hybrids. *Cell* 82:611–620.

Yamada, T. and Dumont, J. N. (1972) Macrophage activity in Wolffian lens regeneration. *J. Morphol.* 136:367–384.

Yamada, T. and McDevitt, D. S. (1974) Direct evidence for transformation of differentiated iris epithelial cels into lens cells. *Dev. Biol.* 38: 104–118.

Yamada, T. and Roesel, M. (1969) Activation of DNA replication in the iris epithelium by lens removal. *J. Exp. Zool.* 171:425–432.

Yamada, Y. and Roesel, M. (1971) Control of mitotic activity in Wolffian lens regeneration. *J. Exp. Zool.* 177:119–128.

Yamada, T., Reese, D. H., McDevitt, D. S. (1973) Transformation of iris into lens in vitro and its dependency on neural retina. *Differentiation* 1:65–82.

Yang, E. V., Gardiner, D. M., and Bryant, S. V. (1999) Expression of *Mmp-9* and related matrix metalloproteinase genes during axolotl limb regeneration. *Dev. Dyn.* 216:2–9.

Yokoyama, H., Ide, H., and Tamura, K. (2001) FGF-10 stimulates limb regeneration ability in *Xenopus laevis*. *Dev. Biol.* 233:72–79.

Zalik, S. E. and Scott, V. (1972) Cell surface changes during dedifferentiation in the metaplastic transformation or iris into lens. *J. Cell Biol.* 55:134–146.

Zalik, S. E. and Scott, V. (1973) Sequential disappearance of cell surface components during dedifferentiation in lens regeneration. *Nat. (N. Biol.)* 244:212–214.

9 Stem Cells in Dermal Wound Healing

WILLIAM J. LINDBLAD, PhD

Repair of damaged dermal tissue is accomplished by dynamic cell-cell and cell-matrix interactions involving activation of resident cells as well as cells that migrate into the lesion. Immediately after injury there is a hemostatic response that activates platelets. Platelets release transforming growth factor-β (TGF-β) and platelet-derived growth factor (PDGF), which begin the recruitment of inflammatory cells and activate several resident dermal cell types. The most notable early inflammatory response is the infiltration of neutrophils. These are necessary to control infection and begin clearing the damaged tissue but are not critical to the healing process. Wound healing begins a few days later with the infiltration of monocyte/macrophages at the site. These cells release a number of cyotokines, synthesize extracellular matrix (ECM), and serve as a source of stem cells. The ECM becomes vascularized by angiogenesis and forms granulation tissue. Neovascularization is accomplished using both resident endothelial cells and circulating progenitor cells from the blood. Scar formation proceeds with the activation, migration, and proliferation of fibroblasts and the formation of a three-dimensional scaffold of ECM. A hematologically derived monocyte/stem cell may be the major contributor to early matrix synthesis. These blood-derived stem cells are able to produce a wide variety of proinflammatory cytokines (IL-1β, TNF-α, MIP-1α, MIP-1β, PDGF-A and TGF-β), which make major contributions to the healing process. Contraction of the wound is driven by activation and differentiation of mesenchymal cells into myofibroblasts that are able to draw the edges of the damaged tissue over the deficit. Migration of viable keratinocytes within the wound and from the edges of it stimulated by types I and IV collagen, fibronectin and vitronectin, as well as serum factors is responsible for reepithelialization of the surface. The cells include both the basal stem cells of the epidermis and the transit-amplifying cells. If epidermal stem cells are lost, migrating stem cells from surviving hair follicles are able to supply epidermal cells.

9.1. INTRODUCTION

The processes involved with repair of damaged dermal tissue require the activation of numerous cell types with subsequent migration, proliferation, and reconstitution of biological structures (Cohen et al., 1992). These dynamic cell–cell and cell–matrix interactions have largely been viewed in terms of activation of terminally differentiated cells. In this paradigm, the resident cells, largely quiescent, retain the capacity to enter the cell cycle given the appropriate environmental cues and repopulate the damaged area. However, recent studies show that for some of these cell types, activation of a stem cell compartment clearly occurs. Although these studies have largely focused on the epidermis, it is not unreasonable to suggest that other wound-healing processes require the contribution of a stem cell component. This chapter highlights discussion of a putative stem cell population that may be recruited to the site of injury where it contributes to the reconstitution of the extracellular matrix (ECM) (Table 1).

From: *Stem Cells Handbook*
Edited by: S. Sell © Humana Press Inc., Totowa, NJ

9.2. WOUND HEALING EVENTS

Tissue repair is initiated with injury to the tissue and consists of a series of highly coordinated, overlapping processes. An initial hemostatic response that staunches blood loss also activates platelets to release growth factor–filled α-granules and synthesize a series of bioactive lipid mediators (Clark, 1996). Of the growth factors released, transforming growth factor-β (TGF-β) and platelet-derived growth factor (PDGF) are central for the prompt recruitment of inflammatory cells and activation of several resident dermal cell types. Within 24 h the wounds become richly populated with neutrophils recruited to the wound site by these soluble mediators (Clark, 1996). Although it does not appear that this initial influx of activated cells is crucial to the overall healing process, it is clear that these cells are required to control infection and clear damaged tissue (Simpson and Ross, 1972). Recent studies suggest that an exaggerated or prolonged influx of the neutrophil can alter the proteolytic environment sufficiently to convert an acute healing response into a chronic site of tissue degradation (Mast and Schultz, 1996; Yager and Nwomeh, 1999).

Table 1
Confirmed and Putative Stem Cell Populations Contributing to Dermal Wound Healing

Process	*Cell type*	*Source*	*Function*
Reepithelialization	Epidermal stem cell and transient-amplifying cells	Resident	Cells provide source of cells for resurfacing skin when significant epidermal cell loss occurs
Angiogenesis	Bone marrow–derived multipotent adult progenitor cell	Blood-borne	May provide a source of angioblasts and mature endothelial cells during wound angiogenesis
ECM synthesis	Bone marrow derived– multipotent adult progenitor cell	Blood-borne	May provide cells to deposit collagenous matrix in provisional ECM

Subsequent to the initial influx of neutrophils, but with significant overlap, there is a directed accumulation of monocyte/macrophages at the wound site. These cells have been repeatedly shown to be central to the development of a complete wound-healing response (Leibovich and Ross, 1975). However, the function of these cells has been largely explained by their ability to synthesize a number of important soluble mediators including TGF-β and thereby these cells serve as regulatory cells (Rappolee et al., 1988). As detailed below, there is evidence that these cells also possess the ability to synthesize numerous ECM proteins and may therefore be intimately involved in assembling the nascent or provisional ECM of the wound bed. In addition, a subpopulation of the monocyte/macrophage lineage may represent a putative stem cell population.

The provisional ECM represents a complex assemblage of such molecules as fibrin, fibronectin, vitronectin, thrombospondin, and tenascin onto which cells can migrate and receive appropriate cues for selected gene expression (Yamada and Clark, 1996). The formation of the provisional ECM is viewed as a spontaneous event, not requiring the active involvement of cells for its initial structure. This matrix then undergoes a series of changes that is a function of cellular activity, to become granulation tissue and, ultimately, the final collagenous structure of the scar. Alteration in the formation of the matrix or the abnormal removal of the matrix may lead to termination of the healing response (Rogers et al., 1995).

With the formation of a pseudo-stable provisional matrix, revascularization of the damaged tissue will occur through an angiogenic response (Madri and Bell, 1992). This response encompasses several stages of new vessel formation, not only the activation, migration, and proliferation of endothelial cells, but also the formation of a three-dimensional (3D) tubular structure that requires stabilization. Just as important as this formation of neovessels is the subsequent regression of the excess vessels that are formed in the initial angiogenic response. This process has been viewed primarily from the standpoint of resident endothelial cell dynamics, but recent studies argue that endothelial cell progenitors are present in adult bone marrow and may be able to contribute to the angiogenic response in wound healing (Carmeliet et al., 2001; Reyes et al., 2002). Thus, the angiogenic response following injury may represent a complex interaction between the activation of resident endothelial cells and endothelial precursor cells that arrive to complement this local activity. Note that the extent to which tissue repair is dependent on the angiogenic response that is normally elicited in adults is not clear. Recent work suggests that in the mouse, the angiogenic response from an excisional wound may be significantly inhibited using α_v-integrin blocking antibodies, without significant inhibition of wound healing (Jang et al., 1999).

With the formation of granulation tissue from the initial provisional ECM plus the angiogenic response, the formation of scar proceeds with the activation, migration, and proliferation of fibroblasts throughout the provisional ECM/granulation tissue. Driving the cells through these cellular activities are a number of soluble mediators, although TGF-β appears to represent the predominant regulator of matrix deposition (Roberts, 1995). The formation of a permanent ECM progresses through a series of stages, starting with a hyaluronic acid–rich phase through to a final stage in which the ECM is largely collagenous with a concomitant abrupt reduction in hyaluronic acid (Dunphy and Upuda, 1955).

In addition to providing for the structural basis of a final collagenous replacement of the damaged tissue, the cells within the ECM participate in a contraction mechanism that reduces the size of the dermal deficit (Gross, 1996). This wound contraction is driven by the activation and differentiation of mesenchymal cells into myofibroblasts, α-actin-rich cells that are transiently present in the healing dermis (Gabbiani et al., 1971). These cells produce the force required to draw the edges of the damaged tissue over the defect, effectively reducing the volume of injured dermis that requires resurfacing and replacement with ECM.

All of these processes advance through an ordered set of activation steps with increasing cell numbers followed by an equally regulated decrement of the cellular and matrix compartments. Mechanisms that control the elimination of cell populations and downregulation of cell activity are not as clearly defined but appear to be the result of apoptotic events (Squier et al., 1995; Savill, 1997). Clearly, the reduction of neutrophils from the damaged tissue encompasses an apoptotic component, as does the clearance of fibroblasts and myofibroblasts (Desmoulière et al., 1995). However, what cues activate the apoptotic pathways are still unclear.

The above outline of dermal tissue repair does not include reepithelialization, the process that replaces an epidermal covering for the underlying dermis. The magnitude of this response is clearly dependent on the extent of epidermal loss, such that with incisional wounds, the response of the epidermis is minimal, whereas with superficial burn injury, this may represent virtually the entire injury response. However, it is with this tissue that stem cell response to injury has been most widely studied.

9.3. REEPITHELIALIZATION AND STEM CELLS

Closure of the epidermis following injury is essential to reestablishing a functional homeostatic barrier over the entire body surface. The injury response of the epidermis initially represents a migratory response by keratinocytes from islands of viable cells within the wound and from around the wound's circumference (Odland and Ross, 1968). This migration represents an intricate interplay between ECM molecules present in the provisional ECM and the keratinocytes with little contribution from cell proliferation at this early time (Clark et al., 1982). The provisional ECM contains several matrix molecules that have been shown to stimulate keratinocytes' motility, including types I and IV collagen, fibronectin and vitronectin, as well as other serum factors (Stenn and Dvoretzky, 1979; Woodley et al., 1990).

Cell proliferation to replace lost epidermal cells occurs within two compartments: a stem cell compartment and a transient-amplifying cell compartment. The stem cell compartment consists of islands of cells located in the basal cell layer adjacent to the epidermal–dermal basement membrane (Jones and Watt, 1993; Jones et al., 1995). These cells stain brightly for $\alpha_2\beta_1$ and $\alpha_3\beta_1$ integrins, through which they would possess high affinity for the underlying basement membrane matrix molecules. The transient-amplifying cells derived from these basal stem cells show two to three times less integrin staining and presumably are able to migrate away from the basal layer and contribute to the expansion of keratinocytes required to repopulate a denuded section of skin. Although the staining for integrins has identified a stem cell population in the epidermal basal layer, there is a discrepancy between the proportions of these cells that are stem cells based on the kinetics of cell division and this integrin staining. Whereas by staining approximately 40% of the basal cells are stem cells, by cell kinetic analysis only 10% are stem cells (Jones et al., 1995).

In thick skin, the stem cells of the basal cell layer are located at the tips of the rete ridges, and from this region transit-amplifying cells differentiate and migrate to fill the rete ridge space (Jensen et al., 1999). These cells then undergo differentiation as they move toward the surface of the epidermis, including the further loss of integrin staining. Consequently, in tangential excisional wounds to the skin or in deep partial-thickness or full-thickness burns, the loss of these basal cells results in an inability of that section of epidermis to be resurfaced, except through the migration of keratinocytes from the wound edges. However, there is another population of epidermal stem cells from which keratinocytes can be obtained for repopulating the epidermis—cells from the hair follicle shaft.

Epidermal stem cells have been identified in human hair follicles in the midregion of the follicle (Rochat et al., 1994). These cells respond to morphogenetic signals to produce hair follicles, sebaceous glands, and epidermis (Oshima et al., 2001). Therefore, residual hair follicles within a damaged region of the skin would be able to supply the cells required to reconstitute a functional epidermis in the adjacent region.

9.4. REPLACEMENT OF COLLAGENOUS STRUCTURES

All of the events of tissue repair occur on or within an appropriate 3D ECM that imparts crucial spatial and regulatory information to the cells. Beginning with the deposition of a provisional ECM via the conversion and self-assembly of fibrinogen to a fibrin network and binding of various soluble attachment proteins, each cell type provides for the deposition of matrix proteins and proteoglycans that are required for the proper migration and/or cellular phenotypic behavior. Ultimately, a greatly expanded mesenchymal cell population synthesizes the massive deposits of collagenous ECM termed a *scar*.

The first 3D scaffold that must be formed in a wound is the provisional ECM. As noted previously, this structure allows for keratinocytes and endothelial cells to migrate into the damaged area and based on its matrix components, induces the appropriate phenotypic behavior of these cells. In addition to the passively accumulated ECM proteins, there is a component of the matrix that results from the active synthesis of collagen molecules that are added to the provisional structure. Current research suggests that a hematologically derived monocyte that may express stem cell surface epitopes contributes partially, if not completely, to this early matrix synthesis.

Speculation about the origin of cells responsible for the deposition of collagen in tissue repair has spanned the past 140 yr (Cohnheim, 1867; Metchnikoff, 1893). At that time it was proposed that blood-borne cells populated the injury site and transformed into connective-tissue fibrocytes. Specifically the monocyte/macrophage was identified as the putative progenitor cell. This concept was supported by Shelton and Rice (1959), who presented evidence that mouse mononuclear cells within a diffusion chamber could transform into fibroblast-like cells within 3 d. Additional studies using autoradiographic techniques also implicated transformation of blood-borne mononuclear cells into collagen biosynthetic–capable cells (Gillman and Wright, 1966). Finally, several investigators examined cells of hematological origin and showed the presence of collagen biosynthetic enzymes within the cells (Goldberg and Green, 1968; Chen-Kiang et al., 1977; Myllyla and Seppa, 1979). Therefore, there are considerable data suggesting that some component of the ECM deposited within a wound area arises from hematological cells migrating to the site of injury and then phenotypically becoming connective-tissue-synthesizing cells.

However, this line of evidence is not without contradictory studies, and one could cite an equal number of studies providing data supporting the concept that the collagen-producing cells of the wound were purely cells of endogenous tissue origin. One of the more influential studies in this latter group was by Ross et al. (1970). These investigators used a parabiotic animal preparation to show that cells from the parabiotic animal that had been radiolabeled with ^{3}H-thymidine did infiltrate a wound; however, all the labeled cells were inflammatory cells (neutrophils, monocytes, and lymphocytes), as shown by light and electron microscope autoradiography. This influential study inhibited further study in this area for a number of years.

With the advent of immunological techniques to identify cell populations by cell-surface antigens and their biosynthetic products, examination of the origins of ECM-producing cells in tissue repair could be reexamined. Several years ago, we identified the deposition of types I, III, and IV collagen within the provisional ECM of rat dermal excisional wounds during the first 24–36 h postinjury (Graham et al., 1984). This finding was surprising because it was doubtful that resident mesenchymal cells would have been able to migrate into the provisional ECM within this time frame. In addition, pan-leukocyte immunohistochemical staining indicated that the only cell type present in this region was hematological cells. Dual immunohistochemical staining for type IV collagen and Mac-1, a macrophage-selective cell-surface antigen, confirmed that the area of collagen deposition corresponded to areas with a large number of Mac-1-positive cells. These results supported the concept that the initial synthesis and deposition of collagenous matrix in wound healing occurs in the provisional matrix and is produced by cells of hematogenous origin.

Subsequent studies using human blood cells confirmed that a circulating population of cells contained the obligate collagen biosynthetic enzyme prolyl hydroxylase and was able to induce the expression of this enzyme within 18–24 h postisolation (Lindblad et al., 1987). Of note, this population of cells was not an adherent cell population, but the cells were sensitive to exposure to adherent cell products and platelet-derived products. Further studies have shown that induction of mRNA for types I and III procollagen by the cells occurs within the first 24–36 h of culture. Therefore, the cells possess all of the characteristics that would be consistent with the in vivo immunohistochemical analysis in rat dermal wounds.

These results were extended to a mouse model in which macrophages were isolated by implanting Nucleopore® membranes in the peritoneum, the membranes removed, and the cells cultured under standard conditions (Vaage and Lindblad, 1990). Using a variety of techniques including phagocytosis of yeast and positivity for anti-Mac-1 immunofluorescence, the cells were identified as 95+% macrophages. These cultures of cells expressed prolyl hydroxylase, type I procollagen mRNA, and using metabolic labeling were able to express complete collagenous proteins. By dual immunnofluorescent staining it was shown that the expression of type I collagen did in fact occur in the Mac-1-positive cells. More recently, *in situ* hybridization and immunohistochemistry extended these results to show that the macrophages contain type I procollagen mRNA (Zhai et al., 1996; Lindblad, 1998).

Another research group has identified a circulating cell type, termed a *fibrocyte*, that is localized to sites of injury (Bucala et al., 1994). These cells have been cell-surface phenotyped as collagen$^+$/CD13$^+$/CD34$^+$/CD45$^+$ and have also been isolated from the wound chamber model of wound healing in mice (Chesney et al., 1998). While these cells are able to express type I collagen, this expression is not observed for several days in vivo, or in vitro it may take up to 2 wk to observe significant collagen expression. In addition, these cells are able to produce an array of proinflammatory cytokines including interleukin-1β, tumor necrosis factor-α, macrophage inflammatory protein-1α (MIP-1α), MIP-1β, PDGF-A, and TGF-β (Chesney et al., 1998). Of note, TGF-β may function in an autocrine fashion within these cells, because these cells are responsive to this growth factor with enhanced cellular activity (Abe et al., 2001).

Currently, the monocyte/macrophage cell population identified in our laboratory with collagen biosynthetic capabilities has not been thoroughly phenotyped for cell-surface markers. However, preliminary results are consistent for a putative stem cell subpopulation. Although it is conceivable that two circulating blood cell subpopulations are involved in the initial deposition of collagen in the provisional ECM, this seems unlikely. Rather, it may be that some of the differences in biochemical activity, e.g., expression of type I and IV collagen within 24–36 h vs 1 to 2 wk, reported are results of experimental conditions. *In toto*, the data are consistent with the initial deposition of collagen in tissue repair resulting from the action of a blood-borne monocyte subpopulation with stem cell markers infiltrating the provisional ECM and depositing a collagenous matrix to support the subsequent processes of wound healing.

9.5. CONCLUSION

Understanding of the cellular and biochemical processes of dermal wound healing has rapidly advanced over the past decade, with many of the key mediators and events now characterized. With these advances, the concept that stem cells may play an important role in contributing pools of cells to repopulate damaged tissue or to play transient roles has emerged. Although the stem cells of the epidermis have received the most attention, studies in angiogenesis and matrix deposition are beginning to add the stem cell to the constellation of cells involved in these processes as well. Along with these putative roles, stem cells may well be involved in expansion of the mesenchymal pool of cells for the reconstitution of the collagenous dermis and scar.

REFERENCES

Abe, R., Donnelly, S. C., Peng, T., Bucala, R., and Metz, C. N. (2001) Peripheral blood fibrocytes: differentiation pathway and migration to wound sites. *J. Immunol.* 166:7556–7562.

Bucala, R., Spiegel, L. A., Chesney, J., Hogan, M., and Cerami, A. (1994) Circulating fibrocytes define a new leukocyte subpopulation that mediates tissue repair. *Mol. Med.* 1:71–81.

Carmeliet, P. and Luttun, A. (2001) The emerging role of the bone marrow-derived stem cells in (therapeutic) angiogenesis. *Thromb. Haemost.* 86:289–297.

Chen-Kiang, S., Cardinale, G. J., and Udenfriend, S. (1977) Expression of collagen biosynthetic activities in lymphocytic cells. *Proc. Natl. Acad. Sci. USA* 75:1379–1383.

Chesney, J., Metz C., Stavitsky, A. B., Bacher, M., and Bucala, R. (1998) Regulated production of type I collagen and inflammatory cytokines by peripheral blood fibrocytes. *J. Immunol.* 160:419–425.

Clark, R. A. F. (1996) Overview of wound repair. In: *The Molecular and Cellular Biology of Wound Repair* (Clark, R. A. F., ed.), Plenum, New York, pp. 3–50.

Clark, R. A. F., Lanigan, J. M., DellaPelle, P., Manseau, E., Dvorak, H. F. and Colvin, R. B. (1982) Fibronecin and fibrin provide a provisional matrix for epidermal cell migration during wound reepithelialization. *J. Invest. Dermatol.* 79:264–269.

Cohen, I. K., Diegelmann, R. F., and Lindblad, W. J., eds. (1992) *Wound Healing: Biochemical and Clinical Aspects*, Saunders, Philadelphia, PA.

Cohnheim, J. (1867) Ueber entzundung und eiterung. *Virchows Arch.* 40:1.

Desmoulière, A., Redard, M., Darby, I., and Gabbiani, G. (1995) Apoptosis mediates the decrease in cellularity during the transition between granulation tissue and scar. *Am. J. Pathol.* 146:56–66.

Dunphy, J. E. and Upuda, K. N. (1955) Chemical and histochemical sequences in the normal healing of wounds. *N. Engl. J. Med.* 253: 847–851.

Gabbiani, G., Ryan, G. B., and Majno, G. (1971) Presence of modified fibroblasts in granulation tissue and their possible role in wound contraction. *Experientia* 27:549–550.

Gillman, T. and Wright, L. J. (1966) Autoradiographic evidence suggesting in vivo transformation of some blood mononuclears in repair and fibrosis. *Nature* 209:1086–1090.

Goldberg, B. and Green, H. (1968) The synthesis of collagen and protocollagen hydroxylase by fibroblastic and nonfibroblastic cell lines. *Pathology* 59:1110–1116.

Graham, M. F., Diegelmann, R. F., Lindblad, W. J., Gay, S., Gay, R., and Cohen, I. K. (1984) Effects of inflammation on wound healing: in vitro studies and in vivo studies. In: *Soft and Hard Tissue Repair: Biological and Clinical Aspects* (Hunt, T. K., Heppenstall, R. B., Pines, E., Rovee, D., eds.), Praeger Scientific, New York, pp. 361–379.

Gross, J. (1996) Getting to mammalian wound repair and amphibian limb regeneration: A mechanistic link in the early events. *Wound Rep. Reg.* 4:190–202.

Jang, Y.-C., Arumugam S., Gibran, N. S., and Isik, F. F. (1999) Role of α_v integrins and angiogenesis during wound repair. *Wound Rep. Reg.* 7:375–380.

Jensen, U., Lowell, S., and Watt, F. M. (1999) The spatial relationship between stem cells and their progeny in the basal layer of human epidermis: a new view based on whole-mount labeling and lineage analysis. *Development* 126:2409–2418.

Jones, P. H. and Watt, F. M. (1993) Separation of human epidermal stem cells from transit amplifying cells on the basis of differences in integrin function and expression. *Cell* 73:713–724.

Jones, P. H., Harper, S., and Watt, F. M. (1995) Stem cell patterning and fate in human epidermis. *Cell* 80:83–93.

Leibovich, S. J. and Ross, R. (1975) The role of the macrophage in wound repair: a study with hydrocortisone and antimacrophage serum. *Am. J. Pathol.* 78:71–100.

Lindblad, W. J., French, J. A., Redford, K. S., Buenaventura, S. K., and Cohen, I. K. (1987) Induction of prolyl hydroxylase activity in a nonadherent population of human leukocytes. *Biochem. Biophys. Res. Commun.* 147:486–493.

Lindblad, W. J. (1998) Collagen expression by novel cell populations in the dermal wound environment. *Wound Rep. Reg.* 6:186–193.

Madri, J. A. and Bell, L. (1992) Vascular cell responses to injury: Modulation by extracellular matrix and soluble factors. In: *Ultrastructure, Membranes and Cell Interactions in Atherosclerosis* (Robenek, H. and Severs, N., eds.), CRC, Boca Raton, FL, pp. 167–181.

Mast, B. A. and Schultz, G. S. (1996) Interactions of cytokines, growth factors, and proteases in acute and chronic wounds. *Wound Rep. Reg.* 4:411–420.

Metchnikoff, E. (1893) *Lectures on the Comparative Pathology of Inflammation*, Keegan, Paul, and Trubner, London, UK.

Myllyla, R. and Seppa, H. (1979) Studies on the enzymes of collagen biosynthesis and the synthesis of hydroxyproline in macrophages and mast cells. *Biochem. J.* 182:311–316.

Odland, G. and Ross, R. (1968) Human wound repair. I. Epidermal regeneration. *J. Cell Biol.* 39:135–151.

Oshima, H., Rochat, A., Kedzia, C., Kobayashi, K., and Barrandon, Y. (2001) Morphogenesis and renewal of hair follicles from adult multipotent stem cells. *Cell* 104:233–245.

Rappolee, D. A., Mark, D., Banda, M. J., and Werb, Z. (1988) Wound macrophages express TGF-α and other growth factors in vivo: analysis by mRNA phenotyping. *Science* 241:708–712.

Reyes, M., Dudek, A., Jahagirdar, B., Koodie, L., Marker, P. H., and Verfaillie, C. M. (2002) Origin of endothelial progenitors in human postnatal bone marrow. *J. Clin. Invest.* 109:337–315.

Roberts, A. B., (1995) Transforming growth factor-β: activity and efficacy in animal models of wound healing. *Wound Rep. Reg.* 3: 408–418.

Rochat, A., Kobayashi, K., and Barrandon, Y. (1994) Location of stem cells of human hair follicles by clonal analysis. *Cell* 76:1063–1073.

Rogers, A. A., Burnett, S., Moore, J. C., Shakespeare, P. G., and Chen, W. Y. J. (1995) Involvement of proteolytic enzymes (plasminogen activators and matrix metalloproteinases) in the pathophysiology of pressure ulcers. *Wound Rep. Reg.* 3:273–283.

Ross, R., Everett, N. B., and Tyler, R. (1970) Wound healing and collagen formation. VI. The origin of the wound fibroblast studied in parabiosis. *J. Cell Biol.* 44:645–654.

Savill, J. (1997) Apoptosis in the resolution of inflammation. *J. Leukoc. Biol.* 61:375–380.

Shelton, E., and Rice, M. E. (1959) Growth of normal peritoneal cells in diffusion chambers: a study in cell modulation. *Am. J. Anat.* 105: 281–303.

Simpson, D. M. and Ross, R. (1972) The neutrophilic leukocyte in wound repair: a study with antineutrophil serum. *J. Clin. Invest.* 51: 2009–2023.

Squier, M. K., Sehnert, A. J., and Cohen, J. J. (1995) Apoptosis in leukocytes. *J. Leukoc. Biol.* 57:2–10.

Stenn, K. S. and Dvoretzky, I. (1979) Human serum and epithelial spread in tissue culture. *Arch. Dermatol. Res.* 246:3–15.

Vaage, J. and Lindblad, W. J. (1990) Production of collagen type I by mouse peritoneal macrophages. *J. Leukoc. Biol.* 48:274–280.

Woodley, D. T., Wynn, K. C., and O'Keefe, E. J. (1990) Type IV collagen and fibronectin enhance human keratinocyte thymidine incorporation. *J. Invest. Dermatol.* 94:139–143.

Yager, D. R. and Nwomeh, B. C. (1999) The proteolytic environment of chronic wounds. *Wound Rep. Reg.* 7:433–441.

Yamada, K. M. and Clark, R. A. F. (1996) Provisional matrix. In: *The Molecular and Cellular Biology of Wound Repair* (Clark, R. A. F., ed.), Plenum, New York, pp. 51–94.

Zhai Y., Vaage, J., and Lindblad, W. J. (1996) Expression of type I procollagen by cultured mouse peritoneal macrophages detected by in situ hybridization and immunohistochemical staining. *Wound Rep. Reg.* 4:185 (abstract).

10 Bone Marrow Mesenchymal Stem Cells

JAMES E. DENNIS, PhD AND ARNOLD I. CAPLAN, PhD

The term *mesenchymal stem cell* (MSC) refers to adult mesenchymal progenitor cells with the potential to produce progeny that differentiate to produce a variety of mesenchymal cell types (e.g., fibroblasts, muscle, bone, tendon, ligament adipose tissue). It is not known if these cells actually have the capacity to self-renew, which is a property of stem cells. MSCs may be found in muscle, skin, and adipose tissue, as well as in the bone marrow. MSCs in the bone marrow may be identified by colony-forming units that produce fibroblasts and make up a very small percentage of the total marrow population. The ability of MSCs in the bone marrow to form bone and cartilage has been known for more than 100 yr. MSCs or their progeny in the bone marrow provide a stromal microenvironment for hematopoiesis. During development, MSCs in the bone marrow may derive from the developing vessels (pericytes) or from circulating precursors. MSCs also produce osteoclasts and osteoblasts responsible for remodeling of bone and adipocytes, which make up a major portion of the bone marrow. MSCs may be isolated from bone marrow, peripheral blood, fat, skin, vasculature, and muscle, where they most likely are responsible for normal tissue renewal, as well as for a response to injury. Bone marrow MSCs are negative for primitive hematopoietic cell markers but express antibody-defined markers: SH2 (type III TGF receptor), SH3 and SH4 (ecto-5'-nucleotidase), and STRO-1. Individual clones of cell lines derived from MSCs have different potentials for differentiation, indicating different stages of determination and levels of plasticity. Transplanted MSCs have been shown to enhance bone, tendon, cartilage, and nerve repair in experimental models. Systemic transplantation of MSCs has not always led to functional results in tissue repair but has tremendous potential. The use of MSCs for gene therapy for hematopoietic, metabolic, and neurological disorders is currently under investigation.

10.1. DEFINITIONS

There are a number of different terms that have been ascribed to the cell type that may be, or to the population that may contain, a mesenchymal stem cell (MSC). Therefore, some clarification of cell terminology is needed for the use of the term MSC in this chapter. The formal definition of the term *stem* includes the ability of a progenitor cell to "self-renew," i.e., to divide and retain a daughter cell that does not differentiate (or produce progeny that differentiate) while producing another daughter cell that differentiates, or divides and then differentiates. To date, no one has rigorously proved that the cells designated "MSCs" have this unique stem cell renewal capability. For this reason, some have termed the expandable population of mesenchymal cells with differentiation potential as *mesenchymal progenitor cells*, while others have used terminology based on conventional terminology in their field of study. For example, Friedenstein refered to cells isolated from bone marrow as colony-forming unit-fibroblastic (CFU-f), a terminology originating from that used by hematologists to describe colony growth in hematopoietic stem cell assays. Similarly, hematologists have referred to the marrow cells that fabricate the unique connective-tissue scaffold that supports hematopoietic cells as "stromal" cells, and these cell preparations could contain MSCs.

In this review, MSCs refers to adult mesenchymal progenitor cells with differentiation potential to form a variety of mesenchymal tissues. These cells may be identical to, or be found in, cell preparations that have been termed *mesenchymal progenitor cells, marrow stromal cells, MSCs*, and *CFU-f*. Indeed, MSCs may also be found in a variety of nonhematopoietic tissues, such as muscle, skin (Pettis et al., 1990; Mizuno and Glowacki, 1996), and adipose tissue (Zuk et al., 2001). The origin of these MSCs in different tissues could be the smooth muscle cell lineage pericytes (Galmiche et al., 1993) associated with vasculature, which have been shown to be both osteogenic (Brighton et al., 1992) and chondrogenic (Doherty et al., 1998).

From: *Stem Cells Handbook*
Edited by: S. Sell © Humana Press Inc., Totowa, NJ

10.2. HISTORY

The presence of mesenchymal progenitor cells within the bone marrow has been documented from the late nineteenth century by the works of Goujon (1869), who was the first to show the osteogenic potential in heterotopic transplants of rabbit marrow, and which was later confirmed in transplantation experiments by Biakow (1870). Danis (1960) showed that whole marrow itself was osteogenic, and not just an inductive factor or chemoattractant for osteogenic cells, by placing marrow within a diffusion chamber (two semipermeable membranes separated by a plastic ring), implanting this within a host animal, and observing the contents by histological methods. Similar confirmatory experiments were conducted by other groups (Petrakova et al., 1963; Friedenstein et al., 1966; Bruder et al., 1990), wherein the formation of cartilage and bone within the diffusion chamber was observed, demonstrating that bone marrow had, at a minimum, the potential to form bone and cartilage. Friedenstein and co-workers (Friedenstein, 1973) showed that the osteogenic potential of bone marrow was a feature of a specific subgroup of cells termed the *CFU-f*, which made up a very small percentage of the total marrow cell population. Friedenstein (1980) later showed that CFU-fs were formed from single cells and that some of these CFU-fs were able to form both bone and the microenvironment necessary for the formation of hematopoitic elements. These cells that form a hematopoietic microenvironment (Friedenstein et al., 1974; Trentin, 1976) can be considered a third phenotype for which the bone marrow has the potential to differentiate into, although other laboratories have categorized bone marrow stromal cells (hematopoietic supportive cells) as early osteogenic progenitors, a debate that has yet to be resolved.

The first formal presentation of the concept of a stem cell residing in bone marrow was in a publication by Owen in 1978, wherein the marrow stroma was hypothesized to consist of a lineage analogous to the hematopoietic lineage (Till and McCulloch, 1980). Owen (1985) expanded on this hypothesis and proposed a model for the stromal lineage that contained "stem cell," "committed progenitor," and "maturing cell" compartments, and included a lineage diagram for "stromal stem cells" that included "reticular," "fibroblastic," "adipocytic," and "osteogenic" cells as end-stage phenotypes. In 1991, Caplan (Caplan, 1991) proposed the existence of an MSC having the capacity to differentiate into multiple mesenchymal phenotypes including adipose, tendon, ligament, muscle, and dermis (Fig. 1) (Caplan, 1994). Formal proof of the multipotentiality of MSCs or CFU-fs came from experimentation on clonal populations. It was demonstrated that approx 30% of isolated CFU-f colonies were able to form bone alone or bone with marrow in open transplants (Friedenstein, 1980) and that about 40% of rabbit marrow cells isolated by limiting dilution or by cloning rings were able to form bone (Bennett et al., 1991). Multipotential mesenchymal cells were also identified in a studies using MSCs isolated from the marrow of transgenic mice containing a gene for conditional immortality (Dennis and Caplan,m 1996; Dennis et al., 1999). In that study, multiple colonies were isolated by cloning rings and limiting dilution while under immortalized culture conditions and then tested for differentiation potential under standard culture conditions or in vivo. For mouse marrow–derived mesenchymal progenitor cells, it was determined that these cells contain a mixture of cells having various differentiation potentials (Dennis and Caplan, 1996) ranging from monopotential to quadripotential (Dennis et al., 1999). Similar results were reported for MSCs isolated from human marrow, wherein three of six isolated and expanded MSC colonies were able to differentiate into bone, adipose, and cartilage while all six were able to differentiate into bone (Pittenger et al., 1999).

10.3. THE EMBRYONIC ORIGIN OF MSCS

While MSCs have an essential role for the differentiation of all the mesenchymal tissues of an organism, it is the function of MSCs in the adult organism that is the focus of this essay. The developing organism seeds within different tissues the progenitor cells necessary for tissue maintainence turnover and repair in the adult. One example of this is the satellite cells in striated muscle located within the basement membrane of myotubes (Mauro, 1961), which do not express any muscle proteins, except when stimulated to expand and differentiate in response to injury. Satellite cells are distinct from the muscle fibroblasts that are found outside the muscle basement membrane, which, interestingly, may be another source of progenitor cells within muscle *tissue* that have the capacity to differentiate into other mesenchymal phenotypes (Lee et al., 2000). Following muscle injury, satellite cells mitotically expand; migrate into the damaged muscle; fuse; and then form fully functional, innervated myotubes. The satellite cells are thought to arise from residual migrating myogenic cells during development that do not fuse into primary or secondary myotubes. For MSCs, the origin of MSCs that eventually reside in adult marrow is less clear.

During limb development, the primitive limb bud first forms a cartilage anlage from condensing mesenchymal cells. Bone then forms around the cartilage anlage at the middiaphysis and expands bidirectionally toward the distal and proximal epiphyses. Just at the time that bone is forming at the middiaphysis, the cartilage beneath the bone becomes hypertrophic, and soon after the initial bone matrix has mineralized, the cartilage of the middiaphysis is invaded by vasculature (Pechak et al., 1986). The invading vasculature infiltrates the region once occupied by hypertrophic cartilage and eventually the entire cartilage anlage is replaced by vasculature and marrow elements. Hematopoietic stem cells migrate to the nascent bone marrow from their preceding embryonic location within the liver. The MSCs in bone marrow can arrive by three different mechanisms: (1) They can enter along with the vasculature; (2) they can migrate into the space after vascularization along the vessel paths, i.e., from the periosteum, which has documented multipotentiality (Nakahara et al., 1992; Yoo and Johnstone, 1998); or (3) they can arrive via the blood proper, indicating the existence of a circulating MSCs. The first two mechanisms differ only with respect to the timing of MSC entrance into the marrow, not by the path of arrival or migration. The third mechanism differs from the other two only with respect to which side of the capillary basement membrane the MSCs use, the luminal or abluminal. However, this mechanism of MSC migration or circulation has important implications for adult tissue repair in that it may be possible to deliver reparative MSCs via the circulation as opposed to only localized applications, especially if MSC docking sites exist on the endothelium lining the vascular network.

The first and second mechanisms of MSC arrival (along with the vasculature) to the bone marrow space are supported by two distinct observations and experiments with the vascular-associated cells (pericytes), and on smooth muscle cells from marrow.

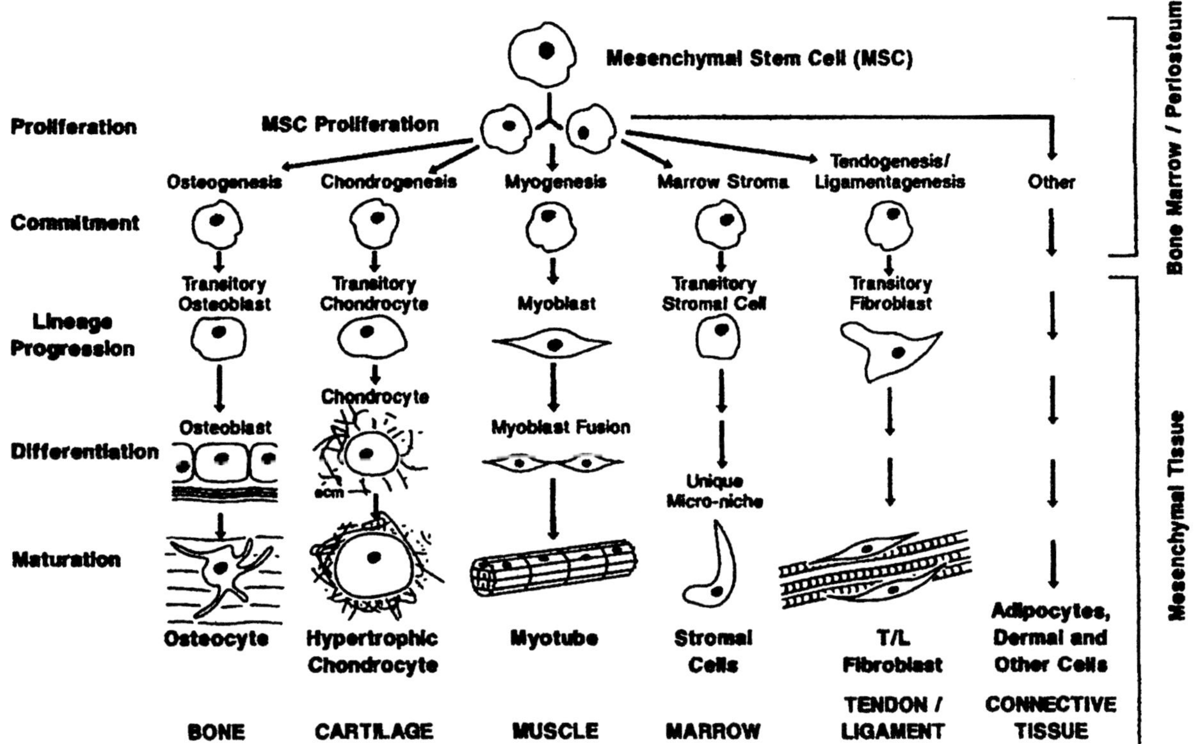

Fig. 1.

Pericytes are cells that are closely associated with capillaries (Rouget, 1873), which express smooth muscle markers (Meyrick et al., 1981; herman and D'Amore, 1985) and have the potential to differentiate into osteoblasts (Brighton et al., 1992). Although often categorized as being a cell type distinct from that of smooth muscle cells, some data indicate that pericytes have the potential to differentiate into smooth muscle cells (Meyrick et al., 1981). The pericytes of the periosteum are one candidate cell type for the origin of marrow MSCs because they are in the correct anatomical region for migration into the nascent marrow space, and also because pericytes have been shown to have the potential to differentiate into chondrocytes and adipocytes (Rhodin, 1968; Doherty et al., 1998). Another set of data addresses the differentiation of MSCs in the marrow from the perspective of the smooth muscle lineage, and these data show that MSCs analyzed for their expression of smooth muscle cell markers can be separated into distinct steps along a proposed smooth muscle cell lineage (Rémy-Martin et al., 1999). Based on this model, both MSCs and pericytes would be part of a smooth muscle cell lineage that includes the multipotential MSCs of marrow. In addition, pericytes have been shown to express the STRO-1 antigen (Doherty et al., 1998), which is a known marker for marrow colony-forming cells (Simmons and Torok-Storb, 1991) and is expressed in multipotential human MSCs (Dennis et al., 2001). While this circumstantial evidence suggests a lineage relationship between pericytes and MSCs that further implies a logical source of marrow MSCs, to date, there is no direct evidence as to the origin of the osteogenic cells in the marrow.

The observation that vascular-associated cells (pericytes) have osteogenic potential adds some modern flavor to an old argument concerning the origin of osteoblastic cells in bone and in bone repair (as reviewed by Keith, 1919). Duhamel du Monceau proposed, in 1741, that the periosteum was the primary agent for the repair of bone. The word *agent* is used here instead of *cell source* because this work precedes the Cell Theory of Schleiden and Schwann by about a century. By contrast, Albrecht von Haller argued that bone was a product of arteries and that periosteum was simply a source of nourishment. It would seem that this scientific argument is relevant even today—von Haller could take the position that the well-documented osteogenic potential of periosteum could be traced back to the presence of pericytes within the periosteal vasculature.

The role of MSCs within the marrow of an adult organism can be as varied as the phenotypes that MSCs can express. (The focus of this discussion is the role of MSCs in the adult, not the developing organism.) If one makes the reasonable, yet debatable, position that MSCs which migrate into marrow, and then differentiate into stromal support cells—i.e., the cells that make up the hematopoietic microenvironment (Trentin, 1976)—then MSCs have an essential role in setting up the adult site of hematopoiesis. Whether MSCs perform the task of hematopoietic support directly, or whether this function is performed by their "committed" and expressive progeny, are issues of lineage progression and phenotype plasticity that are discussed in detail later. Our interpretation of the data from many laboratories on marrow MSCs or stromal cells (which are probably identical) is that some or all stromal hematopoietic support cells have the ability to differentiate into other mesenchymal phenotypes, and that these cells may reassume a quiescent multipotential MSC phenotype when the physiological demand for hematopoietic output diminishes.

MSCs within marrow also function as a reservoir of stem cells for the repair of bone fractures and for the natural turnover of bone. The sequence of cellular events in fracture repair includes the formation of a clot, migration of macrophages and "mesenchymal" cells into the site, formation of cartilage in the fracture callus, and the bridging of the fracture via endochondral ossification (Ham and harris, 1971). The MSCs are the cells that form both the cartilage callus and the bone that bridges the repair site. MSCs also play an osteogenic role in normal bone remodeling or turnover. During natural bone turnover, hematopoietic lineage-derived osteoclasts resorb portions of bone in "resorption pits," the osteoclasts expire, and the excavated pits are then refilled with new bone by osteoblasts that differentiate from MSCs. The MSCs are the reservoir for osteoblastic cells that are needed for the turnover process that occurs throughout an organism's lifetime. It is hypothesized that diminished numbers of MSCs or the restricted access to MSC reservoirs is responsible for some age-related bone loss.

A more controversial role for MSCs has to do with the expression of the adipocyte phenotype. A number of laboratories have documented the ability of MSCs to differentiate into adipocytes, but the precise role of adipocytes within the marrow remains unknown. There is evidence that rat bone marrow adipocytes produce an uncoupling protein that is consistent with marrow adipocytes functioning in the regulation of heat (Marko et al., 1995). Another theory is that adipocytes serve as an expandable and retractable space-regulatory cell within the confines of the bone marrow (Tavassoli, 1974). When there is high hematopoietic demand, adipocytes dwindle to make room for hematopoiesis, and when hematopoietic demand declines, the adipocytes expand to fill in the unused space. This hypothesis of a space-filling role is supported by the fact that marrow adipocytes do not mobilize lipids in response to starvation, as do extramedullary adipocytes (Tavassoli, 1974). A variety of other functions have been ascribed to adipocytes (reviewed by Gimble, 1990), including hematopoietic support, which may be more relevant to discuss within the context of differentiation plasticity (*see* below).

In summary, MSCs within adult bone marrow have five primary roles: as progenitor cells for bone formation during turnover or repair, in cartilage formation during repair, in vascular support, in hematopoietic support, and as progenitors for adipocytes whose precise function has yet to be defined.

10.4. HARVESTING, ISOLATION, AND CHARACTERIZATION

The primary source for MSCs is the bone marrow, although recent reports indicate that MSCs can be isolated from other sources, such as peripheral blood (Kuznetsov et al., 2001), fat (Zuk et al., 2001), skin (Pettis et al., 1990; Mizuno and Glowacki, 1996), vasculature (Brighton et al., 1992), and muscle (Lee et al., 2001). Interestingly, the demonstration of a progenitor cell population in vasculature fits with data suggesting that MSCs are part of a smooth muscle cell lineage (Galmiche et al., 1993; Rémy-Martin et al., 1999), and that the source of multipotential cells in fat may actually be the vasculature. Whether derived from animals or humans, the selection of fetal bovine serum (FBS) is critical to the success of expansion and maintenance of differentiation potential (Lennon et al., 1996). Human MSCs are isolated from marrow aspirates as described originally by Haynesworth et al. (1992b), with some modifications. Marrow aspirates are diluted with phosphate-buffered saline (PBS) at 3.5 parts PBS to 1 part marrow aspirate, centrifuged at 900*g* for 10 min at room temperature. The pellets are combined, resuspended in PBS, and counted, and the volume is adjusted to a cell concentration of 2.0×10^7 cells/mL. The cells are then layered onto 1.073 g/mL of Percoll (e.g., 5–10 mL of marrow cells is layered onto 25 mL of Percoll in a sterile 50-mL tube). The tubes are centrifuged for 30 min at 1100*g* with the brake off, and the top cell band is removed with a 5-mL pipet and combined with 25 mL of PBS and centrifuged, and the pellet is taken up in 10 mL of hMSC medium (Dulbecco's modified Eagle's medium [DMEM] [low glucose] plus 10% FBS). Cells are plated 7.5 – 25.0 $\times 10^4$ cells/cm^2 and cultured at 37°C in humidified air containing 5% carbon dioxide; the medium is exchanged twice weekly.

The procedure for nonhuman bone marrow samples or for human samples obtained directly from marrow space is slightly different than for aspirates. Plugs of marrow are suspended in DMEM and dispersed by passage through 18-, 20-, and 22-gage needles. The dispersed marrow is centrifuged at 900*g* for 5 min, resuspended in DMEM, counted, and plated at 7.5 – 25.0 $\times 10^4$ cells/cm^2 in DMEM containing a selected lot of FBS. Between 12 and 18 days later, the cultures have formed colonies and are ready for passaging. The MSCs are washed once in Tyrode's salt solution, incubated for 5 min in 0.25% Trypsin-EDTA (Gibco), and a one-half volume of calf serum is added to inhibit digestion. The cells are then centrifuged and replated at approx 7.5 – 25.0 $\times 10^4$ cells/cm^2. A significant number of contaminating hematopoietic cells, especially monocytes, remain attached to the primary culture plates.

Numerous laboratories have published methods to isolate and expand MSCs from different species, including human (Kuznetsov et al., 1997; Colter et al., 2000), mouse (Krebsbach et al., 1997), rabbit (Wakitani, 1994). (Bone marrow cultures for all of these species and of guinea pig have been described by a pioneer in the studies, A. J. Friedenstein [Friedenstein et al., 1974, 1976].) Most of these methods of culturing MSCs rely on differential adherence to plastic and the use of serum concentrations lower than that used in hematopoietic cultures to eliminate most nonadherent hematopoietic cells. Differences in preparations among laboratories are primarily in the use of base medium (usually αMEM or DMEM), the lot of serum, and whether or not cells have been partially purified by density gradient centrifugation.

A number of markers have been identified to distinguish MSCs from hematopoietic cells. Three of the earliest markers identified are SH2, SH3, and SH4, which were shown to bind to MSCs and not to hematopoietic cells (Haynesworth et al., 1992a). Subsequent studies have revealed that the antibody SH2 binds endoglin (Barry et al., 1999), a type III transforming growth factor β receptor found on mesenchymal tissues and on macrophages and endothelial cells, indicating that this marker is not specific for MSCs. In addition, the antibodies SH3 and SH4 have been shown to specifically bind an ecto-5'-nucleotidase, CD73. In combination, SH2, SH3, and SH4 are used to characterize individual preparations of human MSCs. Markers for hematopoietic cells, such as anti-CD31, anti-CD1, and anti-CD14, have been used to negatively select (deplete) hematopoietic cells from MSC cultures (Rickard et al., 1996).

The antibody STRO-1 (Simmons and Torok-Storb, 1991) has, so far, been the most useful antibody for identifying and positively selecting for MSCs in bone marrow. STRO-1 has been successfully used to isolate the CFU-f cells from marrow

(Simmons et al., 1994), and STRO-1-selected cells have been shown to be osteogenic (Gronthos et al., 1994), chondrogenic, adipogenic, and hematopoiesis supportive (Dennis et al., 2001). Unlike the other antibodies to date, STRO-1 can be used not only to stain and characterize MSCs, but also to sort cells by magnetic activated cell sorting (Gronthos et al., 1994; Tamayo et al., 1994; Dennis et al., 2001).

10.5. CELL LINES VS PRIMARY CULTURES

The use of cell lines for research carries several important caveats. One primary criticism concerning the use of cell lines is that the entire basis of generating or selecting for a cell line renders that cell line nonphysiological in that most of the cells are transformed and some are tumorigenic. Indeed, the base characteristic of a cell line is that it will expand in culture indefinitely. This alone makes these cells different from native cells. Additionally, the details of the origin and selection of each individual cell line is important for assessing how each cell's phenotypic characteristics may, or may not, be representative of native cells. For example, many cell lines, especially from rodents, are derived from long-term cultures that spontaneously transform. This spontaneous transformation generally follows a time period in culture ("crisis") during which cells cease dividing, many of them die, or they show features of senescence. The cells that break out of this crisis represent a rare subpopulation of cells and usually contain abnormal numbers of chromosomes, which is a likely cause of their transformation. Some cell lines are derived from cancerous lesions, wherein transformation has occurred in vivo instead of in vitro.

More recently, cell lines have been formed by introducing genes into cells that can impart immortality, such as the SV-40 large T-antigen (Takahashi et al., 1995; Houghton et al., 1998). (The transformation of a cell into a cell line is often termed *immortalization*, to reflect the cells' ability to divide indefinitely.) The advantage of this method is that the mechanism of immortalization is known and is not such a highly selective and rare event. The disadvantage to this method is that the transforming gene is always expressed, which is non-physiologic, and the gene is usually randomly introduced into the host cell, which can have other unknown effects on the physiology of the cell line.

A more elegant method of immortalization has been devised, wherein cells are transduced with a immortalization gene that is "conditional"; that is, the culture conditions can be adjusted to stably express the immortalizing gene or gene product. One example of this is the temperature-sensitive gene for the SV-40 large T-antigen. At low temperatures (i.e., 33°C), the large T-antigen remains stable, whereas at higher temperatures (37–39°C) the large T-antigen is very labile and cannot function to derepress mitosis. A useful addition to this method is the inclusion of a promoter region that can increase the expression of large T-antigen after the addition of exogenous molecules. One example of this is the class I histocompatability antigen promotor with a γ-interferon-sensitive element that has been combined with the temperature-sensitive gene for large T-antigen and inserted into the mouse genome (Jat et al., 1991). This transgenic, H-2K^b tsA58 mouse, or "immortomouse," contains this construct that has the potential to confer conditional immortality in each of it's cells, although not all cell isolates behave identically. The advantages of cell lines isolated from these transgenic mice are that there is a single copy of a defined gene, that the gene is conditionally expressed, and that the presence of the inactivated gene does not seriously affect the growth and development of the organism. These cells are not identical to native cells because they are induced to express the immortalization gene while the cells are being propogated, and the effect of the abnormal expression of this gene on cell physiology (even when the gene is later turned off) is not known.

An extensive number of cell lines have been made from cultures containing MSCs. Many of these cell lines have been derived from hematopoietic cultures as a method of analyzing the contribution of "stromal" cells to the support and differentiation of hematopoietic stem cells (reviewed by Deryugina and Muller-Sieburg, 1993). Many of these "stromal" cell lines have since been found to have characteristics of MSCs. For example, the mouse-derived stomal line MBA-15 (Benayahu et al., 1989) has been shown to be osteogenic, as has the cell line MBA-1, which is also chondrogenic (Benayahu et al., 1994). The stromal-derived cell line BMS2 exhibits an adipocytic morphology along with the ability to support lymphocytes in culture (Pietrangeli et al., 1988; Dorheim et al., 1993). It is beyond the scope of this chapter to cover all of the cell lines developed for other applications (besides MSCs) that have MSC-like characteristics. Interestingly, many of these cell lines may have originated from MSCs, although the laboratories that have studied them have been concentrating on other applications. There are, however, some laboratories that have specifically set out to obtain cell-lines composed of MSCs (Dennis and Caplan, 1996; Ragab et al., 1998).

In our laboratory, we set out to isolate, clone, and analyze the differentiation potential of MSCs isolated from the marrow of the H-2K^b tsA58 transgenic mouse, which contains a temperature-sensitive gene for large T-antigen under the control of the class I histocompatibility antigen promotor (Jan et al., 1991). Individual marrow-derived clones from this transgenic mouse were shown to vary in their differentiation potential to include clones that were mono-, bi-, and tripotential (with respect to osteogenic, adipogenic, and hematopoietic support phenotypes) (Dennis and Caplan, 1996). A later study identified another conditionally immortalized clone, BMC9, which has the potential to differentiate into at least four phenotypes: osteoblasts, chondrocytes, adipocytes, and osteoclast-supportive cells (Dennis et al., 1999).

A temperature-sensitive construct of large T-antigen has also been used to immortalize human marrow–derived stromal cells (Hicok et al., 1998), which were shown to express both osteoblastic and adipocytic markers.

10.6. DIFFERENTIATION STUDIES

10.6.1. PLASTICTY *Determination* is a well-known concept in embryology (Hopper and Hart, 1985) that describes the progression of a totipotent oocyte into a collection of cells that is more restricted in the number of cell types into which they can differentiate. Cells become determined to differentiate along one of three germ layers and later become determined to differentiate into specific organs or cells within that layer. While it has generally been accepted that cells retained some *plasticity* (the ability to transdifferentiate or dedifferentiate and redifferentiate into other cell types), especially in the developing embryo, recent results on the differentiation of some mesenchymal phenotypes suggest that plasticity may be more common than previously thought. Mesenchymal tissues may be particularly prone to phenotype plasticity than the other germinal layers, although recent results suggest that even neural tissues may exhibit plasticity toward the hematopoi-

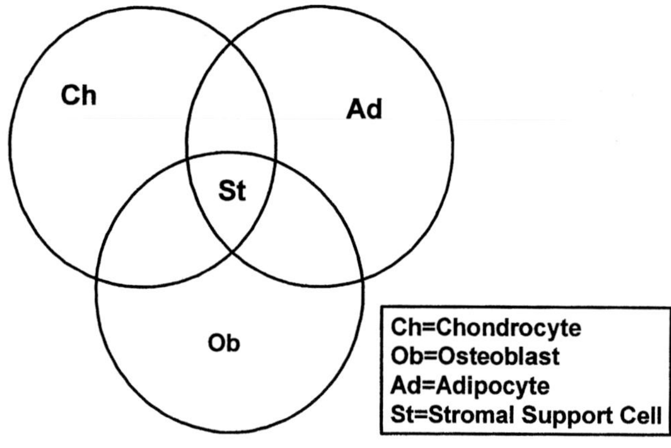

Fig. 2.

etic lineage (Bjornson et al., 1999) and hematopoietic stem cells may differentiate into hepatic oval cells (Petersen et al., 1999).

One line of evidence of mesenchymal plasticity comes from analysis of the expression of markers for differentiation of specific cell phenotypes. For example, adipocytes express a particular set of genes that are associated with the accumulation of lipid. These adipocyte gene include, AP2; adipsin; and a set of transcription factors such as PPARγ2, CEBP/α and CEBP/β. Interestingly, there is evidence that differentiated adipocytes, cells that have accumulated lipid droplets, have the ability to transdifferentiate into ostoblasts (Bennett et al., 1991). Another characteristic of mesenchymal cells from marrow is that they often have overlapping functions and phenotypes. Put another way, there is often a blurring of the boundary between phenotypes. For example, cells that support the formation of osteoclasts are often considered to be preosteoblasts. But isn't the ability to express specific signaling molecules in response to physiological stimuli to promote osteoclast formation a property of a differentiated cell? Are all osteoclast-supportive cells osteoblastic? Data from our laboratory indicate that some mouse-derived marrow clones are osteoclast supportive yet nonosteogenic, while other clones are chondrogenic, osteogenic, and adipogenic but do not support osteoclast formation (unpublished results). Indeed, when reviewing the literature on hematopoietic, preadipocyte, and preosteoblastic cell lines, there seem to be cells that fit every combination and permutation of different mesenchymal markers and phenotypes (although some phenotypes are more common that others). For example, there are preadipocytes that can support myeloid hematopoietic cells (Udagawa et al., 1989; Gimble et al., 1992), and the murine marrow-derived cell line, BMS2, can support osteoclast formation either as a preadipocyte or after induction and expression of mature adipocyte markers (Kelly et al., 1998). There are nonadipogenic cells, such as ST2 cells, that also support osteoclast formation. In a study by Takahashi et al. (1994), clones of a parental cell line, MCHT-1, were analyzed for their response to adipogenic and osteogenic conditions and for their ability to support osteoclast formation. They found that all the clones supported osteoclast formation (at different rates and levels), but varied widely in their expression of osteoblast or adipocyte markers (Takahashi et al., 1994).

To account for these data on the multiple differentiation potentials of MSCs, several different models have been proposed, including detailed lineage models (Minguell et al., 2001) or models that try to account for overlapping phenotypic markers, such as the diagram in Fig. 2, which is modified version of one of the models proposed by Gimble et al. (1996). Within the model proposed in Fig. 2, the stromal support cell is considered an early progenitor of other mesenchymal phenotypes, which does not fit the unpublished data from our laboratory, nor does it account for the published phenotypes of various mouse cell lines. In addition, there is no accounting for plasticity of phenotypes after cells express phenotypic markers. To account for plasticity, lineage progression, and multipotentiality, the Stochastic Activation/Repression Model has been proposed (Dennis and Charbord, 2002) and is illustrated in Fig. 3. In this model, the multipotential events are considered to occur within the MSC compartment, which is not physical compartment but is more accurately defined as a genetic state of differentiation potential. Cells within the MSC compartment can exhibit a range of multiple or monopotentials, while cells in the differentiation compartment are expressing those potentials in response to external stimuli. Plasticity occurs when a cell exits the MSC compartment with multiple differentiation potentials and switches from one phenotype to another. The primary advantage of the Stochastic Activation/Repression Model is that it accounts for the (apparent) randomness of the multiple potentials observed for MSCs and cell lines. However, the mechanism for the formation of this random mixed population of MSCs with different potentials remains unknown.

10.6.2. TRANSPLANTATION STUDIES The ability of MSCs to be expanded in culture, coupled with their potential to differentiate into end-stage mesenchymal phenotypes, makes MSCs a likely source of cells for tissue repair strategies. One advantage of the use of MSCs is that cells are easily obtained by a standard, outpatient bone marrow harvest. The cells can then be expanded and used for autologous transplantation. The key obstacles to overcome are expansion while retaining differentiation potential, delivery of cells to the site of repair, promotion of differentiation toward the appropriate mesenchymal phenotype, and the integration of the differentiated tissue into the host repair site. Although advances have been made in different repair models, several significant obstacles have yet to be overcome.

10.6.3. BONE REPAIR The utility of using MSCs to improve bone healing was demonstrated in a critical-sized bone defect in rat (Kadiyala et al., 1997), wherein calcium phosphate ceramics showed significantly more bone fill (43%) than was found in bone marrow–loaded ceramics (19%) or cell-free ceramics (10%). MSCs were further tested for their ability to repair bone in a femoral gap model in a dog segmental defect model (Bruder et al., 1998). Again, radiological and histomorphometic analysis demonstrated greater bone fill in implants loaded with MSCs than in empty control implants. In a study in rabbits, the use of PGLA plymer fleece or fibrin beads cultured in periosteum-derived MSCs also showed superior bone healing by histological and radiological criteria than did matching empty control implants (Perka et al., 2000). In another approach to enhancing bone repair, calcium phosphate scaffolds and metallic plates were coated with MSCs and cultured for 1 wk in osteogenic conditions and then implanted into nude mice. The precoating and incubation in osteogenic conditions promoted the formation of bone around the implants (de Bruijn et al., 1999). A similar method was employed to "jump start" the bone healing process in a dog segmental bone defect. In this case, perestium was used as a source of MSCs (or osteogenic cells), which were then coated onto ceramic scaffolds,

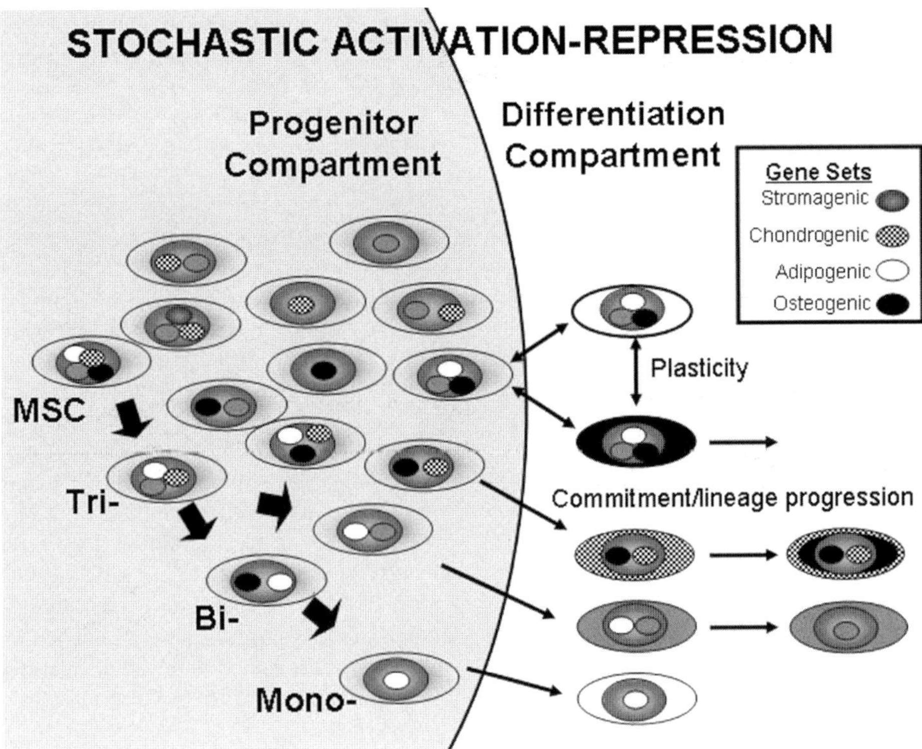

Fig. 3.

and implanted for 1 mo intramuscularly (Cong et al., 2001). The implants were transferred to the segmental defect site, whereupon the implants rapidly fused to the ends of the defect. These studies have demonstrated the feasibility of using MSCs to augment the repair of bone and to augment the integration of scaffold materials into bone sites.

10.6.4. TENDON REPAIR MSCs have also been tested in tendon repair models, which showed that MSCs co-implanted in a collagen matrix carrier were able to integrate into the neotissue and that the implanted MSCs became aligned in parallel with the fibers and native cells. In a rabbit Achillies tendon repair model, it was shown that MSCs combined with a type I collagen regained nearly 65% of the normal tendon maximum force compared with only 31% of maximum force in the cell-free type I collagen controls (Young et al., 1998; Butler and Awad, 1999). In a similar study of MSCs combined with type I collagen gels, MSCs again showed improved biomechnical properties at 4 wk (the end of the study) over matched cell-free controls (Awad et al., 1999). It would be interesting to determine how MSCs would perform in comparison to type I collagen loaded with some other source of cells, such as dermal fibroblasts.

10.6.5. CARTILAGE REPAIR MSCs are chondrogenic (Yoo et al., Dennis et al., 1999, 2001), so another tissue engineering application is for the repair of cartilage, a tissue that is notorious for its inability to heal, especially in adult animals. A number of studies have been conducted on the repair of articular cartilage defects that have been implanted with MSCs contained within different polymers, such as hyluronic acid–based polymers (Solchaga et al., 1999, 2000), type I collagen (Wakitani et al., 1994), and polylactic acid (Dounchis et al., 2000). None of the polymer-cell constructs to date have produced a consistent repair of articular cartilage defects, even though the cells used for these studies have the potential to express a chondrogenic phenotype. The application of MSCs to cartilage repair will not be achieved until methods are devised that promote a consistent healing of defects without the formation of fibrous tissue. Another major obstacle to the application of MSCs in cartilage repair is improving the integration of neocartilage matrix with the surrounding native cartilage matrix.

10.6.6. NERVOUS SYSTEM MSCs have been tested as therapeutic agents for the repair of brain injury. Mice injected with bromodeoxyuridine-tagged MSCs were shown to express NruN, a neuronal-specific protein and glial fibrillary acidic protein, and functional tests indicated significant recovery from cerebral artery occlusion as compared to vehicle-only controls (Li et al., 2000). In a drug-induced model of Parkinson's disease, mice injected with MSCs expressed tyrosine hydroxylase and exhibited significant improvement in functional testing (Li et al., 2001b), and in a model for stroke, MSC-injected rats showed improvement in neurological severity scores (Li et al., 2001a). MSCs have also been shown to promote peripheral nerve repair. Rat MSCs cultured in the presence of forskolin, basic fibroblast growth factor, platelet-derived growth factor, and heregulin were transplanted into the cut ends of sciatic nerves and were able to differentiate into cells expressing myelinating proteins (Dezawa et al., 2001).

10.6.7. SYSTEMIC DELIVERY Another potential application for MSCs is the systemic introduction of osteogenic cells to counteract age-related osteoporosis or as a potential cure for osteogenesis imperfecta (Caplan and Bruder, 2001). To effectively treat these diseases, it is necessary to devise a method for the systemic introduction of MSCs. The literature on the subject of circulating MSCs contains conflicting results. Early studies on the reconstitution of bone marrow after ablation indicated that the stromal component of marrow was derived from the local environment, not from a systemic source. In rabbits with ablated marrow of the femur, the stromal component of the marrow was recovered in animals that had received total body irradiation with the femur shielded, while animals that received just femoral irradiation did not recover the marrow stroma, indicating that recovery of marrow stroma is local (Maloney and Patt, 1968). In another study, male–female parabionts were formed and one partner was irradiated while the other was shielded, and the chromosomes of the CFU-f were determined 30 d later. The results showed some chimerism of the stromal cell population, at a level of approx 1 CFU per femur (Maloney et al., 1985), indicating that systemic movement of MSCs was possible, but that it occurred very rarely.

Studies in human have also yielded conflicting results. One study of the Y chromosomes in sex-mismatched chimeras in human indicated that most of the stromal cell population was of donor origin (Keating et al., 1982). Two subsequent studies, both of which used the more accurate method of *in situ* hybridisation to identify Y chromosomes, as opposed to the identification of "Y bodies" by differential fluorescent staining, showed that the stromal cell population in sex-mismatched chimeras were entirely of host origin (Simmons et al., 1987; Athanasou et al., 1990). A more recent study showed evidence of engraftment of MSCs injected into mice with the use of polymerase chain reaction (PCR) to detect a mutated human minigene for the proα1 (I) chain of procollagen I (Pereira et al., 1995). Based on the ratio of the normal gene to the mutated gene, up to 12% of the spleen cells were reported to be of donor origin by d 30, although the percentage (for all tissues assayed) decreased by 150 d postinjection. In a similar study, transgenic mice carrying a defective gene for type I collagen were shown to express donor MSC DNA in a variety of tissues by PCR analysis and by fluorescent *in situ* hybridization (Pereira et al., 1998). Another study in mouse demonstrated the expression of a marker gene (chloramphenicol acetyltransferase) linked to the osteocalcin promotor in the bones of mice injected with MSCs carrying the marker gene (Hou et al., 1999).

Further evidence of the transplantability of MSCs was provided by a study of the fate of human MSCs infused into fetal sheep, wherein human cells were detected in a variety of tissues, such as muscle, cartilage, and thymus, using *in situ* hybridization to ALU sequences and staining with human-specific antibody to β-2-microglobulin (Hou et al., 1999). These more recent studies show the potential for MSCs to engraft and express different phenotypes after systemic introduction. However, the ability to detect implanted MSCs is more likely a function of improved detection methods of the more recent reports, and not improved implantation efficiencies, which are still quite low. For clinical applications of MSCs to diseases, such as osteogenesis imperfecta, much more efficient methods of MSC engraftment will be needed to cure the disease. For some applications, such as the introduction of a genetically modified cell producing a soluble factor, a low level of engraftment may be sufficient for some therapeutic applications.

10.6.8. GENE THERAPY The potential for applications of gene therapy through the use of MSCs are enormous, based on the potential for MSCs to differentiate into multiple phenotypes and the ease with which MSCs can be harvested and expanded in culture. Limitations to the application of gene therapy with MSCs is primarily a molecular biological question of introducing and expressing the genes of interest. It is beyond the scope of this review to address all the advances and prospects for gene therapy in general. With respect to the cellular issues that relate to MSCs, there are specific advancements and limitations of MSC use in gene therapy that are relevant to this review.

The attractiveness of MSCs as targets for gene therapy lies in three primary considerations: (1) MSCs are easy to obtain by bone marrow aspiration, with little or no morbidity to the patient; (2) MSCs are a homogeneous population of cells that can be easily expanded; and (3) MSCs have the potential to differentiate into multiple cell phenotypes, which may make them useful for targeting specific tissues. Most of these advantages have already been recognized with respect to their application to tissue engineering. The fact that MSCs divide quickly has made them appealing targets for retroviral transfer, although, with advancements in the use of lentiviruses, a specific subset of retoviruses that efficiently transfect nondividing cells, the high mitotic rate of MSCs is no longer of critical importance for cell transduction. The primary hurdles in MSC applications for gene therapy with respect to the cellular issues and not the molecular biological are the reintroduction of MSCs into the host and engraftment and expression of the gene of interest.

In one of the earliest reports on the use of MSCs as targets for gene therapy, Allay et al. (1997) showed expression of bacterial *LacZ* reporter gene in human MSCs implanted in athymic mice that were expressing an osteoblastic and osteocytic pheonotype. In addition, it was demonstrated that hMSCs transfected with a gene for human interleukin-3 (IL-3) and then implanted into nonobese diabetic/severe combined immunodeficiency disease (NOD/SCID) mice showed detectable levels of secreted hIL-3 for up to 12 wk postimplantation. Evidence was also presented showing that mouse MSCs transfected with the gene for human factor IX and then implanted into syngeneic mice had detectable levels of protein in plasma (Gordon et al., 1997). Another study demonstrated the presence of human growth hormone and factor IX expression in dogs that had been injected with genetically modified autologous MSCs (Hurwitz et al., 1997). Therapeutic levels of human factor VIII were detected in NOD/SCID mice injected with hMSCs transduced with a retroviral vector containing the B domain-deleted fragment of human factor VIII (Chuah et al., 2000), although the level of expression in injected mice dropped back to basal levels within 3 wk. Baboon MSCs transduced with a gene for human erythropoietin were shown to express detectable levels of erythropoietin in recipient mice for up to 28 d, and when implanted in immunoisolation devices in allogeneic baboons, human erythropoietin was detected for up to 137 d (Bartholomew et al., 2001).

In addition to the potential application of MSCs to defects in hematopoiesis, MSCs were shown to be effective targets for the introduction of the gene for insulin-like growth factor-1 (IGF-1), via adenovirus transduction, with the specific goal of improving cartilage repair (Nixon et al., 2000). MSCs showed a 1000 multiplicities of infection at optimal viral doses and >100 ng/mL of IGF-1 was detected in the medium after 48 h. Although MSCs have an inherent ability to form bone, some laboratories have augmented this bone-forming ability through introduction of the gene for bone morphogenetic protein-2 (BMP-2). BMP-2-transduced MSCs were also used to repair fractures in nonunion animal models (Lieberman et al., 1999; Turgeman et al., 2001) and in a spine fusion model in rabbits (Riew et al., 1998).

Because MSCs have been shown to have potential for neural differentiation (Chuah et al., 2000), they were tested for their potential for gene therapy for Parkinson's disease. Rat MSCs expressing transfected genes of two enzymes (typrosine hydroxylase and guanosine 5'-triphosphate cyclohydrolase I) necessary for production of L-DOPA (3,4-dihydroxyphenylalanin) were injected into syngeneic rats. Although transgene expression ceased by 9 d, there was a physiological reduction in symptoms, as shown by a transient reduction in apomorphin-induced rotation compared to controls (Schwarz et al., 1999). Mice injected with MSCs transduced with glial cell line–derived neurotrophic factor showed greater numbers of typrosine hydroxylase–positive neurons than mice injected with nontransduced control MSCs (Park et al., 2001).

MSCs were also chosen as a target cell for introduction of the gene α-L-iduronidase, which is deficient in Hurler syndrome. Cultures of MSCs transduced with α-L-iduronidase were shown to secrete this enzyme into the extracellular space and could cross-correct this defect in neighboring cells by protein transfer (Baxter et al., 2002).

Clearly, MSCs are a viable target for the delivery of therapeutic genes for the reconstruction of different tissues and for the potential correction of different metabolic syndromes. Just as clear is the fact that the number of target genes used for MSC genetic therapy can be expected to increase dramatically in the next few years.

10.7. PROSPECTS

Applications for MSCs to correct a variety of disorders in the musculoskeletal system and for systemic delivery of genetically engineered products is still in the early stages of development. The use of MSCs for bone repair is probably the most advanced with respect to potential for clinical applications. The use of MSCs to repair tendon is continuing to be developed while the application of MSCs to repair cartilage has been fraught with difficulties, perhaps related to the propensity of MSCs to differentiate into hypertrophic cartilage, as shown by the expression of type X collagen (Yoo et al., 1998; Dennis et al., 2001). Interestingly, little effort has been made in the application of MSCs to ligament repair, which may be an even more important clinical problem than tendon repair. Applications of MSCs to cardiac repair and to nerve regeneration are still relatively recent and could develop into clinical studies in the near future. The potential for these applications is exciting, as is the potential for MSCs as delivery vehicles of therapeutic genes. Certainly the next decade will be an interesting and challenging one for the advancement of these potential applications into actual therapies.

REFERENCES

Allay, J. A., Dennis, J. E., Haynesworth, S. E., et al. (1997) LacZ and interleukin-3 expression *in vivo* after retroviral transduction of marrow-derived human osteogenic mesenchymal progenitors. *Hum. Gene Ther.* 8:1417–1427.

Athanasou, N. A., Quinn, J., Brenner, M. K., et al. (1990) Origin of bone marrow stromal cells and haemopoietic chimerism following bone marrow transplantation determined by in situ hybridization. *Br. J. Cancer* 61:385–389.

Awad, H., Butler, D., Boivin, G., et al. (1999) Autologous mesenchymal stem cell-mediated repair of tendon. *Tissue Eng.* 5:267–277.

Baikow, A. (1870) Uber transplantation von knochenmark. *Centralbl. F. D. Med. Wiss.* 8:371–373.

Barry, F., Boynton, R., Haynesworth, S., Murphy, J., and Zaia J. (1999) The monoclonal antibody SH-2, raised against human mesenchymal stem cells, recognizes an epitope on endoglin (CD105). *Biochem. Biophys. Res. Commun.* 265:134–139.

Bartholomew, A., Patil, S., Mackay, A., et al. (2001) Baboon mesenchymal stem cells can be genetically modified to secrete human erythropoietin in vivo. *Hum. Gene Ther.* 12:1527–1541.

Baxter, M. A., Wynn, R. F., Deakin, J. A., et al. (2002) Retrovirally mediated correction of bone marrow-derived mesenchymal stem cells from patients with mucopolysaccharidosis type I. *Blood* 99: 1857–1859.

Benayahu, D., Kletter, Y., Zipori, D., and Wientroub, S. (1989) Bone marrow-derived stromal cell line expressing osteoblastic phenotype in vitro and osteogenic capacity in vivo. *J. Cell. Physiol.* 140:1–7.

Benayahu, D., Gurevitz, O. A., and Shamay A. (1994) Bone-related matrix proteins expression *in vitro* and *in vivo* by marrow stromal cell line. *Tissue Cell* 26:661–666.

Bennett, J. H, Joyner, C. J., Triffitt, J. T., and Owen, M. E. (1991) Adipocytic cells cultured from marrow have osteogenic potential. *J. Cell. Sci.* 99:131–139.

Bjornson, C. R., Rietze, R. L., Reynolds, B. A., Magli, M. C., and Vescovi, A. L. (1999) Turning brain into blood: a hematopoietic fate adopted by adult neural stem cells in vivo. *Science* 283:534–537.

Brighton, C., Lorich, D., Kupcha, R., Reilly, T., Jones, A., and Woodbury, R. N. (1992) The pericyte as a possible osteoblast progenitor cell. *Clin. Orthop. Rel. Res.* 275:287–299.

Bruder, S., Kraus, K., Goldberg, V., and Kadiyala, S. (1998) The effect of implants loaded with autologous mesenchymal stem cells on the healing of canine segmental bone defects. *J. Bone Joint Surg.* 80: 985–996.

Bruder, S. P., Gazit, D., Passi-Even, L., Bab, I., and Caplan, A. I. (1990) Osteochondral differentiation and the emergence of stage-specific osteogenic cell-surface molecules by bone marrow cells in diffusion chambers. *Bone Min.* 11:141–151.

Butler, D. and Awad, H. (1999) Perspectives on cell and collagen composites for tendon repair. *Clin. Orthop. Rel. Res.* 367:S324–S332.

Caplan, A. I. (1991) Mesenchymal stem cells. *J. Orthop. Res.* 9:641–650.

Caplan, A. I. (1994) The mesengenic process. *Clin. Plastic Surg.* 21: 429–435.

Caplan, A. I. and Bruder, S. P. (2001) Mesenchymal stem cells: building blocks for molecular medicine in the 21st century. *Trends Mol. Med.* 7:259–264.

Chuah, M. K., Van, D. A., Zwinnen, H., et al. (2000) Long-term persistence of human bone marrow stromal cells transduced with factor VIII-retroviral vectors and transient production of therapeutic levels of human factor VIII in nonmyeloablated immunodeficient mice. *Hum. Gene Ther.* 11:729–738.

Colter, D. C., Class R., DiGirolami, C. M., and Prockop, D. J. (2000) Rapid expansion of recycling stem cells in cultures of plastic-adherent cells from human bone marrow. *Proc. Natl. Acad. Sci. USA* 97: 3213–3218.

Cong, Z., Jianxin, W. U., Huaizhi, F., Bing, L., and Xingdong Z. (2001) Repairing segmental bone defects with living porous ceramic cylinders: an experimental study in dog femora. *J. Biomed. Mater. Res.* 55:28–32.

Danis, A. (1960) Après une greffe de tissu squelettique ostéogène, est à partir des cellules tranplantées que se constitue l'os denouvelle formation. *Bull. Soc. Int. Chirgurie* 6:647–652.

de Bruijn, J. D., van den Brink, I., Mendes, S., Dekker, R., Bovell, Y. P., and van Blittersvinjk, C. A. (1999) Bone induction by implants coated with cultured osteogenic bone marrow cells. *Adv. Dent. Res.* 13:74–81.

Dennis, J. E. and Caplan, A. I. (1996) Differentiation potential of conditionally immortalized mesenchymal progenitor cells from adult marrow of a H-2Kb-tsA58 transgenic mouse. *J. Cell. Physiol.* 167: 523–538.

Dennis, J. E. and Charbord, P. (2002) Origin and differentiation of human and murine stroma. *Stem Cells* 20:205–214.

Dennis, J. E., Merriam, A., Awadallah, A., Yoo J.U., Johnstone, B., and Caplan, A. I. (1999) A quadripotential mesenchymal progenitor cell isolated from the marrow of an adult mouse. *J. Bone Miner. Res.* 14: 700–709.

Dennis, J. E., Carbillet, J.-P., Caplan, A., and Charbord, P. (2002) The STRO-1+ marrow cell population is multi-potential. *Cells Tissues Organs* 170:73–82.

Deryugina, E. I. and Muller-Sieburg, C. E. (1993) Stromal cells in long-term cultures: keys to the elucidation of hematopoietic development? *Crit. Rev. Immunol.* 13:115–150.

Dezawa, M., Takahashi, I., Esaki, M., Takano, M., and Sawada, H. (2001) Sciatic nerve regeneration in rats induced by transplantation of in vitro differentiated bone-marrow stromal cells. *Eur. J. Neurol.* 14:1771–1776.

Doherty, M., Ashton, B., Walsh, S., Beresford, J., Grant, M., and Canfield, A. (1998) Vascular pericytes express osteogenic potential in vitro and in vivo. *J. Bone Miner. Res.* 13:828–838.

Dorheim, M. A., Sullivan, M., Dandapani, V., et al. (1993) Osteoblastic gene expression during adipogenesis in hematopoietic supporting murine bone marrow stromal cells. *J. Cell. Physiol.* 154:317–328.

Dounchis, J., Bae, W., Chen, A., Sah, R., Coutts, R., and Amiel, D. (2000) Cartilage repair with autogenic perichondrium cell and polylactic acid grafts. *Clin. Orthop. Rel. Res.* 337:248–264.

Friedenstein, A. J. (1973) Determined and inducible osteogenic precursor cells. In: *Hard Tissue Growth, Repair and Remineralization* (Elliott K, Fitzsimons D, eds.), Elsevier-Exerpta Medic, North Holland, Amsterdam, pp. 169–185.

Friedenstein, A. J. (1980) Stromal mechanisms of bone marrow: cloning in vitro and retransplantation in vivo. *Hamatol. Bluttrans.* 25:19–29.

Friedenstein, A. J., Piatetzky-Shapiro, I. I., and Petrakova, K. V. (1966) Osteogenesis in transplants of bone marrow cells. *J. Embryol. Exp. Morph.* 16:381–390.

Friedenstein, A. J., Chailakhyan, R. K., Latsinik, N. V., Panasyuk, A. F., and Keiliss-Borok, I. V. (1974) Stromal cells responsible for transferring the microenvironment of the hemopoietic tissues. Cloning in vitro and retransplantation in vivo. *Transplantation* 17:331–340.

Friedenstein, A. J., Gorskaja, J. F., and Kulagina, N. N. (1976) Fibroblast precursors in normal and irradiated mouse hematopoietic organs. *Exp. Hematol.* 4:267–274.

Galmiche, M., Koteliansky, V., Briere, J., Herve, P., and Charbord, P. (1993) Stromal cells from human long-term marrow cultures are mesenchymal cells that differentiate following a vascular smooth muscle differentiation pathway. *Blood* 82:66–76.

Gimble, J., Youkhana, K., Hua, X., et al. (1992) Adipogenesis in a myeloid supporting bone marrow stromal cell line. *J. Cell. Biochem.* 50: 73–82.

Gimble, J. M. (1990) The function of adipocytes in the bone marrow stroma. *N. Biol.* 2:304–312.

Gimble, J. M., Robinson, C. E., Wu, X., and Kelly, K. A. (1996) The function of adipocytes in the bone marrow stroma: an update. *Bone* 19: 421–428.

Gordon, E. M., Skotzko, M., Kundu, R. K., et al. (1997) Capture and expansion of bone marrow-derived mesenchymal progenitor cells with a transforming growth factor-β1-von Willebrand's factor fusion protein for retrovirus-mediated delivery of coagulation factor IX. *Hum. Gene Ther.* 8:1385–1394.

Goujon, E. (1869) Recherches experimentales sur les proprietes physiologiques de la moelle des os. *J. L'Anat. Physiol.* 6:399–412.

Gronthos, S., Graves, S. E., Ohta, S., and Simmons, P. J. (1994) The STRO-1+ fraction of adult human bone marrow contains the osteogenic precursors. *Blood* 84:4164–4173.

Ham, A. W. and Harris, W. R. (1971) Repair and transplantation of bone. In: *The Biochemistry and Physiology of Bone: Development and Growth* (Bourne, G. H., ed.), Academic, New York, pp. 337–399.

Haynesworth, S. E., Baber, M. A., and Caplan, A. I. (1992a) Cell surface antigens on human marrow-derived mesenchymal cells are detected by monoclonal antibodies. *Bone* 13:69–80.

Haynesworth, S. E., Goshima, J., Goldberg, V. M., and Caplan, A.I. (1992b) Characterization of cells with osteogenic potential from human marrow. *Bone* 13:81–88.

Herman, I. and D'Amore, P. (1985) Microvascular pericytes contain muscle and nonmuscle actins. *J. Cell. Biol.* 101:43–52.

Hicok, K. C., Thomas, T., Gori, F., Rickard, D. J., Spelsberg, T. C., and Riggs, B. L. (1998) Development and characterization of conditionally immortalized osteoblast precursor cell lines from human bone marrow stroma. *J. Bone Miner. Res.* 13:205–217.

Hopper, A. F. and Hart, N. H. (1985) *Foundations of Animal Development*, 2 ed. Oxford University Press, New York.

Hou, Z., Nguyen, Q., Frenkel, B., et al. (1999) Osteoblast-specific gene expression after transplantation of marrow cells: Implications for skeletal gene therapy. *Proc. Natl. Acad. Sci. USA* 96:7294–7299.

Houghton, A., Oyajobi, B., Foster, G., Russell, R., and Stringer, B. (1998) Immortalization of human marrow stromal cells by retroviral transduction with a temperature sensitive oncogene: identification of bipotential precursor cells capable of directed differentiation to either an osteoblast or adipocyte phenotype. *Bone* 22:7–16.

Hurwitz, D. R., Kirchgesser, M., Merrill, W., et al. (1997) Systemic delivery of human growth hormone or human factor IX in dogs by reintroduced genetically modified autologous bone marrow stromal cells. *Hum. Gene Ther.* 8:137–156.

Jat, P. S., Noble, M. D., Ataloitis, P., et al. (1991) Direct derivation of conditionally immortal cell lines from an *H-2K^b-tsA58* transgenic mouse. *PNAS* 88:5096–5100.

Kadiyala, S., Jaiswal, N., and Bruder, S. P. (1997) Culture-expanded, bone marrow-derived mesenchymal stem cells can regenerate a critical-sized segmental bone defect. *Tissue Eng.* 3:173–185.

Keating, A., Singer, J. W., Killen, P. D., et al. (1982) Donor origin of the in vitro haematopoietic microenvironment after marrow transplantation in man. *Nature* 298:280–283.

Keith A., Sir. (1919) *Menders of the Maimed: The Anatomical & Physiological Principles Underlying the Treatment of Injuries to Muscles, Nerves, Bones & Joints*, Frowde, Hodder & Stoughton, London, UK.

Kelly, K. A., Tanaka, S., Baron, R., and Gimble, J. M. (1998) Murine bone marrow stromally derived BMS2 adipocytes support differentiation and function of osteoclast-like cells in vitro. *Endocrinology* 139:2092–2101.

Krebsbach, P. H., Kuznetsov, S. A., Satomura, K., Emmons, R. V., Rowe, D. W., and Robey, P. G. (1997) Bone formation in vivo: comparison of osteogenesis by transplanted mouse and human marrow stromal fibroblasts. *Transplantation* 63:1059–1069.

Kuznetsov, S., Mankani, M., Gronthos, S., Satomura, K., Bianco, P., and Robey, P. (2001) Circulating skeletal stem cells. *J. Cell. Biol.* 153: 1133–1140.

Kuznetsov, S. A., Krebsbach, P. H., Satomura, K., et al. (1997) Single-colony derived strains of human marrow stromal fibroblasts form bone after transplantation in vivo. *J. Bone Miner. Res.* 12:1335–1347.

Lee, J., Qu-Peterson, Z., Cao, B., et al. (2000) Clonal isolation of muscle-derived cells capable of enhancing muscle regeneration and bone healing. *J. Cell. Biol.* 150:1085–1100.

Lee, J., Musgrave, D., Pelinkovic, D., et al. (2001) Effect of bone morphogenetic protein-2-expressing muscle-derived cells on healing of critical-sized bone defects in mice. *J. Bone Joint Surg.* 83-A:1032–1039.

Lennon, D. P., Haynesworth, S. E., Bruder, S. P., Jaiswal, N., and Caplan, A. I. (1996) Human and animal mesenchymal progenitor cells from bone marrow: identification of serum for optimal selection and proliferation. *In Vitro Cell Dev. Biol.* 32:602–611.

Li, Y., Chopp, M., Chen, J., et al. (2000) Intrastriatal transplantation of bone marrow nonhematopoietic cells inproves functional recovery after stroke in adult mice. *J. Cer. Blood Flow Metab.* 20:1311–1319.

Li, Y., Chen, J., Wang, L., Lu, M., and Chopp, M. (2001a) Treatment of stroke in rat with intracarotid administration of marrow stromal cells. *Neurology* 56:1661–6672.

Li, Y., Chen, J., Wang, L., Zhang, L., Lu, M., and Chopp, M. (2001b) Intracerebral transplantation of bone marrow stromal cells in a 1-methyl-4-phenyl-1,2,3,6-tetrahydropyridine mouse model of Parkinson's disease. *Neurol. Lett.* 316:67–70.

Lieberman, J. R., Daluiski, A., Stevenson, S., et al. (1999) The effect of regional gene therapy with bone morphogenetic protein-2-producing bone-marrow cells on the repair of segmental femoral defects in rats. *J. Bone Joint Surg.* 81:905–917.

Maloney, M. and Patt, H. (1968) Origin in repopulating cells after localized bone marrow depletion. *Science* 165:71–73.

Maloney, M., Lamela, R., and Patt H. (1985) The question of bone marrow stromal fibroblast traffic. *Ann. NY Acad. Sci.* 459:190–197.

Marko, O., Cascieri, M. A., Ayad, N., Strader, C. D., and Candelore, M. R. (1995) Isolation of a preadipocyte cell line from rat bone marrow and differentiation into adipocytes. *Endocrinology* 136: 4582–4588.

Mauro, A. (1961) Satellite cells of skeletal muscle fibers. *Biophys. Biochem. Cytol.* 9:493–495.

Meyrick, B., Fujiwara, K., and Reid, L. (1981) Smooth muscle myosin in precursor and mature smooth muscle cells in normal pulmonary arteries and the effect of hypoxia. *Exp. Lung Res.* 2:303–313.

Minguell, J. J., Erices, A., and Conget, P. (2001) mesenchymal stem cells. *Exp. Biol. Med.* 226:507–520.

Mizuno, S. and Glowacki, J. (1996) Chondroinduction of human dermal fibroblasts by demineralized bone in three-dimensional culture. *Exp. Cell Res.* 227:89–97.

Nakahara, H., Goldberg, V. M., and Caplan, A. I. (1992) Culture-expanded periosteal-derived cells exhibit osteochondrogenic potential in porous calcium phosphate ceramics in vivo. *Clin. Orthop. Rel. Res.* 276:291–298.

Nixon, A. J., Brower-Toland, B. D., Bent, S. J., et al. (2000) Insulin-like growth factor-I gene therapy applications for cartilage repair. *Clin. Orthop. Rel. Res.* 379:S201–S213.

Owen, M. (1978) Histogenesis of bone cells. *Calcif. Tissue Res.* 25: 205–207.

Owen, M. (1985) Lineage of osteogenic cells and their relationship to the stromal system. In: *Bone and Mineral Research* Elsevier, Amsterdam, pp. 1–25.

Park, K., Eglitis, M., and Miuradian, M. (2001) Protection of nigral neurons by GDNF-engineered marrow cell transplantation. *Neurol. Res.* 40:315–323.

Pechak, D. G., Kujawa, M. J., and Caplan, A. I. (1986) Morphological and histochemical events during first bone formation in embryonic chick limbs. *Bone* 7:441–458.

Pereira, R., Halford, K., O'Hara, M. D., et al. (1995) Cultured adherent cells from marrow can serve as long-lasting precursor cells for bone, cartilage, and lung in irradiated mice. *Proc. Natl. Acad. Sci. USA* 92: 4857–4861.

Pereira, R., O'Hara, M. D., Laptev A., et. al. (1998) Marrow stromal cells as a source of progenitor cells for nonhematopoietic tissues in transgenic mice with a phenotype of osteogenesis imperfecta. *Proc. Natl. Acad. Sci. USA* 95:1142–1147.

Perka, C., Schultz, O., Spitzer, R. S., Lindenhayn, K., Burmester, G. R., and Sittinger, M. (2000) Segmental bone repair by tissue-engineered periosteal cell transplants with bioresorbable fleece and fibrin scaffolds in rabbits. *Biomaterials* 21:1145–1153.

Petersen, B., Bowen, W., Patrene, K., et al. (1999) Bone marrow as a potential source of hepatic oval cells. *Science* 284:1168–1170.

Petrakova, K. V., Tolmacheva, A. A., and Friedenstein, A. J. (1963) Bone formation occurring in bone marrow transplantation in diffusion chambers [in Russian]. *Bull. Exp. Biol. Med.* 56:87–91.

Pettis, G. Y., Kaban, L. B., and Glowacki, J. (1990) Tissue response to composite ceramic hydroxyapatite/demineralized bone implants. *J. Oral Maxillo. Surg.* 48:1068–1074.

Pietrangeli, C. E., Hayashi, S.-I., and Kincade, P. W. (1988) Stromal cell lines which support lymphocyte growth: characterization, sensitivity to radiation and responsiveness to growth factors. *Eur. J. Immunol.* 18:863–872.

Pittenger, M., Mackay, A., Beck, S., et al. (1999) Multilineage potential of adult human mesenchymal stem cells. *Science* 284:143–147.

Ragab, A., Lavish, S., Banks, M., Goldberg, V., and Greenfield, E. (1998) Osteoclast differentiation requires ascorbic acid. *J. Bone Miner. Res.* 13:970–977.

Rémy-Martin, J., Marandin, A., Challier, B., et al. (1999) Vascular smooth muscle differentiation of murine stroma: a sequential model. *Exp. Hematol.* 27:1782–1795.

Rhodin, J. (1968) Ultrastructure of mammalian venous capillaries, venules and small collecting veins. *J. Ultrastruct. Res.* 25:452–500.

Rickard, D. J., Kassem, M., Hefferan, T. E., Sarkar, G., Spelsberg, T. C., and Riggs, B. L. (1996) Isolation and characterization of osteoblast precursor cells from human bone marrow. *J. Bone Miner. Res.* 11: 312–324.

Riew, K. D., Wright, N. M., Cheng, S., Avioli, L. V., and Lou, J. (1998) Induction of bone formation using a recombinant adenoviral vector carrying the human BMP-2 gene in a rabbit spinal fusion model. *Calcif. Tissue Int.* 63:357–360.

Rouget, C. (1873) Memoire sur le developpement, la structure et les proprietes physiogiques des capillaires sanguins et lymphatiques. *Arch. Physiol. Norm. Pathol.* 5:603–661.

Schwarz, E., Alexander, G., Prockop, D., and Azizi, S. (1999) Multipotential marrow stromal cells transduced to produce L-DOPA: engraftment in a rat model of Parkinson disease. *Hum. Gene Ther.* 10: 2539–2549.

Simmons, P., Gronthos, S., Zannettino, A., Ohta, S., and Graves, S. (1994) Isolation, characterization and functional activity of human marrow stromal progenitors in hemopoiesis. *Prog. Clin. Biol. Res.* 389: 271–280.

Simmons, P. J. and Torok-Storb, B. (1991) Identification of stromal cell precursors in human bone marrow by a novel monoclonal antibody, STRO-1. *Blood* 78:55–62.

Simmons, P. J., Przepiorka, D., Thomas, E. D., and Tork-Storb, B. (1987) Host origin of marrow stromal cells following allogeneic bone marrow transplantation. *Nature* 328:429–432.

Solchaga, L., Dennis, J., Goldberg, V., and Caplan, A. (1999) Hyaluronic acid-based polymers as cell carriers for tissue-engineered repair of bone and cartilage. *J. Orthop. Res.* 17:205–213.

Solchaga, L., Yoo, J., Lundberg, M., et al. (2000) Hyaluronan-based polymers in the treatment of osteochondral defects. *J. Orthop. Res.* 18: 773–780.

Takahashi, H., Matsuishi, T., and Yoshizato, K. (1994) Establishment and characterization of stromal cell lines that support differentiation of murine hematopoietic blast cells into osteoclast-like cells. *In Vitro Cell Dev. Biol.* 30A:384–393.

Takahashi, S., Reddy, S. V., Dallas, M., Devlin, R., Chou, J. Y., and Roodman, G. D. (1995) Development and characterization of a human marrow stromal cell line that enhances osteoclast-like cell formation. *Endocrinology* 136:1441–1449.

Tamayo, E., Charbord, P., Li, J., and Herve, P. (1994) A quantitative assay that evaluates the capacity of human stromal cells to support granulomonopoiesis in situ. *Stem Cells* 12:304–315.

Tavassoli, M. (1974) Differential response of bone marrow and extramedullary adipose cells to starvation. *Experientia* 30:424–425.

Tavassoli, M. (1976) Marrow adipose cells. *Arch. Pathol. Lab. Med.* 100: 16–18.

Till, J. and McCulloch, E. (1980) Hemopoietic stem cell differentiation. *Biochem. Biophys. Acta* 605:431–459.

Trentin, J. J. (1976) Hemopoietic inductive microenvironments. In: *Stem Cells* (Cairnie, A. B., Lala, P. K., and Osmond, D. G., eds.), Academic, New York, pp 255-261.

Turgeman, G., Pittman, D. D., Muller, R., et al. (2001) Engineered human mesenchymal stem cells: a novel platform for skeletal cell mediated gene therapy. *J. Gene Med.* 3:240–251.

Udagawa, N., Takahashi, N., Akatsu, T., et al. (1989) The bone marrow-derived stromal cell lines MC3T3-G2/PA6 and ST2 support osteoclast-like cell differentiation in cocultures with mouse spleen cells. *Endocrinology* 125:1805–1813.

Wakitani, S., Goto, T., Pineda, S. J., et al. (1994) Mesenchymal cell-based repair of large, full-thickness defects of articular cartilage. *J. Bone Joint Surg.* 76:579–592.

Yoo, J. and Johnstone B. (1998) The role of osteochondral progenitor cells in fracture repair. *Clin. Orthop. Rel. Res.* 355:S73–S81.

Yoo, J., Barthel, T., Nishimura, K., et al. (1998) The chondrogenic potential of human bone-marrow-derived mesenchymal progenitor cells. *J. Bone Joint Surg.* 80:1745–1757.

Young, R., Butler, D., Weber, W., Caplan, A., Gordon, S., and Fink, D. (1998) Use of mesenchymal stem cells in a collagen matrix for Achilles tendon repair. *J. Orthop. Res.* 16:406–413.

Zuk, P., Zhu, M., Mizuno, H., et al. (2001) Multilineage cells from human adipose tissue: implications for cell-based therapies. *Tissue Eng.* 7:211–228.

11 Normal and Leukemic Hematopoietic Stem Cells and Lineages

ERNEST A. MCCULLOCH, MD, FRS

The spleen colony assay in which single cells injected into heavily irradiated mice produce colonies (clones) that contain cells representing all three myelopoietic lineages (plurpiotent), developed by the author and J. E. Till, was a milestone in the study of hematopoietic stem cells (HSCs) and lineages. Hematopoietic development requires interaction of stem cell factor with its receptor c-kit on stem cells. A hierarchy of stem and progenitor cells results in different lines of differentiaton by transit-amplifying cells belonging to different sublineages and responding to different growth factors. A more primitive stem cell than the spleen colony-forming cell was identified by cell sorting, and a set of cell markers may now be used to identify cells at different stages of the hematopoietic lineage. Bone marrow transplantation studies show that HSCs can give progeny that contribute to other organs, such as liver, muscle, and brain, and "stem" cells from other tissues, such as muscle and brain can produce blood precursor cells. Malignancies of the hematopoietic system are manifested by either an increased growth fraction (acute leukemias) or decreased death fraction (maturation arrest and failure of cells to die). In very simplistic terms, cure of acute leukemia requires elimination of the most primitive stem cell, which harbors the genetic change; cure of chronic leukemias requires ways to force cells to differentiate and die.

11.1. INTRODUCTION

Adult organs and tissues vary in the mechanisms used to maintain integrity and function. The neurons of the central nervous system, once established, are maintained throughout life; if lethal damage is inflicted, recovery does not occur, although adjacent cells may proliferate to form a scar. Other organs, such as liver and many endocrine glands, consist of functional cells that do not divide; if such organs are damaged, the functional cells can proliferate, restoring cellular number and function. Such organs are considered to be conditionally self-renewing. The functional cells of other systems, such as the skin, gastrointestinal tract, and hematopoiesis, are both short-lived and incapable of proliferation. These organs are maintained by a small population of cells with extensive capacity for proliferation together with the ability to have daughters either that are functional or, during a limited number of divisions, whose progeny acquire the special systems that are needed for function. Such systems are obligatory renewal systems; the cells that maintain them are stem cells. Recent evidence suggests that this standard classification of tissues may not be complete. Satellite cells in skeletal muscle, oval cells in the liver, and ependymal cells in brain have stem properties, including the capacity to give rise to hematopoietic descendants. This stem cell plasticity is discussed later.

Tumors are obligatory renewal systems; each is a clone, a cellular population derived from a single cell (Fialkow, 1976). Like normal obligatory renewal systems, many tumor cells are not capable of proliferation; the tumor expands because of the divisions in a population of malignant stem cells. These, like normal stem cells, can give rise both to new stem cells ("birth") or to daughters with limited or no growth potential ("death"). The generation of inert tumor cells limits the growth of the tumor. Without "death" tumors would double in 12–14 h; cancer would be an acute fatal disease, rather than a chronic condition, often responsive to treatment. Death in tumors may be analogous to normal differentiation, as both deprive cells of growth potential. Alternatively, death may be genetically programmed (apoptosis) (Cotter et al., 1990). Apoptosis has characteristic morphological features; these changes are often seen in tumors after injury but are less frequent in untreated controls. It seems possible that both a differentiation-like process and apoptosis occur in malignant clones.

Hematopoiesis has many features that facilitate the experimental study of obligatory renewal. The cells of the system may circu-

From: *Stem Cells Handbook*
Edited by: S. Sell © Humana Press Inc., Totowa, NJ

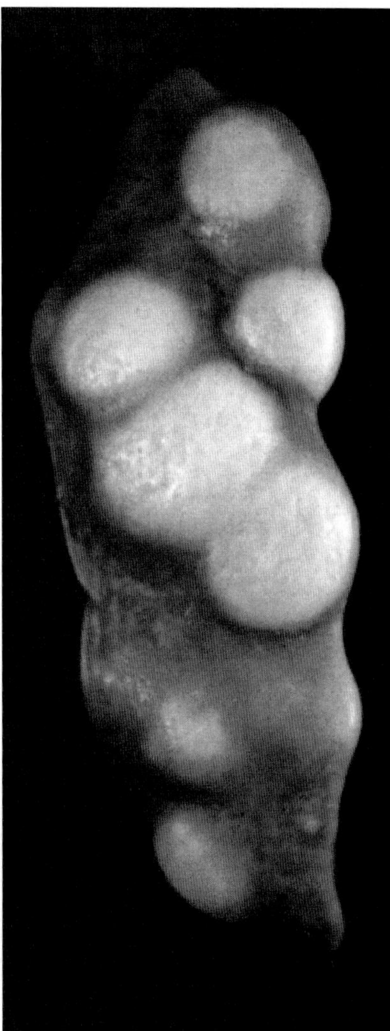

Fig. 1. Mouse spleen taken 12 d after lethal irradiation and injection of 8×10^4 normal mouse marrow cells. The spleen was fixed in Bouin's solution; such fixation enhances distinction of the colonies. (Reprinted from The origin of the cells of the blood. In: *The Physiological Basis of Medical Practice*, 8th ed. (Best, C. H. and Taylor, N. B., eds.), Williams & Wilkins, Baltimore, with permission.

late in the blood or be held loosely by nonhematopoietic stroma in bone marrow or spleen. Thus, they are obtained as cell suspensions, ready for assay. In the human, blood and marrow are routinely safely sampled for diagnosis. Usually, enough cells are taken to allow experimental studies when these are approved by the appropriate committee.

The morphological appearance of maturing and mature cells has been well established, using staining methods designed for the study of blood cells. Functional assays for stem cells and precursors have been devised and well characterized. Hematopoietic malignancies, particularly the leukemias, retain the experimental advantages of the normal. In addition, leukemias have often been the system in which therapeutic methods are developed and tested. Thus, the experimental techniques may be used for therapeutic research as well as studies of disease mechanisms.

This chapter provides an account of some of the questions that have arisen about normal and leukemic hematopoiesis. In particular, the roles of stem cells are considered. Emphasis is placed on how information about the normal has helped in the understanding of the malignant.

11.2. NORMAL STEM CELLS

11.2.1. COLONIES IN SPLEEN In the late nineteenth century, staining methods were developed that made it possible to visualize blood cell structures. Work on human cells led to a classification of blood cells based on their origin. Cells from marrow were considered to be myelopoietic. Lineages were described for red cells, granular leukocytes, and platelets. For each, nuclear and cytoplasmic structures were used to find the earliest recognizable precursor; daughter cells with increasing evidence of the characteristic of each cell type were described and named. Cells from lymphatic organs, called lymphocytes, were morphologically easily distinguished from myeloid lineages, although both circulated together in the peripheral blood. The division of hematopoiesis into myeloid and lymphoid cells was retained when experimental animals were studied, although often both had a common organ of origin in the spleen.

Morphology did not lead further up the lineage, although earlier cells, perhaps stem cells, were postulated. Lack of data often leads to hypothesis and controversy. Two schools of thought were soon established; members of one taught that a single stem cell was the origin of all the cells of the blood. Others were equally committed to the doctrine that each lineage was headed by its own stem cell. Either point of view could be supported by the leukemias, because both acute and chronic leukemias had distinct morphological and clinical characteristics that allowed them to be defined as either lymphoid or myeloid. The issue could be resolved only if new experimental evidence was found. In 1961, J. E. Till and E. A. McCulloch reported experiments in which small numbers of normal mouse marrow cells were transplanted by iv injection into heavily irradiated mice; when these recipients were killed after 10–14 d, their spleens were seen to contain nodules that could easily be counted. A linear relationship was established between the number of marrow cells injected and the number of nodules found in the recipients' spleens. When the spleens were examined histologically, the nodules were found to consist of maturing and mature blood cells. Histological examination had a further advantage: fixation in the commonly used Bouin's solution made the nodules more distinct and the counting easier (Till and McCulloch 1961) (Fig. 1).

The linear relationship between cells injected and nodules, now known to be hematopoietic colonies, made some quantitative experiments possible. For example, many marrow radiobiological parameters were determined. The value of the procedure was limited, because it was not clear whether each colony was derived from a single cell or a closely adherent cell clump. This uncertainty was acknowledged in the nomenclature used to describe the assay. The term *colony-forming unit*, or CFU, was developed; the results could then be given as CFUs per cell number injected; usually, for normal mice, the value was about 1 CFU per 10^3 marrow cells.

The spleen colony assay became much more important when it was shown that each colony was a clone, derived from a single cell. This result was achieved using irradiated marrow cells to form colonies. Sometimes surviving CFUs had unique chromosomal abnormalities that served as clonal makers. In colonies with marker chromosomes, the marker was found in all dividing cells; normal karyotypes were not seen. This was proof that all the cells in such colonies were derived from single cells that had survived irradiation but with a radiation-induced chromosomal change (Becker et al., 1963). Because colonies were readily shown to contain all three myelopoietic lineages, it followed that CFUs were pluripotent. Their identity as hematopoietic stem cells (HSCs) was settled further by the demonstration that spleen colonies contained not only differentiating cells but also cells that retained the capacity to form spleen colonies. The presence of such cells was shown by dissecting out well-separated colonies, making cell suspensions from them, and injecting them into secondary irradiated hosts; these were killed after 10 d and were often found to have spleens with colonies (Siminovitch et al., 1963). The generation of new CFUs during colony formation showed that the cells of origin of primary colonies were able to make cells like themselves, the property of self-renewal. These experiments provided an experimentally based definition of a stem cell as a cell with extensive proliferative capacity, including self-renewal and the ability to give rise to differentiated progeny. The long controversy was settled, at least for myelopoiesis; the three lineages had a common origin.

Later work showed that colony formation in the spleen was complex, consisting of at least three waves. First, about 6 d after transplantation small colonies are seen, consisting only of erythropoietic cells. The cells of origin of the early colonies are under the control of the flexed- tail locus in the mouse; they are committed to erythropoiesis and are active only in embryogenesis or marrow regeneration (Fowler et al., 1967; Gregory et al., 1975a, b). Because the progenitors of these colonies lack self-renewal, they are transient. About 4 d later, larger colonies are seen that contain the three myelopoietic lineages but have little capacity for self-renewal. These also are replaced and by d 14 larger colonies are seen derived from pluripotent, self-renewing stem cells (Magli et al., 1982).

The waves of colonies seen as hematopoietic regeneration studied in the spleens of irradiated marrow-transplanted mice show that hematopoietic lineages are not well represented by the simple linear diagrams that are the most common way in which they are described. Rather than simple and direct transitions from one stage of differentiation to the next, a three-dimensional (3D) process unfolds, as committed progenitors undergo their limited divisions and stem cells either renew themselves or enter into a state of rest.

11.2.2. STEM CELL HETEROGENEITY AND REGULATION CFU self-renewal was shown in experiments in which cells from colonies were transplanted into secondary irradiated recipients. The outcome showed much more than self-renewal; the distribution of new CFUs among spleen colonies was very skewed; many colonies had few, if any, new CFUs, while stem cells were very common in a few. Information was available about such skewed distributions and possible mechanisms by which they might be generated (Arley, 1943; Feller, 1957). An attractive possibility was a stochastic process in which "birth" (self-renewal) and "death" (differentiation) occurred at random, governed only by set probabilities. A computer simulation was constructed, and the outcome was in excellent agreement with the experimental data. It seemed likely that the marked colony-to-colony variation was the outcome of a stochastic process occurring during clonal expansion (Till et al., 1964). Similar variation in other systems has also been attributed to stochastic processes (Nakahata et al., 1982; Kobayashi and Nakahata, 1989; Hayman et al., 1993).

The lax regulation of hematopoiesis inherent in the stochastic model seemed counterintuitive to many, who saw the blood formation as capable of both sustaining normal levels of functional cells in the circulation and responding promptly to infection or after bleeding. More precise control mechanisms were evident in studies of genetically anemic mice. Mutations in two loci proved particularly informative. Mice with mutations of both alleles at the *W* (genotype W/W^v) or the Steel locus (genotype Sl/Sl^d) are phenotypically similar; they are anemic, sterile, with white skin coat and markedly increased sensitivity to the lethal effects of radiation. The spleen colony method disclosed a marked difference between them. Mice of genotype W/W^v have defective stem cells. When their marrow was injected into irradiated recipients, spleen colonies were not observed. By contrast, marrow from Sl/Sl^d mice formed colonies normally; however, when these animals were irradiated and used as recipients, their spleens and marrow spaces were found to be incapable of supporting hematopoietic growth (McCulloch et al., 1964; McCulloch et al., 1965). It was clear that the *W* locus coded for an intrinsic regulator and the Steel locus for an extrinsic regulator. It is now known that the gene at *W* is *c-kit*; the gene product is a transmembrane receptor. The product of the *Sl* is the ligand, now called Steel factor or stem cell factor (Geissler et al., 1988, 1991; Giebel and Spritz, 1991; McCulloch and Minden, 1993). Binding of kit by its ligand is an example of a very common regulatory mechanism. Many cell functions, including both mitogenic and apoptotic events, are based on genetically determined proteins, as either surface receptors or ligands; ligand binding initiates a series of signaling events, usually based on phosphorylation or dephosphorylation. Multiple proteins are involved, sometimes with overlapping function. Bray (1995) has made the suggestion that the protein-based signaling system in cells is highly analogous to the binary language used by computers.

The data support the coexistence of stochastic processes and genetically controlled protein signaling networks. The latter are complex and interactive; yet often the outcome of a signal is binary; examples are differentiation or self-renewal; apoptosis or survival. Perhaps there is a stochastic element to signaling, based on fixed probabilities of certain interactions in a 3D network.

11.2.3. HIERARCHIES AFTER SPLEEN CFU Stem cells, capable of self-renewal, may have daughters that undergo limited divisions while acquiring the cellular molecules or activities required for function. Useful clonogenic assays in culture have been developed for these lineage-committed progenitors. The first such were discovered independently in Australia by Bradley and Metcalf (1966) and Pluznick and Sachs (1966) in Israel; both techniques required that hematopoietic cell suspensions be immobilized in viscid medium (agar or methylcellulose) and supported by either "feeder cells" or medium derived from such cells. The first culture assays detected progenitors of macrophages, granulocytes, and cells with progeny of both lineages. The CFU terminology was adapted to these new assays by the addition of suffixes. The original spleen colony-forming cells became CFU-S; the progenitors of granulocytes and macrophages were called CFU-G and CFU-M, respectively, while the bipotential cells for both were CFU-GM. Similar assays were soon found for both immature and maturing erythropoietic cells(Stephenson et al., 1971; McLeod et al., 1974; Heath et al., 1976); the assay for immediate red cell precursors came first, and, using the then-growing nomenclature, these were called CFU-E. As earlier cells were recognized as clusters of

CFU-E, a new title had to be found; it was BFU-E, for burst-forming unit, erythroid. For both classes, the hormone erythropoietin was required. Assays for megakaryocytic cells were also developed (Sauvage et al., 1994; Wendling et al., 1994). The culture methods were used to seek earlier committed progenitors, with capacity for giving rise to several lineages. An example is the generation in culture of colonies with macrophages, and granulocytic, erythropoietic, and megakayoctic cells; the progenitor of such colonies was called CFU-GEMM (Fauser and Messner, 1979). Lymphoid cells also proved to be capable of colony formation in culture (Radnay et al. 1979; Izaguirre et al., 1980).

With these assays available, it was a requirement that they be ordered in their lineages. Two general methods have been used. In the first, clonogenic cells could be shown to be different if they could be separated physically. A number of techniques were used including density gradients and separation at unit gravity (Turner et al., 1967; Miller and Phillips, 1969; Worton et al., 1969). A second different, indirect, but satisfying method was based on the stochastic model. The model supposes that heterogeneity is generated during clonal expansion and increases with time of growth. Thus, if two clonal methods were detecting the same cell, the results of such assays would be perfectly correlated when assessed in individual spleen colonies, and the correlation would be retained as duration of colony -formation increased. By contrast, if two assays were detecting different cells, the results of the assays should not be perfectly correlated and the correlation should decrease with time. Moreover, correlation would be a function of the degree of difference between two cell classes, each measured by its own assay. This method was used first to compare cells forming macrophage-granulocytic colonies in culture (CFU-GM) with CFU-S. The result was the measurement of a correlation that was high but significantly different from unity. The conclusion was that CFU-S and CFU-GM were not identical but closely related in the myelopoietic lineage, with CFU-S considered the most primitive of the two, based on its self-renewal potential (Wu et al., 1968). As individual spleen colonies were characterized in detail, differing correlations were found between committed progenitors detected using culture assays, CFU-S, and differentiated cells recognized by morphology. The data on correlations were used to construct a lineage map of murine myelopoiesis (Gregory and Henkelman, 1977); this map remains the basis for the present understanding of myelopoiesis.

The colony assays for committed precursors were essential for the construction of lineages. They had equal importance for two other issues in hematopoiesis. The first was that the requirement for feeder cells or media conditioned by such cells served to identify important proteins regulating hematopoiesis and to provide bioassays for known regulators such as erythropoietin. Some colony-stimulating factors (CSF), recognized by their specificity for colony assays, were named by that relationship. For example, the factor required for granulopoiesis in culture was called G- CSF (Avalos et al., 1987). Later, CSF molecules were seen as members of the large family of regulators called interleukins; names were chosen for this relationship. An example is interleukin-3 (IL-3), a potent stimulator of early committed progenitors (Delwel et al., 1987). At the time these biological activities were observed, advances in molecular biology made it possible to clone the genes responsible for their production; with such genes, pure recombinant molecules were made (Fung et al., 1984; Gough et al., 1984; Clark and Kamen, 1987). Growth factors were soon found to be ligands; the genes encoding their receptors were cloned (Avalos et al., 1987; DiPersio et al., 1988; D'Andrea et al., 1989; Gorman et al., 1990; Itoh et al., 1990; Jones et al., 1990; Sherr, 1990). It became clear that binding of growth factors to their receptors initiated a signaling process that governed cellular responses, the same mechanism described earlier in relation to the binding of c-kit to its ligand. With these cloned molecules, responses to growth factors could be described with specificity. The outcomes were found to be complex, with many examples of synergism or other interactions (Bartelmez and Stanley, 1985; Ikebuchi et al., 1987; Donahue et al., 1988; Warren and Moore, 1988; Bot et al., 1989; Caracciolo et al., 1989; Ferrero et al. 1989). Donald Metcalf (2002), who, with his colleagues at the Walter and Eliza Hall Institute, has contributed greatly to research on growth factors, has written an informative book on the subject.

The second important consequence of culture assays for hematopoietic precursor cells was that the technology was immediately applicable to human cells. All of the culture methods for committed precursors proved effective when human marrow or peripheral blood was tested. It followed that diseases of bone marrow could be examined for the involvement of precursors. A limitation remained; there was no exact human equivalent of the spleen colony assay. Investigators sought a way to detect pluripotent stem cells in culture. One surrogate assay was that for CFU-GEMM (Fauser and Messner, 1979), as the colonies contained at least four lineages. Self-renewal was not convincingly demonstrated, because replating of multilineage colonies usually gave rise to secondary colonies with fewer lineages. Another surrogate for CFU-S was a method that gave rise to small colonies of cells without evidence of differentiation; these appeared at varying times after the cells were plated. Colonies arising late could often be replated (Nakahata and Ogawa, 1982; Suda et al., 1985; Ikebuchi et al., 1987). While useful, the method was not widely adopted; the principle contribution to knowledge came from the Ogawa laboratory where the method was developed.

11.2.4. STEM CELLS EARLIER THAN CFU-S Cells earlier than CFU-S must exist, since, eventually, all mature cells are derived from embryonic stem cells (ES). CFU-S can readily be detected in fetal liver; cells from this source have greater self-renewal capacity than CFU-S from adult tissues. Whether or not HSCs exist that are earlier than CFU-S remains controversial. The questions are posed in two related ways. First, is their an earlier cell that gives rise to stem cells committed to myelopoiesis (CFU-S) and others that give rise to both B- and T-lymphocytes? Second, repopulation of irradiated hosts, leading to their survival, is almost certainly a function of CFU-S, but, are earlier cells with greater proliferative capacity required for long-term reconstitution of hematopoiesis? The usual methods of cell separation and purification were used to study these issues. The most unequivocal results were used when the unique integration sites of retroviruses were used to mark marrow cells (Williams et al., 1984). It was then possible to follow in detail the fate of transplanted marrow as a function of time. Early polyclonal hematopoiesis in grafts was soon replaced by clonal populations. During the first year, these clones were found to be unstable; often both myeloid and lymphoid cells were present, although one lineage was usually dominant. Clones with only a single lineage were also seen, but, with time, a second lineage might appear. At much later times, only a single clone was found in each recipient and this clone always contained both lymphoid and myeloid cells

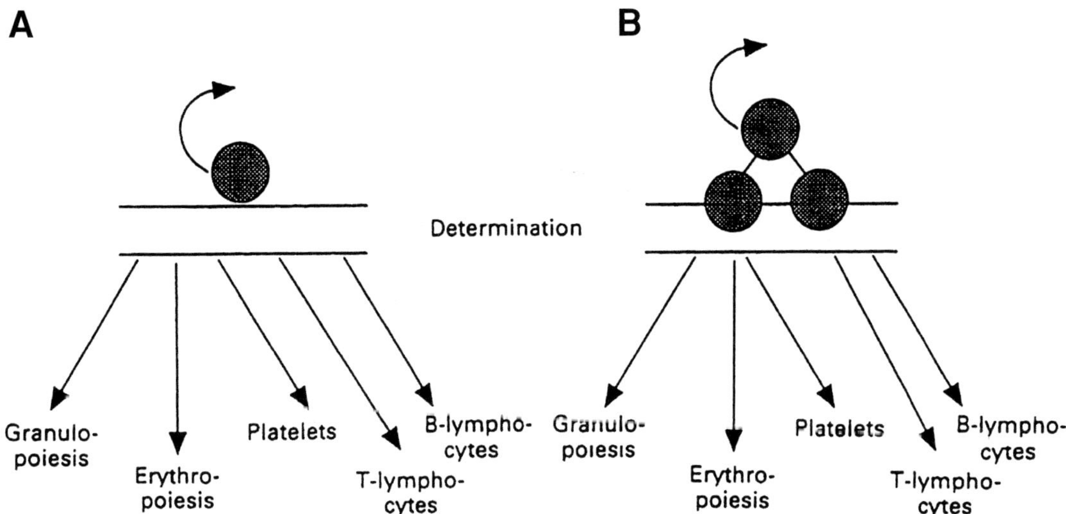

Fig. 2. Diagrams of two alternative ideas about early stem cell differentiation. **(A)** Shows all of the lineages derived from a common stem cell. **(B)** Shows a stem cell giving rise to daughter stem cells restricted to either myelopoietic or lymphopoietic differentiation. (Reprinted from McCulloch, 1993, with permission.)

(Lemischka et al., 1986; Jordan and Lemischka, 1990). This last result is proof that a stem cell exists with both lymphopoictic and myelopoietic potential. Enough clones with only a single lineage were seen at earlier times to provide support for a model that supposes a pluripotent stem gives rise to stem cells committed to either lymphoid or myeloid differentiation. Yet, these clones were sufficiently unstable, varying with time regarding their cellular composition, that the data are also compatible with the view that a single stem cell gives rise to the three lineages of myelopoiesis and two lineages of lymphopoiesis (Fig. 2).

Much of the uncertainty about the earliest HSCs arises from limitations to the colony assays. These assays are both selective and quantitative, yet they are indirect in that they measure what a cell has done, rather than what a cell might do. Determination of cell phenotypes by measuring surface or cytoplasmic markers, defined immunologically or by other criteria, when linked to functional assays, might show if there is more than one class of stem cell. Flow cytometry was sufficiently developed that it provided the means for both measuring and separating cells on the basis of markers. Weissman and colleagues pioneered this approach (Muller-Sieburg et al., 1988; Spangrude et al., 1988, 1991). They separated from marrow a tiny subpopulation with a defined phenotype (thy1^{lo}Sca^{+}); using a variety of assays, they found evidence compatible with the conclusion that lymphopoietic, myelopoietic, and spleen colony-forming capacities resided in each cell of the purified population. This tiny population, enriched 1000-fold from the starting material, meant that their functional characterization was limited; for example, capacity for spleen colony formation rested on findings from a single animal. Because of these limitations, many researchers attempted to repeat their finding and extend their phenotypic characterization, including looking for the phenotypes of human cells. Markers other than those defined immunologically were proposed. Chaudhary and Ronninson (1991) have shown that P-glycoprotein, a cell membrane protein that serves to pump poisons out of cells, was expressed on stem cells. Mulligan and colleagues have suggested that the ability to efflux Hoechst dyes may be a stem cell property (Goodell et al., 1996, 1997); dye exclusion using rhodamine also proved useful as a phenotypic marker (Bertoncello et al., 1885).

Iscove and colleagues took a different tack, one that made use of the powerful tools of molecular biology. They plated marrow in culture with appropriate growth factors and watched growth carefully; when four cell colony "starts" were seen, the cells were collected. DNA was extracted from a single cell and the others were replated. Complementary DNA was then prepared as a series of 150–400 base strands. These were amplified further. Then, an expression profile of the original single cell was obtained by probing the cDNA with cloned sequences for known significant proteins, such as growth factors. This could be compared with the phenotypes of the other members of the "start," and expression pattern of the cell used as a DNA source could be determined from their culture characteristics (Billia et al., 2001). These approaches use cell function, rather than markers, as a way of describing a phenotype.

Even with the work of many groups, there is no widely accepted stem cell phenotype; in part this may be because expression of membrane or cytoplasmic protein may vary during a cell's lifetime, without affecting its proliferative properties (Nolta and Joran, 2001).

Phenotypes could be used to find and measure cells at varying differentiation stages only when their in vivo potential was known. The spleen colony method and the culture techniques for committed progenitors were useful but did not permit a search for cells earlier than CFU-S. To seek earlier stem cells, assays were developed in which the end point was permanent reconstitution of both lymphoid and myeloid cell populations of heavily irradiated mice. A popular model developed by Harrison was based on competition between two cell populations. One of these was from the same strain as the recipient; the other was compatible with the recipient but with a marker, usually a hemoglobin with an electrophoretic mobility difference. A mixture of the two populations was injected after irradiation; the peripheral blood was then sampled as a function of time. Repopulating cells in the marked population were then measured as a function of the number of such cells that competed successfully with normal donor marrow, to remain over a long term as the peripheral blood population (Harrison and Zong, 1992). Using this method, it was possible to design experiments to determine whether cells with capacity for long-term reconstitution differed from CFU-S.

The experiments of Weissman and colleagues, described earlier, could be interpreted to mean that CFU-S had properties that would permit long-term reconstitution of all lineages. Others, using separation procedures, have reported a high degree of separation of cells detected by the two assays. An example is a separation experiment using centrifugal elutriation reported by Jones et al. (1990). They obtained two fractions, one containing cells that formed colonies in culture or in spleen; this fraction protected animals after irradiation, but spleen, marrow, and lymph nodes of long-term survivors were populated principally with host cells. The long-term survival of irradiated mice receiving their second fraction was achieved by mixing the separated cells of the two fractions, but prepared from mice that differed at a sex- linked locus. Using this marker, Jones et al. (1990) found that tissues taken from mice 60 days after irradiation contained principally cells from the fraction without progenitors. The data were consistent with the separation of CFU-S from long-term reconstituting cells, but the degree of purification was sufficiently limited to cast doubt on this conclusion.

A combination of methods, such as physical separation and sorting on the basis of phenotypes associated with long-term reconstituting ability, regularly gave extensive separation of cells with long-term reconstituting ability and CFU-S. In all of this work, fractions greatly depleted of colony-forming cells and enriched for long-term reconstituting cells regularly contained small numbers of CFU-S (Muller-Sieburg et al., 1988; Spangrude et al., 1988, 1991; Trevisan and Iscove, 1995; Trevisan et al., 1996).

Apart from these separation experiments, there is a striking difference between CFU-S and long-term repopulating cells; CFU-S are present in marrow with a frequency of about $1/10^3$, while long-term repopulating cells are 10- to a 100-fold less frequent (Harrison et al., 1988; Szilvassy et al., 1990). Among CFU-S, much heterogeneity can be shown by transplantation experiments; these indicate variation in self-renewal capacity and, hence, in the size of clones that may be generated. Furthermore, heterogeneity in renewal capacity changes as cells are passaged. For example, when secondary colonies are generated by transplantation of cells from primary colonies, new CFU-S cannot be detected by transplantation into tertiary hosts. This loss of renewal capacity as a function of growth history, called "decline," may not be differentiation, since the colonies do not lose the capacity to give rise to the three myelopoietic lineages (Siminovitch et al., 1964). It may be that the small numbers of CFU-S that are regularly found in populations greatly enriched for long-term repopulating ability may be cells that, because of a short previous proliferative history, have the extensive proliferative capacity required for long-term reconstitution (McCulloch, 1993).

11.2.5. STEM CELL PLASTICITY Description of cellular hierarchies usually implies a unidirectional differentiation from stem cells to their descendants, and that stem cells committed to an organ system, such as blood formation or intestine, can only provide new differentiated cells for that system. These long-held ideas have now been seriously challenged by experiments showing that HSCs can give rise to liver cells (Peterson et al., 1999), muscle (Ferrari et al., 1998; Jackson et al., 1999; Goodel et al., 2001), and brain (Bjornson et al., 1999; Brazelton et al., 2000); furthermore, subsets of cells from these organs can repopulate blood and marrow. The design and interpretation of the work required understanding normal regeneration in each tissue. Liver is usually quiescent, but cell proliferation starts rapidly in hepatocytes in response to injury. When hepatocyte cell division is inhibited, a special class of liver cells, called oval cells, divide; these cells may be liver stem cells that can give rise to hepatocytes. Muscle stem cells, sometimes called satellite cells, are present between muscle fibers; these both proliferate and differentiate to maintain muscle mass. Satellite cells are not present in cardiac muscle; after a myocardial infarction, the heart heals by fibrosis, rather than by muscle regeneration. Adult neurons are unable to proliferate. Brain does contain a population of glial cells that can divide, giving rise to progeny with long, branching extensions. Such cells are most readily seen in the olfactory bulb, which is commonly used for their study.

Differentiation of HSCs in other organs is studied after marrow transplantation into heavily irradiated recipients. A genetic marker, often the Y chromosome or one of its gene products, is used to identify donor cells in female recipients. Usually, it is necessary to induce a proliferative state in a potential recipient organ other than marrow. Liver may be damaged to promote oval cell proliferation. Skeletal muscle regeneration is required to show differentiation of marrow cells into muscle; cardiac muscle is prepared for engraftment by tying a coronary artery leading to infarction. For brain, the olfactory bulb is the site were marrow-derived neural cells could be most readily identified.

Showing that muscle cells could differentiate into a hematopoietic population required a modified approach. A technique originally designed to enrich for satellite cells was used to make the donor population. Heavily irradiated mice were recipients; to ensure their survival, a competitive assay was used in which differing numbers of muscle cell preparation were added to a genetically distinguishable marrow source (see above). Then, after time, cells of peripheral blood were examined to identify red cells or leukocytes derived from the muscle preparation.

The ability of neural stem cells to repopulate marrow was shown by injecting them into irradiated hosts; here, survival of the animals was evidence of marrow repopulation. Confirmation was obtained by culturing marrow from neural stem cell recipients in methylcellulose with appropriate cytokines (see above). Colonies of differentiating blood cells were readily identified with the marker of the donor neural stem cells.

These experiments, while strongly supporting the concept that stem cells of adult tissues are not entirely tissue determined, left unanswered questions. Perhaps the most serious was the possibility that reconstitution of marrow by cells from other organs might be explained by contamination of the donor population with blood-derived stem cells. Alternatively, stem cells of different tissues might coexist, explaining the results without requiring a stem cell, irreversibly committed to a certain tissue, to differentiate to the characteristic population of a different organ. Both of these objectives might be met if highly purified stem cells were available. Lagasse et al. (2000) used the flow cytometry methods described earlier to prepare populations of highly purified HSCs with the immunophenotype of c-kit^{high}Thy^{low}Lin^{-}Sca^{+}; their previous experiments were consistent with the view that such cells were 1000-fold enriched for HSCs and might be a nearly homogeneous population. These HSCs were than transplanted into mice with a hereditary lethal liver disease, fumarylacetoacetate hydrolase (FAH) deficiency; the animals can be kept alive by treatment with 2(2-nitro-4-trifluoro-methylbenzyol)-1,3-cyclohexanedione (NTBC). Lethally irradiated female FAH-deficient mice were given varying numbers of purified stem cells mixed with a constant number of protective, unfractionated marrow cells that could be readily distinguished from the purified populations. The animals were maintained with NTBC in their drinking water. Two months later, peripheral blood was tested; engraftment with purified stem cells was proportional to the number of such cells in the transplanted innoculum. Later, NTBC was withdrawn in order to provide a selection pressure in favor of hepatocytes derived from the transplant. After two such selections the animals were killed. Nodules of proliferating cells were found in their livers. Analysis showed conclusively that the nodules were derived from purified stem cells. The results are strong evidence that, at least in this system, HSCs were able to differentiate into liver cells and ameliorate the disease in FAH-deficient animals. These experiments made contamination a very unlikely explanation for the observations.

The capacity of stem cells to repopulate an organ other than that of their origin is now called stem cell plasticity.

The positive results from different laboratories make it very unlikely that a trivial explanation will be found for stem cell plasticity. It follows that at least some aspects of stem cell theory must be reexamined. For example, there is evidence that stem cells from various sources have phenotypic characteristics in common. Satellite cells from muscle and oval cells from liver have surface markers, such as c-kit and Sca, that are regularly found on marrow. Adult neural cells can be cloned in culture; when these are cocultured with embryonal cells, a population of neural- derived cells can be isolated that has the capacity to differentiate along many different lines, although not to hematopoietic cells (Clarke et al., 2000). Perhaps the most radical suggestion advanced is the idea that organ specificity is derived from stromal cells, and that functional parenchyma is provided by a common population of circulating stem cells (Lagasse et al., 2000). Much further work is required to test such hypotheses.

11.3. STEM CELLS IN LEUKEMIA

11.3.1. LEUKEMIC SPLEEN COLONIES AND A CELLULAR BASIS FOR CHEMOTHERAPY An experimental link was made between normal stem cells and their leukemic counterparts when Bruce and colleagues showed that murine leukemia cells could form colonies in the spleens of genetically identical recipients (Bruce and van der Gass, 1963). These colonies were found to contain cells that were easily shown to be very similar to the leukemic cells from which they were derived. Bruce used his method to measure dose-response curves for radiation and drugs used in cancer treatment. The discovery that was to have the greatest impact was made when both normal and leukemic spleen colony assays were used to study very rare stem cells that could not be detected by any other means. The question was, could CFUs exist in a resting state? For common cell populations, radioautography, using tritiated thymidine (^{3}HTdR), allowed cells in DNA synthesis to be recognized and, using kinetic procedures, other components of the cell cycle to be measured. Incorporation of ^{3}HTdR into DNA might be used to measure the S phase of the cell cycle of stem cells if a method other than morphological was found to detect the isotope. Becker hit on the idea of using very high specific activity ^{3}HTdR; when this was incorporated into a cell's DNA, the radiation was sufficient to kill the cell. Thus, loss of colony- formation after incubation with high specific activity ^{3}HTdR was a measure of cells in DNA synthesis. With this "suicide" technique Becker showed that normal CFUs could exist in either a state of rest (Go) or rapid proliferation (Becker et al., 1965).

Bruce knew that his lymphoma studies used cells that were growing rapidly as transplantable tumor. He saw that the difference between rapidly growing cancer cells and resting normal stem cells might be the basis of a differential effect of certain anticancer drugs on tumor and normal blood formation. In a series of experiments comparing normal and leukemic cells, he and his group showed that this was indeed the case. His work allowed him to propose a classification of chemotherapeutic drugs, based on their method of toxicity. He described how a therapeutic advantage could be obtained by using drugs that acted on DNA; such agents killed tumor cells while sparing resting populations (Bruce and Bergsagel, 1968). These observations, of great practical importance for cancer treatment, could only be made with functional assays, such as spleen colony formation, that detect minority populations of stem cells.

11.3.2. HETEROGENEITY IN ACUTE MYELOBLASTIC LEUKEMIA The heterogeneity of normal HSCs was stressed earlier. Variation is also found in the leukemias, especially in acute myeloblastic leukemia (AML). In this clonal hemopathy, great patient variation is seen in the karyotypes of the clones, in response to treatment and in almost every biological parameter tested (Fialkow et al., 1981; McCulloch et al., 1988; Second MIC Cooperative Study Group, 1988). An obvious link was present between the malignant and the normal; in early studies Moore and colleagues showed that the culture methods, including the requirement for colony-stimulating activity, developed for normal cells could be applied successfully to marrow from AML patients. Several growth patterns were seen, varying from no growth to profuse proliferation; many colonies sometimes contained cells with apparently normal differentiation but often with abnormal cellular maturation. An association was found between such growth patterns and clinical outcome (Moore et al., 1974). Although this association was not always seen, it was important to look for a mechanism generating variation.

Lan, working in Till's laboratory, used an assay in which AML cells were immobilized in methylcellulose and stimulated with factors that produce either granulopoietic or erythropoietic colonies in cultures of normal marrow. He found colonies very similar to the normal, but a marked patient-to-patient variation was seen for each colony type. In cultures from many patient samples, colonies were small in number or absent, but profuse proliferation was seen in cultures of other samples. Since each patient harbors a dominant single AML clone, the similarity between the distribution of new CFU-S among spleen colonies and committed hematopoietic progenitors among leukemic patient samples raised the possibility that the same mechanism was operating in both conditions. For the normal, the stochastic model was a good fit with the data. If stochastic processes were also operative in AML, growth patterns would not be repeated if leukemic clones were reduced in size and allowed to expand again. Alternatively, if the frequency of a given type of progenitor was an inherent property of each clone, the same growth pattern would be seen whenever clonal expansion occurred. A test to distinguish between these two outcomes was possible, since AML patients are regularly

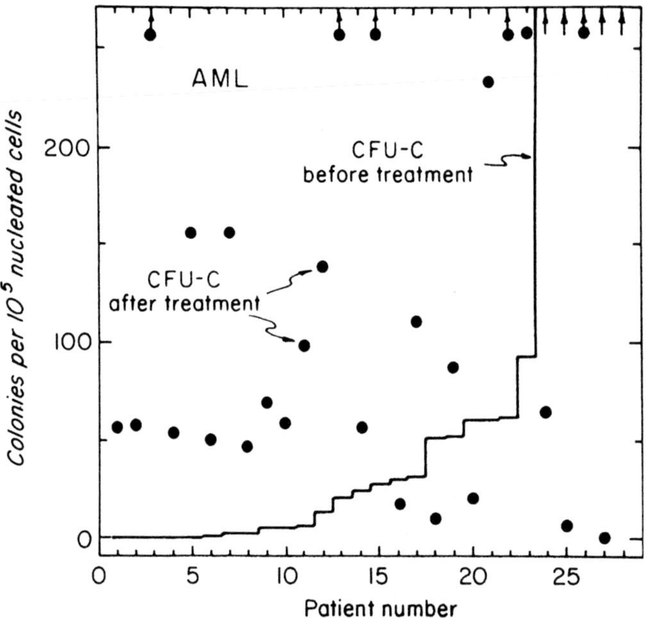

Fig. 3. Distributions of colony-forming cells (CFUs in culture of CFU-C) obtained by plating marrow from patients with AML. Columns show the values obtained at presentation. Closed circles show the values obtained after chemotherapy and clonal reexpansion. For a description, see the text. (Reprinted from Lan et al., 1978.)

treated with cytoreductive chemotherapy. When Lan was working, cure was rare; after chemotherapy, leukemic clones recovered from their drug- induced damage. A comparison was made between the culture results seen before treatment and after recovery. The result was clear: no correlation was seen between the colony numbers found at the first observation and those at the second (Lan et al., 1978) (Fig. 3). It was reasonable to think that some of the variation among patients in AML was generated by a stochastic mechanism, similar to that seen in other types of clonal expansion.

11.4. THE BLAST LINEAGE IN AML

Because representation of apparently normal progenitors in AML clones appeared to have a stochastic basis, there was little point in using assays for such cells to look for a mechanism. Rather, attention was given to the dominant and diagnostic population, consisting of cells with little or no evidence of differentiation; these have been called blast cells, and, traditionally, they were considered to be the result of blocked differentiation. Culture methods were applied to the work on AML blasts; three groups described assays for a progenitor cell capable of giving rise to cells with a morphology very similar to that of the blasts in marrow or blood used to start the cultures (Dicke et al., 1976; Buick et al., 1977; Lowenberg et al., 1980). With assays on hand, the properties of the cell of origin of blasts cells were determined. Replating experiments showed that the blast colonies contained new clonogenic cells, although these were usually few and always variable from population to population (Buick et al., 1979, 1981). It followed that clonogenic blast cells were stem cells that were capable both of self-renewal and, through a process analogous to normal differentiation, of giving rise to cells that retained the morphology of blasts but had lost proliferative potential. The idea that events might occur in blasts that were like differentiation was supported by experiments in which immunologically detected surface markers were seen to increase during growth of blasts in culture (Marie et al., 1985). The marked patient-to-patient variation in blast stem cells again raised the possibility of a stochastic basis for heterogeneity. However, studies of patients at two sequential relapses showed very similar growth patterns, a finding consistent with a clone- intrinsic basis for the frequency of clonogenic blast stem cells (McCulloch et al., 1981).

These data formed the basis of a model of AML blast cells. It was proposed that the blast population was a lineage, headed by blast-committed self-renewing stem cells that could also give rise to nonproliferative blasts. Furthermore, it was proposed that the blast lineage was derived from leukemic pluripotent stem cells, independently of the committed stem cells that differentiated into normal-appearing myelopoietic lineages. In the model, the predominant blast cell population coexisted with minor, but detectable, cells of the granulocytic, megakaryocytic, and lymphoid lineages. This latter conclusion was strengthened by the demonstration that separation of T-lymphocytes from blast cells was necessary for blast colony formation (Minden et al., 1979). The model also provided the impetus for the development of a culture system, in which blasts were cultured in suspension and clonogenic blast number was measured as a function time. Increases in blast stem cells in suspension were considered as self-renewal and could be used to measure this important stem cell function (Nara and McCulloch, 1985).

The discovery and cloning of factors that are required for colony formation by normal cells in culture provided another link between the normal and leukemic. With pure cloned growth factors, it was soon evident that colony formation by blast stem cells required the same stimulation as normal progenitors. As in other features of blast population, great heterogeneity was seen. Responses to factors varied greatly among clones; usually there was a response to stimulators that act early in the normal lineages, such as IL-3 and granulocyte macrophage-CSF, as well as factors with a later and more restrictive normal response, such as G-CSF and CSF-1. Striking synergism was often seen, especially when early-acting and late-acting factors are supplied together. In addition, the specific factor cocktail supplied influenced the balance between "birth" and "death" in the blast stem cell population (Kelleher et al., 1987; Miyauchi et al., 1987, 1988; Nara et al., 1987; Mirro et al., 1993).

11.5. SEARCH FOR ASSOCIATION BETWEEN BLAST PROPERTIES AND OUTCOME

In heterogeneous AML, response to treatment and duration of survival vary greatly from patient to patient. Much clinical research had been directed toward finding AML disease characteristics that predict response. One, karyotype, stands out in every study. Examples of favorable cytogenetics are Inv16 and t8;21, whereas trisomy 8 and deletions of chromosomes 5 and 7 are unfavorable (Estey et al., 1989; Freireich, 1990).

The great heterogeneity found in the properties of blast stem cells in culture gave room to look for associations between culture measurements and clinical outcome. Three properties seemed to be candidates: response to growth factors, capacity for self-renewal, and sensitivity to chemotherapeutic agents. Any positive results had to be compared with the strong association between karyotype and treatment outcome.

When growth factors, including Kit ligand, were found to be active in vivo as well as in culture (Ganser et al., 1989; Teshima et al., 1989; Ottmann et al., 1990; Andrews et al., 1991), it became reasonable to look for a way to use them in treatment or relate blast cell responses to outcome. Support was obtained from experiments in which blast growth was evaluated by measuring uptake of ^{3}HTdR. With this assay, about half of blast cell populations did not need factors to incorporate ^{3}HTdR into DNA; an association was found between clinical outcome and this stem cell autonomy (Lowenberg et al., 1993). By contrast, when colony formation in culture was used, very few autonomous populations were seen, and no association emerged between response and survival (McCulloch et al., 1996).

Growth factors might prove useful in treatment if combined with chemotherapy; the idea was that administration of factor might increase the number of cells in the S phase of the cycle and hence their sensitivity to cycle-dependent drugs. Further, growth factors might reduce the time in aplasia following drug treatment and thus decrease the number of early deaths from infection or bleeding. Many trials were done and conflicting results obtained (summarized in Buchner et al., 1998); a consensus is emerging that adding growth factors to treatment protocols does not significantly improve treatment outcome. The major clinical use of growth factors is to mobilize stem cells from marrow to peripheral blood, where they may be collected with ease for supportive care or transplantation (Sheridan et al., 1992).

Blast stem cell self-renewal was tested for association with clinical outcome because it was thought that high self-renewal would be associated with rapid growth and poor prognosis. Levels of colony formation by cells harvested from primary blast colonies (the secondary plating efficiency, or PE2) were measured; as expected, there was an association between high PE2 values and poor outcome (Buick et al., 1981), a result repeated in other laboratories (Marie et al., 1987; Delmer et al., 1989). Although the association was significant, it was not strong enough to be useful clinically.

Sensitivity of clonogenic blast cells to cytotoxic drugs can be measured readily in culture by determining the decrease in colony formation after exposure to increasing drug concentrations. Often the resulting dose-response curve can be fitted with a simple negative exponential, although more complex relationships, sometimes with a plateau, may sometimes be encountered (Minkin 1991; Minkin and Chow, 1995). Regardless, sensitivity may be described satisfactorily as the dose of drug required to reduce survival of colony formation to 10% of control (D10). Associations could then be sought between D10 and response to drug. Such associations were often found, although, as in the case of the self-renewal measurements,

they were never strong enough to be used to predict drug response or select chemotherapy (Dow et al., 1986; Griffin and Lowenberg, 1986; Nara et al., 1986; McCulloch, 1990).

It was disappointing when neither measurements of self renewal nor response to drug in culture were found to be useful clinically. Perhaps an explanation of the weak associations might be that the culture end points were only readily measured responses to complex regulatory networks, in which other consequences of perturbation might be more important. For example, it may be reasonable to think that the cell culture measurement of PE2 may only be an imprecise surrogate for signaling events more tightly linked to treatment outcome. In the context of measurements of drug sensitivity, there is evidence to support the idea that the culture measurements might be the outcome of complex intracellular mechanisms. The response of cancer cells to chemotherapy is not necrosis but apoptosis (Gunji et al., 1991). Drug sensitivity may be enhanced or reduced by such regulators as growth factors and retinoic acid. These changes have been shown to be responses to alterations in regulators of apoptosis such as Bcl-2. Usually, post-translational changes, such as phosphorylation, alter protein activity. The result is a change in the balance between apoptosis and cell recovery; the effect is seen as a change in drug sensitivity (Hu et al., 1996, 1998; Hedley and McCulloch, 1998). These findings are an example of how measurement of a phenotype may reflect not only the function of that phenotype but also, and perhaps more important, regulatory mechanisms that are the source of the measured attribute.

11.5.1. DIAGNOSTIC KARYOTYPES Blast stem cell attributes measured in culture never approached the prognostic power of abnormal karyotypes. Among these, two are not only important for prognosis, but also define specific leukemic diseases and determine their treatment. The Philadelphia chromosome, now known to be a reciprocal translocation between chromosomes 9 and 22 (Rowley, 1973), is diagnostic of chronic myelogenous leukemia (CML). The break point on chromosome *22* is limited to a small region, termed the *break point cluster region*, or bcr (Groffen et al., 1984). An oncogene, ABL, originally isolated from Abelson murine leukemia virus, is part of the translocation from *9* to *22*, where it encodes a fusion protein, *bcr/abl*. This protein is a tyrosine kinase, with enhanced activity. Transfer of the *bcr/abl* gene either in vivo or in vitro causes abnormal stem cell proliferation (Groffen et al., 1984; Gishizky and Witte, 1992). These findings make it highly likely that bcr/abl either is the immediate cause of CML or is very close to it.

Acute promyelocytic leukemia (APL) has a characteristic chromosomal abnormality, a *15:17* translocation (Rowley, 1973; Rowley et al., 1977). The effect of the translocation is to bring the retinoic acid receptor α (RARα) close to the gene for a protein, myl, which may be a transcription factor. The result is a novel fusion protein (de The et al., 1990), which confers on the cells a markedly increased sensitivity to all *trans* retinoic acid (ATRA); APL cells exposed to ATRA in culture lose clonogenicity and appear to differentiate to mature granulocytes (Chomienne et al., 1990). A remarkable finding was that APL patients responded to treatment with ATRA, often going into complete remission, where clinical evidence of the disease could not be found (Meng-er et al., 1988; Castaigne et al., 1990). This early evidence was conferred in large trials. Whereas often APL patients treated with ATRA alone relapsed, the addition of an anthracycline drug to the regimen often resulted in long-term survival (Fenaux et al., 1993; Warrell et al., 1994).

The idea of using ATRA may have come from the involvement of RARα in the translocation, fortified by previous studies of the induction of differentiation, using model systems (Breitman et al., 1980). A treatment for CML based on the molecular pathology required developing a drug that bound to an important component of the fusion protein, *bcr-abl*. A large program was undertaken looking for tyrosine kinase inhibitors. A compound, 2- phenylaminopyrimidine, with some capacity to inhibit protein kinase C was chosen for further work, analyzing structure/function relationships between chemically related compounds. Eventually one, STI571, was found that was an active tyrosine kinase inhibitor; culture experiments showed that STI571 was very active in decreasing colony formation by CML cells with the 9:21 translocation while normal cells were not affected. Its mechanism appeared to be binding to the adenosine triphosphatase (ATP)–binding site on the bcr-abl protein. This mechanism was confirmed by 3D structural studies that showed the position of ATP-binding cleft and the localization of STI571 to it (Druker and Lydon, 2000).

The preclinical development of STI571 quickly led to phase I/II clinical trials, based on patients who had failed while on the standard treatment with interferon. In this early work, many of these end-stage patients entered remission, and, in some, the Philadelphia chromosome–positive cell population was eliminated. Quickly, large trials were organized; these demonstrated the value of the drug in chronic phase, accelerate phase, and blast phase of CML. In contrast to cytotoxic chemotherapy, very few side effects were encountered, and these were not serious enough to require that treatment be stopped. With time, relapses were seen; this experience, like that with ATRA in the treatment of APL, makes it reasonable that STI567 will be more effective if combined with other antileukemic drugs such as cytosine arabinoside or an anthracyline. Tissue culture experiments with these drugs have already shown both additive and synergistic effects (Druker et al., 2001).

The genetic lesion in both APL and CML was found by looking at dividing cells. Nonetheless, the abnormal chromosomal translocations in both diseases are clonal markers. These must, therefore, be present in the stem cells that maintain the abnormal clones. Since treatment often eliminates all cells with the marker, if also follows that ATRA and STI567 can destroy stem cells.

11.6. CONCLUSION

This chapter describes the development of knowledge about normal and leukemic HSCs. Fundamental stem cell properties, self-renewal, and differentiation are maintained after malignant transformation. It follows that concepts in the normal may often be applied in the leukemic. Examples are the skewed distributions that are generated by stochastic processes and the sensitivity of leukemic blast cells to myelopoietic regulators of normal progenitors.

As knowledge grew, social, ethical, and economic issues arose that were only marginally connected with the scientific and medical motivations of investigators. The finding of stem cell plasticity impinged directly on plans to use ES cells in therapy. ES cells are considered in detail elsewhere in this volume. Here, it is sufficient to note that their great capacity for proliferation and ability to differentiate into the parenchyma of many organs have led to the hope that transplantation of such cells might improve the functions of damaged organs. Neurological diseases, such as Alzheimer and Parkinson diseases, are mentioned as targets for stem cell therapy, along with many musculoskeletal disorders, such as rheumatoid arthritis. The ethical problem arises from the source of such cells—human embryos. Many people find it repugnant to sacrifice embryos to obtain cells for therapy. Governments have been moved by such concerns to consider the regulation of ES research. The plasticity of adult stem cells contributes to the debate since it might be argued that these cells could be used in therapy without conflict with ethical or religious beliefs. The popular press covers these issues, suggesting, characteristically, that stem cell therapy could be started almost at once. Scientists understand the need for much better information and the long research process that will be needed (Civin, 2000, 2001). Regardless, the controversy helps to ensure that stem cell plasticity will be given a high priority. The scientific ideas about developmental biology that arise from the present knowledge of stem cell plasticity certainly need to be tested, perhaps modified, and then included in the basic knowledge bank that is the resource required for further practical developments.

The early success in the treatment of CML with STI571 proved a proof of principle. Investigators have long argued, particularly in grant applications, that knowledge about the basic mechanisms of disease could be applied in therapy; the development of STI571 showed convincingly that understanding the proximate molecular lesion in a disease could be used in successful drug development. The striking clinical finding in trials of STI571 led to its approval in record time. It is now marked by Novartis under the trade name Gleevec; wisely, the company's literature requires that Gleevec be used only by fully trained medical oncologists.

It is most unlikely that the success of STI671 will be an isolated example. Indeed, many drug companies are now using the plan that proved successful to search for other drugs (Druker and Lydon, 2000). The problem faced by the field is not only scientific; it is also economic. Drugs based on molecular pathology are likely to be specific. Because most cancers have either unique or multiple genetic abnormalities, drug development may yield many compounds that are useful only to a small number of patients, or, perhaps, even to a single patient. Under the present system, such drugs will not be economical and there will be little motivation to find them. The challenge, then, is not only to use molecular pathology for drug development but also to make the process simple and quick, so that it can be applied, where necessary, to individual patients or very small groups. Fortunately, history would support the belief that once a principle is established, research and development will ensure that its results will become available to those who need them.

REFERENCES

Andrews, R. G., Knitter, G. H., Bartelmez, S. H., et al. (1991) Recombinant human stem cell factor, a c-kit ligand, stimulates hematopoiesis in primates. *Blood* 78:1975–1980.

Arley, N. (1943) *Stochastic Processes and Their Application to the Theory of Cosmic Radiation*, Wiley, New York.

Avalos, B. R., Hedzat, C., Baldwin, G. C., Golde, D. W., Gason, J. C., and DiPersio, J. F. (1987) Biological activities of human G-CSF and characterization of the human G-CSF receptor. *Blood* 70 (Suppl. 1):165a.

Bartelmez, S. H. and Stanley, E. R. (1985) Synergism between hemopoietic growth factors (HGFs) detected by their effects on cells bearing receptors for a lineage specific HGF: assay of hemopoietin-1. *J. Cell Physiol.* 122:370–378.

Becker, A., McCulloch, E., Siminovitch, L., and Till, J. (1965) The effect of differing demands for blood cell production on DNA synthesis of by hemopoietic colony forming cells of mice. *Blood* 26: 296–308.

Becker, A. J., McCulloch, E. A., and Till, J. E. (1963) Cytological demonstration of the clonal nature of spleen colonies derived from transplanted mouse marrow cells. *Nature* 197:452.

Bertoncello, I., Hodgson, G. S., and Bradley, T. R. (1985) Multiparameter analysis of transplantable hemopoietic stem cells: I. The separation and enrichment of stem cells homing to marrow and spleen on the basis of rhodamine-123 fluorescence. *Exp. Hematol.* 13:999.

Billia, F., Barbara, M., McEwan, J., Trevisan, M., and Iscove, N. N. (2001) Resolution of pluripotential intermediates in murine hematopoietic differentiation by global complementary DNA amplification from single cells: confirmation of assignments by expression profiling of receptor transcripts. *Blood* 97:2257–2268.

Bjornson, C. R. R., Rietze, R. L., Reynolds, B. A. Magli, M. C., and Vescovi, A. L. (1999) Turning brain into blood: a hematopoietic fate adopted by neural stem cells in vivo. *Science* 283:534–536.

Bot, F. J., Eijk, L. V., Broeders, L., Aarden, L. A. and Lowenberg, B. (1989) Interleukin-6 synergizes with M-CSF in the formation of macrophage colonies from purified human marrow progenitor cells. *Blood* 73:435–437.

Bradley, T. R. and Metcalf, D. (1966) The growth of mouse bone marrow cells in vitro. *Aust. J. Exp. Biol. Med. Sci.* 44:287–299.

Bray, D. (1995) Protein molecules as computational elements in living cells. *Nature* 376:307–312.

Brazelton, T. R., Rossi, F. M. V., Kekshet, G. I., and Blau, H. M. (2000) From marrow to brain: expression of neuronal phenotypes in adult mice. *Science* 290:1775–1779.

Breitman, T. R., Selonick, S. E., and Collins, S. J. (1980) Induction of differentiation of the human promyelocytic leukemia cell line (HL-60) by retinoic acid. *Proc. Natl. Acad. Sci. USA* 77:2936–2940.

Bruce, W. R. and Bergsagel, D. E. (1968) On the application of results from a model system to the treatment of leukemia in man. *Cancer Res.* 27:501–511.

Bruce, W. R. and van der Gass, H. (1963) Quantitative assay of the number of murine lymphoma cells capable of proliferation *in vivo*. *Nature* 199:79–80.

Buchner, T., Hiddeman, W., Zuhlsddorf, M., et al. (1998) Hematopoietic growth factors: supportive and proiming effects in AML. In: *Acute Leukemias VII: Experimental Approaches and Novel Therapies*, Springer-Verlag, Münster, Germany.

Buick, R. N., Till, J. E., and McCulloch, E. A. (1977) Colony assay for proliferative blast cells circulating in myeloblastic leukaemia. *Lancet* 1:862–863.

Buick, R. N., Minden, M. D., and McCulloch, E. A. (1979) Self-renewal in culture of proliferative blast progenitor cells in acute myeloblastic leukemia. *Blood* 54:95–104.

Buick, R. N., Chang, L. J.-W., Messner, H. A., Curtis, J. E., and McCulloch, E. A. (1981) Self renewal capacity of leukemic blast progenitor cells. *Cancer Res.* 41:4849–4852.

Caracciolo, D., Clark, S. C., and Rovera, G. (1989) Human interleukin-6 supports granulocytic differentiation of hemopoietic progenitor cells and acts synergistically with GM-CSF. *Blood* 73:666–670.

Castaigne, S., Chomienne, C., Daniel, M. T., et al. (1990) All-trans retinoic acid as a differentiation therapy for acute promyelocytic leukemia I. Clinical results. *Blood* 76:1704–1709.

Chaudhary, P. and Roninson, I. (1991) Expression and activity of p-glycoprotein, a multidrug efflux pump, in human hematopoietic stem cells. *Cell* 66:85–94.

Chomienne, C., Ballerini, P., Balitrand, N., et. al. (1990) All-trans retinoic acid as a differentiation therapy for acute promyelocytic leukemia. II. In vitro studies: structure-funcion relationship. *Blood* 76:1710–1717.

Civin, C. I. (2000) Pluripotent stem cells: science fiction poses no immediate dangers. *Stem Cells* 6:iv–v.

Civin, C. I. (2001) Editorial: stem cell research: back to the future. *Stem Cells* 19:356–357.

Clark, S. C. and Kamen, R. (1987) The human hematopoietic colony-stimulating factors. *Science* 236:1229–1237.

Clarke, D. L., Johansson, C. B., Johannes, W., et al. (2000) Generalized potential of adult neural stem cells. *Science* 2888:1660–1663.

Cotter, T. G., Lennon, S. V., Glynn, J. G. and Martin, S. J. (1990) Cell death via apoptosis and its relationship to growth: development and differentiation of both tumor and normal cells. *Anticancer Res.* 10: 1153–1160.

D'Andrea, A. D., Lodish, H. F., and Wong, G. G. (1989) Expression cloning of the murine erythropoietin receptor. *Cell* 57:277–285.

Delmer, A., Marie, J.-P., Thevenin, D., Cadiou, M., Viguie, F. and Zittoun, R. (1989) Multivariate analysis of prognostic factors in acute myeloid leukemia: value of clonogenic leukemic cell properties. *J. Clin. Oncol.* 7:738–746.

Delwel, R., Dorssers, L., Touw, I., Wagemaker, E. R. and Lowenberg, B. (1987) Human recombinant multilineage colony stimulating factor (interleukin-3): stimulator of acute myelocytic leukemia progenitor cells in vitro. *Blood* 70:333–336.

de The, H., Chomienne, C., Lanotte, M., Degos, L., and Dejean, D. (1990) The t(15:17) translocation of acute promyelocytic leukemia fuses the retinoic acid receptor a gene to a novel transcribed locus. *Nature* 347: 558–561.

Dicke, K. A., Spitzer, G. and Ahearn, M. J. (1976) Colony-formation in vitro in acute myelogenous leukemia with phytohaemagglutinin as stimulating factor. *Nature* 259:129–130.

DiPersio, J., Billing, P., Kaufman, S., Eghtesady, P., Williams, R. E., and Gasson, J. C. (1988) Characterization of the human granulocyte-macrophage colony-stimulating factor receptor. *J. Biol. Chem.* 263: 1834–1841.

Donahue, R. E., Metzger, M., Rock, B., et al. (1988) Human Il-3 and GM-CSF act synergistically in stimulating hematopoiesis in primates. *Science* 241:1820–1823.

Dow, L. W., Dahl, G. V., Kalwinsky, D. K., Mirro, J., Nash, M. B., and Roberson, P. K. (1986) Correlation of drug sensitivity in vitro with clinical responses in childhood acute myeloid leukemia. *Blood* 68: 400–405.

Druker, B. J. and Lydon, M. B. (2000) Lessons learned from the development of an abl tyrosine kinase inhibitor fror chronic myelogenous leukemia. *J. Clin. Invest.* 105:3–7.

Druker, B. J., Talpaz, M., Resta, D. J., et al. (2001) Efficacy and safety of a specific inhibitor of the bcr-abl tyrosine kinase in chronic myeloid leukemia. *N. Engl. J. Med.* 344:1031–1037.

Estey, E., Smith, T. L., Keating, N. J., McCredie, K. B., Gehan, E. A. and Freireich, E. J. (1989) Prediction of survival during induction therapy in patients with newly diagnosed acute myeloblastic leukemia. *Leukemia* 3:257–263.

Fauser, A. A. and Messner, H. A. (1979) Identification of megakaryocytes, macrophages and eosinophils in colonies of human bone marrow containing neutrophilic granulocytes and erythroblasts. *Blood* 53:1023–1027.

Feller, W. (1957) *An Introduction to Probability Theory and Its Applications*, Wiley, New York.

Fenaux, P., M. LeDeley, C., Castaigne, S., et al. (1993) Effect of all trans retinoic acid in newly diagnosed acute promyelocyctic leukemia: results of a multicentre randomized trial. *Blood* 82:3241–3249.

Ferrari, G., Cusella-De Angelis, G., Coleta, M., et al. (1998) Muscle regeneration by bone marrow-derived myogenic progenitors. *Science* 279:1528–1530.

Ferrero, D., Tarella, C., Badoni, R., et al. (1989) Granulocyte-macrophage colony-stimulating factor requires interaction with accessory cells or granulocyte-colony stimulating factor for full stimulation of human myeloid progenitors. *Blood* 73:402–405.

Fialkow, P. J. (1976) Clonal origin of human tumors. *Biochim. Biophys. Acta* 456:283–321.

Fialkow, P. J., Singer, J. W., Adamson, J. W., et al. (1981) Acute nonlymphocytic leukemia: heterogeneity of stem cell origin. *Blood* 57:1068–1073.

Fowler, J. H., Till, J. E., McCulloch, E. A. and Siminovitch, L. (1967) The cellular basis for the defect in haemopoiesis in flexed-tailed mice. II. The specificity of the defect for erythropoiesis. *Br. J. Haematol.* 13:256–264.

Freireich, E. J. (1990) The impact of cytogenetics and molecular genetics on diagnosis and treatment. In: *New Approaches to the Treatment of Leukemia* (Freirecih, E. J. ed.), Springer-Verlag, New York.

Fung, M. C., Hapel, A. J., Ymer, S., et al. (1984) Molecular cloning of cDNA for murine interleukin-3. *Nature* 307:233–237.

Ganser, A., Volkers, B., Greher, J., et al. (1989) Recombinant human granulocyte-macrophage colony-stimulating factor in patients with myelodysplastic syndromes—a phase I/II trial. *Blood* 73:31–37.

Geissler, E. N., Ryan, M. A., and Housman, D. E. (1988) The dominant white-spotting (W) locus of the mouse encodes the c-kit proto-oncogene. *Cell* 55:185–192.

Geissler, E. N., Liao, M., Brook, J. D., et al. (1991) Stem cell factor (SCF), a novel hematopoietic growth factor and ligand for c-kit tyrosine kinase receptor, maps on human chromosome 12 between 12q14.3 and 12qter. *Somatic Cell Mol. Genet.* 17:207–214.

Giebel, L. B. and Spritz, R. A. (1991) Mutation of the KIT (mast/stem cell growth factor receptor) protooncogene in human piebaldism. *Proc. Natl. Academ. Sci. USA* 88:8696–8699.

Gishizky, M. and Witte, O. N. (1992) Initiation of deregulated growth of multipotent progenitor cells by bcr-abl in vitro. *Science* 256:836–839.

Goodell, M. A., Brose, K., Paradis, G., Conner, A. S., and Mulligan, R. C. (1996) Isolation and functional properties of murine hematopoietic stem cells that are replicating *in vivo*. *J. Exp. Med.* 183:1797–1806.

Goodell, M. A., Rosenzweig, M., Kim, H., et al. (1997) Dye efflux studies suggest that hematopoietic stem cells expressing low or undetectable levels of CD34 antigen exist in many species. *Nat. Med.* 3:1337–1345.

Goodell, M. A., Jackson, K. A., Majka, S. M., et al. (2001) Stem cell plasticity in muscle and bone marrow. *Ann. NY Acad. Sci.* 938:208–218.

Gorman, D. M., Itoh, N., Kitamura, T., et al. (1990) Cloning and expression of a gene encoding an interleukin 3 receptor-like protein: identification of another member of the cytokine receptor gene family. *Proc. Natl. Academ. Sci. USA* 87:5459–5463.

Gough, N. M., Gough, J., Metcalf, D., et al. (1984) Molecular cloning of cDNA encoding a murine hematopoietic growth regulator, granulocyte-macrophage colony stimulating factor. *Nature* 309:763–767.

Gregory, C. J. and Henkelman, R. M. (1977) Relationships between early hemopoietic progenitor cells determined by correlation analysis of their numbers in individual spleen colonies. In: *Experimental Hematology Today*, Springer-Verlag, New York, pp. 93–101.

Gregory, C. J., McCulloch, E. A. and Till, J. E. (1975a) The cellular basis for the defect in flexed- tailed mice. III. Restriction of the defect to erythropoietic progenitors capable of transient colony formation in vivo. *Br. J. Haematol.* 30:401–410.

Gregory, C. J., McCulloch, E. A., and Till, J. E. (1975b) Transient erythropoietic spleen colonies: Effects of erythropoietin in normal and genetically anemic W/W^v mice. *J. Cell. Physiol.* 86:1–8.

Griffin, J. D. and Lowenberg, B. (1986) Clonogenic cells in acute myeloblastic leukemia. *Blood* 68:1185–1195.

Groffen, J., Stephenson, J. R., Heistrkamp, N., de Klein, A., Bartram, C. R. and Grosveld, G. (1984) Philadephia chromosome breakpoints are clustered within a limited region, bcr, on chromosome 22. *Cell* 36:93–99.

Gunji, H., Kharbanda, S., and Kufe, D. (1991) Induction of internucleosomal DNA fragmentation in human myeloid leukemia cells by 1-β arabinofuranosylcytosine. *Cancer Res.* 51:741–743.

Harrison, D. E. and Zong, R.-K. (1992) The same exhaustible multilineage precursor produces both myeloid and lymphoid cells as early as 3–4 weeks after marrow transplantation. *Proc. Natl. Acad. Sci. USA* 89:10,134–10,138.

Harrison, D. E., Astle, C. M., and Lerner, C. (1988) Number and continuous proliferative pattern of transplanted primitive immunohematopoietic stem cells. *Proc. Natl. Acad. Sci. USA* 85:822–826.

Hayman, M. J., Meyer, S., Martin, F., Steinlein, P., and Beug, M. (1993) Self-renewal and differentiation of normal avian erythroid progenitor cells: regulatory roles of the TGFalpha/cErbB and SCF/cKit receptors. *Cell* 74:157–169.

Heath, D. S., Axelrad, A. A., and Mcleod, D. L. (1976) Separation of erythropietin-responsive progenitors BFU-E and CFU-E in mouse bone marrow by unit gravity sedimentation. *Blood* 47:777–792.

Hedley, D. W. and McCulloch, E. A. (1998) Physiological events during ara-C toxicity mapped using multiparametric flow cytometry. In: *Acute Leukemias VII: Experimental Approaches and Novel Therapies* Springer-Verlag, Münster, Germany.

Hu, Z.-B., Minden, M. D. and McCulloch, E. A. (1996) Regulation of synthesis of bcl-2 protein by growth factors. *Leukemia* 10:1925–1929.

Hu, Z.-B., M. D. Minden and E. A. McCulloch (1998) Phosphorylation of bcl-2 after exposure of human leukemic cells to retinoic acid. *Blood* 92:1768-1775.

Ikebuchi, K., Wong, G. G., Clark, S. C., Ihle, J. N., Hirai, Y., and Ogawa, M. (1987) Interleukin 6 enhancement of interleukin 3-dependent proliferation of multipotential hemopoietic progenitors. *Proc. Natl. Acad. Sci. USA* 84:9035–9039.

Itoh, N., Schreurs, J., Gorman, D. M., et al. (1990) Cloning of an interleukin-3 receptor gene: a member of a distinct receptor gene family. *Science* 247:324–327.

Izaguirre, C. A., Minden, M. D., Howatson, A. F., and McCulloch, E. A. (1980) Colony formation by normal and malignant B Lymphocytes. *Br. J. Cancer* 42:430–437.

Jackson, K. A., Mi, T., and Goodel, M. A. (1999) Hematopoietic potential of stem cells isolated from murine skeletal muscle. *Proc. Natl. Acad. Sci. USA* 96:14,482–14,486.

Jones, R. I., Wagner, J. E., Celano, P., Zicha, M. S., and Sharkis, S. J. (1990) Separation of of pluripotent haematopoietic stem cells from spleen colony- forming cells. *Nature* 347:188–189.

Jones, S. S., D'Andrea, A. D., Haines, L. L., and Wong, G. G. (1990) Human erythropoietin receptor: cloning, expression and biologic characterization. *Blood* 76:1–35.

Jordan, C. T. and Lemischka, I. R. (1990) Clonal and systemic analysis of long term hemopoiesis in the mouse. *Genes Dev.* 4:220–232.

Kelleher, C., Miyauchi, J., Wong, G., Clark, S., Minden, M. D., and McCulloch, E. A. (1987) Synergism between recombinant growth factors, GM-CSF and G-CSF, acting on the blast cells of acute myeloblastic leukemia. *Blood* 69:1498–1503.

Kobayashi, T. and Nakahata, T. (1989) Stochastic model of mast cell proliferation in culture of murine peritoneal cells. *J. Cell. Physiol.* 138:24–28.

Lagasse, E., Connors, H., Al-Dhalimy, M., et al. (2000) Purified hematopoietic stem cells can differentiate into hepatocyes in vivo. *Nat. Med.* 6:1229–1234.

Lan, S., McCulloch, E. A., and Till, J. E. (1978) Cytodifferentiation in the acute myeloblastic leukemias of man. *J. Natl. Cancer Inst.* 60: 265–269.

Lemischka, I. R., Raulet, D., and Mulligan, R. C. (1986) Developmental potential and dynamic behaviour of hematopoietic stem cells. *Cell* 45: 917–927.

Lowenberg, B., Swart, K., and Hagemeijer, A. (1980) PHA-induced colony-formation in acute nonlymphocytic and chronic myeloid leukemia. *Leukemia Res.* 4:143–149.

Lowenberg, B., van Putten, W. L. J., Touw, I. P., Delwel, R., and Santini, V. (1993) Autonomous proliferation of leukemic cells in vitro as a determinant of prognosis in adult acute myeloid leukemia. *N. Engl. J. Med.* 328:614–619.

Magli, M. C., Iscove, N. N., and Odartchenko, N. (1982) Transient nature of early haematopoietic spleen colonies. *Nature* 295:527–529.

Marie, J. P., Izaguirre, C. A., Civin, C. I., Mirro, J., and McCulloch, E. A. (1985) Granulopoietic differentiation in AML blasts in culture. *Blood* 58:670–674.

Marie, J.-P., Zittoun, R., Deimer, A., Thevenin, D., and Suberville, A.-M. (1987) Prognostic value of clonogenc assay for induction and duration of complete remission in acute myelogenous leukemia. *Leukemia* 1:121–136.

McCulloch, E. A. (1990) Biological characteristics of acute myeloblastic leukemia contributing to management strategy. In: *New Approaches to the Treatment of Leukemia: A Monograph of the European School of Oncology* (Freireich, E. J. ed.), Springer-Verlag, Berlin, pp. 87–116.

McCulloch, E. A. (1993) Stem cell renewal and determination during clonal expansion in normal and leukemic haemopoiesis. The 1993 Lewis Schiffer Memorial lecture to the Cell Kinetics Society. *Cell Prolif.* 26:399–425.

McCulloch, E. A., Siminovitch, L., and Till, J. E. (1964) Spleen colony formation in anemic mice of genotype W/W^v. *Science* 144:844–846.

McCulloch, E. A., Russell, E. S., Siminovitch, L., Till, J. E., and Bernstein, S. E. (1965) The cellular basis of the genetically determined hemopoietic defect in anemic mice of genotype Sl/Sl^d. *Blood* 26:399–410.

McCulloch, E. A., Buick, R. N., Curtis, J. E., Messner, H. A., and Senn, J. S. (1981) The heritable nature of clonal characteristics in acute myeloblastic leukemia. *Blood* 58:105–109.

McCulloch, E. A., Kelleher, C. A., Miyauchi, J., et al. (1988) Heterogeneity in acute myeloblastic leukemia. *Leukemia Suppl.* 2:38S– 49S.

McCulloch, E. A. and Minden, M. D. (1993) *The Cell Surface Receptor Encoded by the Proto- Oncogene KIT and its Ligand*, Klewer Academic, Boston, MA.

McCulloch, E. A., Minkin, S., Curtis, J. E., and Minden, M. D. (1996) Response of the blast stem cells of acute myelogenous leukemia to G-CSF, GM-CSF or the ligand for c-KIT, alone or in combination. *Hematopathol. Mol. Hematol.* 10(3):111–122.

McLeod, D. L., Shreeve, M. M., and Axelrad, A. A. (1974) Improved plasma culture system for production of erythrocytic colonies in vitro: quantitative assay method for CFU-E. *Blood* 44:517–534.

Meng-er, H., Yu-chen, Y., Shu-rong, C., et al. (1988) Use of all-trans retinoic acid in the treatment of acute promyelocytic leukemia. *Blood* 72:567–572.

Metcalf, D. (2000) *Summon up the Blood: In Dogged Pursuit of Blood Cell Regulators*, Alphamed, Miamisburg.

Miller, R. G. and Phillips, R. A. (1969) Separation of cells by velocity sedimentation. *J. Cell. Physiol.* 73:191–202.

Minden, M. D., Buick, R. N., and McCulloch, E. A. (1979) Separation of blast cell and T- lymphocyte progenitors in the blood of patients with acute myeloblastic leukemia. *Blood* 54:186–195.

Minkin, S. (1991) A statistical model for in-vitro assessment of patient sensitivity to cytotoxic drugs. *Biometrics* 47:1581–1591.

Minkin, S. and Chow, A. (1995) Assessing sensitivity in suspension to cytosine arabinoside: Statistical analysis, associations with clinical outcome and experimental design. *Stat. Med.* 14:1081–1095

Mirro, J., Hurwitz, C. A., Behm, F. G., et al. (1993) Effects of recombinant human hematopoietic growth factors on leukemic blasts from children with acute myeloblastic or lymphoblastic leukemia. *Leukemia* 7:1026–1033.

Miyauchi, J., Kelleher, C., Yang, Y.-C., et al. (1987) The effects of three recombinant growth factors, IL-3, GM-CSF and G- CSF, on the blast cells of acute myeloblastic leukemia maintained in short term suspension culture. *Blood* 70:657–663.

Miyauchi, J., Kelleher, C. A., Wong, G. G., et al. (1988) The effects of combinations of the recombinant growth factors GM-CSF, G-CSF, IL-3 and CSF-1 on leukemic blast cells in suspension culture. *Leukemia* 2:382–387.

Moore, M. A. S., Spitzer, G., Williams, N., Metcalf, D., and Buckely, J. (1974) Agar culture studies in 127 cases of untreated acute leukemia: the prognostic value of reclassification of leukemia according to in vitro growth characteristics. *Blood* 44:1–18.

Muller-Sieburg, C. E., Townsend, K., Weissman, I. L., and Rennick, D. (1988) Proliferation and differentiation of highly enriched mouse hematopoietic stem cells and progenitor cells in response to defined growth factors. *J. Exp. Med.* 167:1825–1840.

Nakahata, T. and Ogawa, M. (1982) Identification in culture of a class of hemopoietic colony- forming units with extensive capacity to self-renew and generate multipotential hemopoietic colonies. *Proc. Natl. Acad. Sci. USA* 79:3843–3847.

Nakahata, T., Gross, A. J., and Ogawa, M. (1982) A stochastic model of self-renewal and committment to differentiation of the primitive hemopoietic stem cells in culture. *J. Cell. Physiol.* 113:455–458.

Nara, N. and McCulloch, E. A. (1985) The proliferation in suspension of the progenitors of the blast cells in acute myeloblastic leukemia. *Blood* 65:1484–1493.

Nara, N., Curtis, J. E., Senn, J. S., Tritchler, D. L., and McCulloch, E. A. (1986) The sensitivity to cytosine arabinoside of the blast progenitors of acute myeloblastic leukemia. *Blood* 67:762–769.

Nara, N., Murohashi, I., Suzuki, T., et al. (1987) Effects of recombinant human granulocyte colony-stimulating factor (G-CSF) on blast progenitors from acute myeloblastic leukaemia patients. *Br. J. Cancer* 56:49–51.

Nolta, J. A. and Joran, C. T. (2001) Spotlight on hematopoietic stem cells: looking beyond dogma: introduction. *Leukemia* 15:1677–1680.

Ottmann, O. G., Ganser, A., Seipelt, G., Eder, M., Schulz, G., and Hoelzer, D. (1990) Effects of recombinant human interleukin-3 on human hematopoetic progenitor and precursor cells in vivo. *Blood* 76:1494–1502.

Peterson, B. E., Bowen, W. C., Patrene, K. D., et al. (1999) Bone marrow as a potential source of hepatic oval cells. *Science* 284: 1168–1170.

Pluznik, D. H. and Sachs, L. (1966) The cloning of normal "mast" cells in tissue culture. *J. Cell. Comp. Physiol.* 66:319–324.

Radnay, J., Goldman, I., and Rozenszajn, L. A. (1979) Growth of human B-lymphocyte colonies in vitro. *Nature* 278:351–353.

Rowley, J. D. (1973) A new consistent chromosomal abnormality in chronic myelogenous leukemia identified by quinacrine fluorescence and giemsa staining. *Nature* 243:290–293.

Rowley, J. D., Golomb, H. M., and Dougherty, C. (1977) 15/17 translocation, a consistent chromosomal change in acute promyelocytic leukemia. *Lancet* i:549–550.

Sauvage, F. J. D., Spencer, S. D., Gurney, A. L., et al. (1994) Stimulation of megakaryocytopoiesis and thrombopoiesis by the c-Mpl ligand. *Nature* 369:533–538.

Second MIC Cooperative Study Group. (1988) Morphologic, immunologic and cytogenetic (MIC) working classification of the acute myeloid leukemias. *Cancer Genet. Cytogenet.* 30:1–15.

Sheridan, W. P., Begley, C. G., Juttner, C. A., et al. (1992) Effects of peripheral blood progenitor cells mobilized by G-CSF (Filgastrin) on platelet recovery after high-dose chemotherapy. *Lancet* 339:640.

Sherr, C. J. (1990) Colony-stimulating factor-1 receptor. *Blood* 75: 1–12.

Siminovitch, L., McCulloch, E. A., and Till, J. E. (1963) The distribution of colony-forming cells among spleen colonies. *J. Cell. Comp. Physiol.* 62:327–336.

Siminovitch, L., Till, J. E., and McCulloch, E. A. (1964) Decline in colony-forming ability of marrow cells subjected to serial transplantation into irradiated mice. *J. Cell. Comp. Physiol.* 64:23–31.

Spangrude, G. J., Heimfeld, S., and Weissman, I. L. (1988) Purification and characterization of mouse hematopoietic stem cells. *Science* 241: 58–62.

Spangrude, G. J., Smith, L., Uchida, N., et al. (1991) Mouse hematopoietic stem cells. *Blood* 78:1395–1402.

Stephenson, J. R., Axelrad, A. A., McLeod, D. C., and Shreeve, M. (1971) Induction of colonies of hemoglobin-synthesizing cells by erythropietin in vitro. *Proc. Natl. Acad. Sci. USA* 68:1542–1546.

Suda, T., Suda, J., Ogawa, M., and Ihle, J. N. (1985) Permissive role of interleukin 3 (IL-3) in proliferation and differentiation of multipotential hemopoietic progenitors in culture. *J. Cell. Physiol.* 124:182–190.

Szilvassy, S., Humphries, R. K., Landsdorph, P. M., Eaves, A. C., and Eaves, C. (1990) Quantitative assay of totipotent reconstituting hematopoietic stem cells by a competitive repopulation strategy. *Proc. Natl. Acad. Sci. USA* 87:8736–8740.

Teshima, H., Kitayam, H., Yamagami, T., et al. (1989) Clinical effects of recombinant human granulocyte colony-stimulating factor in leukemia patients. *Exp. Hematol.* 17:853–858.

Till, J. E. and McCulloch, E. A. (1961) A direct measurement of the radiation sensitivity of normal mouse bone marrow cells. *Radiat. Res.* 14:213–222.

Till, J. E., McCulloch, E. A., and Siminovitch, L. (1964) A stochastic model of stem cell proliferation, based on the growth of spleen colony forming cells. *Proc. Natl. Acad. Sci. USA* 51:29–36.

Trevisan, M. and Iscove, N. N. (1995) Phenotypic analysis of murine long-term hemopoietic reconstituting cells quantitated competitively in vivo and comprison with more advanced colony- forming progeny. *J. Exp. Med.* 181:93–103.

Trevisan, M., Yan, X.-Q., and Iscove, N. N. (1996) Cycle initiation and colony-formation in culture by murine marrow cells with long-term reconstituting potential *in vivo*. *Blood* 88:4149– 4158.

Turner, W. A., Siminovitch, L., McCulloch, E. A., and Till, J. E. (1967) Density gradient centrifugation of hemopoietic colony forming cells. *J. Cell. Physiol.* 69:73–81.

Warrell, J. R., Maslak, P., Eardley, A., Heller, G., Miller, J. R., and Frankel, S. R. (1994) Treatment of acute promyelocytic leukemia with all-trans retinoic acid: an update of the New York experience. *Leukemia* 8:929–933.

Warren, D. J. and Moore, M. A. S. (1988) Synergism among interleukin-1, interleukin-3 and interleukin-5 in the production of eosinophils from primitive hemopoietic stem cells. *J. Immunol.* 140:94–99.

Wendling, F., Maraskovsky, E., Debili, N., et al. (1994) c-Mpl ligand is a humoral regulator of megakaryoctopoiesis. *Nature* 369:571–574.

Williams, D. A., Lemischka, I. R., Nathan, D. G., and Mulligan, R. C. (1984) Introduction of new genetic material into pluripotent hematopoietic stem cells of the mouse. *Nature (Lond.)* 310:476–480.

Worton, R. G., McCulloch, E. A., and Till, J. E. (1969) Physical separation of hemopoietic stem cells differing in their capacity for self-renewal. *J. Exp. Med.* 130:91–103.

Wu, A. M., Siminovitch, L., Till, J. E., and McCulloch, E. A. (1968) Evidence for a relationship between mouse hemopoietic stem cells and cells forming colonies in culture. *Proc. Natl. Acad. Sci. USA* 59:1209–1215.

12 Developmental Origin of Murine Hematopoietic Stem Cells

LORRAINE ROBB, MD, BS, PhD AND KYUNGHEE CHOI, PhD

Where do hematopoietic stem cells (HSCs) come from? In the early embryo, prior to development of the fetal liver, two independent sites of hematopoiesis are generally accepted: the yolk sac and the paraaortic splanchnopleure. In the yolk sac, blood islands are first detected at embryonic day (E) 7.5. Blood islands consist of primitive erythroblasts, surrounded by endothelial cells. The yolk sac endothelial network fuses to form vascular channels and primitive erythrocytes circulate after E 8.5 when the heart starts to beat. These first blood cells are thought to arise from a precursor cell known as the hemangioblast. Hemangioblasts, which can be derived by culture of embryonic stem cells in vitro, give rise to both blood cells and endothelial cells. In contrast to the yolk sac, the paraaortic splanchnopleure, later known as the aorta–gonadal–mesoneohros (AGM) region, does not generate primitive hematopoietic cells. It is, however, a major site of formation of HSCs, which seed fetal liver. The liver becomes the major blood-forming organ at E 11–12, generating definitive hematopoietic cells and HSC, which subsequently colonize spleen and bone marrow. The receptor tyrosine kinase *Flk-1* (VEGFR2) and the transcription factor *Scl* have been shown to be critical for hemangioblast development. Further knowledge of the key signals for growth and differentiation of HSCs is necessary for learning how to control hematopoietic malignancies as well as congenital and acquired disorders of blood cell production.

12.1. INTRODUCTION

During mouse embryogenesis, blood cells are sequentially generated from several distinct anatomical sites. Morphologically distinctive primitive blood cells are first identifiable in the blood islands of the yolk sac at E 7.5 of gestation. The liver anlage is colonized by hematopoietic cells on E 10.5 of the 20-day murine gestation period and thereafter becomes the principal fetal hematopoietic organ (Houssaint, 1981). The fetal thymus is also seeded by hematopoietic stem cells (HSCs) on E10.5 and the fetal spleen on E12.5 (Owen and Ritter, 1969; Fontaine-Perus et al., 1981; Godin et al., 1999). Bone marrow is colonized by fetal liver–derived HSCs around E15 (Zanjani et al., 1993; Clapp et al., 1995; Delassus and Cumano, 1996) and, shortly after birth, becomes the major site of blood production throughout adult murine life. The term *primitive hematopoiesis* is given to the first, yolk sac–derived, erythroid lineage and definitive hematopoiesis is applied to all lineages other than primitive erythroid (Keller et al., 1999).

While the yolk sac is widely accepted to be the source of primitive hematopoietic cells, identification of the initial site of definitive hematopoiesis remains controversial. The origin of the definitive HSCs that initially colonize fetal liver and subsequently bone marrow is also a subject of intense debate. Initially, the yolk sac was thought to be the source of definitive progenitor cells that seeded the fetal liver (Moore and Owen, 1967; Moore and Metcalf, 1970). However, it is now accepted that there is a second, intraembryonic site of *de novo* generation of multipotential hematopoietic cells in the mammalian embryo known as the paraaortic splanchnopleure or aorta-gonad-mesonephros (AGM) region (Godin et al., 1993; Medvinsky et al., 1993; Godin et al., 1995; Medvinsky and Dzierzak, 1996). In recent years, a large number of studies of developmental hematopoiesis have been published, indicating an ongoing pledge by researchers to better understand the origins, commitment, and potential of embryonic stem (ES) cells (reviewed in Robb, 1997; Dzierzak et al., 1998; Medvinsky and Dzierzak, 1998; Keller et al., 1999; Cumano and Godin, 2001; Palis et al., 2001; Choi, 2002).

This chapter reviews the biology of blood development in preliver murine embryo and examines evidence for the origin of the stem cells that give rise to definitive hematopoiesis. In addition, it describes experiments characterizing hematopoietic progenitor cells arising from in vitro differentiation of ES, which have contributed to the understanding of the embryological origin of HSCs.

12.2. BLOOD DEVELOPMENT IN THE YOLK SAC

The yolk sac is initially a bilaminar structure composed of an inner layer of mesoderm and an outer visceral endoderm layer. Embryonic mesoderm migrates from the primitive streak to the yolk sac during early gastrulation (Lawson et al., 1991; Kinder et al., 1999). Blood islands are first detectable in yolk sac mesoderm at E 7.5. They consist of clusters of large, nucleated, basophilic cells surrounded by a layer of endothelial cells (angioblasts)

From: *Stem Cells Handbook*
Edited by: S. Sell © Humana Press Inc., Totowa, NJ

(Haar and Ackerman, 1971). The visceral endoderm is thought to serve as a source of inductive signals for development of blood cells and endothelial cells (Palis et al., 1995; Belaoussoff et al., 1998). The yolk sac endothelial cell network fuses to form vascular channels, and when the heart starts to beat, at about E 8.5 (the eight-somite stage), primitive erythrocytes circulate throughout the embryo. The close association of vasculogenesis with embryonic blood cell development, both temporally and spatially, has led to the hypothesis that the two lineages may derive from a common progenitor cell, dubbed the hemangioblast (Murray, 1932).

Primitive erythroblasts arise exclusively in the yolk sac and are found in the embryonic circulation as soon as it is established, in what appears morphologically to be a developmentally synchronous wave. These cells are indispensable for the survival of the embryo until the liver generates the first circulating definitive erythrocytes at E 12 of gestation (Copp, 1995). The circulating primitive erythroblasts continue to divide until E 13 and then gradually differentiate, with progressive nuclear condensation and increasing accumulation of hemoglobin (Steiner and Vogel, 1973). Some enucleated primitive erythroblasts remain in the embryonic circulation until E18 (Bethlenfalvay and Block, 1970). Primitive erythroblasts initially synthesize embryonic hemoglobins and later synthesize adult globins (Brotherton et al., 1979).

In addition to its role in the generation of primitive erythroid cells, the yolk sac has the capacity to generate multiple definitive hematopoietic lineages. Early studies documented the presence of erythroid and myeloid progenitors in the yolk sac prior to their detection in the embryo proper (Wong et al., 1986). Yolk sacs isolated from precirculation embryos generated both erythroid and myeloid precursors when cultured for 48 h (Cumano et al., 1996). A recent study of hematopoietic progenitor cell numbers in yolk sac and early embryo by Palis et al. (1999) combined careful staging of embryos with the use of culture conditions optimized for growth and maturation of the primitive erythroid lineage. This demonstrated that primitive erythroid colony-forming cells (EryP-CFCs) are present in the yolk sacs of midstreak stage (E 7) embryos, within hours of the establishment of extra-embryonic mesoderm. The numbers of EryP-CFCs in yolk sac increased up to early somite stages, disappearing by E 9. EryP-CFCs were never found in cultures of the embryo proper. The earliest definitive erythroid progenitor detectable in cytokine-stimulated semisolid cultures is the burst-forming unit, erythroid (BFU-E). BFU-Es are large colonies composed of hemoglobinized cells, detectable after 7–10 d of culture. More mature erythroid progenitors form colonies of smaller cells known as colony-forming unit, erythroid (CFU-Es) after 2 to 3 d in culture. In addition to EryP-CFCs, BFU-E and macrophage progenitors were also cultured from the yolk sac of precirculation embryos. CFU-Es were first detectable in the yolk sac of older, 26–29 somite stage embryos. Very small numbers of BFU-Es and CFU-Es were detectable in cultures of embryo proper prior to the onset of fetal liver hematopoiesis. The lower numbers of CFU-Es found in the embryo proper compared with yolk sac indicated that most definitive erythroid progenitors were maturing in the yolk sac. In contrast with the sequential development of BFU-Es followed by CFU-Es in the yolk sac, both BFU-E and CFU-E numbers were high in the fetal liver rudiment at E10.5. This suggested migration, rather than *de novo* generation, of the first fetal liver hematopoietic progenitor cells (Palis et al., 1999).

In a second study, multipotential hematopoietic precursor cells, which give rise to high proliferative potential (HPP)-CFCs, were first detected in the yolk sac just prior to the onset of circulation. On replating, these colonies gave rise to definitive erythroid and macrophage colonies. The yolk sac was demonstrated to be the predominant site of HPP-CFCs (100-fold higher than in the embryo proper) until the onset of hematopoiesis in fetal liver (Palis et al., 2001). Other myeloid progenitors, including mast cell and granulocyte-macrophage CFC, could be cultured from yolk sacs of 10–15 somite stage concepti (Palis et al., 1999). Lymphoid progenitors were first detectable in yolk sac after E 8.5 (Godin et al., 1995; Cumano et al., 1996).

The results of these studies suggest the following: The yolk sac initially generates a wave of primitive erythroid progenitors that mature within the embryonic vasculature. The yolk sac also generates definitive erythroid and myeloid progenitors. However, definitive erythrocytes are not detectable in the embryonic circulation until they emerge from fetal liver at E 12, because the embryonic yolk sac environment is unable to support the differentiation of the definitive erythroid progenitors it generates. This notion is supported by studies showing that yolk sac explants from 20 to 15 somite stage embryos only gave rise to a transient population of primitive erythroid progenitors, but, when cultured with liver primordium, the explants gave rise to an additional wave of definitive erythroid progenitors (Cudennec et al., 1981). Erythropoietin signaling may be responsible for the timing of the maturation of definitive erythroid progenitors. In a recent study, homologous recombination in ES cells was used to replace the endogenous erythropoietin receptor locus with a cDNA encoding a constitutively active erythropoietin receptor. In E8.5 concepti injected with ES cells expressing the constitutively activated erythropoietin receptor, ES cell–derived mature, enucleated, definitive erythroid progenitors were found in the chimeric yolk sac (Lee et al., 2001).

The development of both primitive and definitive hematopoietic progenitor cells in the blood islands of the yolk sac raises the question, Is there a common progenitor cell that gives rise to both populations? A common precursor that gives rise to primitive and definitive hematopoietic progeny has been identified in embryoid bodies (EBs) derived from murine ES cells in culture (Kennedy et al., 1997), and EB-derived clonal cell lines displaying primitive and definitive hematopoietic potential have been established (Keller et al., 1997; Perlingeiro et al., 2001). Whether such a progenitor exists in vivo remains to be determined.

12.3. BLOOD DEVELOPMENT IN P-SP/AGM

Experiments using avian models unequivocally demonstrated that, while the yolk sac produced blood cells during embryonic stages, definitive hematopoiesis originated from an intraembryonic site (Dieterlen-Lievre, 1975; Lassila et al., 1982). This led to the identification of an intraembryonic site of hematopoiesis in the mouse conceptus. A region known in early somite stage embryos as the paraaortic splanchnopleure (P-Sp) and in older embryos as the AGM was shown, at E 8.5, to harbor hematopoietic cells that gave rise to multiple lineages (Godin et al., 1995) and at the 27 somite stage (E 9.5) to contain d 8 spleen colony-forming units (CFU-S8) (Medvinsky et al., 1993). At E10.5, this region was found to contain stem cells capable of engrafting lethally irradiated adult recipient mice (*see* below) (Muller et al., 1994).

In the E 7 to 8 embryo, the P-Sp comprises the caudal splanchnic mesoderm, gut endoderm, and endoderm of the paired dorsal aortae and omphalomesenteric artery. After E 9, the region includes the fused dorsal aorta and the urogenital ridge. The latter includes the pro/mesonephros and the developing gonads. In contrast with the yolk sac, the P-Sp/AGM region is not morphologically overtly hematopoietic. However, light microscopy of P-Sp/AGM sections reveals clusters of hematopoietic cells in certain regions. In the chick, small clusters of hematopoietic cells are found in close association with the wall of the dorsal aorta at E 3 and then as paraaortic clusters at E 6–8 (Dieterlen-Lievre and Martin, 1981). DiI labeling and retroviral marking studies have shown that endothelial cells in the prehematopoietic chick aorta can give rise to the intraaortic hematopoietic cells and that these cells migrate to form the paraaortic hematopoietic foci (Jaffredo et al., 1998, 2000). In the murine embryo, intraarterial clusters of hematopoietic cells are found in the lumina of the omphalomesenteric and umbilical arteries and dorsal aorta between E 9.5 and E 11 (Garcia-Porrero et al., 1995; Wood et al., 1997). In birds and mice, the intravascular hematopoietic cell aggregates are restricted to arteries, and their distribution within the arteries is confined to one aspect of the arterial wall. The results of avian lineage–tracing studies imply that, as in yolk sac, there is a close ontological relationship between endothelial cells and hematopoietic progenitors in the P-Sp/AGM (reviewed in Cumano and Godin, 2001; Nishikawa, 2001).

Because the embryonic lymphoid program does not initiate until fetal liver and thymus develop, assays that detect lymphoid progenitors in the pre–fetal liver embryo are generally accepted to be a measure of definitive, multipotential hematopoietic precursors. This was formally proven in experiments by Godin et al. (1995), who demonstrated cells in the early embryo with the potential to give rise to erythromyeloid, T-, and B-cell progenitors. Single cells were expanded for 10 to 14 d on a stromal layer with appropriate cytokines. Individual clones were tested in fetal thymic organ culture, for the generation of mature T-cells, in semisolid cultures stimulated with cytokines to detect erythroid and myeloid CFCs, and in cultures containing stromal cells and interleukin-7 to permit B-cell generation. Small numbers of multipotential hematopoietic precursors were found in the P-Sp of the 12–15 somite stage embryo. Thereafter, progenitor numbers increased rapidly in both yolk sac and the embryo proper (Godin et al., 1995). The AGM was also shown to harbor large numbers of multipotent precursors, the absolute number and the frequency (1 in 12 cells tested) being maximal between E 10.5 and E 11.5 (Godin et al., 1999). The progenitors were found in aorta, gonad, and mesentery of the AGM of E 10 embryos but were most frequent in the aorta (Godin et al., 1999). To address the question of the origin of definitive hematopoietic progenitors, yolk sac and P-Sp were collected prior to the establishment of circulation. The intact tissues were cultured for 2 d prior to assay. Progenitor cells were found in both yolk sac and P-Sp from the same embryo. However, splanchnopleure-derived progenitors showed a wider differentiation potential, with only the P-Sp containing cells giving rise to lymphoid progeny (Cumano et al., 1996). This result was confirmed in a study of the lymphopoietic potential of CD45-positive hematopoietic progenitor cells from yolk sac and embryo proper (Nishikawa et al., 1998a). Allowing for caveats concerning the selective nature of in vitro assays (Keller et al., 1999), the data strongly support the concept of two independent origins for hematopoiesis in the precirculation embryo.

12.4. HSCs IN PRE–FETAL LIVER MURINE EMBRYO

During embryogenesis, the liver anlage is colonized by exogenous blood cells (Johnson and Moore, 1975; Houssaint, 1981), and HSCs capable of engrafting adult mice are present in the fetal liver from E 11 to 12. Although the experiments described earlier point to dual origins of definitive multipotential hematopoietic precursors, they do not reveal the source of the circulating HSCs that seed the liver primordium.

The classic assay for the detection of HSCs is to measure the ability of genetically distinguishable test cells to provide long-term repopulation of the multiple hematopoietic lineages of a lethally irradiated, adult recipient mouse (the LTR-HSC assay). Donor cells detectable in this assay have fulfilled two of the criteria of the HSC—they are multipotent, and as they produce these lineages indefinitely posttransplant, they are capable of self-renewal. The CFU-S assay is often employed as a surrogate stem cell assay, although this detects a more mature hematopoietic progenitor cell. In a CFU-S assay, donor cells are injected into lethally irradiated recipients that are sacrificed 6–14 d later (the day of sacrifice being appended to the abbreviation CFU-S), and the number of macroscopically visible colonies on the surface of the spleen is counted. Rare CFU-S6s were first detected in E9 yolk sac by Perah and Feldman (1977), who also showed that CFU-S6 numbers increased if the yolk sac was cultured prior to transplantation. CFU-S8 numbers in the yolk sac and AGM of 26 to 27 somite embryos were found to be similar, although the number of CFU-S8s in the AGM, but not in yolk sac, increased markedly over the next gestational day. CFU-S8s were first detectable in liver primordium at the 30 to 40 somite stage (Medvinsky et al., 1993, 1996).

The frequency and location of LTR-HSC in the pre–liver murine embryo was studied initially by direct transplantation of cells from staged embryos into irradiated adult recipient mice. Small numbers of stem cells capable of repopulating adult recipients were first identified in the AGM of E10.5 embryos (Muller et al., 1994). LTR-HSCs were not found in yolk sac until E11. This result has since been confirmed in studies in which the donor AGM cells have been fractionated according to cell-surface markers (Sanchez et al., 1996; Ohmura et al., 2001). The addition of an organotypic culture period, prior to transplantation, dramatically increased the numbers of CFU-Ss and LTR-HSCs recoverable from the AGM region at E 10 (35–38 somite pairs). Whereas the direct transplantation assay demonstrated reconstitution by AGM cells in only 3 of 97 recipients (with each receiving a little more than the number of cells present in a single AGM), 24 of 27 recipients showed engraftment if the AGM was cultured prior to transplantation. Organotypic culture might permit either expansion or maturation of P-Sp/AGM LTR-HSCs. However, even after culture, HSC activity was not present in any E 9 tissues and was not found in yolk sac until 24 hours after it became detectable in the AGM, suggesting that the AGM is a major contributing source of the LTR-HSCs that seed fetal liver (Medvinsky and Dzierzak, 1996). In an extension of these studies, the anatomical components of the AGM region of E 11 to 12 embryos were individually assayed for LTR-HSC activity. At E 11, stem cell activity was found in the aorta and the omphalomesenteric and vitelline arteries. By E 12, stem cell activity was also found in the urogenital ridge (de Bruijn et al., 2000).

Although several investigators had conclusively shown the failure of yolk sac–derived hematopoietic cells to engraft in conditioned adult recipients, an intriguing study by Toles et al. (1999) showed long-term engraftment of yolk sac cells administered to immunocompromised host embryos via the placental circulation. In addition, Weissman et al. (1978) found that E 8–10 yolk sac cells injected into E 8 to E 9 recipient embryos produced low levels of lymphoid cell engraftment. This led to the development of a newborn recipient model of transplantation (Yoder et al., 1996). Pregnant dams were treated late in gestation with busulfan and the anemic newborn pups received donor cells either intravenously or via direct intrahepatic injection. The donor cells could colonize bone marrow, and/or the liver, which is an active hematopoietic organ in the murine neonatal period. Using this assay, cells from E 9 and E 10 yolk sacs were shown to provide long-term multilineage reconstitution of busulfan-conditioned neonatal pups (Yoder and Hiatt, 1997; Yoder et al., 1997). These cells, which were CD34 and c-kit positive, were found in both yolk sac and AGM at E9, with a 37-fold preponderance in yolk sac (Yoder et al., 1997). The progeny of transplanted yolk sac stem cells, harvested from bone marrow of 4–6-mo old recipients, were capable of engrafting lethally irradiated secondary adult recipients. This demonstrated that although yolk sac stem cells prior to E11 are unable to directly repopulate adult conditioned mice, once engrafted in newborn recipients, the stem cells can migrate to the bone marrow and there function long term (Yoder and Hiatt, 1997; Yoder et al., 1997). This result reinforced the notion that adult and embryonic HSCs differ and that assays of stem cell activity are influenced by the properties of stem cell population, such as their colonization capacity. The inability of the stem cells from E 9 embryonic tissues to directly repopulate adult recipients could be due to failure to express appropriate homing molecules, or the cells may not survive and expand in the adult microenvironment because of a requirement for embryonic growth factors.

In two important recent studies, yolk sac and P-Sp explants collected prior to the onset of circulation were assayed for their ability to reconstitute adult recipient mice (Cumano et al., 2001; Matsuoka et al., 2001). In the study by Cumano et al. (2001), yolk sac and P-Sp explants were cultured prior to transplantation into irradiated recipient mice deficient in B- and T-cells (*Rag2*$^{-/-}$) or in T-, B-and natural killer (NK) cells (*Rag2*$\gamma c^{-/-}$). The latter recipient strain was selected in order to circumvent NK-mediated rejection of donor embryonic cells, which have low levels of major histocompatibility complex class I. The use of *Rag2*$\gamma c^{-/-}$ recipients improved engraftment capacity of all tissues tested, including AGM from E 9.5 to 10 (28–32 somite) embryos, indicating that NK activity may be important in rejection of donor embryonic tissues. The study revealed that while cultured E 8 yolk sac explants gave short-term myeloid reconstitution of *Rag2*$\gamma c^{-/-}$ recipients, no long-term engraftment occurred. By contrast, cultured E 8 P-Sp explants contained precursors that generated multilineage hematopoietic progeny in the recipients. The detection of LTR-HSCs in precirculation P-Sp, but not in yolk sac, is in keeping with the previous studies by this group demonstrating the different in vitro hematopoietic capacity of the extraembryonic and intraembryonic sites of blood development in the precirculation murine embryo.

In the study by Matsuoka et al. (2001), yolk sac and P-Sp cells were cocultured with a stromal cell line (AGM-S3), derived from the E 10.5 AGM region, for 4–6 d prior to transfer to irradiated adult recipient mice. After coculture, LTR-HSCs were present in E 8.5 yolk sac and P-Sp and were also detectable in tissues collected from E 8.0 embryos, before the onset of circulation. Time course studies showed that the kinetics of LTR-HSC development was similar in the cultured cells of both populations (Matsuoka et al., 2001). The difference between these results and those of Cumano et al. (2001), raises intriguing questions. What are the properties of the AGM-S3 cell line that enable it to induce stem cell activity in yolk sac cells? Are these properties found in vivo in the P-Sp/AGM, implying that it is the site of maturation of stem cells? Answers to these questions will give further insights into the biology of embryonic hematopoiesis and could conceivably translate into tools for the generation and expansion of stem cells for therapeutic use.

12.5. TRANSCRIPTIONAL REGULATION OF PRIMITIVE AND DEFINITIVE HEMATOPOIESIS

Key regulators of primitive and definitive blood development in the murine embryo have been revealed by the creation of mouse strains with targeted disruption of genes encoding transcriptional factors. *Scl/Tal-1* is a member of the basic helix-loop-helix family of transcription factors. Embryos with a null mutation of *Scl/Tal-1* die around E10 owing to anemia. Primitive blood cells are completely absent. Vasculogenesis occurs normally in the early yolk sac, but vitelline vessels fail to develop (Robb et al., 1995; Shivdasani et al., 1995; Visvader et al., 1998). Disruption of LMO2, which encodes a member of the LIM domain protein family that interacts with *Scl/Tal-1*, also abrogates primitive erythropoiesis, and, like *Scl/Tal-1*, *LMO2* has been implicated in regulation of vasculogenesis (Warren et al., Yamada et al., 2000). Studies of the differentiation potential of *Scl/Tal-1* and *LMO2* null ES cells in chimeric mice have revealed that these genes are also absolutely required for definitive hematopoiesis (Porcher et al., 1996; Robb et al., 1996; Yamada et al., 1998). GATA transcription family members bind a DNA consensus sequence that is found in the regulatory region of all erythroid-specific genes (Weiss and Orkin, 1995). Mice with a null mutation of *GATA-1* have a profound disruption of primitive hematopoiesis (Fujiwara et al., 1996). In in vitro cultures of *GATA-1* null ES cells, erythroid differentiation is arrested at the proerythroblast stage (Pevny et al., 1995). *GATA-2* null mice have quantitative defects in primitive erythropoiesis and die of severe anemia (Tsai et al., 1994). The abnormalities in primitive hematopoiesis in *GATA-1* null embryos are phenocopied in embryos with a null mutation of *FOG-1*, which encodes a transcriptional cofactor of GATA-1 (Tsang et al., 1998). Definitive hematopoiesis is also defective in *FOG-1*, *GATA-1*, *GATA-2* null mice.

The transcription factors identified to date that regulate primitive hematopoiesis are also required for regulation of definitive hematopoiesis. By contrast, gene-targeting studies have identified several transcription factors that regulate definitive, but not primitive, blood development. The first to be discovered was *c-Myb*. *c-Myb* null mice have profound defects in definitive hematopoiesis and die of anemia in the midgestational period (Mucenski et al., 1991). *Cbfa2* (*AML1*, *Runx1*) and *Cbfb* encode CBFα and CBFβ, the two subunits of a core binding factor that is required for definitive hematopoiesis. Null mutations of the genes encoding either subunit ablate definitive hematopoiesis but do not affect primitive hematopoiesis (Okuda et al., 1996; Sasaki et al., 1996; Wang et al., 1996a,1996b; Niki et al., 1997). The null embryos contain no definitive progenitors within fetal liver or in circulating blood. Furthermore, *Cbfa2* and *Cbfb* null ES cells are unable to contribute to definitive hematopoiesis in chimeric mice. *Cbfa2* is expressed in the endothelium of the aorta, vitelline, and umbilical arteries and in intraarterial hematopoietic cell clusters and is required for the formation of these clusters (North et al., 1999). Homozygous *Cbfa2* null yolk sacs and P-Sp do not contain LTR-HSCs (Cai et al., 2000).

12.6. THE HEMANGIOBLAST

As discussed earlier, the hemangioblast concept was introduced by the observations that blood cells and the vascular system develop closely together in the yolk sac and in the P-Sp/AGM (Sabin, 1920; Murray, 1932; Wagner, 1980). Support for the existence of the hemangioblast comes from the observations that hematopoietic and endothelial cell lineages share the expression of a number of different genes. For example, receptor tyrosine kinases expressed in endothelial cells, such as *Flk-1*, *Tie-1*, and *Tie-2* (also known as *Tek*), are also expressed in early hematopoietic progenitors (Yano et al., 1997). The transcription factor *Scl* is expressed in both hematopoietic and endothelial cells (Kallianpur et al., 1994; Elefanty et al., 1999). The erythropoietin receptor, which is expressed in erythroid progenitors, is also expressed in endothelial cells (Anagnostou et al., 1994). CD34, which is used as a marker for early hematopoietic progenitor cells, is also expressed in endothelial cells as well (Fina et al., 1990). Furthermore, mouse AGM cells expressing podocalyxin-like protein 1 (PCLP-1) but not CD45 (PCLP-1^+CD45$^-$) can clonally give rise to both hematopoietic and endothelial cells (Hara et al., 1999). Additionally, the zebrafish *cloche* mutants lack both hematopoietic and endothelial cell differentiation, suggesting the presence of a common progenitor (Stainier et al., 1995).

12.7. THE IDENTIFICATION OF BLAST-COLONY-FORMING CELLS FROM IN VITRO DIFFERENTIATED ES CELLS

Even though there has been a great interest in identifying the hemangioblast in the developing embryo, the use of embryo-derived cells has proven difficult since the developmental sequence occurs rapidly, the tissues are difficult to access, and only a small number of cells can be obtained. An alternate source of embryonic cells for the studies of early embryonic events is the in vitro differentiated progeny of ES cells. ES cells differentiate efficiently in vitro and give rise to a three-dimensional, differentiated cell mass called EBs (Fig. 1) (reviewed in Keller et al., 1999; Choi, 2002). ES cells can also be differentiated on stromal cells or type IV collagen without intermediate formation of the EB structure (Nakano et al., 1994; Nishikawa et al., 1998). Many different lineages have been reported to develop within EBs, including neuronal, muscular, endothelial, and hematopoietic lineages (reviewed in Choi, 2002). Of these, the hematopoietic lineage has been most extensively characterized. The development of hematopoietic and endothelial cells within EBs mimics in vivo events such that yolk sac blood island-like structures with vas-cular channels containing hematopoietic cells can be found with-in cystic EBs (Doetschman et al., 1985). As in the developing embryo, the primitive erythroid cells develop prior to definitive hematopoietic populations (Keller et al., 1993; Palis et al., 1999). The developmental kinetics of various hematopoietic lineage precursors within EBs and molecular and cellular studies of these cells have demonstrated that the sequence of events leading to the onset of hematopoiesis within EBs is similar to that found within the mouse embryo. In addition, EBs provide a large number of- cells representing an early/primitive stage of development that are otherwise difficult to access in the embryo (Choi et al., 1998). Therefore, the ES in vitro differentiation model is an ideal system for obtaining and studying primitive progenitors of all cell lineages.

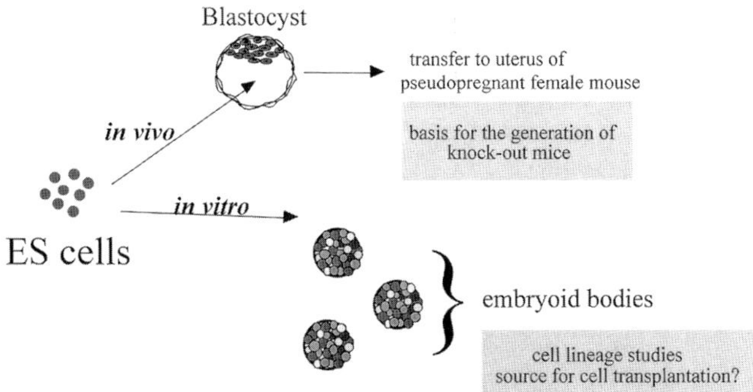

Fig. 1. In vivo and in vitro application of ES cells.

Using the in vitro ES differentiation model system, the blast colony-forming cell (BL-CFC) population present within day (d) 2.5–3.5 EBs has been shown to represent the hemangioblast (Kennedy et al., 1997; Choi et al., 1998). BL-CFCs are transient and develop prior to the primitive erythroid population (Kennedy et al., 1997; Choi et al., 1998) (Fig. 2). BL-CFCs form blast colonies in response to vascular endothelial growth factor (VEGF), a ligand for the receptor tyrosine kinase Flk-1 (Matthews et al., 1991; Millaer et al., 1993), in semisolid media such as methylcellulose cultures. Gene expression analysis indicated that cells within the blast colonies (blast cells) express a number of genes common to both hematopoietic and endothelial lineages, including *Scl*; CD34; and the VEGF receptor, *Flk-1* (Kennedy et al., 1997). Blast cells give rise to primitive, definitive hematopoietic, and endothelial cells when replated in medium containing both hematopoietic and endothelial cell growth factors (Kennedy et al., 1997; Choi et al., 1998). Most importantly, the hematopoietic and endothelial precursors within the blast colonies are clonal, as demonstrated by the mixing studies of two different ES lines (Choi et al., 1998).

12.8. MOLECULES CRITICAL FOR HEMANGIOBLAST DEVELOPMENT

12.8.1. Flk-1 The VEGF receptor Flk-1 (VEGFR2) is known to play a key role in the regulation of embryonic vascular and hematopoietic development. In the mouse embryo, *Flk-1* expression is detected in the presumptive mesodermal yolk sac blood-island progenitors as early as E7.0 (Yamaguchi et al., 1993; Dumont et al., 1995). As has been described (Shalaby et al., 1995), mice deficient in Flk-1 do not develop blood vessels or yolk sac blood islands, and die between E8.5 and 9.5. In chimeric aggregation studies with wild-type embryos, *Flk-1*$^{-/-}$ ES cells fail to participate in vessel formation, or to contribute to primitive or definitive hematopoiesis in vivo, suggesting that Flk-1 inactivation results in a cell autonomous endothelial and hematopoietic defect (Shalaby et al., 1997).

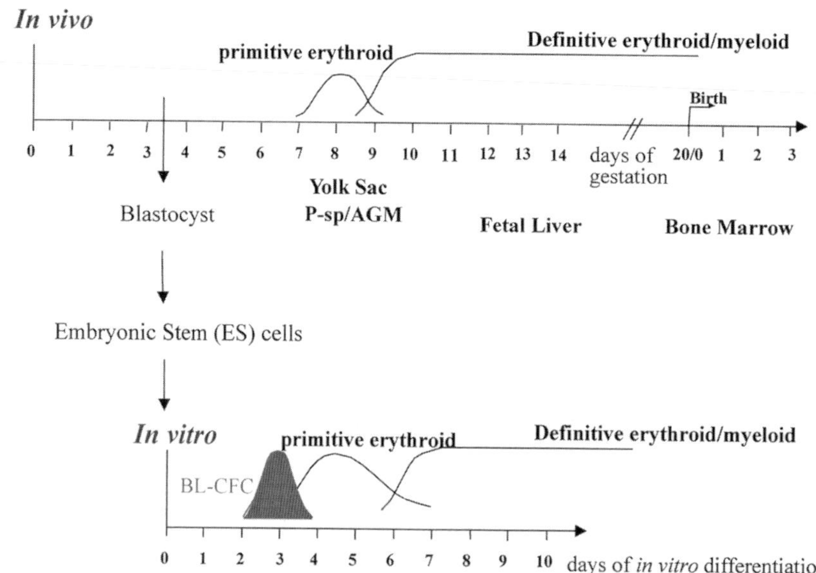

Fig. 2. In vivo vs in vitro hematopoietic development. To date, BL-CFCs have only been identified in vitro. Both BL-CFCs and primitive erythroid progenitors are transient. Definitive erythroid and myeloid lineages develop shortly after and persist throughout EB differentiation. Only the developmental kinetics, not the actual numbers, of each hematopoietic cell population is illustrated.

The findings that BL-CFCs form blast colonies in response to VEGF and that their blast cell progeny express the *Flk-1* gene argue that Flk-1 is expressed on the hemangioblast/BL-CFC (Choi et al., 1998). Indeed, blast colonies develop from a sorted Flk-1$^+$, but not Flk-1$^-$, cell population from d 2.75–3 EBs, when replated with VEGF (Faloon et al., 2000). Consistent with the findings that the hemangioblast expresses Flk-1, VEGFR2$^+$ (Flk-1$^+$) cells from the mesoderm of chicken embryos give rise to both hematopoietic and endothelial cells (Eichmann et al., 1997). In this system, the differentiation of hematopoietic cells from the VEGFR2$^+$ cells occurred in the absence of VEGF, while endothelial cell generation required VEGF.

Several studies suggest a role for *Flk-1* in the hemangioblast migration. For example, Flk-1-deficient cells from *Flk-1*$^{-/-}$/+/+ chimeric embryos accumulate in the amnion, even though Flk-1 (β-galactosidase)-expressing cells are normally found in the yolk sac of *Flk-1*$^{+/-}$ embryos (Shalaby et al., 1997). In addition, in vitro differentiated *Flk-1*$^{-/-}$ ES cells normally give rise to hematopoietic and endothelial cells and E 7.5 *Flk-1*$^{-/-}$ embryos contain hematopoietic progenitors, even though E 8.5 *Flk-1*$^{-/-}$ embryos are known to be profoundly deficient in such cells (Shalaby et al., 1995; Schuh et al., 1999). Future studies will need to address the precise role of VEGF/Flk-1 interaction in hemangioblast development, namely hemangioblast migration, survival and/or proliferation, and their subsequent differentiation.

12.8.2. SCL There are many studies supporting a view that SCL is critical for hemangioblast development. As discussed earlier, *Scl* is required for primitive and definitive hematopoiesis and for endothelial cell development. Furthermore, SCL-deficient ES cells fail to give rise to blast colonies, an in vitro measure of the hemangioblasts (Faloon et al., 2000; Robertson et al., 2000). Recently, Drake and Fleming (2000) have examined early mouse embryos for Flk-1 and SCL expression. At E6.5, Flk-1$^+$SCL$^+$ cells were already present, but dispersed, in the extraembryonic yolk sac. They argue that Flk-1$^+$SCL$^+$ cells are angioblasts since primary vascular networks become evident in the regions where Flk-1$^+$SCL$^+$ cells were initially detected. However, because of the transient nature of the BL-CFC/hemangioblast, it is still possible that these Flk-1$^+$SCL$^+$ cells represent the hemangioblast.

In zebrafish, *Scl* is expressed in the lateral mesoderm, which gives rise to hematopoietic, endothelial, and pronephric lineages (Gering et al., 1998). As discussed earlier, the zebrafish *cloche* (*clo*) mutation affects both hematopoietic and endothelial differentiation (stainier et al., 1995). Scl expression is greatly reduced in *clo* mutants (Liao et al., 1998). More importantly, ectopically expressed *Scl* can rescue, although incompletely, the hematopoietic and endothelial cell defects in these mutants. Rescued embryos develop red blood cells and express the GATA-1 and *Flk-1* genes (Liao et al., 1998), suggesting that the function of SCL lies downstream of clo and upstream of Flk-1 and GATA-1. Cells expressing both Flk-1 and SCL appear to give rise to SCL$^+$Flk-1$^-$ hematopoietic and SCL$^-$Flk-1$^+$ endothelial cells (Gering et al., 1998). These observations predict that cells expressing both SCL and Flk-1 represent hemangioblasts. Indeed, cells expressing both SCL and Flk-1 increase dramatically when SCL is overexpressed in zebrafish embryos (Gering et al., 1998). As a result, both hematopoietic and endothelial cells also increase. A role for SCL in the specification of the hemangioblast is suggested by the observation that the expansion of SCL$^+$Flk-1$^+$ cells occurs at the expense of somitic and pronephric duct tissues (Gering et al., 1998).

Elefanty et al. (1998, 1999) knocked-in a lacZ reporter gene into the *Scl* locus to follow Scl-expressing cells. Histochemical staining of *Scl*$^{+/lacZ}$ embryos for β-galactosidase activity showed that lacZ was expressed in hematopoietic and endothelial cells, and in the developing brain. Cell sorting and replating studies of β-galactosidase$^+$ cells from the fetal liver showed that erythroid and myeloid progenitors are present within the β-galactosidase$^+$ population. Furthermore, the β-galactosidase$^+$ fraction of the bone marrow is enriched for erythroid, myeloid, lymphoid, and CFU-

S12. $Scl^{+/LacZ}$ mice should prove useful for further understanding hemangioblast, hematopoietic, and endothelial cell development.

12.9. IN VIVO HEMATOPOIETIC POTENTIAL OF ES DERIVED CELLS

The ability of ES cells to generate many different somatic cells in vitro argues for their usage as a source for cell transplantation, provided that they can function in vivo. Accumulating studies demonstrate that in vitro–generated hematopoietic progenitors can function in vivo, although the generation of HSCs from ES cells has not been firmly established. Muller and Dzierzak (1993) have utilized in vitro–differentiated ES cells as donor cells in cell transfer studies using newborn W^v/W^v and SCID mice as recipients. In these studies, donor-derived cells have been found only within the lymphoid cell lineages, although donor-derived cells persisted for greater than 6 months. This study utilized d 13 total EB cells without further enrichment of hematopoietic progenitors. Therefore, the finding that ES-derived cells had limited lymphoid potential could reflect the rarity of HSCs present within d-13 EB cells (the time point from which EB cells were generated and used). Potocnik et al. (1997) have isolated AA4.1$^+$B220$^+$ and AA4.1$^+$B220$^-$ cell populations from in vitro–differentiated ES cells (d 15 EBs) and tested for in vivo reconstitution potential in 4- to 6-wk-old *Rag-1*-deficient mice. Both cell populations engrafted to the recipients, but AA4.1$^+$B220$^+$ cells had a limited life-span (up to 8 wk) and limited potential; that is, they could only give rise to B-cell lineage. However, AA4.1+B220- derived cells survived longer in the recipients and could be found 24 weeks after the transplantation. Donor AA4.1$^+$B220$^-$-originated B- and T-cells could be found in these recipients and donor-derived B220$^+$c-Kit$^+$ cells were also found in the recipient bone marrow. These studies suggest that AA4.1$^+$B220– cells contained more primitive hematopoietic progenitors. However, HSC activity was not pursued in this study.

Palacios et al. (1995) reported that E14.1 ES cells differentiated on the RP.0.10 bone marrow cell line (H-2d) and sorted for PgP-1 (CD44)$^+$Lin$^-$ (H-2b) could reconstitute all the hematopoietic cell lineages, including lymphoid, myeloid, and erythroid, of sublethally irradiated 6- to 12-wk-old CB17SCID (H-2d) or C3HSCID (H-2k) mice. Furthermore, bone marrow from the primary recipients could reconstitute the hematopoietic system of secondary recipients, arguing that ES-derived HSCs could self-renew. Perlingeiro et al. (2001) transduced EB-derived blast cells with the *Bcr/Abl* oncogene. Transformed blast cells were cultured on OP9 cells, previously shown to support HSCs (Nakano et al., 1994), and then used to repopulate sublethally irradiated, 8-wk-old 129Sv/Ev or NOD/SCID mice. Consistent with the previous studies demonstrating that EB-derived blast cells contain both primitive and definitive erythroid, and myeloid progenitors (Kennedy et al., 1997; Choi et al., 1998), the *Bcr/Abl*-transduced blast cells gave rise to primitive and definitive erythroid, myeloid, and lymphoid cell lineages in the recipients. The caveat of this study is that the cell population used for transplantation harbors an oncogene. The recipients ultimately developed myeloproliferative disorders between 5 and 9 wk after transplantation. In summary, in vivo hematopoietic reconstitution studies utilizing ES-derived cells argue that HSCs can be derived from in vitro–differentiated ES cells. However, because the reconstitution rate is still low, there is a need to develop an optimal protocol for utilizing ES-derived cells for hematopoietic reconstitution.

12.10. FUTURE DIRECTIONS

Future work in developmental hematopoiesis should clearly address the precise relationship between the primitive and definitive hematopoiesis, the yolk sac and AGM hematopoiesis, and the hematopoietic and endothelial cell lineages. The availability of BL-CFCs, the most primitive progenitors committed to become hematopoietic and endothelial cell lineages identified to date, should be valuable for the studies of hematopoietic specification. For example, the molecular mechanisms leading to BL-CFC development can unveil the pathway of hematopoietic specification in the developing embryo. More specifically, determining the transcription factors and their target(s) expressed in BL-CFCs could provide major insights into hematopoietic and endothelial cell lineage commitment. Furthermore, investigation of growth factors, other than VEGF/Flk-1, and their signals involved in BL-CFC development could also provide insights into early hematopoietic development. It will also be important to determine whether in vitro–derived BL-CFCs can generate both hematopoietic and endothelial cells in in vivo. Such demonstration should facilitate utilization of in vitro–differentiated ES cells for cell therapy as well as cell-mediated gene therapy applications. Ultimately, the in vivo equivalent progenitors of BL-CFCs should be identified and characterized to further understand hematopoietic and vascular development.

ACKNOWLEDGMENTS

This work was supported by the National Health and Medical Research Council of Australia, the Cooperative Research Centre for Cellular Growth Factors, and The Sylvia and Charles Viertel Charitable Foundation; and by National Institutes of Health grants R0155337 and R01 HL63736 and American Cancer Society grant RPG-00-062-01-CCE.

REFERENCES

Anagnostou, A., Liu, Z., Steiner, M., et al. (1994) Erythropoietin receptor mRNA expression in human endothelial cells. *Proc. Natl. Acad. Sci. USA* 91:3974–3978.

Belaussof, M., Farrington, S. M., and Baron, M. H. (1998) hematopoietic induction and respecificaiotn of A-P identity by visceral endoderm signaling in the mouse embryo. *Development* 125: 5009–5018.

Bethlenfalvay, N. C. and Block, M. (1970) Fetal erythropoiesis: maturation in megaloblastic (yolk sac) erythropoiesis in the C57 B1-6J mouse. *Acta Haematol.* 44: 240–245.

Brotherton, T. W., Chui, D. H., Gauldie, J., and Patterson, M. (1979) hemoglobin ontogeny during normal mouse fetal development. *Proc. Natl. Acad. Sci. USA* 76:2853–2857.

Cai, Z., de Bruijn, M., Ma, X., et al. (2000) Haploinsuffieciency of AML1 affects the temporal and spatial generation of hematopoietic stem cells in the mouse embryo. *Immunity* 13:423–431.

Choi, K. (2002) The hemangioblast: a common porgenitor of hematopoietic and endothelial cells. *J. Hematother. Stem Cell Res.* 11:91–101.

Choi, K., Kennedy, M., Kazarov, A., Papadimitriou, J., and Keller, G. (1998) A common precursor for hematopoietic and endothelial cells. *J. Hematother. Stem Cell Res.* 11: 91–101.

Clapp, D. W., Freie, B., Lee, W. H., and Zhang, Y. Y. (1995) Molecular evidence that in situ-transduced fetal liver hematopoietic stem/progenitor cells give rise to medullary hematopoiesis in adult rats. *Blood* 86;2113–2122.

Copp, A. J. (1995) Death before birth: clues from gene knockouts and mutations. *Trends Genet.* 11: 87–93.

Cudennec, C. A., Thiery, J. P., and Le Douarin, N. M. (1981) In vitro induction of adult erythropoiesis in early mouse yolk sac. *Proc. Natl. Acad. Sci. USA* 78:2412–2416.

Cumano, A. and Godin, I. (2001) Pluripotent hematopoietic stem cell development during embryogenesis. *Curr. Opin. Immunol.* 13:166–171.

Cumano, A., Dieterlen-Lievre, F., and Godin, I. (1996) Lymphoid potential, probed before circulation in mouse, is restricted to caudal intraembryonic splanchnopleura. *Cell* 86:907–916.

Cumano, A., Ferraz, J. C., Klaine, M., Di Santo, J. P., and Godin, I. (2001) Intraembryonic, but not yolk sac hematopoietic precursors, isolated before ccirculation , provide long-term multilineage reconstitution. *Immunity* 15;477–485.

De Bruijn, M. F., Speck, N. A., Peeters, M. C., and Dzierzak, E. (2000) Definitive hematopoietic stem cells first develop within the major arterial regions of the mouse embryo. *EMBO. J.* 19:2465–2474.

Delassus, S. and Cumano, A. (1966) Circulation of hematopoietic progenitors in the mouse embryo. *Immunity* 4:97–106.

Dieterlen-Lievre, F. (1975) On the origin of haemopoietic stem cells in the avian embryo: an experimental approach. *J. Embryol. Exp. Morphol.* 33: 607–619.

Dieterlen-Lievre, F. (1977) Intraembryonic hematopoietic stem cells. *Aplastic Anemia Stem Cell Biol.* 11:1149–1171.

Dieterlen-Lievre, F. and Martin, C. (1981) Diffuse intraembryonic hemopoiesis in normal and chimeric avian development. *Dev. Biol.* 88:180–191.

Doetschman, T. C., Eistetter, H., Katz, M., Schmidt, W., and Kemler, R. (1985) The in vitro development of blastocyst-derived embryonic stem cell lines: formation of visceral yolk sack, blood islands and myocardium. *J. Embryol. Exp. Morphol.* 87:27–45.

Drake, C. J. and Fleming, P. A. (2000) Vasculogenesis in the day 6.5 to 9.5 mouse embryo. *Blood* 95:1671–1679.

Dzierzak, E., Medvinsky, A., and de Bruijn, M. (1998) Qualitative and quantitative aspects of haematopoietic cell development in the mammalian embryo. *Immunol. Today* 19:228–236.

Dumont, D. J., Fong, G. H., Puri, M. C., Gradwohl, G., Alitalo, K., and Breitman, M. L. (1995) Vascularization of the mouse embryo: a study of flk-1, tek, tie, and vascular endothelial growth factor expression during development. *Dev. Dyn.* 203:80–92.

Eichmeann, A., Corbel, C., Nataf, V., Vaigot, P., Breant, C., and Le Douarin, N. M. (1997) Ligand-dependent development of the endothelial and hemopoietic lineages from embryonic mesodermal cells expressing vascular endothelial growth factor receptor 2. *Proc. Natl. Acad. Sci. USA* 94:5141–5146.

Elefanty, A. G., Begley, C. G., Metcalf, D., Barnett, L., Kontgen, F., and Robb, L. (1998) Characterization of hematopoietic progenitor cells that express the transcription factor SCL, using a lacZ "knock-in" strategy. *Proc. Natl. Acad. Sci. USA* 95:11,897–11,902.

Elefanty, A. G., Begley, C. G., Hartley, L., Papaevangeliou, B., and Robb, L. (1999) SCL expression in the mouse embryo detected with a targeted lacZ reporter gene demonstrates its localization to hematopoietic, vascular, and neural tissues. *Blood* 94:3754–3763.

Faloon, P., Arentson, E., Kazarov, A., et al. (2000) Basic fibroblast growth factor positively regulates hematopoietic development. *Development* 127:1931–1941.

Fina, L., Molgaard, H. V., Robertson, D., et al. (1990) Expression of the CD34 gene in vascular endothelial cells. *Blood* 75:2417–2426.

Fontaine-Perus, J. C., Calman, F. M., Kaplan, C., and Le Douarin, N. M. (1981) Seeding of the 10-day mouse embryo thymic rudiment by lymphocyte precursors in vitro. *J. Immunol.* 126:2310–2316.

Fujiwara, Y., Browne, C. P., Cunniff, K., Goff, S. C., and Orkin, S. H. (1996) Arrested development of embryonic red cell precursors in mouse embryos lacking transcription factor GATA-1. *Proc. Natl. Acad. Sci. USA* 93:12,355–12,358.

Garcia-Porrero, J. A., Godin, I. E., and Dieterlen-Lievre, F. (1995) potential intraembryonic hemogenic sites at pre-liver stages in the mouse. *Anat. Embryol. (Berl.)* 192:425–435.

Gering, M., Rodaway, A. R. F., Gottgens, B., Patient, R. K., and Green, A. R. (1998) The SCL gene specifies haemangioblast development from early mesoderm. *EMBO J.* 17:4029–4045.

Godin, I., Dieterlen-Lievre, F., and Cumano, A. (1995) Emergence of multipotent hematopoietic cells in the yolk sac and paraaortic splanchnopleura of 8.5 dpc mouse embryos. *Proc. Natl. Acad. Sci. USA* 92:773–777.

Godin, I., Garcia-Oorrero, J. A., Dieterlen-Lievre, F., and Cumano, A. (1999) Stem cell emergence and hemopoietic activity are incompatible in mouse intraembryonic sites. *J. Exp. Med.* 190:43–52.

Godin, I. E., Garcia-Porrero, J. A., Goutinho, A., Dieterlen-Lievre, F., and Marcos, M. A. (1993) Para-aortic splanchnopleura from early mouse embryos contains B1a cell progenitors. *Nature* 364:67–70.

Haar, J. L. and Ackerman, G. A. (1971) Ultrastructural changes in mouse yolk sac associated with the initiation of vitelline circulation. *Anat. Rec.* 170:437–455.

Hara, T., Nakano, Y., Tanaka, M., et al. (1999) Identificaiton of podocalyxin-like protein 1 as a novel cell surface marker for hemangioblasts in the murine aorta-gonad-mesonephros region. *Immunity* 11:567–578.

Houssaint, E. (1981) Differentiation of the mouse hepatic primordium. II. Extrinsic origin of the haemopoietic cell line. *Cell Differ.* 10:243–252.

Jaffredo, T., Gautier, R., Eichmann, A., and Dieterlen-Lievre, F. (1998) Intraaortic hemopoietic cells are derived from endothelial cells during ontogeny. *Development* 125:4575–4583.

Jaffredo, T., Gautier, R., Brajeul, V., and Dieterlen-Lievre, F. (2000) Tracing the progeny of the aortic hemangioblast in the avian embryo. *Dev. Biol.* 224:204–214.

Johnson, G. R. and Moore, M. A. (1975) Role of stem cell migration in initiation of mouse foetal liver haemopoiesis. *Nature* 258:726–728.

Kallianpur, A. R., Jordan, J. E., and Brandt, S. J. (1994) The SCL/TAL-1 gene is expressed in progenitors of both the hematopoietic and vascular systems during embryogenesis. *Blood* 83:1200–1208.

Keller, G., Kennedy, M., Papayannopoulou, T., and Wiles, M. V. (1993) hematopoietic commitment during embryonic stem cell differentiation in culture. *Mol. Cell Biol.* 13:473–486.

Keller, G., Wall, C., Fong, A., Hawley, T., and Hawley, R. (1997) Overexpression of HOX11 leads to immortalization of embryonic precursors with both primitive and definitive hematopoietic potential. *Blood* 90:877–887.

Keller, G., Lacaud, G., and Robertson, S. (1999) Development of the hematopoietic system in the mouse. *Exp. Hematol.* 27:777–787.

Kennedy, M., Firpo, M., Choi, L., et al. (1997) A common precursor for primitive erythropoiesis and definitive haematopoiesis. *Nature* 386: 488–493.

Kinder, S. J., Tsang, T. E., Quinlan, G. A., Hadjantonakis, A., Nagy, A., and Tam, P. P. (1999) The orderly allocation of mesodermal cells to the extraembryonic structures and the anteroposterior axis during gastrulation of the mouse embryo. *Development* 126:4691–4701.

Lassila, O., Martin, C., Toivanen, P., and Dieterlen-Lievre, F. (1982) Erythropoiesis and lymphopoiesis in the chick yolk-sac-embryo chimeras: contribution of yolk sac and intraembryonic stem cells. *Blood* 59:377–381.

Lawson, K. A., Meneses, J. J., and Pedersen, R. A. (1991) Clonal analysis of epiblast fate during germ layer formation in the mouse embryo. *Development* 113:891–911.

Lee, R., Kertesz, N., Joseph, S. B., Jegalian, A., and Wu, H. (2001) Erythropoietin (Epo) and EpoR expression and 2 waves of erythropoiesis. *Blood* 98:1408–1415.

Liao, E. C., Paw, B. H., Oates, A. C., Pratt, S. J., Postlethwait, J. H., and Zon, L. I. (1998) SCL/Tal-1 transcription factor acts downstream of cloche to specify hematopoietic and vascular progenitors in zebrafish. *Genes Dev.* 12:621–626.

Matsuoka, S., Tsuji, K., Hisakwaw, H., et al. (2001) Generation of definitive hematopoietic stem cells from murine early yolk sac and paraaortic splanchnopleures by aorta-gonad-mesonephros region-derived stromal cells. *Blood* 98:6–12.

Matthews, W., Jordan, C. T., Gavin, M., Jenkins, N. A., Copeland, N. G., and Lemischka, I R. (1991) A receptor tyrosine kinase cDNA isolated from a population of enriched primitive hematopoietic cells and exhibiting close genetic linkage to c-kit. *Proc. Natl. Acad. Sci. USA* 88:9026–9030.

Medvinsky, A. and Dzierzak, E. (1996) Definitive hematopoiesis is autonomously initiated by the AGM region. *Cell* 86:897–906.

Medvinsky, A. L. and Dzierzak, E. A. (1998) Development of the definitive hematopoietic hierarchy in the mouse. *Dev. Comp. Immunol.* 22:289–301.

Medvinsky, A. L., Samoylina, N. L., Muller, A. M., and Dzierzak, E. A. (1993) An early pre-liver embryonic source of CFU-S in the developing mouse. *Nature* 364:64–67.

Medvinsky, A. L., Gan, O. I., Semenova, M. L., and Samoylina, N. L. (1996) Development of day-8 colony-forming unit-spleen hematopoietic progenitors during early murine embryogenesis: spatial and temporal mapping. *Blood* 87:557–566.

Millauer, B., Wizigmann-Voos, S., Schnurch, H., et al. (1993) High affinity VEGF binding and developmental expression suggest Flk-1 as a major regulator of vasculogenesis and angiogenesis. *Cell* 72: 835–846.

Moore, M. A. and Metcalf, D. (1970) Ontogeny of the haemopoietic system: yolk sac origin of *in vivo* and *in vitro* colony forming cells in the developing mouse embryo. *Br. J. Haematol.* 18:279–296.

Moore, M. A. S. and Owen, J. J. T. (1967) Stem-cell migration in developing myeloid and lymphoid systems. *Lancet* 11:658–659.

Mucenski, M. L., McLain, K., Kier, A. B., et al. (1991) A functional c-*myb* gene is required for normal murine fetal hepatic hematopoiesis. *Cell* 65:677–689.

Muller, A. M. and Dzierzak, E. A. (1993) ES cells have only a limited lymphopoietic potential after adoptive transfer into mouse recipients. *Development* 118:1343–1351.

Muller, A. M., Medvinsky, A., Strouboulis, J., Grosveld, F., and Dzierzak, E. (1994) Development of hematopoietic stem cell activity in the mouse embryo. *Immunity* 1:291–301.

Murray, P. (1932) *The development 'in vitro' of blood of the early chick embryo*. Strangeways Research Labatory, Cambridge, 497–521.

Nakano, T., Kodama, H., and Honjo, T. (1994) Generation of lymphohematopoietic cells from embryonic stem cells in culture. *Science* 265:1098–1101.

Niki, M., Okada, H., Takano, H., et al. (1997) Hematopoiesis in the fetal liver is impaired by targeted mutagenesis of a gene encoding a non-DNA binding subunit of the transcription factor, polyomavirus enhancer binding protein 2/core binding factor. *Proc. Natl. Acad. Sci. USA* 94: 5697–5702.

Nishikawa, S. I. (2001) A complex linkage in the developmental pathway of endothelial and hematopoietic cells. *Curr. Opin. Cell Biol.* 13:673–678.

Nishikawa, S. I., Nishikaws, S., Kawamoto, H., et al. (1998) In vitro generation of lymphohematopoietic cells from endothelial cells purified from murine embryos. *Immunity* 8:761–769.

North, T., Gu, T. L., Stacy, T., et al. (1999) Cbfa2 is required for the formation of intra-aortic hematopoietic clusters. *Development* 126: 2563–2575.

Ohmura, K., Kawamoto, H., Lu, M., et al. (2001) Immature multipotent hemopoietic progenitors lacking long-term bone marrow-reconstituting activity in the aorta-gonad-mesonephros region of murine day 10 fetuses. *J. Immunol.* 166:3290–3296.

Okuda, T., van Deursen, J., Hiebert, S. W., Grosveld, G., and Downing J. R. 91996) AML1, the target of multiple chromosomal translocations in human leukemia, is essential for normal fetal liver hematopoiesis. *Cell* 84:321–330.

Owen, J. J. and Ritter, M. A. (1969) Tissue interaction in the development of thymus lymphocytes. *J. Exp. Med.* 129:431–442.

Palacios, R., Golunski, E., and Samaridis, J. (1995) In vitro generation of hematopoietic stem cells from an embryonic stem cell line. *Proc. Natl. Acad. Sci. USA* 92:7530–7534.

Palis, J., McGrath, K. E., and Kingsley, P. D. (1995) Initiation of hematopoiesis and vasculaogenesis in murine yolk sac explants. *Blood* 86:156–163.

Palis, J., Robertson, S., Kennedy, M., Wall, C., and Keller, C. (1999) Development of erythroid and myeloid progenitors in the yolk sac and embryo proper of the mouse. *Development* 126:5073–5084.

Palis, J., Chan, R. J., Koniski, A., Patel, R., Starr, M., and Yoder, M. C. (2001) Spatial and temporal emergence of high proliferative potential hematopoietic precursors during murine embryogenesis. *Proc. Natl. Acad. Sci. USA* 98:4528–4533.

Perah, G. and Feldman, M. (1977) In vitro activation of the in vivo colony-forming units of the mouse yolk sac. *J. Cell. Physiol.* 91: 193–199.

Perlingeiro, R. C., Kyba, M., and Daley, G. Q. (2001) Clonal analysis of differentiating embryonic stem cells reveals a hematopoietic progenitor with primitive erythroid and adult lymphoid-myeloid potential. *Development* 128:4597–4604.

Pevny, L., Lin, C.-S., D'Agati, V., Simon, M. C., Orkin, S. H., and Costantini, F. (1995) Development of hemopoietic cells lacking transcription factor GATA-1. *Development* 121:163–172.

Porcher, C., Swat, W., Rockwell, K., Fujiwara, Y., Alt, F. W., and Orkin, S. H. (1996) The T cell leukemia oncoprotein SCL/tal-1 is essential for development of all hematopoietic lineages. *Cell* 86:47–57.

Potocnik, A. J., Kohler, H., and Eichmann, K. 91997) Hemato-lymphoid in vivo reconstitution potential of subpopulations derived from in vitro differentiated embryonic stem cells. *Proc. Natl. Acad. Sci. USA* 94:10,295–10,300.

Robb, L. (1977) Hematopoiesis: origin pinned down at last? *Curr. Biol.* 7:10–12.

Robb, L., Rasko, J. E. J., Bath, M. L., Strasser, A., and Begley, C. G. (1995) scl, a gene frequently activated in human T cell leukaemia, does not induce lymphomas in transgenic mice. *Oncogene* 10: 205–209.

Robb, L., Elwood, N. J., Elefanty, A. G., Kontgen, F., Li, R., Barnett, L. D., and Begley, C. G. (1996) The *scl* gene product is required for the generation of all hematopoietic lineages in the adult mouse. *EMBO J.* 15:4123–4129.

Robertson, S. M., Kennedy, M., Shannon, J. M. and Keller, G. (2000) A transitional stage in the commitment of mesoderm to hematopoiesis requiring the transcription factor SCL/tal-1. *Development* 127:2447–2459.

Sabin, F. R. (1920) Studies on the origin of blood vessels and of red corpuscles as seen in the living blastoderm of the chick during the second day of incubation. *Contrib. Embryol.* 9:213–262.

Sanchez, M. J., Holmes, A., Miles, C., and Dzierzak, E. (1996) Characterization of the first definitive hematopoietic stem cells in the AGM and liver of the mouse embryo. *Immunity* 5:513–525.

Sasaki, K., Yagi, H., Bronson, R. T., et al. (1996) Absence of fetal liver hematopoiesis in mice deficient in transcriptional coactivator core binding factor beta. *Proc. Natl. Acad. Sci. USA* 93:12,359–12,363.

Schuh, A. C., Faloon, P., Hu, Q. L., Bhimani, M., and Choi, K. (1999) In vitro hematopoietic and endothelial potential of flk-1(–/–) embryonic stem cells and embryos. *Proc. Natl. Acad. Sci. USA* 96: 2159–2164.

Shalaby, F., Rossant, J., Yamaguchi, T. P., et al. (1995) Failure of blood-island formation and vasculogenesis in Flk-1 deficient mice. *Nature* 376:62–66.

Shalaby, F., Ho, J., Stanford, W. L., et al. (1997) A requirement for Flk1 in primitive and definitive hematopoiesis and vasculogenesis. *Cell* 89:981–990.

Shivdasani, R. A., Mayer, E. L., and Orkin, S. H. (1995) Absence of blood formation in mice the T-cell leukaemia oncoprotein tal-1/SCL. *Nature* 373:432–434.

Stainier, D. Y., Weinstein, B. M., Detrich, H. W. 3rd, Zon, L. I., and Fishman, M. C. (1995) Cloche, an early acting zebrafish gene, is required by both the endothelial and hematopoietic lineages. *Development* 121:3141–3150.

Steiner, R. and Vogel, H. (1973) On the kinetics of erythroid cell differentiation in fetal mice. I. Microspectrophotometric determination of the hemoglobin content in erythroid cells during gestation. *J. Cell. Physiol.* 81:323–338.

Tavin, M., Coulombel, L., Luton, D., Clemente, H. S., Dieterlen-Lievre, F., and Peault, B. (1996) Aorta-associated CD34+ hematopoietic cells in the early human embryo. *Blood* 87:67–72.

Toles, J. F., Chui, D. H., Belbeck, L. W., Starr, E., and Barker, J. E. (1989) Hemopoietic stem cells in murine embryonic yolk sac and peripheral blood. *Proc. Natl. Acad. Sci. USA* 86:7456–7459.

Tsai, F. Y., Keller, G., Kuo, F. C., et al. (1994) An early haematopoietic defect in mice lacking the transcription factor GATA-2. *Nature* 371:221–226.

Tsang, A. P., Fujiwara, Y., Hom, D. B., and Orkin, S. H. (1998) Failure of megakaryopoiesis and arrested erythropoiesis in mice lacking the GATA-1 transcriptional cofactor FOG. *Genes Dev.* 12:1176–1188.

Visvader, J. E., Fujiwara, Y., and Orkin, S. H. (1998) Unsuspected role for the T-cell leukemia protein SCL/tal-1 in vascular development. *Genes Dev.* 12:473–479.

Wagner, R. C. (1980) Endothelial cell embryology and growth. *Adv. Microcirc.* 9:45–75.

Wang, Q., Stacy, T., Miller, J. D., et al. (1996a) The CBFbeta subunit is essential for5 CBFalpha2 (AML1) function in vivo. *Cell* 87:697–708.

Wang, Q., Stacy, T., Binder, M., Marin-Padilla, M., Sharpe, A. H., and Speck, N. A. (1996b) Disruption of the Cbfa2 gene causes necrosis and hemorrhaging in the central nervous system and blocks definitive hematopoiesis. *Proc. Natl. Acad. Sci. USA* 93:3444–3449.

Warren, A. J., Colledge, W. H., Carlton, M. B. L., Evans, M. J., Smith, A. J. H., and Rabbitts, T. H. (1994) The oncogenic cysteine-rich LIM domain protein Rbtn2 is essential for erythroid development. *Cell* 78:45–57.

Weiss, M. J. and Orkin, S. H. (1995) GATA transcription factors: key regulators of hemopoiesis. *Exp. Hematol.* 23:99–107.

Weissman, I., Papaioannou, V., and Gardner, R. (1978) Fetal hematopoietic origins of the adult hematolymphoid system. In: *Differentiation of Normal and Neoplastic Hematopoietic Cells.* (Calarkson, B., Marks, P. A., and Tills, J. E., eds.), Cold Spring Harbor Laboratory Press, Cold Spring Harbor, NY, pp. 33–47.

Wong, P. M., Chungm S. W., Chui, D. H., and Eaves, C. J. (1986) Properties of the earliest clonogenic hemopoietic precursors to appear in the developing murine yolk sac. *Proc. Natl. Acad. Sci. USA* 83: 3851–3854.

Wood, H. B., May, G., Healy, L., Enver, T., and Morriss-Kay, G. M. (1997) CD34 expression patterns during early mouse development are related to modes of blood vessel formation and reveal additional sites of hematopoiesis. *Blood* 90:2300–2311.

Yamada, Y., Warren, A. J., Dobson, C., Forster, A., Pannell, R., and Rabbitts, T. H. (1998) The T cell leukemia LIM protein Lmo2 is necessary for adult mouse hematopoiesis. *Proc. Natl. Acad. Sci. USA* 95:3890–3895.

Yamada, Y., Pannell, R., Forster, A., and Rabbitts, T. H. (2000) The onogenic LIM-only transcription factor Lmo2 regulates angiogenesis but not basculogenesis in mice. *Proc. Natl. Acad. Sci. USA* 97: 320–324.

Yamaguchi, T. P., Dumont, D. J., Conlon, R. A., Breitman, M. L., Rossant, J. (1993) *flk-1*, and *flt*-related receptor tyrosine kinase is an early marker for endothelial cell precursors. *Development* 118:489–498.

Yano, M., Iwama, A., Nishio, H., Suda, J., Takada, G., and Suda, T. (1997) Expression and function of murine receptor tyrosine kinases, TIE and TEK, in hematopoietic stem cells. *Blood* 89:4317–4326.

Yoder, M. C. and Hiatt, K. (1997) Engraftment of embryonic hematopoietic cells in conditioned newborn recipients. *Blood* 89:2176–2183.

Yoder, M. C., Cumming, J. G., Hiatt, K., Mukherjee, P., and Williams, D. A. (1996) A novel method of myeloablation to enhance engraftment of adult bone marrow cells in newborn mice. *Biol. Blood Marrow Tansplant* 2:59–67.

Yoder, M. C., Hiatt, K., and Mukherjee, P. (1997) In vivo repopulating hematopoietic stem cells are present in the murine yolk sac at day 9.0 postcoitus. *Proc. Natl. Acad. Sci. USA* 94:6776–6780.

Zanjani, E. D., Ascensao, J. L., and Tavassoli, M. (1993) Liver-derived fetal hematopoietic stem cells selectivelya nd preferentially home to the fetal bone marrow. *Blood* 81:399–404.

13 Stromal Support of Hematopoiesis

PIERRE CHARBORD, MD, DR INSERM

Bone marrow stromal cells make up the microenvironment for hematopoiesis. The prototype for stem cell and lineages is that of the hematopoietic stem cell (HSC), as demonstrated by clonal analysis. The fate of the HSC is determined largely by the microenvironment (niche) provided by the mesenchymal stem cells (MSC). MSCs give rise to cells constituting the hematopoietic microenvironnement—i.e., stromal cells, endothelial cells, vascular smooth muscle cells (VSMCs), adipocytes, and osteoblasts—but also to other cell types—i.e., chondrocytes; sarcomeric muscle; and, in certain reports, neuroectoderm or endodermal cells. Transdifferentiation of MSC progeny seems more likely than in other cell lineages, and stromal cells from the different hematopoietic sites (bone marrow, fetal liver, aorta-gonad-mesonephros region) express a variety of mesenchymal cell–type markers, including markers for VSMCs, adipocytes, and osteoblasts. Stromal cells act through a number of mediators, cytokines, adhesion molecules, peptides, hormones, and other molecules such as wnts and eicosanoids. Cytokines and chemokines include the colony-stimulating factors, interleukin-6 (IL-6), leukemia inhibitory factor, IL-1, IL-7, IL-8, stem cell factor, flt3 ligand, hepatocyte growth factor, thrombopoietin, insulin-like growth factor-1, transforming growth factor-β (TGF-β), γ-interferon-inducible protein-10 (IP-10), monocyte chemoattractant protein-1 (MCP-1), and stromal-derived factor-1 (*see* Table 2 in this chapter). IL-1 is a major inducer of the production of the other cytokines. TGF-β is the most potent inhibitor of hematopoiesis. It induces expression of inhibitory chemokines, such as MCP-1 and IP-10, and decreases expression of stromal adhesion molecules. Extracellular matrix collagens, laminins, and fibronectins each appear to have a role in hematopoiesis through binding to a number of cell adhesion molecules (*see* Table 3 of this chapter). This indicates a highly controlled, multifactorial, and redundant regulation of hematopoiesis by mesenchymal stromal cells.

13.1. INTRODUCTION

Stromal cells constitute the cell population that assists the hematopoietic stem cell (HSC) and its progeny, i.e., the set of cells modulating quiescence, self-renewal and commitment of HSCs and the proliferation, maturation, and apoptosis of more mature hematopoietic cells. Stromal cells are easily defined in culture, forming the non-hematopoietic adherent cell component from long-term cultures. In vivo they make up the microenvironment of hematopoiesis, comprising the set of nonhematopoietic cells from the different hematopoietic sites.

The development of the concept of hematopoietic stroma is intimately associated with that of HSC and stem cells in general. Major hallmarks are given on Table 1. In 1924, Maximow developed the idea of a common precursor cell for hematopoiesis and mesenchyme, basing his hypothesis on morphological studies. Experimental works made later provided evidence for two types of stem cells in the hematopoietic sites, one giving rise to the hematopoietic lineages, the HSCs, and the other yielding the lineages of the hematopoietic microenvironment, the mesenchymal stem cell (MSC). However, the concept of a common hematopoietic/mesenchymal stem cell (H/MSC) lingered, supported from time to time by controversial experimental evidence. In 1987, Singer et al. showed that stromal cell clones transformed by the SV-40 virus gave rise to cells of both mesenchymal and hematopoietic nature, which suggested that the infection had reactivated a cell acting early on in development. Such cells might indeed be operative in some patients with myeloproliferative syndrome where stromal cells from long-term cultures expressed the same glucose-6-phosphate dehydrogenase isotype as the neoplastic cells of the hematopoietic lineages, while equal amounts of the two allelic isoforms were expressed in normal tissues (Singer et al., 1984). In 1992, Huang and Terstappen published the phenotype of the presumptive H/MSC but retracted the data 2 yr later. In 1995, Waller et al. followed suit, distinguishing the phenotype of fetal marrow stromal precursors from that of hematopoietic precursors. Recently Huss et al. (1995, 2000) provided some evidence that hematopoietic cells could derive from marrow fibroblasts in dogs and mice.

From: *Stem Cells Handbook*
Edited by: S. Sell © Humana Press Inc., Totowa, NJ

Table 1
Historical Hallmarks

Year	*Author*	*Concept/Data*
1924	Maximow	Hematopoietic/mesenchymal common stem cell
1951	Lorenz et al.	Experimental bone marrow transplantation
1963	Till et al.	Stochastic model for HSC
1968	Wolf and Trentin	Hematopoietic-inductive microenvironments
1974	Friedenstein	CFU-f transplantation
1976	Dexter et al.	Long-term marrow culture (mouse)
1977	Dexter and Moore	In vitro duplication of steel defects
1978	Schofield	Stem cell niche
1980	Gartner and Kaplan	Long-term marrow culture (human)
1984	Whitlock and Witte	Lymphoid long-term marrow culture
1988	Owen	Stromal stem cell
1989	Andrews, R. G. et al.	Pre-CFUs
1989	Sutherland et al.	LTC-ICs
1991	Caplan et al.	MSCs
1991	Ploemacher et al.	CAFCs
1996	Wineman et al.; Friedrich et al.	Stromal heterogeneity at clonal level
2002	Hackney et al.	Molecular characterization of HSC niche

Ten years after the first experimental transplantation of HSCs (Lorenz et al., 1951), Till and McCulloch (1961) described the colony-forming unit in the spleen (CFU-S) of transplanted animals. This team developed the stochastic model of CFU-S differentiation, whereby the fate of a given cell (commitment toward a lineage or self-renewal) appeared to be random (Till et al., 1963). In the 1970s, Wolf and Trentin challenged the stochastic model by considering that the stem cell behavior was determined by external factors, cellular and molecular, and posited the existence of "hematopoietic inductive microenvironments" (reviewed in Trentin, 1989). The major feature of the model was the opposition between spleen territories, most of which were favorable to erythropoiesis, and marrow territories inducing granulopoiesis. In the meantime, Friedenstein (reviewed in Friedenstein, 1980), working on guinea pig and mouse fibroblastic clonal colonies (CFU-f), provided other data in favor of the microenvironmental model. A third of the marrow CFU-f implanted under the kidney capsule gave rise to both bone and hematopoiesis. Since hematopoietic cells were of recipient origin, it appeared probable that a specific subpopulation of donor fibroblasts had recruited recipient circulating HSCs and favored the generation of hematopoietic foci.

These experiments in vivo suggested the potential role of microenvironmental cells for stem cell behavior. Demonstration of the existence and functional role of such "stromal" cells was provided in 1976 when Dexter et al. established long-term bone marrow cultures from mice. The generation of an adherent layer proved mandatory for CFU-S maintenance, and microenvironmental and stem cell defects observed in mice with steel and white-spotting mutations could be duplicated in the culture system (Dexter and Moore, 1977). Subsequent experiments provided further evidence for the HSC-supportive role of stromal cells. The adherent layer was where CFU-S with sustained self-renewal capacity and HSCs with long-term competitive repopulating ability were maintained, and where few HSC clones proliferated (Mauch et al., 1980; Harrison et al., 1987; Fraser et al., 1992). HSCs under Dexter conditions, although generating myeloid cells, remained bipotential, being able to provide B-lymphocytes after switch to Witte–Whitlock conditions (Whitlock et al., 1984; Lemieux et al., 1995). A property of HSCs growing onto stromal layers, probably due to their expression of Edg2 lysophospholipid receptors (Yanai et al., 2000), is their propensity to migrate under the stromal cells, before subsequent proliferation and differentiation giving rise to cobblestone areas. Ploemacher et al. (1991) showed that the time of appearance of cobblestone areas depended on the time of entry into cycle of the cobblestone area–forming cells (CAFCs) and found highly significant correlations among different classes of HSCs as assessed in vivo and the sequential generation of CAFCs. These data suggested that the long-term culture system may serve as a useful surrogate procedure for investigating stem cell behavior, being able, to a certain extent, to substitute for several in vivo assays.

In 1980, Gartner and Kaplan adapted long-term marrow cultures to humans. The fact that progenitors (revealed by hematopoietic CFUs grown in semisolid medium) were generated for only several weeks (instead of months, as in murine cultures) entertained doubts concerning the ability of the human culture system to allow the maintenance of HSCs. A set of data proved to the contrary. Two waves of progenitor production were observed; the second wave occurring after 4 to 5 wk would reflect the influx of CFUs from a more immature compartment of pre-CFU cells. This compartment was made patent by showing that marrow samples depleted in progenitors and cultured onto stromal layers yielded progenitors in increasing number with time in culture (Andrews, R. G. et al., 1989). A quantitative evaluation of the pre-CFU cells was provided with the development of the long-term culture-initating cell (LTC-IC) assay and the adaptation of the CAFC assay to humans (Sutherland et al., 1989; Breems et al., 1994). As for the murine system, these data indicated that stromal cells were critically important for HSCs, preventing apoptosis and providing, at times adequate with the internal clock of the primitive cell, signals for proliferation and determination.

The concept of stem cell niche was first expounded by Schofield in 1978. Schofield posited that when HSCs are fixed by surrounding microenvironmental cells, they would maintain a stable potential of self-renewal. In 1994, Weissman suggested that commitment to the different hematopoietic lineages would also be controlled in specific niches. The generation of continuous cloned murine marrow lines has confirmed the existence of niches. Some lines are able to maintain either myelopoiesis or B-lymphopoiesis (Friedrich et al., 1996), while others show a remarkable capacity to sustain erythropoiesis (Yanai et al., 1989). Most important, very few lines prove able to maintain the HSC potential in vitro, either in syngeneic (culture of murine HSCs on murine lines) or in xenogeneic assays (culture of human HSCs on murine lines) (Issaad et al., 1993; Wineman et al., 1993, 1996; Szilvassy et al., 1996). One of these lines, AFT024, derived from 12.5 d postcoitus mouse fetal liver, has been extensively studied by Moore et al. (1997). The study of transcripts differentially expressed by these cells may serve for the molecular characterization of the niche with HSC-supportive ability. The recent study of *Drosophila* germinal stem cells (Spradling et al., 2001) clearly indicates that the stem cells' cardinal balance between self-renewal and differentiation is preserved when they are in physical contact with stromal cells and associated extracellular matrix (ECM), which fully validates Schofield's hypothesis. Any type of stem cells is probably controlled by associated stromal cells. One extreme hypothesis is that the stroma of a given tissue would specify stem cells that would otherwise remain totipotent.

13.2. PHENOTYPE OF STROMAL CELLS

MSCs are multipotential progenitors that give rise to the different cells of the hematopoietic microenvironment, i.e., stromal cells, endothelial cells, adipocytes, and osteoblasts (Owen, 1988; Caplan, 1991; Prockop, 1997; Reyes et al., 2001). These cells also generate lineages that are not part of the microenvironment, chondrocytes and sarcomeric muscle cells (cardiac and skeletal), and, as reported recently, nonmesodermal lineages, cells from the neuroectoderm (neurons, astrocytes, oligodendrocytes) or the endoderm (hepatocytes) (Azizi et al., 1998; Oh et al., 2000). Whether cells with larger differentiation potential might be totipotent stem cells residing in the bone marrow (Weissman, 2000) remains an open question.

MSCs reside in bone marrow and have recently been found in human fetal liver (Campagnoli et al., 2001). Whether the fetal liver is the site of MSC expansion, as it is for HSCs, remains to be studied. MSCs have not yet been described in the sites of embryonic hematopoiesis, the aorta–gonad–mesonephros (AGM) region, and the yolk sac. However, indirect arguments suggest their presence in yolk sac. In the absence of vascular endothelial growth factor (VEGF) signaling, mesodermal precursors appear unable to home at the only site, that of yolk sac, where they can give rise to hematopoietic precursors, but commit to alternative cell fates such as the amnion; these data suggest a deep anomaly of mesodermal precursor and not only hemangioblastic development (Shalaby et al., 1997). Moreover, we have observed that murine yolk sac stromal cell lines yielded vascular smooth muscle–like cells, suggesting that mesodermal precursors giving rise to yolk sac endothelial cells in vivo may differentiate toward other mesodermal lineage in vitro (Rémy-Martin et al., 1999).

Some characteristics clearly distinguish MSCs from HSCs (Bianco and Robey, 2000). In the hematopoietic system there is a strict requirement for perennial cell renewal due to a daily loss of 10^9 to 10^{11} cells/kg of body weight (in humans). In the mesenchymal system in the adult, under stationary conditions such steady-state loss is nonexistent. For example, the turnover of bone is usually a local event, involving the coordinated excavation and refilling of bone by hematopoietic lineage–derived osteoclasts and mesenchymal-derived osteoprogenitor cells, but systemic changes in bone may occur in certain disease states, such as osteogenesis imperfecta (Pereira et al., 1995). Lineages from MSCs are not so strictly delineated as lineages from HSCs. This property of plasticity is exemplified in cell clones where a switch from the adipocytic to the osteoblastic lineage or from hypertrophic chondrocytes to osteoblasts is induced (Bennett et al., 1991; Gentili et al., 1993). Such plasticity might be explained by a stochastic model in which the plastic nature of the committed cells entering the differentiation compartment would be the result of multiple random activation and repression events affecting transcription factor genes within MSCs (Dennis and Charbord, 2002). Such plasticity should remove the theoretical need for stem cells with indefinite self-renewal capacity because some cells in the progeny of the mother cell would express, anew, genes that were transitorily suppressed. Therefore, the MSC system appears to be distinct from other classic stem cell systems (hematopoietic, intestinal, or epidermal), wherein the existence of a stem cell compartment is required by the irreversible differentiation of the progeny.

We have shown that stromal cells from human long-term marrow cultures expressed a number of markers specific for vascular smooth muscle cells (VSMCs) : cytoskeletal proteins (α-SM actin, SM-actinin, calponin, h-caldesmon, metavinculin, SM22-α and smooth muscle myosin heavy chain SM-1) and ECM glycoproteins (fibronectins comprising CS-1, EDa, and EDb domains; laminin-β-2; thrombospondin-1; and tropo-elastin) (Galmiche et al., 1993; Li et al., 1995; Rémy-Martin et al., 1999). VSMC-like stromal cells are a major component of murine Dexter and Whitlock–Witte cultures (Funk and Witte, 1992; Penn et al., 1993). Moreover, by studying a number of murine lines established not only from cultures of bone marrow, but also from spleen, fetal liver, AGM, and yolk sac, we have found a clear-cut VSMC differentiation for most, with expression of desmin in addition to the markers just described (Rémy-Martin et al., 1999). The existence of several steps along the VSMC differentiation pathway, where differentiation is arrested in some lines, indicates the heterogeneity of the stromal population. Such heterogeneity is also a characteristic of mesenchymal cells that acquire VSMC characteristics in physiological or pathological conditions (Desmoulière and Gabbiani, 1995). Stromal cells are not only essential for hematopoietic support, but are also involved in the trafficking of hematopoietic cells within bone marrow since they modulate the tightness of endothelial junctions in cells lining marrow sinuses (Lichtman, 1981). The contractility of the cells may therefore serve a specific funtional role.

Cells with smooth muscle differentiation are detected in vivo at hematopoietic sites: myoid cells, barrier cells, and pericytes in bone marrow from mice and humans; cells undergoing an epithelial-to-mesenchymal transition in fetal liver; early differentiated smooth muscle cells underneath hematopoietic foci of embryonic aorta (Weiss and Geduldig, 1991; Galmiche et al., 1993; Kiassov et al., 1995; Charbord et al., 1996; Tavian et al., 1999). Barrier cells have also been described in the spleen of rodents

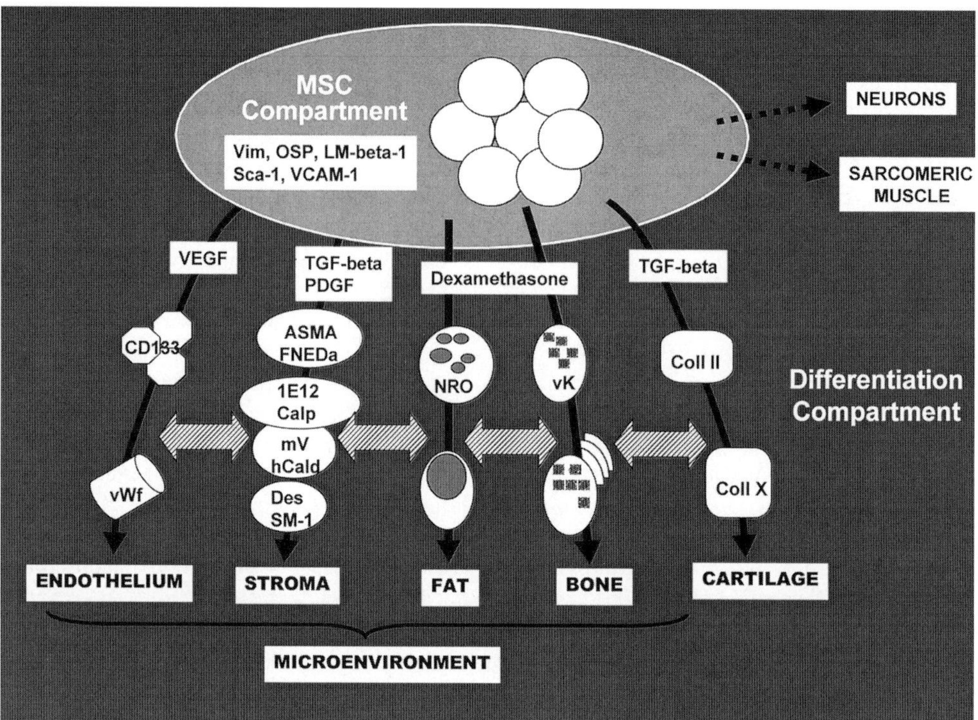

Fig. 1. Diagram of MSC. Vim, vimentin; OSP, osteopontin; LM, laminin; Sca, stem cell antigen; ASMA, α-SM actin; FNEDa, fibronectin EDa⁺; mV, metavinculin; hcald, h-caldesmon; 1E12, SM-actinin recognized by the 1E12 antibody; Calp, calponin; Des, desmin; SM-1, SM myosin heavy chain-1; vWf, von Willebrand factor; NRO, Nile red O; vK, von Kossa stain; Coll, collagen.

(Toccanier-Pelte et al., 1987) and kidneys of salmonids, the major hematopoietic sites in these fishes (Alvarez et al., 1996). It therefore appears probable that MSCs would differentiate following a VSMC differentiation pathway specific for hematopoietic-supportive tissues. Certain cytokines, such as transforming growth factor-β (TGF-β) and platelet-derived growth factor (PDGF), implicated in VSMC recruitment and differentiation and known to be growth factors for stromal precursors, may modulate, at the blood-forming sites, stromal growth and differentiation in coordination with ECM molecules.

The VSMC pathway is the major pathway of differentiation observed in stromal cells, but not the exclusive one. Certain features relate stromal cells to other mesenchymal lineages. Stromal cells with VSMC differentiation express alkaline phosphatase and collagen I, considered early markers of the osteoblastic lineage. Some human stromal cells that have the morphology of an adipocytic cell at one pole (lipid-filled vesicles) and a VSMC morphology at the other pole, featuring bundles of myofilaments, caveolae, and fibronexus, may be considered as cell isoforms intermediate between adipocytes and VSMCs (Dennis et al., 2002). The existence of such isoforms is in agreement with the cellular plasticity of the mesenchymal lineages.

Summary of the data on the phenotype is given in Fig. 1. In addition to the markers for the stromal lineage following a VSMC differentiation pathway, markers for other lineages are shown: CD133 characterizes early endothelial cells and von Willebrand factor is found in Weibel–Palade bodies of mature endothelial cells, Nile red O stains lipid vesicles, bone deposits are revealed using von Kossa stain, collagen II is specific for chondrocytes, and collagen X is induced in hypertrophic cartilage. Major cytokines or hormones required for each pathway are indicated: VEGF for endothelial cells, PDGF and TGF-β for stroma, dexamethasone for adipocytes and osteoblasts, and TGF-β for chondrocytes. Double arrows between pathways indicate the existence of intermediate isoforms and shifts between pathways characterizing the plasticity of the system. Finally, some markers of the MSCs are shown: intracellular proteins, such as the intermediate filament component vimentin and the adhesive glycoproteins osteopontin and laminin-β-1, or cell membrane antigens such as the stem cell antigen-1 and the vascular cell adhesion molecule-1 (VCAM-1). These membrane antigens have been used for the isolation from bone marrow of stromal precursors (van Vlasselaer et al., 1994; Simmons et al., 2001). Other membrane molecules have also been used for this purpose : antigens recognized by the Stro-1 monoclonal antibody, Thy-1, MUC18/CD146, α-1 integrin subunit, and endoglin (Simmons and Torok-Storb, 1991; Guerriero et al., 1997; Filshie et al., 1998; Deschaseaux and Charbord, 2000; Majumdar et al., 2000).

13.3. MEDIATORS INVOLVED IN CONTROL OF HEMATOPOIESIS BY STROMAL CELLS

Almost three decades of study of murine and human long-term cultures has disclosed that the generation or maintenance of HSCs and progenitors was regulated by a large number of mediators, cytokines, adhesion molecules, peptides, hormones, molecules involved in organogenesis, and lipid mediators. Evidence for specific molecules has already been reviewed (Charbord, 2001). In this chapter, I therefore emphasize recent data.

13.3.1. CYTOKINES The macrophage colony-stimulating factor (M-CSF) was the first cytokine detected in stromal cells. It is constitutively expressed by stromal cells because there is no need to stimulate the cells (irradiation, treatment with mitogens, or inflammatory cytokines) to detect the cytokine. Different isoforms of M-CSF are generated by alternative splicing; one isoform, slowly released from the membrane, may serve as adhesion molecule. Moreover, M-CSF binds to proteoglycans, resulting in high-molecular-weight ECM-bound molecule. Experiments using embryonic stem cells cultured onto stromal cells from M-CSF-null mice indicate that the cytokine promotes the macrophagic lineage while impeding the generation of other lineages (Nakano et al., 1994). M-CSF may also serve other purposes in long-term cultures, because it is a growth factor for murine stromal precursors that express its tyrosine kinase receptor (Tanaka-Douzono et al., 2001). Detection of the granulocyte macrophage CSF (GM-CSF) has been less easy than that of M-CSF, usually requiring stimulation of stromal cells. While stromal cell production of GM-CSF probably explains most of the granulomonocytic progenitor output in long-term marrow cultures (Charbord et al., 1991), the study of stromal feeders engineered to synthesize and release continuously GM-CSF suggests that this cytokine does not exert a significant role on more immature precursors (Sutherland et al., 1991). Finally, GM-CSF is a critical factor for microenvironmental cell homeostasis because it modifies the macrophage–to–stromal cell ratio (Charbord et al., 1991). Granulocyte CSF (G-CSF) is detected only in stimulated stromal cells. G-CSF appears to be active on growth of immature hematopoietic precursors in long-term marrow cultures, as indicated by the study of G-CSF-engineered stromal feeders (Hogge et al., 1991). Multi-CSF or interleukin-3 (IL-3) has not been detected in stromal cells, except for subliminal levels in yolk sac mesodermal or endodermal stromal cell lines (Yoder et al., 1994).

Many cytokines from the gp 130 receptor family are constitutively expressed by stromal cells but greatly enhanced by inflammatory cytokines. IL-6, which has been cloned from a murine stromal cell line (Chiu et al., 1988), in long-term marrow cultures does not appear to be effective on immature hematopoietic progenitors (Hauser et al., 1997), but it remains possible that other stromal-derived members of the family substitute for IL-6 in neutralizing studies. In addition, IL-6 is a growth factor for stromal cells (Sensebé et al., 1995; Hauser et al., 1997). The downregulation of leukemia inhibitory factor (LIF) under conditions of osteogenesis and its upregulation under conditions favorable for hematopoiesis suggest its potential role in the equilibrium between bone formation and hematopoietic support by marrow MSCs (Haynesworth et al., 1996). IL-11, cloned from a nonhuman primate stromal cell line (Paul et al., 1990), may play a role on the differentiation of hematopoietic progenitors and on stromal formation.

The stem cell factor (SCF) or c-kit ligand has been cloned by one group (Williams et al., 1990) using a stromal cell line that proved inactive in direct coculture assays of stromal cells and hematopoietic precursors, which confirmed its synergistic activity. Stromal cells express not only the soluble isoform, but also the transmembrane isoform that proved more efficient for hematopoietic progenitor cell maintenance. Several reports suggest that the factor is critical for myelopoiesis, but not essential for the survival of the most primitive precursors (Kodama et al., 1992; Wineman et al., 1993; Sutherland et al., 1993). The flt3 ligand (FL) (Lisovsky et al., 1996) and c-met- ligand (hepatocyte growth factor [HGF]) (Takai et al., 1997; Weimar et al., 1998) are other related cytokines whose receptor has tyrosine kinase activity, produced by stromal cells and with synergistic activity on HSCs/progenitors.

Constitutive expression of thrombopoietin (TPO), or c-mpl ligand, is observed in stromal cells, which is in agreement with the observation that bone marrow parasinusal cells express the cytokine (Sungaran et al., 1997). On the contrary, it is unlikely that stromal cells produce erythropoietin (EPO) efficiently because EPO has to be added to Dexter cultures for erythropoiesis maintenance (Yanai et al., 1989).

IL-7, which has been cloned from murine marrow stromal cells, is constitutively expressed by stromal cells from Witte–Whitlock cultures, where it is one of the factors responsible for B-lymphopoiesis. While the expression of IL-7 is constitutive, its release depends on tight contact between stromal cells and B-lymphocytes (Stephan et al., 1998).

IL-8, a constitutive stromal chemokine whose secretion is greatly increased by IL-1 or tumor necrosis factor-α (TNF-α), is an inhibitor of hematopoiesis, as are the stromal chemokines γ-interferon-inducible protein-10 (IP-10) and monocyte chemoattractant protein-1 (MCP-1) (Gautam et al., 1995). On the contrary, stromal cell–derived factor (SDF-1) cloned from murine stromal cell lines (Nagasawa et al., 1994 ; Bleul et al., 1996) is a stimulating factor for B-cell precursors and a potent chemoattractant of HSCs. Macrophage inflammatory protein-1α (MIP-1-α), another constitutive stromal chemokine enhanced by IL-1 or TNF-alpha, may improve, in association with other growth factors, the survival of HSCs (Gupta et al., 1996).

IL-1 is rarely detected without stimulation in stromal cells. It is a major inducer or enhancer of the production of other cytokines, GM-CSF, G-CSF, IL-6, IL-8, LIF, MIP-1-α, SCF, TNF-α, and IL-1 itself. IL-1 is also involved in stromal cell formation and phenotype. It is a growth factor for stromal precursors (Yan et al., 1990; Sensebé et al., 1995), it downregulates collagen I expression (Andrews, D. F. et al., 1989) and upregulates synthesis and release of plasminogen activator (Hannocks et al., 1992), it inhibits adipocyte formation (Delikat et al., 1993), and it affects the pattern of expression of ECM and cell adhesion molecules. Like IL-1, TNF-α, detected in stimulated stromal cells, is a potent inducer or enhancer of other cytokines and may affect stromal formation and phenotype.

Probably the most potent HSC inhibitor constitutively expressed by stromal cells is TGF-β. The addition of neutralizing anti-TGF-β1 antibodies to long-term cultures induces the expansion of hematopoietic cells. Addition of TGF-β1 to Witte–Whitlock cultures causes inhibition of lymphoid progenitors due, in part, to downregulation of stromal secretion of IL-7 (Tang et al., 1997). TGF-β1 also inhibits stromal cell release of SCF, induces the synthesis of MCP-1 or IP-10 inhibitory chemokines, and decreases the expression of stromal cell adhesion molecules, critical for HSC/progenitor cell adherence. Eventually, TGF-β1 is a potent modulator of stromal cells, increasing colony formation (Sensebé et al., 1995; Kuznetsov et al., 1997; Andrades et al., 1999), modifying the pattern of expression of ECM components (Nemunaitis et al., 1989), and increasing stromal cell proteolytic activity (Hannocks et al., 1992).

Many reports indicate the expression of insulin-like growth factor (IGF-1) and its binding proteins by human and murine stromal cells (Abboud et al., 1991; Grellier et al., 1995; Thomas

Table 2
Major Stromal Cytokines[a]

	Stromal cell		*ECM/stromal*	*Hematopoietic target*			*Stromal*
	Constitutive	*Inducible*	*reservoir*	*pre-CFU*	*CFU*	*Mature cell*	*target*
M-CSF	+	–	+	?	+ (M)	+ (M)	+
GM-CSF	±	+	+	–	+ (GM)	+ (M)	+
G-CSF	±	+	?	+	+ (GM)	+	?
IL-3	?	–	+	+	+	+ (mastocyte)	?
IL-6	+, cloned	+	+	?	+	+ (GM)	+
LIF	+	+	+	?			?
IL-11	+, cloned	+		?	+		+
IL-7	+, cloned	+	+		+ (B)		
TPO	+	+		?	+ (Meg)	+ (Meg)	
EPO	?	?			+ (E)	+ (E)	
SCF	+, cloned	?	+	+	+ (B, E)		+
FL	+, cloned	+	+	?			
HGF	+		+		+		
IL-8	+	+			+		
SDF-1	+, cloned	+		+	+ (B)		
IP-1-α	+	+		?			
MCP-1	+				+ (GM)		
IL-1	?	+		indirect	indirect	indirect	+
TNF-α	–	+	+	indirect	indirect	indirect	+
TGF-β	+	–	+	+	+		+
IGF-1	+		+			+ (B, E)	+
PDGF	+		+				+
FGF-2	+		+				+

[a] M, monocytic; GM, granulomonocytic; E, erythroblastic; B, B-lymphocytic; Meg, megakaryocytic; CFU, colony-forming unit; indirect, indirect effect; ECM, extra-cellular matrix; ?, restricted data; cloned, cloned from stromal cells.

et al., 1999). It has been shown that IGF-1 is a growth factor for erythroblasts and pre-B cells in stromal-based assays (Landreth et al., 1992; O'Prey et al., 1998), and some reports indicate that it is also a growth factor for stromal cells (Jia and Heersche, 2000). PDGF BB, epidermal growth factor and fibroblast growth factors (FGFs) are potent mitogens for stromal cells (Brunner et al., 1991; Gronthos and Simmons, 1995; Sensebé et al., 1995; Kuznetsov et al., 1997) that express their tyrosine kinase receptors (Satomura et al., 1998).

In conclusion, long-term marrow cultures have proven to be a crucial model for the study of cytokines (Table 2) effective on HSCs. Two salient points emerge. First, constitutive expression of soluble cytokine by stromal cells usually results in low concentrations that may be highly effective due to their continuous release and to the proximity of stromal cells to their hematopoietic target cells. Moreover, cytokines might be stored at different sites, resulting in very high local concentrations, owing to their binding to ECM or stromal cell membrane components or related to the existence of transmembrane isoforms. Second, many cytokines are induced or enhanced (by 10- to 1000-fold) by other, usually inflammatory, cytokines, provided by macrophages, T-cells, and HSCs/progenitors. Cell interactions therefore appear to be of critical importance for hematopoietic cytokine synthesis: hematopoietic cells produce mediators effective on stromal cells and stromal cells express cytokines effective on hematopoiesis.

13.3.2. ADHESION MOLECULES In 1985, Zuckerman et al. showed that hematopoiesis maintained in long-term marrow cultures provided the presence of an ECM where collagens served as the framework for other extracellular components. The large representation of collagens in long-term cultures (collagens I, III–VI, XII, and XIV) (Koeningsmann et al., 1992; Klein et al., 1995, 1998), their diffuse distribution in the bone marrow, and the fact that collagen I gene expression is downregulated in cases of anchorage-independant growth of SV40-transformed stromal cells confirm a structural role for these molecules; by contrast, evidence for a role in HSC adhesion is more restricted.

Laminins (LMs) are another set of cytoadhesive glycoproteins largely represented in long-term marrow cultures. Recent studies have shown that the major laminins expressed in bone marrows and stromal cells contain α-4, α-5, β-1, and γ-1 polypeptide chains and are therefore LM-8 (α-4/β-1/γ-1) and LM-10 (α-5/β-1/γ-1) (Gu et al., 1999; Siler et al., 2000). These LMs are detected in the marrow cord and, in humans, in myoid cells. HSCs/progenitors adhere to LM-10 via β-1 integrins. LMs are also involved in the control of HSC migration (Strobel et al., 1997).

Fibronectins are probably the adhesive glycoproteins most studied so far. Stromal cells are able not only to synthesize the molecule and secrete it in the culture supernatant, but also to assemble it in the ECM via α-5/β-1 (VLA-5) integrins (Lerat et al., 1993; van der Velde-Zimmermann et al, 1997). HSCs bind to one or several domains of fibronectin: the CS-1 peptide containing the LDV sequence and ligand for VLA-4 integrins, the III 10 repeat containing RGDS and ligand for VLA-5, and the high-affinity C-terminal heparin-binding IDAPS domain, ligand for VLA-4 integrins, chondroitin sulfate proteoglycan, and CD44. Immature hematopoietic cells may bind to different domains often in a more efficient way than to the whole molecule. Several cytokines, such as IL-3 and SCF, modulate adhesion and migration of HSCs to cell-binding domains by eliciting intracellular signals (inside-out signaling) that modify HSC expression of VLA-4 integrins. Adhesion of HSCs/progenitors to fibronectin leads to increased growth in the presence of cytokines (Schofield et al., 1998), inhibition of proliferation (Hurley et al., 1995), or induction of apoptosis (Sugahara et al., 1994). Such conflictory results may be due to differing hematopoietic target cells, different fibronectin fragments, or the presence of ECM proteins (such as fibulins) (Gu et al., 2000) or cytokines bound to fibronectin. For example, trace amounts of active TGF-β and TNF-α bound to fibronectin (Taipale and Keski-Oja, 1997) may substantially modify the experimental outcome.

Other adhesive glycoproteins described in murine or human cultures comprise thrombospondin-1 and tenascin. Thrombospondin-1 presents binding sites for HSCs and may serve as a storage site for IL-3, SCF, and active TGF-β, possibly via heparan sulfate proteoglycan (Gupta et al., 1996). Hematopoietic output in cultures from tenascin-C-deficient mice is inferior to that of control mice (Ohta et al., 1998). Addition of tenascin-C fully restores the output, while that of fibronectin partially corrects the deficit, which shows a functional overlap between distinct ECM networks.

Proteoglycans have been described early on in long-term marrow cultures. While the role in hematopoiesis of chondroitin sulfate proteoglycans appears to be controversial, that of heparan sulfate proteoglycans (HSPGs) is well recognized. Different protein cores have been detected: the major ECM HSPG is perlecan (Klein et al., 1994) while syndecan-3, syndecan-4, glypican-1, glypican-4 and βglycan are represented on the stromal cell membrane (Drzeniek et al., 1997; Siebertz et al., 1999; Schofield et al., 1999). HSPGs serve as a reservoir for a variety of cytokines (IL-3, IL-6, G-CSF, GM-CSF, LIF, SCF, HGF, FGFs, IL-8, IP-10, MCP-1, and MIP-1α) inducing high local concentrations and/or modifying the cytokine conformation (Gupta et al., 1996). Large HSPGs, heavily O-sulfated on glucosamine residues, appear specific for marrow stromal cells with HSC-supportive ability (Gupta et al., 1998). Being at the crossroad between the two major adhesive and cytokine regulation pathways, HSPGs are essential components of the HSC niche.

The major cell adhesion molecules (CAMs) of the immunoglobulin family expressed by stromal cells are the VCAM-1 and the hematopoietic cell antigen/activated leukocyte cell adhesion molecule (HCA/ALCAM). VCAM-1 is modulated by diverse cytokines, upregulation being observed with IL-1, IL-4, and TNF-α and downregulation with TGF-β1. VCAM-1 is operative in the heterotypic VLA-4: VCAM-1 adhesion of B-lymphoid and erythroblastic progenitor to stromal cell. VCAM-1$^+$ cells are found in murine bone marrow as perivascular reticular cells similar to myoid cells (Jacobsen et al., 1996). HCA/ALCAM, constitutively expressed by stromal cells, is involved in both homophilic HCA:HCA and heterophilic HCA:CD6 adhesion. It is operative in stromal cell homotypic adhesion, in the heterotypic adhesion of CD6$^+$ lymphocyte to stromal cell, and in the homotypic or heterotypic adhesion of HSCs (expressing both HCA and CD6) to stromal cell (Cortès et al., 1999). Remarkably, this adhesion molecule is detected during ontogeny at all the primary sites of hematopoiesis, in the mesoderm underneath clusters of hematopoietic cells in embryonic aortas, in hepatoblasts of fetal liver, in subsets of epithelial cells of fetal thymus, and in presumably myoid cells of fetal bones (Cortès et al., 1999). HCA appears also to play a role in the differentiation of hemangioblastic precursors into hematopoietic and endothelial cells (Ohneda et al., 2001).

CD44 is intensely expressed by stromal cells. Antibodies to CD44 inhibit the initial establishment of lymphoid and myeloid cultures but have no effect when added at a later stage. However, other anti-CD44 antibodies have a stimulating effect on adhesion of human hematopoietic precursor cells to human stroma, probably by modulating the availability of the epitope to different potential ligands (hyaluronic acid, laminins, fibronectins, collagens) (Oostendorp et al., 1997).

Sialomucins, heavily glycosylated proteins with extended extracellular configuration, have recently been implicated in stromal-to-HSC interaction. Expression by HSCs/progenitors of CD34 appears to promote adhesive interaction to stroma (Healy et al., 1995), while that of CD164 partially blocks this interaction (Zannettino et al., 1998). CD34 and CD164 are also expressed by stromal cells, but there are no data suggesting a role in homotypic adhesion.

Stromal cells express VLA-1, -2, -3, -5, and -6 integrins. As already indicated, VLA-5 is involved in cell binding and fibronectin assembly. Other receptors are probably involved in laminin and collagen binding and assembly. Integrins linked to the actin-based cytoskeleton play a major role in the organization of the ECM. They may do so in association with tetraspanins such as CD9. CD9 has been described as associated with β-1 integrin subunits in stromal cells. The addition of anti-CD9 antibodies to Dexter culture at culture inception abrogates myelopoiesis, which may be owing to the disruption of the hematopoietic niche (Oritani et al., 1996). However, other mechanisms may be involved since antibody ligation to HSCs/progenitors impedes differentiation toward myeloid lineages (Aoyama et al., 1999).

In conclusion, long-term marrow cultures have proven to be a crucial model for the study of adhesive pathways between HSCs/progenitors and stromal cells. A summary of the pathways (including ECM molecules, cell adhesion molecules, and cytokine receptors) is presented in Table 3. Adhesion of HSCs/progenitors to the microenvironment appears to be the result of a number of molecular interactions: First, interactions between adhesive molecules (collagens, fibulins and fibronectins, HSPGs, and thrombospondins) may modify the adhesive capacity of single constituents. Second, cytokines modulate adhesive pathways. Third, antiadhesive pathways are also at play: perlecan may reverse chemotaxis of HSCs/progenitors (Klein et al. 1994), and antiadhesive SC-1 protein has been isolated from a marrow stromal cell line (Oritani and Kincade, 1996). Four, stromal cells synthesize and secrete a number of proteases that may degrade the ECM, free cytokines from their reservoirs, activate cytokines from latent forms, and make patent specific domains within ECM glycoproteins.

To add to the overall complexity of HSCs/microenvironmental interactions other pathways/molecules have recently been described. Diverse wnts, molecules involved in organogenesis, are expressed by stromal cells, as well as wnt receptors (frizzled-1, -3, 8). wnts show diverse effects on hematopoiesis, depending on the molecule (wnt-3a, -5a, -10b), the hematopoietic target (fetal liver vs bone marrow), and the stromal cell type (fetal liver vs bone marrow lines) (Austin et al., 1997; Brandon et al., 2000). Stromal cells respond to wnt-3a (proliferation via stabilization and translocation of β-catenin?) (Yamane et al., 2001), and

Table 3
Major Stromal Adhesion Molecules[a]

	Type	*Adhesive pathway stroma:hematopoietic cell*	*Hematopoietic target*
Collagen I	ECM gp		HSC, E
Collagen VI	ECM gp	Collagen VI:HSPG	HSC (lines)
Collagen XIV	ECM gp	Collagen XIV:HSPG	HSC (lines)
Laminin-10 (LM-10)	ECM gp	LM-10:β-1 integrin	HSC (lines)
Fibronectin (FN)	ECM gp	FN (RGD):VLA-5 integrin	HSC, E
		FN (LDV):VLA-4 integrin	HSC, L
		FN (IDAPS):CD44, VLA-4, CSPG	HSC
Thrombospondin-1 (TSP-1)	ECM gp	TSP-1 : CD36	Meg
Tenascin	ECM gp	Tenascin:HSPG	HSC
Fibulin	ECM gp	Fibulin:β-3 integrin	Meg
Perlecan	HSPG	(Antiadhesion)	HSC (lines)
HSPG	HSPG	HSPG:pil protein	HSC
VCAM-1	IgCAM	VCAM-1:VLA-4 integrin	E, L
HCA/ALCAM	IgCAM	HCA:HCA	HSC
		HCA:CD6	HSC, L
CD44	CAM	CD44:hyaluronate	HSC, E, L
CD34	Sialomucin		HSC
CD164	Sialomucin		HSC
M-CSF β	Cytokine	M-CSF β:c-fms	M
tm SCF	Cytokine	tm SCF:c-kit	HSC, E
tm FL	Cytokine	tm FL:flt-3	HSC

[a]IgCAM, immunoglobulin cell adhesion molecule; ECM gp, extracellular matrix glycoprotein; HSPG, heparan sulfate proteoglycan; CSPG, chondroitin sulfate proteoglycan; pil, phosphatidyl inositol linked; HSC, hematopoietic stem cell; L, lymphoid precursor; M, macrophage precursor; E, erythroid precursor; Meg, megakaryocytic precursor; tm, transmembrane.

wnt-10b is a molecular switch for MSCs (inhibition of adipocytic pathway and stimulation of myogenesis) (Ross et al., 2000). Eicosanoids, lipid mediators also synthesized by stromal cells, may be involved in the regulation of hematopoiesis. Recent studies have shown that prostaglandin-E_2 (PGE_2) released by stromal cells reduces IL-8 and GM-CSF stromal output while increasing that of IL-6 (Denizot et al., 1998; 1999). PGE_2 also affects the growth of stromal cell precursors (Dobson et al., 1999). Moreover, in mice deficient in cyclooxygenase 2 (the PGH_2 synthase activated by inflammatory cytokines) the recruitment, in case of stress, of quiescent HSCs is delayed, leading to delayed hematopoietic recovery (Lorenz et al., 1999).

13.4. CONCLUSION

Stromal control of HSCs' activity appears to involve a great number of probably redundant regulatory loops. Such highly coordinated, multifactorial, and degenerate regulation is probably required by the perennial and intense cell renewal observed in the hematopoietic system (Lemischka, 2001).

REFERENCES

Abboud, S. L., Bethel, R., and Aron, D. C. (1991) Secretion of insulin like growth factor I and insuline like growth factor-binding proteins by murine bone marrow stromal cells. *J. Clin. Invest.* 88:470–475.

Alvarez, F., Flano, E., Castillo, A., Lopez-Fierro, P., Razquin, B., and Villena, A. (1996) Tissue distribution and structure of barrier cells in the hematopoietic and lymphoid organs of salmonids. *Anat. Rec.* 245: 17–24.

Andrades, J. A., Han, B., Becerra, J., Sorgente, N., Hall, F. L., and Nimni, M. E. (1999) A recombinant human TGF-beta-1 fusion protein with collagen-binding domain promotes migration, growth, and differentiation of bone marrow mesenchymal cells. *Exp. Cell. Res.* 250:485–498.

Andrews, D. F., Nemunaitis, J. J., and Singer, J. W. (1989) Recombinant tumor necrosis factor -alpha and interleukin 1-alpha increase expression of c-abl protooncogene mRNA in cultured human marrow stromal cells. *Proc. Natl. Acad. Sci. USA* 86:6788–6792.

Andrews, R. G., Singer, J. W., and Bernstein, I. D. (1989) Precursors of colony-forming cells in humans can be distinguished from colony-forming cells by expression of the CD33 and CD34 antigens and light scatter properties. *J. Exp. Med.* 169:1721–1731.

Aoyama, K., Oritani, K., Yokota, T., et al. (1999) Stromal cell CD9 regulates differentiation of hematopoietic stem/progenitor cells. *Blood* 93: 2586–2594.

Austin, T. W., Solar, G. P., Ziegler, F. C., Liem, L., and Matthews, W. (1997) A role for the Wnt gene family in hematopoiesis: expansion of multilineage progenitor cells. *Blood* 89:3624–3635.

Azizi, S. A., Stokes, D., Augelli, B. J., DiGirolamo, C., and Prockop, D. J. (1998) Engraftment and migration of human bone marrow stromal cells implanted in the brains of albino rats—similarities to astrocyte grafts. *Proc. Natl. Acad. Sci. USA* 95:3908–3913.

Bennett, J. H., Joyner, C. J., Triffitt, J. T., and Owen, M. E. (1991) Adipocytic cells cultured from marrow have osteogenic potential. *J. Cell Sci.* 99:131–139.

Bianco, P. and Robey, P. G. (2000) Marrow stromal stem cells. *J. Clin. Invest.* 105:1663–1668.

Bleul, C. C., Fuhlbrigge, R. C., Casanovas, J. M., Aiuti, A., and Springer, T. A. (1996) A highly efficacious lymphocyte chemoattractant, stromal cell-derived factor 1 (SDF-1). *J. Exp. Med.* 184:1101–1109.

Brandon, C., Eisenberg, L. M., and Eisenberg, C. A. (2000) WNT signaling modulates the diversification of hematopoietic cells. *Blood* 96: 4132–4141.

Breems, D. A., Blokland, E. A. W., Neben, S., and Ploemacher, R. E. (1994) Frequency analysis of human primitive haematopoietic stem cell subsets using a cobblestone area forming cell assay. *Leukemia* 8:1095–1104.

Brunner, G., Gabrilove, J., Rifkin, D. B., and Wilson, E. L. (1991) Phospholipase C release of basic fibroblast growth factor from human bone marrow cultures as a biologically active complex with a phosphatidylinositol-anchored heparan sulfate proteoglycan. *J. Cell Biol.* 114:1275–1283.

Campagnoli, C., Roberts, I. A., Kumar, S., Bennett, P. R., Bellantuono, I., and Fisk, N. M. (2001) Identification of mesenchymal stem/progenitor cells in human first-trimester fetal blood, liver, and bone marrow. *Blood* 98:2396–2402.

Caplan, A. I. (1991) Mesenchymal stem cells. *J. Orthop. Res.* 9:641–650.

Charbord, P. (2001) Mediators involved in the control of hmatopoiesis by the microenvironment. In: *Hematopoiesis. A developmental approach.* Zon, L., ed., Oxford University Press, New York, pp. 702–717.

Charbord, P., Tamayo, E., Saeland, S., Duvert, V., Poulet, J., Gown, A. M., and Hervé, P. (1991) Granulocyte-macrophage colony-stimulating factor (GM-CSF) in human long-term bone marrow cultures: endogeneous production in the adherent layer and effect of exogeneous GM-CSF on granulomonopoiesis. *Blood* 78:1230–1236.

Charbord, P., Tavian, M., Humeau, L., and Péault, B. (1996) Early ontogeny of the human marrow from long bones: an immunohistochemical study of hematopoiesis and its microenvironment. *Blood* 87:4109–4119.

Chiu, C. P., Moulds, C., Coffman, R. L., Rennick, D., and Lee, F. (1988) Multiple biological activities are expressed by a mouse interleukin 6 cDNA clone isolated from bone marrow stromal cells. *Proc. Natl. Acad. Sci. USA* 85:7099–7103.

Cortès, F., Deschaseaux, F., Uchida, N., et al. (1999) HCA, an immunoglobulin-like adhesion molecule present on the earliest human hematopoietic precursor cells, is also expressed by stromal cells in blood-forming tissues. *Blood* 93:1–14.

Delikat, S., Harris, R. J., and Galvani, D. W. (1993) IL-1 alpha inhibits adipocyte formation in human long-term bone marrow culture. *Exp. Hematol.* 21:31–37.

Denizot, Y., Trimoreau, F., and Praloran, V. (1998) Effects of lipid mediators on the synthesis of leukemia inhibitory factor and interleukin 6 by human bone marrow stromal cells. *Cytokine* 10:781–785.

Denizot, Y., Godard, A., Raher, S., Trimoreau, F., and Praloran, V. (1999) Lipid mediators modulate the synthesis of interleukin 8 by human bone marrow stromal cells. *Cytokine* 11:606–610.

Dennis, J. E. and Charbord, P. (2002) Origin and differentiation of human and murine stroma. *Stem Cells* 20:205–214.

Dennis, J. E., Carbillet, J. P., Caplan, A., and Charbord, P. (2002) The STRO-1+ marrow cell population is multopotential. *Cell. Tiss. Org.* 170:73–82.

Deschaseaux, F. and Charbord, P. (2000) Human marrow stromal cell precursors are alpha1 integrin subunit-positive. *J. Cell. Physiol.* 184: 319–325.

Desmoulière, A. and Gabbiani, G. (1995) Smooth muscle cell and fibroblast biological and functional features : similarities and differences. In: *The Vascular Smooth Muscle Cell.* Schwartz, S. M. and Mecham, R. P., eds., Academic Pres, San Diego, CA, pp. 329–359.

Dexter, T. M. and Moore, M. A. S. (1977) In vitro duplication and "cure" of haemopoietic defects in genetically anaemic mice. *Nature* 269:412–414.

Dexter, T. M., Allen, T. D., and Lajtha, L. G. (1976) Conditions controlling the proliferation of haemopoietic stem cells in vitro. *J. Cell. Physiol.* 91:335–344.

Dobson, K., Reading, L., and Scutt, A. (1999) A cost-effective method for the automatic quantitative analysis of fibroblastic colony-forming units. *Calcif. Tissue Int.* 65:166–172.

Drzeniek, Z., Siebertz, B., Stocker, G., et al. (1997) Proteoglycan synthesis in haematopoietic cells: isolation and characterization of heparan sulphate proteoglycans expressed by the bone-marrow stromal cell line MS-5. *Biochem. J.* 327:473–480.

Filshie, R. J. A., Zannettino, A. C. W., Makrynikola, V., et al. (1998) MUC18, a member of the immunoglobulin superfamily, is expressed on bone marrow fibroblasts and a subset of hematological malignancies. *Leukemia* 12:414–421.

Fraser, C. C., Szilvassy, S. J., Eaves, C. J., and Humphries, R. K. (1992) Proliferation of totipotent hematopoietic stem cells in vitro with retention of long-term competitive in vivo reconstituting ability. *Proc. Natl. Acad. Sci. USA* 89:1968–1972.

Friedenstein, A. J. (1980) Stromal mechanisms of bone marrow cloning in vitro and retransplantation in vivo. In: *Immunology of Bone Marrow Transplantation.* Thienfelder, S., ed., Springer Verlag, Berlin, pp. 19–28.

Friedenstein, A. J., Chailakhjan, R. K., Latsinik, N. V., Panasyuk, A. F., and Keiliss-Borok, V. (1974) Stromal cells responsible fro transferring the microenvironment of the hemopoietic tissues. *Transplantation* 17:331–340.

Friedrich, C., Zausch, E., Sugrue, S. P., and Gutierrez-Ramos, J. C. (1996) Hematopoietic supportive functions of mouse bone marrow and fetal liver microenvironment: dissection of granulocyte, B-lymphocyte, and hematopoietic progenitor support at the stroma cell clone level. *Blood* 87:4596–4606.

Funk, P. E. and Witte, P. L. (1992) Enrichment of primary lymphocyte-supporting stromal cells and characterization of associated B lymphocyte progenitors. *Eur. J. Immunol.* 22:1305–1313.

Galmiche, M. C., Koteliansky, V. E., Hervé, P., and Charbord, P. (1993) Stromal cells from human long-term marrow cultures are mesenchymal cells that differentiate following a vascular smooth muscle differentiation pathway. *Blood* 82:66–76.

Gartner, S. and Kaplan, H. S. (1980) Long-term culture of human bone marrow cells. *Proc. Natl. Acad. Sci. USA* 77:4756–4759.

Gautam, S. C., Noth, C. J., Janakiraman, N., Pindolia, K. R., and Chapman, R. A. (1995) Induction of chemokine mRNA in bone marrow stromal cells: modulation by TFG-beta-1 and IL-4. *Exp. Hematol.* 23:482–491.

Gentili, C., Bianco, P., Neri, M., et al. (1993) Cell proliferation, extracellular matrix mineralization, and ovotransferrin transient expression during in vitro differentiation of chick hypertrophic chondrocytes into osteoblast-like cells. *J. Cell Biol.* 122:703–712.

Grellier, P., Yee, D., Gonzalez, M., and Abboud, S. L. (1995) Characterization of insulin-like growth factor binding proteins (IGFBP) and regulation of IGFBP-4 in bone marrow stromal cells. *Br. J. Haematol.* 90:249–257.

Gronthos, S. and Simmons, P. J. (1995) The growth factor requirements of STRO-1—positive human bone marrow stromal precursors under serum-deprived conditions in vitro. *Blood* 85:929–940.

Gu, Y., Sorokin, L., Durbeej, M., Hjalt, T., Jonsson, J. I., and Ekblom, M. (1999) Characterization of bone marrow laminins and identification of alpha-5-containing laminins as adhesive proteins for multipotent hematopoietic FDCP-Mix cells. *Blood* 93:2533–2542.

Gu, Y. C., Nilsson, K., Eng, H., and Ekblom, M. (2000) Association of extracellular matrix proteins fibulin-1 and fibulin-2 with fibronectin in bone marrow stroma. *Br. J. Haematol.* 109:305–313.

Guerriero, A., Worford, L., Holland, H. K., Guo, G. R., Sheehan, K., and Waller, E. K. (1997) Thrombopoietin is synthesized by bone marrow stromal cells. *Blood* 90:3444–3455.

Gupta, P., McCarthy, J. B., and Verfaillie, C. M. (1996) Stromal fibroblast heparan sulfate is required for cytokine-mediated ex vivo maintenance of human long-term culture-initiating cells. *Blood* 87: 3229–3236.

Gupta, P., Oemega, T. R., Brazil, J. J., Dudek, A. Z., Slungaard, A., and Verfaillie, C. M. (1998) Structurally specific heparan sulfates support primitive human hematopoiesis by formation of a multimolecular stem cell niche. *Blood* 92:4641–4651.

Hackney, J. A., Charbord, P., Brunk, B. P., Stoeckert, C. J., Lemischka, I. R., and Moore, K. A. (2002) A molecular profile of a hematopoietic stem cell niche. *Proc. Natl. Acad. Sci. USA* 99:13,061–13,066.

Hannocks, M. J., Oliver, L., Gabrilove, J. L., and Wilson, E. L. (1992) Regulation of proteolytic activity in human bone marrow stromal cells by basic fibroblast growth factor, interleukin-1, and transforming growth factor beta. *Blood* 79:1178–1184.

Harrison, D. E., Lerner, C. P., and Spooncer, E. (1987) Erythropoietic repopulating ability of stem cells from long-term marrow culture. *Blood* 69:1021–1025.

Hauser, S. P., Kajkenova, O., and Lipschitz, D. A. (1997) The pivotal role of interleukin 6 in formation and function of hematopoietically active murine long-term bone marrow cultures. *Stem Cells* 15:125–132.

Haynesworth, S. E., Baber, M. A., and Caplan, A. I. (1996) Cytokine expression by human marrow-derived mesenchymal progenitor cells in vitro: effects of dexamethasone and IL-1-alpha. *J. Cell. Physiol.* 166:585–592.

Healy, L., May, G., Gale, K., Grosveld, F., Greaves, M., and Enver, T. (1995) The stem cell antigen CD34 functions as a regulator of hemopoietic cell adhesion. *Proc. Natl. Acad. Sci. USA* 92:12240–12244.

Hogge, D. E., Cashman, J. D., Humphries, R. K., and Eaves, C. J. (1991) Differential and synergistic effects of human granulocyte-macrophage colony-stimulating factor and human granulocyte colony-stimulating factor on hematopoiesis in human long-term marrow cultures. *Blood* 77:493–499.

Huang, S. and Terstappen, L. X. (1994) Correction to Huang and Terstappen (1992). *Nature* 368:664.

Huang, S. and Terstappen, L. X. (1992) Formation of haematopoietic microenvironment and haematopoietic stem cells from single human bone marrow stem cells. *Nature* 360:745–749.

Hurley, R. W., McCarthy, J. B., and Verfaillie, C. M. (1995) Direct adhesion to bone marrow stroma via fibronectin receptors inhibits hematopoietic progenitor proliferation. *J. Clin. Invest.* 96:511–519.

Huss, R., Hong, D. S., McSweeney, P. A., Hoy, C. A., and Deeg, H. J. (1995) Differentiation of canine bone marrow cells with hemopoietic characteristics from an adherent stromal cell precursor. *Proc. Natl. Acad. Sci. USA* 92:748–752.

Huss, R., Lange, C., Weissinger, E. M., Kolb, H. J., and Thalmeier, K. (2000) Evidence of peripheral blood-derived, plastic-adherent CD34(-/low) hematopoietic stem cell clones with mesenchymal stem cell characteristics. *Stem Cells* 18:252–260.

Issaad, C., Croisille, L., Katz, A., Vainchenker, W., and Coulombel, L. (1993) A murine stromal cell line allows the proliferation of very primitive human CD34++/CD38– progenitor cells in long-term cultures and semi-solid assays. *Blood* 81:2916–2924.

Jacobsen, K., Kravitz, J., Kincade, P. W., and Osmond, D. G. (1996) Adhesion receptors on bone marrow stromal cells : in vitro expression of vascular cell adhesion molecule-1 by reticular cells and sinusoidal endothelium in normal and gamma-irradiated mice. *Blood* 87: 73–82.

Jia, D. and Heersche, J. N. (2000) Insulin-like growth factor-1 and -2 stimulate osteoprogenitor proliferation and differentiation and adipocyte formation in cell populations derived from adult rat bone. *Bone* 27:785–794.

Kiassov, A. P., Van Eyken, P., van Pelt, J. F., et al. (1995) Desmin expressing nonhematopoietic liver cells during rat liver development: an immunohistochemical and morphometric study. *Differentiation* 59:253–258.

Klein, G., Conzelmann, S., Beck, S., Timpl, R., and Muller, C. A. (1994) Perlecan in human bone marrow : a growth-factor-presenting, but anti-adhesive, extracellular matrix component for hematopoietic cells; *Matrix Biol.* 14:457–465.

Klein, G., Muller, C. A., Tillet, E., Chu, M., and Timpl, R. (1995) Collagen type VI in the human bone marrow microenvironment : a strong cytoadhesive component. *Blood* 86:1740–1748.

Klein, G., Kibler, C., Schermutzki, F., Brown, J., Muller, C. A., and Timpl, R. (1998) Cell binding properties of collagen type XIV for human hematopoietic cells. *Matrix Biol.* 16:307–317.

Kodama, H., Nose, M., Yamaguchi, Y., et al. (1992) In vitro proliferation of primitive hemopoietic stem cells supported by stromal cells: evidence for the presence of a mechanism(s) other than that involving c-kit receptor and its ligands. *J. Exp. Med.* 176:351–361.

Koenigsmann, M., Griffin, J. D., DiCarlo, J., and Cannistra, S. A. (1992) Myeloid and erythroid progenitor cells from normal bone marrow adhere to collagen type I. *Blood* 79:657–665.

Kuznetsov, S., Friedenstein, A. J., and Robey, P. G. (1997) Factors required for bone marrow stromal fibroblast colony formation in vitro. *Brit. J. Haematol.* 97:561–570.

Landreth, K. S., Narayanan, R., and Dorshkind, K. (1992) Insulin-like growth factor-I regulates pro-B cell differentiation. *Blood* 80: 1207–1212.

Lemieux, M. E., Rebel, V. I., Lansdorp, P. M., and Eaves, C. J. (1995) Characterization and purification of a primitive hematopoietic cell type in adult mouse marrow capable of lymphomyeloid differentiation in long-term marrow "switch" cultures. *Blood* 86:1339–1347.

Lemischka, I. L. (2001). Regulation of hematopoietic stem cells : some conceptual and practical considerations. In: *Hematopoiesis: A Developmental Approach.* Zon, L., ed., Oxford University Press, New York, pp. 48–60.

Lerat, L., Lissitzky, J. C., Singer, J. W., Keating, A., Hervé, P., and Charbord, P. (1993) The role of stromal cells and macrophages in fibronectin biosynthesis and matrix assembly in human long-term marrow cultures. *Blood* 82:1480–1492.

Li, J., Sensebé, L., Hervé, P., and Charbord, P. (1995) Non-transformed colony-derived stromal cell lines from normal human marrows. II. Phenotypic characterization and differentiation pathway. *Exp. Hematol.* 23:133–141.

Lichtman, M. A. (1981) The ultrastructure of the hemopoietic environment of the marrow : a review. *Exp. Hematol.* 9:391–410.

Lisovsky, M., Braun, S. E., GE, Y., et al. (1996) Flt3-ligand production by human bone marrow stromal cells. *Leukemia* 10:1012–1018.

Lorenz, E., Uphoff, D., Reid, T. R., and Shelton, E. (1951) Modification of irradiation injury in mice and guinea pigs by bone marrow injections. *J. Natl. Cancer Inst.* 12:197–201.

Lorenz, M., Slaughter, H. S., Wescott, D. M., et al. (1999) Cyclooxygenase-2 is essential for normal recovery from 5-fluorouracil-induced myelotoxicity in mice. *Exp. Hematol.* 27:1494–1502.

Majumdar, M. K., Banks, V., Peluso, D. P., and Morris, E. A. (2000) Isolation, characterization, and chondrogenic potential of human bone marrow-derived multipotential stromal cells. *J. Cell. Physiol.* 185:98–106.

Mauch, P., Greenberger, J. S., Botnick, L., Hannon, E., and Hellman, S. (1980) Evidence for structured variation in self-renewal capacity within long-term bone marrow cultures. *Proc. Natl. Acad. Sci. USA* 77:2927–2930.

Maximow, A. A. (1924) Relation of blood cells to connective tissues and endothelium. *Physiol. Rev.* 4:533–563.

Moore, K. A., Ema, H., and Lemischka, I. R. (1997) In vitro maintenance of highly purified, transplantable hematopoietic stem cells. *Blood* 89:4337–4347.

Nagasawa, T., Kikutani, H., and Kishimoto, T. (1994) Molecular cloning and structure of a pre-B-cell growth-stimulating factor. *Proc. Natl. Acad. Sci. USA* 91:2305–2309.

Nakano, T., Kodama, H., and Honjo, T. (1994) Generation of lymphohematopoietic cells from embryonic stem cells in culture. *Science* 265:1098–1101.

Nemunaitis, J., Tompkins, C., Andrews, F., Rullian, M., and Singer, J. (1989) Marrow stromal cells: hematopoietic growth factors and extracellular matrix proteins. In: *Experimental Hematology Today*, Springer-Verlag, New York, pp. 53–57.

Oh, S. H., Miyazaki, M., Kouchi, H., et al. (2000) Hepatocyte growth factor induces differentiation of adult rat bone marrow cells into a hepatocyte lineage in vitro. *Biochem. Biophys. Res. Commun.* 279: 500–504.

Ohneda, O., Ohneda, K., Arai, F., et al. (2001) ALCAM (CD166): its role in hematopoietic and endothelial development. *Blood* 98:2134–2142.

Ohta, M., Sakai, T., Saga, Y., Aizawa, S. I., and Saito, M. (1998) Suppression of hematopoietic activity in tenascin-C-deficient-mice. *Blood* 91:4074–4083.

Oostendorp, R. A. J., Spitzer, E., Reisbach, G., and Dormer, P. (1997) Antibodies to the beta-1-integrin chain, CD44, or ICAM-3 stimulate adhesion of blast colony-forming cells and may inhibit their growth. *Exp. Hematol.* 25:345–349.

O'Prey, J., Leslie, N., Itoh, K., Ostertag, W., Bartholomew, C., and Harrison, P. R. (1998) Both stroma and stem cell factor maintain long-term growth of ELM erythroleukemia cells, but only stroma prevents erythroid differentiation in response to erythropoietin and interleukin-3. *Blood* 91:1548–1555.

Oritani, K. and Kincade, P. W. (1996) Identification of stromal cell products that interact with pre-B cells. *J. Cell Biol.* 134:771–782.

Oritani, K., Wu, X., Medina, K., et al. (1996) Antibody ligation of CD9 modifies production of myeloid cells in long-term cultures. *Blood* 87:2252–2261.

Owen, M. (1988) Marrow stromal stem cells. *J. Cell Sci.* (Suppl. 10): 63–76.

Paul, S. R., Bennett, F., Calvetti, J. A., et al. (1990) Molecular cloning of a cDNA encoding interleukin 11, a stromal cell-derived lymphopoietic and hematopoietic cytokine. *Proc. Natl. Acad. Sci. USA* 87:7512–7516.

Penn, P. E., Jiang, D. Z., Fei, R. G., Sitnicka, E., and Wolf, N. S. (1993) Dissecting the hematopoietic microenvironment. IX. Further characterization of murine bone marrow stromal cells. *Blood* 81:1205–1213.

Pereira, R. F., Halford, K. W., O'Hara, M. D., et al. (1995) Cultured adherent cells from marrow can serve as long-lasting precursor cells for bone, cartilage, and lung in irradiated mice. *Proc. Natl. Acad. Sci. USA* 92:4857–4861.

Ploemacher, R. E., Van der Sluijs, J. P., Van Beurden, C. A. J., Baert, M. R. M., and Chan, P. L. (1991) Use of limiting-dilution type long-term marrow cultures in frequency analysis of marrow-repopulating and spleen colony-forming hematopoietic stem cells in the mouse. *Blood* 78:2527–2533.

Prockop, D. J. (1997) Marrow stromal cells as stem cells for nonhematopoietic tissues. *Science* 276:71–74.

Rémy Martin, J. P., Marandin, A., Challier, B., et al. (1999) The vascular smooth muscle differentiation of murine stroma. A sequential model. *Exp. Hematol.* 27:1782–1795.

Reyes, M., Lund, T., Lenvik, T., Aguiar, D., Koodie, L., and Verfaillie, C. M. (2001) Purification and ex vivo expansion of postnatal human marrow mesodermal progenitor cells. *Blood* 98:2615–2625.

Ross, S. E., Hemati, N., Longo, K. A., et al. (2000) Inhibition of adipogenesis by Wnt signaling. *Science* 289:950–953.

Satomura, K., Derubeis, A. R., Fedarko, N. S., et al. (1998) Receptor tyrosine kinase expression in human bone marrow stromal cells. *J. Cell. Physiol.* 177:426–438.

Schofield, K. P., Humphries, M. J., de Wynter, E., Testa, N., and Gallagher, J. T. (1998) The effect of alpha-4 beta-1-integrin binding sequences of fibronectin on growth of cells from human hematopoietic progenitors. *Blood* 91:3230–3238.

Schofield, K. P.,Gallagher, J. T., and David, G. (1999) Expression of proteoglycan core proteins in human bone marrow stroma. *Biochem. J.* 343:663–668.

Schofield, R. (1978) The relationship between the spleen colony-forming cell and the haemopoietic stem cell. A hypothesis. *Blood Cells* 4:7–25.

Sensebé, L., Li, J., Lilly, M., et al. (1995). Non-transformed colony-derived stromal cell lines from normal human marrows. I. Growth requirement and myelopoiesis supportive ability. *Exp. Hematol.* 23:507–513.

Shalaby, F., Ho, J., Stanford, W. L., et al. (1997) A requirement for Flk1 in primitive and definitive hematopoiesis and vasculogenesis. *Cell* 89: 981–990.

Siebertz, B.,Stocker, G.,Drzeniek, Z.,Handt, S., Just, U., and Haubeck, H.-D. (1999) Expression of glypican-4 in haemopoietic-progenitor and bone-marrow-stromal cells. *Biochem. J.* 344:937–943.

Siler, U., Seiffert, M., Puch, S., et al. (2000) Characterization and functional analysis of laminin isoforms in human bone marrow. *Blood* 96: 4194–4203.

Simmons, P. J. and Torok-Storb, B. (1991) Identification of stromal cell precursors in human bone marrow by a novel monoclonal antibody, Stro-1. *Blood* 78:55–62.

Simmons, P. J., Gronthos, S., and Zannettino, A. C. W. (2001) The development of stromal cells. In: *Hematopoiesis. A Developmental Approach.* Zon, L., ed., Oxford University Press, New York, pp. 718–726.

Singer, J. W., Keating, A., Cuttner, J., et al. (1984) Evidence for a stem cell common to hematopoiesis and its in vitro microenvironment : studies of patients with clonal hematopoietic neoplasia. *Leuk. Res.* 8: 535–545.

Singer, J. W., Charbord, P., Keating, A., et al. (1987) Simian virus 40-transformed adherent cells from human long-term marrow cultures: cloned cell lines produce cells with stromal and hematopoietic characteristics. *Blood* 70:464–474.

Spradling, A., Drummond-Barbosa, D., and Kai, T. (2001) Stem cells find their niche. *Nature* 414:98–104.

Stephan, R. P., Reilly, C. R., and Witte, P. L. (1998) Impaired ability of bone marrow stromal cells to support B-lymphopoiesis with age. *Blood* 91:75–88.

Strobel, E. S., Moebest, D., von Kleist, S., et al. (1997) Adhesion and migration are differentially regulated in hematopoietic progenitor cells by cytokines and extracellular matrix. *Blood* 90:3524–3532.

Sugahara, H., Kanakura, Y., Furitsu, T., et al. (1994) Induction of programmed cell death in human hematopoietic cell lines by fibronectin via its interaction with very late antigen 5. *J. Exp. Med.* 179:1757–66.

Sungaran, R., Markovic, B., and Chong, B. H. (1997) Localization and regulation of thrombopoietin mRNA expression in human kidney, liver, bone marrow, and spleen using in situ hybridization. *Blood* 89: 101–107.

Sutherland, H. J., Eaves, C. J., Eaves, A. C., Dragowska, W., and Lansdorp, P. M. (1989) Characterization and partial purification of human marrow cells capable of initiating long-term hematopoiesis in vitro. *Blood* 74:1563–1570.

Sutherland, H. J., Eaves, C. J., Lansdorp, P. M., Thacker, J. D., and Hogge, D. E. (1991) Differential regulation of primitive human hematopoietic cells in long-term cultures maintained on genetically engineered murine stromal cells. *Blood* 78:666–672.

Sutherland, H. J., Hogge, D. E., Cook, D., and Eaves, C. J. (1993) Alternative mechanisms with and without Steel factor support primitive human hematopoiesis. *Blood* 81:1465–1470.

Szilvassy, S. J., Weller, K. P., Lin, W., et al. (1996) Leukemia inhibitory factor upregulates cytokines expression by a murine stromal cell line enabling the mainteance of highly enriched competitive repopulating stem cells. *Blood* 87:4618–4628.

Taipale, J. and Keski-Oja, J. (1997) Growth factors in the extracellular matrix. *FASEB J.* 11:51–59.

Takai, K., Hara, J., Matsumoto, K., et al. (1997) Hepatocyte growth factor is constitutively produced by human bone marrow stromal cells and indirectly promotes hematopoiesis. *Blood* 89:1560–1565.

Tanaka-Douzono, M., Suzu, S., Yamada, M., et al. (2001) Detection of murine adult bone marrow stroma-initiating cells in Lin(−)c-fms(+)c-kit(low)VCAM-1(+) cells. *J. Cell. Physiol.* 189:45–53.

Tang, J., Nuccie, B. l., Ritterman, I., Liesveld, J. L., Abboud, C. N., and Ryan, D. H. (1997) TGF-beta down-regulates stromal IL-7 secretion and inhibits proliferation of human B cell precursors. *J. Immunol.* 157: 117–125.

Tavian, M., Hallais, M. F., and Péault, B. (1999) Emergence of intraembryonic hematopoietic precursors in the pre-liver human embryo. *Development*126:793–803.

Thomas, T., Gori, F., Spelsberg, T. C., Khosla, S., Riggs, B. L., and Conover, C. A. (1999) Response of bipotential human marrow stromal cells to insulin-like growth factors: effect on binding protein production, proliferation, and commitment to osteoblasts and adipocytes. *Endocrinol.* 140:5036–5044.

Till, J. E. and McCulloch, E. A. (1961) A direct measurement of the radiation sensitivity of normal mouse bone marrow cells. *Rad. Res.*, 14:213–222.

Till, J. E., McCulloch, E. A., and Siminovitch, L. (1963) A stochastic model of stem cell proliferation, based on the growth of spleen colony-forming cells. *Proc. Natl. Acad. Sci. USA* 51:29–36.

Toccanier-Pelte, M. F., Skalli, O., Kapanci, Y., and Gabbiani, G. (1987) Characterization of stromal cells with myoid features in lymph nodes and spleen in normal and pathologic conditions. *Am. J. Pathol.* 129:109–118.

Trentin, J. J. (1989) Hemopoietic microenvironments. Historical perspectives, status and projections. In: *Handbook of the Hemopoietic Microenvironment.* Tavassoli, M., ed., Humana Press, Totowa, NJ, 1–87.

van der Velde-Zimmermann, D., Verdaasdonk, M. A. M., Rademakers, L. H. P. M., De Weger, R. A., Van den Tweel, J. G., and Joling, P. (1997) Fibronectin distribution in human bone marrow stroma : matrix assembly and tumor cell adhesion via alpha-5-beta-1 integrin. *Exp. Cell. Res.* 230:111–120.

van Vlasselaer, P., Falla, N., Snoeck, H., and Mathieu, E. (1994) Characterization and purification of osteogenic cells from murine bone marrow by two-color cell sorting using anti-Sca-1 monoclonal antibody and wheat germ agglutinin. *Blood* 84:753–763.

Waller, E. K., Olweus, J., Lund-Johansen, F., et al. (1995) The "common stem cell" hypothesis reevaluated : human fetal bone marrow contains separate populations of hematopoietic and stromal progenitors. *Blood* 85:2422–2435.

Weimar, I. S., Miranda, N., Muller, E. J., et al. (1998) Hepatocyte growth factor/scatter factor (HGF/SF) is produced by human bone marrow stromal cells and promotes proliferation, adhesion and survival of human hematopoietic progenitor cells (CD34+). *Exp. Hematol.* 26: 885–894.

Weiss, L. and Geduldig, U. (1991) Barrier cells : stromal regulation of hematopoiesis and blood cell release in normal and stressed murine bone marrow. *Blood* 78:975–990.

Weissman, I. L. (1994) Developmental switches in the immune system. *Cell* 76:207–218.

Weissman, I. L. (2000) Stem cells: units of development, units of regeneration, and units in evolution. *Cell* 100:157–168.

Whitlock, C. A., Robertson, D., and Witte, O. N. (1984) Murine B cell lymphopoiesis in long term culture. *J. Immunol. Methods.* 67: 353–369.

Williams, D. E., Eisenman, J., Baird, A., et al. (1990) Identification of a ligand for the c-kit proto-oncogene. *Cell* 63:167–174.

Wineman, J., Moore, K., Lemischka, I., and Muller-Sieburg, C. (1996) Functional heterogeneity of the hematopoietic microenvironment: rare stromal elements maintain long-term repopulating stem cells. *Blood* 87:4082–4090.

Wineman, J. P., Nishikawa, S. I., and Muller-Sieburg, C. E. (1993) Maintenance of high levels of pluripotent hematopoietic stem cells in vitro : effect of stromal cells and c-kit. *Blood* 81:365–372.

Wolf, N. S. and Trentin, J. J. (1968) Hemopoietic colony studies. V. Effect of hemopoietic organ stroma on differentiation of pluripotent stem cells. *J. Exp. Med.* 127:205–214.

Yamane, T., Kunisada, T., Tsukamoto, H., et al. (2001) Wnt signaling regulates hemopoiesis through stromal cells. *J. Immunol.* 167:765–772.

Yan, Z. J., Wang, Q. R., McNiece, I. K., and Wolf, N. S. (1990) Dissecting the hematopoietic microenvironment. VII. The production of an autostimulatory factor as well as a CSF by unstimulated murine marrow fibroblasts. *Exp. Hematol.* 18:348–354.

Yanai, N., Matsuya, Y., and Obinata, M. (1989) Spleen stromal cell lines selectively support erythroid colony formation. *Blood* 74:2391–2397.

Yanai, N., Matsui, N., Furusawa, T., Okubo, T., and Obinata, M. (2000) Sphingosine-1-phosphate and lysophosphatidic acid trigger invasion of primitive hematopoietic cells into stromal cell layers. *Blood* 96: 139–144.

Yoder, M. C., Papaioannou, V. E., Breitfeld, P. P., and Williams, D. A. (1994) Murine yolk sac endoderm- and mesoderm-derived cell lines support in vitro growth and differentiation of hematopoietic cells. *Blood* 83:2436–2443.

Zannettino, A. C., Buhring, H. J., Niutta, S., Watt, S. M., Benton, M. A., and Simmons, P. J. (1998) The sialomucin CD164 (MGC-24v) is an adhesive glycoprotein expressed by human hematopoietic progenitors and bone marrow stromal cells that serves as a potent negative regulator of hematopoiesis. *Blood* 92:2613–2628.

Zuckerman, K. S., Rhodes, R. K., Goodrum, D. D., et al. (1985) Inhibition of collagen deposition in the extracellular matrix prevents the establishment of a stroma supportive of hematopoiesis in long-term murine bone marrow cultures. *J. Clin. Invest.* 75:970–975.

14 Hematopoietic Stem Cells

Identification, Characterization, and Assays

Ian Ponting, PhD, Yi Zhao, MD, and W. French Anderson, MD

The small population of pluripotent hematopoietic stem cells (PHSCs) in the bone marrow consists of short-term reconstituting cells (STRCs) and long-term reconstituting cells (LTRCs), based on how quickly the transplanted cells can produce progeny in an irradiated recipient. They can be "purified" using a combination of cell size; density; fluorescent dye uptake; resistance to cytotoxic chemicals; and cell-surface markers including Thy 1.1 (T), Sca-1 (S), c-kit (K), lineage (L), CD38 (38), and CD34 (34). Using five-color fluorescence-activated cell sorting the long term, very primitive mouse LTRCs are $L^{-/lo}$, S^+, K^+, 38^+, 34^-, and appear to mature to $L^{-/lo}$, S^+, K^+, 38^+, 34^+ cells and then to $L^{-/lo}$, S^+, K^+, 38^-, 34^+; thus, STRCs acquire CD34 and lose CD38 on maturation from $CD34^-$ LTRCs. CD34 has been used to isolate PHSCs for human transplantation studies; therefore, the LTRC may be lost during this procedure. Experimental transplantation studies indicate that the best reconstitution occurs when both cell populations are present, the more mature cells activating the immature cells after myeloablation, whereas the mature cells provide negative control in normal animals. Functionally the type of assay that has been most widely used for the quantitation of mouse stem cells is the in vivo repopulating assay. Different numbers of donor cells are combined with a standard number of normal bone marrow cells. The normal cells protect against the immediate effects of myeloablation and compete with the donor stem cells. The proportions of mature cells derived from the donor stem cells are determined by the detection of a donor-specific marker, such as an isoenzyme, Y-chromosome, or congenic antigen. Similarly, using limiting dilution transplant of a donor test population of cells and a standard number of stem cell–compromised serially transplanted cells, the relative contribution of the donor cells is measured as a competitive repopulating unit. Finally, the repopulating stem cell unit assay using complete myeloablation and busulfan-treated bone marrow radioprotective support cells provides comparatively rapid and sensitive detection of the very small numbers of LTRCs present in limiting dilution transplants. This procedure utilizes busulfan because it appears preferentially stem cell toxic, and it provides radioprotective support cells that are unable to compete effectively with normal donor stem cells in the population under investigation. Stem cells are selected based on their ability to produce both lymphoid and myeloid repopulation in severely ablated mice, rather than competitive ability.

14.1. INTRODUCTION

Pluripotent hematopoietic stem cells (PHSCs) have been defined as a population of cells that can reconstitute and maintain all lymphoid and myeloid lineage cells over the long term. It has been estimated, based on limiting dilution assays in vivo, that the frequency of long-term reconstituting hematopoietic stem cells (HSCs) in murine bone marrow is about 1 in 100,000 whole bone marrow nucleated cells (Harrison et al., 1993). Classically, according to their characteristics after bone marrow transplantation (BMT), cells in murine bone marrow, which have multilineage repopulating ability, can be divided into two groups: short-term reconstituting cells (STRCs) and long term reconstituting cells (LTRCs) (Harrison et al., 1992; Spangrude et al., 1995; Zhong et al., 1996). STRCs can supply radioprotection and reconstitute the blood system transiently shortly after BMT. LTRCs do not have radioprotection ability and do not appear to reconstitute the blood system at an early stage after BMT, but they can reconstitute at later stages after BMT and provide multilineage cells for the life-span of the animal. This classification is based on two properties: (1) whether the cells can or cannot protect recipients from lethal irradiation, and (2) the time of appearance of progeny cells in the blood after BMT. The hypothesis is that the STRCs are more mature and rapid cycling progenitors that can respond to irradiation damage more quickly, whereas the LTRCs are more primitive and slower cycling cells that need time to reconstitute the blood system long term.

From: *Stem Cells Handbook*
Edited by: S. Sell © Humana Press Inc., Totowa, NJ

14.2. IDENTIFICATION OF HSCs

Identification and isolation of PHSCs is a first step in order to study function. Because of the limited frequency of the PHSC in the bone marrow, major efforts have been made to obtain "pure" stem cells for both clinical and research purposes. Clinically, transplantation of pure stem cells should eliminate graft-vs-host disease as well as the reintroduction of malignancies; from a research viewpoint, pure stem cells would facilitate the study of stem cell regulation in vivo and in vitro, which would ultimately benefit clinical applications.

There have been three major approaches to stem cell identification: (1) stem cell physiological characteristics—cell size and density (Jones et al., 1990); (2) cell specific metabolic or cycling stages—low uptake of the fluorescence dye (Rhodamine 123 or Hoechst 33342) and/or resistance to cytotoxic reagents (5-fluorouracil [S-FU]) (Shapiro, 1981; Van Zant, 1984; Harrison and Lerner, 1991; Belloc et al., 1994; Zijlmans, 1995); (3) cell-specific biochemical characteristics—expression of cell-specific proteins (cell-surface markers). The combination of fluorescence-activated cell sorting (FACS) and multiple cell-surface marker identification is one of the most efficient methods used in both clinical and research settings for stem cell purification.

14.2.1. PHYSIOLOGICAL CHARACTERISTICS Using counterflow centrifugal elutriation (CCE), which sort cells on the basis of size and density, Jones et al. (1990) separated a bone marrow subset (CCE25) of small, dense cells, showing delayed, but long-term repopulating ability. CCE25 cells cannot rescue lethally irradiated animals when transplanted alone. They require the presence of a second population of cells. Reconstitution from the CCE25 subset can be detected 2 mo posttransplantation. This observation led to the concept of the coexistence of long-term and short-term repopulating cells in bone marrow and the different roles of these two subsets in BMT.

14.2.2. METABOLIC CHARACTERISTICS The level of uptake of Rhodamine 123 by cells can be used as a means to isolate PHSCs. Rhodamine 123 binds to the mitochondrial DNA, and a low level of Rhodamine 123 suggests an active outflow pump that has been shown to be highly active in PHSCs. Another DNA-binding dye, Hoechst 33342, has also been used in PHSC identification. With split spectrum analysis, Goodell et al. (1996) discovered side population cells (SP cells) and demonstrated their LTRC properties. It appears that a low level of Hoechst 33342 results from a more activated multi drug- resistant gene product, as evidenced by the fact that verapamil will block Hoechst efflux from SP cells. Both Rhodamine 123 and Hoechst 33342 also have been used in combination with surface markers to isolate PHSCs. A so-called pocket cell has been identified using a combination of Rhodamine 123, Hoechst 33342, Sca-1, and c-kit simultaneously, a method first described by Sitnicka et al. (1996).

14.2.3. CELL-SURFACE MARKERS Although both CCE and fluorescence dye uptake have their advantages, a more commonly used method is to identify cells via recognition of surface-specific expressed proteins (surface markers are abbreviated herein by the first letter of the name in bold). Spangrude et al. (1988) showed that PHSCs in murine bone marrow can be enriched 1000- to 2000-fold in a population of cells that express low levels of Thy-1.1 (**T**^{lo}), high levels of Sca-1 (**S**^{hi}), and no (or low) lineage markers (**T**^{lo}, **S**^{hi}, **L**$^{-/lo}$: TSL cells). As few as 30 cells with these markers could rescue 50% of lethally irradiated mice and repopulate the hematopoietic system of the host (Spangrude et al., 1988). The receptor of the stem cell factor, c-kit, has also been used as one of the markers for recognizing stem cells (Okada et al., 1991). Cells with high Sca-1and c-kit (**S** and **K**), but no or low expression of lineage markers (**L**$^{-/lo}$, **S**$^{+}$, **K**$^{+}$) have been demonstrated to have self-renewal and long-term repopulating ability (Osawa et al., 1996a). It has also been reported that CD34^{+} (hereafter abbreviated **34**) cells in murine bone marrow could repopulate lethally irradiated mice (Krause et al., 1994). However, with monoclonal antibody RAM34, Osawa et al. reported that cells with low or no expression of CD34 (**L**$^{-/lo}$, **S**$^{+}$, **K**$^{+}$, **34**$^{-}$) were LTRCs in the mouse. With a competitive repopulation assay, they demonstrated that one **L**$^{-/lo}$, **S**$^{+}$, **K**$^{+}$, **34**$^{-}$ cell could reconstitute the hematopoietic system in lethally irradiated mice (Osawa et al., 1996b). At the same time, Randall et al. (1996) confirmed that high CD38 (hereafter abbreviated **38**) expression in **L**$^{-/lo}$, **S**$^{+}$, **K**$^{+}$ cells are LTRC (**L**$^{-/lo}$, **S**$^{+}$, **K**$^{+}$, **38**$^{+}$).

3. CHARACTERIZATION OF HSCs

14.3.1. MATURATION PATHWAY OF MURINE PHSCs To characterize more fully the surface markers of murine PHSCs, our laboratory developed a five-color FACS sorting technique and analyzed murine PHSCs by testing different subsets of fractionated bone marrow cells with the competitive repopulation assay (Zhao et al., 2000). The very primitive PHSCs in adult murine bone marrow harbor the characteristic surface marker profile of **L**$^{-/lo}$, **S**$^{+}$, **K**$^{+}$, **38**$^{+}$, **34**$^{-}$. These cells appear to mature into **L**$^{-/lo}$, **S**$^{+}$, **K**$^{+}$, **38**$^{+}$, **34**$^{+}$ and then into **L**$^{-/lo}$, **S**$^{+}$, **K**$^{+}$, **38**$^{-}$, **34**$^{+}$ cells. The LTRC resides in the **L**$^{-/lo}$, **S**$^{+}$, **K**$^{+}$, **38**$^{+}$, **34**$^{-}$ subpopulation.

14.3.2. ROLE OF CD34 IN MURINE SYSTEM Since several lines of evidence suggest that the **34**$^{-}$ cells are LTRCs, the observations from Sato et al. (1999) must be taken into account. Their results suggest that **34** expression in bone marrow cells can be influenced by the kinetic state of the cells, such as by treatment with 5-FU. Thus, both **34**$^{+}$ and **34**$^{-}$ cells can be LTRCs under different conditions (Sato et al., 1999). Furthermore, **34** expression in the stem cell is ontologically dependent: mice before 7 wk of age have primitive stem cells that express **34** (**34**$^{+}$), while the stem cells of older mice (>7 weeks old) are **34**$^{-}$ (Ito et al., 2000). Another surface marker, **38** expression, like **34**, has been shown to be expressed in a reciprocal fashion over time (Tajima et al., 2001). In 5-FU-treated adult mice, the majority of PHSCs are **38**$^{-}$ **34**$^{+}$, while after recovery in steady state they are **38**$^{+}$ **34**$^{-}$.

14.3.3. CD34^{-} HUMAN PHSCs In humans, **34**$^{+}$ bone marrow cells have long been considered as PHSCs. Clinical stem cell transplantation relies on **34**$^{+}$ selection together with other surface markers, such as **38**$^{-}$ and **T**^{lo} (Terstappen et al., 1991; Cardoso et al.,1993; Huang and Tersapper, 1994). **34**$^{+}$-selected cells reconstitute lethally irradiated baboons (Berenson, 1988). The discovery that **34**$^{-}$ cells in adult mice are the true PHSCs raised the question, could the real stem cells in preparations of human blood cells be discarded when using **34**$^{+}$ selection? Using different animal models, both Zanjani (1998), using human/sheep chimeras, and Bhatia et al. (1998), using human/severe combined immunodeficiency disease (SCID) mice, demonstrated that **34**$^{-}$ human bone marrow cells could develop into **34**$^{+}$ cells and give rise to multilineage cells in long term. Goodell et al. (1997) also showed that SP cells from rhesus bone marrow are largely **34**$^{-}$ cells. Efforts to identify other surface molecules that are specifically expressed in **34**$^{-}$ PHSCs are being carried out by several groups, including our laboratory.

It has been shown that anti-KDR receptor antibody (Ziegler et al., 1999) as well as AC133 antibody recognizes subsets from

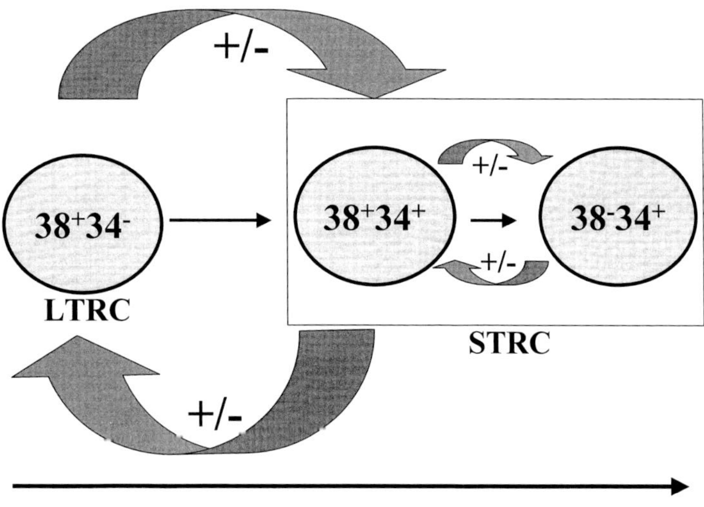

Fig. 1. Schematic representation of suggested regulation among subsets in the mouse stem cell compartment. Both positive and negative feedback probably exists between the two major subsets (LTRC and STRC). The small straight arrows are the differentiation/maturation pathway; the large curved arrows suggest the regulation between the LTRC and STRC.

34^- bone marrow cells that appear to have the LTRC property in the human/SCID mouse model (Gallacher et al., 2000). In our laboratory, we are using differential display polymerase chain reaction to identify genes specifically expressed in mouse **$L^{-/lo}$**, **S^+**, **K^+**, **38^+**, **34^-** cells to identify surface-expressing proteins that might be used for stem cell recognition. Several candidates have been identified and follow-up studies are under way. Human homologs to promising mouse surface antigens will be sought.

14.3.4. REGULATORY PATHWAYS If the **34^-** cells in human are the more primitive stem cells, and if the technology becomes available to isolate these cells, will they be a better choice for clinical transplantation than the **34^+** cells? From our studies in the murine model, the answer may not be yes. To understand stem cell transplantation kinetics, we examined stem cell transplantation regulation. It has been assumed that the LTRC cannot give rapid engraftment in lethally irradiated mice since these cells cycle slowly. When we examined the three subsets from the mouse stem cell compartment (because all three are **$L^{-/lo}$**, **S^+**, **K^+**, we refer to them as **38^+ 34^-**, **38^+ 34^+**, and **38^- 34^+**), we found that all three subsets can give rise to early (8 and 14 d) engraftment when cotransplanted with whole bone marrow cells (Zhao et al., 2000). However, none of the subsets by itself could rescue lethally irradiated mice. However, when the **38^+ 34^-** cells were cotransplanted with either of the other two more mature subsets, the animals survived with a significant majority of the LTRCs arising from the **38^+ 34^-** cells.

We then tested the hypothesis that the **34^+** populations (**38^+ 34^+** and **38^- 34^+**) can facilitate the early engraftment of **38^+ 34^-** cells. We transplanted the **34^+** cells 7 d before **38^+ 34^-** cells and then checked the engraftment of **38^+ 34^-** cells in bone marrow 7 d after transplantation. We observed **38^+ 34^-** cell early engraftment under these conditions. This result suggests that in the presence of **34^+** cells, the **38^+ 34^-** cell can proliferate in the bone marrow immediately after transplantation.

Although a general supportive function of **34^+** cells cannot be excluded, a specific facilitative role (with an unknown mechanism) appears the more likely mechanism. This result also suggests that a regulatory mechanism exists in the stem cell compartment between the three subsets. With the maturation from **38^+ 34^-** cells to **38^+ 34^+** cells and then to **38^- 34^+** cells, the downstream cells appear to provide feedback regulation on the most primitive cells. This regulation is positive when there has been myeloablation, and it is negative in normal animals in order to control **38^+ 34^-** cell proliferation (Fig. 1). This regulation could be via cell-cell interactions and/or through systemic factors. On the other hand, the **38^+ 34^-** cells may also play an important role in the regulation of the stability of the stem cell compartment, as suggested by gene expression analysis indicating that several important regulatory proteins (growth factors, cytokines, chemokines) are uniquely expressed in **38^+ 34^-** cells.

These results suggest that transplantation with pure primitive (**34^-**) stem cells may not be the best approach. Successful stem cell transplantation probably needs both LTRCs and STRCs. Because the **34^+** human stem cells, at least as presently purified, clearly have LTRC and STRC activity, only an incremental improvement in clinical outcome would be expected by adding additional **34^-** cells. If the regulation of human stem cell transplantation becomes better understood, clinical use of pure PHSC populations could be beneficial in reducing malignant cell contamination when BMT is done as a part of cancer therapy.

14.4. MOUSE HSC ASSAYS

The term *hematopoietic stem cell* has been a much abused and misunderstood term that has been used in its loosest form to describe any hematopoietic progenitor cell. A major reason for the frequent incorrect use of the term in the literature is that generally stem cells have been operationally defined based on the assay used to detect and enumerate them. Over time stem cell assays have changed considerably, especially those for the most commonly used hematopoietic model, the mouse. We now know that previous mouse stem cell assays, as well as some of those in vitro assays currently in use, either do not detect stem cells at all, or more usually detect other types of hematopoietic progenitors, as well as stem cells (Till and McCulloch, 1961; Gan et al., 1997).

The biological characteristics that, when taken together, distinguish stem cells from other types of hematopoietic cells include self-renewal ability, extensive proliferative potential, and developmental multipotentiality, this being the ability to produce all the different cells of the lymphoid and myeloid lineages. In addition, it has recently become apparent that to be a functional stem cell in vivo, it is also necessary to have chemokine receptors and cell-surface adhesion molecules, which enable it to efficiently home and remain in an environment capable of supporting the expression of the stem cell's proliferative and developmental potential (Nagasawa et al., 1996; Zou et al., 1998; Berrios et al., 2001; Szilvassy et al., 2001; Voermans et al., 2001).

It is important to note that certain types of hematopoietic progenitors that are not stem cells have some of these characteristics, particularly extensive proliferative potential. However, only stem cells have them all. The detection and quantitation of HSCs therefore requires assays that can identify those cells with the characteristics of a stem cell, and that can ideally discriminate them from similar hematopoietic progenitors. We next briefly discuss the various in vivo mouse assays currently in use and also describe a new assay that we have developed that, by comparison, provides substantial advantages in the quantitation of repopulating HSCs.

14.4.1. IN VIVO STEM CELL REPOPULATING ASSAYS The type of assay that has been most widely used for the quantitation of stem cells is the in vivo repopulating cell assay. The assays that use this approach detect those cells capable of repopulating the hematopoietic system, following their injection into mice that have been treated with a potentially lethal dose of radiation. These assays are very stem cell specific because to achieve this goal it is necessary to have all of the stem cell characteristics.

The assay that has been the standard for over two decades is the long-term competitive repopulating assay (Harrison, 1980, Harrison et al., 1993). In this assay, different numbers of donor cells, containing an unknown frequency of stem cells, are cotransplanted with a standard number of normal bone marrow cells. This latter population of cells serves two functions. First, it provides radioprotective support cells that quickly produce the mature cells necessary to ensure the survival of the mice. This function is particularly important during the initial 6 wk that it takes for the donor stem cells to develop into a sufficient number of mature hematopoietic cells. Second, the normal bone marrow cell population contains stem cells that compete with the donor stem cells, and act as a standard with which to compare repopulating ability. The proportions of mature cells derived from stem cells of donor origin are then determined by the detection of one of several markers that differ between the donor and the competitor/recipient cells, such as isoenzyme, Y chromosome or congenic antigen (usually CD45) (Spangrude et al., 1988; Hampson et al., 1989; Harrison, 1980). The donor cell proportions obtained are used to statistically determine the frequency of repopulating units in the transplanted donor population using a binomial model.

This assay has been extremely useful in the study of stem cell biology and transplantation; however, its use has been restricted by the fact that it usually requires at least 3–6 months to complete (Harrison et al., 1993, Zhong et al., 1996). This long period of time is necessary, at least in part, due to the presence of substantial numbers of relatively short-lived progenitors, including radioprotective cells, colony-forming cells and colony-forming units-spleen, that either survive the radiation treatment or are present in the donor and competitor cell populations. The initial production of mature hematopoietic cells of all lineages is predominantly from these developmentally later progenitors, which have to be largely exhausted and replaced by progeny of the competing stem cells before the results of the assay can be determined.

The development of the competitive repopulating unit (CRU) assay provided an alternative approach to in vivo stem cell quantitation (Szilvassy et al., 1990; Szilvassy and Cory, 1993). It utilizes a limiting dilution stem cell transplant, Poisson statistics, and a stem cell–compromised population of serially transplanted cells, rather than normal bone marrow, as a source of competitive/radioprotective cells. The use of these transplant conditions enables stem cell detection and quantitation only 10 wk after injection. Unfortunately, however, when using the CRU assay there is no overall reduction in the time taken to perform the assay, when compared to the long-term competitive repopulation assay, since it takes a minimum of nine additional weeks to prepare the population of stem cell–compromised radioprotective cells.

14.4.2. REPOPULATING STEM CELL UNIT ASSAY With the objective of further understanding the requirements for optimal stem cell repopulation, the effects of the different variables involved in this process were investigated. These variables included radiation dose, recipient hematopoiesis, radioprotective support cells, repopulation kinetics, and stem cell competition (Ponting et al., 2000). It was determined that the optimal conditions for stem cell repopulation following transplantation were effectively no competition and maximum stem cell repopulation pressure. These conditions were achieved first, by ablating recipient stem cells and progenitors with an unusually high dose of radiation, 1400 cGy as compared with 900–1100 cGy for other in vivo stem cell assays (Harrison, 1980; Szilvassy et al., 1990), and second, by performing a limiting dilution cotransplant of congenic donor stem cells with busulfan-treated bone marrow radioprotective support cells (BUS-BM). Poisson analysis of the proportion of mice not showing donor lymphoid (B- and/or T-lymphocytes) and myeloid cells only 8 wk after transplant enables quantitation of the repopulating ability of the cell population under investigation. Stem cell specificity is therefore dependent on their multipotentiality and the ability to proliferate extensively for a period of at least 8 weeks when exposed to intense repopulation pressure. The transplantation system and quantitative stem cell assay resulting from this work has been called the repopulating stem cell unit (RESCU) assay to distinguish it from the other in vivo assays that use different procedures (Fig. 2).

The high dose of radiation used in the RESCU assay provides a virtual "clean slate" with which to study hematopoietic repopulation and is critical to the sensitive detection of the very small numbers of donor stem cells present in the limiting dilution transplant. As a result, the proportion of mature cells derived from donor stem cells, as well as the proportion of mice that show donor reconstitution of both lymphoid and myeloid cells, is substantially higher than seen when using the traditional lower doses of radiation. Of course, donor reconstitution will still be observed at these lower doses, but more donor cells must be transplanted to obtain the same effect.

Busulfan has been used extensively in the clinic as a part of conditioning regimens that ablate patient's endogenous stem cells prior to BMT (Parkman et al., 1984; Hassan 1999). Until the development of the RESCU assay, however, busulfan had not

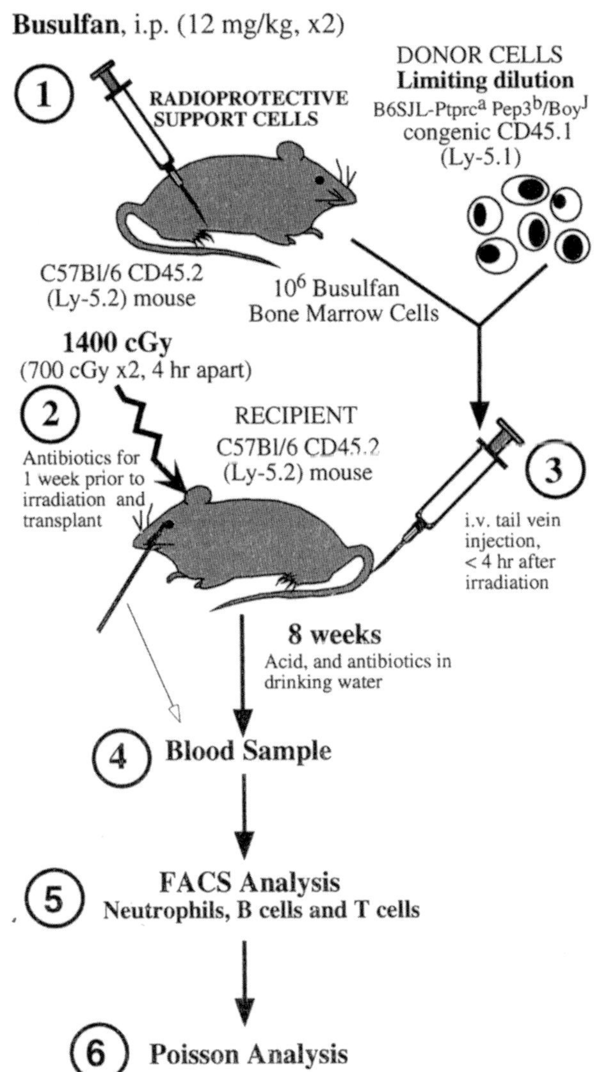

Fig. 2. Schematic representation of in vivo stem cell RESCU assay procedure. Ten mice/donor cell dilution is optimal, with five or six different donor cell dilution groups/experiment. The proportion of mice that did not show any donor repopulation is determined, with a 37% negative response being equivalent to the RESCU frequency.

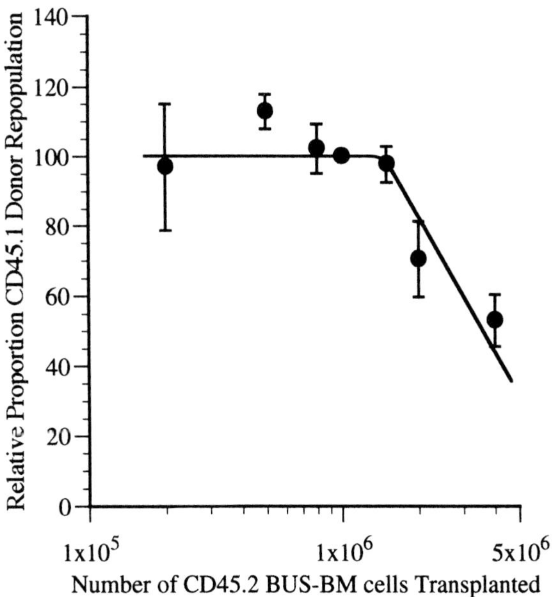

Fig. 3. BUS-BM cotransplants are non-competitive. Data were obtained 8 wk after the irradiation (1400 cGy) and transplantation of 2×10^5 donor CD45.1 (0,1,2) 5-FU-BM stem cells together with increasing numbers of CD45.2 BUS-BM radioprotective support cells. The results are shown as a percentage of the CD45.1 repopulation obtained when 10^6 BUS-BM cells were cotransplanted (100% represents 42.5 ± 0.4% mature donor CD45.1 cells). Data are expressed as the mean ± SD of three experiments, with 10 mice/group.

been used to produce radioprotective support cells for stem cell transplantation studies. The novel aspect of cotransplanting BUS-BM cells is the ability to provide a sufficient number of radioprotective cells to ensure survival, while also enabling more efficient donor stem cell repopulation, due to the lack of effective stem cell competition. The BUS-BM support cells are obtained by treating mice with two 12 mg/kg ip injections of Busulfex™, 24 h apart; the bone marrow cells are harvested 24 h after the last injection. As can be seen in Fig. 3, the cotransplant of 1.5×10^6 or fewer CD45.2 BUS-BM support cells results in the production of the same proportion of mature CD45.1 donor cell reconstitution. This result indicates that when cotransplanted with 10^6 BUS-BM cells, there is effectively no stem cell competition encountered by the donor stem cells. However, this number of BUS-BM cells when transplanted alone is able to maintain long-term survival in 80% of animals. This finding indicates that a small number of stem cells are present in the BUS-BM cells, but they are of poor quality due to the busulfan treatment, and are therefore unable to effectively compete with normal donor stem cells.

The apparent explanation for this observation is that busulfan is preferentially stem cell toxic, producing a decreased stem cell:radioprotective cell ratio for the population of BUS-BM cells, as compared to normal bone marrow. In support of this conclusion, busulfan has been shown to be more toxic toward primitive murine hematopoietic progenitor cells (cobblestone area–forming cell [CAFC] d 28–35) as compared with developmentally more mature progenitors (CAFC d 5–18) (Down and Ploemacher, 1993). As already discussed in relation to the use of a high dose of radiation, the reduction in stem cell competition produces a substantial increase in the sensitivity of detecting the small number of stem cells present in a limiting dilution transplant. The fact that it takes 2 d to prepare the BUS-BM population of cells and only 8 wk to complete the assay makes the RESCU assay by far the quickest in vivo stem cell assay currently available.

As a result of the special transplant conditions described, the RESCU assay is extremely sensitive in detecting repopulating stem cells. This sensitivity is strikingly revealed by the finding that 62% of repopulating units in a population of 5-FU-treated bone marrow (5-FU-BM) cells are not detected when the BUS-BM cells are replaced by as few as 10^5 competitive NBM cells (Fig. 4). This result can be explained by a relative increase in the number of mature CD45.2 cells when cotransplanting normal bone marrow cells, which produces a corresponding decrease in the proportion of mature CD45.1 cells. As a result, those mice that were previously identified as being repopulated by a low proportion of mature donor cells when using BUS-BM cells would then become negative for donor stem cell repopulation when using normal bone

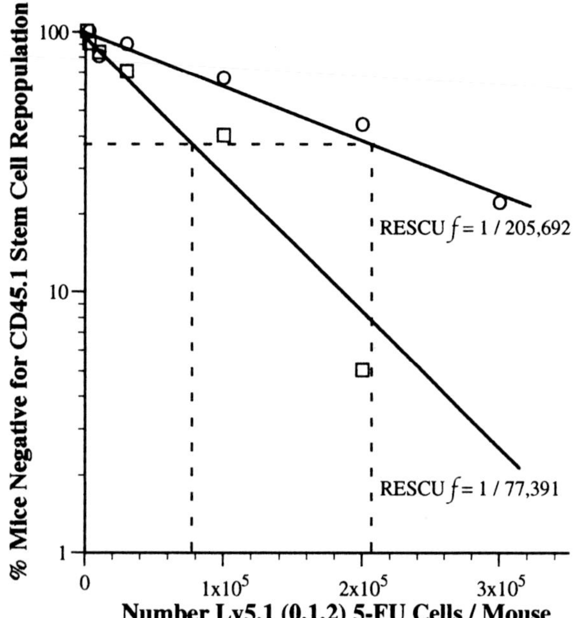

Fig. 4. Determination of RESCU frequency in (0,1,2) 5-FU-BM using either BUS-BM or normal bone marrow support cells. Recipient CD45.2 mice were irradiated with 1400 cGy and transplanted with a range (1×10^4 to 3×10^5) of CD45.1 (0,1,2) 5-FU-BM cells, together with either 10^6 CD45.2 BUS-BM (□) or 10^5 CD45.2 normal bone marrow (○) support cells. After 8 wk the proportions of mice that did not show any CD45.1 donor repopulation were determined. Data were obtained from three experiments, with a total of 30 mice/data point.

marrow cells. It is therefore apparent that stem cell competition results in the selection of those repopulating units that produce the greatest proportion of mature donor cells. This can be explained either by the selection of those repopulating units that for intrinsic or microenvironmental reasons comprise stem cells with the greatest proliferative potential or, alternatively, simply those repopulating units comprising the largest number of stem cells.

Current dogma supports the possibility that stem cell competition selects for those repopulating units comprising of "better quality" stem cells that have the greatest intrinsic proliferative potential. However, from experiments performed using the RESCU assay, it appears more likely that competition selects for those repopulating units that contain a larger number of stem cells and/or stem cells that have seeded optimal microenvironments. As competition increases, there will therefore be a corresponding increase in the number of stem cells comprising a repopulating unit and a decrease in the frequency of repopulating units. A direct result of this conclusion is that repopulating units will comprise the smallest number of stem cells when there is effectively no competition, whereas, e.g., mice that have not been irradiated will require the transplant of a comparatively much larger number of stem cells before a donor repopulating unit is detected (Stewart et al., 1998).

REFERENCES

Belloc, F., Dumain, P., Boisseau, M. R., et al. (1994) A flow cytometric method using Hoechst 33342 and propidium iodide for simultaneous cell cycle analysis and apoptosis determination in unfixed cells. *Cytometry* 17:59–65.

Berenson, R. J., Andrews, R. G., Bensinger, W. I., et al. (1988) Antigen CD34+ marrow cells engraft lethally irradiated baboons. *J. Clin. Invest.* 81:951–955.

Berrios, V. M., Dooner, G. J., Nowakowski, G., et al. (2001) The molecular basis for the cytokine-induced defect in homing and engraftment of hematopoietic stem cells. *Exp. Hematol.* 29:1326–1335.

Bhatia, M., Bonnet, D., Murdoch, B., Gan, O. I., and Dick, J. E. (1998) A newly discovered class of human hematopoietic cells with SCID-repopulating activity. *Nat. Med.* 4:1038–1045.

Cardoso, A. A., Li, M. L., Batard, B., et al. (1993) Release from quiescence of CD34+CD38- human umbilical cord blood cells reveals their potentiality to engraft adults. *Proc. Natl. Acad. Sci. USA* 90:8707–8711.

Down, J.D. and Ploemacher, R.E. (1993) Transient and permanent engraftment potential of murine hematopoietic stem cell subsets: differential effects of host conditioning with gamma radiation and cytotoxic drugs. *Exp. Hematol.* 21:913–921.

Gallacher, L., Murdoch, B., Wu, D. M., Karanu, F. N., Keeney, M., and Bhatia, M. (2000) Isolation and characterization of human CD34 (–)Lin(–) and CD34(+)Lin(–) hematopoietic stem cells using cell surface markers AC133 and CD7. *Blood* 95:2813–2820.

Gan, O. I., Murdoch, B., Larochelle, A., and Dick, J. E. (1997) Differential maintenance of primitive human SCID-repopulating cells, clonogenic progenitors, and long-term culture-initiating cells after incubation on human bone marrow stromal cells. *Blood* 90:641–650.

Goodell, M. A., Brose, K., Paradis, G., Conner, A. S., and Mulligan, R. C. (1996) Isolation and functional properties of murine hematopoietic stem cells that are replicating in vivo. *J. Exp. Med.* 183:1797–1806.

Goodell, M. A., Rosenzweig, M., Kim, H., et al. (1997) Dye efflux studies suggest that hematopoietic stem cells expressing low or undetectable levels of CD34 antigen exist in multiple species. *Nat. Med.* 3:1337–1345.

Hampson, I. N., Spooncer, E. and Dexter, T. M. (1989) Evaluation of a mouse Y chromosome probe for assessing marrow transplantation. *Exp. Hematol.* 17:313–315.

Harrison, D. E. (1980) Competitive repopulation: a new assay for long-term stem cell functional capacity. *Blood* 55:77–81.

Harrison, D. E. and Lerner, C. P. (1991) Most primitive hematopoietic stem cells are stimulated to cycle rapidly after treatment with 5-fluorouracil. *Blood* 78:1237–1240.

Harrison, D. E. and Zhong, R.-K. (1992) The same exhaustible multilineage precursor produces both meyloid and lymphoid cells as early as 3–4 weeks after marrow transplantation. *Proc. Natl. Acad. Sci. USA* 89:10,134–10,138.

Harrison, D. E., Jordan, C. T., Zhong, R. K., and Astle, C. M. (1993) Primitive hemopoietic stem cells: direct assay of most productive populations by competitive repopulation with simple binomial, correlation and covariance calculations. *Exp. Hematol.* 21:206–219.

Hassan, M. (1999) The role of busulfan in bone marrow transplantation. *Med. Oncol.* 16:166–176.

Huang, S. and Terstappen, L. W. (1994) Lymphoid and myeloid differentiation of single human CD34+, HLA-DR+, CD38– hematopoietic stem cells. *Blood* 83:1515–1526

Ito, T., Tajima, F., and Ogawa, M. (2000) Developmental changes of CD34 expression by murine hematopoietic stem cells. *Exp. Hematol.* 28:1269–1273.

Jones, R. J., Wagner, J. E., Celano, P., Sicha, M. S., and Sharkis, S. J. (1990) Separation of pluripotent haematopoietic stem cells from spleen colony-forming cells. *Nature* 347:188–189.

Krause, D. S., Ito, T., Fackler, M. J., et al. (1994) Characterization of murine CD34, a marker for hematopoietic progenitor and stem cells. *Blood* 84:691–701.

Nagasawa, T., Hirota, S., Tachibana, K., et al. (1996) Defects of B-cell lymphopoiesis and bone-marrow myelopoiesis in mice lacking the CXC chemokine PBSF/SDF-1. *Nature* 382:635–638.

Okada, S., Nakauchi, H., Nagayoshi, K., et al. (1991) Enrichment and characterization of murine hematopoietic stem cells that express c-kit molecule. *Blood* 78:1706–1712.

Osawa, M., Nakamura, K., Nishi, N., Takahashi, N., Tokuomoto, Y., Inoue, H., and Nakauchi, H. (1996) In vivo self-renewal of c-kit$^+$ Sca-1$^+$ Lin$^{low/-}$ hemopoietic stem cells. *J. Immunol.* 156:3207–3214.

Osawa, M., Hanada, K., Hamada, H., and Nakauchi, H. (1996b) Long-term lymphohematopoietic reconstitution by a single CD34-low/negative hematopoietic stem cell. *Science* 273:242–245.

Parkman, R., Rappeport, J. M., Hellman, S., et al. (1984) Busulfan and total body irradiation as antihematopoietic stem cell agents in the preparation of patients with congenital bone marrow disorders for allogenic bone marrow transplantation. *Blood* 64:852–857.

Ponting, I., Wang, H.-M., Chiu, L., Shpaner, A., and Shin, F. (2000) Excessive stem cell competition in the competitive long-term repopulation assay—development of a solution. *Exp. Hematol.* 28:39 (abstract).

Randall, T. D., Lund, F. E., Howark, M. C., and Weissman, I. L. (1996) Expression of murine CD38 defines a population of long-term reconstituting hematopoietic stem cells. *Blood* 87:4057–4067.

Sato, T., Laver, J. H., and Ogawa, M. (1999) Reversible expression of CD34 by murine hematopoietic stem cells. *Blood* 94:2548–2553.

Shapiro, H. M. (1981) Flow cytometric estimation of DNA and RNA content in intact cells stained with Hoechst 33342 and Pyronin Y. *Cytometry* 2:143–150.

Sitnicka, E., Ruscetti, F. W., Priestley, G. V., Wolf, N. S., and Bartelmez, S. H. (1996) Transforming growth factor beta 1 directly and reversibly inhibits the initial cell divisions of long-term repopulating hematopoietic stem cells. *Blood* 88:82–88.

Spangrude, G. J., Heimfeld, S., and Weissman, I.L. (1988) Purification and characterization of mouse hematopoietic stem cells [published erratum appears in *Science* (1989) 244(4908):1030]. *Science* 241:58–62.

Spangrude, G. J., Brooks, D. M., and Tumas, D. B. (1995) Long-term repopulation of irradiated mice with limiting numbers of purified hematopoietic stem cells: in vivo expansion of stem cell phenotype but not function. *Blood* 85:1006–1016.

Stewart, F. M., Zhong, S., Wuu, J., Hsieh, C., Nilsson, S. K., and Quesenberry, P. J. (1998) Lymphohematopoietic engraftment in minimally myeloablated hosts. *Blood* 91:3681–3687.

Szilvassy, S. J. and Cory, S. (1993) Phenotypic and functional characterization of competitive long-term repopulating hematopoietic stem cells enriched from 5-fluorouracil-treated murine marrow. *Blood* 81: 2310–2320.

Szilvassy, S. J., Humphries, R. K., Lansdorp, P. M., Eaves, A. C., and Eaves, C. J. (1990) Quantitative assay for totipotent reconstituting hematopoietic stem cells by a competitive repopulation strategy. *Proc. Natl. Acad. Sci. USA* 87:8736–8740.

Szilvassy, S. J., Meyerrose, T. E., Ragland, P. L., and Grimes, B. (2001) Homing and engraftment defects in ex vivo expanded murine hematopoietic cells are associated with downregulation of beta1 integrin. *Exp. Hematol.* 29:1494–1502.

Tajima, F., Deguchi, T., Laver, J. H., Zeng, H., and Ogawa, M. (2001) Reciprocal expression of CD38 and CD34 by adult murine hematopoieitc stem cells . *Blood* 97: 2618–2624.

Terstappen, L. W., Huang, S., Safford, M., Lansdorp, P. M., and Loken, M. R. (1991) Sequential generations of hematpoietic colonies derived from single nonlineage-committed CD34+CD38– progenitor cells. *Blood* 77:1218–1227.

Till, J. E. and McCulloch, E. A. (1961) A direct measurement of the radiation sensitivity of normal mouse bone marrow cells. *Radiat. Res.* 14:213–222.

Van Zant, G. (1984) Studies of hematopoietic stem cells. *J. Exp. Med.* 159:679–690.

Voermans, C., van Hennik, P. B., and van der Schoot, C. E. (2001) Homing of human hematopoietic stem and progenitor cells: new insights, new challenges? *J. Hematother. Stem Cell Res.* 10:725–738.

Zanjani, E. D, Almeida-Porada, G., Livingston, A. G., Flake, A. W., and Ogawa, M. (1998) Human bone marrow CD34– cells engraft in vivo and undergo multilineage expression that includes giving rise to CD34+ cells. *Exp. Hematol.*26:353–360.

Zhao, Y., Lin, Y., Zhan, Y., et al. (2000) Murine hematopoietic stem cell characterization and its regulation in BM transplantation. *Blood* 96: 3016–3022.

Zhong, R. K., Astle, C. M., and Harrison, D. E. (1996) Distinct developmental patterns of short-term and long-term functioning lymphoid and myeloid precursors defined by competitive limiting dilution analysis in vivo. *J. Immunol.* 157:138–145.

Ziegler, B. L., Valtieri, M., Porada, G. A., et al. (1999) KDR receptor: a key marker defining hematopoietic stem cells. *Science* 285: 1553–1558.

Zijlmans, J. M., Visser, J. W. M., Kleiverda, K., Kluin, P. M., Willemze, R., and Fibbe, W. E. (1995) Modification of rhodamine staining allows identification of hematopoietic stem cells with preferential short-term or long-term bone marrow-repopulating ability. *Proc. Natl. Acad. Sci. USA* 92:8901–8905.

Zou, Y. R., Kottmann, A. H., Kuroda, M., Taniuchi, I., and Littman, D. R. (1998) Function of the chemokine receptor CXCR4 in haematopoiesis and in cerebellar development. *Nature* 393:595–599.

15 Hematopoietic Stem Cells in Leukemia and Lymphoma

STEPHEN M. BAIRD, MD

Chemotherapy for acute leukemia acts on both malignant and proliferating transit-amplifying cells and produces a marked drop in white cells and platelets in 1 to 2 wk. Resting cells, such as bone marrow stem cells and memory T-cells, are relatively unaffected. When chemotherapy is discontinued, red cells, white cells, and platelets recover in about 10–14 d. The most critical cell at this point is the neutrophil, and failure of these to recover may result in fatal infections. Neutrophil recovery can be stimulated by administration of granulocyte colony-stimulating factor. Recovery is mediated by generation of a new population of transit-amplifying white and red cell precursors from the chemotherapy-resistant stem cell. The phenotype of the cell that could accomplish this is $CD34^+$, Thy1 lo and lineage negative. An even more primitive stem cell may actually be $CD34^-$. The defining features of the human bone marrow stem cell are that it can give rise to all hematopoietic colonies in vitro, repopulate the marrow, or can give rise to hematopoiesis in the SCID-Hu mouse. CD markers representative of different stages of white cell poiesis can be used to define different leukemias and lymphomas, the response to chemotherapy, and the differentiating populations during recovery. Cyclic administration of chemotherapy may be curative of some leukemias, suggesting that the most primitive bone marrow stem cell is not affected. Expression of the multiple-drug resistance gene product may partially explain this resistance. For some cases of leukemia, the malignant genetic event occurs in the most primitive stem cell, so that ablative irradiation or chemotherapy must be given to destroy all malignant stem cells. The stem cells are then replaced by bone marrow or circulating blood stem cell transplantation. Although cytokine-liberated autologous blood stem cells are often used to avoid graft vs host reactions owing to the presence of T-cells in the blood or marrow of allogeneic donors, a mild graft-versus-host disease reaction after transplantation has been shown to be effective in eliminating residual host tumor cells. The results of bone marrow transplantation clearly show that this is an effective treatment for acute and some chronic leukemias and that malignant transformation therefore occurs either in the most primitive bone marrow stem cell or in an early transit-amplifying cell.

15.1. INTRODUCTION

When a physician administers chemotherapy to a patient for acute leukemia, there is a general pattern of response. Most combination drug regimens produce a nadir of white blood cell (WBC) and platelet counts in 1–2 wk. Among the white cells, the granulocytes (neutrophils, eosinophils, and basophils) drop the most, while some lymphocytes remain. This is understandable because we know that granulocytes typically circulate <1 d before emigrating into the tissues, where they die in a few days. There are no long-lived neutrophils. By contrast, memory T-lymphocytes may live for decades, and if they are not stimulated, they rest in G0, making them resistant to many chemotherapeutic agents.

When chemotherapy is stopped, most patients recover normal red blood cell (RBC), WBC and platelet counts in 10–14 d. Administration of the cytokine granulocyte colony-stimulating factor (GCSF) speeds up neutrophil recovery by a few days. This is the most important cell for response to acute bacterial infection. The recovery time indicates the approximate time required for hematopoietic cells to go from the postulated resting stem cell that survived chemotherapy to mature cells circulating in the blood.

Figure 1 is a standard chart of hematopoiesis with the site of action of some important cytokines listed. A few selected CD markers are also listed beside cells at specific stages of differentiation. Data on where cytokines act in the differentiation pathways and where CD markers are acquired and lost have been generated in many different ways, mostly in vitro, so Fig. 1 provides all the strengths and weaknesses of any model illustrating a complex process.

What we have learned from chemotherapy and recovery is that the marrow contains drug-resistant cells capable of reconstituting all the circulating cells in the bloodstream. Since Till and McCulloch's seminal observations in 1961 (Till and McCulloch, 1961) on spleen colonies formed by hematopoietic cells in the mouse, the nature of the cells responsible for recovery from che-

From: *Stem Cells Handbook*
Edited by: S. Sell © Humana Press Inc., Totowa, NJ

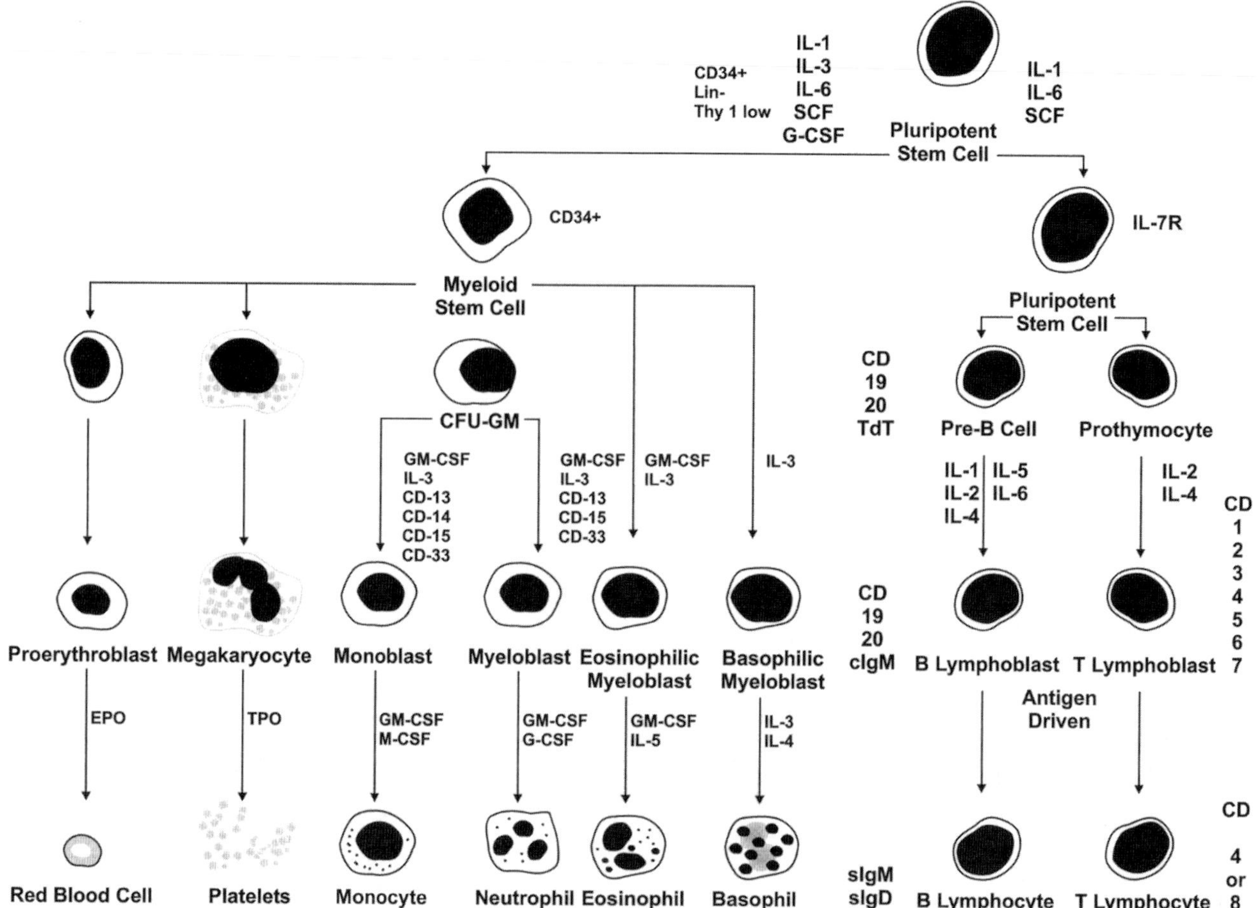

Fig. 1. Brief chart of hematopoiesis. Some CD markers characteristic of various maturational states are indicated as are some cytokines responsible for driving cells through the maturational process.

motherapy and those that successfully engraft after bone marrow transplantation (BMT) have been described. In 1988, Spangrude et al. reported the phenotype of murine cells capable of completely reconstituting hematopoiesis in a lethally irradiated mouse. In their "best" experiment, as few as seven cells would reconstitute an animal (Spangrude et al., 1988). Subsequent work has shown that one cell can be sufficient to repopulate an entire animal (Baum et al., 1992). This cell expressed CD34, Thy 1 lo, and was lineage negative, meaning that the investigators sorted to remove cells that expressed CD markers characteristic of mature cells of any hematopoietic lineage. Microscopically, the cell resembled a small lymphocyte. Subsequent work has shown that cells that have not yet expressed CD34 also have stem cell capability (Craig et al., 1993; Brandt et al., 1988). Similar cells have been isolated from human marrow, but their capacities have not been demonstrated with the same definitive experiments used in mice. CD patterns of mouse and human hematopoietic stem cells (HSCs) are slightly different. The earliest human stem cell is $CD34^-$ but in practice $CD34^+$ cells are quantitated (Uchida et al., 1996; Goodell et al., 1997). Bear in mind that the definitions are not exactly the same. The stem cell in the mouse is the cell that can repopulate the entire hematopoietic system after transplantation into a lethally irradiated recipient. The stem cell in the human can give rise to colonies in tissue culture, can repopulate the marrow after BMT, or can give rise to hematopoiesis in the SCID-Hu mouse (Berenson et al., 1988; Negrin et al., 2000). The properties of normal HSCs that are critical to understand as they relate to the treatment of leukemias and lymphomas are that they are capable of repopulating all the cells normally produced in the marrow and that a large percentage of them appears to be quiescent at any given time. This second feature allows recovery of normal hematopoiesis after chemotherapy directed against dividing cells.

Figures 2–6 show a typical case of early pre-B acute lymphoblastic leukemia (ALL) from diagnosis through a few cycles of chemotherapy and recovery of the white cell count in the blood. Flow cytometry analysis of WBCs with fluorescent antibodies against selected CD markers is shown. The results were generated by monitoring a patient through clinical diagnosis and therapy, not in a research program.

Figure 2A shows the distribution of cell sizes and complexity at diagnosis when the blood was filled with lymphoblasts. The *X*-axis shows cell size and the *Y*-axis shows complexity such as multilobed nuclei or cytoplasmic granules. The cells in the gate labeled R1 range from small to large and are of low internal complexity. These are lymphocytes and lymphoblasts. Very few other cells such as neutrophils or monocytes were present in the blood at diagnosis.

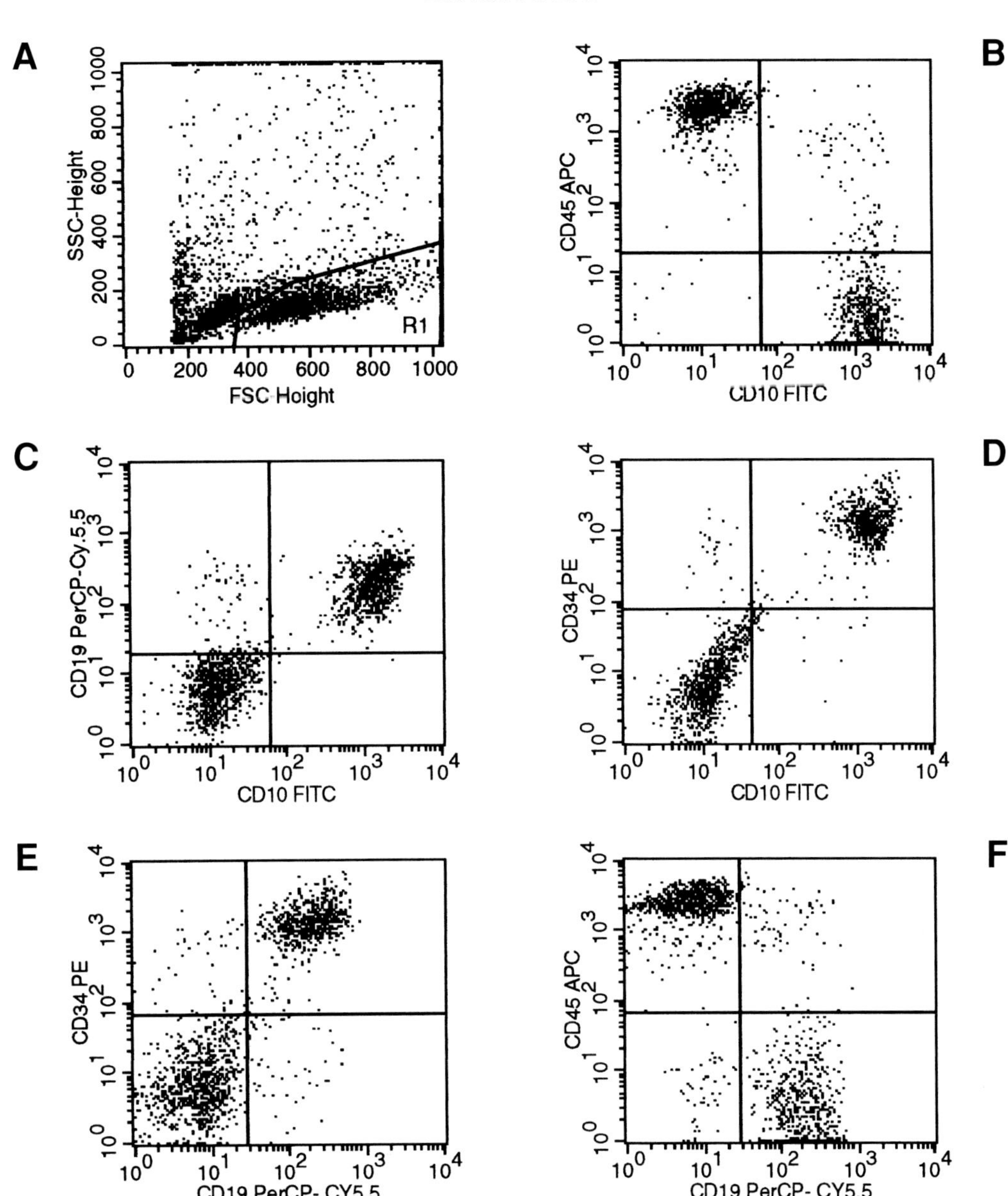

Fig. 2. Bone marrow was aspirated from the iliac crest and prepared for histological and flow cytometric analysis. For flow cytometry, red cells were lysed, and then nucleated cells were concentrated by ficoll gradient centrifugation and put into aliquots. They were incubated with fluorescent antibodies to the indicated CD markers, washed, and analyzed on a Beckman FACScan using three-color discrimination. Results are discussed in text. FSC, forward scatter; FITC, fluorescein isothiocyanate; PE, phycoerythrin; APC, allo phycocyanin.

Figure 2B shows two populations of cells as defined by expression of CD45, the common leukocyte antigen, and CD10, the common ALL antigen. Figures 2C–E shows that CD10 is coexpressed with CD34 and CD19. Two different sets of labeled antibodies were used in Fig. 2D,E. These results and others defined a leukemic cell expressing CD19, CD10, CD34, TdT, and low levels of the myeloid antigens CD13 and CD33, an example of lineage infidelity. The bright CD45 cells seen in Fig. 2B,F were shown to be T-lymphocytes. Essentially no granulocytes were present.

Figure 3, after the hypoplasia induced by chemotherapy, shows that essentially all cells expressing CD10, CD19, and CD34 are gone. The small number of events emerging from the negative pool of cells at the lower left is conventionally interpreted as nonspecific antibody adherence to cells since it runs up at a 45° angle.

Figure 4 shows the distribution of the patient's WBCs in the marrow 16 d after the hypoplasia seen in Fig. 3. Figure 4A demonstrates the presence of small to large cells of low internal complexity—small and large lymphoid cells. Two other populations are easily seen: relatively large cells of moderate internal complexity and somewhat smaller cells of high internal complexity. These were demonstrated to be monocytes and neutrophils, respectively.

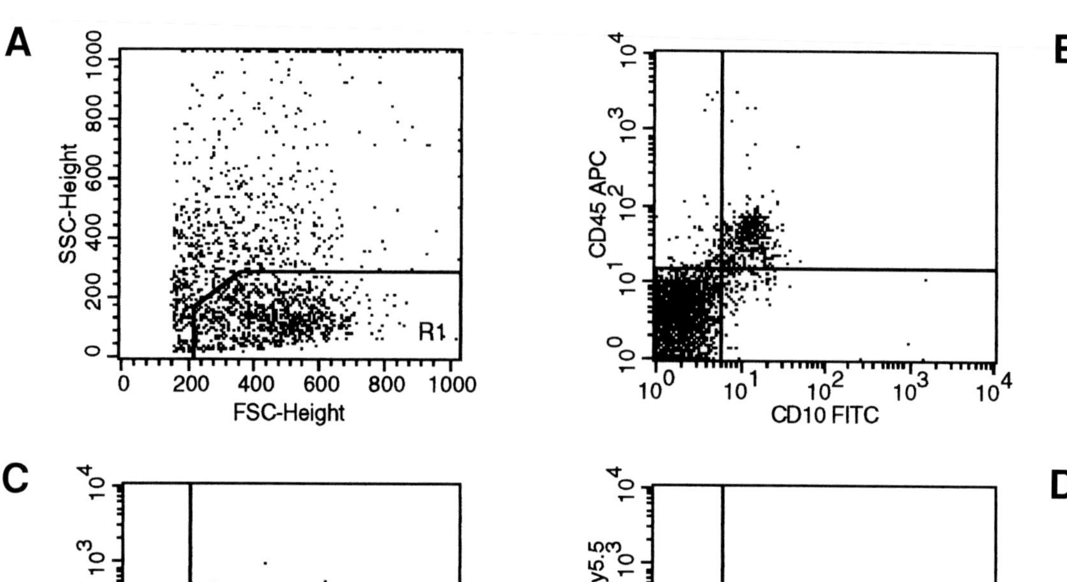

Fig. 3. Bone marrow from the iliac crest on d 51 after treatment was analyzed as in Fig. 2.

The CD45 expression seen in Fig. 4B shows the appearance of cells ranging from negative through weak positive to bright positive. This is the pattern of expression of CD45 in normal maturing leukocytes. Figure 4C–E shows the regrowth of the malignant clone of cells first demonstrated at diagnosis in Fig. 2. In addition, best seen in Fig. 4D, some cells not in the malignant clone are beginning to express dim to bright CD34. Figure 4F shows that the cells expressing CD10 (and, by inference, coexpressing CD19 and CD34 as well) are not part of the normal population expressing CD45. Recall that the original malignant clone was CD45 negative, a common finding in ALL.

Figure 4G,H shows some early expression of CD13 and CD33 on cells that are CD19 negative (normal, nonlymphoid cells) and on a uniform cluster of CD19-positive cells (the original malignant clone that is reexpressing its lineage infidelity). Figure 4H shows CD13 and CD33 expression on cells ranging from CD45 negative through rather bright positive. The very bright CD45-positive cells that are CD13 and CD33 negative are T-lymphocytes. In summary, Fig. 4 shows recovery of normal leukocytes and persistence of the malignant clone, a common problem early in chemotherapy.

Figure 5 shows the patient 1 mo after the findings depicted in Fig. 4. Note the continued production of normal leukocytes as well as persistence of the malignant clone. As assessed by CD13 and CD33 expression, normal granulocytes have continued to recover. CD34 cells that are not part of the malignant clone seem to have declined in number, perhaps indicating maturation to normal "adult" leukocytes that have lost their CD34.

Figure 6, depicting findings about 6 wk after Fig. 5, represents recovery from an additional round of chemotherapy. Figure 6A shows abundant lymphocytes, monocytes (confirmed in Fig. 6F,G), and neutrophils. Figure 6B shows three distinct groups of CD45 positive cells that are *not* monocytes (CD14 positive). Figure 6C,D shows the development of CD33, CD13, and CD15 positive cells, which are markers for granulocytes. Figure 6E again shows the emergence of CD34 dim to bright cells that are not part of the malignant clone (shown as the CD34/CD10 coexpressing cells) .

This series of figures shows the normal sequence of recovery of leukocytes after bone marrow aplasia induced by chemotherapy. Extensive research has confirmed that the "stem" cell responsible for recovery is the CD34-positive, lineage negative cell seen especially well in Figs. 4 and 6. This case also illustrates the problem of persistence of leukemic clones after conventional chemotherapy in many patients. In such cases, cure may be attempted by complete myeloablation followed by rescue with either a bone marrow transplant or stem cell transplantation.

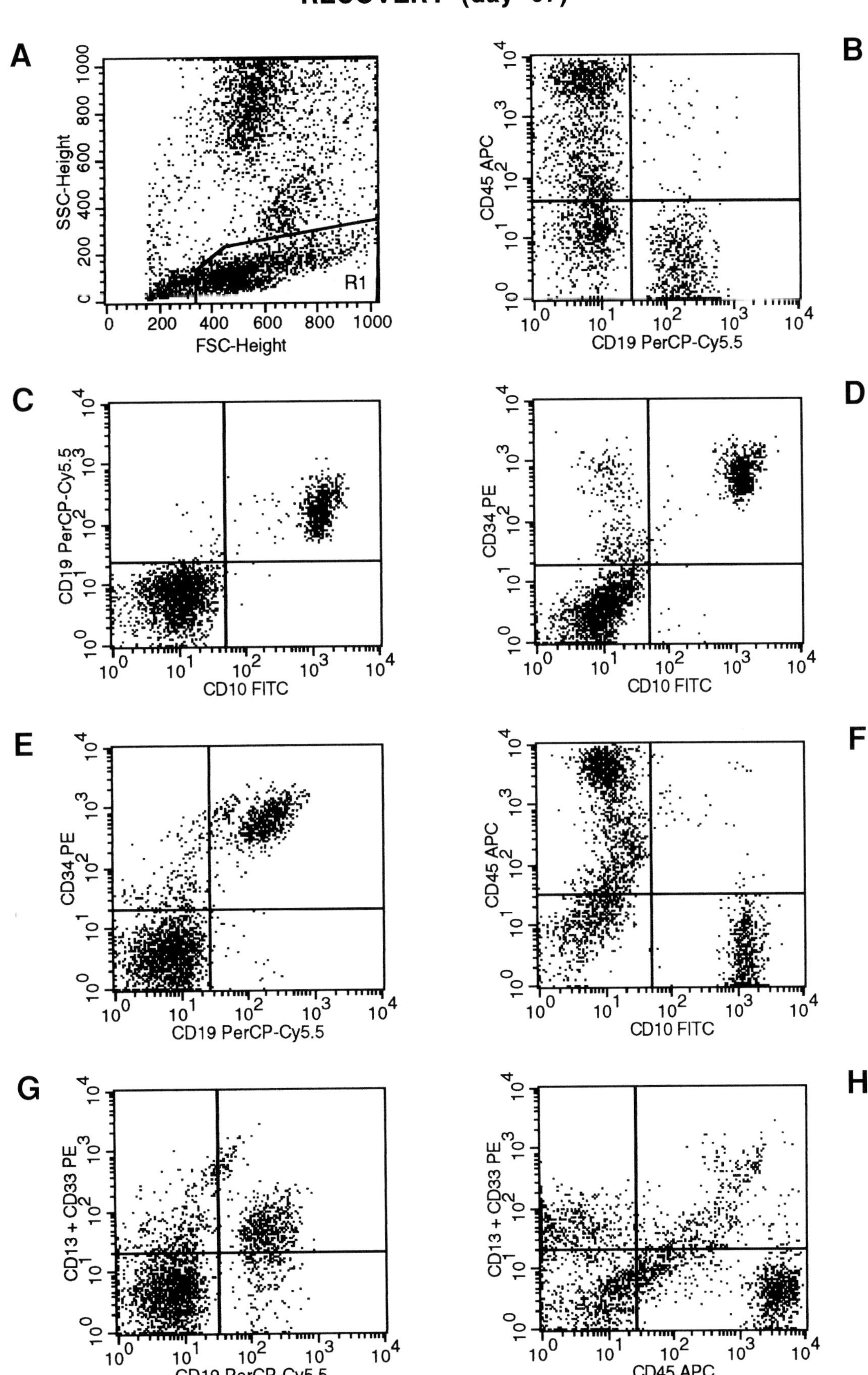

Fig. 4. Bone marrow obtained on d 67 after treatment was analyzed as in Fig. 2.

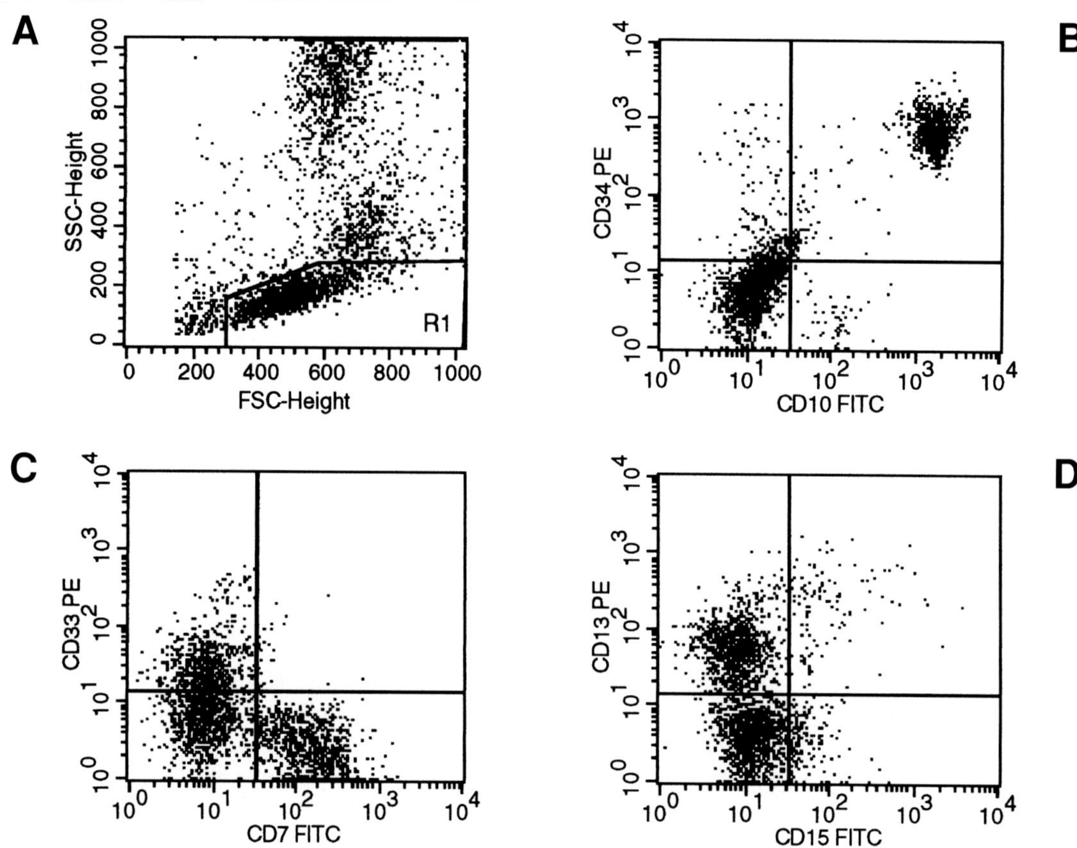

Fig. 5. Bone marrow from d 102 after treatment was analyzed as in Fig. 2.

15.2. HSC TRANSPLANTATION IN TREATMENT OF LEUKEMIA AND LYMPHOMA

Historically, the first source of cells to restore hematopoiesis was bone marrow. In 1951, Brecher and Cronkite showed that rats could be rescued from lethal irradiation by parabiosis, which we now know to be owing to circulating stem cells from the nonirradiated rat. In 1965, a patient being treated for ALL demonstrated engraftment of donor marrow after being infused with marrow harvested from six different donors (Mathe et al., 1965). Children with severe combined immune deficiency were first successfully treated by marrow transplantation in 1968 (Gatti et al., 1968; Hong et al., 1968). By 1975, many patients who suffered from leukemia had been treated with myeloablative chemotherapy and rescued with bone marrow transplants from human lymphocyte antigen (HLA) identical twins or siblings (Buckner et al., 1970; Thomas et al., 1975; Fefer et al., 1982; O'Reilly, 1983).

Marrow transplantation, however, is arduous for the donor. It usually requires general anesthesia and the aspiration of several hundred milliliters of marrow from the iliac crests. While this can be done safely, it is painful. Adult marrow, besides being a source of self-renewing HSCs, is also a source of mature T effector cells and therefore a risk for inducing graft-versus-host disease (GVHD). This is usually dealt with by choosing histocompatible marrow from siblings matched at HLA A, B, C, and DR loci and by various methods to purge mature T-cells before transplantation (Martin et al., 1985; Storek et al., 1997).

Both allogeneic and autologous marrow may be used to restore hematopoiesis. In the case of autologous donation, the patient is harvested during periods of remission; the marrow is stored frozen, then reinfused after a round of what would be lethal myeloablative therapy. Although GVHD is not a problem, another potential difficulty was demonstrated in our previous case: residual malignant cells in the marrow. When the antigenic phenotype of the malignant cell has been determined and appropriate antibodies or chemotherapeutic agents are available, attempts to purge the marrow of these undesirable cells in vitro are made before the marrow is infused. (Rowley et al., 1989; Gorin et al., 1990; Gribben et al., 1991; Shpall et al., 1991; Linker et al., 1998) This technology is not yet completely successful. The problems of obtaining marrow, GVHD, and purging of unwanted cells led to investigations of other sources of HSCs in the last decade.

Cytokine-mobilized stem cells from the peripheral blood are now the preferred cell type to restore hematopoiesis in patients being treated for leukemia and lymphoma (Siena et al., 1989; Korbling, et al., 1995). The success of this approach depends on

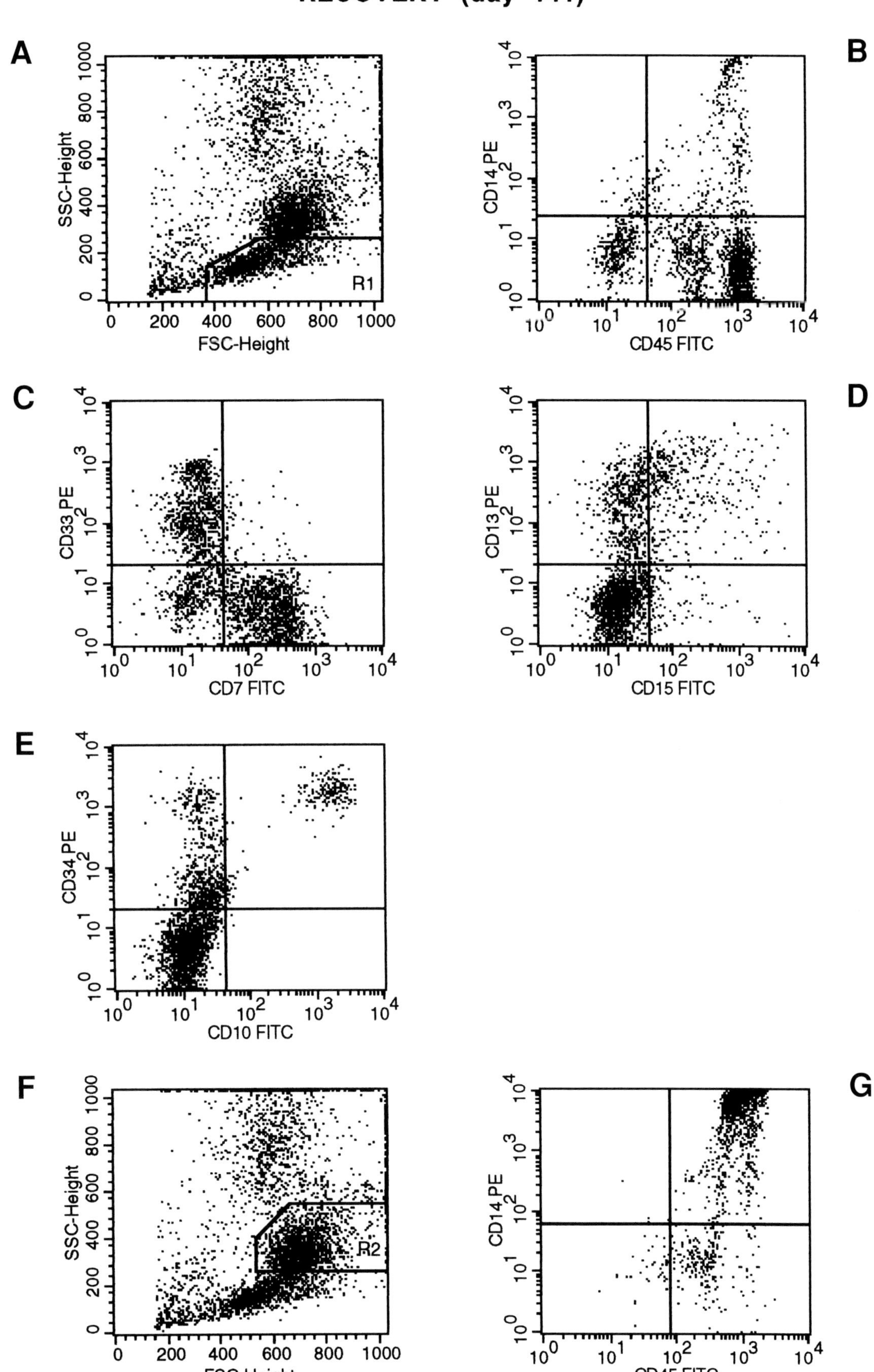

Fig. 6. Bone marrow from d 141 after treatment was analyzed as in Fig. 2.

the observation that self-renewing HSCs circulate in the blood at a frequency of about 10^{-4} per nucleated cell (Spangrude et al., 1988). This frequency rises after treatment with some chemotherapeutic agents and after treatment with cytokines such as GCSF or granulocyte-monocyte CSF (Socinski et al., 1988; Chao et al., 1993; Bensinger et al., 1994). The magnitude of the increase may be as much as 100-fold (Kosatec et al., 1992; Lane, 1995; Gillespie et al., 1996). This makes peripheral blood a useful source of stem cells for both autologous and allogeneic transplantation.

Collection is still irksome for the donor. Apheresis for several hours is required, often for more than one session. Stimulation of the donor with cytokines for several days before the apheresis procedure may produce side effects such as fever and flulike symptoms. It should be obvious without stating that leukemia and lymphoma patients who are acting as autologous donors may receive both chemotherapy and cytokines to mobilize stem cells whereas allogeneic donors receive only cytokines. Optimal restoration of red cell, granulocyte and platelet production is achieved at about 10×10^6 CD34$^+$ cells per kg of recipient after myeloablative chemotherapy (Shpall et al., 1998). The minimum satisfactory dose is about 1×10^6 CD34$^+$ cells/kg (Shpall et al., 1998). Most patients receive $2–5 \times 10^6$ CD34$^+$ cells/kg. This a larger number of cells per kilogram than is required in various murine experiments to restore hematopoiesis after lethal irradiation. The two systems are not strictly comparable, though, because in the murine experiment, preparative cell sorting to obtain and infuse the most active population of cells is usually done, while in humans CD34$^+$ cells in the apheresis specimen are simply quantitated and usually not further purified.

Umbilical cord blood has also been found to be an excellent source of CD34$^+$, self-renewing stem cells, but this source has not found wide application to date (Gluckeman et al., 1997).

Typical treatment courses and some general data on engraftment, complications, and survival are discussed next. Two patients are presented to illustrate the general process of stem cell transplantation in lymphoma and leukemia. The first is a 70-yr-old male with non-Hodgkin's lymphoma diagnosed 2 yr before the transplant. He underwent conventional chemotherapy and monoclonal antibody (MAb) treatment with anti-CD20 (he had a B-cell lymphoma expressing CD20). After 8 mo of chemotherapy and antibody therapy, he was judged to be in complete remission. Three months later, however, he developed a rash that was biopsied and showed infiltration by B-cell lymphoma, compatible with a recurrence of his original disease. He received more chemotherapy and MAb therapy and was judged to be in complete remission 5 mo later. A bone marrow biopsy showed no evidence of disease.

At this point the patient was judged to be a candidate for further high-dose chemotherapy and autologous stem cell transplantation because the statistical likelihood that he was really free of disease was very low. He had no suitable allogeneic donors, so peripheral stem cell mobilization with autologous transplantation was deemed appropriate.

The patient's hospital course with reference to his white cell, red cell, and platelet counts is indicated in Tables 1 and 2.

Table 1 shows the patient's response to conditioning and mobilization of stem cells with GCSF at 10 μg/kg for 4 d. It is typical for the WBC count to rise to several fold the normal value in 4 or 5 d. The "blast" count as determined morphologically and the CD34$^+$ cell quantity as determined by flow cytometry typically peak at about d 4 as can be seen in this patient.

Table 1
Preparation of Patient for "Stem Cell" Harvest[a]

Day	*WBCs* ($\times 10^{-3}$/μL)	*RBCs* ($\times 10^{-6}$/μL)	*Platelets* ($\times 10^{-3}$/μL)	*pmn* (%)	*blast* (%)
1	3.5	3.5	119	47	0
2					
3					
4	25	3.4	94	44	9
5	35	3.4	50	74	4
6	40	3.3	50	85	—

[a] The patient received GCSF at 10 μg/kg from d 1 to d 4 and then underwent cytopheresis to harvest peripheral CD34$^+$ "stem cells."

Table 2
Myeloablative Chemotherapy and Autologous "Stem Cell" Infusion[a]

Day	*WBCs* ($\times 10^{-3}$/μL)	*RBCs* ($\times 10^{-6}$/μL)	*Platelets* ($\times 10^{-3}$/μL)	*pmn* (%)
1	29	3.4	106	53
2	26	3.5	94	35
3	27	3.0	66	51
4	41	3.2	70	70
10	3.1	3.5	155	55
11	2.4	3.1	141	50
12	8.3	3.4	189	89
13	10.0	3.3	193	
14	7.0	3.2	182	
15	5.3	3.2	164	
16	5.6	3.1	141	
17	5.6	2.9	132	98
18	2.3	3.0	95	77
19	0.7	3.0	80	56
20	0.7	3.2	50	8
21	0.2	2.8	23	
22	0.2	2.6	7	
23	0.2	2.4	16	
24	0.3	2.7	39	
25	0.4	2.6	15	
27	0.5	2.3	10	26
28	2.1	2.9	16	52
29	2.7	3.0	31	69
30	2.1	3.0	23	58
31	1.6	3.2	41	56
33	1.7	3.2	25	41
35	2.5	3.2	20	39
38	3.2	3.3	16	36

[a] The patient received myeloablative chemotherapy and then was infused with 1.6×10^6/kg of CD34$^+$ autologous stem cells on d 17. WBC, RBC, and platelet counts reached their nadir on d 21 after therapy was begun. The patient was supported with platelets when counts fell below 10×10^3/uL. Peripheral WBCs began to recover on d 27.

The patient underwent cytopheresis and sufficient CD34$^+$ cells were harvested from the blood to permit an infusion of 1.6×10^6 CD34$^+$ cells/kg. These cells were stored frozen and the patient was prepared for stem cell transplantation by myeloablative chemotherapy for 2 wk. It is common to harvest rather low numbers of CD34$^+$ cells from patients who have had extensive chemotherapy.

Table 3
Myeloablative Chemotherapy and Allogeneic "Stem Cell" Infusion[a]

Day	*WBCs* ($\times 10^{-3}/\mu L$)	*RBCs* ($\times 10^{-6}/\mu L$)	*Platelets* ($\times 10^{-3}/\mu L$)	*pmn* (%)
1	3.3	3.0	71	80
2	1.9	2.9	80	64
3				
4	3.3	3.0	100	84
5				
6	2.1	2.9	89	77
7				
8	2.0	2.5	56	83
9	1.9	2.4	54	92
10	1.8	2.3	53	98
11	1.0	2.3	43	92
12	0.3	3.3	30	—
13	0.1	3.0	16	—
14	0.1	2.8	6	—
15	0.1	2.7	34	—
16	0.1	2.5	21	—
17	0.1	3.3	8	—
18	0.2	2.8	20	—
19	0.4	2.6	38	—
20	0.8	2.7	25	10
21	1.6	2.5	20	29
22	4.6	3.7	16	46
23				
24	3.2	3.9	34	29

[a] The patient received myeloablative chemotherapy followed by infusion of HLA-matched allogeneic "stem cells" at 4.4×10^6/kg on d 9. The nadir of peripheral blood counts was reached on d 14–15. The patient experienced bleeding and fever on d 17 and was treated with platelets and antibiotics. Neutrophils appeared in the blood on d 20; she defervesced and was discharged on d 22.

Table 2 shows the patient's response through chemotherapy, stem cell infusion, complications, and discharge over 28 d. The patient's sole complication was bacteremia with staphyloccus epidermidis, for which he received appropriate antibiotics. He was afebrile for 5 d before discharge.

On admission for chemotherapy, the patient's WBC count was $29 \times 10^3/\mu L$, reflecting residual effects of the stimulation with GCSF. His RBC count was $3.43 \times 10^3/\mu L$ and platelet count was $106 \times 10^3/\mu L$. With chemotherapy preceding transplantation, these counts fell to a nadir of $0.1 \times 10^3/\mu L$ for WBCs, $2.3 \times 10^6/\mu L$ for RBCs, and $7 \times 10^3/\mu L$ for platelets. These nadirs occurred on different days because of different circulating half-lives of different cell types. Recognizable neutrophils disappeared from the peripheral blood smear on d 20 after the initiation of chemotherapy (and d 4 after the infusion of stem cells). Neutrophils returned on d 9 after the infusion, making up 26% of a total count of $0.5 \times 10^3/\mu L$. The patient was discharged on d 12 after the infusion with a WBC count of $2.1 \times 10^3/\mu L$, 58% of which were neutrophils, an RBC count of $3.01 \times 10^6/\mu L$, and platelets of $23 \times 10^3/\mu L$. Platelet recovery is often the slowest, but achieving an absolute neutrophil count of $>1 \times 10^3/\mu L$ is critical to prevent infection. Patients are often supported with platelet transfusions when their counts fall below $10 \times 10^3/\mu L$. GCSF is also used to hasten recovery of neutrophils.

The second patient is a 22-yr-old female with acute myelogenous leukemia (AML). She achieved complete remission with her first chemotherapy regimen and was in bone marrow remission 8 mo later. Because of the statistical likelihood of relapse, she underwent further consolidation chemotherapy and allogeneic peripheral blood stem cell transplantation harvested from an HLA-matched sibling.

Chemotherapy was begun 8 d before transplantation. The patient was infused with 4.4×10^6/kg of $CD34^+$ cells on d 9. The nadir of her WBC, RBC, and platelet counts was on d 14, followed by recovery of recognizable blood neutrophils on d 20 and discharge on d 22.

Complications were bleeding, probably menstrual, on d 17, treated with a platelet transfusion, and fever on the same day, treated with iv antibiotics. No organisms were ever cultured from the blood, and the patient defervesced in 4 d. The results are presented in Table 3.

15.3. LONG-TERM FOLLOW-UP IN STEM CELL TRANSPLANTATION

As one would suppose, the most common complications in stem cell transplantation are infections and bleeding, as illustrated by the two presented cases. For this reason, it is common to prophylactically "cover" the patient with antibiotics at the first sign of fever, even if no specific organism has been identified. The two cases illustrated successful antibiotic treatment of both cases, one in which an organism was identified and one in which it was not. Bleeding is managed with platelet transfusions. Because bleeding is so likely at peripheral counts below $10 \times 10^3/\mu L$, platelets are given prophylactically to keep counts above this level during the aplastic period. The second case also illustrated this problem managed successfully. Failure to engraft is still a problem in a minority of patients. It is seen most frequently when less than 2×10^6 $CD34^+$ cells/kg are transfused (Shpall et al., 1998). As illustrated by the first case patient, patients receiving lower numbers of cells may have successful outcomes. Transfusing lower numbers of cells is often required when using autologous stem cells in leukemia and lymphoma patients. Higher numbers of $CD34^+$ cells are regularly harvested from allogeneic donors but this, too, does not guarantee engraftment, since occasional failures will be seen even with $CD34^+$ cell counts exceeding 5×10^6 cells/kg.

GVHD is not a problem when autologous cells are used for transplantation. However, it is a continuing problem in allogeneic transplantation. For stem cell transplantation, HLA A, B, and C (class I) and HLA DR (class II) are typed. Because class I antigens are closely linked on chromosome 6 and the class II genes are also closely linked at a different location on the same chromosome, all siblings have a 25% chance of sharing exactly the same alleles. Because many alleles are very common in the population, e.g., HLA-A2, and because crossing over can take place between class I and class II, or even within class I or II, the likelihood of various siblings being a "perfect match" is complicated. Different types of mismatch also produce GVHD of different severity. Mismatch at HLA DR is the worst (Anasetti et al., 1989).

Even with a so-called perfect match between siblings who are not identical twins, some GVHD is possible because of the mechanism by which T-cells recognize antigen. T-cells recognize a complex of class I or II molecules containing a peptide in an external groove of the histocompatibility molecule. In the case of class I molecules, this peptide is usually derived from the metabolism of internal proteins in the cell that are transported to the surface with the class I molecule. Thus, CD8 T-cells from the donor, which recognize peptides in the context of class I, will recognize as foreign any host peptide with any amino acid variation from the donor. Since we type potential donors only for class I and II molecules and not for every protein they make, there will always be some variation between host and donor that can be recognized by donor T-cells (unless the donor and recipient are monozygotic twins).

GVHD, paradoxically, is not always detrimental to the recipient. Because mature T-cells can recognize slight differences in mutant proteins on both normal and malignant cells, the T-cells from the donor could theoretically be expected to recognize any residual lymphoma or leukemia cells in the recipient. If such cells were susceptible to T killer cells or other mechanisms of graft rejection initiated by T-cells, then one might expect to see a "graft vs leukemia" effect in allogeneic bone marrow or stem cell recipients (Biggs et al., 1902a). This is, in fact, what is observed. Patients receiving allogeneic bone marrow or stem cells that have not been purged of T-cells and who experience *mild* GVHD have a better outcome than those who do not have any GVHD at all (MacKinnon et al., 1995).

Table 4 shows the ranges of disease-free survival that have been reported in multiple studies for patients with leukemia and lymphoma after autologous stem cell transplantation (Thomas et al., 1979; Blume et al., 1980; Powles et al., 1980; Bacigalupo et al., 1985; Clift et al., 1987; Feig et al., 1987; McGlave et al., 1988; Forman et al., 1991; Biggs et al., 1992). The ranges are large and studies are not directly comparable. As we have seen, varying amounts of stem cells may be harvested and infused. In general, patients with leukemia in second remission or with lymphomas in advanced stage do more poorly. The age of the patient is also a factor as can be seen in Table 5. The study by Hansen et al. (1998) compared probability of survival ranging from 5 to 10 yr with varying numbers of patients in each age group. Although this study also used allogeneic bone marrow and not peripheral "stem cells," the findings that younger patients have better survival than older patients is generalizable to most therapeutic modalities.

In the case of autologous marrow transplantation, the question arises, should the harvested marrow be purged when the patient is in remission of any putative residual leukemia or lymphoma cells? This can be done with chemotherapeutic agents or MAbs, or both (Rowely et al., 1989; Gribben et al., 1991). No prospective study has been conducted to answer this question although the process of peripheral stem cell mobilization and harvesting of CD34$^+$ cells in the blood is often called "in vitro" purging. One hopes that leukemia or lymphoma cells are not mobilized as efficiently as normal stem cells. The efficacy of this "purging" technique has also not been proven. It is generally accepted that chemotherapy of acute leukemia followed by allogeneic BMT fails most frequently from problems associated with the transplant (engraftment, infections, bleeding, GVHD) while patients receiving autologous marrow most often fail owing to relapse of leukemia. This suggests that further work on purging needs to be done and that the previously described "graft vs tumor" effect of allogeneic marrow is real.

Leukemic genotypes and inherent sensitivity to chemotherapeutic agents also affect the outcome of studies in which patients receive either allogeneic or autologous marrow or peripheral stem cell transplants. Past experience has shown that acute leukemia

Table 4
Disease-Free Survival After "Stem Cell" Transplantation

Disease		*Five-year disease-free survival (%)*
CML	Chronic phase	50–75
	Accelerated phase	30–40
	Blast crisis	5–15
AML	First remission	40–70
	Second remission	30
CLL		50
ALL		30–50
Non-Hodgkin's lymphoma	First relapse	40–50
Hodgkin's disease	First remission, standard treatment	40–70
	High-stage disease	15–30

Table 5
Effect of Age on Allogeneic Unrelated BMT in CML

Probability of survival (5 yr)	*Age range (yr)*
0.8	<21
0.7	21–36
0.65	37–50
0.3	>50

with the karyotypes t(15:17), t(8:21), and inv 16 do very well with chemotherapy alone, so many centers do not transplant these patients in first remission (Maruyomi et al., 1994; Schoch et al., 1996; Wang et al., 1997; Lo Coco et al., 1998). By contrast, patients with leukemic cells having abnormalities of chromosomes 5, 7, or 8 do poorly with chemotherapy alone and are often transplanted in the first remission (Cuneo et al., 1995; Vendetti et al., 1997; Villamor et al., 1998). Obviously, the mixture of leukemic karyotypes in any prospective study of allogeneic or autologous marrow or stem cell transplantation would affect the likelihood of residual cells in unpurged marrow or blood and therefore would also affect event-free and possibly total survival.

Table 6 presents the results of three large studies of AML reported in the late 1990s (Hewlett et al., 1995; Harousseau et al., 1997; Cassileth et al., 1998). Disease-free survival at 5 yr and relapse rates were studied comparing allogeneic BMT, autologous bone marrow transplantation, and standard chemotherapy. The range of disease-free survival with comparable treatment was wide. However in all three studies, as noted previously, the relapse rate was lowest for patients receiving allogeneic bone marrow transplants. This was true even when disease-free survival was not improved. There is also a suggestion in two of the three studies that autologous marrow transplantation offers some protection against relapse compared to chemotherapy alone. This may suggest that the concept of "in vivo purging" has merit, although it does not appear as efficacious as the putative graft vs tumor effect of allogeneic marrow transplantation. However desirable such an effect may be, it may never be available to all because all patients do not have nor will they have a histocompatible sibling to serve as a marrow donor.

Table 6
Allogeneic BMT vs Autologous Marrow vs Chemotherapy Alone for AML in First Remission

Study	*Treatment*[a]	*No. of patients*	*Disease-free survival (%)*	*Relapse (%)*
EORTC/GIMEMA, Hewlett et al. (1995)	allo BMT	168	55	27
	ABMT	128	48	41
	Chemo	126	30	57
GEOLAM, Harousseau, et al. (1997)	allo BMT	126	45	38
	ABMT	67	47	44
	Chemo	67	53	43
US Intergroup, Cassileth et al. (1998)	allo BMT	113	43	29
	ABMT	116	34	48
	Chemo	117	34	62
	allo BMT	92	47	
	ABMT	63	48	

[a]allo BMT, allogeneic BMT; ABMT, autologous BMT; Chemo, standard chemotherapy.

15.4. CONCLUSION

It is now clear that stem cells capable of reconstituting all hematopoietic function are present in the bone marrow of mammals. They are present in the circulating blood and can be mobilized by cytotoxic agents and certain cytokines to increase their concentration 100-fold. These cells are responsible for survival from chemotherapy because a large fraction of them may be resting in G0 and perhaps because they are able to exclude some supravital dyes, which may indicate the ability to exclude some cytotoxic agents (Chaudhary and Roninson, 1991). Such cells, obtained either autologously or allogeneically (and HLA matched), are capable of restoring hematopoiesis after myeloablative treatment (whole-body irradiation in mice) or therapy (various treatments in humans).

In ALL there is statistically significant evidence that allogeneic BMT can rescue patients who have relapsed from chemotherapy and that it is superior to additional rounds of chemotherapy alone (Dopfer et al., 1991). Autologous stem cell transplantation is also beneficial in high-risk patients when performed during the first complete remission (Simonsson et al., 1989; Blaise et, al, 1990). When patients with a specific genetic marker have been studied by polymerase chain reaction (PCR) while in "remission" and before transplant, absence of PCR-detectable disease was clearly associated with a more favorable outcome following transplantation (Knechtti et al., 1998).

In chronic myelogeneous leukemia (CML), transplantation of allogeneic marrow clearly has a beneficial effect compared to conventional chemotherapy (Gale et al., 1998). Survival after transplantation is critically dependent on two factors: phase of the disease and age of the patient (Goldman et al., 1986; McGlave 1990; Biggs et al., 1992; Hansen et al., 1998). Patients with disease in the chronic phase do better than patients in accelerated phase who do better than patients in blast crisis. Relapse rates are higher in autologous than in allogeneic transplants, suggesting the graft vs leukemia effect (McGlave et al., 1994; Carella et al., 1999). Recent studies also suggest that allogeneic grafts from HLA matched, unrelated donors may be almost as good as those from HLA-matched siblings (Hansen et al., 1998). The ongoing trials with STI 571 (Gleevec), which inhibits the tyrosine kinase fusion protein produced by the 9:22 translocation in CML, may change the role of stem cell transplantation in this disease.

The role of stem cell transplantation in chronic lymphocytic leukemia (CLL) has not been as clearly demonstrated, mostly because most of the patients are well over 50 yr of age (Khouri et al., 1994). This is an age group known to do poorly after attempts to do BMT in other diseases.

Non-Hodgkin lymphoma is a very heterogeneous group of diseases. In humans, about 85% are B-cell disorders, while T-cell and "null cell" make up the rest. Genetic abnormalities vary from chromosomal translocations such as t14:18, leading to high levels of Bcl-2 production and apoptosis resistance, to t8:14, leading to rapid cell division. Normal karyotypes and complex, aneuploid karyotypes are also seen. Age ranges at onset are wide, association with various viruses has been demonstrated, and susceptibility or resistance to various forms of chemotherapy vary as well. Thus, studies on the efficacy of stem cell transplantation, whether allogeneic or autologous, must be interpreted with care.

The PARMA study of 215 patients randomized 109 patients who had responded to chemotherapy to further chemotherapy or autologous stem cell rescue after myeloablation. After 5 yr event-free survival was 46% in the stem cell group and 12% in the chemotherapy group, a highly significant result. Overall survival was 53% for the stem cell patients and 32% for the chemotherapy patients, some of whom were rescued with stem cells after relapse (Philip et al., 1995).

In a separate study, 916 patients were stratified by factors such as histology, stage, age, and performance status into four risk groups, high to low. Patients who had achieved complete remission by chemotherapy were randomized to further chemotherapy or autologous stem cell transplantation. Benefit was noted mainly in the highest risk group, which had a 5-yr survival with no evidence of disease of 59% for stem cell recipients compared with 39% for chemotherapy patients (Stockerl-Goldstein et al., 1996).

In Hodgkin's disease, studies of the beneficial effect of autologous stem cell transplantation are ongoing. In patients in whom standard radiation and chemotherapeutic treatments have failed to induce remission, salvage rates of 35–50% have been reported using autologous "stem cells" (Chopra et al., 1993; Reece et al., 1995). These are not prospective studies.

In the near future, the next major improvement is likely to be preparative cell sorting of mobilized stem cells so that very low numbers of cells can be used to rescue patients from myeloablative treatment. This should drastically reduce the likelihood of contaminating malignant cells being reinjected into the patient—still a recognized problem in autologous transplants.

REFERENCES

Anasetti, C., Amos, D., Beatty, P., et al. (1989) Effect of HLA compatibility on engraftment of bone marrow transplants in patients with leukemia or lymphoma. *N. Engl. J. Med.* 320:197.

Bacigalupo, A., Frassoni, F., Van Lint, M., et al. (1985) Bone marrow transplantation (BMT) for acute nonlymphoid leukemia (ANLL) in first remission. *Acta Haematol.* 74:23.

Baum, C., Weissman, I., Tsukamoto, A., Buckle, A., and Peault, B. (1992) Isolation of a candidate human hematopoietic stem-cell population. *Proc. Natl. Acad. Sci. USA* 89:2804.

Bensinger, W., Longin, K., Appelbaum, F., et al. (1994) Peripheral blood stem cells (PBSCs) collected after recombinant granulocyte colony stimulating factor (rhG-CSF): an analysis of factors correlating with the tempo of engraftment after transplantation. *Br. J. Haematol.* 87:824.

Berenson, R., Andrews, R., Bensinger, W., et al. (1988) Antigen $CD34^+$ marrow cells engraft lethally irradiated baboons. *J. Clin. Invest.* 81:951.

Biggs, J., Szer, J., Crilley, P., et al. (1992) Treatment of chronic myeloid leukemia with allogeneic bone marrow transplantation after preparation with BuCy2. *Blood* 80:1352.

Biggs, J., Horowitz, M., Gale, R., et al. (1992) Bone marrow transplants may cure patients with acute leukemia never achieving remission with chemotherapy. *Blood* 80:1090.

Blaise, D., Gaspard, M., Stoppa, A., et al. (1990) Allogeneic or autologous bone marrow transplantation for acute lymphoblastic leukemia in first complete remission. *Bone Marrow Transplant.* 5:7.

Blume, K., Beutler, E., Bross, K., et al. (1980) Bone-marrow ablation and allogeneic marrow transplantation in acute leukemia. *N. Engl. J. Med.* 302:1041.

Brandt, J., Baird, N., Lu, L., Srour, E., and Hoffman. R. (1988) Characterization of a human hemopoietic progenitor cell capable of blast cell containing colonies in vitro. *J. Clin. Invest.* 82:1017.

Brecher, G. and Cronkite, E. (1951) Post radiation parabiosis and survival in rats. *Proc. Soc. Exp. Biol. Med.* 77:292.

Buckner, C., Epstein, R., Rudolph, R., et al. (1970) Allogeneic marrow engraftment following whole body irradiation in a patient with leukemia. *Blood* 35:741.

Carella, A., Lerma, E., Corsetti, M., et al. (1999) Autografting with Philadelphia chromosome negative mobilized hematopoietic progenitor cells in chronic myelogenous leukemia. *Blood* 93:1534.

Cassileth, P., Harrington, D., Appelbaum, F., et al. (1998) Chemotherapy compared with autologous or allogenic bone marrow transplantation in the management of acute myeloid leukemia in first remission. *N. Engl. J. Med.* 339:1649.

Chao, N., Schriber, J., Grimes, K., et al. (1993) Granulocyte colony-stimulating factor "mobilized" peripheral blood progenitor cells accelerate granulocyte and platelet recovery after high-dose chemotherapy. *Blood* 81:2031.

Chaudhary, P. and Roninson, I. (1991) Expression and activity of P-glycoprotein: a multi-drug efflux pump in human hematopoietic stem cells. *Cell* 66:85.

Chopra, R., McMillan, A., Linch, D., et al. (1993) The place of high-dose BEAM therapy and autologous bone marrow transplantation in poor-risk Hodgkin's disease: A single-center eight-year study of 155 patients. *Blood* 81:1137.

Clift, R. A., Buckner, C. D., Thomas, E. D., et al. (1987) The treatment of acute non-lymphoblastic leukemia by allogeneic marrow transplantation. *Bone Marrow Transplant.* 2:243–258.

Craig, W., Kay, R., Cutler, R., and Lansdorp, P. (1993) Expression of thy-1 on human hematopoietic progenitor cells. *J. Exp. Med.* 177:1331.

Cuneo, A., Ferrant, A., Michaux, J., et al. (1995) Cytogenetic profile of minimally differentiated (FAB M0) acute myeloid leukemia: correlation with clinicobiologic findings. *Blood* 85:3688.

Dopfer, R., Henze, G., Bender-Gotze, C., et al. (1991) Allogeneic bone marrow transplantation for childhood acute lymphoblastic leukemia in second remission after intensive primary and relapse therapy according to the BFM- and CoALL-protocols: results of the German Cooperative Study. *Blood* 78:2780.

Fefer, A., Cheever, M., Greenberg, P., et al. (1982) Treatment of chronic granulocytic leukemia with chemoradiotherapy and transplantation of marrow from identical twins. *N. Engl. J. Med.* 306:63.

Feig, S. A., Nesbit, M. E., Buckley, J., et al. (1987) Bone marrow transplantation for acute non-lymphocytic leukemia: a report from the Childrens Cancer Study Group of sixty-seven children transplanted in first remission. *Bone Marrow Transplant.* 2:365–374.

Forman, S., Schmidt, G., Nademanee, A., et al. (1991) Allogeneic bone marrow transplantation as therapy for primary induction failure for patients with acute leukemia. *J. Clin. Oncol.* 9:1570.

Gale, R., Horowitz, M., Ash, R., et al. (1994) Twin bone marrow transplants for leukemia. *Ann. Intern. Med.* 120:646.

Gale, R., Hehlmann, R., Zhang, M., et al. (1998) Survival with bone marrow transplantation versus hydroxyurea or interferon for chronic myclogenous leukemia: The German CML Study Group. *Blood* 91:1810.

Gatti, R., Meuwissen, H., Allen, H., Hong, R., and Good, R. (1968) Immunological reconstitution of sex-linked lymphopenic immunological deficiency. *Lancet* 2:1366.

Gillespie, T. and Hillyer, C. (1996) Peripheral blood progenitor cells for marrow reconstitution: mobilization and collection strategies. *Transfusion* 36:611.

Gluckman, E., Rocha, V., Boyer-Chammard, A., et al. (1997) Outcome of cord-blood transplantation from related and unrelated donors. Eurocord Transplant Group and the European Blood and Marrow Transplantation Group. *N. Engl. J. Med.* 337:373–381.

Goldman, J., Apperley, J., Jones, L., et al. (1986) Bone marrow transplantation for patients with chronic myeloid leukemia. *N. Engl. J. Med.* 314:202.

Goodell, M., Rosenzweig, M., Kim, H., et al. (1997) Dye efflux studies suggest that hematopoietic stem cells expressing low or undetectable levels of CD34 antigen exist in multiple species. *Nat. Med.* 3:1337.

Gorin, N., Aegerter, P., Auvert, B., et al. (1990) Autologous bone marrow transplantation for acute myelocytic leukemia in first remission: a European survey of the role of marrow purging. *Blood* 75:1606.

Gribben, J. G., Freedman, A. S., Neuberg, D., et al. (1991) Immunologic purging of marrow assessed by PCR before autologous bone marrow transplantation for B-cell lymphoma. *N. Engl. J. Med.* 325:1525.

Hansen, J., Gooley, T., Martin, P., et al. (1998) Bone marrow transplants from unrelated donors for patients with chronic myeloid leukemia. *N. Engl. J. Med.* 338:962.

Harousseau, J., Cahn, J., Pignon, B., et al. (1997) Comparison of autologous bone marrow transplantation and intensive chemotherapy as postremission therapy in adult acute myeloid leukemia: The Groupe Ouest Est Leucemies Aigues Myeloblastiques (GOELAM). *Blood* 90:2978.

Hewlett, J., Kopecky, K., Head, D., et al. (1995) A prospective evaluation of the roles of allogeneic marrow transplantation and low-dose monthly maintenance chemotherapy in the treatment of adult acute myelogenous leukemia (AML): A Southwest Oncology Group study. *Leukemia* 9:562.

Hong, R., Cooper, M., Allan, M., et al. (1968) Immunological restitution in lymphopenic immunological deficiency syndrome. *Lancet* 1:503.

Khouri, I., Keating, M., Vriesendorp, H., et al. (1994) Autologous and allogeneic bone marrow transplantation for chronic lymphocytic leukemia: preliminary results. *J. Clin. Oncol.* 12:748.

Knechtli, C., Goulden, N., Hancock, J., et al. (1998) Minimal residual disease status before allogeneic bone marrow transplantation is an important determinant of successful outcome for children and adolescents with acute lymphoblastic leukemia. *Blood* 92:4072.

Korbling, M., Przepiorka, D., Huh, Y., et al. (1995) Allogeneic blood stem cell transplantation for refractory leukemia and lymphoma: Potential advantage of blood over marrow grafts. *Blood* 85:1659.

Kosatec, D., Shepherd, K., Sage, R., et al. (1992) Factors affecting blood stem cell collections following high dose cyclophosphamide mobilization in lymphoma myeloma, and solid tumors. *Int. J. Cell Cloning* (Suppl. 1):35.

Lane, T. (1995) Mobilization of hematopoietic progenitor cells In: *Processing, Standards and Practice*, (Brecher, M. E., Lasky, L. C., Sacher, R. A., and Issit, L. A., eds.), American Association of Blood Banks, pp. 59–108.

Linker, C., Ries, C., Damon, L., Rugo, H., and Wolf, J. (1998) Autologous bone marrow transplantation for acute myeloid leukemia using 4-hydroperoxycyclophosphamide-purged bone marrow and the busulfan/etoposide preparative regimen: a follow-up report. *Bone Marrow Transplant.* 22:865.

Lo Coco, F., Nervi, C., Avvisati, G., and Mandelli, F. (1998) Acute promyelocytic leukemia: a curable disease. *Leukemia* 12:1866.

MacKinnon, S., Papadoupoulous, E., Carabasi, M., et al. (1995) Adoptive immunotherapy evaluating doses of donor leukocytes for relapse of chronic myeloid leukemia after bone marrow transplantation: separation of graft-versus-leukemia responses from graft-versus-host disease. *Blood* 86(4): 1261.

Martin, P., Hansen, J., Bucker, C., et. al (1985) Effects of in vitro depletion of T cells in HLA-identical allogeneic marrow grafts. *Blood* 66:664.

Maruyami, F., Stass, S., Estey, E., et al. (1994) Detection of AMLI/ETO fusion transcript as a tool for diagnosing t(8;21) positive acute myelogenous leukemia. *Leukemia* 8:40.

Mathe, G., Amiel, J., Schwartzberg, L., et al. (1965) Successful allogeneic bone marrow transplantation in man: chimerism, induced specific tolerance and possible anti leukemic effects. *Blood* 25:179.

McGlave, P. B., Haake, R. J., Bostrom, B. C., et al. (1988) Allogeneic bone marrow transplantation for acute nonlymphocytic leukemia in first remission. *Blood* 72:1512–1517.

McGlave, P. (1990) Bone marrow transplants in chronic myelogenous leukemia: an overview of determinants of survival. *Semin. Hematol.* 27:23.

McGlave, P., DeFabritiis, P., Deisseroth, A., et al. (1994) Autologous transplants for chronic myelogenous leukemia: Results from eight transplant groups. *Lancet* 343:1486.

Negrin, R., Atkinson, K., Leemhuis, T., et al. (2000) Transplantation of highly purified $CD34^{+}Thy\text{-}1^{+}$ hematopoietic stem cells in patients with metastatic breast cancer. *Biol. Blood Marrow Transplant.* 6:262.

O'Reilly, R. (1983) Allogenic bone marrow transplantation: current status and future directions. *Blood* 62:941.

Philip, T., Guglielmi, C., Hagenbeek, A., et al. (1995) Autologous bone marrow transplantation as compared with salvage chemotherapy in relapses of chemotherapy-sensitive non-Hodgkin's lymphoma. *N. Engl. J. Med.* 333:1540.

Powles, R., Morgenstern, G., Clink, H., et al. (1980) The place of bone-marrow transplantation in acute myelogenous leukemia. *Lancet* 1:1047.

Reece, D., Barnett, M., Shepherd, J., et al. (1995) High-dose cyclophosphamide, carmustine (BCNU), and etoposide (VP16-213) with or without cisplatin (CBV (P) and autologous transplantation for patients with Hodgkin's disease who fail to enter a complete remission after combination chemotherapy. *Blood* 86:451.

Rowley, S., Jones, R., Piantadosi, S., et al. (1989) Efficacy of ex vivo purging for autologous bone marrow transplantation in the treatment of acute nonlymphoblastic leukemia. *Blood* 74:501.

Santos, G., Sensenbrenner, L., Burke, P., et al. (1974) Allogeneic marrow grafts in man using cyclophosphamide. *Transplant. Proc.* 6:345.

Schoch, C., Haase, D., Haferlach, T., et al. (1996) Fifty-one patients with acute myeloid leukemia and translocation t(8;21) (q22; q22): an additional deletion in 9q is an adverse prognostic factor. *Leukemia* 10:1288.

Shpall, E., Bast, R., Jones, W., et al. (1991) Immunomagnetic purging of breast cancer for bone marrow for autologous transplantation. *Bone Marrow Transplant.* 7:145.

Shpall, E., Champlin, R., and Glaspy, J. (1998) Effect of $CD34^{+}$ peripheral blood progenitor cell dose on hematopoietic recovery. *Biol. Blood Marrow Transplant.* 4:84.

Siena, S., Bregni, M., Brando, B., et. al (1989) Circulation of $CD34^{+}$ hematopoietic progenitor cells in the peripheral blood of high-dose cyclophosphamide-treated patients: Enhancement by intravenous human granulocyte-macrophage colony-stimulating factor. *Blood* 74:1905.

Simonsson, B., Burnett, A., Prentice, H., et al. (1989) Autologous bone marrow transplantation with monoclonal antibody purged marrow for high risk acute lymphoblastic leukemia. *Leukemia* 3:631.

Socinski, M., Elias, A., Schnipper, L., et al. (1988) Granulocyte-macrophage colony stimulating factor expands the circulating haematopoietic progenitor cell compartment in man. *Lancet* 1:1194.

Spangrude, G., Heimfeld, S., and Weissman, I. (1988) Purification and characterization of mouse hematopoietic stem cells. *Science* 241:58.

Stockerl-Goldstein, K., Horning, S., Negrin, R., et al. (1996) Influence of preparatory regimen and source of hematopoietic cells on outcome of autotransplantation for non-Hodgkin's lymphoma. *Biol. Blood Marrow Transplant.* 2:76.

Storek, J., Gooley, T., Siadek, M., et al. (1997) Allogeneic peripheral blood stem cell transplantation may be associated with a high risk of chronic graft-versus-host disease. *Blood* 90:4705.

Thomas, E., Storb, R., Clift, R., et al. (1975) Bone-marrow transplantation. *N. Engl. J. Med.* 292:832.

Thomas, E., Buckner, C., Clift, R., et al. (1979) Marrow transplantation for acute nonlymphoblastic leukemia in first remission. *N. Engl. J. Med.* 301:597.

Till, J. and McCulloch, E. (1961) A direct measurement of the radiation sensitivity of normal mouse bone marrow cells. *Radiat. Res.* 14: 213–222.

Uchida, N., Combs, J., Chen, S., et al. (1996) Primitive human hematopoietic cells displaying differential efflux of the rhodamine-123 dye have distinct biological activities. *Blood* 88:1297.

Venditti, A., Del Poeta, G., Buccisano, F., et al. (1997) Minimally differentiated acute myeloid leukemia (AML MO): comparison of 25 cases with other French-American-British subtypes. *Blood* 89:621.

Villamor, N., Zarco, M. A., Rozman, M., et al. (1998) Acute myeloblastic leukemia with minimal myeloid differentiation: phenotypical and ultrastructural characteristics. *Leukemia* 12:1071.

Wang, J., Wang, M., and Liu, J. (1997) Transformation properties of the ETO gene fusion partner in t(8;21) leukemias. *Cancer Res.* 57:2951.

16 Neurons, Stem Cells and Potential Therapies

FIONA C. MANSERGH, PhD, MICHAEL A. WRIDE, PhD, AND DERRICK E. RANCOURT, PhD

Contrary to long held theory it has been shown recently that there are self-renewing stem cells in the adult central nervous system and that neuronal precursor cells can be derived in vitro from embryonic stem (ES) cells as well as other types of adult stem cells. Differentiation can be controlled by growth factors and culture conditions, such as, epidermal growth factor for the astrocyte lineage, and basic fibroblast growth factor for neuronal phenotypes, presumably by selection of stem cells with different potentials. Cell lines derived from different stages of neuronal development may provide cell sources with more directed potential to different types of neuronal cells. For example, lines restricted to glial differentiation may be useful therapeutically for transplantation for remyelination and other cell lines for replacement of lost neurons. On the other hand, adult neural stem cells appear to have multi–cell type plasticity and can give rise to progeny of many different tissue cell types, and ES cells can give rise to different types of neuronal cells. ES cells show great promise for replacement of injured nerve tissues, but technical and ethical problems need to be resolved.

16.1. INTRODUCTION

The mammalian central nervous system (CNS) is incredibly complex and possesses only a limited ability to recover from damage. These characteristics have made treatments for neurological disorders and injuries hard to devise. The challenge is made all the greater by the aging population profiles of most Western societies; for example, the annual cost of dementia to the US economy is estimated at $100 billion annually (www.alzheimers.org/pubs/biomed95.html). These costs are likely to increase. Moreover, estimates of the economic burden of dementia, psychiatric disorders, cognitive dysfunction, stroke, cancer, injury, and other such disorders leave the emotional costs to those affected and their families unaccounted for.

Fortunately, the discovery of self-renewing stem cell populations within the fetal and adult CNS has opened promising lines of inquiry (Reynolds and Weiss 1992; Richards et al., 1992). It is also becoming increasingly clear that neurons can be derived from numerous sources of progenitor cell lines in vitro. Many cell types, including neurons and glia, can be obtained from differentiation of both mouse and human embryonic stem (ES) cells in vitro (Mansergh et al., 2000). ES cells hold out particular promise, because treatments involving ES cells have the potential to sidestep the thorny problem of immunological rejection. This could conceivably be achieved by using genetic modification of preexisting ES cell lines, or by creating ES cell lines identical to the patient. However, while such technology would undoubtedly have therapeutic potential, its potential use in humans presents ethical issues. Nevertheless, it is likely that the potential of stem cell research will soon be clinically realized.

From: *Stem Cells Handbook*
Edited by: S. Sell © Humana Press Inc., Totowa, NJ

16.2. NATURALLY OCCURRING STEM CELLS

Cells of the very early embryo can divide indefinitely and are also capable of differentiation into all cell types present in the adult. Standard developmental theory has determined that cell fate becomes progressively more restricted as development proceeds. However, as recent studies have demonstrated, the developmental potential of early embryonic cells appears to be retained by small cell populations within most adult tissues (Rao and Mattson, 2001). This would suggest that transplantation of manipulated stem cells or precise stimulation of endogenous precursor populations may allow treatment of many diseases, including those involving neural or glial problems. Culture of a wide range of adult and fetal stem cells that vary in the extent of their developmental potential is becoming more and more feasible. As these cells are untransformed, their therapeutic use is less fraught with concerns about their oncogenic potential. It may also be possible to expand stem cells from a given individual in vitro, while encouraging them to adopt specific cell fates. This approach avoids the risks of immune rejection. The primary disadvantage of these cells is that, owing to finite life-span and/or pleuripotency, fresh sources of stem cells may be continually required. In addition, amelioration of genetic defects might be tricky, because stem cells derived from the patient would require correction of deleterious mutation(s) prior to reintroduction.

16.2.1. MULTIPOTENT NEURAL STEM CELLS Fetal tissue has long been used as a source of multipotent neural progenitors. However, until recently, the adult CNS was considered incapable of self-renewal or repair. The discovery of stem cells within the adult brain (Reynolds and Weiss 1992; Richards et al., 1992) rendered this concept, at least in part, obsolete. These cells are capable of self-renewal and can give rise to highly differentiated progeny (neurons and glia), thus conforming to the standard definition of a stem cell (Watt and Hogan, 2000). Adult neural precursors were initially obtained from the forebrain subventricular zone (SVZ) and have also since been isolated from the hippocampus, spinal cord, striatum, septum, cortex, and optic nerve (Gage, 2000; Reynolds and Weiss, 1992; Weiss et al., 1996; Palmer et al., 1999). Both adult stem cells and those of fetal origin are isolated via dissection of brain regions previously shown to be mitotically active in vivo. The SVZ and hippocampus are possibly the most commonly used source of adult precursors, whereas the fetal brain contains a large number of such regions (Gage, 2000). Cells can be cultured either as floating aggregates (neurospheres) or in monolayers. They are maintained in a pluripotent state using various mitogens (most commonly epidermal growth factor [EGF] or basic fibroblast growth factor [bFGF]) and can be expanded by dissociation of primary neurospheres to form secondary spheres (Mansergh et al., 2000).

Induction toward various different lineages can be achieved by varying the mitogens used. Precursor cell lines grown in the presence of EGF seem biased toward an astrocytic lineage, whereas those exposed to bFGF appear to have greater neuronal potential (Rao, 1999; Whittemore et al., 1999). This may reflect different origins; EGF-responsive stem cells cannot be isolated from rat embryos prior to embryonic day (E) 14.5, whereas bFGF-dependent stem cells (i.e., neuroepithelial stem cells [NEPs]) can be isolated as early as E10.5. Both can be isolated after E14.5 (Weiss et al., 1996; Rao 1999). The question as to whether these cells share a common lineage is still somewhat unclear. Transplantation studies have shown that both can contribute to neurogenesis. However, bromodeoxyuridine pulse labeling following infusion of bFGF or EGF demonstrates that EGF-dependent stem cells mainly give rise to glia. Furthermore, studies of EGF receptor knockout mice suggest that EGF-dependent stem cells do not contribute to early neurogenesis in vivo (Kuhn et al., 1997; Craig et al., 1999; Rao 1999). EGF-responsive stem cells have been shown to appear transiently in mouse spinal cord between E12 and E14 (a period encompassing the onset of gliogenesis and when neurogenesis is at its peak). By contrast, FGF-dependent stem cells are present before, during, and following this interval, and are also present in adult spinal cord. It has been suggested that the FGF-responsive stem cells may be more primitive and may give rise to EGF-responsive progenitors (Represa et al., 2001). Both types of stem cell can be persuaded to differentiate along neural lineages in vitro, however.

Depending on requirements, an ever-increasing number of factors can be used either to encourage further proliferation of stem cell cultures or for differentiation. Ciliary neurotropic factor has been shown to inhibit lineage restriction of EGF-responsive stem cells to glial progenitors in vitro, resulting in a rise in cell number and maintenance of pluripotency (Shimazaki et al., 2001). Erythropoietin is known to have neurotrophic capabilities in the nervous system (Nagai et al., 2001). The presence of insulin-like growth factor (IGF) has been shown to be necessary for the proliferation of E14 mouse striatal cells to neurospheres (Arsenijevic et al., 2001a). Differentiation of neural stem cells along specific lineages is usually achieved by withdrawal of mitogens or by exposure of the cells to differentiative factors (Gage, 2000). Neurons have been derived from EGF-dependent stem cells using retinoic acid (RA), IGF, and brain-derived neurotrophic factor (Arsenijevic and Weiss, 1998; Wohl and Weiss, 1998). Fetal rat neural stem cells can be immunoselected for polysialylated neural cell adhesion molecule (PSA-NCAM)–positive precursors. Culturing these in the presence of thyroid hormone and bFGF induces predominantly glial differentiation (Keirstead et al., 1999). Cortical precursor cells, most of which give rise to neurons, can be grown using FGF-2, FGF-4, and FGF-8b; FGF-8b also gives rise to astroglia. Using 24-h exposure to bFGF in the presence of glial cell–conditioned media, the percentage of forebrain dopaminergic neurons has been increased (Daadi and Weiss 1999). However, although these studies show promise, stem cell–derived neural cultures are still mixed cell populations. Further study will be of value in identifying additional factors or differentiation "cocktails" that allow selective enrichment for specific cell types.

Markers specific to certain lineages would also be useful; currently, neural stem cells are known to express nestin (an intermediate filament and marker of NEPs). Cells with morphology characteristic of neurons are known to express mitogen-activated protein-2 (MAP-2), neurofilament, β tubulin (type III), γ-aminobutyric acid (GABA), and substance P. Cells with glial characteristics express the cell-surface protein O4 (oligodendrocytes) and glial fibrillary acidic protein (astrocytes) (Reynolds et al., 1992; Richards et al., 1992; Reynolds and Weiss, 1996; Weiss et al., 1996; Weiss and van der Kooy, 1998). Current advances in genomics and microarray technology will probably facilitate the isolation of additonal factors that can be used to direct the differentiation of precursors for specific neuronal subtypes.

16.2.1.1. Neural Crest Stem Cells During development, neural crest cells migrate through the embryo, giving rise to a wide variety of cell types. Multipotent neural crest stem cells (NCSCs) have been isolated in vitro, are antigenically and morphologically distinct from brain-derived neural stem cells, and can give rise to multipotent progeny (Stemple and Anderson, 1992; Rao and Anderson, 1997; Mujtaba et al., 1998). NCSCs can give rise to a variety of cells types on differentiation, including cholinergic parasympathetic and noradrenergic sympathetic neurons, smooth muscle, Schwann cells, cartilage, melanocytes, neurons, and glia (Baroffio et al., 1991; Stemple and Anderson, 1992; White et al., 2001). Transforming growth factor-β1, glial growth factor-2, and bone morphogenetic protein-2 (BMP-2) can be used to induce differentiation along smooth muscle, glial, and neural lineages in vitro (Shah and Anderson, 1997). The relative concentration of BMP-2 appears to be important in whether a parasympathetic or sympathetic neural fate is chosen; cholinergic neurons in vitro differentiate at lower BMP-2 concentrations than do noradrenergic neurons (White et al., 2001). NCSCs are more responsive to some of these growth factors than others when used in combination, suggesting some degree of intrinsic bias toward certain cell fates. Transplantation studies involving rat NCSCs into chick embryos have shown that E10.5 NCSCs have a greater ability to generate neurons than do those derived from E14.5 embryos, suggesting cell intrinsic changes in sensitivity to external cues with time (White et al., 2001). NCSCs have a wide developmental potential; however, this advantage is somewhat offset by the fact that maintenance of their pluripotency in culture becomes progressively more difficult to maintain after six cell divisions (Rao and Anderson, 1997). Derived immortalized cell lines are thought to be nontumorigenic, thus circumventing that problem (Rao and Anderson, 1997). However, neuronal progeny of NCSCs are thought to be restricted to peripheral nervous system (PNS)-type derivatives; NCSCs are possibly committed to a PNS-like fate by E11 and are incapable of generating CNS-like derivatives (Kalyani et al., 1997; Mujtaba et al., 1998).

16.2.1.2. Neural and Glial Restricted Precursor Cells NCSCs and adult and fetal neural stem cells have relatively wide developmental potential. In addition to these stem cells, however, more restricted types of stem cells have been isolated, namely neuron, glial, astrocyte, and oligodendrocyte restricted precursor cells (NRPs, GRPs, APCs and OPCs [or O2A cells], respectively). Each of these cell types are morphologically distinct from each other and possess different potentials (Mayer-Proschel et al., 1997; Kalyani and Rao, 1998; Rao et al., 1998; Rao, 1999). NRPs can give rise to multiple neuronal subtypes; synapses and the presence of excitatory, inhibitory, and cholinergic neurons have been demonstrated in culture. NRPs were originally isolated from spinal cord, but similar, NRP-like cells have also been isolated from other regions of the developing brain (Rao, 1999). GRPs, APCs, and OPCs (or O-2As) are all glial precursors. APCs are EGF dependent and were isolated from E16 mouse spinal cord (Seidman et al., 1997; Pringle et al., 1998). When differentiated, they give rise primarily to type 1 astrocytes. O-2As were initially isolated from postnatal rat optic nerve, cortex, cerebellum, brain stem, and spinal cord; however, similar cells have been isolated from human fetal and adult rodent tissue (Lee et al., 2000). They can be maintained in a proliferative state using FGF and platelet-derived growth factor (PDGF) and give rise to oligodendrocytes and type 2 astrocytes under differentiating conditions (Collarini et al., 1991; Rao, 1999; Lee et al., 2000). Myelinating oligodendrocytes have been obtained after transplantation. GRPs possess the widest potential of all the glial precursor types; these can be persuaded to give rise to oligodendrocytes and types 1 and 2 astrocytes. These cells were initially isolated from embryonic mouse spinal cord; similar cell types have also been derived from human fetal tissue (Rao et al., 1998; Lee et al., 2000).

Restricted glial precursors have promise therapeutically, as they have been involved in transplantation experiments in which remyelination in myelin-deficient backgrounds has been achieved successfully (Keirstead et al., 1999; Zhang et al., 1999; Herrera et al., 2001). In one case, the precursors were shown to generate astrocytes, oligodendrocytes, and Schwann cells. This was notable because Schwann cells had not been obtained from in vitro differentiation of the GRPs; factors at the site of injury may have widened the differentiation potential of these cells (Keirstead et al., 1999).

Spinal cord NRPs have also been shown to migrate to some, but not all, regions of the forebrain after transplantation, and to differentiate into region-specific neuronal subtypes (Yang et al., 2000). Notably, the migration pattern and neuronal phenotypes obtained from these cells after transplantation, while specific to their environment, were different from NRPs obtained from anterior forebrain SVZ-derived NRPs (Yang et al., 2000).

16.2.1.3. Human Neural Stem Cells Human neural stem cells can be isolated from various regions of adult brain, including the SVZ, temporal and frontal cortex, amygdala, hippocampus, spinal cord, and lateral ventricle wall (Frisen et al., 1998; Johansson et al., 1999a, 1999b; Arsenijevic et al., 2001b). Human fetal stem cells can be isolated from embryos left over after in vitro fertilization or from aborted material (Shamblott et al., 1998; Thomson et al., 1998a; Vescovi et al., 1999b). They share many characteristics with rodent stem cells; they can differentiate into electrophysiologically active neurons and glia, and they express appropriate cell type–specific markers (Vescovi and Snyder, 1999). Fetal neural stem cells have been shown to be stable over multiple passages and therefore can be expanded for transplantation (Vescovi and Snyder, 1999; Vescovi et al., 1999a). There are some differences between human and rodent neural stem cells; prolonged culture of the human cells requires both EGF and FGF2, and some protocols also include leukemia inhibitory factor (LIF) (Fricker et al., 1999). Laminin is required for oligodendrocyte generation, while bFGF and PDGF can inhibit the generation of oligodendrocytes and neurons. Adult human and fetal progenitors differ from each other in that adult cells seem not to require withdrawal of mitogen for differentiation (Arsenijevic et al., 2001b).

Transplantation experiments involving human fetal neural stem cells have been shown to be successful in a number of instances. They have been shown to incorporate into all major brain compartments of adult rodents and to give rise to neurons, astrocytes, and oligodendrocytes (Brustle et al., 1998; Fricker et al., 1999; Vescovi et al., 1999a; Rubio et al., 2000; Uchida et al., 2000). They have also been shown to take the routes normally used by endogenous rodent stem cells (Fricker et al., 1999) and to take part in normal mouse brain development (following transplantation into newborn mouse or embryonic rat brain) (Brustle et al., 1998; Flax et al., 1998). Notably, human fetal neural stem cells have been shown to incorporate extensively into aged rat brain, with resulting improvements in cognitive function, implying potential in treatments for age-related disorders (Qu et al., 2001). Interestingly, in rodent recipients of human stem cells, smaller transplants (approx 200,000 cells) extended more neuronal fibers and were less likely to provoke immunological rejection than transplants of 2 million cells (Ostenfeld et al., 2000). There is also potential for differentiation of fetal human stem cells into specific types of neuron for therapeutic purposes; human mesencephalic neural precursors have been differentiated to yield a 1% culture of dopaminergic neurons (Storch et al., 2001).

Transfection and expression of transgenes has also been shown to occur, suggesting that fetal human stem cells could be used as vehicles for gene therapy (Flax et al., 1998). The same could be true of adult neural stem cells; lentiviral vectors have been shown to integrate efficiently into their genomic DNA (Arsenijevic et al., 2001b). A big boost for stem cells in gene therapy comes from rodent studies involving gene therapy for glioblastoma, an intractable type of primary brain tumor. Neural stem cells derived from newborn mice were retrovirally transfected with interleukin-4 (IL-4) cDNA and transplanted into syngeneic established glioblastomas. This led to greatly enhanced survival rates and tumor regression, as did IL-4-transfected rat neural progenitors when similarly transplanted (Benedetti et al., 2000). Notably, the stem cells were more effective than retroviral vectors in terms of their therapeutic effect (Benedetti et al., 2000).

16.2.2. STEM CELL LINEAGE RELATIONSHIPS AND PLURIPOTENCY The interrelationships among the aforementioned types of neural and/or glial precursors are still far from clear. As mentioned previously, it has been suggested that FGF-responsive neural stem cells (NEPs) may be more primitive and may give rise to EGF-responsive progenitors (Weiss et al., 1996; Rao 1999; Represa et al., 2001). NEPs have been shown to give rise to PNS derivatives similar to NCSCs, which would suggest that a lineage relationship between these two types of stem cell is possible (Mujtaba et al., 1998). NEPs and EGF-responsive stem cells have also been shown to produce glial precursors (Mayer-Proschel et al., 1997; Zhang et al., 1999). GRP cells, in turn, may be able to generate O-2A cells (Lee et al., 2000). Neural and glial stem cells can also be isolated from ES cells, implying lineage relationships (Okabe et al., 1996; Li et al., 1998; Mujtaba et al., 1999). Human neural stem cells implanted into cerebral germinal zones of monkey fetuses have been shown to contribute both to differentiated neurons and glia and to a pool of undifferentiated cells within the SVZ (Ourednik et al., 2001). This would imply a direct lineage relationship between adult and fetal neural stem cells.

"Differentiated" glia have also been shown to act as neural stem cells (Doetsch et al., 1999; Alvarez-Buylla et al., 2001). Embryonic neural tube stem cells do not show glial characteristics, but radial glia in the developing brain are thought to be able to give rise to both cortical astrocytes and cortical neurons (Alvarez-Buylla et al., 2001). Almost all radial glia have been found to proliferate throughout neurogenesis and to be divisible into three subsets based on immunostaining for different antigens: RC2, astrocyte-specific glutamate transporter (GLAST), and brain lipid-binding protein. The GLAST-negative subset is specific to neurogenesis, while multipotential fetal or adult neurosphere cultures are reactive for all three (Hartfuss et al., 2001). The role of radial glia has been somewhat redefined. Their previously characterized role in development was thought to involve provision of support to migrating and differentiating neurons and provision of signals involved in neural precursor migration, differentiation, and survival. Following neuronal migration, these cells were thought to differentiate along astrocytic lineages. To this list of vital functions, we should add that of precursor cells (Chanas-Sacre et al., 2000). The finding that glia can behave as progenitors is exciting from a therapeutic perspective, especially because it may also be possible to reverse glial cell fate decisions. Recent findings suggest that oligodendrocyte precursor cells can be persuaded to revert to stem cells with developmental potential that includes neurons and astrocytes as well as oligodendrocytes (Kondo and Raff, 2000). Similarities in differentiation among mouse, rat, chick, and human stem cells are also promising, suggesting that results in animal models may be readily translated to therapies (Kalyani and Rao, 1998; Duff and Rao, 2001). However, ascertainment of definite relationships among different stem cell types is somewhat complicated by the fact that methods of stem cell isolation and culture differ among investigators (Mansergh et al., 2000).

This issue is further complicated by the recent discovery that cell fate assignments in the adult are more blurred than previously assumed. Until recently, prevalent theories suggested that stem cells from a given tissue are restricted to producing cell types that form part of that tissue. Research using embryos showed that, e.g., adult hematopoietic cells injected into mouse blastocysts went on to form part of the hematopoietic system (Geiger et al., 1998). However, transplants of adult stem cells into adult recipients suggest a greater flexibility. Bone marrow–derived stem cells can generate liver, blood, endothelial, bone, cartilage, fat, tendon, lung, muscle, marrow stroma, and brain cell types (Ferrari et al., 1998; Petersen et al., 1999; Magli et al., 2000). Mesenchymal stem cells (multipotent cells present in the adult human bone marrow) can be induced to develop along adipocytic, chondrocytic, tenocytic, myotubular, neural, and osteocytic lineages, in addition to hematopoietic stroma (Pittenger et al., 1999; Minguell et al., 2001). These cells are easy to isolate and culture and have a high expansive capacity, potentially making them a very attractive therapeutic tool (Minguell et al., 2001).

Adult murine neural stem cells can give rise to all germ layers in chicks and chimeric mice (Clarke et al., 2000). Neural stem cells can give rise to skeletal myotubes in vitro and in vivo, provided that the neural stem cells were directly exposed to preexisting myotubes and were not clustered together (Galli et al., 2000). Similarly, they have been shown to generate myeloid, lymphoid, and hematopoietic cell types after transplantation into irradiated hosts (Eglitis and Mezey, 1997; Bjornson et al., 1999; Kopen et al., 1999). Endothelial and hematopoietic receptors have been found on the surface of human embryonic neural stem cells; these are partially maintained during development (Parati et al., 2002). Notably, many types of neural stem cell can be shown to reconstitute the hematopoietic system after transplantation into immunocompromised irradiated mice (Shih et al., 2001). This would suggest that these cells have a wider potential than previously assumed and may point to a link between neural and hematopoietic stem cells (Parati et al., 2002).

Human umbilical cord blood cells treated with RA and nerve growth factor (NGF) express molecular markers associated with neurogenesis and extend long cytoplasmic processes, suggesting that these cells could also form a readily accessible source of stem cells (Ha et al., 2001; Sanchez-Ramos et al., 2001). Similarly, amniotic epithelial cells have been shown to express MAP-2 and nestin, and to repopulate areas of the brain damaged by ischemia (Okawa et al., 2001).

Differentiation of stem cells from rodent dermis and adult human scalp can produce neural and mesodermal derivatives and may produce a readily accessible source of stem cells for transplantation (Toma et al., 2001). The feasibility of expansion and long-term culturing of various types of somatic stem cell, along with their versatility, makes them attractive therapeutically (Magli et al., 2000). This, combined with the possibility of dedifferentiation, may soon result in plentiful sources of committed stem cells to treat neurodegenerative disorders (Singh, 2001). Using either of these strategies, stem cells for transplantation could conceivably be obtained from the individual for whom they are intended, thus avoiding the use of immunosuppression.

However, this process would require advances in culturing conditions beyond what is possible today. These scenarios may also be complicated by suggestions that "reprogramming" of stem cells may be explained by the presence of rare stem cells, present in some or all tissues, that are capable of greater plasticity than their neighbours. This may have implications for the ease with which different tissue types can be obtained from, say, skin-derived stem cells and also for the quantity obtainable from in vitro culture methods (Temple, 2001).

16.3. ALTERNATIVE SOURCES OF NEURAL PRECURSORS

16.3.1. IMMORTALIZED NEURAL PRECURSOR CELLS Stem cells derived from sources in vivo have many advantages, however, some of the drawbacks include the time-limited nature of expansion protocols and hence finite limits to the number of cells that can eventually be obtained from any given source. Ethical issues therefore arise regarding the use of large amounts of fetal material, meaning that access to any such treatments may be limited to those nations that permit both elective abortion and medical use of any ensuing fetal tissue. Such problems do not arise in the case of adult stem cells, but these tend to be more of a challenge to expand and give lower yields. Immortalized stem cell lines can be expanded indefinitely and therefore would provide a consistent source of similar cellular material for transplantation. Some lines are also easy to transfect and may be of use in cell-based gene therapy. The challenge would be to ensure that these are not oncogenic post-transplantation.

16.3.1.1. C17.2 Cells C17.2 cells are an immortalized, multipotent cell line derived from neonatal mouse cerebellum. C17.2 cells have been shown to integrate into many different brain structures. Notably, they can integrate into regions of adult mouse neocortex that had previously been subjected to apoptotic degeneration (Snyder et al., 1997a). In addition, C17.2 cells have been shown to engraft widely and to partially correct the cerebellar defects present in meander mice (Rosario et al., 1997). C17.2 transplantation and engraftment (Yandava et al., 1999) ameliorated defective myelination present in shiverer mice. C17.2 cells are also capable of genetic complementation after transplantation (Snyder et al., 1997b). Transfection with β-glucuronidase and transplantation into an animal model of Sly disease (mucopolysaccharidosis VII) resulted in engraftment and partial correction of the lysosomal storage disorder in neurons and glia of affected mice (Snyder and Macklis, 1995). Transfection with human HexA and transplantation into Tay-Sachs mice resulted in amelioration of neurodegeneration (Lacorazza et al., 1996).

16.3.1.2. Embryonic Carcinoma Cells More pluripotent cell lines offer the advantage of greater flexibility on differentiation. Embryonic carcinoma (EC) cell lines were the first stem cell lines recognized as being widely pluripotent. These are derived from gonadal tumors that contain a mixture of undifferentiated and a wide variety of differentiated cell types. EC cells can give rise to cell types derived from all three germ layers (endoderm, mesoderm, and ectoderm). The therapeutic potential of both human and mouse EC lines has been investigated since the mid-1970s (Donovan and Gearhart, 2001; Lovell-Badge, 2001). The P19 line, isolated from a tumor generated artificially following implantation of a 7-d embryo under the testis capsule of an adult male mouse, has proved one of the most valuable models of neural in vitro differentiation (Jones-Villeneuve et al., 1982).

P19 cells resemble primitive ectoderm cells, have a broad developmental potential, and can contribute to a wide variety of somatic cell types in murine chimeras. P19 cells are most easy to differentiate along neural and muscular lineages in vitro (Jones-Villeneuve et al., 1982; McBurney et al., 1982). Production of neurons from P19 cells requires culture of cellular aggregates, followed by culture in the presence of $>5 \times 10^{-7}$ *M* RA. The cells are then desegregated and plated out on adhesive substrates. A mixed cell population containing neuron, glia, and fibroblast-like cells is then obtained (Jones-Villeneuve et al., 1982; McBurney et al., 1982; McBurney, 1993; Bain et al., 1994).

Investigation of the neuron and glia-like cells has proved promising. P19-derived neuron-like cells are postmitotic; express neural markers including neurotransmitters and their receptors; possess cell-surface carbohydrate antigens characteristic of neurons; and form dendrites, axons, and functional synapses (Levine and Flynn, 1986; McBurney et al., 1988; Bain et al., 1994; MacPherson et al., 1997). Neurotransmitters expressed in mature cultures include GABA, GABA transaminase, glutamic acid decarboxylase (GAD), neuropeptide Y, somatostatin, tyrosine hydroxylase, DOPA decarboxylase, serotonin, calcitonin gene–related peptide, galanin, substance P, and enkephalin (Staines et al., 1994), suggesting the presence of a wide variety of neuronal subtypes. Transplantation studies have also been carried out; RA-treated P19 cells have been shown to produce neurons and glia when transplanted into neonatal rat brains (Magnuson et al., 1995a; Staines et al., 1996). Transplanted oligodendrocytes appeared to be myelinating host axons, while transplanted neurons expressed GABA, glycine, and glutamate and exhibited the neurophysiological characteristics of neurons 2–4 wk posttranslation (Magnuson et al., 1995a, 1995b; Staines et al., 1996).

Taken together, these results would imply a huge therapeutic potential for cells of this type. P19 cells have disadvantages, however. They lose and/or inactivate transgenes at a high rate, which has precluded attempts at their use as a gene therapy vehicle (Schmidt-Kastner et al., 1996, 1998). EC cells also have the potential to become oncogenic. Human EC lines are tumor derived and often aneuploid (Lovell-Badge, 2001), while P19 cells have been shown to have an inactivated p53 gene (Schmidt-Kastner et al., 1996, 1998). The risk of their oncogenic potential has so far prevented the use of immortalized cell lines in clinical trials. Mice may be an imperfect model for transplantation studies in this instance; studies of numerous knockout mice that failed to recapitulate the cancers seen in human patients with similar gene loss (Rb, p53) imply that oncogenic pathways may differ substantially between mouse and human (Jacks, 1996).

16.3.2. ES CELLS Perhaps the most valuable contribution of EC cells was that their analysis led to the discovery of ES cells. ES cells were derived from mouse blastocysts that had been prevented from implanting and that were then cultured in conditions optimized from previous EC research (Evans and Kaufman, 1981; Lovell-Badge, 2001). ES cells are derived from the inner cell mass (ICM) of mouse blastocysts, possess normal karyotypes, and can be cultured indefinitely (Evans and Kaufman, 1981; Martin, 1981). Prevention of differentiation requires culture on a feeder cell layer and/or culture in the presence of LIF. ES cells, when aggregated or microinjected into early mouse embryos, can give rise to all cell types in the resulting chimera, including the germline. The discovery that they can be easily, stably, and precisely genetically manipulated has led to the production of more than 1000 different mouse lines via targeted mutagenesis. This provides a revolutionary accuracy, both to the production of animal models for human disease and to the analysis of murine (and, hence, by extrapolation, mammalian) gene function (Capecchi, 2000). ES-like cells have since been isolated from various species, including humans and various primates (Thomson et al., 1998a, 1998b; Reubinoff et al., 2000; Suemori et al., 2001). Similarly, pluripotent embryonic germ (EG) cells have also been isolated. These cells are derived from primitive germ cells colonizing the developing gonadal ridge and have been isolated from various species, including humans and mice (Shamblott et al., 1998, 2001). Mouse EG cells can also demonstrate germline transmission in chimeric mice but have been less widely used than mouse ES cells (Shamblott et al., 1998).

Extensive parallels with EC cell research led quickly to the probability that ES cells would be differentiated in vitro in similar ways. The extensive pluripotency of ES cells would also imply that, given advances in culturing techniques, all cell types may eventually be derivable from undifferentiated ES cells (Mansergh et al., 2000). ES cells cultured in the absence of feeder layers and LIF form floating aggregates (embryoid bodies [EBs]) and can be cultured for extended periods in this state. ES cells cultured for a long time as EBs will spontaneously differentiate into various cell types, including myocardium, blood islands, and visceral yolk sac (Doetschman et al., 1985; Risau et al., 1988).

Differentiation protocols have been developed that influence these cultures to produce a larger proportion of specific cell types. Dimethyl sulfoxide induces the formation of skeletal, cardiac, and smooth muscle cells (Dinsmore et al., 1996). Overexpression of IGF-2 can also be used to enhance myogenic differentiation in vitro (Prelle et al., 2000). Culture on stromal cell layers with hematopoietic growth factors induces hematopoietic differentiation (Keller et al., 1993; Nakano et al., 1994; Nakayama et al., 1998). Culture as EBs, followed by plating on a collagen matrix and addition of angiogenic growth factors, can recapitulate vasculogenesis and angiogenesis (Feraud et al., 2001). ES cell–derived hematopoietic progenitors can be shown to give rise to cells of the lymphoid, myeloid, and embryonic erythroid systems in vivo (Perlingeiro et al., 2001). RA-treated ES cells can be cultured with ascorbic acid, β-glycerophosphate, and then BMP-2 or compactin, resulting in the formation of bone nodules in culture (Phillips et al., 2001). Osteoclast formation can be obtained by culture on a stromal cell line and enhanced by the addition of macrophage colony-stimulating factor and osteoprotegerin-ligand (Hemmi et al., 2001). Cells expressing insulin and other other pancreatic endocrine hormones have also been obtained by differentiation of ES cells (Lumelsky et al., 2001).

Based on P19 neural differentiation methods, a protocol for differentiation of murine ES cells into neurons and glia was first developed by Bain et al. (1995). Because RA causes extensive cell death in undifferentiated ES cells (Robertson, 1987), ES cells were cultured as aggregates in standard serum containing medium without LIF for 4 d. The resulting EBs were then cultured in the presence of 5×10^{-7} *M* all-*trans* RA for a further 4 d. EBs were next desegregated and plated out on adhesive substrates. Following 4 to 5 d of growth under these conditions, approx 40% of the resulting cells possessed neuronal or glial (astrocytic and oligodendrocytic) morphology (Bain and Gottlieb, 1998; Bain et al., 1995).

Variations on this protocol have since been developed, yielding similar results (Fraichard et al., 1995; Strubing et al., 1995; Dinsmore et al., 1996; Okabe et al., 1996; Brustle et al., 1997; Li et al., 1998). A number of studies have used FGF-2 or insulin transferrin selenium fibronectin medium as alternatives to RA, with similar results (Okabe et al., 1996; Brustle et al., 1997). Markers of neurogenesis are expressed in a manner that strongly resemble temporal patterns of gene expression in vivo. Subsets of ES-derived neurons are thought to differentiate into dopaminergic, noradrenergic, cholinergic, and GABAergic neurons, since expression of tyrosine hydroxylase, choline acetyl transferase, vesicular inhibitory amino acid transporter, and GAD have been noted in mature cultures (Bain et al., 1995, 2000; Strubing et al., 1995; Fraichard et al., 1995; Dinsmore et al., 1996; Okabe et al., 1996; Bain and Gottlieb, 1998; Kawasaki et al., 2000; Westmoreland et al., 2001). Markers characteristic of somatic motor neurons, cranial motor neurons, and interneurons have also been found, implying high levels of differentiation along very specific neuronal pathways (Renoncourt et al., 1998). Voltage clamp studies have demonstrated the presence of K^+, Ca^+, and Na^+ voltage-dependent channels within cell membranes in ES cell neural cultures (Bain et al., 1995; Fraichard et al., 1995; Strubing et al., 1995). Cells were also capable of action potential generation and were sensitive to excitatory and inhibitory agonists (kainate, *N*-methyl-D-aspartate, GABA, and glycine) (Bain et al., 1995).

The aforementioned neural differentiation protocols typically yield 30–40% neurons and 10% glia (Bain et al., 1995; Li et al., 1998). The identity of the remaining 50% of the cells is unknown. Some may represent multipotent or lineage-restricted neuroepithelial, neuronal, or glial precursors, but this remains speculative (Li et al., 1998; Mujtaba et al., 1999). Derivation of purer neural precursor cultures and/or more restricted neuronal and glial subtypes may be facilitated by different culture methods. Uncovering the links between ES cell–derived neuronal precursors and various types of neural stem cells may also aid this process. Successful attempts have been made to enrich the proportion of neurons and/or glia by passaging through media containing different mitogens, the use of transgenic methods, and growth on specific feeder layers (Li et al., 1998; Brustle et al., 1999; Kawasaki et al., 2000).

A modified RA-based protocol has been used to produce oligodendrocyte cultures (Liu et al., 2000). Use of PA6 stromal cells to culture mouse ES cells resulted in differentiation of a high proportion of dopaminergic neurons without the use of RA or EBs (Kawasaki et al., 2000). Splitting ES cells into single cells prior to EB formation, followed later by addition of a mitogenic cocktail and withdrawal of HEPES from the medium, also resulted in a higher proportion of dopaminergic neurons (Lee et al., 2000). Growth of ES cells at low density, with LIF, but in the absence of serum, has been used to generate low percentages of neurospheres from ES cells (Tropepe et al., 2001; Xian and Gottlieb 2001). These neurospheres have the potential to generate cultures composed entirely of nestin-expressing neural precursors, mature neurons, and glia under differentiating conditions. However, ES-derived neurospheres possessed a wider developmental potential than those isolated from fetal or adult brain; they could contribute extensively to all embryonic tissues present at E9.5 (Tropepe et al., 2001).

A selection method involving the integration of a β-galactosidase-neo cassette into the Sox2 gene, a marker of neuroepithelial precursors, has also been useful. Using this system, treatment with G418 allows selection of only neuroepithelial precursors. Enrichment of neuroepithelial precursors led to a concomitant increase in neurons and glia after differentiation (Li et al., 1998).

ES cell–derived neurons show great promise in vivo. Transplantation of mouse ES–derived neural precursors into embryonic rat brain, lesioned rat striatum, or injured spinal cord has shown migration, differentiation, and integration of ES cell–derived neurons and glia into the host neural architechture (Dinsmore et al., 1996; Brustle et al., 1997; McDonald et al., 1999; Liu et al., 2000). Resulting neurons and glia have shown survival times of at least 6 wk (Dinsmore et al., 1996). Transplantation-related improvements in locomotor function have also been noted, implying that the grafted cells are fully functional (McDonald et al., 1999). Furthermore, transplanted ES cell–derived glia have been shown to be capable of remyelination in a myelin-deficient rat model of Pelizaeus–Merzbacher disease, in artificially demyelinated spinal cord, and in myelin-deficient shiverer mice (Brustle et al., 1999; Liu et al., 2000). Transplantation of ES-derived dopaminergic neurons resulted in integration into mouse striatum and retention of thyroid hormone expression (Kawasaki et al., 2000), suggesting utility in the treatment of Parkinson disease. Mature genetically altered neural grafts have been obtained from transplantation of EB-derived cells; expression of a marker gene demonstrated both the feasibility of cell-based gene transfer into the nervous system and the derivation of mature neurons, microglial cells, oligodendrocytes, and astrocytes (Benninger et al., 2000).

Transfection of an enhanced green fluorescent protein (EGFP) construct under the control of the thymidine kinase promoter/nestin second intron has been used successfully to visualize the integration of ES-derived neural precursors posttransplantation (Andressen et al., 2001). Similar strategies may be useful for selecting correctly differentiating neural precursors for transplantation by cell sorting (Andressen et al., 2001). Also of interest is the discovery that ES-derived neural precursors (as opposed to ES cells prior to differentiation) can be transfected with plasmids using lipofection or infected with retroviruses (Westmoreland et al., 2001).

Transplantation of differentiated ES cell derivatives has great therapeutic promise. However, certain challenges need to be overcome. The presence of mixed cell types in cultures of ES-derived precursors leads to the worry that some undifferentiated cell types may still remain. Undifferentiated ES cells can give rise to teratocarcinomas if transplanted in an undifferentiated state (Robertson, 1987). Tumorigenesis has not resulted from transplantation studies carried out to date (Dinsmore et al., 1996; Brustle et al., 1997). However, as previously stated, human ES cells may differ from mouse cells in terms of their tumorigenic potential.

16.3.2.1. Human ES Cells and Therapeutic Cloning Taken together, the results of transplant studies in mice would imply that therapeutic benefits could result from the transplantation of ES cell–derived neural and glial precursors. These results are especially promising given the aforementioned isolation of human and primate ES and EG cell lines (Thomson et al., 1998a, 1998b; Reubinoff et al., 2000; Shamblott et al., 2001; Suemori et al., 2001). Human ES cells, like their murine equivalents, have been obtained from the ICM of blastocysts. They can be propagated indefinitely in culture, maintain high levels of telomerase, and possess normal karyotypes (Amit et al., 2000; Eiges et al., 2001; Odorico et al., 2001). Like mouse ES cells, human ES cells express Oct4 (a POU domain transcription factor expressed only in pluripotent cells) and can be cultured on feeder cell layers (normally mouse embryonic fibroblasts) (Amit et al., 2000; Reubinoff et al., 2000).

Human ES cell differentiation in vitro has yielded promising results to date. Simple culture of EBs in suspension yields a cell population containing approx 8% cardiomyocytes (Kehat et al., 2001). Coculture with bone marrow or yolk sac endothelial cell lines can be used to derive hematopoietic precursors from human ES cells (Kaufman et al., 2001). RA and βNGF have been shown to act as enhancers of neural differentiation in human ES cell cultures; resulting cells expressed neuronal markers and showed extensive process outgrowth (Schuldiner et al., 2001). Neurospheres have been isolated from human ES cells via isolation of NCAM-expressing cells from multilayered aggregates (Reubinoff et al., 2000). Culture of selected cells in serum-free medium resulted in the formation of nestin-expressing spherical structures, which could be either expanded in culture or plated on a substrate. Following substrate plating, differentiated cells displayed morphological characteristics of neurons and expressed markers of neuronal differentiation, including neurofilament, β-tubulin, synaptophysin, glutamate, GAD, and GABA receptors (Reubinoff et al., 2000; Pera, 2001). EBs derived from human EG cells can be used to generate cultures of precursor cells capable of robust proliferation in culture. These cultures contain a variety of cell types but seem to express markers of the neural lineage most strongly (Shamblott et al., 2001).

Potentially, human ES cells could provide a promising source of neural precursors for transplantation, with reduced requirements for repeated harvesting of fetal tissue. Protocols for the genetic alteration of human ES and EG cells have been established and the stable persistence of marker genes noted (Eiges et al., 2001; Shamblott et al., 2001), further expanding their therapeutic potential. DNA can be introduced into human EG cells via lipofection or lentiviral or retroviral transduction (Shamblott et al., 2001). Murine ES cells are usually transfected with foreign DNA via electroporation, but this technique resulted in poor survival in human ES cells. EGFP was introduced into human ES cells using ExGen 500 (a transfection reagent produced by Fermentas), under the control of a promoter specific to germ cells, ES cells, and trophoblast (Rex1). In this case, the marker gene was used to track undifferentiated cells for ease of maintenance of pluripotency (Eiges et al., 2001). However, scenarios in which EGFP could be turned on in response to early neuronal differentiation have already been tested in mouse (Andressen et al., 2001) and may be feasible in human cells. In such cases, EGFP-positive cells could potentially be used to purify neural progenitors for transplantation by fluorescence-activated cell sorting. A different selection strategy could also be envisaged in which tissue-specific promoters could be used to activate an antibiotic resistance gene or a similar selectable marker (Odorico et al., 2001).

A further potential method of tracking transplanted stem cells in humans involves the use of magnetodendrimers. These have been developed as a versatile class of magnetic tags that are capable of labeling mammalian cells and would be traceable in vivo using magnetic resonance imaging (Bulte et al., 2001). A potential drawback with these strategies, however, is the fact that these cells would express a foreign protein or contain a foreign material that might trigger immunological rejection. Insertion of a transgene in an unfavorable position could also result in malignancies (Odorico et al., 2001).

The successful transfection of human ES cells means that they could be used as cell-based vectors for gene therapy. ES cells could be used for production of tissue immunologically compatible with the donor, a desirable outcome because currently available immunosuppressive drugs have a number of undesirable side effects. One strategy would involve alteration of the histocompatibility antigens in existing ES cell lines to match the patient at least partially, reducing (or even eliminating) the need for immunosuppression (Gearhart, 1998; Solter and Gearhart, 1999). However, techniques involving precise genetic manipulations may be costly in terms of both time and money.

A second strategy would involve the use of somatic nuclear transfer protocols to create an embryo identical to the donor. Derivation of ES cell lines from embryos created by somatic nuclear transfer has been achieved in mice (Wakayama et al., 1999; Munsie et al., 2000; Rideout et al., 2000). Somatic nuclear transfer involves the isolation of a somatic cell nucleus from an individual. This is then injected into an enucleated egg. The resulting zygote can then be cultured to the blastocyst stage, at which point ES cell lines can be isolated from the ICM. The advantage of this technique is that any cell type–specific precursors differentiated from the resulting ES cell lines would be identical to the donor of the original somatic cell nucleus, thus, it is hoped, eliminating risk of immunological rejection. This technique is popularly known as therapeutic cloning. In terms of cytotherapeutics, it has great potential. Patient-specific ES cell lines could also be genetically altered prior to transplantation, differentiated to the required precursor type, and used as vectors for gene therapy and/or cytotherapeutics. Use of homologous recombination techniques may even allow targeted replacement of deleterious dominant or recessive alleles in patient-specific ES cell lines; targeted murine ES cell nuclei have been shown to give rise to healthy cloned adult mice (Rideout et al., 2000).

However, the use of therapeutic cloning in humans is fraught with ethical and moral concerns. Ethical dilemmas include the fact that a cloned human embryo is first created and then destroyed, doubly abhorrent in the eyes of many. Widespread use of the technique may eventually result in its acquisition by the unscrupulous, resulting in the increasing likelihood of reproductive cloning being attempted by someone, somewhere. As a result, existing or forthcoming legislation in many jurisdictions may preclude therapeutic cloning. Other drawbacks include feasibility. ES cell lines derived from a cloned embryo have yet to be isolated in humans. Human oocytes are not available in high numbers, the nuclear transfer procedure is highly inefficient, and human ES cells have a doubling time three times that of murine ES cells (Odorico et al., 2001). One can add to this the issue of cost and most of the theoretical risks associated with transplantation of differentiated ES cells in general.

One significant problem is the relatively poor characterization of ES cell differentiation to date. Most ES differentiation protocols result in enrichment for the desired cell type or types. However, as previously mentioned, many techniques still result in mixed cell populations. In addition, despite the use of identical growth conditions, there is considerable variability among cultures (Odorico et al., 2001). More effort should be expended in the positive identification of unknown cell types resulting from ES neural differentiation. Further development of improved culture methods may be aided by the wealth of information emerging from various genome and expressed sequence tag (EST) sequencing projects.

16.4. GENOMICS, GENE DISCOVERY, AND STEM CELLS

Bioinformatic analysis of genes implicated in neural differentiation could be important in unraveling stem cell biology (Bain et al., 2000; Mansergh et al., 2000). The advantage of in vitro systems such as ES neural differentiation is that large amounts of material undergoing synchronous differentiation can be made available for analysis. Such systems have, to date, been invaluable in gene discovery projects intended to uncover novel regulators of differentiation. Subtractive libraries and microarray experiments have been used in ES cell neural, extraembryonic endoderm, and hematopoietic differentiation and adult neural and haematopoietic stem cell systems with success (Bain et al., 2000; Kelly and Rizzino, 2000; Baird et al., 2001; Geschwind et al., 2001; Terskikh et al., 2001). Notably, production of an ES neural differentiation library resulted in a high proportion of uncharacterized or novel sequences, in addition to a large number of genes involved in development and/or function of the nervous system (Bain et al., 2000). This would suggest that some important regulators of neurogenesis may remain as-yet undiscovered. Array-based studies of hematopoietic and neural stem cells suggest that these two systems may have overlapping patterns of gene expression, implying the existence of a common set of genes expressed in stem cells (Geschwind et al., 2001; Terskikh et al., 2001). The use of two-dimensional electrophoresis and peptide mass fingerprinting has revealed 24 proteins that are differentially regulated during RA-mediated neural differentiation of ES cells. Nine of the genes encoding these proteins are known to be involved in neural differentiation or survival (Guo et al., 2001). Similar studies, possibly also including large-scale proteomics approaches, will undoubtedly be of use in uncovering hitherto unknown regulators of neural differentiation and stem cell maintenance. We therefore anticipate that elucidation of the function of further genes involved in neurogenesis will allow improvement of culture systems and isolation of more restricted cell types from cell culture systems.

16.5. INTRINSIC BRAIN REPAIR

The discovery of genes whose therapeutic use or expression could promote endogenous nervous system repair may be another result of the genomics revolution (Weiss, 1999; Mansergh et al., 2000). The brain is now known to be capable of the intrinsic generation of new cells, although some of these persist only transiently (Gould et al., 2001). Physical and/or mental stimulation of rodents or primates can result in the birth of new cells in brain regions associated with learning and memory (Gould et al., 1997, 1998; Gould et al., 1999a, 1999b; van Praag et al., 1999a, 1999b) There is now evidence supporting the involvement of *de novo* neurogenesis in learning and memory; associative learning enhances the survival of newborn hippocampal neurons, while depletion of new neurons adversely affects hippocampal-dependent trace conditioning (Shors et al., 2001). This lends credence to the belief that physical and/or mental stimulation could be used to ameliorate the neurodegenerative consequences of old age. Avoidance of stress may also help, because high circulating glucocorticoid hormone levels inhibit the proliferation of precursors in the dentate gyrus. Chronic stress results in long-term inhibition of proliferation and changes in the structure of the dentate gyrus; stress may therefore alter and possibly impair hippocampal function (Gould and Tanapat, 1999).

There is also the possibility that intrinsic brain repair could be triggered therapeutically (Weiss, 1999). The injection of EGF or FGF can stimulate cell division and the production of new neurons and glia in the SVZ and olfactory tract (Craig et al., 1999; Wagner et al., 1999). Similar administration of neurogenic substances or the use of exercise regimes could be used to stimulate the proliferation of endogenous precursors without the use of more invasive transplantation procedures (Weiss, 1999).

16.6. CONCLUSIONS

Results from clinical trials involving the transplantation of fetal neural tissue are promising. Follow-up studies of patients with Huntington or Parkinson disease who had received fetal neural transplants have shown improvements in symptoms over time (Spencer et al., 1992; Widner et al., 1992; Lopez-Lozano et al., 1995, 1997; Bachoud-Levi et al., 2000a, 2000b). Histological studies performed on one patient following death (from cardiovascular disease) 18 mo post-transplantation demonstrated surviving graft-derived neurons that had integrated into the host neural architecture and were unaffected by the Huntington disease process (Freeman et al., 2000). However, the use of human fetal tissue remains problematic owing to the relatively large amount of fetal tissue required for transplantation and the ethical issues involved.

Stem cells have potential as a replacement for fetal tissues in therapies for neurodegenerative disorders. Increasing numbers of stem cell types can be cultured in vitro, producing specifically defined progeny. Transplant studies in animal models provide evidence of graft integration into host neural structures and, in some cases, amelioration of symptoms. Given that human patients show analogous improvements on receipt of fetal neurons, one could hypothesize that they would respond in a similar way to stem cell transplantation. Stem cells show promise as vehicles for gene therapy, demonstrating expression of transfected genes in host tissues. Novel techniques may permit the production of neural progenitors of which the transplant recipient is immunotolerant.

However, for realization of stem cell promise, improvements still need to be made to existing technology. More stringent tests of posttransplant oncogenesis in animal models (preferably including nonrodents) and reliable production of a variety of well-defined, pure progenitor cultures from human stem cell sources are probably the most urgent developments required. Improvements in culturing techniques will undoubtedly be aided by the data emerging from genome sequencing, EST discovery, and global expression studies involving stem cells. Given the speed with which this field is moving, the wait for stem cells in clinical trials probably will not be long.

ACKNOWLEDGMENTS

FCM was supported by postdoctoral fellowships from the Alberta Cancer Board, the Alberta Heritage Foundation for Medical Research, and the Canadian Institutes of Health Research. MAW was supported by a postdoctoral fellowship from the Alberta Heritage Foundation for Medical Research. DER is a scholar of the Alberta Heritage Foundation for Medical Research.

REFERENCES

Alvarez-Buylla, A., Garcia-Verdugo, J. M., and Tramontin, A. D. (2001) A unified hypothesis on the lineage of neural stem cells. *Nat. Rev. Neurosci.* 2:287–293.

Amit, M., Carpenter, M. K., Inokuma, M. S., et al. (2000) Clonally derived human embryonic stem cell lines maintain pluripotency and proliferative potential for prolonged periods of culture. *Dev. Biol.* 227:271–278.

Andressen, C., Stocker, E., Klinz, F. J., et al. (2001) Nestin-specific green fluorescent protein expression in embryonic stem cell-derived neural precursor cells used for transplantation. *Stem Cells* 19:419–424.

Arsenijevic, Y. and Weiss, S. (1998) Insulin-like growth factor-I is a differentiation factor for postmitotic CNS stem cell-derived neuronal precursors: distinct actions from those of brain-derived neurotrophic factor. *J. Neurosci.* 18:2118–2128.

Arsenijevic, Y., Weiss, S., Schneider, B., and Aebischer, P. (2001a) Insulin-like growth factor-I is necessary for neural stem cell proliferation and demonstrates distinct actions of epidermal growth factor and fibroblast growth factor-2. *J. Neurosci.* 21:7194–7202.

Arsenijevic, Y., Villemure, J. G., Brunet, J. F., et al. (2001b) Isolation of multipotent neural precursors residing in the cortex of the adult human brain. *Exp. Neurol.* 170:48–62.

Bachoud-Levi, A. C., Remy, P., Nguyen, J. P., et al. (2000a) Motor and cognitive improvements in patients with Huntington's disease after neural transplantation. *Lancet* 356:1975–1979.

Bachoud-Levi, A., Bourdet, C., Brugieres, P., et al. (2000b) Safety and tolerability assessment of intrastriatal neural allografts in five patients with Huntington's disease. *Exp. Neurol.* 161:194–202.

Bain, G. and Gottlieb, D. I. (1998) Neural cells derived by in vitro differentiation of P19 and embryonic stem cells. *Perspect. Dev. Neurobiol.* 5:175–178.

Bain, G., Ray, W. J., Yao, M., and Gottlieb, D. I. (1994) From embryonal carcinoma cells to neurons: the P19 pathway. *Bioessays* 16:343–348.

Bain, G., Kitchens, D., Yao, M., Huettner, J. E., and Gottlieb, D. I. (1995) Embryonic stem cells express neuronal properties in vitro. *Dev. Biol.* 168:342–357.

Bain, G., Mansergh, F. C., Wride, M. A., et al. (2000) ES cell neural differentiation reveals a substantial number of novel ESTs. *Funct. Integr. Genomics* 1:127–139.

Baird, J. W., Ryan, K. M., Hayes, I., et al. (2001) Differentiating embryonal stem cells are a rich source of haemopoietic gene products and suggest erythroid preconditioning of primitive haemopoietic stem cells. *J. Biol. Chem.* 276:9189–9198.

Baroffio, A., Dupin, E., and Le Douarin, N. M. (1991) Common precursors for neural and mesectodermal derivatives in the cephalic neural crest. *Development* 112:301–305.

Benedetti, S., Pirola, B., Pollo, B., et al. (2000) Gene therapy of experimental brain tumors using neural progenitor cells. *Nat.. Med.* 6:447–450.

Benninger, Y., Marino, S., Hardegger, R., Weissmann, C., Aguzzi, A., and Brandner, S. (2000) Differentiation and histological analysis of embryonic stem cell-derived neural transplants in mice. *Brain Pathol.* 10:330–341.

Bjornson, C. R., Rietze, R. L., Reynolds, B. A., Magli, M. C., and Vescovi, A. L. (1999) Turning brain into blood: a hematopoietic fate adopted by adult neural stem cells in vivo. *Science* 283:534–537.

Brustle, O., Spiro, A. C., Karram, K., Choudhary, K., Okabe, S., and McKay, R. D. (1997) In vitro-generated neural precursors participate in mammalian brain development. *Proc. Natl. Acad. Sci. USA* 94: 14,809–14,814.

Brustle, O., Choudhary, K., Karram, K., et al. (1998) Chimeric brains generated by intraventricular transplantation of fetal human brain cells into embryonic rats. *Nat. Biotechnol.* 16:1040–1044.

Brustle, O., Jones, K. N., Learish, R. D., et al. (1999) Embryonic stem cell-derived glial precursors: a source of myelinating transplants. *Science* 285:754–756.

Bulte, J. W., Douglas, T., Witwer, B., et al. (2001) Magnetodendrimers allow endosomal magnetic labeling and in vivo tracking of stem cells. *Nat. Biotechnol.* 19:1141–1147.

Capecchi, M. R. (2000) How close are we to implementing gene targeting in animals other than the mouse? *Proc. Natl. Acad. Sci. USA* 97:956–957.

Chanas-Sacre, G., Rogister, B., Moonen, G., and Leprince, P. (2000) Radial glia phenotype: origin, regulation, and transdifferentiation. *J. Neurosci. Res.* 61:357–363.

Clarke, D. L., Johansson, C. B., Wilbertz, J., et al. (2000) Generalized potential of adult neural stem cells. *Science* 288:1660–1663.

Collarini, E. J., Pringle, N., Mudhar, H., et al. (1991) Growth factors and transcription factors in oligodendrocyte development. *J. Cell. Sci.* (Suppl.)15:117–123.

Craig, C. G., Tropepe, V., Morshead, C. M., Reynolds, B. A., Weiss, S., and van der Kooy, D. (1999) In vivo growth factor expansion of endogenous subependymal neural precursor cell populations in the adult mouse brain. *J. Neurosci.* 16:2649–2658.

Daadi, M. M., and Weiss, S. (1999) Generation of tyrosine hydroxylase-producing neurons from precursors of the embryonic and adult forebrain. *J. Neurosci.* 19:4484–4497.

Dinsmore, J., Ratliff, J., Deacon, T., et al. (1996) Embryonic stem cells differentiated in vitro as a novel source of cells for transplantation. *Cell Transplant.* 5:131–43.

Doetsch, F., Caille, I., Lim, D. A., Garcia-Verdugo, J. M., and Alvarez-Buylla, A. (1999) Subventricular zone astrocytes are neural stem cells in the adult mammalian brain. *Cell* 97:703–16.

Doetschman, T. C., Eistetter, H., Katz, M., Schmidt, W., and Kemler, R. (1985) The in vitro development of blastocyst-derived embryonic stem cell lines: formation of visceral yolk sac, blood islands and myocardium. *J. Embryol. Exp. Morphol.* 87:27–45.

Donovan, P. J. and Gearhart, J. (2001) The end of the beginning for pluripotent stem cells. *Nature* 414:92–97.

Duff, K. and Rao, M. V. (2001) Progress in the modeling of neurodegenerative diseases in transgenic mice. *Curr. Opin. Neurol.* 14: 441–447.

Eglitis, M. A. and Mezey, E. (1997) Hematopoietic cells differentiate into both microglia and macroglia in the brains of adult mice. *Proc. Natl. Acad. Sci. USA* 94:4080–4085.

Eiges, R., Schuldiner, M., Drukker, M., Yanuka, O., Itskovitz-Eldor, J., and Benvenisty, N. (2001) Establishment of human embryonic stem cell-transfected clones carrying a marker for undifferentiated cells. *Curr. Biol.* 11:514–518.

Evans, M. J. and Kaufman, M. H. (1981) Establishment in culture of pluripotential cells from mouse embryos. *Nature* 292:154–156.

Feraud, O., Cao, Y., and Vittet, D. (2001) Embryonic stem cell-derived embryoid bodies development in collagen gels recapitulates sprouting angiogenesis. *Lab. Invest.* 81:1669–1681.

Ferrari, G., Cusella-De Angelis, G., Coletta, M., et al. (1998) Muscle regeneration by bone marrow-derived myogenic progenitors. *Science* 279:1528–1530.

Flax, J. D., Aurora, S., Yang, C., et al. (1998) Engraftable human neural stem cells respond to developmental cues, replace neurons, and express foreign genes. *Nat. Biotechnol.* 16:1033–1039.

Fraichard, A., Chassande, O., Bilbaut, G., Dehay, C., Savatier, P., and Samarut, J. (1995) In vitro differentiation of embryonic stem cells into glial cells and functional neurons. *J. Cell. Sci.* 108:3181–3188.

Freeman, T. B., Cicchetti, F., Hauser, R. A., et al. (2000) Transplanted fetal striatum in Huntington's disease: phenotypic development and lack of pathology. *Proc. Natl. Acad. Sci. USA* 97:13,877–13,882.

Fricker, R. A., Carpenter, M. K., Winkler, C., Greco, C., Gates, M. A., and Bjorklund, A. (1999) Site-specific migration and neuronal differentiation of human neural progenitor cells after transplantation in the adult rat brain. *J. Neurosci.* 19:5990–6005.

Frisen, J., Johansson, C. B., Lothian, C., and Lendahl, U. (1998) Central nervous system stem cells in the embryo and adult. *Cell. Mol. Life Sci.* 54:935–945.

Gage, F. H. (2000) Mammalian neural stem cells. *Science* 287:1433–1439.

Galli, R., Borello, U., Gritti, A., et al. (2000) Skeletal myogenic potential of human and mouse neural stem cells. *Nat. Neurosci.* 3:986–991.

Gearhart, J. (1998) New potential for human embryonic stem cells. *Science* 282:1061–1062.

Geiger, H., Sick, S., Bonifer, C., and Muller, A. M. (1998) Globin gene expression is reprogrammed in chimeras generated by injecting adult hematopoietic stem cells into mouse blastocysts. *Cell* 93:1055–1065.

Geschwind, D. H., Ou, J., Easterday, M. C., et al. (2001) A genetic analysis of neural progenitor differentiation. *Neuron.* 29:325–339.

Gould, E. and Tanapat, P. (1999) Stress and hippocampal neurogenesis. *Biol. Psychiatry* 46:1472–1479.

Gould, E., McEwen, B. S., Tanapat, P., Galea, L. A., and Fuchs, E. (1997) Neurogenesis in the dentate gyrus of the adult tree shrew is regulated by psychosocial stress and NMDA receptor activation. *J. Neurosci.* 17:2492–2498.

Gould, E., Tanapat, P., McEwen, B. S., Flugge, G., and Fuchs, E. (1998) Proliferation of granule cell precursors in the dentate gyrus of adult monkeys is diminished by stress. *Proc. Natl. Acad. Sci. USA* 95:3168–3171.

Gould, E., Beylin, A., Tanapat, P., Reeves, A., and Shors, T. J. (1999a) Learning enhances adult neurogenesis in the hippocampal formation. *Nat. Neurosci.* 2:260–265.

Gould, E., Reeves, A. J., Graziano, M. S., and Gross, C. G. (1999b) Neurogenesis in the neocortex of adult primates. *Science* 286:548–552.

Gould, E., Vail, N., Wagers, M., and Gross, C. G. (2001) Adult-generated hippocampal and neocortical neurons in macaques have a transient existence. *Proc. Natl. Acad. Sci. USA* 98:10,910–10,917.

Guo, X., Ying, W., Wan, J., et al. (2001) Proteomic characterization of early-stage differentiation of mouse embryonic stem cells into neural cells induced by all-trans retinoic acid in vitro. *Electrophoresis* 22: 3067–3075.

Ha, Y., Choi, J. U., Yoon, D. H., et al. (2001) Neural phenotype expression of cultured human cord blood cells in vitro. *Neuroreport* 12:3523–3527.

Hartfuss, E., Galli, R., Heins, N., and Gotz, M. (2001) Characterization of CNS precursor subtypes and radial glia. *Dev. Biol.* 229:15–30.

Hemmi, H., Okuyama, H., Yamane, T., et al. (2001) Temporal and spatial localization of osteoclasts in colonies from embryonic stem cells. *Biochem. Biophys. Res. Commun.* 280:526–534.

Herrera, J., Yang, H., Zhang, S. C., et al. (2001) Embryonic-derived glial-restricted precursor cells (GRP cells) can differentiate into astrocytes and oligodendrocytes in vivo. *Exp. Neurol.* 171:11–21.

Jacks, T. (1996) Tumor suppressor gene mutations in mice. *Annu. Rev. Genet.* 30:603–636.

Johansson, C. B., Momma, S., Clarke, D. L., Risling, M., Lendahl, U., and Frisen, J. (1999a) Identification of a neural stem cell in the adult mammalian central nervous system. *Cell* 96:25–34.

Johansson, C. B., Svensson, M., Wallstedt, L., Janson, A. M., and Frisen, J. (1999b) Neural stem cells in the adult human brain. *Exp. Cell. Res.* 253:733–736.

Jones-Villeneuve, E. M., McBurney, M. W., Rogers, K. A., and Kalnins, V. I. (1982) Retinoic acid induces embryonal carcinoma cells to differentiate into neurons and glial cells. *J. Cell. Biol.* 94:253–262.

Kalyani, A., Hobson, K., and Rao, M. S. (1997) Neuroepithelial stem cells from the embryonic spinal cord: isolation, characterization, and clonal analysis. *Dev. Biol.* 186:202–223.

Kalyani, A. J. and Rao, M. S. (1998) Cell lineage in the developing neural tube. *Biochem. Cell. Biol.* 76:1051–1068.

Kaufman, D. S., Hanson, E. T., Lewis, R. L., Auerbach, R., and Thomson, J. A. (2001) Hematopoietic colony-forming cells derived from human embryonic stem cells. *Proc. Natl. Acad. Sci. USA* 98:10,716–10,721.

Kawasaki, H., Mizuseki, K., Nishikawa, S., et al. (2000) Induction of midbrain dopaminergic neurons from ES cells by stromal cell-derived inducing activity. *Neuron* 28:31–40.

Kehat, I., Kenyagin-Karsenti, D., Snir, M., et al. (2001) Human embryonic stem cells can differentiate into myocytes with structural and functional properties of cardiomyocytes. *J. Clin. Invest.* 108:407–414.

Keirstead, H. S., Ben-Hur, T., Rogister, B., O'Leary, M. T., Dubois-Dalcq, M., and Blakemore, W. F. (1999) Polysialylated neural cell adhesion molecule-positive CNS precursors generate both oligodendrocytes and Schwann cells to remyelinate the CNS after transplantation. *J. Neurosci.* 19:7529–7536.

Keller, G., Kennedy, M., Papayannopoulou, T., and Wiles, M. V. (1993) Hematopoietic commitment during embryonic stem cell differentiation in culture. *Mol. Cell. Biol.* 13:473–486.

Kelly, D. L., and Rizzino, A. (2000) DNA microarray analyses of genes regulated during the differentiation of embryonic stem cells. *Mol. Reprod. Dev.* 56:113–123.

Kondo, T. and Raff, M. (2000) Oligodendrocyte precursor cells reprogrammed to become multipotential CNS stem cells. *Science* 289: 1754–1757.

Kopen, G. C., Prockop, D. J., and Phinney, D. G. (1999) Marrow stromal cells migrate throughout forebrain and cerebellum, and they differentiate into astrocytes after injection into neonatal mouse brains. *Proc. Natl. Acad. Sci. USA* 96:10,711–10,716.

Kuhn, H. G., Winkler, J., Kempermann, G., Thal, L. J., and Gage, F. H. (1997) Epidermal growth factor and fibroblast growth factor-2 have different effects on neural progenitors in the adult rat brain. *J. Neurosci.* 17:5820–5829.

Lacorazza, H. D., Flax, J. D., Snyder, E. Y., and Jendoubi, M. (1996) Expression of human beta-hexosaminidase alpha-subunit gene (the gene defect of Tay-Sachs disease) in mouse brains upon engraftment of transduced progenitor cells. *Nat. Med.* 2:424–429.

Lee, J. C., Mayer-Proschel, M., and Rao, M. S. (2000) Gliogenesis in the central nervous system. *Glia* 30:105–121.

Levine, J. M. and Flynn, P. (1986) Cell surface changes accompanying the neural differentiation of an embryonal carcinoma cell line. *J. Neurosci.* 6:3374–3384.

Li, M., Pevny, L., Lovell-Badge, R., and Smith, A. (1998) Generation of purified neural precursors from embryonic stem cells by lineage selection. *Curr. Biol.* 8:971–974.

Liu, S., Qu, Y., Stewart, T. J., et al. (2000) Embryonic stem cells differentiate into oligodendrocytes and myelinate in culture and after spinal cord transplantation. *Proc. Natl. Acad. Sci. USA* 97:6126–6131.

Lopez-Lozano, J. J., Bravo, G., Brera, B., et al. (1995) Long-term follow-up in 10 Parkinson's disease patients subjected to fetal brain grafting into a cavity in the caudate nucleus: the Clinica Puerta de Hierro experience. CPH Neural Transplantation Group. *Transplant. Proc.* 27:1395–1400.

Lopez-Lozano, J. J., Bravo, G., Brera, B., et al. (1997) Long-term improvement in patients with severe Parkinson's disease after implantation of fetal ventral mesencephalic tissue in a cavity of the caudate nucleus: 5-year follow up in 10 patients. Clinica Puerta de Hierro Neural Transplantation Group. *J. Neurosurg.* 86:931–942.

Lovell-Badge, R. (2001) The future for stem cell research. *Nature* 414: 88–91.

Lumelsky, N., Blondel, O., Laeng, P., Velasco, I., Ravin, R., and McKay, R. (2001) Differentiation of embryonic stem cells to insulin-secreting structures similar to pancreatic islets. *Science* 292:1389–1394.

MacPherson, P. A., Jones, S., Pawson, P. A., Marshall, K. C., and McBurney, M. W. (1997) P19 cells differentiate into glutamatergic and glutamate-responsive neurons in vitro. *Neuroscience* 80:487–499.

Magli, M. C., Levantini, E., and Giorgetti, A. (2000) Developmental potential of somatic stem cells in mammalian adults. *J. Hematother. Stem Cell Res.* 9:961–969.

Magnuson, D. S., Morassutti, D. J., Staines, W. A., McBurney, M. W., and Marshall, K. C. (1995a) In vivo electrophysiological maturation of neurons derived from a multipotent precursor (embryonal carcinoma) cell line. *Brain Res. Dev. Brain Res.* 84:130–141.

Magnuson, D. S., Morassutti, D. J., McBurney, M. W., and Marshall, K. C. (1995b) Neurons derived from P19 embryonal carcinoma cells develop responses to excitatory and inhibitory neurotransmitters. *Brain Res. Dev. Brain Res.* 90:141–150.

Mansergh, F. C., Wride, M. A., and Rancourt, D. E. (2000) Neurons from stem cells: implications for understanding nervous system development and repair. *Biochem. Cell. Biol.* 78:613–628.

Martin, G. R. (1981) Isolation of a pluripotent cell line from early mouse embryos cultured in medium conditioned by teratocarcinoma stem cells. *Proc. Natl. Acad. Sci. USA* 78:7634–7638.

Mayer-Proschel, M., Kalyani, A. J., Mujtaba, T., and Rao, M. S. (1997) Isolation of lineage-restricted neuronal precursors from multipotent neuroepithelial stem cells. *Neuron* 19:773–785.

McBurney, M. W. (1993) P19 embryonal carcinoma cells. *Int. J. Dev. Biol.* 37:135–140.

McBurney, M. W., Jones-Villeneuve, E. M., Edwards, M. K., and Anderson, P. J. (1982) Control of muscle and neuronal differentiation in a cultured embryonal carcinoma cell line. *Nature* 299:165–167.

McBurney, M. W., Reuhl, K. R., Ally, A. I., Nasipuri, S., Bell, J. C., and Craig, J. (1988) Differentiation and maturation of embryonal carcinoma-derived neurons in cell culture. *J. Neurosci.* 8:1063–1073.

McDonald, J. W., Liu, X. Z., Qu, Y., et al. (1999) Transplanted embryonic stem cells survive, differentiate and promote recovery in injured rat spinal cord. *Nat. Med.* 5:1410–1412.

Minguell, J. J., Erices, A., and Conget, P. (2001) Mesenchymal stem cells. *Exp. Biol. Med. (Maywood)* 226:507–520.

Mujtaba, T., Mayer-Proschel, M., and Rao, M. S. (1998) A common neural progenitor for the CNS and PNS. *Dev. Biol.* 200:1–15.

Mujtaba, T., Piper, D. R., Kalyani, A., Groves, A. K., Lucero, M. T., and Rao, M. S. (1999) Lineage-restricted neural precursors can be isolated from both the mouse neural tube and cultured ES cells. *Dev. Biol.* 214:113–127.

Munsie, M. J., Michalska, A. E., O'Brien, C. M., Trounson, A. O., Pera, M. F., and Mountford, P. S. (2000) Isolation of pluripotent embryonic stem cells from reprogrammed adult mouse somatic cell nuclei. *Curr. Biol.* 10:989–992.

Nagai, A., Nakagawa, E., Choi, H. B., Hatori, K., Kobayashi, S., and Kim, S. U. (2001) Erythropoietin and erythropoietin receptors in human CNS neurons, astrocytes, microglia, and oligodendrocytes grown in culture. *J. Neuropathol. Exp. Neurol.* 60:386–392.

Nakano, T., Kodama, H., and Honjo, T. (1994) Generation of lymphohematopoietic cells from embryonic stem cells in culture. *Science* 265:1098–1101.

Nakayama, N., Fang, I., and Elliott, G. (1998) Natural killer and B-lymphoid potential in CD34+ cells derived from embryonic stem cells differentiated in the presence of vascular endothelial growth factor. *Blood* 91:2283–2295.

Odorico, J. S., Kaufman, D. S., and Thomson, J. A. (2001) Multilineage differentiation from human embryonic stem cell lines. *Stem Cells* 19: 193–204.

Okabe, S., Forsberg-Nilsson, K., Spiro, A. C., Segal, M., and McKay, R. D. (1996) Development of neuronal precursor cells and functional postmitotic neurons from embryonic stem cells in vitro. *Mech. Dev.* 59:89–102.

Okawa, H., Okuda, O., Arai, H., Sakuragawa, N., and Sato, K. (2001) Amniotic epithelial cells transform into neuron-like cells in the ischemic brain. *Neuroreport* 12:4003–4007.

Ostenfeld, T., Caldwell, M. A., Prowse, K. R., Linskens, M. H., Jauniaux, E., and Svendsen, C. N. (2000) Human neural precursor cells express low levels of telomerase in vitro and show diminishing cell proliferation with extensive axonal outgrowth following transplantation. *Exp. Neurol.* 164:215–226.

Ourednik, V., Ourednik, J., Flax, J. D., et al. (2001) Segregation of human neural stem cells in the developing primate forebrain. *Science* 293: 1820–1824.

Palmer, T. D., Markakis, E. A., Willhoite, A. R., Safar, F., and Gage, F. H. (1999) Fibroblast growth factor-2 activates a latent neurogenic program in neural stem cells from diverse regions of the adult CNS. *J. Neurosci.* 19:8487–8497.

Parati, E. A., Bez, A., Ponti, D., et al. (2002) Human neural stem cells express extra-neural markers. *Brain Res.* 925:213–221.

Pera, M. F. (2001) Scientific considerations relating to the ethics of the use of human embryonic stem cells in research and medicine. *Reprod. Fertil. Dev.* 13:23–29.

Perlingeiro, R. C., Kyba, M., and Daley, G. Q. (2001) Clonal analysis of differentiating embryonic stem cells reveals a hematopoietic progenitor with primitive erythroid and adult lymphoid-myeloid potential. *Development* 128:4597–4604.

Petersen, B. E., Bowen, W. C., Patrene, K. D., et al. (1999) Bone marrow as a potential source of hepatic oval cells. *Science* 284:1168–1170.

Phillips, B. W., Belmonte, N., Vernochet, C., Ailhaud, G., and Dani, C. (2001) Compactin enhances osteogenesis in murine embryonic stem cells. *Biochem. Biophys. Res. Commun.* 284:478–484.

Pittenger, M. F., Mackay, A. M., Beck, S. C., et al. (1999) Multilineage potential of adult human mesenchymal stem cells. *Science* 284:143–147.

Prelle, K., Wobus, A. M., Krebs, O., Blum, W. F., and Wolf, E. (2000) Overexpression of insulin-like growth factor-II in mouse embryonic stem cells promotes myogenic differentiation. *Biochem. Biophys. Res. Commun.* 277:631–638.

Pringle, N. P., Guthrie, S., Lumsden, A., and Richardson, W. D. (1998) Dorsal spinal cord neuroepithelium generates astrocytes but not oligodendrocytes. *Neuron* 20:883–893.

Qu, T., Brannen, C. L., Kim, H. M., and Sugaya, K. (2001) Human neural stem cells improve cognitive function of aged brain. *Neuroreport* 12: 1127–1132.

Rao, M. S. (1999) Multipotent and restricted precursors in the central nervous system. *Anat. Rec.* 257:137–148.

Rao, M. S., and Anderson, D. J. (1997) Immortalization and controlled in vitro differentiation of murine multipotent neural crest stem cells. *J. Neurobiol.* 32:722–746.

Rao, M. S. and Mattson, M. P. (2001) Stem cells and aging: expanding the possibilities. *Mech. Ageing Dev.* 122:713–734.

Rao, M. S., Noble, M., and Mayer-Proschel, M. (1998) A tripotential glial precursor cell is present in the developing spinal cord. *Proc. Natl. Acad. Sci. USA* 95:3996–4001.

Renoncourt, Y., Carroll, P., Filippi, P., Arce, V., and Alonso, S. (1998) Neurons derived in vitro from ES cells express homeoproteins characteristic of motoneurons and interneurons. *Mech. Dev.* 79:185–197.

Represa, A., Shimazaki, T., Simmonds, M., and Weiss, S. (2001) EGF-responsive neural stem cells are a transient population in the developing mouse spinal cord. *Eur. J. Neurosci.* 14:452–462.

Reubinoff, B. E., Pera, M. F., Fong, C. Y., Trounson, A., and Bongso, A. (2000) Embryonic stem cell lines from human blastocysts: somatic differentiation in vitro. *Nat. Biotechnol.* 18:399–404.

Reynolds, B. A. and Weiss, S. (1992) Generation of neurons and astrocytes from isolated cells of the adult mammalian central nervous system. *Science* 255:1707–1710.

Reynolds, B. A. and Weiss, S. (1996) Clonal and population analyses demonstrate that an EGF-responsive mammalian embryonic CNS precursor is a stem cell. *Dev. Biol.* 175:1–13.

Reynolds, B. A., Tetzlaff, W., and Weiss, S. (1992) A multipotent EGF-responsive striatal embryonic progenitor cell produces neurons and astrocytes. *J. Neurosci.* 12:4565–4574.

Richards, L. J., Kilpatrick, T. J., and Bartlett, P. F. (1992) De novo generation of neuronal cells from the adult mouse brain. *Proc. Natl. Acad. Sci. USA* 89:8591–8595.

Rideout, W. M. 3rd, Wakayama, T., Wutz, A., Eggan, K., Jackson-Grusby, L., Dausman, J., Yanagimachi, R., and Jaenisch, R. (2000) Generation of mice from wild-type and targeted ES cells by nuclear cloning. *Nat. Genet.* 24:109–110.

Risau, W., Sariola, H., Zerwes, H. G., Sasse, J., Ekblom, P., Kemler, R., and Doetschman, T. (1988) Vasculogenesis and angiogenesis in embryonic-stem-cell-derived embryoid bodies. *Development* 102: 471–478.

Robertson, E. J. (1987) In: *Teratocarcinomas and Embryonic Stem Cells: A Practical Appraoch* (Robertosn, E. J., ed.), IRL, Oxford.

Rosario, C. M., Yandava, B. D., Kosaras, B., Zurakowski, D., Sidman, R. L., and Snyder, E. Y. (1997) Differentiation of engrafted multipotent neural progenitors towards replacement of missing granule neurons in meander tail cerebellum may help determine the locus of mutant gene action. *Development* 124:4213–4224.

Rubio, F. J., Bueno, C., Villa, A., Navarro, B., and Martinez-Serrano, A. (2000) Genetically perpetuated human neural stem cells engraft and differentiate into the adult mammalian brain. *Mol. Cell. Neurosci.* 16:1–13.

Sanchez-Ramos, J. R., Song, S., Kamath, S. G., Zigova, T., Willing, A., Cardozo-Pelaez, F., Stedeford, T., Chopp, M., and Sanberg, P. R. (2001) Expression of neural markers in human umbilical cord blood. *Exp. Neurol.* 171:109–115.

Schmidt-Kastner, P. K., Jardine, K., Cormier, M., and McBurney, M. W. (1996) Genes transfected into embryonal carcinoma stem cells are both lost and inactivated at high frequency. *Somat. Cell. Mol. Genet.* 22:383–392.

Schmidt-Kastner, P. K., Jardine, K., Cormier, M., and McBurney, M. W. (1998) Absence of p53-dependent cell cycle regulation in pluripotent mouse cell lines. *Oncogene* 16:3003–3011.

Schuldiner, M., Eiges, R., Eden, A., et al. (2001) Induced neuronal differentiation of human embryonic stem cells. *Brain Res.* 913:201–205.

Seidman, K. J., Teng, A. L., Rosenkopf, R., Spilotro, P., and Weyhenmeyer, J. A. (1997) Isolation, cloning and characterization of a putative type-1 astrocyte cell line. *Brain Res.* 753:18–26.

Shah, N. M., and Anderson, D. J. (1997) Integration of multiple instructive cues by neural crest stem cells reveals cell-intrinsic biases in relative growth factor responsiveness. *Proc. Natl. Acad. Sci. USA* 94: 11,369–11,374.

Shamblott, M. J., Axelman, J., Wang, S., et al. (1998) Derivation of pluripotent stem cells from cultured human primordial germ cells. *Proc. Natl. Acad. Sci. USA* 95:13,726–13,731.

Shamblott, M. J., Axelman, J., Littlefield, J. W., et al. (2001) Human embryonic germ cell derivatives express a broad range of developmentally distinct markers and proliferate extensively in vitro. *Proc. Natl. Acad. Sci. USA* 98:113–118.

Shih, C. C., Weng, Y., Mamelak, A., LeBon, T., Hu, M. C., and Forman, S. J. (2001) Identification of a candidate human neurohematopoietic stem-cell population. *Blood* 98:2412–2422.

Shimazaki, T., Shingo, T., and Weiss, S. (2001) The ciliary neurotrophic factor/leukemia inhibitory factor/gp130 receptor complex operates in the maintenance of mammalian forebrain neural stem cells. *J. Neurosci.* 21:7642–7653.

Shors, T. J., Miesegaes, G., Beylin, A., Zhao, M., Rydel, T., and Gould, E. (2001) Neurogenesis in the adult is involved in the formation of trace memories. *Nature* 410:372–376.

Singh, G. (2001) Sources of neuronal material for implantation. *Neuropathology* 21:110–114.

Snyder, E. Y. and Macklis, J. D. (1995) Multipotent neural progenitor or stem-like cells may be uniquely suited for therapy for some neurodegenerative conditions. *Clin. Neurosci.* 3:310–316.

Snyder, E. Y., Yoon, C., Flax, J. D., and Macklis, J. D. (1997a) Multipotent neural precursors can differentiate toward replacement of neurons undergoing targeted apoptotic degeneration in adult mouse neocortex. *Proc. Natl. Acad. Sci. USA* 94:11,663–11,668.

Snyder, E. Y., Park, K. I., Flax, J. D., et al. (1997b) Potential of neural "stem-like" cells for gene therapy and repair of the degenerating central nervous system. *Adv. Neurol.* 72:121–132.

Solter, D. and Gearhart, J. (1999) Putting stem cells to work. *Science* 283: 1468–1470.

Spencer, D. D., Robbins, R. J., Naftolin, F., et al. (1992) Unilateral transplantation of human fetal mesencephalic tissue into the caudate nucleus of patients with Parkinson's disease. *N. Engl. J. Med.* 327: 1541–1548.

Staines, W. A., Morassutti, D. J., Reuhl, K. R., Ally, A. I., and McBurney, M. W. (1994) Neurons derived from P19 embryonal carcinoma cells have varied morphologies and neurotransmitters. *Neuroscience* 58: 735–751.

Staines, W. A., Craig, J., Reuhl, K., and McBurney, M. W. (1996) Retinoic acid treated P19 embryonal carcinoma cells differentiate into oligodendrocytes capable of myelination. *Neuroscience* 71:845–853.

Stemple, D. L. and Anderson, D. J. (1992) Isolation of a stem cell for neurons and glia from the mammalian neural crest. *Cell* 71:973–985.

Storch, A., Paul, G., Csete, M., et al. (2001) Long-term proliferation and dopaminergic differentiation of human mesencephalic neural precursor cells. *Exp. Neurol.* 170:317–325.

Strubing, C., Ahnert-Hilger, G., Shan, J., Wiedenmann, B., Hescheler, J., and Wobus, A. M. (1995) Differentiation of pluripotent embryonic stem cells into the neuronal lineage in vitro gives rise to mature inhibitory and excitatory neurons. *Mech. Dev.* 53:275–287.

Suemori, H., Tada, T., Torii, R., et al. (2001) Establishment of embryonic stem cell lines from cynomolgus monkey blastocysts produced by IVF or ICSI. *Dev. Dyn.* 222:273–279.

Temple, S. (2001) The development of neural stem cells. *Nature* 414, 112–117.

Terskikh, A. V., Easterday, M. C., Li, L., et al. (2001) From hematopoiesis to neuropoiesis: evidence of overlapping genetic programs. *Proc. Natl. Acad. Sci. USA* 98:7934–7939.

Thomson, J. A., Itskovitz-Eldor, J., Shapiro, S. S., et al. (1998a) Embryonic stem cell lines derived from human blastocysts. *Science* 282: 1145–1147.

Thomson, J. A., Marshall, V. S., and Trojanowski, J. Q. (1998b) Neural differentiation of rhesus embryonic stem cells. *APMIS* 106:149–157.

Toma, J. G., Akhavan, M., Fernandes, K. J., et al. (2001) Isolation of multipotent adult stem cells from the dermis of mammalian skin. *Nat. Cell. Biol.* 3:778–784.

Tropepe, V., Hitoshi, S., Sirard, C., Mak, T. W., Rossant, J., and van der Kooy, D. (2001) Direct neural fate specification from embryonic stem cells: a primitive mammalian neural stem cell stage acquired through a default mechanism. *Neuron* 30:65–78.

Uchida, N., Buck, D. W., He, D., et al. (2000) Direct isolation of human central nervous system stem cells. *Proc. Natl. Acad. Sci. USA* 97: 14,720–14,725.

van Praag, H., Christie, B. R., Sejnowski, T. J., and Gage, F. H. (1999a) Running enhances neurogenesis, learning, and long-term potentiation in mice. *Proc. Natl. Acad. Sci. USA* 96:13,427–13,431.

van Praag, H., Kempermann, G., and Gage, F. H. (1999b) Running increases cell proliferation and neurogenesis in the adult mouse dentate gyrus. *Nat. Neurosci.* 2:266–270.

Vescovi, A. L., and Snyder, E. Y. (1999) Establishment and properties of neural stem cell clones: plasticity in vitro and in vivo. *Brain Pathol.* 9:569–598.

Vescovi, A. L., Parati, E. A., Gritti, A., et al. (1999a) Isolation and cloning of multipotential stem cells from the embryonic human CNS and establishment of transplantable human neural stem cell lines by epigenetic stimulation. *Exp. Neurol.* 156:71–83.

Vescovi, A. L., Gritti, A., Galli, R., and Parati, E. A. (1999b) Isolation and intracerebral grafting of nontransformed multipotential embryonic human CNS stem cells. *J. Neurotrauma* 16:689–693.

Wagner, J. P., Black, I. B., and DiCicco-Bloom, E. (1999) Stimulation of neonatal and adult brain neurogenesis by subcutaneous injection of basic fibroblast growth factor. *J. Neurosci.* 19:6006–6016.

Wakayama, T., Rodriguez, I., Perry, A. C., Yanagimachi, R., and Mombaerts, P. (1999) Mice cloned from embryonic stem cells. *Proc. Natl. Acad. Sci. USA* 96:14,984–14,989.

Watt, F. M. and Hogan, B. L. (2000) Out of Eden: stem cells and their niches. *Science* 287:1427–1430.

Weiss, S. (1999) Pathways for neural stem cell biology and repair. *Nat. Biotechnol.* 17:850–851.

Weiss, S. and van der Kooy, D. (1998) CNS stem cells: where's the biology (a.k.a. beef)? *J. Neurobiol.* 36:307–314.

Weiss, S., Dunne, C., Hewson, J., et al. (1996) Multipotent CNS stem cells are present in the adult mammalian spinal cord and ventricular neuroaxis. *J. Neurosci.* 16:7599–7609.

Westmoreland, J. J., Hancock, C. R., and Condie, B. G. (2001) Neuronal development of embryonic stem cells: a model of GABAergic neuron differentiation. *Biochem. Biophys. Res. Commun.* 284:674–680.

White, P. M., Morrison, S. J., Orimoto, K., Kubu, C. J., Verdi, J. M., and Anderson, D. J. (2001) Neural crest stem cells undergo cell-intrinsic developmental changes in sensitivity to instructive differentiation signals. *Neuron* 29:57–71.

Whittemore, S. R., Morassutti, D. J., Walters, W. M., Liu, R. H., and Magnuson, D. S. (1999) Mitogen and substrate differentially affect the lineage restriction of adult rat subventricular zone neural precursor cell populations. *Exp. Cell. Res.* 252:75–95.

Widner, H., Tetrud, J., Rehncrona, S., et al. (1992) Bilateral fetal mesencephalic grafting in two patients with parkinsonism induced by 1-methyl-4-phenyl-1,2,3,6-tetrahydropyridine (MPTP). *N. Engl. J. Med.* 327:1556–1563.

Wohl, C. A. and Weiss, S. (1998) Retinoic acid enhances neuronal proliferation and astroglial differentiation in cultures of CNS stem cell–derived precursors. *J. Neurobiol.* 37:281–290.

Xian, H. Q. and Gottlieb, D. I. (2001) Peering into early neurogenesis with embryonic stem cells. *Trends Neurosci.* 24:685–686.

Yandava, B. D., Billinghurst, L. L., and Snyder, E. Y. (1999) "Global" cell replacement is feasible via neural stem cell transplantation: evidence from the dysmyelinated shiverer mouse brain. *Proc. Natl. Acad. Sci. USA* 96:7029-34.

Yang, H., Mujtaba, T., Venkatraman, G., Wu, Y. Y., Rao, M. S., and Luskin, M. B. (2000) Region-specific differentiation of neural tube-derived neuronal restricted progenitor cells after heterotopic transplantation. *Proc. Natl. Acad. Sci. USA* 97:13,366–13,371.

Zhang, S. C., Ge, B., and Duncan, I. D. (1999) Adult brain retains the potential to generate oligodendroglial progenitors with extensive myelination capacity. *Proc. Natl. Acad. Sci. USA* 96:4089–4094.

17 Neural Stem Cells

From In Vivo to In Vitro and Back Again—Practical Aspects

MICHAEL A. MARCONI, BS, KOOK I. PARK, MD, DMSc,
YANG D. TENG, MD, PhD, JITKA OUREDNIK, PhD,
VACLAV OUREDNIK, PhD, ROSANNE M. TAYLOR, DMV, PhD,
ALEKSANDRA E. MARCINIAK, BS, MARCEL M. DAADI, PhD,
HEATHER L. ROSE, MA, ERIN B. LAVIK, PhD, ROBERT LANGER, ScD,
KURTIS I. AUGUSTE, MD, MAHESH LACHYANKAR, PhD,
CURT R. FREED, MD, D. EUGENE REDMOND, MD,
RICHARD L. SIDMAN, MD, AND EVAN Y. SNYDER, MD, PhD

The application of neural stem cell transplantation for cellular repair of lesions of the brain and spinal cord appears to have much greater promise than bone marrow transplantation, viral-mediated gene therapy, or systemic enzyme replacement. The initial approach for treatment of Parkinson's disease by transfer of dopamine-producing cells provides a prototype for cell transplantation therapy that can be extended to the use of multipotent neural stem cells, which not only have the ability to self-renew and to differentiate into cells of all glial and neuronal lineages, but also can migrate to areas of CNS disease or injury. How transplanted cells and injured/diseased brain communicate with each other in what appears to be directed migration and differentiation is a subject of current investigation. The abililty to isolate and culture cells in vitro that have the migration and differentiation properties of neural stem cells is a major advance in obtaining cells for transplantation therapy. Such cells are relative easy to inject into the ventricles, migrate across the blood–brain barrier, and integrate as different CNS cell type into damaged brain, yet do not give rise to inappropriate cell type or neoplasms. In some proof-of-principle experimental models, transplanted neural stem cells have been used to treat a mouse model of neurogenic lysosomal storage disease, mutant mice with congenital anatomic abnormalities, myelination disorders, hypoxic–ischemic injury, amyloid plaques, and brain tumors by delivery of oncolysis-promoting cytosine deaminase.

17.1. INTRODUCTION

"Think globally and act locally." While this phrase may have been coined for environmental issues, we employ it here to describe the effectiveness of some therapies for neurological diseases. The pathological lesions of many neurological disorders are often globally dispersed throughout the brain and spinal cord. Such diseases include not only the inherited neurodegenerative diseases of the pediatric age group (e.g., lysosomal storage diseases, leukodystrophies, inborn errors of metabolism, hypoxic–ischemic encephalopathy), but also such adult maladies as Alzheimer's disease (AD), Huntington's disease, multiinfarct dementia, multiple sclerosis, amyotrophic lateral sclerosis, and brain tumors (especially glioblastomas). These "global" problems have typically been treated by attempting to deliver gene products via viral vectors, pumps, or even grafts of fetal tissue, but, as discussed below, these treatments often have only "local" effects. Conversely, attempts to provide a somewhat more "global" effect through the systemic administration of pharmacological agents or growth factors are often incapable of sufficiently restricting their influence to just the nervous system or to the neural cell types of interest and, hence, are accompanied by unacceptable side effects. Thus, these latter treatments may be "too global."

Cell-based therapies provide another avenue with which to pursue treatments of central nervous system (CNS) disorders. Until recently, therapies such as neural transplantation have been reserved for more regionally restricted neurological diseases than those mentioned. Parkinson's disease fits this description (Kordower et al., 1995; Nikkhah et al., 1995a, 1995b; Freed et al., 2001), and treatment regimens have been designed that centered on the engraftment and enhanced survival of dopamine-producing cells within the striatum, or the delayed degeneration of dopaminergic neurons within the substantia nigra (SN). However, with the recognition that neural progenitor or neural stem cells (NSCs) might integrate appropriately *throughout* the mam-

From: *Stem Cells Handbook*
Edited by: S. Sell © Humana Press Inc., Totowa, NJ

malian CNS following transplantation (e.g., Snyder et al., 1995; Brüstle et al., 1998; Flax et al., 1998; Yandava et al., 1999; Ourednik et al., 2001a), a new road for neural transplantation and gene therapy in treating global manifestations of CNS diseases has been opened for exploration.

Multipotent NSCs are operationally defined by their ability to self-renew, to differentiate into cells of all glial and neuronal lineages throughout the neuraxis, and to populate developing or degenerating CNS regions (reviewed in Snyder and Flax, 1995; Fisher, 1997; McKay, 1997; Alvarez-Buylla and Temple, 1998; Vescovi and Snyder, 1999; Gage, 2000). Thus, their use as graft material can be considered analogous to hematopoietic stem cell–mediated reconstitution and gene transfer. Studies have demonstrated that NSCs are capable of engrafting in a cytoarchitecturally and developmentally appropriate manner within the normal or abnormal CNS following neural transplantation. Furthermore, these cells are capable of migrating to areas of the CNS affected by disease or injury; in other words, they have the ability to "home in" on pathology, even over great distances. With their relative ease of propagation and manipulation in vitro, NSCs may represent a readily available and replenishable source for cell-based therapies for the aforementioned conditions.

How NSCs perform these functions is the current focus of intensive study, and models of NSC–host tissue interactions are constantly evolving. The phenomenon of the host CNS environment instructing the fate of transplanted NSCs has been observed. This has been evidenced by NSCs migrating to areas of pathology and assuming the phenotype(s) of host cells that are affected by disease or injury. However, a model has been emerging in which "communication" between the injured host and the transplanted NSC is not one-way but, rather, a genuine dialogue: the transplanted NSC appears equally capable of instructing and altering the fate of *host* CNS tissue. In some animal models of degeneration and injury that have received NSC transplantation, we have noted "reconstitution" of CNS regions or systems targeted by the insult. On further investigation, many of the cells participating in this "recovery" have actually been found to be of host, not donor, origin, a phenomenon never seen in untransplanted animals. In other words, the donor NSCs are interacting with host tissue in a regenerative or restorative manner; in some cases, the new host neural processes seem to be elaborated and in others preexisting host cells seem to be rescued (as opposed to promoting new host neurogenesis). Illuminating the mechanisms behind these phenomena is now a major focus of our efforts.

17.2. WHY SOME CURRENT THERAPIES FOR CNS DISORDERS MAY NOT BE ENOUGH

Current therapies for diseases of the CNS include bone marrow transplantation (BMT) and systemic enzyme replacement, viral delivery of gene products, grafting of synthetic biocompatible materials and nonneural tissue, and neural transplantation. Each of these strategies has achieved some measure of success in treating various conditions; however, each has inherent limitations that we will enumerate here.

BMT or enzyme replacement has been employed to treat a number of heritable metabolic diseases, especially those in which a single gene product is deficient. These interventions have been successful in treating the peripheral manifestations of such diseases but have had disappointing results in ameliorating damage to the CNS in most cases. This is due to the yeoman work that the blood–brain barrier (BBB) performs in restricting entry of nonneural cells or novel molecules into the brain from the vasculature, even if such molecules have therapeutic value. (Umbilical cord stem cells may ultimately prove to be an exception to this rule, but understanding of their biology is still in its infancy.) Typically, BMT also involves irradiation and massive immunosuppression as a preconditioning regimen, with unfortunate deleterious effects on the developing CNS.

Given the difficulty of crossing the BBB, direct gene transfer to the CNS has been used to deliver genetic material that is either missing or damaged in the recipient. Several types of viral vectors have been engineered to deliver genes of interest. These include retrovirus, lentivirus, and adenoassociated virus. Although these vectors are quite effective in delivering genetic material to certain cell types or localized areas of tissue, they may not possess the complete arsenal needed to treat the whole condition. Retroviral vectors infect only mitotic cells, which are less abundant in the postdevelopmental CNS. Although lentiviral and adenoassociated virus vectors are capable of infecting postmitotic tissue, they often do not target the widespread lesions and multiple cell types that are characteristic of many neurogenetic diseases.

Since the ultimate goal of gene therapy is to supply a missing gene product, synthetic "pumps," or genetically modified donor cells, have been implanted into host CNS tissue to deliver exogenous factors to host cells. Genetically engineered nonneural cells (e.g., fibroblasts) can be used for delivery of discrete molecules to the CNS and can be implanted autologously (Chen and Gage, 1995; Tuszynski and Gage, 1995; Grill et al., 1997). However, the range of tissue positively affected by this approach is limited. In addition, these nonneural grafts are not incorporated into host tissue in a functional manner and therefore are not subject to the control of feedback loops or regulated release (inherent in neurally derived cells) so critical in biological systems. For some conditions and substances, unregulated release may actually be harmful to the host (e.g., nerve growth factor [NGF] [Chen and Gage, 1995] and dopamine [Freed et al., 2001]). In addition to these concerns, downregulation of the engineered neural genes in these nonneural cells may leave them ineffective in the host tissue.

Even if the concerns mentioned thus far were somehow reduced or eliminated, there is still another factor to consider regarding the host tissue in many neurogenetic diseases: there can be widespread, often progressive, degeneration of the neuronal and/or glial cell populations. This may be due to the presence of certain toxins or lack of trophic factors in the milieu, or to pathological processes intrinsic to the diseased cells. Whatever the cause, the neural substrate is damaged in these conditions; simply to add new genetic material or gene products to this environment might be akin to handing a new bucket to the captain of a sinking ship. The challenge is to provide permissive substrates (including creating new substrates) on which these therapeutic genes can operate.

Neurons derived from the CNS could be used as a graft material to help create new substrates in diseased tissue. Mature neurons are limited in this role because of their limited mitotic capacity. This limits their ability to be transduced by ex vivo viral vectors, and the logistics of expanding these cells into adequate numbers for grafting are quite difficult. Although some very impressive connections with host tissue have been demonstrated in some grafts of neurons that have matured from implanted fetal tissue (Bjorklund and Lindvall, 2000), their ability to form extensively new connections to and from all the necessary target and projection regions may be limited. Nevertheless, primary fetal neuronal tissue has historically been the most successful donor tissue for CNS grafting and has shown promise for the amelioration of certain neurological conditions (reviewed in Dunnett and Bjorklund, 2000). However, the use of fetal tissue involves significant concerns, including ethical considerations, the availability of sufficient amounts of suitable disease-free material, the need to ensure survival of desired cells in tissue that is typically heterogeneous and contains nonneural cells, the inability to augment the expression of biological molecules by genetically manipulating donor fetal tissue, and the restricted or very focal integration of the fetal graft into the host brain.

17.3. WHY NSCs MAY BE THERAPEUTICALLY USEFUL IN THE CNS

The ability to isolate NSCs, propagate them in culture, and successfully reimplant them into mammalian brain (Snyder et al., 1992) has highlighted their use as a potential vehicle for CNS gene therapy and repair. A number of studies over the past decade (reviewed in Vescovi and Snyder, 1999) have established that NSCs isolated from disparate regions and developmental stages could be maintained and perpetuated in vitro, both epigenetically and genetically. These methods include the transduction of NSCs with genes interacting with cell-cycle proteins (e.g., myc, telomerase) and by mitogen stimulation (e.g., epidermal growth factor [EGF] and/or basic fibroblast growth factor [bFGF]). Despite using varied culture and propagation methods, progenitor cell lines of different origins have largely behaved quite similarly in their ability to reintegrate into the CNS, suggesting the involvement of common downstream cellular mechanisms. Some of these NSC lines are sufficiently plastic to participate in normal CNS development from germinal zones of multiple regions along the neuraxis and at multiple stages of development from embryo to old age (Snyder et al., 1995, 1997; Suhonen et al., 1996, Flax et al., 1998; Fricker et al., 1999; Ourednik et al., 2001a; Reubinoff et al., 2001; Zhang et al., 2001). In addition, they appear to model the in vitro and in vivo behavior of some endogenous fetal and adult neural cells (Weiss et al., 1996; Park et al., 1998; Gould et al., 1999; Magavi et al., 2000).

NSCs display several properties that make them attractive as potential constituents in the treatment of metabolic, degenerative, or other widespread lesions in the brain. They are relatively easy to administer (often directly into the cerebral ventricles) and remain quite viable throughout any manipulation required prior to transplantation. Once transplanted, NSCs are capable of crossing the BBB and readily engraft into neural tissue. NSCs are apparently able to integrate in a cytoarchitecturally and functionally appropriate manner (Zlomanczuk et al., 2002) throughout many regions of the host brain as neurons; astrocytes; oligodendrocytes; and even undifferentiated, quiescent progenitors. The latter group represents a sort of "minutemen brigade," ready to migrate and differentiate as the need arises. Thus, a given NSC clone can give rise to multiple cell types within the same region. This is important in the likely situation in which return of function may require the reconstitution of the whole milieu of a given region—e.g., not just the neurons but also the glia and support cells required to nurture, detoxify, and/or myelinate the neurons. NSCs appear to respond in vivo to neurogenic signals not only when they occur appropriately during development, but even when induced at later stages by certain neurodegenerative processes, such as during apoptosis (Fig. 1) (Snyder et al., 1997; Doering and Snyder, 2000). NSCs may be attracted to regions of neurodegeneration in the young (Park et al., 1999) as well as in the adult (Aboody et al., 2000) and aged (Ourednik et al., 1999).

NSCs apparently have the intrinsic ability to know where to go, and what to become, in the diseased CNS. This reduces the necessity of obtaining donor cells of specific cell types from specific regions and of the precise targeting of the implant. NSCs are known to express many neurotrophic factors intrinsically, or they can be engineered ex vivo to deliver other agents as needed. Delivery of these factors to the host CNS can be done in a direct and stable manner (Snyder et al., 1995; Lacorazza et al., 1996; Yandava et al., 1999; Aboody et al., 2000). While NSCs can migrate and integrate widely throughout the brain particularly well when implanted into *germinal* zones, allowing reconstitution of enzyme or cellular deficiencies in a global manner (Snyder et al., 1995; Lacorazza et al., 1996; Yandava et al., 1999), this extensive migratory ability is present even in the *parenchyma* of the diseased adult (Snyder et al., 1997; Aboody et al., 2000) and aged (Ourednik et al., 1999) brain, conventionally regarded as nonsupportive of migration. Despite their extensive plasticity, NSCs never give rise to cell types inappropriate to the brain (e.g., muscle, bone, teeth) or yield neoplasms.

17.4. HOW NSCs STAND UP TO TESTS OF THERAPEUTIC POTENTIAL

The therapeutic potential of NSCs was first tested experimentally in mouse models of genetically based neurodegeneration. As proof of principle, a mouse model of a neurogenetic lysosomal storage disease was approached. The naturally occurring MPS VII mutant mouse, characterized by a frameshift mutation in the β-*glucuronidase* gene, emulates the neurodegenerative condition mucopolysaccharidosis type VII. This mutation leads to a deficiency in the secreted enzyme β-glucuronidase (GUSB), which in turn results in lysosomal accumulation of undegraded glycosaminoglycans in the brain and other tissues. Treatments for MPS VII and most other lysosomal storage diseases are designed to provide a source of normal enzyme for uptake by diseased cells, a process termed *cross-correction* (Neufeld and Fratantoni, 1970).

An intracerebroventricular injection technique was devised for the delivery of GUSB-expressing NSCs (Snyder et al., 1995). By injecting the NSCs into the cerebral ventricles, they were placed in close proximity to the subventricular germinal zone (SVZ), the home of endogenous progenitor cells and presumably a permissive environment for donor cells. The engraftment and widespread integration of NSCs throughout the newborn MPS VII mutant brain succeeded in providing a sustained, lifelong, widespread source of cross-correcting enzyme. While MPS VII may be regarded as "uncommon," the broad category of diseases that it models (neurogenetic conditions) afflicts as many as 1 in 1500 children and serves as a model for many adult neurodegenerative processes of genetic origin (e.g., AD). Even in the adult brain, there are routes of relatively extensive migration followed by both endogenous and transplanted NSCs (Luskin, 1993; Lois and Alvarez-Buylla, 1994; Goldman and Luskin, 1998; Gritti et al., 2002). If injected into the cerebral ventricles of normal adult mice, NSCs will integrate into the SVZ and migrate long distances, such as to the olfactory bulb, where they differentiate into interneurons, and occasionally into subcortical parenchyma, where they become glia (Suhonen et al., 1996; Flax et al., 1998; Snyder, 1998; Fricker et al., 1999; Uchida et al., 2000).

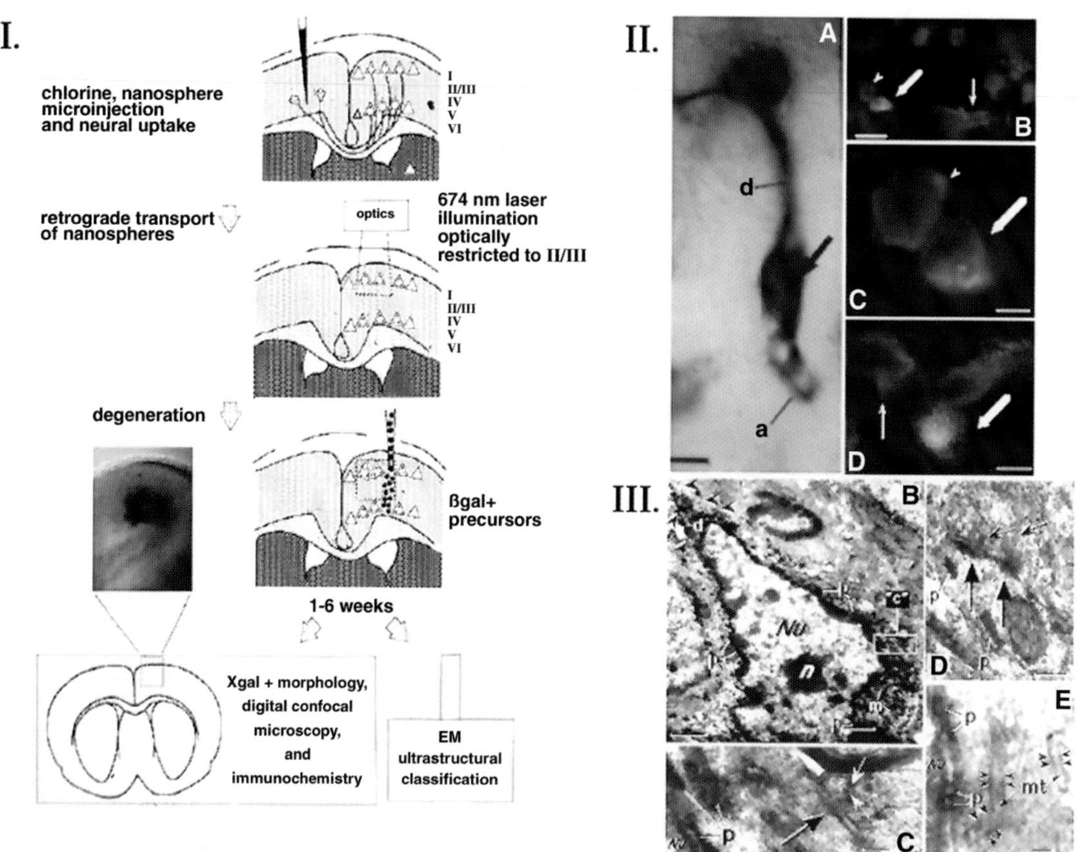

Fig. 1. Multipotent NSCs acquire a neuronal phenotype in regions of adult neocortex subjected to targeted apoptotic neuronal degeneration (modified from Snyder et al., 1997). **(I)** Schematic of procedure by which pyramidal neurons in layer II of adult mouse neocortex are induced to die an apoptotic death by injecting nanospheres into their target region of their transcallosal projections, allowing them to be retrogradely transported back to their somata, and then energizing them via laser illumination such that they promote degeneration of those neurons. NSCs were then implanted into the degenerated region. **(II)** NSCs differentiate appropriately into only glia (or remain undifferentiated) in intact adult cortex, where neurogenesis has normally ceased but gliogenesis persists. However, 15% of engrafted cells (identified by their *lacZ* expression) in regions of apoptotic neurodegeneration developed neuronal morphology, resembling pyramidal neurons within layer II/III 6 wk following transplantation, at 12 wk of age **(A)** and immunocytochemical properties consistent with a neuronal phenotype **(B–D)**. (A) Donor-derived cell (which stains blue following X-gal histochemistry to detect *lacZ* expression) with pyramidal neuronal morphology and size (large arrow) under bright-field microscopy: apical dendrites (d); with descending axons (a). These features are readily confirmed by ultrastructural criteria under EM (III) where donor-derived cells are noted to receive both axosomatic and axodendritic synaptic input. Donor-derived cells in control intact adult cortices had only morphological, ultrastructural, and immunocytochemical features of glia (astrocytes and oligodendrocytes). (Not shown here but pictured in Snyder et al., 1997). (B–D) Immunocytochemical analysis. Donor-derived *lacZ*-expressing cells (identified by an anti-β-galactosidase (β-gal) antibody conjugated to Texas red) are immunoreactive for NeuN (fluorescein, green), a marker for mature neurons. In (B) and at higher magnification in (C), a β-gal$^+$ cell (large arrow) double labels for NeuN. Other small β-gal$^+$ cells with nonneuronal morphology are NeuN$^-$ (small arrow). Remaining host neurons (NeuN$^+$) are β-gal$^-$ (arrowhead). (D) Donor-derived neuron (β-gal$^+$, NeuN$^+$; large arrow) adjacent to two NeuN$^-$ donor-derived cells (small arrow). Bars: (A,B), 25 µm; (C,D), 10 µm. **(III)** Ultrastructure of donor-derived neurons in regions of adult cortex subjected to targeted apoptotic neuronal degeneration. Donor-derived neurons were restricted to the cortex within the degenerated region of layer II/III. In all panels, X-gal precipitate (p) is visible in the nuclear membrane, cytoplasmic organelles, and processes. **(B)** EM characteristics suggestive of layer II/III pyramidal neurons are present 6 wk following transplantation at 12 wk of age: large somata (20–30 µm), large nuclei (Nu), prominent nucleoli (n), abundant ER and mitochondria (m), and apical dendrites (d). An afferent synapse is indicated on the donor-derived neuron in (B) (box c). **(C)** Higher magnification of the axosomatic synapse boxed in (B): presynaptic vesicles (white arrows) and postsynaptic specialization (large black arrow); the postsynaptic region is in continuity with the donor cell nucleus via uninterrupted cytoplasm. Both the cytoplasm and the nuclear membrane contain precipitate (p). **(D)** Axodendritic synapse on dendrite of a donor-derived neuron: crystalline, linear X-gal precipitate (p) in the postsynaptic region of the dendrite confirms its donor origin; postsynaptic specialization with a hazy, nonlinear, noncrystalline appearance immediately under the membrane (large arrows); presynaptic vesicles clustered near the synaptic densities (small arrows). **(E)** Microtubules (arrows, mt) of 20–26 nm diameter (outlined by arrowheads) in a donor-derived neuron near precipitate (p) in nuclear membrane. Bars = 1 µm (B), 100 nm, (C), 200 nm, (D), 40nm.

The amount of enzyme required to restore normal metabolism in many inherited metabolic diseases may be quite small. Thus, even normal NSCs, unengineered but constitutively expressing physiological levels of the enzyme in question, may be used to cross-correct diseased cells. Each disease, each animal model, each enzyme needs to be evaluated individually to ask whether the amount of gene product inherently produced by the NSC is sufficient to restore metabolism or reverse pathology for the case at hand, or whether the NSC will need to be further engineered to enhance its synthesis, processing, and secretion of the required gene product. Using these observations as a blueprint, neural progenitors and stem cells have been used for the expression of other classes of gene products, including neurotrophic factors (e.g., neurotrophin-3 [NT-3], brain-derived nsurotrophic factor [BDNF], and/or glial cell line–derived neurotrophic factor [GDNF] within the rat spinal cord [Liu et al., 1999; Himes et al., 2001; Lu et al., 2000]), NGF and BDNF within the septum [Martinez-Serrano et al., 1995a, 1995b; Rubio et al., 1999), GDNF within mesostriatum (Akerud et al., 2001); synthetic enzymes for neurotransmitters (e.g., tyrosine hydroxylase [TH] [Anton et al., 1994; Sabate et al., 1995]), antiapoptotic proteins (e.g., Bcl-2 [Anton et al., 1995] to the striatum), and antitumor proteins (e.g., cytosine deaminase [Aboody et al, 2000]). NSCs have even been converted to engraftable, migratory packaging lines for viral vectors in order to optimize and maximize their gene delivery capabilities within the CNS and avoid the "halo" effect often seen when vectors are administered as a simple inoculation (Lynch et al., 1999).

17.5. WHEN MORE THAN GENES NEED REPLACING

If the integrity of the CNS cytoarchitecture is compromised, then mere replacement of gene products will have little effect. Such is the case with many neurological diseases that are characterized by degeneration of cells or circuits. Early experiments with NSC clones in various rodent mutants and injury models have provided evidence that NSCs may be able to replace some degenerated or dysfunctional neural cells. Although it may be a daunting task to attempt even the slightest restructuring of CNS tissue, studies of classic fetal transplants demonstrate that even modest anatomical reconstruction may sometimes have an unexpectedly beneficial functional effect (Dunnett and Bjorklund, 2000).

17.5.1. CEREBELLAR MODELS The mutant mouse models *meander tail* (*mea*) and *reeler* (*rl*) have been used to study the ability of donor NSCs to replace cells and restore normal cytoarchitecture, respectively. In the *mea* mouse, which is characterized by a deficiency of cerebellar granule cell (GC) neurons, NSCs implanted at birth were capable of "repopulating" large portions of the GC-poor internal granular layer with neurons (Rosario et al., 1997) (a phenomenon that was subsequently duplicated with human NSCs [hNSCs] [Flax et al., 1998]). One of the most important observations from this work is the apparent receptiveness of donor NSCs to cues from the host tissue, and the capability to become cell types in the abnormal brain that they might not become based on the prevailing developmental cues of the normal brain. Indeed, in the typically "nonneurogenic" milieu of the adult neocortex, it was learned that NSCs would "choose" to differentiate into neurons under conditions in which host neurons were experimentally eliminated by a targeted apoptotic process (Snyder et al., 1997).

The *rl* mouse presents the additional challenge of restoring a cytoarchitecture gone awry. This mutant is characterized by an abnormal laminar assignment of neurons due to a mutation in a gene encoding for the secreted extracellular matrix (ECM) molecule Reelin. As in the *mea*, NSCs were implanted into the cerebella of newborn *rl* mice. NSCs appeared not only to replace missing granule cells, but also to promote a more wild-type-laminated appearance in engrafted regions by "rescuing" aspects of the abnormal migration, positioning, and survival of host neurons. The underlying mechanism is most likely the provision of molecules (including Reelin) at the cell surface to guide proper histogenesis (Auguste et al., 2000).

17.5.2. WHITE MATTER DISEASE Mutants characterized by CNS-wide white matter disease provided ideal models for testing whether NSCs might be useful for widespread cell replacement (Yandava et al., 1999). The *shiverer* (*shi*) and *twitcher* (*twi*) mouse mutants both exhibit white matter degeneration, albeit from somewhat different pathological mechanisms. The oligodendroglia of the dysmyelinated *shi* mouse are dysfunctional because they lack myelin basic protein (MBP), which is essential for proper myelination. The dysmyelination seen in the *twi* mouse is the result of dysfunction and degeneration of oligodendrocytes caused by the loss of galactocerebrosidase-α (GalC) activity, which in turn allows the accumulation of the glycolipid toxin pyschosine. Using these models to test the ability of NSCs to generate myelinating cells is important because disordered myelination plays a critical role in many genetic and acquired (injury, infection) neurodegenerative processes; oligodendroglial pathology is prominent in stroke, spinal cord injury, head trauma, and ischemia and may account for a significant proportion of the neurological handicap seen in asphyxiated and premature newborns.

Newborn *shi* mice were transplanted with NSCs using the same intracerebroventricular implantation technique employed in previous models. Widespread engraftment throughout the *shi* brain was observed with repletion of significant amounts of MBP. Accordingly, of the many donor-derived oligodendroglia (NSCs indeed "shifted" to yield a higher proportion of such cells), a subgroup myelinated up to 52% of host neuronal processes with better compacted myelin. Some animals experienced a decrease in their symptomatic tremor (Yandava et al., 1999).

The *twitcher* mouse is an authentic model of the childhood disease Krabbe's or globoid cell leukodystrophy (GLD). Children afflicted with GLD exhibit inexorable psychomotor deterioration and early death, owing to failing oligodendroglia and subsequent loss of myelin. Onset of symptoms is early (before weaning) in the affected *twi* mouse, and murine NSCs (mNSCs) have been transplanted into the brains of both neonatal and symptomatic juvenile *twi* mice. In preliminary studies, the NSCs engrafted well throughout the brains, particularly in newborns but also in end-stage young adult animals, and differentiated extensively into healthy oligodendrocytes that appeared to remyelinate up to 30–50 host axons. Success in these white matter disease mutants represents another step forward in the use of donor NSCs in host CNS tissue. Whereas in some disease models the NSCs are only "asked" essentially to become gene product factories, in the *shi* and *twi* models the NSCs are required to interact with host tissue in a more intimate manner, i.e., to wrap host axons with donor myelin.

Interestingly, this degree of extensive cell replacement was unable to remediate symptoms of the *twi* mouse as it was in the *shi* mouse. The differences clearly derive from the different etiologies of the two diseases. In the *shi* mouse, the disease process is purely cell autonomous. In the *twi* mouse, the disease process may have certain cell nonautonomous components that will require attention. That cell replacement could proceed at all, however, despite the likely toxic environment of the *twi* brain, is significant and instructive. Wild-type NSCs appear, at least in preliminary studies, to be more resistant to psychosine than Krabbe cells and hence differentiate into mature, effectively myelinating oligodendrocytes. Whether host cells also need to be protected remains an area of active investigation. The replacement of both cells and molecules, e.g., with NSCs that can both become oligodendrocytes and serve as pumps for secreted cross-corrective GalC, remains a feasible and promising basis for a future multidisciplinary strategy. Interventions in this complex leukodystrophy may also call for implantation of NSCs at *multiple* locations as well as at *multiple* time points in the evolution of the disease, possibly beginning *presymptomatically* and continuing after the disease is established. Interventions will likely need to combine NSC implantation in the brain with strategies to address extracranial manifestations (e.g., BMT) that together may prolong life. Such strategizing represents how stem cell investigators will need to problem solve through each disease and not glibly extrapolate from one disease to another.

17.6. NSCs CAN COVER A LOT OF GROUND

Given the ability of NSCs to pursue alternative differentiation paths in response to certain types of neurodegeneration (Rosario et al., 1997; Snyder et al., 1997) and their ability to migrate and express foreign genes, can NSCs be selectively targeted to the regions most in need of such? Evidence (some published, some preliminary) in various models suggests that NSCs may in fact be able to follow the "bread crumbs" laid down by actively degenerating areas. These "bread crumbs" are as-yet-unidentified factors transiently elaborated by troubled tissue, which may be a target for NSCs. We have speculated that some of those signals are cytokines and chemokines for which, we have determined, some NSCs have receptors and to which they can migrate.

17.6.1. HYPOXIC-ISCHEMIC BRAIN INJURY In pilot studies, when NSCs are transplanted into brains of young mice subjected to unilateral hypoxic–ischemic (HI) brain injury (a model for cerebral palsy), donor-derived cells migrate preferentially to and integrate extensively within the large ischemic areas that typically span the injured hemisphere. A subpopulation of donor NSCs, particularly in the penumbra of the infarct, "shift" their differentiation fate toward becoming neurons and oligodendrocytes, the neural cell types typically damaged following asphyxia/stroke. Furthermore, there appears to be an optimal window of time following injury (3–7 d) during which signals are elaborated within the degenerating region and to which NSCs respond with migration and reconstitution of lost neural cells (similar observations have been noted by Hodges et al. [2000]).

To determine whether this "natural" shift by NSCs toward neuronal replacement could be further augmented, a subclone of the same murine NSCs was engineered via retroviral transduction to overexpress NT-3, a factor known to play a role in inducing neuronal differentiation. This subclone was then implanted into asphyxiated mouse brains. The percentage of donor-derived neurons increased from 5 to 20% in the infarction cavity and to >80% in the penumbra. While it remains far from clear that one would even desire so many neurons, this observation suggests that a naturally occurring NSC-based process in a degenerative environment can be augmented via genetic engineering.

17.6.2. BRAIN TUMORS Brain tumors such as gliomas are very migratory, expanding and infiltrating the surrounding normal tissue. This attribute renders them "elusive" to effective resection, irradiation, chemotherapy, or gene therapy. Aboody et al. (2000) demonstrated that murine and hNSCs, when implanted into intracranial gliomas in adult rodents, distributed quickly and extensively throughout the tumor bed. Furthermore, when implanted intracranially at distant sites from the tumor bed in adult brain (e.g., into normal tissue, into the contralateral hemisphere, or into the cerebral ventricles), the donor NSCs crossed from one hemisphere to the other, migrating through normal parenchyma to the tumor on the opposite side. NSCs could also deliver the oncolysis-promoting enzyme cytosine deaminase, resulting in a reduction in tumor cell burden. With their ability to track down migrating tumor cells across relatively large distances, and to deliver therapeutic molecules, NSCs have warranted further investigation as a possible treatment tool for brain tumors. It is quite conceivable that NSCs are so effective in "shadowing" tumor cells because there is a common biology between the two; tumor cells could conceivably be NSCs that "have gone bad." It will be important to illuminate the basic mechanisms by which NSCs are driven to bypass stereotypical migratory routes in their pursuit of tumor cells.

17.6.3. AMYLOID PLAQUES Although the amyloid deposits characteristic of AD may not have the aggressively invasive nature of gliomas, they can be quite widespread and their extent and distribution seems to mirror the degree of dementia displayed by patients. In preliminary studies in our laboratory, Tate et al. (2000) used NSCs to address lesions present in an adult animal model of AD-like pathology. Tate previously demonstrated that chronically infused human amyloid will cause an inflammatory response in the rat brain. Subsequently, it was observed that mNSCs placed in the opposite lateral ventricle will migrate to and surround areas of amyloid infusion. While, as noted above, the signals that stimulate migration of NSCs are not yet identified, inflammatory molecules are among the likely candidates. Inflammation has recently been recognized to characterize the pathology seen in both transgenic mouse models of AD and AD patients. Accordingly, and reassuringly, similar results have been observed in pilot studies involving transgenic mouse models of AD and the homing response of implanted foreign gene–expressing NSCs. Hence, as reagents come to be identified that might be of help in reducing amyloid plaques or other aspects of AD pathology, NSCs may play a role in helping to deliver those therapeutic molecules.

17.7. NSC-HOST INTERACTIONS: A TWO-WAY STREET

There is growing evidence to support the notion that donor NSCs and host tissue communicate in a reciprocal fashion. While it has been known that a given host CNS environment is instructive of donor stem cells, influencing their migration and differentiation patterns, the donor NSCs are apparently equally capable of manipulating their milieu. While it seems evident that NSCs that have been engineered to express foreign gene products may influence the host CNS into which they have been implanted, it is now becoming evident that even unengineered NSCs, through their unanticipated constitutive secretion of many endogenous growth factors (or perhaps by as-yet-unknown other mechanisms), may instruct the host CNS.

Ourednik et al. (2002) and Teng et al. (2002) have accumulated data suggesting that this may be true. In Ourednik et al.'s (2002) article, aged mice received repeated systemic administrations of 1-methyl-4-phenyl-1,2,3,6-tetrahydropyridine (MPTP), a neurotoxin that produces persistent impairment of mesencephalic dopaminergic neurons and their striatal projections (Fig. 2 I[A,B]). mNSCs were unilaterally implanted into the SN 1–4 wk following MPTP treatment. Unilaterally implanted NSCs were found to have migrated and integrated extensively within *both* hemispheres and were associated with dramatic reconstitution of TH and dopamine (DA) transporter (DAT) expression and function throughout the mesostriatal system (Fig. 2 I[C,D]). While there was spontaneous conversion of NSCs to TH^+ cells in DA-depleted areas, and while TH^+ cells of donor origin contributed to

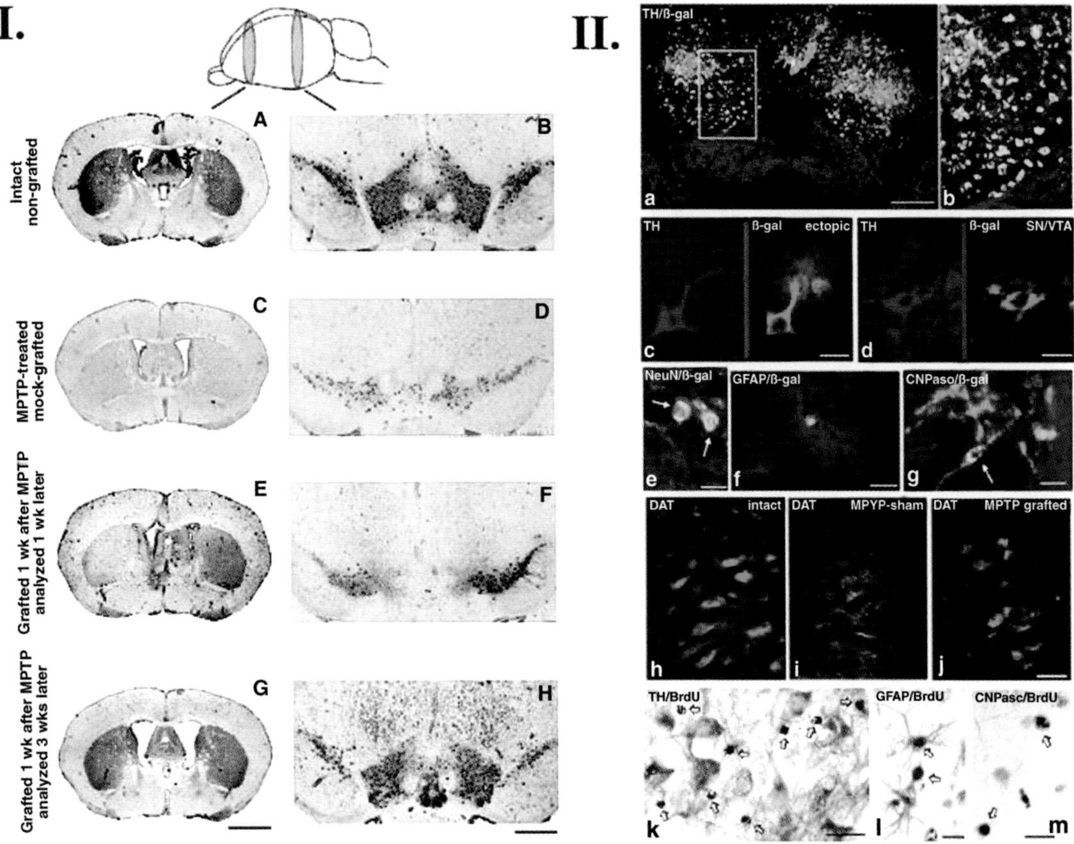

Fig. 2. NSCs possess inherent mechanism for rescuing dysfunctional neurons: evidence from effects of NSCs in restoration of mesencephalic dopaminergic function (modified from Ourednik et al., 2002). **(I)** TH expression in mesencephalon and striatum of aged mice following MPTP lesioning and unilateral NSC engraftment into SN/ventral tegmental area (SN/VTA). A model that emulates the slow dysfunction of aging dopaminergic neurons in SN was generated by giving aged mice repeated high doses of MPTP. The schematic at top indicates the levels of the analyzed transverse sections along the rostrocaudal axis of the mouse brain. Representative coronal sections through the striatum are presented in the left column **(A,C,E,G)** and through the SN/VTA area in the right column **(B,D,F,H)**. (A,B) Immunodetection of TH (black cells) shows the normal distribution of DA-producing TH$^+$ neurons in coronal sections in the intact SN/VTA (B) and their projections to the striatum (A). (C,D) Within 1 wk, MPTP treatment caused extensive and permanent bilateral loss of TH immunoreactivity in both the mesostriatal nuclei (C) and the striatum (D), which lasted life long. Shown in this example, and matching the time point in (G,H), is the situation in a mock-grafted animal 4 wk after MPTP treatment. (E,F) Unilateral (right side) stereotactic injection of NSCs into the nigra is associated, within 1 wk after grafting, with substantial recovery of TH synthesis within the ipsilateral DA nuclei (F) and their ipsilateral striatal projections (E). By 3 wk posttransplant, however (G,H), the asymmetric distribution of TH expression disappeared, giving rise to TH immunoreactivity in the midbrain (H) and striatum (G) of both hemispheres that approached that of intact controls (A,B) and gave the appearance of mesostriatal restoration. Similar observations were made when NSCs were injected 4 wk after MPTP treatment (not shown). Bars: 2 mm (left), 1 mm (right). Note the ectopically placed TH$^+$ cells in (H). These are analyzed in greater detail, along with the entire SN, in (II). **(II)** Immunohistochemical analyses of TH, DAT, and bromodeoxyuridine (BrdU)-positive cells in MPTP-treated and grafted mouse brains. The initial presumption was that the NSCs had replaced the dysfunctional TH neurons. However, examination of the reconstituted SN with dual β-gal (green) and TH (red) immunocytochemistry showed that **(a,c)** 90% of the TH$^+$ cells in the SN were *host*-derived cells that had been rescued and only 10% donor derived (**d**). Most NSC-derived TH$^+$ cells were actually just above the SN ectopically (blocked area in [**a**], enlarged in [**b**]). These photomicrographs were taken from immunostained brain sections from aged mice exposed to MPTP, transplanted 1 wk later with NSCs, and sacrificed after 3 wk. The following combinations of markers were evaluated: **(a–d)** TH (red) with β-gal (green); **(e)** NeuN (red) with β-gal (green); **(f)** glial fibrillary acid protein (GFAP) (red) with β-gal (green); **(g)** CNPase (green) with β-gal (red); as well as **(k)** TH (brown) and BrdU (black); **(l)** GFAP (brown) with BrdU (black); and **(m)** CNPase (brown) with BrdU (black). Anti-DAT-stained areas are revealed in green in the SN of intact **(h)**, mock-grafted **(i)**, and NSC-grafted **(j)** brains. Three different fluorescence filters specific for Alexa Fluor 488 (green), Texas Red (red), and a double filter for both types of fluorochromes (yellow) were used to visualize specific antibody binding; **(c,d,h–j)** single-filter exposures; **(a,b,e–g)** are double-filter exposures. In (**a**) a low-power overview of the SN + VTA of both hemispheres, similar to the image in Fig. 2IH, is shown. The majority of TH$^+$ cells (red cells in [**a**]) within the nigra are actually of *host* origin (approx 90%) with a much smaller proportion there being of donor derivation (green cells) (approx 10%) (representative close-up of such a donor derived TH$^+$ cell in [**d**]). Although a significant proportion of NSCs did differentiate into TH$^+$ neurons, many of these actually resided ectopically, dorsal to the SN (boxed area in [**a**], enlarged in [**b**]; high-power view of donor-derived (green) cell that was also TH$^+$ (red) in [**c**]), where the ratio of donor-to-host cells was inverted: ~90% donor derived compared with ~10% host derived. Note the almost complete absence of a green β-gal-specific signal in the SN + VTA while, ectopically, many of the TH$^+$ cells were double labeled and thus NSC derived (appearing yellow-orange in higher power under a red/green double filter in [**b**]). (c–g) NSC-derived *non*-TH neurons (NeuN$^+$) (**e**, arrows), astrocytes (GFAP$^+$) (**f**), and oligodendrocytes (CNPase$^+$) (**g**, arrow) were also seen, both within the mesencephalic nuclei and dorsal to them.

(Caption continued on page 198)

nigral reconstitution, the *majority* (~90%) of TH^+ cells in the reconstituted SN were actually *host* cells "rescued" by constitutively produced NSC-derived factors (Fig. 2 II). One of these factors appeared to be GDNF, produced by an undifferentiated or glial-differentiated subpopulation of donor NSC-derived cells. GDNF is a molecule known to be protective of DA neurons. However, GDNF is probably just one of a number of factors elaborated by NSCs in various states of differentiation that may play a role in host neuronal cell rescue. NSCs, e.g., are known to produce a broad range of other neurotrophic factors (including NT-3, NT-4/5, NGF, BDNF), adhesion molecules (e.g., L1), ECMs (e.g., reelin), and lysosomal enzymes. Hence, host cells of various types in various regions and in various disease conditions may similarly benefit from other of the many intrinsically expressed NSC-derived molecules.

Ourednik et al. (2001b) observed this phenomenon of donor-host interaction in another set of preliminary experiments. Three mouse mutants were chosen that exhibit degeneration of the cerebellar Purkinje cells (PCs). The *nervous* (*nr*) mouse, the *Purkinje cell degeneration* (*pcd*) mouse, and the *Lurcher* (*Lc*) mouse, however, differ in the severity; onset; degree of secondary loss of granule cell neurons; and, most significantly, the etiology of their PC loss. Nevertheless, intriguingly, in all three mutants a protective influence was exerted on the vulnerable PC population by engrafted NSCs. While a small proportion of donor NSCs did, indeed, differentiate into PCs (a phenomenon observed only in the face of PC degeneration, given that PCs are normally born only prenatally), the *majority* of NSCs actually differentiated into *nonneurons*—glia and quiescent, undifferentiated neural progenitors, which, in turn, appeared to nurture, protect, and/or rescue host PCs. In all three mutants, the PC layer appeared to be significantly reconstituted, but the vast majority of the PCs in that layer were of host origin. This strong potential by NSCs to preserve neuronal populations in the mutant cerebella seemed most prominent if NSCs were implanted before or at the onset of PC degeneration, specifically postnatal (P) d 0–10, rather than during more advanced stages of PC loss (P14 d to 6 mo). This positive effect of the grafts was reflected in both an impressively improved cerebellar cytoarchitecture and the motor behavior of the host animals, the latter seen most prominently in the *Lc* mouse. The molecular mechanism underlying this "rescue" remains unclear but might be related to the exuberant production of neurotrophic factors by NSCs when in their undifferentiated or glial differentiated state.

17.7.1. ENHANCING NSC-HOST COMMUNICATION—A THIRD MECHANISM OF STEM CELL–MEDIATED REPAIR

In previously mentioned studies, implanted NSCs were generally afforded the luxury of somewhat intact cyto-architecture in which to migrate and engraft, even if the environment was degenerative or toxic. However, some conditions or injuries present the conundrum of extensively degraded cytoarchitecture, and even the most migratory and engraftable of donor cells have little chance of surviving in, let alone repairing, that environment. Both Park et al. (2002a) and Teng et al. (2002), in the brain and spinal cord, respectively, hypothesized that three-dimensional highly porous "scaffolds" composed of polyglycolic acid (PGA) may act as a third party to optimize communication between donor and host tissue in these models. PGA is a synthetic biodegradable polymer used widely in clinical medicine. Highly hydrophilic, PGA loses its mechanical strength rapidly over 2–4 wk in the body, after which time it is degraded, obviating concerns over long-term biocompatibility. This is enough time, however, to provide an initial matrix to guide cellular organization and growth, allow diffusion of nutrients to the transplanted cells, and become vascularized. This last point is important because large volumes of cells will not survive if situated greater than a few hundred microns from a capillary.

To first test this hypothesis, mNSCs were cocultured with PGA in vitro (the details of which are outlined below). The mNSCs adhered well to the scaffold and migrated throughout the matrix [Park et al., 2002a; Teng et al., 2002]. Spontaneous differentiation of the mNSCs resulted in neuronal and glial cell formation, with many of the neurons sending out long, complex processes along the synthetic fibers of the scaffold. With NSCs and PGA apparently working well together, this new complex was then used to effect repair in a rodent model of induced HI brain injury (Park et al., 2002a). After coculturing for 4 d in vitro, the NSC–PGA complex was transplanted into the cerebral infarction cavity 4–7 d following induction of an experimental HI insult. The NSCs not only completely impregnated the PGA matrix, but the NSC-PGA complex seemed to refill the infarction cavity, even becoming vascularized by the host. The NSCs seeded on polymers displayed robust engraftment, foreign gene expression, and differentiation into neurons and glia within the region of HI injury. Many long neuronal processes of host and donor-derived neurons enwrapped the polymer fibers and ran along the length of the fibers, often interconnecting with each other (Fig 3 III). Donor-derived neurons appeared to extend many very long, complex processes along the length of the disappearing matrix, extending ultimately into host parenchyma to as far as presumably appropriate target regions in the opposite intact hemisphere (Fig. 3 IV). Host neuronal processes, in a reciprocal manner, appeared to enter the matrix, possibly making contact with donor-derived neurons. Indeed, neuronal tracing studies with DiI and biotinylated dextrose amine (BDA) showed that the long-distance neuronal circuitry between donor-derived and host neurons in both cerebral hemispheres may have been reformed through the corpus callosum in some instances (Fig. 3 IV).

These findings (Park et al., 2002a) suggest that NSCs—when fixed in space by an artificial ECM, such as the scaffold, in order to maximize the reciprocal interaction between stem cell and injured CNS—may facilitate even further the differentiation of

Fig. 2. *(Caption continued from page 197)* **(h–j)** The green DAT-specific signal in **(j)** suggests that the reconstituted mesencephalic nuclei in the NSC-grafted mice (as in Figs. 2IIa and 2IH) were functional DA neurons comparable with those seen in intact nuclei **(h)** but not in MPTP-lesioned sham-engrafted controls **(i)**. This further suggests that the TH^+ mesostriatal DA neurons affected by MPTP are, indeed, functionally impaired. (Note that sham-grafted animals *(i)* contain only punctate residual DAT staining within their dysfunctional fibers, while DAT staining in normal **(h)** and, similarly, in engrafted **(j)** animals was normally and robustly distributed both within processes and throughout their cell bodies.) **(k–m)** Any proliferative $BrdU^+$ cells after MPTP insult and/or grafting were confined to glial cells while the TH^+ neurons **(k)** were $BrdU^-$. This finding suggested that the reappearance of TH^+ host cells was not the result of neurogenesis but, rather, the recovery of extant host TH^+ neurons. Bars: 90 **(a)**, 20 **(c–e)**, 30 **(f)**, 10 **(g)**; 20 **(h–j)**, 25 **(k)**, 10 **(l)**, and 20 μm **(m)**.

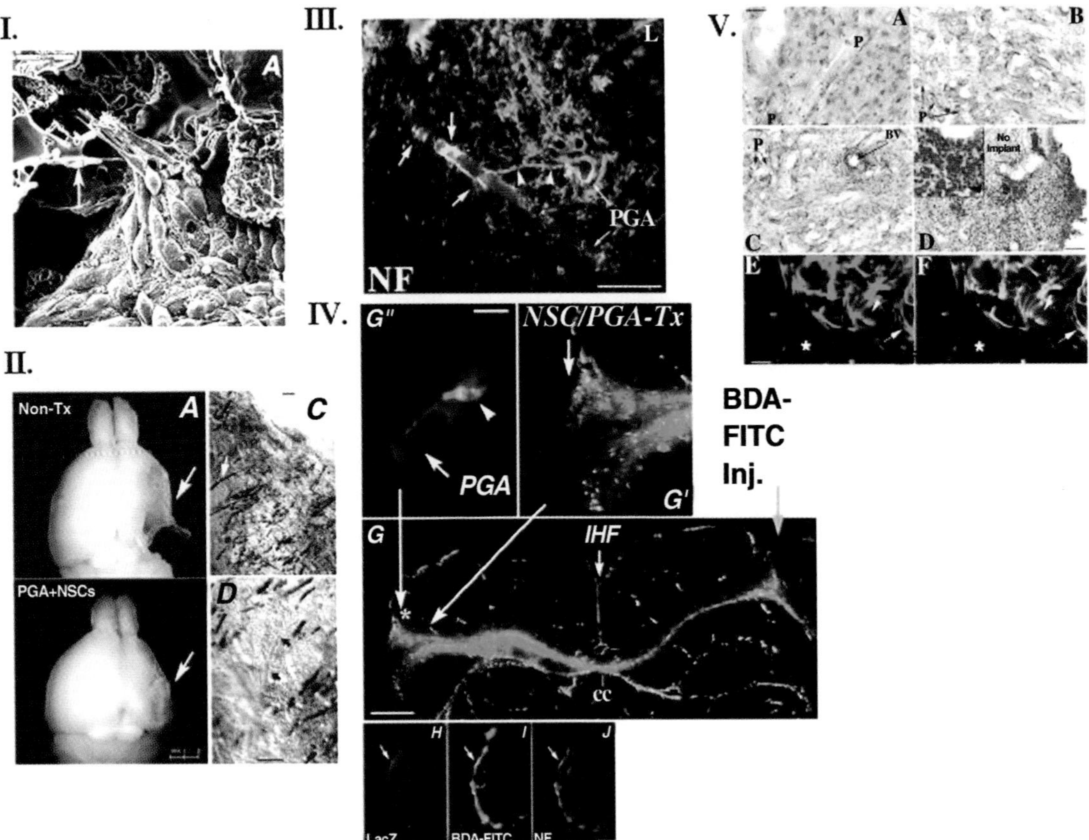

Fig. 3. The injured brain interacts reciprocally with NSCs supported by scaffolds to reconstitute lost tissue—evidence from HI injury (modified from Park et al., 2002a). **(I)** Characterization of NSCs in vitro when seeded on PGA scaffold. Cells, seen with scanning electron microscopy at 5 d after seeding, were able to attach to, impregnate, and migrate throughout a highly porous PGA matrix (arrow). The NSCs differentiated primarily into neurons (>90%) that sent out long, complex processes that adhered to, enwrapped, and interconnected the PGA fibers. **(II)** Implantation of NSC/PGA complexes into region of cavity formation following extensive HI brain injury and necrosis. **(A)** Brain of untransplanted (non-Tx) mouse subjected to right HI injury with extensive infarction and cavitation of ipsilateral right cortex, striatum, thalamus, and hippocampus (arrow). By contrast, **(B)** shows the brain of a similarly injured mouse implanted with an NSC/PGA complex (PGA+NSCs) (generated in vitro as per (I) into the infarction cavity 7 d after induction of HI (arrow) ($n = 60$). At maturity (age matched to the animal in [A]), the NSC/scaffold complex appears, in this whole mount, to have filled the cavity (arrow) and become incorporated into the infarcted cerebrum. Representative coronal sections through that region are seen at higher magnification in **(C)** and **(D)**, in which parenchyma appears to have filled in spaces between the dissolving black polymer fibers (white arrow in [C]) and, as seen in [D], even to support neovascularization by host tissues (blood vessel indicated by closed black arrow in [D]; open arrow in [D] points to degrading black polymer fiber). Bars: 100 μm (C,D). **(III)** Characterization in vivo of the neural composition of NSC/PGA complexes within HI-injured brain. At 2 wk following transplantation of the NSC/PGA complex into the infarction cavity, donor-derived cells showed robust engraftment within the injured region. An intricate network of multiple long, branching neurofilament⁺ (Nf⁺) (green) processes were present within the NSC/PGA complex and its parenchyma enwrapping the PGA fibers (orange autofluorescent tubelike structures under a Texas Red filter), adherent to and running along the length of the fibers (arrows), often interconnecting and bridging the fibers (arrowheads). Those NF⁺ processes were of both host and donor derivation. In other words, not only were donor-derived neural cells present, but also *host*-derived cells seemed to have entered the NSC/PGA complex, migrating along and becoming adherent to the PGA matrix. In a reciprocal manner, donor-derived (*lacZ*⁺) neurons (NF⁺ cells) within the complex appeared to send processes along the PGA fibers out of the matrix into host parenchyma as seen in (IV). Bars: 100 μm. **(IV)** Long-distance neuronal connections extend from the transplanted NSC/PGA complexes in the HI-injured brain toward presumptive target regions in the intact contralateral hemisphere. By 6 wk following engraftment, donor-derived *lacZ*⁺ cells appeared to extend many exceedingly long, complex NF⁺ processes along the length of the disappearing matrix, apparently extending into host parenchyma. To confirm the suggestion that long-distance processes projected from the injured cortex into host parenchyma, a series of tract tracing studies was performed. **(G–G")** BDA-fluorescein isothiocyanate (FITC) was injected (G) into the contralateral intact cortex and external capsule (green arrow) at 8 wk following implantation of the NSC/PGA complex into the infarction cavity (NSC/PGA-Tx). Axonal projections (labeled green with fluorescein under an FITC filter) are visualized (via the retrograde transport of BDA) leading back to (across the interhemispheric fissure [IHF] via the corpus callosum [cc]) and emanating from cells in the NSC/PGA complex within the damaged contralateral cortex and penumbra (seen at progressively higher magnification in [G']—region indicated by arrow to [G]—and [G"]—region indicated by arrow and asterisk in [G]). In (G"), the retrogradely BDA-FITC-labeled perikaryon of a representative neuron adherent to a dissolving PGA fiber is well visualized. That such cells are neurons of donor derivation is supported by their triple labeling **(H–J)** for *lacZ* (H) (β-gal), BDA-FITC (I), and the neuronal marker NF (J); arrows in (H–J) indicate same cell in all three panels. Such neuronal clusters were never seen under control conditions—i.e., in untransplanted cases or when vehicle or even an NSC suspension *unsupported* by scaffolds was injected into the infarction cavity. Bars: 500 μm (G), 20 μm (G), 30 μm (H–J).

(Caption continued on page 200)

host and donor neurons and enhance the growth of such cells to help promote reformation of cortical tissue, both structurally and functionally. An additional interesting observation to emerge from these studies was the fact that the otherwise prominent infiltration by inflammatory mononuclear cells and the development of astroglial scarring following typical HI injury seemed to be inhibited by the NSC–PGA complex (Fig. 3 V), most likely minimizing a factor that tends to impede brain repair and, hence, abetting spontaneous regenerative processes in the injured brain.

In a parallel study by Teng et al. (2002), it was observed in the hemisectioned adult rat spinal cord that implantation of NSCs seeded on a PGA-based scaffold (Fig. 4I) significantly improved functional recovery and preservation of the parenchyma of the injured cord (Fig. 4II,III) (Teng et al., 2002). In these experiments, although donor *murine* NSCs were present in abundance as undifferentiated nestin-immunopositive progenitors in the *rat* spinal cord (Fig. 4IV), the parenchyma and regenerated neurites (confirmed by species-specific markers) were *not* derived from donor cells, but rather from the *host*. Indeed, the NSCs did not simply differentiate into astrocytes that might have contributed to glial scar formation; in fact, scarring and evidence of secondary injury were significantly diminished compared to untreated controls (Fig. 4IV), a situation similar to that observed by Park et al. (2002a). In fact, there was evidence of a significant regenerative process (as suggested by robust GAP-43 immunoreactivity and the typical morphological hallmarks of regenerating axons [Fig. 4V]). Apparently, *host* tissue preservation and regeneration—as stimulated by the NSCs using the scaffold as a template—were responsible for the functional improvement [Fig. 4-III]. Indeed, tract tracing of host-derived fibers coursing from rostral to caudal within the cord through the injury epicenter confirmed the anatomical substrate to account for such functional recovery.

Thus, donor and host cells are apparently engaged in communication that is instructive to both. The broader implications for CNS repair are that host structures may benefit not only from NSC-derived replacement of missing neurons but also from the "chaperone" effect of undifferentiated or glial-differentiated NSCs. While NSCs have been touted for cell and gene therapy, these findings suggest a *third mechanism* by which therapeutic outcomes might be achieved: an inherent capacity of NSCs to create host environments sufficiently rich in trophic and/or neuroprotective support to promote the recovery of damaged *endogenous* cells.

17.8. HUMAN NSCs

After mounting studies with rodent NSCs began to elucidate their potential as therapeutic vehicles, questions naturally arose about the potential of *human* NSCs. With the identification of NSCs of human derivation (Flax et al., 1998; Pincus et al., 1998; Vescovi et al., 1999; Roy et al., 2000; Uchida et al., 2000; Villa et al., 2000), it could be ascertained that many of the biological principles first gleaned from examining rodent cells had been conserved with the human CNS. Although important differences also existed—principally attributable to the much-prolonged duration of the NSC cell cycle in humans—stem cell–based strategies seemed a reasonable consideration for human neurodegenerative conditions.

Lines of engraftable hNSCs have been isolated from normal human fetuses that, in many ways, emulate their rodent counterparts (reviewed in Vescovi and Snyder, 1999). Their fundamental biology seems to cut across species and therefore makes them an attractive tool for CNS therapy, as suggested by studies using rodent NSCs. For example, hNSCs can participate in CNS development, including migration from germinal zones along migratory streams, to widely disseminated CNS regions (Flax et al., 1998; Fricker et al., 1999; Rubio et al., 2000; Uchida et al., 2000), and respond to local and temporal developmental cues. Like the rodent NSCs, hNSCs can be transduced by viral vectors ex vivo and express these transgenes in vivo. Secretory products from hNSCs can cross-correct genetic metabolic defects, and hNSCs can differentiate into neuronal subtypes that are deficient in mouse mutants (Flax et al., 1998). In models of brain tumor, hNSCs migrate to areas of pathology when transplanted contralaterally to those areas in the adult rodent brain (Aboody et al., 2000). In pilot studies involving the contused adult rat spinal cord, hNSCs can yield neurons (including motor neurons) that can make long-distance connections both rostral (as far as the thalamus) and caudal to the lesion, have the ability to conduct corticospinal signals (as assessed by cortical evoked potentials), and result in apparent functional improvement (Park et al., 2002b).

To make the transition from rodents to humans, experiments with subhuman primates have been undertaken to bridge the spe-

Fig. 3. *(Caption continued from page 199)* **(V)** Adverse secondary events that typically follow injury (e.g., monocyte infiltration and astroglial scar formation) are minimized by and within the NSC/PGA complex. **(A–D)** Photomicrographs of hematoxylin and eosin (H&E)–stained sections prepared to visualize degree of monocyte infiltration in relation to NSC/PGA complex and the injured cortex 3 wk following implantation into infarction cavity. Monocytes are classically recognized under H & E as very small cells with small round nuclei and scanty cytoplasm (e.g., inset in [D], arrowhead). While some very localized monocyte infiltration was present immediately surrounding a blood vessel (BV in [C], arrow) that grew into the NSC/PGA complex from the host parenchyma, there was little or no monocyte infiltration either in the center of the NSC/PGA complex (B) or at the interface between the NSC/PGA complex and host cortical penumbra (A)—in stark contrast to the excessive monocyte infiltration seen in an untransplanted infarct of equal duration, age, and extent (D), the typical histopathological picture otherwise seen following HI brain injury (see inset, a higher magnification of the region indicated by the asterisk in [D]; a typical monocyte is indicated by the arrowhead). While neural cells (nuclei of which are seen in [A–C]) adhere exuberantly to the many polymer fibers (P in [A–C]), monocyte infiltration was minimal compared with that in (D). **(E,F)** Astroglial scarring (another pathological condition confounding recovery from ischemic CNS injury) is also much constrained and diminished following implantation of the NSC/PGA complex. While $GFAP^+$ cells (astrocytes) were among the cell types into which NSCs differentiated when in contact with the PGA fibers, *away* from the fibers (*), there was minimal astroglial presence of either donor or host origin. (E) GFAP immunostaining recognized by a fluorescein-conjugated secondary antibody (green) is observed. Note little scarring in the regions indicated by the asterisk. Under a Texas red filter (F) (merged with the fluorescein filter image), the tubelike PGA fibers (arrowhead in both panels) becomes evident (as an autofluorescent orange), and most of the donor-derived astrocytes (arrows) (yellow because of their dual *lacZ* and GFAP immunoreactivity) are seen to be associated with these fibers, again leaving most regions of the infarct (*) astroglial scar free (arrows in [E] and [F] point to the same cells). Far from creating a barrier to the migration of host- or donor-origin cells or to the ingrowth/outgrowth of axons of host- or donor-origin neurons (as per [III and IV]), NSC-derived astrocytes may actually have helped provide a facilitating bridge. Bars: 10 (A) and (C–F) 20 μm.

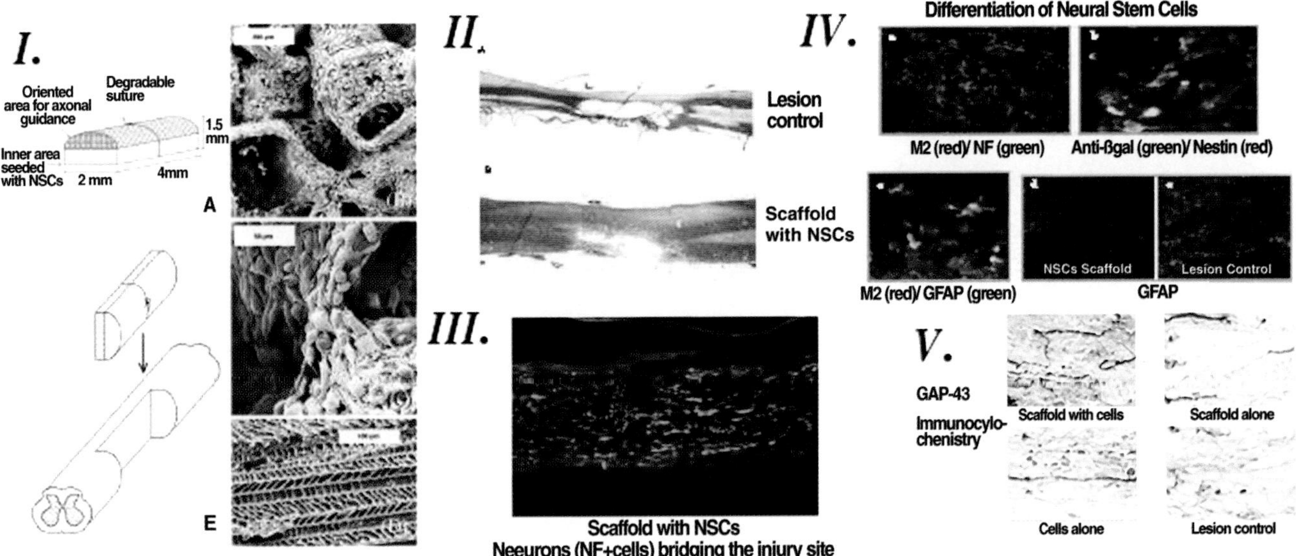

Fig. 4. Functional recovery following traumatic spinal cord injury mediated by unique polymer scaffold seeded with NSCs (modified from Teng et al., 2002). **(I)** Schematics of scaffold design showing inner and outer portions. **(B,C)** Inner scaffold seeded with NSCs. **(D)** Outer scaffold created to have long, axially oriented pores for axonal guidance & radial pores to allow fluid transport and inhibit ingrowth of cells. **(E)** Schematic of surgical insertion of implant into SC. **(II)** Based on Basso–Bresnahan–Beatie open-field walking scores, the scaffold with NSCs group showed significant improvement in open-field locomotion compared with lesion control groups ($p < 0.007$). Histology (H&E) of longitudinal sections from **(A)** lesioned untreated and **(B)** scaffold with NSCs groups was revealing. Note greater integrity of parenchyma in the latter. **(III)** Examination of composition of the tissue at the lesion site demonstrated numerous NF$^+$ cells and processes. However, as illustrated in (IV), the neurons were host NSC derived. **(IV)** The neurons were host and *not* donor NSC derived. The mNSCs were identified with M2, a mouse-specific marker that works reliably in the rat host SC. The mNSCs were neither NF$^+$ **(A)** nor GFAP$^+$, the latter finding suggesting that they did not contribute to the glial scar. In fact, glial scarring was *diminished* in NSC+scaffold SCs **(D)** compared with lesion control SCs **(E)** based on GFAP immunoreactivity. Most mNSCs remained undifferentiated nestin$^+$ cells **(B)**. **(V)** Scaffold with cells implantation significantly increased the presence of GAP-43$^+$ fibers relative to other controls, a marker for regenerating neurites. Following administration of BDA for antegrade tracing, BDA$^+$ axons (not shown here but pictured in Teng et al., 2002) were coursed through the lesion epicenter (as in [III]) to reach areas caudal to the lesion in the scaffold-containing groups, suggesting an anatomical substrate for the functional improvement seen in those animals (a mean of 14 on the 21-point Basso–Bresnahan–Beatie scale).

cies gap. hNSCs transplanted into normal fetal Old World monkeys via *in utero* intracerebroventricular injections that allowed the cells access to the VZ integrated throughout the developing brain, yielding neurons and glia appropriate to given cortical laminae as well as contributing to such secondary germinal zones as the SVZ (Ourednik et al., 2001a) (Fig. 5). Given the successful integration of hNSCs into fetal subhuman primates, this intervention bears significance in the realm of possible *in utero* therapy of human neurogenetic disorders. Such prenatal treatments could be directed not only at congenital disorders but also, theoretically, at neurodegenerative diseases that are not expressed until adulthood or middle age but whose antenatal genetic diagnosis is possible (e.g., Huntington disease).

Experiments have begun involving hNSCs in lesioned nonhuman primates, often the primate equivalent of some of the rodent models that we have described. These experiments will provide insights into the performance of hNSCs in an environment more relevant to humans. These include the cells' safety and efficacy, as well as their response to a neurodegenerative milieu. They also offer an opportunity to fine-tune the many parameters involved in the practical administration of cells such as total cell number to inject, placement of injection, number of injections, and rate of delivery. One experiment has analyzed the fate and impact of hNSCs in the MPTP-induced model of DA depletion and parkinsonism in Old World St. Kitts African Green Monkeys, probably the most authentic animal model of true Parkinson disease in humans. In encouraging pilot studies, hNSCs appeared to survive in the SN and some spontaneously converted to TH$^+$ cells. Given that hNSCs, like murine NSCs, intrinsically produce a great many neurotrophic and neuroprotective factors, improvement in DA activity in some recipient pilot monkeys (as assessed by single-photon emission computed tomography and an improvement in parkinsonian symptom scores) is likely to be the combined effect of not only DA cell replacement by NSCs but also the provision by NSCs of factors promoting the survival and enhanced function of host DA neurons and their connections. These double-barreled mechanisms, if corroborated, will likely be therapeutically significant.

17.9. PRACTICAL ASPECTS: IN VITRO CONSIDERATIONS FOR NSCs

17.9.1. mNSCs—ISOLATION AND PROPAGATION The most widely used line of mNSCs in our laboratory and many others internationally is clone C17.2. A prototypical NSC clone, C17.2, was originally derived from neonatal mouse cerebellum (Ryder et al., 1990; Snyder et al., 1992).

NSC lines can most readily be derived by dissecting the primary structure of interest—usually a germinal zone such as the

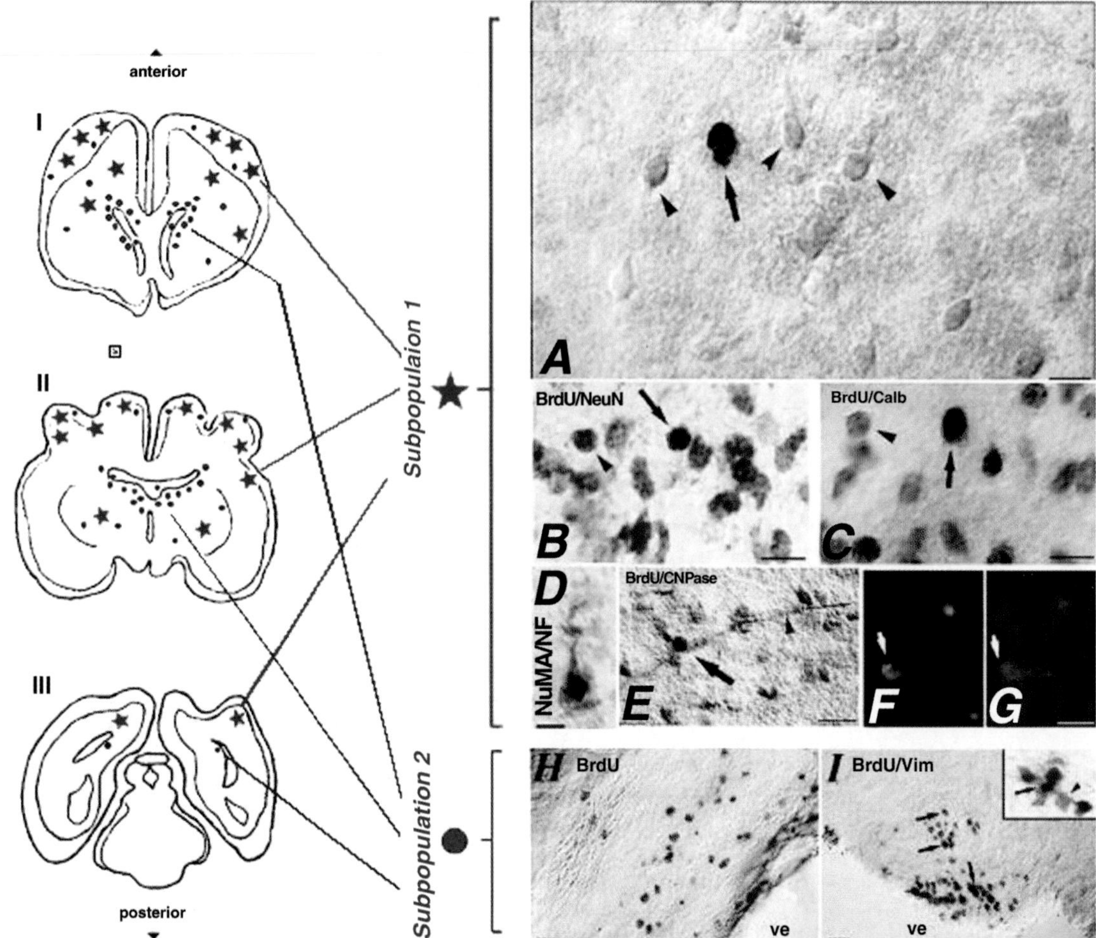

Fig. 5. Segregation of the fates of hNSCs and their progeny into two subpopulations following implantation into the ventricular zones (VZs) of developing Old World monkey brains (modified from Ourednik et al., 2002). Schematics (left) and photomicrographs (right) illustrate the cells. Coronal sections from representative levels throughout the telencephelon are illustrated on the left. hNSCs (prelabeled with BrdU) and implanted dispersed throughout the ventricular system and integrated into the VZ. From there, clonally related hNSC-derived cells pursued one of two fates, as shown by immunocytochemical analysis (**A–I**). Those donor cells that migrated outward from the VZ along radial glial fibers into the developing neocortex constituted one pool or subpopulation. The differentiated phenotype of cells in this subpopulation 1 (red stars in the schematic) (particularly in layers II/III) are pictured in (A–G). (A) A human NSC-derived cell (black nucleus, arrow)—likely a neuron according to its size, morphology, large nucleus, and location—is visualized (under Nomarski optics) intermingled with the monkey's own similar neurons (arrowheads) in neocortical layers II/III. The neuronal identity of such donor-derived cells is confirmed in (B–D). (B,C,E–G) High-power photomicrographs of human donor-derived cells integrated into monkey cortex double-stained with antibodies against BrdU and cell type–specific markers: (B) NeuN and (C) calbindin (Calb) for neurons (arrows = donor-derived cells; arrowheads = host-derived cells); (E) CNPase for oligodendroglia (arrow = BrdU$^+$ black nucleus in CNPase$^+$ brown cell; arrowhead indicates long process emanating from the soma); (F,G) GFAP for astroglia (anti-BrdU$^+$ via fluorescein in [F]; anti-GFAP via Texas Red in [G]). The human origin of the cortical neurons is further independently confirmed in (D), where the human-specific nuclear marker NuMA (black nucleus) is colocalized in the same cell with NF immunoreactivity (brown). Progeny from this same hNSC clone were also allocated to a second cellular pool—subpopulation 2 (blue dots in the schematic and pictured in [H,I, arrows])—which remained mainly confined to the SVZ and stained only for an immature neural marker (vimentin [brown] colocalized with BrdU [black nucleus] in inset, arrows; arrowhead indicates host vimentin + cell). Some members of subpopulation 2 were identified within the developing neocortex intermixed with differentiated cells ([F] and [G] employ immunofluorescence; the other immunostains use a DAB-based color reaction). (The photomicrographs were taken from different animals as representative of all animals.) ve, lateral cerebral ventricle; arrow, BrdU$^+$ donor-derived cell; arrowhead, BrdU$^-$, host-derived cell. Bars: 30 (A–C) and 20 μm (D–I).[Modified from Ourednik et al. (2001a).]

SVZ, VZ, subgranular layer of the hippocampus, central canal of the spinal cord—but also parenchymal regions (e.g., olfactory bulb, cortex, striatum) from immature (either fetal or newborn) but also adult rodents (with somewhat less efficiency). The tissue is then incubated in 0.1% trypsin (with EDTA) in phosphate-buffered saline (PBS) at 37°C for approx 30 min. If the tissue is abstracted from adult animals, dissociation will probably require collagenase as well as longer incubation times. The tissue is then triturated well and centrifuged for 3 min at 1000 rpm. The supernatant is discarded and the cell pellet is resuspended in serum-containing medium, which is followed by centrifugation again. Serum both inactivates the trypsin and ensures the health of the dissociated cells. This washing step is repeated once. After discarding the supernatant, the cells can be resuspended in serum-free medium (e.g., Dulbecco's minimal essential medium [DMEM] plus F12 base medium plus a series of hormones and trace elements (Snyder and Kim, 1979; Bottenstein and Sato, 1980) with the added mitogens EGF (10–20 ng/mL), bFGF (10–20 ng/mL), heparin, as a cofactor for bFGF (8 μg/mL) and leukemia inhibitory factor (LIF) (10 ng/mL). Cells can then be plated onto uncoated, tissue culture–treated dishes.

Stem cells occupy a small population of the CNS (or of any solid organ, for that matter) and, hence, must be expanded ex vivo to a useful graft material. As an alternative to the use of mitogens, cells can be propagated by transducing them with a gene that works downstream of these mitogens directly on the cell cycle, such as *vmyc*. To employ this genetic approach most effectively, the primary culture just described is infected 24–48 h after plating, by incubation with a retrovirus encoding the gene of interest (clone C17.2 was transduced with MLV-derived vector [Kaplan et al., 1987] called PK-VM-2 encoding avian *myc* or viral *myc* [v*myc*] plus a selectable antibiotic [neomycin] resistance gene [neo]). Successfully transduced cells will be neomycin resistant and, hence, are used to select infectants. Cells are then cultured in DMEM + 10% fetal bovine serum (FBS) + 5% horse serum (HS) + 2 m*M* glutamine for 3–7 d, until the cultures have undergone at least two doublings. At that point, cultures are trypsinized and seeded at 3–10% confluence in DMEM + 10% FBS, along with 0.3 mg/mL of G418 (a neomycin analog) for selection. Neomycin-resistant colonies are typically observed within 7–10 d. Chosen colonies are then isolated by brief exposure to trypsin within plastic cloning cylinders at that time. Individual colonies are replated and expanded in separate uncoated 24-well Costar plates. At confluence, these cultures are further passaged and serially expanded first to 35-mm dishes, then 6-cm, and ultimately 10-cm uncoated tissue culture dishes. Cells infected with control vectors without propagating genes do not survive beyond this passage in serum-containing medium.

NSC lines can then be transfected with additional genes, such as *lacZ*, the gene encoding *Escherichia coli* β-Gal. Such a reporter gene is stably integrated into clone C17.2. This gene product can be detected in vivo and in vitro by X-gal histochemistry or by anti-β-Gal immunohistochemistry, making such cells ideal for transplantation studies. The procedure for transfecting with *lacZ* begins with plating a recent passage of the target cells onto 60-mm tissue culture dishes. Twenty-four to 48 h after plating, the cells are incubated with a *lacZ*-encoding retroviral vector (e.g., BAG) plus 8 μg/mL of polybrene for 1–4 h (a procedure similar to that used for v*myc*). The polycation polybrene enhances the effectiveness of retroviral vectors by permeabilizing the cell membrane. Cells are then cultured in fresh feeding medium (DMEM + 10%FCS + 5%HS + 20 m*M* glutamate) for approx 3 d until they appear to have undergone at least two doublings. Cultures are trypsinized and seeded at low density, 50–5000 cells, on a 10-cm tissue culture dish. If the newly inserted transgene contains a selection marker distinct from the first selection marker (a nonneomycin-based selection marker such as hygromycin or puromycin), then one can select for the new infectants as described above. After about 3 d, well-separated colonies are isolated by brief exposure to trypsin within plastic cloning cylinders. If the new transgene does not have a new selection marker or also employs *neo*, identification of infectants is still possible. The low seeding density allows for relatively easy isolation of colonies with cloning cylinders, which can then be plated separately into 24-well plates. When confluent, these cultures can be expanded into increasingly larger tissue culture dishes. A representative dish from each clone can then be stained histochemically for the new gene product of interest. In the case of l*acZ* infectants, β-Gal expression is examined via X-gal histochemistry. The percentage of X-gal$^+$ blue cells is assessed microscopically, and the clones with the highest percentage of blue cells are then used for future studies. Although these procedures were described specifically for *lacZ*, they are applicable to almost any transgene of interest, whether an alternative reporter gene such as green fluorscent protein, or a therapeutic gene, such as TH.

With the identification of viable, transgene-expressing clones, the primary concerns become the expansion and cryopreservation of cell lines. Preferred mNSC lines are grown in DMEM + 10% FBS + 5% HS + 2 m*M* glutamine on coated tissue culture–treated dishes. In the past, we have grown C17.2 cells on dishes precoated with poly-L-lysine. However, either through curiosity or laziness, we have discovered that this cell line grows quite readily on uncoated dishes. The rate of growth typically seen in mNSCs is on the order of an 18 to 24-h doubling time. Normally growing cultures can be fed twice per week, with the cells being passaged during one of the feedings. The feeding alone consists of merely removing old medium and adding fresh medium. It may be beneficial to retain approx 25% of the old medium because it has been "conditioned"; however, because these cells grow fairly quickly they have no trouble "conditioning" fresh medium. Passaging consists of trypsinizing and transferring cells to new dishes. After removal of the medium, PBS should be added to the dish and then aspirated to remove any traces of serum. A small amount of 0.05% trypsin/EDTA is added to the dish to just cover the cell layer—0.5 mL works well for a 10-cm dish. The cultures are then incubated at 37°C for 3 min. On removal from the incubator, several milliliters of normal feeding medium should be added to the dish to inactivate the trypsin and allow for easy trituration. Cells are then transferred to a centrifuge tube and spun at 1000 rpm for 3 min. The supernatant is discarded and the cells are resuspended in 10 mL of fresh medium. We generally passage mNSCs at a 1:10 ratio, which can be accomplished in a couple of ways. For example, if one is growing a single 10-cm dish of cells, and wishes to maintain this scale, then 1 mL of the cell suspension can be transferred to a new dish and the remaining 9 mL can be discarded. Nine milliliters of fresh medium should be subsequently added to the dish. However, since the main concern here is to expand cells to be frozen, the second method of obtaining a 1:10 passage should be employed. Simply increasing the available growing surface 10-fold will accomplish this. Ten-centimeter dishes have a growing surface of approx 75cm^2 and will support about 5 million cells when confluent. Alternatively, T150 flasks (with a 150-cm^2 growing surface) may be used. These large flasks are ideal for safely growing large numbers of cells to be frozen. When these cultures have reached no more than 90% confluence, the cells can be harvested for freezing. Ten 10-cm dishes, or five T150 flasks, will yield roughly 40–50 million cells. The freezing medium consists of 70% normal growth medium with an additional 20% FBS, plus 10% dimethyl sulfoxide (DMSO). Cells should be trypsinized as described above; however, following centrifugation the cells should be resuspended in prechilled freezing medium. The desired concentration is 1×10^6 cells/mL, with 1 mL of this suspension placed into each prelabeled cryovial. The cryovials are kept on ice until they are transferred to a controlled-rate freezing device that is put into a –80°C freezer. This device allows the cells to cool at approximately 1°C/min. This transfer should be done as quickly as possible. After 24–48 h, and no longer than 2 wk, the cryovials should be transferred to a –140°C mechanical freezer or a liquid nitrogen freezer, where they can be maintained indefinitely.

It is always advisable to thaw a representative vial from each freeze to check for viability. To thaw these cells, it is best initially to seed them into a small dish, such as a 35-mm dish, containing 2 to 3 mL of prewarmed growth medium. Quickly thaw the cells and transfer them immediately to the dish. Centrifugation is not recommended, because the cells are extremely fragile when coming out of freeze. Over the course of 8 hours the cells will settle and attach to the dish, at which point the medium should be aspirated and replaced with fresh medium. Given that the cells successfully recover from freezing, one can feel secure in knowing that no matter what happens to growing cultures, many vials of early passage cells are stored safely.

17.9.2. PREPARATION OF mNSCs FOR TRANSPLANTATION Just as with freezing, cultures selected for transplantation studies should be no more than 80–90% confluent and still undifferentiated and in a log growth phase. Passaging at a 1:5 ratio 48 h before a transplantation procedure generally yields the proper density. If cells are allowed to become too confluent for more than 48 h prior to transplantation, they begin to elaborate an ECM that causes them to become clumpy and renders poor engraftment and often-autonomous clusters of cells in vivo. Confluent cells also begin to exit the cell cycle and differentiate spontaneously, which also predisposes them to poor engraftment. The more immature the NSCs, the better the engraftment, integration, and migration. Sister cultures of those chosen for transplantation can be allowed to grow to confluence and analyzed in vitro for their ability to differentiate into mature neural phenotypes and the efficiency of reporter transgene expression (e.g., *lacZ*). If the cells do not express *lacZ* well in a dish, they are even less likely to be identifiable by such a gene in the brain. In such a case, the experiments either should be aborted or an alternative marker should be used—e.g., preincubating the cells in bromodeoxyuridine (BrdU) ex vivo prior to transplantation (Flax et al., 1998; Ourednik et al., 2001b) or using a species-specific marker in xenograft paradigms (*see* below for more details).

For NSCs that have been engineered with transgenes via retrovirus, there is another criterion to consider when selecting a clone for transplantation. "Helper virus-free" status needs to be affirmed prior to transplantation; otherwise, host-derived cells may be falsely interpreted as being donor derived. (Helper virus refers to a replication-competent virus that emerges from a replication-incompetent vector owing to unintended recombination events in which enabling genes get placed before the transgene). This status can be confirmed by growing naive cells such as 3T3 fibroblasts with the supernatant from the NSC line in question. If these conditions fail to produce neomycin-resistant colonies or X-gal$^+$ colonies in the naive cells, then the NSCs are considered "helper virus free" and can be legitimately used for transplantation studies.

Cells to be used for transplantation are trypsinized well and gently triturated to yield a single-cell suspension. Cells are then washed in medium and finally resuspended in PBS or Hank's balanced salt solution (HBSS) for injection. A cellular concentration of 5–10 × 10^4cells/μL is ideal for most injections, tending more to the lower end of this range. Concentrations above this level can lead to obstruction of the needle. Although we use finely drawn glass micropipets, one can also employ a Hamilton syringe. The tip should be as thin as possible while still allowing for the ready flow of a well-dissociated single-cell suspension. We color the tip black to help with its visualization during penetration, especially in young animals whose brains can be transilluminated. A few microliters of concentrated trypan blue (0.04% [w/v]) can also be added to the cell suspension to permit ready visualization of the suspension during injection as well as to calculate the percentage of viable cells (by trypan blue exclusion). Cells should be kept on ice until injection to minimize clumping. However, this alone is not sufficient; the cells will settle out of suspension within minutes and should be repeatedly triturated to maintain a uniform distribution. Even within the needle or micropipet the cells can settle; therefore, it is advisable to draw up only enough cells to implant one or two animals at a time.

A typical intraventricular injection into a newborn mouse consists of a 2-μL NSC suspension injected into each ventricle. The subject is placed against a fiberoptic light source to visualize the ventricles and the NSCs are injected over 1 to 2 s. The best engraftment of cells is usually found when implanted into mice that are P0–P3.

Once NSCs are implanted, the focus switches to the identification of donor cells in vivo. While the most efficient method for detecting cells in vivo is via the presence of a transgene (e.g., *lacZ*), the risk of downregulation of the transgene warrants the simultaneous use of other donor cell detection strategies. Even if the transgene is downregulated, it can still be detected through fluorescent *in situ* hybridization (FISH) against the transgene itself. In addition, by using NSCs derived from a male animal, and implanting them into a female animal, one can perform FISH against the Y-chromosome (Gussoni et al., 1999). NSCs can also be preincubated in BrdU 48 h prior to transplantation at a concentration of 10 μ*M* in the culture medium, although if a transplant had to be done sooner than expected, 24 h of BrdU preincubation will suffice. BrdU is taken up by mitotic cells and incorporated into their DNA. This presents another reason for using subconfluent cells for transplantation, because they will most likely be dividing normally, giving a high yield of BrdU$^+$ cells. These cells can subsequently be identified in vivo through the use of an anti-BrdU antibody that will indicate donor-derived cells via their immunopositive nucleus (Flax et al., 1998; Ourednik et al., 2001b). Most NSCs undergo zero to two cell divisions in vivo following transplantation, usually not enough to "dilute" the BrdU marker (which remains stably detectable for as many as six murine cell divisions).

17.9.3. HUMAN NSCs-ISOLATION AND PROPAGATION The isolation and propagation of NSCs from the human CNS followed a path well trod by prior experience with the successful mNSC clone C17.2. A suspension of primary dissociated neural cells (5 × 10^5 cells/mL) prepared from the periventricular region of the telencephalon of an early second-trimester human fetal cadaver (as detailed elsewhere [Flax et al., 1998]) was plated on uncoated tissue culture dishes. Cells were first plated in a serum-containing medium for approx 1 wk and then switched to a serum-free medium. This latter medium consists of essentially the same serum-free medium described earlier for mouse. The following growth factors were added at the time of feeding: bFGF (10–20 ng/mL); heparin, as a cofactor for bFGF (8 μg/mL); EGF

(10–20 μg/mL); and LIF (10 ng/mL). Cultures were then put through a "growth factor" selection process, under the assumption—based on our much earlier prior work—that true NSCs bear a receptor to both FGF and EGF simultaneously. Cells were transferred to serum-free medium containing bFGF and heparin alone for 2 to 3 wk, transferred to serum-free medium containing EGF alone for 2 to 3 wk, and then returned to serum-free medium containing bFGF and heparin alone for 2 to 3 wk. Finally, they were maintained in serum-free medium containing bFGF, heparin, and LIF. This selection process eliminates most cells, and the only ones left surviving, dividing, and capable of being passaged at its conclusion were immature, proliferative cells that were responsive to both FGF and EGF. These cells also fulfilled our operational definition of a neural stem cell, capable of giving rise to mature progeny of all neural cell types in vitro and in vivo throughout the neuraxis as well as to new stemlike cells.

On completion of the selection process, cells were maintained on a schedule of medium changes every 5–7 d. Cell aggregates were dissociated in 0.05% trypsin/EDTA (when the size was >10 cells in diameter) and replated in growth medium at 2×10^5 cells/mL. Some dissociated NSCs were set aside for characterization and plated on poly-L-lysine–coated slides in DMEM + 10% FBS. In these conditions, most cells differentiated spontaneously and could be processed with immunocytochemistry to search for neuronal and glial markers to verify multipotency.

Polyclonal populations of the hNSCs were then separated into single clonal lines either by serial dilution alone (i.e., one cell per well) or by first infecting the cells with a retrovirus and then performing serial dilution. The proviral integration site also provides a molecular marker of clonality. The transgenes transduced via retrovirus were either nontransforming propagation enhancement genes such as v*myc*, or purely reporter genes such as *lacZ*, or both. Two xenotropic, replication-incompetent retroviral vectors were used to infect hNSCs. A vector encoding *lacZ* was similar to BAG except for bearing a PG13 xenotropic envelope. An amphotropic vector encoding v*myc* was generated using the ecotropic vector described for generating mNSC clone C17.2 to infect the GP+envAM12 amphotropic packaging line (Flax et al., 1998). Helper virus-free status was confirmed using the technique described for mNSCs. The titer used for either vector was 4×10^5 CFU, and the procedure for infection was similar to that described for murine cells.

For cloning of hNSCs, cells were dissociated and diluted to 1 cell/15 μL of bFGF-containing growth medium and plated at 15 μL/well of a 96-well plate. Low-density cultures grow best when the medium is supplemented with 25–50% conditioned medium from previously grown dense cultures. Wells plated with single cells were identified and expanded. The monoclonality of these colonies was confirmed by the presence of a single and identical genomic insertion site on Southern analysis for either the *lacZ*- or the v*myc*-encoding provirus (Flax et al., 1998).

As with the mNSCs, once viable, transgene-expressing hNSC populations or clones are identified, expansion and cryopreservation of cells becomes all important. For standard culture conditions, hNSCs are normally maintained in DMEM + F12 medium, supplemented with N2 medium, bFGF, heparin, and LIF. Because of their relatively slow cell cycle, anywhere from 3-6 days or longer, hNSCs may not need to be passaged every week; however, the medium should be changed every 5–7 d. It is advisable to retain 25% of the old conditioned growth medium at each change, supplemented with 75% fresh medium. Determining the need to passage hNSCs may not be as straightforward as with many cell lines. They tend to grow as large cell aggregates—sometimes floating, sometimes attaching to the surface of the culture dish and separated but with many processes traversing between and connecting the aggregates. This often gives the culture a "honeycomb" appearance. If this "honeycomb" covers the entire surface of the vessel, the culture is most likely ready to be passaged, even though it is not confluent in the traditional sense. Indeed, many of the cells will have already differentiated and these will not passage well. It will be the phase bright, nonprocessing bearing, round cells that will likely be immature and proliferative enough to passage as stemlike cells. It makes sense to tap the flask on a daily basis to dislodge as many attached cells as possible each day and forestall their differentiation. Another criterion for passaging is if the aggregates have grown to >10–20 cells in diameter. Beyond this point cells in the center are unable to obtain necessary nutrients and gas exchange is compromised. The procedure for trypsinizing the hNSCs is outlined below in Subheading 17.9.4.

hNSCs are cryopreserved by resuspending cells in a freezing solution composed of 10% DMSO, 50% FBS, and 40% bFGF-containing growth medium. It is best not to trypsinize hNSCs immediately prior to freezing; twenty-four to forty-eight hours before the freeze is ideal. To harvest cells for freezing, most can be dislodged from the culture vessel with trituration alone. Cells are normally concentrated at 1×10^6/mL of freezing solution. The cells should be transferred to a controlled-rate freezing device as quickly as possible. The cells are then stored at –80°C for 24–48 h and transferred to ultracold storage, either a –140°C mechanical freezer or a liquid nitrogen freezer. They can be maintained in the latter conditions indefinitely.

Thawing hNSCs can be tricky. Like the mNSCs, hNSCs are very fragile after thawing and should not be centrifuged right away. In the past, we have thawed hNSCs into prewarmed medium in a small dish, 35–60 mm, and waited 8 h before changing the medium. Unlike mNSCs, however, many hNSCs do not attach to the dish quickly, and in order to retrieve the suspended cells when the medium is changed 8 h later, they must be centrifuged. Moreover, many of the cells that do attach will differentiate because of the high serum content of the freezing medium. Some cells will survive and proliferate, but the yield is generally low. Although this technique still has merit, we have recently tried a new thawing technique designed to counteract these problems. When hNSCs are thawed, they are placed into a 15-mL centrifuge tube with approx 9 mL of prewarmed feeding medium. The tube is placed vertically in an incubator at 37°C for approx 1–2 h. During this time, the cells will settle, but they will not differentiate due to the nonpermissive surface of the centrifuge tube. After 1–2 h the tube should be centrifuged at 700 rpm for 1 min. The mixture of freezing and feeding medium can then be safely aspirated. Trying to aspirate the medium without centrifugation will result in the loss of many cells, and since the centrifugation is done at a low rate for a short period, the cells tolerate it well. The pellet is then resuspended in a mixture of 50% fresh feeding medium and 50% conditioned feeding medium. This conditioned medium should be collected from previously growing cultures, filtered, and stored at –80°C. After 1 or 2 wk in culture, the percentage of conditioned medium may be reduced to 25%. Although the yield of viable and immature cells appears to be higher with the latter technique, it can still take several weeks after a thaw to amass usable numbers of cells.

17.9.4. PREPARATION OF hNSCs FOR TRANSPLANTATION While most hNSCs tend to grow attached to the vessel, many viable cells remain suspended and all should be collected. Adherent cells are mechanically dislodged by trituration of the medium, which is then transferred to a centrifuge tube. Cells are spun for 3 min at 1000 rpm. The supernatant is discarded and 0.7 mL of trypsin/EDTA (0.05% trypsin, 0.53 m*M* EDTA) is added. For larger cell pellets containing more than about 5 million cells, a larger volume of trypsin/EDTA should be used. The cells are again triturated briefly before a 5-min incubation at 37°C. Trypsinization is then terminated by adding an equal volume of soybean trypsin inhibitor (0.25 mg/mL in PBS) into the tube, and triturating the mixture thoroughly. Cells are spun once again for 3 min at 1000 rpm, the supernatant is removed, and the cells are resuspended in medium. This wash step can be repeated if there is a sufficient supply of cells for the given transplantation, but if cells are scarce, it is best to omit the second wash because cells are inevitably lost with each centrifugation. Following the final centrifugation, cells are resuspended in a small volume of PBS or HBSS and counted, and the volume is adjusted to achieve a concentration of 5–10 × 10^4 cells/μL for injections. Trypan blue (0.04% [w/v]) can be added to permit visualization of the injected suspension, if desired, as well as to calculate the percentage of viable cells by trypan blue exclusion. As described earlier for mNSCs, hNSCs may be injected with either a finely drawn glass micropipet or a Hamilton syringe. Cells are kept on ice until injection and should be triturated repeatedly to maintain a uniform density and avoid clumping. Like mNSCs, these cells settle quickly so only enough cells to inject one or two animals should be drawn into the needle or micropipet at a time.

Transplantation procedures for hNSCs are identical to those used for mNSCs. Many of the methods for detecting donor cells in vivo are also similar. One notable exception is the BrdU prelabeling technique. Since the hNSCs divide slower than their murine counterparts, BrdU (10 μ*M*) should be added at least 48 h prior to transplantation, and preferably 72 h, in order to catch more cells undergoing mitosis. Cells cannot tolerate BrdU indefinitely, however, and the yield of BrdU$^+$ cells in vitro may be unacceptably low. With this in mind, we have tried other prelabeling techniques using such substances as DiI. This dye does not need to be added to the medium of growing cultures beforehand. It can be included in the transplant preparation protocol according to manufacturers' specifications, while only slightly lengthening the time it takes to prepare cells. We are currently testing the efficacy of a new substance as a prelabeling dye, carboxyfluorescein, which may have better retention within the cell than other dyes. A very powerful technique for detecting hNSC-derived cells in vivo in the rodent or monkey CNS is to use human-specific antibodies such as NuMA, human EGF receptor, human nuclear antigen, or human mitochondrial antigen (Aboody et al., 2000; Ourednik et al., 2001a). In addition, human-specific neural cell–type-specific antibodies can perform the dual task of identifying hNSC-derived cells while ascertaining their phenotype, (e.g., human-specific tau and neurofilament for neurons, human-specific GFAP for astrocytes) (Flax et al., 1998).

17.10. INTO THE FUTURE

We have touched on the many possibilities for NSCs as therapeutic tools for the future. Will these multipotent cells truly be a "godsend" or just a "devil in a trypan blue dress"? Time and further research will be the judge, but the evidence is mounting that clearly points to their place among the armamentarium of therapeutic weapons. Nevertheless, many aspects concerning the clinical potential of NSCs remain speculative and much work needs to be done (Snyder and Park, 2002). There are many unanswered questions about the intrinsic properties of these cells. We have outlined the cell culture conditions that have worked for us, but surely these can be refined, or even redefined, to give the highest possible yield of multipotent and engraftable NSCs. In vivo, we have only touched on the many parameters involved in successfully delivering cells to areas of pathology and allowing them the optimal opportunity to integrate functionally into host tissue. Are they best left to perform these functions spontaneously, or should they be instructed and preinduced to do so each step of the way? Such a fundamental question as this remains a topic of future investigations. The field of NSC biology is quite young and promising and deserves the highest scrutiny that basic and applied scientific approaches can provide to optimize both efficacy and safety. Thus, we are left with perhaps a new phrase, at least as far as CNS therapies are concerned: "Think globally and act locally, with global effects, but not too global." It is not very catchy, but if NSCs could be employed in such a manner to effectively aid in the treatment of neurological diseases, it would sound very nice indeed.

ACKNOWLEDGMENTS

We thank our many collaborators, some of whom are listed here: Drs. Mellitta Schachner, Jonathan D. Flax, John H. Wolfe, David Wenger, Itzhak Fischer, Alan Tessler, Tim Himes, Mark Tuszynksi, Paul Lu, Xandra Breakefield, Ralph Weisleder, Ulrich Herrlinger. Some of the work described here was supported in part by grants from the National Institute of Neurologic Diseases and Stroke, March of Dimes, Project ALS, Brain Tumor Society, Hunter's Hope, Canavan Research Fund, Late Onset Tay Sachs Foundation, A-T Children's Project, International Organization for Glutaric Acidemia, and Parkinson's Action Network/Michael J. Fox Foundation; grant no. 981-0713-097-2 from the Basic Research Program and BDRC of the Korean Science and Engineering Foundation; and grant HMP-98-N-1-0003 from the Ministry of Health & Welfare, R. O. Korea.

REFERENCES

Aboody, K. S., Brown, A., Rainov, N. G., er al. (2000) Neural stem cells display extensive tropism for pathology in the adult brain: evidence from intracranial gliomas. *Proc. Natl. Acad. Sci. USA* 97:12,846–12,851.

Akerud, P., Canals, J. M., Snyder, E. Y., and Arenas, E. (2001) Neuroprotection through delivery of GDNF by neural stem cells in a mouse model of Parkinson's disease. *J. Neurosci.* 21(20):8108–8118.

Alvarez-Buylla, A. and Temple, S. (1998) Neural stem cells. *J. Neurobiol.* 36:105–314.

Anton, R., Kordower, J. H., Maidment, N. T., et al. (1994) Neural-targeted gene therapy for rodent and primate hemiparkinsonism, *Exp. Neurol.* 127:207–218.

Anton, R., Kordower, J. H., Kane, D. J., Markahma, C. H., and Bredesen, D. E. (1995) Neural transplantation of cells expressing the anti-apoptotic gene bcl-2. *Cell Transplant.* 4:49–54.

Auguste, K. I., Nakajima, K., Miyata, T., et al. (2000) Neural progenitor transplantation into *reeler* cerebellum complements mutant lamination and neuronal survival by Reelin-and non-Reelin-producing processes. *J. Neurosci.*, in revision.

Bjorklund, A. and Lindvall, O. (2000). Cell replacement therapies for central nervous system disorders. *Nat. Neurosci.* 3:537–544.

Bottenstein, J. E. and Sato, G. H. (1980) Fibronectin and polylysine requirement for proliferation of neuroblastoma cells in defined medium. *Exp. Cell. Res.* 129(2):361–366.

Brüstle, O., Choudhary, K., Karram, K., et al. (1998) Chimeric brains generated by intarventricular transplantation of fetal human brain cells into embryonic rats. *Nat. Biotechnol.* 11:1040–1049.

Chen, K. S. and Gage, F. H. (1995) Somatic gene transfer of NGF to the aged brain: behavioral and morphologic amelioration. *J. Neurosci.* 15 (4):2819–2825.

Doering, L. and Snyder, E. Y. (2000) Cholinergic expression by a neural stem cell line grafted to the adult medial septum/diagonal band complex. *J. Neurosci. Res.* 61:597–604.

Dunnett, S. B. and Bjorklund, A. (2000) *Functional Neural Transplantation*, Raven, New York.

Fisher, L. J. (1997) Neural precursor cells: applications for the study and repair of the central nervous system. *Neurobiol. Dis.* 4:1–22.

Flax, J. D., Aurora, S., Yang, C., et al. (1998) Engraftable human neural stem cells respond to developmental cues, replace neurons, and express foreign genes. *Nat. Biotechnol.* 16:1033–1039.

Freed, C. R., Greene P. E., Breeze, R. E., et al. (2001) Transplantation of embryonic dopamine neurons for severe Parkinson's disease. *N. Engl. J. Med.* 344:763–765.

Fricker, R. A., Carpenter, M. K., Winkler, C., Greco, C., Gates, M. A., and Bjorklund, A. (1999) Site-specific migration and neuronal differentiation of human neural progenitor cells after transplantation in the adult rat brain. *J. Neurosci.* 19(14):5990–6005.

Gage, F. H. (2000) Mammalian neural stem cells. *Science* 287:1433–1438.

Goldman, S. A. and Luskin, M. B. (1998) Strategies utilized by migrating neurons of the postnatal vertebrate forebrain. *Trends Neurosci.* 21(3): 107–14.

Gould, E., Reeves, A. J., Graziano, M. S., and Gross, C. G. (1999) Neurogenesis in the neocortex of adult primates. *Science* 286:548–552.

Grill, R., Murai, K., Blesch, A., Gage, F. H., and Tuszynski, M. H. (1997) Cellular delivery of neurotrophin-3 promotes corticospinal axonal growth and partial functional recovery after spinal cord injury. *J. Neurosci.* 17:5560–5572.

Gritti, A., Bonfanti, L., Doetsch, F., et al. (2002) Multipotent neural stem cells reside into the rostral extension and olfactory bulb of adult rodents. *J. Neurosci.* 22(2):437–45.

Gussoni, E., Soneoka, Y., Strickland, C. D., et al. (1999) Dystrophin expression in the mdx mouse restored by stem cell transplantation. *Nature* 401(6751):390–394.

Himes BT, Liu Y, Solowska JM, Snyder EY, Fischer I, Tessler A (2001) Transplants of cells genetically modified to express neurotrophin-3 rescue axotomized Clarke's nucleus neurons after spinal cord hemisection in adult rats. *J. Neurosci. Res.* 65(6):549–564.

Hodges, H., Veizovic, T., Bray, N., et al. (2000) Conditionally immortal neuroepithelial stem cell grafts reverse age-associated memory impairments in rats. *Neuroscience* 101:945–955.

Kaplan, P. L., Simon, S., Cartwright, C. A., and Eckhart, W. (1987) cDNA cloning with a retrovirus expression vector: generation of a pp60c-src cDNA clone. *J. Virol.* 61:1731–1734.

Kordower, J. H., Freeman, T. B., Snow, B. J., et al. (1995) Neuropathological evidence of graft survival and striatal reinnervation after the transplantation of fetal mesencephalic tisssue in a patient with Parkinson's disease. *N. Engl. J. Med.* 332(17):1118–1124.

Lacorazza, H. D., Flax, J. D., Snyder, E. Y., and Jendoubi, M. (1996) Expression of human β-hexosaminidase α-subunit gene (the gene defect of Tay-Sachs disease) in mouse brains upon engraftment of transduced progenitor cells. *Nat. Med.* 4:424–429.

Liu, Y., Himes, B. T., Solowska, J., et al. (1999) Intraspinal delivery of neurotrophin-3 using neural stem cells genetically modified by recombinant retrovirus. *Exp. Neurol.* 158:9–26.

Lois, C. and Alvarez-Buylla, A. (1994) Long distance neuronal migration in the adult mammalian brain. *Science* 264:1145–1148.

Lu, P., Jones, L., Park, K. I., Snyder, E. Y., and Tuszynski, M. (2000) Neural stem cells secrete BDNF and GDNF, and promote axonal growth after spinal cord injury [abstract]. *Soc. Neurosci. Abstr.* 26:332.

Luskin, M. B. (1993) Restricted proliferation and migration of postnatally generated neurons derived from the forebrain subventricular zone. *Neuron* 11:173–189.

Lynch, W. P., Sharpe, A. H., and Snyder, E. Y. (1999) Neural stem cells as engraftable packaging lines optimize viral vector-mediated gene delivery to the CNS: evidence from studying retroviral *env*-related neurodegeneration. *J. Virol.* 73:6841–6851.

Magavi, S. S., Leavitt, B. R., and Macklis, J. D. (2000) Induction of neurogenesis in the neocortex of adult mice. *Nature* 405:951–955.

Martinez-Serrano, A., Lundberg, C., Horellou, P., et al. (1995a) CNS-derived neural progeniutor cells for gene transfer of nerve growth factor to the adult rat brain: complete rescue of axotomized cholinergic neurons after transplantation into the septum. *J. Neurosci.* 15(8): 5668–5680.

Martinez-Serrano, A., Fischer, W., and Bjorklund, A. (1995b) Reversal of age-dependent cognitive impairments and cholinergic neuron atrophy by NGF-secreting neural progenitors grafted to the basal forebrain. *Neuron* 15:473–484.

McKay, R. (1997) Stem cells in the central nervous system. *Science* 276:66–71.

Neufeld, E. F. and Fratantoni, J. C. (1970) Inborn errors of mucopolysaccharide metabolism. *Science* 169:141–146.

Nikkhah, G., Cunningham, M. G., Cenci, M. A., McKay, R. D., and Bjorklund, A. (1995a) Dopaminergic microtransplants into the substantia nigra of neonatal rats with bilateral 6-OHDA lesions. I. Evidence for anatomical reconstruction of the nigrostriatal pathway. *J. Neurosci.* 15(5):3548–3561.

Nikkhah, G., Cunningham, M. G., McKay, R., and Bjorklund, A. (1995b) Dopaminergic microtransplants into the substantia nigra of neonatal rats with bilateral 6-OHDA lesions. II. Transplant-induced behavioral recovery. *J. Neurosci.* 15(5):3562–3570.

Ourednik, J., Ourednik, V., Lynch, W. P., Snyder, E. Y., and Schachner, M. (1999) Massive regeneration of substantia nigra neurons in aged parkinsonian mice after transplantation of neural stem cells overexpressing L1 [abstract]. *Soc. Neurosci. Abstr.* 25:1310.

Ourednik, V., Ourednik, J., Flax, J. D., et al. (2001a) Segregation of human neural stem cells in the developing primate forebrain. *Science* 293(5536):1820–1824.

Ourednik, V., Ourednik, J., Kosaras, B., Sidman, R. L., and Snyder, E. Y. (2001b) Nerve cell rescue and behavioral changes induced by grafted clonal neural stem cells in ataxic cerebellar mouse mutants [abstract]. *Soc. Neurosci. Abstr.* 27:371.3.

Ourednik, V., Ourednik, J., Lynch, W. P., Snyder, E. Y., and Schachner, M. (2002) Neural stem cells display an inherent mechanism for rescuing dysfunctional neurons. *Nat. Biotechnol.* 20(11):1103–1110.

Park, K. I., Jensen, F. E., Stieg, P. E., and Snyder, E. Y. (1998) Hypoxic-ischemic (HI) injury may direct the proliferation, migation, and differentiation of endogenous neural progenitors [abstract]. *Soc. Neurosci. Abstr.* 24:1310.

Park, K. I., Liu, S., Flax, J. D., Nissim, S., Stieg, P. E., and Snyder, E. Y. (1999) Transplantation of neural progenitor and stem-like cells: developmental insights may suggest new therapies for spinal cord and other CNS dysfunction. *J. Neurotrauma* 16(8):675–687.

Park, K. I., Teng, Y. D., and Snyder, E. Y. (2002) The injured brain interacts reciprocally with neural stem cells supported by scaffolds to reconstitute lost tissue. *Nat. Biotechnol.* 20(11):1111–1117.

Park, K. I., Lee, K. H., Lee, B. W., Teng, Y. D., and Snyder, E. Y. (2002) Transplanted human neural stem cells replace lost neural cells and promote neuronal reinnervation and functional recovery in injured rat spinal cord [abstract]. *Soc. Neurosci. Abstr.* Program 825.9.

Pincus, D. W., Keyoung, H. M., Harrison-Restelli, C., et al. (1998). FGF2/BDNF-associated maturation of new neurons generated from adult human subependymal cells. *Ann. Neurol.* 43:576–585.

Reubinoff, B. E., Itsykson, P., Turetsky, T., et al. (2001) Neural progenitors from human embryonic stem cells: derivation, expansion, and characterization of their developmental potential *in vitro* and *in vivo*. *Nat. Biotechnol.* 19:1134–1140.

Rosario, C. M., Yandava, B. D., Kosaras, B., Zurakowski, D., Sidman, R. L., and Snyder, E. Y. (1997) Differentiation of engrafted multipotent neural progenitors toward replacement of missing granule neurons in meander tail cerebellum may help determine the locus of mutant gene action. *Development* 124:4213–4224.

Roy, N. S., Wang, S., Jiang, L., et al. (2000) *In vitro* neurogenesis by progenitor cells isolated from the adult human hippocampus. *Nat. Med.* 6:271–277.

Rubio, F. J., Kokai, Z., del Arco, A., et al. (1999) BDNF gene transfer to the mammalian brain using CNS-derived neural precursors. *Gene Ther.* 6:1851–1866.

Rubio, F. J., Bueno, C., Villa, A., Navarro, B., and Martinez-Serrano, A (2000) Genetically perpetuated human neural stem cells engraft and differentiate into the adult mammalian brain. *Mol. Cell Neurosci.* 16:1–13.

Ryder, E. F., Snyder, E. Y., and Cepko, C. L. (1990) Establishment and characterization of multipotent neural cell lines using retrovirus vector mediated oncogene transfer. *J. Neurobiol.* 21:356–375.

Sabate, O., Horellou, P., Vigne, E., et al. (1995) Transplantation to the rat brain of human neural progenitors that were genetically modified using adenoviruses. *Nat. Genet.* 9:256–260.

Snyder, E. Y. (1998) Neural stem-like cells: developmental lessons with therapeutic potential. *Neuroscientist* 4(6):408–425.

Snyder, E. Y. and Flax, J. D. (1995) Transplantation of neural progenitors and stem-like cells as a strategy for gene therapy and repair of neurodegenerative diseases. *Ment. Retard. Dev. Dis. Res. Rev.* 1:27–38.

Snyder, E. Y. and Kim, S. U. (1979) Hormonal requirements for neuronal survival in culture. *Neurosci. Lett.* 13(3):225–230.

Snyder, E. Y. and Park, K. I. (2002) Limitations in brain repair. *Nat. Med.* 8(9):928–930.

Snyder, E. Y., Deitcher, D. L., Walsh, C., Arnold-Aldea, S., Hartwieg, E. A., and Cepko, C. L. (1992) Multipotent neural cell lines can engraft and participate in development of mouse cerebellum. *Cell* 68:33–55.

Snyder, E. Y., Taylor, R. M., and Wolfe, J. H. (1995) Neural progenitor cell engraftment corrects lysosomal storage throughout the MPS VII mouse brain. *Nature* 374:367–370.

Snyder, E. Y., Yoon, C., Flax, J. D., and Macklis, J. D. (1997) Multipotent neural precursors can differentiate toward replacement of neurons undergoing targeted apoptotic degeneration in adult mouse neocortex. *Proc. Natl. Acad. Sci. USA* 94(21):11,663–11,668.

Suhonen JO, Peterson DA, Ray J, Gage FH (1996) Differentiation of adult hippocampus-derived progenitors into olfactory neurons in vivo. *Nature* 383:624–627.

Tate, B. A., Werzanski, D., Marciniak, A., and Snyder, E. Y. (2000) Migration of neural stem cells to Alzheimer-like lesions in an animal model of AD [abstract]. *Soc. Neurosci Abstr.* 26:496.

Teng, Y. D., Lavik, E. B., Qu, X., et al. (2002) Functional recovery following traumatic spinal cord injury mediated by a unique polymer scaffold seeded with neural stem cells. *Proc. Natl. Acad. Sci. USA* 99: 3024–3029.

Tuszynski, M. H. and Gage, F. H. (1995) Bridging grafts and transient nerve growth factor infusions promote long-term central nervous system neuronal rescue and partial functional recovery. *Proc. Natl. Acad. Sci. USA* 92(10):4621–4625.

Uchida, N., Buck, D. W., He, D., et al. (2000) Direct isolation of human central nervous system stem cells. *Proc. Natl. Acad. Sci. USA* 97: 14,720–14,725.

Vescovi, A. L., Parati, E. A., Gritti, A., et al. (1999) Isolation and cloning of multipotential stem cells from the embryonic human CNS and establishment of transplantable human neural stem cell lines by epigenetic stimulation. *Exp. Neurol.* 156:71–83.

Vescovi, A. L. and Snyder, E. Y. (1999) Establishment and properties of neural stem cell clones: plasticity *in vitro* and *in vivo*. *Brain Pathol.* 9:569–598.

Villa, A., Snyder, E. Y., Vescovi, A., and Martinez-Serrano, A. (2000) Establishment and properties of a growth factor-dependent, perpetual neural stem cell line from the human CNS. *Exp. Neurol.* 161: 67–84.

Weiss, S., Reynolds, B. A., Vescovi, A. L., Morshead, C., and van der Kooy, D. (1996) Is there a neural stem cell in the mammalian forebrain? *Trends Neurosci.* 19:387–393.

Yandava, B., Billinghurst, L., and Snyder, E. (1999) "Global" cell replacement is feasible via neural stem cell transplantation: evidence from the dysmyelinated shiverer mouse brain. *Proc. Natl. Acad. Sci. USA* 96:7029–7034.

Zhang, S. C., Wernig, M., Duncan, I. D., Brustle, O., and Thomson, J. A. (2001) In vitro differentiation of transplantable neural precursors from human embryonic stem cells. *Nat. Biotechnol.* 19:1129–1133.

Zlomanczuk, P., Mrugala, M., de la Iglesia, H. O., et al. (2002). Transplanted clonal neural stem-like cells respond to remote photic stimulation following incorporation within the suprachiasmatic nucleus. *Exp. Neurol.* 174(2):162–168.

18 Molecular Genetic Approaches in the Study of Retinal Progenitor Cells and Stem Cells

Till Marquardt, PhD and Peter Gruss, PhD

Understanding how to take advantage of the special nature of multipotential stem cells responsible for renewal of retinal neurons in the mammalian eye could lead to more effective treatment of retinal blindness. The retina develops from precursor cells in the anterior neural plate through a highly conserved histogenic order. The differentiating cells migrate vertically from multipotent retinal progenitor cells in the pseudostratified neuroepithelium to form the inner layer of the optic cup. The differentiation of retinal cell types is promoted or inhibited by a number of secreted factors. Like other lineage systems, the developing retina has a few stem cells and larger numbers of transit-amplifying cells that are more restricted, presumably through neurogenic transcription factors of the bHLH class, including Pax6, Rx1, Six3/6, and Lhx2. In the adult, the peripheral rim of the neuroretina of amphibians and fish contribute to growth of the eye throughout adult life. In adult mammals, rare progenitor cells may be found in the pigmented ciliary margin of the eye. Under specific conditions, donor progenitor cells may be incorporated into the adult eye after injection into the subretinal space, but differentiation into functioning retinal neurons has not been achieved. This might be accomplished by the appropriate control of the required transcription factors.

18.1. INTRODUCTION

The success in reconstituting retinal function by introducing heterogeneous neuronal stem cells has so far been limited, underlining the highly specialized nature of retinal cells. These findings stress the importance of understanding the endogenous mechanisms governing cellular diversification from retinal progenitor cells (RPCs) in vivo. Recent advances have led to the exciting discovery of retinal stem cells in the adult mammalian eye and have provided valuable insights into the molecular mechanisms underlying theying theenic potential of RPCs. In addition to providing background information on retinal specification and the generation of retinal cell diversity, this chapter provides an overview of recent approaches and concepts for the study of mammalian retinal progenitor and stem cells in vivo.

18.1.1. CELLULAR DIVERSIFICATION IN VERTEBRATE RETINA The detection, processing, and transmission of visual input to the brain is achieved by six prinicipal classes of retinal neurons, each in turn comprising several subtypes. In outline, visual stimuli are detected by the rod and cone photoreceptor cells and are either directly or indirectly transmitted via bipolar interneurons to retinal ganglion cells, which eventually relay the signals to the brain (Dowling, 1987; Wassle and Boycott, 1991). Visual functions such as movement detection and contrast enhancement essentially depend on the modulation of intraretinal transmission, requiring two types of interneurons, horizontal and amacrine cells (Wassle and Boycott, 1991; Taylor et al., 2000).

During development, this array of vastly different cell types derives from a common pool of multipotent RPCs residing in the pseudostratified neuroepithelium that forms the inner layer of the optic cup (Fig. 1A). The multipotent nature of RPCs was elucidated by a series of seminal lineage tracing experiments in mammals and amphibia (Turner and Cepko, 1987; Holt et al., 1988; Wetts and Fraser, 1988; Turner et al., 1990). These studies revealed that RPCs generally retain their ability to generate different cell types up to the final cell division.

The differentiation of the distinct retinal cell types from RPCs proceeds in a highly conserved histogenetic order (Fig. 1B). Ganglion cells are generated first, followed in overlapping phases by horizontal, cone photoreceptor, amacrine, rod photoreceptor, bipolar and, finally, Müller glia cells (Fig. 1B) (Young, 1985a, 1985b). Retinal cell-cell differentiation is initiated in the central optic cup, close to the optic nerve head, and progresses toward the periphery in a wavelike fashion until reaching the region of the presumptive iris (Prada et al., 1991). Clonal analysis in mice and the claw frog *Xenopus laevis* revealed that newly generated retinal cells migrate vertically to their point of settlement, i.e., along the vitreoscleral axis, but not laterally, along the proximodistal (optic nerve iris) axis (Wetts and Fraser, 1988; Goldowitz et al., 1996). During later retinogenesis, with increasing proportion of postmitotic cells, new cells are generated in a

From: *Stem Cells Handbook*
Edited by: S. Sell © Humana Press Inc., Totowa, NJ

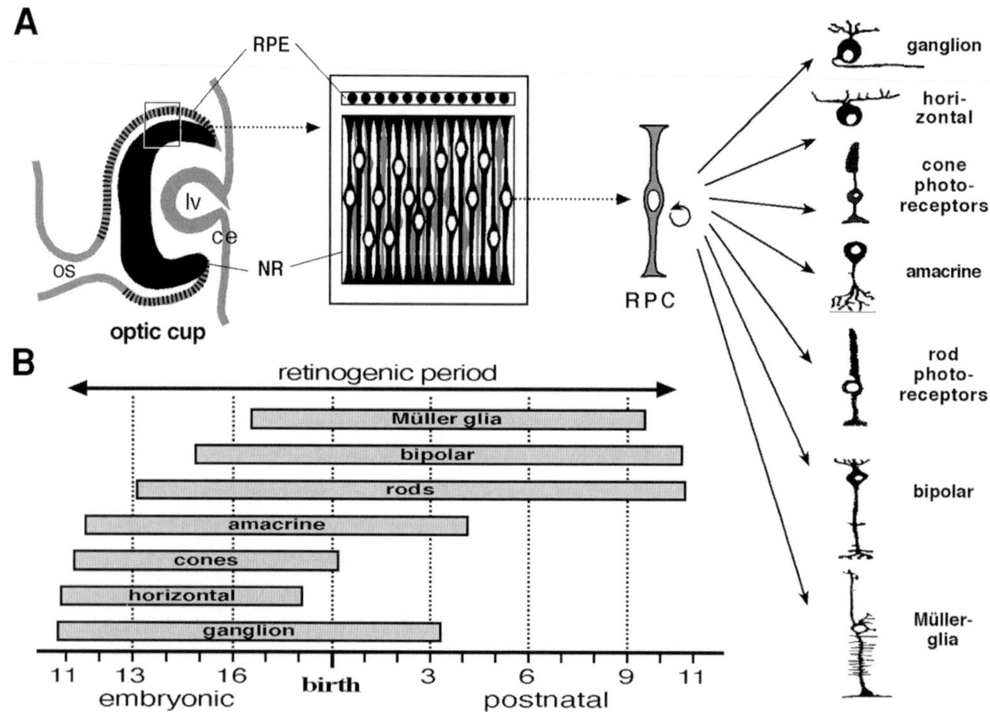

Fig. 1. Generation of cellular diversity during retinal development. **(A)** A common population of multipotent RPCs in the inner layer of the optic cup gives rise to the seven cell classes of the neuroretina. **(B)** The distinct cell classes are generated in a defined temporal sequence. The scheme is derived from cell birth dating studies in the developing mouse retina (after Young, 1985a, 1985b). *See* text for details. ce, corneal ectoderm; lv, lens vesicle; OS, optic stalk; RPE, retinal pigment epithelium; NR, neuroretina.

proliferative zone (by most authors referred to as the neuroblast layer) and migrate vertically to their sites of settlement (Fig. 2B).

In mice, retinogenesis extends over a period of nearly 3 wk (Young, 1985a) (Fig. 1), while in amphibians and fish this process is extremely temporally compressed, obviously reflecting the adaptive necessity of the pelagic larva for immediate visual function. Histogenesis in the central retina of *Xenopus*, e.g., is completed within 24 h (Holt et al., 1988). However, despite this radical temporal compression, the general histogenetic order is conserved in lower vertebrates, with retinal ganglion cells being born first and Müller glia born last. In these species, this initial phase of retinogenesis is followed by a second phase, which is further discussed in Subheading 18.1.4.1.

18.1.1.1. Intrinsic Changes in RPC Potential The differentiation of particular retinal cell types from RPCs can be promoted or inhibited by secreted factors, such as transforming growth factors (α/β, epidermal growth factor (EGF), Sonic hedgehog (Shh), nerve growth factor, leukemia inhibitory factor, and ciliary neurotrophic factor (reviewed in Reh and Levi, 1998; Cepko, 1999). These factors mostly act on mitotic RPCs but, in some instances, can also redirect the fate of postmitotic precursor cells (Ezzeddine et al., 1997). Shh, e.g., is secreted by newly postmitotic ganglion cells and promotes further production of ganglion cells by adjacent RPCs, which in turn start secreting Shh (Neumann and Nuesslein-Volhard, 2000). The production of appropriate numbers of particular cell types appears to be controlled by negative feedback signaling by postmitotic cells (Waid and McLoon, 1998; McCabe et al., 1999). During retinogenesis, the continuous increase in the number of postmitotic cells of different types automatically results in alterations of the signaling environment, which in turn seems to be a motor of cellular diversification.

In addition to these extrinsic signaling systems, RPCs display significant intrinsic changes in their potential to generate different cell types with progression of retinogenesis, as well as differential changes in their response to inductive and mitogenic factors (reviewed in: Lillien, 1998; Cepko, 1999). These intrinsic changes are reflected by the observation that, e.g., embryonic retinal cells cultured with an excess of postnatal retinal cells do largely not adopt cell fates preferred by postnatal RPCs (Alexiades and Cepko, 1997; Belliveau and Cepko, 1999; Belliveau et al., 2000). The molecular mechanisms that drive these intrinsic shifts are largely unclear, although a possible mechanism involves changes in the expression level of cell-surface receptors, such as the EGF receptor (Lillien, 1995, 1998).

18.1.1.2. Multipotency and Lineage Restriction in the Retinal Progenitor Cell Pool When cultured in vitro, small numbers of multipotent RPCs with stem cell characteristics (i.e., passagability and neurosphere formation) can be retrieved from the late embryonic retina, while the majority of RPCs apparently undergo immediate differentiation (Kelley et al., 1995; Jensen and Raff, 1997). Therefore, similar to other regions of the developing central nervous system (CNS), the RPC pool appears to comprise a complex population of few stem cells and distinct lineage-restricted progenitor cell populations (Lillien, 1998; Anderson, 2001). A subpopulation of RPCs selectively expressing the epitopes syntaxin and VC1.1 was isolated and preferentially gave rise to amacrine cells and, later, rod photoreceptor cells (Alexiades and Cepko, 1997).

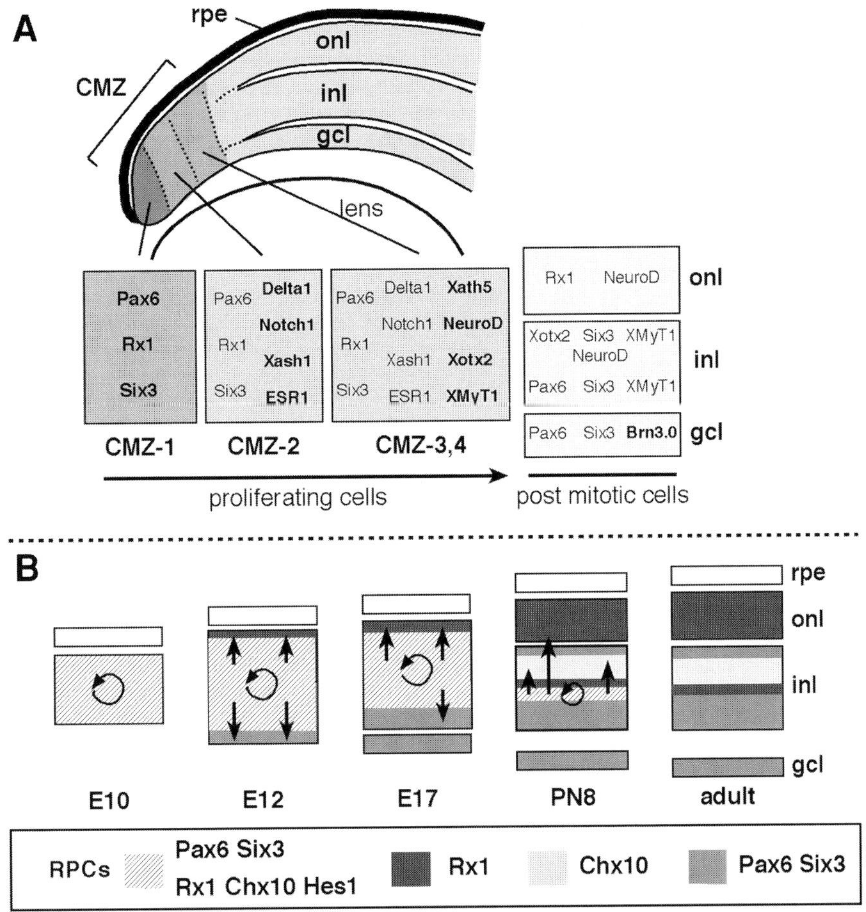

Fig. 2. Spatiotemporal dynamics of transcription factor expression during retinogenesis in mammals and peripheral retinogenesis in amphibians. **(A)** In the CMZ of the *X. laevis* retina, new cells are generated from multipotential stem cells residing in the peripheralmost CMZ, which coexpress Pax6, Rx1, and Six3. Cells in the course of retinogenesis become progressively localized toward the central CMZ and display characteristic switches in their expression profiles of transcription factors involved in neurogenesis and cell fate specification. Postmitotic cells display a segregation of the factors initially coexpressed in the CMZ stem cells. **(B)** Retinal lamination during retinogenesis in mice. Similar to the CMZ stem cells in (A), RPCs characteristically coexpress a common set of transcription factors. Their expression pattern rapidly segregates after cell-cycle exit. onl, inl, gcl: outer nuclear, inner nuclear, and ganglion cell layer, respectively; rpe, retinal pigment epithelium.

Good candidates for mediating intrinsic cell fate biases of distinct lineage-restricted RPC subpopulations are neurogenic transcription factors of the basic helix-loop-helix (bHLH) class (Guillemot, 1999). A number of bHLH factors are localized to both overlapping and nonoverlapping subsets of RPCs and are potent promoters of particular retinal cell fates (Brown et al., 1998; Morrow et al., 1999; Perron et al., 1999; Hojo et al., 2000; Wang et al., 2001). The expression of these factors generally occurs in a mosaic-like pattern and is restricted to only subsets of RPCs presumably via the action of Notch/Delta signaling–mediated lateral inhibition (Artavanis-Tsakonas et al., 1999; Perron and Harris, 2000), in addition to repressive interactions among the factors themselves (reviewed in Marquardt and Gruss, 2002) because neurogenic bHLH factors can drive neural progenitor cells out of the cell cycle (Farah et al., 2000), this mechanism potentially prevents the premature depletion of the RPC pool (Tomita et al., 1996; Dorsky et al., 1997).

Next to restricting the numbers of cell cycles available to RPCs, the transient action of these factors in RPCs appears to promote the expression of other transcription factors that act as terminal differentiation factors and continue to be present postmitotically. The bHLH factor Math5, e.g., appears to directly activate expression of the POU domain transcription factor Brn3b (Liu et al., 2001; Wang et al., 2001), which is required for the terminal differentiation of retinal ganglion cells (Xiang, 1998; Gan et al., 1999). However, the regulatory relationships between retinogenic bHLH factors and the extrinsic signaling systems operating during retinogenesis remain unresolved.

18.1.2. ORIGIN OF RETINAL PROGENITOR CELLS Formation of the retina is initiated with the evagination of the anterior neural plate in the region of the presumptive diencephalon. These evaginations, the optic sulci, eventually form to the optic vesicles, which ultimately give rise to three principal ocular tissues: neuroretina, retinal pigment epithelium, and optic nerve. The neuroepithelium of the optic vesicle distally invaginates to form the optic cup. The outer layer of the optic cup eventually differentiates into the retinal pigment epithelium, while the inner layer contains the RPCs that give rise to the different cell types that assemble to the functional architecture of the neuroretina.

What are the molecular players that determine retinal fate? The homeodomain and Pax transcription factors Pax6, Rx1 (rax), Six3, Six6 (Optx2), Lhx2, and Otx2 become expressed in the retinal anlage prior to the formation of the optic sulci (Walther and Gruss, 1991; Li et al., 1994; Oliver et al., 1995; Furukawa et al., 1997a, 1997b; Mathers et al., 1997; Jean et al., 1999; Crossley, 2001; Crossley et al., 2001; Kenyon et al., 2001) and are implicated in a conserved regulatory network that initiates retinal development (Chow et al., 1999; Gehring and Ikeo, 1999; Zuber et al., 1999; Bernier et al., 2000; Ashery-Padan and Gruss, 2002). Their central role in this process is impressively demonstrated by the finding that forced expression of either Pax6, Six3, Six6, or Rx1 in fish and frog embryos promotes the formation of ectopic retinal tissue (Chow et al., 1999; Loosli et al., 1999; Zuber et al., 1999; Bernier et al., 2000; Kenyon et al., 2001). Therefore, each of these factors appears to be capable of triggering the events that ultimately lead to retinal differentiation. Ectopic expression of any of these factors can trigger the expression of the others, an observation that highlights the network character of their regulatory interrelationships. The central function in initiating retinal development is furthermore reflected by failure of eye development in null mutants for the genes encoding these factors. Rx1 null mutant mice display a complete failure to form the optic sulci, while in Pax6 and Lhx2 null mutants the optic vesicle is formed but arrests at a stage prior to the initiation of retinal neurogenesis (Grindley et al., 1995; Mathers et al., 1997; Porter et al., 1997).

The intriguing regulatory network of these factors might pave the way to retinal development via mutually stabilizing and thereby coordinating their expression. However, the mechanisms that initiate and concisely restrict these retinal determinants to the initial retinal primordium largely remain to be elucidated. In addition, these factors appear to be capable of promoting ectopic retinal tissue only in certain areas, indicating that only particular tissues are competent to be guided to the retinal fate. Otx2 was implicated in marking the area of this "retinal competetence field" (Chow et al., 1999; Loosli et al., 1999; Bernier et al., 2000; Kenyon et al., 2001). However, the actual requirement of Otx2 in this process remains obscure, since in null mutant mice the development of head structures is severely compromised (Boncinelli and Morgan, 2001).

18.1.3. MOLECULAR SETUP OF THE UNCOMMITTED RPC STATE In all vertebrate species analyzed so far, the same set of transcription factors (Pax6, Rx1, Six3/6, and Lhx2) that act in initiating vertebrate eye development continues to be present during the ensuing steps of retinal neurogenesis (Fig. 2A,B) (Walther and Gruss, 1991; Belecky-Adams et al., 1997; Furukawa et al., 1997a; Perron et al., 1998). Pax6, e.g., is localized in virtually all RPCs throughout murine retinogenesis (Marquardt and Gruss, 2002). With increasing proportion of postmitotic cells, these factors display a segregation in their expression patterns, being downregulated in some cell lineages, while being maintained in others (Fig. 2B). Therefore, the actual coexpression of these factors appears to be a defining feature of RPCs. Given their coordinate function in the very initiation of retinal development, this observation strongly suggests a continued requirement for the combined action of these factors in retinogenesis.

The inactivation of Pax6 specifically in the RPCs of the distal optic cup of mouse embryos by using the Cre-loxP approach (*see* Subheading 18.2.3.) led to the complete restriction in the retinogenic potential of RPCs toward the production of only one of the different cell fates normally available to RPCs—amacrine cells (Marquardt et al., 2001). Pax6 appears to function, at least in part, by mediating the activation of retinogenic bHLH transcription factors such as Math5 and Ngn2 (Marquardt et al., 2001; Marquardt and Gruss, 2002). The presence of other RPC factors (i.e., Rx1, Six3/6, and Lhx2) was maintained, probably accounting for the restricted retinogenesis still observed from Pax6-deficient RPCs. These results therefore confirm the continued requirement for these transcription factors for the full retinogenic potential of RPCs.

The observation that a single factor, active in all RPCs throughout retinogenesis, is essential for the formation of nearly all retinal cell types, affecting early and late born types alike, suggests the existence of a common molecular framework underlying RPCs during all stages of retinogenesis. In this respect, the intrinsic changes in the retinogenic potential of RPCs at particular stages of retinal development (*see* Subheading 18.1.1.1.) might merely be superimposed on a more "primitive" RPC state. The coexpression of this set of retinal determinants therefore appears to represent the molecular setup of the principal RPC state and might serve as a useful tool in the identification of retinal stem cells in vivo and in vitro.

18.1.4. ADULT RETINAL STEM CELLS

18.1.4.1. Retinal Regeneration and Retinal Stem Cells in Lower Vertebrates In amphibians and fish the retina continues to grow postembryonically and throughout adult life, keeping pace with increase in body size of larva and adult (Hollyfield, 1971; Johns, 1978). After a general phase of retinogenesis, which proceeds in the usual proximodistal pattern (*see* above), new cells are now added laterally from a specialized proliferative zone in the peripheral rim of the neuroretina, the ciliary marginal zone (CMZ) (Hollyfield, 1971; Johns, 1978; Wetts et al., 1989). In the CMZ, the least determined cells, "retinal stem cells," reside closest to the periphery, and cells in the process of differentiation progressively localize to the central portion of the CMZ, therefore representing a spatial recapitulation of the sequential events of retinogenesis (Wetts et al., 1989; Perron et al., 1998). In analogy to the findings in the murine retina (Subsection 18.1.3.), the distalmost CMZ stem cells were shown to coexpress Rx1, Six3, and Pax6 prior to and during the ensuing steps of retinogenesis (Perron et al., 1998).

CMZ stem cells are multipotent and can give rise to all retinal cell types, including retinal pigment epithelial cells. The proliferative activity of the CMZ stem cells appears to depend on their localization to the CMZ, since transfer to the central retina leads to immediate cell-cycle exit and differentiation (Wetts et al., 1989). A most intriguing property of the CMZ cells is their apparent capability to respond to stimuli from the central retina with the production of particular classes of cells. In the *Xenopus* retina, ablation of neurons of a particular neurotransmitter phenotype has been reported to stimulate the production of this cell type by the CMZ (Reh, 1987).

The eyes of urodele amphibians display a remarkable regenerative capacity. After complete retinectomy, the RPE of salamanders can transdifferentiate into retinal neuroepithelium (Stone, 1950). Similarly, following lens ablation, the urodele RPE can also transdifferentiate to form a lens (Reyer, 1977). However, most anuran species, like *Xenopus*, lack such a regenerative capacity, although some retinal tissue is replaced after partial retinectomy, but the source of the new cells might mainly be the CMZ (Reh, 1987; Reh and Nagy, 1987).

18.1.4.2. Retinal Stem Cells in Adult Mammals Like other regions of the CNS, the retina of adult mammals generally lacks regenerative capacity. Recently, however, retinal stem cells could be retrieved from the pigmented ciliary margin (PCM) of the adult mouse and human eye (Tropepe et al., 2000). On culture, a number of cells of this region were passageable and gave rise to sphere colonies containing various types of retinal neurons and glia. Retinal stem cells could only be retrieved from the PCM, not from the neuroretina and other regions of the iris or RPE. Interestingly, the proportion of PCM cells with retinal stem cell properties dramatically increased postnatally, while the embryonic retinal neuroepithelium contains only a few stem cells (*see* Subheading 18.1.1.2.).

The PCM cells are speculated to constitute an evolutionary "remnant" of the CMZ of lower vertebrates. However, the region of the PCM in the mouse does not precisely correspond to that of the CMZ. The nonpigmented CMZ of amphibian represents the distal margin of (and is continous with) the neuroretina and approximately corresponds to the area around the ora serrata in mammals. The latter appears to be devoid of retinal stem cells (Tropepe et al., 2000). Whether the retinal stem cells of the PCM actually play a role in the adult mammalian retina in vivo remains unclear.

In the retina of the chick, a limited number of proliferative cells in the nonpigmented ciliary margin have been observed, bearing closer resemblance to the CMZ cells of lower vertebrates (Fischer and Reh, 2000). Evidence has been provided that mild injury can induce a certain regenerative potential exerted by Müller glia cells in the adult chick retina (Fischer and Reh, 2001). The production of new cells by Müller glia was preceded by the concomitant upregulation of Chx10 and Pax6, which normally are only coexpressed in RPCs (*see* Fig. 2). Furthermore, a rare population of stem cells in the inner nuclear layer of the adult teleost retina (possibly Müller glia) that express Pax6 could be retrieved, (Otteson, 2001). In this context it might be of interest to note that the mammalian PCM cells likewise express Pax6 (*see* Marquardt et al., 2001). Therefore, it will be highly significant to analyze the contribution of pivotal retinal factors such as Pax6, Six3, and Rx1 in mediating the retinogenic potential of these stem cells. These findings might be an important step in developing therapies for retinal trauma or degeneration in humans by cell replacement strategies using PCM stem cells. However, it still has to be elucidated in animal models whether the PCM cells are capable of eliciting their full retinogenic potential on introduction into the retina of an adult host (*see* Subheading 18.3.).

18.2. MOLECULAR GENETIC APPROACHES IN THE IN VIVO STUDY OF RETINAL STEM CELLS AND PROGENITOR CELLS

18.2.1. CELL FATE TRACING AND GAIN-OF-FUNCTION EXPERIMENTS USING RETROVIRUSES The use of retroviral vectors in the study of retinal cell fate was pioneered in the laboratory of C. Cepkos (Price et al., 1987; Turner and Cepko, 1987). These experiments were the first to demonstrate unanimously the multipotent nature of mammalian RPCs. The retroviral vector is injected into the eyes of neo- or early postnatal rats via the vitreal cavity, between pigment epithelium and neuroretina (for a technical account, *see* Cepko et al., 1998). The injected animals are dissected after a few weeks, and the retina is histochemically analyzed for the presence of transfected cells. Infection of cells is followed by integration of a single copy of the retrovirus into the host genome, which is passed on to its daughter cells. Most retroviruses require the host cell to be in the M phase of the cell cycle for succesful genomic integration (Roe et al., 1993). Therefore, only mitotic cells are infected, resulting in the selective tracing of progenitor cells. In addition, replication-incompetent retroviruses are used to prevent secondary infections, allowing the tracing of the actual progeny of a given RPC.

Using a modified vector containing the gene of interest "*geneX*" (e.g., encoding a transcription factor implicated in retinal cell specification), as well as a repoter gene seperated by an internal ribosome entry site (IRES) sequence, allowed the expression of both genes encoded by a single bicistronic mRNA. Comparing the cell type composition of the labeled cell progeny after transfection with the *geneX/reporter* and the control vector should reveal the potential of the respective factor in influencing cell fate (*see* Furukawa et al., 1997b; Morrow et al., 1999).

Although powerful in revealing cell fate of single or few labeled RPCs in vivo, the delivery by direct injection into the eye is technically largely limited to postnatal stages of rats. The lineage of embryonic mouse embryos could be followed using application of retroviral vectors via *ex utero* surgery (Turner et al., 1990). However, this microsurgical approach requires considerable technical effort, results in low survival rates, and appears to be applicable only to some genetic backgrounds (Cepko et al., 1998; Ngo-Muller and Muneoka, 2000). These difficulties can be partially circumvented by using infection of retinal explant cultures, in which retinal maturation and lamination proceeds to a remarkable degree normally (Sparrow et al., 1990). Recently, techniques for gene delivery to the CNS of mouse embryos *in utero* were successfully deployed, in which mouse embryos are injected with retroviral vectors by puncturing the uterus of anesthetsized pregnant mice (Gaiano et al., 1999; Fukuchi-Shimogori and Grove, 2001; Saito and Nakatsuji, 2001; Tabata and Nakajima, 2001). It remains to be seen whether these methods offer the possibility of routinely transfecting the retina.

18.2.2. POWER AND LIMITS OF CLASSICAL GENE TARGETING IN MICE "Classic" gene targeting to create null mutations for a particular gene in mice by homologous recombination in embryonic stem (ES) cells has for years been a powerful tool to study the requirement for a particular gene in vivo (Thomas and Capecchi, 1987; for a technical account, *see* Mansouri, 2001). However, the role of a particular gene during later stages of development or in a certain tissue can be severely compromised by earlier and/or more global functions. Often such earlier roles lead to premature lethality, thereby preventing the study of an involvment in later processes, such as retinogenesis.

For the retina, the problem of prenatal lethality can partially be overcome by culturing retinal explants prior to the onset of lethality. In such retinal explant cultures, cell differentiation and lamination proceeds to a remarkable degree autonomously, somewhat resembling the in vivo situation (Sparrow et al., 1990). Briefly, late embryonic retinas are dissociated from vitreal, iris, optic nerve, and lens tissue; cut into smaller pieces; and cultured on laminin-, fibronectin-, or collagen-coated dishes or cover slips (Sparrow et al., 1990). The presence of serum in the medium appears to be a prerequisite for proper retinal lamination. This method has been succesfully employed for targeted mutations that lead to perinatal lethality (*see* Morrow et al., 1999) and can be readily combined with retrovirally mediated lineage tracing or reexpression.

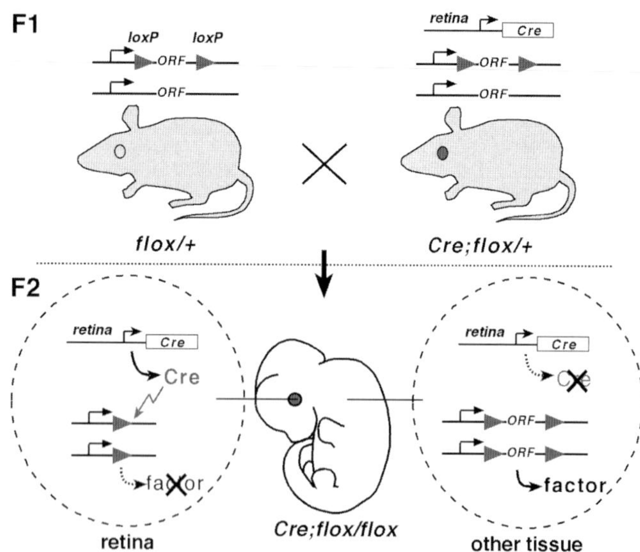

Fig. 3. Outline of Cre-loxP-mediated gene inactivation strategy in mouse retina. **(F1)** A mouse heterozygous for a *floxed* version of a particular gene (in which an essential portion of the open reading frame [ORF] has been flanked by two loxP sites) is mated with a mouse heterozygous for the same *floxed* allele, in addition to carrying a transgene that drives retina-specific Cre expression. The latter has been previously generated by interbreeding. **(F2)** Approximately one-eighth of the offspring will be homozygous for the *floxed* allele in addition to carrying the retina-specific Cre transgene. In this genotype, the gene will be deleted only in the retina, where Cre is expressed, but not in other tissues.

18.2.3. GENETIC LOSS-OF-FUNCTION, GAIN-OF-FUNCTION AND CELL FATE TRACING EXPERIMENTS USING THE CRE-LOXP RECOMBINASE SYSTEM

18.2.3.1. Outline of Cre-loxP System Loss-of-Function Studies The Cre-loxP genetic system takes advantage of the properties of the sequence-specific Cre recombinase of the bacteriophage *P1* (Rajewsky et al., 1996). Cre can efficiently catalyze the circularization and excision of DNA segments that are flanked by two of its asymmetric recognition sequences, the 34-bp loxP sites, in the same orientation. Cre, coupled to a nuclear localization signal, readily functions in the environment of the mammalian cell nucleus (Rajewsky et al., 1996; Tsien et al., 1996), making this system a uniquely suitable tool to study gene function in the mouse.

Cre-loxP-mediated inactivation of a given *geneX* gene utilizes two components each carried by a separate mouse line (Fig. 3). The first mouse line carries a manipulated ("floxed") version of the gene of interest, *geneX^{flox}*, which is flanked by two loxP sites. The second mouse line carries a transgene, *retina-Cre*, expressing the Cre recombinase in the desired tissue by utilizing a tissue-specific promoter or regulatory element. Tissue-specific inactivation is finally achieved by the interbreeding of these mouse lines to generate *retina-Cre*; *geneX$^{flox/flox}$* animals, which are homozygous for the targeted locus (Fig. 3). Stable transgenic mouse lines expressing Cre are generated either by microinjection into fertilized mouse oocytes followed by random genomic integration (for a technical account, *see* Hogan et al., 1994) or by targeted insertion into a known gene by homologous recombination in ES cells (Mansouri, 2001). The "floxed" allele is generated by standard gene-targeting procedures (*see* Subheading 18.2.2.).

Cre-loxP-based strategies are particularly powerful because, once established, they principally allow the combination of preexisting reporter strains with different Cre lines, allowing fate tracing and manipulation of distinct progenitor cell populations. However, for the same reason, this strategy is currently limited by the still rarely available suitable retina-specific Cre lines or retina-specific promoter elements (Table 1).

18.2.3.2. Cre-loxP System-Mediated Cell Fate Tracing Cre-loxP-mediated fate mapping utilizes two components, each carried by a separate transgenic mouse line. One transgene, *retina-Cre*, utilizes a tissue-specific promoter or regulatory element that drives expression of the Cre recombinase in RPCs. The other mouse line carries a Cre reporter transgene, *lox/stop/lox-reporter* (Fig. 4), in which a stop signal (usually one or more *polyA* signals that prevent transcription to proceed farther downstream) is flanked by two loxP sites and precedes a constitutively expressed reporter gene. Interbreeding produces *retina-Cre; lox/stop/lox-reporter* double transgenic mouse embryos in which Cre activity can now lead to the excision of the stop signal, thereby allowing reporter gene expression (Fig. 4). If this recombination event occurs in a given progenitor cell, the recombined constellation of the reporter transgene eventually will be passed on to all of its cell progeny, which then can be identified by the expression reporter gene (Fig. 4).

This strategy can be combined with a gain-of-function approach (*see* Sharma et al., 2000). In outline, the Cre reporter vector is modified to encode a bicistronic mRNA by using an IRES sequence, allowing expression of both the reporter gene and the gene of interest, *geneX*. Crossing this hypothetical transgene, *lox/stop/lox-geneX-reporter*, into a particular retina-specific Cre line, *retina-Cre*, ideally results in efficient transgene activation in all or a subset of RPCs. Comparing the cellular compostion of labeled clones in the mature retina of *retina-Cre; lox/stop/lox-reporter* and *retina-Cre; lox/stop/lox-geneX-reporter* double transgenic mice should reveal the potential influence of *geneX* expression on retinal cell fate decisions (*see* Fig. 4). The outcome will critically depend on the population of RPCs in which Cre is active and the timing of Cre activity, since these parameters will determine the intrinsic potential (i.e., presence of other cell fate factors in the respective RPCs).

18.2.3.3. Combining Genetic Cell Fate Tracing with a Cre-loxP-Mediated Loss of Function Combining genetic cell fate tracing with a Cre-loxP-mediated loss of function, although potentially very powerful, suffers from some technical contraints, which require more elaborate efforts. The concomitant insertion of reporter genes into the targeted allele potentially leads to undesired effects on the expression of the gene prior to the Cre-mediated recombination event. Alternatively, the *lox/stop/lox-reporter* transgene can in principle be crossed together with *retina-Cre* and the *floxedgeneX* parallel to generate *retina-Cre; lox/stop/lox-reporter; lox/geneX/lox* (homozygous) offspring. This strategy of crossing involves considerable effort and time and necessarily requires the largely independent segregation of all genetic components. Nevertheless, the strategy could be succesfully deployed to trace the progeny of RPCs in which the *Pax6* gene was inactivated via the Cre-loxP approach (Marquardt et al., 2001). Another potentially powerful approach is to utilize retroviruses expressing Cre to directly trace the fate of cell progeny after loss of gene function in the infected progenitor cell. Of course, the same technical limitations apply regarding the applicability of retroviruses in utero (*see* Subheading 18.2.1.).

Table 1
Examples of Established Transgenic Mouse Lines with Retinal Cre Expression[a]

Retina-expressing Cre lines	*Cre reporter lines*	*Cre transgenic database*	*Transgenic/targeted mutation database*
Optic vesicle, all RPCs: *Six3-Cre* (Furuta et al., 1000) All distal RPCs: α-*(Pax6)-Cre-Gfp* (Marquardt et al., 2001) Postmitotic ganglio cells: *c-kit-Cre* (Eriksson et al., 2000) Postmitotic photoreceptors: *Ret-P1-Cre* (Marszalek et al., 2000) Postanatal retina, all layers OHT (hydroxy tamoxifen) inducible: *PrP-CreER(T)* (Weber et al., 2001)	Ubiquitous *lacZ-hPLAP* double reporter: *Z/AP* (Lobe et al., 1999)	[http://www.mshri.on.ca/nagy/cre.htm] A highly useful, noncomprehensive source for available Cre lines, Cre reporter strains, and ES cell lines	[http://tbase.jax.org] A useful collection of published mouse mutants, transgenes, and mouse mutant–related research tools

Unfortunately, the number of retina-specific promoters is still limited. The web sites given might serve as a useful tool in searching for suitable Cre lines or promoters.

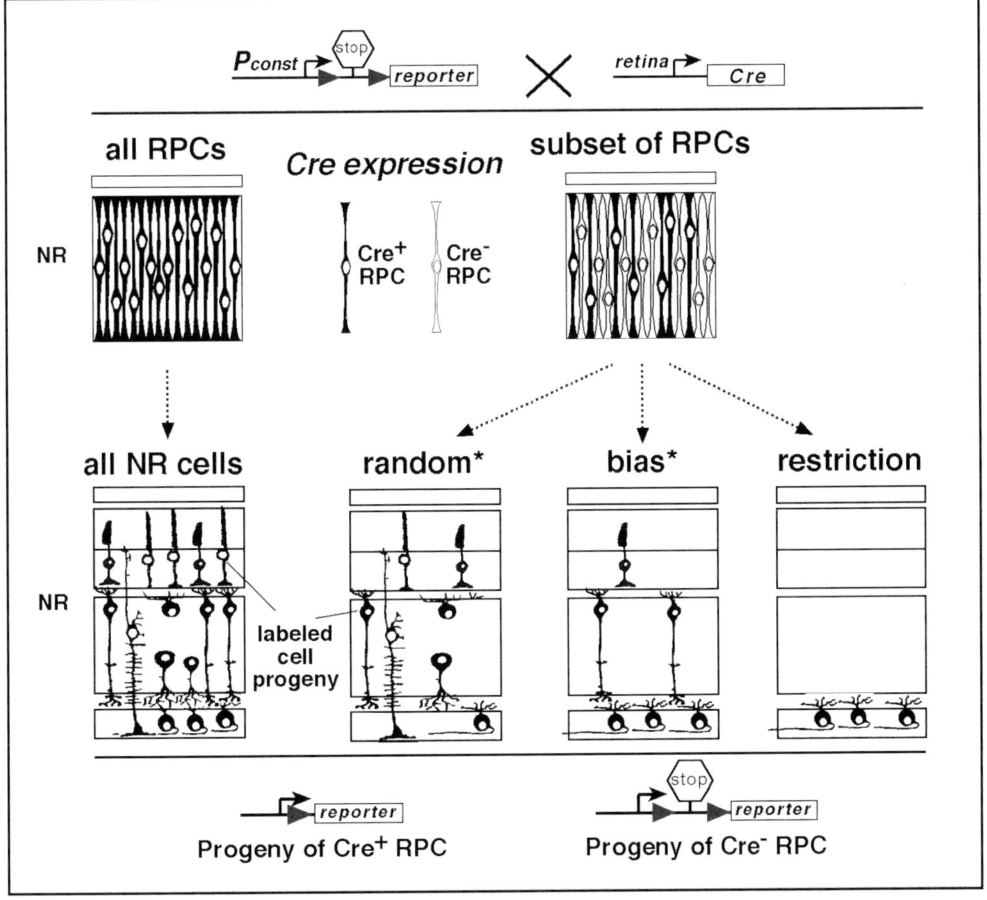

Fig. 4. Cre-loxP-mediated cell fate tracing in mouse retina. **(A)** A Cre reporter transgene includes an ubiquitously active promoter followed by a loxP-flanked polyA (stop) signal and a reporter gene. The other transgene expresses Cre under control of a promoter driving expression in RPCs. **(B)** Interbreeding creates double transgenic mice. Cre activity in all RPCs results in constitutive reporter gene expression and histochemical labeling in all postmitotic retinal cells. Cre in distinct RPC subpopulations results in either (1) random labeling of different cell types, (2) a bias toward the labeling of certain cell types, or (3) a restriction in the labeling of only one cell type; the latter two indicate the presence of a lineage-restricted RPC population. *However, it has to be taken into account that, because of the sequential birth order of retinal cell types, a certain bias might also be created by the timing of Cre expression in RPCs. NR, neuroretina.

Table 2
Retinal Cell–Type Specific Antibody Markers[a]

		Brn3b	Thy-1	Isl1	PKC	Grk-1	Recoverin	Glut.Synth.	NF_{165kDA}(2H3)	NeuN	βIII-tubulin	Syntaxin	VC1.1	GABA	GlyT1	Glycine	Calretinin	ChAT	TH	Substance P	NR1	GAD_{65}	GluR2	Pax6
Source		*Santa Cruz Biotechnology*	*Chemicon*	*DSHB[b,c]*	*Sigma*	*Santa Cruz Biotechnology*	*Dr. K. W. Koch, Kiel[c]*	*Transduction Laboratory*	*DSHB*	*Chemicon*	*Babco*	*Sigma*	*Sigma*	*Sigma*	*Chemicon*	*Chemicon*	*Chemicon*	*Chemicon*	*Chemicon*	*Pharmingen*	*Chemicon*	*Roche*	*Chemicon*	*DSHB, CAPPEL*
Cell type	Ganglion	+	+	+	–	–	–	–	±	+	+	–	–	–	–	–	±	–	–	–	–	–	–	+
	Horizontal	–	–	–	–	–	–	–	+	±	–	±	±	–	–	–	–	–	–	–	±	–	–	±
	Amacrine	–	–	±	–	–	–	–	–	±	+	+	+	+	+	+	±	±	±	±	+	+	+	+
	Bipolar	–	–	+	+	–	±	–	–	–	–	–	–	–	–	–	–	–	–	–	–	–	–	–
	Müller glia	–	–	–	–	–	–	+	–	–	–	–	–	–	–	–	–	–	–	–	–	–	–	–
	Photoreceptor	–	–	–	±	+	+	–	–	–	–	–	–	–	–	–	–	–	–	–	–	–	–	–

(Column headings: *Antigen/antibody*; row "Source" gives the *Source* of each antibody.)

[a]Note that βIII-tubulin is present at low levels in virtually all retinal neurons, but at much higher levels in ganglion and amacrine cells. Grk-1 (rhodopsin kinase) labels rod outer segments, while PKC labels rod outer segments, in addition to bipolar cells, and recoverin labels all photoreceptor cell bodies. Syntaxin, NR1, GluR2, GAD_{65}, and VC1.1 label virtually all amacrine cells, the latter also additional cells in the inner nuclear layer. GABA, GlyT1, and glycine each label 30–40% of amacrine cells, while calretinin, TH, and ChAT label only small amacrine subpopulations. Also note that polyclonal antisera might be subject to considerable variations in their quality and/or working conditions.

[b]DSHB Developmental Studies Hybridoma Bank.

[c]Non-commercial antibody, courtesy of Dr. K. W. Koch.

18.2.4. Identification of Retinal Cell Types A considerable number of suitable antibody markers are available for the identification of particular retinal cell types *in situ* or in cultured cells (Table 2) (*see* Haverkamp and Wassle, 2000). In addition, in retinal sections the identity of a cell can in part be revealed by the position of its cell body in the outer, inner, or ganglion cell layer, which is stereotypic for the different cell types. However, when using such markers it is important to be aware that they often label more than one particular cell type. Syntaxin, e.g., labels cell bodies and processes of amacrine cells, but also horizontal cells (Barnstable et al., 1985), and Islet1, which is sometimes referred to as a ganglion cell marker, is also strongly expressed in bipolar cells and cholinergic amacrine cells (Galli-Resta et al., 1997). This is especially a concern when working with cultured cells in which the type of a cell is not additionally revealed by its position within the retinal laminae. In addition, some care must be taken when transferring results from one species to another, because some seemingly conserved markers can display unexpected species specificity regarding their cell type–specific expression (Haverkamp and Wassle, 2000).

18.3. CURING RETINOPATHIES WITH STEM CELLS: ANIMAL MODELS

Donor cells can be delivered to the retina by injection of usually 10^5–10^6 cells into the subretinal space (i.e., between RPE and neuroretina), similar to the administration of retroviral vectors described under Subheading 18.2.1. Usually, entry into the vitreal cavity is achieved by directly inserting a thin (1-mm) glass micropipet connected to a Hamilton syringe through a self-sealing wound at the corneoscleral junction (*see* Warfvinge et al., 2001). The entry via the vitreous cavity appears to minimize reflux of cells by the intraocular pressure.

Injection of adult hippocampal (Takahashi et al., 1998) or transformed embryonic cerebellar– and medulla (C17-2, RN33B) (Warfvinge et al., 2001)–derived progenitor cells into the retina of neonatal or adult rats resulted, with varying efficiency, in the integration of donor cells into the host retina. Many of these cells appear to be incorporated into the host retina, where they survive and differentiate. However, the vast majority of grafted cerebellar- and medulla-derived progenitor cells cells did not differentiate neurons but, rather, microglia-like cells situated in the inner and outer plexiform layers of the host retina (Warfvinge et al., 2001).

Interestingly, adult hipocampal progenitor cells were only integrated into the host retina of neonatal, not adult, rats (Takahashi et al., 1998). However, when these progenitor cells were injected into mechanically injured or dystrophic retina of adult rats, they were incorporated, suggesting that injury facilitates integration and differentiation of donor cells (Young et al., 2000). These cells displayed efficient integration into the host retina and neuronal differentiation. However, the cells did not display marker profiles typical for mature retinal neurons, indicating that terminal differentiation into retinal neurons did not occur. Injection of immortalized rat RPCs (E1A-line) (Seigel, 1996) was reported to lead to the succesful incorporation into the retina of neonatal and adult rats (Seigel et al., 1998). While some of these cells were observed to acquire marker characteristics of mature photoreceptor cells, Müller glia differentiation was not observed.

The limited success of these transplantation studies in differentiating donor cells into mature retinal cells is probably due to two main factors. First, the heterogeneous neuronal stem cells that were used appear to be restricted in their cytogenic potential. Apart from giving rise to "generic" neurons and glia, they appear to lack the potential to generate highly specialized retinal neurons, such as photoreceptors. Intrinsic limitations might also be attributed to embryonic retinal stem cells, which reflect stage-specific alterations in the retinogenic potential of RPCs (*see* Subheading 18.1.1.1.). Second, in addition to these intrinsic factors, the limited success in differentiating into mature retinal cells is also likely due to the altered signaling environment in the postnatal and adult retina, which might be insufficient to promote the generation of all retinal cell types.

18.4. CONCLUSION

Given the limitations experienced with using heterogeneous donor cells for retinal transplantation, it will be highly significant to test whether the grafting of the recently identified adult RPCs, which in vitro display a potential to give rise to most of the retinal cell types (Tropepe et al., 2000), are able to overcome these constraints. However, it still has to be elucidated in animal models whether the PCM cells are capable of eliciting their full retinogenic potential on introduction into the retina of an adult host. In an ideal therapeutic strategy, PCM cells would be derived from the patient's eye to minimize the risk of rejection, expanded in vitro, and reintroduced. However, while iridectomy seems to be relatively uncomplicated, biopsy from the ciliary body (containing the PCM cells) carries the risk of severe trauma to the patient's eye. In this respect, two recent studies are of considerable interest. The first found that forced expression of the homeodomain transcription factor Crx1 in cultured iris tissue, which normally seems to lacks intrinsic stem cell potential (Tropepe et al., 2000), induces the differentiation of cells expressing markers of mature photoreceptors (Haruta et al., 2001). The other study observed that the bHLH transcription factor Ngn2 appears to promote the differentiation of retinal neurons from cultured RPE cells (Yan et al., 2001). Together, these studies further signify that a deeper understanding of the mechanisms acting during normal retinal development in vivo will be instrumental in designing successful cell replacement strategies aiming at the reconstitution of retinal function.

REFERENCES

Alexiades, M. R. and Cepko, C. L. (1997) Subsets of retinal progenitors display temporally regulated and distinct biases in the fates of their progeny. *Development* 124:1119–1131.

Anderson, D. J. (2001) Stem cells and pattern formation in the nervous system: the possible versus the actual. *Neuron* 30:19–35.

Artavanis-Tsakonas, S., Rand, M. D., and Lake, R. J. (1999) Notch signaling: cell fate control and signal integration in development. *Science* 284:770–776.

Ashery-Padan, R. and Gruss P. (2002). Pax6 lights-up the way for eye development. *Curr. Opin. Cell. Biol.* 13:706–714.

Barnstable, C. J., Hofstein, R., and Akagawa, K. (1985) A marker of early amacrine cell development in rat retina. *Brain Res.* 352:286–290.

Belecky-Adams, T., Tomarev, S., Li, H. S., et al. (1997) Pax-6, Prox 1, and Chx10 homeobox gene expression correlates with phenotypic fate of retinal precursor cells. *Invest. Ophthalmol. Vis. Sci.* 38:1293–1303.

Belliveau, M. J. and Cepko, C. L. (1999) Extrinsic and intrinsic factors control the genesis of amacrine and cone cells in the rat retina. *Development* 126:555–566.

Belliveau, M. J., Young, T. L., and Cepko, C. L. (2000) Late retinal progenitor cells show intrinsic limitations in the production of cell types and the kinetics of opsin synthesis. *J. Neurosci.* 20:2247–2254.

Bernier, G., Panitz, F., Zhou, X., Hollemann, T., Gruss, P., and Pieler, T. (2000) Expanded retina territory by midbrain transformation upon overexpression of Six6 (Optx2) in *Xenopus* embryos. *Mech. Dev.* 93: 59–69.

Boncinelli, E. and Morgan, R. (2001) Downstream of Otx2, or how to get a head. *Trends Genet.* 17:633–666.

Brown, N. L., Kanekar, S., Vetter, M. L., Tucker, P. K., Gemza, D. L., and Glaser, T. (1998) Math5 encodes a murine basic helix-loop-helix transcription factor expressed during early stages of retinal neurogenesis. *Development* 125:4821–4833.

Cepko, C. L. (1999) The roles of intrinsic and extrinsic cues and bHLH genes in the determination of retinal cell fates. *Curr. Opin. Neurobiol.* 9:37–46.

Cepko, C. L., Ryder, E., Austin, C., Golden, J., Fields-Berry, S., and Lin, J. (1998) Lineage analysis using retroviral vectors. *Methods* 14: 393–406.

Chow, R. L., Altmann, C. R., Lang, R. A., and Hemmati-Brivanlou, A. (1999) Pax6 induces ectopic eyes in a vertebrate. *Development* 126: 4213–4222.

Crossley, P. H., Martinez, S., Ohkubo, Y., and Rubenstein, J. L. (2001) Coordinate expression of Fgf8, Otx2, Bmp4, and Shh in the rostral prosencephalon during development of the telencephalic and optic vesicles. *Neuroscience* 108:183–206.

Dorsky, R. I., Chang, W. S., Rapaport, D. H., and Harris, W. A. (1997) Regulation of neuronal diversity in the *Xenopus* retina by Delta signalling. *Nature* 385:67–70.

Dowling, J. E. (1987) *The Retina: An Approachable Part of the Brain.* Belknap Press of Harvard University Press, Cambridge, MA.

Eriksson, B., Bergqvist, I., Eriksson, M., and Holmberg, D. (2000) Functional expression of Cre recombinase in sub-regions of mouse CNS and retina. *FEBS Lett.* 102:106–110.

Ezzeddine, Z. D., Yang, X., DeChiara, T., Yancopoulos, G., and Cepko, C. L. (1997) Postmitotic cells fated to become rod photoreceptors can be respecified by CNTF treatment of the retina. *Development* 124: 1055–1067.

Farah, M. H., Olson, J. M., Sucic, H. B., Hume, R. I., Tapscott, S. J., and Turner, D. L. (2000) Generation of neurons by transient expression of neural bHLH proteins in mammalian cells. *Development* 127: 693–702.

Fischer, A. J. and Reh, T. A. (2000) Identification of a proliferating marginal zone of retinal progenitors in postnatal chickens. *Dev. Biol.* 220:197–210.

Fischer, A. J. and Reh, T. A. (2001) Muller glia are a potential source of neural regeneration in the postnatal chicken retina. *Nat. Neurosci.* 4:247–252.

Fukuchi-Shimogori, T. and Grove, E. A. (2001) Neocortex patterning by the secreted signaling molecule FGF8. *Science* 128:3585–3594.

Furukawa, T., Kozak, C. A., and Cepko, C. L. (1997a) rax, a novel paired-type homeobox gene, shows expression in the anterior neural fold and developing retina. *Proc. Natl. Acad. Sci. USA* 94:3088–3093.

Furukawa, T., Morrow, E. M., and Cepko, C. L. (1997b) Crx, a novel otx-like homeobox gene, shows photoreceptor-specific expression and regulates photoreceptor differentiation. *Cell* 14:531–541.

Furuta, Y., Lagutin, O., Hogan, B. L., and Oliver, G. C. (2000) Retina- and ventral forebrain-specific Cre recombinase activity in transgenic mice. *Genesis* 26:130–132.

Gaiano, N., Kohtz, J. D., Turnbull, D. H., and Fishell, G. (1999) A method for rapid gain-of-function studies in the mouse embryonic nervous system. *Nat. Neurosci.* 2:812–819.

Galli-Resta, L., Resta, G., Tan, S. S., and Reese, B. E. (1997) Mosaics of islet-1-expressing amacrine cells assembled by short-range cellular interactions. *J. Neurosci.* 17:7831–7838.

Gan, L., Wang, S. W., Huang, Z., and Klein, W. H. (1999) POU domain factor Brn-3b is essential for retinal ganglion cell differentiation and survival but not for initial cell fate specification. *Dev. Biol.* 210:469–480.

Gehring, W. J. and Ikeo, K. (1999) Pax 6: mastering eye morphogenesis and eye evolution [*see* comments]. *Trends Genet.* 15:371–377.

Goldowitz, D., Rice, D. S., and Williams, R. W. (1996) Clonal architecture of the mouse retina. *Prog. Brain Res.* 108:3–15.

Grindley, J. C., Davidson, D. R., and Hill, R. E. (1995) The role of Pax-6 in eye and nasal development. *Development* 121:1433–1442.

Guillemot, F. (1999) Vertebrate bHLH genes and the determination of neuronal fates. *Exp. Cell. Res.* 253:357–364.

Haruta, M., Kosaka, M., Kanegae, Y., et al. (2001) Induction of photoreceptor-specific phenotypes in adult mammalian iris tissue. *Nat. Neurosci.* 4:1163–1164.

Haverkamp, S. and Wassle, H. (2000) Immunocytochemical analysis of the mouse retina. *J. Comp. Neurol.* 424:1–23.

Hogan, B., Beddington, R., Costantini, F., and Lacy, E. (1994) *Manipulating the Mouse Embryo*, 2nd ed. Cold Spring Harbor Laboratory Press, Cold Spring Harbor, NY.

Hojo, M., Ohtsuka, T., Hashimoto, N., Gradwohl, G., Guillemot, F., and Kageyama, R. (2000). Glial cell fate specification modulated by the bHLH gene Hes5 in mouse retina. *Development* 127:2515–2522.

Hollyfield, J. G. (1971). Differential growth of the neural retina in Xenopus laevis larvae. *Dev. Biol.* 24:264–286.

Holt, C. E., Bertsch, T. W., Ellis, H. M., and Harris, W. A. (1988). Cellular determination in the Xenopus retina is independent of lineage and birth date. *Neuron* 1:15–26.

Jean, D., Bernier, G., and Gruss, P. (1999) Six6 (Optx2) is a novel murine Six3-related homeobox gene that demarcates the presumptive pituitary/hypothalamic axis and the ventral optic stalk. *Mech. Dev.* 84:31–40.

Jensen, A. M. and Raff, M. C. (1997) Continuous observation of multipotential retinal progenitor cells in clonal density culture. *Dev. Biol.* 188:267–279.

Johns, P. R. (1978) Growth of the adult goldfish eye. III. Source of the new retinal cells. *J. Comp. Neurol.* 176:178–198.

Kelley, M. W., Turner, J. K., and Reh, T. A. (1995) Regulation of proliferation and photoreceptor differentation in fetal human retinal cell cultures. *Invest. Ophthalmol. Vis. Sci.* 36:1280–1289.

Kenyon, K. L., Zaghloul, N., and Moody, S. A. (2001) Transcription factors of the anterior neural plate alter cell movements of epidermal progenitors to specify a retinal fate. *Dev. Biol.* 240:77–91.

Li, H. S., Yang, J. M., Jacobson, R. D., Pasko, D., and Sundin, O. (1994) Pax-6 is first expressed in a region of ectoderm anterior to the early neural plate: implications for stepwise determination of the lens. *Dev. Biol.* 162:181–194.

Lillien, L. (1995) Changes in retinal cell fate induced by overexpression of EGF receptor. *Nature* 377:158–162.

Lillien, L. (1998) Neural progenitors and stem cells: mechanisms of progenitor heterogeneity. *Curr. Opin. Neurobiol.* 8:37–44.

Liu, W., Mo, Z., and Xiang, M. (2001) The Ath5 proneural genes function upstream of Brn3 POU domain transcription factor genes to promote retinal ganglion cell development. *Proc. Natl. Acad. Sci. USA* 98: 1649–1654.

Lobe, C. G., Koop, K. E., Kreppner, W., Lomeli, H., Gertsenstein, M., Nagy, A. (199) Z/AP, a double reporter for cre-mediated recombination. *Dev. Biol.* 208:281–292.

Loosli, F., Winkler, S., and Wittbrodt, J. (1999) Six3 overexpression initiates the formation of ectopic retina. *Genes Dev.* 13:649–654.

Mansouri, A. (2001) Determination of gene function by homologous recombination using embryonic stem cells and knockout mice. *Methods Mol. Biol.* 175:397–413.

Marquardt, T. and Gruss, P. (2002). Generating neuronal diversity in the retina: one for nearly all. *Trends Neurosci.* 25:32–38.

Marquardt, T., Ashery-Padan, R., Andrejewski, N., Scardigli, R., Guillemot, F., and Gruss, P. (2001). Pax6 is required for the multipotent state of retinal progenitor cells. *Cell* 105:43–55.

Marszalek, J. R., Liu, X., Roberts, E. A., et al. (2000) Genetic evidence for selective transport of opsin and arrestin by kinesin-II in mammalian photoreceptors. *Cell* 102:175–187.

Mathers, P. H., Grinberg, A., Mahon, K. A., and Jamrich, M. (1997) The Rx homeobox gene is essential for vertebrate eye development. *Nature* 387:603–760.

McCabe, K. L., Gunther, E. C., and Reh, T. A. (1999) The development of the pattern of retinal ganglion cells in the chick retina: mechanisms that control differentiation. *Development* 126:5713–5724.

Morrow, E. M., Furukawa, T., Lee, J. E., and Cepko, C. L. (1999) NeuroD regulates multiple functions in the developing neural retina in rodent. *Development* 126:23–36.

Neumann, C. J. and Nuesslein-Volhard, C. (2000) Patterning of the zebrafish retina by a wave of sonic hedgehog activity. *Science* 289: 2137–2139.

Ngo-Muller, V. and Muneoka, K. (2000) Exo utero surgery. *Methods Mol. Biol.* 135:481–492.

Oliver, G., Mailhos, A., Wehr, R., Copeland, N. G., Jenkins, N. A., and Gruss, P. (1995) Six3, a murine homologue of the sine oculis gene, demarcates the most anterior border of the developing neural plate and is expressed during eye development. *Development* 121:4045–4055.

Otteson, D. C. (2001) Putative stem cells and the lineage of rod photoreceptors in the mature retina of the goldfish. *Dev. Biol.* 232:62–76.

Perron, M. and Harris, W. A. (2000) Determination of vertebrate retinal progenitor cell fate by the Notch pathway and basic helix-loop-helix transcription factors. *Cell. Mol. Life Sci.* 57:215–223.

Perron, M., Kanekar, S., Vetter, M. L., and Harris, W. A. (1998) The genetic sequence of retinal development in the ciliary margin of the Xenopus eye. *Dev. Biol.* 199:185–200.

Perron, M., Opdecamp, K., Butler, K., Harris, W. A., and Bellefroid, E. J. (1999) X-ngnr-1 and Xath3 promote ectopic expression of sensory neuron markers in the neurula ectoderm and have distinct inducing properties in the retina. *Proc. Natl. Acad. Sci. USA* 96:14,996–15,001.

Porter, F. D., Drago, J., Xu, Y., et al. (1997) Lhx2, a LIM homeobox gene, is required for eye, forebrain, and definitive erythrocyte development. *Development* 124:2935–2944.

Prada, C., Puga, J., Perez-Mendez, L., Lopez, R., and Ramirez, G. (1991) Spatial and temproal patterns of neurogenesis in the chick retina. *Eur. J. Neurosci.* 3:559–569.

Price, J., Turner, D., and Cepko, C. (1987) Lineage analysis in the vertebrate nervous system by retrovirus-mediated gene transfer. *Proc. Natl. Acad. Sci. USA* 84:156–160.

Rajewsky, K., Gu, H., Kuhn, R., Betz, U. A., Muller, W., Roes, J., and Schwenk, F. (1996) Conditional gene targeting. *J. Clin. Invest.* 98:600–603.

Reh, T. (1987) Cell-specific regulation of neuronal production in the larval frog retina. *J. Neurosci.* 7:3317–3324.

Reh, T. A. and Levine, E. M. (1998) Multiopotential stem cells and progenitor cells in the vertebrate retina. *J. Neurobiol.* 36:206–220.

Reh, T. A. and Nagy, T. (1987) A possible role for the vascular membrane in retinal regeneration in *Rana catesbienna* tadpoles. *Dev. Biol.* 122:471–482.

Reyer, R. W. (1977) Repolarization of reversed, regenerating lenses in adult newts, *Notophtalmus viridescens*. *Exp. Eye Res.* 24:501–509.

Roe, T., Reynolds, T. C., Yu, G., and Brown, P. O. (1993) Integration of murine leukemia virus DNA depends on mitosis. *EMBO J.* 12: 2099–2108.

Saito, T. and Nakatsuji, N. (2001) Efficient gene transfer into the embryonic mouse brain using in vivo electroporation. *Dev. Biol.* 240:237–246.

Seigel, G. M. (1996) Establishment of an E1A-immortalized retinal cell culture. *In Vitro Cell Dev. Biol. Anim.* 32:66–68.

Seigel, G. M., Takahashi, M., Adamus, G., and McDaniel, T. (1998) Intraocular transplantation of E1A-immortalized retinal precursor cells. *Cell Transplant.* 7:559–566.

Sharma, K., Leonard, A. E., Lettieri, K., and Pfaff, S. L. (2000) Genetic and epigenetic mechanisms contribute to motor neuron pathfinding. *Nature* 406:515–519.

Sparrow, J. R., Hicks, D., and Barnstable, C. J. (1990) Cell commitment and differentiation in explants of embryonic rat neural retina: comparison with the developmental potential of dissociated retina. *Brain Res. Dev. Brain Res.* 51:69–84.

Stone, L. S. (1950) The role of retinal pigment cells in regenerating neural retinae of adult salamander eyes. *J. Exp. Zool.* 113:9–31.

Tabata, H. and Nakajima, K. (2001) Efficient in utero gene transfer system to the developing mouse brain using electroporation: visualization of neuronal migration in the developing cortex. *Neuroscience* 103:865–872.

Takahashi, M., Palmer, T. D., Takahashi, J., and Gage, F. H. (1998) Widespread integration and survival of adult-derived neural progenitor cells in the developing optic retina. *Mol. Cell. Neurosci.* 12: 340–348.

Taylor, W. R., He, S., Levick, W. R., and Vaney, D. I. (2000) Dendritic computation of direction selectivity by retinal ganglion cells [*see* comments]. *Science* 289:2347–2350.

Thomas, K. R. and Capecchi, M. R. (1987) Site-directed mutagenesis by gene targeting in mouse embryo-derived stem cells. *Cell* 51:503–512.

Tomita, K., Ishibashi, M., Nakahara, K., et al. (1996) Mammalian hairy and Enhancer of split homolog 1 regulates differentiation of retinal neurons and is essential for eye morphogenesis. *Neuron* 16:723–734.

Tropepe, V., Coles, B. L., Chiasson, B. J., et al. (2000) Retinal stem cells in the adult mammalian eye. *Science* 287:2032–2036.

Tsien, J. Z., Chen, D. F., Gerber, D., et al. (1996) Subregion- and cell type–restricted gene knockout in mouse brain [*see* comments]. *Cell* 87:1317–1326.

Turner, D. L. and Cepko, C. L. (1987) A common progenitor for neurons and glia persists in rat retina late in development. *Nature* 328: 131–136.

Turner, D. L., Snyder, E. Y., and Cepko, C. L. (1990) Lineage-independent determination of cell type in the embryonic mouse retina. *Neuron* 4:833–845.

Waid, D. K. and McLoon, S. C. (1998) Ganglion cells influence the fate of dividing retinal cells in culture. *Development* 125:1059–1066.

Walther, C. and Gruss, P. (1991) Pax-6, a murine paired box gene, is expressed in the developing CNS. *Development* 113:1435–1449.

Wang, S. W., Kim, B. S., Ding, K., et al. (2001) Requirement for math5 in the development of retinal ganglion cells. *Genes Dev.* 15:24–29.

Warfvinge, K., Kamme, C., Englund, U., and Wictorin, K. (2001) Retinal integration of grafts of brain-derived precursor cell lines implanted subretinally into adult, normal rats. *Exp. Neurol.* 169:1–12.

Wassle, H. and Boycott, B. B. (1991) Functional architecture of the mammalian retina. *Physiol. Rev.* 71:447–480.

Weber, P., Metzger, D., Chambon, P. (2001) Temporally controlled targeted somatic mutagenesis in the mouse brain. *Eur. J. Neurosci.* 14: 1777–8173.

Wetts, R. and Fraser, S. E. (1988) Multipotent precursors can give rise to all major cell types of the frog retina. *Science* 239:1142–1145.

Wetts, R., Serbedzija, G. N., and Fraser, S. E. (1989) Cell lineage analysis reveals multipotent precursors in the ciliary margin of the frog retina. *Dev. Biol.* 136:254–263.

Xiang, M. (1998) Requirement for Brn-3b in early differentiation of postmitotic retinal ganglion cell precursors. *Dev. Biol.* 197:155–169.

Yan, R. T., Ma, W. X., and Wang, S. Z. (2001) Neurogenin2 elicits the genesis of retinal neurons from cultures of nonneural cells. *Proc. Natl. Acad. Sci. USA* 98:15,014–15,019.

Young, M. J., Ray, J., Whiteley, S. J. O., Klassen, H., and Gage, F. H. (2000) Neuronal differentiation and morphological integration of hippocampal progenitor cells transplanted to the retina of immature and mature dystrophic rats. *Mol. Cell. Neurosci.* 16:197–205.

Young, R. W. (1985a) Cell differentiation in the retina of the mouse. *Anat. Rec.* 212:199–205.

Young, R. W. (1985b) Cell proliferation during postnatal development of the retina in the mouse. *Brain Res.* 353:229–239.

Zuber, M. E., Perron, M., Philpott, A., Bang, A., and Harris, W. A. (1999) Giant eyes in *Xenopus laevis* by overexpression of XOptx2. *Cell* 98: 341–352.

19 Endothelial Progenitor Cells

TAKAYUKI ASAHARA, MD, PhD, AND JEFFREY M. ISNER, MD, PhD

Differentiation of organs depends on signals derived from developing vasculature. Embryonic endothelial progenitor cells (EPCs), angioblasts, arise from migrating mesodermal cells and have a precursor in common with hematopoietic stem cells (hemangioblasts, HSCs). These cells appear together during formation of blood islands and the yolk sac capillary network with the EPCs located peripherally to the HSCs. EPCs respond to fibroblast growth factor-2, and vascular endothelial growth factor (VEGF). Activation of vasculoneogenesis in the adult in response to hyperplasia, injury, or tumor growth involves both endothelial cells *in situ* and circulating EPCs from the bone marrow. Bone marrow–derived EPCs may be mobilized by growth factors such as granulocyte macrophage colony-stimulating factor and VEGF. The therapeutic use of EPCs became feasible when it was shown that statins activate EPCs and enhance angiogenesis in vivo. In addition, mobilized EPCs may be expanded in vitro and used for transplantation enhancement of angiogenesis. Gene therapy to enhance circulation in premature atherosclerosis (Buerger disease) may be accomplished using phVEGF. Decreased neoangiogenesis in older animals may be corrected by transplantation of bone marrow from young animals. However, there is a major hurdle to overcome in obtaining enough EPCs for human use. There is preliminary evidence that this may be overcome by genetic modification of EPCs to overexpress angiogenic growth factors.

19.1. STEM AND PROGENITOR CELLS FOR REGENERATION

Tissue replacement in the body takes place by two mechanisms. One is the replacement of differentiated cells by newly generated populations derived from residual cycling stem cells. Blood cells are a typical example for this kind of regeneration. Whole hematopoietic lineage cells are derived from a few self-renewal stem cells by regulated differentiation under the influence of appropriate cytokines and/or growth factors. The second mechanism is the self-repair of differentiated functioning cells preserving their proliferative activity. Hepatocytes, endothelial cells, smooth muscle cells, keratinocytes, and fibroblasts are considered to possess this ability. Following physiological stimulation or injury, factors secreted from surrounding tissues stimulate cell replication and replacement. However, these fully differentiated cells are still limited in terms of infinite proliferation by senescence, and their inability to incorporate into remote target sites.

While most cells in adult organs are composed of differentiated cells, which express a variety of specific phenotypic genes adapted to each organ's environment, quiescent stem or progenitor cells are maintained locally or in the systemic circulation and are activated by environmental stimuli for physiological and pathological tissue regeneration. These reserved quiescence stem or progenitor cells are mobilized in response to environmental stimuli when an emergent regenerative process is required, while during a minor event, neighboring differentiated cells are relied on. Researchers have defined in the past decade the stem or progenitor cells from various tissues, including bone marrow, peripheral blood, brain, liver and reproductive organs, in both adult animals and humans.

Among these stem/progenitor cells, the endothelial progenitor cell (EPC) has been identified recently and investigated to elucidate its biology for therapeutic applications. Recent reports demonstrate that endothelial lineage cells play a critical role in the early stage of liver or pancreatic differentiation, (Lammert et al., 2001; Matsumoto et al., 2001), and the significance of vascular development in organogenesis has become a crucial issue in regenerative medicine. In this review, we introduce the profiles of embryonic and postnatal EPCs and their designs of therapeutic application.

19.2. EMBRYONIC ENDOTHELIAL PROGENITOR CELLS

Embryonic EPCs, so-called angioblast, for blood vessel development arise from migrating mesodermal cells. EPCs have the capacity to proliferate, migrate, and differentiate into endothelial lineage cells but have not yet acquired characteristic mature endothelial markers. Available evidence suggests that hema-

From: *Stem Cells Handbook*
Edited by: S. Sell © Humana Press Inc., Totowa, NJ

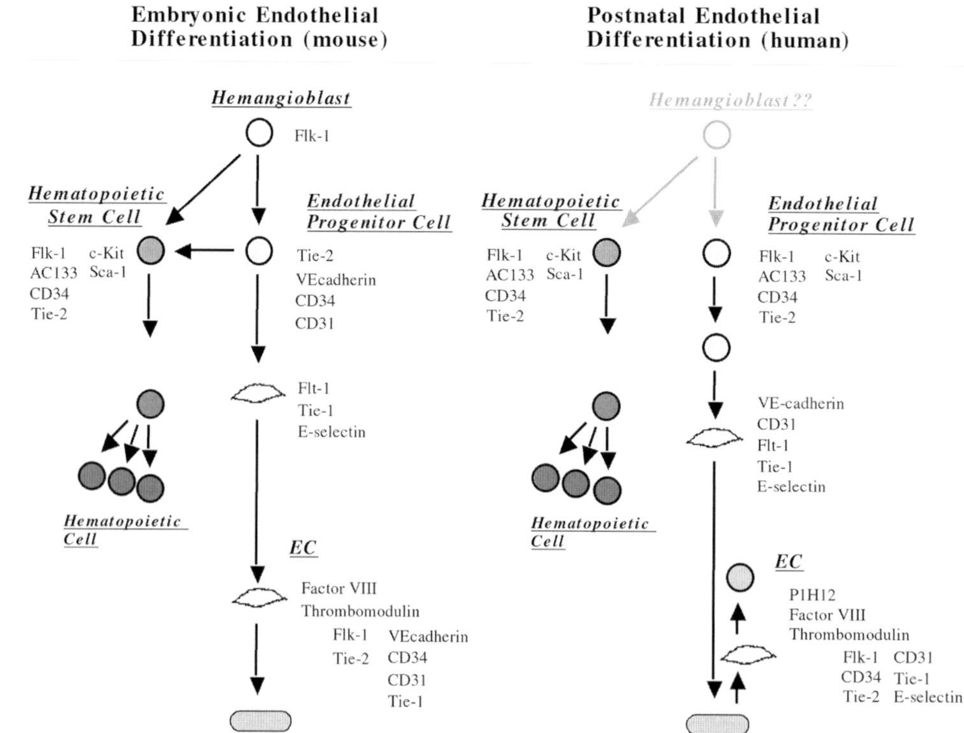

Fig.1. Embryonic vs postnatal differentiation profiles of endothelial lineage cell. EC, endothelial cell.

toporetic stem cells (HSCs) and EPCs (Risau et al., 1988; Pardanaud et al., 1989) are derived from a common precursor (hemangioblast) (His, 1900; Flamme and Risau, 1992; Weiss and Orkin, 1996). Growth and fusion of multiple blood islands in the yolk sac of the embryo ultimately give rise to the yolk sac capillary network (Risau and Flamme, 1995); after the onset of blood circulation, this network differentiates into an arteriovenous vascular system (Risau et al., 1988). The integral relationship between the elements that circulate in the vascular system, the blood cells, and the cells that are principally responsible for the vessels themselves, endothelial cells, is implied by the composition of the embryonic blood islands. The cells destined to generate hematopoietic cells are situated in the center of the blood island and are termed HSCs. EPCs are located at the periphery of the blood islands.

The key molecular players determining the fate of the hemangioblast are not fully elucidated. However, several factors have been identified that may play a role in this early event. Studies in quail/chick chimeras have shown that the fibroblast growth factor-2 mediates the induction of EPCs from the mesoderm (Poole et al., 2001). These embryonic EPCs express flk-1, the receptor-2 for vascular endothelial growth factor (VEGFR-2), and respond to a pleiotropic angiogenic factor, VEGF, for proliferation and migration. Deletion of the flk-1 gene in mice results in embryonic lethality, lacking both hematopoietic and endothelial lineage development, supporting the critical importance of flk-1 at that developmental stage, although not defining the steps regulating differentiation into endothelial vs hematopoietic cells.

The Flk-1-expressing mesodermal cell has also been defined as an embryonic common vascular progenitor that differentiates into endothelial and smooth muscle cells (Yamashita et al., 2000). The vascular progenitors differentiated to endothelial cells in response to VEGF, whereas they developed into smooth muscle cells in response to PDGF-BB.

It remains to be determined whether embryonic EPCs or vascular progenitor cells persist with an equivalent capability during adult life and whether these cells contribute to postnatal vessel growth (*see* below).

19.3. IDENTIFICATION OF ADULT EPCs

Embryonic HSCs and EPCs share certain antigenic determinants, including Flk-1, Tie-2, c-Kit, Sca-1, CD133, and CD34. These progenitor cells have consequently been considered to derive from a common precursor, putatively termed a *hemangioblast* (His, 1900; Flamme and Risau, 1992; Weiss and Orkin, 1996) (Fig. 1).

The identification of putative HSCs in peripheral blood and bone marrow and the demonstration of sustained hematopoietic reconstitution with these HSC transplants have constituted inferential evidence for HSCs in adult tissues (Kessinger and Armitage, 1991; Sheridan et al., 1992; Shpall et al., 1994; Brugger et al., 1995). Recently, the related descendents— EPCs—have been isolated along with HSCs in hematopoietic organs. Flk-1 and CD34, shared by embryonic EPCs and HSCs, were used to detect putative EPCs from mononuclear cell fraction of peripheral blood (Asahara et al., 1997). In vitro, these cells differentiated into endothelial lineage cells, and in animal models of ischemia, heterologous, homologous, and autologous EPCs were shown to incorporate into sites of active neovascularization. This finding was followed by diverse identifications of EPCs using equivalent or different methodologies by several groups (Shi et al., 1998; Gehling et al., 2000; Gunsilius et al., 2000; Lin et al., 2000; Peichev et al., 2000). EPCs

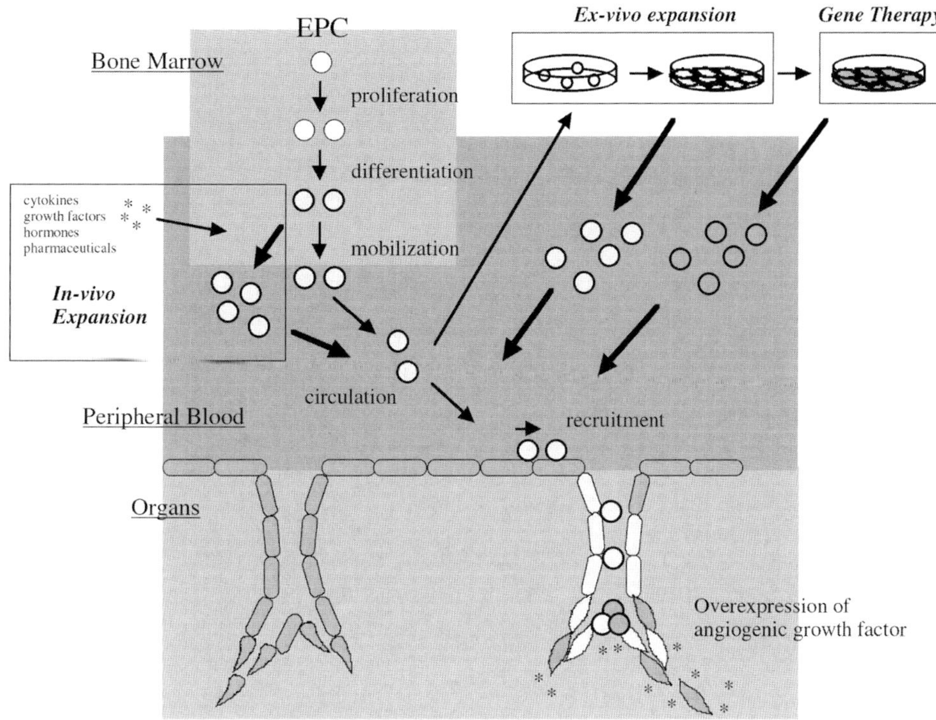

Fig.2. Endothelial progenitor cell kinetics in postnatal body.

were subsequently shown to express VE-cadherin, a junctional molecule, and AC133, an orphan receptor that is specifically expressed on EPCs but whose expression is lost once they differentiate into more mature endothelial cells (Peichev et al., 2000). Their high proliferation rate distinguishes circulating marrow-derived EPCs in the adult from mature endothelial cells shed from the vessel wall (Lin et al., 2000). Thus far, a bipotential common vascular progenitor, giving rise to both endothelial and smooth muscle cells, has not been documented in the adult.

19.4. KINETICS OF EPCs

More recently, bone marrow transplantation (BMT) experiments have demonstrated the incorporation of bone marrow–derived EPCs into foci of physiological and pathological neovascularization (Asahara et al., 1999a). Wild-type mice were lethally irradiated and transplanted with bone marrow harvested from transgenic mice in which constitutive LacZ expression is regulated by an endothelial cell--specific promoter, Flk-1 or Tie-2. The tissues in growing tumor, healing wound, ischemic skeletal and cardiac muscles, and cornea micropocket surgery have shown localization of Flk-1 or Tie-2 expressing endothelial lineage cells derived from bone marrow in blood vessels and stroma around vasculatures. Similar incorporation was observed in physiological neovascularization in uterus endometrial formation following induced ovulation as well as administration of estrogen (Asahara et al., 1999a).

Previous investigators have shown that wound trauma causes mobilization of hematopoietic cells, including pluripotent stem or progenitor cells in spleen, bone marrow, and peripheral blood. Consistent with EPC/HSC common ancestry, recent data from our laboratory have shown that mobilization of bone marrow–derived EPCs constitutes a natural response to tissue ischemia (Takahashi et al., 1999). The former murine BMT model presented the direct evidence of enhanced bone marrow–derived EPC incorporation into foci of corneal neovascularization following the development of hind-limb ischemia. Light microscopic examination of corneas excised 6 d after micropocket injury and concurrent surgery to establish hind-limb ischemia demonstrated a statistically significant increase in cells expressing β-galactosidase in the corneas of mice with vs those without an ischemic limb (Takahashi et al., 1999). This finding indicates that circulating EPCs are mobilized endogenously in response to tissue ischemia, following which they may be incorporated into neovascular foci to promote tissue repair. This was comfirmed by clinical findings of EPC mobilization in patients of coronary artery bypass grafting, burns (Gill et al., 2001), and acute myocardial infarction (Shintani et al., 2001).

Having demonstrated the potential for endogenous mobilization of bone marrow–derived EPCs, we considered that iatrogenic expansion and mobilization of this putative endothelial cell precursor population might represent an effective means to augment the resident population of ECs that is competent to respond to administered angiogenic cytokines. Such a program might thereby address the issue of endothelial dysfunction or depletion that may compromise strategies of therapeutic neovascularization in older, diabetic, and/or hypercholesterolemic animals and patients. Granulocyte macrophage colony-stimulating factor (GM-CSF), which stimulates hematopoietic progenitor cells and myeloid lineage cells, as well as nonhematopoietic cells including bone marrow stromal cells and endothelial cells, has been shown to exert a potent stimulatory effect on EPC kinetics (Takahashi et al., 1999). Such cytokine-induced EPC mobilization could enhance neovascularization of severely ischemic tissues as well as *de novo* corneal vascularization (Takahashi et al., 1999) (Fig. 2).

The mechanisms whereby these EPCs are mobilized to the peripheral circulation are in their early definitions. Among other growth factors, VEGF is the most critical factor for vasculogenesis and angiogenesis (Shalaby et al., 1995; Carmeliet et al., 1996; Ferrara et al., 1996). Recent data indicate that VEGF is an important factor for the kinetics of EPC as well. Our studies performed first in mice (Asahara et al., 1999b) and subsequently in patients undergoing VEGF gene transfer for critical limb ischemia (Kalka et al., 2000a) and myocardial ischemia (Kalka et al., 2000c) established that a previously unappreciated mechanism by which VEGF contributes to neovascularization is via mobilization of bone marrow–derived EPCs. Similar EPC kinetics modulation has been observed in response to other hematopoietic stimulators, such as granulocyte-colony stimulating factor (G-CSF) (Gehling et al., 2000), angiopoietin-1 (Hattori et al., 2001), and stroma-derived factor-1 (Peichev et al., 2000).

This therapeutic strategy of EPC mobilization has recently been implicated not only by natural hematopoietic or angiogenic stimulants but also by recombinant pharmaceuticals. The 3-hydroxy-3-methylglutaryl coenzyme A (HMG-CoA) reductase inhibitors—statins—inhibit the activity of HMG-CoA reductase, which catalyzes the synthesis of mevalonate, a rate-limiting step in cholesterol biosynthesis. The statins rapidly activate Akt signaling in endothelial cells, and this stimulates endothelial cell bioactivity in vitro and enhances angiogenesis in vivo (Kureishi et al., 2000). Recently, we and Dimmeler et al. (2001) demonstrated a novel function for HMG-CoA reductase inhibitors that contributes to postnatal neovascularization by augmented mobilization of bone marrow–derived EPCs through stimulation of the Akt signaling pathway (Vasa et al., 2000; Llevadot et al., 2001). Regarding its pharmacological safety and effectiveness on hypercholesterolemia, one of the risk factors for atherogenesis, the statin might be a potent medication against atherosclerotic vascular diseases.

19.5. THERAPEUTIC VASCULOGENESIS BY EPC TRANSPLANTATION

Recently, the regenerative potential of stem cells has been under intense investigation. In vitro, stem and progenitor cells possess the capability of self-renewal and differentiation into organ-specific cell types. In vivo, transplantation of these cells may reconstitute organ systems, as shown in animal models of diseases (Anklesaria et al., 1987; Lindvall et al., 1990; Asahara et al., 1997; Flax et al., 1998; Evans et al., 1999; Kalka et al., 2000b). By contrast, differentiated cells do not exhibit such characteristics. Human EPCs have been isolated from the peripheral blood of adult individuals, expanded in vitro, and committed into an endothelial lineage in culture (Asahara et al., 1997). Transplantation of these human EPCs has been shown to facilitate successful salvage of limb vasculature and perfusion in athymic nude mice with severe hind-limb ischemia, while differentiated endothelial cells (human microvascular endothelial cells) failed to accomplish limb-saving neovascularization (Kalka et al., 2000b) (Fig. 2).

These experimental findings call into question certain fundamental concepts regarding blood vessel growth and development in adult organisms. Postnatal neovascularization has previously been considered synonymous with proliferation and migration of preexisting, fully differentiated endothelial cells resident within parent vessels (i.e., angiogenesis) (Folkman, 1971). The finding that circulating EPCs may home to sites of neovascularization and differentiate into endothelial cells *in situ* is consistent with "vasculogenesis" (Risau et al., 1988), a critical paradigm for establishment of the primordial vascular network in the embryo. While the proportional contributions of angiogenesis and vasculogenesis to postnatal neovascularization remain to be clarified, our findings, together with recent reports from other investigators (Hatzopoulos et al., 1998; Shi et al., 1998), suggest that growth and development of new blood vessels in the adult is not restricted to angiogenesis but encompasses both embryonic mechanisms. As a corollary, augmented or retarded neovascularization—whether endogenous or iatrogenic—likely includes enhancement or impairment of vasculogenesis.

We therefore considered a novel strategy of EPC transplantation to provide a source of robust endothelial cells that might supplement fully differentiated endothelial cells thought to migrate and proliferate from preexisting blood vessels according to the classic paradigm of angiogenesis developed by Folkman and colleagues (Folkman, 1993). Our studies indicated that ex vivo cell therapy, consisting of culture-expanded EPC transplantation, successfully promotes neovascularization of ischemic tissues, even when administered as "sole therapy," i.e., in the absence of angiogenic growth factors. Such a "supply-side" version of therapeutic neovascularization in which the substrate (endothelial cells as EPCs) rather than ligand comprises the therapeutic agent was first demonstrated in the hind-limb ischemia model of immunodeficient mouse, using donor cells from human volunteers (Kalka et al., 2000b). These findings provided novel evidence that exogenously administered EPCs augment naturally impaired neovascularization in an animal model of experimentally induced critical limb ischemia. Not only did heterologous cell transplantation improve neovascularization and blood flow recovery, but important biological consequences—notably limb necrosis and autoamputation—were reduced by 50% in comparison with mice receiving differentiated endothelial cells or control mice receiving media in which harvested cells were expanded ex vivo prior to transplantation. A similar strategy applied in a model of myocardial ischemia in the nude rat demonstrated that transplanted human EPCs incorporated into rat myocardial neovascularization, differentiated into mature endothelial cells in ischemic myocardium, enhanced neovascularization preserved left ventricular (LV) function, and inhibited myocardial fibrosis (Kawamoto et al., 2001).

Recently, Kocher et al. (2001) attempted *intravenous* infusion of freshly isolated (not cultured) human $CD34^+$ mononuclear cells (EPC-enriched fraction) into nude rats with myocardial ischemia. This strategy resulted in preservation of LV function associated with inhibition of cardiomyocyte apoptosis. These experimental findings using immunodeficient animals suggest that both cultured and freshly isolated human EPCs have therapeutic potential in peripheral and coronary artery diseases.

19.6. THE LIMITATION OF PRIMARY EPC TRANSPLANTATION

Despite promising potential for regenerative applications, the fundamental scarcity of EPC populations in the hematopoietic system and the possible functional impairment of EPCs associated with a variety of phenotypes, such as aging, diabetes, hypercholesterolemia, and homocysteinemia (vide infra) constitute important limitations of primary EPC transplantation. Ex vivo expansion of EPCs cultured from the peripheral blood of healthy human volunteers typically yields approx 5.0×10^6 cells/100 mL of blood. Our animal studies (Kalka et al., 2000b) suggest that heterologous transplantation required 0.5–2.0 × 10^4 human EPCs/g weight (of the recipient mouse) to achieve satisfactory reperfusion of the ischemic hind limb. Rough extrapolation of this experience to humans suggests that a volume of as much as 12 L of peripheral blood may be necessary to harvest the EPCs required to treat critical limb ischemia. Even with the integration of certain technical improvements, the adjustment of species compatibility by autologous transplantation, and adjunctive strategies (e.g., cytokine supplements) to promote EPC mobilization (Asahara et al., 1999b; Takahashi et al., 1999), the primary scarcity of a viable and functional EPC population constitutes a potential limitation of therapeutic vasculogenesis based on the use of ex vivo expansion alone.

19.7. IMPACT OF CLINICAL PHENOTYPE ON EPCs

Preliminary clinical findings in patients with critical limb ischemia indicated that the response to phVEGF gene transfer was most robust and expeditious in young patients with premature atherosclerosis involving the lower extremities, so-called Buerger's disease (Isner et al., 1998). This clinical observation was supported by experiments performed in live animal models, specifically young (6–8 mo) vs old (4 to 5 yr) rabbits and young (8 wk) vs old (2 yr) mice. In both cases, native neovascularization of the ischemic hind limb was markedly retarded in old vs young animals. Retardation of neovascularization in old animals appeared in part to result from reduced expression of VEGF in tissue sections harvested from the ischemic limb (Rivard et al., 1999a).

Endogenous cytokine expression, however, is not the only factor contributing to impaired neovascularization. Older, diabetic and hypercholesterolemic animals—like human subjects (Drexler et al., 1991; Johnstone et al., 1993; Luscher and Tshuci, 1993; Taddei et al., 1995; Chauhan et al., 1996; Gerhard et al., 1996; Tschundi et al., 1996; Cosentino and Luscher, 1998)—also exhibit evidence of age-related endothelial dysfunction. While endothelial dysfunction does not necessarily preclude a favorable response to cytokine replacement therapy, indices of limb perfusion fail to reach ultimate levels recorded in wild-type animals, reflecting limitations imposed by a less responsive endothelial cell substrate (Van Belle et al., 1997; Couffinhal et al., 1999; Rivard et al., 1999a, 1999b).

It is then conceivable that unfavorable clinical situations (such as aging) might be associated with dysfunctional EPCs, defective vasculogenesis, and thus impaired neovascularization. Indeed, preliminary results from OUR laboratory indicated that replacement of native bone marrow (including its compartment of progenitor cells) of young mice with bone marrow transplanted from old animals led to a marked reduction in neovascularization following corneal micropocket injury, in comparison with young mice transplanted with young bone marrow (Rivard et al., 1998). These studies thus established evidence of an age-dependent impairment in vasculogenesis (as well as angiogenesis) and the origin of progenitor cells as a critical parameter influencing neovascularization. Moreover, analysis of clinical data from older patients at our institution disclosed a significant reduction in the baseline levels and the population of EPCs mobilized in response to VEGF165 gene transfer (Kalka et al., 1999); specifically, the number of EPCs in the systemic circulation of young patients with critical limb ischemia was five times more than the number circulating in old individuals. Impaired EPC mobilization and/or activity in response to VEGF may thus contribute to the age-dependent defect in postnatal neovascularization.

19.8. GENE THERAPY OF EPCs

Given the presented findings, together with the limited quantity of EPCs available even under healthy, physiological conditions, one must consider a strategy that addresses this shortfall and mitigates the possibility of dysfunctional EPCs for therapeutic vasculogenesis in ischemic disorders complicated by aging, diabetes, hypercholesterolemia, and/or hyperhomocysteinemia. Genetic modification of EPCs to overexpress angiogenic growth factors, enhance signaling activity of the angiogenic response, and rejuvenate the bioactivity and/or extend the life-span of EPCs constitutes one potential strategy that might address these limitations of EPC transplantation and thereby optimize therapeutic neovascularization (Fig. 2).

Our recent findings provide the first evidence that exogenously administered gene-modified EPCs augment naturally impaired neovascularization in an animal model of experimentally induced limb ischemia (Iwaguro et al., 2002). Transplantation of heterologous EPCs transduced with adenovirus encoding VEGF improved not only neovascularization and blood flow recovery, but also had meaningful biological consequences: limb necrosis and autoamputation were reduced by 63.7% in comparison with controls. The dose of EPCs used in the current in vivo experiments was subtherapeutic; that is, the dose of EPCs was 30 times less than that required in previous experiments to improve the rate of limb salvage above that seen in untreated controls. Adenoviral VEGF EPC-gene transfer, however, accomplished a therapeutic effect, as evidenced by a functional outcome, despite a subtherapeutic dose of EPCs. Thus, VEGF EPC-gene transfer constitutes one option to address the limited number of EPCs that can be isolated from peripheral blood prior to ex vivo expansion and subsequent autologous readministration.

REFERENCES

Anklesaria, P., Kase, K., Glowacki, J., et al. (1987) Engraftment of a clonal bone marrow stromal cell line in vivo stimulates hematopoietic recovery from total body irradiation. *Proc. Natl. Acad. Sci. USA* 84:7681–7685.

Asahara, T., Murohara, T., Sullivan, A., et al. (1997) Isolation of putative progenitor endothelial cells for angiogenesis. *Science* 275:964–967.

Asahara, T., Masuda, H., Takahashi, T., et al. (1999a) Bone marrow origin of endothelial progenitor cells responsible for postnatal vasculogenesis in physiological and pathological neovascularization. *Circ. Res.* 85:221–228.

Asahara, T., Takahashi, T., Masuda, H., et al. (1999b) VEGF contributes to postnatal neovascularization by mobilizing bone marrow–derived endothelial progenitor cells. *EMBO J.* 18:3964–3972.

Brugger, W., Heimfeld, S., Berenson, R. J., Mertelsmann, R., and Kanz, L. (1995) Reconstitution of hematopoiesis after high-dose chemotherapy by autogous progenitor cells generated ex vivo. *N. Engl. J. Med.* 333:283–287.

Carmeliet, P., Ferreira, V., Breier, G., et al. (1996) Abnormal blood vessel development and lethality in embryos lacking a single VEGF allele. *Nature* 380:435–439.

Chauhan, A., More, R. S., Mullins, P. A., Taylor, G., Petch, M. C., and Schofield, P. M. (1996) Aging-associated endothelial dysfunction in humans is reversed by L-arginine. *J. Am. Coll. Cardiol.* 28:1796–1804.

Cosentino, F. and Luscher, T. F. (1998) Endothelial dysfunction in diabetes mellitus. *J. Cardiovasc. Pharmacol.* 32:S54–S61.

Couffinhal, T., Silver, M., Kearney, M., et al. (1999). Impaired collateral vessel development associated with reduced expression of vascular endothelial growth factor in $ApoE^{-/-}$ mice. *Circulation* 99:3188–3198.

Dimmeler, S., Aicher, A., Vasa, M., et al. (2001) HMG-CoA-reductase inhibitors (statins) increase endothelial progenitor cells via the P13 kinase/Akt pathway. *J. Clin. Invest.* 108:391–397.

Drexler, H., Zeiher, A.M., Meinzer, K., and Just, H. (1991) Correction of endothelial dysfunction in coronary microcirculation of hypercholesterolemic patients by L-arginine. *Lancet* 338:1546–1550.

Evans, J. T., Kelly, P. F., O'Neill, E., and Garcia, J. V. (1999) Human cord blood CD34+CD38– cell transduction via lentivirus-based gene transfer vectors. *Hum. Gene Ther.* 10:1479–1489.

Ferrara, N., Carver-Moore, K., Chen, H., et al. (1996) Heterozygous embryonic lethality induced by targeted inactivation of the VEGF gene. *Nature* 380:439–442.

Flamme, I. and Risau, W. (1992) Induction of vasculogenesis and hematopoiesis in vitro. *Development* 116:435–439.

Flax, J. D., Aurora, S., Yang, C., et al. (1998) Engraftable human neural stem cells respond to developmental cues, replace neurons, and express foreign genes. *Nat. Biotechnol.* 16:1033–1039.

Folkman, J. (1971) Tumor angiogenesis: therapeutic implications. *N. Engl. J. Med.* 285:1182–1186.

Folkman, J. (1993) Tumor angiogenesis. In: *Cancer Medicine* (Holland, J. F., Frei, E. III, Bast, R. C. Jr., Kute, D. W., Morton, D. L., and Weichselbaum, R. R., eds.), Lea & Febiger, Philadelphia, PA, pp. 153–170.

Gehling, U. M., Ergun, S., Schumacher, U., et al. (2000) In vivo differentiation of endothelial cells from AC133-positive progenitor cells. *Blood* 95:3106–3112.

Gerhard, M., Roddy, M.-A., Creager, S. J., and Creager, M. A. (1996) Aging progressively impairs endothelium–dependent vasodilation in forearm resistance vessles of humans. *Hypertension* 27:849–853.

Gill, M., Dias, S., Hattori, K., et al. (2001) Vascular trauma induces rapid but transient mobilization of VEGFR(+)AC133(+) endothelial precursor cells. *Circ. Res.* 88:167–174.

Gunsilius, E., Duba, H. C., Petzer, A. L., et al. (2000) Evidence from a leukaemia model for maintenance of vascular endothelium by bone-marrow-derived endothelial cells. *Lancet* 355:1688–1691.

Hattori, K., Dias, S., Heissig, B., et al. (2001) Vascular endothelial growth factor and angiopoietin-1 stimulate postnatal hematopoiesis by recruitment of vasculogenic and hematopoietic stem cells. *J. Exp. Med.* 193: 1005–1014.

Hatzopoulos, A. K., Folkman, J., Vasile, E., Eiselen, G. K., and Rosenberg, R. D. (1998) Isolation and characterization of endothelial progenitor cells from mouse embyros. *Development* 125:1457–1468.

His, W. (1900). Leoithoblast und angioblast der wirbelthiere. *Abhandl. KSGes. Wiss. Math. Phys.* 22:171–328.

Isner, J. M., Baumgartner, I., Rauh, G., et al. (1998) Treatment of thromboangiitis obliterans (Buerger's disease) by intramuscular gene transfer of vascular endothelial growth factor: preliminary clinical results. *J. Vasc. Surg.* 28:964–975.

Iwaguro, H., Yamaguchi, J., Kalka, C., et al. (2002) Endothelial progenitor cell vascular endothelial growth factor gene transfer for vascular regeneration. *Circulation* 105:732–738.

Johnstone, M. D., Creager, S. J., Scales, K. M., Cusco, J. A., Lee, B. K., and Creager, M. A. (1993) Impaired endothelium-dependent vasodilation in patients with insulin-dependent diabetes mellitus. *Circulation* 88:2510–2516.

Kalka, C., Masuda, H., Gordon, R., Silver, M., and Asahara, T. (1999) Age dependent response in mobilization of endothelial progenitor cells (EPC) to VEGF gene therapy in human subjects. *Circulation* 100:I–40.

Kalka, C., Masuda, H., Takahashi, T., et al. (2000a) Vascular endothelial growth factor 165 gene transfer augments circulating endothelial progenitor cells in human subjects. *Circ. Res.* 86:1198–1202.

Kalka, C., Masuda, H., Takahashi, T., et al. (2000b) Transplantation of ex vivo expanded endothelial progenitor cells for therapeutic neovascularization. *Proc. Natl. Acad. Sci. USA* 97:3422–3427.

Kalka, C., Tehrani, H., Laudenberg, B., et al. (2000c) Mobilization of endothelial progenitor cells following gene therapy with VEGF165 in patients with inoperable coronary disease. *Ann. Thorac. Surg.* 70: 829–834.

Kawamoto, A., Gwon, H.-C., Iwaguro, H., et al. (2001). Therapeutic potential of *ex vivo* expanded endothelial progenitor cells for myocardial ischemia. *Circulation* 103:634–637.

Kessinger, A. and Armitage, J. O. (1991) The evolving role of autologous peripheral stem cell transplantation following high-dose therapy for malignancies. *Blood* 77:211–213.

Kocher, A. A., Schuster, M. D., Szabolcs, M. J., et al. (2001) Neovascularization of ischemic myocardium by human bone marrow–derived angioblasts prevents cardiomyocyte apoptosis, reduces remodeling and improves cardiac function. *Nat. Med.* 7:430–436.

Kureishi, Y., Luo, Z., and Shiojima, I. (2000) The HMG-CoA reductase inhibitor simvastatin activates the protein kinase Akt and promotes angiogenesis in normocholesterolemic animals. *Nat. Med.* 6:1004–1010.

Lammert, E., Cleaver, O., and Melton, D. (2001) Induction of pancreatic differentiation by signals from blood vessels. *Science* 294: 564–567.

Lin, Y., Weisdorf, D. J., Solovey, A., and Hebbel, R. P. (2000) Origins of circulating endothelial cells and endothelial outgrowth from blood. *J. Clin. Invest.* 105:71–77.

Lindvall, O., Brundin, P., Widner, H., et al. (1990) Grafts of fetal dopamine neurons survive and improve motor function in Parkinson's disease. *Science* 247:574–577.

Llevadot, J., Murasawa, S., Kureishi, Y., et al. (2001) HMG-CoA reductase inhibitor mobilizes bone-marrow derived endothelial progenitor cells. *J. Clin. Invest.* 108:399–405.

Luscher, T. F. and Tshuci, M. R. (1993) Endothelial dysfunction in coronary artery disease. *Ann. Rev. Med.* 44:395–418.

Matsumoto, K., Yoshitomi, H., Rossant, J., and Zaret, K. S. (2001) Liver organogenesis promoted by endothelial cells prior to vascular function. *Science* 294:559–563.

Pardanaud, L., Altman, C., Kitos, P., and Dieterien-Lievre, F. (1989) Relationship between vasculogenesis, angiogenesis and haemopoiesis during avian ontogeny. *Development* 105:473–485.

Peichev, M., Naiyer, A. J., Pereira, D., et al. (2000) Expression of VEGFR-2 and AC133 by circulating human CD34+ cells identifies a population of functional endothelial precursors. *Blood* 95:952–958.

Poole, T. J., Finkelstein, E. B., and Cox, C. M. (2001) The role of FGF and VEGF in angioblast induction and migration during vascular development. *Dev. Dyn.* 220:1–17.

Risau, W. and Flamme, I. (1995) Vasculogenesis. *Ann. Rev. Cell Dev. Biol.* 11:73–91.

Risau, W., Sariola, H., Zerwes, H.-G., et al. (1988) Vasculogenesis and angiogenesis in embryonic stem cell-derived embryoid bodies. *Development* 102:471–478.

Rivard, A., Asahara, T., Takahashi, T., Chen, D., and Isner, J. M. (1998) Contribution of endothelial progenitor cells to neovascularization (vasculogenesis) is impaired with aging. *Circulation* 98: 1–39.

Rivard, A., Fabre, J.-E., Silver, M., et al. (1999a) Age-dependent impairment of angiogenesis. *Circulation* 99:111–120.

Rivard, A., Silver, M., Chen, D., et al. (1999b) Rescue of diabetes related impairment of angiogenesis by intramuscular gene therapy with adeno-VEGF. *Am. J. Pathol.* 154:355–364.

Shalaby, F., Rossant, J., Yamaguchi, T. P., et al. (1995) Failure of blood-island formation and vasculogenesis in Flk-1 deficient mice. *Nature* 376:62–66.

Sheridan, W. P., Begley, C. G., and Juttener, C. (1992) Effect of peripheral-blood progenitor cells mobilised by filgrastim (G-CSF) on platelet recovery after high-dose chemotherapy. *Lancet* 339:640–644.

Shi, Q., Rafii, S., Wu, M. H.-D., et al. (1998) Evidence for circulating bone marrow–derived endothelial cells. *Blood* 92:362–367.

Shintani, S., Murohara, T., Ikeda, H., et al. (2001) Augmentation of postnatal neovascularization with autologous bone marrow transplantation. *Circulation* 103:897–903.

Shpall, E. J., Jones, R. B., and Bearman, S. I. (1994) Transplantation of enriched CD34-positive autologous marrow into breast cancer patients following high-dose chemotherapy. *J. Clin. Oncol.* 12:28–36.

Taddei, S., Virdis, A., Mattei, P., et al. (1995) Aging and endothelial function in normotensive subjects and patients with essential hypertension. *Circulation* 91:1981–1987.

Takahashi, T., Kalka, C., Masuda, H., et al. (1999) Ischemia- and cytokine-induced mobilization of bone marrow–derived endothelial progenitor cells for neovascularization. *Nat. Med.* 5:434–438.

Tschudi, M. R., Barton, M., Bersinger, N. A., et al. (1996) Effect of age on kinetics of nitric oxide release in rat aorta and pulmonary artery. *J. Clin. Invest.* 98:899–905.

Van Belle, E., Rivard, A., Chen, D., et al. (1997). Hypercholesterolemia attenuates angiogenesis but does not preclude augmentation by angiogenic cytokines. *Circulation* 96:2667–2674.

Vasa, M., Breitschopf, K., Zeiher, A. M., and Dimmeler, S. (2000) Nitric oxide activates telomerase and delays endothelial cell senescence. *Circ. Res.* 87:540–542.

Weiss, M. and Orkin, S. H. (1996) In vitro differentiation of murine embryonic stem cells: new approaches to old problems. *J. Clin. Invest.* 97:591–595.

Yamashita, J., Itoh, H., Hirashima, M., et al. (2000) Flk1-positive cells derived from embryonic stem cells serve as vascular progenitors. *Nature* 408:92–96.

20 Development of the Cardiovascular System in Embryoid Bodies Derived from Embryonic Stem Cells

HEINRICH SAUER, PhD, MARIA WARTENBERG, PhD,
AGAPIOS SACHINIDIS, PhD, AND JÜRGEN HESCHELER, PhD

The vertebrate heart and circulatory system is the first embryonic organ to develop from cells in the embryoid body. Understanding how this happens has been examined through the use of reporter genes by expression of green fluorescent protein under the control of cardiac- and endothelial cell–specific promoters. The cardiac primordial may be recognized as bilaterally symmetric strands of mesenchymal cells derived from the lateral plate mesoderm, which forms separate layers of myocardium and endocardium separated by extracellular matrix (cardiac jelly). The endoderm derives from ingressing cells during gastrulation and provides signals (factors such as bone morphogenetic protein-2 (BMP-2), fibroblast growth factor-1,2,4 (FGF-1,2,4)) that influence the adjacent mesoderm. The endocardium is in turn influenced by transforming growth factor-β and vascular endothelial growth factor (VEGF). Other factors such as cardiotrophin-1 and leukemia inhibitory factor also control cardiac development. First there is a simple linear heart tube, which goes on to form modular elements (atria, ventricles, septa, and valves). By folding on itself and fusing, the four-chambered heart is formed. These structural events are triggered by specific signaling molecules and involve reactive oxygen species and characteristic action potentials correlated with specialized types of ion channels, as well as a developmentally controlled expression pattern of the cardiac-specific genes encoding atrial natriuretic factor and sarcomeric proteins (i.e., α- and β-cardiac myosin heavy chain, myosin light chain isoform 2V, titin [Z-disk], titin [M-band], α-actinin, myomesin, sarcomeric α-actin, and troponin T), followed by M-protein. Whereas in early stage embryonic stem (ES) cell–derived cardiomyocytes cell contraction is triggered by Ca^{2+} transients arising from intracellular Ca^{2+} stores, contraction of terminally differentiated cardiac cells is dependent on an evoked action potential leading to opening of L-type voltage-dependent Ca^{2+} channels that appear several days before "beating" of cardiomyocytes. The vascular structures of the embryo develop from angioblasts in the paraaxial and lateral plate mesoderm as well as in the yolk sac extraembryonic mesoderm, where they form the outer layer of blood islands. Vasculogenesis and angiogenesis are strictly regulated by the pericellular oxygen tension of the tissue through hypoxia inducible factor-1. Many of the genes involved in cardiovascular differentiation are directly or indirectly regulated by hypoxia, with VEGF being a principle factor required for vasculogenesis. Therapeutic use of ES cells would be greatly enhanced if we knew how to direct ES cells to specific cell lineages in the cardiovascular or other organ systems.

20.1. BACKGROUND

Embryonic stem (ES) cells have been established from the inner cell mass (ICM) of blastocysts of several species, including humans. ES cells display the potential to generate all embryonic germ layers and cell lineages when they undergo differentiation within spheroidal cell aggregates termed *embryoid bodies* (EBs). ES cells can be maintained for infinite times in their pluripotential state when cultured in the presence of leukemia inhibitory factor (LIF), which inhibits their differentiation. The ability to grow mass cultures of EBs and the possibility of cultivating vast amounts of tissue have raised the hope of using ES cells differentiated to specific cell types for cell replacement therapies. Among others, ES cells differentiate a primitive cardiovascular system comprising spontaneously contracting cardiac cell clusters as well as hollow tube–forming blood vessel–like structures. The vertebrate heart and circulatory system is the first embryonic organ to develop and is therefore difficult to investigate in vivo owing to the small size of very early embryos. Furthermore, studies of the function of genes in knockout animals is frequently hampered by early death of embryos *in utero*.

The easy availability of early precursors of cardiac and vascular cells from ES cells has therefore initiated intensive scientific investigations to analyze the mechanisms of the initial stages of cardiovascular development. By the use of ES cells, basic signaling cascades governing the differentiation of precursor cells of the

From: *Stem Cells Handbook*
Edited by: S. Sell © Humana Press Inc., Totowa, NJ

cardiovascular system to terminally differentiated cardiac and vascular cells have been unraveled and have significantly contributed towards our understanding of the molecular events occurring during early embryogenesis of the heart and circulatory system.

20.2. INTRODUCTION

ES cells derived from the ICM of mouse blastocysts display self-renewal and pluripotent capacity and have been demonstrated to differentiate into phenotypes characteristic of derivatives of all three germ layers. The cardiovascular system is the first organ system developed during embryonic development and has been investigated in detail by the use of murine ES cells that are differentiated within the three-dimensional (3D) tissue of EBs. Numerous studies by us and others have revealed that the time course of development as well as gain of function of ES cell–derived cell types of the cardiovascular system is nearly indistinguishable in many aspects from the developmental events occurring in vivo (for reviews, *see* Wobus et al., 1995; Hescheler et al., 1997, 1999, 2002). This makes the ES cell system a valuable and reliable tool for the analysis of physiological and biochemical differentiation patterns of cardiac and/or vascular cells that arise as yet poorly defined stem cells in the early mesoderm and develop toward terminally differentiated cardiomyocytes with tightly regulated autorhythmic activity, as well as endothelial cells lining functional blood vessel–like structures.

The detailed knowledge of the physiological properties of cells of the cardiovascular system of early developmental stages has received particular interest; recent studies have demonstrated that cardiomyocytes of failing hearts tend to dedifferentiate toward an embryonic phenotype (Townsend et al., 1995; Sack et al., 1997). This may also imply that dysfunctions in adult cardiac cells may arise from the activation of dormant cellular pathways that have been developed earlier during cellular ontogenesis. With the availability of human pluripotent stem cells derived from primordial germ cells (Shamblott et al., 2001) and ES cells derived from human blastocysts (Thomson et al., 1998), the very early steps of human embryonic cell differentiation are now available for scientific investigation. Differentiation capacity for the cardiovascular system (e.g. for cardiomyocytes) (Kehat et al., 2001) as well as endothelial cells (Levenberg et al., 2002) has recently been found to be present in ES cells of human origin, which has raised the possibility of using human ES cell–derived cardiac as well as endothelial cells for cell replacement therapies. This approach is even more promising because ES cells can be cultivated in mass cultures (Wartenberg et al., 1998; Hescheler et al., 2002), and genetic approaches based on ES cell clones that carry resistance genes toward antibiotics under the control of either cardiac-specific (Klug et al., 1996) or vascular endothelium–derived promotors (Marchetti et al., 2002) allow the selection and enrichment of specific cell lineages.

The ability to inactivate virtually any gene in ES cells by gene targeting has significantly improved our understanding of the roles played by specific genes during cellular and organismic development and has vastly improved our scientific knowledge of how specific genes are turned on and turned off during embryonic development. The potency of using ES cells differentiated in specific cell lineages for regenerative medicine has raised a so-far unknown general public and political interest. In the near future, the rapidly growing scientific knowledge that can be achieved with ES cells will increasingly shape our understanding of organ-specific stem cells, mammalian development, and cell replacement therapy.

20.3. ORGANOGENESIS OF MURINE HEART

The circulatory system is the first functioning organ system of the embryo. The very early development of the cardiovascular system is essentially important for the demands of nutrition and excretion of the developing embryo because neither the fertilized egg nor the yolk sac contains sufficient amounts of nutritive substances. Furthermore, the catabolic waste products must be eliminated and oxygen must be supplied to avoid impairment of growth and development of the embryo.

Development of the vertebrate heart can be considered an additive process, in which additional layers of complexity have been put together throughout the evolution of a simple structure (linear heart tube) in the form of modular elements (atria, ventricles, septa, and valves) (for a review, *see* (Olson and Srivastava, 1996; Fishman and Chien, 1997; Epstein and Buck, 2000; Sedmera et al., 2000). Each modular element confers an added capacity to the vertebrate heart and can be identified as individual structures patterned in a sophisticated manner (Bruneau, 2002). In this respect, formation of the heart requires the precise integration of cell type–specific gene expression to regulate morphological development.

The mouse embryo at the commencement of gastrulation is made up of two tissue layers: the epiblast, which gives rise to all fetal tissue types, and the hypoblast (i.e., the visceral and parietal endoderm), which contributes to the yolk sac endoderm. Early in gestation (d 7.5 in mouse), cardiac primordia are evident as bilaterally symmetric strands of mesenchymal cells derived from the lateral plate mesoderm (Yutzey and Bader, 1995). From d 8.5 (mouse) on, the cardiac primordia migrate medially and fuse to form a single heart tube. This primitive cardiac tissue comprises the myocardium and the endocardium, which are separated by the extracellular matrix (ECM) of cardiac jelly. Despite the already functional differentiation (Kamino, 1991), there is minimal variation in structure along the cardiac tube. The first morphological myocardial diversification can be perceived in the circumferential arrangement of actin and fibronectin (chick embryo) (Shiraishi et al., 1995). Subsequently, cardiac looping occurs; that is, the midline heart tube folds on itself, which results in the formation of discrete regions corresponding to future heart chambers (Taber et al., 1995). By d 10.5 (mouse), localized swellings in the cardiac jelly are invaded by underlying endothelial cells to initiate the formation of the endocardial cushions that will later condense to form the mature cardiac valves. Subsequently, the cardiac chambers are formed through a series of complex septation events. As revealed in chick embryos, endocardial and myocardial lineages are distinct within the cardiogenic mesoderm prior to the fusion of the heart tube and do not derive from a common precursor cell (Cohen-Gould and Mikawa, 1996). By d 11.5 (mouse), neural crest cells have migrated from the dorsal neural tube along aortic arches 3, 4, and 6 and have invaded the outflow tract of the heart. By this process, the septation of the single great vessel emerging from the embryonic heart (the truncus arteriosus) is initiated, resulting in the formation of the aorta and the pulmonary artery. Neural crest cells contribute to the mesenchymal elements of the ductus arteriosus and great vessels and induce remodeling of the aortic arches. Furthermore, cardiac neural crest differentiates into a subpopulation of arterial smooth muscle cells (Kirby and Waldo,

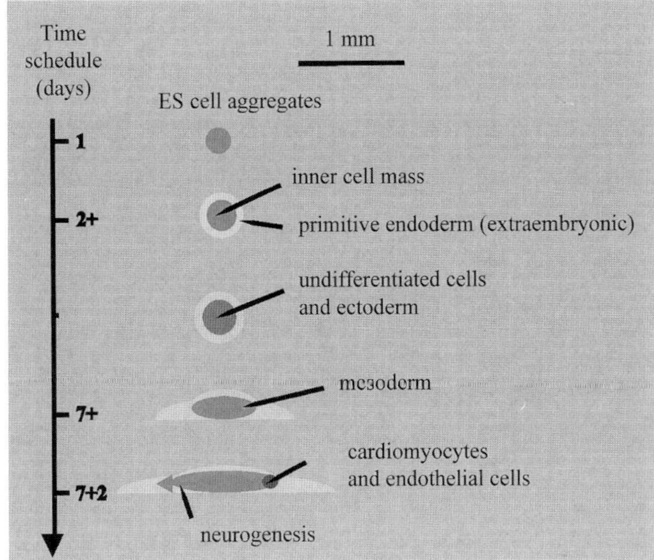

Fig. 1. Prospective time course of ES cell differentiation in EB. The first differentiation occurring from d 2+ on is the development of cell layers of extraembryonic primitive endoderm that further differentiate toward parietal and visceral endoderm. After 7 d in suspension culture, EBs are plated, which probably induces further structural changes. In outgrowing EBs mesoderm develops at the border zone between extraembryonic visceral and parietal endoderm and deeper layers of ectoderm. Mesodermal cells differentiate toward a variety of tissues, including cardiac cells, endothelial cells, hematopoietic cells, skeletal and smooth muscle, as well as adipocytes and bone/cartilage.

1990; Olson and Srivastava, 1996). At birth, the ductus arteriosus closes, which completes the development of separate pulmonary and systemic circulations.

20.4. CARDIOMYOGENESIS OF ES CELLS

Studies in the in vitro EB model have revealed a strictly regulated program of cardiomyogenesis that is triggered by specific signaling molecules and mediated by tissue-specific transcription factors. This program controls the specification of cardiomyocytes from mesodermal stem cells and the subsequent activation of genes involved in cardiac force generation. During EB differentiation toward functional cardiomyocytes, a number of transcription factors known to play a critical role in early cardiac development, including GATA-4, dHAND, eHAND, and Nkx-2.5, are expressed (Czyz and Wobus, 2001) (*see* Chapter 21).

However, organogenesis of heart structures is absent when ES cells are differentiated within the tissue of EBs, indicating that a strict structural organization of the cell lineages that give rise to cell types of the cardiovascular system is not required for cellular differentiation. The induction of cardiomyogenesis of ES cells requires the 3D tissue context of EBs that are grown either for 2 d in hanging drops attached to the lid of a Petri dish and subsequently for 5 d in suspension culture (Wobus et al., 1991) or in spinner flasks, which allows the performance of mass cultures (Wartenberg et al., 1998) (Fig. 1). Within these spherical cell aggregates, a tissue microenvironment comparable to the embryo *in utero* (e.g., hypoxic conditions, gradients of nutrients) prevails and may be at least a cofactor in the induction of specific cell differentiations. After plating of EBs on day 7 of cell culture, cardiomyocytes appear as spontaneously beating cell clusters that increase their beating frequency during subsequent days (Fig. 2). Commitment and differentiation of cardiomyocytes in EBs closely recapitulates murine cardiomyogenesis *in utero*. Cardiac development in EBs has been observed as early as 7 d after formation of the cell aggregates, which correlates well with the murine embryo in which first beating is seen on embryonic day (E) 8.5–9.5 (i.e., 8.5–9.5 d postcoitum).

The differentiation of cardiac cells within EBs is critically dependent on the intracellular redox state of the tissue, which is defined by the balance of reactive oxygen intermediate generation and the antioxidative capacity of the cells (Sauer et al., 2001b). We have previously shown that EBs transiently express an NADPH-oxidase-like enzyme that is regulated by the phosphatidylinositol 3-kinase and robustly generates reactive oxygen species (ROS) (Sauer et al., 2000). Incubation with free radical scavengers severely inhibited cardiomyogensis in EBs, indicating that ROSs are utilized as signaling molecules in signal transduction cascades involved in cardiomyocyte differentiation and/or proliferation (Sauer et al., 2000).

Despite the absence of regulated organogenesis, ES cell–derived cardiomyocytes have been shown to differentiate into sinusnodal-, atrium-, and ventricle-like cells (Maltsev et al., 1993, 1994). Furthermore, the expression of cardiac-specific genes and ion channels, as well as the occurrence of action potentials, was developmentally controlled, which indicates that a localized formation of cardiac chambers and outflow tracts is not necessary for the differentiation of cardiac cell types. In EBs action potential development in mammalian differentiating cardiomyocytes can be studied in vitro in all consecutive stages. ES cells develop into mesodermal progenitors, and early cardiac progenitor cells. Within one spontaneously contracting EB, action potentials characteristic for different cardiac cell types (i.e., pacemaker-, Purkinje-, atrial- and ventricular-like) are present (Maltsev et al., 1993; Hescheler et al., 1997).

During the early stages of cardiomyogenesis predominently pacemaker-like action potentials are found, whereas in the late developmental stage cardiomyocytes of atrial and ventricular origin could be demonstrated. The characteristic shapes of action potentials in ES cell–derived cardiomyocytes are well correlated with the expression of specialized types of ion channels (Maltsev et al., 1994). Terminally differentiated cardiomyocytes responded with characteristic positive or negative chronotropic responses to cardioactive agents comparably with the responses of cardiomyocytes from mammalian species. The β-adrenoceptor agonists (–)isoprenaline and clenbuterenol, the adenylate cyclase activator forskolin, and the phosphodiesterase inhibitor isobutylmethylxanthine, as well as the α1-adrenoceptor agonist (–)phenylephrine, induced a positive chronotropic response, whereas the muscarinic cholinoceptor agonist carbachol and L-type voltage-dependent Ca^{2+} channel (VDCC) inhibitors exerted a negative chronotropic response (Wobus et al., 1991). Interestingly, β-adrenergic modulation of the L-type Ca^{2+} current (I_{CaL}) was insensitive to isoproterenol, forskolin, and 8-bromo-cAMP in early developmental stage cardiomyocytes but highly stimulated by these substances in late stages, indicating that all signaling cascade components became functionally coupled during cardiomyocyte differentiation (Maltsev et al., 1999). While in early developmental stage cardiomyocytes the primitive pacemaker action potentials are generated by VDCC and transient K^+ channels ($I_{K,to}$), terminally differentiated cardiomyocytes of late

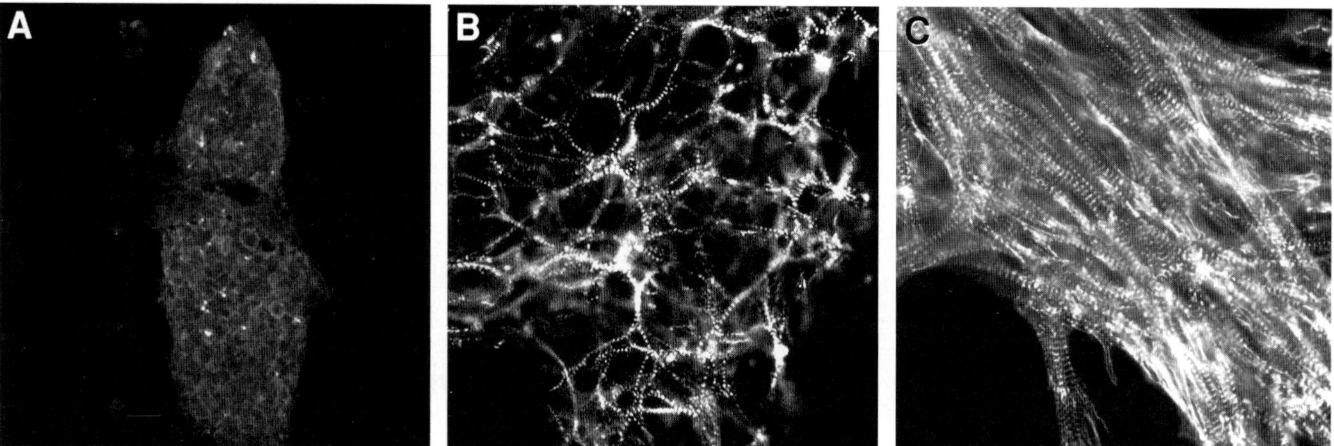

Fig. 2. Structural organization of clusters of developing cardiomyocytes in plated whole mount EBs. The EBs were immunostained with a monoclonal antibody directed against sarcomeric α-actinin. Shown are representative 6-d-old **(A)**, 9-d-old **(B)**, and 19-d-old **(C)** EBs. Note that no sarcomeric organization was visible in 6-d-old cardiac precursor cells. The bar in (A) represents 45 μm. The bars in (B) and (C) represent 15 μm. (From Sauer et al., 2001a, with permission).

developmental stages express various additional ion channels: voltage-dependent Na^+ channels (I_{Na}), delayed outward rectifying K^+ channels, inward rectifying K^+ channels (I_{KI}), muscarinic acetylcholine-activated K^+ channels ($I_{K,Ach}$), and hyperpolarization-activated pacemaker channels (Maltsev et al., 1993).

Besides the establishment of patterns of cardiac specific ion channels, a developmentally controlled expression pattern of the cardiac-specific genes encoding α- and β-cardiac myosin heavy chain, atrial natriuretic factor and myosin light chain isoform 2V has been observed by reverse transcriptase polymerase chain reaction (Fassler et al., 1996). Furthermore, during cardiogenesis of ES cells in vitro, organization of sarcomeric proteins occurred in a strictly controlled manner in the sequence: titin (Z-disk), α-actinin, myomesin, titin (M-band), myosin heavy chain (sarcomeric MHC), sarcomeric α-actin, and troponin T at early cardiac differentiation stages, followed by M-protein at a terminal differentiation stage (Guan et al., 1999).

20.4.1. ROLE OF PRIMITIVE ENDODERM AND LIF IN CARDIOMYOGENESIS OF ES CELLS During embryogenesis, the first differentiated cell lineage emerging from the ICM of a blastocyst is the primitive endoderm that belongs to the hypoblast and gives rise to the extraembryonic visceral and parietal endoderm. The embryonic endoderm first appears with the initial ingression cohort of gastrulating cells (Lough and Sugi, 2000). During the further time course of gastrulation, the embryonic endoderm migrates anteriorly and laterally, thus outwardly replacing the previously formed hypoblast. The relationship between embryonic endoderm cells and the development of cardiac tissue from mesodern has been known for decades. However, the precise molecular mechanisms and the endoderm-derived molecules that are inductive for these processes are just now emerging (Lough and Sugi, 2000).

During recent years, increasing evidence has accumulated for a cross talk between embryonic endoderm and mesoderm during the specification of cardiac precursor cells, which is mediated by specific endoderm-derived factors that are secreted in the vicinity of this tissue. In this respect, it has been evidenced that the specification of cardiac myocytes is induced by molecules such as bone morphogenetic protein-2 and fibroblast growth factor-1 (FGF-1), FGF-2, and FGF-4, whereas endocardiocytes are induced by transforming growth facor-β2 (TGF-β2) TGF-β3, TGF-β4, and vascular endothelial growth factor (VEGF). Besides the embryonic endoderm, differentiation-inducing factors may be likewise secreted from extraembryonic primitive endoderm. In the mouse embryo, cardiac inductive activity resides in the anterior visceral endoderm, which has raised the possibility that embryonic endoderm and the extraembryonic endoderm layers may participate in the cardiogenic process because both tissues are close together (Lough and Sugi, 2000). It has been suggested that the hypoblast intitiates a cascade of inductive events during which the epiblast forms mesoderm and embryonic endoderm, whereupon the latter induces a subset of cells in the former to enter the cardiogenic pathway (Lough and Sugi, 2000).

In EBs, cardiomyogenesis is regulated by different cytokines, e.g., members of the interleukin-6 family of cytokines such as cardiotrophin-1, which is known to stimulate cardiomyocyte proliferation, inhibit cardiomyocyte apoptotic cell death, and induce hypertrophic cell growth in adult cardiomyocytes (Pennica et al., 1995; Sheng et al., 1996), or LIF. LIF is used in ES cell culture to keep ES cells in an undifferentiated state (Williams et al., 1988; Gough et al., 1989). During in vivo embryogenesis, LIF is expressed in the uterine endometrial glands. This burst of expression is under the maternal control and always precedes implantation of the blastocyst (Murray and Edgar, 2001). It has been demonstrated that transient expression of LIF in mice is essential for implantation. Females lacking a functional LIF gene are fertile, but their blastocysts fail to implant and do not develop (Stewart et al., 1992). Before gastrulation LIF is downregulated to nearly undetectable levels, which may be necessary to induce commitment of the mesodermal cell lineage (Conquet et al., 1992). In this respect, it has been evidenced that overexpression of the ECM-bound isoform of LIF (M-LIF) in mice resulted in inhibition of cardiomyogenesis and impaired gastrulation (Conquet et al., 1992). On the other hand, it has been conclusively demonstrated in the EB that LIF is necessary for cardiomyocyte differentiation, since cardiomyogenesis was severely hampered in *lif*$^{-/-}$ EBs and could be restored at very low (picomolar) concentrations of diffusible LIF (D-LIF) (Bader et al., 2001). Furthermore, D-LIF was shown to act as a proliferative stimulus on fully differentiated cardiomyocytes and increased their longevity (Bader et al., 2000). LIF receptor mRNA has been detected in yolk sac visceral endoderm (Passavant et al., 2000), which is thought to be involved in cardiogenesis.

In EBs differentiation of primitive extraembryonic endoderm to visceral and parietal endoderm has been studied in detail (Boller et al., 1987; Sauer et al., 1998; Bader et al., 2001) and has been demonstrated to influence cardiomygenesis in EBs (Bader et al., 2001). In parietal and visceral endoderm cell layers, spontaneous Ca^{2+} oscillations occur and have been demonstrated by our group to trigger transcytotic transport pathways toward the underlying cell layers. It is thought that Ca^{2+} signals in concert with factors secreted by the extraembryonic endoderm may trigger cardiomyogenesis in the underlying mesodermal cell layers (Sauer et al., 1998).

The development of the extraembryonic visceral endoderm is critically dependent on the transcription factor GATA-4. Targeted mutagenesis of the transcription factor GATA-4 gene in mouse ES cells resulted in inhibition of visceral endoderm differentiation (Soudais et al., 1995). In GATA4$^{-/-}$ mice, heart formation was severely impaired, which was owing to a failure of GATA4$^{-/-}$ procardiomyocytes to migrate to the ventral midline to form a linear heart tube, resulting in the formation of aberrant cardiac structures in the anterior and dorsolateral regions of the embryo (Kuo et al., 1997). This points toward the notion that growth factors and/or cytokines secreted by extraembryonic tissues may promote specification of the cardiac cell lineage. In murine GATA4$^{-/-}$ ES cell–derived EBs, however, cardiomyogenic differentiation was not totally abolished, although cardiomyocytes were observed more often in wild-type than in mutant EBs (Narita et al., 1997). Hence, GATA-4 either may not be essential for terminal differentiation of cardiomyocytes or additional GATA-binding proteins known to be in cardiac tissue, such as GATA-5 or GATA-6, may compensate for a lack of GATA-4 (Narita et al., 1997).

20.4.2. Ca2+ SIGNALS AND CONTRACTILITY OF ES CELL–DERIVED CARDIOMYOCYTES In the classic view of excitation–contraction coupling, contraction of cardiomyocytes is initiated by an evoked action potential leading to the opening of VDCCs. Ca^{2+} entering into the cell via VDCCs then triggers Ca^{2+} release from intracellular ryanodine-sensitive stores via a mechanism termed *Ca^{2+}-induced Ca^{2+} release* (CICR). Several years ago, localized discrete Ca^{2+} release events, called Ca^{2+} sparks, were discovered in cardiac cells with the use of the fluorescent Ca^{2+} probe fluo-3 and confocal laser scanning microscopy. A spark event occurs when Ca^{2+} stored in the sarcoplasmic reticulum is released by one or more ryanodine receptors. The whole-cell intracellular Ca^{2+} transient elicited by a strong depolarization is thought to represent the recruitment and summation of many Ca^{2+} sparks after an increase in opened L-type Ca^{2+} channels (Lopez-Lopez et al., 1995; Cleemann et al., 1998).

In EBs, the first cardiac-specific transcripts of the gene coding for the α_1-subunit of the VDCC are already detectable at d 3 of EB culture, i.e., 5 d before the onset of spontaneous beating (Fassler et al., 1996). Slow Ca^{2+} fluctuations with largely variable frequency and sometimes irregular shape were observed in cardiac precursor cells (5–7 d old) that did not yet exert spontaneous contractions. In these cardiomyocytes, the ryanodine receptor agonist caffeine induced Ca^{2+} release, indicating the presence of

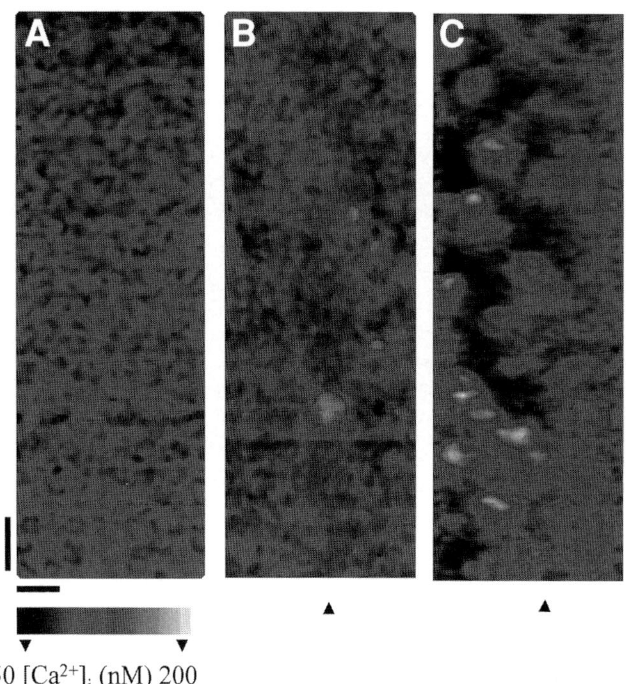

Fig. 3. Occurrence of Ca^{2+} sparks in ES cell–derived cardiomyocytes. Representative false color line scan images of nonstimulated **(A)** cardiac precursor cells, **(B)** early developmental stage, and **(C)** late developmental stage cardiomyocytes differentiated from ES cells and loaded with the Ca^{2+} indicator fluo-3. The line scan images were recorded after spontaneous global $[Ca^{2+}]_i$ transients had declined to baseline values. The position along the cell is presented horizontally, and successive scan lines in time are displayed vertically. Each image includes 512 lines scanned at 250 Hz. The time is running from top to bottom. Note that localized Ca^{2+} sparks were absent in noncontracting cardiac precursor cells, whereas typical, spatially restricted sparks were observed in contracting early developmental stage and late developmental stage cardiomyocytes. Vertical bar = 0.2 s. Horizontal bar = 5 μm. (From Sauer et al., 2001a, with permission).

ryanodine-sensitive Ca^{2+} stores. However, in these cells no Ca^{2+} sparks were observed, which may be owing at least in part to the low expression levels of ryanodine receptors in cardiac precursor cells as well as the reduced filling state of the sarcoplasmic reticulum (Sauer et al., 2001a) (Fig. 3). From d 8 on, spontaneous contractions occur in EBs which increase in frequency during subsequent days of cell culture. In 8 to 11-d-old EBs, these contractions were shown to be independent of transmembrane currents (e.g., of VDCC channels) since contractions persisted in high extracellular K^+ solution as well as in the absence of extracellular Ca^{2+}. Apparently, the contractions of early stage EBs were driven by intracellular Ca^{2+} oscillations that could be inhibited by the Ca^{2+}–adenosine triphosphatase inhibitor thapsigargin (Viatchenko-Karpinski et al., 1999).

These findings strongly argue for an autonomic rhythm generation in early stage EBs that is independent on the classic CICR and relies on the Ca^{2+} release from intracellular stores, whereas VDCCs may serve primarily for store refilling. In terminally differentiated late developmental stage EBs (16–22 d old), rhythm generation is switched to CICR and driven by the well-characterized interplay of different ion channels (e.g., VDCC,

K^+, and sarcoplasmic reticulum). Consequently, in late developmental stage cardiomyocytes derived from ES cells, spontaneous contractions are stopped in cardioplegic solution comparably with cardiomyocytes of adult mice. In late developmental stage cardiomyocytes, a prominent expression of ryanodine receptors was observed whereas the expression of the inositol 3-phosphate receptor was downregulated (unpublished observations). Furthermore, the Ca^{2+} content within the sarcoplasmic reticulum was increased as compared to early stages of cardiomyocyte differentiation, and Ca^{2+} sparks underlying the rhythmic global Ca^{2+} transients occurred with a characteristic that is very similar to elementary intracellular Ca^{2+} events observed in adult murine cardiomyocytes (Sauer et al., 2001a) (*see* Fig. 3).

20.5. VASCULARIZATION OF EBs

Vascularization of the embryo is closely linked to the development of the heart and to the differentiation of the hematopoietic cell lineage. The hematopoietic and vascular system of the mouse develops from the extraembryonic splanchnopleuric mesoderm from which cells migrate through the primitive streak to the presumptive yolk sac between d 6.5 and 7.0 of gestation. During murine embryogenesis, angioblasts arise from differentiation of mesodermal cells shortly after gastrulation. Angioblasts are first detected within paraxial and lateral plate mesoderm as early as stage E7.0, and within the yolk sac extraembryonic mesoderm between E7.5 and E8.5, where they constitute the outer layer of blood islands (Coffin et al., 1991; Risau, 1991). These blood islands are organized into an outer layer of endothelial precursors that surrounds an inner cluster of primitive erythroblasts that comprise large nucleated erythrocytes producing embryonic globins (Barker, 1968). The developing angioblasts of the blood islands rapidly form the first vascular network in the yolk sac through a process termed *vasculogenesis*. Hematopoiesis within yolk sac declines between d 10 and 12 of gestation with the onset of intraembryonic hematopoiesis, which is initiated in the paraaortic splanchnopleura aorta-gonad-mesonephros region and shortly thereafter in the developing fetal liver (Godin et al., 1993; Medvinsky et al., 1993; Muller et al., 1994). The transition from yolk sac to fetal liver defines the switch from the single-lineage primitive erythroid program to multilineage hematopoiesis, which includes definitive erythropoiesis, myelopoiesis, and lymphopoiesis (Robertson et al., 2000).

The appearance of blood island–like structures consisting of immature hematopoietic cells that were surrounded by endothelial cells has been reported to occur in EBs (Doetschman et al., 1985; Risau et al., 1988; Wang et al., 1992; Wartenberg et al., 1998), indicating that molecular mechanisms comparable to in vivo embryonic development occur in ES cell differentiation. In EBs, a "transitional stage" in the commitment of mesoderm to hematopoiesis was identified (Robertson et al., 2000), which is followed by "blast cell colonies" that generate both endothelial and hematopoietic precursors (Choi et al., 1998). Several endothelial markers have been evidenced to be expressed in EBs, including the growth factor receptors Flk-1, tie-1, and tie-2; adhesive molecules such as vascular endothelial cadherin and platelet endothelial cell adhesion molecule-1 (PECAM-1); as well as other endothelial-specific antigens such as MECA-32 and MEC-14.7 (Vittet et al., 1996).

The process of vascularization and angiogenesis is strictly regulated by the pericellular O_2 tension of the tissue. During embryogenesis, the natural progression of organogenesis involves hypoxia since the diffusion of within the embryo is limited by its size. The primary molecular mechanism of gene activation during hypoxia is through hypoxia-inducible factor-1 (HIF-1). HIF-1 is activated during development, as well as during the normal homeostasis of most tissues, since organs naturally generate a gradient of O_2. Mice with HIF mutations (HIF-1$\alpha^{-/-}$, ARNT$^{-/-}$) display severe cardiovascular abnormalities, such as, intact endothelial cell differentiation but aberrant blood vessel maturation, and exhibit lethality by E10.5 (Maltepe et al., 1997; Carmeliet et al., 1998; Iyer et al., 1998a; Ryan et al., 1998; Kotch et al., 1999). Many genes involved in cardiovascular differentiation are directly or indirectly regulated by hypoxia. These include erythropoietin, transferrin, transferrin receptor, VEGF, Flk-1, Flt-1, platelet-derived growth factor-β, basic FGF, and other genes affecting glycolysis (Rolfs et al., 1997; Iyer et al., 1998; Wenger et al., 1998; Wood et al., 1998; Pugh et al., 1999).

Although several candidate O_2-sensing molecules have been discussed, the molecular basis of how cells sense O_2 levels is poorly characterized (Zhu and Bunn, 2001; Zhu et al., 2002). A key role as a mediator of O_2 homeostasis has recently been ascribed to the von Hippel–Lindau tumor suppressor protein (pVHL) (Maxwell and Ratcliffe, 2002). pVHL acts as the recognition component of an ubiquitin E3 ligase complex that binds HIF-1α, thereby targeting it for ubiquitin-mediated degradation (Kallio et al., 1999; Cockman et al., 2000; Tanimoto et al., 2000). HIF-1α is stabilized under hypoxic conditions, associates with the aryl hydrocarbon receptor nuclear translocator protein ARNT, and forms a functional transcription factor complex.

Since embryos mutant in either HIF-1 or ARNT die *in utero*, the EB in vitro model has been used to corroborate in vivo studies on O_2 regulation of gene expression. It was observed that ARNT$^{-/-}$ ES cells are defective in generating different hematopoietic lineage progenitors, which indicates that hypoxic responses are critical for the proliferation and/or survival of hematopoietic progenitor cells (Adelman et al., 1999). Furthermore, the role of VEGF signaling, which is inevitably essential for both endothelial cell division and morphogenesis, has been studied by the use of the ES cell model. Knockout mice that lack one or two copies of the *VEGF* gene die *in utero* owing to vascular defects. Moreover, embryonic development is disrupted by even modest increases in VEGF gene expression (Miquerol et al., 2000), indicating a fine-tuned regulation of the location and duration of VEGF expression (Kearney et al., 2002). In differentiating EBs, vasculogenesis is severely impaired in *VEGF-A*$^{+/-}$ and *VEGF-A*$^{-/-}$ ES cells. It is thought that the location and duration of VEGF expression provide the first level of vascularization control, but other components are likely involved in the fine-tuning of the signal that is transduced to the cell through the receptor tyrosine kinases flk-1 (VEGF receptor 2) and flt-1 (VEGF receptor 1). It has previously been shown that in mice or ES cells lacking flk-1, blood vessel formation is dramatically reduced, suggesting that the majority of the downstream effects of VEGF, especially the mitogenic effect of VEGF on endothelial cell proliferation, are transduced through flk-1. By contrast, mice lacking flt-1 die of vascular overgrowth, which could be likewise observed in the ES cell–derived EB model and was attributed to an aberrant division of angioblasts and/or endothelial cells since in EBs the vascular overgrowth phenotype could be abolished by the cell-cycle inhibitor mitomycin C (Kearney et al., 2002).

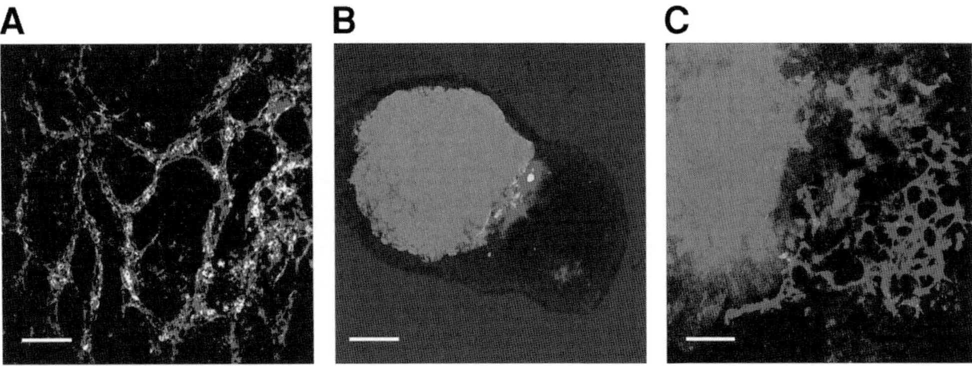

Fig. 4. ES cells to study tumor-induced angiogenesis. **(A)** PECAM-1 immunostaining in a 10-d-old outgrown EB. The bar represents 50 μm. **(B)** Confrontation culture established between an EB (right) and a multicellular prostate tumor spheroid (left) stained with long-term cell tracker CMFDA. After 24–48 h of confrontation culture, endothelial cell structures (red) appear in the contact region of both objects and start to invade into the tumor tissue. The bar represents 150 μm. **(C)** Invasion of ES cell–derived endothelial cells (red) into the tumor tissue. The bar represents 50 μm.

Recently, we have evidenced a transient expression of HIF-1α in ES cell–derived EBs that paralleled the expression of VEGF and the formation of capillary-like structures (Wartenberg et al., 2001). In avascular EBs (3 d old) that expressed prominent levels of HIF-1α, steep O_2 gradients were observed whereas with the onset of vascularization O_2 gradients dissipated and both HIF-1α and VEGF expression were downregulated. VEGF is now known to be a key requirement for tumor growth. VEGF expression has been observed to be increased in avascular hypoxic regions of solid tumors, which may induce directed blood vessel growth during tumor-induced angiogenesis. Because growth and metastasis of malignant neoplasms require the presence of an adequate blood supply, pharmacological inhibition of tumor-induced angiogenesis represents a promising target for antineoplastic therapy (Folkman, 1997; Kieran and Billett, 2001; Tosetti et al., 2002).

We have recently established a confrontation culture system comprising multicellular tumor spheroids and EBs the latter representing the vascularized host tissue (Wartenberg et al., 2001) (Fig. 4). It was observed that endothelial cell structures occurred first in the confrontation area of tumor spheroid and EBs where increased levels of VEGF expression were observed. After several days of confrontation culture, an invasion of PECAM-1-positive cell structures was observed that resulted in a pronounced growth stimulation of the tumor spheroid on vascularization of the tumor tissue. Incubation with known antiangiogenic agents inhibited blood vessel formation in the confrontation cultures as well as in EBs, indicating that the ES cell–derived in vitro models are suitable for the testing of antiangiogenic agents that may be used for the treatment of cancer patients (Wartenberg et al., 1998, 2001).

20.6. FUTURE PROSPECTS

Pluripotent ES cells cultivated within EBs recapitulate the development of the cardiovascular system from primitive precursor cells to highly specialized cardiac and endothelial cell types. The "mesodermal stem cell" has not yet been identified owing to the current inavailability of the earliest developmental markers for the diversification of cell lineages from the pluripotent ICM. During past years, reporter gene-based approaches, such as expression of green fluorescent protein under the control of cardiac- and endothelial cell–specific genes, have been elaborated to identify early precursor cells of the cardiac (Kolossov et al., 1998) as well as endothelial cell lineage (Marchetti et al., 2002) and make them available for biochemical and physiological analysis. With the future discovery of the earliest markers for precursor cells of the mesodermal cell lineage, signaling cascades will be unraveled that direct the diversification of the mesodermal cell lineage toward specialized cell types. Besides the uncontestable impact of ES cells on the investigation of signaling cascades established during very early development, the EB has been successfully used as an in vitro test system to evaluate neurotoxic (Arnhold et al., 1998), cardiotoxic (Bremer et al., 2001), as well as antiangiogenic (Wartenberg et al., 1998, 2001) agents and has significantly contributed to the attempts to reduce animal experiments (for a review, *see* Genschow et al., 2000; Rohwedel et al., 2001).

Most public and scientific interest has, however, raised the possibility of using ES cells for cell replacement therapies. Unlike adult stem cells, the enormous renewal capacity of ES cells allows mass cultures of EBs that could potentially be used for cell transplantation (for a review, *see* Jones and Thomson, 2000; Watt and Hogan, 2000). Future research will focus on the investigation of cell culture procedures that allow direct cell differentiation toward specific cell lineages, the establishment of cell purification protocols based on antibiotic resistance genes under the control of tissue-specific promotors, and/or the elaboration of magnetic bead separation and differential centrifugation techniques. With the availability of large amounts of cardiac and vascular tissue derived from ES cells, therapeutic approaches based on cell transplantation will initiate a revolutionary era for the treatment of cardiovascular diseases that are currently worldwide the primary cause of death.

REFERENCES

Adelman, D. M., Maltepe, E., and Simon, M. C. (1999) Multilineage embryonic hematopoiesis requires hypoxic ARNT activity. *Genes Dev.* 13:2478–2483.

Arnhold, S., Andressen, C., Hescheler, J., and Addicks, K. (1998) Microcinematographic studies on neurodifferentiation and neurotoxicity in vitro using mouse embryonic stem cells. *ALTEX* 15:59–66.

Bader, A., Al Dubai, H., and Weitzer, G. (2000) Leukemia inhibitory factor modulates cardiogenesis in embryoid bodies in opposite fashions. *Circ. Res.* 86:787–794.

Bader, A., Gruss, A., Hollrigl, A., Al Dubai, H., Capetanaki, Y., and Weitzer, G. (2001) Paracrine promotion of cardiomyogenesis in embryoid bodies by LIF modulated endoderm. *Differentiation* 68: 31–43.

Barker, J. E. (1968) Development of the mouse hematopoietic system. I. Types of hemoglobin produced in embryonic yolk sac and liver. *Dev. Biol.* 18:14–29.

Boller, K., Kemler, R., Baribault, H., and Doetschman, T. (1987) Differential distribution of cytokeratins after microinjection of anti-cytokeratin monoclonal antibodies. *Eur. J. Cell Biol.* 43:459–468.

Bremer, S., Worth, A. P., Paparella, M., et al. (2001) Establishment of an in vitro reporter gene assay for developmental cardiac toxicity. *Toxicol. In Vitro* 15:215–223.

Bruneau, B. G. (2002) Transcriptional regulation of vertebrate cardiac morphogenesis. *Circ. Res.* 90 :509–519.

Carmeliet, P., Dor, Y., Herbert, J. M., et al. (1998) Role of HIF-1alpha in hypoxia-mediated apoptosis, cell proliferation and tumour angiogenesis. *Nature* 394:485–490.

Choi, K., Kennedy, M., Kazarov, A., Papadimitriou, J. C., and Keller, G. (1998) A common precursor for hematopoietic and endothelial cells. *Development* 125:725–732.

Cleemann, L., Wang, W., and Morad, M. (1998) Two-dimensional confocal images of organization, density, and gating of focal Ca^{2+} release sites in rat cardiac myocytes. *Proc. Natl. Acad. Sci. USA* 95: 10,984–10,989.

Cockman, M. E., Masson, N., Mole, D. R., et al. (2000) Hypoxia inducible factor-alpha binding and ubiquitylation by the von Hippel-Lindau tumor suppressor protein. *J. Biol. Chem.* 275:25,733–25,741.

Coffin, J. D., Harrison, J., Schwartz, S., and Heimark, R. (1991) Angioblast differentiation and morphogenesis of the vascular endothelium in the mouse embryo. *Dev. Biol.* 148:51–62.

Cohen-Gould, L. and Mikawa, T. (1996) The fate diversity of mesodermal cells within the heart field during chicken early embryogenesis. *Dev. Biol.* 177:265–273.

Conquet, F., Peyrieras, N., Tiret, L., and Brulet, P. (1992) Inhibited gastrulation in mouse embryos overexpressing the leukemia inhibitory factor. *Proc. Natl. Acad. Sci. USA* 89:8195–8199.

Czyz, J. and Wobus, A. (2001) Embryonic stem cell differentiation: the role of extracellular factors. *Differentiation* 68:167–174.

Doetschman, T. C., Eistetter, H., Katz, M., Schmidt, W., and Kemler, R. (1985) The in vitro development of blastocyst-derived embryonic stem cell lines: formation of visceral yolk sac, blood islands and myocardium. *J. Embryol. Exp. Morphol.* 87:27–45.

Epstein, J. A. and Buck, C. A. (2000) Transcriptional regulation of cardiac development: implications for congenital heart disease and DiGeorge syndrome. *Pediatr. Res.* 48:717–724.

Fassler, R., Rohwedel, J., Maltsev, V., et al. (1996) Differentiation and integrity of cardiac muscle cells are impaired in the absence of beta 1 integrin. *J. Cell. Sci.* 109(Pt. 13):2989–2999.

Fishman, M. C. and Chien, K. R. (1997) Fashioning the vertebrate heart: earliest embryonic decisions. *Development* 124:2099–2117.

Folkman, J. (1997) Angiogenesis and angiogenesis inhibition: an overview. *EXS* 79:1–8.

Genschow, E., Scholz, G., Brown, N., et al. (2000) Development of prediction models for three in vitro embryotoxicity tests in an ECVAM validation study. *In Vitro Mol. Toxicol.* 13:51–66.

Godin, I. E., Garcia-Porrero, J. A., Coutinho, A., Dieterlen-Lievre, F., and Marcos, M. A. (1993) Para-aortic splanchnopleura from early mouse embryos contains B1a cell progenitors. *Nature* 364:67–70.

Gough, N. M., Williams, R. L., Hilton, D. J., et al. (1989) LIF: a molecule with divergent actions on myeloid leukaemic cells and embryonic stem cells. *Reprod. Fertil. Dev.* 1:281–288.

Guan, K., Furst, D. O., and Wobus, A. M. (1999) Modulation of sarcomere organization during embryonic stem cell–derived cardiomyocyte differentiation. *Eur. J. Cell Biol.* 78:813–823.

Hescheler, J., Fleischmann, B. K., Lentini, S., et al. (1997). Embryonic stem cells: a model to study structural and functional properties in cardiomyogenesis. *Cardiovasc. Res.* 36:149–162.

Hescheler, J., Fleischmann, B. K., Wartenberg, M., et al. (1999). Establishment of ionic channels and signalling cascades in the embryonic stem cell–derived primitive endoderm and cardiovascular system. *Cells Tissues Organs* 165:153–164.

Hescheler, J., Wartenberg, M., Fleischmann, B. K., Banach, K., Acker, H., and Sauer, H. (2002). Embryonic stem cells as a model for the physiological analysis of the cardiovascular system. *Methods Mol. Biol.* 185:169–187.

Iyer, N. V., Kotch, L. E., Agani, F., et al. (1998) Cellular and developmental control of O2 homeostasis by hypoxia-inducible factor 1 alpha. *Genes Dev.* 12:149–162.

Jones, J. M. and Thomson, J. A. (2000) Human embryonic stem cell technology. *Semin. Reprod. Med.* 18:219–223.

Kallio, P. J., Wilson, W. J., O'Brien, S., Makino, Y., and Poellinger, L. (1999) Regulation of the hypoxia-inducible transcription factor 1 alpha by the ubiquitin-proteasome pathway. *J. Biol. Chem.* 274: 6519–6525.

Kamino, K. (1991) Optical approaches to ontogeny of electrical activity and related functional organization during early heart development. *Physiol. Rev.* 71:53–91.

Kearney, J. B., Ambler, C. A., Monaco, K. A., Johnson, N., Rapoport, R. G., and Bautch, V. L. (2002) Vascular endothelial growth factor receptor Flt-1 negatively regulates developmental blood vessel formation by modulating endothelial cell division. *Blood* 99: 2397–2407.

Kehat, I., Kenyagin-Karsenti, D., Snir, M., et al. (2001) Human embryonic stem cells can differentiate into myocytes with structural and functional properties of cardiomyocytes. *J. Clin. Invest.* 108:407–414.

Kieran, M. W. and Billett, A. (2001) Antiangiogenesis therapy: current and future agents. *Hematol. Oncol. Clin. North Am.* 15:835–51; viii, Review.

Kirby, M. L. and Waldo, K. L. (1990) Role of neural crest in congenital heart disease. *Circulation* 82:332–340.

Klug, M. G., Soonpaa, M. H., Koh, G. Y., and Field, L. J. (1996) Genetically selected cardiomyocytes from differentiating embronic stem cells form stable intracardiac grafts. *J. Clin. Invest.* 98:216–224.

Kolossov, E., Fleischmann, B. K., Liu, Q., et al. (1998) Functional characteristics of ES cell–derived cardiac precursor cells identified by tissue-specific expression of the green fluorescent protein. *J. Cell. Biol.* 143:2045–2056.

Kotch, L. E., Iyer, N. V., Laughner, E., and Semenza, G. L. (1999) Defective vascularization of HIF-1alpha-null embryos is not associated with VEGF deficiency but with mesenchymal cell death. *Dev. Biol.* 209:254–267.

Kuo, C. T., Morrisey, E. E., Anandappa, R., et al. (1997). GATA4 transcription factor is required for ventral morphogenesis and heart tube formation. *Genes Dev.* 11:1048–1060.

Levenberg, S., Golub, J. S., Amit, M., Itskovitz-Eldor, J., and Langer, R. (2002) Endothelial cells derived from human embryonic stem cells. *Proc. Natl. Acad. Sci. USA* 99:4391–4396.

Lopez-Lopez, J. R., Shacklock, P. S., Balke, C. W., and Wier, W. G. (1995) Local calcium transients triggered by single L-type calcium channel currents in cardiac cells. *Science* 268:1042–1045.

Lough, J. and Sugi, Y. (2000) Endoderm and heart development. *Dev. Dyn.* 217:327–342.

Maltepe, E., Schmidt, J. V., Baunoch, D., Bradfield, C. A., and Simon, M. C. (1997) Abnormal angiogenesis and responses to glucose and oxygen deprivation in mice lacking the protein ARNT. *Nature* 386:403–407.

Maltsev, V. A., Rohwedel, J., Hescheler, J., and Wobus, A. M. (1993) Embryonic stem cells differentiate in vitro into cardiomyocytes representing sinusnodal, atrial and ventricular cell types. *Mech. Dev.* 44: 41–50.

Maltsev, V. A., Wobus, A. M., Rohwedel, J., Bader, M., and Hescheler, J. (1994) Cardiomyocytes differentiated in vitro from embryonic stem cells developmentally express cardiac-specific genes and ionic currents. *Circ. Res.* 75:233–244.

Maltsev, V. A., Ji, G. J., Wobus, A. M., Fleischmann, B. K., and Hescheler, J. (1999) Establishment of beta-adrenergic modulation of L-type Ca^{2+} current in the early stages of cardiomyocyte development. *Circ. Res.* 84:136–145.

Marchetti, S., Gimond, C., Iljin, K., et al. (2002) Endothelial cells genetically selected from differentiating mouse embryonic stem cells incorporate at sites of neovascularization in vivo. *J. Cell. Sci.* 115: 2075–2085.

Maxwell, P. H. and Ratcliffe, P. J. (2002) Oxygen sensors and angiogenesis. *Semin. Cell. Dev. Biol.* 13:29–37.

Medvinsky, A. L., Samoylina, N. L., Muller, A. M., and Dzierzak, E. A. (1993) An early pre-liver intraembryonic source of CFU-S in the developing mouse. *Nature* 364:64–67.

Miquerol, L., Langille, B. L., and Nagy, A. (2000) Embryonic development is disrupted by modest increases in vascular endothelial growth factor gene expression. *Development* 127:3941–3946.

Muller, A. M., Medvinsky, A., Strouboulis, J., Grosveld, F., and Dzierzak, E. (1994) Development of hematopoietic stem cell activity in the mouse embryo. *Immunity* 1:291–301.

Murray, P. and Edgar, D. (2001) The regulation of embryonic stem cell differentiation by leukaemia inhibitory factor (LIF). *Differentiation* 68:227–234.

Narita, N., Bielinska, M., and Wilson, D. B. (1997) Cardiomyocyte differentiation by GATA-4-deficient embryonic stem cells. *Development* 124:3755–3764.

Olson, E. N. and Srivastava, D. (1996) Molecular pathways controlling heart development. *Science* 272:671–676.

Passavant, C., Zhao, X., Das, S. K., Dey, S. K., and Mead, R. A. (2000) Changes in uterine expression of leukemia inhibitory factor receptor gene during pregnancy and its up-regulation by prolactin in the western spotted skunk. *Biol. Reprod.* 63:301–307.

Pennica, D., King, K. L., Shaw, K. J., et al. (1995) Expression cloning of cardiotrophin 1, a cytokine that induces cardiac myocyte hypertrophy. *Proc. Natl. Acad. Sci. USA* 92:1142–1146.

Pugh, C. W., Chang, G. W., Cockman, M., et al. (1999) Regulation of gene expression by oxygen levels in mammalian cells. *Adv. Nephrol. Necker Hosp.* 29:191–206.

Risau, W. (1991) Embryonic angiogenesis factors. *Pharmacol. Ther.* 51:371–376.

Risau, W., Sariola, H., Zerwes, H. G., et al. (1988) Vasculogenesis and angiogenesis in embryonic-stem-cell-derived embryoid bodies. *Development* 102:471–478.

Robertson, S. M., Kennedy, M., Shannon, J. M., and Keller, G. (2000) A transitional stage in the commitment of mesoderm to hematopoiesis requiring the transcription factor SCL/tal-1. *Development* 127: 2447–2459.

Rohwedel, J., Guan, K., Hegert, C., and Wobus, A. M. (2001) Embryonic stem cells as an in vitro model for mutagenicity, cytotoxicity and embryotoxicity studies: present state and future prospects. *Toxicol. In Vitro* 15:741–753.

Rolfs, A., Kvietikova, I., Gassmann, M., and Wenger, R. H. (1997) Oxygen-regulated transferrin expression is mediated by hypoxia-inducible factor-1. *J. Biol. Chem.* 272:20,055–20,062.

Ryan, H. E., Lo, J., and Johnson, R. S. (1998) HIF-1 alpha is required for solid tumor formation and embryonic vascularization. *EMBO J.* 17:3005–3015.

Sack, M. N., Disch, D. L., Rockman, H. A., and Kelly, D. P. (1997) A role for Sp and nuclear receptor transcription factors in a cardiac hypertrophic growth program. *Proc. Natl. Acad. Sci. USA* 94:6438–6443.

Sauer, H., Hofmann, C., Wartenberg, M., Wobus, A. M., and Hescheler, J. (1998) Spontaneous calcium oscillations in embryonic stem cell–derived primitive endodermal cells. *Exp. Cell. Res.* 238:13–22.

Sauer, H., Rahimi, G., Hescheler, J., and Wartenberg, M. (2000) Role of reactive oxygen species and phosphatidylinositol 3-kinase in cardiomyocyte differentiation of embryonic stem cells. *FEBS Lett.* 476:218–223.

Sauer, H., Theben, T., Hescheler, J., Lindner, M., Brandt, M. C., and Wartenberg, M. (2001a) Characteristics of calcium sparks in cardiomyocytes derived from embryonic stem cells. *Am. J. Physiol. Heart Circ. Physiol.* 281:H411–H421.

Sauer, H., Wartenberg, M., and Hescheler, J. (2001b) Reactive oxygen species as intracellular messengers during cell growth and differentiation. *Cell. Physiol. Biochem.* 11:173–186.

Sedmera, D., Pexieder, T., Vuillemin, M., Thompson, R. P., and Anderson, R. H. (2000) Developmental patterning of the myocardium. *Anat. Rec.* 258:319–337.

Shamblott, M. J., Axelman, J., Littlefield, J. W., et al. (2001) Human embryonic germ cell derivatives express a broad range of developmentally distinct markers and proliferate extensively in vitro. *Proc. Natl. Acad. Sci. USA* 98:113–118.

Sheng, Z., Pennica, D., Wood, W. I., and Chien, K. R. (1996) Cardiotrophin-1 displays early expression in the murine heart tube and promotes cardiac myocyte survival. *Development* 122:419–428.

Shiraishi, I., Takamatsu, T., and Fujita, S. (1995) Three-dimensional observation with a confocal scanning laser microscope of fibronectin immunolabeling during cardiac looping in the chick embryo. *Anat. Embryol. (Berl.)* 191:183–189.

Soudais, C., Bielinska, M., Heikinheimo, M., et al. (1995) Targeted mutagenesis of the transcription factor GATA-4 gene in mouse embryonic stem cells disrupts visceral endoderm differentiation in vitro. *Development* 121:3877–3888.

Stewart, C. L., Kaspar, P., Brunet, L. J., et al. (1992) Blastocyst implantation depends on maternal expression of leukaemia inhibitory factor. *Nature* 359:76–79.

Taber, L. A., Lin, I. E., and Clark, E. B. (1995) Mechanics of cardiac looping. *Dev. Dyn.* 203:42–50.

Tanimoto, K., Makino, Y., Pereira, T., and Poellinger, L. (2000) Mechanism of regulation of the hypoxia-inducible factor-1 alpha by the von Hippel–Lindau tumor suppressor protein. *EMBO J.* 19:4298–4309.

Thomson, J. A., Itskovitz-Eldor, J., Shapiro, S. S., et al. (1998) Embryonic stem cell lines derived from human blastocysts. *Science* 282: 1145–1147.

Tosetti, F., Ferrari, N., De Flora, S., and Albini, A. (2002) Angioprevention: angiogenesis is a common and key target for cancer chemopreventive agents. *FASEB J.* 16:2–14.

Townsend, P. J., Barton, P. J., Yacoub, M. H., and Farza, H. (1995) Molecular cloning of human cardiac troponin T isoforms: expression in developing and failing heart. *J. Mol. Cell. Cardiol.* 27:2223–2236.

Viatchenko-Karpinski, S., Fleischmann, B. K., Liu, Q., et al. (1999). Intracellular Ca^{2+} oscillations drive spontaneous contractions in cardiomyocytes during early development. *Proc. Natl. Acad. Sci. USA* 96:8259–8264.

Vittet, D., Prandini, M. H., Berthier, R., et al. (1996) Embryonic stem cells differentiate in vitro to endothelial cells through successive maturation steps. *Blood* 88:3424–3431.

Wang, R., Clark, R., and Bautch, V. L. (1992) Embryonic stem cell–derived cystic embryoid bodies form vascular channels: an in vitro model of blood vessel development. *Development* 114:303–316.

Wartenberg, M., Gunther, J., Hescheler, J., and Sauer, H. (1998) The embryoid body as a novel in vitro assay system for antiangiogenic agents. *Lab. Invest.* 78:1301–1314.

Wartenberg, M., Donmez, F., Ling, F. C., Acker, H., Hescheler, J., and Sauer, H. (2001) Tumor-induced angiogenesis studied in confrontation cultures of multicellular tumor spheroids and embryoid bodies grown from pluripotent embryonic stem cells. *FASEB J.* 15: 995–1005.

Watt, F. M. and Hogan, B. L. (2000). Out of Eden: stem cells and their niches. *Science* 287:1427–1430.

Wenger, R. H., Kvietikova, I., Rolfs, A., Camenisch, G., and Gassmann, M. (1998) Oxygen-regulated erythropoietin gene expression is dependent on a CpG methylation-free hypoxia-inducible factor-1 DNA-binding site. *Eur. J. Biochem.* 253:771–777.

Williams, R. L., Hilton, D. J., Pease, S., et al. (1988) Myeloid leukaemia inhibitory factor maintains the developmental potential of embryonic stem cells. *Nature* 336:684–687.

Wobus, A. M., Wallukat, G., and Hescheler, J. (1991) Pluripotent mouse embryonic stem cells are able to differentiate into cardiomyocytes expressing chronotropic responses to adrenergic and cholinergic agents and Ca2+ channel blockers. *Differentiation* 48:173–182.

Wobus, A. M., Rohwedel, J., Maltsev, V., and Hescheler, J. (1995) Development of cardiomyocytes expressing cardiac-specific genes, action potentials, and ionic channels during embryonic stem cell–derived cardiogenesis. *Ann. NY Acad. Sci.* 752:460–469.

Wood, S. M., Wiesener, M. S., Yeates, K. M., et al. (1998) Selection and analysis of a mutant cell line defective in the hypoxia-inducible factor-1 alpha-subunit (HIF-1alpha). Characterization of hif-1alpha-dependent and -independent hypoxia-inducible gene expression. *J. Biol. Chem.* 273:8360–8368.

Yutzey, K. E. and Bader, D. (1995) Diversification of cardiomyogenic cell lineages during early heart development. *Circ. Res.* 77:216–219.

Zhu, H. and Bunn, H. F. (2001) Signal transduction: how do cells sense oxygen? *Science* 292:449–451.

Zhu, H., Jackson, T., and Bunn, H. F. (2002) Detecting and responding to hypoxia. *Nephrol. Dial. Transplant.* 17(Suppl. 1):3–7.

21 Transcription Factors, Growth Factors, and Signal Cascades Capable of Priming Cardiogenesis

AGAPIOS SACHINIDIS, PhD, HEINRICH SAUER, PhD,
MARIA WARTENBERG, PhD, AND JÜRGEN HESCHELER, MD

Similar signaling molecules and factors are involved in heart development in *Drosophila* and vertebrates. Heart development is regulated by factors and signals secreted by the anterior primitive endoderm and by the neuronal tube. Bone morphogenic proteins (BMPs) are expressed in endoderm and ectoderm and play a central role in induction of heart formation in vertebrate embryos. They elicit expression of a number of cardiac transcription factors and structural genes and induce full cardiac differentiation in medial mesoderm. They act with signals from the anterior endoderm, whereas signals from the axial tissues repress heart formation. The zinc finger GATA proteins regulate both hematopoiesis and cardiogenesis; GATA-4 transcription factor is required for ventral morphogenesis and heart tube formation. Leukemia inhibitory factor at high concentrations inhibits differentiation of embryonic stem (ES) cells into embryoid bodies, whereas other factors, such as retinoic acid (neuronal and smooth muscle), TGF-β (cardiomyogenesis), IL-3 (white blood cells), insulin growth factor-1 (pericardium and heart), fibroblast growth factor (heart), BMP (heart), IL-6-related factor CT-1 (heart), erythropoietin (heart and red blood cells), and dynorphin B, induce selective differentiation through reaction with cell-surface receptors and activation of intracellular kinase signal transduction. Wnt family members inhibit cardiogenesis, and the Wnt inhibitor, Crescent, blocks this effect. The ability to control heart differentiation in vitro could provide the means to direct ES cell therapy to replacement of damaged or defective heart tissue.

21.1. HEART DEVELOPMENT IN *DROSOPHILA*

Basic mechanisms of heart development (cardiogenesis) are conserved between vertebrates and invertebrates. Remarkably, although the *Drosophila* heart is linear and thus totally different in the morphology of the looped and chambered heart of vertebrates, outstanding similarities exist concerning the embryonic origins and genes involved in the cardiac specification (Harvey, 1996; Olson and Srivastava, 1996; Bodmer and Venkatesh, 1998; Su et al., 1999). In this context, the homeobox gene *tinman* in *Drosophila* (similar to Nkx-2.5 in vertebrate) is essential for the specification of myocardial progenitors in the fly (Bodmer, 1993; Azpiazu and Frasch, 1993; Lyons et al., 1995; Fu et al., 1998; Grow and Krieg, 1998; Su et al., 1999; Tanaka et al., 1999). Similar members of the transforming growth family (TGF) family genes are involved in cardiogenesis in both vertebrates and *Drosophila* (Frasch et al., 1987; Staehling-Hampton et al., 1994; Su et al., 1999).

According to a recent model (Bodmer and Frasch, 1999), cardiogenesis in *Drosophila* is initiated by the basic helix–loop–helix (bHLH) protein Twist and the zinc finger protein Snail; both of which are essential for the formation of mesoderm (Kosman et al., 1991; Leptin, 1991; Bodmer and Frasch, 1999). Downstream targets of *twist* that are expressed in the early mesoderm are *Dmef2* (Bour et al., 1995; Lilly et al., 1995), *tinman* (Bodmer et al., 1990), *heartless* (codes for the Fibroblast growth factor [FGF] receptor) (Shishido et al., 1993, 1997; Beiman et al., 1996; Gisselbrecht et al., 1996), and *zfh-1* (Lai et al., 1993; Su et al., 1999). All these genes participate in the further differentiation of the mesodermal cells. According to this model (Bodmer and Frasch, 1999; Su et al., 1999), mesoderm subdivides into four major groups of cells with restricted cell fates: the somatic mesoderm (forms the skeletal body wall muscle), the visceral mesoderm (forms the gut muscles), the fat body/gonadal mesoderm, and the cardiac mesoderm (forms the heart). Gene expression studies indicated that the mesoderm is organized in parasegmental units similar to ectoderm (Azpiazu et al., 1996; Riechmann et al., 1997; Su et al., 1999). According to Su et al. (1999), the early expression in transverse stripes of the pair rule genes *even skipped* (*eve*), *sloppy-paired* (*slp*) as well as of the segment polarity genes *wingless* (*wg*) and *hedgehog* (*hh*) is crucial for the anteroposterior patterning of the developing mesoderm (Wu et al., 1995; Azpiazu et al., 1996; Riechmann et al., 1997; Bodmer and Frasch, 1999; Su et al., 1999). In contrast, TGF-β encoded by *dpp*, and the homeobox gene *tinman*, are

From: *Stem Cells Handbook*
Edited by: S. Sell

expressed perpendicularly to these four genes in a broad dorsal domain along the anteroposterior axis. Dpp is secreted from the dorsal ectoderm and is essential to maintain *tinman* expression in the mesoderm (Su et al., 1999). These findings demonstrate that both *tinman* and *dpp* are required for subdividing the mesoderm and for specifying dorsal mesodermal fates including cardiogenesis (Azpiazu and Frasch, 1993; Bodmer, 1993; Frasch, 1995; Su et al., 1999). Opposite to *dpp* and *tinman*, which are necessary for heart and visceral mesoderm formation, *wg* is needed for cardiac but not the visceral mesodermal precursors (Baylies et al., 1995; Su et al., 1999; Wu et al., 1995). Because *wg*, *dpp*, and *tinman* are necessary for all aspects of heart formation in *Drosophila*, they are unlikely to be sufficient for specifying among different cardiac cell types (Su et al., 1999). Like *tinman*, the zinc finger– *and homeobox-containing gene zfh-1* is widely expressed in the early mesoderm and later in the pericardial cells of the forming heart tube (Lyons et al., 1995; Su et al., 1999).

The vertebrate homolog of *zfh-1*, *dEF1* is also involved in some aspects of mesodermal differentiation (Takagi et al., 1998). In contrast to *zfh-1*, the homeobox gene *eve* is expressed in small subsets in the dorsal mesoderm within the cardiac primordium (Frasch et al., 1987). Two *eve*-expressing mesodermal progenitor cells undergo asymmetric divisions under the control of *numb* to give rise to a subset of pericardial cells (endothelial progenitor cells [EPCs]) and the founder cells of the dorsal muscles (DA1) (Carmena et al., 1998a, 1998b; Park et al., 1998; Su et al., 1999). In *zfh-1* mutants, the EPC subset of pericardial cells is missing without affecting formation of the EPC progenitors or the *eve*-expressing DA1 muscles (Su et al., 1999). Cardiac and other pericardiac cells of the heart are formed, but abnormalities in the morphology of heart tube have been observed in *zfh-1* mutants (Su et al., 1999). Overexpession of Eve in *zfh-1* mutants partially restores EPC formation, suggesting that *eve* acts downstream of *zfh-1*.

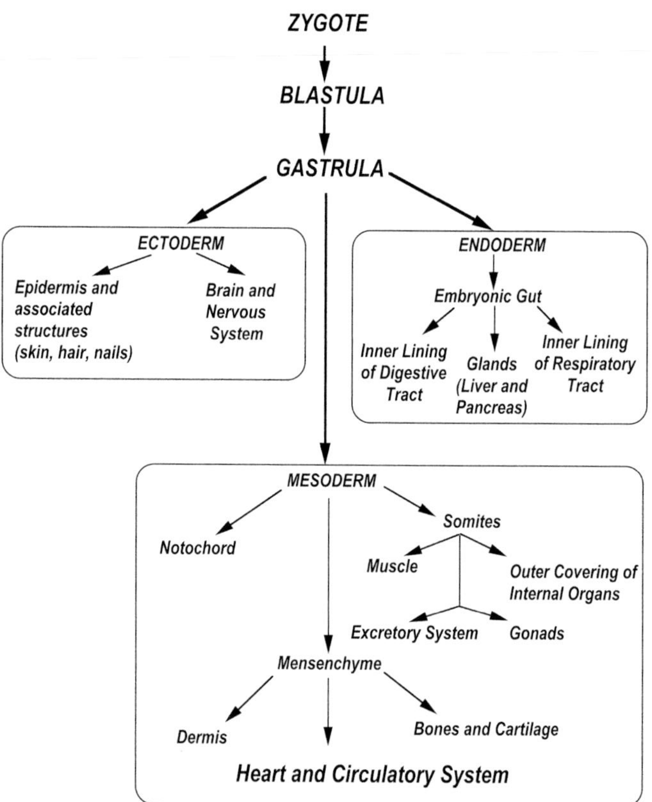

Fig. 1. Development of different organs from germ layers, ectoderm, endoderm, and mesoderm after gastrulation process.

21.2. FACTORS AND SIGNALS PROMOTING CARDIOGENESIS IN VERTEBRATES

Identification of soluble growth factors, transcription factors, and signal cascades capable of priming cardiogenesis is a crucial issue for understanding heart development. In recent years, through applying the embryonic stem (ES) cell and knockout animal model, several soluble factors and intracellular transcription factors as well as intracellular signal transduction pathways have been recognized to be involved in heart formation.

In mammalians, the heart is the first mesoderm-derived functional embryonic organ that is developed after gastrulation in amniotic embryos (Fig. 1). Heart development is a complex process involving proliferation and differentiation as well as tissue organization into specific architecture. In early gastrulation, cells migrating to the anterior and lateral to the primitive streak contribute to the development of heart tissue (Garcia-Martinez and Schoenwolf, 1993). There is accumulating evidence that development of uncommitted mesodermal precardiac cells to early cardiac cells is regulated by signals secreted by the anterior primitive endoderm (Antin et al., 1994; Sugi and Lough, 1994; Montgomery et al., 1994; Schultheiss et al., 1997; 1995). In addition, various signals from the lateral regions of the embryo are also essential for heart formation (Schultheiss et al., 1995, 1997). Many studies in the past demonstrated that factors secreted by the neuronal tube suppress cardiogenesis in adjacent mesoderm (Jacobson, 1960; Climent et al., 1995; Schultheiss et al., 1997; Raffin et al., 2000). This has been proved by the observation that extirpation of the endoderm blocked heart formation in gastrula-stage newt embryos, whereas extirpation of both endoderm and neural plates restored heart formation (Jacobson, 1960). Furthermore, anterior endoderm has heart-inducing properties, as demonstrated by the ability of this tissue to promote heart formation in coculture with posterior primitive streak, a tissue normally fated to form blood (Schultheiss et al., 1995). These findings suggest that development of the heart is regulated by signals secreted by the surrounding tissues.

21.2.1. MORPHOGENETIC PROTEINS AND TRANSCRIPTION FACTORS In general, bone morphogenetic proteins (BMPs) are expressed in lateral endoderm, and ectoderm and BMP signaling plays a central role in the induction of heart formation in vertebrate embryos (Schultheiss et al., 1997; Andree et al., 1998; Schlange et al., 2000). Application of BMP-2 in vivo elicits ectopic expression of the cardiac transcription factors Nkx-2.5; GATA-4; GATA-5; GATA-6; MEF2; eHAND; and dHAND and the cardiac structure gene, ventricular myosin heavy chain (vMHC) (Schultheiss et al., 1997; Schlange et al., 2000; Tzahor and Lassar, 2001). Furthermore, administration of soluble BMP-2 or BMP-4 to explant cultures induces full cardiac differentiation in stage 5–7 anterior medial mesoderm, a tissue that is normally not cardiogenic (Schultheiss et al., 1997). However, when the adjacent neural tube and notochord were included in these explants, BMP-2 administration only induced Nkx-2.5 and failed to induce expression of either GATA-4 or vMHC (Schultheiss et al., 1997). The competence to undergo cardiogenesis in response to BMPs is restricted to mesoderm located in the anter-ior regions of gastrula- to neurula-stage embryos.

Signals from the anterior endoderm work to promote heart formation in concert with BMP signals in the anterior lateral mesoderm, and signals from the axial tissues repress heart formation in the more dorsomedial anterior paraxial mesoderm (Tzahor and Lassar, 2001). The heart-promoting activities of the anterior endoderm and BMPs are antagonized by repressive signals from the axial tissues that block cardiogenesis in the anterior paraxial mesoderm (Tzahor and Lassar, 2001). The secreted protein noggin, which binds to BMPs and antagonizes BMP activity, completely inhibits differentiation of the precardiac mesoderm, indicating that BMP activity is required for myocardial differentiation in this tissue (Schultheiss et al., 1997).

GATA transcription factors are important regulators of developmental processes in worms, flies and mammals (Orkin, 1992; Merika and Orkin, 1993; Simon, 1995; Zhu et al., 1997; Svensson et al., 1999). Transcriptional activities of the GATA proteins are modulated by their interactions with other transcriptional coactivators and repressors. All six proteins (GATA1–6) contain zinc finger DNA-binding domains, thereby binding to the consensus motif, WGATAR, as well as to the related sequences such as CGATGG and AGATTA (Ko and Engel, 1993; Merika and Orkin, 1993; Svensson et al., 1999). GATA regulates both hematopoiesis (GATA-1, -2, -3) (Pevny et al., 1991; Tsai et al., 1994; Ting et al., 1996) and cardiogenesis (GATA-4) (Olson and Srivastava, 1996; Fishman and Chien, 1997a) in mammals. GATA-1, -2, -3 are expressed in the hematopoietic cell lineages, critical roles in the development of the erythroid, hematopoietic stem cell, and T-cell lineages (Pevny et al., 1991; Tsai et al., 1994; Ting et al., 1996). GATA-4, -5, -6 are expressed in heart, gut, urogenital system, and smooth muscle cell lineages (Heikinheimo et al., 1994; Laverriere et al., 1994; Morrisey et al., 1996; Svensson et al., 1999). GATA-4 plays a key role in cardiac muscle development and regulates the expression of several cardiac-specific genes including α-MHC, cardiac troponin C, and atrial natriuretic peptide (Wang et al., 1998; Svensson et al., 1999; Ventura and Maioli, 2000). GATA-4 is a regulator of early cardiogenesis, expressed before the formation of a linear heart tube (Heikinheimo et al., 1994; Svensson et al., 1999). Disruption of the GATA-4 gene in mice leads to early embryonic lethality because of a specific defect in ventral heart tube formation (Kuo et al., 1997; Molkentin et al., 1997; Svensson et al., 1999). These findings suggest that GATA-4 transcription factor is required for ventral morphogenesis and heart tube formation.

The transcriptional activity of the GATA proteins can be regulated by other transcriptional factors and by transcriptional coactivators and repressors. In this context, it has been shown that C-terminal zinc finger of GATA-1 interacts specifically with the SP1 and EKLF transcription factors to synergistically activate erythroid-specific gene expression (Merika and Orkin, 1995; Svensson et al., 1999). In addition, Nkx-2.5 and NFAT3 associate with the C-terminal zinc finger of GATA-4 and coactivate the transcription of cardiac-restricted genes (Durocher et al., 1997; Lee et al., 1998; Molkentin et al., 1998; Sepulveda et al., 1998). Remarkably, mutation in Nkx-2.5 causes congenital human heart diseases (Schott et al., 1998) and impairs normal heart architecture by affecting cardiac developmental pathways in a wide population of cardiopathic patients (Benson et al., 1999; Ventura and Maioli, 2000).

Similarly, two related zinc finger proteins, U-shaped (USH) and Friend of GATA-1 (FOG) interact with the GATA protein Pannier of the *Drosophila* and GATA-1, respectively (Haenlin et al., 1997; Tsang et al., 1997). The interactions of USH with Pannier repress the transcriptional activity of Pannier. By contrast, FOG in conjunction with GATA-1 synergistically activates expression of the erythroid-specific NF-E2 promoter and, like GATA-1, is required for normal erythropoiesis in the mouse (Tsang et al., 1997, 1998; Svensson et al., 1999).

Recently it has been demonstrated that FOG-2, an 1151 amino acid nuclear protein, contains eight zinc finger motifs, structurally related to both FOG and USH. FOG-2 is expressed in mouse embryonic heart, neuroepithelium, and urogenital ridge. In the adult, FOG-2 is expressed in the heart, brain, and testis. FOG-2 associates physically with the N-terminal zinc finger of GATA-4 in vitro and in vivo. Interactions modulate specifically the transcriptional activity of GATA-4. From these findings, it has been concluded that FOG-2 is a novel modulator of GATA-4 function during cardiac development and represents a paradigm in which tissue-specific interactions between different FOG and GATA proteins regulate the differentiation of distinct mesodermal cell lineages (Svensson et al., 1999).

21.2.2. GROWTH FACTORS AND SIGNALS PROMOTING HEART DEVELOPMENT ES cells isolated from the inner cell mass of the early mammalian blastocyst-stage embryo are pluripotent (Sugi and Lough, 1994; Thomson et al., 1998; Odorico et al., 2001). Murine ES cells can serve as an in vitro model for in vivo differentiation. In the presence of a relatively high concentration of leukemia inhibitory factor (LIF), ES cells remain undifferentiated (Smith et al., 1988). In the absence of LIF, murine ES cells spontaneously differentiate into multicellular aggregates, termed *embryoid bodies* (EBs), which resemble early postimplantation embryos.

In general, it is believed that formation of EBs initiates signaling and spontaneous differentiation of ES cells to the three embryonic germ layer cells: the ectoderm, mesoderm, and endoderm (for a review, *see* (Jacobson, 1960; Azpiazu et al., 1996; Thomson et al., 1998; Hescheler et al., 1999; Itskovitz-Eldor et al., 2000). In the last few years, it has been shown that several soluble factors are able to induce selective differentiation of ES cells. In this context, retinoic acid (RA) induces neuronal (Bain et al., 1995; Brustle et al., 1999; Su et al., 1999) and smooth muscle cell formation (Drab et al., 1997), and TGF-β induces cardiomyogenesis (Rohwedel et al., 1994), whereas interleukin-3 (IL-3) promotes differentiation of ES cells to macrophages and neutrophils (Wiles and Keller, 1991).

These findings demonstrate that a selective differentiation of ES cells is regulated by several growth factors via their respective receptors as well as by the cell culture conditions. During differentiation, ES cells are exposed to various extracellular stimuli, resulting in a regulation of differentiation. In most cases, the initial mediators of cellular responses are cell-surface receptors that transmit differentiation signals into the cell after binding to their respective receptors. TGF-β itself was discovered as an autocrine growth factor required for tumor cell growth. However, TGF-β is also able to inhibit growth of several cell types including epithelial, endothelial neuronal, and immune system cells. Furthermore, TGF-β stimulates the formation of extracellular matrix (ECM) components as well as suppresses the degradation of ECM by inhibiting the expression of proteases and increasing protease inhibitors. These effects emphasize the importance of TGF-β for tissue maintenance and repair. Recently, many other members of the TGF-β superfamily were discovered to act as morphogens, thereby playing an important role in embryogenesis. This has been clearly shown in *Drosophila*, demonstrating that TGF-β encoded by *dpp* is required for mesoderm formation and cardiogenesis (Azpiazu and Frasch, 1993; Bodmer, 1993; Frasch, 1995; Su et al., 1999). Binding of TGF-β to its transmembrane receptor, which possesses cytoplasmic serine/threonine kinase domains, initiates signaling of TGF-β via stabilization and activation of heterotetrameric TGF-β receptor complex (Whitman, 1998; for a review *see* Wakefield and Roberts, 2002). This complex activates a member of the Smad family of transcription factors via phosphorylation. The Smad proteins are classified into receptor-regulated Smads (R-Smads) which become phosphorylated by the TGF-β receptor complex and common Smads (Co-Smads) that heterooligomerize with the R-Smads and finally the inhibitory Smads. After phosphorylation of R-Smad by the TGF-β receptor, it migrates to the nucleus and heterooligomerizes with the Co-Smad. In the nucleus, the heterooligomeric complex binds to DNA in a site-specific fashion and interacts with a variety of transcriptional factors, coactivators, and corepressors to induce the expression of TGF-β-responsive genes. Although stimulation of the Smad pathway is the major pathway, activation of the mitogen activated protein kinase (MAPK) pathway including the extracellular response kinases 1 and 2 (ERK1/2) and the c-jun N-terminal kinases (JNKs) is also possible by TGF-β in some cell types. In contrast to the Smad pathway, which is probably involved in developmental processes, activation of ERK1/2 and JNKs appears to be required for motility and epithelial-mesenchymal transformation induced by TGF-β (Whitman, 1998; Wakefield and Roberts, 2002).

Insulin growth factor 1 (IGF-1) has been shown to be essential for normal embryonic growth in mice (Powell-Braxton et al., 1993), and mouse ES cells express receptors for IGF-1 (Shi et al., 1995). In this context, avian pericardial mesoderm development is regulated by IGF-1 (Antin et al., 1996). Furthermore, IGF-1 has been shown to be essential for the formation of the functional heart (for a review *see* Fishman and Chien, 1997). The early development of avian pericardial mesoderm was found to be regulated by FGF-2, insulin, and IGFs (Antin et al., 1996).

Opposite to receptor types such as the FGF receptor, which are activated by autophosphorylation, activation of the insulin receptor (IR)/IGFR family by insulin and IGF leads to the tyrosine phosphorylation of a variety of intracellular IR substrates (IRSs) (for a review *see* Siddle et al., 2001). Phosphorylated IRSs recruit and activate class 1a phosphoinositide 3-kinase (PI-3-kinase) via Src homology-2 (SH2) domains leading to synthesis of membrane-associated phosphatidylinositol 4,5-bisphosphate (PIP_2). Phosphorylated IRSs are also able to phosphorylate and thus activate other kinases such as phospholipase c-γ1 (PLC-γ1), Akt/protein kinase B, $p70^{rsk}$, and protein kinase C (Siddle et al., 2001). Activation of PLC-γ1 results in hydrolysis of PIP_2 to diacylglycerol and $InsP_3$, which releases Ca^{2+} from endoplasmic reticulum. A second signaling pathway involves recruitment to both IRS-1 and Shc of the guanine-nucleotide-exchange factor Sos, via the SH2 domain of the adapter Grb2. This leads to activation of the small G-protein Ras, which in turn activates the protein serine kinase Raf and the MAPK cascade including the ERK1/2 (Siddle et al., 2001).

Recently, it has been demonstrated that the specific inhibitor of PI-3-kinase, LY294002, induced a massive loss of α-actinin-stained cardiomyocytes in murine EBs (Klintz et al., 1999). In parallel, we observed a strong decrease in the number of EBs containing area(s) with beating cardiomyocytes. The specific action of the PI-3-kinase inhibitor on development of cardiomyocytes was demonstrated by the observation that formation of endothelial cells was not affected in the same EBs. Our results provided the first evidence that signal transduction via the PI-3-kinase pathway is essential for development of mammalian early cardiomyocytes.

A heart-inducing growth factor in combination with BMP-2 is the FGF. In this context, it has been observed that zebrafish bearing a mutation for FGF, normally expressed in the cardiogenic mesoderm, show a decreased expression of cardiac markers. This finding suggests that FGF signaling is necessary for heart formation (Reifers et al., 2000). Up until now, 20 FGFs have been identified belonging to the FGF family of small polypeptide growth factors (for a review *see* Powers et al., 2000). FGFs propagate their growth and developmental signals synergistically with heparan-like glycosaminoglycans through dimerization and tyrosine phosphorylation of FGF receptors, thereby activating several tyrosine kinases via their phosphorylation on tyrosine residues. On activation of FGF receptors, PLC-γ1 and Src tyrosine kinases bind to the autophosphorylated receptor via their SH2 domain and become tyrosine phosphorylated. Furthermore, activation of FGF receptors results in an activation of the MAPK cascade including ERK1/2 via the Ras/Raf pathway (Powers et al., 2000).

It is well established that activation of serum response factor (SRF) occurs by growth factors. SRF mediates growth factor–stimulated transcriptional induction of immediate-early genes. The transcription of immediate-early genes appears to be regulated by SRF and Elk-1 protein. Both bind to serum response element promoters of immediate-early genes. Transcription of IEGs occurs only after phosphorylation of Elk-1. Phosphorylation of Elk-1 occurs by the mitogen-activated kinase cascades (especially by ERK1/2) stimulated by serum and growth factors (Treisman et al., 1992). Recently, there is increasing evidence that SRF is essential for murine embryogenesis (Weinhold et al., 2000). It has been shown that mice lacking Srf ($Srf^{-/-}$) stop developing at the onset of gastrulation, lacking detectable mesoderm and mesodermal marker genes such as BMP-2. Furthermore, the ability of $Srf^{-/-}$ ES cells to differentiate in vitro to mesodermal cells was impaired, but this impairment could be modulated by external cell-independent factors such as RA, which induced expression of mesodermal marker genes (Weinhold et al., 2000). In this context, it has been shown that RA accelerates differentiation of ES cells to cardiac differentiation (Wobus et al., 1997). Mesodermal marker gene expression such as T(Bra) was also observed when SRF was expressed in $Srf^{-/-}$ ES cells. These findings suggest that SRF contributes to mesodermal gene expression of ES cells and that $Srf^{-/-}$ ES cells possess a nonautonomous defect in differentiation toward mesoderm (Weinhold et al., 2000).

Cardiogenesis is a multistep process regulated by a hierarchy of factors defining each developmental stage of the heart (Bader et al., 2000). One of these factors is the LIF that belongs to the IL-6 family of cytokines. In general, LIF is a growth factor for hematopoiesis, bone, and neuroectodermal tissue as well as a differentiation factor with pleiotropic properties (Auernhammer and Melmed, 2000). Furthermore, transient expression of LIF in mice is essential for implantation of blastocyst (Stewart et al., 1992). LIF-deficient mice derived by gene-targeting techniques have dramatically decreased numbers of stem cells in spleen and bone

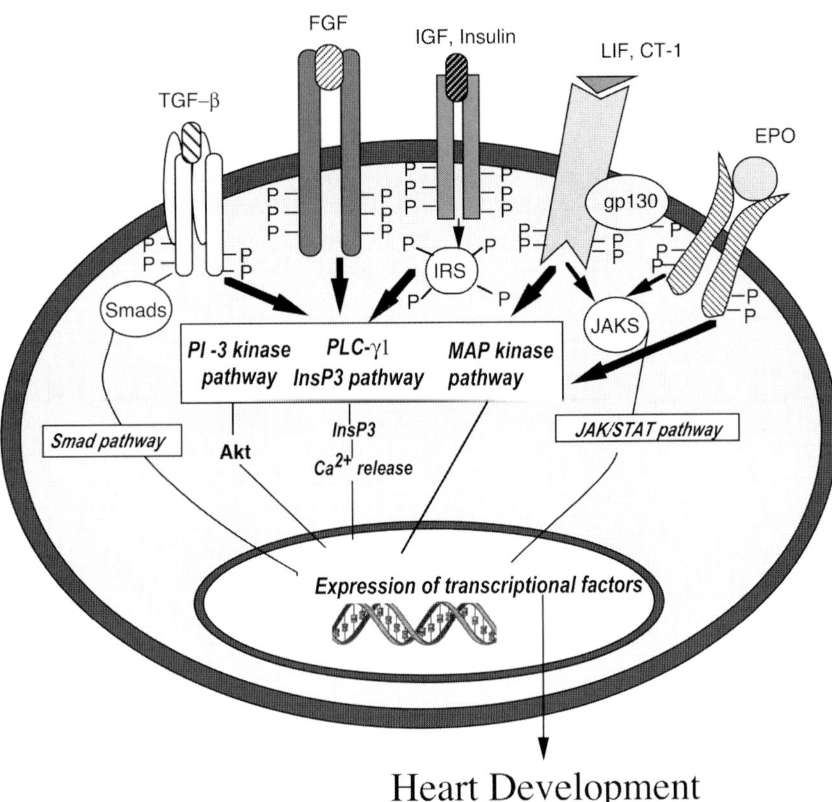

Fig. 2. Induction of heart development by classic growth factors and cytokines via activation of MAPK, PI-3-kinase, and PLC-γ1 pathways as well as via Smad and JAK/STAT pathway (*see* text for more details).

marrow, and defects in stem cell number can be compensated by exogenous LIF, suggesting that LIF is also required for survival of the normal pool of stem cells (Escary et al., 1993). LIF exists in two different forms: a diffusible form (D-LIF) and a matrix-bound form (M-LIF) (Rathjen et al., 1990; Conquet et al., 1992). Embryos overexpressing the D-LIF cDNA looked essentially normal. Chimerae expressing LIF associated with the ECM cDNA showed an abnormal proliferation of tissues and the absence of differentiated mesoderm (Conquet et al., 1992).

lif$^{-/-}$ EBs lacking of both diffusible and matrix-bound forms of LIF (*lif*$^{-/-}$) EBs) showed a severe suppression of early differentiation of cardiomyocytes (Bader et al., 2000). Furthermore, onset of differentiation could be rescued by a very low concentration of D-LIF, but consecutive differentiation was attenuated in a concentration-dependent manner by increasing D-LIF concentration in wild-type and *lif*$^{-/-}$ cardiomyocytes. At fully differentiated state, paracrine and autocrine LIF promoted proliferation and longevity of cardiomyocytes. Bader et al. (2000) suggested that both D-LIF and M-LIF contribute to the modulation of cardiogenesis in a subtile, opposite manner and control proliferation and maintenance of the differentiated state of cardiomyocytes.

LIF signaling is mediated mainly by the janus kinase signal transducer and activator of transcription (JAK/STAT) pathway (Auernhammer and Melmed, 2000). LIF binds to a receptor complex consisting of LIFR and to signal transducer gp130. Heterodimerization of the LIFR-gp130 complex by LIF activates Jak kinase activity followed by phosphorylation of gp130 and LIFR. Phosphorylated tyrosine residues on LIFR and gp130 provide specific docking sites for the SH2 domains of STAT proteins causing receptor association and subsequent phosphorylation of STATs that are transcriptional factors (Auernhammer and Melmed, 2000). In addition to the STAT pathway, binding of IL-6-related cytokines to LIF/gp130 complex can also activate the p21^{ras}/Raf pathway, thereby stimulating the MAPK pathway including ERK1/2 (Kumar et al., 1994). However, it is still not clear whether the LIF/gp130 complex–mediated activation of the MAPK pathway is more relevant than the JAK/STAT pathway for developmental processes.

Recently, cardiotrophin-1 (CT-1), another LIFR/gp130-binding protein, has been isolated using a cDNA expression library from murine ES cells (Wollert and Chien, 1997). CT-1 is structurally similar to the IL-6 related cytokines. It activates several features of cardiomyocyte hypertrophy in vitro, including sarcomeric organization and embryonic gene expression (Wollert and Chien, 1997). Furthermore, receptor-binding studies and functional studies reveal that CT-1 shares the signal transducing receptor components gp130 and LIFR. CT-1 rapidly activates gp130 and LIFR tyrosine phosphorylation in cultured cardiac myocytes, resulting in cardiomyocyte hypertrophy (Wollert and Chien, 1997). Knockout gp130-lacking mice (gp130$^{-/-}$) show a severe ventricular hypoplasia and fail to form intact ventricular chamber (Yoshida et al., 1996). These findings suggest that cytokines signaling through gp130 are also potent regulators of embryonic heart development (Wollert and Chien, 1997).

In summary, several classic growth factors—TGF-β, IGF, insulin, and FGF—as well as cytokines have been identified to promote heart development via activation of the MAPK pathway, the PI-3-kinase pathway, and the PLC-γ1 pathway (Fig. 2).

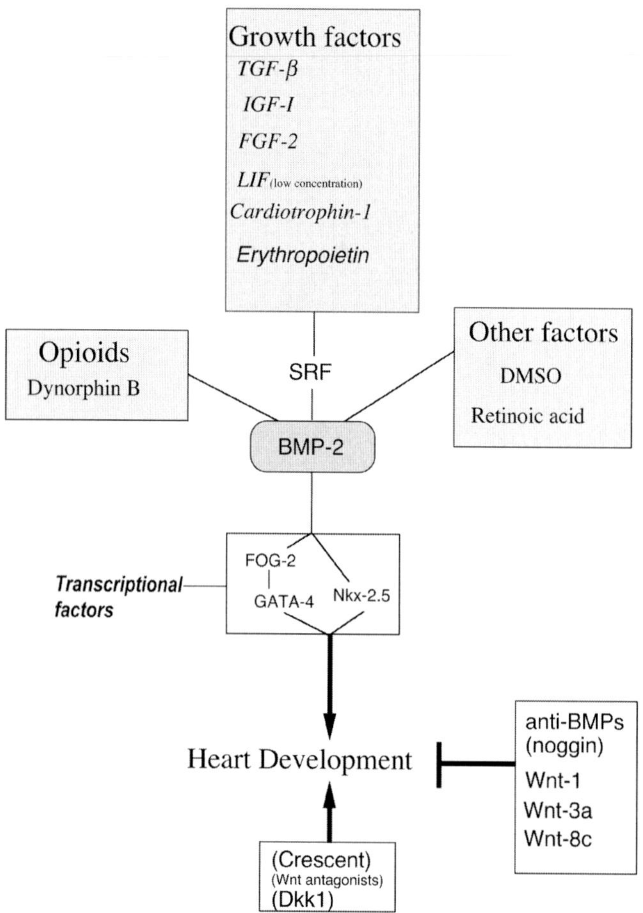

Fig. 3. Possible induction pathway of heart development by classic growth factors as well as by other factors via activation of SRF and transcription factors such as FOG-2, GATA-4, and Nkx-2.5 by BMP-2. Wnt signaling as well as BMP antagonists such as noggin inhibit heart development in mammalians. Wnt protein antagonists such as Crescent and Dkk1 promote heart development (*see* text for more details).

Erythropoietin (EPO) is a growth factor inducing formation of erythrocytes and proliferation and/or differentiation of megakaryocytes (Ishibashi et al., 1987; Krantz, 1991; Broudy et al., 1995; Papayannopoulou et al., 1996; Wu et al., 1999). Furthermore, EPO receptors (EPORs) are expressed in nonhematopoietic cells such as umbilical cord and placental endothelial cells as well as in neurons (Wu et al., 1999). EPO also induces proliferation of endothelial cells (Krantz, 1991; Anagnostou et al., 1994). The EPOR belongs to the cytokine receptor superfamily. Similar to the IL-6 receptor, after binding of EPO to its receptor, homodimerization and autophosphorylation of the EPOR occurs following stimulation of the JAK/STAT, PI-3-kinase, and the MAPK pathways (for a review *see* Yoshimura and Misawa, 1998; Cheung and Miller, 2001). Studies on mice lacking EPO$^{-/-}$ and EPOR$^{-/-}$ demonstrate that EPO is involved in the control of proliferation, survival, and irreversible terminal differentiation of erythroid progenitors (Wu et al., 1999). However, novel findings from EPO$^{-/-}$ and EPOR$^{-/-}$ animals suggest that EPO and EPOR are also in involved in heart development (Wu et al., 1999). In this context, it has been shown that embryonic heart development in EPO$^{-/-}$ and EPOR$^{-/-}$ lacking mice suffers from ventricular hyperplasia, which is coupled to defects in the intraventricular septum. Cardiac abnormalities are likely due to a reduction in proliferation, specific to the heart, since proliferation in other organs including brain in EPOR null animals is indistinguishable from that of wild mice (Wu et al., 1999). After the creation of chimeric animals by injecting EPOR$^{-/-}$ into wild-type blastocysts, it has been shown that EPOR is essential for normal structure and development of cardiomyocytes. It has been suggested that EPO triggers proliferation of cardiomyocytes in a non-cell-autonomous manner, thereby promoting cardiac morphogenesis (Wu et al., 1999).

Remarkably, there is increasing evidence that opioid peptides are regulators of cardiogenesis. In this context, it has been shown that dimethyl sulfoxide (DMSO) promotes the differentiation of ES cells to embryonic cardiac cells (McBurney et al., 1982), which express GATA-4 and Nkx-2.5 (Skerjanc et al., 1998; Ventura and Maioli, 2000). Inhibition of GATA-4 blocked DMSO-induced cardiogenesis whereas transfection of Nkx-2.5 to ES cells in the absence of DMSO led to the appearance of myocardial cells (Skerjanc et al., 1998; Ventura and Maioli, 2000). In this context, adult cardiac myocytes express the prodynorphin gene and are able to synthesize and secrete dynorphin B, a natural κ-opioid. P19 cells and murine ES cells synthesize and secrete dynorphin B, a biologically active end product of the prodynorphin gene (Ventura and Maioli, 2000). DMSO-induced expression of GATA-4 and Nkx-2.5 gene expression was preceded by a marked increase in prodynorphin gene expression and dynorphin B synthesis and secretion (Ventura and Maioli, 2000). In the absence of DMSO, dynorphin B triggered GATA-4 and Nkx-2.5 gene and led to the appearance of both α-MHC and myosin light chain 2V transcripts (Ventura and Maioli, 2000). Opioid receptor antagonists and antisense oligonucleotides blocked DMSO-induced cardiogenesis. These findings suggest the implication of an autocrine opioid gene in the development of heart (Ventura and Maioli, 2000).

In addition to classic growth factors, other factors such as DMSO and RA, as well as opioids, promote heart development probably by activation of SRF; formation of BMP-2; and activation of transcriptional factors such as FOG-2, GATA-4, and Nkx-2.5. Finally, interactions of Nkx-2.5 with GATA-4 result in the transcription of cardiac-restricted genes (Fig. 3).

21.2.3. INHIBITORY SIGNALS FOR HEART FORMATION

Wingless (Wg) and Wnt are secreted signaling molecules that regulate key developmental processes in *Drosophila* and vertebrates, respectively (Miller and Moon, 1996; Orsulic and Peifer, 1996; Moon et al., 1997). Wnts are a family of cysteine-rich glycosylated ligands (more than 16 mammalian family members) that mediate cell–cell communication in diverse developmental processes. The loss or inappropriate activation of Wnt expression has been shown to alter cell fate, morphogenesis, and mitogenesis (for a review *see* Dale, 1998). The Wg/Wnt signaling pathway was first discovered during genetic studies in *Drosophila* (Orsulic and Peifer, 1996). After binding of Wg or Wnt to its seven-transmembrane domain receptor, the frizzled receptor 2 (fz-2) or Frizzled, respectively, a signal transduction cascade is stimulated resulting first in the hyperphosphorylation of cytoplasmic protein known as Disheveled (Bhanot et al., 1996; Dale, 1998; Yanagawa et al., 1995) or Dsh (Fig. 4). Activated Disheveled inhibits the serine/threonine kinases Zeste-White-3 (ZW3) (in *Drosophila*) or glycogen synthase kinase-β (GSK-β) (in vertebrates). Inactivation of

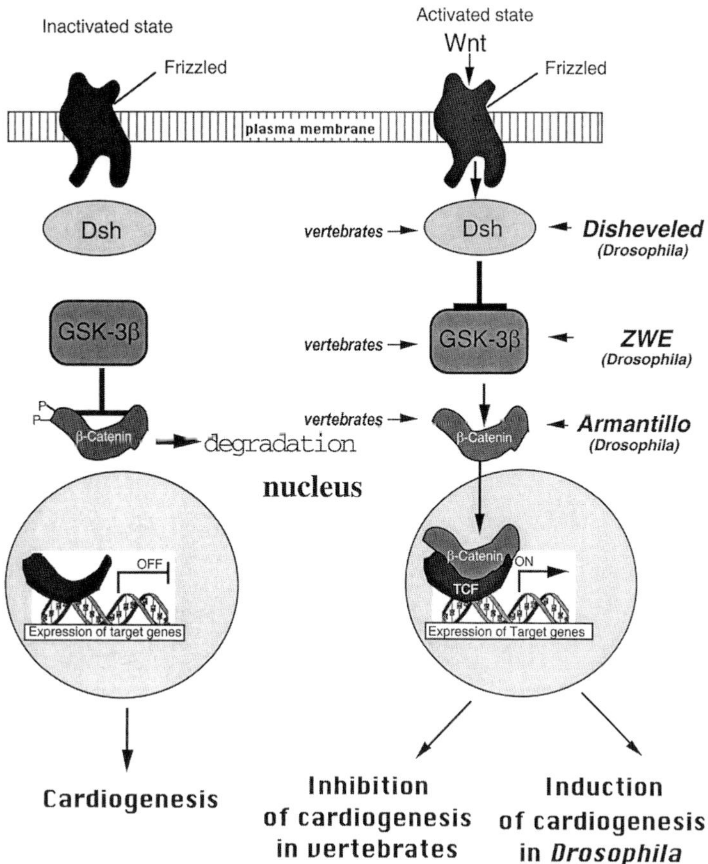

Fig. 4. Wnt signaling transduction pathway. Binding of Wnt or Wg to Frizzled-activated Dsh (vertebrates) or Disheveled (*Drosophila*) inhibits the serine/threonine kinases ZW3 (*Drosophila*) or GSK-β (in mammalians). Inactivation of ZW3 or GSK-β leads to an accumulation of cytoplasmic Armantillo or β-catenin, respectively, and its translocation to the nucleus. Armantillo or β-catenin interacts with dTCF/pangolin or the TCF/Lef transcription factor family, resulting in an expression of target genes promoting or inhibiting cardiac development in *Drosophila* or mammalians, respectively. In the absence of Wg or Wnt signaling, ZW3 or GSK-β degrade Armantillo or β-catenin, thereby inhibiting expression of target developmental genes.

ZW3 or GSK-β leads to an accumulation of cytoplasmic Armantillo or β-catenin, respectively, and its translocation to the nucleus. Armantillo or β-catenin interacts with dTCF/pangolin or the T-cell factor (TCF)/Lef trancription factor, resulting in the expression of developmental target genes promoting or inhibiting cardiac development in *Drosophila* or vertebrates, respectively. In the absence of Wg or Wnt signalling, ZW3 or GSK-β targets Armantillo or β-catenin for degradation, thereby inhibiting expression of target developmental genes.

More recently, it has been shown that activation of Frizzled receptors and subsequently Disheveled proteins relieves β-catenin from inhibition by a protein complex including GSK-3β, Axin, and adenomatous polyposis coli tumor suppressor. This allows β-catenin to form nuclear complexes with TCF and transcriptional repressor proteins such as CtBP, Rpd3, and TLE to counteract promoter repression and accomplish target gene activation. In vertebrates, Wnt stimulation requires not only TCF-binding elements but also sequences bound by Smad4, which is an essential mediator of signaling events trigged by members of the TGF-β subfamily (for a review *see* Hecht and Kemler, 2000).

Inhibitory signals that block heart formation in anterior paraxial mesoderm include Wnt family members expressed in dorsal neural tube (Wnt-1 and Wnt-3a) and anti-BMPs expressed in the axial tissues (i.e., noggin in the notochord). The sum of these positive and negative signals determines the medial–lateral boarders of the heart-forming region (Marvin et al., 2001; Tzahor and Lassar, 2001).

Signal from the neural tube that blocks cardiogenesis in mesoderm of stage 8 to 9 chick embryos can be mimicked by ectopic expression of either Wnt-3a or Wnt-1, both of which are expressed in the dorsal neural tube (Tzahor and Lassar, 2001). Repression can be overcome by ectopic expression of Wnt antagonist (Tzahor and Lassar, 2001). These findings suggest that heart formation is cued by a combination of positive and negative signals from surrounding tissues.

Although in vertebrates BMP signals play a positive role in promoting expression of the NK homeobox gene Nkx-2.5 and subsequent heart formation, these findings suggest that Wnt signals paradoxically repress heart formation in vertebrates (Tzahor and Lassar, 2001). By contrast, in *Drosophila*, the BMP family members dpp (Frasch, 1995) and the Wnt family members wingless are required to maintain the expression of the NK homeobox gene *tinman* and for subsequent cardiogenesis (Tzahor and Lassar, 2001; Wu et al., 1995).

Recently, it has been demonstrated that chick heart mesoderm is induced by signals from the anterior endoderm (Marvin et al., 2001). BMP-2, one candidate that is expressed in the anterior endoderm, is necessary but not sufficient for heart formation. One or more additional factors from anterior endoderm are required. There is increasing evidence that Crescent, a Wnt (Wnt-8) antagonist, promotes cardiogenesis (Marvin et al., 2001). Crescent is a Frizzled-related protein that is expressed in anterior endoderm during gastrulation (Schultheiss et al., 1995). The importance of Crescent for heart development has been proved by experiments showing that coculture of posterior lateral plate mesoderm with anterior endoderm or overexpression of Cresent using a replication-competent retrovirus (RCAS) induces formation of beating heart muscle and represses formation of blood (Marvin et al., 2001). Posterior lateral plate mesoderm normally forms derivatives of blood cells.

Dickkopf1 (Dkk1) is another class of Wnt antagonist that inhibits both Wnt-8 and Wnt-3a and induces heart-specific gene expression in posterior lateral plate mesoderm (Krupnik et al., 1999; Kazanskaya et al., 2000). Moreover, ectopic Wnt signals can repress heart formation from anterior mesoderm in vitro and in vivo. Expression of either Wnt-3a or Wnt-8c promotes development of primitive erythrocytes from the precardiac region (Marvin et al., 2001). From these findings it has been concluded that inhibition of Wnt signaling promotes heart formation in the anterior lateral mesoderm, whereas active Wnt signaling in the posterior lateral mesoderm promotes blood development (Marvin et al., 2001). According to this model two orthogonal gradients—Wnt activity and BMP signals—promote cardiogenesis. Wnt activity along the anterior–posterior axis and BMP signals along the dorsal–ventral axis intersect in the heart-forming region to induce cardiogenesis in the region of high BMP and low Wnt activity (Marvin et al., 2001). In summary, in striking contrast to *Drosophila*, activation of the Wnt signaling in vertebrates induces an inhibition of the heart development. In contrast to BMP-2 antagonists such as noggin, Wnt antagonists such as Cresent and Dkk1 appear to induce cardiogenesis (Fig. 3).

21.3. SUMMARY, CONCLUSIONS AND PROSPECT OF THE FUTURE

Basic mechanisms of heart development are conserved between vertebrates and invertebrates. In vertebrates, the heart is the first mesoderm-derived functional embryonic organ that is developed after gastrulation in amniotic embryos. Heart development is a complex process involving proliferation and differentiation as well as tissue organization into specific architecture. Heart development is regulated by factors and signals secreted by the anterior primitive endoderm and by neuronal tube. In the last few years, several soluble factors and intracellular transcription factors as well as intracellular signal cascades have been recognized to be involved in the development of heart. Classic growth factors such as IGF, TGF-β, FGF, LIF (at low concentration), CT-1, and EPO; other factors such as DMSO and RA; as well as opioids promote heart development, probably by activation of SRF, formation of BMP-2 and activation of the transcriptional factors such as FOG-2, GATA-4, and Nkx-2.5. Activation of the Wnt signaling in vertebrates induces an inhibition of heart development. BMP-2 antagonists such as noggin and Wnt antagonists such as Cresent or Dkk1 appear to promote cardiac development.

Many efforts have been made to develop new cellular therapy concepts for treatment of numerous diseases. ES cells grown in vitro can be selectively differentiated under appropriate culture conditions into the cell type required for transplantation in numerous diseases and for biomedical engineering. Therefore, identification of growth factors, as well as signal cascades capable of priming selective differentiation of ES to cardiac cells, and in vitro generation of a great amount of cardiac cells, remain crucial issues for the future in order to develop cellular therapy concepts for treatment of severe cardiac diseases.

REFERENCES

Anagnostou, A., Liu, Z., Steiner, M., et al. (1994) Erythropoietin receptor mRNA expression in human endothelial cells. *Proc. Natl. Acad. Sci. USA* 91:3974–3978.

Andree, B., Duprez, D., Vorbusch, B., Arnold, H. H., and Brand, T. (1998) BMP-2 induces ectopic expression of cardiac lineage markers and interferes with somite formation in chicken embryos. *Mech. Dev.* 70:119–131.

Antin, P. B., Taylor, R. G., and Yatskievych, T. (1994) Precardiac mesoderm is specified during gastrulation in quail. *Dev. Dyn.* 200:144–154.

Antin, P. B., Yatskievych, T., Dominguez, J. L., and Chieffi, P. (1996) Regulation of avian precardiac mesoderm development by insulin and insulin-like growth factors. *J. Cell Physiol.* 168:42–50.

Auernhammer, C. J. and Melmed, S. (2000) Leukemia-inhibitory factor-neuroimmune modulator of endocrine function. *Endocr. Rev.* 21:313–345.

Azpiazu, N. and Frasch, M. (1993) tinman and bagpipe: two homeo box genes that determine cell fates in the dorsal mesoderm of Drosophila. *Genes Dev.* 7:1325–1340.

Azpiazu, N., Lawrence, P. A., Vincent, J. P., and Frasch, M. (1996) Segmentation and specification of the *Drosophila* mesoderm. *Genes Dev.* 10:3183–3194.

Bader, A., Al-Dubai, H., and Weitzer, G. (2000) Leukemia inhibitory factor modulates cardiogenesis in embryoid bodies in opposite fashions. *Circ. Res.* 86:787–794.

Bain, G., Kitchens, D., Yao, M., Huettner, J. E., and Gottlieb, D. I. (1995) Embryonic stem cells express neuronal properties in vitro. *Dev. Biol.* 168:342–357.

Baylies, M. K., Martinez Arias, A., and Bate, M. (1995) wingless is required for the formation of a subset of muscle founder cells during *Drosophila* embryogenesis. *Development* 121:3829–3837.

Beiman, M., Shilo, B. Z., and Volk, T. (1996) Heartless, a Drosophila FGF receptor homolog, is essential for cell migration and establishment of several mesodermal lineages. *Genes Dev.* 10:2993–3002.

Benson, D. W., Silberbach, G. M., Kavanaugh-McHugh, A., et al. (1999) Mutations in the cardiac transcription factor NKX2.5 affect diverse cardiac developmental pathways. *J. Clin. Invest.* 104:1567–1573.

Bhanot, P., Brink, M., Samos, C. H., et al. (1996) A new member of the frizzled family from *Drosophila* functions as a Wingless receptor. *Nature* 382:225–230.

Bodmer, R. (1993) The gene tinman is required for specification of the heart and visceral muscles in *Drosophila*. *Development* 118:719–729.

Bodmer, R. and Frasch, M. (1999) Genetic determination of *Drosophila* heart development. In: *Heart development* (Rosenthal, N. and Harvey, R., eds.) Academic Press, San Diego, London, New York, pp. 65–90.

Bodmer, R. and Venkatesh, T. V. (1998) Heart development in *Drosophila* and vertebrates: conservation of molecular mechanisms. *Dev. Genet.* 22:181–186.

Bodmer, R., Jan, L. Y., and Jan, Y. N. (1990) A new homeobox-containing gene, msh-2, is transiently expressed early during mesoderm formation of *Drosophila*. *Development* 110:661–669.

Bour, B. A., O'Brien, M. A., Lockwood, W. L., et al. (1995) *Drosophila* mef2, a transcription factor that is essential for myogenesis. *Genes Dev.* 9:730–741.

Broudy, V. C., Lin, N. L., and Kaushansky, K. (1995) Thrombopoietin (c-mpl ligand) acts synergistically with erythropoietin, stem cell factor, and interleukin-11 to enhance murine megakaryocyte colony growth and increases megakaryocyte ploidy in vitro. *Blood* 85:1719–1726.

Brustle, O., Jones, K. N., Learish, R. D., et al. (1999) Embryonic stem cell-derived glial precursors: a source of myelinating transplants. *Science* 285:754–756.

Carmena, A., Gisselbrecht, S., Harrison, J., Jimenez, F., and Michelson, A. M. (1998a) Combinatorial signaling codes for the progressive determination of cell fates in the *Drosophila* embryonic mesoderm. *Genes Dev.* 12:3910–3922.

Carmena, A., Murugasu-Oei, B., Menon, D., Jimenez, F., and Chia, W. (1998b) Inscuteable and numb mediate asymmetric muscle progenitor cell divisions during *Drosophila* myogenesis. *Genes Dev.* 12: 304–315.

Cheung, J. Y. and Miller, B. A. (2001) Molecular mechanisms of erythropoietin signaling. *Nephron* 87:215–222.

Climent, S., Sarasa, M., Villar, J. M., and Murillo-Ferrol, N. L. (1995) Neurogenic cells inhibit the differentiation of cardiogenic cells. *Dev. Biol.* 171:130–148.

Conquet, F., Peyrieras, N., Tiret, L., and Brulet, P. (1992) Inhibited gastrulation in mouse embryos overexpressing the leukemia inhibitory factor. *Proc. Natl. Acad. Sci. USA* 89:8195–8199.

Dale, T. C. (1998) Signal transduction by the Wnt family of ligands. *Biochem. J.* 329 :209–223.

Drab, M., Haller, H., Bychkov, R., et al. (1997) From totipotent embryonic stem cells to spontaneously contracting smooth muscle cells: a retinoic acid and db-cAMP in vitro differentiation model. *FASEB J.* 11:905–915.

Durocher, D., Charron, F., Warren, R., Schwartz, R. J., and Nemer, M. (1997) The cardiac transcription factors Nkx2-5 and GATA-4 are mutual cofactors. *EMBO J.* 16:5687–5696.

Escary, J. L., Perreau, J., Dumenil, D., Ezine, S., and Brulet, P. (1993) Leukaemia inhibitory factor is necessary for maintenance of haematopoietic stem cells and thymocyte stimulation. *Nature* 363: 361–364.

Fishman, M. C. and Chien, K. R. (1997) Fashioning the vertebrate heart: earliest embryonic decisions. *Development* 124:2099–2117.

Frasch, M. (1995) Induction of visceral and cardiac mesoderm by ectodermal Dpp in the early *Drosophila* embryo. *Nature* 374:464–467.

Frasch, M., Hoey, T., Rushlow, C., Doyle, H., and Levine, M. (1987) Characterization and localization of the even-skipped protein of *Drosophila. EMBO J.* 6:749–759.

Fu, Y., Yan, W., Mohun, T. J., and Evans, S. M. (1998) Vertebrate tinman homologues XNkx2-3 and XNkx2-5 are required for heart formation in a functionally redundant manner. *Development* 125:4439–4449.

Garcia-Martinez, V. and Schoenwolf, G. C. (1993) Primitive-streak origin of the cardiovascular system in avian embryos. *Dev. Biol.* 159: 706–719.

Gisselbrecht, S., Skeath, J. B., Doe, C. Q., and Michelson, A. M. (1996) heartless encodes a fibroblast growth factor receptor (DFR1/DFGF-R2) involved in the directional migration of early mesodermal cells in the *Drosophila* embryo. *Genes Dev.* 10:3003–3017.

Grow, M. W. and Krieg, P. A. (1998) Tinman function is essential for vertebrate heart development: elimination of cardiac differentiation by dominant inhibitory mutants of the tinman-related genes, XNkx2-3 and XNkx2-5. *Dev. Biol.* 204:187–196.

Haenlin, M., Cubadda, Y., Blondeau, F., et al. (1997) Transcriptional activity of pannier is regulated negatively by heterodimerization of the GATA DNA-binding domain with a cofactor encoded by the u-shaped gene of *Drosophila. Genes Dev.* 11:3096–3108.

Harvey, R. P. (1996) NK-2 homeobox genes and heart development. *Dev. Biol.* 178:203–216.

Hecht, A. and Kemler, R. (2000) Curbing the nuclear activities of beta-catenin. Control over Wnt target gene expression. *EMBO Rep.* 1:24–28.

Heikinheimo, M., Scandrett, J. M., and Wilson, D. B. (1994) Localization of transcription factor GATA-4 to regions of the mouse embryo involved in cardiac development. *Dev. Biol.* 164:361–373.

Hescheler, J., Fleischmann, B. K., Wartenberg, M., et al. (1999) Establishment of ionic channels and signalling cascades in the embryonic stem cell–derived primitive endoderm and cardiovascular system. *Cells Tissues Organs* 165:153–164.

Ishibashi, T., Koziol, J. A., and Burstein, S. A. (1987) Human recombinant erythropoietin promotes differentiation of murine megakaryocytes in vitro. *J. Clin. Invest.* 79:286–289.

Itskovitz-Eldor, J., Schuldiner, M., Karsenti, D., et al. (2000) Differentiation of human embryonic stem cells into embryoid bodies compromising the three embryonic germ layers. *Mol. Med.* 6:88–95.

Jacobson, A. G. (1960) Influences of ectoderm and endoderm on heart differentiation in the newt. *Dev. Biol.* 2:138–154.

Kazanskaya, O., Glinka, A., and Niehrs, C. (2000) The role of Xenopus dickkopf1 in prechordal plate specification and neural patterning. *Development* 127:4981–4992.

Klintz, F., Bloch, W., Addicks, K., and Hescheler, J. (1999) Inhibition of phosphatidylinositol-3-kinase blocks development of functional embryonic cardiomyocytes. *Exp. Cell. Res.* 247:79–83.

Ko, L. J. and Engel, J. D. (1993) DNA-binding specificities of the GATA transcription factor family. *Mol. Cell. Biol.* 13:4011–4022.

Kosman, D., Ip, Y. T., Levine, M., and Arora, K. (1991) Establishment of the mesoderm-neuroectoderm boundary in the *Drosophila* embryo. *Science* 254:118–122.

Krantz, S. B. (1991) Erythropoietin. *Blood* 77:419–434.

Krupnik, V. E., Sharp, J. D., Jiang, C., et al. (1999) Functional and structural diversity of the human Dickkopf gene family. *Gene* 238:301–313.

Kumar, G., Gupta, S., Wang, S., and Nel, A. E. (1994) Involvement of Janus kinases, p52shc, Raf-1, and MEK-1 in the IL-6- induced mitogen-activated protein kinase cascade of a growth-responsive B cell line. *J. Immunol.* 153:4436–4447.

Kuo, C. T., Morrisey, E. E., Anandappa, R., et al. (1997) GATA4 transcription factor is required for ventral morphogenesis and heart tube formation. *Genes Dev.* 11:1048–1060.

Lai, Z. C., Rushton, E., Bate, M., and Rubin, G. M. (1993) Loss of function of the *Drosophila zfh-1* gene results in abnormal development of mesodermally derived tissues. *Proc. Natl. Acad. Sci. USA* 90:4122–4126.

Laverriere, A. C., MacNeill, C., Mueller, C., Poelmann, R. E., Burch, J. B., and Evans, T. (1994) GATA-4/5/6, a subfamily of three transcription factors transcribed in developing heart and gut. *J. Biol. Chem.* 269:23,177–23,184.

Lee, Y., Shioi, T., Kasahara, H., et al. (1998) The cardiac tissue-restricted homeobox protein Csx/Nkx2.5 physically associates with the zinc finger protein GATA4 and cooperatively activates atrial natriuretic factor gene expression. *Mol. Cell. Biol.* 18:3120–3129.

Leptin, M. (1991) twist and snail as positive and negative regulators during Drosophila mesoderm development. *Genes Dev.* 5:1568–1576.

Lilly, B., Zhao, B., Ranganayakulu, G., Paterson, B. M., Schulz, R. A., and Olson, E. N. (1995) Requirement of mads domain transcription factor d-mef2 for muscle formation in drosophila. *Science* 267: 688–693.

Lyons, I., Parsons, L. M., Hartley, L., et al. (1995) Myogenic and morphogenetic defects in the heart tubes of murine embryos lacking the homeo box gene Nkx2-5. *Genes Dev.* 9:1654–1666.

Marvin, M. J., Di Rocco, G., Gardiner, A., Bush, S. M., and Lassar, A. B. (2001) Inhibition of Wnt activity induces heart formation from posterior mesoderm. *Genes Dev.* 15:316–327.

McBurney, M. W., Jones-Villeneuve, E. M., Edwards, M. K., and Anderson, P. J. (1982) Control of muscle and neuronal differentiation in a cultured embryonal carcinoma cell line. *Nature* 299:165–167.

Merika, M. and Orkin, S. H. (1993) DNA-binding specificity of GATA family transcription factors. *Mol. Cell. Biol.* 13:3999–4010.

Merika, M. and Orkin, S. H. (1995) Functional synergy and physical interactions of the erythroid transcription factor GATA-1 with the Kruppel family proteins Sp1 and EKLF. *Mol. Cell. Biol.* 15:2437–2447.

Miller, J. R. and Moon, R. T. (1996) Signal transduction through beta-catenin and specification of cell fate during embryogenesis. *Genes Dev.* 10:2527–2539.

Molkentin, J. D., Lin, Q., Duncan, S. A., and Olson, E. N. (1997) Requirement of the transcription factor GATA4 for heart tube formation and ventral morphogenesis. *Genes Dev.* 11:1061–1072.

Molkentin, J. D., Lu, J. R., Antos, C. L., et al. (1998) A calcineurin-dependent transcriptional pathway for cardiac hypertrophy. *Cell* 93: 215–228.

Montgomery, M. O., Litvin, J., Gonzalez-Sanchez, A., and Bader, D. (1994) Staging of commitment and differentiation of avian cardiac myocytes. *Dev. Biol.* 164:63–71.

Moon, R. T., Brown, J. D., and Torres, M. (1997) WNTs modulate cell fate and behavior during vertebrate development. *Trends Genet.* 13: 157–162.

Morrisey, E. E., Ip, H. S., Lu, M. M., and Parmacek, M. S. (1996) GATA-6: a zinc finger transcription factor that is expressed in multiple cell lineages derived from lateral mesoderm. *Dev. Biol.* 177:309–322.

Odorico, J. S., Kaufman, D. S., and Thomson, J. A. (2001) Multilineage differentiation from human embryonic stem cell lines. *Stem Cells* 19: 193–204.

Olson, E. N. and Srivastava, D. (1996) Molecular pathways controlling heart development. *Science* 272:671–676.

Orkin, S. H. (1992) GATA-binding transcription factors in hematopoietic cells. *Blood* 80:575–581.

Orsulic, S. and Peifer, M. (1996) An in vivo structure-function study of armadillo, the beta-catenin homologue, reveals both separate and overlapping regions of the protein required for cell adhesion and for wingless signaling. *J. Cell. Biol.* 134:1283–1300.

Papayannopoulou, T., Brice, M., Farrer, D., and Kaushansky, K. (1996) Insights into the cellular mechanisms of erythropoietin-thrombopoietin synergy. *Exp. Hematol.* 24:660–669.

Park, M., Yaich, L. E., and Bodmer, R. (1998) Mesodermal cell fate decisions in *Drosophila* are under the control of the lineage genes numb, Notch, and sanpodo. *Mech. Dev.* 75:117–126.

Pevny, L., Simon, M. C., Robertson, E., Klein, W. H., Tsai, S. F., D'Agati, V., Orkin, S. H., and Costantini, F. (1991) Erythroid differentiation in chimaeric mice blocked by a targeted mutation in the gene for transcription factor GATA-1. *Nature* 349:257–260.

Powell-Braxton, L., Hollingshead, P., Warburton, C., Dowd, M., Pitts-Meek, S., Dalton, D., Gillett, N., and Stewart, T. A. (1993) IGF-I is required for normal embryonic growth in mice. *Genes Dev.* 7: 2609–2617.

Powers, C. J., McLeskey, S. W., and Wellstein, A. (2000) Fibroblast growth factors, their receptors and signaling. *Endocr. Relat. Cancer* 7:165–197.

Raffin, M., Leong, L. M., Rones, M. S., Sparrow, D., Mohun, T., and Mercola, M. (2000) Subdivision of the cardiac Nkx2.5 expression domain into myogenic and nonmyogenic compartments. *Dev. Biol.* 218:326–340.

Rathjen, P. D., Toth, S., Willis, A., Heath, J. K., and Smith, A. G. (1990) Differentiation inhibiting activity is produced in matrix-associated and diffusible forms that are generated by alternate promoter usage. *Cell* 62:1105–1114.

Reifers, F., Walsh, E. C., Leger, S., Stainier, D. Y., and Brand, M. (2000) Induction and differentiation of the zebrafish heart requires fibroblast growth factor 8 (fgf8/acerebellar). *Development* 127:225–235.

Riechmann, V., Irion, U., Wilson, R., Grosskortenhaus, R., and Leptin, M. (1997) Control of cell fates and segmentation in the *Drosophila* mesoderm. *Development* 124:2915–2922.

Rohwedel, J., Maltsev, V., Bober, E., Arnold, H. H., Hescheler, J., and Wobus, A. M. (1994) Muscle cell differentiation of embryonic stem cells reflects myogenesis in vivo: developmentally regulated expression of myogenic determination genes and functional expression of ionic currents. *Dev. Biol.* 164:87–101.

Schlange, T., Andree, B., Arnold, H. H., and Brand, T. (2000) BMP2 is required for early heart development during a distinct time period. *Mech. Dev.* 91:259–270.

Schott, J. J., Benson, D. W., Basson, C. T., et al. (1998) Congenital heart disease caused by mutations in the transcription factor NKX2-5. *Science* 281:108–111.

Schultheiss, T. M., Xydas, S., and Lassar, A. B. (1995) Induction of avian cardiac myogenesis by anterior endoderm. *Development* 121: 4203–4214.

Schultheiss, T. M., Burch, J. B., and Lassar, A. B. (1997) A role for bone morphogenetic proteins in the induction of cardiac myogenesis. *Genes Dev.* 11:451–462.

Sepulveda, J. L., Belaguli, N., Nigam, V., Chen, C. Y., Nemer, M., and Schwartz, R. J. (1998) GATA-4 and Nkx-2.5 coactivate Nkx-2 DNA binding targets: role for regulating early cardiac gene expression. *Mol. Cell. Biol.* 18:3405–3415.

Shi, C. Z., Dhir, R. N., Kesavan, P., Zhang, S. L., Matschinsky, F. M., and Heyner, S. (1995) Mouse embryonic stem cells express receptors of the insulin family of growth factors. *Mol. Reprod. Dev.* 42:173–179.

Shishido, E., Higashijima, S., Emori, Y., and Saigo, K. (1993) Two FGF-receptor homologues of *Drosophila*: one is expressed in mesodermal primordium in early embryos. *Development* 117:751–761.

Shishido, E., Ono, N., Kojima, T., and Saigo, K. (1997) Requirements of DFR1/Heartless, a mesoderm-specific *Drosophila* FGF-receptor, for the formation of heart, visceral and somatic muscles, and ensheathing of longitudinal axon tracts in CNS. *Development* 124 :2119–2128.

Siddle, K., Urso, B., Niesler, C. A., et al. (2001) Specificity in ligand binding and intracellular signalling by insulin and insulin-like growth factor receptors. *Biochem. Soc. Trans.* 29:513–525.

Simon, M. C. (1995) Gotta have GATA. *Nat. Genet.* 11:9–11.

Skerjanc, I. S., Petropoulos, H., Ridgeway, A. G., and Wilton, S. (1998) Myocyte enhancer factor 2C and Nkx2-5 up-regulate each other's expression and initiate cardiomyogenesis in P19 cells. *J. Biol. Chem.* 273:34,904–34,910.

Smith, A. G., Heath, J. K., Donaldson, D. D., et al. (1988) Inhibition of pluripotential embryonic stem cell differentiation by purified polypeptides. *Nature* 336:688–690.

Staehling-Hampton, K., Hoffmann, F. M., Baylies, M. K., Rushton, E., and Bate, M. (1994) dpp induces mesodermal gene expression in *Drosophila*. *Nature* 372:783–786.

Stewart, C. L., Kaspar, P., Brunet, L. J., et al. (1992) Blastocyst implantation depends on maternal expression of leukaemia inhibitory factor. *Nature* 359:76–79.

Su, M. T., Fujioka, M., Goto, T., and Bodmer, R. (1999) The *Drosophila* homeobox genes zfh-1 and even-skipped are required for cardiac-specific differentiation of a numb-dependent lineage decision. *Development* 126:3241–3251.

Sugi, Y. and Lough, J. (1994) Anterior endoderm is a specific effector of terminal cardiac myocyte differentiation of cells from the embryonic heart forming region. *Dev. Dyn.* 200:155–162.

Svensson, E. C., Tufts, R. L., Polk, C. E., and Leiden, J. M. (1999) Molecular cloning of FOG-2: a modulator of transcription factor GATA-4 in cardiomyocytes. *Proc. Natl. Acad. Sci. USA* 96:956–961.

Takagi, T., Moribe, H., Kondoh, H., and Higashi, Y. (1998) DeltaEF1, a zinc finger and homeodomain transcription factor, is required for skeleton patterning in multiple lineages. *Development* 125:21–31.

Tanaka, M., Chen, Z., Bartunkova, S., Yamasaki, N., and Izumo, S. (1999) The cardiac homeobox gene Csx/Nkx2.5 lies genetically upstream of multiple genes essential for heart development. *Development* 126: 1269–1280.

Thomson, J. A., Itskovitz-Eldor, J., Shapiro, S. S., Waknitz, M. A., Swiergiel, J. J., Marshall, V. S., and Jones, J. M. (1998) Embryonic stem cell lines derived from human blastocysts. *Science* 282:1145–1147.

Ting, C. N., Olson, M. C., Barton, K. P., and Leiden, J. M. (1996) Transcription factor GATA-3 is required for development of the T-cell lineage. *Nature* 384:474–478.

Treisman, R., Marais, R., and Wynne, J. (1992) Spatial flexibility in ternary complexes between SRF and its accessory proteins. *EMBO J.* 11:4631–4640.

Tsai, F. Y., Keller, G., Kuo, F. C., Weiss, M., Chen, J., Rosenblatt, M., Alt, F. W., and Orkin, S. H. (1994) An early haematopoietic defect in mice lacking the transcription factor GATA-2. *Nature* 371:221–226.

Tsang, A. P., Visvader, J. E., Turner, C. A., et al. (1997) FOG, a multitype zinc finger protein, acts as a cofactor for transcription factor GATA-1 in erythroid and megakaryocytic differentiation. *Cell* 90:109–119.

Tsang, A. P., Fujiwara, Y., Hom, D. B., and Orkin, S. H. (1998) Failure of megakaryopoiesis and arrested erythropoiesis in mice lacking the GATA-1 transcriptional cofactor FOG. *Genes Dev.* 12:1176–1188.

Tzahor, E. and Lassar, A. B. (2001) Wnt signals from the neural tube block ectopic cardiogenesis. *Genes Dev.* 15:255–260.

Ventura, C. and Maioli, M. (2000) Opioid peptide gene expression primes cardiogenesis in embryonal pluripotent stem cells. *Circ. Res.* 87:189–194.

Wakefield, L. M. and Roberts, A. B. (2002) TGF-beta signaling: positive and negative effects on tumorigenesis. *Curr. Opin. Genet. Dev.* 12:22–29.

Wang, G. F., Nikovits, W. Jr., Schleinitz, M., and Stockdale, F. E. (1998) A positive GATA element and a negative vitamin D receptor-like element control atrial chamber-specific expression of a slow myosin heavy-chain gene during cardiac morphogenesis. *Mol. Cell. Biol.* 18:6023–6034.

Weinhold, B., Schratt, G., Arsenian, S., Berger, J., Kamino, K., Schwarz, H., Ruther, U., and Nordheim, A. (2000) Srf(–/–) ES cells display non-cell-autonomous impairment in mesodermal differentiation. *EMBO J.* 19:5835–5844.

Whitman, M. (1998) Smads and early developmental signaling by the TGFbeta superfamily. *Genes Dev.* 12:2445–2462.

Wiles, M. V. and Keller, G. (1991) Multiple hematopoietic lineages develop from embryonic stem (ES) cells in culture. *Development* 111:259 267.

Wobus, A. M., Kaomei, G., Shan, J., et al. (1997) Retinoic acid accelerates embryonic stem cell–derived cardiac differentiation and enhances development of ventricular cardiomyocytes. *J. Mol. Cell. Cardiol.* 29:1525–1539.

Wollert, K. C. and Chien, K. R. (1997) Cardiotrophin-1 and the role of gp130-dependent signaling pathways in cardiac growth and development. *J. Mol. Med.* 75:492–501.

Wu, H., Lee, S. H., Gao, J., Liu, X., and Iruela-Arispe, M. L. (1999) Inactivation of erythropoietin leads to defects in cardiac morphogenesis. *Development* 126:3597–3605.

Wu, X., Golden, K., and Bodmer, R. (1995) Heart development in *Drosophila* requires the segment polarity gene wingless. *Dev. Biol.* 169: 619–628.

Yanagawa, S., van Leeuwen, F., Wodarz, A., Klingensmith, J., and Nusse, R. (1995) The dishevelled protein is modified by wingless signaling in Drosophila. *Genes Dev.* 9:1087–1097.

Yoshida, K., Taga, T., Saito, M., et al. (1996) Targeted disruption of gp130, a common signal transducer for the interleukin 6 family of cytokines, leads to myocardial and hematological disorders. *Proc. Natl. Acad. Sci. USA* 93:407–411.

Yoshimura, A. and Misawa, H. (1998) Physiology and function of the erythropoietin receptor. *Curr. Opin. Hematol.* 5:171–176.

Zhu, J., Hill, R. J., Heid, P. J., et al. (1997) end-1 encodes an apparent GATA factor that specifies the endoderm precursor in *Caenorhabditis elegans* embryos. *Genes Dev.* 11:2883–2896.

22 Strategies Using Cell Therapy to Induce Cardiomyocyte Regeneration in Adults with Heart Disease

SILVIU ITESCU, MD

Congestive heart failure remains a major public health problem and is frequently the end result of cardiomyocyte apoptosis and fibrous replacement after myocardial infarction (MI), a process referred to as *left ventricular remodeling*. Cardiomyocytes undergo terminal differentiation soon after birth and are generally considered to withdraw irreversibly from the cell cycle. In response to ischemic insult, adult cardiomyocytes undergo cellular hypertrophy, nuclear ploidy, and a high degree of apoptosis. A small number of human cardiomyocytes retain the capacity to proliferate and regenerate in response to ischemic injury. However, whether these cells are derived from a resident pool of cardiomyocyte stem cells or from a renewable source of circulating bone marrow–derived stem cells that home to the damaged myocardium is at present not known. Replacement and regeneration of functional cardiac muscle after an ischemic insult to the heart could be achieved either by stimulating proliferation of endogenous mature cardiomyocytes or resident cardiac stem cells, or by implanting exogenous donor-derived or allogeneic cells such as fetal or embryonic cardiomyocyte precursors, bone marrow–derived mesenchymal stem cells, or skeletal myoblasts. The newly formed cardiomyocytes must integrate precisely into the existing myocardial wall in order to augment synchronized contractility and avoid potentially life-threatening alterations in the electrical conduction of the heart. A major impediment to survival of the implanted cells is altered immunogenicity by prolonged ex vivo culture conditions. In addition, concurrent myocardial revascularization is required to ensure viability of the repaired region and prevent further scar tissue formation. Human adult bone marrow contains endothelial precursors, which resemble embryonic angioblasts and can be used to induce infarct bed neovascularization after experimental MI. This results in protection of cardiomyocytes against apoptosis, induction of cardiomyocyte proliferation and regeneration, long-term salvage and survival of viable myocardium, prevention of left ventricular remodeling, and sustained improvement in cardiac function. It is reasonable to anticipate that cell therapy strategies for ischemic heart disease will need to incorporate (1) a renewable source of proliferating, functional cardiomyocytes, and (2) angioblasts to generate a network of capillaries and larger-size blood vessels for supply of oxygen and nutrients to both the chronically ischemic endogenous myocardium and to the newly implanted cardiomyocytes.

22.1. INTRODUCTION

Congestive heart failure (CHF) remains a major public health problem and is frequently the end result of cardiomyocyte apoptosis and fibrous replacement after myocardial infarction (MI), a process referred to as left ventricular remodeling. Cardiomyocytes undergo terminal differentiation soon after birth and are generally considered to withdraw irreversibly from the cell cycle. In response to ischemic insult, adult cardiomyocytes undergo cellular hypertrophy, nuclear ploidy, and a high degree of apoptosis. A small number of human cardiomyocytes retain the capacity to proliferate and regenerate in response to ischemic injury. However, whether these cells are derived from a resident pool of cardiomyocyte stem cells or from a renewable source of circulating bone marrow–derived stem cells that home to the damaged myocardium is at present not known. Replacement and regeneration of functional cardiac muscle after an ischemic insult to the heart could be achieved either by stimulating proliferation of endogenous mature cardiomyocytes or resident cardiac stem cells, or by implanting exogenous donor-derived or allogeneic cells such as fetal or embryonic cardiomyocyte precursors, bone marrow–derived mesenchymal stem cells, or skeletal myoblasts. The newly formed cardiomyocytes must integrate precisely into the existing myocardial wall in order to augment synchronized contractility and avoid potentially life-threatening alterations in the electrical conduction of the heart. A major impediment to survival of the implanted cells is altered immunogenicity by prolonged ex vivo culture conditions. In addition, concurrent myocardial revascularization is required to ensure viability of the repaired region and prevent further scar tissue formation.

From: *Stem Cells Handbook*
Edited by: S. Sell © Humana Press Inc., Totowa, NJ

Human adult bone marrow contains endothelial precursors that resemble embryonic angioblasts and can be used to induce infarct bed neovascularization after experimental MI. This results in protection of cardiomyocytes against apoptosis, induction of cardiomyocyte proliferation and regeneration, long-term salvage and survival of viable myocardium, prevention of left ventricular remodeling, and sustained improvement in cardiac function. It is reasonable to anticipate that cell therapy strategies for ischemic heart disease will need to incorporate (1) a renewable source of proliferating, functional cardiomyocytes, and (2) angioblasts to generate a network of capillaries and larger-size blood vessels for supply of oxygen and nutrients to both the chronically ischemic endogenous myocardium and to the newlyimplanted cardiomyocytes.

22.2. ISCHEMIC HEART DISEASE AND HEART FAILURE: A PROBLEM OF EPIDEMIC PROPORTION

CHF remains a major public health problem, with recent estimates indicating that end-stage heart failure with 2yr mortality rates of 70–80% affects more than 60,000 patients in the United States each year (Adachi et al., 2001). In Western societies heart failure is primarily the consequence of previous MI (Agocha et al., 1997). As new modalities have emerged that have enabled significant reduction in early mortality from acute MI, affecting more than 1 million new patients in the United States annually, there has been a paradoxical increase in the incidence of postinfarction heart failure among the survivors. Current therapy of heart failure is limited to the treatment of already established disease and is predominantly pharmacological in nature, aiming primarily to inhibit the neurohormonal axis that results in excessive cardiac activation through angiotensin- or norepinephrine-dependent pathways. For patients with end-stage heart failure, treatment options are extremely limited, with less than 3000 patients being offered cardiac transplants annually owing to the severely limited supply of donor organs ("Annual Report," 1990; Andres and Walsh, 1996), and to implantable left ventricular assist devices being expensive, not proven for long-term use, and associated with significant complications (Asahara et al., 1997, 1999; Anversa and Nadal-Ginard, 2002). Clearly, development of approaches that prevent heart failure after MI would be preferable to those that simply ameliorate or treat already established disease.

22.3. HEART FAILURE AFTER MYOCARDIAL INFARCTION RESULTS FROM PROGRESSIVE VENTRICULAR REMODELING

Heart failure after MI occurs as a result of a process termed *myocardial remodeling*. This process is characterized by myocyte apoptosis, cardiomyocyte replacement by fibrous tissue deposition in the ventricular wall (Atkins et al., 1999; Beauchamp et al., 1999; Beltrami et al., 2001), progressive expansion of the initial infarct area, and dilation of the left ventricular lumen (Colucci, 1997; Choi et al., 1998). Another integral component of the remodeling process appears to be the development of neoangiogenesis within the myocardial infarct scar ("Effects of Enalapil," 1987; Elefanty et al., 1997), a process requiring activation of latent collagenase and other proteinases (Field, 1998). Under normal circumstances, the contribution of neoangiogenesis to the infarct bed capillary network is insufficient to keep pace with the tissue growth required for contractile compensation and is unable to support the greater demands of the hypertrophied, but viable, myocardium. The relative lack of oxygen and nutrients to the hypertrophied myocytes may be an important etiological factor in the death of otherwise viable myocardium, resulting in progressive infarct extension and fibrous replacement. Because late reperfusion of the infarct vascular bed in both humans and experimental animal models significantly benefits ventricular remodeling and survival (Frazier et al., 1992; Halevy et al., 1995; Folkman, 1998), we have postulated that methods to successfully augment vascular bed neovascularization might improve cardiac function by preventing loss of hypertrophied, but otherwise viable, cardiac myocytes.

22.4. INABILITY OF DAMAGED MYOCARDIUM TO UNDERGO REPAIR DUE TO CELL-CYCLE ARREST OF ADULT CARDIOMYOCYTES

Cardiomyocytes undergo terminal differentiation soon after birth and are thought by most investigators to withdraw irreversibly from the cell cycle. Analysis of cardiac myocyte growth during early mammalian development indicates that cardiac myocyte DNA synthesis occurs primarily *in utero*, with proliferating cells decreasing from 33% at midgestation to 2% at birth (Hescheler et al., 1999). While ventricular karyokinesis and cytokinesis are coupled during fetal growth, resulting in increases in mononucleated cardiac myocytes, karyokinesis occurs in the absence of cytokinesis for a transient period during the postnatal period, resulting in binucleation of ventricular myocytes without an overall increase in cell number. A similar dissociation between karyokinesis and cytokinesis characterizes the primary adult mammalian cardiac response to ischemia, resulting in myocyte hypertrophy and an increase in nuclear ploidy rather than myocyte hyperplasia (Hill and Singal, 1997; Heymans et al., 1999). Moreover, in parallel with an inability to progress through cell cycle, ischemic adult cardiomyocytes undergo a high degree of apoptosis.

When cells proliferate, progression of the mitotic cycle is tightly regulated by an intricate network of positive and negative signals. Progress from one phase of the cell cycle to the next is controlled by the transduction of mitogenic signals to cyclically expressed proteins known as cyclins and subsequent activation or inactivation of several members of a conserved family of serine/threonine protein kinases known as the cyclin-dependent kinases (cdks) (Hochman and Choor, 1987). Growth arrest observed with such diverse processes as DNA damage, terminal differentiation, and replicative senescence is due to negative regulation of cell-cycle progression by two functionally distinct families of cdk inhibitors—the Ink4 and Cip/Kip families (Hescheler et al., 1999). The cell-cycle inhibitory activity of p21Cip1/WAF1 is intimately correlated with its nuclear localization and participation in quaternary complexes of cell-cycle regulators by binding to G1 cyclin–cdk through its N-terminal domain and to proliferating cell nuclear antigen (PCNA) through its C-terminal domain (Hognes, 1991; Jaffredo et al., 1998; Hodgetts et al., 2000; Jackson et al., 2001). The latter interaction blocks the ability of PCNA to activate DNA polymerase, the principal replicative DNA polymerase (Kajstura et al., 1998). For a growth-arrested cell to subsequently enter an apoptotic pathway, signals provided by specific apoptotic stimuli in concert with cell-cycle regulators are required. For example, caspase-mediated cleavage of p21, together with upregulation of cyclin A–associated cdk2 activity, has been shown to be critical for induction of cellular apoptosis by either deprivation of growth factors (Kalka et al., 2000) or hypoxia of cardiomyocytes (Kalkman et al., 1996).

Throughout life, a mixture of young and old cells is present in the normal myocardium. Although most myocytes seem to be terminally differentiated, there is a fraction of younger myocytes (15–20%) that retains the capacity to replicate (Kawana et al., 2000). Moreover, recent observations have suggested that some human ventricular cardiomyocytes also have the capacity to proliferate and regenerate in response to ischemic injury (Kellerman et al., 1992; Kehat et al., 2001). The dividing myocytes can be identified on the basis of immunohistochemical staining of proliferating nuclear structures such as Ki67 and cell-surface expression of specific surface markers, including c-kit (CD117). Whether these cells are derived from a resident pool of cardiomyocyte stem cells or are derived from a renewable source of circulating bone marrow–derived stem cells that home to the damaged myocardium remains to be determined. More important, the signals required for homing, *in situ* expansion, and differentiation of these cells are, at present, unknown. Gaining an understanding of these issues would open the possibility of manipulating the biology of endogenous cardiomyocytes in order to augment the healing process after myocardial ischemia.

22.5. STRATEGIES FOR USE OF CELLULAR THERAPY TO IMPROVE MYOCARDIAL FUNCTION

Replacement and regeneration of functional cardiac muscle after an ischemic insult to the heart could be achieved either by stimulating proliferation of endogenous mature cardiomyocytes or resident cardiac stem cells, or by implanting exogenous donor-derived or allogeneic cardiomyocytes. The newly formed cardiomyocytes must integrate precisely into the existing myocardial wall in order to augment contractile function of the residual myocardium in a synchronized manner and avoid alterations in the electrical conduction and syncytial contraction of the heart, potentially resulting in life-threatening consequences. In addition, whatever the source of the cells used, it is likely that concurrent myocardial revascularization must also occur in order to ensure viability of the repaired region and prevent further scar tissue formation. The following section discusses various methods of using cellular therapies to replace damaged myocardium or reinitiate mitosis in mature endogenous cardiomyocytes, including transplanted bone marrow–derived cardiomyocyte or endothelial precursors, fetal cardiomyocytes, and skeletal myoblasts.

22.6. POTENTIAL ROLE FOR BONE MARROW–DERIVED OR EMBRYONIC CARDIOMYOCYTE LINEAGE STEM CELLS IN MYOCARDIAL REPAIR/REGENERATION

Over the past several years, a number of studies have suggested that stem cells can be used to generate cardiomyocytes ex vivo for potential use in a range of cardiovascular diseases (Klug et al., 1996; Kennedy et al., 1997; Lebastie et al., 1998; Kocher et al., 2001, 2002). Multipotent bone marrow–derived mesenchymal stem cells (MSCs) have been identified in adult murine and human bone marrow functionally by their ability to differentiate to lineages of diverse mesenchymal tissues, including bone, cartilage, fat, tendon, and both skeletal and cardiac muscle (Kocher et al., 2002), and phenotypically by their expression of specific surface markers and lack of hematopoietic lineage markers such as CD34 or CD45 (Kocher et al., 2001). It is well established that murine embryonic stem (ES) cells can give rise to cardiomyocytes in vitro and in vivo (Li et al., 1994; Levkau et al., 1998). Recently, Kehat et al. (2001) were able to demonstrate that human ES cells can also differentiate in vitro into cells with characteristics of cardiomyocytes. However, there are striking differences in the human and murine stem cell models, and this needs to be taken into account when extrapolating results of mouse experiments to the human condition. For example, human ES cells have a very low efficiency of differentiation to cardiomyocytes compared with murine ES cells, and a considerably slower time course (a median of 11 vs 2 d). Whether these differences reflect true variations between species, or differences in the experimental protocols, remains to be determined.

Irrespective of whether the cardiomyocyte lineage stem cell precursors are obtained from adult bone marrow or embryonic sources, the newly generated cardiomyocytes appear to resemble normal cardiomyocytes in terms of phenotypic properties, such as expression of actinin, desmin, and troponin I; and function, including positive and negative chronotropic regulation of contractility by pharmacological agents and production of vasoactive factors such as atrial and brain natriuretic peptides. However, in vivo evidence for functional cardiac improvement following transplantation of adult bone marrow–derived or ES-derived cardiomyocytes has been exceedingly difficult to show to date. In part, this may be because the signals required for cardiomyocyte differentiation and functional regulation are complex and poorly understood. For example, phenotypic and functional differentiation of MSCs to cardiomyocyte lineage cells in vitro requires culture with exogenously added 5-azacytidine (Klug et al., 1996; Kennedy et al., 1997). Alternatively, the poor functional data obtained to date may reflect immune-mediated rejection of cells that have been modified during the ex vivo culture process or poor viability owing to the lack of a sufficient vascular supply to the engrafted cells (*see* below).

22.7. POTENTIAL ROLE FOR AUTOLOGOUS SKELETAL MYOBLASTS IN MYOCARDIAL REPAIR

An alternative approach to replacing damaged myocardium involves the use of autologous skeletal myoblasts (Liechty et al., 2000). The procedure involves harvesting a patient's skeletal muscle cells, expanding the cells in a laboratory, and reinjecting the cells into the patient's heart. Perceived advantages of the approach include ease of access to the cellular source, the fact that immunosuppression is not needed, and the lack of ethical dilemmas associated with the use of allogeneic or embryonic cells. It has also been argued that using relatively ischemia-resistant skeletal myoblasts rather than cardiomyocytes might enable higher levels of cell engraftment and survival in infarcted regions of the heart, where cardiomyocytes would probably perish (Lin et al., 2000).

Successful engraftment of autologous skeletal myoblasts into injured myocardium has been reported in multiple animal models of cardiac injury. These studies have demonstrated survival and engraftment of myoblasts into infarcted or necrotic hearts (Liechty et al., 2000), differentiation of the myoblasts into striated cells within the damaged myocardium (Liechty et al., 2000), and improved myocardial functional performance (Mahon et al., 1999; Liechty et al., 2000; MacLellan and Schneider, 2000). Other studies have shown that the survival of transplanted myoblasts can be improved by heat-shock pretreatment (Maier et al., 2001) and have confirmed that the benefits of skeletal myoblast transfer are additive with those of conventional therapies, such as angiotensin-converting enzyme inhibition (Makino et al., 1999). More recently, the procedure has been reported anecdotally to result in improved myocardial function in humans (McCarthy et al., 1991). On the basis of these preliminary results, clinical trials have begun in both Europe and in the United States. In addition to the need to demonstrate functional improvement in large, prospective series, questions remain to be addressed such as the following: Can skeletal myoblasts make meaningful electromechanical connections to the surrounding endogenous cardiomyocytes through gap junctions? Will the cellular mass contract in concert? Will electrical impulses be transmitted to the myoblast tissue without inducing significant tachyarrhythmias?

22.8. POOR SURVIVAL OF CELLS TRANSPLANTED INTO DAMAGED MYOCARDIUM AFTER EX VIVO CULTURE

A major limitation to successful cellular therapy in animal models of myocardial damage has been the inability of the introduced donor cells to survive in their host environment, whether such transplants have been congenic (analogous to the autologous scenario in humans) or allogeneic. It has become clear that a major impediment to survival of the implanted cells is the alteration of their immunogenic character by prolonged ex vivo culture conditions. For example, whereas myocardial implantation of skeletal muscle in the absence of tissue culture does not induce any adverse immune response and results in grafts showing excellent survival for up to a year, injection of cultured isolated (congenic) myoblasts results in a massive and rapid necrosis of donor myoblasts, with more than 90% dead within the first hour after injection (Murry et al., 1996; Nelissen-Vrancken et al., 1996; McEwan et al., 1998; Menasche et al., 2001). This rapid myoblast death appears to be mediated by host natural killer (NK) cells (Nelissen-Vrancken et al., 1996) that respond to immunogenic antigens on the transplanted myoblasts altered by exposure to tissue culture conditions (Menasche et al., 2001). It seems likely that a similar mechanism of host NK cell–mediated rejection will apply also to transplanted ES-derived, cultured cardiomyocytes (Nidorf et al., 1992) since massive death of injected donor cells is recognized as a major problem with transplanted cardiomyocytes, especially in the inflammatory conditions that follow infarction (Ogawa et al., 1999). In this regard, the report that cultured MSCs obtained from adult human bone marrow are not rejected on transfer to other species is intriguing (Kocher et al., 2002), needs confirmation in humans, and requires detailed investigation into possible tolerogenic mechanisms.

22.9. CONCOMITANT INDUCTION OF VASCULAR STRUCTURES AUGMENTS SURVIVAL AND FUNCTION OF CARDIOMYOCYTE PRECURSORS

An additional explanation for the poor survival of transplanted cardiomyocytes or skeletal myoblasts may be that viability and prolonged function of transplanted cells requires an augmented vascular supply. Whereas many transplanted cardiomyocytes die by apoptosis, cultured cardiomyocytes that incorporate more vascular structures in vivo demonstrate significantly greater survival (Ogawa et al., 1999). Moreover, in situations in which transplanted cardiomyocyte precursors contained an admixture of cells also giving rise to vascular structures, survival and function of the newly formed cardiomyocytes has been significantly augmented.

In one study, direct injection of whole rat bone marrow into a cryodamaged heart resulted in neovascularization, cardiomyocyte regeneration, and functional improvement (Klug et al., 1996). More recently, systemic delivery of highly purified bone marrow–derived hematopoietic stem cells in lethally irradiated mice contributed to the formation of both endothelial cells and long-lived cardiomyocytes in ischemic hearts (Orlic et al., 2001). Most strikingly, significant improvement in cardiac function of mice that had previously undergone left anterior descending (LAD) ligation was demonstrated after direct myocardial injection of syngeneic bone marrow–derived stem cells, defined on the basis of c-kit (CD117) expression (Oz et al., 1997). This population of cells contains a mixture of cellular elements in addition to cardiomyocyte precursors, including CD117-positive endothelial progenitors (*see* below). These cells were found to proliferate and differentiate into myocytes, smooth muscle cells, and endothelial cells, resulting in the partial regeneration of the destroyed myocardium and prevention of ventricular scarring. Together, these findings raise the intriguing possibility that for long-term in vivo viability and functional integrity of stem cell–derived cardiomyocytes, it may be necessary to induce neovascularization by coadministration of endothelial cell progenitors (*see* below).

22.10. ENDOTHELIAL PRECURSORS AND FORMATION OF VASCULAR STRUCTURES DURING EMBRYOGENESIS

To develop successful methods for inducing neovascularization of the adult heart, one needs to understand the process of definitive vascular network formation during embryogenesis. In the prenatal period, hemangioblasts derived from the human ventral aorta give rise to cellular elements involved in both vasculogenesis, or formation of the primitive capillary network, and hematopoiesis (J. M. Pfeffer et al., 1991; M. A. Pfeffer et al., 1992). In addition to hematopoietic lineage markers, embryonic hemangioblasts are characterized by expression of the vascular endothelial cell growth factor receptor-2 (VEGFR-2), and have high proliferative potential with blast colony formation in response to VEGF (Rafii et al., 1994; Pitt et al., 1997; Pittenger et al., 1999; Pouzet et al., 2001). Under the regulatory influence of various transcriptional and differentiation factors, embryonic hemangioblasts mature, migrate, and differentiate to become endothelial lining cells and create the primitive vasculogenic network. The differentiation of embryonic hemangioblasts to pluripotent stem cells and to endothelial precursors appears to be related to coexpression of the GATA-2 transcription factor, since GATA-2 knockout ES cells have a complete block in definitive hematopoiesis and seeding of the fetal liver and bone marrow (Ravichandran and Puvanakrishnan, 1991). Moreover, the earliest precursor of both hematopoietic and endothelial cell lineage to have diverged from embryonic ventral endothelium has been shown to express VEGFRs as well as GATA-2 and α4-integrins (Shi et al., 1998). Subsequent to capillary tube formation, the newly created vasculogenic vessels undergo sprouting, tapering, remodeling, and regression under the direction of VEGF, *angiopoietins*, and other factors, a process termed angiogenesis. The final component required for definitive vascular network formation to sustain embryonic organogenesis is influx of mesenchymal lineage cells to form the vascular supporting structures such as smooth muscle cells and pericytes.

22.11. CHARACTERIZATION OF ENDOTHELIAL PROGENITORS, OR ANGIOBLASTS, IN HUMAN ADULT BONE MARROW

In studies using various animal models of peripheral ischemia, several groups have shown the potential of adult bone marrow–derived elements to induce neovascularization of ischemic tissues (Steinman et al., 1994; Tavian et al., 1996; Soonpaa and Field, 1997; Takahashi et al., 1999; Smythe and Grounds, 2000; Suzuki et al., 2000; Smythe et al., 2001). In the most successful of these studies (Smythe and grounds, 2000), bone marrow–derived cells injected directly into the thighs of rats that had undergone ligation of the left femoral artery and vein induced neovascularization and augmented blood flow in the ischemic limb, as documented by laser Doppler and immunohistochemical analyses. Although the nature of the bone marrow–derived endothelial progenitors was not precisely identified in these studies, the cumulative reports indicated that this site may be an important source of endothelial progenitors that could be useful for augmenting collateral vessel growth in ischemic tissues, a process termed *therapeutic angiogenesis*.

In more recent studies, our group has identified such endothelial progenitors in human adult bone marrow (Taylor et al., 1998). By employing both in vitro and in vivo experimental models, we have sought to precisely identify the surface characteristics and biological properties of these bone marrow–derived endothelial progenitor cells. Following granulocyte colony-stimulating factor (G-CSF) treatment, mobilized mononuclear cells were harvested and $CD34^+$ cells were separated using anti-CD34 monoclonal antibody (MAb) coupled to magnetic beads. Ninety to ninety-five percent of $CD34^+$cells coexpressed the hematopoietic lineage marker CD45, 60–80% coexpressed the stem cell factor receptor CD117, and <1% coexpressed the monocyte/macrophage lineage marker CD14. By quadruple parameter analysis, the VEGFR-2-positive cells within the $CD34^+CD117^{bright}$ subset displayed phenotypic characteristics of endothelial progenitors, including coexpression of Tie-2, as well as AC133, but not markers of mature endothelium such as ecNOS, vWF, E-selectin, and intracellular adhesion molecule. Sorting $CD34^+$ cells on the basis of CD117 bright or dim expression demonstrated that GATA-2 mRNA and protein levels were significantly higher in the $CD117^{bright}$ population, indicating that human adult bone marrow contains cells with an angioblast-like phenotype.

Because the frequency of circulating endothelial cell precursors in animal models has been shown to be increased by either VEGF (Tomita et al., 1999) or regional ischemia (Steinman et al., 1994; Soonpaa and Field, 1997; Smythe and Grounds, 2000; Smythe et al., 2001), phenotypically defined angioblasts were examined for proliferative responses to VEGF and to factors in ischemic serum. $CD117^{bright}GATA\text{-}2^{hi}$ angioblasts demonstrated significantly higher proliferative responses relative to $CD117^{dim}GATA\text{-}2^{lo}$ bone marrow–derived cells from the same donor following culture for 96 h with either VEGF or ischemic serum. The expanded angioblast population consisted of large blast cells, defined by forward scatter, which continued to express immature markers, including GATA-2 and $CD117^{bright}$, but not markers of mature endothelial cells, including eNOS or E-selectin, indicating blast proliferation without differentiation under these culture conditions. However, culture on fibronectin with endothelial growth medium resulted in outgrowth of monolayers with endothelial morphology and functional and phenotypic features characteristic of endothelial cells, including uniform uptake of acetylated low-density lipoprotein, and coexpression of CD34, factor VIII, and eNOS. Thus, G-CSF treatment of adult humans mobilizes into the peripheral circulation a bone marrow–derived population with phenotypic and functional characteristics of embryonic angioblasts, as defined by specific surface phenotype, high proliferative responses to VEGF and cytokines in ischemic serum, and the ability to differentiate into endothelial cells by culture in medium enriched with endothelial growth factors.

22.12. HUMAN ANGIOBLASTS INDUCE NEOVASCULARIZATION OF MYOCARDIAL INFARCT ZONE

Intravenous injection of freshly obtained human angioblasts into athymic nude rats that had undergone ligation of the LAD coronary artery resulted in infarct zone infiltration within 48 h. Few human cells were detected in unaffected areas of hearts with regional infarcts or in myocardium of sham-operated rats. Histological examination at 2 wk postinfarction revealed that injection of human angioblasts was accompanied by a significant increase in infarct zone microvascularity, cellularity, and numbers of capillaries, and by reduction in matrix deposition and fibrosis in comparison to controls. Neovascularization was significantly increased both within the infarct zone and in the periinfarct rim in rats receiving angioblasts compared with controls receiving saline or other cells that infiltrated the heart (e.g., $CD34^-$ cells or saphenous vein endothelial cells [SVECs]). The neovascularization induced by human angioblasts was due to both an increase in capillaries of human origin as well as of rat origin, as defined by MAbs with specificity for human or rat CD31 endothelial markers. Capillaries of human origin, defined by coexpression of DiI fluorescence and human CD31, but not rat CD31, accounted for 20–25% of the total myocardial capillary vasculature, which was located exclusively within the central infarct zone of collagen deposition. By contrast, capillaries of rat origin, as determined by expression of rat, but not human, CD31, demonstrated a distinctively different pattern of localization, being absent within the central zone of collagen deposition and abundant both at the periinfarct rim between the region of collagen deposition and myocytes, and between myocytes.

22.13. HUMAN ANGIOBLASTS PROTECT HYPERTROPHIED ENDOGENOUS CARDIOMYOCYTES AGAINST APOPTOSIS

By concomitantly staining rat tissues for the myocyte-specific marker desmin and performing DNA end labeling using the TUNEL technique, temporal examinations demonstrated that the infarct zone neovascularization induced by injection of human angioblasts prevented an eccentrically extending proapoptotic process evident in saline controls. Thus, at 2 wk postinfarction, myocardial tissue of LAD-ligated rats that received saline demonstrated sixfold higher numbers of apoptotic myocytes compared with that from rats receiving iv injections of human angioblasts. Moreover, these myocytes had a distorted appearance and an irregular shape. By contrast, myocytes from LAD-ligated rats that received human angioblasts had a regular oval shape and were significantly larger than myocytes from control rats.

22.14. HUMAN ANGIOBLASTS INDUCE SUSTAINED REGENERATION/PROLIFERATION OF ENDOGENOUS CARDIOMYOCYTES

In addition to protection of hypertrophied myocytes against apoptosis, human angioblast-dependent neovascularization resulted in a striking induction of regeneration/proliferation of endogenous rat cardiomyocytes at the periinfarct rim (Tsai et al., 1994). At 2 wk after LAD ligation, rats receiving human angioblasts demonstrated numerous "fingers" of cardiomyocytes of rat origin, as determined by expression of rat major histocompatibility complex class I molecules, extending from the periinfarct region into the infarct zone. The islands of cardiomyocytes at the periinfarct rim in animals receiving human angioblasts contained a high frequency of rat myocytes with DNA activity, as determined by dual staining with MAbs reactive against cardiomyocyte-specific troponin I and rat Ki67. By contrast, in animals receiving saline, there was a high frequency of cells with fibroblast morphology and reactivity with rat Ki67, but not troponin I, within the infarct zone. The number of cardiomyocytes progressing through the cell cycle at the periinfarct region of rats receiving human angioblasts was 40-fold higher than that at sites distal to the infarct, 20-fold higher than that found in noninfarcted hearts, and fivefold higher than that at the periinfarct rim of animals receiving saline (Tsai et al., 1994).

22.15. NEOVASCULARIZATION OF ACUTELY ISCHEMIC MYOCARDIUM BY HUMAN ANGIOBLASTS PREVENTS VENTRICULAR REMODELING AND CAUSES SUSTAINED IMPROVEMENT IN CARDIAC FUNCTION

By 15 wk postinfarction, rats receiving human angioblasts demonstrated markedly smaller scar sizes together with increased mass of viable myocardium within the anterior free wall. Whereas collagen deposition and scar formation extended almost through the entire left ventricular wall thickness in controls, with aneurysmal dilatation and typical electrocardiogram abnormalities, the infarct scar extended only to 20–50% of the left ventricular wall thickness in rats receiving $CD34^+$ cells. Moreover, pathological collagen deposition in the noninfarct zone was markedly reduced in rats receiving $CD34^+$ cells. At 15 wk, the mean proportion of scar/normal left ventricular myocardium was 13% in rats receiving $CD34^+$ cells compared with 36–45% for each of the other groups studied (saline, $CD34^-$, SVEC).

Remarkably, by 2 wk after injection of human angioblasts, and in a parallel time frame with the observed neovascularization, left ventricular ejection fraction (LVEF) recovered by a mean of 22%. This effect was long-lived, with LVEF recovering by a mean of 34% at the end of follow-up, 15 wk postinjection. Neither $CD34^-$ cells nor SVECs demonstrated similar effects. At 15 wk postinfarction, the mean cardiac index in rats injected with human angioblasts was only reduced by 26% relative to normal rats, whereas for each of the other groups it was reduced by 48–59%. Together, these results indicate that the neovascularization, reduction in periinfarct myocyte apoptosis, and increase in cardiomyocyte regeneration/proliferation observed at 2 wk prevented myocardial replacement with fibrous tissue and caused sustained improvement in myocardial function.

22.16. POTENTIAL USE OF ANGIOBLASTS IN COMBINATION WITH CARDIOMYOCYTE PROGENITORS FOR REPAIR AND REGENERATION OF ISCHEMIC MYOCARDIUM

While increasing capillary density through angioblast-dependent neovascularization is a promising approach for preventing apoptotic death and inducing regeneration of endogenous cardiomyocytes following acute MI, the role of angioblast therapy for the treatment of CHF following chronic ischemia is at present unknown. Nevertheless, it is reasonable to anticipate that cellular therapies for CHF owing to ischemic cardiomyopathy will need to address two interdependent processes: (1) a renewable source of proliferating, functional cardiomyocytes; and (2) development of a network of capillaries and larger-size blood vessels for supply of oxygen and nutrients to both the chronically ischemic, endogenous myocardium and the newly implanted cardiomyocytes. To achieve these end points, it is likely that coadministration of angioblasts and mesenchymal stem cells will be needed in order to develop regenerating cardiomyocytes, vascular structures, and supporting cells such as pericytes and smooth muscle cells. Future studies will need to address the timing, relative concentrations, source and route of delivery of each of these cellular populations in animal models of acute and chronic myocardial ischemia.

In addition to synergistic cellular therapies, it is likely that optimal regimens for the treatment of acute and chronically ischemic hearts will require a combined approach employing additional pharmacological strategies. For example, augmentation in myocardial function might be achieved by combining infusion of human angioblasts and cardiomyocyte progenitors together with β blockade, angiotensin-converting-enzyme inhibition, or AT1-receptor blockade to reduce angiotensin II–dependent cardiac fibroblast proliferation and collagen secretion (White et al., 1987, 1994; Tsurimoto, 1999; Zhang et al., 2001). Understanding the potential of defined lineages of stem cells or undifferentiated progenitors, and their interactions with pharmacological interventions, will lead to better and more focused clinical trial designs using each cell type independently or in combination, depending on which particular clinical indication is being targeted.

REFERENCES

Adachi, S,, Ito, H.,m Tamamori-Adochi, M., et al. (2001) Cyclin A/cdk2 activation is involved in hypoxia-induced apoptosis in cardiomyocytes. *Circ. Res.* 88:408–4140.

Agocha, A., Lee, H. W., and Eghali-Webb, M. (1997) Hypoxia regulates basal and induced DNA synthesis and collagen type I production in human cardiac fibroblasts: effects of TGF-beta, thyroid hormone, angiotensis II and basic fibroblast growth factor. *J. Mol. Cell. Cardiol.* 29:2233H–2244H.

Andres, V. and Walsh, K. (1996) Myogenin expression, cell cycle withdrawal and phenotypic differentiation are temporally separable events that precedes cell fusion upon myogenesis. *J. Cell Biol.* 132:657–666.

US Department of Health and Human Services (1990) *Annual Report of the US Scientific Registry for Organ Transplantation and the Organ Procurement and Transplantation Network*. Washingtion, DC.

Anversa, P. and Nadal-Ginard, B. (2002) Myocyte renewal and ventricular remodeling. *Nature* 415:240–243.

Asahara, T., Murohara, T., Sullivan, A., et al. (1997) Isolation of putative progenitor cells for endothelial angiogenesis. *Science* 275: 964–967.

Asahara, T., Takahashi, T., Masuda, H., et al. (1999) VEGF contributes to postnatal neovascularization by mobilizing bone marrow-derived endothelial progenitor cells. *EMBO J.* 18:3964–3972.

Atkins, B. Z, Hueman, M. T., Meuchel, J. M., Cottman, M. J., Hutcheson, K. A., Taylor, D. A. (1999) Myogenic cell transplantation improves in vivo regional performance in infarcted rabbit myocardium. *J. Heart Lung Transplant.* 18:1173–1180.

Beauchamp, J. R., Morgan, J. E., Pagel, C. N., and Partridge, T. A. (1999) Dynamics of myoblast transplantation reveal a discrete minority of precursors with stem cell-like properties as the myogenic source. *J. Cell. Biol.* 144:1113–1122.

Beltrami, A. P., Urbarek, K., Kajstura, J., et al. Evidence that human cardiac myocytes divide after myocardial infarction. *N. Engl. J. Med.* 344:1750–1757 .

Choi, K., Kennedy, M., Kazarov, A., Papadimitriou, J. C., and Keller, G. (1998) A common precursor for hematopoietic and endothelial cells. *Development* 125:725–732.

Colucci, W. S. (1997) Molecular and cellular mechanisms of myocardial failure. *Am. J. Cardiol.* 80:15L–25L.

Effects of enalapril on mortality in severe congestive heart failure: results of the Cooperative North Scandinavian Enalapril Survival Study (CONSENSUS) Trial Study Group. (1987) *N. Engl. J. Med.* 316(23):1429–1435.

Elefanty, A. G., Robb, L., Birner, R., and Begley, C. G. (1997) Hematopoietic-specific genes are not induced during in vitro differentiation of scl-null embryonic stem cells. *Blood* 90:1435–1447.

Field, L. (1998) Future therapy for cardiovascular disease. In: *Proceedings of the NHLBI Workshop Cell Transplantation: Future Therapy for Cardiovascular Disease?* Columbia, MD.

Folkman, J. (1998) Therapeutic angiogenesis in ischemic limbs. *Circulation* 97:108–110.

Frazier, O. H., Rose, E. A., Macmanus, Q., et al. (1992) Multicenter clinical evaluation of the Heartmate 1000 IP left ventricular assist device. *Ann. Thorac. Surg.* 102:578–587.

Halevy, O., Novitch, B. G., Spicer, D. B., et al. (1995) Correlation of terminal cell cycle arrest of skeletal muscle with induction of p21 by MyoD. *Science* 267:1018–1021.

Hescheler, J., Fleischmann, B. K., Wartenberg, M., et al. (1999) Establishment of ionic channels and signalling cascades in the embryonic stem cell-derived primitive endoderm and cardiovascular system. *Cells Tissues Organs* 165:153–164.

Heymans, S., Luutun, A., Nuyens, D., et al. (1999) Inhibition of plasminogen activators or matrix metalloproteinases prevents cardiac rupture but impairs therapeutic angiogenesis and causes cardiac failure. *Nat. Med.* 5:1135–1142.

Hill, M. F. and Singal, P. K. (1997) Right and left myocardial antioxidant responses during heart failure subsequent to myocardial infarction. *Circulation* 96:2414–2420.

Hochman, J. S. and Choo, H. (1987) Limitation of myocardial infarct expansion by reperfusion independent of myocardial salvage. *Circulation* 75:299–306.

Hodgetts, S. I., Beilharz, M. W., Scalzo, T., and Grounds, M. D. (2000) Why do cultured transplanted myoblasts die in vivo? DNA quantification shows enhanced survival of donor male myoblasts in host mice depleted of CD4+ and CD8+ or NK1.1+ cells. *Cell Transplant.* 9: 489–502.

Hognes, J. R. (1991) *The Artificial Heart: Prototypes, Policies and Patients*, National Academy Press, Washingtion, DC.

Jackson, K. A., Majka, S. M., Wang, H., et al. (2001) Regeneration of ischemic cardiac muscle and vascular endothelium by adult stem cells. *J. Clin. Invest.* 107:1395–1402.

Jaffredo, T., Gautier, R., Eichmann, A., and Dieterlen-Lievre, F. (1998) Intraaortic hemopoietic cells are derived from endothelial cells during ontogeny. *Development* 125:4575–4583.

Kajstura, J., Leri, A., Finato, N., di Loreto, N., Beltramo, C. A., and Anversa, P. (1998) Myocyte proliferation in end-stage cardiac failure in humans. *Proc. Natl. Acad. Sci. USA* 95:8801–8805.

Kalka, C., Masuda, H., Takahashi, T., et al. (2000) Transplantation of ex vivo expanded endothelial progenitor cells for therapeutic neovascularization. *Proc. Natl. Acad. Sci. USA* 97:3422–3427.

Kalkman, E. A. J., Bilgin, Y. M., van Haren, P., van Suylen, R.-J., Saxena, P. R., and Schoemaker, R. G. (1996) Determinants of coronary reserve in rats subjected to coronary artery ligation or aortic banding. *Cardiovasc. Res.* 32:1088–1095.

Kawano, H., Do, Y. S., Kawano, Y., et al. (2000) Angiotensin II has multiple profibrotic effects in human cardiac fibroblasts. *Circulation* 101:1130–1137.

Kehat, I., Kenyagin-Karsenti, D., Snir, M., et al. (2001) Human embryonic stem cells can differentiate into myocytes with structural and functional properties of cardiomyocytes. *J. Clin. Invest.* 108: 407–414.

Kellerman, S., Moore, J. A., Zierhut, W., Zimmer, H. G., Campbell, J., and Gerdes, A. M. (1992) Nuclear DNA content and nucleation patterns in rat cardiac myocytes from different models of cardiac hypertrophy. *J. Mol. Cell. Cardiol.* 24:497–505.

Kennedy, M., Firpo, M., Choi, K., et al. (1997) A common precursor for primitive erythropoiesis and definitive haematopoiesis. *Nature* 386:488–493.

Klug, M. G., Soonpaa, M. H., Koh, G. Y., and Field, L. J. (1996) Genetically selected cardiomyocytes from differentiating embryonic stem cells form stable intracardiac grafts. *J. Clin. Invest.* 98:216–224

Kocher, A. A., Schuster, M. D., Szabolcs, M. J., et al. (2001) Neovascularization of ischemic myocardium by human bone-marrow-derived angioblasts prevents cardiomyocyte apoptosis, reduces remodeling and improves cardiac function. *Nat. Med.* 7:430–436.

Kocher, A., Schuster, M., Szabolcs, M., and Itescu, S. (2002) Cardiomyocyte regeneration after neovascularization of ischemic myocardium by human bone marrow-derived angioblasts. *Nat. Med.*, in press.

Labastie, M.-C., Cortes, F., Romeo, P.-H., Dulac, C., and Peault, B. (1998) Molecular identity of hematopoietic precursor cells emerging in the human embryo. *Blood* 92:3624–3635.

Levkau, B., Koyama, H., Raines, E. W., et al. (1998) Cleavage of p21cip1/waf1 and p27 kip1 mediates apoptosis in endothelial cells through activation of cdk2: role of a caspase cascade. *Mol. Cell.* 1:553–563.

Li, Y., Jenkins, C. W., Nichols, M. A., and Xiong, Y. (1994) Cell cycle expression and p53 regulation of the cyclin-dependent kinase inhibitor p21. *Oncogene* 9:2261–2268.

Liechty, K. W., MacKenzie, T. C., Shaaban, A. F., et al. (2000) Human mesenchymal stem cells engraft and demonstrate site-specific differentiation after in utero transplantation in sheep. *Nat. Med.* 6: 1282–1286.

Lin, Y., Weisdorf, D. J., Solovey, A., and Hebbel, R. P. (2000) Origins of circulating endothelial cells and endothelial outgrowth from blood. *J. Clin. Invest.* 105:71–77.

MacLellan, W. R. and Schneider, M. D. (2000) Genetic dissection of cardiac growth control pathways *Annu. Rev. Physiol.* 62:289–320.

Mahon, N. G., O'Roke, C., Codd, M. B., et al. (1999) Hospital mortality of acute myocardial infarction in the thrombolytic era. *Heart* 81: 478–82.

Maier, S., Tertilt, C., Chambron, N., et al. (2001) Inhibition of natural killer cells results in acceptance of cardiac allografts in CD28–/– mice. *Nat. Med.* 7:557–562.

Makino, S., Fukuda, K., Miyoshi, S., et al. (1999) Cardiomyocytes can be generated from marrow stromal cells in vitro. *J. Clin. Invest.* 103: 697–705.

McCarthy, P. M., Rose, E. A., Macmanus, Q., et al. (1991) Clinical experience with the Novacor ventricular assist system. *J. Thorac. Cardiovasc. Surg.* 102:578–587.

McEwan, P. E., Gray, G. A., Sherry, L., Webb, D. J., and Kenyon, C. J. (1998) Differential effects of angiotensin II on cardiac cell proliferation and intramyocardial perivascular fibrosis in vivo. *Circulation* 98: 2765–2773.

Menasche, P., Hagege, A. A., Scorsin, M., et al. (2001) Myoblast transplantation for heart failure. *Lancet* 347:279–280.

Murry, C. E., Wiseman, R. W., Schwartz, S. M., and Hauschka, S. D. (1996) Skeletal myoblast transplantation for repair of myocardial necrosis. *J. Clin. Invest.* 98:2512–2523.

Nelissen-Vrancken, H., Debets, J., Snoeckx, L., Daemen, M., and Smits, J. (1996) Time-related normalization of maximal coronary flow in isolated perfused hearts of rats with myocardial infarction. *Circulation* 93:349–355.

Nidorf, S. M., Siu, S. C., Galambos, G., Weyman, A. E., and Picard, M. H. (1992) Benefit of late coronary reperfusion on ventricular morphology and function after myocardial infarction. *J. Am. Coll. Cardiol.* 20:307–313.

Ogawa, M., Kizumoto, M., Nishikawa, S., Fujimoto, T., Kodama, H., and Nishikawa, S. I. (1999) *Blood* 93:1168–1177.

Orlic, D., Kajstura, J., Chimenti, S., et al. (2001) Bone marrow cells regenerate infarcted myocardium. *Nature* 410:701–705.

Oz, M.C., Argenziano, M., Catanese, K. A., et al. (1997) Bridge experience with long-term implantable left ventricular assist devices. Are they an alternative to transplantation? *Circulation* 95:1844–1852.

Pfeffe,r J. M., Pfeffer, M. A., Fletcher, P. J., and Braunwald, E. (1991) Progressive ventricular remodeling in rat with myocardial infarction. *Am. J. Physiol.* 260:H1406–H1414.

Pfeffer, M. A., Braunwald, E., Moye, L. A., et al.(1992) Effect of captopril on mortality and morbidity in patients with left ventricular dysfunction after myocardial infarction. Results of the survival and ventricular enlargement trial. The SAVE investigators. *N. Engl. J. Med.* 327:669–677.

Pitt, B, Segal, R., Martinez, F. A., et al. (1997) Randomised trial of losartan versus captopril in patients over 65 with heart failure (Evaluation of Losartan in the Elderly Study, ELITE). *Lancet* 349:747–752.

Pittenger, M. F., Mackay, A. M., Beck, S. C., et al. (1999) Multilineage potential of adult human mesenchymal stem cells. *Science* 284: 143–147.

Pouzet, B., Ghostine, S., Vilquin, J.-T., et al. (2001) Is skeletal myoblast transplantation clinically relevant in the era of angiotensin-converting enzyme inhibitors? *Circulation* 104(Suppl 1):I223–228.

Rafii, S., Shapiro, F., Rimarachin, J., et al. (1994) Isolation and characterization of human bone marrow microvascular endothelial cells: hematopoietic progenitor cell adhesion. *Blood* 84:10–19.

Ravichandran, L. V. and Puvanakrishnan, R. (1991) In vivo labeling studies on the biosynthesis and degradation of collagen in experimental myocardial myocardial infarction. *Biochem. Intl.* 24:405–414.

Shi, Q., Rafii, S., Wu, M. H.-D., et al. (1998) Evidence for circulating bone marrow-derived endothelial cells. *Blood* 92:362–367.

Smythe, G. M. and Grounds, M. D. (2000) Exposure to tissue culture conditions can adversely affect myoblast behaviour in vivo in whole muscle grafts: implications for myoblast transfer therapy. *Cell Transplant.* 9:379–393.

Smythe, G. M., Hodgetts, S. I., and Grounds, M. D. (2001) Problems and solutions in myoblast transfer therapy. *J. Cell. Mol. Med.* 5:33–47.

Soonpaa, M. H. and Field, L. J. (1997) Assessment of cardiomyocyte DNA synthesis in normal and injured adult mouse hearts. *Am. J. Physiol.* 272:H220–H226.

Steinman, R. A., Hoffman, B., Iro, A., Guillouf, C., Liebermann, D. A., and El-Houseini, M. E. (1994) Induction of p21 (WAF1/CIP1) during differentiation. *Oncogene* 9:3389–3396.

Suzuki, K., Smolenski, R. T., Jayakumar, J., Murtuza, B., Brand, N. J., and Yacoub, M. H. (2000) Heat shock treatment enhances graft cell survival in skeletal myoblast transplantation to the heart. *Circulation* 102(Suppl. 19):III216–221.

Takahashi, T., Kalka, C., Masuda, H., et al. (1999) Ischemia- and cytokine-induced mobilization of bone marrow-derived endothelial progenitor cells for neovascularization. *Nat. Med.* 5:434–438.

Tavian, M., Coulombel, L., Luton, D., San Clemente, H., Dieterlen-Lievre, F., and Peault, B. (1996) Aorta-associated CD34 hematopoietic cells in the early human embryo. *Blood* 87:67–72.

Taylor, D. A., Atkins, B. Z., Hungspreugs, P., et al. (1998) Regenerating functional myocardium: improved performance after skeletal myoblast transplantation. *Nat. Med.* 4:929–933.

Tomita, S., Li, R.-K., Weisel, R. D., et al. (1999) Autologous transplantation of bone marrow cells improves damaged heart function. *Circulation* 100(Suppl. 19):II247–256.

Tsai, F. Y., Keller, G., Kuo, F. C., et al. (1994) An early hematopoietic defect in mice lacking the transcription factor GATA-2. *Nature* 371:221–225.

Tsurimoto, T. (1999) PCNA Binding Proteins. *Frontiers Biosci.*4:849–858.

White, H. D., Norris, R. M., Brown, M. A., Brandt PWT, Whitlock, R. M. L., and Wild, C. J. (1987) Left ventricular end systolic volume as the major determinant of survival after recovery from myocardial infarction. *Circulation* 76:44–51.

White, H. D., Cross, D. B., Elliot, J. M., et al. (1994) Long-term prognostic importance of patency of the infarct-related coronary artery after thrombolytic therapy for myocardial infarction. *Circulation* 89:61–67.

Zhang, M., Methot, D., Poppa, V., Fujio, Y., Walsh, K., and Murry, C. E. (2001) Cardiomyocyte grafting for cardiac repair: graft cell death and anti-death strategies. *J. Mol. Cell. Cardiol.* 33:907–921.

23 Generation and Stem Cell Repair of Cardiac Tissue

KATHYJO A. JACKSON, PhD AND MARGARET A. GOODELL, PhD

Cellular-based therapies may be a future strategy for treatment of cardiac injury. Full differentiation of the heart leaves an organ without a stem cell population to respond to injury. This can be addressed by providing cardiomyocyte precursors from embryonic or adult stem cells. The coronary vasculature derives from both vasculogenesis from the putative hemangioblast and angiogenesis by sprouting from newly formed vessels from the proepicardium, which is formed from villi-like protrusions of the fetal liver. Interruption of the vascular supply leads to infarction of myocardium and death of myocardiocytes. Normally there is little or no regeneration of this tissue, which heals by the process of inflammation and scarring. Although some myocardiocytes may proliferate after injury, these represent <0.01% and are insufficient to mediate any useful replacement. The endothelial cells of the coronary arteries are able to proliferate in response to the inflammation following an infarct and establish a collateral circulation to support the remaining heart and scar tissue. Cell transplantation experiments have explored the use of cardiomyocytes derived from embryonic stem cells or fetal hearts for cardiac repair with mixed success. Adult muscle stem cells (satellite cells) may populate ischemic heart tissue and potentially improve function, but they do not establish electromechanical coupling with cardiac myocytes. Mesenchymal stem cells (MSCs) from the bone marrow differentiate into multiple cell types in vitro including myocardial-like cells, which may contract in vitro. These cells can also be grafted onto ischemic heart tissue, but their degree of functional integration is not yet clear. Bone marrow–derived cells can be found in small numbers in recipient hearts after transient cardiac ischemia, either after bone marrow replacement of depleted mice or following direct injection into areas adjacent to the injured myocardium. Cytokines, which liberate autogenous bone marrow stem cells, may improve the outcome of experimental cardiac injury. Both bone marrow and MSCs may contribute to revascularization after injury. In selected models, transplanted stem cells may replace or restore lost gene function in the heart, and bone marrow progeny have been identified in the hearts of humans with heart transplants. Adult stem cells may be expected to play an increasing role in the treatment of many forms of heart disease, but extensive validation and development research still needs to be undertaken.

23.1. INTRODUCTION

In Western societies, heart disease is associated with a high rate of morbidity and mortality. In 1999, heart disease was the number one cause of death in the United States (Hoyert et al., 2001). The high rate of mortality associated with diseases of the heart is due to the inability of this organ to repair following injury. New approaches for the treatment of heart disease are necessary to overcome the lack of regenerative potential of this organ. Cellular-based therapies employing stem cells could be the next strategy for the clinical treatment of heart disease.

An understanding of the origin of cardiomyocytes during development is essential for comprehending the decreased regenerative capacity of the heart compared with other organs. Terminal differentiation of cardiomyocyte precursors leaves the heart without a stem cell population, resulting in only fully differentiated cells with a limited proliferation potential.

To this end, much work has been done to attempt to identify cell populations that could be used as donor cells in heart injury models. Fetal and adult cardiomyocytes as well as embryonic stem (ES) cells have been tested for their ability to engraft into injured myocardium. Cellular therapies using adult stem cells are receiving an increasing amount of attention because of the potential for autologous transplants, thereby preventing graft rejection.

Adult stem cells from multiple tissues have been explored for repair of myocardial cells and vascular endothelium. Studies have shown that in animal models, grafting of stem cells into infarcted hearts may improve cardiac function (Orlic et al., 2001a, 2001b). Recent evidence suggests that stem cells may also participate in heart regeneration and remodeling in human hearts (Quaini et al., 2002).

From: *Stem Cells Handbook*
Edited by: S. Sell © Humana Press Inc., Totowa, NJ

23.2. DEVELOPMENT, GROWTH, AND REPAIR OF THE HEART

23.2.1. DEVELOPMENTAL ORIGIN OF THE HEART

23.2.1.1. Heart The heart is the first functional organ in the developing vertebrate embryo. Formation of this organ requires complex interactions from multiple cell types including lateral plate mesoderm and neural crest. Lateral plate mesoderm is induced to differentiate into cardiogenic mesenchyme by factors produced by the anterior endoderm (Sugi and Lough, 1994; Schultheiss et al., 1995). Cardiogenic mesenchyme gives rise to the first specified cardiogenic lineages: myocardium and endocardium. Cells of the myocardium are precursors for cardiomyocytes, which terminally differentiate. Although the myocardium is mainly thought of as a contractile tissue, it also plays an important role in the conduction system, which allows the heart to maintain a coordinated wave of contraction (Farrell and Kirby, 2001).

This developmental program of myocardium, which gives rise to terminally differentiated cells, results in a decreased regenerative capacity for cardiac tissue during adult life. For example, during skeletal muscle development, myoblasts fuse to form myotubes, which mature into myofibers (Schultz, 1989). A population of muscle stem cells—satellite cells—adheres to the myotubes and resides under the basal lamina of the mature myofiber. Following skeletal muscle injury, satellite cells become activated, proliferate, and participate in tissue regeneration. By contrast, cardiomyocyte precursors are thought to terminally differentiate during development, resulting in the absence of a stem cell pool in the adult heart.

The endocardium of the fetal heart acts as a source of mesenchymal cells that form the endocardial cushions. The endocardial cushions function as primitive valves in the embryonic heart. This tissue will ultimately give rise to the atrioventricular valves and the membranous portion of the interventricular septum of the adult heart (Markwald et al., 1977).

Extracardiac sources of cells also participate in the development of the four-chambered heart. The neural crest lies at the dorsal aspect of the neural tube. Neural crest cells migrate to the primitive heart and contribute mesenchymal cells to the septum of the outflow tract, neuronal cells to the cardiac ganglia, and smooth muscle cells to the great arteries (Kirby et al., 1983). In addition, cells derived from cardiac neural crest also participate in the remodeling of the cardiovascular tissues, maturation of the excitation–contraction coupling apparatus of the myocardium, and muscularization of the outflow tract septum (Farrell and Kirby, 2001).

23.2.1.2. Vasculature Formation of blood vessels during embryonic development takes place via two processes: vasculogenesis and angiogenesis. Vasculogenesis results in the *de novo* formation of blood vessels via the putative hemangioblast, an intraembryonic stem cell population that gives rise to both endothelial precursors and hematopoietic precursors. By contrast, angiogenesis involves the sprouting of new blood vessels from preexisting vessels. During cardiac development, both vasculogenesis and angiogenesis play a role in the formation of the coronary vasculature.

The coronary vasculature arises from an extracardiac primordium called the proepicardium. The proepicardium forms as villi-like protrusions of the primordial liver (Viragh et al., 1993). Epicardial cells from these protrusions proliferate and migrate cranially to envelop the entire heart from the sinus venosus to the outflow tract. The proepicardial-derived cells then undergo an epithelial–mesenchymal transformation that gives rise to the three coronary vessel–associated cell populations: coronary smooth muscle, endothelial cells, and perivascular connective tissue (Mikawa and Gourdie, 1996; Dettman et al., 1998).

Development of the coronary vasculature begins when the sinus venosus plexus extends branches into the proepicardium toward the dorsal side of the atrioventricular sulcus (Farrell and Kirby, 2001). Vessels then grow toward the ventral side and apex of the heart. After the coronary vessels penetrate the aorta, a tunica media of smooth muscle cells is established (Waldo et al., 1994). In addition, fibroblasts form an adventia around the coronary arteries. Finally, the coronary vasculature undergoes remodeling to form the final network of vessels, which supply the heart with oxygen and nutrients.

23.2.2. POSTNATAL GROWTH OF THE HEART

23.2.2.1. Heart After birth, the heart continues to grow proportional to the growth of the body. This growth occurs either by an increase in the number of cardiomyocytes (hyperplasia) or by an increase in the size of preexisting cardiomyocytes (hypertrophy). During the first postnatal months of human life, growth of the heart occurs via hyperplasia, with the total number of cardiomyocytes doubling in the first few months (Rakusan, 1984). Thereafter, the numbers of cardiomyocytes remain relatively constant. Hyperplasia plays an even more significant role in growth of the rat heart postnatally. In rats, cardiomyocyte numbers double within 11 d of birth and quadruple over a 2yr period (Anversa et al., 1980; Sasaki et al., 1968). Hypertrophy of cardiomyocytes also occurs during postnatal development, resulting in an overall increase in heart size (Anversa et al., 1980; Rakusan, 1984).

External stimuli are also capable of increasing the size of the adult heart. High-altitude hypoxia induces enlargement of the right ventricle by both hyperplasia and hypertrophy of cardiomyocytes (Bartels et al., 1979; Pietschmann and Bartels, 1985). This hypoxia-induced increase in heart size was found to be greater in young animals than in adult animals. In addition, exercise training results in cardiac growth by hypertrophy of cardiomyocytes, with the extent of heart growth varying depending on the type of exercise (Hudlika and Brown, 1996).

23.2.2.2. Vasculature With the rapid amount of cardiac growth occurring postnatally, it is natural that the supplying blood vessels must increase as well. This increase in the coronary vasculature occurs mainly at the level of the capillaries by the process of angiogenesis. Although capillary numbers increase, the diameter of capillaries decreases, resulting in a greater area for oxygen transport to cardiomyocytes (Van Groningen et al., 1991). Capillary growth is not accompanied by a proportional increase in the number of smaller arterioles (Rakusan et al., 1994). In rats, there is a decrease in the length of smaller arterioles accompanied by a decrease in arteriole numbers with aging (Wiest et al., 1992; Vitullo et al., 1993; Rakusan et al., 1994). In humans, large coronary arteries increase in diameter and elongate postnatally, but their numbers are fixed at birth (Rakusan et al., 1992; Hudlika and Brown, 1996).

23.2.3. REPAIR OF THE HEART FOLLOWING MYOCARDIAL INFARCTION

23.2.3.1. Heart Coronary artery occlusion and cardiomyocyte hypertrophy result in decreased blood flow and oxygen levels in the myocardium, producing an ischemic region in the heart.

Severe ischemia results in myocardial infarction (MI), a focal loss of cardiomyocytes by necrosis. Repair of the injured region begins quickly in order to repair the damaged myocardium and maintain structural integrity of the ventricle (Sun and Weber, 2000). Although inflammation can increase myocardial injury, an inflammatory response is a prerequisite for cardiac healing and scar formation.

The complement cascade is activated by ischemic myocardial injury in a rat model of MI (Hill and Ward, 1971). Cardiomyocyte necrosis results in the release of mitochondria and mitochondrial membrane fragments that trigger the early acting components of the complement cascade (Pinckard et al., 1975; Rossen et al., 1994). Activation of the complement system may play an important role in mediating the recruitment of neutrophils and monocytes to the injured myocardium (Frangogiannis et al., 2002).

Infiltration of neutrophils into reperfused myocardial infarcts may be responsible for a significant amount of cardiomyocyte injury. Studies in which neutrophils were depleted in animals undergoing MI/reperfusion injury led to a decrease in the infarct size (Romson et al., 1983). Neutrophils participate in cardiac injury by infiltrating the ischemic region, adhering to the capillary endothelium, and plugging capillaries, preventing reperfusion of the infarcted area (Engler et al., 1986; Engler, 1989). In a canine model of ischemia/reperfusion injury, interleukin-8 (IL-8) induced neutrophils to adhere to cardiomyocytes, resulting in direct cytotoxicity (Kukielka et al., 1995). In addition, neutrophils may also directly injure parenchymal cells through the release of toxic products (Jaeschke and Smith, 1997).

Evidence suggesting that the inflammation process could extend myocardial injury led to a clinical study using methylprednisolone, a corticosteroid, with catastrophic results. Patients with acute MI given this drug had increased incidence of ventricular arrhythmias and extended infarct size (Roberts et al., 1976). Later investigations demonstrated that this drug inhibited the inflammatory process by decreasing the number of infiltrating leukocytes and delaying healing and collagen deposition (Kloner et al., 1978). Further studies demonstrated that an inflammatory response is vital for tissue repair.

In the first few hours following reperfusion, monocytes infiltrate the infarcted myocardium, where they eventually differentiate into macrophages. Macrophages may function in scar healing by providing cytokines and growth factors as well as regulating extracellular matrix metabolism (Ganz, 1993; Weihrauch et al., 1995; Frangogiannis et al., 2000). Mast cells also participate in scar healing in damaged myocardium. Mast cell numbers increase in canine myocardial infarcts during the healing phase (Frangogiannis et al., 1998). Furthermore, degranulation products of mast cells induce fibroblast proliferation (Ruoss et al., 1991).

Myofibroblasts appear at the infarct site and participate in fibrogenesis at sites where rebuilding and remodeling are occurring (Sun and Weber, 1996). Myofibroblasts are localized to the heart valve leaflets of normal cardiac tissue but appear at the infarct site following MI. These cells appear to arise from interstitial fibroblasts or adventitial fibroblasts (Skalli et al., 1989; Sappino et al., 1990). In a rat model of MI, myofibroblasts appear at the infarct site by d 3 and remain for months (Sun and Weber, 1996, 2000). Myofibroblasts appear in the infarct of humans 4–6 days postinfarction and have been found to persist in the infarct scar for years (Willems et al., 1994). These cells express α–smooth muscle actin and develop ultrastructural and phenotypic characteristics of smooth muscle cells (Willems et al., 1994). In addition, myofibroblasts express the embryonal isoform of smooth muscle myosin heavy chain (MHC), which may reflect their dedifferentiation and phenotypic plasticity (Richard et al., 1995). These cells are responsible for increased production of type I/III fibrillar collagens that function in tissue fibrosis (Cleutjens et al., 1995; Sun et al., 1998).

It has long been thought that cardiomyocytes do not proliferate following MI to replace necrotic cardiac tissue. Recent research has shown that human cardiomyocytes may replicate following injury. Beltrami et al. (2001) have found Ki-67-positive cardiomyocytes in the hearts of patients with MI, as well as microtubules in the mitotic spindles indicative of dividing cells. Although these researchers show that cardiomyocytes may divide following injury, the percentage of cells that undergo division is probably not enough to result in sufficient repair of the damaged tissue. In mouse heart following injury, <0.01% of cardiomyocytes undergo DNA synthesis, which is lower than the amount needed for improvement of damaged hearts (Soonpaa and Field, 1997).

23.2.3.2. Vasculature The endothelial and smooth muscle cells of normal coronary arteries are mitotically inactive (Schaper and Ito, 1996). However, during ischemia, hypoxia, or inflammation, these cells begin to migrate and divide to form collateral vessels in order to supply oxygen and nutrients to the affected region. Angiogenesis is the predominant mechanism for increasing collateral vessels in the ischemic heart. Early studies demonstrated that collateral vessels are better developed in hearts with a longer duration of ischemic symptoms and/or with more severe coronary arterial narrowing (Sasayama and Fujita, 1992).

Inflammation may be the major factor regulating angiogenesis, rather than ischemia (Schaper and Ito, 1996). Experiments have shown that the presence of macrophages and neutrophils is sufficient for angiogenesis to occur, but ischemia is not required (More and Sholley, 1985; Sunderkotter et al., 1991). MI results in the infiltration of inflammatory cells into the infarcted region including macrophages, mast cells, and neutrophils. Inflammatory cells release cytokines and growth factors including transforming growth factor-β and IL-6, which increase expression of vascular endothelial growth factor (VEGF), thereby stimulating angiogenesis (More and Sholley, 1985; Sunderkotter et al., 1991). The importance of VEGF expression in myocardial angiogenesis was demonstrated in a mouse model in which two isoforms of VEGF, $VEGF_{164}$ and $VEGF_{188}$, were knocked out (Carmeliet et al., 1999). These animals exhibited a decreased capillary density resulting in myocardial ischemia as well as atrophy and degeneration of cardiomyocytes.

23.3. EXPERIMENTAL REPAIR OF DAMAGED HEART BY ADULT STEM CELLS

With the high morbidity and mortality associated with cardiac ischemia, researchers have been searching for a cell source to replace dying cardiomyocytes for many years. Much work has been done in animal models using adult cardiomyocytes with moderate success (Leor et al., 1996). Even more promising is the use of fetal stem cells including ES cells, fetal cardiomyocytes, and fetal skeletal myoblasts (Leor et al., 1996; El Oakley et al., 2001; Min et al., 2002). In the following sections, the discussion is limited to experimental models using a defined population of adult stem cells.

23.3.1. HEART

23.3.1.1. Skeletal Muscle Stem Cells (Satellite Cells) Satellite cells are skeletal muscle stem cells, which reside under the basal lamina of muscle fibers in the adult. On skeletal muscle injury, these cells become activated, proliferate and fuse to repair the damaged myofiber. With the inability of cardiomyocytes to regenerate following injury, satellite cells have become widely used as a stem cell source in cardiac regeneration models. Studies have shown that satellite cells implanted in the infarcted region differentiate into striated muscle (Chiu et al., 1995; Yoon et al., 1995; Atkins et al., 1999). Reports indicate that new muscles within the infarcted region appear as both multinucleated myotubes (Murry et al., 1996; Atkins et al., 1999) and cells with centrally located nuclei (Chiu et al., 1995; Taylor et al., 1998). These newly formed striated muscles showed evidence of intercalated disks (Chiu et al., 1995; Taylor et al., 1998).

Although satellite cells do form new muscle in the infarcted region, gap junctions must be formed with preexisting cells in order for communication to occur. In vitro studies indicate that myotubes from rat satellite cells form gap junctions when cocultured with neonatal or adult cardiomyocytes (Reinecke et al., 2000). When cocultured, approx 10% of skeletal myotubes contracted in synchrony with adjacent cardiomyocytes. N-Cadherin and connexin 43 were expressed at the junctions between myotubes and cardiomyocytes, indicating the formation of electromechanical junctions.

Recent work seems to suggest that satellite cells do not transdifferentiate into cardiomyocytes in vivo. Reinecke et al. (2002) found that rat satellite cell grafts into the heart formed multinucleated, cross-striated myofibers that expressed fast skeletal MHC, a marker of mature fast-twitch skeletal muscle. Satellite cell grafts did not express α-MHC, cardiac troponin I, nor atrial natriuretic peptide, markers of differentiated cardiomyocytes. Furthermore, myofibrils of satellite cell grafts express prominent Z-bands but do not express other sarcomeric repeats such as A- and M-bands (Taylor et al., 1998). Although gap junctions were formed in vitro, grafts expressed neither N-cadherin nor connexin 43, suggesting that electromechanical coupling does not occur in vivo (Reinecke et al., 2002).

Transdifferentiation of satellite cells into cardiomyocytes may not occur in vivo, but this does not necessarily mean that they are not useful for clinical therapies. It still remains to be determined whether satellite cells have the potential to differentiate into modified slow-twitch skeletal muscle that can develop resistance to fatigue. Murry et al. (1996) found that satellite cell grafts 7 wk postinjection express β-MHC, a hallmark of slow-twitch muscle fibers. In addition, they found that wounds containing satellite cell grafts contract when stimulated ex vivo (Murry et al., 1996). Furthermore, these grafts could alternate between tetanus and relaxation simulating a cardiac like cycle suggesting that they may be useful for cardiac regeneration. Further evidence supporting this theory comes from in vivo experiments in which satellite cell injections improved myocardial performance in a rabbit model of myocardial injury (Taylor et al., 1998; Hutcheson et al., 2000).

Delivery of cells into the heart for regenerative purposes should be considered when planning experiments. In most studies, cells are directly injected into the infarcted region of the heart, resulting in a concentration of cells localized to the injection site with cell numbers decreasing as distance from the injection site increases (Kessler and Byrne, 1999). Infusion of satellite cells into the left coronary circulation results in discrete loci of cells found in the epicardium, myocardium, and endocardium of the left ventricle (Robinson et al., 1996; Taylor et al., 1997). Dispersal of cells throughout the heart might be advantageous for gene therapy strategies in which local delivery of the gene of interest may not be enough to ameliorate the symptoms of the disease.

23.3.1.2. Mesenchymal Stem Cells Mesenchymal stem cells (MSCs) reside in the bone marrow and provide supporting stromal cells that support hematopoiesis. Human MSCs (hMSCs) differentiate into multiple cell types including adipocytes, osteocytes, chondrocytes, and smooth muscle cells (Pittenger et al., 1999; Deans and Moseley, 2000). This differentiation potential of MSCs have led researchers to examine these cells' ability to differentiate into myocardium.

Fukuda (2001) demonstrated that immortalized mouse stromal cells when treated with 5-azacytidine would spontaneously beat in culture. Furthermore, these cells expressed atrial natriuretic peptide and brain natriuretic peptide along with a cardiomyocyte-like ultrastructure including typical sarcomeres and atrial granules. In vivo experiments with rat MSCs demonstrated that these cells have growth potential in the myocardial environment and express sarcomeric MHC and organized contractile proteins (Wang et al., 2000, 2001). Gap junctions were detected by positive staining for connexin 43 (Wang et al., 2000).

hMSCs have been shown to differentiate into cardiomyocytes when transplanted into multiple animal models including mice and sheep (Liechty et al., 2000; Toma et al., 2002). In mice, MSCs engrafted and began to hypertrophy over time (Toma et al., 2002). These cells stained positive for desmin; α-actinin; β-MHC; cardiac troponin T' and phospholamban, a phosphoprotein that plays a role in regulating the cardiac sarcoplasmic reticulum Ca^{2+}–adenosine triphosphatase (ATPase). *In utero* transplants into sheep were performed using hsMSC and resulted in human-derived cardiomyocytes (Liechty et al., 2000). Furthermore, in this xenogeneic system, hMSCs were able to evade the immune system and survive for as long as 13 mo posttransplant.

23.3.1.3. Hematopoietic Stem Cells One of the most exciting discoveries in the stem cell field is the finding that hematopoietic stem cells (HSCs) may be able to participate in regeneration of the ischemic heart. This phenomenon has been demonstrated by two groups using HSCs isolated by different methods.

In the first study, lethally irradiated mice were transplanted with highly enriched HSCs, side population cells, from Rosa 26 mice, which express the *lacZ* gene ubiquitously (Goodell et al., 2001; Jackson et al., 2001). The side population cells ($CD34^-$/low, c-Kit^+, Sca-1^+) of murine bone marrow have previously been shown to give rise to all hematopoietic lineages in the mouse (Goodell et al., 1996, 1997). Following stable hematopoietic engraftment, recipient hearts were rendered ischemic by coronary artery occlusion for 60 min followed by reperfusion. Two or four weeks after injury, *lacZ*-positive cardiomyocytes were found primarily within the periinfarct region of the heart and stained positive for α-actinin, a marker of cardiomyocytes. Approximately 0.02% of cardiomyocytes was derived from HSCs or their progeny. This study did not determine whether the newly formed cardiomyocytes formed gap junctions with surrounding cells nor whether they were functional.

Orlic et al., (2001a) also provided evidence for the engraftment of cells from a population enriched for HSCs into injured myocardium. In this study, MI was induced, and 3–5 hours postinfarction

Lin$^-$, c-Kit$^+$ hematopoietic cells expressing green fluorescent protein (GFP) were injected into the myocardium bordering the infarct. Nine days after surgery, ventricular function was measured with animals treated with HSCs having a left ventricular end-diastolic pressure 36% lower than that of controls. These animals also had GFP-positive myocytes that stained positive for cardiac markers, myocyte enhancer factor, GATA-4, and Csx/Nkx2.5, as well as connexin 43, demonstrating electrical coupling with surrounding cells.

Orlic et al. (2001b) have also shown that administration of growth factors capable of recruiting HSCs into the circulation followed by experimentally induced ischemia resulted in superior cardiac function relative to control mice. They hypothesize that this is due to recruitment of HSCs from the circulation to repair injured tissue. The study showed that treatment with stem cell factor and granulocyte colony-stimulating factor may improve the outcome of ischemic heart injury by decreasing mortality and infarct size. In addition, these investigators reported new cardiomyocytes in the injured hearts. Although it is known that these growth factors mobilize HSCs, the experiments did not show whether the newly formed cardiomyocytes were actually derived from HSCs. Because there are multiple tissue-specific stem cell populations in an organism, it is possible that other stem cell types may also be mobilized by growth factor treatment and participate in cardiac regeneration. Alternatively, the results may be due to some unrelated aspect of growth factor treatment, such as higher hematocrit, which may have a protective effect against ischemia.

23.3.2. VASCULATURE Increases in vasculature structures following ischemia were thought to take place by sprouting of preexisting vessels. In light of new information demonstrating that stem cells are present in the adult and participate in the formation of new vessels proves that vasculogenesis is not restricted to embryogenesis.

Studies have shown that both MSCs and HSCs participate in neovascularization in the ischemic heart. Following coronary arterial delivery of MSCs after MI, capillaries of MSC origin were found within and outside the infarcted scar area (Wang et al., 2001). In addition, HSCs gave rise to vascular endothelial (Jackson et al., 2001; Orlic et al., 2001a) and smooth muscle cells (Orlic et al., 2001a) following MI.

More extensive work has been done examining the role of bone marrow–derived endothelial progenitors in neovascularization of ischemic tissues (Asahara et al., 1999). Endothelial progenitor cells have been isolated from peripheral blood and bone marrow based on their expression of CD34 (Masuda et al., 2000) Recent work has shown that IV injection of human CD34$^+$ endothelial precursor cells into rats after left anterior descending artery ligation resulted in an increase in infarct zone microvascularity (Kocher et al., 2001). Neoangiogenesis was increased in the periinfarct region with increased survival of hypertrophied cardiomyocytes and improved myocardial function (Kocher et al., 2001).

In humans, circulating CD34$^+$ endothelial progenitor cells increased following acute MI (Shintani et al., 2001). This increase in circulating progenitors coincided with an increase in the plasma levels of VEGF. Increased VEGF mRNA levels have also been detected in ventricular biopsies from patients undergoing coronary bypass surgery, suggesting the importance of this growth factor in ischemic injury (Lee et al., 2000). Studies have shown that administration of VEGF in vivo induced mobilization of endothelial precursor cells into the peripheral blood (Asahara et al., 1999)

23.4. GENE THERAPY

The use of stem cells is a potential method of delivering genes of interest into the heart for therapeutic purposes. In the rat, heart-specific expression of the β-galactosidase gene has been shown demonstrating the feasibility of this approach for clinical therapies (Griscelli et al., 1998).

23.4.1. HEART One potential application for gene therapy is preventing ischemic myocardial damage by protecting tissue from repeated ischemia/reperfusion injury. To this end, Melo et al. (2002) delivered the cytoprotective *human heme oxygenase-1* gene (*hHO-1*) into rat myocardium using a recombinant adenoassociated virus (rAAV). After MI, animals expressing *hHO-1* experienced a >75% reduction in left ventricular MI accompanied by a decrease in myocardial lipid peroxidation. Levels of Bax and IL-1β, proapoptotic and proinflammatory proteins, respectively, decreased while the antiapoptotic Bcl-2 protein increased.

Gene therapy can also be used as a strategy to replace defective genes. TO-2 hamsters exhibit congenital dilated cardiomyopathy as a result of defects in the δ-sarcoglycan (δ-SG) gene similar to that seen in humans. An rAAV vector was used to introduce the δ-SG gene under a cytomegalovirus promoter in TO-2 hamsters (Kawada et al., 2002). Expression of the transgene preserved sarcolemmal permeability, normalized myocardial contractility and hemodynamics, and increased the survival period of hamsters.

Failing cardiomyocytes have abnormal calcium handling, negative force-frequency relationship, and decreased sarcoplasmic reticulum Ca^{2+} ATPase activity. In vitro adenoviral gene transfer of antisense phospholamban into human cardiomyocytes from patients with end-stage heart failure resulted in a decrease in phospholamban expression over 48 hours (del Monte et al., 2002). The decrease in phospholamban expression increased the velocity of contraction and relaxation to normal levels. Frequency response returned to normal, resulting in enhanced sarcoplasmic reticulum Ca^{2+} release and contraction.

23.4.2. VASCULATURE Gene therapy using angiogenic growth factors for coronary artery disease is likely to be a viable therapy in the near future. Early studies in animals and clinical trials in patients suggest that gene therapy with VEGF can promote neovascularization in ischemic tissue (Kalka et al., 2000a; Isner, 2002). Intramuscular injection of naked plasmid DNA encoding the VEGF gene resulted in increased plasma levels of VEGF and increased numbers of circulating endothelial precursor cells (Kalka et al., 2000a). Patients receiving intramyocardial injections of the VEGF-165 gene had increased VEGF levels with increased mobilization of circulating endothelial precursor cells (Kalka et al., 2000b). This increase in endothelial precursor cells continued for 9 wk after treatment followed by a subsequent decrease.

23.5. HUMAN STUDIES

Recent work by Quaini et al. (2002) suggests that adult stem cells may take part in the recovery of human hearts. Y-chromosome expression was detected in female donor hearts when transplanted into male recipients suffering from heart failure. The male chromosome was found in myocytes (9 ± 4%), arterioles

(10 ± 3%), and capillaries (7 ± 1%) at similar percentages. The highest levels of chimerism were found in hearts 4 and 28 days after transplantation, with chimerism decreasing as days posttransplantation increased. This raises the question, Will the Y-chromosome-positive cells take part in heart function long term or are they just a transient population of cells?

23.6. CONCLUSION

Despite the substantial advances made in the treatment of cardiac ischemia over the years, heart disease continues to be the number one cause of death in the United States. Much more work is needed to overcome the mortality associated with this condition. Adult stem cells provide a potential therapy for the future treatment of many forms of heart disease, but considerable developmental work must occur prior to the development of therapies applicable to the clinic.

REFERENCES

Anversa, P., Olivetti, G., and Loud, A. V. (1980) Morphometric study of early postnatal development in the left and right ventricular myocardium of the rat. I. Hypertrophy, hyperplasia, and binucleation ofmyocytes. *Circ. Res.* 46:495–502.

Asahara, T., Masuda, H., Takahashi, T., et al. (1999) Bone marrow origin of endothelial progenitor cells responsible for postnatal vasculogenesis in physiological and pathological neovascularization. *Circ. Res.* 85:221–228.

Atkins, B.Z., Lewis, C.W., Kraus, W.E., Hutcheson, K. A., Glower, D. D., and Taylor, D. A. (1999) Intracardiac transplantation of skeletal myoblasts yields two populations of striated cells in situ. *Ann. Thorac. Surg.* 67:124–129.

Bartels, H. Bartels, S., Rathschlag-Schaefer, A.M., Robbel, H., and Ludders, S. (1979) Acclimatization of newborn rats and guinea pigs to 3000 to 5000m simulated altitudes. *Respir. Physiol.* 36:375–389.

Beltrami, A.P., Urbanek, K., Kajstura, J., et al. (2001) Evidence that human cardiac myocytes divide after myocardial infarction. *N. Engl. J. Med.* 344:1750–1757.

Carmeliet, P., Ng, Y. S., Nuyens, D., et al. (1999) Impaired myocardial angiogenesis and ischemic cardiomyopathy in mice lacking the vascular endothelial growth factor isoforms $VEGf_{164}$ and $VEGF_{188}$. *Nat. Med.* 5:495–502.

Chiu, R. C. J., Zibaitis, A., and Kao, R. L. (1995) Cellular cardiomyoplasty: myocardial regeneration with satellite cell implantation. *Ann. Thorac. Surg.* 60:12–18.

Cleutjens, J. P. M., Verluyten, M. J. A., Smits, J. F. M., and Daemen, M. J. A. P. (1995) Collagen remodeling after myocardial infarction in the rat heart. *Am. J. Pathol.* 147:325–338.

Deans R. J. and Moseley, A.B. (2000) Mesenchymal stem cells: biology and potential clinical uses. *Exp. Hematolo.* 28:875–884.

del Monte, F., Harding, S. E., Dec, G. W., Gwathmey, J. K., and Hajjar, R. J. (2002) Targeting phospholamban by gene transfer in human heart failure. *Circulation* 105:904–907.

Dettman, R.W., Denetclaw, W., Jr., Ordahl, C. P., and Bristow, J. (1998) Common origin of coronary vascular smooth muscle, perivascular fibroblasts, and intermyocardial fibroblasts in the avian heart. *Dev. Biol.* 193:169–181.

El Oakley, R. M., Ooi, O. C., Bongso, A., and Yacoub, M. H. (2001) Myocyte transplantation for myocardial repair: a few good cells can mend a broken heart. *Ann. Thorac. Surg.* 71:1724–1733.

Engler, R. L. (1989) Free radical and granulocyte-mediated injury during myocardial ischemia and reperfusion. *Am. J. Cardiol.* 63: 19E–23E.

Engler, R. L., Dahlgren, M. D., Morris, D. D., Peterson, M. A., and Schmidt-Schoenbein, G. W. (1986) Role of leukocytes in response to acute myocardial ischemia and reflow in dogs. *Am. J. Physiol.* 251: H314–H323.

Farrell, M. J. and Kirby, M. L. (2001) Cell biology of cardiac development. *Int. Rev. Cytol.* 202:99–158.

Frangogiannis, N. G., Perrard, J L., Mendoza, L., et al. (1998) Stem cell factor induction is associated with mast cell accumulation after canine myocardial ischemia and reperfusion. *Circulation* 98:687–698.

Frangogiannis, N. G., Mendoza, L. H., Lindsey, M. L., et al. (2000) IL-10 is induced in the reperfused myocardium and may modulate the reaction to injury. *J. Immunol.* 165:2798–2808.

Frangogiannis, N. G., Smith, C. W., and Entman, M. L. (2002) The inflammatory response in myocardial infarction. *Cardiovasc. Res.* 53:31–47.

Fukuda, K. (2001) Development of regenerative cardiomyocytes from mesenchymal stem cells for cardiovascular tissue engineering. *Artif. Organ* 25:187–193.

Ganz, T. (1993) Macrophage function. *N. Horiz.* 1:23–27.

Goodell, M. A., Brose, K., Paradis, G., Conner, A., and Mulligan, R. (1996) Isolation and functional properties of murine hematopoietic stem cells that are replication in vivo. *J. Exp. Med.* 183:1797–1806.

Goodell, M. A., Rosenzweig, M., Kim, H., et al. (1997) Dye efflux studies suggest the existence of CD34-negative/low hematopoietic stem cells in multiple species. *Nat. Med.* 3:1337–1345.

Goodell, M.A., Jackson, K.A., Majka, S.M., et al. (2001) Stem cell plasticity in muscle and bone marrow. *Ann. NY Acad. Sci.* 938:208–220.

Griscelli, F., Gilardi-Hebenstreit, P., Hanania, N., et al. (1998) Heart-specific targeting of beta-galactosidase by the ventricle-specific cardiac myosin light chain 2 promoter using adenovirus vectors. *Hum. Gene Ther.* 9:1919–1928.

Hill, J. H. and Ward, P. A. (1971) The phlogistic role of C3 leukotactic fragment in myocardial infarcts of rats. *J. Exp. Med.* 133:885–890.

Hoyert, D. L., Arias, E., Smith, B. L., Murphy, S. L., and Kochanek, K. D. (2001) Deaths: final data for 1999. *Natl. Vital Stat. Rep.* 49:1–113.

Hudlika, O. and Brown, M. D. (1996) Postnatal growth of the heart and its blood vessels. *J. Vasc. Res.* 33:266–287.

Hutcheson, K. A., Atkins, B. Z., Hueman, M. T., Hopkins, M. B., Glower, D. D., and Taylor, D. A. (2000) Comparison of benefits on myocardial performance of cellular cardiomyoplasty with skeletal myoblasts and fibroblasts. *Cell Transplant* 9:359–368.

Isner, J. M. (2002) Myocardial gene therapy. *Nature* 415:234–239.

Jackson, K. A., Majka, S. M., Wang, H., et al. (2001) Regeneration of ischemic cardiac muscle and vascular endothelium by adult stem cells. *J. Clin. Invest.* 107:1395–1402.

Jaeschke, H. and Smith C. W. (1997) Mechanisms of neutrophil-induced parenchymal cell injury. *J. Leukoc. Biol.* 61:647–653.

Kalka, C., Masuda, H., Takahashi, T., et al. (2000a) Vascular endothelial growth factor(165) gene transfer augment circulating endothelial progenitor cells in human subjects. *Circ. Res.* 86:1198–1202.

Kalka, C., Tehrani, H., Laudenberg, B., et al. (2000b) VEGF gene transfer mobilizes endothelial progeitor cells in patients with inoperable coronary disease. *Ann. Thorac. Surg.* 70:829–834.

Kawada, T., Nakazawa, M., Nakauchi, S., et al. (2002) Rescue of hereditary form of dilated cardiomyopathy by rAAV-mediated somatic gene therapy: amelioration of morphological findings, sarcolemmal permeability, cardiac performances, and the prognosis of TO-2 hamsters. *PNAS* 99:901–906.

Kessler, P. D. and Byrne, B. J. (1999) Myoblast cell grafting into heart muscle: cellular biology and potential applications. *Annu. Rev. Physiol.* 61:219–242.

Kirby, M. L., Gale, T. F., and Stewart, D. E. (1983) Neural crest cells contribute to normal aorticopulmonary septation. *Science* 220: 1059–1061.

Kloner R. A., Fishbein, M.C ., Lew, H., Maroko, P. R., and Braunwald, E. (1978) Mummification of the infarcted myocardium by high dose corticosteroids. *Circulation* 57:56–63.

Kocher, A. A., Schuster, M. D., Szabolcs, M. J., et al. (2001) Neovascularization of ischemic myocardium by human bone-marrow-derived angioblasts prevents cardiomyocytes apoptosis, reduces remodeling and improves cardiac function. *Nat .Med.* 7:430–436.

Kukielka, G. L., Smith, C. W., LaRosa, G. J., et al. (1995) IL-8 gene induction in the myocardium after ischemia nad reperfusion in vivo. *J Clin. Invest.* 95:89–103.

Lee, S. H., Wolf, P. L., Escudero, R., Deutsch, R., Jamieson, S. W., and Thistlethwaite, P. A. (2000) Early expression of angiogenesis factors in acute myocardial ischemia and infarction. *N. Engl. J. Med.* 342:626–633.

Leor, J., Patterson, M., Quinones, M. J., Kedes, L. H., and Kloner, R. A. (1996) *Circulation* 94:II332–II336.

Liechty, K. W., MacKenzie, T. C., Shaaban, A. F., et al. (2000) Human mesenchymal stem cells engraft and demonstrate site-specific differentiation after *in utero* transplantation in sheep. *Nat. Med.* 6:1282–1286.

Markwald, R. R., Fitzharris, T. P., and Manasek, F. J. (1977) Structural development of endocardial cushions. *Am. J. Anat.* 148:85–119.

Masuda, H., Kalka, C., and Asahara, T. (2000) Endothelial progenitor cells for regeneration. *Hum. Cell* 13:153–160.

Melo, L. G., Agrawal, R., Zhang, L., et al. (2002) Gene therapy strategy for long-term myocardial protection using adeno-associated virus-mediated delivery of heme oxygenase gene. *Circulation* 105:602–607.

Mikawa, T. and Gourdie, R. G. (1996) Pericardial mesoderm generates a population of coronary smooth muscle cells migrating into the heart along with ingrowth of the epicardial organ. *Dev. Biol.* 174:221–232.

Min, J.Y., Yang, Y., Converso, K.L., et al. (2002) Transplantation of embryonic stem cells improves cardiac function in postinfarcted rats. *J. Appl. Physiol.* 92:288–296.

More, J. W. I. and Sholley, M. M. (1985) Comparison of the neovascular effects of stimulated macrophages and neutrophils in autologous rabbit corneas. *Am. J. Pathol.* 120:87–98.

Murry, C. E., Wiseman, R. W., Schwartz, S. M., and Hauschka, S. D. (1996) Skeletal myoblast transplantation for repair of myocardial necrosis. *J. Clin. Invest.* 98:2512–2523.

Orlic, D., Kajstura, J., Chimenti, S., et al. (2001a) Bone marrow cells regenerate infarcted myocardium. *Nature* 410:701–705.

Orlic, D., Kajstura, J., Chimenti, S., et al. (2001b) Mobilized bone marrow cells repair the infarcted heart, improving function and survival. *PNAS* 98:10344–10349.

Pietschmann, M. and Bartels, H. (1985) Cellular hyperplasia and hypertrophy, capillary proliferation and myoglobin concentration n the heart of newborn and adult rats at high altitude. *Respir. Physiol.* 59: 347–360.

Pinckard, R. N., Olson, M. S., Giclas, P. C., Terry, R., Boyer, J. T., and O'Rourke, R. A. (1975) Consumption of classical complement components by heart subcellular membranes in vitro and in patients after acute myocardial infarction. *J. Clin. Invest.* 56: 740–750.

Pittenger, M. F., Mackay, A. M., Beck, S. C., et al. (1999) Multilineage potential of adult human mesenchymal stem cells. *Science* 284: 143–147.

Quaini, F., Urbanek, K., Beltrami, A. P., et al. (2002) Chimerism of the transplanted heart. *N. Engl. J. Med.* 346:5–15.

Rakusan, K. (1984) Cardiac growth, maturation, and ageing. In: *Growth of the Heart in Health and Disease* (Zak, R., ed.), Raven, New York, pp. 131–164.

Rakusan, K., Flanagan, M. F., Geva, T., Southern, J., and Vanpraagh, R. (1992) Morphometry of human coronary capillaries during normal growth and the effect of age in left ventricular pressure-overload hypertrophy. *Circulation* 86:38–46.

Rakusan, K., Cicutti, M., and Flanagan, M. F. (1994) Changes in the microvascular network during cardiac growth, development, and aging. *Cell. Mol. Biol. Res.* 40:117–122.

Reinecke, H., MacDonald, G. H., Hauschka, S. D., and Murry, C. E. (2000) Electromechanical coupling between skeletal and cardiac muscle: implications for infarct repair. *J. Cell. Biol.* 149:731–740.

Reinecke, H., Poppa, V., and Murry, C. E. (2002) Skeletal muscle stem cells do not transdifferentiate into cardiomyocytes after cardiac grafting. *J. Mol. Cell. Cardiol.* 34:241–249.

Richard, V., Murry, C. E., and Reimer, K. A. (1995) Healing of myocardial infarcts in dogs: effects of late reperfusion. *Circulation* 92:1891–1901.

Roberts, R., DeMello, V., and Sobel, B. E. (1976) Deleterious effects of methylprednisolone in patients with myocardial infarction. *Circulation* 53:I204–I206.

Robinson, S. W., Cho, P. W., and Levitsky, H. I. (1996) Arterial delivery of genetically labeled skeletal myoblasts to the murine heart: long-term survival and phenotypic modification of implanted myoblasts. *Cell Transplant* 5:77–91.

Romson, J. L., Hook, B. G., Kunkel, S. L., Abrams, G. D., Schork, M. A., and Lucchesi, B. R. (1983) Reduction of the extent of ischemic myocardial injury by neutrophil depletion in the dog. *Circulation* 67:1016–1023.

Rossen, R. D., Michael, L. H., Hawkins, H. K., et al. (1994) Cardiolipin-protein complexes and initiation of complement activation after coronary artery occlusion. *Circ. Res.* 75:546–555.

Ruoss, S. J., Hartmann, T., and Caughey, G. H. (1991) Mast cell tryptase is a mitogen for cultured fibroblasts. *J. Clin. Invest.* 88:493–499.

Sapino, A. P., Schurch, W., Gabbiani, G. (1990) Differentiation repertoire of fibroblastic cells: expression of cytoskeletal proteins as marker of phenotypic modulations. *Lab. Invest.* 63: 144–161.

Sasaki, R., Watanabe, Y., Morishita, T., and Yamagata, S. (1968) Estimation of the cell number of heart muscles in normal rats. *Tohoku J. Exp. Med.* 95:177–184.

Sasayama, S. and Fujita, M. (1992) Recent insights into coronary collateral circulation. *Circulation* 85:1197–1204.

Schaper, W. and Ito, W. D. (1996) Molecular mechanisms of coronary collateral vessel growth. *Circ. Res.* 79:911–919.

Schulthesis, T. M., Xydas, S., and Lassar, A. B. (1995) Induction of avian cardiogenic myogenesis by anterior endoderm. *Development* 121: 4203–4214.

Schultz, E. (1989) Satellite cell behavior during skeletal muscle growth and regeneration. *Med. Sci. Sports Exerc.* 21:S181–S186.

Shintani, S., Murohara, T., Ikeda, H., et al. (2001) Mobilization of endothelial progenitor cells in patients with acute myocardial infarction. *Circulation* 103:2776–2779.

Skalli, O., Schurch, W., Seemayer, T., et al. (1989) Myofibroblasts from diverse pathologic settings are heterogeneous in their content of actin isoforms and intermediate filament proteins. *Lab. Invest.* 60:275–285.

Sugi, X. and Lough, J. (1994) Anterior endoderm is a specific effector of terminal cardiac myocyte differentiation of cells from embryonic heart forming region. *Dev. Dyn.* 200:155–162.

Soonpaa, M. H. and Field, L. J. (1997) Assessment of cardiomyocyte DNA synthesis in normal and injured adult mouse hearts. *Am. J. Physiol.* 272:H220–H226.

Sun, Y., and Weber, K.T. (1996) Angiotensin converting enzyme and myofibroblasts during tissue repair in the rat heart. *J. Mol. Cell. Cardiol.* 28:851-858.

Sun, Y. and Weber, K. T. (2000) Infarct scar: a dynamic tissue. *Cardiovasc. Res.* 46:250–256.

Sun, Y., Zhang, J. Q., Zhang, J., and Ramires, F. J. A. (1998) Angiotensin II transforming growth factor-β1 and repair in the infarcted heart. *J. Mol. Cell. Cardiol.* 30:1559–1569.

Sunderkotter, C., Goebeler, M., Schulze-Osthoff, K., Bhardwaj, R., and Sorg, C. (1991) Macrophage-derived angiogenesis factors. *Pharmacol. Ther.* 51:195–216.

Taylor, D. A., Silvestry, S. C., Bishop, S. P., et al. (1997) Delivery of primary autologous skeletal myoblasts into rabbit heart by coronary infusion; a potential approach to myocardial repair. *PNAS* 109: 245–253.

Taylor, D. A., Atkins, B. Z., Hungspreugs, P., et al. (1998) Regenerating functional myocardium: improved performance after skeletal myoblast transplantation. *Nat. Med.*4:929–933.

Toma, C., Pittenger, M. F., Cahill, K. S., Byrne, B. J., and Kessler, P. D. (2002) Human mesenchymal stem cells differentiate to a cardiomyocytes phenotype in the adult murine heart. *Circulation* 105:93–98.

Van Gronigen, J. P., Wenink, A. C., and Testers, L. H. (1991) Myocardial capillaries: Increase in number by splitting of existing vessels. *Anat. Embryol.* 184:65–70.

Viragh, S. C., Gittenberger-deGroot, R. E., Poelmann, F., and Kalman, F. (1993) Early development of quail heart epicardium and associated vascular and glandular structures. *Anat. Embryol.* 188: 381–393.

Vitullo, J. C., Penn, M. S., Rakusan, K., and Wicker, P. (1993) Effects of hypertension and aging on coronary arteriolar density. *Hypertension* 21:406–414.

Waldo, K. L., Kumiski, D. H., and Kirby M. L. (1994) Association of the cardiac neural crest with development of the coronary arteries in the chick embryo. *Anat. Rec.* 239:315–331.

Wang, J. S., Shum-Tim, D., Galipeau, J., Cjedrawy, E., Eliopoulos, N., and Chiu, R. C. (2000) Marrow stromal cells for cellular cardiomyoplasty: feasibility and potential clinical advantages. *J. Thorac. Cardiovasc. Surg.* 120:999–1005.

Wang, J. S., Shum-Tum, D., Chedrawy, E., and Chiu, R. C. (2001) The coronary delivery of marrow stromal cells for myocardial regeneration: pathophysiologic and therapeutic implications. *J. Thorac. Cardiovasc. Surg.* 122:699–705.

Weihrauch, D., Arras, M., Zimmerman, R., and Schaper, J. (1995) Importance of monocytes/macrophages and fibroblasts for healing of micronecroses in porcine myocardium. *Mol. Cell. Biochem.* 147: 13–19.

Wiest, G., Gharehbaghi, H., Ammann, K., Simon, T., Mattfeldt, T., and Mall, G. (1992) Physiological growth of arteries in the rat heart parallels the growth of capillaries, but not of myocytes. *J. Mol. Cell. Cardiol.* 24:1423–1431.

Willems, I. E. M. G., Havenith, M. G., DeMey, J. G. R., and Daemen, M. J. A. P. (1994) The α-smooth muscle actin-positive cells in healing human myocardial scars. *Am. J. Pathol.* 145:868–875.

Yoon, P. D., Kau, R. L., and Macgovern, G. J. (1995) Myocardial regeneration. Transplanting satellite cells into damaged myocardium. *Tex. Heart Inst. J.* 22:119–125

24 Stem Cells In Kidney Morphogenesis

EMMA M. A. BALL, PhD AND GAIL P. RISBRIDGER, PhD

Development of the permanent mammalian kidney (metanephros) involves complex processes such as tissue induction, branching morphogenesis of epithelia, and mesenchyme-to-epithelium transition, which give rise to the two main components of the renal system: the collecting ducts and the nephrons. While renal stem cell or embryonic progenitor cell populations have not been definitively identified, their existence is implied throughout development, during maintenance of the mature organ and in disease states. The issue of whether renal stem cells are epithelial or mesenchymal in nature needs to be resolved, as well as whether there is a single progenitor or multiple stem cell populations in the kidney. Here we examine the development of the kidney and the evidence for renal stem cells during this process. We also review the data in support of a single renal cell progenitor or multiple stem cell populations. Drawing analogies with other branched organs, we discuss where these cells are likely to be located and we examine different models for stem cell lineage. Because renal stem cells are considered targets for therapeutic drug discovery and development, their identification is critical to both oncology and regenerative medicine in nephrology.

24.1. INTRODUCTION

Development of the permanent mammalian kidney (metanephros) involves complex processes, such as tissue induction, branching morphogenesis of epithelia, and mesenchyme-to-epithelium transition, which give rise to the two main components of the renal system: the collecting ducts and the nephrons. Although renal stem cell or embryonic progenitor cell populations have not been definitively identified, their existence is implied throughout development, during maintenance of the mature organ, and in disease states. The issue of whether renal stem cells are epithelial or mesenchymal in nature needs to be resolved, as well as whether there is a single progenitor or multiple stem cell populations in the kidney. Here, we examine the development of the kidney and evidence for renal stem cells during this process. We also review the data in support of a single renal cell progenitor or multiple stem cell populations. Drawing analogies with other branched organs, we discuss where these cells are likely to be located and examine different models for stem cell lineage. Because renal stem cells are considered targets for therapeutic drug discovery and development, their identification is critical to both oncology and regenerative medicine in nephrology.

24.2. OVERVIEW OF KIDNEY DEVELOPMENT

The embryonic permanent kidney follows a pattern of development similar to that of other glandular organs, such as the pancreas, lung, and prostate, in the mammal, but there are distinct characteristics specific to the kidney. The kidney undergoes branching morphogenesis, a repetitive process that follows a tightly regulated spatial and temporal sequence of events, resulting in a three-dimensional (3D) treelike structure. Given the recognizable pattern unique to each branched organ, geometric features such as the length of ducts, the angle of branching, and the number of branches are all highly coordinated, requiring complex cell–cell signals involving epithelial–mesenchymal interactions (reviewed in Ball and Risbridger, 2001).

24.2.1. BRANCHING OF URETERIC BUD AND COLLECTING DUCTS The embryonic mammalian kidney originates from the intermediate mesoderm, developing sequentially as three distinct organs. Components of the pronephros (forekidney) and the mesonephros (midkidney) undergo involution or contribute to other structures in the urogenital system, while the metanephros (hindkidney) persists and develops into the adult kidney. Remnants of each kidney structure are utilized and incorporated by the next organ as it develops. Metanephric kidney development is induced by signaling between the metanephric mesenchyme and the ureteric bud epithelium, two mesodermally derived components. The ureteric bud forms as an outgrowth from the wolffian duct (embryonic d [E] 11 in the mouse and wk 4 to 5 of human gestation) (Saxen, 1987; Barasch et al., 1997), signaling to promote survival of the nephrogenic mesenchyme (Fig. 1). The bud dilates, growing toward and penetrating the adjacent metanephrogenic mesenchyme, which, in turn, signals back to

From: *Stem Cells Handbook*
Edited by: S. Sell © Humana Press Inc., Totowa, NJ

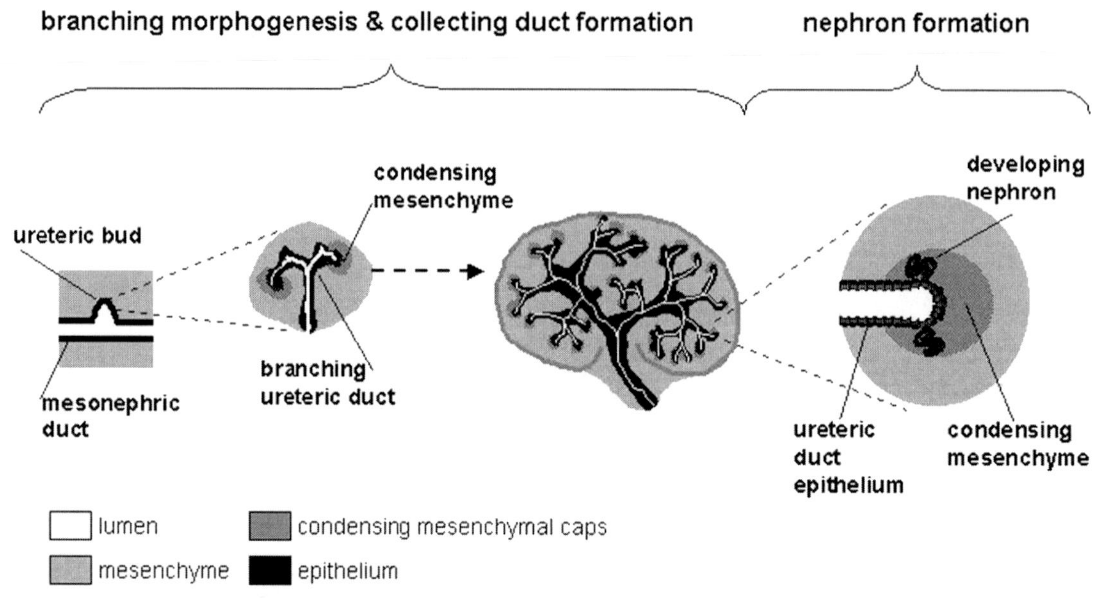

Fig. 1. Diagram of kidney development showing initial bud stage, branching morphogenesis creating collecting ducts, and nephron formation by mesenchyme-to-epithelium transition. The ureteric bud forms as an outgrowth from the wolffian (mesonephric) duct. The bud then penetrates the metanephric mesenchyme, subdivides, and induces the mesenchyme to condense around each tip. Next, higher-order branching events give rise to the entire collecting duct system. Finally, the invading epithelial tips of the collecting duct system induce the cells at the lateral edges of the mesenchymal caps to aggregate and eventually form into nephrons.

promote growth of the ureteric bud. The bud subdivides and higher-order branches form the major calyces followed by the minor calyces of the renal pelvis, undergoing further tree like branching to eventually form the epithelial collecting ducts. As branching occurs, the canalized ducts form a continuous lumen. Signaling events between the ureteric bud epithelium and surrounding mesenchyme promote the formation of a cap of condensed mesenchyme surrounding the epithelial tips (Fig. 1) (Dudley et al., 1999). For recent reviews on collecting duct morphogenesis, *see* Al-Awqati and Goldberg (1998) and Davies and Davey (1999); also visit http://cpmcnet.columbia.edu/dept/genetics/kidney/ for time-lapse movies of ureteric bud growth of in vitro organ cultures (Srinivas et al., 1999).

24.2.2. DIFFERENTIATION OF NEPHRONS In contrast to the collecting ducts that form by branching morphogenesis of the ureteric bud epithelium, the epithelial excretory tubules that give rise to nephrons originate from transition of mesenchyme to epithelium (reviewed in Davies, 1996; Horster et al., 1997; Sariola, 2002). This process has not been reported during development of other glandular organs, such as the prostate, that also undergo branching morphogenesis, and it is unique to the kidney in this regard. The invading epithelial tips of the collecting duct system induce the cells at the lateral edge of the mesenchymal caps surrounding the distal end of each tip to aggregate (Fig. 1). These aggregates differentiate into the excretory epithelial nephrons, which then differentiate into segmented renal vesicles and join to the collecting ducts themselves, creating an epithelial tubule (uriniferous tubule) with a continuous lumen. The segment of the nephron closest to the collecting duct forms the distal tubule, and the end of the renal vesicle farthest from the collecting duct becomes the glomerulus, linking the kidney to the vasculature, while the proximal tubule lies between the distal tubule and the glomerulus. In the human, nephron formation begins around wk 11 or 12 of gestation, continuing until approximately wk 36. Within 10 d from the initiation of kidney development on E11 (in mouse), at least 26 recognizable terminally differentiated cell types have formed from the ureteric bud and metanephric mesenchyme, with at least 14 epithelial cell types descended from the metanephric mesenchyme alone (Herzlinger et al., 1992; Al-Awqati and Goldberg, 1998).

24.1. REGULATION OF TEMPOROSPATIAL EVENTS IN KIDNEY DEVELOPMENT

Reciprocal stromal–epithelial interactions regulate kidney development. The metanephric mesenchyme is a powerful inducer of renal epithelium (Grobstein, 1953), although the nature of the epithelial induction depends on the type of mesenchyme. For example, kidney mesenchyme (Grobstein, 1953) or spinal cord mesenchyme (Grobstein, 1956) can induce branching of the kidney ureter in vitro, yet mesenchyme from seminal vesicle and lung induces the ureteric bud to grow like seminal vesicle and lung epithelium, respectively (Lipschutz et al., 1996; Lin et al., 2001). Using a combination of knockout models, organ culture experiments, and 3D cell culture systems, a number of factors have been identified that are necessary for or modulate kidney development. They are briefly reviewed here, but bear in mind that this account is by no means exhaustive.

24.3.1. FACTORS REGULATING URETERIC BUDDING AND BRANCHING OF COLLECTING SYSTEM Several factors have been identified as autocrine or paracrine regulators of branching and collecting duct formation. For example, the transcription factor Pax2 is expressed in both developing renal epithelium and mesenchyme, where it is essential for initiation of kidney development (Dressler et al., 1990; Torres et al., 1995). Signaling from the mesenchyme regulates Wnt-11 expression in the ureteric tip, where it acts as an autocrine growth factor on ureteric epithelium (Kispert et al., 1996). Other factors such as hepatocyte growth factor (HGF), glial cell line–derived neurotrophic factor, the enzyme heparin sulfate 2-sulfotransferase, and various matrix metalloproteinases have a positive effect on branching morphogenesis in the developing kidney collecting system (Bullock et al., 1998; Maeshima et al., 2000; Pohl et al., 2000b; for a review, *see* Sariiola, 2001). By contrast, many factors that constrain branching morphogenesis belong to the transforming growth factor-β (TGF-β) superfamily such as TGF-β itself (Ritvos et al., 1995), bone morphogenetic proteins (BMPs) (Piscione et al., 1997), and activin (Hilden et al., 1994; Roberts and Barth, 1994; Tuuri et al., 1994; for a review, *see* Ball and Risbridger, 2001).

24.3.2. FACTORS REGULATING NEPHROGENESIS A different subset of factors is involved in the mesenchyme-to-epithelium transition and nephron formation. For example, basic fibroblast growth factor (bFGF or FGF-2), is secreted from the ureteric bud to promote survival of metanephric mesenchyme, which, in turn, signals back to the ureter to allow it to branch (Perantoni et al., 1995; Barasch et al., 1997). Other factors such as proteoglycans; cytokines such as BMP-7, TGFβ-2, and leukemia inhibitory factor; the secreted glycoprotein Wnt-4; as well as protein phosphatases also promote survival and maintain the competence of metanephric mesenchyme, required for subsequent nephron formation (Stark et al., 1994; Svennilson et al., 1995; Godin et al., 1998; Plisov et al., 2001).

24.3.3. FACTORS REGULATING MULTIPLE STEPS IN KIDNEY DEVELOPMENT Some factors are involved in both phases of kidney development, adding to the complexity of the system. For example, the cell adhesion molecule α8β1-integrin is required for initial budding of the ureteric bud and subsequent branching as well as for the mesenchyme-to-epithelium transition (Muller et al., 1997). The transcription factor Pax2 has also been implicated in both initial budding and the mesenchyme-to-epithelium transition (Rothenpieler and Dressler, 1993; Torres et al., 1995). The Winged Helix transcription factor, BF-2, is expressed in stromal cells. BF-2 knockout mice highlight the importance of the metanephric stroma on kidney development; as these mice exhibit reduced growth and branching of ureteric structures as well as reduced differentiation of metanephric mesenchyme into nephrons (Hatini et al., 1996).

As stated, the description of factors involved in kidney development is not comprehensive; it merely provides a snapshot of examples known to contribute to the overall milieu of signaling molecules, transcription factors, and matrix proteins that regulate kidney development. For more detailed descriptions of factors involved in kidney development, readers are directed to these review articles: Horster et al. (1997), Vainio and Muller (1997), Wallner et al. (1998), Davies and Davey (1999), Clark and Bertram (2000) and Pohl et al. (2000a). This section serves to illustrate the complex nature of renal branching morphogenesis and nephrogenesis, both of which require precise temporospatial regulation of numerous factors acting in both autocrine and paracrine modes between cell compartments. Following development of the kidney, the organ must be maintained by achieving a balance between natural attrition and division of cells. Therefore, it is assumed that in the kidney, as in other branched organs, stem cells are involved in normal morphogenic processes. As well, the preservation of stem cells in the adult is assumed to contribute to maintenance of the normal organ and to development of disease states. Given the central role of "stem cells" (in development and in the etiology of disease), definitive identification of these cells is crucial to the understanding of basic biological processes and advances in therapeutics.

24.4. DEFINITION OF THE RENAL STEM CELL

Several criteria must be satisfied if a particular cell type is to be considered a true organ-specific stem cell (reviewed in Watt and Hogan, 2000; Al-Awqati and Oliver, 2002). In the case of the kidney, as in other organs, the putative stem cell must be able to generate a number of different cell types found in that organ, a characteristic known as *multipotency*, and the stem cell must be capable of self-renewal. It has been hypothesized that when a mammalian stem cell divides, its daughter cells fall into two populations, becoming either stem cells or committed progenitors that undergo differentiation, a process known as *asymmetric cell division*.

During kidney development, it is commonly considered that there are two types of stem cells: one epithelial cell type involved in the formation of the collecting ducts and a stem cell of mesenchymal origin involved in nephrogenesis. It has not been determined whether there is a single progenitor for all kidney cell types. In comparison with other branched organs, it has been suggested that there are adult stem cells in the kidney. Although it is a relatively quiescent organ, the kidney has the potential to regenerate following toxicity or ischemia (Cuppage et al., 1969; Humes et al., 1989, 1996; Witzgall et al., 1994). As with other organs, it is not known if the putative adult stem cells are the same as the stem cells involved in embryonic development. Furthermore, it is not known if there is one adult stem cell for the whole organ or if multiple stem cell types contribute to maintenance and regeneration of different compartments in the mature organ.

Very little is known about putative renal stem cells and their resulting lineages during development or in adulthood. The following sections examine the evidence for renal stem cells and consider the localisation and potential markers for these cell populations within the kidney. Where appropriate, comparisons are made with other organs that undergo branching morphogenesis.

24.5. EVIDENCE FOR RENAL STEM CELLS

In epithelia during collecting duct formation, the existence of renal stem cells is assumed, but the number of distinct populations in the kidney is still a contentious issue. Historically, it was believed that the two major epithelial compartments in the kidney are derived from different cell populations; that is, the ureteric bud epithelium gives rise to the collecting duct system while the metanephric mesenchyme gives rise to the nephrons. According to this hypothesis, cells isolated from the collecting system would not be able to differentiate into cells forming nephrons and vice versa. As such, the formation of an entire kidney would require at least two populations of progenitor stem cells and their resulting cell lineages. Because the cell lineages themselves are poorly defined, identification of renal stem cells has been limited. Early studies demonstrated that when isolated ureteric bud was cocultured with metanephric mesenchyme, an entire collecting system developed (Grobstein, 1953). Since then, further experiments have been conducted allowing observation of the different components of the kidney in vitro, in an attempt to elucidate gross cell lineages.

It is noteworthy that many of the studies described in the following sections are based on specific in vitro culture conditions using developing kidneys, often with the addition of specific growth factors to induce correct cell differentiation. Furthermore, several of the studies used transformed cells; therefore, extrapolation of these artificial experimental situations to the in vivo situation, in either the developing or mature kidney, may be limited or flawed.

24.5.1. STEM CELLS IN EPITHELIA Suzuki et al. (1994) cultured transformed rabbit collecting duct cells and cloned specific cell types, which they termed *principal cells*, *nonprincipal cells*, and *intercalated cells*, based on their ability to bind peanut agglutinin. They demonstrated that in the presence of HGF, nonprincipal cells exhibited stemcell–like characteristics, differentiating into principal cells and intercalated-like cells, and formed collecting ducts (Suzuki et al., 1994). None of the cell types differentiated into cells exhibiting nephron-like phenotypes.

Experiments conducted with cultured proximal tubule cells isolated from adult rabbit nephrons demonstrated that under certain culture conditions, kidney tubules formed from clonal expansion of a single epithelial cell (Humes et al., 1996). The tubules consisted of a central lumen surrounded by a polarized epithelial layer separated from the collagen matrix by basal cells (Humes et al., 1996).

Together, these two studies support the hypothesis that both epithelial compartments can be regenerated from their own individual subset of adult stem cells. It is not certain, however, if these stem cells are the same as the embryonic progenitors that give rise to epithelia during kidney development.

24.5.2. STEM CELLS IN MESENCHYME Ekblom et al. (1981) isolated metanephric mesenchyme from E11 mice and cocultured it with embryonic spinal cord, which, like the ureteric bud in vivo, transmits inductive signals to the mesenchyme, initiating the mesenchyme-to-epithelium transition. Using specific segment markers, they demonstrated that only nephronic epithelia (the glomerulus, the proximal and distal tubules) developed (Ekblom et al., 1981). More important, no collecting duct structures formed, implying that cells from both the ureteric bud and metanephric mesenchyme are required for complete formation of the kidney. Herzlinger et al. (1992) provided further evidence for mesenchymal stem cells (MSCs) by demonstrating that a single metanephric mesenchymal cell could give rise to nephron epithelia. In support of the previous studies, they did not observe the formation of cells associated with the collecting tubule (Herzlinger et al., 1992). Because these experiments were performed with embryonic tissues, they provide evidence for an MSC population during development. Whether this population is maintained into adulthood is currently unresolved.

24.5.3. A SINGLE PROGENITOR In gathering evidence for stem cells in the mesenchyme or epithelia, the prevailing hypothesis is that individual subsets of stem cells from both the epithelium and the mesenchyme are required to generate collecting duct as well as nephron epithelial compartments. The idea that there is a single progenitor for the kidney was proposed by Qiao et al. (1995), who used tagged mesenchymal cells isolated from E13 rats cocultured with ureteric bud epithelia. They demonstrated that cells from the ureteric bud were able to give rise to the collecting system as well as differentiating nephron epithelia. Additionally, they were also able to show that metanephric mesenchyme gave rise to both nephrons and collecting duct epithelia. While they could not conclusively rule out cross contamination of cell populations, by comparing cell division rates in each compartment, they demonstrated that the number of potential contaminating mesenchymal cells was not sufficient to account for the quantity of nephron-like cells generated by the isolated ureteric buds. Assuming then that contamination was highly unlikely, they hypothesized that the collecting system forms, at least in part, from recruitment of metanephric mesenchymal cells. In contrast to the previous studies, they postulated that the metanephric mesenchymal cells are the progenitors of the entire kidney (Qiao et al., 1995). They proposed that this occurs via a series of mesenchyme-to-epithelium and epithelium-to-mesenchyme transitions whereby the metanephric mesenchyme gives rise to ureteric bud epithelium, which then switches back to metanephric mesenchyme, which, in turn, eventually converts again to nephron epithelia (Fig. 2). The exact sequence of events is as yet unknown. Simplistically, the metanephric mesenchyme is an intermediate through which both types of epithelia arise; therefore, it is likely that the metanephric mesenchymal cells are the true renal stem cell population, challenging the previous hypothesis that two stem cell populations are required for kidney development. The implication that there is a single mesenchymal progenitor may be important for other organs such as prostate, lung, and breast, for which only epithelial stem cells have been identified.

Further studies have examined the possibility that the ureteric bud is the source of mesenchymal cells destined to become nephrons. Herzlinger et al. (1993) hypothesize that ureteric bud cells are nephron progenitors. This theory is based on the observation that epithelial cells from cultured ureteric bud explants isolated from E12.5 rats dissociate from the epithelium and integrate with the mesenchyme, later becoming nephrons. Electron microscopy coupled with cell tagging ensured that contaminating mesenchymal cells were not the source of the nephron structures. Mice lacking RET kinase activity exhibit variable phenotypes associated with defects in ureteric bud growth; however, mesenchyme isolated from these animals is able to differentiate to nephron epithelia in vitro (Schuchardt et al., 1994, 1996). Taken together, these investigations demonstrate that although the ureteric bud cells can physically convert to nephron epithelium via a mesenchymal intermediate, they are not necessary for nephron formation other than as a source of inductive signals

The data supporting a single progenitor of mesenchymal origin come mainly from a single laboratory, and, ideally, these studies require validation by other investigators, preferably without the use of retroviral transformation, before strong conclusions can be drawn. Nevertheless, the phenomenon of epithelium-to-mesenchyme transitions is observed in a number of developmental processes involving cell migration and scattering as well as tumor invasion and metastasis (reviewed in Boyer et al., 2000; Savagner, 2001). There is further evidence that this may be a process involved in kidney development. In vitro, the epithelium-to-mesenchyme transition has been observed in cultured tubular epithelial cells that can be induced with TG-β1 to differentiate into myofibroblasts (Yang and Liu, 2001). In addition, Humes and Cieslinski (1992) demonstrated that tubule cells grown in monolayers were also able to give rise to mesenchymal-like cells surrounding tubule-like structures. It is difficult to ascertain how much significance to attach to these studies because epithelial cells commonly acquire mesenchymal characteristics in vitro. Nevertheless, it is conceivable that this process, in addition to the mesenchyme-to-epithelium transition associated with nephrogenesis, may be occurring during normal kidney development in vivo.

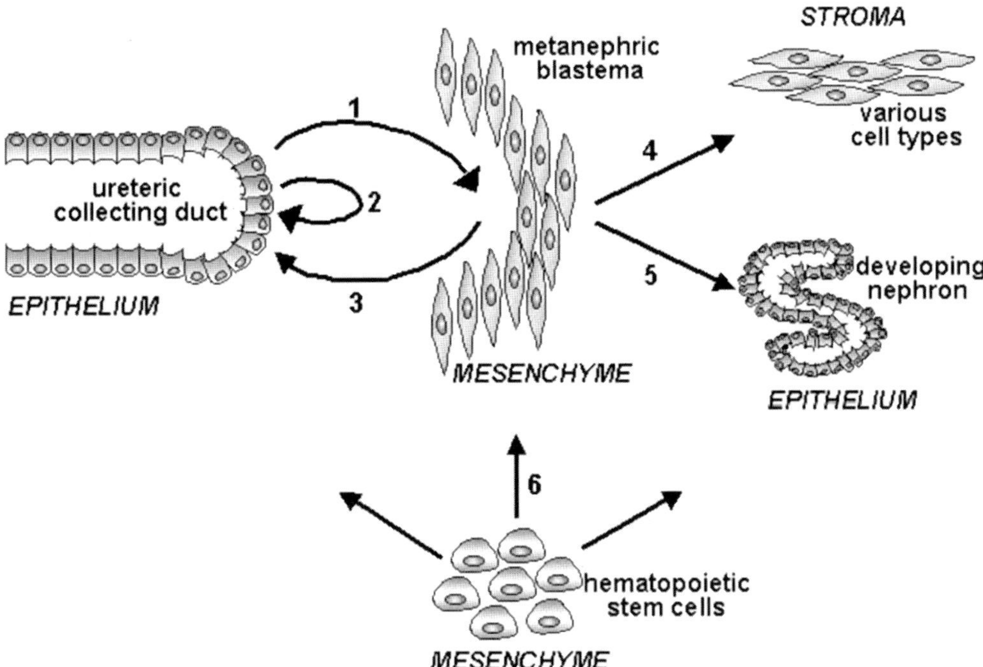

Fig.2. Possible cell lineage relationships during kidney development. 1, Epithelium-to-mesenchyme transition from ureteric bud epithelium to condensing metanephric mesenchyme; 2, self-renewal of ureteric epithelium; 3, mesenchyme-to-epithelium transition from condensing metanephric mesenchyme to ureteric duct epithelium; 4, mesenchyme differentiation into various stromal cell types; 5, mesenchyme-to-epithelium transition from condensing metanephric mesenchyme to nephron epithelia; 6, transdifferentiation of HSCs into multiple renal cell types.

24.5.4. STEM CELLS OF NONEPITHELIAL CELL TYPES Most of the studies that we have described focused on the origin of renal epithelia. Metanephric mesenchyme has also been demonstrated to be the source of other cell types in the kidney. In transfilter experiments in which the isolated embryonic metanephric mesenchyme and the source of inducing signals—in this case spinal cord—were physically separated, the mesenchymal cells gave rise to stromal cells in addition to epithelia associated with tubules (Aufderheide et al., 1987). Additionally, in cell lineage studies using a combination of immunohistochemical markers and lacZ-tagged kidneys, it was demonstrated that juxtaglomerular cells, which later acquire smooth muscle–like characteristics, originate from metanephric mesenchyme (Sequeira Lopez et al., 2001). In similar studies in which metanephric mesenchyme cultures were grafted into anterior eye chambers, angioblasts and endothelial cells arose (Robert et al., 1996; St John et al., 2001). There is also evidence to suggest that interstitial cells, consisting of fibroblasts and cells from the immune system, are descended directly from the metanephric mesenchyme (reviewed in Alcorn et al., 1999). Taken together with the data on epithelial generation, these studies demonstrate that the putative MSC has the potential to contribute to many of the cell types required for complete kidney formation.

24.5.5. ROLE OF TRANSDIFFERENTIATION IN KIDNEY DEVELOPMENT A further complication to the identification of renal stem cells arises from the possibility of transdifferentiation of resident hematopoietic stem cells (HSCs). Transdifferentiation of HSCs was considered because they are detected in urogenital ridge explants and mesonephri isolated at E12 as well in cultures of mesonephric explants from E11 (de Bruijn et al., 2000; Oostendorp et al., 2002). Evidence that HSCs can give rise to kidney-like structures in the adult is derived from studies in both humans and mice. In female mice receiving bone marrow cells from male recipients, Y-chromosome-positive epithelial cells were observed in many organs such as the lung, gastrointestinal tract, and skin (Krause et al., 2001). In the kidney, donor-derived bone marrow cells gave rise to tubular epithelia and myofibroblasts in host animals (Poulsom et al., 2001). In male humans receiving kidney transplants from female donors, the recipient's cells, presumably derived from circulating bone marrow cells, populated the grafted organ, forming some of the tubular epithelia and endothelial cells (Poulsom et al., 2001). Although transdifferentiation of HSCs is a very low-frequency event, its involvement in normal kidney development and regeneration in vivo cannot be excluded. HSC markers such as Sca-1, CD34, and c-kit are available to identify and sort out HSCs and may resolve whether or not transdifferentiation of HSCs results in kidney structures in in vitro cell lineage studies (Sanchez et al., 1996). Elucidation of cell lineage within the kidney using appropriate stem cell markers must resolve this area of question if the true renal stem cell population(s) is to be identified. Further evidence is required to establish whether transdifferentiation is crucial in kidney development and maintenance.

24.5.1. CONCLUSIONS FROM PRELIMINARY CELL LINEAGE STUDIES Based on the previously discussed studies, it is clear that the identity of the renal stem cell(s) remains uncertain. Early studies suggested that two stem cell populations in the kidney gave rise to collecting duct and tubule epithelia, respectively. However, more recent experiments have challenged this idea, suggesting that signaling from the ureteric bud causes the mesenchyme to acquire a "stem cell phenotype" and that metanephric mesenchymal cells are capable of giving rise to the nephron epithelia, collecting duct epithelia, and stromal cell types (reviewed in Davies, 1996). Much of the controversy arises during consideration of kidney development, but it appears that differentiated tubule cells may still remain multipotent, giving rise to a number of epithelial cell types in the nephron as well as mesenchymal cells.

The additional roles of mesenchyme-to-epithelium and epithelium-to-mesenchyme transitions and the contribution of HSC transdifferentiation to kidney development add other layers of complexity to determining renal cell lineage (Fig. 2). Given the need to identify stem cell populations within the kidney, it is imperative that the intricate sequences of events determining cell lineages are documented. It is necessary to follow these lineages to determine categorically whether there are single or multiple progenitors of renal cells.

24.6. IDENTIFICATION OF RENAL STEM CELLS

Given the difficulty of identifying renal stem cells, we can only postulate on aspects such as the likely location(s) of the stem cells and resultant cell lineages. Therefore, the aim of the following sections is to draw analogies with other organs, in pursuit of the renal stem cell population(s).

24.6.1. LOCATION It is not known where renal stem cells are likely to reside either during development or in adulthood. In other organs, stem cells are usually located in a specific position ("niche") where adjacent cells secrete factors to prevent rapid proliferation of the stem cells, where they are protected from environmental harm, and where they have plentiful blood supply (reviewed in Watt and Hogan, 2000; Al-Awqati and Oliver, 2002). The location of this hypothetical niche in the kidney is uncertain, but evidence from other organs might yield some clues. For example, putative prostatic stem cells are located in the basal epithelial layer, nestled between the stroma and secretory epithelium (De Marzo et al., 1998; Collins et al., 2001; Wang et al., 2001). Mammary and lung stem cells also reside within the epithelium (reviewed in Li et al., 1998; Warburton et al., 1998).

24.6.2. MARKERS The lack of known markers hampers the identification of stem cell population(s) in the kidney, and if the field is to progress, specific markers for the renal stem cells must be identified. A133-2 (an isoform of CD133) is one recently proposed marker. Originally identified as HSC marker, its pattern of expression in a number of putative stem cell niches and its rapid downregulation during differentiation suggests that A133-2 may be a marker for stem cells in a number of organs (Yu et al., 2002). A133-2 transcripts have been detected in whole kidney although its exact cellular localization is as yet unknown (Yu et al., 2002). Another potential marker is telomerase. Telomerase prevents telomeric shortening of chromosomes, thus delaying senescence. Pluripotent embryonic stem (ES) cells express telomerase, which is then downregulated at an early stage of development, although in highly regenerative tissues, expression is retained in small subsets of cells thought to be stem cells (reviewed in Forsyth et al., 2002). Telomerase detection is used as a marker for cancer diagnosis (reviewed in Hiyama and Hiyama, 2002); therefore, it is possible that it may also be useful in identifying renal stem cells.

Markers for stem cells in other branching organs may provide some insight into potential markers for renal stem cells. For example, in the prostate, $\alpha 2\beta 1$-integrin, p27Kip1, P-cadherin, and connexin 43 have been used to either identify or isolate cell populations containing prostatic stem cells (Jarrard et al., 1997; Soler et al., 1997; De Marzo et al., 1998; Collins et al., 2001; Habermann et al., 2002). When used in conjunction with the cytokeratin (CK) group of intermediate filament proteins that can distinguish among different cell types in this compartment (Wang et al., 2001), it is hypothesized that the true prostatic stem cell can be identified. Similar subsets of markers may prove to be useful in the identification of renal stem cells.

24.6.3. STEM CELL LINEAGE IN COLLECTING DUCT EPITHELIA Generally, epithelial stem cells divide slowly, but rapid proliferation of their progeny is achieved by an intermediate cell type known as the transit-amplifying cell (reviewed in Watt and Hogan, 2000; Al-Awqati and Oliver, 2002). Both of these populations remain unidentified in the kidney. Analogies can be drawn among other organs in which putative stem cell lineages have been identified.

In the polarized epithelium of the kidney collecting ducts, as in the prostate, it has long been assumed that stem cells of normal kidney epithelia undergo asymmetric cell division to produce one copy of the stem cell and one amplifying intermediate cell that in turn terminally differentiates (reviewed in Al-Awqati and Oliver, 2002). However, this concept was recently revisited in the prostate, providing new insights into prostatic epithelial cell lineages. This may serve as a template on which to build an alternative model of renal stem cell lineages. As in the kidney, the most common prostatic stem cell model is one in which prostatic stem cells in the undifferentiated state are located in the basal layer and give rise to intermediate and then secretory cells, in a direct precursor-progeny relationship (Bonkhoff and Remberger, 1996; De Marzo et al., 1998; van Leenders and Schalken, 2001) (Fig. 3). More recently, Wang et al. (2001) comprehensively mapped the pattern of CK expression in murine and human prostatic tissues during development and compared that with the pattern of expression in the mature organ in order to identify the adult stem cell. Interestingly, their interpretation of the results led to an alternate perspective of the lineage of cells in the prostatic epithelium (Fig. 3). They suggested that progenitor/stem cells of the prostate were cells of intermediate phenotype expressing the full range of luminal and basal epithelial differentiation markers (including CK8, CK18, CK14, CK5) as distinct from the classic basal cells (that express CK14 and CK5) or fully differentiated luminal cells (that express CK8 and CK18). This suggestion was based on the hypothesis that embryonic progenitor cells, initially located in the urogenital sinus epithelium, are retained in the mature epithelium as adult stem cells. Only cells of the intermediate phenotype were found in both urogenital sinus epithelium and in adult epithelium. Interestingly, the basal cells, previously thought to be stem cells, were detected only in the mature organ, suggesting that they represent a later stage of differentiation (Wang et al., 2001).

In the prostate, it is widely accepted that the ES cell population is retained during adulthood and contributes to malignancy. Furthermore, in the mammary gland, the highest concentration of stem cell is at the branch tips, which is the area most susceptible to malignancy (reviewed in Li et al., 1998). Therefore, tremendous efforts are directed at definitively identifying and characterizing these cell populations in an attempt to elucidate how these cells give rise to disease. These findings have the potential to greatly enhance our understanding and assist identification of the renal stem cell population(s).

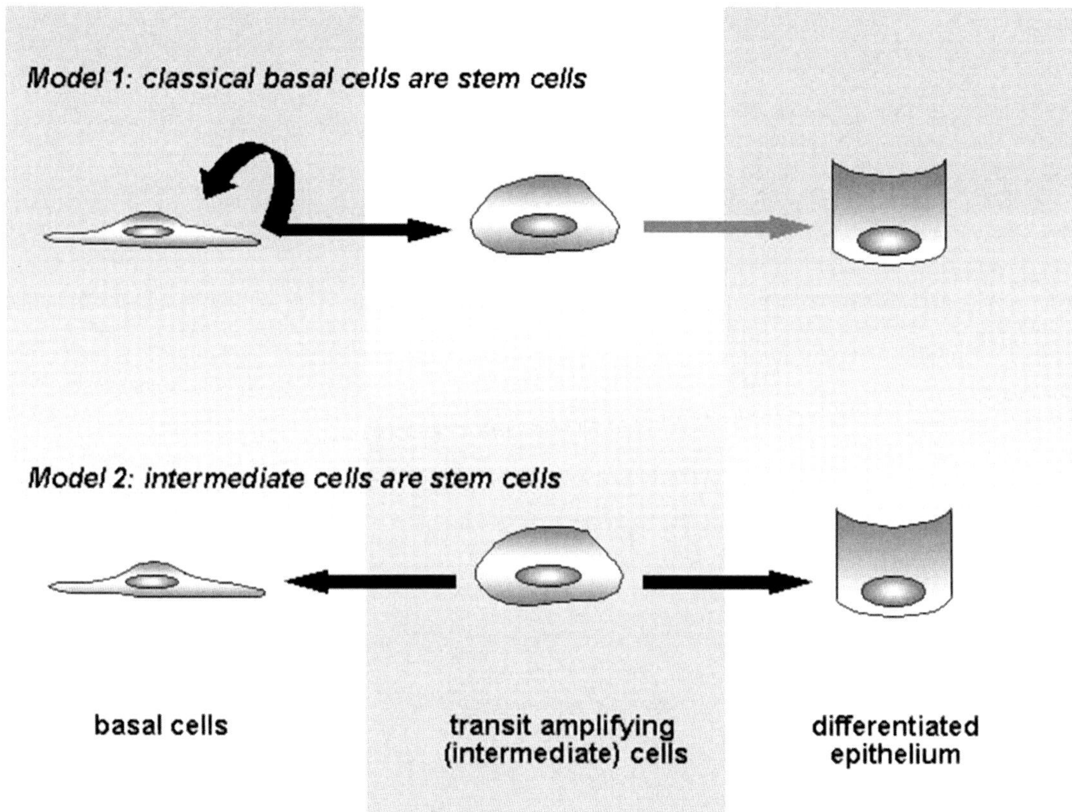

Fig. 3. Alternative models of epithelial cell lineage in the collecting ducts. Model 1 shows the classic precursor-progeny relationship between stem cells and other epithelial cells representing different stages of differentiation. Model 2 shows an alternative model of cell lineage in which the intermediate cells are stem cells that give rise to epithelial cell types in a nonlinear fashion.

24.6.4. STEM CELL LINEAGE IN NEPHROGENESIS The exact lineages of cells contributing to nephron formation are unclear. They may involve mesenchyme-to-epithelium transitions, epithelium-to-mesenchyme transitions, or both. Comparisons between prostate and kidney are limited when considering the nephrogenic stem cell, however, because renal epithelia are derived from mesoderm whereas prostatic glandular epithelia are derived from endoderm. In addition, in organs such as the prostate and mammary gland, the mesenchyme plays simple inductive and structural roles but, unlike the kidney mesenchyme, does not supply cells to the proliferating epithelium. Therefore, perhaps an equally valid system with which to draw comparisons to the kidney stem cell is the mesodermally derived bone marrow in which there resides a population of nonhematopietic cells known as MSCs. Bone marrow stem cells can undergo multiple differentiation pathways, progressively losing their lineage potential as they differentiate (Muraglia et al., 2000). Cell-surface markers used to isolate these MSCs may prove useful in the search for markers to recognize the renal stem cell population(s) (Majumdar et al., 1998; Pittenger et al., 1999).

24.7. RENAL STEM CELLS AND POTENTIAL THERAPEUTICS—A CHALLENGE FOR THE FUTURE

While the identification of both embryonic and renal stem cells is not conclusive, any discussion on the contribution of stem cells to renal disease remains speculative. It is easy to imagine how our understanding of stem cells might be used in future treatments for renal agenesis or degenerative diseases to restore normal kidney architecture and function. Undoubtedly, the adult stem cell populations will also be used as therapeutic targets for proliferative diseases such as Wilms' tumor and renal cell carcinoma.

Several unresolved issues are currently constraining our attempt to understand renal cell lineages and design appropriate therapies to treat developmental and adult renal disease. First, the issue of whether a single or multiple renal stem cell populations contribute to kidney development must be resolved. Second, it must be determined whether these are the same cells that persist through to adulthood. Third, the contribution of HSC transdifferentiation in kidney morphogenesis and regeneration must be assessed. Cell lineages must be mapped in order to identify where there is overlap and interaction between these pathways.

The issue of the adult stem cell is central to the development of therapeutic agents. The confusion that currently surrounds the issue of the ES cell will greatly impact our understanding of the biology of the mature organ. Using the prostate as an example, we may assume that the renal ES cell is retained during adulthood and contributes to disease. The exact lineage of renal stem cells cannot be assumed to operate in a direct precursor-progeny relationship, whether they be epithelial stem cells, MSCs, or a common progenitor of both cell types. Further work using inductive factors to differentiate tagged pluripotent ES cells into kidney will allow tracking of renal cell lineages by comparing subsets of markers at each stage of development. Furthermore, it must be determined whether mesenchyme-to-epithelium or epithelium-to-mesenchyme transitions occur at maturity and whether these phenomena contribute to pathogenesis. Certainly, preliminary studies draw a link between mechanisms operating during renal cell carcinoma and developmental mesenchymal-epithelial transition (reviewed in Walker, 1998). Furthermore, TGF-β-dependent epithelial-mesenchymal transition is associated with renal fibrosis (Jinde et al., 2001; Yang and Liu, 2001). Elucidation of these pathways will be a significant conceptual advance in our understanding of kidney morphogenesis and will identify specific cell types as important targets for the design and development of therapeutic agents. Undoubtedly, real breakthroughs will come in terms of the identification of stem cells in the adult organ and understanding how aberrant changes contribute to disease.

ACKNOWLEDGMENTS

We thank Assoc. Prof. Martin Pera, Monash Institute of Reproduction & Development, and Prof. John Bertram, Department of Anatomy & Cell Biology, both of Monash University, Victoria, Australia; and Dr. Alan Perantoni, National Cancer Institute, Frederick, MD, for critically evaluating this review.

REFERENCES

Al-Awqati, Q. and Goldberg, M. R. (1998) Architectural patterns in branching morphogenesis in the kidney. *Kidney Int.* 54:1832–1842.

Al-Awqati, Q. and Oliver, J. A. (2002) Stem cells in the kidney. *Kidney Int.* 61:387–395.

Alcorn, D., Maric, C., and McCausland, J. (1999) Development of the renal interstitium. *Pediatr. Nephrol.* 13:347–354.

Aufderheide, E., Chiquet-Ehrismann, R., and Ekblom, P. (1987) Epithelial-mesenchymal interactions in the developing kidney lead to expression of tenascin in the mesenchyme. *J. Cell. Biol.* 105:599–608.

Ball, E. M. and Risbridger, G. P. (2001) Activins as regulators of branching morphogenesis. *Dev. Biol.* 238:1–12.

Barasch, J., Qiao, J., McWilliams, G., Chen, D., Oliver, J. A., and Herzlinger, D. (1997) Ureteric bud cells secrete multiple factors, including bFGF, which rescue renal progenitors from apoptosis. *Am. J. Physiol.* 273:F757–F767.

Bonkhoff, H. and Remberger, K. (1996) Differentiation pathways and histogenetic aspects of normal and abnormal prostatic growth: a stem cell model. *Prostate* 28:98–106.

Boyer, B., Valles, A. M., and Edme, N. (2000) Induction and regulation of epithelial-mesenchymal transitions. *Biochem. Pharmacol.* 60: 1091–1099.

Bullock, S. L., Fletcher, J. M., Beddington, R. S., and Wilson, V. A. (1998) Renal agenesis in mice homozygous for a gene trap mutation in the gene encoding heparan sulfate 2-sulfotransferase. *Genes Dev.* 12:1894–1906.

Clark, A. T. and Bertram, J. F. (2000) Advances in renal development. *Curr. Opin. Nephrol. Hypertens.* 9:247–251.

Collins, A. T., Habib, F. K., Maitland, N. J., and Neal, D. E. (2001) Identification and isolation of human prostate epithelial stem cells based on alpha(2)beta(1)-integrin expression. *J. Cell. Sci.* 114: 3865–3872.

Cuppage, F. E., Cunningham, N., and Tate, A. (1969) Nucleic acid synthesis in the regenerating nephron following injury with mercuric chloride. *Lab. Invest.* 21:449–457.

Davies, J. A. (1996) Mesenchyme to epithelium transition during development of the mammalian kidney tubule. *Acta Anat.* 156:187–201.

Davies, J. A. and Davey, M. G. (1999) Collecting duct morphogenesis. *Pediatr. Nephrol.* 13:535–541.

de Bruijn, M. F., Speck, N. A., Peeters, M. C., and Dzierzak, E. (2000) Definitive hematopoietic stem cells first develop within the major arterial regions of the mouse embryo. *EMBO J.* 19:2465–2474.

De Marzo, A. M., Meeker, A. K., Epstein, J. I., and Coffey, D. S. (1998) Prostate stem cell compartments: expression of the cell cycle inhibitor p27Kip1 in normal, hyperplastic, and neoplastic cells. *Am. J. Pathol.* 153:911–919.

Dressler, G. R., Deutsch, U., Chowdhury, K., Nornes, H. O., and Gruss, P. (1990) Pax2, a new murine paired-box-containing gene and its expression in the developing excretory system. *Development* 109: 787–795.

Dudley, A. T., Godin, R. E., and Robertson, E. J. (1999) Interaction between FGF and BMP signaling pathways regulates development of metanephric mesenchyme. *Genes Dev.* 13:1601–1613.

Ekblom, P., Miettinen, A., Virtanen, I., Wahlstrom, T., Dawnay, A., and Saxen, L. (1981) In vitro segregation of the metanephric nephron. *Dev. Biol.* 84:88–95.

Forsyth, N. R., Wright, W. E., and Shay, J. W. (2002) Telomerase and differentiation in multicellular organisms: turn it off, turn it on, and turn it off again. *Differentiation* 69:188–197.

Godin, R. E., Takaesu, N. T., Robertson, E. J., and Dudley, A. T. (1998) Regulation of BMP7 expression during kidney development. *Development* 125:3473–3482.

Grobstein, C. (1953) Inductive epithelio-mesenchymal interaction in cultured organ rudiments of the Mouse. *Science* 118:52–55.

Grobstein, C. (1956) Transfilter induction of tubules in mouse metanephrogenic mesenchyme. *Exp. Cell. Res.* 10:424–440.

Habermann, H., Ray, V., Habermann, W., and Prins, G. S. (2002) Alterations in gap junction protein expression in human benign prostatic hyperplasia and prostate cancer. *J. Urol.* 167:655–660.

Hatini, V., Huh, S. O., Herzlinger, D., Soares, V. C., and Lai, E. (1996) Essential role of stromal mesenchyme in kidney morphogenesis revealed by targeted disruption of Winged Helix transcription factor BF-2. *Genes Dev.* 10:1467–1478.

Herzlinger, D., Koseki, C., Mikawa, T., and al-Awqati, Q. (1992). Metanephric mesenchyme contains multipotent stem cells whose fate is restricted after induction. *Development* 114:565–572.

Herzlinger, D., Abramson, R., and Cohen, D. (1993) Phenotypic conversions in renal development. *J. Cell. Sci.* (Suppl. 17):61–64.

Hilden, K., Tuuri, T., Eramaa, M., and Ritvos, O. (1994). Expression of type II activin receptor genes during differentiation of human K562 cells and cDNA cloning of the human type IIB activin receptor. *Blood* 83:2163–2170.

Hiyama, E. and Hiyama, K. (2002). Clinical utility of telomerase in cancer. *Oncogene* 21:643–649.

Horster, M., Huber, S., Tschop, J., Dittrich, G., and Braun, G. (1997). Epithelial nephrogenesis. *Pflügers Arch.* 434:647–660.

Humes, H. D. and Cieslinski, D. A. (1992). Interaction between growth factors and retinoic acid in the induction of kidney tubulogenesis in tissue culture. *Exp. Cell. Res.* 201:8–15.

Humes, H. D., Cieslinski, D. A., Coimbra, T. M., Messana, J. M., and Galvao, C. (1989). Epidermal growth factor enhances renal tubule cell regeneration and repair and accelerates the recovery of renal function in postischemic acute renal failure. J. *Clin. Invest.* 84: 1757–1761.

Humes, H. D., Krauss, J. C., Cieslinski, D. A., and Funke, A. J. (1996). Tubulogenesis from isolated single cells of adult mammalian kidney: clonal analysis with a recombinant retrovirus. *Am. J. Physiol.* 271:F42–F49.

Jarrard, D. F., Paul, R., van Bokhoven, A., et al. (1997). P-Cadherin is a basal cell–specific epithelial marker that is not expressed in prostate cancer. *Clin. Cancer Res.* 3:2121–2128.

Jinde, K., Nikolic-Paterson, D. J., Huang, X. R., et al. (2001). Tubular phenotypic change in progressive tubulointerstitial fibrosis in human glomerulonephritis. *Am. J. Kidney Dis.* 38:761–769.

Kispert, A., Vainio, S., Shen, L., Rowitch, D. H., and McMahon, A. P. (1996) Proteoglycans are required for maintenance of Wnt-11 expression in the ureter tips. *Development* 122:3627–3637.

Krause, D. S., Theise, N. D., Collector, M. I., et al. (2001) Multi-organ, multi-lineage engraftment by a single bone marrow-derived stem cell. *Cell* 105:369–377.

Li, P., Barraclough, R., Fernig, D. G., Smith, J. A., and Rudland, P. S. (1998) Stem cells in breast epithelia. *Int. J. Exp. Pathol.* 79:193–206.

Lin, Y., Zhang, S., Rehn, M., et al. (2001) Induced repatterning of type XVIII collagen expression in ureter bud from kidney to lung type: association with sonic hedgehog and ectopic surfactant protein C. *Development* 128:1573–1585.

Lipschutz, J. H., Young, P., Taguchi, O., and Cunha, G. R. (1996) Urothelial transformation into functional glandular tissue in situ by instructive mesenchymal induction. *Kidney Int.* 49:59–66.

Maeshima, A., Zhang, Y. Q., Furukawa, M., Naruse, T., and Kojima, I. (2000) Hepatocyte growth factor induces branching tubulogenesis in MDCK cells by modulating the activin-follistatin system. *Kidney Int.* 58:1511–1522.

Majumdar, M. K., Thiede, M. A., Mosca, J. D., Moorman, M., and Gerson, S. L. (1998) Phenotypic and functional comparison of cultures of marrow-derived mesenchymal stem cells (MSCs) and stromal cells. *J. Cell. Physiol.* 176:57–66.

Muller, U., Wang, D., Denda, S., Meneses, J. J., Pedersen, R. A., and Reichardt, L. F. (1997) Integrin alpha8beta1 is critically important for epithelial-mesenchymal interactions during kidney morphogenesis. *Cell* 88:603–613.

Muraglia, A., Cancedda, R., and Quarto, R. (2000) Clonal mesenchymal progenitors from human bone marrow differentiate in vitro according to a hierarchical model. *J. Cell. Sci.* 113:1161–1166.

Oostendorp, R. A., Medvinsky, A. J., Kusadasi, N., et al. (2002) Embryonal subregion-derived stromal cell lines from novel temperature-sensitive SV40 T antigen transgenic mice support hematopoiesis. *J. Cell. Sci.* 115:2099–2108.

Perantoni, A. O., Dove, L. F., and Karavanova, I. (1995) Basic fibroblast growth factor can mediate the early inductive events in renal development. *Proc. Natl. Acad. Sci. USA* 92:4696–4700.

Piscione, T. D., Yager, T. D., Gupta, I. R., et al. (1997) BMP-2 and OP-1 exert direct and opposite effects on renal branching morphogenesis. *Am. J. Physiol.* 273:F961–F975.

Pittenger, M. F., Mackay, A. M., Beck, S. C., et al. (1999) Multilineage potential of adult human mesenchymal stem cells. *Science* 284:143–147.

Plisov, S. Y., Yoshino, K., Dove, L. F., Higinbotham, K. G., Rubin, J. S., and Perantoni, A. O. (2001) TGF beta 2, LIF and FGF2 cooperate to induce nephrogenesis. *Development* 128:1045–1057.

Pohl, M., Stuart, R. O., Sakurai, H., and Nigam, S. K. (2000a) Branching morphogenesis during kidney development. *Annu. Rev. Physiol.* 62: 595–620.

Pohl, M., Sakurai, H., Bush, K. T., and Nigam, S. K. (2000b) Matrix metalloproteinases and their inhibitors regulate in vitro ureteric bud branching morphogenesis. *Am. J. Physiol. Renal Physiol.* 279: F891–F900.

Poulsom, R., Forbes, S. J., Hodivala-Dilke, K., et al. (2001) Bone marrow contributes to renal parenchymal turnover and regeneration. *J. Pathol.* 195:229–235.

Qiao, J., Cohen, D., and Herzlinger, D. (1995) The metanephric blastema differentiates into collecting system and nephron epithelia in vitro. *Development* 121:3207–3214.

Ritvos, O., Tuuri, T., Eramaa, M., et al. (1995) Activin disrupts epithelial branching morphogenesis in developing glandular organs of the mouse. *Mech. Dev.* 50:229–245.

Robert, B., St John, P. L., Hyink, D. P., and Abrahamson, D. R. (1996) Evidence that embryonic kidney cells expressing flk-1 are intrinsic, vasculogenic angioblasts. *Am. J. Physiol.* 271:F744–F753.

Roberts, V. J. and Barth, S. L. (1994) Expression of messenger ribonucleic acids encoding the inhibin/activin system during mid- and late-gestation rat embryogenesis. *Endocrinology* 134:914–923.

Rothenpieler, U. W. and Dressler, G. R. (1993) Pax-2 is required for mesenchyme-to-epithelium conversion during kidney development. *Development* 119:711–720.

Sanchez, M. J., Holmes, A., Miles, C., and Dzierzak, E. (1996) Characterization of the first definitive hematopoietic stem cells in the AGM and liver of the mouse embryo. *Immunity* 5, 513–525.

Sariola, H. (2001) The neurotrophic factors in non-neuronal tissues. *Cell. Mol. Life Sci.* 58:1061–1066.

Sariola, H. (2002) Nephron induction revisited: from caps to condensates. *Curr. Opin. Nephrol. Hypertens.* 11:17–21.

Savagner, P. (2001) Leaving the neighborhood: molecular mechanisms involved during epithelial-mesenchymal transition. *Bioessays* 23: 912–923.

Saxen, L. (1987) *Organogenesis of the Kidney*, Cambridge University Press, Cambridge, UK.

Schuchardt, A., D'Agati, V., Larsson-Blomberg, L., Costantini, F., and Pachnis, V. (1994) Defects in the kidney and enteric nervous system of mice lacking the tyrosine kinase receptor Ret. *Nature* 367: 380–383.

Schuchardt, A., D'Agati, V., Pachnis, V., and Costantini, F. (1996) Renal agenesis and hypodysplasia in ret-k– mutant mice result from defects in ureteric bud development. *Development* 122:1919–1929.

Sequeira Lopez, M. L., Pentz, E. S., Robert, B., Abrahamson, D. R., and Gomez, R. A. (2001) Embryonic origin and lineage of juxtaglomerular cells. *Am. J. Physiol. Renal Physiol.* 281:F345–F356.

Soler, A. P., Harner, G. D., Knudsen, K. A., et al. (1997) Expression of P-cadherin identifies prostate-specific-antigen-negative cells in epithelial tissues of male sexual accessory organs and in prostatic carcinomas: implications for prostate cancer biology. *Am. J. Pathol.* 151:471–478.

Srinivas, S., Goldberg, M. R., Watanabe, T., D'Agati, V., al-Awqati, Q., and Costantini, F. (1999) Expression of green fluorescent protein in the ureteric bud of transgenic mice: a new tool for the analysis of ureteric bud morphogenesis. *Dev. Genet.* 24:241–251.

Stark, K., Vainio, S., Vassileva, G., and McMahon, A. P. (1994). Epithelial transformation of metanephric mesenchyme in the developing kidney regulated by Wnt-4. *Nature* 372:679–683.

St John, P. L., Wang, R., Yin, Y., Miner, J. H., Robert, B., and Abrahamson, D. R. (2001) Glomerular laminin isoform transitions: errors in metanephric culture are corrected by grafting. *Am. J. Physiol. Renal Physiol.* 280:F695–F705.

Suzuki, M., Nakamura, T., Ikeda, M., Hayashi, T., Kawaguchi, Y., and Sakai, O. (1994) Cloned cells develop renal cortical collecting tubules. *Nephron* 68:118–124.

Svennilson, J., Durbeej, M., Celsi, G., et al. (1995) Evidence for a role of protein phosphatases 1 and 2A during early nephrogenesis. *Kidney Int.* 48:103–110.

Torres, M., Gomez-Pardo, E., Dressler, G. R., and Gruss, P. (1995) Pax-2 controls multiple steps of urogenital development. *Development* 121:4057–4065.

Tuuri, T., Eramaa, M., Hilden, K., and Ritvos, O. (1994) The tissue distribution of activin beta A- and beta B-subunit and follistatin messenger ribonucleic acids suggests multiple sites of action for the activin-follistatin system during human development. *J. Clin. Endocrinol. Metab.* 78:1521–1524.

Vainio, S. and Muller, U. (1997) Inductive tissue interactions, cell signaling, and the control of kidney organogenesis. *Cell* 90:975–978.

van Leenders, G. J. and Schalken, J. A. (2001) Stem cell differentiation within the human prostate epithelium: implications for prostate carcinogenesis. *BJU Int.* 88(Suppl. 2):35–42; discussion 49–50.

Walker, C. (1998) Molecular genetics of renal carcinogenesis. *Toxicol. Pathol.* 26:113–120.

Wallner, E. I., Yang, Q., Peterson, D. R., Wada, J., and Kanwar, Y. S. (1998) Relevance of extracellular matrix, its receptors, and cell adhesion molecules in mammalian nephrogenesis. *Am. J. Physiol.* 275: F467–F477.

Wang, Y., Hayward, S., Cao, M., Thayer, K., and Cunha, G. (2001) Cell differentiation lineage in the prostate. *Differentiation* 68:270–279.

Warburton, D., Wuenschell, C., Flores-Delgado, G., and Anderson, K. (1998) Commitment and differentiation of lung cell lineages. *Biochem. Cell. Biol.* 76:971–995.

Watt, F. M. and Hogan, B. L. (2000) Out of Eden: stem cells and their niches. *Science* 287:1427–1430.

Witzgall, R., Brown, D., Schwarz, C., and Bonventre, J. V. (1994) Localization of proliferating cell nuclear antigen, vimentin, c-Fos, and clusterin in the postischemic kidney. Evidence for a heterogenous genetic response among nephron segments, and a large pool of mitotically active and dedifferentiated cells. *J. Clin. Invest.* 93: 2175–2188.

Yang, J. and Liu, Y. (2001) Dissection of key events in tubular epithelial to myofibroblast transition and its implications in renal interstitial fibrosis. *Am. J. Pathol.* 159:1465–1475.

Yu, Y., Flint, A., Dvorin, E. L., and Bischoff, J. (2002). AC133-2, a Novel Isoform of Human AC133 Stem Cell Antigen. *J. Biol. Chem.* 277: 20,711–20,716.

25 Nephroblastoma

A Metanephric Caricature

ALAN O. PERANTONI, PhD

The histological appearance of nephroblastoma, or Wilms' tumor, closely resembles the histodifferentiation of the kidney from primitive metanephrogenic tissue. This is classically one of the tumors for which pathologists first recognized the possible role of stem cells in cancer as the blastemal cells from the nephroblastoma behave as pluripotent stem cells differentiating along multiple defined pathways. Wilms' tumors may appear predominantly blastemal, epithelial, or stromal; are associated with persistence of metanephric mesenchyme postnatally (embryonic/nephrogenic rests); and arise from a clonal expansion, as all cells carry the same genetic change, when present. About 10% of cases are associated with inheritance of inactivating mutations in the Wilms' tumor suppressor gene, *Wt1*. A second locus in Wilms' tumor on chromosome 11 is characterized by overexpression of insulin-like growth factor-2 or mutation of $p57^{KIP2}$ and is associated with the Beckwith-Wiedmann syndrome, but only 4% of patients with this syndrome develop Wilms' tumors. However, the relative infrequency of defined genetic abnormalities (others include *Wnt*, *N-myc*, *β-catenin*, and *bcl-2*), and early appearance suggest developmental control problems. Premature expression of neurotrophins (nerve growth factor and NT-3) and their receptors in the blastemal component may also contribute to tumorigenicity. The most promising approach for these types of cancers is to develop techniques for the early detection of lesions and perhaps differentiation therapies to induce neoplastic cells to a nonmalignant phenotype.

25.1. INTRODUCTION

Often described as a disease of differentiation, the nephroblastoma, or Wilms' tumor, in humans displays a remarkable histological phenotype as a result of its ability to mimic metanephric development (Fig. 1). Similar to the multipotent embryonal carcinoma, blastemal cells from the nephroblastoma behave as pluripotent stem cells, differentiating along multiple defined pathways. As described in Chapter 24, the metanephric mesenchymal stem cell, which arises from the intermediate meso-derm in the nephrogenic cord, has the potential to develop into interstitial stroma and nephrogenic mesenchyme, which, in turn, is specified to form the epithelial structures of the nephron, including the podocytes of the glomeruli and the proximal and distal tubular epithelia. The presence in nephroblastomas of primitive tubular epithelial structures, stroma, and blastemal elements, yielding a characteristic triphasic histologic phenotype, provides compelling evidence that these tumors arise from uninduced stem cells of metanephric mesenchyme and that the neoplastic counterpart has retained some ability to differentiate, albeit to another neoplastic cell with characteristics of a more advanced cellular phenotype.

From: *Stem Cells Handbook*
Edited by: S. Sell © Humana Press Inc., Totowa, NJ

25.2. WILMS' TUMOR:A PEDIATRIC PROBLEM

Wilms' tumor is predominantly a pediatric neoplasm, arising in some 1 in 10,000 children by age 15, although tumors may occur during fetal development or even well into adulthood (Bennington and Beckwith, 1975; Breslow et al., 1988). The vast majority of tumors (more than 75%), however, are actually diagnosed by age 5. The relatively high frequency (i.e., 6% of all childhood neoplasms) makes it one of the most common solid pediatric cancers. Unlike the majority of adult tumors, though, the number of chromosomal abnormalities, such as rearrangements, duplications, or deletions in Wilms' tumor, and the frequency of loss of heterozygosity (LOH) throughout the genome are limited. In fact, the apparent lack of genomic instability in neoplasms (Maw et al., 1992), their early appearance, and absence of any linkage with external factors seem to argue against environmental causation, further suggesting that delineation of the pathogenetic process and genetic mechanism responsible for Wilms' tumor should be attainable.

Wilms' tumors generally occur unilaterally but may arise bilaterally in a limited number of cases (5–10%). Additionally, their appearance is most often sporadic, although a familial predisposition is observed in 1–2% of Wilms' tumor cases (Breslow et al., 1993). Because familial tumors are more often bilateral and

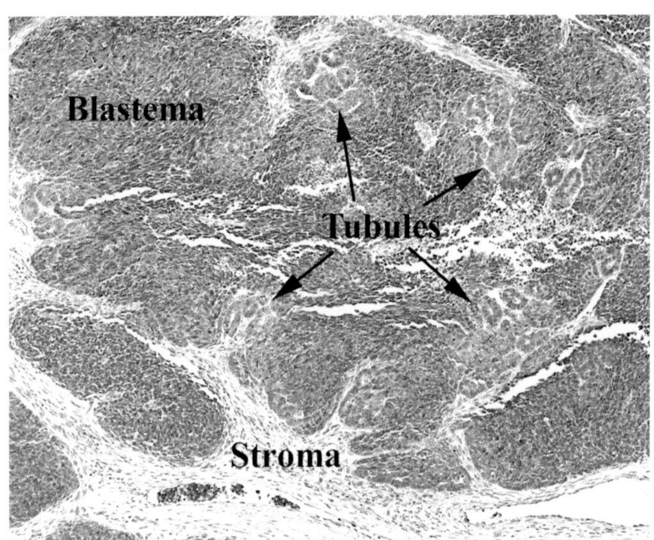

Fig. 1. Section from a Wilms' tumor showing the characteristic histology of blastemal, tubular, and stromal elements.

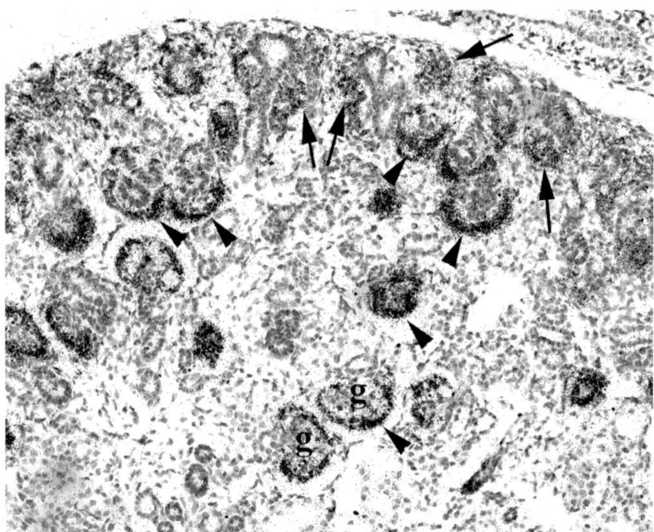

Fig. 2. Expression of Wilms' tumor suppressor gene-1 (Wt1) in fetal rat metanephros by ISH using ^{35}S-labeled riboprobe. Wt1 is localized to the podocytes of glomeruli (g), S-shaped bodies (arrowheads), and condensed metanephric mesenchyme (arrows).

exhibit an earlier onset, there may be a cancer predisposition involving a germline mutation, which could accelerate the tumorigenic process. The familial mutation, however, shows incomplete penetrance, because the disease is passed infrequently from parent to child (Matsunaga, 1981). Although other congenital anomalies, such as aniridia, a rare malformation of the iris, or a variety of genitourinary deficiencies (ureteral or urethral malformations), are occasionally observed in Wilms' cases, the vast majority of familial tumors are not associated with any apparent phenotype other than the tumor itself.

In small percentages of Wilms' tumors, specific constitutional syndromes have been identified. Of particular note are the WAGR (Wilms' tumor, aniridia, genitourinary anomalies, and mental retardation) syndrome, which is found in about 1% of patients with Wilms' tumor, and the Denys–Drash syndrome (DDS) (Wilms' tumor, genitourinary anomalies, early onset renal failure, and predisposition to germ cell tumors), which is reported in a similarly small percentage (3%) of patients with Wilms' tumor (Drash et al., 1970; Pendergrass, 1976). For both WAGR syndrome and DDS, germline mutations of chromosome 11p13 have been identified. In addition, demonstration of somatic mutations in chromosome 11p13 of some 40% of patients with Wilms' tumor who lacked germline mutations and/or LOH at 11p have further implicated this sequence in tumorigenesis (Koufos et al., 1984; Orkin et al.,1984; Kaneko et al., 1991).

25.2.1. Wt1: A PREDISPOSITION GENE FOR WILMS' TUMOR Following identification of the Wilms' tumor predisposing locus, a gene encoding a transcription factor was subsequently cloned and implicated in Wilms' tumorigenesis (Call et al., 1990). The Wilms' tumor suppressor gene *Wt1* encodes a Krüppel-like zinc finger DNA-binding protein. It is localized to the nucleus and can either suppress or activate transcription depending on its target and/or interactive partner (reviewed in Lee and Haber, 2001). During metanephric development (Fig. 2), *Wt1* expression is upregulated in induced metanephric mesenchyme, and expression remains high in condensates and subsequently in newly formed epithelia (Armstrong et al., 1993). With nephron segmentation, however, expression becomes restricted to podocytes, where it remains into adulthood. *Wt1* null homozygotes manifest a severe renal phenotype, showing no metanephric development as the mesenchyme undergoes apoptosis (Kreidberg et al., 1993). Thus, *Wt1* plays a critical role in renal stem cell maintenance and differentiation, although it is unclear whether the effect is direct or not.

25.2.2. MUTATIONS IN THE Wt1 LOCUS FROM WILMS' TUMORS The detection and distribution of *Wt1*-inactivating mutations in a number of studies has implicated the *Wt1* locus in the pathogenesis of a significant portion of Wilms' tumors (Huff et al., 1991, 1995; reviewed in Huff, 1998). The vast majority of tumors from patients with the WAGR syndrome contain a germline loss of chromosomal locus 11p13, causing deletion of a constitutional *Wt1* allele. Under these circumstances, the second allele has frequently been reported to contain a somatic point mutation (Baird et al., 1992a). Thus, loss of both alleles may be necessary for Wilms' tumor development, suggesting that Wt1 functions as a tumor suppressor in these neoplasms. In the case of DDS, tumor formation is associated with specific point mutations in *Wt1* that frequently involve the zinc finger DNA-binding domain of the encoded protein. Most often the mutation results in an Arg residue replacement for Trp in codon 394 and within the third zinc finger (Baird et al., 1992b). Of more than 50 cases examined, about 50% contain missense mutations in this codon or the adjacent codon Asp396.

While it is thought that these mutant proteins behave as dominant negatives, the remaining wild-type allele is apparently also lost in tumors that have been evaluated (Pellitier et al., 1991), so tumor formation could simply be dependent on the loss of normal *Wt1* function. A third nephropathy associated with *Wt1*, called the Frasier syndrome, resembles DDS in that germline alterations involving intronic point mutations interfere with splicing between zinc fingers 3 and 4 of *Wt1*; however, these alterations do not result in Wilms' tumors. Thus, mutations are not necessarily sufficient for renal tumorigenesis, although patients with this syndrome are

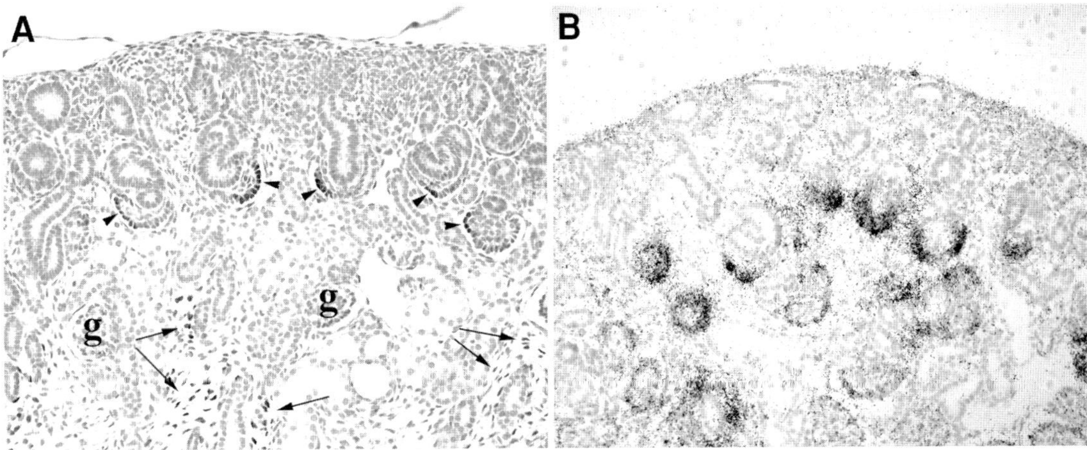

Fig. 3. Expression of p57^{KIP2} in fetal rat metanephros by immunohistochemistry (**A**) or ISH with ^{35}S-labeled riboprobe (**B**). Induced metanephric mesenchyme and interstitial stroma (arrows) show moderate expression, while podocytes in S-shaped bodies (arrowheads) and newly formed glomeruli (g) express high levels of p57^{KIP2}.

predisposed to gonadal tumors. In any case, study of mutations in *Wt1* from WAGR and DDS supports the "two-hit" model of carcinogenesis described by Knudson and colleagues (1972, 1993).

Familial predispostions to Wilms' tumor are autosomal dominant (Matsunaga, 1981), and tumor formation with no associated congenital abnormalities is often the only indication of a predisposition. In the rare familial form of the disease, *Wt1* mutations are infrequent. In one study, only 1 of 26 tumors exhibited a mutation; in this case a germline deletion passed from parent to child (Huff et al., 1995). Such observations suggest the involvement of other genes (perhaps downstream of *Wt1*) either individually or in combination with *Wt1* in tumorigenesis (Huff et al., 1997).

Wt1 mutations are considerably more prevalent in sporadic tumors. Of more than 600 neoplasms now analyzed, roughly 10% carry *Wt1* mutations. These have included somatic inactivating deletions, insertions, or missense mutations. The bilateral tumors examined in these studies all contain germline mutations, while unilateral tumors generally carried somatic *Wt1* alterations (Huff et al., 1995), a finding again consistent with a "two-hit" model. However, not all Wilms' tumors follow this formula. In some cases, germline mutations are detectable in one allele, but the remaining allele appears normal, which would seem to indicate that other genes may cooperate with *Wt1* to silence the normal allele. Finally, a large fraction of Wilms' tumors yield no *Wt1* mutations and, in fact, express high levels of normal functional Wt1 protein, showing clearly that other mechanisms mediate tumor formation. In this regard, abnormalities implicating chromosomes 1, 6–8, 11 and 12, 16–18, and 22 have been described in cytogenetic analyses of Wilms' tumors (Wang-Wuu et al., 1990; Maw et al., 1992; Miozzo et al., 1996; Steenman et al., 1997; Brown et al., 2002). Furthermore, LOH is observed in chromosomal loci of 1p, 4q, 7p, 11p15, 16q, and 17p.

25.2.3. BECKWITH–WIEDEMANN SYNDROME INVOLVES A SECOND LOCUS IN CHROMOSOME 11 AND IMPRINTED GENES IN THIS LOCUS A Wilms' tumor-associated syndrome that is also distinct from the *Wt1* locus has been identified through characterization of a pathological process involving umbilical hernia, organomegaly, and predisposition to malignancies including Wilms' tumor, rhabdomyosarcoma, and hepatoblastoma. Called Beckwith–Wiedemann syndrome (BWS), this disorder affects some 1 in 14,000 live births and has been linked to chromosomal locus 11p15.5, through an examination of familial cases of BWS. Familial predisposition occurs in approx 15% of patients with BWS, and, interestingly, the associated genetic locus encompasses a cluster of imprinted genes (Koufos et al., 1989), i.e., genes that are normally expressed from a single allele and for which expression is parent specific. Candidate imprinted genes that are also implicated in cell proliferation and/or metanephric differentiation include fetal-specific *insulin-like growth factor-2* (*IGF-2*) and cell-cycle cyclin-dependent kinase inhibitor *p57^{KIP2}*. In patients with BWS, *IGF-2* expression is increased due to a loss of imprinting in some cases (Weksberg et al., 1993), and the genetic lesion seems to depend on the presence of two copies of the paternal chromosome 11 through either duplication or selective inheritance. Since *IGF-2* is carried in this locus and its expression is imprinted and restricted to the paternal allele, its overexpression may contribute to organ enlargement and tumor formation. This hypothesis is supported in several mouse models in which *IGF-2* levels have been manipulated. By removing controls for imprinting and allowing both parental copies of *IGF-2* to be expressed, organomegaly occurs; however, other characteristics of BWS are not observed (Leighton et al., 1995; Eggenschwiler et al., 1997), suggesting the involvement of other sequences in the deleted locus.

Expression of the cyclin-dependent kinase inhibitor *p57^{KIP2}* from 11p15.5 is also imprinted but restricted to the maternal allele. Its loss could therefore facilitate deregulation of growth control and also contribute to the accumulation of blastemal populations and their inability to differentiate. During metanephric development in the mouse, *p57^{KIP2}* expression is localized to interstitial stroma, condensing mesenchyme, and developing proximal tubule by immunohistochemistry (Westbury et al., 2001). We have observed moderate expression in the mesenchyme of the nephrogenic zone but very pronounced staining in putative podocytes of S-shaped bodies by both immunohistochemistry and *in situ* hybridization (ISH) (Fig. 3). Thus, its pattern of expression is perhaps more consistent with that of *Wt1*, although expression is not sustained into adulthood. However, similar populations of cells should be at risk of tumorigenesis during metanephric development as a result of lesions in these genes.

In Wilms' tumors, maternally inherited mutations of $p57^{KIP2}$ are demonstrable in sporadic (approx 10%) and familial (approx 50%) cases of BWS (Hatada et al., 1996; Lam et al., 1999), and null homozygotes for this gene have numerous characteristics of BWS, including renal dysplasia and adrenal cortical hyperplasia, but without the organomegaly associated with *IGF-2* overexpression. (Zhang et al., 1997). To determine whether the composite disorder could be duplicated in a mouse model, studies were performed in which both *IGF-2* and $p57^{KIP2}$ were dysregulated as in BWS. Double mutants manifested a significantly more severe placental and renal dysplasia relative to the $p57^{KIP2}$ knockout (KO) animals, leading the investigators to conclude that the two genes function antagonistically to one another (Caspary et al., 1999). While it is quite satisfying to connect imprinting of these two genes to BWS and the BWS-like phenotype in double mutants, the relationship between these genes and Wilms' tumor remains unclear. Despite the renal dysplasia in $p57^{KIP2}$ KO mice, these animals do not develop nephroblastomas. While this may simply reflect species differences, as described later in this chapter, similar mutations in patients with BWS are infrequently observed in tumors. Only 4% of BWS patients eventually develop Wilms' tumor (Wiedemann, 1964; Beckwith, 1969), although, in a small number of patients, normal tissues show a mosaic pattern of 11p15 LOH with somatic duplication of the paternal locus and loss of imprinted maternal sequences (Chao et al., 1993). The percentage of cells with LOH exceeded 75% in several normal tissues, suggesting that these cells undergo selection during tissue differentiation and maintenance. Since tumors also arose from cells with LOH, it is conceivable that these abnormalities provide a growth advantage during tumorigenesis. This is further supported by more recent findings in which imprinted genes were found to be perturbed in Wilms' preneoplastic lesions (Cui et al., 1997).

25.3. NEPHROBLASTOMA: A STEM CELL TUMOR?

As described above, Wilms' tumors generally exhibit a triphasic histological phenotype with blastemal, stromal, and epithelial cell populations, although there is a continuum in the quantitative representation of each population. Thus, tumors may appear predominantly blastemal, epithelial, or stromal. While the blastemal and epithelial elements resemble the structures observed in the differentiating metanephros, the stroma can undergo heterologous differentiation to yield cartilage, adipose tissue, striated smooth muscle, and even bone, suggesting either that cells of metanephric mesenchyme have such a pluripotential, or that tumor cells are derived from an even more primitive stem cell population.

A variety of markers have been utilized in comparative studies of Wilms' tumors and normal metanephric populations, but staining for these markers generally demonstrates similarity in expression profiles between specific neoplastic elements and the normal counterpart. For several commonly applied lectins, blastemal populations in Wilms' tumor and in fetal kidney bind only concanavalin A with strong affinity, while tubular elements from either tumor or kidney react in a similar pattern and with comparable affinities (Yeger et al., 1987).

25.3.1. INTERMEDIATE FILAMENTS AND EXTRACELLULAR MATRIX COMPONENTS In addition to lectin distribution, patterns of intermediate filament expression similarly reflect the nature of the histological components of Wilms' tumor. Blastemal cells in either Wilms' tumor or the metanephros express vimentin, while primitive normal and neoplastic epithelia show staining for cytokeratins and stromal tissues for vimentin (Denk et al., 1985). Integrin and extracellular matrix (ECM) transitions have been implicated in metanephric differentiation, and, again, patterns observed in Wilms' tumors resemble expression profiles of normal tissue components (Sariola et al., 1985; Yeger et al., 1985; Peringa et al., 1994). Specifically, integrins $\alpha2$, $\alpha3$, and $\alpha6$ as well as laminin and type IV collagen are all elevated in the epithelial component of Wilms' tumors and normal primitive tubules, while $\alpha1$, $\alpha4$, and $\alpha5$ with fibronectin were detected in stromal populations. Blastemal elements in Wilms' tumors, however, were deficient for expression of most ECM protein examined but did stain for $\alpha3$ and $\alpha6$ integrins and showed variable levels of fibronectin. Neural cell adhesion molecule and the transcription factor N-myc have also proven useful as markers of both neoplastic and nonneoplastic renal blastema (Nisen et al., 1986; Roth et al., 1988; Hirvonen et al., 1989; Satoh et al., 2000).

25.3.2. NEUROTROPHINS AND AUTOCRINE SIGNALING For neurotrophins, there is evidence of autocrine ligand signaling in tumorigenesis. Nerve growth factor (NGF) was identified in the metanephros only in interstitial stroma, and NT-3 was expressed by epithelia of the glomeruli (Huber et al., 1996). For receptors, the low-affinity p75 receptor and trk lined the early epithelial structures of the nephron (i.e., comma-shaped bodies) and subsequently committed to either mesangial or mature tubular cells, respectively. *TrkC* expression also occurred in glomerular structures and tubules, while *TrkB* was localized to interstitial stroma. The shift of expression patterns in Wilms' tumors involves the reported premature appearance of both neurotrophins (NGF and NT-3) and their receptors (p75) in the blastemal component (Donovan et al., 1994), an alteration that could influence tumorigenicity, although autocrine signaling by neurotrophins occurs later in development. Notably, in an examination of neurotrophin receptors as prognostic indicators of unfavorable outcome for Wilms' tumors, TrkBfull was associated with tumors having the worst outcome despite its localization to stromal elements in tumors (Eggert et al., 2001). This may suggest that stromal components are secreting critical neurotrophin-induced factors for blastemal cell maintenance.

Use of the various tissue-specific markers effectively demonstrates the consistent nature of marker expression among Wilms' tumor tissue components relative to normal embryonic blastemal, epithelial, and stromal tissues and additionally serves to verify the classification of those tissues. Furthermore, the markers implicate several developmentally regulated molecules in Wilms' tumor pathogenesis and provide evidence for autocrine shifts that may contribute to or enhance tumor formation. Finally, they have been shown in some cases to serve as prognostic indicators of clinical outcome.

25.3.3. EACH WILMS' TUMOR CELL COMPONENT CARRIES THE SAME GENETIC POLYMORPHISM That these populations are indeed derived from tumor tissues and not simply recruited from normal surrounding structures has been demonstrated through microdissection of the various tissue elements. In all tumors analyzed with LOH for the *Wt1* locus, each histological element contained the identical polymorphism (Zhuang et al., 1997), indicating a clonal origin for each tumor component despite the histological differences. Thus, stromal elements, including striated muscle, show the same loss of *Wt1* as the blastemal populations, suggesting that all heterologous elements are neoplastic.

25.3.4. PRECURSOR LESIONS IN WILMS' TUMOR PATHOGENESIS In humans, metanephric development is completed around 36 wk of gestation, so the presence of metanephric mesenchyme postnatally is abnormal. In a retrospective study of more than 1000 necropsies of newborns, almost 1 in 100 carried persistent blastemal cells in the kidney, which is 100 times the incidence of Wilms' tumors. Furthermore, more than 40% of patients with Wilms' tumor with unilateral disease and 100% with bilateral disease have readily demonstrable preneoplastic lesions. The term *nephrogenic rest* has been applied to these Wilms' tumor precursors and includes all lesions from dormant to maturing, hyperplastic, or neoplastic. Nephroblastomatosis then describes a condition characterized by multiple or diffuse nephrogenic rests. Beckwith and colleagues (1990, 1998) have characterized these lesions based on their distribution in the renal lobe, which also denotes their differentiation status. Because differentiation occurs along a proximal–distal axis of nephron layering from medulla to cortex, the most medullary structures arise first and those distributed cortically occur at the termination of organogenesis. Lesions are therefore classified either as perilobar (PLNR), which are distributed at the periphery of the renal lobe, or as intralobar (ILNR), which appear within the lobe (Beckwith et al., 1990). Morphologically, the PLNRs occur principally as spherical blastemal foci, although later lesions are more epithelial in character. Kidneys often contain several, well-defined lesions or, less frequently, more diffuse margins. ILNRs, on the other hand, are predominantly stromal but with some blastemal and epithelial elements. These occur randomly within the renal lobe as single foci with irregular margins. In the 40% of unilateral Wilms' tumors with nephrogenic rests, the incidence of either PLNR or ILNR is roughly equivalent. In bilateral disease, however, tumors with PLNR vastly outnumber those with ILNR by 3 to 1. Similarly, in Wilms' tumor patients with BWS, PLNR also predominates. Conversely, in cases of Wilms' tumor associated with WAGR syndrome or DDS, the primary lesion is ILNR, which arises in most tumors evaluated (84 and 91%, respectively), whereas PLNR is observed in <20% of cases.

Beckwith's (1998) descriptive studies suggest a varied outcome to the presence of nephrogenic rests. Clearly all rests do not progress to Wilms' tumors as only 1 in 100 patients with rests develops neoplastic disease. In fact, the rests may remain dormant for years, mature and sclerose, become cystic, regress, or form hyperplastic nodules. These pathways are not unlike those described in tumor progression involving preneoplastic lesions in other tissues (e.g., liver or skin). However, a clear association between rests and tumors has not been forthcoming as in other tissues. One might predict from Knudson's model that the rest would carry the first "hit," e.g., a mutation in one allele of *Wt1*, with the second alteration arising during neoplastic conversion. Thus far, this has not been clearly demonstrated. In the case of *IGF-2*, constitutively high levels of expression have been reported for both nephrogenic rests and Wilms' tumors, and even epithelial structures occasionally showed sustained expression (Yun et al., 1993). On the other hand, *Wt1* mutations have been observed in nephrogenic rests but a progression has not (Park et al., 1993). This may simply reflect the complexity of nephroblastoma development in that all of the genes involved have not been identified. This is supported by genetic studies linking high frequencies of PLNR with either trisomy 13, which is associated with hyperplastic nephromegaly and Wilms tumor (Keshgegian and Chatten, 1979), or trisomy 18, which also predisposes to Wilms tumor in surviving patients (Bove et al., 1976; Olson et al., 1995).

25.4. OTHER GENES ASSOCIATED WITH WILMS TUMORIGENESIS

In addition to *Wt1*, *IGF-2*, and $p57^{KIP2}$, other genes have been implicated in the neoplastic process of renal stem cells. While some of these genes are normally expressed during metanephric development, their misexpression due to constitutive activation or inappropriate tissue distribution could involve them in Wilms' tumor pathogenesis. Thus, the autocrine loop created by IGF-2 secretion and expression of the type 1 IGF receptor, which interacts with and is activated by IGF-2, in Wilms' tumor tissues could contribute to tumorigenesis. An exhaustive study of expression during renal organogenesis, however, suggests otherwise (Lindenbergh-Kortleve et al.,1997). In fact, both the receptor and IGF-2 are highly expressed in normal mesenchymal cells prior to condensation, with condensation, and with epithelial conversion, so the patterns observed in tumor tissues seem to reflect the expression profiles of the metanephros, suggesting that they may be important for sustaining the stem cell populations but probably not responsible for tumorigenesis. Furthermore, efforts to establish IGF-2 as a prognostic factor in Wilms' tumor development have been unsuccessful (Little et al.,1987).

25.4.1. BONE MORPHOGENETIC PROTEIN-7 IN TUMORIGENESIS For bone morphogenetic protein-7 (BMP-7), a member of the transforming growth factor-β (TGF-β) family that signals through Smad1, patterns of expression may differ from those of normal rudiments. BMP-7 was initially thought to function as an inductive factor for epithelial conversion (Vukicevic et al., 1996), which was consistent with an observed reduction in nephron formation in null homozygotes (Dudley et al., 1995). However, it was later shown to function in vitro as a cell maintenance factor but not as a morphogen (Dudley et al., 1999). Studies of rat nephroblastomas, on the other hand, demonstrated deficient *BMP-7* expression in tumors, which differed markedly from the high levels of expression in induced blastemal populations observed during nephrogenesis (Higinbotham et al., 1998). Although it was originally thought that the deficiency was responsible for a block in differentiation and accumulation of stem cells in the tumors, the inability of BMP-7 to function in this capacity suggests instead that tumor cell survival is independent of BMP-7 signaling. Perhaps tumor cells have undergone a necessary adaptation to overcome the absence of this maintenance factor or the stem cells are blocked in their differentiation at a stage prior to *BMP-7* expression, which occurs at induction.

25.4.2. AUTOCRINE TGF-α/EPIDERMAL GROWTH FACTOR RECEPTOR Other autocrine mechanisms that may contribute to cell proliferation and/or tumorigenesis include TGF-α and its binding partner, the epidermal growth factor receptor (EGFR), and hepatocyte growth factor (HGF) and its receptor c-met. TGF-α functions as a growth factor to stimulate cell proliferation through mitogen-activated protein kinase activation or as a regulator of apoptosis by inhibiting Fas-mediated cell death. In a recent examination of patients with Wilms' tumor, elevated expression of *TGF-α* and *TGF-α/EGFR* correlated with tumor phenotype and prognosis, suggesting that autocrine expression may promote cell transformation or progression during tumorigenesis (Ghanem et al., 2001a). This is supported by efforts to localize ligand and receptor, which indicate a paracrine relationship in normal tissues derived from metanephric mesenchyme (Bernardini et al., 1996).

25.4.3. AUTOCRINE HGF/c-met For HGF and c-met, conversion to an autocrine mechanism is also supported in tumor studies. While *HGF* is expressed primarily in mesenchymal cells, which are more characteristic of stroma than nephrogenic mesenchyme (Karavanova et al., 1996) during nephrogenesis, in Wilms' tumors, it is also detected in blastemal and epithelial populations (Alami et al., 2002). *c-met* expression, on the other hand, occurs in blastema and epithelia during normal metanephric differentiation and in Wilms' tumors. Thus, misexpression of HGF in tissues bearing the cognate receptor could contribute to tumor formation in those tissues. Because constitutive activation of the c-met receptor through mutation conveys tumorigenic potential in certain papillary renal cell carcinomas (Schmidt et al., 1997), it is reasonable to hypothesize a mechanism involving activation of this signaling pathway for Wilms' tumor as well, but such a mechanism remains to be proven.

25.4.4. DYSREGULATION OF N-myc In addition to aberrant extracellular signaling, there are indications for nuclear complications. For example, N-myc, a transcription factor and oncogene, is localized to blastemal populations in the metanephros, and loss of expression results in hypoplastic kidneys with markedly reduced numbers of glomeruli (Bates et al., 2000). Similarly, elevated levels of expression have been reported in the blastemal component of most Wilms' tumors (Nisen et al., 1986; Shaw et al.,1988). This, however, corresponds to patterns of expression identified in normal fetal kidney (Hirvonen et al., 1989). Most interestingly, normal Wt1 protein represses transcription from the *N-myc* promoter; however, a mutant form of *Wt1* was incapable of downregulating *N-myc*, suggesting that Wt1 involvement may, in part, explain the relaxed expression of this transcription factor (Hewitt et al., 1995;Zhang et al., 1999).

25.4.5. Wnt SIGNALING As described in a previous chapter, Wnt-4 signaling plays a critical role in the epithelial conversion of metanephric mesenchyme to form the epithelial structures of the nephron and, as such, functions in a pivotal position to regulate differentiation of renal stem cells/blastema (Stark et al., 1994). Wnts interact with a Frizzled membrane receptor to indirectly stabilize a secondary messenger β-catenin through inactivation of a glycogen synthase kinase-3β (GSK-3β)-mediated ubiquitination pathway responsible for β-catenin degradation. Accumulated β-catenin can then translocate to the nucleus for transcriptional activation of Wnt-dependent targets via canonical cooperation with a member of the T-cell factor family of DNA-binding proteins. While blocking activity in this pathway inhibits differentiation, it also impacts metanephric mesenchymal cell survival. Thus, cells in which Wnt signaling is specifically targeted do not survive even in the presence of an inductive factor (Plisov et al., 2001), indicating that this pathway may also be critical to cell maintenance and/or proliferation. Indeed, the ability of the Wnt pathway to activate transcription of cell-cycle regulator cyclin D1 is consistent with this role (Tetsu and McCormick, 1999). β-Catenin behaves as an oncogene in a variety of tumors (Morin et al., 1997), and mutations represent dominant lesions, requiring single changes for dysregulation. In an analysis of multiple Wilms' tumors, 15% contained mutations in a domain of β-catenin targeted by GSK-3β for phosphorylation, resulting in stabilization of the protein (Koesters et al., 1999; Maiti et al., 2000). Mutations were observed in sporadic and DDS- or WAGR- associated neoplasms. Furthermore, Wilms' tumors exhibiting mutations in β-catenin consistently also carried *Wt1* mutations (Maiti et al., 2000). While these alterations directly impact separate signaling pathways, it is possible that cross talk may exacerbate the effects of a single lesion. Regardless, further study of the Wnt signaling pathway components could prove quite fruitful due to its critical role in stem cell differentiation.

25.4.6. APOPTOTIC ANTAGONIST bcl-2 Finally, apoptosis is commonly affected in tumorigenesis, and evidence suggests a deregulation of programmed cell death in some Wilms' tumors as well. bcl-2, a protein of nuclear and ER membranes, antagonizes apoptosis, and studies reveal that Wilms' tumors frequently express it in blastemal cells and that levels of expression increase with pathological staging (Ghanem et al., 2001b). Thus, it functions as a significant prognostic indicator for clinical progression. Additionally, elevated expression is detected in nephrogenic rests whether or not they are associated with Wilms' tumor, so such a lesion may represent an early stage in tumor development, allowing for the selective expansion of this altered cell population. Because Wt1 can also repress *bcl-2* expression and presumably effect apoptosis, conversely, the loss of Wt1 may promote tumorigenesis through deregulation of *bcl-2* and inhibition of programmed cell death.

The list of implicated factors is certainly much longer than those just described; however, for others, such as *ras* family members or *p53* (Reeve et al., 1984; Waber et al., 1993; Bardeesy et al., 1994), the evidence of involvement is considerably less compelling primarily because studies are somewhat contradictory. That is not to say that these other molecules are less important but, rather, require clarification. With the advent of array analysis, we can expect a rapid evolution of molecular and prognostic markers for Wilms' tumors as sufficient numbers and types of tumors are evaluated. Already, efforts have identified a series of expressed genes that are differentially regulated in Wilms' tumors vs fetal and adult kidneys, some of which may provide insight into mechanisms of tumorigenesis as well as provide diagnostic tools for the identification of susceptible children (Rigolet et al., 2001).

25.5. ANIMAL MODELS FOR NEPHROBLASTOMA

Nephroblastoma occurs spontaneously in a variety of animals. In surveys of slaughterhouses, it is the most common tumor observed in swine, and both unilateral (80%) and bilateral disease (20%) have been noted (Migaki et al., 1971). Classic nephroblastomas arise frequently in chickens (Feldman and Olson, 1965), and tumors are readily induced with a myeloblastosis virus (Walter et al., 1962). While tumors have been reported in cattle, the incidence is relatively low (1 of 302 renal tumors) (Sandison and Anderson, 1968), and there is anecdotal evidence for nephroblastomas in dogs (Wick et al., 1986; Simpson et al., 1992).

Considerably more information is available on nephroblastomas of rodents, because they have provided useful models for the study of carcinogenesis. Most striking is the apparent inability of certain rodent species such as the mouse or hamster to develop spontaneous or chemically induced nephroblastomas. Mice, e.g., are susceptible only to adult-type adenomas and adenocarcinomas regardless of age at carcinogen exposure, including transplacental induction (Vesselinovitch et al., 1979). Rats and rabbits, however, are susceptible both spontaneously and with chemical treatment to nephroblastoma induction (Hard and Fox, 1983). These tumors manifest the same triphasic histology as Wilms' tumors and resemble the human neoplasms (Fig. 4).

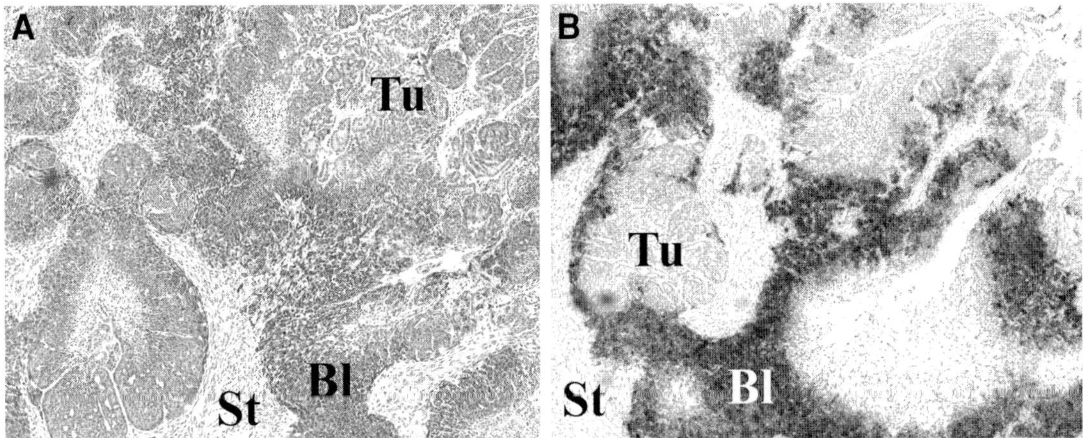

Fig. 4. Section from rat nephroblastoma **(A)** showing blastemal (Bl), tubular (Tu), and stromal (St) components. The same tumor **(B)** was probed with a marker specific for blastemal cells, demonstrating the gross expansion of this tissue component.

Nephroblastomas are rapidly and readily induced at high frequency in both rats and rabbits following transplacental exposure to direct-acting alkylating agents such as *N*-nitrosoethylurea (NEU). In rats, spontaneous and chemically induced nephroblastomas are strain dependent. While tumors have been reported in both Sprague-Dawley (Ohaki, 1989; Chandra et al.,1993) and Noble (Hard and Noble, 1981) rat strains, F344 rats are insensitive to blastemal tumor induction but instead develop mesenchymal tumors (Sukumar et al., 1986), which are stromal in nature, localized to the outer cortex, and resemble histologically the congenital mesoblastic nephroma. Induction of nephroblastomas in Noble rats is both chemical and age dependent. While NEU is a potent initiator of nephroblastoma transplacentally, causing about a 50% tumor incidence with a single exposure at 18 d of gestation, it produces renal mesenchymal tumors predominantly with neonatal treatment (Diwan and Rice, 1995). Similarly, dimethylnitrosamine, a carcinogen requiring metabolic activation, causes renal mesenchymal tumors in neonatal Noble rats but no nephroblastomas regardless of exposure age (Hard, 1985). In crosses between Noble and F344 rats, susceptibility to NEU-induced nephroblastomas is inherited as an incomplete dominant trait, suggesting multilocus involvement (B. Diwan, personal communication). For Sprague-Dawley rats, preneoplastic lesions were reported and described as intralobar nephroblastomatosis, resembling ILNR in humans (Mesfin, 1999).

Finally, *Wt1* mutations may also contribute to nephroblastoma development in rodent tumors. In a small series of NEU-induced neoplasms, point mutations were detected in about 40% of tumors; however, half of the tumors with mutations were renal mesenchymal tumors (Sharma et al., 1994). In our own series of NEU-induced nephroblastomas and renal mesenchymal tumors, we have observed no mutations in any of the hot spots reported for Wilms' tumors (unpublished observation). Thus, it is unclear if the rat tumors depend upon similar mechanisms for tumor formation.

25.6. IN VITRO MODELS OF WILMS' TUMOR

Efforts to develop a renewable source of nephroblastoma tissue for studies of the biochemical and molecular pathogenesis of Wilms' tumor have been limited and largely unsuccessful. These have included the serendipitous immortalization of cultured tumor cells, the use of tissues from heterotransplants of spontaneous or chemically induced tumors, primary cultures of dissociated Wilms' tumors, or dissociated tissues transduced with a transforming virus. Of these approaches, none has provided a particularly effective model that has been embraced by others nor has any approach been validated in attempting to provide such a model. At present, other than a few poorly characterized cell lines, the principal source of Wilms' tumor, or nephroblastoma tissues other than primary material, has been transplanted and passaged tumors.

25.6.1 CELL LINES The primary problem with the in vitro propagation of cells from Wilms' tumor tissues is their limited life span in cell culture, surviving for no more than 10–15 passages before cell crisis. In rare cases, however, cell lines have been cultivated successfully and apparently immortalized. The human SK-NEP-1 line was isolated from a malignant pleural effusion that was passaged in RPMI-1640 medium with 10% fetal bovine serum (Trempe and Old, 1975). The cells in this line resemble metanephric blastema cells, and subsequent efforts to characterize them have demonstrated cellular coexpression of TGF-α and EGFR (Mattii et al., 2001), formation of "tubule-like" structures in a matrix, and suppression of tumorigenicity with introduction of normal human chromosome 11 (Nagashima et al., 1992). This line has also been useful as a model for regulation of tumor aggressiveness with heterotransplantation (Lovvorn et al., 2000). More recently, it has been applied to cDNA expression array studies for the identification of possible prognostic markers (Rigolet et al., 2001). A rat nephroblastoma cell line designated ENU-T-1 was derived from a chemically induced tumor that had been maintained through xenotransplantation (Nagashima et al., 1989). This line retained its ability to form tumors in nude mice, and subclones of differing histological phenotype could be isolated, suggesting that the originating cell had remained pluripotent (Sumino et al., 1992). The most frequently studied cell line, G401, has a questionable derivation based on subsequent characterization. It is now thought to originate from a rhabdoid tumor of the kidney following comparison with Wilms' tumors (Garvin et al.,1993). Other cell lines derived from Wilms' tumors but currently having limited application include HFWT (Ishiwata et al., 1991), which has been passaged in tissue culture for more than 95 passages and which forms an anaplastic tumor in nude mice, or RM1 (Haber et al., 1993), that is growth suppressed following transfection of Wt1. Finally, the tumor cell line HEK293, which was derived from human embryonic kidney by immortalization with adenovirus type 5 (Graham et al., 1977), has been used extensively in studies of signal transduction mechanisms. The cells express many properties of metanephric mesenchyme despite their transformation. While not generated from a Wilms' tumor and apparently incapable of producing a triphasic phenotype, the cells may still serve as a valid model for transformed renal stem/blastemal cells.

25.6.2. TISSUE CULTURE OF WILMS' TUMOR HETEROTRANSPLANTS Another option for in vitro study is the application of culture techniques to Wilms' tumor heterotransplants, i.e., tumors that are propagated subcutaneously by serial passage in immunocompromised or syngeneic hosts. These tissues presumably have undergone a level of growth selection in the host and, therefore, may be more easily propagated in vitro. The most comprehensive efforts in this area have involved heterotransplantation of Wilms' tumors into Balb/c nude mice. Dissociation and subsequent culture of cells in the heterotransplants using a serum-free medium supplemented with several growth factors, bovine pituitary extract, and conditioned medium from a proximal tubule cell line (Garvin et al., 1987) allows for the serial passage in vitro and apparent selection for blastemal cells in these tumors. Under slightly different culture conditions, it is possible to select instead for stromal (Garvin et al., 1985) or epithelial (Hazen-Martin et al., 1993) elements as well. Selective outgrowth of the epithelial populations, however, requires the direct culture of tumorous tissue into a complex growth medium. In a similar study involving direct outgrowth from primary Wilms' tumor tissues, different cell fractions were obtained that individually expressed features of epithelia, stroma, or blastema (Velasco et al., 1993). It is worth noting that human (Yeger et al., 1985; Garvin et al., 1988) as well as rat tumors can be maintained with serial transplantation in the mouse (human and rat tumors) or in rat hosts (Tomashefsky et al., 1972; Hard and Noble, 1981). While heterotransplantation selects for blastemal populations in human neoplasms, in the rat, the tumors become somewhat more epithelial in character with serial passage. Several cell clones have been established from transplanted rat nephroblastomas, and these also manifest a varied histological profile despite their lengthy passage in vivo (Murphy et al., 1987).

25.6.3 "IMMORTALIZATION" OF WILMS' TUMOR CELLS WITH A TUMOR VIRUS Immortalization with SV-40 transforming genes has also been attempted in cultures of Wilms' tumor tissues in order to stabilize the growth of tumor cells in vitro. This method is not without its risks in interpretating results, but the approach is effective and can be relevant if results are verified in primary tissues. In SV40 large T antigen–transformed Wilms' tumor cells, cultures could be maintained for up to 35–40 passages, far longer than nontransformed cultures, but certainly not to the point of bona fide immortalization. Cells eventually did undergo a crisis from which they invariably failed to recover (Maitland et al., 1989). On the other hand, cells showed anchorage-independent growth, although they failed to form tumors in nude mice. Since the time this work was performed, several conditional transforming/immortalizing constructs have become available and may prove more effective than the method described here.

25.7. CONCLUSION

The nephroblastoma, or Wilms' tumor, provides a compelling paradigm of a tumor originating in a defined pluripotent stem cell population. This has been aptly demonstrated histologically and confirmed biochemically. The expression patterns of most blastemal markers resemble the profiles observed for stem cells of putative preneoplastic lesions (i.e., the nephrogenic rests) and of the Wilms' tumors themselves. Most striking, however, is the genetic evidence that despite morphological differences, each tissue component, whether primitive epithelia or stroma, is derived from the same stem cell clone in any given Wilms' tumor. While certain marker expression patterns are consistent in Wilms' tumors and metanephroi, others indicate shifts from paracrine to autocrine signaling, suggesting movement toward cellular autonomy and loss of growth control. Such shifts may contribute significantly to the pathogenetic process that directs tumor formation in metanephric tissues, as has been demonstrated in other tissues. Furthermore, the genetic lesions associated with tumorigenesis in genes such as *Wt1*, *IGF-2*, and *p57*, which are relevant and in some cases critical to normal blastemal differentiation, provide strong evidence that Wilms' tumor is indeed a disease of differentiation and that an understanding of the events responsible for the accumulation of these blastemal cells may eventually lead to therapies that could reregulate and commit cells to a nonneoplastic and differentiated phenotype, which is, of course, the ultimate goal of stem cell research involving any tumor.

REFERENCES

Alami, J., Williams, B. R. G., and Yeger, H. (2002) Expression and localization of HGF and met in Wilms' tumours. *J. Pathol.* 196: 76–84.

Armstrong, J. F., Pritchard-Jones, K., Bickmore, W. A., Hastie, N. D., and Bard, J. B. (1993) The expression of the Wilms' tumour gene, *WT1*, in the developing mammalian embryo. *Mech. Dev.* 40:85–97.

Baird, P. N., Groves, N., Haber, D. A., Housman, D. E., and Cowell, J. K. (1992a) Identification of mutations in the *WT1* gene in tumours from patients with the WAGR syndrome. *Oncogene* 7:2141–2149.

Baird, P. N., Santos, A., Groves, N., Jadresic, L., and Cowell, J. K. (1992b) Constitutional mutations in the *WT1* gene in patients with Denys-Drash syndrome. *Hum. Mol. Genet.* 1:301–305.

Bardeesy, N., Falkoff, D., Petruzzi, M. J., et al. (1994) Anaplastic Wilms' tumour, a subtype displaying poor prognosis, harbours *p53* gene mutations. *Nat. Genet.* 7:91–97.

Bates, C. M., Kharzai, S., Erwin, T., Rossant, J., and Parada, L. F. (2000) Role of N-myc in the developing mouse kidney. *Dev. Biol.* 222:317–325.

Beckwith, J. B. (1969) Macroglossia, omphalocele, adrenal cytomegaly, gigantism, and hyperplastic visceromegaly. In: *The Clinical Delineation of Birth Defects: Part II*, National Foundation—March of Dimes, New York, pp. 188–190.

Beckwith, J. B. (1998) Nephrogenic rests and the pathogenesis of Wilms tumor: developmental and clinical considerations. *Am. J. Med. Genet.* 79:268–273.

Beckwith, J. B., Kiviat, N. B., and Bonadio, J. F. (1990) Nephrogenic rests, nephroblastomatosis, and the pathogenesis of Wilms' tumor. *Pediatr. Pathol.* 10:1–36.

Bennington, J. and Beckwith, J. (1975) Tumors of the kidney, renal pelvis, and ureter. In: *Atlas of Tumor Pathology*, Series 2, Fascile 12, Armed Forces Institute of Pathology, Washington, DC.

Bernardini, N., Bianchi, F., Lupetti, M., and Dolfi, A. (1996) Immunohistochemical localization of the epidermal growth factor, transforming growth factor alpha, and their receptor in the human mesonephros and metanephros. *Dev. Dyn.* 206:231–238.

Bove, K. E., Soukup, S., Ballard, E. T., and Ryckman, F. (1996) hepatoblastoma in a child with trisomy 18: cytogenetics, liver anomalies, and literature review. *Pediatr. Pathol. Lab. Med.* 16: 253–262.

Breslow, N., Beckwith, J. B., Ciol, M., and Sharples, K. (1988) Age distribution of Wilms' tumor: report from the National Wilms' Tumor Study. *Cancer Res.* 48:1653–1657.

Breslow, N., Olshan, A., Beckwith, J. B., and Green, D. M. (1993) Epidemiology of Wilms tumor. *Med. Pediatr. Oncol.*21:172–181.

Brown, N., Cotterill, S. J., Roberts, P., et al. (2002) Cytogenetic abnormalities and clinical oucome in Wilms tumor: a study by the U.K. Cancer Cytogenetics Group and the U.K. Children's Cancer Study Group. *Med. Pediatr. Oncol.* 38:11–21.

Call, K. M., Glaser, T., Ito, C. Y., et al. (1990) Isolation and characterization of a zinc finger polypeptide gene at the human chromosome 11 Wilms' tumor locus. *Cell* 60:509–520.

Caspary, T., Cleary, M. A., Perlman, E. J., Zhang, P., Elledge, S. J., and Tilghman, S. M. (1999) Oppositely imprinted genes $p57^{Kip2}$ and *Igf2* interact in a mouse model for Beckwith-Wiedemann syndrome. *Genes Dev.* 13:3115–3124.

Chandra, M., Riley, M. G., and Johnson, D. E. (1993) Spontaneous renal neoplasms in rats. *J. Appl. Toxicol.* 13:109–116.

Chao, L.-Y., Huff, V., Tomlinson, G., Riccardi, V. M., Strong, L. C., and Saunders, G. F. (1993) Genetic mosaicism in normal tissues of Wilms' tumour patients. *Nat. Genet.* 3:127–131.

Cui, H., Hedborg, F., He, L., et al. (1997) Inactivation of H19, an imprinted and putative tumor repressor gene, is a preneoplastic event during Wilms' tumorigenesis. *Cancer Res.* 57:4469–4473.

Denk, H., Weybora, W., Ratschek, M., Sohar, R., and Franke, W. W. (1985) Distribution of vimentin, cytokeratins, and desmosomal-plaque proteins in human nephroblastoma as revealed by specific antibodies: co-existence of cell groups of different degrees of epithelial differentiation. *Differentiation* 29:88–97.

Diwan, B. A. and Rice, J. M. (1995) Effect of stage of development on frequency and pathogenesis of kidney tumors induced in Noble (Nb) rats exposed prenatally or neonatally to *N*-nitrosoethylurea. *Carcinogenesis* 16:2023–2028.

Donovan, M. J., Hempstead, B., Huber, L. J., et al. (1994) Identification of the neurotrophin receptors p75 and trk in a series of Wilms' tumors. *Am. J. Pathol.* 145:792–801.

Drash, A., Sherman, F., Hartmann, W. H., and Blizzard, R.M. (1970) A syndrome of pseudohermaphroditism, Wilms' tumor, hypertension, and degenerative renal disease. *J. Pediatr.* 76:585–593.

Dudley, A. T., Lyons, K. M., and Robertson, E. J. (1995) A requirement for bone morphogenetic protein-7 during development of the mammalian kidney and eye. *Genes Dev.* 9:2795–2807.

Dudley, A. T., Godin, R. E., and Robertson, E. J. (1999) Interaction between FGF and BMP signaling pathways regulates development of metanephric mesenchyme. *Genes Dev.* 13:1601–1613.

Eggenschwiler, J., Ludwig, T., Fisher, P., Leighton, P. A., Tilghman, S. M., and Efstratiadis, A. (1997) Mouse mutant embryos overexpressing IGF-II exhibit phenotypic features of the Beckwith-Wiedemann and Simpson-Golabi-Behmel syndromes. *Genes Dev.* 11:3128–3142.

Eggert, A., Grotzer, M. A., Ikegaki, N., et al. (2001) Expression of the neutrophin receptor TrkB is associated with unfavorable outcome in Wilms' tumor. *J. Clin. Oncol.* 19:689–696.

Feldman, W. H. and Olson, C. (1965) Neoplastic diseases of the chicken. In: *Diseases of Poultry* (Biester, H. E. and Schwarte, L. H., eds.), Iowa State University Press, Ames, pp. 913–924.

Garvin, A. J., Surrette, F., Hintz, D. S., Rudisill, M. T., Sens, M. A., and Sens, D. A. (1985) The *in vitro* growth and characterization of the skeletal muscle component of Wilms' tumor. *Am. J. Pathol.* 121: 298–310.

Garvin, A. J., Sullivan, J. L., Bennett, D. D., Stanley, W. S., Inabnett, T., and Sens, D. A. (1987) The in vitro growth, heterotransplantation, and immunohistochemical characterization of the blastemal component of Wilms' tumor. *Am. J. Pathol.* 129:353–363.

Garvin, A. J., Congleton, L., Inabnett, T., Gansler, T., and Sens, D. A. (1988) Growth characteristics of human Wilms' tumor in nude mice. *Pediatr. Pathol.* 8:599–615.

Garvin, A. J., Re, G. G., Tarnowski, B. I., Hazen-Martin, D. J., and Sens, D. A. (1993) The G401 cell line, utilized for studies of chromosomal changes in Wilms' tumor, is derived from a rhabdoid tumor of the kidney. *Am. J. Pathol.* 142:375–380.

Ghanem, M. A., Van Der Kwast, T. H., Den Hollander, J. C., et al. (2001a) Expression and prognostic value of epidermal growth factor receptor, transforming growth factor-α, and c-erb B-2 in nephroblastoma. *Cancer* 92:3120–3129.

Ghanem, M. A., Van der Kwast, T. H., Den Hollander, J. C., et al. (2001b) The prognostic significance of apoptosis-associated proteins BCL-2, BAX and BCL-X in clinical nephroblastoma. *Br. J. Cancer* 85:1557–1563.

Graham, F. L., Smiley, J., Russel, W. C., and Nairn, R. (1977) Characteristics of a human cell line transformed by DNA from human adenovirus type 5. *J. Gen. Virol.* 36:59–74.

Haber, D. A., Park, S., Maheswaran, S., et al. (1993) WT1-mediated growth suppression of Wilms tumor cells expressing a WT1 splicing variant. *Science* 262:2057–2059.

Hard, G. C. (1985) Differential renal tumor response to N-ethylnitrosourea and dimethylnitrosamine in the Nb rat: basis for a new rodent model of nephroblastoma. *Carcinogenesis* 6:1551–1558.

Hard, G. C. and Fox, R. R. (1983) Histologic charcterization of renal tumors (nephroblastomas) induced transplacentally in IIIVO/J and WH/J rabbits by N-ethylnitrosourea. *Am. J. Pathol.* 113:8–18.

Hard, G. C. and Noble, R. L. (1981) Occurrence, transplantation, and histologic characteristics of nephroblastoma in the Nb hooded rat. *Invest. Urol.* 18:371–376.

Hatada, I., Ohashi, H., Fukushima, Y., et al. (1996) An imprinted gene $p57^{KIP2}$ is mutated in Beckwith-Wiedemann syndrome. *Nat. Genet.* 14:171–173.

Hazen-Martin, D. J., Garvin, A. J., Gansler, T., Tarnowski, B. I., and Sens, D. A. (1993) Morphology and growth characteristics of epithelial cells from classic Wilms' tumors. *Am. J. Pathol.* 142:893–905.

Hewitt, S. M., Hamada, S., McDonnell, T. J., Rauscher, F. J. 3rd, and Saunders, G. F. (1995) Regulation of the proto-oncogenes bcl-2 and c-myc by the Wilms' tumor suppressor gene *WT1*. *Cancer Res.* 55: 5386–5389.

Higinbotham, K. G., Karavanova, I. D., Diwan, B. A., and Perantoni, A. O. (1998) Deficient expression of mRNA from the putative inductive factor bone morphogenetic protein-7 in chemically initiated rat nephroblastomas. *Mol. Carcinog.* 23:53–61.

Hirvonen, H., Sandberg, M., Kalimo, H., et al. (1989) The N-myc proto-oncogene and IGF-II growth factor mRNAs are expressed by distinct cells in human fetal kidney and brain. *J. Cell Biol.* 108: 1093–1104.

Huber, L. J., Hempstead, B., and Donovan, M. J. (1996) Neurotrophin and neurotrophins receptors in human fetal kidney. *Dev. Biol.* 179:369–381.

Huff, V. (1998) Wilms tumor genetics. *Am. J. Med. Genet.* 79:260–267.

Huff, V., Miwa, H., Haber, D. A., et al. (1991) Evidence for WT1 as a Wilms tumor (WT) gene: intragenic germinal deletion in bilateral WT. *Am. J. Hum. Genet.* 48:997–1003.

Huff, V., Jaffe, N., Saunders, G. F., Strong, L. C., Villalba, F., and Ruteshouser, E. C. (1995) WT1 exon 1 deletion/insertion mutations in Wilms tumor patients associated with di- and trinucleotide repeats and deletion hotspot consensus sequences. *Am. J. Hum. Genet.* 56:84–90.

Huff, V., Amos, C. I., Douglass, E. C., et al. (1997) Evidence for genetic heterogeneity in familial Wilms' tumor. *Cancer Res.* 57:1859–1862.

Kaneko, Y., Homma, C., Maseki, N., Sakurai, M. and Hata, J. (1991) Correlation of chromosome abnormalities with histological and clinical features in Wilms' and other childhood renal tumors. *Cancer Res.* 51:5937–5942.

Ishiwata, I., Ono, I., Ishiwata, C., et al. (1991) Carcinoembryonic proteins produced by Wilms' tumor cells in vitro and in vivo. *Exp. Pathol.* 41: 1–9.

Karavanova, I. D., Dove, L. F., Resau, J. H., and Perantoni, A. O. (1996) Conditioned medium from a rat ureteric bud cell line in combination with bFGF induces complete differentiation of isolated metanephric mesenchyme. *Development* 122:4159–4167.

Keshgegian, A. A. and Chatten, J. (1979) Nodular renal blastema in trisomy 13. *Arch. Pathol. Lab. Med.* 103:73–75.

Knudson A. G. and Strong, L. C. (1972) Mutation and Cancer: a model for Wilms' tumour of the kidney. *J. Natl. Cancer Inst.* 48:313–324.

Knudson, A. G. Jr. (1993) Introduction to the genetics of primary renal tumors in children. *Med. Pediatr. Oncol.* 21:193–198.

Koesters, R., Ridder, R., Kopp-Schneider, A., et al. (1999) Mutational activation of the beta-catenin proto-oncogene is a common event in the development of Wilms' tumors. *Cancer Res.* 59:3880–3882.

Koufos, A., Hansen, M. F., Lampkin, B. C., et al. (1984) Loss of alleles at loci on human chromosome 11 during genesis of Wilms' tumor. *Nature* 309:170–172.

Koufos, A., Grundy, P., Morgan, K., et al. (1989) Familial Wiedemann-Beckwith syndrome and a second Wilms' tumor locus both map to 11p15.5. *Am. J. Hum. Genet.* 44:711–719.

Kreidberg, J. A., Sariola, H., Loring, J. M., et al. (1993) WT-1 is required for early kidney development. *Cell* 74:679–691.

Lam, W. W., Hatada, I., Ohishi, S., et al. (1999) Analysis of germline *CDKN1C* (*p57KIP2*) mutations in familial and sporadic Beckwith-Wiedemann syndrome (BWS) provides a novel genotype/phenotype correlation. *J. Med. Genet.* 7:518–523.

Lee, S. B. and Haber, D. A. (2001) Wilms tumor and the *WT1* gene. *Exp. Cell Res.* 264:74–99.

Leighton, P. A., Ingram, R. S., Eggenschwiler, J., Efstratiadia, A., and Tilghman, S. M. (1995) Disruption of imprinting caused by deletion of the H19 gene region in mice. *Nature* 375:34-39.

Lindenbergh-Kortleve, D. J., Rosato, R. R., van Neck, J. W., et al. (1997) Gene expression of the insulin-like growth factor system during mouse kidney development. *Mol. Cell. Endocrinol.* 132:81–91.

Little, M. H., Ablett, G., and Smith, P. J. (1987) Enhanced expression of insulin-like growth factor II is not a necessary event in Wilms' tumour progression. *Carcinogenesis* 8:865–868.

Lovvorn, H. N. 3rd, Savani, R. C., Ruchelli, E., Cass, D. L., and Adzick, N. S. (2000) Serum hyaluronan and its association with unfavorable histology and aggressiveness, of heterotransplated Wilms' tumor. *J. Pediatr. Surg.* 35:1070–1078.

Maiti, S., Alam, R., Amos, C. I., and Huff, V. (2000) Frequent association of beta-catenin and WT1 mutations in Wilms tumors. *Cancer Res.* 60: 6288–6292.

Maitland, N. J., Brown, K. W., Poirier, V., Shaw, A. P. W., and Williams, J. (1989) Molecular and cellular biology of Wilms' tumor. *Anticancer Res.* 9:1417–1426.

Matsunaga, E. (1981) Genetics of Wilms' tumor. *Hum. Genet.* 57:231–246.

Mattii, L. Bianchi, F., Da Prato, I., Dolfi, A., and Bernardini, N. (2001) Renal cell cultures for the study of growth factor interactions underlying kidney organogenesis. *In Vitro Cell Dev. Biol. Anim.* 37:251–258.

Maw, M., Grundy, P. E., Millow, L. J., et al. (1992) A third Wilms' tumor locus on chromosome 16q. *Cancer Res.* 52:3094–3098.

Mesfin, G. M. (1999) Intralobar nephroblastometosis: precursor lesions of nephroblatoma in the Sprague-Dawley rat. *Vet. Pathol.* 36:379–390.

Migaki, G., Nelson, L. W., and Todd, G. C. (1971) Prevalence of embryonal nephroma in slaughtered swine. *J. Am. Vet. Med. Assoc.* 159: 441–442.

Miozzo, M., Perotti, D., Minoletti, F., et al. (1996) Mapping of a putative tumor suppressor locus to proximal 7p in Wilms tumors. *Genomics* 37:310–315.

Morin, P. J., Sparks, A. B., Korinek, V., et al. (1997) Activation of beta-catenin-Tcf-signaling in colon cancer by mutations in beta-catenin or APC. *Science* 275:1752–1753.

Murphy, G. P., Kawinski, E., and Horoszewicz, J. S. (1987) Cell cultures derived from Wilms' tumour animal model. *Anticancer Res.* 6: 717–720.

Nagashima, Y., Ohaki, Y., Umeda, M., Oshimura, M., and Misugi, K. (1989) Establishment and characterization of an immature epithelial cell line (ENU-T-1) derived from a rat nephroblastoma. *Virchows Arch. B Cell. Pathol. Incl. Mol. Pathol.* 57:383–392.

Nagashima, Y., Miyagi, Y., Sumino, K., et al. (1992) Characterization of experimental rat nephroblastoma and its cell line. *Tohoku J. Exp. Med.* 168:303–305.

Nisen, P. D., Zimmerman, K. A., Cotter, S. V., Gilbert, F., and Alt, F. W. (1986) Enhanced expression of the N-myc gene in Wilms' tumors. *Cancer Res.* 46:6217–6222.

Ohaki, Y. (1989) Renal tumors induced transplacentally in the rat by N-ethylnitrosourea. *Pediatr. Pathol.* 9:19–33.

Olson, J. M., Hamilton, A., and Breslow, N. E. (1995) Non-11p constitutional chromosomal abnormalities in Wilms' tumor patients. *Med. Pediatr. Oncol.* 24:305–309.

Orkin, S. H., Goldman, D. S., and Salan, S. E. (1984) Development of homozygosity of chromosome 11p markers in Wilms' tumour. *Nature* 309:172–174.

Park, S., Bernard, A., Bove, K. E., et al. (1993) Inactivation of WT1 in nephrogenic rests, genetic precursors to Wilms' tumour. *Nat. Genet.* 5:363–367.

Pelletier, J. Bruening, W., Kashtan, C. D., et al. (1991) Germline mutations in the Wilms' tumor suppressor gene are associated with abnormal urogenital development in Denys-Drash syndrome. *Cell* 67: 437–447.

Pendergrass, T. (1976) Congenital anomalies in children with Wilms' tumor: a new survey. *Cancer* 37:403–408.

Peringa, J., Molenaar, W. M., and Timens, W. (1994) Integrins and extracellular matrix-proteins in the different components of the Wilms' tumor. *Virchows Arch.* 425:113–119.

Plisov, S. Y., Yoshino, K., Dove, L. F., Higinbotham, K. G., Rubin, J. S., and Perantoni, A. O. (2001) TGF beta 2, LIF and FGF2 cooperate to induce nephrogenesis. *Development* 128:1045–1057.

Reeve, A. E., Housiaux, P. J., Gardner, R. J., Chewings, W. E., Grindley, R. M., and Millow, L. J. (1984) Loss of a Harvey ras allele in sporadic Wilms' tumour. *Nature* 309:174–176.

Rigolet, M., Faussillon, M., Baudry, D., Junien, C., and Jeanpierre, C. (2001) Profiling of differential gene expression in Wilms tumor by cDNA expression array. *Pediatr. Nephrol.* 16:1113–1121.

Roth, J., Zuber, C., Wagner, P., Blaha, I., Bitter-Suermann, D., and Heitz, P. U. (1988) Presence of the long chain form of polysialic acid of the neural cell adhesion molecule in Wilms' tumor: Identification of a cell adhesion molecule as an oncodevelopmental antigen and implications for tumor histogenesis. *Am. J. Pathol.* 133:227–240.

Sandison, A. T. and Anderson, L. J. (1968) Tumors in the kidney in cattle, sheep and pigs. *Cancer* 21:727–742.

Sariola, H., Ekblom, P., Rapola, J., Vaheri, A., and Timpl, R. (1985) Extracellular matrix and epithelial differentiation of Wilms' tumor. *Am. J. Pathol.* 118:96–107.

Satoh, F., Tsutsumi, Y., Yokoyama, S., and Osamura, R. Y. (2000) Comparative immunohistochemical analysis of developing kidneys, nephroblastomas and related tumors: consideration on their histogenesis. *Pathol. Int.* 50:458–471.

Schmidt, L., Duh, F.-M., Chen, F., et al. (1997) Germline and somatic mutations in the tyrosine kinase domain of the MET proto-oncogene in papillary renal carcinomas. *Nat. Genet.* 16:68–73.

Sharma, P. M., Bowman, M., Yu, B.-F., and Sukumar, S. (1994) A rodent model for Wilms tumors: embryonal kidney neoplasms induced by *N*-nitroso-*N'*-methylurea. *Proc. Natl. Acad. Sci. USA* 91:9931–9935.

Shaw, A. P., Poirier, V., Tyler, S., Mott, M., Berry, J., and Maitland, N. J. (1988) Expression of the N-myc oncogene in Wilms' tumour and related tissues. *Oncogene* 3:143–149.

Simpson, R. M., Gliatto, J. M., Casey, H. W., and Henk, W. G. (1992) The histologic, ultrastructural, and immunohistochemical features of a blastema-predominant canine nephroblastoma. *Vet. Pathol.* 29:250–253.

Stark, K., Vainio, S., Vassileva, G., and McMahon, A. P. (1994) Epithelial transformation of metanephric mesenchyme in the developing kidney regulated by Wnt-4. *Nature* 372:679–683.

Steenman, M., Redeker, B., de Meulemeester, M., et al. (1997) Comparative genomic hybridization analysis of Wilms tumors. *Cytogenet. Cell. Genet.* 77:296–303.

Sukumar, S., Perantoni, A., Reed, C., Rice, J. M., and Wenk, M. L. (1986) Activated K-ras and N-ras oncogenes in primary renal mesenchymal tumors induced in F344 rats by methyl(methoxymethyl)nitrosamine. *Mol. Cell. Biol.* 6:2716–2720.

Sumino, K., Nagashima, Y., Ohaki, Y., and Umeda, M. (1992) Isolation of subclones with different tumorigenicity and metastatic ability from rat nephroblastoma cell line, ENUT. *Acta Pathol. Japan* 42:166–176.

Tetsu, O. and McCormick, F. (1999) Beta-catenin regulates expression of cyclin D1 in colon carcinoma cells. *Nature* 398:422–426.

Tomashefsky, P., Furth, J., Lattimer, J. K., Tannenbaum, M., and Priestley, J. (1972) The Furth-Columbia rat Wilms tumor. *J. Urol.* 107:348–354.

Trempe, G. and Old, L. J. (1975) New human tumor cell lines. In: *Human Tumor Cells in Vitro.* (Fogh, J., ed.) Plenum, New York, pp. 115–159.

Velasco, S., D'Amico, D., Schneider, N. R., et al. (1993) Molecular and cellular heterogeneity of Wilms' tumor. *Int. J. Cancer* 53:672–679.

Vesselinovitch, S. D., Rao, K. V., and Mihailovich, N. (1979) Neoplastic response of mouse tissues during perinatal age periods and its significance in chemical carcinogenesis. *Natl. Cancer Inst. Monogr.* 51:239–250.

Vukicevic, S., Kopp, J. B., Luyten, F. P., and Sampath, T. K. (1996) Induction of nephrogenic mesenchyme by osteogenic protein 1 (bone morphogenetic protein 7). *Proc. Natl. Acad. Sci. USA* 93: 9021–9026.

Waber, P. G., Chen, J., and Nisen, P. D. (1993) Infrequency of ras, p53, WT1, or RB gene alterations in Wilms tumors. *Cancer* 72: 3732–3738.

Walter, W. G., Burmester, B. R., and Cunningham, C. H. (1962) Studies on the transmission and pathology of a viral-induced avian nephroblastoma (embryonal nephroma). *Avian Dis.* 6:455–477.

Wang-Wuu, S., Soukup, S., Bove, K., Gotwals, B., and Lampkin, B. (1990) Chromosome analysis of 31 Wilms' tumors. *Cancer Res.* 50:2786–2793.

Weksberg, R., Shen, D. R., Fei, Y. L., Song, Q. L., and Squire, J. (1993) Disruption of insulin-like growth factor 2 imprinting in Beckwith-Wiedemann syndrome. *Nat. Genet.* 5:143–150.

Westbury, J., Watkins, M., Ferguson-Smith, A. C., and Smith, J. (2001) Dynamic temporal and spatial regulation of the cdk inhibitor $p57^{KIP2}$ during embryo morphogenesis. *Mech. Dev.* 109:83–89.

Wick, M. R., Manivel, C., O'Leary, T. P., and Cherwitz, D. L. (1986) Nephroblastoma: a comparative immunocytochemical and lectin-histochemical study. *Arch. Pathol. Lab. Med.* 110:630–635.

Wiedemann, H. R. (1964) Complexe malformatif familial avec hernie ombilicale et macroglossie—un "syndrome nouveau"? *J. Génét Hum.* 13:223–232.

Yeger, H., Baumal, R., Bailey, D., Pawlin, G., and Phillips, M. J. (1985) Histochemical and immunohistochemical characterization of surgically resected and heterotransplanted Wilms' tumor. *Cancer Res.* 45:2350–2357.

Yeger, H., Baumal, R., Harason, P., and Phillips, M. J. (1987) Lectin histochemistry of Wilms' tumor: comparison with normal adult and fetal kidney. *Am. J. Clin. Pathol.* 88:278–285.

Yun, K., Molenaar, A. J., Fiedler, A. M., et al. (1993) Insulin-like growth factor II messenger ribonucleic acid expression in Wilms tumor, nephrogenic rest, and kidney. *Lab. Invest.* 69:603–615.

Zhang, P., Liegeios, N. J., Wong, C., et al. (1997) Altered cell differentiation and proliferation in mice lacking $p57^{KIP2}$ indicates a role in Beckwith-Wiedemann syndrome. *Nature* 387:151–158.

Zhang, X., Xing, G., and Saunders, G. F. (1999) Proto-oncogene N-myc promoter is down regulated by the Wilms' tumor suppressor gene WT1. *Anticancer Res.* 19:1641–1648.

Zhuang, Z., Merino, M. J., Vortmeyer, A. O., et al. (1997) Identical genetic changes in different histologic components of Wilms' tumors. *J. Natl. Cancer Inst.* 89:1148–1152.

26 Stem Cells in Nonmelanoma Skin Cancer

WENDY C. WEINBERG, PhD AND STUART H. YUSPA, MD

Mosaic pattern analysis and genetic mutations common to all cells of a cancer show that squamous cell carcinoma (SCC) and basal cell carcinoma (BCC), as well as squamous dysplasias, are clonal, whereas focal hyperplasias are polyclonal. One compartment of putative stem cells in the skin is located in the bulge of the hair follicle. Cells in this compartment are multipotent and can give rise to progeny that differentiate into any of the epidermal cells or adnexal organs. The interfollicular epidermal proliferative unit (EPU) in normal skin is a columnar group of differentiating cells overlying 10–12 basal cells and is believed to be derived from a single, centrally located stem cell with a more limited potential than the follicular stem cell. Stem cells in the skin cycle slowly and are identified by retaining a pulsed DNA marker for extended periods. Other markers include increased expression of β1 or β4 integrins; decreased expression of the transferrin receptor or connexin 43; and unique expression of keratins 15, 17, and 19. BCCs appear to arise from follicular bulge stem cells and are associated with genetic changes in the Sonic Hedgehog developmental pathway. SCCs can arise from stem cells in the interfollicular EPU and infundibulum of the hair follicle as well as the bulge. Benign squamous neoplasms may also arise from the more differentiated cell populations. Alterations in the *ras* pathway have been implicated in both experimental and human squamous cell carcinogenesis. Genetic or epigenetic changes in stem cell markers that have been associated with squamous cell neoplasms include alterations in integrins, telomerase, c-*myc*, and p63.

26.1. INTRODUCTION

The evaluation of stem cells as targets for the clonal origin of nonmelanoma cutaneous cancers (primarily squamous cell carcinoma [SCC] and basal cell carcinoma [BCC]) must encompass morphological, biochemical, and molecular criteria that link stem cell characteristics to those of developing neoplasms. Defining those characteristics for stem cells in normal skin has not been without controversies. Stem cells would be expected to cycle slowly, have extended or unlimited self-renewal capacity, and be multipotential regarding commitment to a spectrum of lineage outcomes required to reconstitute the epithelial component of the integument. However, with the exception of the self-renewal capacity, skin tumors generally behave differently, with rapid cycling and limited lineage commitment. Thus, the dilemma arises as to how to recognize stemness in the cancer cell while also understanding the fundamental changes that cause tumor cells to divert from stem cell characteristics. To begin, it is worthwhile to review the evidence indicating that cutaneous cancers and their precursors arise from a single cell.

From: *Stem Cells Handbook*
Edited by: S. Sell © Humana Press Inc., Totowa, NJ

26.2. EVIDENCE FOR CLONALITY OF CUTANEOUS TUMORS

The cellular origin and evolution of epithelial neoplasms within a target tissue can be delineated by mosaic pattern analysis. Spontaneous mosaicism in human and mouse models (e.g., via X-chromosome inactivation) and experimental chimerism in rodent models have been exploited to trace stem cell lineage and dissect the derivation of evolving neoplastic lesions (Iannaccone et al., 1987; Weinberg et al., 1992). In these models, genetic variants are present equally in a random distribution within the normal tissue, and patches of single variants represent a blend of clonal growth and cellular migration patterns. On this background, lesions arising from multiple cells would include a random distribution of the genetic markers, provided that the variegation pattern in the normal tissue is small relative to the sample size evaluated. By contrast, a monoclonal origin requires that only one marker be present.

The clonal origin of murine skin tumors was evaluated in experimental chimeras utilizing electrophoretic variants of glucose phosphate isomerase (Iannaccone et al., 1978), as well as in spontaneous genetic mosaics resulting from random inactivation of X-chromosome-linked phosphoglycerate kinase-1 (Pgk-1) isoenzymes (Reddy et al., 1983; Taguchi et al., 1984). Two-stage carcinogenesis studies revealed a nonrandom distribution of markers within both papillomas and carcinomas, consistent with a clonal origin of these tumors (Iannaccone et al., 1978; Reddy et al., 1983; Taguchi et al., 1984; Deamant and Iannaccone, 1987). Carcinomas arising at the site of prior papillomas were of the same genotype, in support of intraclonal conversion (Taguchi et al., 1984; Deamant and Iannaccone, 1987). By contrast, focal hyperplasias are polyclonal (Winton et al., 1989). Skin carcinomas induced by ultraviolet (UV) radiation in Pgk-1 mosaic mice are also clonally derived (Burnham et al., 1986).

The clonal origin and evolution within a tumor can also be evaluated by following the presence of a consistent genetic mutation. One set of genetic alterations identified at high frequency in both SCC and BCC, as well as in putative precursor lesions, occurs in the p53 gene. The characteristic occurrence of p53 mutations at dipyrimidine sites is evidence for the contribution of UV radiation to these lesions. p53 mutations are observed in SCC following UV exposure of both human and mouse skin (Kanjilal et al., 1993; Ren et al., 1996). Positive immunostaining for p53 in UV-irradiated mouse skin, indicative of mutations in the p53 gene, has also been observed in clusters of macroscopically normal but putative preneoplastic epidermal cells (Berg et al., 1996). Similarly, alterations in p53 have been noted in a large proportion of microscopically normal human sun-exposed skin and actinic keratoses (Ren et al., 1996). Multiple samplings encompassing different histopathologies from individuals with simultaneous presentation of SCC representing various degrees of differentiation revealed the identical mutation across these lesions, supporting their derivation from the same cellular clone; the identical mutation could also be traced from dysplasia through carcinoma (Ren et al., 1996). Distinct mutations were observed in morphologically normal epidermis that displayed p53 immunoreactivity (Ren et al., 1996). The mutations in these sharply demarcated areas may provide an enhanced proliferative advantage, but these cell clusters are not believed to be direct precursors to carcinomas due to their distinct p53 genotype (Ren et al., 1996, 1997).

The epidermal proliferative unit (EPU) in mouse skin is a clonal structure defined as a columnar organization of differentiating cells overlying 10–12 basal keratinocytes and derived from an individual centrally located stem cell (Mackenzie, 1997). The relationship of p53 mutant clusters to the EPU cannot be definitively evaluated owing to lack of suitable stem cell markers. However, recent studies in the mouse have suggested that the EPU provides a barrier to clonal expansion that must be breached before autonomous cell expansion can ensue (Zhang et al., 2001).

The heterogeneous nature of BCC has called into question the clonal origin of this tumor type. However, the nonrandom inactivation pattern of the X-linked HUMARA microsatellite marker gene in sporadic BCCs is consistent with a monoclonal origin (Walsh et al., 1998). Multiple sampling of individual BCCs by microdissection and DNA sequencing of the *p53* gene as a genetic marker revealed the presence of a common mutation across all samples. Within these tumors, additional genetic changes were observed with some overlap across samples. Thus, a single cell gives rise to multiple BCC phenotypes, with alterations in *p53* reflecting further clonal selection and expansion within a cancer (Ponten et al., 1997).

26.3. EVIDENCE FOR EPIDERMAL STEM CELLS AND LOCATION

The location of the stem cell population in skin has been a subject of much controversy but might provide clues to the origin of cutaneous tumors. A source of disparate information initially came from contrasting results in the analyses of human and mouse skin, but recent, more sophisticated, studies have resolved the differences and confirmed the similarities of the two models (Morris and Potten, 1999; Taylor G. et al., 2000; Oshima H. et al., 2001). This is an important advance because much of the experimental skin carcinogenesis data derive from mouse studies while the clinical analyses are from human tumors. Prevailing data suggest that cells of the hair follicle bulge can fulfill all of the biological characteristics of a stem cell population, including multipotentiality to recapitulate the entire epidermis and adnexal structures (Oshima et al., 2001). However, interfollicular skin contains a basal population of adhesive, clonogenic cells with the capacity to renew the epidermis and form autonomous proliferative units (EPUs) that persist from the same progenitor lineage (Jones and Watt, 1993; Mackenzie, 1997; Tani et al., 2000; Watt, 2001). Thus, one has to conclude that stem cells or stem cell compartments may come in several flavors and could give rise to separate lineages of tumor phenotypes.

The earliest phenotypic marker of a stem cell population in the skin came with the recognition of a population of slow cycling cells that retained a pulsed DNA marker for extended periods (Mackenzie and Bickenbach, 1985). Surmising that such "label-retaining" cells must be long-lived, investigators showed that clonogenic cells in vitro were derived from this population (Morris et al., 1990). Using clonogenic capacity or follicle bulge location as markers of stemness, recent studies have associated several cell-surface and internal molecular markers with stem cell capacity. Among these, increased expression of β1 or β4 integrins and decreased expression of the transferrin receptor or connexin 43 have provided quantitative markers, while unique expression of keratin 15, 17, and 19 have provided qualitative markers for stem cell isolation (Jones and Watt, 1993; Michel et al., 1996; Lyle et al., 1998; McGowan and Coulombe, 1998; Tani et al., 2000). What is missing from these analyses is molecular evidence for the contribution of any marker to stem cell characteristics. Since several of these markers are typical of cutaneous tumors, understanding their contribution to cell cycling, self-renewal capacity and multipotentiality would improve insight into the pathogenesis of these lesions. Recently, a functional marker based on the exclusion of the vital dye Hoechst 33342 by label-retaining, clonogenic mouse epidermal cells has provided evidence for the multipotentiality of this population and a potential pathway for such cells to regulate their internal signaling by excluding certain extracellular signaling molecules (Liang and Bickenbach, 2002). In addition, gene targeting of the epidermis by transgenic or knockout techniques has revealed several internal signaling pathways that have been associated with stem cell populations and contribute to tumor formation (*see* below).

26.4. EVIDENCE FOR STEM CELL ORIGIN OF CUTANEOUS TUMORS BASED ON LOCATION OF TARGET CELLS

A long-standing hypothesis is that tumors developing from stem cells will display the maximal range of histopathologies and malignant potential compared with those derived from transient amplifying cells or differentiating compartments. The persistence of the initiated cell (Yuspa, 1994) points to the stem cell as the target cell type for carcinogenesis; however, cells in the differentiating compartment are capable of neoplastic transformation (Bailleul et al., 1990; Greenhalgh et al., 1993). Careful histological dissection of progressing lesions and direct oncogene targeting have been used to evaluate systematically the contribution of specific cell populations to skin carcinogenesis (Bailleul et al., 1990; Greenhalgh et al., 1993; Hansen and Tennant 1994; Binder et al., 1997; Brown et al., 1998; Frame and Balmain 1999).

Focal hyperplastic (papilloma precursor) lesions arising during two-stage carcinogenesis were traced to the infundibular region of a single hair follicle that clonally expanded (Binder et al., 1997). Likewise, papillomas arising in TG-AC mice in which oncogenic *ras* is the under control of the ζ-globin promoter can be traced histologically to the permanent portion of the follicle where the mRNA for the *ras* transgene was initially localized. This indicates that the potential to form squamous tumors persists throughout the hair cycle. However, the transient proliferation of bulge cells during early anagen imparts enhanced responsiveness of the anagen follicle to two-stage carcinogenesis compared to the telogen follicle (Miller et al., 1993). When hair follicle cells and basal interfollicular keratinocytes are isolated in vitro and subjected to transformation by a *ras* oncogene, they produce squamous tumors in vivo that are indistinguishable, regarding both histopathology and malignant conversion frequency (Weinberg et al., 1991). The contribution of these cell populations to two-stage carcinogenesis was also evaluated by selective removal of the interfollicular epidermis in vivo following 7,12 dimethylbenz[a]anthracene (DMBA) initiation (Morris et al., 2000). Mice were promoted with 12-*0*-tetradecanoylphorbol-13-acetate (TPA) after the remaining hair follicles were allowed to reconstitute the epidermis. Those with intact initiated epidermis developed twice as many papillomas as mice with abraded skin, demonstrating that both cell populations are targets in these carcinogenesis protocols. Incidence of carcinoma, however, was similar between the groups, suggesting that hair follicles harbor initiated cells with increased malignant potential (Morris et al., 2000).

In vivo targeting of oncogenic *ras* to different epithelial compartments in transgenic skin also influences the tumorigenic outcome. For example, targeting the *Ha-ras* oncogene primarily to cells of the outer root sheath of mouse hair follicles using a truncated K5 promoter yielded spontaneous papillomas and keratoacanthomas that were refractory to the tumor promoter TPA (Brown et al., 1998). These tumors expressed keratin 13 and had a high probability of progressing to invasive carcinoma. However, targeting oncogenic *ras* to the suprabasal cell compartment by K1 or K10 promoters results in hyperkeratosis, and papillomas develop only at sites of wounding or following TPA treatment after a long latency (Bailleul et al., 1990; Greenhalgh et al., 1993). These lesions are self-limited with low malignant potential, possibly reflecting their derivation from more differentiated cells. Definitive evidence that stem cells impart the difference in autonomy and malignant potential among these cell types awaits elucidation of stem cell–specific proteins and more selective promoter targeting.

26.5. EVIDENCE FOR STEM CELL ORIGIN OF CUTANEOUS TUMORS BASED ON CHARACTERISTICS OF TARGET CELLS

A subpopulation of "dark cells" has been recognized within the infundibulum of the hair follicle and in the basal layer of interfollicular epidermis (Raick 1973; Klein-Szanto et al., 1980). The abundance of these cells during embryonic development, and their induction following carcinogenic doses of DMBA, suggested that they are both stem cells and target cells for tumor-initiating agents (Slaga and Klein-Szanto, 1983). The proportion of these cells increases up to 11-fold following treatment of mouse skin with TPA (Klein-Szanto et al., 1980). While squamous papillomas are enriched in dark cells (Raick, 1974), the lack of a molecular explanation for the dark phenotype has prevented their identification as true stem cells.

The contribution of label-retaining cells to carcinogenesis is implied by the specific retention of carcinogen–DNA adducts in these cells in vivo, consistent with the slow cycling of this population and the irreversible nature of the initiation event in two-stage carcinogenesis protocols (Morris, 2000). Carcinogen-retaining cells are found in the bulge region of the hair follicle and the central region of the EPU (Morris et al., 1986; Morris, 2000). Further support for a slow-cycling population as the target for initiation came from studies demonstrating that initiated cells persisted in skin exposed to 5-fluorouracil after carcinogen exposure and prior to tumor promotion by TPA (Morris et al., 1997). Many basal cells undergo differentiation in response to TPA, while label-retaining cells persist in the basal layer and are induced to proliferate (Morris, 2000). Cells resistant to TPA-induced differentiation in vitro were detected in basal keratinocytes that were purified by Percoll density gradients. This subpopulation of basal cells was capable of long-term proliferation and may represent the label-retaining cells observed in vivo. DMBA initiation alone in vivo did not alter the number of clonogenic epidermal cells identified in vitro by colony formation assays. However, in vivo promotion with TPA expanded the number of clonogenic cells isolated from uninitiated skin and, to a greater degree, from DMBA-initiated skin (Morris et al., 1988). Together, these results imply that clonogenic cells are derived from the quiescent, differentiation-resistant population and are targets for the mutagenic activity of chemical initiators.

26.6. EVIDENCE FOR STEM CELL ORIGIN OF CUTANEOUS TUMORS BASED ON MOLECULAR PATHWAYS

26.6.1. INTEGRINS High expression of β1 and β4 integrins has been associated with putative stem cells in mouse and human models. In human SCCs, β1 integrins are reduced or lost while α6β4 increases along with malignant progression (Rossen et al., 1994; Savoia et al., 1994). Similar changes are associated with SCC development in mouse skin in which a splice variant of α6 is also detected in more malignant lesions (Tennenbaum et al., 1993, 1995). By contrast, β1 integrins persist in BCC while α6β4 is decreased or lost (Rossen et al., 1994; Tuominen et al., 1994). How can these changes provide clues to the stem cell origin of these biologically and genetically distinct lesions? Genetic ablation of β1 integrins in mice is not compatible with hair follicle formation (Brakebusch et al., 2000; Raghavan et al., 2000), a trait shared with sonic hedgehog (SHH) ablation (Chiang et al.,

1999). Conversely, overexpression of β1 integrins in human keratinocytes suppresses differentiation (Levy et al., 2000). Absence of differentiation into follicular or interfollicular structures is characteristic of BCC. Activation of α6β4 integrin receptors initiates signals through the ras-mitogen-activated protein kinase pathway and AP-1 transcription factors to influence cellular function, and these pathways are essential to SCC development (Mainiero et al., 1997). Furthermore, overexpression of α3β1 integrins in the suprabasal compartment of mouse skin suppresses malignant conversion of squamous papillomas (Owens and Watt, 2001). Together, these findings suggest that mutations in the stem cell population characterized by bright staining of β1 integrins are more likely to yield BCC while α6β4 bright stem cells could be precursors to SCC.

26.6.2. β-CATENIN Integrin β1–positive human keratinocytes are also rich in non-cadherin-associated β-catenin, a downstream effector of the Wnt signaling pathway (Zhu and Watt, 1999). Consistent with the inability to form hair follicles in the β1 null mouse genotype, mice ablated of the β-catenin gene also fail to form hair follicles and lose hair follicles when β-catenin is deleted postnatally (Huelsken et al., 2001). Instead, they form intradermal cysts that express markers of keratinocyte differentiation but retain a β1/keratin 15–positive stem cell population in close contact. This indicates that β-catenin is required for stem cell potentiality in the formation of hair follicle structures. That such a conclusion is related to skin cancer development was confirmed by the targeting of a nondegradable mutant form of β-catenin to the skin of transgenic mice (Gat et al., 1998). These mice not only develop a high density of hair follicles, but also develop hair follicle tumors of the pilomatricoma phenotype. The power of the mouse model was confirmed when β-catenin mutations were detected in a high proportion of human pilomatricomas (Chan et al., 1999). These studies closely link developmental processes with certain forms of cutaneous cancer and suggest that pathways that regulate development, such as Wnt/β-catenin/Lef-1/Tcf, are important in both stem cell function and tumor development.

26.6.3. SONIC HEDGEHOG Activation of the SHH developmental pathway is fundamental to the pathogenesis of several skin tumor types including BCC, trichoepithelioma, and sebaceous nevi (Bale and Yu, 2001). Since activation of SHH is detected in BCC in all of its varied phenotypic presentations, the target cell for these tumors is likely to be multipotential. In *Drosophila*, Hedgehog is identified as a somatic stem cell factor (Zhang and Kalderon, 2001), and its downstream effector Gli 1 is localized in the mesenchyme surrounding the hair follicle bulge in mice (Ghali et al., 1999). Genetic ablation of SHH in mice produces a hairless phenotype (Chiang et al., 1999). Together, these data support the hair follicle bulge stem cells as the target for BCC formation. Further support comes from a model for BCC development in which mice heterozygous for inactivating mutations in the SHH receptor *patched* upregulate the promoter for *patched* in the hair follicle bulge and develop BCC lesions originating from hair follicles after skin irradiation (Aszterbaum et al., 1999).

26.6.4. TELOMERASE When one considers the long-lived capacity ascribed to stem cells, it would be logical to consider that such cells must maintain telomere length through multiple generations presumably through telomerase activity. While initial reports suggested that human and mouse skin contained a subpopulation of telomerase-positive cells, detailed analyses concluded that cells with stem cell properties did not have telomerase activity nor was telomerase activity high in the hair follicle bulge (Ramirez et al., 1997; Bickenbach et al., 1998). By contrast, telomerase activity was highest in the transit-amplifying population. Nevertheless, virtually all skin tumors examined have high telomerase activity (Taylor et al., 1996; Wu et al., 1999; Chen et al., 2001), and reconstitution of cultured human keratinocytes with hTERT extends the culture life-span without altering differentiation potential (Dickson et al., 2000). Furthermore, targeting mTERT to the basal cells of transgenic mouse epidermis enhances tumor formation (Gonzalez-Suarez et al., 2001) whereas ablating the mTERT gene reduces skin tumor susceptibility (Gonzalez-Suarez et al., 2000). From these studies, it would appear that telomeres must be maintained by an alternative mechanism in stem cells or that the slow-cycling cells do not shorten telomeres sufficiently until called on to proliferate, e.g., as an incipient tumor cell. If this is correct, then telomerase could be a good target for therapy of skin tumors without the additional concern of stem cell targeting.

26.6.5. c-*myc* The protooncogene c-*myc* is downstream of β-catenin/Tcf in the Wnt pathway and thus would be a good candidate to mediate tumor formation arising from the stem cell pool. In fact c-*myc* transcripts are elevated in BCC, the c-*myc* gene is amplified in a subset of SCC, and targeting c-*myc* to the basal epidermis of transgenic mice enhances chemical carcinogenesis (Pelisson et al., 1996; Bonifas et al., 2001; Rounbehler et al., 2001). However targeting of c-*myc* to the skin of transgenic mice appears to deplete the stem cell pool and induce progression of stem cells into the transit-amplifying population (Arnold and Watt, 2001; Waikel et al., 2001). Interestingly, if c-*myc* is targeted suprabasally in transgenic mice, benign squamous tumors develop that regress when the overexpressed c-*myc* is repressed (Pelengaris et al., 1999). When combined, this information suggests that skin tumors may evolve from keratinocytes outside the stem cell compartment, but such tumors have a different biological potential (Brown et al., 1998).

26.6.6. p63 Considerable excitement in the field of cutaneous stem cells arose with the discovery that mice genetically deleted for p63, a member of the p53 tumor suppressor family, had arrested limb development and failed to form a stratified epidermis and appendages (Mills et al., 1999; Yang et al., 1999). The p63 gene encodes a variety of transcriptional species, some of which share overlap with p53 transcriptional activity, while others can act as dominant negatives for p63 or p53 transcriptional activation (Yang et al., 1998). In fact, the predominant species expressed in proliferating keratinocytes is a truncated form with dominant-negative activity (Parsa et al., 1999). p63 is associated with proliferation in vivo and in vitro, and high expression correlates with the anaplastic phenotype in oral SCC (Parsa et al., 1999). However, in vivo studies have also suggested that expression is restricted to a subpopulation of epidermal basal cells reminiscent of the columnar organization of the EPU. Furthermore, in vitro studies suggest that p63 may serve as a marker to distinguish the stem cell and transit-amplifying cell populations, since expression is high in stem cell–derived holoclones but lost in paraclones derived from transit-amplifying cells (Pellegrini et al., 2001). Thus, this very promising pathway is undeciphered with respect to stem cell functioning or tumor induction and awaits clarification by ongoing studies.

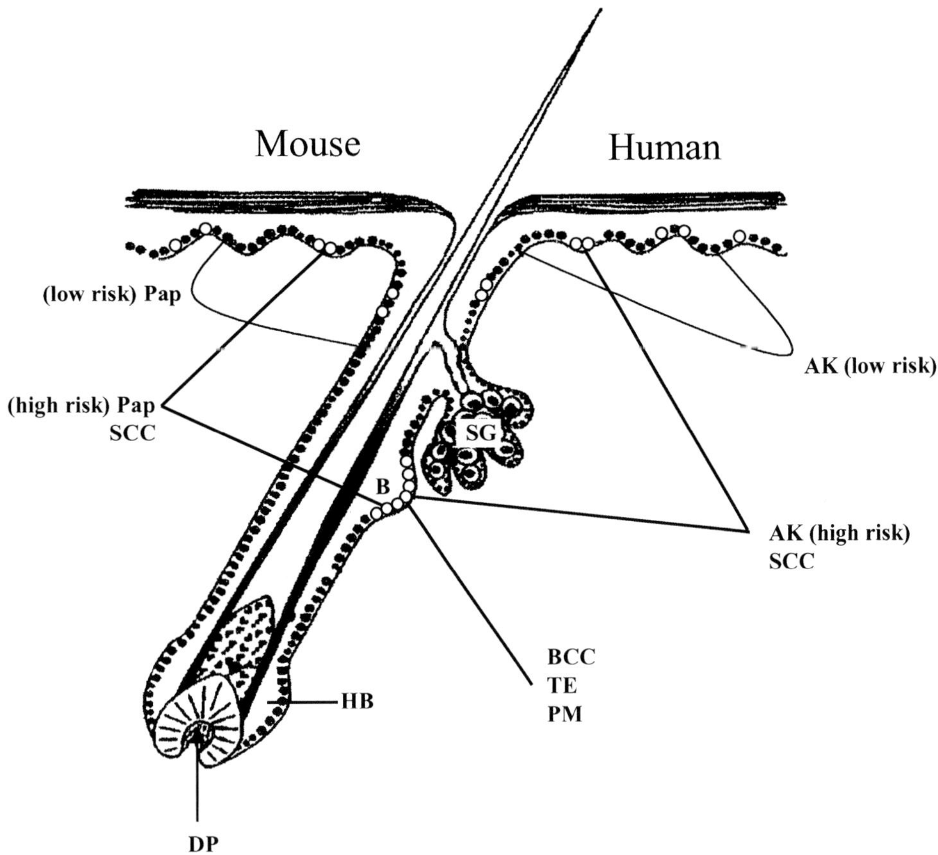

Fig. 1. Potential target cells for origin of cutaneous tumors in rodents and humans. Open circles represent integrin (-β1 or -β4) bright putative stem cells and solid circles are integrin dull cells. DP, dermal papilla; HB, hair bulb; B, bulge; SG, sebaceous gland; AK, actinic keratosis; SCC, squamous cell carcinoma; BCC, basal cell carcinoma; TE, trichoepithelioma; PM, pilomatricoma; Pap, squamous papilloma.

26.7. CONCLUSIONS

It is likely that stem cells are the target for at least a subpopulation of cutaneous tumors since a variety of phenotypic manifestations can evolve from similar genetic lesions. Perhaps this is best seen in BCC in which nodular, papillary, basosquamous and adenomatous lesions all have defects in SHH signaling. Furthermore, this pathway may contribute to trichoepithelioma and sebaceous nevus development and could be downstream from the pilomatricomas induced by mutations in β-catenin (Gat et al., 1998). This spectrum of tumors suggests that BCC evolves from mutational lesions in hair follicle stem cells presumably from the bulge area (Fig. 1). Since complete interruption of SHH signaling is incompatible with hair follicle formation, it is logical to conclude that the SHH pathway is integral to the regeneration of the hair follicle in each anagen cycle, and persistent activity of the pathway would select cells that behave autonomously. What is not clear is why mice do not develop BCC when exposed to carcinogenic stimuli unless they have a partially disrupted SHH pathway, as in the case of the *patched* mouse. One explanation might be the common use of hairless mice for UV carcinogenesis studies. These mice have bulge cells, but they are not in contact with the inductive stroma of the dermal papilla that is required for hair follicle morphogenesis (Cotsarelis et al., 1990; Panteleyev et al., 1999). It has been proposed that BCC cells must also remain in contact with inductive stroma to survive (Miller, 1995). Additionally, in chemical carcinogenesis or transgenic mouse studies, the use of phorbol ester tumor promoters may convert label-retaining cells into transit-amplifying cells (Morris, 2000). Alternatively, the relatively short anagen period and rapid onset of catagen in mice may eliminate occasional cells migrating from the bulge that harbor heterozygous mutations of the SHH pathway.

Experimentally, squamous papillomas induced in mice by chemical carcinogens have the capacity to differentiate into keratinous or sebaceous cysts, depending on the microenvironment (Andrews, 1974). This is an interesting example of neoplastic reversion. It is increasingly clear that transit-amplifying or perhaps even differentiation-committed cells have the capacity to form benign tumors, as in the example of c-*myc* or oncogenic *ras* targeted suprabasally to the skin of transgenic mice (Greenhalgh et al., 1993; Brown et al., 1998; Pelengaris et al., 1999). Since tumors from cells in these compartments rarely develop into SCC, they may be analogous to actinic keratoses in sun-damaged human skin (Fig. 1). This raises the issue of the consequence of stem cell vs committed cell targeting for cancer. Many of the tumors most likely of follicle stem cell origin (BCC and hair follicle tumors) are induced through alterations in developmental pathways and are relatively benign lesions. Likewise, most precursor lesions for SCC (actinic keratoses in human and papillomas in mice) are terminally benign and may spontaneously regress. However, a subpopulation of SCC precursor lesions in both species is at high risk for premalignant progression, and in mouse studies these lesions are predetermined and recognizable from their initial phenotype (Tennenbaum et al., 1993; Rehman et al., 1994). Could the determining factors reside in the inherent properties of the target cell, the disturbed genetic pathway initiating the tumor host-modifying factors, or the location of the incipient tumor? One logical explanation would be that high-risk SCC precursors evolve as a result of oncogenic activation of nondevelopmental pathways in epidermal or follicle stem cells. These populations would lose multipotentiality by virtue of the somatic oncogenic pathway but retain self-renewal by virtue of the stem cell origin. With increasing knowledge of the biochemistry of the stem cell and the impact of tumor-initiating mutations, such questions will be answered in the near future.

REFERENCES

Andrews, E. J. (1974) The morphological, biological, and antigenic characteristics of transplantable papillomas and keratinous cysts induced by methylcholanthrene. *Cancer Res.* 34:2842–2851.

Arnold, I. and Watt, F. M. (2001) c-Myc activation in transgenic mouse epidermis results in mobilization of stem cells and differentiation of their progeny. *Curr. Biol.* 11:558–568.

Aszterbaum, M., Epstein, J., Oro, A., et al. (1999) Ultraviolet and ionizing radiation enhance the growth of BCCs and trichoblastomas in patched heterozygous knockout mice. *Nat. Med.* 5:1285-1291.

Bailleul, B., Surani, M. A., White, S., et al. (1990) Skin hyperkeratosis and papilloma formation in transgenic mice expressing a *ras* oncogene from a suprabasal keratin promoter. *Cell* 62:697–708.

Bale, A. E. and Yu, K. P. (2001) The hedgehog pathway and basal cell carcinomas. *Hum. Mol. Genet.* 10:757–762.

Berg, R. J., van Kranen, H. J., Rebel, H. G., et al. (1996) Early p53 alterations in mouse skin carcinogenesis by UVB radiation: immunohistochemical detection of mutant p53 protein in clusters of preneoplastic epidermal cells. *Proc. Natl. Acad. Sci. USA* 93:274–278.

Bickenbach, J. R., Vormwald-Dogan, V., Bachor, C., Bleuel, K., Schnapp, G., and Boukamp, P. (1998) Telomerase is not an epidermal stem cell marker and is downregulated by calcium. *J. Invest. Dermatol.* 111: 1045–1052.

Binder, R. L., Gallagher, P. M., Johnson, G. R., et al. (1997) Evidence that initiated keratinocytes clonally expand into multiple existing hair follicles during papilloma histogenesis in SENCAR mouse skin. *Mol. Carcinog.* 20:151–158.

Bonifas, J. M., Pennypacker, S., Chuang, P. T., et al. (2001) Activation of expression of hedgehog target genes in basal cell carcinomas. *J. Invest. Dermatol.* 116:739–742.

Brakebusch, C., Grose, R., Quondamatteo, F., et al. (2000) Skin and hair follicle integrity is crucially dependent on beta 1 integrin expression on keratinocytes. *EMBO J.* 19:3990–4003.

Brown, K., Strathdee, D., Bryson, S., Lambie, W., and Balmain, A. (1998) The malignant capacity of skin tumours induced by expression of a mutant H-ras transgene depends on the cell type targeted. *Curr. Biol.* 8:516–524.

Burnham, D. K., Gahring, L. C., and Dayes, R. A. (1986) Clonal origin of tumors induced by ultraviolet radiation. *J. Natl. Cancer Inst.* 76: 151–157.

Chan, E. F., Gat, U., McNiff, J. M., and Fuchs, E. (1999) A common human skin tumour is caused by activating mutations in beta-catenin. *Nat. Genet.* 21:410–413.

Chen, Z., Smith, K. J., Skelton, H. G. III, Barrett, T. L., Greenway, H. T. Jr., and Lo, S. C. (2001) Telomerase activity in Kaposi's sarcoma, squamous cell carcinoma, and basal cell carcinoma. *Exp. Biol. Med. (Maywood)* 226:753–757.

Chiang, C., Swan, R. Z., Grachtchouk, M., et al. (1999) Essential role for Sonic hedgehog during hair follicle morphogenesis. *Dev. Biol.* 205:1–9.

Cotsarelis, G., Sun, T. T., and Lavker, R. M. (1990) Label-retaining cells reside in the bulge area of pilosebaceous unit: implications for follicular stem cells, hair cycle, and skin carcinogenesis. *Cell* 61: 1329–1337.

Deamant, F. D. and Iannaccone, P. M. (1987) Clonal origin of chemically induced papillomas: separate analysis of epidermal and dermal components. *J. Cell Sci.* 88(Pt. 3):305–312.

Dickson, M. A., Hahn, W. C., Ino, Y., et al. (2000) Human keratinocytes that express hTERT and also bypass a p16(INK4a)-enforced mechanism that limits life span become immortal yet retain normal growth and differentiation characteristics. *Mol. Cell. Biol.* 20:1436–1447.

Frame, S. and Balmain, A. (1999) Target genes and target cells in carcinogenesis. *Br. J. Cancer* 80(Suppl. 1):28–33.

Gat, U., DasGupta, R., Degenstein, L., and Fuchs, E. (1998) De novo hair follicle morphogenesis and hair tumors in mice expressing a truncated beta-catenin in skin. *Cell* 95:605–614.

Ghali, L., Wong, S. T., Green, J., Tidman, N., and Quinn, A. G. (1999) Gli1 protein is expressed in basal cell carcinomas, outer root sheath keratinocytes and a subpopulation of mesenchymal cells in normal human skin. *J. Invest. Dermatol.* 113:595–599.

Gonzalez-Suarez, E., Samper, E., Flores, J. M., and Blasco, M. A. (2000) Telomerase- deficient mice with short telomeres are resistant to skin tumorigenesis. *Nat. Genet.* 26:114–117.

Gonzalez-Suarez, E., Samper, E., Ramirez, A., et al. (2001) Increased epidermal tumors and increased skin wound healing in transgenic mice overexpressing the catalytic subunit of telomerase, mTERT, in basal keratinocytes. *EMBO J.* 20:2619–2630.

Greenhalgh, D. A., Rothnagel, J. A., Quintanilla, M. I., et al. (1993) Induction of epidermal hyperplasia, hyperkeratosis, and papillomas in transgenic mice by a targeted v-Ha-*ras* oncogene. *Mol. Carcinog.* 7: 99–110.

Hansen, L. A. and Tennant, R. W. (1994) Follicular origin of epidermal papillomas in v- Ha-*ras* transgenic TG.AC mouse skin. *Proc. Natl. Acad. Sci. USA* 91:7822–7826.

Huelsken, J., Vogel, R., Erdmann, B., Cotsarelis, G., and Birchmeier, W. (2001) beta- Catenin controls hair follicle morphogenesis and stem cell differentiation in the skin. *Cell* 105:533–545.

Iannaccone, P. M., Gardner, R. L., and Harris, H. (1978) The cellular origin of chemically induced tumors. *J. Cell Biol.* 29:249–269.

Iannaccone, P. M., Weinberg, W. C., and Deamant, F. D. (1987) On the clonal origin of tumors: a review of experimental models. *Int. J. Cancer* 39:778–784.

Jones, P. H. and Watt, F. M. (1993) Separation of human epidermal stem cells from transit amplifying cells on the basis of differences in integrin function and expression. *Cell* 73:713–724.

Kanjilal, S., Pierceall, W. E., Cummings, K. K., Kripke, M. L., and Ananthaswamy, H. N. (1993) High frequency of *p53* mutations in ultraviolet radiation-induced murine skin tumors: evidence for strand bias and tumor heterogeneity. *Cancer Res.* 53:2961–2964.

Klein-Szanto, A. J., Major, S. K., and Slaga, T. J. (1980) Induction of dark keratinocytes by 12-O-tetradecanoylphorbol-13-acetate and mezerein as an indicator of tumor-promoting efficiency. *Carcinogenesis* 1:399–406.

Levy, L., Broad, S., Diekmann, D., Evans, R. D., and Watt, F. M. (2000) beta1 integrins regulate keratinocyte adhesion and differentiation by distinct mechanisms. *Mol. Biol. Cell* 11:453–466.

Liang, L. and Bickenbach, J. R. (2002) Somatic epidermal stem cells can produce multiple cell lineages during development. *Stem Cells* 20:21–31.

Lyle, S., Christofidou-Solomidou, M., Liu, Y., Elder, D. E., Albelda, S., and Cotsarelis, G. (1998) The C8/144B monoclonal antibody recognizes cytokeratin 15 and defines the location of human hair follicle stem cells. *J. Cell Sci.* 111(Pt. 21):3179–3188.

Mackenzie, I. C. (1997) Retroviral transduction of murine epidermal stem cells demonstrates clonal units of epidermal structure. *J. Invest. Dermatol.* 109:377–383.

Mackenzie, I. C. and Bickenbach, J. R. (1985) Label-retaining keratinocytes and Langerhans cells in mouse epithelia. *Cell Tissue Res.* 242:551–556.

Mainiero, F., Murgia, C., Wary, K. K., et al. (1997) The coupling of alpha6beta4 integrin to Ras-MAP kinase pathways mediated by Shc controls keratinocyte proliferation. *EMBO J.* 16:2365–2375.

McGowan, K. M. and Coulombe, P. A. (1998) Onset of keratin 17 expression coincides with the definition of major epithelial lineages during skin development. *J. Cell Biol.* 143:469–486.

Michel, M., Torok, N., Godbout, M. J., et al. (1996) Keratin 19 as a biochemical marker of skin stem cells in vivo and in vitro: keratin 19 expressing cells are differentially localized in function of anatomic sites, and their number varies with donor age and culture stage. *J. Cell Sci.* 109(Pt. 5):1017–1028.

Miller, S. J. (1995) Etiology and pathogenesis of basal cell carcinoma. *Clin. Dermatol.* 13:527–536.

Miller, S. J., Wei, Z. G., Wilson, C., Dzubow, L., Sun, T. T., and Lavker, R. M. (1993) Mouse skin is particularly susceptible to tumor initiation during early anagen of the hair cycle: possible involvement of hair follicle stem cells. *J. Invest. Dermatol.* 101:591–594.

Mills, A. A., Zheng, B., Wang, X. J., Vogel, H., Roop, D. R., and Bradley, A. (1999) p63 is a p53 homologue required for limb and epidermal morphogenesis. *Nature* 398:708–713.

Morris, R. J. (2000) Keratinocyte stem cells: targets for cutaneous carcinogens. *J. Clin. Invest.* 106:3–8.

Morris, R. J. and Potten, C. S. (1999) Highly persistent label-retaining cells in the hair follicles of mice and their fate following induction of anagen. *J. Invest. Dermatol.* 112:470–475.

Morris, R. J., Fischer, S. M., and Slaga, T. J. (1986) Evidence that a slowly cycling subpopulation of adult murine epidermal cells retains carcinogen. *Cancer Res.* 46:3061–3066.

Morris, R. J., Tacker, K. C., Fischer, S. M., and Slaga, T. J. (1988) Quantitation of primary in vitro clonogenic keratinocytes from normal adult murine epidermis, following initiation, and during promotion of epidermal tumors. *Cancer Res.* 48:6285–6290.

Morris, R. J., Fischer, S. M., Klein-Szanto, A. J., and Slaga, T. J. (1990) Subpopulations of primary adult murine epidermal basal cells sedimented on density gradients. *Cell Tissue Kinet.* 23:587–602.

Morris, R. J., Coulter, K., Tryson, K., and Steinberg, S. R. (1997) Evidence that cutaneous carcinogen-initiated epithelial cells from mice are quiescent rather than actively cycling. *Cancer Res.* 57:3436–3443.

Morris, R. J., Tryson, K. A., and Wu, K. Q. (2000) Evidence that the epidermal targets of carcinogen action are found in the interfollicular epidermis of infundibulum as well as in the hair follicles. *Cancer Res.* 60:226–229.

Oshima, H., Rochat, A., Kedzia, C., Kobayashi, K., and Barrandon, Y. (2001) Morphogenesis and renewal of hair follicles from adult multipotent stem cells. *Cell* 104:233–245.

Owens, D. M. and Watt, F. M. (2001) Influence of beta1 integrins on epidermal squamous cell carcinoma formation in a transgenic mouse model: alpha3beta1, but not alpha2beta1, suppresses malignant conversion. *Cancer Res.* 61:5248–5254.

Panteleyev, A. A., Botchkareva, N. V., Sundberg, J. P., Christiano, A. M., and Paus, R. (1999) The role of the hairless (hr) gene in the regulation of hair follicle catagen transformation. *Am. J. Pathol.* 155:159–171.

Parsa, R., Yang, A., McKeon, F., and Green, H. (1999) Association of p63 with proliferative potential in normal and neoplastic human keratinocytes. *J. Invest. Dermatol.* 113:1099–1105.

Pelengaris, S., Littlewood, T., Khan, M., Elia, G., and Evan, G. (1999) Reversible activation of c-Myc in skin: induction of a complex neoplastic phenotype by a single oncogenic lesion. *Mol. Cell* 3: 565–577.

Pelisson, I., Soler, C., Chardonnet, Y., Euvrard, S., and Schmitt, D. (1996) A possible role for human papillomaviruses and c-myc, c-Ha-ras, and p53 gene alterations in malignant cutaneous lesions from renal transplant recipients. *Cancer Detect. Prev.* 20:20–30.

Pellegrini, G., Dellambra, E., Golisano, O., et al. (2001) p63 identifies keratinocyte stem cells. *Proc. Natl. Acad. Sci. USA* 98:3156–3161.

Ponten, F., Berg, C., Ahmadian, A., et al. (1997) Molecular pathology in basal cell cancer with p53 as a genetic marker. *Oncogene* 15:1059–1067.

Raghavan, S., Bauer, C., Mundschau, G., Li, Q., and Fuchs, E. (2000) Conditional ablation of beta1 integrin in skin. Severe defects in epidermal proliferation, basement membrane formation, and hair follicle invagination. *J. Cell Biol.* 150:1149–1160.

Raick, A. N. (1973) Ultrastructural, histological, and biochemical alterations produced by 12-O-tetradecanoyl-phorbol-13-acetate on mouse epidermis and their relevance to skin tumor promotion. *Cancer Res.* 33:269–286.

Raick, A. N. (1974) Cell differentiation and tumor-promoting action in skin carcinogenesis. *Cancer Res.* 34:2915–2925.

Ramirez, R. D., Wright, W. E., Shay, J. W., and Taylor, R. S. (1997) Telomerase activity concentrates in the mitotically active segments of human hair follicles. *J. Invest. Dermatol.* 108:113–117.

Reddy, A. L. and Fialkow, P. J. (1983) Papillomas induced by initiation-promotion differ from those induced by carcinogen alone. *Nature* 304: 69–71.

Rehman, I., Quinn, A. G., Healy, E., and Rees, J. L. (1994) High frequency of loss of heterozygosity in actinic keratoses, a usually benign disease. *Lancet* 344:788–789.

Ren, Z. P., Hedrum, A., Ponten, F., et al. (1996) Human epidermal cancer and accompanying precursors have identical p53 mutations different from p53 mutations in adjacent areas of clonally expanded non-neoplastic keratinocytes. *Oncogene* 12:765–773.

Ren, Z. P., Ahmadian, A., Ponten, F., et al. (1997) Benign clonal keratinocyte patches with p53 mutations show no genetic link to synchronous squamous cell precancer or cancer in human skin. *Am. J. Pathol.* 150:1791–1803.

Rossen, K., Dahlstrom, K. K., Mercurio, A. M., and Wewer, U. M. (1994) Expression of the alpha 6 beta 4 integrin by squamous cell carcinomas and basal cell carcinomas: possible relation to invasive potential? *Acta Derm. Venereol.* 74:101–105.

Rounbehler, R. J., Schneider-Broussard, R., Conti, C. J., and Johnson, D. G. (2001) Myc lacks E2F1's ability to suppress skin carcinogenesis. *Oncogene* 20:5341–5349.

Savoia, P., Cremona, O., Trusolino, L., Pepino, E., and Marchisio, P. C. (1994) Integrins and basement membrane proteins in skin carcinomas. *Pathol. Res. Pract.* 190:950–954.

Slaga, T. J. and Klein-Szanto, A. J. (1983) Initiation-promotion versus complete skin carcinogenesis in mice: importance of dark basal keratinocytes (stem cells). *Cancer Invest.* 1:425–436.

Taguchi, T., Yokoyama, M., and Kitamura, Y. (1984) Intraclonal conversion from papilloma to carcinoma in the skin of Pgk-1a/Pgk-1b mice treated by a complete carcinogenesis process or by an initiation-promotion regimen. *Cancer Res.* 44:3779–3782.

Tani, H., Morris, R. J., and Kaur, P. (2000) Enrichment for murine keratinocyte stem cells based on cell surface phenotype. *Proc. Natl. Acad. Sci. USA* 97:10,960–10,965.

Taylor, G., Lehrer, M. S., Jensen, P. J., Sun, T. T., and Lavker, R. M. (2000) Involvement of follicular stem cells in forming not only the follicle but also the epidermis. *Cell* 102:451–461.

Taylor, R. S., Ramirez, R. D., Ogoshi, M., Chaffins, M., Piatyszek, M. A., and Shay, J. W. (1996) Detection of telomerase activity in malignant and nonmalignant skin conditions. *J. Invest. Dermatol.* 106:759–765.

Tennenbaum, T., Weiner, A. K., Belanger, A. J., Glick, A. B., Hennings, H., and Yuspa, S. H. (1993) The suprabasal expression of $\alpha 6 \beta 4$ integrin is associated with a high risk for malignant progression in mouse skin carcinogenesis. *Cancer Res.* 53:4803–4810.

Tennenbaum, T., Belanger, A. J., Glick, A. B., Tamura, R., Quaranta, V., and Yuspa, S. H. (1995) A splice variant of $\alpha 6$ integrin is associated with malignant conversion in mouse skin tumorigenesis. *Proc. Natl. Acad. Sci. USA* 92:7041–7045.

Tuominen, H., Junttila, T., Karvonen, J., and Kallioinen, M. (1994) Cell-type related and spatial variation in the expression of integrins in cutaneous tumors. *J. Cutan. Pathol.* 21:500–506.

Waikel, R. L., Kawachi, Y., Waikel, P. A., Wang, X. J., and Roop, D. R. (2001) Deregulated expression of c-Myc depletes epidermal stem cells. *Nat. Genet.* 28:165–168.

Walsh, D. S., Peacocke, M., Harrington, A., James, W. D., and Tsou, H. C. (1998) Patterns of X chromosome inactivation in sporadic basal cell carcinomas: evidence for clonality. *J. Am. Acad. Dermatol.* 38: 49–55.

Watt, F. M. (2001) Stem cell fate and patterning in mammalian epidermis. *Curr. Opin. Genet. Dev.* 11:410–417.

Weinberg, W. C., Morgan, D., George, C., and Yuspa, S. H. (1991) A comparison of interfollicular and hair follicle derived cells as targets for the v-ras^{Ha} oncogene in mouse skin carcinogenesis. *Carcinogenesis* 12:1119–1124.

Weinberg, W. C., Ng, Y. K., and Iannaccone, P. M. (1992) Clonal analysis of hepatic neoplasms by mosaic pattern. In: *The Role of Cell Types in Hepatocarcinogenesis*. (Sirica, A. E., ed.), CRC Press, Boca Raton, FL, pp. 29–53.

Winton, D. J., Blount, M. A., and Ponder, B. A. (1989) Polyclonal origin of mouse skin papillomas. *Br. J. Cancer* 60:59–63.

Wu, A., Ichihashi, M., and Ueda, M. (1999) Correlation of the expression of human telomerase subunits with telomerase activity in normal skin and skin tumors. *Cancer* 86:2038–2044.

Yang, A., Kaghad, M., Wang, Y., Gillett, E., Fleming, M. D., Dotsch, V., Andrews, N. C., Caput, D., and McKeon, F. (1998) p63, a p53 homolog at 3q27-29, encodes multiple products with transactivating, death-inducing, and dominant-negative activities. *Mol. Cell* 2:305–316.

Yang, A., Schweitzer, R., Sun, D., Kaghad, M., Walker, N., Bronson, R. T., Tabin, C., Sharpe, A., Caput, D., Crum, C., and McKeon, F. (1999) p63 is essential for regenerative proliferation in limb, craniofacial and epithelial development. *Nature* 398:714–718.

Yuspa, S. H. (1994) The pathogenesis of squamous cell cancer: lessons learned from studies of skin carcinogenesis—Thirty-third G.H.A. Clowes Memorial Award Lecture. *Cancer Res.* 54:1178–1189.

Zhang, W., Remenyik, E., Zelterman, D., Brash, D. E., and Wikonkal, N. M. (2001) Escaping the stem cell compartment: sustained UVB exposure allows p53-mutant keratinocytes to colonize adjacent epidermal proliferating units without incurring additional mutations. *Proc. Natl. Acad. Sci. USA* 98:13,948–13,953.

Zhang, Y. and Kalderon, D. (2001) Hedgehog acts as a somatic stem cell factor in the *Drosophila* ovary. *Nature* 410:599–604.

Zhu, A. J. and Watt, F. M. (1999) beta-catenin signalling modulates proliferative potential of human epidermal keratinocytes independently of intercellular adhesion. *Development* 126:2285–2298.

27 The Stem Cell Plasticity of Aggressive Melanoma Tumor Cells

MARY J. C. HENDRIX, PhD, ELISABETH A. SEFTOR, BS, PAUL S. MELTZER, MD, PhD, ANGELA R. HESS, PhD, LYNN M. GRUMAN, HT, BRIAN J. NICKOLOFF, MD, PhD, LUCIO MIELE, MD, PhD, DON D. SHERIFF, PhD, GINA C. SCHATTEMAN, PhD, MARIO A. BOURDON, PhD, AND RICHARD E.B. SEFTOR, PhD

Microarray analysis of melanoma cell lines suggests that aggressive melanomas have a pluripotent, embryonic-like phenotype, implying the possibility of a stem cell origin for tumor components. Aggressive melanoma cells also form vascular structures and express endothelial-associated genes (including vascular endothelial-cadherin and EphA2) critical for vessel formation, indicating that the tumor cells have the plasticity to generate progeny, which express multiple cellular phenotypes with additional biological potential. This does not occur in poorly aggressive tumors, and, thus, expression of these genes is a predictor of biological behavior of the tumor. The aggressive tumor cells were able to participate in the neovascularization of ischemic tissue and produce factors that influence poorly aggressive tumor cells to assume a vascular phenotype. Understanding the molecular underpinnings of the plasticity of melanoma cells may lead to more effective diagnosis, treatment, and prevention measures for aggressive melanoma tumors.

27.1. INTRODUCTION

Like many types of cancers, cutaneous and uveal melanoma could benefit tremendously from the development of valid predictors of the aggressive potential of these diseases in patients. Recent studies aimed at characterizing the molecular signature of melanoma tumor cells generated a classification scheme for malignant cutaneous melanoma (Bittner et al., 2000), in addition to a molecular profile for uveal melanoma (Seftor et al., 2002b). These molecular tools demonstrated how global transcript analysis could contribute to the classification of melanoma, which may hold great promise in improving diagnosis and treatment.

The microarray analysis studies of differential gene expression of highly aggressive vs poorly aggressive human cutaneous and uveal melanoma cell lines (Bittner et al., 2000; Seftor et al., 2002b) revealed the coexpression of multiple phenotype-specific genes by the aggressive tumor cells, including those of endothelial, epithelial, fibroblastic, hematopoietic, kidney, neuronal, muscle, and several precursor cell types. These observations were quite intriguing and suggested that aggressive melanoma cells may undergo a genetic reversion to a pluripotent, embryonic-like phenotype. However, the biological significance of the unexpected expression of multiple molecular phenotypes by melanoma cells remains enigmatic. Indeed, these findings have prompted further investigation regarding the potential relevance of a "plastic" tumor cell phenotype and challenge our current thinking of identifying and targeting tumor cells that may masquerade as other cell types.

27.1.1. TUMOR CELL PLASTICITY ALLOWS FOR VASCULOGENIC MIMICRY BY AGGRESSIVE MELANOMA

Many of the biological properties germane to embryogenesis are also important in tumor growth. For example, during embryonic development, the formation of primary vascular networks occurs by the process of vasculogenesis—the *in situ* differentiation of mesodermal progenitor cells (angioblasts or hemangioblasts) to endothelial cells that organize into a primitive network (for a review, *see* Risau, 1997; Carmeliet, 2000) (Fig.1). The subsequent remodeling of the vasculogenic network into a more refined complex of vasculature occurs through angiogenesis—the sprouting of new capillaries from preexisting networks. Similarly, it is widely accepted that during cancer progression, tumors require a blood supply for growth (Folkman, 1995; Rak and Kerbel, 1996; Kumar and Fidler, 1998). Based on the molecular profile of the aggressive melanoma cells, together with novel in vitro observations and correlative histopathology findings, our laboratory and collaborators have introduced the concept of vasculogenic mimicry to describe the plasticity and unique ability of aggressive melanoma tumor cells to express endothelial-associated genes and form tubular structures and patterned networks in three-dimensional (3D) culture that mimic embryonic vasculogenic networks (Maniotis et al., 1999; Hendrix et al., 2001; Hess et al., 2001; Seftor et al., 2001). In animal and patient tumors, many of

From: *Stem Cells Handbook*
Edited by: S. Sell © Humana Press Inc., Totowa, NJ

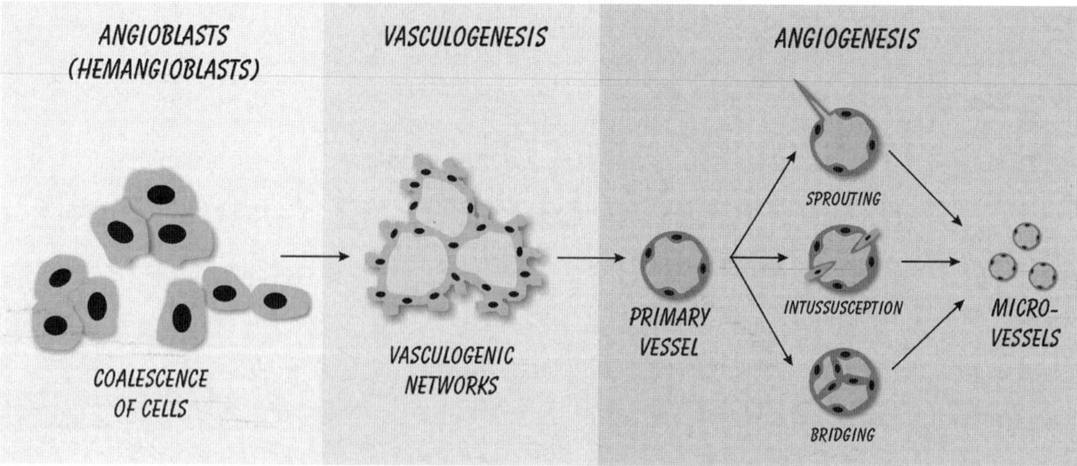

Fig. 1. Overview of vasculogenesis and angiogenesis.

these networks are lined by tumor cells adjacent to channels, some containing red blood cells (RBCs) and plasma, and possibly providing a perfusion mechanism and dissemination route within the tumor compartment that functions either independently of or simultaneously with angiogenesis. Tumors exhibiting vasculogenic mimicry have a poor prognosis (Maniotis et al., 1999); however, the biological relevance of this potential perfusion pathway remains to be elucidated (Hendrx et al., 2003).

A noteworthy example of melanoma tumor cell plasticity is illustrated in Fig. 2, which shows the coalescence of several spheroids of aggressive human cutaneous melanoma cells after 1 mo in suspension culture. A histological section through this spheroidal mass revealed that the outermost layer resembled a stratified squamouse-like epithelium, while the inner mass was organized like a mesoderm with evidence of vasculogenic-like networks. This seemingly simplistic observation is an excellent illustration of the phenotypic diversity exhibited by aggressive melanoma cells, which is greatly influenced by microenvironmental factors.

The molecular underpinnings of melanoma vasculogenic mimicry are beginning to unfold. Recent experimental evidence has shown the importance of several key molecules in the formation of vasculogenic-like networks by melanoma cells, including vascular endothelial (VE)-cadherin (CD144 or cadherin 5), EphA2 (epithelial cell kinase), and laminin (Hendrix et al., 2001; Hess et al., 2001; Seftor et al., 2001). These molecules (and their binding partners) are also essential in the formation and maintenance of blood vessels (Risau, 1997; Carmeliet, 2000). VE-cadherin is an adhesive protein, previously considered to be endothelial cell-specific, belonging to the cadherin family of transmembrane proteins promoting homotypic cell-to-cell interaction (Hynes, 1992, Kemler, 1992; Lampugnani et al., 1992; Gumbiner, 1996). EphA2 is a receptor protein tyrosine kinase that is part of a large family of ephrin receptors (Pasquale, 1997). Binding of EphA2 to its ligand ephrin-A1 results in the phosphorylation of EphA2; however, there is evidence indicating that EphA2 can also be constitutively phosphorylated in unstimulated cells (Rosenburg et al., 1997). Strong expression of EphA2 and ephrin-A1 has been associated with increased melanoma thickness and decreased survival, and the EphA2/ephrin-A1 pathway has been linked to tumor cell proliferation (Straume and Akslen, 2001). Laminins are major components of basement membranes and play an active role in neurite outgrowth, tumor metastasis, cell attachment and migration, and angiogenesis (Malinda and Kleinman, 1996; Colognato and Yurchenco, 2000; Straume and Akslen, 2001). Proteolytic cleavage of laminin, particularly the laminin 5 $\gamma 2$ chain, can alter and regulate the integrin-mediated migratory behavior of certain cells (Giannelli et al., 1997; Malinda et al., 1999; Koshikawa et al., 2000; Straume and Akslen, 2001), illustrating its potential importance as a molecular trigger in the microenvironment.

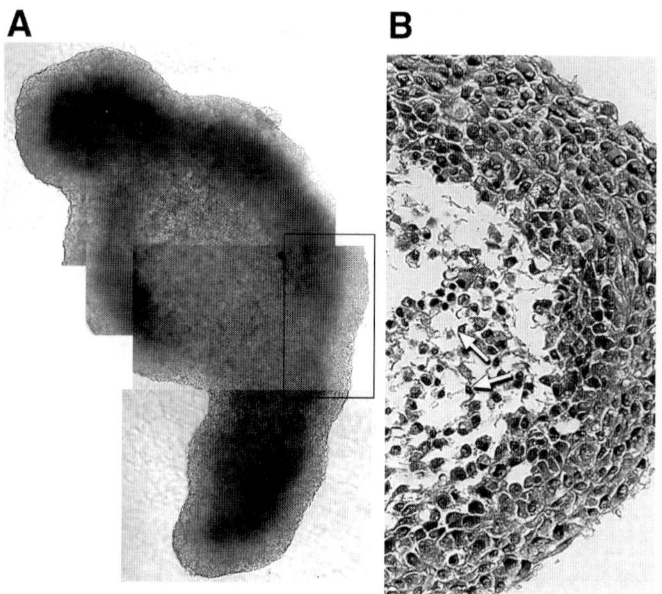

Fig. 2. Example of melanoma tumor cell plasticity. **(A)** Microscopic low-power view of the coalescence of several large spheroids of aggressive human cutaneous C8161 melanoma cells after 1 mo in suspension culture; **(B)** a histological section through an area delineated by rectangle in (A), where outermost layer appears similar to stratified squamous epithelium and inner mass shows vasculogenic-like networks (arrows). Magnification: (A) ×15; original magnification: (B) ×40.

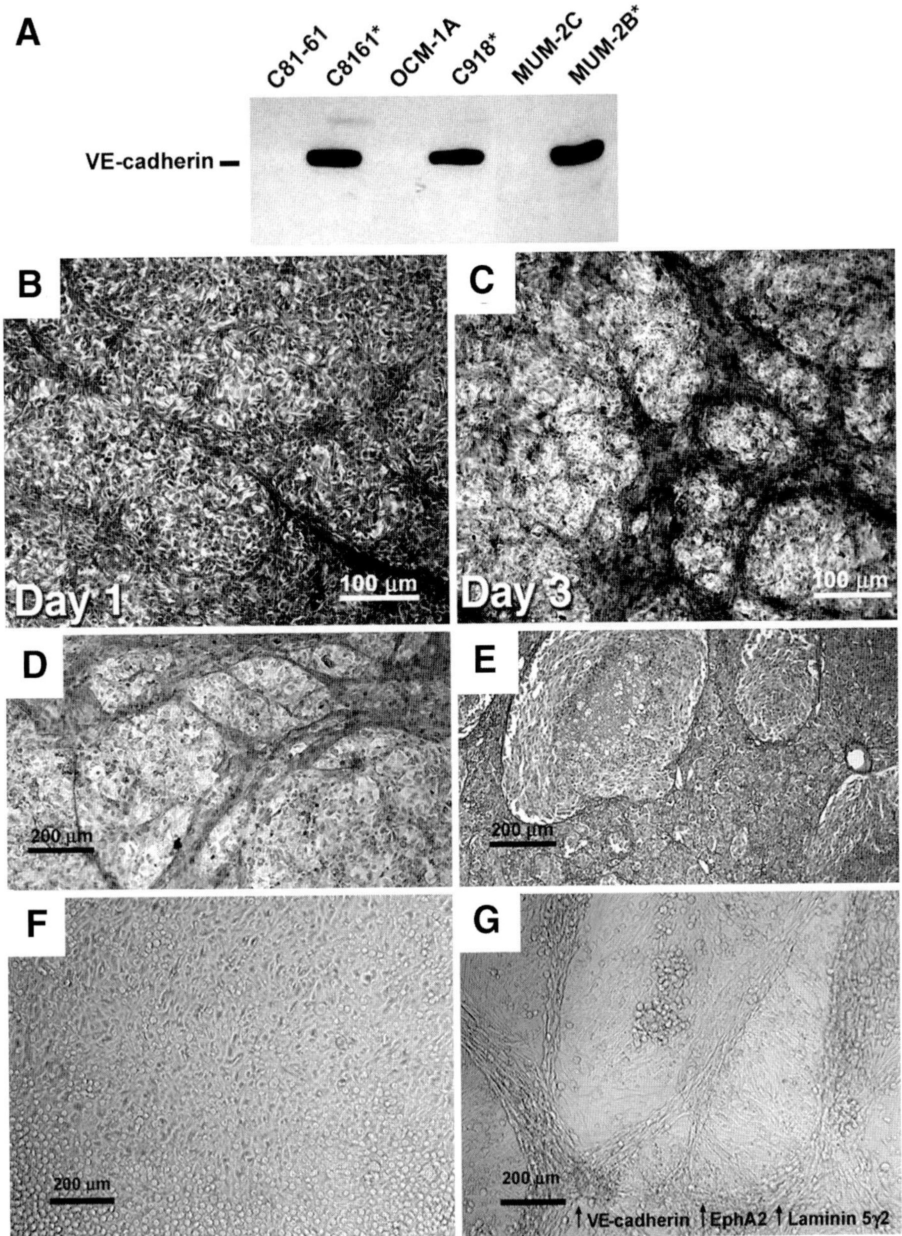

Fig. 3. Vasculogenic mimicry in aggressive human uveal and cutaneous melanoma. **(A)** Western blot analysis of VE-cadherin by aggressive* human cutaneous (C8161) and uveal (C918, MUM-2B), but not poorly aggressive, cutaneous (C81-61) and uveal (OCM-1A, MUM-2C) melanoma cells. **(B,C)** phase contrast microscopy of developing vasculogenic networks, at d 1 and 3, respectively, formed by aggressive MUM-2B cells in 3D culture. **(D,E)** immunohistochemical localization of laminin 5γ2 chain (dark staining highlighting vasculogenic-like networks) in aggressive C8161 cells in 3D culture (D), similar to laminin staining (dark staining surrounding spheroidal nests of tumor cells) of network patterns in a patient tissue section with aggressive uveal melanoma. **(F)** poorly aggressive C81-61 cells cultured on a collagen I gel for 4 d (showing no vasculogenic networks) and on a gel preconditioned for 2 d **(G)** by aggressive C8161 cells, which were subsequently removed before addition of poorly aggressive C81-61 cells—which now form vasculogenic-like networks and express VE-cadherin, EphA2, and laminin 5γ2 chain.

The importance of VE-cadherin, EphA2, and laminin 5 γ2 chain is summarized in Fig. 3. Molecular analyses revealed that these three molecules, among others, were dramatically overexpressed in aggressive human cutaneous and uveal melanoma cells (Bittner et al., 2000; Hendrix et al., 2001; Hess et al., 2001; Seftor et al., 2001, 2002b). At the protein level, these molecules were expressed only by aggressive tumor cells, not by poorly aggressive melanoma cells. The biological relevance of these molecules in vasculogenic mimicry was further demonstrated by downregulating their expression independently and measuring the consequences on their ability to form vasculo-genic-like networks. This approach revealed that downregulation of VE-cadherin, EphA2, or laminin 5 γ2 chain resulted in the complete inability of aggressive melanoma cells to engage in the formation of vasculogenic-like networks in 3D culture (Hendrix et al., 2001; Hess et al., 2001; Seftor et al., 2001). Additional experiments focused on tumor cell–

associated extracellular matrix (ECM) led to the discovery of the inductive potential of the microenvironment preconditioned by aggressive melanoma cells and shown to influence poorly aggressive melanoma cells to assume a vasculogenic phenotype (Seftor et al., 2001). These data, along with subsequent observations of vascular-associated gene expression (summarized in Fig. 3G), suggest that highly aggressive melanoma cells deposit molecular signals in their environment that have the potential to induce a vasculogenic phenotype in poorly aggressive cells, with respect to the formation of vasculogenic-like networks and the concomitant expression of vascular-associated genes (VE-cadherin, EphA2, and laminin 5 $\gamma 2$ chain).

Based on the intriguing immunohistochemical colocalization pattern of VE-cadherin and EphA2, together with immunoprecipitation data, we have proposed a novel hypothetical model for signaling during vasculogenic mimicry (Seftor et al., 2002). This model highlights the cooperative interactions of VE-cadherin and EphA2 and suggests downstream signaling involving phosphatidylinositol-3 kinase and focal adhesion kinase, initiated by ephrin-A1 binding. These molecules are the focus of ongoing investigations to elucidate important signaling pathways involved in vasculogenic mimicry. Collectively, these findings may provide the basis of new therapeutic and diagnostic methods for cancer detection and mangement.

27.1.2. BIOLOGICAL RELEVANCE OF VASCULOGENIC PHENOTYPE IN VIVO Studies addressing the biological significance of the melanoma vasculogenic phenotype in animal models have focused on three major questions: (1) Is there a fundamental difference in the vascularization of highly aggressive vs poorly aggressive tumor cell masses? (2) Can a morphological connection be identified between tumor cell-lined vascular channels within aggressive tumors and endothelial-lined vasculature? (3) Is it possible for aggressive tumor cells to provide a vascular function when challenged to an ischemic environment?

To address the first question regarding potential differences in the vascularization of highly vs poorly aggressive melanoma cell masses, we utilized a dorsal skinfold window chamber in a nude mouse model. Into the chambers were placed either aggressive human cutaneous melanoma cells (in 3D collagen gels) or poorly aggressive human cutaneous melanoma cells under similar conditions, and the data were recorded at various time intervals over 2 wk with videomicroscopy and histological analyses. As summarized in Fig. 4, the aggressive melanoma 3D culture showed the presence of RBCs and plasma in tumor-lined channels and networks—in the absence of angiogenesis within the tumor compartment. At the periphery of the tumor mass in the stromal compartment, blood vessels were evident, suggesting a possible connection between tumor cell–lined channels and endothelial-lined vasculature. By contrast, the poorly aggressive cutaneous melanoma cells in 3D collagen gels induced an angiogenic response in the mouse model. These data demonstrate a fundamental difference in the vascularization of highly aggressive vs poorly aggressive melanoma tumor cell masses.

The second question sought to identify a potential biological connection between tumor-lined vascular channels within aggressive melanoma tumors and endothelial-lined vasculature. Using a nude mouse model injected subcutaneously with red fluorescent protein (RFP)–labeled aggressive human cutaneous melanoma cells, primary tumors were allowed to develop over 42 d. As summarized in Fig. 5, a combination of routine histological analysis and microsphere perfusion of the mouse vasculature and confocal microscopy with 3D reconstruction revealed the presence of microspheres (derived from the perfused vasculature) within the human tumor compartment. The endothelial-lined mouse vasculature was seen at the periphery of the human tumor, where fluorescein isothiocyanate (FITC)-labeled vasculature overlapped with RFP-labeled tumor cells forming vasculo-genic-like networks and channels. The aggressive human tumors demonstrated vasculogenic mimicry and no necrosis within the tumor compartment which coalesced with angiogenic mouse vessels at the human–mouse interface. These data are the first demonstration of a physiological connection between tumor cell line vascular channels in aggressive melanoma tumors and mouse endothelial-lined vasculature. However, Shirakawa et al. (2002) have recently used dynamic micromagnetic resonance angiography analysis and laser-capture microdissection to demonstrate a similar vascular connection between vasculogenic mimicry within inflammatory breast cancer xenografts and mouse angiogenic/host vessels at the tumor margin. They used the strategy of laser-capture microdissection to document the expression of endothelial-associated genes by breast cancer cells within the inflammatory breast xenograft that exhibited no central necrosis, no endothelial cells, and no fibrosis.

The third question explored the plasticity of aggressive melanoma tumor cells by challenging them to an ischemic environment and assessing whether the cells would participate in neovascularization and/or form a tumor. Hendrix et al. (2002), injected intramuscularly Ad-RFP (RFP expressing adenovirus)- or enhanced green fluorescent protein (EGFP)-labeled human cutaneous metastatic melanoma cells into the hind limbs of nude mice with surgically induced ischemia (resulting in blood flow <15% of normal). Highlights of this study are presented in Fig. 6. Five days postischemia, the presence of microspheres in limb vasculature (perfused via the aorta) demonstrated reperfusion of the limb as demonstrated by the presence of perfused beads within the vasculature. Mice were perfused via the aorta with FITC-labeled *Bandeira simplicifolia* lectin B4 (BSLB4) and *Ulex europaeus* agglutinin to visualize the luminal side of the vascular wall, and confocal microscopy revealed Ad-RFP-labeled melanoma cells adjacent to the luminal label. To exclude the possibility that Ad-vector targets murine vessels, similar experiments were performed using metastatic melanoma cells stably transfected with EGFP. Five days postischemia, histological cross-sections of muscular tissue stained with mouse endothelial cell–specific BSLB4 (followed by streptavidin–conjugated Alexa dye 594) showed human melanoma cells adjacent to and overlapping with mouse endothelial cells in a linear arrangement. When poorly aggressive EGFP-labeled melanoma cells were introduced at the site of ischemia in separate experiments, they were not found 5 d postischemia. This evidence demonstrated the powerful influence of the microenvironment on the transendothelial differentiation of malignant melanoma cells needed for neovascularization and reperfusion of ischemic limbs.

In our investigation of ischemic limb reperfusion, we chose to examine selected Notch proteins known to promote the differentiation of endothelial cells into vascular networks (Gridley, 2001; Uyttendaele et al., 2001). By immunohistochemistry, Notch 3 and Notch 4 were strongly expressed in the malignant melanoma cells, but not in the nonspecific antibody control nor in the poorly aggressive melanoma cells. Notch signaling molecules are inte-

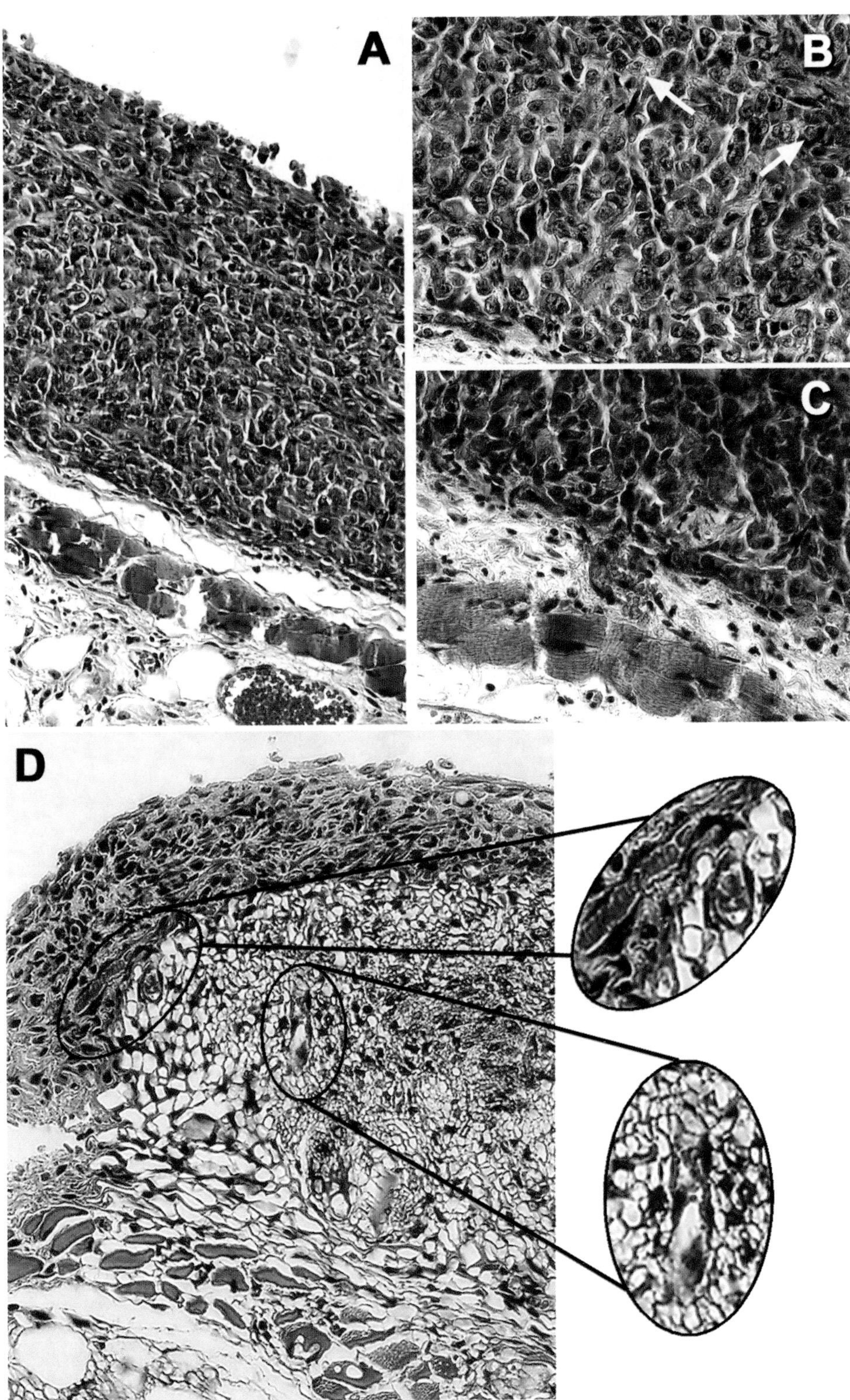

Fig. 4. Microscopic comparison of vasculogenic vs angiogenic human cutaneous melanoma tumor xenografts (stained with hematoxylin and eosin [H&E]) in 3D collagen gels in dorsal skinfold window chambers in nude mice for 10 d. Tumor cell–lined vascular channels are prominent in lower magnification **(A)** and higher magnification **(B,C)** views of C8161 human aggressive melanoma cells in 3D collagen gels in dorsal skinfold window chambers. RBCs (arrows) are seen in some vascular channels (B), and they appear to originate from vasculature in the stromal/muscular region (C) underlying the tumor mass. **(D)** Poorly aggressive human cutaneous A375P melanoma cells in 3D collagen gels in a dorsal skinfold window chamber demonstrate angiogenesis and tissue necrosis associated with the stromal/muscular compartment. Angiogenic neovascularization is prevalent (higher-magnification insets) within the tumor mass. Original magnifications: (A) ×20; (B,C) ×63; (D) ×10.

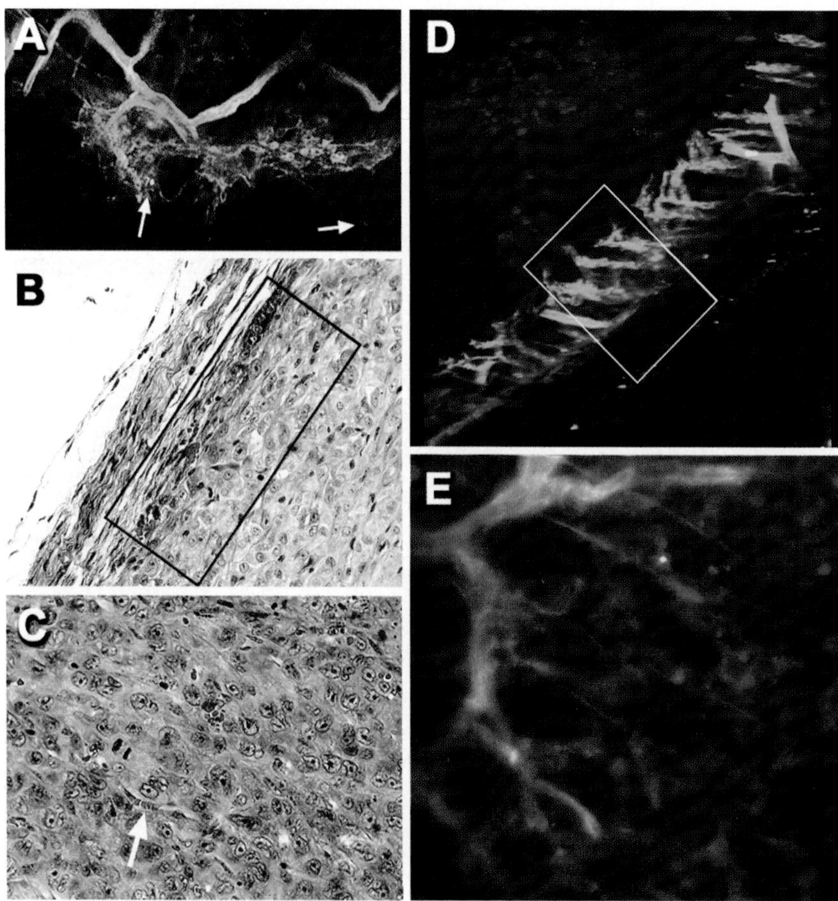

Fig. 5. Morphological assessment of the relationship between tumor cell–lined vascular channels within aggressive human cutaneous melanoma tumors (42 d after sc administration in nude mice) and endothelial-lined vasculature. **(A)** Confocal microscopy of mouse vasculature perfused with F-BSLB$_4$ and *U. europaeus* agglutinin delineating mouse vessels (green) showing red fluorescent microspheres (arrows) delivered from mouse vasculature into the human melanoma tumor interface. **(B)** H&E-stained histological section reveals an area (outlined by the rectangle) containing the mouse vasculature: melanoma interface at the tumor periphery, similar to that shown in (A). **(C)** H&E-stained histological section within the central region of the melanoma tumor shows melanoma cell–lined vascular channels (arrow). **(D)** Confocal microscopy shows the perfused mouse endothelial-lined vasculature (green) at the periphery (rectangle) of the melanoma tumor (containing RFP-labeled melanoma cells), and **(E)** overlaps with the RFP-labeled melanoma cells. Original magnifications: (A) ×20; (B,C,E) ×40; (D) ×10.

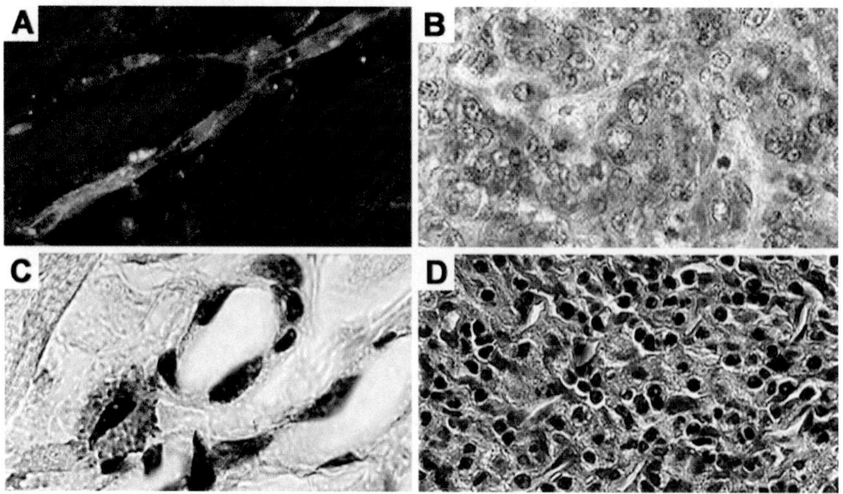

Fig. 6. Microscopic analyses of revascularization of nude mouse ischemic hind limbs 5 **(A,B)** and 20 **(C,D)** d after surgical severing of femoral arterial branches followed by inoculation with 1×10^3 Ad-RFP-labeled human metastatic cutaneous melanoma cells (C8161) intramuscularly. (A) Confocal microscopy of mouse neovasculature at 5 d postischemia perfused with F-BSLB$_4$ and *U. europaeus* agglutinin delineating mouse vessels (green) and Ad-RFP-labeled tumor cells (red) with branched neovasculature containing tumor cells; (B) immunohistochemical localization of Notch 3 protein expression in C8161 melanoma cells in reperfused ischemic muscle 5 d postischemia; (C) cross-section of reperfused ischemic hind limb at 20 d postischemia showing brown immunostaining (for cytokeratins 8 and 18) of one C8161 melanoma cell external to mouse vasculature; (D) microscopic view of H&E-stained tumor (section) formed intramuscularly by C8161 metastatic melanoma cells at 20 d postischemia. Original magnifications: (A,B,D) ×40; (C) ×63.

Table 1
Stem Cell–Associated Genes in Melanoma[a]

Gene name	*Unigene*	*Function*	*Ratio*[b]
Stem cell factor receptor c-kit (KIT)	Hs.81665	Tyrosine kinase receptor; oncogene	2.5
CD34 antigen	Hs.85289	Hematopoietic progenitor cells	2.1
Leukemia inhibitory factor (LIF)	Hs.2250	IL-6 superfamily member	8.3
Leukemia inhibitory factor receptor (LIFR)	Hs.2798	LIF receptor	9.9
Colony-stimulating factor; M-CSF (CSF-1)	Hs.173894	Hematopoietic progenitor cell factor	13.4
Colony-stimulating factor 3 receptor (CSF3R)	Hs.2175	G-CSF receptor	2.6
Hematopoietic-lineage cell-specific protein (HCLS1)	Hs.14601	LYN tyrosine kinase substrate	5.7
Lymphocyte-specific protein 1 (LSP1)	Hs.56729	Signal transduction	2.9
Activated leukocyte cell adhesion molecule (ALCAM)	Hs.10247	CD6 ligand	17.5
Neutral endopeptidase, CALLA (MME)	Hs.1298	Cell adhesion molecule; CD10	4.7
Interleukin-4 receptor (IL-4R)	Hs.75545	Lymphoid stem cell cytokine receptor	2.5

[a]Altered gene expression in human cutaneous melanoma was identified by cDNA microarray analysis.

[b]Selected genes with a differential cutaneous expression of 2.1-fold or greater are reported as a ratio of highly aggressive to poorly aggressive cutaneous melanoma cells.

grally involved in cell fate determination of stem cells and nonterminally differentiated cell types (Gridley, 2001; Uyttendaele et al., 2001). Alterations in Notch expression and signaling have also been implicated in human T-cell leukemia (Ellisen et al., 1991), cervical carcinoma (Zagouras et al., 1995), murine mammary carcinomas (Robbins et al., 1992), and prostatic tumor progression (Shou et al., 2001).

Evaluation of the mouse reperfused limbs 20 d postischemia showed well-formed vasculature with no detectable colocalized metastatic melanoma cells. Rarely, a melanoma cell was found external to the hind-limb vasculature, which had been completely reperfused. By this time period, melanoma tumors had developed in other areas of the musculature, suggesting that the influence of the microenvironmental cues under ischemic conditions ceased to control the transendothelial phenotype function of the aggressive, malignant melanoma cells. Collectively, these observations advance our understanding of the remarkable inductive nature of the microenvironment on aggressive tumor cells that express vasculogenic/angiogenic molecules, together with cell fate determination signaling proteins, associated with a transendothelial function. These findings present new possibilities for therapeutic strategies and novel perspectives on tumor cell plasticity.

27.2. EXPRESSION OF STEM CELL–ASSOCIATED AND PHENOTYPE-SPECIFIC GENES BY AGGRESSIVE MELANOMA CELLS

Highlights of selected differentially expressed genes from the molecular analyses of more than 30 human cutaneous melanoma cell lines are presented in Tables 1 and 2. Highly aggressive vs poorly aggressive melanoma cells, including genetically matched highly and poorly aggressive cell lines from human cutaneous melanoma patients, were hybridized to cDNA microarrays, as previously described (Bittner et al., 2000). This experimental strategy allowed the expression analysis of 6000 genes, simultaneously. Genes that were differentially expressed 2.1-fold and greater (*see* Tables 1 and 2) are accompanied by information regarding their putative function. Many of the listed genes have been confirmed by reverse transcriptase polymerase chain reaction measurement, which has shown an excellent concordance with the differential gene expression profile derived by microarray (Seftor et al., 2002a). The spectrum of genes upregulated and highlighted in Tables 1 and 2 reflects multiple molecular phenotypes, including those of progenitor cells, endothelial, epithelial, fibroblastic, hematopoietic, kidney, neuronal, muscle, and placental cell types. We have previously presented the concept that aggressive melanoma cells exhibit a deregulated genotype reminiscent of an embryonic-like stem cell. Although many of our studies to date have focused on the biological significance of melanoma upregulated genes associated with the vascular phenotype, the potential role (or roles) of other stem cell–related and phenotype-specific genes has not been investigated. Based on the experimental findings demonstrating the transendothelial function of human metastatic melanoma cells challenged to an ischemic environment (Hendrix et al., 2002), it is tempting to speculate that the expression of other genes (listed in Tables 1 and 2) might have functional relevance in the appropriate microenvironment. However, further studies are required to address this speculation.

27.3. CONCLUSION

The exquisite tools currently available to probe the genetic profile of tumor cells and tissues have contributed significantly to our understanding of the molecular underpinnings of tumor progression. The molecular analyses of various cancers, including melanoma, have generated more questions than answers about the phenotype of aggressive tumor cells. Molecular profiling of other tumor types such as breast, prostatic, and renal cell carcinomas, and rhabdomyosarcoma has yielded molecular signatures reflecting multiple phenotypes, thus supporting the speculation of a deregulated genotype (Khan et al., 1998; Perou et al., 2000; Voehringer et al., 2000; Luo et al., 2001). In our current studies with cutaneous and uveal melanoma, we have just begun to address the biological significance of the endothelial- and epithelial-associated phenotypes as they relate to vasculogenic mimicry. Based on the expression of specific genes that are key to vasculogenic and angiogenic events, we have suggested that aggressive melanoma cells have the potential to revert to an embryonic-like phenotype, as demonstrated by the *de novo* formation of vasculogenic-like networks reminiscent of embryological events.

Table 2
Phenotype-Specific Genes in Melanoma[a]

Gene name	*Unigene*	*Function*	*Ratio [b]*
TIE-1 (TIE)	Hs.78824	Endothelial tyrosine kinase	25.0
Epithelial cell kinase (EphA2)	Hs.171596	Receptor tyrosine kinase, ECK	13.0
Vascular endothelial growth factor-C (VEGFC)	Hs.79141	Lymphatics flt-4 ligand	6.5
Neuropilin (NRP1)	Hs.69285	VEGF receptor	5.3
VE-Cadherin, cadherin-5 (CDH5)	Hs.76206	Cell-cell adhesion molecule	11.0
Connective-tissue growth factor (CTGF)	Hs.75511	Growth factor	26.0
Hypoxia-inducible factor 1α (HIF1A)	Hs.197540	bHLH transcription factor	3.1
Podocalyxin-like protein (PODXL)	Hs.16426	Sialoglycoprotein foot process	15.0
Endothelial differentiation receptor (EDG1)	Hs.154210	G-coupled receptor	3.7
Endothelial cell–specific molecule (ESM1)	Hs.41716	Endothelial–specific signaling molecule	41.0
Endothelial differentiation-related factor-1 (EDF-1)	Hs.174050	Endothelial cell differentiation regulator	4.8
Macrophage maturation-associated protein (MMA)	Hs.79889	Differentiation molecule	4.5
Placental Bikunin (SPINT2)	Hs.31439	Serine protease/coagulation inhibitor	11.5
Smoothelin; α actinin (SMTN)	Hs.149098	Colocalizes with actin/desmin	6.1
Desmin (DES)	Hs.171185	Muscle IF	2.7
Fibrillin 1 (FBN1)	Hs.750	Elastic fiber	5.0
Keratin 7 (KRT7)	Hs.23881	Intermediate filament	29.0
Keratin 8 (KRT8)	Hs.230298	Intermediate filament	13.0
Keratin 18 (KRT18)	Hs.65114	Intermediate filament	16.0
Urokinase (PLAU)	Hs.77274	Serine protease	48.0
Urokinase plasminogen activator receptor (PLAUR)	Hs.179657	Protease receptor	9.2
Neuronal tissue–enriched acidic protein (NAP22)	Hs.79516	Intercellular signaling	9.7

[a] Altered gene expression in human cutaneous melanoma was identified by cDNA microarray analysis.
[b] Selected genes with a differential expression of 2.1-fold or greater are reported as a ratio of highly aggressive to poorly aggressive cutaneous melanoma cells.

It has been suggested that tumor plasticity allows vasculogenic mimicry to occur (Bissell, 1999). However, relatively little is known about the molecular switch for this event. Our observations have indicated the importance of VE-cadherin and EphA2 in the formation of vasculogenic-like networks, which may serve as components of a vasculogenic switch (Hendrix et al., 2001; Hess et al., 2001). However, the most exciting data revealing embryonic decision-making genes overexpressed by aggressive melanoma cells have identified Notch proteins as potential molecular switches (Hendrix et al., 2002). Notch proteins are known to promote the differentiation of endothelial cells into vascular networks (Gridley, 2001; Uyttendaele et al., 2001). These findings, along with the melanoma expression of Notch proteins, EphA2, and VE-cadherin, suggest that aggressive melanoma cells utilize a signaling pathway similar to angioblasts in cell fate determination that results in vasculogenic mimicry.

The fundamental question that should be addressed is: How can one influence this embryonic-like stem cell phenotype and convert aggressive melanoma cells to nonaggressive cell types, with the hope of ultimately destroying them? In this regard, two pieces of evidence may guide us in developing new strategies for cancer management based on the tumor microenvironment. First, the observation that aggressive melanoma cells, when placed in an ischemic environment, can be influenced to form normal vasculature, at least transiently, before developing a tumor mass, is remarkable (Hendrix et al., 2002). This key finding signifies the importance of microenvironmental factors in altering the fate of tumor cells. Second, the ability of ECM preconditioned by aggressive melanoma cells to influence poorly aggressive melanoma cells to assume a vasculogenic phenotype suggests that important molecular signals are deposited within the microenvironment that can induce cells to become aggressive (Seftor et al., 2001). If we can decipher these molecular signals, new strategies could be developed to inhibit them.

The field of stem cell research, including cancer stem cells, is rapidly emerging with new information and concepts (Reya et al., 2001). Indeed, recent studies have challenged previously held dogma regarding tissue-restricted differentiation of postnatal stem cells with convincing evidence that demonstrates pluripotency for mesenchymal, neural, and hematopoietic stem cells (for a review, *see* Anderson et al., 2001). Our melanoma transendothelial differentiation results coincide with the recent observation that cardiomyocytes induced endothelial cells to transdifferentiate into cardiac muscle in ischemic hearts (Condorelli et al., 2001). This study may provide clues as to the potential inductive nature of the ischemic skeletal muscle in our experiments. Other studies supporting our unique observations have shown that fibrocytes can induce an angiogenic phenotype in cultured endothelial cells and promote angiogenesis in vivo (Hartlapp et al., 2001), and transformation of fibroblasts into endothelial cells during angiogenesis has also been shown (Kazunori and Fujiwara, 1994). Most intriguing is the potential significance of the expression of CD34, a progenitor cell antigen, normally found in endothelial cells and now reported in invasive malignant melanoma (Hoang et al., 2001). A recent report has provided new information regarding the signaling pathways involved in arterial-venous decisions by angioblast precursor cells, including Notch-Gridlock pathways and EphB2 and EphB4 expression (Zhong et al., 2001). Previous observations from our laboratory have shown the necessity for EphA2 in melanoma vasculogenic mimicry (Hess et al., 2001), which may coincide with some of the findings in the angioblast cell fate determination study.

The plasticity of aggressive tumor cells and their ability to mimic other cell types has been observed in melanoma, breast, prostate and ovarian cancer (Bittner et al., 2000; Hendrix et al., 2000; Sood et al., 2001; Seftor et al., 2002b; Sharma et al., 2002; Shirakawa et al., 2002). These findings represent a significant clinical challenge in the detection and classification of tumor types. Although we and others have focused on the importance of vasculogenic mimicry with respect to providing possible perfusion and dissemination routes to rapidly growing tumors, this mimicry may also provide an evasion mechanism from immune surveillance based on the expression of genes that are associated with normal cell phenotypes (Seftor et al., 2003). This is an intriguing and daunting speculation that requires additional investigation. As these questions are addressed regarding the molecular determinants of tumor cell plasticity, our hope is that new diagnostic markers and therapeutic targets will be identified for clinical use.

ACKNOWLEDGMENTS

We gratefully acknowledge, Dr. Jeff Trent for collaborating on the microarray analyses, and Dr. Dawn Kirschmann at The University of Iowa for contributing the scientific model. This research was supported by National Institutes of Health grants R37CA59702, CA84065, HL46314, DK55965, DK59223, CA52879, CA83137, and CA80318.

REFERENCES

Anderson, D. J., Gage, F. H., and Weissman, L. (2001) Can stem cells cross lineage boundaries? *Nat. Med.* 7(4):393–395.

Bissell, M. (1999) Tumor plasticity allows vasculogenic mimicry, a novel form of angiogenic switch : A rose by any other name? *Am. J. Pathol.* 155(3):675–679.

Bittner, M., Meltzer, P., Chen, Y., et al. (2000) Molecular classification of cutaneous malignant melanoma by gene expression profiling. *Nature* 406(6795):536–540.

Carmeliet, P. (2000) Mechanisms of angiogenesis and arteriogenesis. *Nat. Med.* 6:389–395.

Colognato, H. and Yurchenco, P.D. (2000) Form and function: the laminin family of heterotrimers. *Dev. Dyn.* 218:213–234.

Condorelli, G., Borello, U., De Angelis, L., et al. (2001) Cardiomyocytes induce endothelial cells to transdifferentiate into cardiac muscle: implications for myocardium regeneration. *Proc. Natl. Acad. Sci. USA* 98(19):10,733–10,738.

Ellisen, L.W., Bird, J., West, D.C., et al. (1991) TAN-1, the human homolog of the *Drosophila* notch gene, is broken by chromosomal translocations in T lymphoblastic neoplasms. *Cell* 66:649–661.

Folkman, J. (1995) Seminars in Medicine of the Beth Israel Hospital, Boston. Clinical applications of research on angiogenesis. *N. Engl. J. Med.* 333:1757–1763.

Giannelli, G., Falk-Marzillier, J., Schiraldi, O., Stetler-Stevenson, W. G., and Quaranta, V. (1997) Induction of cell migration by matrix metalloprotease-2 cleavage of laminin-5. *Science* 277:225–228.

Gridley, T. (2001) Notch signaling during vascular development. *Proc. Natl. Acad. Sci. USA* 98:10,733–10,738.

Gumbiner, B. M. (1996) Cell adhesion: the molecular basis of tissue architecture and morphogenesis. *Cell* 4:345–357.

Hartlapp, I., Abe, R., Saeed, R. W., et al. (2001) Fibrocytes induce an angiogenic phenotype in cultured endothelial cells and promote angiogenesis in vivo. *FASEB J.* 15:2215–2224.

Hendrix, M. J. C., Seftor, E. A., Kirschmann, D. A., and Seftor, R. E. B. (2000) Molecular biology of breast cancer metastasis: molecular expression of vascular markers by aggressive breast cancer cells. *Breast Cancer Res.* 2:417–422.

Hendrix, M. J. C., Seftor, E. A., Meltzer, P. S., et al. (2001) Expression and functional significance of VE-cadherin in aggressive human melanoma cells: role in vasculogenic mimicry. *Proc. Natl. Acad. Sci. USA* 98:8018–8023.

Hendrix, M. J. C., Seftor, R. E. B., Seftor, E. A., et al. (2002) Transendothelial function of human metastatic melanoma cells: role of the micro-environment in cell-fate determination. *Cancer Res.* 62: 665—668.

Hendrix, M. J. C., Seftor, E. A., Hess, A. R., and Seftor, R. E. B. (2003) Vasculogenic mimicry and tumour-cell plasticity: Lessons from melanoma. *Nature Rev. Cancer* 3:411–421.

Hess, A. R., Seftor, E. A., Gardner, L. M. G., et al. (2001) Molecular regulation of tumor cell vasculogenic mimicry by tyrosine phosphorylation: role of epithelial cell kinase (Eck/EphA2). *Cancer Res.* 61: 3250–3255.

Hoang, M. P., Selim, M. A., Bentley, R. C., Burchette, J. L., and Shea, C. R. (2001) CD34 expression in desmoplastic melanoma. *J. Cutan. Pathol.* 28:508–512.

Hynes, R. O. (1992) Specificity of cell adhesion in development: the cadherin superfamily. *Curr. Opin. Genet. Dev.* 2:621–624.

Kazunori, K. and Fujiwara, T. (1994) Transformation of fibroblasts into endothelial cells during angiogenesis. *Cell Tissue Res.* 278:625–628.

Kemler, R. (1992) Classical cadherins. *Semin. Cell Biol.* 3:149–155.

Khan, J., Simon, R., Bittner, M., et al. (1998) Gene expression profiling of alveolar rhabdomyosarcoma with cDNA microarrays. *Cancer Res.* 58:5009–5013.

Koshikawa, N., Giannelli, G., Cirulli, V., Miyazaki, K., and Quaranta, V. (2000) Role of cell surface metalloprotease MT1-MMP in epithelial cell migration over laminin-5. J. *Cell Biol.* 148:615–624.

Kumar, R. and Fidler, I. J. (1998) Angiogenic molecules and cancer metastasis. *In Vivo* 12:27–34.

Lampugnani, M. G., Resnati, M., Raiteri, M., et al. (1992) A novel endothelial-specific membrane protein is a marker of cell-cell contacts. *J. Cell Biol.* 118:1511–1522.

Luo, J., Duggan, D. J., Chen, Y., et al. (2001) Human prostate cancer and benign prostatic hyperplasia: molecular dissection by gene expression profiling. *Cancer Res.* 61:4683–4688.

Malinda, K. M. and Kleinman, H. K. (1996) The laminins. *Int. J. Biochem. Cell Biol.* 28(9):957–995.

Malinda, K. M., Motoyoshi, N., Melissa, C., et al. (1999) Identification of laminin α1 and β1 chain peptides active for endothelial cell adhesion, tube formation, and aortic sprouting. *FASEB J.* 13:53–62.

Maniotis, A. J., Folberg, R., Hess, A., et al. (1999) Vascular channel formation by human melanoma cells in vivo and in vitro: vasculogenic mimicry. *Am. J. Pathol.* 155(3):739–752.

Pasquale, E. B. (1997) The Eph family of receptors. *Curr. Opin. Cell. Biol.* 9:608–615.

Perou, C. M., Sorlie, T., Eisen, M. B., et al. (2000) Molecular portraits of human breast tumours. *Nature* 406:747–752.

Rak, J. and Kerbel, R. S. (1996) Treating cancer by inhibiting angiogenesis: new hopes and potential pitfalls. *Cancer Metastasis. Rev.* 15: 231–236.

Reya, T, Morrison, S. J., Clarke, M. F., and Weissman, I. (2001) Stem cells, cancer, and cancer stem cells. *Nature* 414:105–111.

Risau, W. (1997) Mechanisms of angiogenesis. *Nature* 386:671–674.

Robbins, J., Blonel, B. J., Gallahan, D., and Callahan, R. (1992) Mouse mammary tumor gene Int-3: a member of the Notch gene family transforms mammary epithelial cells. *J. Virol.* 66:2594–2599.

Rosenburg, I. M., Goke, M., Kanai, M., Reinecker, H. C., Podolsky, D. K. (1997) Epithelial cell kinase-B-61: an autocrine loop modulating intestinal epithelial migration and barrier function. *Am. J. Physiol.* 273(4 Pt. 1):G824–G832.

Seftor, E. A., Meltzer, P. S., Schatteman, G. C., et al. (2002a) Expression of multiple molecular phenotypes by aggressive melanoma tumor cells: role in vasculogenic mimicry. *Crit. Rev. Oncol. Hematol.* 44: 17–27.

Seftor, E.A., Meltzer, P.S., Kirschmann, D.A., Pe'er, J., Maniotis, A.J., Trent, J.M., Folberg, R., and Hendrix, M.J.C. (2002) Molecular determinants of human uveal melanoma invasion and metastasis. *Clin. Exp. Metastasis.* 19:233–246.

Seftor, R. E. B., Seftor, E. A., Koshikawa, N., et al. (2001) Cooperative interactions of laminin 5 γ2 chain, matrix metalloproteinase-2, and membrane type-1 matrix/metalloproteinase are required for mimicry of embryonic vasculogenesis by aggressive melanoma. *Cancer Res.* 61:6322–6327.

Seftor, R. E. B., Seftor, E. A., Hess, A. R., et al. (2003) The role of the vasculogenic phenotype and its associated extracellular matrix in tumor progression: Implications for immune surveillance. *Clin. Applied Immunol. Rev.* 3:263–276.

Sharma, N., Seftor, R. E. B., Seftor, E. A., et al. (2002) Prostatic tumor cell plasticity involves cooperative interactions of distinct phenotypic subpopulations: role in vasculogenic mimicry. *Prostate* 50: 189–201.

Shirakawa, K., Kobayashi, H. Keike, Y., et al. (2002) Rapid accumulation and internalization of radiolabeled herceptin in an inflammatory breast cancer xenograft with vasculogenic mimicry predicted by the contrast-enhanced dynamic MRI with the macromolecular contrast agent G6-(1B4M-Gd)(256). *Cancer Res.* 62(3):860–866.

Shou, J., Ross, S., Koeppen, H., de Sauvage, F. J., and Gao, W.-Q. (2001) Dynamics of Notch expression during murine prostate development and tumorigenesis. *Cancer Res.* 61:7291–7297.

Sood, A. K., Seftor, E. A., Fletcher, M. S., et al. (2001) Molecular determinants of ovarian cancer plasticity. *Am. J. Pathol.* 158(4):1279–1288.

Straume, O. and Akslen, L. A. (2001) Importance of vascular phenotype by basic fibroblast growth factor, and influence of the angiogenic factors basic fibroblast growth factor/fibroblast growth factor receptor-1 and Ephrin-A1/EphA2 on melanoma progression. *Am. J. Pathol.* 160(3):1009–1019.

Uyttendaele, H., Ho, J., Rossant, J., and Kitajewski, J. (2001) Vascular patterning defects associated with expression of activated Notch4 in embryonic endothelium. *Proc. Natl. Acad. Sci. USA* 98, 5643–5648.

Voehringer, D. W., Hirschberg, D. L., Xiao, J., et al. (2000) Gene microarray identification of redox and mitochondrial elements that control resistance or sensitivity to apoptosis. *Proc. Natl. Acad. Sci. USA* 97: 2680–2685.

Zagouras, P., Stifani, S., Blaumueller, C. M., Carcangiu, M. L., and Artavanis-Tsakonas, S. (1995) Alterations in Notch signaling in neoplastic lesions of the human cervix. *Proc. Natl. Acad. Sci. USA* 92: 6414–6418.

Zhong, T. P., Childs, S., Leu, J. P., and Fishman, M. C. (2001) Gridlock signaling pathway fashions the first embryonic artery. *Nature* 414: 216–220.

28 Stem Cells in Glandular Organs

KARIN WILLIAMS, PhD AND SIMON W. HAYWARD, PhD

The interaction of stroma and glandular tissue determines the phenotype of the glandular organ. The gastrointestinal (GI) stem cell model system is presented as an example of epithelial tissue turnover. In the GI epithelium, there are approximately four to six functional stem cells per crypt. These give rise to transit-amplifying cells that can differentiate into the different mature cells of the GI tract including mucous, absorptive, secretory, and neuroendocrine cells with different proportions at different levels of the GI tract. About 80–90% of the long-lived clones in the GI tract are unipotent, with the remainder being multipotent. Other glandular organs have similar cell lineage arrangements with different differentiation potentials. The differentiation potential of immature epithelial cells is controlled by the mesenchymal stroma through permissive or instructive effects. For example, embryonic mammary mesenchyme can induce mammary gland differentiation of skin epithelium through a process of induction of a new phenotype by the developing transit-amplifying cells, an example of metaplasia. Metaplasia, the change from one type of epithelium to another, is a prominent feature of mature glandular organs; squamous metaplasia of the prostate occurs in response to estrogens, and in the lung as a result of exposure to cigarette smoke. Metaplasia is a process of altered differentiation of the progeny of the tissue progenitor or stem cell; transdifferentiation is a change from one differentiated phenotype to another by the same cells. Thus, transdifferentiation in this context does not involve stem cells. Tissue recombination experiments show that exposure to inductive mesenchyme is sufficient to stimulate a stem cell population to generate daughter populations different from the tissue of origin but is insufficient to force transdifferentiation of committed epithelial cells.

28.1. INTRODUCTION

At the heart of any discussion of stem cell biology is the question of definition: what is a stem cell? Embryonic stem (ES) cells are totipotent, having the plasticity to give rise to all of the cell types required for development and adult function. However, in the adult, it has been considered that stem cells have a more restricted repertoire. Accepted definitions of adult stem cells include the ability to continue to self-replicate in adulthood and multipotentiality (the ability to give rise to daughter cells with more than one phenotype). Stem cells are often described as normally having low proliferative activity with the ability to multiply in response to appropriate stimuli, such as in wound healing. This potential to repopulate a tissue was thought to be restricted to the organ in which the stem cell resides. However, data have accumulated suggesting that the plasticity of adult stem cells may be far wider, perhaps encompassing the repertoire of an entire embryonic germ (EG) layer. Metaplastic changes, in which cells of one tissue type take on characteristics of another, have been recognized by pathologists for many years (Lugo and Putong, 1984). Metaplasia demonstrates that adult epithelia retain an ability to take on a new differentiated phenotype. This may be due to true transdifferentiation (the conversion of one differentiated cell type to another) or may represent a proliferative response of stem cells implying that stem cells have wider specificity than the cells of the organs from which they are derived (Slack, 2000; Slack and Tosh, 2001).

Mammalian cloning experiments have shown that nuclei from established cell lines and from differentiated adult tissues retain the ability to regenerate an entire fetal and adult animal (Campbell et al., 1996; Wilmut et al., 1997). Such phenomena, however, require highly specialized manipulations. Recent investigations have demonstrated that the livers of human bone marrow transplant patients contain hepatocytes derived from their grafts (Alison et al., 2000). This result confirms experiments in which the fate of bone marrow cells transplanted into mice was carefully tracked (Petersen et al., 1999). A series of studies relating to the ability of cells to populate tissues has established that bone marrow cells can contribute to the epithelial parenchyma of the kidney (Poulsom et al., 2001). Cells derived from the central nervous system (CNS) have been reported to contribute to tissues derived from all three EG layers (Clarke et al., 2000; Kondo and Raff, 2000). One recent study has suggested that germ layer shifts in CNS cells could be

From: *Stem Cells Handbook*
Edited by: S. Sell © Humana Press Inc., Totowa, NJ

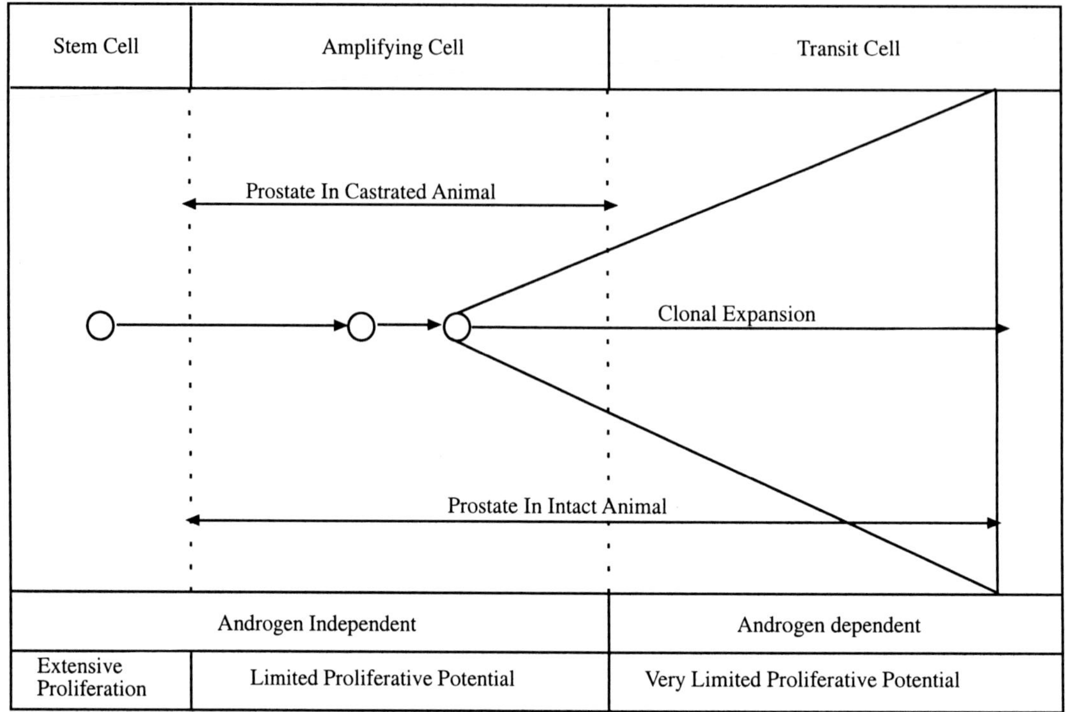

Fig. 1. Model of cell amplification in the prostate. This figure, modified from Isaacs (1985), illustrates a stem-amplifying-transit cell lineage model of organ development. In the case of the prostate, an androgen-dependent gland, Isaacs proposed that there were androgen-independent and androgen-dependent populations. In this model, androgen-independent cells are able to initiate the growth of prostate tissue and retain glandular rudiments following castration that can then undergo clonal expansion to repopulate the gland in response to androgenic stimulation.

attributable to cellular potency changes resulting from spontaneous cell fusion (Ying et al., 2002). Another study suggests that the change from neural to hematopoietic cell types may result from genetic or epigenetic alterations occurring during prolonged cell passaging (D'Amour and Gage, 2002; Morshead et al., 2002). This ongoing debate demonstrates that a major change is taking place in our perception of the nature of stem cells and the nature of cellular differentiation.

Current perceptions of stem cells focus on the plasticity of the differentiated phenotype and on mechanisms that might induce changes from different levels of differentiation. Traditional models of differentiation suggest that there are certain irreversible points of commitment between the totipotent ES cell and the fully differentiated cells in functional adult tissue. By contrast, recent analyses have proposed a continuum of differentiation with cells having less propensity to exhibit stem cell–like behavior as they express more specific characteristic markers (Blau et al., 2001). On a molecular basis, the "developmental commitment" can be considered as being encoded in combinations of transcription factors (Slack and Tosh, 2001). This represents the resurrection of the idea of combinatorial gene control proposed nearly 30 yr ago (Gierer, 1974). Commitment to a specific phenotype is considered reversible, allowing the possibility that individual cells can revert to a less committed phenotype by changing their combination of expressed transcription factors. The caveat to this is that as differentiation proceeds the degree of difficulty involved in reversing the process increases.

In this chapter, we examine the broad topic of stem cells in glandular organs rather than the role of stem cells in a particular organ. We also address the question of what stem cells are and whether a given organ is served by a truly organ-specific stem cell population. In addition, we focus on stem cells in epithelial tissues and discuss the interactions of these cells with their local microenvironment. Finally, we discuss the phenomena of metaplasia and transdifferentiation as they apply to our comprehension of stem cell biology.

28.2. MODELS OF STEM CELL BIOLOGY

Stem cell–based models of development and proliferation have been proposed in a number of organs. These models have generally suggested the presence of a stem cell compartment with the potential to undergo many rounds of division. Divisions of stem cells are proposed to give rise to daughter cells that are stem cells or that join a population of transient-amplifying cells capable of undergoing clonal expansion, finally giving rise to a cell population that is fully differentiated but has limited potential for proliferative activity (Hume and Potten, 1979; Lavker and Sun, 1983). Isaacs (1985) proposed such a stem, amplifying, transit cell model for the prostate, which is summarized in Fig. 1. The prostate is potentially a good model in which to study stem cell biology because it is dependent on androgenic stimulation for its development and for maintenance of its glandular structure in adulthood. If androgens are removed, the prostate will regress, and if they are reintroduced, it will regrow. Stem, amplifying, transit cell models provide a simple basis from which to determine how the growth and maintenance of organs could come about. However, these models do not address the questions: Where are stem cells located? How does this location affect glandular structures? What might be the phenotypic characteristics of stem cells?

The location of stem cell populations in glandular tissues has been studied most closely in the gastrointestinal (GI) tract. Stem cells of the gut epithelium give rise, via asymmetric cell mitosis, to daughter cells with more than one phenotype. These phenotypes include a range of distinctive cell types, including the Paneth, goblet, and endocrine cells. The perpetually self-renewing gut tissue consisting of proliferating, differentiating, migrating, and dying cells is not directly the progeny of the stem cells but rather of stem cell daughters. The stem cell daughters retain proliferative potential giving rise to a clonally expanding population; however, the proliferative potential of each daughter is finite, unlike the potential of the stem cell. Commitment of daughter cells to a specific secretory cell–type fate in the small intestine apparently occurs in a cell type–determined manner at different points between a stem and a differentiated cell. For example cells become committed to the endocrine lineage at an early point but do not commit to one of the nine or so different endocrine cell types until a later point (Rindi et al., 1999).

Gut stem cells all reside within the epithelial units that are known in the stomach as glands and in the intestine as crypts. Both of these units consist of areas in which the stem cells and their proliferating daughters reside, migratory zones in which cells are both migrating and differentiating and a fully differentiated zone in which mature and dying cells are located. Nomura et al. (1998) have identified the existence of at least three to four stem cells located within the isthmus region of the developing/maturing stomach glands and maintenance of one or more stem cells in the mature units. Their elegant study examined cell lineage by use of a lacZ-carrying X-chromosome. Chimeric (mixed blue and white) glands were observed at early time points, indicating the presence of more than one progenitor cell. Two progenitor cells would produce 50% mixed, 25% homochromatic ratios. The ratios noted by these researchers were, however, closer to 75%, indicating that at least three stem cells are involved. Glands are therefore initially seeded by more than one stem cell (Nomura et al., 1998), but mature glands generally are derived from only one stem cell (Canfield et al., 1996; Nomura et al., 1998), possibly as a result of stem cell competition, in which a stem cell must compete for a favored position within the gland (Loeffler et al., 1993), or due to symmetrical/asymmetrical cell divisions removing stem cells from the population. As compared to the stomach, the intestinal crypts progress to monoclonality in a similar but accelerated fashion (Ponder et al., 1985; Griffiths et al., 1988; Schmidt et al., 1988; Winton et al., 1988; Roth et al., 1991; Gordon et al., 1992; Hermiston and Gordon, 1995). While crypts are generally accepted to be monoclonal, it is thought that there are four to six functional stem cells per crypt (Bjerknes and Cheng, 1999). This apparent discrepancy arises in part because the stem cells present within the gland are themselves clonal (i.e., they are daughters of a single stem cell produced by symmetrical division). Such cells cannot be differentiated using the present techniques of following X-chromosome inactivation patterns, chimeric lineage analysis, or retroviral integration.

Repopulation experiments have failed to answer the question of how many stem cells are present because immature first-through third-generation stem cell daughters (which are, of course, themselves monoclonal) may be able to repopulate the entire crypt in times of crisis by reprogramming back into stem cells (Merritt et al., 1994; Potten et al., 1997; Potten and Grant, 1998; Booth et al., 2000;). Targeted ablation studies using transgenic mice carrying reporter genes have confirmed that certain cell lineages of the gut epithelium share a multipotent progenitor intermediate (Rindi et al., 1999). However, Bjerknes and Cheng (1999) found that 80–90% of the long-lived clones in the small intestine are unipotent, with the remainder being multipotent. This finding does not undermine the existence of true stem cells but does indicate the problems that arise when the stem cells and the long-lived but finite first- through third-generation daughters cannot be readily distinguished from each other.

Multiple stem cells are also found within a single niche in the tracheal epithelium and submucosal glands (Engelhardt et al., 1995). Lineage analysis using recombinant retroviruses has demonstrated the existence of both unipotent and multipotent progenitors in submucosal glands (Engelhardt et al., 1995). Borthwick et al. (2001) have further localized the stem cells to specific regions of the glands and trachea.

Studies in the urogenital tract using chimeric Balb/c/C3H mice indicate that glandular organs such as the prostate and seminal vesicles, as well as nonglandular structures such as the epididymis, bladder, ureters, and kidney, were not clonal (Lipschutz et al., 1999). This finding suggests that these structures were derived from more than one progenitor cell. By contrast, in the same study, uterine glands were never found to be chimeric. The investigators suggest that this indicates a monoclonal origin for uterine glands. However, it is entirely possible that these glands are in fact originally polyclonal but undergo sorting procedures similar to the stem cell selection described for intestinal crypts.

The submandibular salivary gland is generated from a diverse array of progenitor cells that contribute to the functional and architectural complexity. It has been implied in many reports in the literature (Toyosawa et al., 1999, and references within) that small cell undifferentiated carcinomas of the salivary gland arise from a ductal stem cell with a multidirectional capacity. This putative stem cell is thought to reside within the intercalated ducts.

There has been a long search for specific cellular markers that will identify stem cells within a tissue. Cytoskeletal proteins such as cytokeratins have been utilized in several studies in an attempt to identify specific cell phenotypes and lineages in proliferative and nonproliferative compartments in glands such as the breast and prostate (Taylor-Papadimitriou and Lane, 1987, 1989, 1992; Bartek et al., 1990; Hudson et al., 2000, 2001; Wang et al., 2001). These studies suggest that there may be specific profiles of cellular marker expression that define the position of a cell within the differentiation process, and as well as that expression of certain combinations of markers is restricted to early or stem cell populations. Other studies suggest that putative stem cells can be identified either by the absence of a specific marker (e.g., $p27^{Kip1}$ in the prostate [De Marzo et al., 1998a, 1998b]) or by the overexpression of certain specific markers (e.g., pp32 in the intestine and prostate [Malek et al., 1990; Walensky et al., 1993]). Such markers of stem cell phenotypes are covered more fully in other chapters.

28.3. POTENTIALITY OF ADULT EPITHELIAL TISSUES

Glandular organs are generally composed of an epithelial parenchyma surrounded by stroma. For a long time the stroma was considered to be a supporting matrix that aided in organ function. For example, the fibromuscular prostatic stroma provides the force needed for ejaculation, while smooth muscle in the gut provides the peristaltic action needed to move food. It is now clear that the stroma plays an active role both in development, in which mesenchyme directs epithelial differentiation, and in the adult, in which the differentiated state of the epithelial cells is maintained by continuous interactions with the adjacent stromal cells (Hayward and Cunha, 2000). This continuous cross talk between tissues in an organ also regulates functions such as proliferation and apoptosis (Kurita et al., 2000a, 2000b, 2001; Risbridger et al., 2001a).

Following gastrulation, the mammalian embryo is composed of cells representing the three germ layers that will give rise to all of the tissues of the body. In a simplistic representation, the external surface is covered with a layer of ectoderm that will give rise to the skin, as well as the sweat, mammary, and preputial glands. The endoderm will give rise to the GI tract and those structures, such as the liver, pancreas, prostate, and bladder, that are derived from it. The mesodermal layer, which occupies the space between these surfaces, will give rise to all of the mesenchymal tissues, including the muscles and connective tissues. As a result of mesenchymal-to-epithelial transitions, mesoderm also gives rise to epithelial structures including the urogenital tract derivatives of the Wolffian and Müllerian ducts (the ureters, epididymis, ductus deferens, and seminal vesicles, and the fallopian tubes and uterus, respectively).

The phenomena of metaplasia and transdifferentiation represent changes in the differentiation of epithelial tissues. Metaplasia represents the presence in a tissue of an epithelial cell phenotype not normally found in the location being examined. Many examples of metaplasia are recognized by pathologists. In many cases, these are protective responses. For example, many epithelial tissues will take on a squamous phenotype following injury or persistent insult, covering and sealing the wounded surface. In other cases, metaplasia may result from a chemical insult. In the prostate, estrogen exposure results in the formation of squamous metaplasia (Triche and Harkin, 1971; Kroes and Teppema, 1972; Helpap and Stiens, 1975; Risbridger et al., 2001a, 2001b;). In human prostate cancer patients, treatment with synthetic estrogens gives rise to squamous metaplasia, a phenomenon now rarely seen in countries where this treatment regimen has been discontinued. *In utero* exposure to female sex hormones, notably diethylstilbestrol, also results in squamous differentiation (Driscoll and Taylor, 1980). Metaplasia can be a benign protective response to a particular insult. However, metaplasia is often seen as a premalignant condition because the insult that gaves rise to a metaplastic response can persist and induce malignant transformation. Thus, e.g., cigarette smoke induces stratified squamous metaplasia in the trachea and bronchi. Continued smoking leads to malignant transformation and a finding of squamous carcinoma in the respiratory tract. It is noteworthy that all of the clinically observed metaplastic changes are apparently restricted to the repertoire of the germ layer from which the epithelium is originally derived.

Metaplasia can be a result of proliferation of a stem cell population to give rise to daughter cells with an inappropriate or abnormal phenotype. Metaplasia includes the more restrictive class of cellular changes known as transdifferentiation. Unlike other forms of metaplasia, in which tissues can result from proliferation of stem or transitional/amplifying cell populations, transdifferentiation is defined as "an irreversible switch of one type of already differentiated cell to another type of normal differentiated cell" (Okada, 1991; Slack and Tosh, 2001). Note, however, that many articles use the word *transdifferentiation* without due regard for its strictly defined meaning. While there are limited documented examples, transdifferentiation occurs during development (Patapoutian et al., 1995), and in some specialized laboratory situations such as in cell culture (Danielpour, 1999). However, in normal and benign adult disease transdifferentiation is not common. In malignant disease, epithelial-to-mesenchymal transitions are well documented. However it can be argued that these fall outside of the definition of transdifferentiation because they do not result in the formation of a normal differentiated cell. In the context of the present article, transdifferentiation is apparently not relevant to stem cell biology since, in contrast to other forms of metaplasia, it involves only previously differentiated adult cells.

A series of tissue recombination experiments has established that epithelial tissues from a range of sources can respond to inductive mesenchyma by changing their pattern of differentiation. These experiments involve the separation of epithelial and mesenchymal cells from different organs and their heterotypic recombination, as shown schematically in Fig. 2. Recombined tissues are then grown as subrenal capsule implants in either syngeneic or immunodeficient rodent hosts or, alternatively, in the case of embryonic birds grafted *in ovo*, and subsequently examined to determine the nature of the resulting recombinant tissue.

Two classes of mesenchymal and stromal interaction with epithelium have been recognized: permissive effects and instructive effects (Haffen et al., 1987). A *permissive* effect supports a previously determined developmental program already specified by the epithelium, e.g., supporting differentiation of an adult tissue. An *instructive* effect elicits a new program of markers in the epithelium specified by the mesenchyme, as found in the process of organogenesis. The first effect can be illustrated in heterospecific but homotypic intestinal recombination experiments, as summarized in Table 1. In these experiments, 14-d fetal rat intestinal mesenchyme and endoderm were recombined with endoderm and mesenchyme from 5.5-d-old embryonic chick intestines. The resultant recombinations were developed as intracolemic grafts *in ovo* and examined in terms of endodermal expression of brush-border enzymes. Recombinants of rat mesenchyme with chick endoderm expressed sucrase but not lactase while the opposite recombination produced the opposite enzyme patterns. This pattern of expression would be expected of the source endoderm because at this stage of development chick intestine expresses sucrase but not lactase while rats have the opposite expression (Kedinger et al., 1981). Thus, the mesenchyme in these experiments supports the differentiation of the gut endoderm and allows the expression of markers characteristic of the species from which the epithelial tissue was derived.

Instructive interactions of mesenchyme with the epithelium have been demonstrated in the development of the gut, using the chicken and the Japanese quail as model systems, as summarized in Table 2. Mesenchyma from the stomach and small intestine of these birds exert instructive influences on the morphogenesis of both allantoic epithelium and epithelium from various levels of the digestive tract (Gumpel-Pinot et al., 1978). Thus, when stomach mesenchyme is recombined with allantoic epithelium, the morphology and expression of markers by the epithelium changes to that of stomach epithelium. Similar experiments have demonstrated glandular induction in ectoderm. For example, embryonic mammary mesenchyme can induce mammary gland differentiation from skin (Cunha et al., 1995), and rabbit corneal

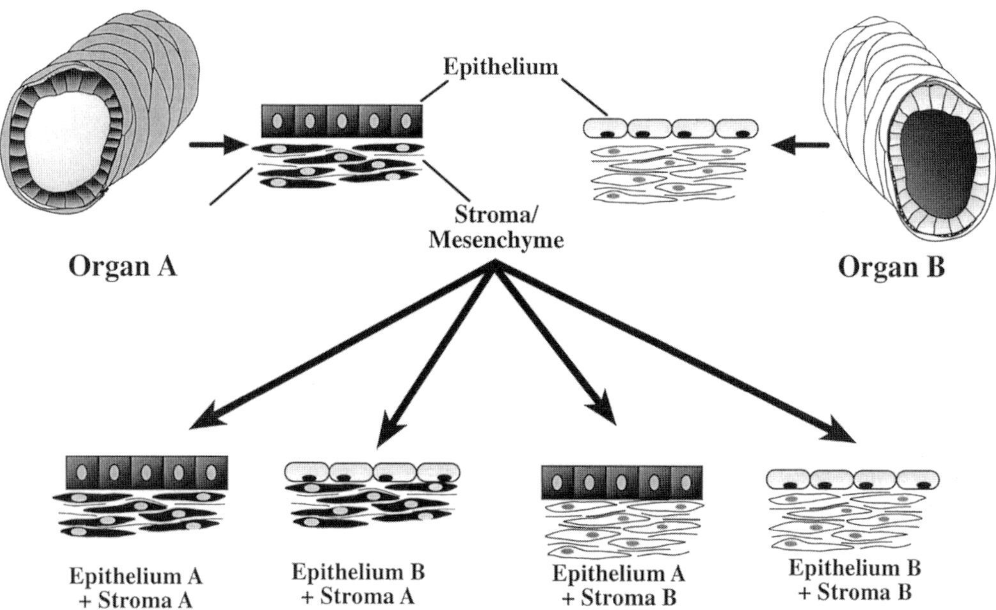

Fig. 2. Schematic representation of tissue recombination. Stromal and epithelial tissues are separated from each other and isolated. The tissues are then recombined in either homotypic (stromal and epithelial cells from the same organ) or heterotypic (stromal and epithelial cells from different organs) combinations.

Table 1
Permissive Mesenchymal Effect on Epithelium in Heterospecific but Homotypic Recombination Experiment[a]

Epithelium source	*Mesenchyme source*	*Marker expression pattern*
Fetal rat intestine	Embryonic chick intestine	Lactase (rat-specific marker)
Embryonic chick intestine	Fetal rat intestine	Sucrase (chick-specific marker)

[a]The mesenchyme supports expression of a set of markers already specified by the epithelium. (Data from Kedinger et al., 1981.)

epithelium likewise has been demonstrated to produce sweat glands or pilosebaceous units when combined with embryonic dermis (Ferraris et al., 2000).

Experiments performed with epithelial and mesenchymal tissues from the endodermal hindgut and Wolffian duct derivatives have shed more light on the role of the germ layer of origin in determining epithelial cell fate. Urogenital sinus mesenchyme (UGM) has been shown to produce prostatic morphogenesis in the epithelium of the urogenital sinus, vagina, adult prostate, and embryonic urinary bladder (Cunha, 1972, 1975; Cunha and Lung, 1978; Cunha et al., 1980; Norman et al., 1986). Stromal–epithelial interactions were once thought to determine irreversibly the developmental fate of the epithelium. However, recombinations of adult bladder epithelium with UGM demonstrated that adult tissue could, in some instances, be made to redifferentiate along another pathway (Cunha et al., 1983; Neubauer et al., 1983). The adult bladder epithelium gives rise to secretory prostatic structure as a result of proliferation and reorganization of the adult bladder basal cells. This process closely resembles the sequence occurring in normal prostatic development. The ductal–acinar structures formed in this experiment resemble prostatic epithelium in terms of histology, histochemistry, expression of androgen receptors, androgen dependency for DNA synthesis, and production of prostate-specific antigens (Cunha et al., 1983; Neubauer et al., 1983). More recent studies have confirmed that adult human bladder urothelium can also be induced to generate prostatic structures by rat UGM (Aboseif et al., 1999). In similar experiments, Li et al. (2000) demonstrated that bladder epithelium can also respond to rectal mesenchyme giving rise to glandular structures with characteristic intestinal histology and secretions. It is noteworthy that their study also documents that a subpopulation of epithelial cells take on a glandular appearance. As in recombinants involving UGM, significant areas of transitional differentiation were still seen in recombinants composed of bladder epithelium and rectal mesenchyme.

Descriptions of bladder epithelium + UGM and bladder epithelium + rectal mesenchyme recombinants strongly suggest that the process that is occurring is not one of transdifferentiation (the change of mature adult cells from one phenotype to another) but, rather, one of induction of a new phenotype. Thus, the bladder epithelium does not uniformly respond to the inductive effects of the mesenchyme by changing its pattern of differentiation, but, rather, a subpopulation of cells within the urothelial layer responds

Table 2
Instructive Mesenchymal Effect on Epithelium in Heterospecific and Heterotypic Recombination Experiment[a]

Source mesenchyme	*Source epithelium*	*Resultant epithelium*
Chick stomach	Quail allantois	Quail stomach
Quail intestine	Chick allantois	Chick intestine

[a]The mesenchyme changes the expression of the markers expressed by the epithelium to those of the epithelial type, which would be associated with the mesenchyme. (Data from Haffen et al., 1987.)

to the inductive mesenchyme and gives rise to new prostatic or intestinal glandular structures. It is noteworthy that in both the bladder and the prostate proliferative rates are naturally extremely low. By contrast, the rectum, which is a part of the GI tract, has a much higher rate of cell turnover. Thus, stem cell populations must have a capacity to engage in a wide range of proliferative activity with that activity being dependent on the organ in which the cell is finally located. Regulation of stem cell proliferation in this context is therefore likely to be controlled by interactions with the local stromal microenvironment.

Like UGM, newborn rat seminal vesicle mesenchyme (SVM) is a powerful inducer of glandular differentiation. In tissue recombination experiments, SVM has been shown to induce seminal vesicle differentiation in the epithelial tissues derived from the seminal vesicle, ureter, and ductus deferens (Higgins et al., 1989a, 1989b; Cunha et al., 1991). Like the seminal vesicle, ureter and ductus deferens are both derived from the Wolffian duct, which is, in turn, a mesodermally derived epithelial tissue.

Thus, epithelial tissues from all three germ layers can be induced to give rise to new glandular tissues by exposure to appropriate mesenchyma. All of the experiments described above demonstrate the ability of mesenchymal cells to induce gland formation within a germ layer of origin. A further series of experiments in the urogenital tract examined the ability of mesenchymal cells to induce changes across germ layer boundaries. Tissue recombinants prepared using UGM with mesodermally derived epithelium gave rise to glandular structures with seminal vesicle morphology and secretions (Cunha et al., 1987; Tsuji et al., 1994). By contrast, SVM recombined with endodermally derived epithelium gave rise to glandular tissues with prostatic phenotype and secretory activity (Donjacour and Cunha, 1995). This series of experiments is summarized in Table 3. It is notable that the epithelial tissues of the bladder, urethra, and ureter all have the same transitional phenotype in vivo and are essentially indistinguishable in terms of appearance and expression of differentiated markers. Yet, in these experiments they give rise to tissues reflecting their EG layer of origin.

Tissue recombination experiments show that there is a subpopulation of cells within embryonic and adult epithelial tissues that can respond to inductive mesenchyma by giving rise to new tissue types. These experiments thus demonstrate either that putative stem cells have the potential to produce a wider variety of daughter cells than are found in their tissue of origin or, alternatively, that tissues contain stem cells for more than one tissue type. On the basis of the experiments described, this second option would suggest that the bladder, e.g., contains a range of stem cell types including prostatic and rectal stem cells, in addition to its native "bladder stem cell" population. Such an explanation is unappealing both intellectually and biologically. In particular, recent findings described in the introduction to this chapter would suggest that developmental plasticity of stem cell populations is a reasonable explanation for this ability to repopulate multiple organs. These tissue recombination data would further suggest that changes in stromal environment are sufficient to change the developmental program executed by stem cells within the confines of their germ layer of origin.

28.4. CONCLUSIONS

In this brief overview, we have highlighted some of the available data suggesting a high degree of plasticity in adult epithelial cell populations. We note that epithelial tissues are capable of changing their pattern of differentiation under both naturally occurring and experimental conditions. These data suggest that the epithelial tissues contain a cell population that is capable of generating epithelial tissues characteristic of multiple organs. Thus, we would support the contention that the idea of organ-specific stem cells is no longer viable. Rather, available data suggest that epithelial stem cells are capable of repopulating many of the organs of their germ layer of origin with relative ease. It is certainly clear that signals from inductive mesenchyma can elicit this response although at present specific molecular pathways that result in the induction of any given gland are unknown.

Recent data would suggest a cellular plasticity that allows cells to repopulate tissues even beyond their germ layer of origin. However, the mechanism by which this phenomenon may occur is presently unclear. Current models of differentiation suggest that fully differentiated and fully plastic states are defined by expression of specific transcription factors by cells (Slack, 2000; Blau et al., 2001). Such models predict that cells have a capacity to move between these states. It is further suggested that the degree of difficulty involved in moving between differentiated and plastic states is a function of the distance moved (*see* Fig. 3). Data from tissue recombination experiments suggest that exposure to inductive mesenchyma is insufficient to force transdifferentiation of committed epithelial cells but is sufficient to stimulate a stem cell population to generate daughter populations different from the tissue of origin. The process of transdifferentiation, although well documented, is uncommon, supporting the idea that moving from one differentiated state to another, although possible, is difficult. Data from tissue recombinants and from naturally occurring metaplastic responses further suggest that crossing the germ layer boundary is a significantly more difficult step than moving differentiation patterns within a germ layer. Thus, the mechanisms required to shift between germ layers are apparently different and perhaps more fundamental than reprogramming within a layer. Therefore, the germ layer might represent a "pinch point" in the continuum from totipotent to committed cells.

We thus return to the question, What is a stem cell? The data presented would suggest that stem cells occurring in glandular organs are simply the least differentiated and most plastic epithelial cell type found in adult glands. Adult epithelial stem cells exist toward one end of a continuum between totipotent ES cells and differentiated adult epithelium. Adult stem cells can easily repopulate a range of organs within the confines of a given germ layer.

Table 3
Role of Epithelial Germ Layer in Response to Inductive Mesenchyme[a]

Epithelium	*Germ layer origin*	*Mesenchyme*	*Resultant tissue*
Ductus veferens	Mesoderm	USM or SVM	Seminal vesicle
Ureter	Mesoderm	USM or SVM	Seminal vesicle
Seminal Vesicle	Mesoderm	USM or SVM	Seminal vesicle
Bladder	Endoderm	USM or SVM	Prostate
Prostate	Endoderm	USM or SVM	Prostate
Urethra	Endoderm	USM or SVM	Prostate

[a]The response of adult urogenital tract epithelia to inductive mesenchyme is limited by the germ layer of origin of the epithelium. Mesodermally derived epithelia give rise to seminal vesicle in response to either USM or SMV. By contrast, endodermally derived epithelia respond to the same inductive influences by generating prostatic tissue.

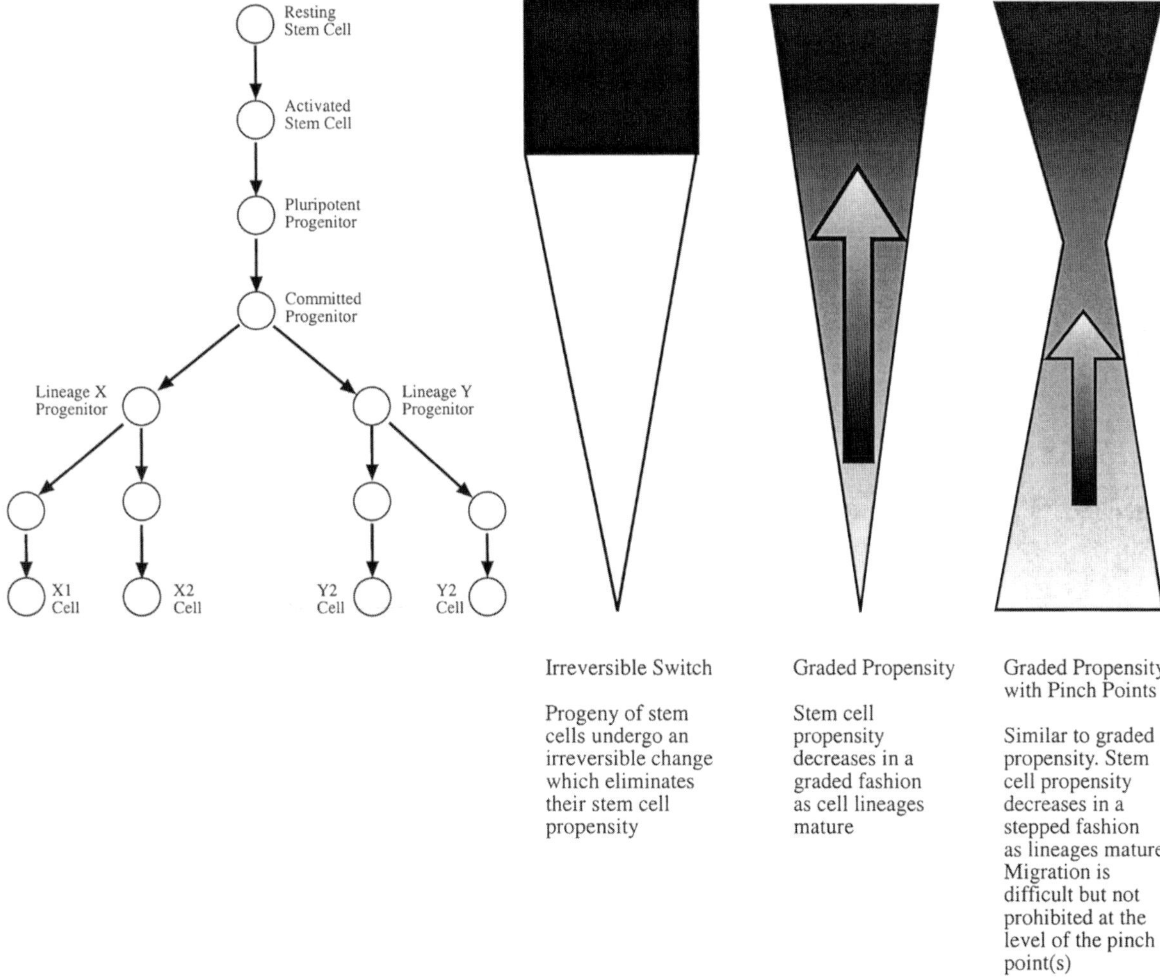

Fig. 3. Concepts of the propensity of cells to migrate between totipotent and committed cell types. Traditional views of cell maturation suggest that at some point on the differentiation pathway, cells undergo an irreversible switch and are no longer able to dedifferentiate. More contemporary views suggest that there is a gradation between totipotent ES cells and fully differentiated adult cells that can be traversed in either direction. Increased commitment is associated with expression of tissue-specific transcription factors. The energetic and biological costs associated with transformations from differentiated to totipotent cell types apparently make this sort of event unlikely. We propose that this graded propensity model contains a discrete pinch point (or points). Movement from below the pinch point to above is more restricted than other forms of change from a committed to stem cell phenotype. An example of such a pinch point might be a change across a germ layer boundary. (Redrawn and modified from Blau et al 2001.)

ACKNOWLEDGMENTS

The topic of stem cell biology has been one of interest and frustration for many years. SWH would like to express thanks to Gerald Cunha and the members of the Cunha laboratory, with whom these ideas have been explored in the past. Our research work is supported by the National Institutes of Health and the Department of Defense Prostate Cancer Research Program in the form of grants DK52721, CA 96403, DAMD 17-02-1-0151, and DAMD 17-01-1-0037.

REFERENCES

Aboseif, S., El-Sakka, A., Young, P., and Cunha, G. (1999) Mesenchymal reprogramming of adult human epithelial differentiation. *Differentiation* 65:113–118.

Alison, M. R., Poulsom, R., Jeffery, R., et al. (2000) Hepatocytes from non-hepatic adult stem cells. *Nature* 406:257.

Bartek, J., Bartkova, J., and Taylor-Papadimitriou, J. (1990) Keratin 19 expression in the adult and developing human mammary gland. *Histochem. J.* 22:537–544.

Bjerknes, M. and Cheng, H. (1999) Clonal analysis of mouse intestinal epithelial progenitors. *Gastroenterology* 116:7–14.

Blau, H. M., Brazelton, T. R., and Weimann, J. M. (2001) The evolving concept of a stem cell: entity or function? *Cell* 105:829–841.

Booth, D., Haley, J. D., Bruskin, A. M., and Potten, C. S. (2000) Transforming growth factor-B3 protects murine small intestinal crypt stem cells and animal survival after irradiation, possibly by reducing stem-cell cycling. *Int. J. Cancer* 86:53–59.

Borthwick, D. W., Shahbazian, M., Krantz, Q. T., Dorin, J. R., and Randell, S. H. (2001) Evidence for stem-cell niches in the tracheal epithelium. *Am. J. Respir. Cell. Mol. Biol.* 24:662–670.

Campbell, K. H., McWhir, J., Ritchie, W. A., and Wilmut, I. (1996) Sheep cloned by nuclear transfer from a cultured cell line. *Nature* 380:64–66.

Canfield, V., West, A. B., Goldenring, J. R. and Levenson, R. (1996) Genetic ablation of parietal cells in transgenic mice: a new model for analyzing cell lineage relationships in the gastric mucosa. *Proc. Natl. Acad. Sci. USA* 93:2431–2435.

Clarke, D. L., Johansson, C. B., Wilbertz, J., et al. (2000) Generalized potential of adult neural stem cells. *Science* 288:1660–1663.

Cunha, G. R. (1972) Tissue interactions between epithelium and mesenchyme of urogenital and integumental origin. *Anat. Rec.* 172: 529–542.

Cunha, G. R. (1975) Age-dependent loss of sensitivity of female urogenital sinus to androgenic conditions as a function of the epithelial-stromal interaction. *Endocrinology* 95:665–673.

Cunha, G. R. and Lung, B. (1978) The possible influences of temporal factors in androgenic responsiveness of urogenital tissue recombinants from wild-type and androgen-insensitive (Tfm) mice. *J. Exp. Zool.* 205:181–194.

Cunha, G. R., Chung, L. W. K., Shannon, J. M., and Reese, B. A. (1980) Stromal-epithelial interactions in sex differentiation. *Biol. Reprod.* 22:19–43.

Cunha, G. R., Fujii, H., Neubauer, B. L., Shannon, J. M., Sawyer, L. M., and Reese, B. A. (1983) Epithelial-mesenchymal interactions in prostatic development. I. Morphological observations of prostatic induction by urogenital sinus mesenchyme in epithelium of the adult rodent urinary bladder. *J. Cell. Biol.* 96:1662–1670.

Cunha, G. R., Donjacour, A. A., Cooke, P. S., et al. (1987) The endocrinology and developmental biology of the prostate. *Endocrine Rev.* 8: 338–362.

Cunha, G. R., Young, P., Higgins, S. J., and Cooke, P. S. (1991) Neonatal seminal vesicle mesenchyme induces a new morphological and functional phenotype in the epithelia of adult ureter and ductus deferens. *Development* 111:145–158.

Cunha, G. R., Young, P., Christov, K., et al. (1995) Mammary phenotypic expression induced in epidermal cells by embryonic mammary mesenchyme. *Acta Anat. (Basel)* 152:195–204.

D'Amour, K. A. and Gage, F. H. (2002) Are somatic stem cells pluripotent or lineage-restricted? *Nat. Med.* 8:213–214.

Danielpour, D. (1999) Transdifferentiation of NRP-152 rat prostatic basal epithelial cells toward luminal phenotype: regulation by glucocorticoid, insulin-like growth factor-I and transforming growth factor-beta. *J. Cell Sci.* 112:169–179.

De Marzo, A. M., Meeker, A. K., Epstein, J. I., and Coffey, D. S. (1998a) Prostate stem cell compartments: expression of the cell cycle inhibitor p27Kip1 in normal, hyperplastic, and neoplastic cells. *Am. J. Pathol.* 153:911–919.

De Marzo, A. M., Nelson, W. G., Meeker, A. K., and Coffey, D. S. (1998b) Stem cell features of benign and malignant prostate epithelial cells. *J. Urol.* 160:2381–2392.

Donjacour, A. A. and Cunha, G. R. (1995) Induction of prostatic morphology and secretion in urothelium by seminal vesicle mesenchyme. *Development* 121:2199–2207.

Driscoll, S. G. and Taylor, S. H. (1980) Effects of prenatal maternal estrogen on the male urogenital system. *Obstet. Gynecol.* 56:537–542.

Engelhardt, J. F., Schlossberg, H., Yankaskas, J. R., and Dudus, L. (1995) Progenitor cells of the adult human airway involved in submucosal gland development. *Development* 121:2031–2046.

Ferraris, C., Chevalier, G., Favier, B., Jahoda, C. A., and Dhouailly, D. (2000) Adult corneal epithelium basal cells possess the capacity to activate epidermal, pilosebaceous and sweat gland genetic programs in response to embryonic dermal stimuli. *Development* 127: 5487–5495.

Gierer, A. (1974) Molecular models and combinatorial principles in cell differentiation and morphogenesis. *Cold Spring Harbor Symp. Quant. Biol.* 38:951–961.

Gordon, J. I., Schmidt, G. H., and Roth, K. A. (1992) Studies of intestinal stem cells using normal, chimeric, and transgenic mice. *FASEB J.* 6: 3039–3050.

Griffiths, D. F., Davies, S. J., Williams, D., Williams, G. T., and Williams, E. D. (1988) Demonstration of somatic mutation and colonic crypt clonality by X-linked enzyme histochemistry. *Nature* 333: 461–463.

Gumpel-Pinot, M., Yasugi, S., and Mizuno, T. (1978) Différenciation d'épitheliums endodermiques associés à du mésoderme splanchnique. *C.R. Acad. Paris* 286:117–120.

Haffen, K., Kedinger, M., and Simon-Assmann, P. (1987) Mesenchyme-dependent differentiation of epithelial progenitor cells in the gut. *J. Pediatr. Gastroenterol. Nutr.* 6:14–23.

Hayward, S. W. and Cunha, G. R. (2000) The prostate: development and physiology. *Radiol. Clin. North Am.* 38:1–14.

Helpap, B. and Stiens, R. (1975) The cell proliferation of epithelial metaplasia in the prostate gland: an autoradiographic in vitro study. *Virchows Arch. B Cell. Pathol.* 19:69–76.

Hermiston, M. L. and Gordon, J. I. (1995) Organization of the crypt-villus axis and evolution of its stem cell hierarchy during intestinal development. *Am. J. Physiol.* 268:G813–G822.

Higgins, S. J., Young, P., Brody, J. R., and Cunha, G. R. (1989a) Induction of functional cytodifferentiation in the epithelium of tissue recombinants. I. Homotypic seminal vesicle recombinants. *Development* 106: 219–234.

Higgins, S. J., Young, P., and Cunha, G. R. (1989b) Induction of functional cytodifferentiation in the epithelium of tissue recombinants. II. Instructive induction of Wolffian duct epithelia by neonatal seminal vesicle mesenchyme. *Development* 106:235–250.

Hudson, D. L., O'Hare, M., Watt, F. M., and Masters, J. R. (2000) Proliferative heterogeneity in the human prostate: evidence for epithelial stem cells. *Lab. Invest.* 80:1243–1250.

Hudson, D. L., Guy, A. T., Fry, P., O'Hare, M. J., Watt, F. M., and Masters, J. R. (2001) Epithelial cell differentiation pathways in the human prostate: identification of intermediate phenotypes by keratin expression. *J. Histochem. Cytochem.* 49:271–278.

Hume, W. J. and Potten, C. S. (1979) Advances in epithelial kinetics—an oral view. *J. Oral. Pathol.* 8:3–22.

Isaacs, J. T. (1985) Control of cell proliferation and death in the normal and neoplastic prostate: a stem cell model. In: *Benign Prostatic Hyperplasia*, vol. 2 (Rogers, C. H., Coffey, D. S., Cunha, G. R., Grayhack,

J. T., Hinman, F. Jr., and Horton, R., eds.), National Institutes of Health, Bethesda, MD pp. 85–94.

Kedinger, M., Simon-Assmann, P. M., Grenier, J. F., and Haffen, K. (1981) Role of epithelial-mesenchymal interactions in the ontogenesis of intestinal brush-border enzymes. *Dev. Biol.* 86:339–347.

Kondo, T. and Raff, M. (2000) Oligodendrocyte precursor cells reprogrammed to become multipotential CNS stem cells. *Science* 289: 1754–1757.

Kroes, R. and Teppema, J. S. (1972) Development and restitution of squamous metaplasia in the calf prostate after a single estrogen treatment. An electron microscopic study. *Exp. Mol. Pathol.* 16: 286–301.

Kurita, T., Lee, K. J., Cooke, P. S., Lydon, J. P., and Cunha, G. R. (2000a) Paracrine regulation of epithelial progesterone receptor and lactoferrin by progesterone in the mouse uterus. *Biol. Reprod.* 62: 831–838.

Kurita, T., Lee, K. J., Cooke, P. S., Taylor, J. A., Lubahn, D. B., and Cunha, G. R. (2000b) Paracrine regulation of epithelial progesterone receptor by estradiol in the mouse female reproductive tract [published erratum appears in *Biol. Reprod.* 2000;63(1):354]. *Biol. Reprod.* 62:821–830.

Kurita, T., Wang, Y. Z., Donjacour, A. A., et al. (2001) Paracrine regulation of apoptosis by steroid hormones in the male and female reproductive system. *Cell Death Differ.* 8:192–200.

Lavker, R. M. and Sun, T. T. (1983) Epidermal stem cells. *J. Invest. Dermatol.* 81:121s–127s.

Li, Y., Liu, W., Hayward, S. W., Cunha, G. R., and Baskin, L. S. (2000) Plasticity of the urothelial phenotype: effects of gastro-intestinal mesenchyme/stroma and implications for urinary tract reconstruction. *Differentiation* 66:126–135.

Lipschutz, J. H., Fukami, H., Yamamoto, M., Tatematsu, M., Sugimura, Y., Kusakabe, M. and Cunha, G. (1999) Clonality of urogenital organs as determined by analysis of chimeric mice. *Cells Tissues Organs* 165: 57–66.

Loeffler, M., Birke, A., Winton, D., and Potten, C. (1993) Somatic mutation, monoclonality and stochastic models of stem cell organization in the intestinal crypt. *J. Theor. Biol.* 160:471–491.

Lugo, M. and Putong, P. B. (1984) Metaplasia: an overview. *Arch. Pathol. Lab. Med.* 108:185–189.

Malek, S. N., Katumuluwa, A. I., and Pasternack, G. R. (1990) Identification and preliminary characterization of two related proliferation-associated nuclear phosphoproteins. *J. Biol. Chem.* 265: 13,400–13,409.

Merritt, A. J., Potten, C. S., Kemp, C. J., et al. (1994) The role of p53 in spontaneous and radiation-induced apoptosis in the gastrointestinal tract of normal and p53-deficient mice. *Cancer Res.* 54:614–617.

Morshead, C. M., Benveniste, P., Iscove, N. N., and van der Kooy, D. (2002) Hematopoietic competence is a rare property of neural stem cells that may depend on genetic and epigenetic alterations. *Nat. Med.* 8:268–273.

Neubauer, B. L., Chung, L. W. K., McCormick, K. A., Taguchi, O., Thompson, T. C., and Cunha, G. R. (1983) Epithelial-mesenchymal interactions in prostatic development. II. Biochemical observations of prostatic induction by urogenital sinus mesenchyme in epithelium of the adult rodent urinary bladder. *J. Cell Biol.* 96:1671–1676.

Nomura, S., Esumi, H., Job, C., and Tan, S. S. (1998) Lineage and clonal development of gastric glands. *Dev. Biol.* 204:124–135.

Norman, J. T., Cunha, G. R., and Sugimura, Y. (1986) The induction of new ductal growth in adult prostatic epithelium in response to an embryonic prostatic inductor. *Prostate* 8:209–220.

Okada, T. (1991) *Transdifferentiation: Flexibility in Cell Differentiation*, Clarendon Press, Oxford, NY.

Patapoutian, A., Wold, B. J., and Wagner, R. A. (1995) Evidence for developmentally programmed transdifferentiation in mouse esophageal muscle. *Science* 270:1818–1821.

Petersen, B. E., Bowen, W. C., Patrene, K. D., et al. (1999) Bone marrow as a potential source of hepatic oval cells. *Science* 284: 1168–1170.

Ponder, B. A., Schmidt, G. H., Wilkinson, M. M., Wood, M. J., Monk, M., and Reid, A. (1985) Derivation of mouse intestinal crypts from single progenitor cells. *Nature* 313:689–691.

Potten, C. S. and Grant, H. K. (1998) The relationship between ionizing radiation-induced apoptosis and stem cells in the small and large intestine. *Br. J. Cancer* 78:993–1003.

Potten, C. S., Wilson, J. W., and Booth, C. (1997) Regulation and significance of apoptosis in the stem cells of the gastrointestinal epithelium. *Stem Cells* 15:82–93.

Poulsom, R., Forbes, S. J., Hodivala-Dilke, K., et al. (2001) Bone marrow contributes to renal parenchymal turnover and regeneration. *J. Pathol.* 195:229–235.

Rindi, G., Ratineau, C., Ronco, A., Candusso, M. E., Tsai, M., and Leiter, A. B. (1999) Targeted ablation of secretin-producing cells in transgenic mice reveals a common differentiation pathway with multiple enteroendocrine cell lineages in the small intestine. *Development* 126:4149–4156.

Risbridger, G., Wang, H., Young, P., et al. (2001a) Evidence that epithelial and mesenchymal estrogen receptor-alpha mediates effects of estrogen on prostatic epithelium. *Dev. Biol.* 229:432–442.

Risbridger, G. P., Wang, H., Frydenberg, M., and Cunha, G. (2001b) The metaplastic effects of estrogen on mouse prostate epithelium: proliferation of cells with basal cell phenotype. *Endocrinology* 142: 2443–2450.

Roth, K. A., Hermiston, M. L., and Gordon, J. I. (1991) Use of transgenic mice to infer the biological properties of small intestinal stem cells and to examine the lineage relationships of their descendants. *Proc. Natl. Acad. Sci. USA* 88:9407–9411.

Schmidt, G. H., Winton, D. J., and Ponder, B. A. (1988) Development of the pattern of cell renewal in the crypt-villus unit of chimaeric mouse small intestine. *Development* 103:785–790.

Slack, J. M. (2000) Stem cells in epithelial tissues. *Science* 287: 1431–1433.

Slack, J. M. and Tosh, D. (2001). Transdifferentiation and metaplasia—switching cell types. *Curr. Opin. Genet. Dev.* 11:581–586.

Taylor-Papadimitriou, J. and Lane, E. B. (1987) Keratin expression in the mammary gland. In: *The Mammary Gland Development, Regulation and Function* (Neville, M. C. and Daniel, C. W., eds.), Plenum, New York, pp. 181–215.

Taylor-Papadimitriou, J., Stampfer, M., Bartek, J., et al. (1989) Keratin expression in human mammary epithelial cells cultured from normal and malignant tissue: relation to in vivo phenotypes and influence of medium. *J. Cell. Sci.* 94:403–413.

Taylor-Papadimitriou, J., Wetzels, R., and Ramaekers, F. (1992). Intermediate filament protein expression in normal and malignant human mammary epithelial cells. *Cancer Treat. Res.* 61:355–378.

Toyosawa, S., Ohnishi, A., Ito, R., et al. (1999). Small cell undifferentiated carcinoma of the submandibular gland: immunohistochemical evidence of myoepithelial, basal and luminal cell features. *Pathol. Int.* 49:887–892.

Triche, T. J. and Harkin, J. C. (1971) An ultrastructural study of hormonally induced squamous metaplasia in the coagulating gland of the mouse prostate. *Lab. Invest.* 25:596–606.

Tsuji, M., Shima, H., Boutin, E., Young, P., and Cunha, G. R. (1994). Effect of mesenchymal glandular inductors on the growth and cytodifferentiation of neonatal mouse seminal vesicle epithelium. *J. Androl.* 15:565–574.

Walensky, L. D., Coffey, D. S., Chen, T. H., Wu, T. C., and Pasternack, G. R. (1993) A novel M(r) 32,000 nuclear phosphoprotein is selectively expressed in cells competent for self-renewal. *Cancer Res.* 53:4720–4726.

Wang, Y. Z., Hayward, S. W., Cao, M., Thayer, K. A., and Cunha, G. R. (2001) Cell differentiation lineage in the prostate. *Differentiation* 68:270–279.

Wilmut, I., Schnieke, A. E., McWhir, J., Kind, A. J., and Campbell, K. H. (1997) Viable offspring derived from fetal and adult mammalian cells. *Nature* 385:810–813.

Winton, D. J., Blount, M. A., and Ponder, B. A. (1988) A clonal marker induced by mutation in mouse intestinal epithelium. *Nature* 333: 463–466.

Ying, Q.-L., Nichols, J., Evans, E., and Smith, A. (2002) Changing potency by spontaneous fusion. *Nature* advance online publication, March 13, 2002

29 Gastrointestinal Stem Cells

Proliferation Kinetics and Differentiation Hierarchies

SHERIF M. KARAM, MD, PhD

In the gastrointestinal (GI) epithelium, cellular differentiation occurs in five stages: stem cells, precursor cells, transit cells, mature cells, and terminal cells. Stem cells are the least differentiated and have the greatest proliferation potential. In the stomach, they are located in the isthmus region of the pit-gland unit and give rise to four main cell lineages through precursor-amplifying cells: pre–pit cells give rise to mucous-secreting pit cells that migrate and reach the gastric luminal surface in 3 d; pre–neck cells differentiate into neck cells that migrate toward the bottom of the gland while changing their phenotype into the longest lived cells in the unit, zymogenic cells; pre–parietal cells complete their differentiation in the isthmus and then undergo bipolar migration to the pit and the base of the gland, where they have a life-span of about 2 mo. The precursors of enteroendocrine cells also originate in the isthmus and follow the bipolar mode of migration. In the small intestine, stem cells are located in the fourth cell layer from the crypt bottom and are derived from a single preceding cell (clonal). An adult mouse has about 1.1 million crypts and each crypt contains about 250 cells, of which about two-thirds go through the cell cycle every 12 h; each crypt produces about 13–16 new cells per hour! They give rise to four main cell lineages: absorptive, goblet, enteroendocrine, and Paneth. While members of the former three lineages differentiate while migrating upward along the crypt-villus axis, those of the latter complete their differentiation and remain at the crypt bottom. M-cell precursors are found in the intestinal crypts around lymphoid follicles and their mature forms migrate to cover the dome-shaped Peyer's patches. In the colon, the lineages are columnar cells (80%), goblet cells (16%), deep crypt secretory cells (3%), and enteroendocrine cells (0.4%). Precursors of caveolated cells are also produced by stem cells throughout the GI tract.

29.1. INTRODUCTION

The lining epithelium of the gastrointestinal (GI) tract is characterized by its continuous and rapid renewal. Pioneering experiments using DNA labeling by radiothymidine and the radioautographic technique have demonstrated that this renewal phenomenon is generated by cells with high proliferative capability anchored in specific locations along the GI epithelium (Leblond et al., 1959). During the last 30 yr, our understanding of the cellular hierarchies of these proliferative "stem" cells has gradually increased, especially with the development of genetically manipulated animal models (Gordon and Hermiston, 1994; Hermiston et al., 1994; Traber and Silberg, 1996; Robine et al., 1997) and laser-capture microdissection (Wong et al., 2000). Thus, the renewal concepts of the GI epithelium have become fundamental to understanding its structure and function in health and disease.

During development, the GI epithelium starts as a single layer of proliferative endodermal stem cells (Maunoury et al., 1992), which then form a pseudostratified epithelium. Eventually, elongation and compartmentalization of the primitive gut occurs with a remarkable increase in the epithelial surface area and production of various cell lineages. This chapter begins with a brief description of the cellular hierarchies in the GI epithelium and the features of stem cells. Then, the main morpho-dynamic features of the various epithelial cell lineages along the GI tract are summarized. Finally, some in vitro studies on these cells are presented.

From: *Stem Cells Handbook*
Edited by: S. Sell © Humana Press Inc., Totowa, NJ

29.2. CELLULAR HIERARCHIES OF GI EPITHELIUM

In the GI epithelium, cells go through five main stages. They are first present as *stem cells*, which divide and produce new stem cells as well as uncommitted and/or committed *precursor cells*, which represent the second stage in GI epithelial cell life. The uncommitted precursor cells exhibit dual-lineage features and eventually become committed precursor cells with features of one lineage. These cells are usually capable of undergoing equivalent mitosis and thus amplifying the population before entering the next stage (Lee and Leblond, 1985a; Karam and Leblond, 1993a).

Transit cells represent the third stage, during which cellular specification gradually occurs by synthesizing new gene products. This may be encountered morphologically by gradual changes in cell structure; immuno- or lectin-cytochemically by expression of new proteins or sugar residues; and biochemically by changes in enzymatic activities, protein composition, and messenger RNA expression. The fourth stage is that of the *mature cells*, which have completed their differentiation and thus become actively functional. In the fifth stage, *terminal cells*, there is a gradual deterioration, and eventually cells undergo death and elimination (Karam et al., 1997b).

29.3. STEM CELLS OF GI EPITHELIUM

Even though specific biomarkers for stem cells of normal gut epithelium are not yet available, these cells can be defined by two major criteria. First, morphologically, they are undifferentiated and exhibit embryonic cell–like features (Fig. 1): high nucleus-to-cytoplasm ratio, a nucleus with much diffuse chromatin and large reticulated nucleoli, and cytoplasm containing a few small organelles but many free ribosomes (Leblond, 1981; Karam, 1995). Second, functionally, they have a high capacity to proliferate so as to ensure their own renewal while producing lineage precursors that differentiate to form transit cells committed to become mature cells (Hall and Watt, 1989; Potten and Loeffler, 1990; Gordon et al., 1992). Therefore, potential stem cells in a population can be tentatively identified by electron microscopy combined with ^{3}H-thymidine radioautography; stem cells would be the least differentiated and most proliferative cells (Karam, 1995).

Recently, the stages of the cell cycle have been redefined in the stem cells of the pyloric antrum (El-Alfy and Leblond, 1987a, 1987b) and duodenum (El-Alfy and Leblond, 1989). It appears that the duration of their mitosis is about 8.4 h including 4.8 h for prophase and 0.2, 0.06, and 3.3 h for metaphase, anaphase and telophase, respectively. Thus, the duration of prophase is quite longer than hitherto believed; that is, chromatin condensation of the prophase begins during the DNA-synthesizing (S) stage, which starts during interphase and lasts for about 5.8 h (El-Alfy and Leblond, 1989).

New insights into the small intestinal stem cells are introduced by cytotoxic radiation damage where the stem cells are defined by their regenerative capabilities. The location of stem cells has been confirmed and precisely defined in the fourth cell stratum from the crypt bottom (Potten et al., 1997).

The powerful technology of mouse aggregation chimeras and transgene expression have provided new insights regarding the clonality and number of the multipotent stem cells in each epithelial unit and their capacity to encode spatial memory or retain a positional address along the cephalocaudal axis of the gut (reviewed in Gordon et al., 1992). The stem cell hierarchy in the small intestinal epithelium is established during crypt formation. This process involves a selection among several multipotential stem cells so that ultimately only one survives to supply descendants to the fully formed crypt. Using genetic mosaic analysis, Wong et al. (2002) found that the selection of multipotent stem cells during morphogenesis of small intestinal crypts is perturbed by stimulation of transcription factor Lef-1/β-catenin signaling. In addition, it has been recently revealed that mutations in the transcription factor GATA-4 disrupt the normal differentiation program of mouse gastric epithelial stem cells (Jacobsen et al, 2002).

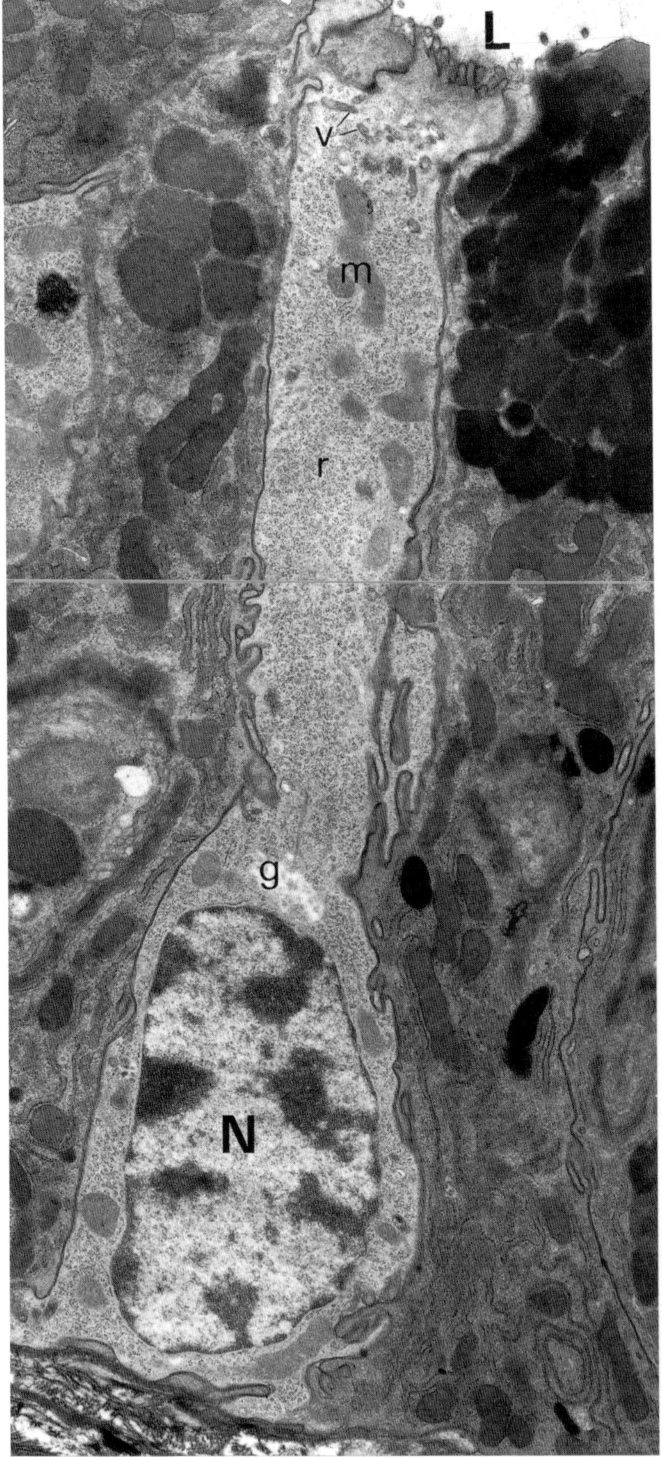

Fig. 1. Electron micrograph showing undifferentiated stem cell of the mouse colonic epithelium. The cell is long columnar with a narrow apex facing the crypt lumen (L), seen at the top, and a wider base sitting on the basement membrane, seen at the bottom. As it appears from the distribution of chromatin in the nucleus (N), the cell is in mitosis. The cytoplasm has many free ribosomes (r), but a small Golgi complex (g) compared with the large Golgi of the neighbor secretory cells. The RER cisternae are small and very few. The mitochondria (m) are few and smaller than those in the neighbor cells. The apical cytoplasm has no secretory granules, but a few small vesicles/tubules (v), which could be endocytic or exocytic (transporting glycocalyx to the cell apex). Note the presence of dark mucous granules in the apical cytoplasm of neighbor cells. Magnification: ×13,000.

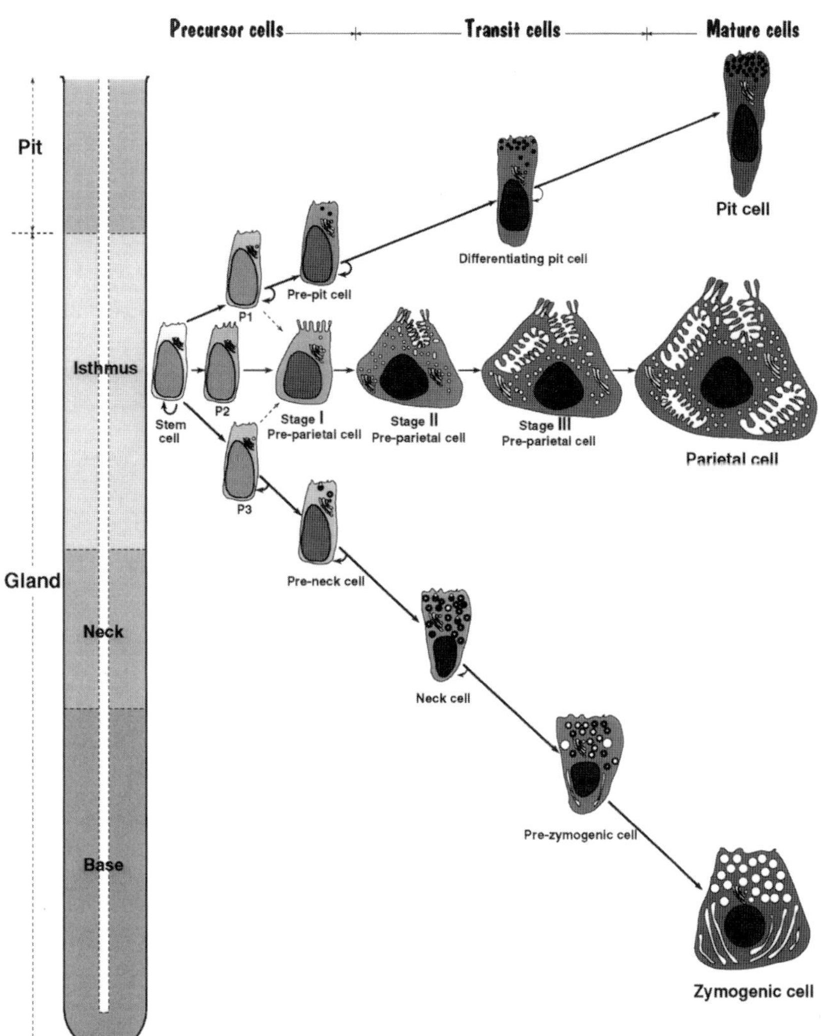

Fig. 2. Diagram of cell differentiation in oxyntic pit-gland unit of stomach. The stem cells are located in the isthmus region, which we enlarged here to accommodate all precursor cells seen at the right. The stem cells give rise to three precursors: pre–pit cell precursor (P1), pre–parietal cell precursor (P2), and pre–neck cell precursor (P3), which, respectively, evolve into pit, parietal, and zymogenic cell lineages. The straight arrows for pit and zymogenic cell lineages indicate the migration pathways, and curved arrows indicate cells capable of mitosis. After their differentiation in the isthmus, parietal cells migrate in both directions.

29.4. CELL LINEAGES IN BODY OF STOMACH

In the body region of the stomach, the oxyntic epithelium invaginates to form short pits continuous with long tubular glands (Helander, 1981; Ito, 1987; Karam and Leblond, 1992). Each gland can be divided into isthmus, neck, and base regions (Fig. 2). In the mouse, the pit-gland unit is lined by a monolayer of about 200 cells (Karam and Leblond, 1992). All cells lining the pit-gland unit of the oxyntic epithelium originate from the stem cells. These cells are stationary, anchored in the isthmus region, and have a turnover time of about 2.5 d (Karam and Leblond, 1993a). They give rise to three main cell lineages (Fig. 2).

29.4.1. PIT CELL LINEAGE

29.4.1.1. Pre–Pit Cell Precursors In the isthmus, about 67% of the progeny of the stem cells produced daily become pre–pit cell precursors (P1 in Fig. 2). They are characterized by a small Golgi apparatus that produces prosecretory vesicles at the *trans*-face. These vesicles vary in density but contain uniformly fine particulate material. These cells are partially committed and have two different progenies. The majority (99%) become pre–pit cells and only 1% become pre–parietal cells with pre–pit cell-like secretory granules. The development of a pre–pit cell precursor into a pre–pit cell is manifested by the maturation of the *trans*-Golgi vesicles into dense secretory granules. In the case of the pre–parietal cell, there is also an elongation of the apical microvilli (Karam and Leblond, 1993a).

29.4.1.2. Pre–Pit Cells Pre–pit cells are located in the upper portion of the isthmus and are characterized by a few 200-nm-wide, dense secretory granules. An average of 10 pre–pit cells are present in each isthmus. Radioautography has revealed that they have two sources of origin. About 57% come from the differentiation of pre–pit cell precursors, the remaining 43% from their own mitosis. After a pulse of ^{3}H-thymidine, 25% of pre–pit cells become labeled. With time, label increases to reach 33% at 6 h, then gradually decreases to 1% at 4 d, and completely disappears thereafter. In continuous ^{3}H-thymidine-labeling experiments, almost all pre–pit cells become labeled by 2 d. Both single injection and continuous-labeling experiments confirm the short turnover time of pre–pit cells (2.5 d). The fate of pre–pit cells is to become pit cells. This occurs as the activity of the cell increases and an increasing number of larger and larger secretory granules are produced and accumulate at the apex before exocytosis (Karam and Leblond, 1993a).

29.4.1.3. Pit Cells Pit cells are located in the pit region and are characterized by a dense apical group of mucous granules. The pit cells migrate outward along the pit wall to reach the gastric luminal surface in a few days (Karam and Leblond, 1993b). During pit cell migration, the apical group of granules enlarges due to an increase in number and size of newly produced secretory granules, from 250 to 400 nm. In addition, cells gradually elongate with tapering of their basal cytoplasm, nucleoli become condensed, the amount of ribosomes diminishes, and the mitochondria decrease in size. Pit cells close to the pit–isthmus border retain some ability to divide. Thus, pit cells are not only developed from maturation of pre–pit cells, but some are also produced by their own mitosis. Even though a pit region may include a few large parietal cells, the migration of pit cells along the pit wall occurs in a fairly regular pipeline manner. It takes about 60 h for a pit cell to reach the surface. At the luminal surface, the transit time is only 12 h. The overall turnover time of pit cells averages 3 d (Karam and Leblond, 1993b).

29.4.2. ZYMOGENIC CELL LINEAGE

29.4.2.1. Pre–Neck Cell Precursors About 24% of the stem cells produced daily become pre–neck cell precursors (P3 in Fig. 2). These precursors are characterized by prosecretory vesicles at the *trans*-face of their Golgi apparatus containing dense irregular material with light periphery. They are partially committed precursors and the fate of 98% of them is to become pre–neck cells; the remaining 2% become pre–parietal cells with cored secretory granules similar to those of pre–neck cells (Karam and Leblond, 1993a).

29.4.2.2. Pre–Neck Cells Pre–neck cells are located in the lower portion of the isthmus; they average 1.8 cells per isthmus. They are characterized by a few 400-nm-wide secretory granules that appear dense with a light core. They are mitotically active (11% become labeled after a radiothymidine pulse), and their division yields new pre–neck cells and cells committed to develop into neck cells. The turnover time of pre–neck cells is about 3 d (Karam and Leblond, 1993a).

29.4.2.3. Neck Cells Neck cells, also called "mucous neck cells," are located in the neck region and are characterized by many dense mucous granules that usually contain a light core made of pepsinogen (Sato and Spicer, 1980). Neck cells close to the isthmus have fewer and smaller granules (430-nm wide) than those close to the base (700 nm). After their production in the isthmus from transformation of pre–neck cells or in the high neck segment from their own mitosis, neck cells migrate inward while completing their differentiation toward the mucous phenotype. Neck cells are not end cells; that is, their fate is not to degenerate and die. They spend from 7 to 14 d in the neck region. Then, at the neck-base border, their phenotype gradually changes from mucous to serous (Karam and Leblond, 1993c).

29.4.2.4. Pre–Zymogenic Cells In the upper segment of the base region of the pit-gland unit of the mouse, there is a group of cells producing secretory granules that appear intermediate between those of neck cells and those of zymogenic cells. These granules contain two different components: electron dense mucus and light pepsinogen. Cells with similar criteria are also described in guinea pigs (Sato and Spicer, 1980) and rats (Suzuki et al., 1983). In the mouse, these cells are classified into subtypes I, II, and III, according to whether the dense mucous component is, respectively, more abundant than, about equal to, or less abundant than the light pepsinogenic component. Moreover, prosecretory vesicles at the Golgi *trans*-face of each of these subtypes exhibit differences parallel to those occurring in the granules. In the basal cytoplasm, rough endoplasmic reticulum (RER) cisternae are more abundant in subtype III than in subtypes I and II. The existence of further intermediates between these subtypes indicates that they transform into one another (I → II → III) and, thus, gradually change their phenotype to become more and more pepsinogenic. The gradual decrease in their mucous production has led to the production of granules that are entirely pepsinogenic (Karam and Leblond, 1993c).

29.4.2.5. Zymogenic Cells Zymogenic cells are pepsinogen-secreting cells that are characterized by spherical zymogen granules with homogeneously light pepsinogenic content. As zymogenic cells migrate inward, their phenotype specificity increases, as suggested by the measurement of zymogen granules, 780-nm wide in the high base vs 1070-nm wide in the low base. The production of larger and larger granules is in line with the increase in the amount of RER cisternae and also with the enlargement of the nucleolus. Zymogenic cells are end cells that eventually acquire signs of degeneration and finally die at the gland bottom after a long turnover time of approx 194 d (Karam and Leblond, 1993c).

29.4.3. PARIETAL CELL LINEAGE

29.4.3.1. Pre–Parietal Cell Precursors Little is known about pre–parietal cell precursors (P2 in Fig. 2). They are defined in a developing transgenic animal model in which the precursors of acid-secreting cell lineage have been amplified. They are characterized by embryonic cell–like features, in addition to having numerous apical microvilli with little glycocalyx (Li et al., 1995; Karam et al., 1997a).

29.4.3.2. Pre–Parietal Cells Pre–parietal cells are characterized by having long apical microvilli and incipient canaliculi. They do not undergo mitosis at any stage of their development (Karam, 1993). Based on the presence or absence of some secretory granules, pre–parietal cells are divided into three variants: (1) pre–parietal cells with no secretory granules that directly develop from pre–parietal cell precursors, (2) pre–parietal cells with a few small dense granules similar to the granules of pre–pit cells that develop from pre–pit cell precursors, and (3) pre–parietal cells carrying a few cored granules similar to those in pre–neck cells that develop from pre–neck cell precursors. Development of pre–parietal cells into parietal cells occurs in four stages. First, an increase in the surface area of the apical plasma membrane forms long numerous microvilli. Second, a few small H,K-ATPase-containing tubules and vesicles appear in the cytoplasm, and the apical membrane invaginates to form an incipient canaliculus at one side of the nucleus. Third, an additional canaliculus appears on the other side of the nucleus and the number and size of mitochondria gradually increase. Finally, expansion of the canaliculi and overall increase in cell size lead to the formation of a fully mature parietal cell (Karam,

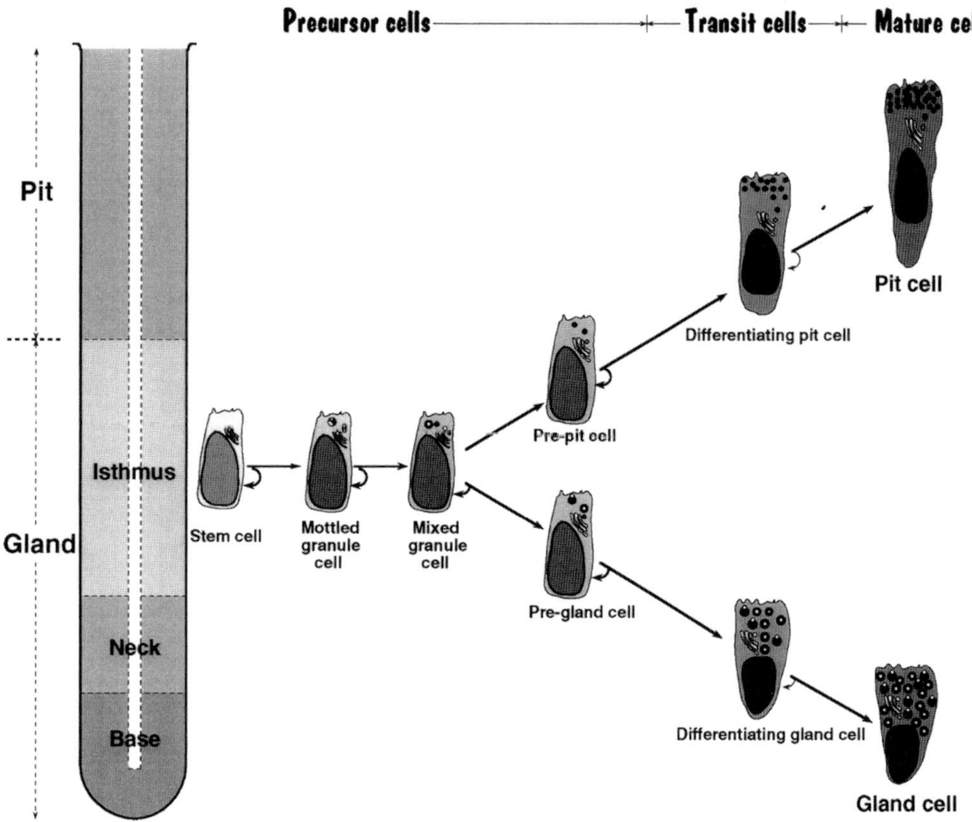

Fig. 3. Diagram of cell differentiation in mucous pit-gland unit of pyloric antrum. The gland includes three equal regions: isthmus (enlarged here to accommodate all precursor cells), neck, and base. Stem cells are located in the isthmus and give rise to mottled granule cells that undergo several rounds of cell division and produce mixed-granule cells. The latter give rise to pre–pit and pre–neck cells that evolve to form pit and gland cell lineages, respectively.

1993; Karam and Forte, 1994; Karam et al., 1997a). The formation of a pre–parietal cell takes about 1 d, and its maturation into a parietal cell requires at least two more days (Karam, 1993).

29.4.3.3. Parietal Cells Parietal cells are produced in the isthmus and migrate bidirectionally along the pit-gland axis. This migration pathway has been visualized by radioautography. Radiolabeled cells are first seen in the isthmus. With time, they appear in the pit in an outward direction and also in the neck in an inward direction until they reach the blind end of the unit. *In situ* hybridization studies and biochemical analysis have demonstrated that the synthetic/secretory activity of parietal cells varies along the pit-gland axis (Karam et al., 1997b). Young parietal cells in the isthmus and neck are more active than old parietal cells in the pit and base regions. The estimated turnover time of parietal cells is about 54 d (Karam, 1993). Ablation of parietal cells in genetically manipulated animal models (Sharp et al., 1995; Canfield et al., 1996; Li et al., 1996; 1998; Karam, 1998) has been associated with a block in the terminal differentiation of zymogenic cells. Thus, in addition to the fact that parietal and zymogenic cell lineages have a common source of origin, it seems that the former produces some regulatory factors necessary for the terminal differentiation of the latter. DNA profiling of isolated parietal cells revealed the identity of various genes that could be responsible for this regulatory function of parietal cells (Mills et al., 2001).

29.5. CELL LINEAGES IN PYLORIC ANTRUM OF STOMACH

In the pyloric antrum of the stomach, the epithelium forms long pits continuous with short glands made of three regions: isthmus, neck, and base (Fig. 3). In the mouse, the pyloric antral pit-gland unit is lined by a monolayer of about 250 cells (Lee et al., 1982). In the isthmus region, stem cells have a turnover time of about 1 d and their immediate descendants are called mottled-granule cells (Lee and Leblond, 1985a). They are characterized by embryonic cell–like features, but they also carry a few small (235 nm) mottled granules in their apical cytoplasm (Fig. 3). These uncommitted precursor cells represent 39% of the isthmal cells; they undergo clonal expansion and divide four times before giving rise to "mixed-granule cells," characterized by a mixture of small dense granules and large cored granules. Mixed-granule cells divide and give rise to dense granule cells (pre–pit cells) and core granule cells (pre–gland cells). The turnover time for each of the mottled- and mixed-granule cells is about 1 d (Lee and Leblond, 1985a).

29.5.1. PIT CELL LINEAGE

29.5.1.1. Pre–Pit Cells Pre–pit cells represent about 17% of all isthmal cells and are usually located near the pit border. Both the morphological features and dynamic behavior of these cells are quite similar to those of the pre–pit cells in the isthmus region of the oxyntic pit-gland units (Karam and Leblond, 1993a; Lee, 1985).

29.5.1.2. Pit Cells Pit cells are located in the pit region and represent about 180 cells per pit-gland unit. The mode of migration and structural features of these cells are similar to those in the oxyntic epithelium (Karam and Leblond, 1993a). They also have the same turnover time (3 d) (Lee, 1985).

29.5.2. GLAND CELL LINEAGE

29.5.2.1. Pre–Gland Cells Pre–gland cells are poorly differentiated cells that are characterized by having a few small (280 nm) cored granules. Pre–gland cells represent about 28% of the isthmal cells and predominate near the neck border. These cells duplicate before their differentiation-associated migration to cross the neck border and become gland cells (Lee and Leblond, 1985b).

29.5.2.2. Gland Cells Gland cells are located in the neck and base regions of the mucous pit-gland unit and represent about 37 cells per unit. Along the neck–base axis, gland cells exhibit more and larger-cored granules toward the base. The granule size varies from 380 nm in the neck to 580 nm in the base. In addition, with the inward migration of gland cells, the amount of ribosomes diminishes, the RER cisternae become numerous, and the Golgi apparatus increases in size (Lee and Leblond, 1985b). The isthmus of each mucous unit produces about 12 gland cells/d by differentiation of pre–gland cells. Gland cells also retain some mitotic activity, which gradually diminishes toward the gland bottom. Thus, a total of 29 gland cells are added daily to the gland cell population. Gland cells migrate inward to the gland bottom in a gradually decreasing rate. The average time spent by a gland cell in the neck region is about 10 h, and in the base region, about 200 h. This pattern of gland cell renewal is known as the "cascade" pattern of cell renewal. Therefore, the overall turnover time of gland cells is highly variable, from 1 to 60 d (Lee and Leblond, 1985b).

29.6. CELL LINEAGES IN THE SMALL INTESTINE

The small intestinal epithelium invaginates to form small crypts continuous with large evaginating villi (Fig. 4). Although the crypts are much smaller than the villi, they are much more numerous (Wright and Alison, 1984). An adult mouse has a total of about 1.1 million crypts; each crypt contains about 250 cells (Li et al., 1994). The crypts include the proliferative stem cells that are arranged in an annulus (Potten and Loeffler, 1990) and immediate precursors. In mice, about two-thirds of the crypt cells pass through the cell cycle every 12 h, and every crypt produces about 13–16 new cells/h that undergo differentiation-associated migration and form several cell lineages (Fig. 4) that supply two to three neighboring villi (Potten et al., 1997).

29.6.1. ABSORPTIVE CELL LINEAGE In mice, there is an overwhelming majority of members of the absorptive cell lineage along the small intestinal crypt–villus unit (>80% of all epithelial cells).

29.6.1.1. Pre–Absorptive Cells Pre–absorptive cells are located in the crypt base. They exhibit stem cell–like features and relatively long microvilli. These cells retain some ability to divide and, thus, are produced by their own mitosis as well as by differentiation of stem cells. Pre–absorptive cells migrate outward along the crypt base to reach the midcrypt, where they gradually differentiate; the microvilli elongate to form brush border. When these cells reach the crypt top, the differentiation is complete as the absorptive cells are formed (Cheng and Leblond, 1974a).

29.6.1.2. Absorptive Cells Absorptive cells are located along the crypt top and the whole villus. They are characterized by the absence of secretory granules and the presence of a prominent apical brush border. The absorptive cells migrate outward along the crypt wall and reach the surface of the villus in a few days (Cheng and Leblond, 1974a). During migration of absorptive cells, they show signs of differentiation: the cell gradually elongates, apical microvilli become long and numerous, lateral membranes interdigitate with neighbor cells, the nucleus gradually becomes heterochromatic and closer to the cell center, and the nucleoli become condensed and small. In the cytoplasm, the amount of free ribosomes diminishes, the Golgi apparatus becomes prominent, mitochondria become abundant, and RER cisternae increase in amount. The migration of members of the absorptive cell lineage is slow in the crypt and gradually increases to become approximately constant in the crypt top and along the villus. The overall turnover time of absorptive cells averages 3 d (Cheng and Leblond, 1974a). It has been recently shown that extracellular matrix (Simon-Assmann et al., 1995); some adhesion molecules, such as N-cadherin and β-catenin (Hermiston and Gordon, 1995; Wong et al., 1998, 2002); and Rac1, a member of the Rho family of guanosine 5'-triphosphate-binding proteins (Stappenbeck and Gordon, 2000) play an important role in maintaining the differentiation program of the absorptive cells and homeostasis of the epithelium. Along the crypt–villus axis, the functional activity of the absorptive cells follows a pattern parallel to their morphological changes. ^{3}H-fucose radioautography has revealed that the turnover rate of the glycoprotein (brush-border enzyme) in the absorptive cells is very rapid. Within 5 min the fucose reaches the Golgi region and by 90 min most of the fucose is on the microvilli. However, the amount of incorporated fucose is variable along the crypt-villus axis. It is low in the crypt, rises toward the villus base, remains high along the midvillus, and then sharply declines at the villus tip (Leblond, 1981).

29.6.2. GOBLET CELL LINEAGE The number of goblet cells increases from the duodenum to the jejunum to the ileum; they respectively constitute 4, 6, and 12% of all epithelial cells (Cheng, 1974a).

29.6.2.1. Pre–Goblet Cells Pre–goblet cells, also called oligomucous cells, are located in the crypt base compartment of the small intestine, where they exhibit stem cell features, and also show, as a sign of early differentiation, a few small mucous granules. These granules are usually homogeneously pale (common oliogomucous cells), but sometimes they include a dense core (granular oligomucous cells). Pre–goblet cells retain some capacity for mitosis, and, therefore, they originate by their own mitosis as well as by differentiation of the stem cells. Then, they migrate outward and reach the midcrypt within 12–24 h. As they migrate, more granules are produced and accumulate in the supranuclear cytoplasm. Thus, they are transformed into goblet cells (Cheng, 1974a).

29.6.2.2. Goblet Cells Goblet cells are scattered along the midcrypt up to the villus' tip. They are characterized by a large group of mucous granules in the supranuclear cytoplasm. From the mid-crypt, the goblet cells migrate outward to reach the villous surface in 1 d (Cheng, 1974a). With their migration, goblet cells producing cored mucous granules (granular mucous cells) gradually change their phenotype and produce homogeneously pale mucous granules. The overall turnover time of goblet cells averages 3 d (Cheng, 1974a).

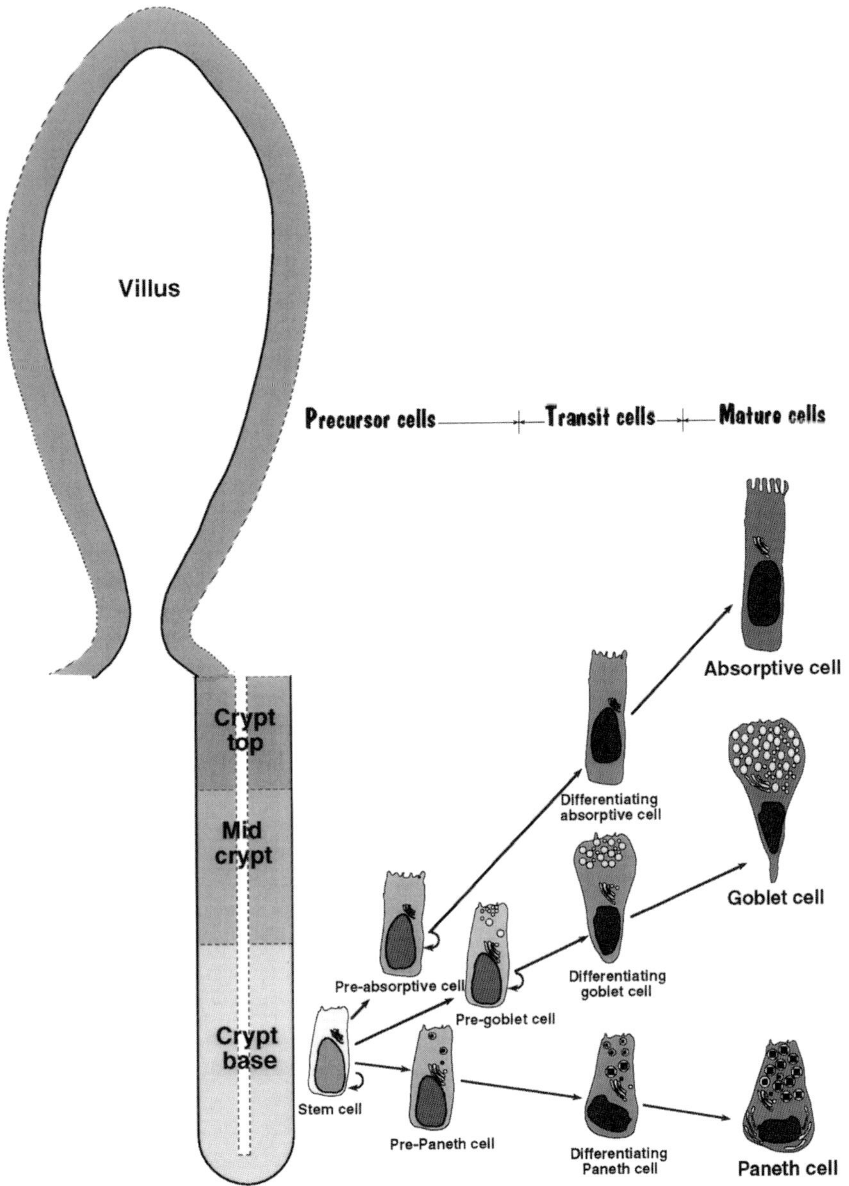

Fig. 4. Diagram of cell differentiation in crypt villous unit of small intestine. The crypt includes three equal regions: crypt top, midcrypt, and crypt base (the crypt base has been enlarged here to fit the precursor and Paneth cells). The stem cells are located in the crypt base and give rise to three main precursors: pre–absorptive, pre–goblet, and pre–Paneth cells, which, respectively, evolve into absorptive, goblet, and Paneth cell lineages.

29.6.3. PANETH CELL LINEAGE The presence, morphology, and number of Paneth cells vary in different species. They are absent in cats and predominate in the crypts of ant-eating Brazilian bears (Creamer, 1967). In mice, they are located in the lower third of the crypts and represent 3.3% of the duodenal crypt cells and, respectively, 7.5 and 6.6% in the jejunum and ileum (Cheng and Leblond, 1974a).

29.6.3.1. Pre–Paneth Cells Pre–paneth cells are located in the crypt base next to their ancestor stem cells. In addition to their stem cell–like features, they carry a few small granules with central dense cores and light halos. These cells migrate toward the crypt bottom while maturing into Paneth cells (Cheng, 1974b).

29.6.3.2. Paneth Cells Paneth cells are characterized by apical secretory granules that exhibit an electron-dense core and a light halo. The basal cytoplasm contains abundant cisternae of RER. Paneth cells exhibit more and larger haloed granules toward the crypt bottom, where they are more numerous. The granule size varies from about 500 nm close from the midcrypt border to about 3000 nm in the bottom of the crypt (Cheng, 1974b). This is due to production of gradually larger granules. In addition, with the inward migration of Paneth cells, the RER cisternae become numerous, and the Golgi increases in size. Paneth cells along the crypt base cannot divide by mitosis. Thus, they are only developed from maturation of the stem cells (Troughton and Trier, 1969; Cheng, 1974b). With time, Paneth cells migrate inward to the crypt bottom. The overall turnover time of Paneth cells is about 15 d (Cheng, 1974b). The location and relatively long residency time of Paneth cells, and the nature

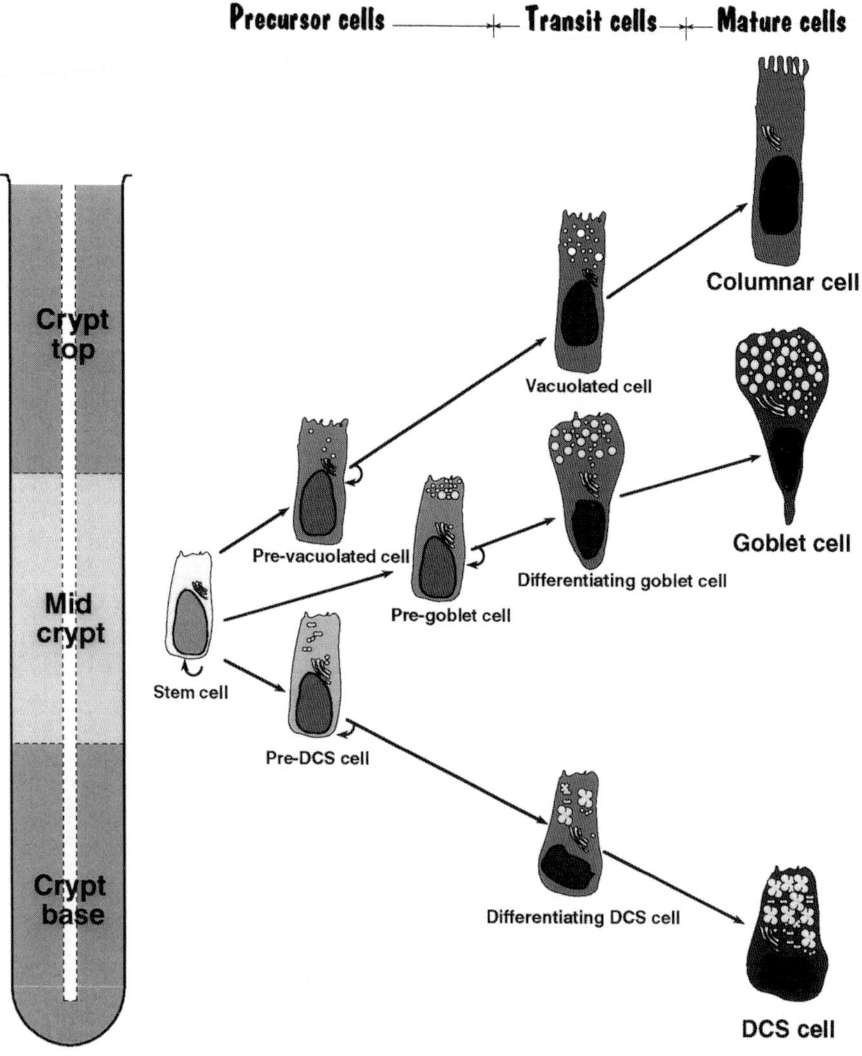

Fig. 5. Diagram of cell differentiation in crypt of ascending colon. The stem cells are anchored in the midcrypt and give rise to three main precursors: pre–vacuolated, pre–goblet, and pre–deep crypt secretory (pre-DCS) cells, which, respectively, differentiate into columnar, goblet, and DCS cell lineages.

of their secreted gene products, suggest that they may influence the structure and/or function of the stem cell niche. However, it has recently been found that ablation of Paneth cells in transgenic mice has no detectable effect on the proliferation/differentiation program in the crypts or on host-microbial interactions (Garabedian et al., 1997).

29.7. CELL LINEAGES IN COLON

The epithelial lining of the colon invaginates to form numerous crypts (Fig. 5). It is estimated that the mouse colon includes about 700,000 crypts (Cheng and Bjerknes, 1985). The size of crypts and their lining cell types vary in the ascending and descending portions of the colon. In the ascending colon, crypts are shorter (about 20 vs 30 cell positions) and are populated by a smaller number of cells (about 300 vs 700 cells/crypt) than in the descending colon. Cell types in the descending colon include vacuolated-columnar, goblet, enteroendocrine, and cavuolated. These cells are all scattered throughout the crypt wall (Chang and Leblond, 1971). In the ascending colon, an additional morphologically distinct cell type, deep crypt secretory cell, is located in the crypt base. Here the vacuolated-columnar and goblet cells are found in the crypt top (Altmann, 1983, 1990; Sato and Ahnen, 1992).

In the mouse colon, each crypt produces 14–21 cells/h (Cheng and Bjerknes, 1982), whereas in the rat, the rate of cell production is slightly lower, 7–11 cells/crypt/h (Maskens and Dujardin-Lorts, 1985). Dynamics of the colonic stem cells vary in the ascending and descending portions of the colon (Chang and Nadler, 1975; Appleton et al., 1980). Whereas stem cells are located in the midcrypt of the ascending colon and there is a bidirectional mode of migration (Fig. 5), they are located in the crypt base of the descending colon and cellular migration occurs in an outward direction (Sato and Ahnen, 1992). Moreover, in the ascending colon, the cell-cycle time is longer (19 vs 15 h), and proliferating cells are less numerous (90 vs 190 cells/crypt) in comparison with the descending colon. The latter explains the higher risk of colon cancer in the descending colon.

29.7.1. VACUOLATED-COLUMNAR CELL LINEAGE Cells of the vacuolated columnar lineage form about 80% of the crypt cell population in the descending colon and include three main members (Fig. 5).

29.7.1.1. Pre–Vacuolated Cells Pre–vacuolated cells are poorly differentiated cells that are located in the midcrypt of the ascending portion of the mouse colon or the crypt base of its descending portion. They exhibit a few small periodic acid-Schiff–negative vacuole-like granules in their apical cytoplasm. Pre–vacuolated cells retain some ability to divide. Thus, they are developed from maturation of stem cells, and also by their own mitosis. With time, pre–vacuolated cells accumulate more and larger granules and become vacuolated cells (Chang and Leblond, 1971).

29.7.1.2. Vacuolated Cells Vacuolated cells are located in the upper two-thirds of the crypts of the ascending portion of the colon and the lower two-thirds of the crypts of the descending colon. They are characterized by the presence of pale vacuole-like granules in the apical cytoplasm. As the cells migrate upward, the granules become large and numerous. The nuclei of vacuolated cells are basal and darkly stained; mitochondria are scanty and free ribosomes are numerous. A few microvilli may project into the crypt lumen (Chang and Leblond, 1971). Vacuolated cells migrate outward along the lower two-thirds of the descending colonic crypts at a constant speed. One to two days after their production, they acquire a striated border characteristic of absorptive cells. Thus, they gradually differentiate toward the absorptive columnar phenotype. This transformation process is completed in about 3 d (Chang and Nadler, 1975).

29.7.1.3. Columnar Cells Columnar cells are found in the crypt top of ascending and descending colons. They are characterized by an apical striated border composed of packed microvilli that are less prominent than those of the small intestinal absorptive cells. Columnar cells maintain the same migration speed as vacuolated cells. In the descending colon, the overall turnover time of vacuolated-columnar cells averages 4.6 d (Chang and Leblond, 1971).

29.7.2. GOBLET CELL LINEAGE In the mouse descending colon, cells of the goblet cell lineage form about 16% of the whole crypt cell population (Chang and Leblond, 1971). They are evenly distributed throughout the crypt except at the crypt bottom and at the surface, where they are very few. They are characterized by various numbers of mucous granules.

29.7.2.1. Pre–Goblet Cells Pre–goblet cells are also called oligomucous cells and are located in the midcrypt of ascending colon and the crypt base of descending colon. These cells are characterized by a few mucous granules. In some cells, the granules are grouped into a narrow theca. These cells are capable of mitosis, and, thus, they are developed from maturation of stem cells and also by their own mitosis (Chang and Leblond, 1971). Following their production, pre–goblet cells migrate upward along the crypt wall and show signs of gradual maturation. Within 1 to 2 d, they accumulate more and more mucous granules and the theca gradually enlarges, and, thus, they become mature goblet cells. This transformation process is completed in about 2 d (Chang and Nadler, 1975).

29.7.2.2. Goblet Cells Goblet cells are located in the upper third (or two-thirds) of the ascending (or descending) colon. The mucous granules are large and numerous, and, thus, the theca is swollen to give a typical goblet appearance. The dark nuclei are squeezed at the base of the goblet. In the descending colon, it takes about 1 d for a goblet cell to migrate along one-third of the crypt and reach the surface. The overall turnover time of goblet cells is similar to that of the vacuolated-columnar cells, averaging 4.6 d (Chang and Leblond, 1971).

29.7.3. DEEP CRYPT SECRETORY CELL LINEAGE Members of the deep crypt secretory cell lineage are found in several species including mice, rats, and humans. They differ from goblet cells in location, histochemistry, and ultrastructure (Altmann, 1983). They lack the goblet-shaped theca but exhibit many mucous granules that appear large and light in mice and rats, but small and dark in humans (Altmann, 1990). These cells are produced from stem cells located in the midcrypt as pre–deep crypt secretory cells and undergo migration-associated differentiation. They reach the crypt bottom as mature deep crypt secretory cells. In mice, they have an average turnover time of about 14–21 d (Altmann, 1990).

29.8. CELL LINEAGES SCATTERED ALONG GI TRACT

29.8.1. ENTEROENDOCRINE CELL LINEAGES The peptide- or polypeptide-producing enteroendocrine cell lineages are scattered throughout the GI epithelium. In mice, they represent about 7% of all cells in the stomach body (Karam and Leblond, 1992) and 3% in the pyloric antrum (Lee, 1985). In the small intestine, they are more abundant in the crypts than in the villi and generally form about 0.5% of all crypt-villus cells (Cheng and Leblond, 1974b). They form about 0.4% of cells in the colon (Tsubouchi and Leblond, 1979). The identification of enteroendocrine cells of different lineages depends on the size, shape, electron density, and immunocytochemical specificity of their secretory granules (Solcia et al., 1987). After a long debate about the neuronal vs epithelial origin of these cells, it has been established that they share a common stem cell with other GI epithelial cell lineages (Cheng and Leblond, 1974b; Inokuchi et al., 1985; Thompson et al., 1990; Karam and Leblond, 1993d). Thus, enteroendocrine cells represent several cell lineages that originate from the GI epithelial stem cells.

29.8.1.1. Pre–Enteroendocrine Cells Pre–enteroendocrine cells are immature cells producing a few small endocrine-like secretory granules. They have been described in the isthmus region of the oxyntic glands of the stomach (Karam and Leblond, 1993d), in the crypt base of the small intestinal epithelium (Cheng and Leblond, 1974b), and in the crypt base of the descending colon (Tsubouchi and Leblond, 1979). Pre–enteroendocrine cells are occasionally seen undergoing mitosis in the stomach (Karam and Leblond, 1993d) and intestine (Cheng and Leblond, 1974b). Thus, they originate mainly by differentiation of the stem cells as well as by their own mitoses. Radioautographic labeling of these immature forms of enteroendocrine cells has revealed that in the stomach body, they mature in the isthmus and then migrate bidirectionally to reach the pit and base regions after about 16 d. In the small intestine, they migrate outward and reach the crypt top by 1 to 2 d, where they produce more and more granules. After an additional 1 to 2 d, they reach the villi, where they are transformed into mature enteroendocrine cells (Cheng and Leblond, 1974b). In the colon, it takes at least 1 d for a pre–enteroendocrine cell to be formed; it differentiates into enteroendocrine cell and reaches the midcrypt by 7 d. Pre–enteroendocrine cells are, thus, left behind by the more rapidly migrating pre–vacuolated and pre–goblet cells (Tsubouchi and Leblond, 1979).

29.8.1.2. Enteroendocrine Cells Enteroendocrine cells are the mature forms of the endocrine cells that are located throughout the gastric pit-gland units, the small intestinal villi, and the colonic crypts (Cheng and Leblond, 1974b; Tsubouchi and Leblond, 1979; Karam and Leblond, 1993d). They are characterized by a large group of granules in the infranuclear cytoplasm and may have bundles of cytoplasmic filaments that appear relatively few owing to increase in cell size. With time, enteroendocrine cells migrate in an inward or outward direction. The overall turnover time of enteroendocrine cells is estimated at about 60 d in the stomach corpus (estimated from Karam and Leblond, 1993d), 4 d in the small intestine (Cheng and Leblond, 1974b; Inokuchi et al., 1985), and 23 d in the descending colon (Tsubouchi and Leblond, 1979).

29.8.2. M-CELL LINEAGES The antigen-sampling M-cells are also called membranous or microfold cells and are found overlying lymphoid follicles in the small as well as large intestines of rodents and humans. They do not show signs of lipid absorption, as do their neighbor absorptive cells, but are able to internalize cationized ferritin (Bye et al., 1984). In rabbits, α1-2-linked fucose- and *N*-acetyl-galactosamine-specific lectins can be used as markers for M-cells (Gebert and Posselt, 1997).

29.8.2.1. Pre–M-cells In rabbits, pre–M-cells are also found in the intestinal crypts around lymphoid follicles and express little α1-2-linked fucose and *N*-acetyl-galactosamine (Gebert and Posselt, 1997). They are characterized by many free ribosomes and numerous microvilli at the cell apex (Bye et al., 1984). These cells are not capable of mitosis, they originate from the crypt-base stem cells. It takes at least 1 d for pre–M-cells to be generated. Then, they migrate outward to cover the dome-shaped lymphoid follicles while differentiating into mature functional M-cells (Gebert and Posselt, 1997).

29.8.2.2. M-Cells M-cells are flattened due to accumulation of lymphoid cells in the underlying connective tissue. They are characterized by central cytoplasmic "pocket"-containing lymphoid cells (Neutra et al., 1996). In M-cells, both free ribosomes and apical microvilli are few. These cells are actively capable of internalizing luminal molecules, particles, or microorganisms and transporting them to the intercellular spaces. In rabbits, they express much α1-2-linked fucose and *N*-acetyl-galactosamine (Neutra et al., 1996).

29.8.3. CAVEOLATED CELL LINEAGES Nabeyama and Leblond (1974) found and described mature members of the caveolated cell lineage throughout the GI epithelium. In the gastric epithelial units, they are most numerous in the pit, isthmus, and neck regions (Karam and Leblond, 1992).

29.8.3.1. Pre–Caveolated Cells Pre–caveolated cells are immature cells that are described in the isthmus regions of the gastric units (Karam and Leblond, 1993d, 1995), and the crypt base regions of the descending colon (Tsubouchi and Leblond, 1979). They are very rare and appear plump with narrow apices and few caveoli. Pre–caveolated cells originate by differentiation of the gut stem cells (Tsubouchi and Leblond, 1979; Karam and Leblond, 1993d). Their maturation is followed by bidirectional migration in the gastric units or by their outward migration in the descending colonic crypts.

29.8.3.2. Caveolated Cells Caveolated cells are mature cells that are characterized by a plump body with a narrow apex projecting microvilli into the luminal surface. The cytoplasm exhibits prominent lysosomes and numerous caveoli separated by bundles of filaments extending from the core of the microvilli deep to the sides of the nucleus (Karam and Leblond, 1993d). The long axis of the ovoid nucleus tends to be parallel to the basement membrane. In the midcrypt regions of the descending colon, caveolated cells exhibit a basal cytoplasmic process that becomes longer in the crypt top, but short at the luminal surface (Tsubouchi and Leblond, 1979). In the stomach, the few data available have shown that caveolated cells follow a bidirectional mode of migration similar to that of enteroendocrine and parietal cells (Karam and Leblond, 1993d). In the descending colon, it takes about 1 d for a caveolated cell to be produced; they migrate outward and spend about 4 d in the crypt base, and 0.5 d in each of the middle and upper thirds of the crypt. The overall turnover time of caveolated cells is about 8 d in the colon (Tsubouchi and Leblond, 1979).

29.9. IN VITRO STUDIES AND PERSPECTIVES

One central question in GI epithelial cell biology is, what factors control proliferation and differentiation of the epithelial stem cells and the uncommitted/commited precursors? Resolution of this question has been difficult because of the lack of an in vitro functioning system that includes the GI epithelial stem cells and lineage precursors. In 1994, Tait et al. reported the isolation, culture, and successful transplantation of small intestinal stem cells of rats. These cells were also isolated from human fetuses and were made conditionally immortalized by using a temperature-sensitive simian virus 40 large tumor antigen. It was possible to manipulate these cells to induce irreversible growth arrest and acquisition of an enterocyte-like phenotype that was found to be due to expression of the cyclin-dependent kinase inhibitor (Quaroni and Beaulieu, 1997). These cells represent a wonderful model for the study of early events in intestinal epithelial cell differentiation. More recently, isolation of sequential fractions of oxyntic epithelial cells along the pit-isthmus axis of the rabbit stomach made it possible to establish a primary culture system enriched in the stem cells and supported their proliferation and differentiation (Karam et al., 2001). These in vitro studies will provide unique opportunities for identifying genes specific for the GI stem cells and exploring mechanisms responsible for (1) keeping balance between their proliferation and differentiation, (2) controlling their commitment program in different regions along the GI tract, and (3) converting of a healthy to a malignant stem cell and thus cancer development.

ACKNOWLEDGMENTS

I am indebted to the research funding of the Faculty of Medicine and Health Sciences, UAE University.

REFERENCES

Altmann, G. G. (1983) Morphological observations on mucus-secreting nongoblet cells in deep crypts of the rat ascending colon. *Am. J. Anat.* 167:95–117.

Altmann, G. G. (1990) Renewal of intestinal epithelium: new aspects as indicated by recent ultrastructural observations. *J. Electron Microscope Technol.* 16:2–14.

Appleton, D. R., Sunter, J. P., and deRodriguez, M. S. B. (1980) Cell proliferation in the mouse large bowel, with details of the analysis of experimental data. In: *Cell Proliferation in the Gastrointestinal Tract* (Appleton, D. R., ed.), Pitman, London, UK.

Bye, W. A., Allan C. H., and Trier, J. S. (1984) Structure, distribution, and origin of M cells in Peyer's patches of mouse ileum. *Gastroenterology* 86:789–801

Canfield, V., West, A. B., Goldenring, J. R., and Levenson, R. (1996) Genetic ablation of parietal cells in transgenic mice: a new model for analyzing cell lineage relationships in the gastric mucosa. *Proc. Nat. Acad. Sci. USA* 93:2413-2435.

Chang, W. W. L. and Leblond, C. P. (1971) Renewal of the epithelium in the descending colon of the mouse. I. Presence of three cell populations: vacuolated-columnar, mucous and argentaffin. *Am. J. Anat.* 131:73–100.

Chang, W. W. L. and Nadler, N. J. (1975) Renewal of the epithelium in the descending colon of the mouse. IV. Cell population kinetics of vacuolated-columnar cells. *Am. J. Anat.* 144:39–58.

Cheng, H. (1974a) Origin, differentiation and renewal of the four main epithelial cell types in the mouse small intestine. II. Mucous cells. *Am. J. Anat.* 141:481–502.

Cheng, H. (1974b) Origin, differentiation and renewal of the four main epithelial cell types in the mouse small intestine. IV. Paneth cells. *Am. J. Anat.* 141:521–536 .

Cheng, H. and Bjerknes, M. (1982) Whole population cell kinetics of mouse duodenal, jejunal, ileal, and colonic epithelia as determined by radioautography and flow cytometry. *Anat. Rec.* 203:251–264.

Cheng, H. and Bjerknes, M. (1985) Whole population cell kinetics and post-natal development of the mouse intestinal epithelium. *Anat. Rec.* 211:420–426.

Cheng, H. and Leblond, C. P. (1974a) Origin, differentiation and renewal of the four main epithelial cell types in the mouse small intestine. I. Columnar cells. *Am. J. Anat.* 141:461–480.

Cheng, H. and Leblond, C.P. (1974b) Origin, differentiation and renewal of the four main epithelial cell types in the mouse small intestine. III. Entero-endocrine cells. *Am. J. Anat.* 141:503–520.

Creamer, B. (1967) Paneth cell function. *Lancet* 1:314–316.

El-Alfy, M. and Leblond, C. P. (1987a) Long duration of mitosis and consequences for the cell cycle concept, as seen in the isthmal cells of the mouse pyloric antrum. I. Identification of early and late steps of mitosis. *Cell Tissue Kinet.* 20:205–213.

El-Alfy, M. and Leblond, C. P. (1987b) Long duration of mitosis and consequences for the cell cycle concept, as seen in the isthmal cells of the mouse pyloric antrum. II. Duration of mitotic phases and cycle stages, and their relation to one another. *Cell Tissue Kinet.* 20: 215–326.

El-Alfy, M. and Leblond, C. P. (1989) An electron microscopic study of mitosis in mouse duodenal crypt cells confirms that prophasic condensation of chromatin begins during the DNA-synthesizing (S) stage of the cycle. *Am. J. Anat.* 186:69–84.

Garabedian, E. M. Roberts, L. J., McNevin, M. S., and Gordon, J. I. (1997) Examining the role of Paneth cells in the small intestine by lineage ablation in transgenic mice. *J. Biol. Chem.* 272:23,729–23,740.

Gebert, A. and Posselt, W. (1997) Glycoconjugate expression defines the origin and differentiation pathway of intestinal M-cells. *J. Histochem. Cytochem.* 45:1341–1350.

Gordon, J. I. and Hermiston, M. L. (1994) Differentiation and self-renewal in the mouse gastrointestinal epithelium. *Curr. Opin. Cell. Biol.* 6:795–803.

Gordon, J. I., Schmidt, G. H., and Roth, K. A. (1992) Studies of intestinal stem cells using normal, chimeric, and transgenic mice. *FASEB J.* 6: 3039–3050.

Hall, P. A. and Watt, F.M. (1989) Stem cells: the generation and maintenance of cellular diversity. *Development* 106:619–633.

Helander, H. (1981) The cells of the gastric mucosa. *Int. Rev. Cytol.* 70: 217–289

Hermiston, M. L. and Gordon, J. I. (1995) In vivo analysis of cadherin function in the mouse intestinal epithelium: essential roles in adhesion, maintenance of differentiation, and regulation of programmed cell death. *J. Cell. Biol.* 129:489–506.

Hermiston, M. L., Simon, T. C., Crossman, M. W., and Gordon, J. I. (1994) Model systems for studying cell fate specification and differentiation in the gut epithelium: from worms to flies to mice. In: *Physiology of the Gastrointestinal Tract* (Johnson, L. R. ed.), Raven, NY.

Inokuchi, H., Fujimoto, S., Hattori, T., and Kawai, K. (1985) Tritiated thymidine radioautographic study on the origin and renewal of secretin cells in the rat duodenum. *Gastroenterology* 89:1014–1020.

Ito, S. (1987) Functional gastric morphology. In: *Physiology of the Gastrointestinal Tract* (Johnson, L. R.) Raven, NY.

Jacobsen, C. M., Narita, N., Bielinska, M., Syder, A. J., Gordon, J. I., Wilson, D. B. (2002) Genetic mosaic analysis reveals that GATA-4 is required for proper differentiation of mouse gastric epithelium. *Dev. Biol.* 241(1):34–46.

Karam, S. M. (1993) Dynamics of epithelial cells in the corpus of the mouse stomach. IV. Bidirectional migration of parietal cells ending in their gradual degeneration and loss. *Anat. Rec.* 236:314–332.

Karam, S. M. (1995) New insights into the stem cells and the precursors of the gastric epithelium. *Nutrition* 11:607–613.

Karam, S. M. (1998) Cell lineage relationship in the stomach of normal and genetically manipulated mice. *Braz. J. Med. Biol. Res.* 31: 271–279.

Karam, S. M. and Leblond, C. P. (1992) Identifying and counting epithelial cell types in the "corpus " of the mouse stomach. *Anat. Rec.* 232: 231–246.

Karam, S. M. and Leblond, C. P. (1993) Dynamics of epithelial cells in the corpus of the mouse stomach. I. Identification of proliferative cell types and pinpointing of the stem cell. *Anat. Rec.* 236: 259–279.

Karam, S. M. and Leblond, C. P. (1993) Dynamics of epithelial cells in the corpus of the mouse stomach. II. Outward migration of pit cells. *Anat. Rec.* 236:280–296.

Karam, S. M. and Leblond, C. P. (1993) Dynamics of epithelial cells in the corpus of the mouse stomach. III. Inward migration of neck cells followed by progressive transformation into zymogenic cells. *Anat. Rec.* 236:297–313.

Karam, S. M. and Leblond, C. P. (1993) Dynamics of epithelial cells in the corpus of the mouse stomach. V. Behavior of entero-endocrine and caveolated cells. General conclusions on cell dynamics in the oxyntic epithelium. *Anat. Rec.* 236:333–340.

Karam, S. M. and Leblond, C. P. (1995) Origin and migratory pathways of the eleven epithelial cell types present in the body of the mouse stomach. *Microscopy Res. Technol.* 31:193–214.

Karam, S. M. and Forte, J. G. (1994) Inhibiting gastric H+-K+-ATPase activity by omeprazole promotes degeneration and production of parietal cells. *Am. J. Physiol.* 266 (*Gastrointest. Liver Physiol.* 29): G745–G758.

Karam, S. M., Yao, X., and Forte, J. G. (1997) Functional heterogeneity of parietal cells along the pit-gland axis. *Am. J. Physiol.* 272 (*Gastrointest. Liver Physiol.* 35):G161–G171.

Karam, S. M., Li, O., and Gordon, J. I. (1997) Gastric epithelial morphogenesis in normal and transgenic mice. *Am. J. Physiol.* 272 (*Gastrointest. Liver Physiol.* 35):G1209–G1220.

Karam, S. M., Alexander, G., Farook, V., Wagdi, A. (2001) Characterization of the rabbit gastric epithelial lineage progenitors in short-term culture. *Cell Tissue Res.* 306(1):65–74.

Leblond, C. P. (1981) Life history of renewing cells. *Am. J. Anat.* 160: 113–158.

Leblond, C. P., Messier, B., and Kopriwa, B. (1959) Thymidine-H3 as a tool for the investigation of the renewal of cell populations. *Lab. Invest.* 8:296–308.

Lee, E. R. (1985) Dynamic histology of the antral epithelium in the mouse stomach: II. Ultrastructure and renewal of pit cells. *Am. J. Anat.* 172: 225–240.

Lee, E. R. and Leblond, C. P. (1985) Dynamic histology of the antral epithelium in the mouse stomach: I. Ultrastructure and renewal of isthmal cells. *Am. J. Anat.* 172:205–224.

Lee, E. R. and Leblond, C. P. (1985) Dynamic histology of the antral epithelium in the mouse stomach: III. Ultrastructure and renewal of gland cells. *Am. J. Anat.* 172:241–259.

Lee, E. R., Trasler, J., Dwivedi, S., and Leblond, C. P. (1982) Division of the mouse gastric mucosa into zymogenic and mucous regions on the basis of gland features. *Am. J. Anat.* 164:187–207.

Li, Q., Karam, S.M., and Gordon, J. I. (1995) Simian Virus 40 T antigen amplification of pre-parietal cells in transgenic mice: effects on other gastric epithelial cell lineages and evidence for a p53-independent apoptotic mechanism that operates in a committed progenitor. *J. Biol. Chem.* 270:15,777–15,788.

Li, Q., Karam, S. M., and Gordon, J. I. (1996) Diphtheria toxin-mediated ablation of parietal cells in the stomach of transgenic mice. *J. Biol. Chem.* 271:3671–3676.

Li, Q., Karam, S. M., Coever, K. A., Matzuk, M. M., and Gordon, J. I. (1998) Stimulation of activin receptor II signaling pathways inhibits differentiation of multiple gastric epithelial lineages. *Mol. Endocrinol.* 12:181–192.

Li, Y. Q., Roberts, S. A., Paulus, U., and Potten, C. S. (1994) The crypt cycle in mouse small intestinal epithelium. *J. Cell. Sci.* 107:3271–3279.

Maskens, A. P. and Dujardin-Lorts, R. (1985) Kinetics of tissue proliferation in colorectal mucosa durinmg postnatal growth. *Cell. Tissue Kinet.* 14:467–477.

Maunoury, R., Robine, S., Pringault, E., Leonard, N., Gaillard, J. A., and Louvard, D. (1992) Developmental regulation of villin gene expression in the epithelial cell lineages of mouse digestive and urogenital tracts. *Development* 115:717–728.

Mills, J. C., Syder, A. J., Hong, C. V., Guruge, J. L., Raaii, F., and Gordon, J. I. (2001) A molecular profile of the mouse gastric parietal cell with and without exposure to Helicobacter pylori. *Proc. Natl. Acad. Sci. USA* 98(24):13,687–13,692.

Nabeyama, A. and Leblond, C. P. (1974) "Caveolated cells" characterized by deep surface invaginations and abundant filaments in mouse gastrointestinal tract. *Am. J. Anat.* 140:147–166.

Neutra, M. R. Frey, A., and Kraehenbuhl, J. P. (1996) Epithelial M cells: gateways for mucosal infection and immunization. *Cell* 86:345–348.

Potten, C. S. and Loeffler, M. (1990) Stem cells: attributes, cycles, spirals, pitfalls and uncertainties: lessons for and from the crypt. *Development* 110:1001–1020.

Potten, C. S., Booth, C., and Pritchard, M. (1997) The intestinal epithelial stem cell: the mucosal governor. *Int. J. Exp. Pathol.* 78:219–243

Quaroni, A. and Beaulieu, J. F. (1997) Cell dynamics and differentiation of conditionally immortalized human intestinal epithelial cells. *Gastroenterology* 113(4):1198–1213.

Robine, S., Jaisser, F., and Louvard, D. (1997) Epithelial cell growth and differentiation. IV. Controlled spatiotemporal expression of transgenes: new tools to study normal and pathological states. *Am. J. Physiol.* 273(*Gastrointest. Liver Physiol.* 36):G759–G762.

Sato, M. and Ahnen, D. (1992) Regional variability of colonocyte growth and differentiation in the rat. *Anat. Rec.* 233:409–414.

Sato, A. and Spicer, S. S. (1980) Ultrastructural cytochemistry of complex carbohydrates of gastric epithelium in the guinea pig. *Am. J. Anat.* 159:307–329.

Sharp, R., Babyatsky, M. W., Takagi, H. et al. (1995) Transforming growth factor a disrupts the normal program of cellular differentiation in the gastric mucosa of transgenic mice. *Development* 121:149–161.

Simon-Assmann, P., Kedinger, M., De Arcangelis, A., Rousseau, V., and Simo, P. (1995) Extracellular matrix components in intestinal development. *Experientia* 51:883–900.

Solcia, E., Capella, C., Buffa, R., Usillini, L., Fiocca, R., and Sessa, F. (1987) Enteroendocrine cells of the digestive system. In: *Physiology of the Gastrointestinal Tract* (Johnson, L. R., ed.), Raven, NY.

Stappenbeck, T. S., Gordon, J. I. (2000) Rac1 mutations produce aberrant epithelial differentiation in the developing and adult mouse small intestine. *Development* 127(12):2629–2642.

Suzuki, S., Tsuyama, S., and Murata, F. (1983) Cells intermediate between mucous neck cells and chief cells in rat stomach. *Cell Tissue Res.* 233: 475–484.

Tait, I. S., Flint, N., Campbell, F. C., and Evans, G. S. (1994) Generation of neomucosa in vivo by transplantation of dissociated rat postnatal small intestinal epithelium. *Differentiation* 56(1-2):91–100.

Thompson, E. M., Fleming, K. A. Evans, D. J., Fundele, R., Surani, M. A., and Wright, N. A. (1990) Gastric endocrine cells share a clonal origin with other gut cell lineages. *Development* 110:477–481.

Traber, P.G. and Silberg, D. G. (1996) Intestine-specific gene transcription. *Ann. Rev. Physiol.* 58:275–297.

Troughton, W. D. and Trier, J. S. (1969) Paneth and goblet cell renewal in mouse duodenal crypts. *J. Cell. Biol.* 41: 251–268.

Tsubouchi, S. and Leblond, C. P. (1979) Migration and turnover of enteroendocrine and caveolated cells in the epithelium of the descending colon, as shown by radioautography after continuous infusion of 3H-thymidine into mice. *Am. J. Anat.* 156:431–452.

Wong, M. H., Huelsken, J., Birchmeier, W., Gordon, J. I. (2002) Selection of multipotent stem cells during morphogenesis of small intestinal crypts of Lieberkuhn is perturbed by stimulation of Lef-1/beta -catenin signaling. *J. Biol. Chem.* 277:15,843–15,850.

Wong, M. H., Rubinfeld, B., and Gordon, J. I. (1998) Effects of forced expression of an NH2-terminal truncated beta-catenin on mouse intestinal epithelial homeostasis. *J. Cell. Biol.* 141:765–777.

Wong, M. H., Saam, J. R., Stappenbeck, T. S., Rexer, C. H., Gordon, J. I. (2000) Genetic mosaic analysis based on Cre recombinase and navigated laser capture microdissection. *Proc. Natl. Acad. Sci. USA* 97(23): 12,601–12,606.

Wright, N. A. and Alison, M. (1984) *The Biology of Epithelial Cell Populations*, Clarendon, Oxford.

30 Stem Cell Origin of Cell Lineages, Proliferative Units, and Cancer in the Gastrointestinal Tract

MAIRI BRITTAN, BSc AND NICHOLAS A WRIGHT, MD, PhD

The gastrointestinal (GI) stem cell has the ability to differentiate into every epithelial lineage in the GI tract and is proposed as the target of GI carcinogenesis. There is considerable difference in the epithelial tissue of the GI tract including the oral cavity, pharynx, esophagus, stomach, small intestine, and colon, yet each appears to have a common stem cell that can give rise to both metaplasia and cancer throughout the GI tract. These cells may be located in the basal layer of the mucosa or in specialized niches at the base or just above the base of glands in the stomach and intestines. The underlying mesenchymal cells of the lamina propria provide an environment, that is responsible for normal differentiation. The crypts are composed of monoclonal units of differentiating cells similar to the epidermal poliferative unit of the skin. However, although the crypts are monoclonal, the villi of the small intestine are polyclonal, indicating that multiple crypts may contribute to the cells of the villi. Bone marrow transplantation studies indicate that marrow stem cells can engraft into the small intestine and colon and differentiate into intestinal subepithelial myofibroblast cells located within the lamina propria. Mucosal stem cells may also derive from transplanted bone marrow cells. There is evidence for both monoclonal and polyclonal origins of epithelial cancer. The presence of multiple, synchronous premalignant foci associated with invasive cancer has led to the field cancerization hypothesis, which predicts that a carcinogenic stimulus leads to transformation of multiple cells in a tissue and that one of these "clones" may grow out because of a mutation that favors its expansion over the others. In this case, a polyclonal expansion may be followed by a clonal proliferation, indicating that the carcinogenic process starts out polyclonal and ends up clonal. The "adenomas' of familial adenomatous polyposis and some animal models of carcinogenesis are multiclonal, whereas the cancers in this condition are monoclonal. Even though the clonal nature of the normal expansion of GI cells is well established, the clonal origin of intestinal epithelial neoplasms is unresolved; data support both possibilities. The genetic and molecular regulatory mechanisms involved in GI tumorigenesis are being unraveled, and it may well be that different molecular mechanisms lead to cellular pathways to cancer. For example, a mutation in *APC* may lead to polyclonal adenoma formation and further mutations to monoclonal expansion.

30.1. INTRODUCTION

From the viewpoint of the maintenance of structural integrity, the gastrointestinal (GI) stem cell is the most important cell in the GI tract, with multipotential ability to differentiate into every epithelial lineage, and regenerative capacity to maintain normal homeostasis in the mucosa to such an extent that a single multipotential stem cell can regenerate entire intestinal crypts after irradiation. Despite the obvious significance of the GI stem cell, it remains elusive and unidentified due to a lack of histological markers, and debate exists as to the number and location of stem cells within the gastric glands and intestinal crypts, although it is generally accepted that stem cells are housed within niches composed of and maintained by myofibroblasts in the adjacent lamina propria. Because of its longevity, the stem cell is the proposed target cell of GI carcinogenesis, and various pathways of morphological progression of such transformed stem cells have been proposed; these include a "top-down" proliferation of stem cells located within intercryptal zones on the mucosal surface, spreading downward into adjacent crypts. Alternatively, a "bottom-up" theory of the upward proliferation of transformed stem cells in the crypt base to produce dysplastic crypts that replicate and expand by crypt fission has also been suggested. The clonal origin of normal and dysplastic gastric glands and intestinal crypts has been a matter of debate for some time. The bulk of evidence suggests that normal intestinal crypts and gastric glands in the mouse are monoclonal, i.e., derived from a single cell. In humans, it appears that intestinal crypts are monoclonal but gastric glands in the fundus are polyclonal. It is not yet clear whether GI epithelial neoplasms are monoclonal in origin, derived from a single transformed stem cell, or are polyclonal and develop in accordance to the "field cancerization" hypothesis.

From: *Stem Cells Handbook*
Edited by: S. Sell © Humana Press Inc., Totowa, NJ

In this chapter, we discuss the stem cell repertoire, stem cell plasticity and models of studying clonality in neoplasms of the GI tract in rodents and humans, and the morphological progression of tumors. We also discuss components of molecular regulatory pathways that appear to play a vital role in the onset and progression of GI tumorigenesis, such as Wnt signaling and transforming growth factor-β (TGFβ) signaling pathways. An important influence on tumorigenesis is the spread of a mutated clone of cells after the initial transformation, initially proposed to be a result of hyperproliferation, although recent growing evidence suggests crypt fission; we also examine these concepts.

30.2. EPITHELIAL CELL LINEAGES IN THE GI TRACT

The GI tract comprises the oral cavity, pharynx, esophagus, stomach, small intestine, and colon. Each organ is histologically distinct, although throughout the tract an epithelial luminal lining with underlying vascular lamina propria forms the GI mucosa. The gastric glands open onto the stomach mucosal surface and are further subdivided into foveolar, isthmus, neck, and base regions. In the fundus, gastric glands contain surface mucous cells and mucous neck cells; acid-secreting parietal (oxyntic) cells; pepsinogen-secreting zymogen (chief or peptic) cells; and numerous endocrine cell families including the histamine-producing enterochromaffin-like cells, and gastrin-producing G-cells. In the pylorus, glands contain many more mucinous cells, no peptic cells, and few parietal cells.

In the small intestine, villi are formed as finger-shaped projections from the intestinal crypts of Lieberkühn, with a central lamina propria. Four main epithelial cell lineages form the epithelium of the numerous crypts of the small intestine and colon and the small intestinal villi. These are the absorptive columnar cells, the most abundant cell type, termed *enterocytes* in the small intestine and *colonocytes* in the large intestine; mucin-secreting goblet cells mainly on the villi and upper half of the crypts; the endocrine, neuroendocrine, or enteroendocrine cells concentrated toward the crypt base; and the Paneth cells, which are almost exclusive to the crypt base of the small intestine and ascending colon. Other less common cell lineages are also present, such as caveolated cells and M- (membranous or microfold) cells (Wright and Alison, 1984).

30.3. ORIGIN OF CELL LINEAGES IN GI EPITHELIUM

It is generally agreed that the epithelial cell lineages of the gastric glands and intestinal crypts derive from one or more stem cells, located within a niche, possibly created and maintained by the underlying mesenchymal cells of the lamina propria (Brittan et al., 2002). The stem cells of the GI tract are unidentified due to a lack of histological markers, but they are thought to reside in the neck/isthmus region of the gastric glands, producing differentiated progeny that migrate bidirectionally to form the mucosa of the foveola and base of the gland. In the small intestine, the stem cells are believed to be located in the base of the crypts of Lieberkühn just superior to the Paneth cells, in the crypt base of the descending colon (Karam, 1999), and in the base or midcrypt of the proximal colon (Kovacs and Potten, 1973).

GI stem cells are identified functionally by their ability to regenerate entire intestinal crypts and villi following radiation treatment, and only surviving cells in the postulated stem cell locations have this regenerative capacity (Withers and Elkind, 1970). The number of stem cells within the gastric glands and intestinal crypts is unknown, although a single surviving crypt cell can regenerate every epithelial cell lineage, creating a monoclonal crypt and demonstrating the multipotentiality of the GI stem cell (Potten and Hendry, 1995). Indeed, the Unitarian Hypothesis, the most popular theory for the derivation of the GI epithelial cell lineages, states that under normal circumstances, all epithelial cells of the intestinal crypts originate from a single progenitor stem cell (Cheng and LeBlond, 1974). The single cell–cloned human colorectal carcinoma cell line, HRA19, forms polarized epithelial monolayers in vitro, which, when engrafted into nude mice, differentiate to form a histologically identical tumor to the original, containing columnar, goblet, and neuroendocrine cells of human origin (Kirkland, 1988). This demonstrates multilineage differentiation of a single epithelial cell, albeit malignant, and thus inconclusive evidence that this process occurs in the normal GI epithelium. Although subsequent studies have amassed evidence substantiating the Unitarian Hypothesis that gastric glands and intestinal crypts are monoclonal populations derived from a single stem cell, it appears that in the human stomach different circumstances occur.

30.3.1. STOMACH Epithelial cell lineages within the mouse antral gastric glands were shown to derive from a common stem cell origin in studies of XX/XY chimeric mice, in which the highly repetitive sequences of the male mouse Y-chromosome were identified using *in situ* hybridization (ISH) with a digoxigenin-labeled probe, pY 353. The male-derived gastric glands were almost exclusively Y-chromosome positive and the female portions of the chimeric stomach were negative (Thompson et al., 1990). This was confirmed in CH3↔Balb/c chimeric mice, in which each gastric gland was composed entirely of either CH3 or Balb/c cells and there were no mixed glands (Tatematsu et al., 1994). A later study, using *lacZ* reporter gene–expressing X-inactivation mosaic mice, demonstrated that gastric glands in the fundic and pyloric regions of the developing mouse stomach initially appear polyclonal with three or four stem cells per gland but become monoclonal during the first 6 wk of life. This 'purification' of the glands is possibly due to an eventual predomination by one stem cell over all other stem cells or, alternatively, by gland fission. Approximately 5–10% of mixed, polyclonal glands persist throughout development and into adulthood, conceivably undetected in studies using aggregation chimeras owing to reduced survival rates of cells of different genotypes (Nomura et al., 1998). The significance of these multi–stem cell–containing glands is not known, although it is possible that they fail to undergo crypt fission or have reduced fission rates, or perhaps have an increased stem cell number during development, and therefore maintain a higher number of stem cells after gland fission.

In the human, the situation appears more complex, perhaps due to regional differences during development of the gastric mucosa. Random methylation of CpG islands on X-chromosome-linked genes including phosphoglycerate kinase (*PGK*), the androgen receptor gene (*HUMARA*), hypoxanthanine phosphoribosyl transferase, and glucose-6-phosphate dehydrogenase (*G6PD*) during embryogenesis results in transcriptional inactivation of the methylated gene providing a mode of distinction between the two female X-chromosomes. In the female human gastric mucosa, the pyloric glands are homotypic for the *PGK* or *HUMARA* locus and thus are monoclonal, although approximately half of the fundic glands studied were heterotypic for *PGK* and *HUMARA* loci and thus appear to be polyclonal (Nomura et al., 1996).

30.3.2. SMALL INTESTINE AND COLON In C57BL/6J Lac (B6)⇔SWR mouse embryo aggregation chimeras, the lectin *Dolichos biflorus* agglutinin (DBA) binds to sites on the B6-derived, but not SWR-derived, cells. Intestinal crypts in these chimeras are either completely positive or negative for DBA and thus are clonal populations, although it was not possible to detect endocrine cells in this study because of their small luminal surface area (Ponder et al., 1985). The endocrine cell was later shown to be of the same clonal origin using the *Dlb* assay in SWR mice lacking a DBA-binding site on intestinal epithelial cells owing to a somatic mutation of the *Dlb-1* locus on chromosome 11. SWR mice are induced to bind DBA by the DNA mutagen ethylnitrosourea (ENU), resulting in wholly DBA-positive or -negative intestinal crypts (Bjerknes and Cheng, 1999). Similarly, mutation of a heterozygously expressed DBA-binding site on intestinal epithelial cells in C57BL/6J⇔SWR F1 chimeric mice either spontaneously or by ENU generates entirely DBA-negative or -positive crypts; thus, mutation of *Dlb-1* in a stem cell produces a clone of cells that cannot bind DBA and remains unstained (Winton and Ponder, 1990). It is important to note that intestinal crypts in C57BL/6J Lac (B6)⇔SWR chimeric mice are polyclonal until postnatal wk 2, and subsequently become monoclonal, possibly due to overgrowth of a single dominant stem cell lineage or by crypt fission, which is highly active throughout this period of development (Schmidt et al., 1988).

To eliminate the *chimeric artifact* hypothesis—the possibility that cells from different parental strains of chimeric animals segregate independently during development to produce monophenotypic crypts and not monoclonal crypts as presumed—mosaic expression of the electrophoretic isoenzymes PGK-1A and PGK-1B was analysed in colonic crypts of mice heterozygous for the X-linked alleles $Pgk\text{-}1^a$ and $Pgk\text{-}1^b$ (Thomas et al., 1988). This confirmed the results of chimeric studies—that crypts were indeed monoclonal—since no mixed crypts were seen. Female mice heterozygous for an X-linked polymorphism causing reduced expression of G6PD are naturally mosaic and provide an *in situ* look at the clonal architecture of many tissues without the use of artificially produced chimeric animals. In these GPDX heterozygous mice, small intestinal crypts are monoclonal and the cells of the villi are polyclonal, as multiple crypts contribute to the cells of the villus (Thomas et al., 1988).

Analogously, the colon carcinogens dimethylhydrazine (DMH) or ENU induce a heterozygous deficiency of the G6PD gene, giving partial or negative expression of G6PD in colonic crypts. It is possible that mutation of a non–stem cell within the proliferative compartment of the crypt produces the observed partial G6PD expression, since these crypts are transient and decrease in frequency parallel to an increase in wholly negative crypts, which will occur when the mutated cell proliferates and is shed into the gut lumen. Wholly negative crypts are likely to be clonal populations, derived from a single mutated stem cell. The time taken for the decrease in partially mutated crypts and emergence of entirely negative crypts to reach a plateau is approx 4 wk in the colon and up to 12 wk in the small intestine (Park et al., 1995). This difference in timing is possibly due to differences in stem cell cycle times of either tissue. Alternatively, in the *stem cell niche* hypothesis, different numbers of stem cells occupy the niche in either tissue with stochastic expansion of a mutated stem cell (Williams et al., 1992). Finally, crypt fission could be responsible, where crypts divide longitudinally, possibly cleansing partially mutated crypts by segregating the mutated and nonmutated cells, and duplicating the wholly mutated monoclonal crypts (Park et al., 1995; Bjerknes and Cheng 1999).

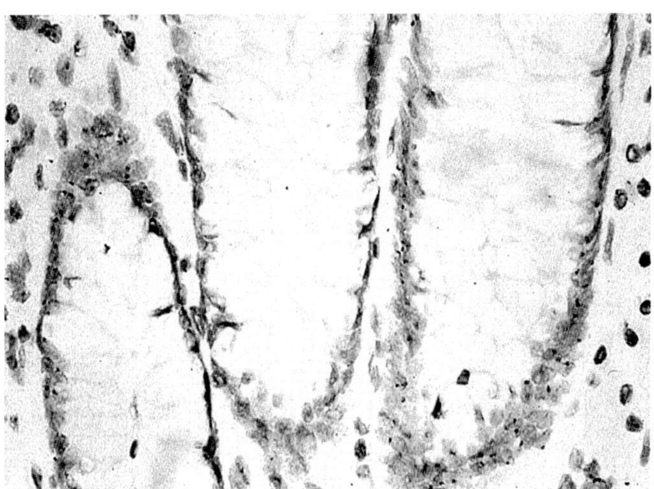

Fig. 1. Monoclonal origin of colonic crypts in XO/XY patient. Normal colonic mucosa in an XO/XY mosaic individual stained by ISH for a Y-chromosome-specific probe showing an XO crypt (left) and an XY crypt is shown. (Courtesy of M. Novelli.)

In the human small intestine and colon, the clonal origin of intestinal epithelial cells appears to be similar to that of the mouse. Approximately 9% of the human Caucasian population has a homozygous genetic mutation in the enzyme *O*-acetyl transferase ($OAT^{-/-}$) (Jass and Robertson, 1994) causing goblet cell secretion of sialic acid lacking in *O*-acetyl substituents in the colon, which is detected using the mild periodic acid-Schiff (mPAS) technique (Sugihara and Jass, 1986). Approximately 42% of the population is heterozygous for this mutation (OAT^-/OAT^+), in accordance with the Hardy–Weinberg law, although in these individuals *O*-acetylation proceeds due to the one active *OAT* gene, and crypts are negative for mPAS. Loss of the remaining active *OAT* gene, producing a homozygous (OAT^-/OAT^-) genotype, gives occasional, random, positively mPAS-stained crypts, with uniform staining of goblet cells from base to luminal surface, an effect that increases with age (Fuller et al., 1990). This could be due to a somatic mutation or nondisjunction in a single crypt stem cell and subsequent formation of a monoclonal crypt, supporting a clonal origin for at least the mucin cell lineages (Wright, 2000).

Moreover, in a rare XO/XY patient with familial adenomatous polyposis (FAP), nonisotopic *in situ* hybridization (NISH) using Y-chromosome-specific probes showed that all indigenous epithelial cell lineages in the colonic crypts, including neuroendocrine cells visualized by immunostaining for neuroendocrine-specific markers, were clonally derived, since each crypt was composed almost entirely of either Y-chromosome-positive or Y-chromosome-negative cells (Fig. 1). In the small intestine, the villous epithelium was a mixture of XO and XY cells, as expected, because villi are believed to derive from stem cells of more than one crypt. Only 4 of the 12,614 crypts examined were composed of mixed XO and XY cells, possibly owing to nondisjunction with loss of the Y-chromosome in a crypt stem cell (Novelli et al.,1996).

Studies of sequence variations in methylation tags of three neutral loci in cells of normal human colonic crypts further support the clonal or "Unitarian" origin of epithelial cells within human colonic crypts. Methylation patterns can predict crypt histories and map the fate of individual cells within a crypt. Sequence differences exist in cells of independent crypts, and in morphologically identical cells within crypts, where closely related methylation tags indicate more closely related cells. It therefore appears that human colonic crypts contain multiple stem cells (i.e., they are quasiclonal), that are lost and replaced several times during life, eventually leading to a bottleneck effect in which all cells within a crypt are related to the closest stem cell ascendant (Yatabe et al., 2001). Thus, substantial evidence strongly suggests that with the exception of the human gastric glands, every epithelial cell lineage of the normal GI mucosa in mice and humans is monoclonally derived from a single multipotent stem cell.

30.4. GI STEM CELL PLASTICITY

Adult stem cells within many tissues are now known to have far higher levels of plasticity than previously thought. The adult hematopoietic bone marrow stem cell differentiates into blood cell lineages and can transdifferentiate into many different cell types including hepatocytes (Petersen et al., 1999; Alison et al., 2000; Lagasse et al., 2000; Thiese et al., 2000), biliary epithelial cells (Thiese et al., 2000), endothelial cells (Gao et al., 2001; Lagaaij et al., 2001), skeletal muscle fibers (Ferrari et al., 1998), cardiomyocytes (Orlic et al., 2001), neuronal cells (Eglitis and Mezey, 1997), and renal tubular epithelial cells (Poulsom et al., 2001). Bone marrow stromal cells, or mesenchymal stem cells (MSCs), produce mesodermal lineages such as osteoblasts, chondrocytes, adipocytes, myocytes, cardiomyocytes and thymic stroma (Beresford et al., 1992; Wakitani et al., 1995; Pittenger et al., 1999; Liechty et al., 2000; Poulsom et al., 2001). Human MSCs, transplanted into rats with cortical brain ischemia, migrate to the infarction site and differentiate into neuroendodermal cell types, which significantly ameliorate neurological defects (Zhao et al., 2002). Similarly, in rats with traumatic brain injury, purified rat MSCs engraft into injury sites and express neuronal and astrocyte antigens and markedly reduce motor and neurological deficits (Mahmood et al., 2001). These differentiation pathways can be bidirectional as muscle (Jackson et al., 1999) and neuronal stem cells (Bjornson et al., 1999) can also form bone marrow, and fully differentiated cells can transdifferentiate into other adult cell types without undergoing cell division; for example, exocrine pancreatic cells can differentiate into hepatocytes in vitro (Shen et al., 2000). Following induction of ischemia by coronary artery occlusion (Jackson et al., 1999) and myocardial infarction (Orlic et al., 2001), bone marrow stem cells differentiate into cardiomyocytes, endothelial cells, and smooth muscle cells. Functional capability of the transdifferentiated cells in repair and regeneration of diseased tissue is demonstrated in fumarylacetoacetate hydrolase (FAH)-deficient mice, which resemble type 1 tyrosinemia in humans and are rescued from liver failure by transplantation of purified hematopoietic stem cells (HSCs) that differentiate into morphologically normal hepatocytes and express the FAH enzyme (Lagasse et al., 2000).

Studies of stem cell plasticity in the GI tract may provide insight into the mechanisms involved in the normal turnover of the GI mucosa and in GI tumorigenesis and disease. In lethally irradiated female mice that are rescued by a whole bone marrow transplant from male donors, the transplanted bone marrow cells were seen to engraft into the small intestine and colon and differentiate into intestinal subepithelial myofibroblast (ISEMF) cells located within the lamina propria underlying the GI epithelia. Indeed, 6 wk after transplantation, these bone marrow–derived ISEMFs were present as columns spanning from crypt base to luminal surface and thus potentially derived from a stem cell in the crypt base, proliferating upward to be shed at the luminal surface. The same study demonstrated bone marrow–derived ISEMFs in duodenal biopsies from human female patients who had developed graft-versus-host disease following bone marrow transplantation from a male donor. In both mouse and human, male donor cells were detected by ISH using a Y-chromosome-specific probe, and the newly differentiated ISMEF cell phenotype was confirmed by positive immunohistochemical reactivity for α-smooth muscle actin (αSMA), and negativity for desmin, the mouse macrophage marker F4/80, and the hematopoietic precursor marker CD34 (Brittan et al., 2002). It is postulated that ISEMFs provide and maintain the intestinal epithelial stem cell niche via epithelial: mesenchymal cross talk and thus influence epithelial cell proliferation and ultimately determine intestinal epithelial cell fate (Powell et al., 1999).

After transplantation of a single purified hematopoietic bone marrow stem cell in the mouse, bone marrow–derived columnar epithelial cells were found in the lung, skin, esophageal lining, a single small intestinal villus, colonic crypt, and gastric pit of the stomach. Engraftment into the pericryptal myofibroblast sheath was not reported, and it is possible that ISEMFs derive from MSC populations within transplanted bone marrow since only a single HSC was transplanted in this study. This point can be addressed by studies of transplantation of defined stem cell populations (Krause et al., 2001). Similarly, in biopsies from human female patients who had received a sex-mismatched hematopoietic bone marrow transplant, 2–7% of epithelial cells in the skin and gastric cardia, and hepatocytes in the liver, expressed a Y-chromosome and therefore were derived from the transplanted donor hematopoietic bone marrow (Korbling et al., 2002). It is possible to engineer small intestinal tissue by grafting a cellular collagen sponge scaffold within the jejunum of dogs, although the regenerated tissue lacks muscularis layers necessary for peristalsis (Hori et al., 2001). Seeding of MSCs from bone marrow onto the collagen sponge scaffold prior to implantation induced initial development of an αSMA-positive muscle layer, although this regressed to a thin muscle layer 16 wk after transplant, and, therefore, it may be necessary to stimulate MSCs toward muscularis development (Hori et al., 2002).

30.5. THERAPEUTIC RELEVANCE OF GI STEM CELL PLASTICITY

The ability of a transplanted bone marrow stem cell to transdifferentiate into intestinal subepithelial myofibroblasts may prove therapeutic in the inflammatory bowel diseases ulcerative colitis and Crohn's disease, which are associated with an increased risk of bowel cancer (Tomlinson et al., 1997). A bone marrow transplant or mobilization of a patient's own bone marrow stem cells to colonize a diseased tissue could potentially lead to regeneration of damaged tissues. Additionally, the use of bone marrow stem cells in gene therapy is conceivable and appears preferential to current possibilities of directly transfecting differentiated cells such as hepatocytes. Manipulation of a bone marrow stem cell in vitro to express a normal copy of a desired gene, such as *APC*, could theoretically provide a means of treatment of a patient with FAP.

30.6. ROLE OF STEM CELL IN EPITHELIAL TUMORIGENESIS

Epithelial cell carcinogenesis is believed to be a multistep process in which an initial carcinogenic stimulus creates an accumulation of aberrant somatic events in specific oncogenes and tumor-suppressor genes eventually resulting in malignant transformation of a cell. Between 6 and 10 genetic events are thought to incite the successive changes in cellular morphology, abnormal cell proliferation capacity and eventual malignant transformation (Renan, 1993). In 1976, Nowell suggested a hypothesis for a clonal evolution of tumors in which a stem cell or epithelial progenitor becomes neoplastic due to an inherited gene defect or exogenous carcinogen and proliferates at an increased rate to produce several genetically variant mutant daughter cells, most of which are eliminated by an immune response or due to their metabolic inferiority. Eventually, a dominant, immune-resistant metastatic clone of cells emerges, which divides to form a solid, invasive malignancy, more genetically variant and unstable with continued proliferation. Overgrowth of the tumor by such a clone could lead to a pseudomonoclonal neoplasm from an initial polyclonal lesion. Studies of clonality in most epithelial cell tumors are difficult to undertake and give conflicting results. The main techniques used to study clonality in human and animal epithelial cell tumors are use of X-chromosome-linked isoenzymes and restriction fragment length polymorphisms (RFLPs) (although these methods are restricted to female tissues), karyotypic and chromosomal markers, aberrant immunoglobulin production, aggregation chimeras or allophenic tetraparental mice, and single-strand conformation polymorphism and direct sequencing analysis of mutations of key genes of tumorigenesis. Early studies of clonality in humans, using G6PD in females with X-chromosome inactivation as a marker, implied that most human neoplasms are clonal (Fialkow, 1976), although there is growing evidence that also suggests a polyclonal origin of epithelial cell tumors.

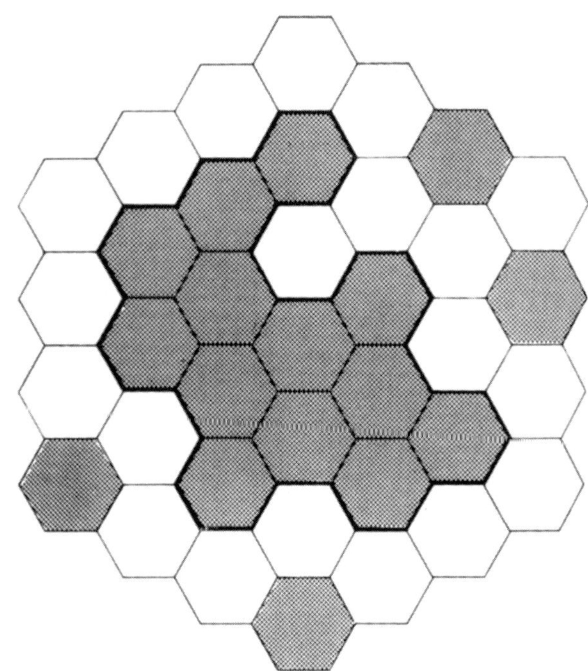

Fig. 2. Patch size. An idealized field of mosaic tissue comprising clear or stipled cells is shown. Clonal expansion in a small hexagonal area can lead to a more complicated patch pattern by coalescence with adjacent clones, which are by chance the same type. The patch size can be determined by examining the variance in the proportion of cell types in many uniform small samples in *pure* tissues. If patch size is small in relation to the sample size, there will be little variation in the proportion of the two cell types from sample to sample. If patch size is large in relation to the sample size, there will be a great deal of variation from sample to sample. If tumors arise from many cells in a field of mosaic tissue with a large patch size, then they may appear to be composed of a single cell type by chance alone. The small size and geometric complexity of the patches make it unlikely that tumors arising from large numbers of cells would comprise a single cell type in either X-chromosome inactivation mosaics or chimeras. (From Iannaccone et al., 1987, with permission.)

30.7. FIELD CANCERIZATION HYPOTHESIS AND PATCH SIZE

In a study of 783 resected head and neck carcinomas by Slaughter et al. (1953), multiple, synchronous premalignant primary foci were present in areas outside and overlapping the original invasive squamous carcinoma. This led to the *field cancerization* hypothesis—that an initial carcinogenic stimulus induces neoplastic transformation in multiple cells within a tissue. Premalignant regions of epithelium adjacent to the initial primary tumor are termed *field cancerizations* and are believed to progress independently to form second primary tumors, secondary or recurrent tumors. It is, however, acknowledged that if these multifocal lesions are monoclonal, they possibly arise from lateral migration and recurrence of the initial tumor clone. Investigations of clonal origins of preneoplastic lesions and tumors of the upper aerodigestive tract and adjacent mucosa have accumulated conflicting evidence that lesions arise both by field cancerization (Gusterson et al., 1991; Yang et al., 1995; Bloching et al., 2000), and monoclonally (Worsham et al., 1995).

In studies concerning clonality, it is important to consider the concept of *patch size* (Schmidt et al., 1990) in which a *patch* is defined as the number of contiguous, functional cells of a single genotype in an area of tissue, derived either from a single clone or by the coalescence of clones of the same lineage, such as cells of the same genotype in chimeric animals or with the same X-chromosome inactivation pattern (Fig. 2). Patch size in normal female human tissues has been suggested to be small, since small pieces of tissue expressed a mixture of cells with either chromosome activated (Nesbitt, 1974). However, heterozygous X-inactivation of the G6PD gene occurs in 17% of Sardinian women, which has permitted histochemical staining of their colonic mucosa and confirmed that patch size in the human colonic mucosa is large, often exceeding 400 crypts (Novelli et al., 2003). Furthermore, identification of X-inactivation patches of the *HUMARA* gene in normal human aorta indicated a large patch size exceeding 4 mm (Chung et al., 1998). If a tumor develops from cells restricted to one patch, it will ultimately be monogenotypic but not necessarily monoclonal because it is derived from an unknown number of cells of the same genotype. Likewise, a polyclonal tumor can only form if it derives from cells at a single patch boundary, and it is therefore important to measure patch size within a tissue before undertaking clonality analysis.

30.8. ANIMAL MODELS OF CLONALITY IN EPITHELIAL TUMORS—IMPORTANT CONSIDERATIONS

Animal models have provided various models of tumorigenesis in many tissues, although they obviously cannot be directly applied to human processes. Chimeric mice can be used to examine clonality, where the presence of both parental lineages in an induced neoplasm indicates a nonclonal origin, whereas restricted expression of either lineage suggests a monoclonal neoplastic proliferation (Iannaccone et al., 1987). Chimeric mice heterozygous for expression of a slow-migrating and fast-migrating electrophoretic variant (A and B, respectively) of the dimorphic autosomal enzyme glucose phosphate isomerase (GPI-1) were chemically induced to develop fibrosarcomas. More than 95% of tumors analyzed displayed only a single variant of GPI-1 enzyme, compared to the normal chimeric animals, which displayed equal activity of both enzyme isoforms even in the smallest tissue samples. However, these tumors were thought to be contaminated by nontumorous host stromal or inflammatory cells of a separate clonal derivation, or formed from the coalescence of two primary tumors, and thus likely to be clonal in origin (Iannaccone et al., 1978). Culturing tissue lysates in medium specific for survival of only the neoplastic cells eliminates inflammatory cell contamination. Remaining neoplastic cells display activity exclusive to one isoenzyme when analyzed at passage 0–2 before selection of cell subpopulations can occur and, therefore, are monoclonal (Deamant and Iannaccone, 1985). High doses of 3-methylcholanthrene (MCA) were used to chemically induce fibrosarcomas in female chimeric mice heterozygous for the X-chromosome-linked alleles $Pgk\text{-}1^a$ and $Pgk\text{-}1^b$ ($Pgk\text{-}1^a/Pgk\text{-}1^b$), which under normal circumstances contain equal numbers of cells that produce either the PGK-1A or PGK-1B isoenzymes. Most fibrosarcomas contained a mixture of cells expressing both enzyme phenotypes, unlikely to be derived from a single cell with both $Pgk\text{-}1^a$ and $Pgk\text{-}1^b$ loci active. Contamination by infiltrating nonneoplastic inflammatory cells was ruled out by histological examination, and cells with double-enzyme expression displayed phenotypes characteristic of transformed cells. The fusion of two cells with different enzyme activity was deemed unlikely because mixed tumors gave rise to PGK-A- and PGK-B-type clones. It is not clear whether the multicellular origin of these fibrosarcomas is a fusion of multiple primary tumors or if they are polyclonal proliferations, although this evidence demonstrates that not all chemically induced tumors appear to be monoclonal growths (Reddy and Fialkow, 1979). A later study showed that fibrosarcomas induced by MCA in $Pgk\text{-}1^a/Pgk\text{-}1^b$ heterozygotes were clonal growths expressing only one PGK-1 isoform. The discrepancy in results of this study and previous studies by Reddy and Fialkow (1979) was explained by an undetected coalescence of multiple primary tumors due to late tissue sampling (Deamant and Iannaccone, 1985).

30.9. TUMORS IN GI TRACT—MONOCLONAL OR POLYCLONAL IN ORIGIN?

The GI tract undergoes constant self-renewal throughout the lifetime of its host, with cell proliferation maintaining normal homeostasis by responding to the increases and decreases in cell death determined by the health of the tissue. The huge numbers of daily mitotic events in this complex homeostatic regulation make the GI tract one of the most common sites of human cancers (Cutler et al., 1974). The life-span of a GI epithelial cell is shorter than the time taken to genetically induce neoplastic change, with the exception of the stem cells, which are therefore believed to be the carcinogen target cells. In normal monoclonally derived structures, such as the intestinal crypts, it could be presumed that neoplasms within this monophenotypic structure would also be monoclonal or homotypic, deriving from a transformed stem cell. However, the clonal origin of intestinal epithelial neoplasms is unresolved, with conflicting data.

RFLPs of inactivations of the X-chromosome-linked *PGK* gene were studied in colorectal adenomas and carcinomas from 50 female patients. In adjacent normal epithelia, X-chromosome inactivation was random and therefore polyclonal, although in the adenomas and carcinomas, a monoclonal pattern of X-chromosome inactivation was observed (Fearon et al., 1987). Conversely, in female patients with Gardner syndrome, a precancerous bowel condition in which patients develop multiple adenomas of the GI mucosa, the adenomas have a multiclonal origin, expressing both forms of G6PD (Hsu et al., 1983). A CAG trinucleotide repeat polymorphism adjacent to methylation sites in the X-chromosome-linked human androgen receptor gene (*HUMARA*) is present in approx 90% of females and provides a means of looking at tumor clonality by polymerase chain reaction (PCR) amplification of a 600-bp DNA fragment encompassing the polymorphic *Bst*XI and methylation-sensitive *Hpa*II sites. In 15 female GI tumors, the majority revealed an unequivocal monoclonal origin with random pattern of X-inactivation in the normal surrounding mucosa, consistent with Southern blot analysis for PGK and the M27β probe (DXS255), which detects X-chromosome tandem repeat polymorphisms in about 90% of females (Kopp et al., 1997). However, it is important to note that patch size was overlooked in these studies, and it is possible that the observed monoclonality may arise by expansion of a preexisting clone of cells, within a large X-linked patch.

30.9.1. CHEMICALLY INDUCED INTESTINAL TUMOR FORMATION IN CHIMERIC MICE Mouse chimeras constructed by aggregation of CBA/Ca (CBA) and C57BL/6J (B6) early embryo strains form colonic epithelial dysplasias and tumors when treated with azoxymethane. Mosaic patches in these mice were identified by immunohistochemical staining for H-2 antigens, since CBA mice are of H-2^K haplotype and B6 mice are of H-2^B haplotype. In the normal chimeric mice, the intestinal crypts are monoclonal; therefore, it is not known whether neoplastic foci within a single crypt form from a single cell or many cells of the same genotype. In 55 dysplastic foci that comprised less than 10 crypts and overlapped patch boundaries, each expressed only a single H-2 antigen; therefore, the foci arose exclusively within a single crypt and expanded by crypt fission, not by recruitment of other independently transformed crypts or the initiation of transformation of cells in neighboring crypts. In 2 of 17 larger adenomas, both H-2 antigens were expressed, but these are proposed to be collision tumors. Therefore, chemically induced epithelial dysplastic foci, the precursors of polypoid adenomas that may develop into invasive carcinomas, form within single crypts, but not necessarily from a single cell (Ponder and Wilkinson, 1986). In GPDX mice, mentioned previously, tumors induced by DMH are also of monocryptal origin (Griffiths et al., 1989).

30.9.2. FAP AND MUTATIONS OF THE *APC* GENE Loss of *APC* gene function is one of the very first, if not the initiating step, in colorectal adenoma formation, preceding mutation of RAS family oncogenes and the p53 tumor suppressor gene (Powell et al., 1992). *APC* is mutated in 85% of human sporadic colorectal tumors, and hundreds of individual mutations of the *APC* gene have been identified, the position of the mutation appearing to dictate the severity and onset of polyposis (Kinzler and Vogelstein, 1996). Patients with FAP have an autosomally dominant inherited germline mutation of the *APC* tumor suppressor gene on chromosome 5q21 (Bodmer et al., 1987), making them susceptible to spontaneous development of hundreds to thousands of colorectal adenomas if the remaining wild-type (WT) *APC* allele is mutated, in concordance with the two-hit loss of suppressor gene hypothesis of Knudson (1993). Novelli et al. (1996) studied the clonal origin of colonic adenomas in an XO/XY patient with FAP confirmed by mutational analysis of the *APC* gene, by NISH using Y-chromosome-specific probes. Normal intestinal crypts were exclusively XY or XO and therefore clonal populations, as mentioned previously. However, 76% of adenomas greater than one crypt in size displayed a mixed karyotype of XO and XY cells and therefore appeared to be polyclonal (Novelli et al., 1996). Alternatively, XY/XO adenomas may be XY adenomas that have lost their Y-chromosome expression, which occurs frequently in other carcinomas, albeit at later stages (Rau et al., 1992; Kovacs et al., 1994). This was deemed unlikely because there was no apparent correlation between the size and stage of tumorigenesis and polyclonality in the adenomas analyzed, and in normal male FAP control individuals there was no loss of Y-chromosome. Random collisions between monoclonal adenomas to produce these mixed-phenotype adenomas was ruled out by statistical analysis, since the occurrence of polyclonal adenomas was too high. It is more conceivable that the adenomas are true polyclonal proliferations caused by clustering of transformed cells, in accordance with the field cancerization hypothesis, or that the early adenomas induce adenomatous growth in neighboring crypts.

Loss of function of the wild-type (WT) *APC* allele in FAP, or loss of heterozygosity (LOH), occurs by either a germline mutation or somatic mutation although both cause truncation of the APC gene product, APC-c (Miyaki et al., 1994). Using an antibody specific for APC-c, 29 colorectal adenomas from seven patients with FAP were shown to express a mixture of APC-c-positive and APC-c-negative cells, implying that at least one mutated stem cell and one normal stem cell were present in each adenoma. This was verified by the discovery that the ratio of positive to negative cells did not change as the adenomas progressed and that APC-c-positive and -negative mitoses were seen throughout the adenomas. Invasive adenocarcinomas contained only APC-c-negative cells, perhaps due to further genetic transformations in APC-c-negative cells providing a selective growth advantage over the normal APC-c-positive stem cell populations (Bjerknes et al., 1997). Approximately 10% of patients with FAP develop desmoids, local aggressive proliferations of fibroblasts, which can cause small bowel obstruction and often mortality. Desmoids behave as neoplastic lesions, although their fibroblastic composition and lack of telomerase expression suggest a nonneoplastic origin. PCR amplification of X-inactivation patterns of the *HUMARA* locus was used to assess clonality in female FAP-associated desmoids (Middleton et al., 2000), demonstrating that desmoids are monoclonal.

30.9.3. Min MOUSE AND *Apc* KNOCKOUT MOUSE MODELS The multiple intestinal neoplasia (Min) mouse, produced by chemical mutation with ENU, has a heterozygous germline mutation of the *Apc* gene, the murine *APC* homologe, and develops intestinal adenomatous polyps, particularly throughout the small intestine, associated with increased LOH of the *Apc* WT allele (Moser et al., 1990). Chimeras of $Apc^{min/+}$ that expressed the ROSA26 β-galactosidase marker were produced with a B6 background (Merritt et al., 1997). Loss of the *Apc* WT allele was confirmed by PCR in all adenomas, in contrast to the partial loss observed in humans by Bjerknes et al. (1997). APC immunohistochemical staining revealed 79% of adenomas with a polyclonal structure, proposed to arise by fusion of *Apc*-negative clones, termed *passive polyclonality*, which suggests that specific regions of intestine develop *Apc*-negative clones, rather than random transformation. Alternatively, it is possible that these clones arise by initiation of *Apc*-negative crypts for adenomatous growth by neighboring adenomas, previously considered by Novelli et al. (1996). Heterozygous *Apc* knockout (KO) mice ($Apc^{\Delta 716}$) are similar to Min mice but develop three to five times more adenomas predominantly in the small intestine following LOH of the WT *Apc* allele (Oshima et al., 1995).

30.9.4. ABERRANT CRYPT FOCI AND COLONIC CANCER Aberrant crypt foci (ACF) are microscopic, preneoplastic lesions in colon tumorigenesis and carcinogenesis of rodents and humans that present as crypts with a thickened epithelium elevated above the normal mucosa. ACF formation is initiated by *APC* mutation and is believed to be the first step in the adenoma–carcinoma sequence in humans (Kinzler and Vogelstein, 1996). PCR amplification of a CAG-repeat methylation site in the *HUMARA* (androgen receptor gene) locus of the X-chromosome allowed a study of X-inactivation patterns in 18 ACF of 12 female patients, although PCR was only successful in 11 of the 18 ACF. Ten of the 11 ACF showed nonrandom inactivation of the X-chromosome regardless of the degree of dysplasia in the tissue, implying that ACF are monoclonal neoplastic precursor lesions of colon cancer (Siu et al., 1999). ACF induced in distal colons of rats by azoxymethane injection showed marked increases in cell number and proliferation rate, demonstrated by incorporation of BrdU primarily in the lower, proliferative regions of the crypts. Increased proliferative activity was maintained at a constant rate for 36 wk, similar to that observed in adenomas and carcinomas, additional evidence that ACF are precursor lesions of colon carcinogenesis (Pretlow et al., 1994). Sunter et al. (1978) proposed a mode of classification of carcinomas in the DMH-treated rat colon. Group I and II carcinomas progress from adenomas in the classic ACF-adenoma-carcinoma pathway, whereas group III carcinomas are poorly differentiated, mucin-secreting carcinomas restricted to the proximal colon. DMH-induced ACF are restricted to the distal colon, possibly indicating the first stage of the ACF-adenoma-carcinoma pathway, but do not occur in the proximal colon, the location of group III carcinomas, which are proposed to develop according to a *de novo* sequence of events lacking intermediate ACF formation. Thus, ACF is a preneoplastic marker for only a subset of colon carcinomas, determined by location and carcinoma histology (Park et al., 1997).

30.9.5. CRYPT FISSION Crypt fission is the process by which intestinal crypts replicate, during embryogenesis and following damage by irradiation or cytotoxic chemicals, wherein a fissure occurs in the base of the parental crypt, causing crypt bifurcation and budding that ascends longitudionally to produce a new daughter crypt, in the crypt cycle. The crypt cycle is believed to occur

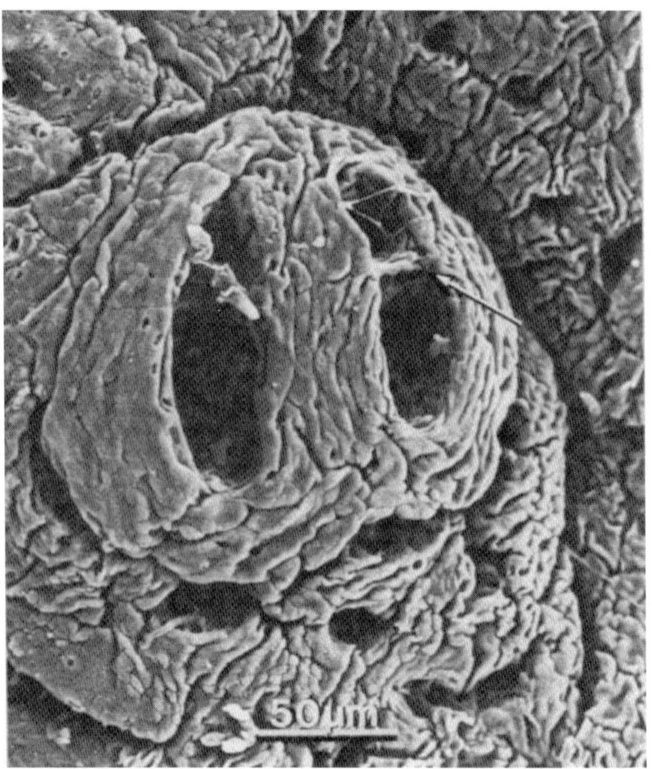

Fig. 3. Crypt fission in aberrant crypts. A scanning electron micrograph of two large elongated aberrant crypts undergoing crypt fission in colonic mucosa from F344 rats treated with DMH is shown. Inside the left aberrant crypt, below a small mucous fragment, a transverse epithelium septum is observed. The transverse septum of the right aberrant crypt has fully divided the larger crypt into two new crypts (arrow). Magnification:×400 (From Paulsen et al., 1994, with permission.)

slowly in the normal GI epithelium in mice (Cheng and Bjerknes, 1985) and in humans (Cheng et al., 1986), giving a gradual expansion of the crypt population. Mechanisms that stimulate a crypt to divide are unknown, although an increased crypt size was initially postulated (Totafurno et al., 1987); subsequent findings that crypt fission is stimulated by DMH with no increase in cell proliferation implies a more complex regulation. In the same study, epidermal growth factor markedly increased cell proliferation in the crypts but reduced the crypt fission index (Park et al., 1995). High crypt fission rates are hypothetically linked with increased stem cell numbers, where resultant elevated numbers of differentiated progeny create a larger crypt more likely to undergo fission (Totafurno et al., 1987). As yet circumstantial, stem cells are believed to be target cells in carcinogenesis of the GI tract, with increased progeny and consequent likelihood to undergo crypt fission, supported by the finding that aberrant crypts and hence their mutant stem cell population undergo a faster rate of replication than normal crypts (Bjerknes, 1996). Crypt fission is therefore a vital component of the ACF-adenoma-carcinoma pathway, giving replication and expansion of mutated crypts. This was demonstrated in a scanning electron microscopic study of DMH-treated rats, showing aberrant colonic crypts with widened luminal openings, which appear to divide by crypt fission, verified by the discovery of dividing septum between crypts (Fig. 3) (Paulsen et al., 1994), and the previous finding that ACF branch to produce clusters of related aberrant crypts in the rat colon (McLellan et al., 1991).

A revealing study by Wasan et al. (1998) determined the effects of *APC* germline mutations in numerous colorectal adenomas in human patients and in the Min mouse on cell proliferation and crypt fission indices (CFIs). Results showed that in patients with FAP mutation of an *APC* WT allele, CFIs in the flat, non-adenomatous mucosa increased 19-fold compared to normal controls, and of 6291 microadenomatous crypts analyzed, 220 were undergoing fission although there was no increase in epithelial cell proliferation or apoptotic counts. Similarly, in Min mice, CFIs increased 1.75-fold in the colon and 1.61-fold in the small intestine with no alteration of mitotic index or apoptosis, and the sites of highest CFI were the locations of largest tumors. Approximately 10.9% of nonadenomatous crypts in fission appeared to be undergoing an abnormal, asymmetrical division, which may represent crypts that have undergone loss of the WT *Apc* allele. This implicates control of crypt commitment to the crypt cycle by APC, possibly via the cadherin–catenin cell signaling pathways described below, likely to occur in the intestinal stem cells, which are largely thought to regulate crypt fission.

30.9.6. MORPHOLOGICAL DEVELOPMENT OF COLORECTAL ADENOMAS As previously mentioned, stem cells are believed to be the target cells for initial genetic alterations in GI carcinogenesis. Initiation of polyp formation is believed to be a spontaneous event in both FAP and sporadic polyps (Bodmer, 1999), and progression from microadenoma to macroscopic adenoma is essentially random. The severity of colonic FAP is primarily influenced by microadenoma number and is independent of the patient's age or disease progression; therefore, some *APC* genotypes are selectively advantageous in tumor initiation, rather than progression (Crabtree et al., 2001). The *bottom-up* theory of spontaneous microadenoma development predicts that mutated stem cells within a niche in the base of the colonic crypts produce neoplastic daughter cells that progress upward toward the intestinal lumen, creating entire dysplastic crypts that divide and spread by crypt fission (Bjerknes and Cheng, 1981). However, neoplastic cells are frequently observed from midway up the crypt to the luminal surface, and not within the crypt base. Histological analysis of 35 small sporadic human adenomas demonstrated dysplastic epithelium restricted to superficial regions of the affected crypts of each specimen, whereas basal crypt regions retained a normal morphology. These dysplastic compartments stained intensely with Ki-67 antibody and β-catenin, indicating abnormal proliferation and LOH of *APC*, respectively. The concept of *top-down* morphogenesis, in which neoplastic epithelial cells originate from stem cells near the mucosal surface, possibly in *intercryptal* zones, and migrate downward, proliferating to produce dysplastic crypts, emerged from this study (Shih et al., 2001). Interestingly, crypt fission cannot be ruled out as the mode of crypt expansion in this theory of adenoma development. Normal epithelial cells near the base of the dysplastic crypts may be transformed cells whose phenotype is only apparent at later stages of progression, or, alternatively, the intercryptal zone in this *top-down morphogenesis* could be the site of LOH of the second *APC* allele. Both bottom-up and top-down concepts identify the initiation of adenomatous growth within a single crypt, thus giving rise to a clonal adenoma, and it is possible that both management structures exist in adenoma morphogenesis, although to what extent is not known (Wright and Poulsom, 2003). However, in Min mice, adenoma formation appears to occur by an entirely different process. Polyps form as initial outpockets of intestinal epithelium at the crypt–villus boundary and protrude into the lacteal side of neighboring villi, where they expand, eventually spreading into adjoining villi. Multiple adenomas within several crypts fuse to form a multivillous polyp, although throughout the process the normal villous epithelium remains intact as the polyps extend into the adjoining villi from the mucosal side (Fig. 4) (Oshima et al., 1997).

30.9.7. MOLECULAR REGULATION OF TUMORIGENESIS IN GI TRACT Advances are being made in understanding the pathways of molecular regulation of epithelial cell proliferation and differentiation in the normal GI tract. The APC protein forms a complex with glycogen synthase kinase-3β (GSK3-β), and axin (Kinzler and Vogelstein, 1996), to maintain low levels of β-catenin by phosphorylation, ubiquitination, and proteasomal degradation. The signaling protein Wnt binds to its receptor Frizzled to activate the cytoplasmic phosphoprotein Dishevelled and inhibit the APC/GSK-3β/axin complex, which in turn increases cytosolic levels of β-catenin. β-Catenin then translocates to the nucleus and combines with members of the T-cell factor/lymphocyte enhancer factor (Tcf/LEF) DNA-binding protein family to activate specific genes and increase cell proliferation (Fig. 5). The apparent downregulation of β-catenin and control of cell proliferation by APC makes the Min mouse an exciting paradigm of Wnt signaling pathways in tumorigenesis. Mutation of *APC* prevents the destabilization of β-catenin, causing increased gene transcription by nuclear β-catenin/Tcf/LEF complexes, believed to lead to intestinal epithelial cell hyperproliferation, and adenoma formation (Willert and Nusse, 1998; Bienz and Clevers, 2000), although several analyses have reported no alterations in cell proliferation in FAP flat mucosa (Nakamura et al., 1993; Wasan et al., 1998).

The inability of mutated APC to downregulate β-catenin is potentially due to loss of the SAMP repeats (Ser, Ala, Met, Pro), which provide the axin-binding site, rather than loss of the β-catenin-binding site (Smits et al., 1999; von Kries et al., 2000). Mutations of the β-catenin gene (*CTNNB1*) affect the serine and threonine residues targeted by GSK-3β, thereby creating a phosphorylation-resistant protein that is not degraded in the Wnt signaling pathway. These mutations are present in 48% of colorectal adenocarcinomas lacking *APC* gene mutations (Sparks et al., 1998). In addition to its role in cell signaling, β-catenin plays an important role in cell adhesion, linking cadherin molecules to the actin cytoskeleton via α-catenin (Barth et al., 1997). Transgenic mice expressing an activated mutant form of β-catenin devoid of GSK-3β and α-catenin-binding sites (ΔN131β-catenin) develop dysplastic lesions in the small intestine with increased cell proliferation and apoptosis, similar to those in the Min mouse, although they appear focal and relatively sparse and do not occur in the colons of these mice (Romagnolo et al., 1999). Harada et al. (1999) report a similar murine model in which the serine/threonine residues on exon 3 that comprise the binding sites of GSK-3β, but not the α-catenin-binding sites, were mutated. These mice developed numerous intestinal tumors, also essentially restricted to the small intestine, further evidence of the involvement of β-catenin in intestinal tumorigenesis regulated by Wnt signaling.

Analysis of Wnt gene expression in normal, adenomatous, and malignant human colorectal tissues showed that Wnt 2 and 5 are markedly upregulated in colorectal adenomas and carcinomas, primarily in the surrounding stromal cells. It is possible that increased β-catenin in *APC*-mutated cells causes an increase in cell signaling by E-cadherin between transformed cells and the surrounding stroma, which consequently increases Wnt 2 transcription (Smith et al., 1999). Interestingly, activation of protein kinase C (PKC) by the phorbol ester PMA in both normal APC-expressing and APC-deficient intestinal cells in vitro causes β-catenin nuclear translocation and gene transcription. This suggests an alternative pathway of β-catenin gene transcription regulated by PKC, independent of APC (Baulida et al., 1999), and a model of PKC involvement in colonic carcinogenesis has previously been suggested (Brasitus and Bissonnette, 1998).

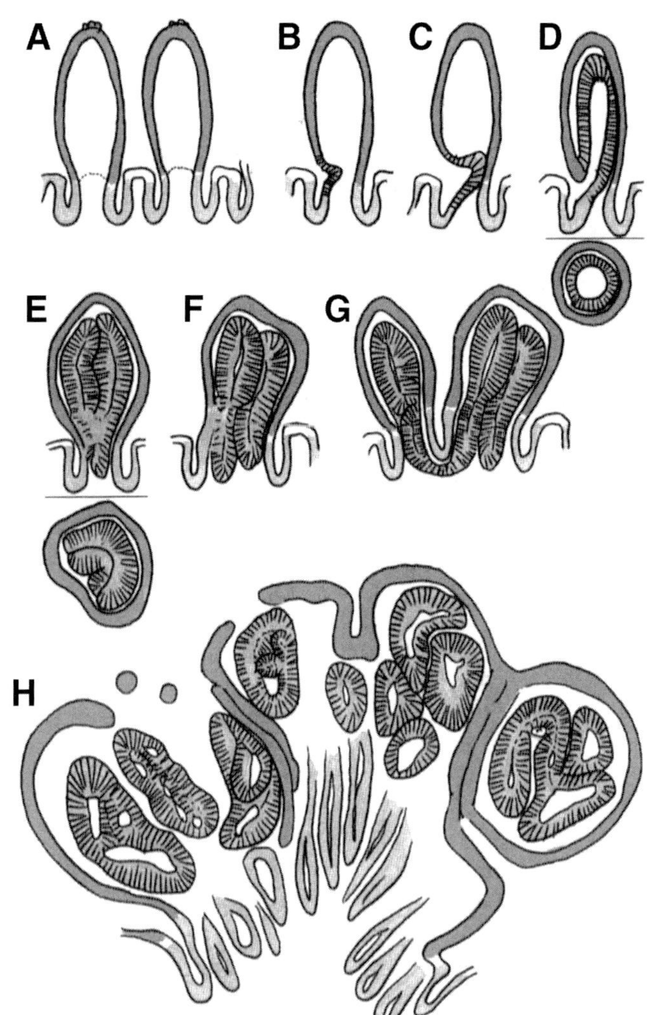

Fig. 4. Intestinal polyp formation in $Apc^{\Delta 716}$ heterozygous mice. **(A)** Normal intestinal epithelium. **(B,C)** Outpocketing pouches formed at the proliferative zone of the crypt develop into the back side of the neighboring villi. **(D)** Double-layer structure of the nascent polyp that consists of microadenoma tissue attaching inside the normal villous epithelium. **(E,F)** Microadenomas growing further inside single villi form folded layers of the adenoma tissue, which remain inside the single villi with the villous epithelium remaining intact. **(G)** Double-villous polyp in which microadenoma tissue moved from the original villus into adjoining one. **(H)** Multivillous polyp consisting of four original villi. The normal villous epithelium remains almost intact. (from Oshima H et al., 1997 with permission).

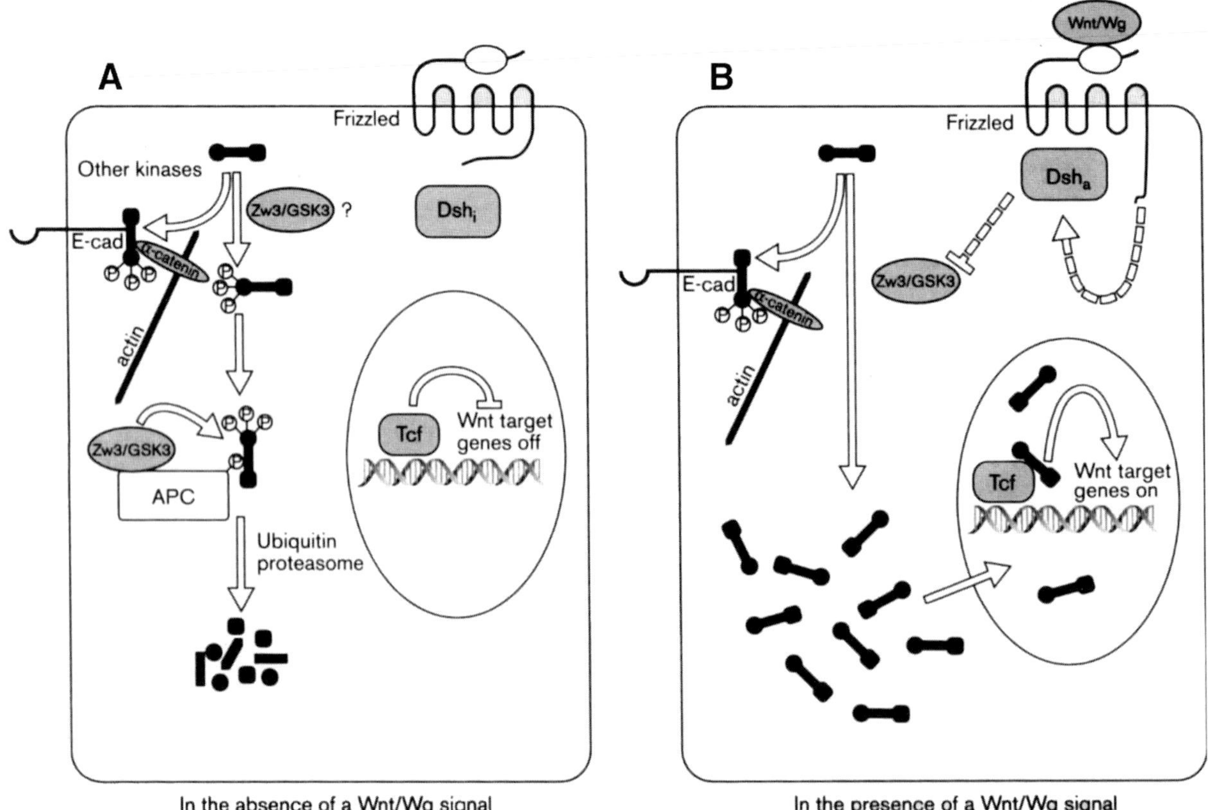

Fig. 5. Wnt signaling pathway. **(A)** In the absence of Wnt signaling, Dishevelled is inactive (Dsh$_i$) and Drosophila Zeste-white 3 or its mammalian homolog glycogen synthase kinase-3 (Zw3/GSK-3) is active. β-Catenin (black dumbbells), via association with the APC-Zw3/GSK-3 complex, undergoes phosphorylation and degradation by the ubiquitin-proteasome pathway. Meanwhile, Tcf is bound to its DNA-binding site in the nucleus, where it represses the expression of genes such as Siamois in Xenopus. **(B)** In the presence of a Wnt signal, Dishevelled is activated (Dsh$_a$), leading to inactivation of Zw3/GSK-3 by an unknown mechanism. β-Catenin fails to be phosphorylated and is no longer targeted into the ubiquitin-proteasome pathway. Instead, it accumulates in the cytoplasm and enters the nucleus by an unknown pathway where it interacts with Tcf, to alleviate repression of the downstream genes and provide a transcriptional activation domain. (From Willert and Nusse, 1998, with permission.)

30.9.8. DOWNSTREAM TRANSCRIPTIONAL TARGETS OF β-CATENIN/Tcf/LEF IN COLON CARCINOGENESIS Several genes including c-myc (He Tet al., 1988; Peifer , 1997), cyclin D1 (Tetsu and McCormick, 1999), c-Jun, Fra-1, urokinase-type plasminogen activator receptor (uPAR) (Mann et al., 1999), fibronectin (Gradl et al., 1999), CD44 (Wielenga et al., 1999), and the matrix metalloproteinase *matrilysin* (Brabletz et al., 1999; Crawford et al., 1999) have been identified as downstream transcriptional targets of the β-catenin/Tcf/LEF nuclear complex, and many have recently been shown to have oncogenic potential in the GI tract. It is well documented that c-myc (Brabletz et al., 2000), cyclin D1 (Arber et al., 1996), and uPAR, possibly via regulation by the oncoproteins c-jun and fra-1 (Mann et al., 1999), are overexpressed in colorectal adenocarcinomas. Differential expression of prostaglandin endoperoxide H synthase-2 is commonly reported in colonic adenomas, believed to be a direct result of increased β-catenin/Tcf/LEF nuclear transcription in *Apc*-mutated cells (Mei, et al., 1999). In mice heterozygous for a defective *Apc* gene, ensuing ACF, adenomas, and carcinomas strongly overexpress CD44. In the same study, Tcf-4 heterozygous mutant mice did not express CD44 in the small intestinal epithelium, indicating that a failure of β-catenin/Tcf/LEF complex formation in these mice prevents CD44 transcription, evidence of the role of CD44 as a target in the intestinal carcinogenesis pathway (Wielenga et al., 1999).

30.9.9. TGF-β AND Smad4 IN GI TUMORIGENESIS Under normal circumstances, the TGF-β family inhibits growth of most epithelial cells, including cells of the GI tract. The TGF-β signaling pathway involves TGF-β ligand binding to the serine/threonine TGF-β type II receptor (TβRII), which forms a heteromeric complex with and activates the TGF-β type I receptor (TβRI), which, in turn, phosphorylates the cytoplasmic proteins, Smad2 and Smad3. Phosphorylated Smad2 and Smad3 form a heteromeric complex with Smad4 and translocate into the nucleus to interact with transcriptional activators and coactivators, such as c-Jun/c-Fos complex and CBP/p300, to give TGF-β target gene transcription (Miyazono et al., 2000). Disruption of the TGF-β/SMAD signaling pathway blocks the normal inhibition of epithelial proliferation by TGF-β and can increase cell proliferation, often leading to tumorigenesis. A study of 11 human colorectal cancer cell lines provided insight into the frequency of these deletions in human colorectal carcinomas, identifying that 70%

of cell lines had a mutation in TβRII (Ijichi et al., 2001). This is in agreement with previous studies demonstrating TβRII mutations associated with microsatellite instability in colorectal carcinoma cell lines (Lu et al., 1995; Markowitz et al., 1995). Inactivation of Smad2 and Smad4 is frequently reported in human cancers, implicating them as tumor-suppressing genes (Hata et al., 1998; Massague, 1998).

An immunohistological study of Smad expression in normal and tumor tissues of the human colon showed an upregulation of receptor-activated Smads, including Smad2 and Smad3, in colorectal cancer cells but not in the normal colonic mucosa, indicating alterations in TGF-β signaling in colorectal carcinogenesis. Interestingly, the same study demonstrated Smad4 expression in surface epithelial cells of colonic crypts, suggesting a role of Smad4 in the normal terminal differentiation and apoptosis of these cells (Korchynskyi et al., 1999). However, heterozygous Smad4 (*Dpc4*) KO mice showed no obvious phenotypic differences from their WT littermates. The same Smad4 mutation, introduced into *Apc* KO mice ($Apc^{\Delta716}$) by meiotic recombination as both mutations are located on chromosome 18, showed that the *cis*-configuration compound heterozygotes (*Cis*-$Apc^{\Delta716}$[+/–] *Dpc4* [+/–]) developed colonic and intestinal polyps owing to LOH of the WT *Apc* allele. LOH of the remaining *Dpc4* allele also occurs in these mice. Further in vitro analysis revealed that the whole of chromosome 18 expressing the WT alleles was inactive in microadenoma cells of the *cis*-configuration heterozygotes and was replaced by a reduplicated chromosome 18 expressing the mutant alleles, thereby producing homozygous mutants. The *cis*-compound heterozygotes developed much larger intestinal lesions with greater proliferation of stromal cells than those seen in $Apc^{\Delta716}$ heterozygote mice. These adenomas progress into adenocarcinomas with a highly malignant nature, indicating that inactivation of the *Dpc4*/*Smad4* gene permits the malignant progression of adenomas in the small intestine and colon that are initiated by LOH of the *Apc* gene (Takaku et al., 1998). This study demonstrates a reciprocal process between the Wnt and TGF-β signaling pathways in malignant progression of colonic and intestinal carcinogenesis, since malignant transformation cannot occur without LOH in genes from both pathways. It would be interesting to investigate whether LOH of other genes in the TGF-β signaling pathway, such as TβRI, TβRII, Smad2, or Smad3, produce the same malignant phenotype in *Apc*-mutated polyps.

30.9.10. HOMEOBOX GENES Cdx-1 AND Cdx-2 IN GI TUMORIGENESIS The mammalian homeobox genes Cdx-1 and Cdx-2 are transcriptional regulators that display specific expression in the small intestine and colon, during embryogenesis and in the adult, and are potentially influential in GI metaplasia. The homozygous Tcf-4 KO mouse lacks Cdx-1 expression in the small intestine, implying that Wnt signaling directly confers Cdx-1 transcription by the β-catenin/Tcf-4 complex during intestinal development (Lickert et al., 2000). Cdx1 protein expression is greatest in nuclei of proliferative epithelial cells of normal intestinal crypts and decreases with the progression of normal colonic epithelium to adenomas to adenocarcinomas (Silberg et al., 1997). Therefore, a regulatory role of stem cell proliferation and differentiation in normal and neoplastic intestinal epithelium by Cdx-1 is possible, although Cdx-1 appears to lack tumor-suppressing properties since colonic tumors were not found in homozygous Cdx-1-deficient mice (Subramaniam et al., 1995).

In the normal colonic epithelium in humans and rodents, Cdx-2 expression is less restricted to proliferative cells, occurring in all crypt cell nuclei in the proximal colon and in cell nuclei in the upper third of the crypts from the descending colon to the rectum. Similar to Cdx-1, immunohistochemical analysis of Cdx-2 expression in human colonic adenoma and carcinoma specimens, and in rat colon treated with DMH, demonstrated a negative correlation between the degree of dysplasia and Cdx-2 expression (Ee et al., 1995). Heterozygous Cdx-2 KO mice ($Cdx2^{+/-}$), develop multiple intestinal polyp-like lesions predominantly in the proximal colon, and occasionally in the small intestine, that contain areas of stratified squamous epithelium histologically similar to that of the forestomach. These lesions were initially believed to be intestinal polyps and tumors, suggesting a role of Cdx-2 as a tumor suppressor gene (Chawengsaksophak et al., 1997), but were later shown to be heterotypic stomach and small intestinal mucosa; thus, Cdx-2 is vital in establishing regional differentiation during GI morphogenesis (Beck et al., 1999).

The colorectal cancer cell line, RKO, does not undergo constitutive β-catenin/Tcf-4-regulated transcription, despite expression of WT *APC*, β-catenin, and Tcf-4 genes. This cell line was subsequently found to express an inactivating mutant Cdx-2 allele, thus linking Cdx-2 to the *APC* tumor suppressor pathway mediated by Wnt signaling. An increase in Cdx2 RNA was observed in an HT29 cell line endogenous for mutant *APC*, stimulated to express WT *APC* following zinc induction. It appears that the normal negative feedback of target genes transcribed by β-catenin/Tcf-4 complexes by *APC* does not apply to Cdx-2, which is increased by *APC*, and therefore works in conjunction with *APC* to enhance the tumor-suppressing effects (da Costa et al., 1999).

It is not known if *APC* mutations alone initiate colon tumorigenesis, although several subsequent mutations have been identified that allow progression of adenomatous growth following *APC* mutation (Tomlinson et al., 1997). The *K-RAS* oncogene becomes activated by a dominant mutation in approx 50% of colorectal tumors, although the exact role of this mutation in tumorigenesis is unclear because *K-RAS* mutations have also been identified in normal colonic epithelium (Minamoto et al., 1995) and in ACF (Pretlow, 1995). *RAS* oncogene activation in colonic cell lines, Caco-2 and HT-29, produced opposing effects on mRNA expression of homeobox genes Cdx-1 and Cdx-2, in which Cdx1 is upregulated via an MEK1-mitogen-activated protein kinase pathway, and Cdx-2, a known tumor suppressor gene, is downregulated via a PKC-activated pathway. Although yet to be demonstrated in vivo, this regulation by *RAS* implicates a distinct signaling pathway between an activated *RAS* oncogene and Cdx-1 and Cdx-2, which may or may not be linked to mutations in the *APC* gene (Lorentz et al., 1999). The tumor suppressor gene p53 becomes inactivated predominantly at the late adenoma stage, probably causing reduced capability to detect DNA damage, karyotypic instability, impaired G1 cell-cycle arrest, and reduced apoptosis (Tomlinson et al., 1997).

30.10. CONCLUSIONS

Lack of a definitive GI stem cell marker has limited studies of clonality in the normal and carcinogenic GI tract, and although continuously vied for, it may be fitting that the postulated primitive, undifferentiated GI stem cell does not express such a marker. It is generally accepted that the intestinal crypts in the adult mouse and human, and the gastric glands in the mouse, are monoclonal populations derived from a single multipotent stem cell located within a mesenchymal niche. The GI stem cell increases and decreases its proliferation rate in response to the demand dictated by the health of the tissue to maintain homeostasis within the GI epithelia. This continual, frequent turnover of epithelial cells in the gut would ensure that a stem cell daughter cell becoming neoplastic subsequent to differentiation would be shed into the gut lumen and lost, and, consequently, the stem cell itself is the proposed target cell in GI carcinogenesis. Clonal analyses have presented evidence of both monoclonal- and polyclonal-derived GI tumors, although many experimental inconsistencies such as patch size have often been overlooked, potentially providing erroneous data. An understanding of the genetic and molecular regulatory mechanisms involved in epithelial cell turnover and the onset and progression of GI tumorigenesis is rapidly becoming established, for example, the role of the Wnt/β-catenin signalling pathway and its downstream components is now well characterized. It is clear that a second-hit mutation of the *APC* gene is one of the initiating events in colorectal adenoma formation, although the mode of morphological progression is a subject of ongoing debate. Recent discoveries of GI stem cell plasticity may provide insight into the cellular mechanisms involved in GI carcinogenesis and increase the potential for therapies of diseases and tumors of the GI tract.

REFERENCES

Alison, M. R., Poulsom, R., Jeffery, R., et al. (2000) Hepatocytes from non-hepatic adult stem cells. *Nature* 406(6793):257.

Arber, N., Hibshoosh, H., Moss, S. F., et al. (1996) Increased expression of cyclin D1 is an early event in multistage colorectal carcinogenesis. *Gastroenterology* 110(3):669–674.

Barth, A. I., Nathke, I. S., Nelson, W. J. (1997) Cadherins, catenins and APC protein: interplay between cytoskeletal complexes and signaling pathways. *Curr. Opin. Cell. Biol.* 9(5):683–690.

Baulida, J., Batlle, E., and Garcia De Herreros, A. (1999) Adenomatous polyposis coli protein (APC)-independent regulation of beta-catenin/Tcf-4 mediated transcription in intestinal cells. *Biochem. J.* 344(Pt. 2):565–570.

Beck, F., Chawengsaksophak, K., Waring, P., et al. (1999) Reprogramming of intestinal differentiation and intercalary regeneration in Cdx2 mutant mice. *Proc. Natl. Acad. Sci. USA* 96(13):7318–7323.

Beresford, J. N., Bennett, J. H., Devlin, C., et al. (1992) Evidence for an inverse relationship between the differentiation of adipocytic and osteogenic cells in rat marrow stromal cell cultures. *J. Cell Sci.* 102: 341–351.

Bienz, M. and Clevers, H. (2000) Linking colorectal cancer to Wnt signaling. *Cell* 103:311–320.

Bjerknes, M. (1996) Expansion of mutant stem cell populations in the human colon. *J. Theor. Biol.* 178(4):381–385.

Bjerknes, M. and Cheng, H. (1981) The stem-cell zone of the small intestinal epithelium. I. Evidence from Paneth cells in the adult mouse. *Am. J. Anat.* 160(1):51–63.

Bjerknes, M. and Cheng, H. (1999) Clonal analysis of mouse intestinal epithelial progenitors. *Gastroenterology* 116:7–14.

Bjerknes, M., Cheng, H., Kim, H., et al. (1997) Clonality of dysplastic epithelium in colorectal adenomas from familial adenomatous polyposis patients. *Cancer Res.* 57(3):355–361.

Bjornson, C. R. R., Rietze, R. L., Reynolds, B. A., et al. (1999) Turning brain into blood: a hematopoietic fate adopted by adult neural stem cells in vivo. *Science* 283(5401):534–537.

Bloching, M., Hofmann, A., Lautenschlager, C., et al. (2000) Exfoliative cytology of normal buccal mucosa to predict the relative risk of cancer in the upper aerodigestive tract using the MN-assay. *Oral Oncol.* 36(6): 550–555.

Bodmer, W. (1999) Familial adenomatous polyposis (FAP) and its gene, APC. *Cytogenet. Cell. Genet.* 86(2):99–104.

Bodmer, W. F., Bailey, C. J., Bodmer, J., et al. (1987) Localization of the gene for familial adenomatous polyposis on chromosome 5. *Nature* 328(6131):614–616.

Brabletz, T., Jung, A., Dag, S., et al. (1999) beta-catenin regulates the expression of the matrix metalloproteinase-7 in human colorectal cancer. *Am. J. Pathol.* 155(4):1033–1038.

Brabletz, T., Herrmann, K., Jung, A., et al. (2000) Expression of nuclear beta-catenin and c-myc is correlated with tumor size but not with proliferative activity of colorectal adenomas. *Am. J. Pathol.* 156(3): 865–870.

Brasitus, T. A. and Bissonnette, M. (1998) PKC isoforms: villains in colon cancer? *Gastroenterology* 115(1):225–227.

Brittan, M., Hunt, T., Jeffery, R. et al. (2002) Bone marrow derivation of pericryptal myofibroblasts in the mouse and human small intestine and colon. *Gut* 50(6):752–757.

Chawengsaksophak, K., James, R., Hammond, V. E., et al. (1997) Homeosis and intestinal tumors in Cdx2 mutant mice. *Nature* 386 (6620):84–87.

Cheng, H. and Bjerknes, M. (1985) Whole population cell kinetics and postnatal development of the mouse intestinal epithelium. *Anat. Rec.* 211(4):420–426.

Cheng, H. and LeBlond, C. P. (1974) Origin, differentiation and renewal of the four main epithelial cell types in the mouse small intestine. V. Unitarian Theory of the origin of the four epithelial cell types. *Am. J. Anat.* 141(4):537–561.

Cheng, H., Bjerknes, M., Amar, J., et al. (1986) Crypt production in normal and diseased human colonic epithelium. *Anat. Rec.* 216(1):44–48.

Chung, I. M., Schwartz, S. M., and Murry, C. E. (1998) Clonal architecture of normal and atherosclerotic aorta: implications for atherogenesis and vascular development. *Am. J. Pathol.* 152(4):913–923.

Crabtree, M. D., Tomlinson, I. P., and Talbot, I. C. (2001) Variability in the severity of colonic disease in familial adenomatous polyposis results from differences in tumor initiation rather than progression and depends relatively little on patient age. *Gut* 49(4):540–543.

Crawford, H. C., Fingleton, B. M., Rudolph-Owen, L. A., et al. (1999) The metalloproteinase matrilysin is a target of beta-catenin transactivation in intestinal tumors. *Oncogene* 18(18):2883–2891.

Cutler, S. J., Scotto, J., Devesa, S. S., et al. (1974) Third National Cancer Survey—an overview of available information. *J. Natl. Cancer Inst.* 53(6):1565–1575.

da Costa, L. T., He, T C., Yu, J., et al. (1999) CDX2 is mutated in a colorectal cancer with normal APC/beta-catenin signaling. *Oncogene* 18(35):5010–5014.

Deamant, F. D. and Iannaccone, P. M. (1985) Evidence concerning the clonal nature of chemically induced tumors: phosphoglycerate kinase-1 isozyme patterns in chemically induced fibrosarcomas. *J. Natl. Cancer Inst.* 74(1):145–150.

Ee, H. C., Erler, T., Bhathal, P. S., et al. (1995) Cdx-2 homeodomain protein expression in human and rat colorectal adenoma and carcinoma. *Am. J. Pathol.* 147(3):586–592.

Eglitis, M. A. and Mezey, E. (1997) Hematopoietic cells differentiate into both microglia and macroglia in the brains of adult mice. *Proc. Natl. Acad. Sci. USA* 94(8):4080–4085.

Fearon, E. R., Hamilton, S. R., and Vogelstein, B. (1987) Clonal analysis of human colorectal tumors. *Science* 238(4824):193–197.

Ferrari, G., Cusella-De Angelis, G., Coletta, M., et al. (1998) Muscle regeneration by bone marrow–derived myogenic progenitors. *Science* 279(5356):1528–1530.

Fialkow, P. J. (1976) Clonal origin of human tumors. *Biochem. Biophys. Acta* 458(3):283–321.

Fuller, C. E., Davies, R. P., Williams, G. T., et al. (1990) Crypt restricted heterogeneity of goblet cell mucus glycoprotein in histologically normal human colonic mucosa: a potential marker of somatic mutation. *Br. J. Cancer* 61(3):382–384.

Gao, Z., McAlister, V. C., and Williams, G. M. (2001) Repopulation of liver endothelium by bone-marrow–derived cells. *Lancet* 357(9260): 932–933.

Gradl, D., Kuhl, M., and Wedlich, D. (1999) The Wnt/Wg signal transducer beta-catenin controls fibronectin expression. *Mol. Cell. Biol.* 19(8):5576–5587.

Griffiths, D. F., Sacco, P., Williams, D., et al. (1989) The clonal origin of experimental large bowel tumors. *Br. J. Cancer* 59(3):385–387.

Gusterson, B. A., Anbazhagan, R., Warren, W, et al. (1991) Expression of p53 in premalignant and malignant squamous epithelium. *Oncogene* 6(10):1785–1789.

Harada, N., Tamai, Y., Ishikawa, T., et al. (1999) Intestinal polyposis in mice with a dominant stable mutation of the beta-catenin gene. *EMBO J.* 18(21):5931–5942.

Hata, A., Shi, Y., and Massague, J. (1998) TGF-beta signaling and cancer: structural and functional consequences of mutations in Smads. *Mol. Med. Today* 4(6):257–262.

He, T. C., Sparks, A. B., Rago, C., et al. (1988) Identification of c-MYC as a target of the APC pathway. *Science* 281(5382):1509–1511.

Hori, Y., Nakamura, T., Matsumoto, K., et al. (2001) Experimental study on in situ tissue engineering of the stomach by an acellular collagen sponge scaffold graft. *ASAIO J.* 47(3):206–210.

Hori, Y., Nakamura, T., Kimura, D., et al. (2002) Experimental study on tissue engineering of the small intestine by mesenchymal stem cell seeding. *J. Surg. Res.* 102(2):156–160.

Hsu, S. H., Luk, G. D., Krush, A. J., et al. (1983) Multiclonal origin of polyps in Gardner syndrome. *Science* 221(4614):951–953.

Iannaccone, P. M., Gasdner, R. L., and Harris, H. (1978) The cellular origin of chemically induced tumors. *J. Cell. Sci.* 29:249–269.

Iannaccone, P. M., Weinberg, W. C., and Deamant, F. D. (1987) On the clonal origin of tumors: a review of experimental models. *Int. J. Cancer* 39(6):778–784.

Ijichi, H., Ikenoue, T., Kato, N., et al. (2001) Systematic analysis of the TGF-beta-Smad signaling pathway in gastrointestinal cancer cells. *Biochem. Biophys. Res. Comm.* 289(2):350–357.

Jackson, K. A., Mi, T., and Goodell, M. A. (1999) Hematopoietic potential of stem cells isolated from murine skeletal muscle. *Proc. Natl. Acad. Sci. USA* 96:14,482–14,486.

Jackson, K. A., Majka, S. M., Wang, H., et al. (2001) Regeneration of ischemic cardiac muscle and vascular endothelium by adult stem cells. *J. Clin. Invest.* 107(11):1395–1402.

Jass, J. R. and Robertson, A M. (1994) Colorectal mucin histochemistry in health and disease: a critical review. *Pathol. Int.* 44:487–504.

Karam, S. M. (1999) Lineage commitment and maturation of epithelial cells in the gut. *Frontiers Biosci.* 4:286–298.

Kinzler, K. W. and Vogelstein, B. (1996) Lessons from hereditary colorectal cancer. *Cell* 87(2):159–170.

Kirkland, S. (1988) Clonal origin of columnar, mucous, and endocrine cell lineages in human colorectal epithelium. *Cancer* 61:1359–1363.

Knudson, A. G. (1993) Antioncogenes and human cancer. *Proc. Natl. Acad. Sci. USA* 90(23):10,914–10,921.

Kopp, P., Jaggi, R., Tobler, A., et al. (1997) Clonal X-inactivation analysis of human tumors using the human androgen receptor gene (HUMARA) polymorphism: a non-radioactive and semiquantitative strategy applicable to fresh and archival tissue. *Mol. Cell. Probes* 11(3):217–228.

Korbling, M., Katz, R L., Khanna, A., et al. (2002) Hepatocytes and epithelial cells of donor origin in recipients of peripheral-blood stem cells. *N. Engl. J. Med.* 346(10):738–746.

Korchynskyi, O., Landstrom, M., Stoika, R., et al. (1999) Expression of Smad proteins in human colorectal cancer. *Int. J. Cancer* 82(2):197–202.

Kovacs, L. and Potten, C. S. (1973) An estimation of proliferative population size in stomach, jejunum and colon of DBA-2 mice. *Cell Tissue Kinet.* 6:125–134.

Kovacs, G., Tory, K., and Kovacs, A. (1994) Development of papillary renal cell tumors is associated with loss of Y-chromosome-specific DNA sequences. *J. Pathol.* 173(1):39–44.

Krause, D. S., Thiese, N. D., Collector, M. I., et al. (2001) Multi-organ, multi-lineage engraftment by a single bone marrow-derived stem cell. *Cell* 105(3):369–377.

Lagaaij, E. L., Cramer-Knijnenburg, G. F., van Kemenade, F. J., et al. (2001) Endothelial cell chimerism after renal transplantation and vascular rejection. *Lancet* 357(9249):33–37.

Lagasse, E., Connors, H., Al-Dhalimy, M., et al. (2000) Purified hematopoietic stem cells can differentiate into hepatocytes in vivo. *Nat. Med.* 6(11):1229–1234.

Lickert, H., Domon, C., Huls, G., et al. (2000) Wnt/(beta)-catenin signaling regulates the expression of the homeobox gene Cdx1 in embryonic intestine. *Development* 127(17):3805–3813.

Liechty, K. W., MacKenzie, T. C., Shaaban, A. F., et al. (2000) Human mesenchymal stem cells engraft and demonstrate site-specific differentiation after in utero transplantation in sheep. *Nat. Med.* 6(11): 1282–1286.

Lorentz, O., Cardoret, A., Duluc, I., et al. (1999) Downregulation of the colon tumor-suppressor homeobox gene Cdx-2 by oncogenic ras. *Oncogene* 18(1):87–92.

Lu, S. L., Akiyama, Y., Nagasaki, H., et al. (1995) Mutations of the transforming growth factor-beta type II receptor gene and genomic instability in hereditary nonpolyposis colorectal cancer. *Biochem. Biophys. Res. Commun.* 216(2):452–457.

Mahmood, A., Lu, D., Wang, L., et al. (2001) Treatment of traumatic brain injury in female rats with intravenous administration of bone marrow stromal cells. *Neurosurgery* 49(5):1196–1203.

Mann, B., Gelos, M., Siedow, A., et al. (1999) Target genes of beta-catenin-T cell-factor/lymphoid-enhancer-factor signaling in human colorectal carcinomas. *Proc. Natl. Acad. Sci. USA* 96(4):1603–1608.

Markowitz, S., Wang, J., Myeroff, L., et al. (1995) Inactivation of the type II TGF-beta receptor in colon cancer cells with microsatellite instability. *Science* 268(5215):1336–1338.

Massague, J. (1998) TGF-beta signal transduction. *Annu. Rev. Biochem.* 67:753–791.

McLellan, E. A., Medline, A., and Bird, R. P. (1991) Sequential analyses of the growth and morphological characteristics of aberrant crypt foci: putative preneoplastic lesions. *Cancer Res.* 51(19):5270–5274.

Mei, J. M., Hord, N. G., Winterstein, D. F., et al. (1999) Differential expression of prostaglandin endoperoxide H synthase-2 and formation of activated beta-catenin-LEF-1 transcription complex in mouse colonic epithelial cells contrasting in Apc. *Carcinogenesis* 20(4):737–740.

Merritt, A. J., Gould, K A., and Dove, W. F. (1997) Polyclonal structure of intestinal adenomas in ApcMin/+ mice with concomitant loss of Apc+ from all tumor lineages. *Proc. Natl. Acad. Sci. USA* 94(25): 13,927–13,931.

Middleton, S. B., Frayling, I. M., and Phillips, R. K. (2000) Desmoids in familial adenomatous polyposis are monoclonal proliferations. *Br. J. Cancer* 82(4):827–832.

Minamoto, T., Yamashita, N., Ochiai, A., et al. (1995) Mutant K-ras in apparently normal mucosa of colorectal cancer patients: its potential as a biomarker of colorectal tumorigenesis. *Cancer* 75(Suppl. 6): 1520–1526.

Miyaki, M., Konishi, M., Kikuchi-Yanoshita, R., et al. (1994) Characteristics of somatic mutation of the adenomatous polyposis coli gene in colorectal tumors. *Cancer Res.* 54(11):3011–3020.

Miyazono, K., ten Dijke, P., and Heldin, C. H. (2000) TGF-beta signaling by Smad proteins. *Adv. Immunol.* 75:115–157.

Moser, A. R., Pitot, H. C., and Dove, W. F. (1990) A dominant mutation that predisposes to multiple intestinal neoplasia in the mouse. *Science* 247(4940):322–324.

Nakamura, S., Kino, I., and Baba, S. (1993) Nuclear DNA content of isolated crypts of background colonic mucosa from patients with familial adenomatous polyposis and sporadic colorectal cancer. *Gut* 34(9):1240–1244.

Nesbitt, M. N. (1974) Chimeras vs X inactivation mosaics: significance of differences in pigment distribution. *Dev. Biol.* 38(1):202–207.

Nomura, S., Kaminishi, M., Sugiyama, K., et al. (1996) Clonal analysis of isolated single fundic and pyloric gland of stomach using X-linked polymorphism. *Biochem. Biophys. Res. Commun.* 226(2):385–390.

Nomura, S., Esumi, H., Job, C., et al. (1998) Lineage and clonal development of gastric glands. *Dev. Biol.* 204(1):124–135.

Novelli, M. R., Williamson. J. A., Tomlinson, I. P., et al. (1996) Polyclonal origin of colonic adenomas in an XO/XY patient with FAP. *Science* 272(5265):1187–1190.

Novelli, M., Cossu, A., Oukrif, D., et al. (2003) X-inactivation patch size in human female tissue confounds the assessment of tumor clonality. *Proc. Natl. Acad. Sci. USA* 100(6):3311–3314.

Nowell, P. C. (1976) The clonal evolution of tumor cell populations. *Science* 196(4260):23–28.

Orlic, D., Kajstura, J., Chimenti, S., et al. (2001) Bone marrow cells regenerate infarcted myocardium. *Nature* 410(6829):701–705.

Oshima, M., Oshima, H., Kitagawa, K., et al. (1995) Loss of Apc heterozygosity and abnormal tissue building in nascent intestinal polyps in mice carrying a truncated Apc gene. *Proc. Natl. Acad. Sci. USA* 92(10):4482–4486.

Oshima, H., Oshima, M., Kobayashi, M., et al. (1997) Morphological and molecular processes of polyp formation in Apc(delta716) knockout mice. *Cancer Res.* 57(9):1644–1649.

Park, H. S., Goodlad, R. A., and Wright, N. A. (1995) Crypt fission in the small intestine and colon: a mechanism for the emergence of G6PD locus-mutated crypts after treatment with mutagens. *Am. J. Pathol.* 147(5):1416–1427.

Park, H. S., Goodlad, R. A., Ahnen, D. J., et al. (1997a) Effects of epidermal growth factor and dimethylhydrazine on crypt size, cell proliferation, and crypt fission in the rat colon: cell proliferation and crypt fission are controlled independently. *Am. J. Pathol.* 151(3), 843–852.

Park, H. S., Goodlad, R. A., and Wright, N. A. (1997b) The incidence of aberrant crypt foci and colonic carcinoma in dimethylhydrazine-treated rats varies in a site-specific manner and depends on tumor histology. *Cancer Res.* 57(20):4507–4510.

Paulsen, J. E., Steffensen, I. L., Namork, E., et al. (1994) Scanning electron microscopy of aberrant crypt foci in rat colon. *Carcinogenesis* 15 (10):2371–2373.

Peifer, M. (1997) Beta-catenin as oncogene: the smoking gun. *Science* 275(5307):1752–1753.

Petersen, B. E., Bowen, W. C., Patrene, K. D., et al. (1999) Bone marrow as a potential source of hepatic oval cells. *Science* 284(5417): 1168–1170.

Pittenger, M. F., Mackay, A. M., Beck, S. C., et al. (1999) Multilineage potential of adult human mesenchymal stem cells. *Science* 284(5411): 143–147.

Ponder, B. A., Schmidt, G. H., Wilkinson, M. M., et al. (1985) Derivation of mouse intestinal crypts from single progenitor cells. *Nature* 313 (6004):689–691.

Ponder, B A. J. and Wilkinson, M. M. (1986) Direct examination of the clonality of carcinogen-induced colonic epithelial dysplasia in chimeric mice. *J. Natl. Cancer Inst.* 77(4), 967–976.

Potten, C. S. and Hendry, J. H. (1995) *Radiation and Gut*, Elsevier, Amsterdam.

Poulsom, R., Forbes, S. J., Hodivala-Dilke, K., et al. (2001) Bone marrow contributes to renal parenchymal turnover and regeneration. *J. Pathol.* 195(2):229–235.

Powell, D. W., Mifflin, R. C., Valentich, J. D., et al. (1999) Myofibroblasts. II. Intestinal subepithelial myofibroblasts. *Am. J. Physiol.* 277:C183–C201.

Powell, S. M., Zilz, N., Beazer-Barclay, Y., et al. (1992) APC mutations occur early during colorectal tumorigenesis. *Nature* 359(6392): 235–237.

Pretlow, T. P. (1995) Aberrant crypt foci and K-ras mutations: earliest recognized players or innocent bystanders in colon carcinogenesis? *Gastroenterology* 108(2):600–603.

Pretlow, T. P., Cheyer, C., and O'Riordan, M. A. (1994) Aberrant crypt foci and colon tumors in F344 rats have similar increases in proliferative activity. *Int. J. Cancer* 56(4):599–602.

Rau, D., Neubauer, S., Koster, A., et al. (1992) Cytogenetic, oncogenetic, and histopathologic characteristics of colorectal carcinomas with 17p abnormalities. *Hum. Genet.* 89(1):64–68.

Reddy, A. L. and Fialkow, P. J. (1979) Multicellular origin of fibrosarcomas in mice induced by the chemical carcinogen 3-methylcholanthrene. *J. Exp. Med.* 150(4):878–887.

Renan, M. J. (1993)How many mutations are required for tumorigenesis? Implications from human cancer data. *Mol. Carcinog.* 7(3): 139–146.

Romagnolo, B., Berrebi, D., Saadi-Keddoucci, S., et al. (1999) Intestinal dysplasia and adenoma in transgenic mice after overexpression of an activated beta-catenin. *Cancer Res.* 59(16):3875–3879.

Schmidt, G. H. and Mead, R. (1990) On the clonal origin of tumors—lessons from studies of intestinal epithelium. *BioEssays* 12(1):37–40.

Schmidt, G., Winton, D. J., and Ponder, B. A. (1988) Development of the pattern of cell renewal in the crypt-villus unit of chimeric mouse small intestine. *Development* 103:785–790.

Shen, C. N., Slack, J. M., and Tosh, D. (2000) Molecular basis of transdifferentiation of pancreas to liver. *Nat. Cell. Biol.* 2(12):879–887.

Shih, I. M., Wang, T. L., Traverso, G., et al. (2001) Top-down morphogenesis of colorectal tumors. *Proc. Natl. Acad. Sci. USA* 98(5): 2640–2645.

Silberg, D. G., Furth, E. E., Taylor, J. K., et al. (1997) CDX1 protein expression in normal, metaplastic, and neoplastic human alimentary tract epithelium. *Gastroenterology* 113(2):478–486.

Siu, I. M., Robinson, D. R., Schwartz, S., et al. (1999) The identification of monoclonality in human aberrant crypt foci. *Cancer Res.* 59(1): 63–66.

Slaughter, D. P., Southwick, H., and Smejkol, W. (1953) "Field cancerization" in oral stratified squamous epithelium: clinical implications of multicentric origin. *Cancer* 6(5):963–968.

Smith, K., Bui, T. D., Poulsom, R., et al. (1999) Up-regulation of macrophage wnt gene expression in adenoma-carcinoma progression of human colorectal cancer. *Br. J. Cancer* 81(3):496–502.

Smits, R., Kielman, M. F., Breukel, C., et al. (1999) Apc1638T: a mouse model delineating critical domains of the adenomatous polyposis coli protein involved in tumorigenesis and development. *Genes Dev.* 13(10):1309–1321.

Sparks, A. B., Morin, P. J., Vogelstein, B., et al. (1998) Mutational analysis of the APC/beta-catenin/Tcf pathway in colorectal cancer. *Cancer Res.* 58(6):1130–1134.

Subramanian, V., Meyer, B. I., Gruss, P., et al. (1995) Disruption of the murine homeobox gene Cdx1 affects axial skeletal identities by altering the mesodermal expression domains of hox genes. *Cell* 83: 641–653.

Sugihara, K. and Jass, J. R. (1986) Colorectal goblet cell sialomucin heterogeneity: its relation to malignant disease. *J. Clin. Pathol.* 39:1088–1095.

Sunter, J. P., Appleton, D. R., Wright, N. A., et al. (1978) Pathological features of the colonic tumors induced in rats by the administration of 1,2-dimethylhydrazine. *Virchows Arch. B Cell. Pathol.* 29(3):211–223.

Takaku, K., Oshima, M., Miyoshi, H., et al. (1998) Intestinal tumorigenesis in compound mutant mice of both Dpc4 (Smad4) and Apc genes. *Cell* 92(5):645–656.

Tatematsu, M., Fukami, H., Yamamoto, M., et al. (1994) Clonal analysis of glandular stomach carcinogenesis in C3H/HeN< == >BALB/c chimeric mice treated with N-methyl-N-nitrosourea. *Cancer Lett.* 83 (1–2):37–42.

Tetsu, O. and McCormick, F. (1999) Beta-catenin regulates expression of cyclin D1 in colon carcinoma cells. *Nature* 398(6726):422–426.

Thiese, N. D., Badve, S., Saxena, R., et al. (2001) Derivation of hepatocytes from bone marrow cells in mice after radiation induced myeloblation. *Hepatology* 31:253–240.

Thiese, N. D., Nimmakalu, M., Gardner, R., et al. (2000) Liver from bone marrow in humans. *Hepatology* 32:11–16.

Thomas, G. A., Williams, D., and Williams, E. D. (1988) The demonstration of tissue clonality by X-linked enzyme histochemistry. *J. Pathol.* 155(2):101–108.

Thompson, M., Fleming, K. A., Evans, D. J., et al. (1990) Gastric endocrine cells share a clonal origin with other gut cell lineages. *Development* 110:477–481.

Tomlinson, I., Ilyas, M., and Novelli, M. (1997) Molecular genetics of colon cancer. *Cancer Metastasis Rev.* 16(1–2):67–79.

Totafurno, J, Bjerknes, M., and Cheng, H. (1987) The crypt cycle: crypt and villus production in the adult intestinal epithelium. *Biophys. J.* 52(2):279–294.

von Kries, J. P., Winbeck, G., Asbrand, C., et al. (2000) Hot spots in beta-catenin for interactions with LEF-1, conductin and APC. *Nat. Struct. Biol.* 7(9):800–807.

Wakitani, S., Saito, T., and Caplan, A. L. (1995) Myogenic cells derived from rat bone marrow mesenchymal stem cells exposed to 5-azacytidine. *Muscle Nerve* 18(12):1417–1426.

Wasan, H. S., Park, H. S., Liu, K. C., et al. (1998) APC in the regulation of intestinal crypt fission. *J. Pathol.* 185(3):246–255.

Wielenga, V. J., Smits, R., Korinek, V., et al. (1999) Expression of CD44 in Apc and Tcf mutant mice implies regulation by the WNT pathway. *Am. J. Pathol.* 154(2):515–523.

Willert, K. and Nusse, R. (1998) Beta-catenin: a key mediator of Wnt signaling. *Curr. Opin. Genet. Dev.* 8:95–102.

Williams, E. D., Lowes, A. P., Williams, D., et al. (1992) A stem cell niche theory of intestinal crypt maintenance based on a study of somatic mutation in colonic mucosa. *Am. J. Pathol.* 141 (4):773–776.

Winton, D. J. and Ponder, B. A. (1990) Stem-cell organization in mouse small intestine. *Proc. Roy. Soc.* 241:13–18.

Withers, H. R. and Elkind, M. M. (1970) Microcolony survival assay for cells of mouse intestinal mucosa exposed to radiation. *Int. J. Radiat. Biol.* 17:261–267.

Worsham, M. J., Wolman, S. R., Carey, T. E., et al. (1995) Common clonal origin of synchronous primary head and neck squamous cell carcinomas: analysis by tumor karyotypes and fluorescence in situ hybridization. *Hum. Pathol.* 26(3):251–261.

Wright, N. A. and Alison, M. R. (1984) *Biology of Epithelial Cell Populations*, vol. 2, Clarendon, Oxford, UK.

Wright, N. A. (2000) Epithelial stem cell repertoire in the gut: clues to the origin of cell lineages, proliferative units and cancer. *Int. J. Exp. Pathol.* 81(2):117–143.

Wright, N. A. and Poulsom, R. (2002) Top down or bottom up? Competing management structures in the morphogenesis of colorectal neoplasms. *Gut* 51(3):306–308.

Yang, H. K., Linnoila, R. I., Conrad, N. K., et al. (1995) TP53 and RAS mutations in metachronous tumors from patients with cancer of the upper aerodigestive tract. *Int. J. Cancer* 64(4):229–233.

Yatabe, Y., Tavare, S., and Shibata, D. (2001) Investigating stem cells in human colon by using methylation patterns. *Proc. Natl. Acad. Sci. USA* 98(19):1187–1190.

Zhao, L. R., Duan, W. M., Reyes, M., et al. (2002) Human bone marrow stem cells exhibit neural phenotypes and ameliorate neurological deficits after grafting into the ischemic brain of rats. *Exp. Neurol.* 174:11–20.

31 Specification of Liver from Embryonic Endoderm

HIDEYUKI YOSHITOMI, MD, PhD AND KENNETH S. ZARET, PhD

During organogenesis, how do endoderm cells acquire their multipotency, become specified to different cell types, and give rise to tissue buds? The liver derives from the definitive gut endoderm, which expresses many genes in common with the visceral endoderm, which give rise to the yolk sac. The gut endoderm forms from epithelial sheets which form the foregut and hindgut, which elongate and converge at the midsection. During determination, different domains of endoderm are dependent on different groups of transcription factors, which appear to be controlled partially by preprogramming and partially by the influence of overlying mesoderm. When progenitor cells become specified, they proliferate more extensively, forming a tissue bud that extends into the surrounding mesenchymal domain. The liver tissue bud interacts with the adjacent septum transversum mesenchyme between the liver tissue bud and the developing heart (cardiogenic mesoderm). Coordinate signaling from both the cardiogenic mesoderm and septum transversum mesenchyme is required to induce liver differentiation in the ventral foregut endoderm. On the other hand, dorsal-posterior mesenchyme inhibits the induction of liver gene expression in endoderm outside the ventral foregut region. Liver progenitor cells in the ventral foregut also have the potential to undergo pancreatic differentiation. At the time of hepatic specification, there is a burst of expression of fibroblast growth factor-1 (FGF-1) and FGF-2 and persistent expression of FGF-8 in the cardiac mesoderm, and the adjacent ventral foregut expresses FGF receptor genes (FGF-R1 and FGF-R4). Inhibition of this signaling results in failure of hepatic gene expression; FGF signaling is sufficient to induce a hepatic fate, while suppressing a pancreatic fate. Bone morphogenetic proteins from the septum transversum, in conjunction with FGF signaling from the cardiogenic mesoderm, pattern the ventral foregut endoderm into liver and pancreatic cell domains. The induction of expression of liver genes is tightly coupled to morphological changes in the cells from cuboidal to columnar and increased proliferation resulting in formation of the liver bud. The bud expands into the surrounding mesenchyme of the septum transversum with subsequent appearance of endothelial cells that will define the sinusoidal paths. Isolation and culture of uncommitted foregut endodermal cells of the mouse between the 2- and 6-somite stage is described.

31.1. INTRODUCTION

A major issue in the area of organogenesis is the extent to which tissues develop from progenitor cells that either are lost in the process of cell differentiation or are from stem cell populations that are self-renewing. For the endoderm-derived organs, such as the liver, lung, thyroid, and pancreas, we presently lack definitive evidence that they arise from self-renewing stem cells in embryonic development, and there is some evidence to the contrary. Under normal conditions, only the intestine has a well-documented, self-renewing stem cell compartment in adults, yet these cells appear to arise as one of several cell types after the initial commitment of the endoderm to an intestinal epithelium (Korinek et al., 1998). Regardless of these distinctions, endoderm cells, like stem cells, are clearly the progenitors of diverse cell types. The means by which endoderm cells acquire their multipotency, by which gut organ cell types are specified, and by which modulations of cell growth lead to the formation of tissue buds seem highly likely to have mechanisms in common with stem cell biology. In addition, such mechanisms could be common to stemlike cells that become activated in endoderm-derived organs during chronic tissue damage, regeneration, and carcinogenesis, as described in other chapters of this book.

In this chapter, we discuss the above issues in relation to early liver development, as well as provide details on the manipulation of the embryonic endoderm as an experimental system. We focus on what has been learned from genetic studies and from isolating and characterizing the endodermal domain that normally gives rise to the liver and part of the pancreas, emphasizing tissue isolation and culture techniques that promote endoderm differentiation and growth. At various points of the review, we indicate how embryological approaches and insights may be useful for understanding and controlling the differentiation of stem cells outside the normal context of liver development.

From: *Stem Cells Handbook*
Edited by: S. Sell © Humana Press Inc., Totowa, NJ

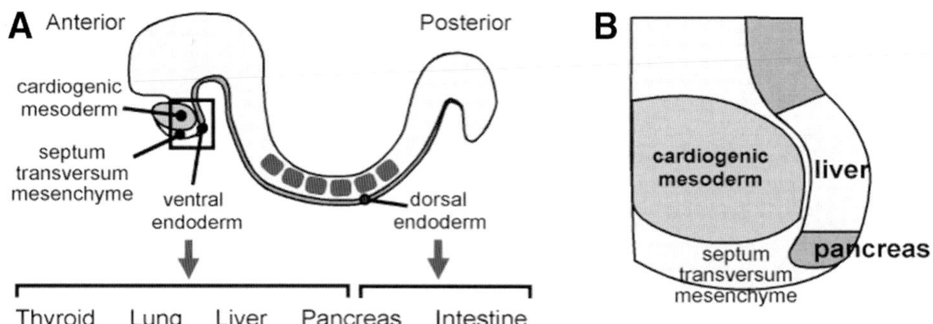

Fig. 1. Patterning of the endoderm into tissues. **(A)** Parasagittal section of mouse embryo at approx 8.25 d of gestation, corresponding to the approx 6-somite stage. Relevant tissue domains and organ derivatives are shown. **(B)** Magnification of boxed area in (A).

31.2. ORIGIN AND INITIAL PATTERNING OF DEFINITIVE ENDODERM, A PROGENITOR CELL POPULATION

During gastrulation, an epithelial layer of the embryo called the epiblast is converted to the definitive gut endoderm and its overlying mesoderm. The definitive endoderm is to be distinguished from the visceral endoderm, the latter being a nonembryonic tissue that gives rise to the yolk sac. The visceral endoderm expresses many genes in common with the liver, such as for serum proteins that help nourish the developing embryo, like the liver does for adults. It is worth noting that one of the classic protocols for differentiating embryonic stem (ES) cells in vitro leads to the creation of visceral endoderm cells, not hepatocytes (Strickland and Mahdavi, 1978). Thus, when intending to generate hepatocytes from ES cells, it is important to distinguish whether in fact visceral endoderm has been created, although in practice this is hard to do and investigators more often rely on diagnostic cell morphologies than marker gene expression. More relevant to the present discussion is the normal means by which the gut endoderm germ layer gives rise to diverse tissues during embryonic development.

In terms of overall morphogenesis of the endoderm into a gut tube, around day 8.0 in mouse gestation (embryonic day 8.0 [E8.0]), the epithelial sheets of endoderm at the anteriormost and posteriormost domains of the embryo fold in to generate the foregut and hindgut (Fig. 1A). As the elongating foregut and hindgut converge at the midsection, by E9.0, the embryo turns from facing outward to facing inward and the gut closes off as a tube. Prior to gut tube formation, progenitors of the thyroid, lung, liver, and ventral rudiment of the pancreas are specified in the ventral foregut endoderm sheet, while progenitors of the dorsal rudiment of the pancreas and the intestinal tract are specified in the dorsal mid- and posterior endoderm.

The two major parameters that govern the ability of the endoderm to differentiate into various tissue types are the heterogeneous pattern of genes expressed in the endoderm itself, across the anterior–posterior axis of the embryo, and the diverse signals from mesoderm cells that overlie the endoderm. Regarding the first parameter, gene inactivation experiments in mice are revealing that the formation of different domains of endoderm are dependent on different groups of transcription factors. For example, the gene for the transcription factor FoxA2 (HNF3β) is essential for the development of the anterior and midgut endoderm (Dufort et al., 1998), whereas the genes for the transcription factors Smad2 (Tremblay et al., 2000) and Sox17 (Kanai-Azuma et al., 2002) are essential for the development of the midgut and hindgut endoderm. Since there is heterogeneity in the requirement for transcription factors across the gut endoderm, it follows that such heterogeneity could govern or limit the developmental potential of different domains of endoderm. This would be an initial "prepatterning" mechanism to facilitate different tissues being specified from endodermal progenitor cell populations, across the anterior–posterior axis of the embryo.

However, recent embryo tissue culture experiments have shown that any such prepatterning must be limited in extent, and thus emphasize a second parameter governing endodermal differentiation: the influence of overlying mesoderm. When Wells and Melton (2000) isolated undifferentiated anterior endoderm from mouse embryos and cultured it with posterior mesoderm, it induced the expression of posterior endodermal markers in the anterior endoderm; similarly, anterior mesoderm induced anterior genes in posterior endoderm. These results in the mouse, and similar studies with *Xenopus* embryos (Horb and Slack, 2001), show that the anterior and posterior domains of endoderm are not irreversibly programmed to one fate or another and that the mesoderm has a primary role in controlling early endoderm differentiation. However, posterior mesoderm was unable to extinguish the induction of anterior genes in cocultures of anterior endoderm, showing that there are limitations to the ability to reprogram the endoderm in response to mesodermal signals.

After progenitor cells of each tissue are specified, each group of cells proliferates more extensively, creating a "bud" of tissue with cells that proliferate into the surrounding mesenchymal domain. Much effort is under way to understand how different mesodermal signals along the anterior-posterior axis of the endoderm lead to the specification of different tissues and the emergence of tissue buds.

31.3. MESODERMAL CELLS PROMOTE OR INHIBIT HEPATOGENESIS IN DIFFERENT DOMAINS OF ENDODERM

Classic developmental studies in the chick identified the specific mesodermal domain required to initiate liver development in the ventral foregut endoderm (Le Douarin, 1964; Fukuda-Taira, 1981). In the chick studies, early somite–stage endoderm from the ventral foregut region was transplanted to the midlateral mesenchyme region of a later-stage embryo (Le Douarin, 1964, 1975). The transplanted endoderm only differentiated into hepatic cells when it was cotransplanted with adjacent cardiogenic mesoderm. Further studies with the chick transplant system showed that after the initial induction of hepatic cells within the endoderm, septum transversum mesenchyme cells, which are mesodermal in origin, promote outgrowth of the liver bud (Le Douarin, 1975). Septum transversum mesenchyme cells surround the developing cardiac and ventral foregut endoderm regions prior to hepatic induction and contribute to both the epicardium and the diaphragm (Fig. 1B). The septum transversum mesenchyme cells apparently elaborate growth factors that promote morphogenesis of the liver bud (*see* below). Taken together, the chick embryo studies showed that sequential interactions with different mesodermal cell types promoted the earliest stages of differentiation and growth of the liver.

Twenty years later, our laboratory adapted this experimental framework by establishing conditions whereby the ventral foregut endoderm could be isolated from mouse embryos and cultured in vitro with or without different mesodermal components or with purified signaling molecules and inhibitors (Gualdi et al., 1996; Jung et al., 1999; Rossi et al., 2001). The details of the experimental system are presented under Subheading 31.7. and Fig.3. With this system, both early hepatic differentiation and growth could be analyzed in detail, allowing a molecular and genetic analysis of relevant signaling mechanisms.

In all studies described hereafter, the developmental stage of the embryo is referred to by the number of somites, which are musculoskeletal precursors, rather than by the number of days of gestation. This provides a far more accurate and reproducible means for predicting the behavior of embryonic tissues in experiments. For example, a single harvest of embryos at E8.5 in the mouse might have a distribution of embryos from the 4- to 8-somite stages, and we now know that prior to the 6-somite stage, the endoderm lacks signs of tissue commitment, whereas by 7–8 somites, genes for various tissues such as the liver are already induced (Gualdi et al., 1996). In addition, using somite stages allows better comparisons between organisms, because the developmental timing of endodermal organogenesis with that of the somites is remarkably conserved in vertebrate evolution.

The prior work with the chick compared favorably with new tissue explant studies in the mouse. Ventral foregut endoderm isolated from the 2- to 6-somite stages and cultured for 2 d in vitro was found to be uncommitted for hepatogenesis, since it failed to express early liver genes (Gualdi et al., 1996). By contrast, when the ventral endoderm from 2- to 6-somite stages was isolated along with its associated cardiogenic mesoderm, it would progress in vitro to express hepatic genes such as for serum albumin and α-fetoprotein. A convenience of this experimental system is that the cardiogenic mesoderm progresses to the beating stage during the culture period; thus, it is easy to identify the cardiac cells from the surrounding endoderm cells. Furthermore, in the presence of cardiogenic mesoderm, the endoderm grows much more vigorously in vitro, providing a simple visual assay for growth promotion analogous to that seen in the liver bud in vivo. As a negative control, when ventral foregut endoderm was cultivated with dorsal–posterior mesoderm, which normally interacts with preintestinal endoderm, it would not induce liver genes in vitro. These studies show that there is specificity for the cardiogenic mesoderm to induce hepatogenesis within the endoderm.

While these findings with the mouse would seem to simply recapitulate those with the chick, the ability to use diverse mouse molecular markers to characterize cell populations, as well as the ability to use genetically altered embryos, has provided deeper perspectives into hepatic specification. For example, molecular marker analysis showed that ventral foregut endoderm explants from the mouse, and thus possibly from the chick, were found to be "contaminated" with mesenchymal cells of the prospective septum transversum (Rossi et al., 2001). Further studies showed that while the septum transversum mesenchyme cells alone are insufficient to induce hepatogenesis in isolated ventral foregut endoderm, signaling from the septum cells is required for the cardiac mesoderm to induce liver genes in the ventral endoderm (Rossi et al., 2001). Thus, the mouse studies showed that coordinate signaling from both the cardiogenic mesoderm and septum transversum mesenchyme is required to induce the liver in the ventral foregut endoderm. To restate, these findings illustrate that the septum transversum mesenchyme cells have an earlier role in endodermal patterning than was previously appreciated from the chick transplant studies.

Additional studies with the tissue explant system provided insight into mesodermal patterning of the endoderm across the embryo. Specifically, dorsal–posterior mesoderm was not only found to lack the hepatic-inducing properties seen in the cardiogenic mesoderm, but it actually inhibited hepatic induction when juxtaposed to explants of ventral foregut endoderm and cardiogenic mesoderm (Gualdi et al., 1996). On the other hand, when dorsal-posterior endoderm, which normally becomes the intestine, was cultured in isolation from dorsal–posterior mesoderm, the dorsal–posterior endoderm induced some early liver genes. Taken together, these studies show that early liver genes are capable of being expressed more or less throughout the endoderm germ layer, and that the dorsal–posterior mesoderm prevents these genes from being induced outside the ventral foregut region (Bossard and Zaret, 1998, 2000). In conclusion, the mesoderm in the mouse is regionalized to both induce and inhibit cell-specific genes in the endoderm. This appears to cause tissues of the liver to emerge at the appropriate position along the anterior–posterior axis of the embryo. These studies emphasize the multipotency of the endoderm germ layer as a progenitor cell population and the importance of mesodermal signals in tissue patterning.

31.4. MULTIPOTENCY OF ENDODERM PROGENITORS OF LIVER

The multipotency of the endoderm is evident not only across the anterior–posterior axis of the embryo; it is also seen within the ventral foregut domain that can give rise to the liver. Initially, we thought that if ventral foregut endoderm was cultivated in the absence of cardiogenic mesoderm, we could expand the progenitor cells and study them in detail in their undifferentiated state. However, when the isolated ventral endoderm was not induced to undergo a hepatic fate, it was found to assume a pancreatic fate as a default (Deutsch et al., 2001). Indeed, the liver and the ventral domain of the pancreas normally arise from adjacent segments of the anteriormost ventral foregut endoderm (Fig. 1B). A sensitive reverse transcriptase polymerase chain reaction (RT-PCR) assay showed that the endoderm closest to the cardiogenic mesoderm, at the 7- to 8-somite stages, expresses serum albumin mRNA, whereas the endoderm slightly more ventral, extending distal to the cardiogenic mesoderm, expresses *pdx*-1 mRNA

(Deutsch et al., 2001). *pdx-1* encodes one of the earliest-acting homeobox transcription factors in the pancreatic lineage (Jonsson et al., 1994; Offield et al., 1996). Further studies showed that in endoderm explants with cardiogenic mesoderm, a *pdx-1*-positive domain sometimes emerged, and it was always distal to the beating cardiac cells in the explant. These findings are consistent with cardiac mesoderm blocking the default pancreatic program in nearby cells, while it induces the liver program.

Further support for the ventral foregut endoderm consisting of a bipotential cell population came from studies to determine whether the cardiogenic mesoderm promotes selective outgrowth of putative "predetermined" hepatic cells and/or the death of putative "prepancreatic" cells (Deutsch et al., 2001). No such evidence was found. Although it is necessary to mark individual progenitor cells to determine definitively whether they are bipotential, the available evidence indicates that the ventral foregut endoderm cells can undergo either a hepatic or pancreatic fate. Because the lung and the thyroid also emerge from slightly more posterior endoderm, it will be interesting in the future to see how these tissues are patterned with respect to the liver and the pancreas.

The apparent bipotentiality of ventral endoderm cells in embryos may be recapitulated in adult tissues, because in certain pathological conditions, pancreatic cells can differentiate into hepatocytes (Rao et al., 1989; Gu and Sarvetnick, 1993; Dabeva et al., 1997; Krakowski et al., 1999). In these cases, it is not yet known whether fully mature exocrine or endocrine cells "transdifferentiate" into hepatocytes or if bi- or multipotent stem cells differentiate *de novo* into hepatocytes. However, a subclone of a pancreatic exocrine cell line has been shown to transdifferentiate directly into an albumin-producing hepatic cell in vitro (Shen et al., 2000; Wang et al., 2001). These studies show that the developmental relationship of the liver and pancreas in embryos can impact our understanding of the plasticity of the differentiated state in adult tissues.

31.5. MOLECULAR MECHANISMS CONTROLLING INITIAL SPECIFICATION OF LIVER

The ability to isolate embryonic liver progenitor cells and study parameters that differentiate them in vitro has provided a powerful system with which to analyze the relevant signaling mechanisms. At the time of hepatic specification, there is a burst of expression of fibroblast growth factor-1 (FGF-1) and FGF-2 and persistent expression of FGF-8 in the cardiac mesoderm (Crossley and Martin, 1995; Zhu et al., 1996; Jung et al., 1999). In addition, the ventral foregut endoderm expresses at least two of the four FGF receptor genes, *FGF-R1* and *FGF-R4* (Stark et al., 1991; Sugi et al., 1995). Mouse homozygous null mutations for these FGF signaling components either are embryonic lethal prior to hepatic induction (Yamaguchi et al., 1992; Deng et al., 1994; Meyers et al., 1998) or have minimal or no phenotype (Dono et al., 1998; Ortega et al., 1998; Weinstein et al., 1998; Miller et al., 2000). To address the issues of prior requirements and redundancy, Jung et al. (1999) treated embryonic tissue explants of cardiogenic mesoderm, ventral foregut endoderm, and septum transversum mesenchyme with general inhibitors of FGF signaling, which resulted in an inhibition of hepatic gene induction.

Further analysis showed that FGF signaling controls a cell fate choice in the ventral endoderm. When ventral foregut endoderm explants, free from cardiac mesoderm, were treated with low doses of purified FGF-2, the cells initiated the expression of liver genes instead of pancreatic genes (Jung et al., 1999; Deutsch et al., 2001). FGF-2 did not appear to affect the selective growth or death of the cells in the population. Taken together, the explant studies show that FGF signaling is sufficient to explain the influence of the cardiogenic mesoderm in inducing a hepatic fate while suppressing a pancreatic fate in the ventral foregut endoderm.

FGF signaling results in the activation of the mitogen-activated protein kinase pathway, which involves signaling via small G proteins such as ras (Kouhara et al., 1997; Casci et al., 1999). Recently, a homozygous null mutation was made in the gene for the enzyme prenylcysteine carboxymethyltransferase (*PCCMT*); this enzyme normally modifies small G proteins so that they stably associate with the plasma membrane (Bergo et al., 2000; Chen et al., 2000). The *PCCMT* null mutation causes a deficiency in ras association with the plasma membrane, but not a complete block (Bergo et al., 2000) and almost certainly affects a variety of other signaling proteins. Interestingly, the *PCCMT* mutation results in a failure in liver development (Lin et al., 2002). Detailed studies revealed a delay in hepatic induction of the endoderm in vivo and inconsistent induction of hepatic genes in cocultures of cardiogenic mesoderm, ventral foregut endoderm, and septum transversum mesenchyme in vitro (Lin et al., 2002). Taken with the above FGF inhibitor studies, there is emerging evidence that growth factor signaling pathways inside the endoderm are crucial to trigger a hepatic fate. Further work is needed to identify the specific signaling pathway components involved and to understand how the pancreatic fate is excluded in this context.

Additional studies with the explant system identified a signaling role for the septum transversum mesenchyme cells. These cells are identifiable by their expression of the genes for the transcription factor *Mrg1* (Dunwoodie et al., 1998) and for the signaling molecule bone morphogenetic protein-4 (*BMP-4*) (Winnier et al., 1995). When cocultures of ventral foregut endoderm, cardiogenic mesoderm, and septum transversum mesenchyme were treated with noggin, an inhibitor of BMP signaling (Zimmerman et al., 1996), the normal expression of albumin mRNA in the explants was suppressed and enhanced the expression of *pdx-1* was enhanced (Rossi et al., 2001). When BMP-2 or BMP-4 was added back to noggin-inhibited explants, albumin expression was restored, and when other non-BMP factors were added back, they failed to work well in this regard. In summary, BMPs from the septum transversum mesenchyme cells, in conjunction with FGF signaling from the cardiogenic mesoderm, pattern the ventral foregut endoderm into liver and pancreatic cell domains. These studies emphasize how multiple signaling factors working together may be required to modulate stem cells to desired cell fates.

31.6. MORPHOGENESIS OF LIVER BUD

The initial induction of liver genes within hepatic endoderm is tightly coupled to changes in cellular morphogenesis and growth. Specifically, the endodermal epithelial cells first undergo a shape transition from cuboidal to columnar, while lining the developing gut tube. Shortly thereafter, by E9.0 in the mouse (approx 15–20 somites), the hepatic endoderm cells start to proliferate more rapidly than the nearby gut endoderm cells. Together, the columnar transition and increased proliferation lead to a thickening of the endoderm, and thus the initial appearance of the liver bud (Fig. 2A,B). By E9.5 (approx 21–25 somites), the hepatic cells then proliferate out of the epithelium and into the surrounding

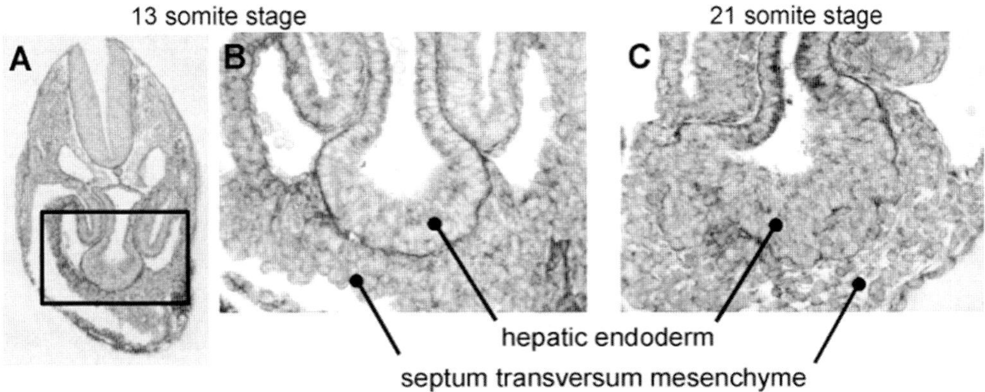

Fig. 2. Morphogenesis of the liver bud. Transverse sections of mouse embryos at designated somite stages, corresponding to 8.5–9.5 d of gestation. The boxed area in **(A)** (×55 original magnification) is magnified to ×220 original in **(B)**. **(C)** Same magnification in (B).

mesenchyme of the septum transversum, causing an expansion of the liver bud (Fig. 2C). Simultaneously, endothelial cells appear and line spaces in the emerging liver bud that will ultimately define the sinusoidal paths through which blood will flow (Matsumoto et al., 2001). The liver is a hematopoietic organ in mammals, and by E10–10.5 in the mouse, hematopoietic cells invade the liver bud. By E11.5, the liver is a defined organ and a primary site of fetal hematopoiesis.

Although there are no genes known whose inactivation leads to a consistent defect in the induction of hepatic cells in the endoderm, there are a variety of genes that are essential for the early steps of liver bud morphogenesis and thus are embryonic lethal when inactivated in mice. For example, inactivation of the genes for the *Hex* or *Prox-1* homeobox-containing transcription factors, which are expressed in the ventral foregut endoderm, leads to a block in liver bud emergence after the initial thickening of the hepatic endoderm (Keng et al., 2000; Martinez-Barbera et al., 2000; Sosa-Pineda et al., 2000). In both of these nulls, the early hepatic genes are induced, but there is a block in liver bud growth. These transcription factors may control the expression of signal transduction pathway components that transduce growth signals from nearby mesoderm.

There is also genetic evidence that the septum transversum mesenchyme is important for morphological outgrowth of the liver bud. Inactivation of the gene for *Hlx*, a homeobox-containing transcription factor expressed in the septum transversum mesenchyme, leads to a block in liver bud growth after hepatic genes are induced (Hentsch et al., 1996). Hlx therefore appears to control the expression of a paracrine signaling factor from the septum transversum mesenchyme to the liver bud.

Explant cultures have been made of ventral foregut endoderm, cardiogenic mesoderm, and septum transversum mesenchyme from homozygous null embryos for *Hex* or *Prox-1*. In each case, the hepatic endoderm domain grew substantially in vitro (B. Sosa-Pineda, R. Bort, and K. Zaret) (unpublished studies). Recall that these mutations block early liver growth after the initial thickening of the hepatic endoderm. By contrast, an inhibitor of FGF signaling markedly prevented the initial outgrowth of ventral foregut tissue explants in culture, and adding FGF-8 back to such inhibited cultures restored outgrowth (Jung et al., 1999). In addition, although a BMP-signaling inhibitor alone failed to affect ventral foregut tissue outgrowth in culture, when the tissue was derived from *BMP-4* heterozygous or homozygous null embryos, the inhibitor strongly inhibited outgrowth (Rossi et al., 2001). Apparently, an endogenous genetic reduction in *BMP* signaling was necessary for the exogenous BMP inhibitor to have an effect. Taken together, these findings indicate that outgrowth of ventral foregut endoderm cells in the embryo tissue explant assay is dependent on the same signals that induce a hepatic fate. Such outgrowth appears to be controlled differently from the morphogenetic signals that control the secondary expansion of the liver bud into an organ.

To test the morphogenetic role of endothelial cells that contribute to the developing liver vasculature, Matsumoto et al. (2001) used mouse embryos homozygous mutant for the *flk-1* gene. *flk-1* encodes the vascular endothelial growth factor receptor-2, which is normally expressed in and required for the development of endothelial cells (Shalaby et al., 1995); null mutants exhibit a failure of maturation of angioblasts and they do not migrate to target sites such as the liver, thus cleanly deleting the cells from the liver bud. Strikingly, loss of endothelial cells caused a failure in liver bud growth after the hepatic induction stage (Matsumoto et al., 2001). Thus, endothelial cells appear to express normally a paracrine factor that promotes liver bud growth. The block in hepatic bud growth in the *flk-1* null embryos is prior to hematopoietic cell invasion, and therefore is not owing to a failure of blood cells, oxygen, or nutrients being delivered to the local region (Matsumoto et al., 2001).

The mutations discussed in this section cause the earliest known defects in liver bud growth and may be germane to issues related to stem cell growth and expansion. Many other mouse mutations affect liver development at subsequent steps of differentiation and morphogenesis and are discussed in reviews elsewhere (Duncan, 2000; Zaret, 2002).

31.7. EMBRYO TISSUE EXPLANT CULTURES THAT RECAPITULATE HEPATOGENESIS

In this section, we describe our methodology for an embryo tissue culture system that supports the viability, growth, and differentiation of ventral foregut endoderm, with or without cardiogenic mesoderm and septum transversum mesenchyme (Fig.3). The single most important parameter in developing this culture

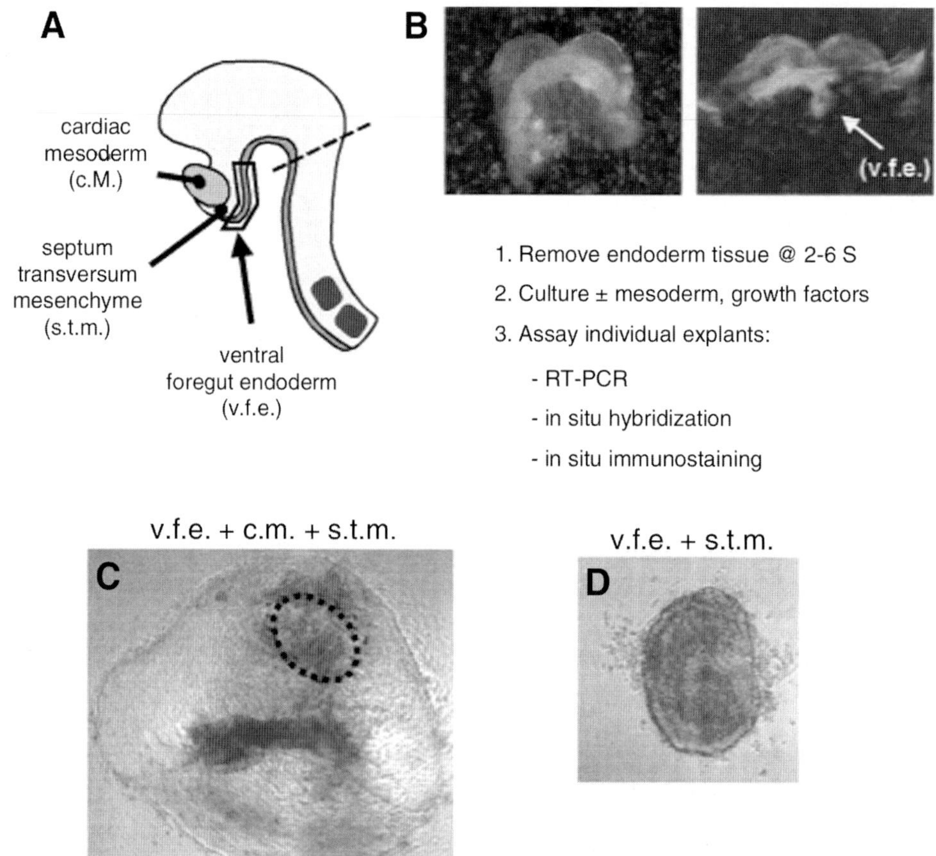

Fig. 3. Isolation of ventral foregut endoderm, including liver progenitor cells, from mouse embryos. **(A)** Parasagittal view of anterior domain of mouse embryo at ~8.25 d of gestation, 2–6 somites. The dashed line indicates the area cut to remove the anterior endoderm domain. **(B)** Dissection of ventral foregut endoderm (v.f.e.) and outline of procedure. **(C,D)** Explants grown for 48 h in vitro. The circled area represents beating cardiac cells, which are present only in the explant in (C). Both images are at the same magnification. (Pictures courtesy of Dr. Jennifer Rossi.)

system was the utilization of 10% calf serum in a standard tissue culture medium. The use of fetal calf sera from any source tested or serum-free media supplemented with growth factors was strongly inhibitory to the survival of the endoderm. It is also critical to assess carefully the somite stage of the isolated embryos, prior to tissue dissection. As described above, the ventral foregut endoderm in the mouse is uncommitted between the 2- and 6-somite stages and therefore must be isolated during this period for cell fate studies. A final general point is that it is worth taking a photograph under the microscope of each embryo, before and during dissection, and of each explant on all days of culture. This greatly facilitates retrospective analysis and troubleshooting.

Explant tissue culture medium consists of Dulbecco's modified Eagle's medium containing 10% calf serum (Hyclone), penicillin (100 U/mL), and streptomycin (100 μg/mL). Place the embryo tissue explants in individual chambers of glass slides containing eight removable chambers mounted on top (Nunc Lab-Tek; cat. no. 177402). The glass substrate facilitates microscopic analysis during the culture period. Prior to use, coat the slides with collagen by adding to each chamber 200 μL of a solution containing 50 μg/mL of collagen type 1 (Collaborative Biomedical; cat. no. 354236) in 0.02 *N* acetic acid in phosphate-buffered saline (PBS) at 37°C for 1 h. Then aspirate the solution and wash the wells twice with PBS and once with medium. Add 400 μL of medium to each well and equilibrate in an incubator with 5% CO_2.

Considering that noon of the day of vaginal plug discovery is taken to be E0.5, embryos are harvested at approx E8.25 (e.g., in the morning) to select embryos prior to hepatic induction, at the 2- to 6-somite stages. Remove the embryos from uteri and transfer them to dishes containing PBS. Dissect the embryos free of decidual tissues and store in PBS in a 6-cm dish on ice. Transfer an individual embryo to a few drops of PBS on black dissecting wax in a dish, under a dissecting microscope. Use electrolytically etched tungsten needles (Hogan et al., 1986) to carefully remove the yolk sac. Because the yolk sac is connected to embryonic tissue at the level of foregut opening, it must be removed near the foregut endoderm. Cut the embryo just below the opening of the foregut (Fig. 3A, dashed line) and save the anterior domain for further dissection (Fig. 3B, left panel).

To isolate the foregut endoderm with cardiac mesoderm, excise and remove the optic lobes and the dorsal (back) part of the embryo fragment. Use a micropipetor to transfer the tissue into a slide chamber well containing medium.

For finer dissection of foregut endoderm free of cardiogenic mesoderm, expose the anterior portion of the embryo to either heat or an enzymatic solution to help separate tissue layers. Prior to dissection, transfer an individual embryo to a dish of PBS on a heat block at 37°C for about 3 min. This loosens the tissues and facilitates dissection of the endoderm. Then transfer the tissue to clean PBS on the dissecting dish and carefully separate the endodermal layer as a sheet from cardiac mesoderm (Fig.3B, right panel, v.f.e.). It is best to try dissecting without enzymes, but if necessary, put the embryo or tissue fragment into a solution of 2.5% pancreatin (Sigma, St. Louis, MO), 0.5% polyvinylpyrrolidone, and 20 m*M* HEPES (pH 7.4) in PBS for a few minutes at room temperature. Excessive enzymatic treatment causes a degradation of tissue. Transfer the embryo or tissue carefully into medium containing 10% serum for 1 min to stop the digestion, and then transfer to clean PBS and dissect as above. Again, it is possible to dissect away the endoderm from the cardiogenic mesoderm without the enzyme treatment, but in this case the several-minute exposure of the embryo to PBS at 37°C is necessary (*see* above).

Embryo tissue fragments are usually incubated at 37°C for 2 to 3 d. Cardiogenic mesoderm cells progress to the beating stage after 8–24 h of culture and thus become easily visible under the microscope (Fig. 3C, 48 h of growth, beating cardiac area is circled by red dashed line). In the absence of cardiogenic mesoderm, the endoderm explants are much smaller and may not attach to the substrate (Fig. 3D, 48 h of growth). Take instant or digital photographs of the explants on each day of culture and carefully circle the beating area of cells in the picture. It is important to score for beating cells in explants immediately after they are removed from the 37°C incubator, because extended periods at room temperature can cause the cells to stop beating. After culturing for an optimal period, the explants are subjected to RNA extraction for RT-PCR, *in situ* hybridization (ISH), or immunohistochemistry. For ISH and immunohistochemistry, fix tissues on the slide in 4% paraformaldehyde in PBS for a few hours at 4°C, and then dehydrate with a series of methanol (for ISH) or ethanol (for immunohistochemistry). They can be stored at –20°C for several months.

31.8. FUTURE PROSPECTS

As discussed at the outset of this chapter, we do not know the extent to which mechanisms of endoderm tissue specification relate to those exhibited by bona fide stem cells in adult tissues in vivo, or by embryonic or adult stem cells that are manipulated in culture. Nonetheless, it is reasonable to assume that the ways that nature normally controls the formation of tissues such as the liver in development are good starting points for thinking about how stem cells might generate such tissues in other contexts.

Two general points emerge from our detailed discussion above on embryo explants and genetic studies of liver specification from the endoderm. The first is that initial cell fate decisions are controlled by multiple, convergent signals from distinct mesodermal cell domains acting on a particular domain of endoderm. Thus, the more we know about regulatory factors that impart hepatogenic competence to the endoderm and signal transduction pathways triggered therein, the better we should be able to predict the behavior of stem cells in other contexts. The second major point is that initial cell fate decisions appear to be intimately linked to growth control; that is, all signaling influences that promote hepatic specification also promote outgrowth of the cells into a nascent liver bud. An interesting aspect of this issue is that enhanced growth of the endoderm occurs when the cells begin to differentiate, which is in contrast to the conventional view that growth and differentiation oppose one another. Indeed, throughout development, and perhaps in stem cell biology, signaling molecules that were originally designated as growth factors typically play critical roles in controlling cell type differentiation. We anticipate that these general points and the specific regulatory molecules discussed here will be useful when considering how stem cells isolated from embryos or residing in adult tissues can be induced to become hepatocytes or other gut organ cell types.

ACKNOWLEDGMENTS

This research was supported by grants from the Honjyo International Foundation and Board of Associates at Fox Chase Cancer Center to H.Y. and grants from the National Institutes of Health (GM36477), National Cancer Institute, Human Frontiers Science Program, W.W. Smith Charitable Trust, and Mathers Charitable Foundation to K.S.Z.

REFERENCES

Bergo, M. O., Leung, G. K., Ambroziak, P., Otto, J. C., Casey, P. J., and Young, S. G. (2000) Targeted inactivation of the isoprenylcysteine carboxyl methyltransferase gene causes mislocalization of K-Ras in mammalian cells. *J. Biol. Chem.* 275:17,605–17,610.

Bossard, P. and Zaret, K. S. (1998) GATA transcription factors as potentiators of gut endoderm differentiation. *Development* 125:4909–4917.

Bossard, P. and Zaret, K. S. (2000) Repressive and restrictive mesodermal interactions with gut endoderm: possible relation to Meckel's Diverticulum. *Development* 127:4915–4923.

Casci, T., Vinos, J., and Freeman, M. (1999) Sprouty, an intracellular inhibitor of Ras signaling. *Cell* 96:655–665.

Chen, Z., Otto, J. C., Bergo, M. O., Young, S. G., and Casey, P. J. (2000) The C-terminal polylysine region and methylation of K-Ras are critical for the interaction between K-Ras and microtubules. *J. Biol. Chem.* 275:41251–41257.

Crossley P. H. and Martin G. R. (1995) The mouse *Fgf8* gene encodes a family of polypeptides and is expressed in regions that direct outgrowth and patterning in the developing embryo. *Development* 121: 439–451.

Dabeva, M. S., Hwang, S. G., Vasa, S. R., et al. (1997) Differentiation of pancreatic epithelial progenitor cells into hepatocytes following transplantation into rat liver. *Proc. Natl. Acad. Sci. USA* 94: 7356–7361.

Deng, C. X., Wynshaw-Boris, A., Shen, M. M., Daugherty, C., Ornitz, D. M., and Leder, P. (1994) Murine FGFR-1 is required for early postimplantation growth and axial organization. *Genes Dev.* 8: 3045–3057.

Deutsch, G., Jung, J., Zheng, M., Lóra, J., and Zaret, K. S. (2001) A bipotential precursor population for pancreas and liver within the embryonic endoderm. *Development* 128:871–881.

Dono, R., Texido, G., Dussel, R., Ehmke, H., and Zeller, R. (1998) Impaired cerebral cortex development and blood pressure regulation in FGF-2-deficient mice. *EMBO J.* 17:4213–4225.

Dufort, D., Schwartz, L., Harpal, K., and Rossant, J. (1998) The transcription factor HNF3β is required in visceral endoderm for normal primitive streak morphogenesis. *Development* 125:3015–3025.

Duncan, S. A. (2000) Transcriptional regulation of liver development. *Dev. Dyn.* 219:131–142.

Dunwoodie, S. L., Rodriguez, T. A., and Beddington, R. S. P. (1998) *Msg1* and *Mrg1*, founding members of a gene family, show distinct patterns of gene expression during mouse embryogenesis. *Mech. Dev.* 72:27–40.

Fukuda-Taira, S. (1981) Hepatic induction in the avian embryo: specificity of reactive endoderm and inductive mesoderm. *J. Embryol. Exp. Morphol.* 63:111–125.

Gu, D. and Sarvetnick, N. (1993) Epithelial cell proliferation and islet neogenesis in IFN-γ transgenic mice. *Development* 118:33–46.

Gualdi, R., Bossard, P., Zheng, M., Hamada, Y., Coleman, J. R., and Zaret, K. S. (1996) Hepatic specification of the gut endoderm in vitro: cell signalling and transcriptional control. *Genes Dev.* 10:1670–1682.

Hentsch, B., Lyons, I., Ruili, L., Hartley, L., Lints, T. J., Adams, J. M., and Harvey, R. P. (1996) Hlx homeo box gene is essential for an inductive tissue interaction that drives expansion of embryonic liver and gut. *Genes Dev.* 10:70–79.

Hogan, B., Costantini, F., and Lacy, E. (1986) *Manipulating the Mouse Embryo: A Laboratory Manual*, Cold Spring Harbor Laboratory Press, Cold Spring Harbor, NY.

Horb, M. E. and Slack, J. M. (2001) Endoderm specification and differentiation in Xenopus embryos. *Dev. Biol.* 236:330–343.

Jonsson, J., Carlsson, L., Edlund, T., and Edlund, H. (1994) Insulin-promoter-factor 1 is required for pancreas development in mice. *Nature* 371:606–609.

Jung, J., Zheng, M., Goldfarb, M., and Zaret, K. S. (1999) Initiation of mammalian liver development from endoderm by fibroblasts growth factors. *Science* 284:1998–2003.

Kanai-Azuma M., Kanai Y., Gad J. M., et al. (2002) Depletion of definitive gut endoderm in Sox17-null mutant mice. *Development* 129: 2367–2379.

Keng V. W., Yagi H., Ikawa M., et al. (2000) Homeobox gene Hex is essential for onset of mouse embryonic liver development and differentiation of the monocyte lineage. *Biochem. Biophys. Res. Commun.* 276:1155–1161.

Korinek V., Barker N., Moerer P., et al. (1998) Depletion of epithelial stem-cell compartments in the small intestine of mice lacking Tcf-4. *Nat. Genet.* 19:379–383.

Kouhara H., Hadari Y. R., Spivak-Kroizman T., et al. (1997) A lipid-anchored Grb2-binding protein that links FGF-receptor activation to the Ras/MAPK signaling pathway. *Cell* 89:693–702.

Krakowski, M. L., Kritzik, M. R., Jones, E. M., et al. (1999) Pancreatic expression of keratinocyte growth factor leads to differentiation of islet hepatocytes and proliferation of duct cells. *Am. J. Pathol.* 154: 683–691.

Le Douarin, N. (1964) Induction de l'endoderme pré-hépatique par le mésoderme de l'aire cardiaque chez l'embryon de poulet. *J. Embryol. Exp. Morphol.* 12:651–664.

Le Douarin, N. M. (1975) An experimental analysis of liver development. *Med. Biol.* 53:427–455.

Lin, X., Jung, J., Zaret, K. S., and Zoghbi, H. (2002) Prenycysteine carboxylmethyltransferase (PCCMT), an essential gene for early liver development in mice. *Gastroenterology* 123:345–351.

Martinez-Barbera, J. P., Clements, M., Thomas, P., et al. (2000) The homeobox gene *hex* is required in definitive endodermal tissues for normal forebrain, liver and thyroid formation. *Development* 127: 2433–2445.

Matsumoto, K., Yoshitomi, H., Rossant, J., and Zaret, K. S. (2001) Liver organogenesis promoted by endothelial cells prior to vascular function. *Science* 294:559–563.

Meyers, E. N., Lewandowski, M., and Martin, G. R. (1998) An *Fgf8* mutant allelic series generated by Cre- and Flp-mediated recombination. *Nat. Genet.* 18:136–141.

Miller, D. L., Ortega, S., Bashayan, O., Basch, R., and Basilico, C. (2000) Compensation by fibroblast growth factor 1 (FGF1) does not account for the mild phenotypic defects observed in FGF2 null mice. *Mol. Cell. Biol.* 20:2260–2268.

Offield, M. F., Jetton, T. L., Labosky, P. A., et al. (1996) PDX-1 is required for pancreatic outgrowth and differentiation of the rostral duodenum. *Development* 122:983–995.

Ortega, S., Ittmann, M., Tsang, S. H., Ehrlich, M., and Basilico, C. (1998) Neuronal defects and delayed wound healing in mice lacking fibroblast growth factor 2. *Proc. Natl. Acad. Sci. USA* 95:5672–5677.

Rao, M. S., Dwivedi, R. S., Yelandi, A., et al. (1989) Role of periductal and ductalar epithelial cells of the adult rat pancreas in pancreatic hepatocyte lineage: a change in the differentiation commitment. *Am. J. Pathol.* 134:1069–1086.

Rossi, J. M., Dunn, N. R., Hogan, B. L. M., and Zaret, K. S. (2001) Distinct mesodermal signals, including BMP's from the septum transversum mesenchyme, are required in combination for hepatogenesis from the endoderm. *Genes Dev.* 15:1998–2009.

Shalaby, F., Rossant, J., Yamaguchi, T. P., et al. (1995) Failure of blood-island formation and vasculogenesis in Flk-1-deficient mice. *Nature* 376:62–66.

Shen, C. N., Slack, J. M., and Tosh, D. (2000) Molecular basis of transdifferentiation of pancreas to liver. *Nat. Cell. Biol.* 2:879–887.

Sosa-Pineda, B., Wigle, J. T., and Oliver, G. (2000) Hepatocyte migration during liver development requires Prox1. *Nat. Genet.* 25:254–255.

Stark, K. L., McMahon, J. A., and McMahon, A. P. (1991) FGFR-4, a new member of the fibroblast growth factor receptor family, expressed in the definitive endoderm and skeletal muscle lineages of the mouse. *Development* 113:641–651.

Strickland, S. and Mahdavi, V. (1978) The induction of differentiation in teratocarcinoma stem cells by retinoic acid. *Cell* 15:393–403.

Sugi, Y., Sasse, J., Barron, M., and Lough, J. (1995) Developmental expression of fibroblast growth factor receptor-1 (cek-1; flg) during heart development. *Dev. Dyn.* 202:115–125.

Tremblay, K. D., Hoodless, P. A., Bikoff, E. K., and Robertson, E. J. (2000) Formation of the definitive endoderm in mouse is a Smad2-dependent process. *Development* 127:3079–3090.

Wang, X., Al-Dhalimy, M., Lagasse, E., Finegold, M., and Grompe, M. (2001) Liver repopulation and correction of metabolic liver disease by transplanted adult mouse pancreatic cells. *Am. J. Pathol.* 158: 571–579.

Weinstein, M., Xu, X., Ohyama, K., and Deng, C.-X. (1998) FGFR-3 and FGFR-4 function cooperatively to direct alveogenesis in the murine lung. *Development* 125:3615–3623.

Wells, J. M. and Melton, D. A. (2000) Early mouse endoderm is patterned by soluble factors from adjacent germ layers. *Development* 127:1563–1572.

Winnier, G., Blessing, M., Labosky, P. A., and Hogan, B. L. M. (1995) Bone morphongenetic protein-4 is required for mesoderm formation and patterning in the mouse. *Genes Dev.* 9:2105–2116.

Yamaguchi, T. P., Conlon, R. A., and Rossant, J. (1992) Expression of the fibroblast growth factor receptor FGFR-1/flg during gastrulation and segmentation in the mouse embryo. *Dev. Biol.* 152:75–88.

Zaret, K. S. (2002) Regulatory phases of early liver development: Paradigms of organogenesis. *Nat. Rev. Genetics* 3:499–512.

Zhu, X., Sasse, J., McAllister, D., and Lough, J. (1996) Evidence that fibroblast growth factors 1 and 4 participate in regulation of cardiogenesis. *Dev. Dyn.* 207:429–438.

Zimmerman, L., De Jesus-Escobar, J., and Harland, R. (1996) The Spemann organizer signal noggin binds and inactivates bone morphogenetic protein 4. *Cell* 86:599–606.

32 Animal Models for Assessing the Contribution of Stem Cells to Liver Development

DOUGLAS C. HIXSON, PhD

A lineage progression from small undifferentiated cells to bipotent cells expressing both biliary and hepatocytic products to unipotent mature hepatocyte or bile duct cells has been identified by a number of laboratories during liver development or in response to injury or carcinogenesis. This leads to a multitiered system of cell renewal designed to provide alternative pathways to liver regeneration that will ensure retention of the reestablishment of liver functionality even when mature hepatocytes and ductal cells have been severely compromised. Recent experiments indicate that the small undifferentiated liver progenitor cell (LPC) may actually be bone marrow derived. Selective activation of LPCs may be accomplished in rodents by combining agents or events that activate proliferation with those that inhibit hepatocyte proliferation. Different injury or carcinogenesis regimens applied to rodents appear to activate the liver lineage cells at different levels including small intraportal undifferentiated cells, bipotent ductal cells, and small hepatocytes. Early lineage (oval) cells also are seen in the livers of humans with chronic liver injury or carcinogen exposure and are a consistent feature of viral hepatocarcinogensis in both animals and humans. Experimental models of chemical hepatocarcinogenesis also implicate LPCs as the cells of origin of hepatocellular carcinoma. Identification of the potential of a cell using transplantation into the liver has been complicated by several confounding factors, but it has been shown that adult hepatocytes have the capacity to replace extensively damaged liver tissue by undergoing a number of cell divisions. This clearly shows that the mature hepatocyte is capable of sustained symmetric divisions, but the proliferation capacity of putative LPCs has not been determined in this fashion. Future directions in using proliferation of endogenous or transplanted stem cells for replacement or gene therapy seem boundless, but much more needs to be known about the liver stem cell lineage and how to control its potential in vivo.

32.1. INTRODUCTION

The concept of bipotent hepatic progenitor cells residing in the cholangioles, the smallest ducts that connect the bile canaliculi to the biliary tree, was postulated nearly 50 yr ago by Wilson and Leduc (1958). This concept was supported by subsequent studies demonstrating the activation and proliferation of small basophilic progenitor cells with ovoid nuclei after physical or chemical injury, or treatment with hepatocarcinogens (Grisham and Porta, 1964; Sell and Salman, 1984; Hixson et al., 1992; Sarraf et al., 1994; Sirica, 1995; Sell, 1998). Because these so-called oval cells were only activated under extreme conditions, they were designated by Grisham and colleagues as "facultative stem cells," a designation suggested by their ability to undergo hepatocyte or ductal cell differentiation (Grisham, 1980). The importance of hepatic stem cells in normal renewal or regenerative processes remains an area of controversy, especially in light of recent studies demonstrating that hepatocytes and bile ductal epithelial cells have an immense capacity for self-renewal that is more than sufficient to regenerate and maintain the size and functional capacity of the liver (Sirica et al., 1994; Overturf et al., 1999). If stem/progenitor cells are involved, their contribution during regeneration and renewal in a healthy liver appears to be relatively small, leading many to question the importance of hepatic stem/progenitor cells since there is no apparent need for them under normal circumstances.

However, under the "extreme conditions" needed for activation (Grisham, 1980), hepatic stem/progenitor cells appear to play an essential role as a fail-safe system that can repair the liver under conditions in which the proliferative capacity of hepatocytes and ductal cells has been compromised. Indeed, recent investigations suggest that there are three, and possibly four, bi- or multipotent cell populations coexisting in the normal adult liver. Primitive periductal progenitors, designated as type 0 by Sell (1998) and as "blast" or "basal" cells by Novikoff (Novikoff and Yam, 1998), are the first cells that undergo DNA synthesis and proliferation following treatment with liver carcinogens (Bisgaard et al., 1996; Novikoff and Yam, 1998; Sell 2001). These cells lack lineage markers and reside either in the portal

From: *Stem Cells Handbook*
Edited by: S. Sell © Humana Press Inc., Totowa, NJ

mesenchyme in close proximity to bile ductules or sequestered inside a basal compartment formed by bile duct epithelial cells. Other bipotent liver progenitors such as those designated type I, type II (classic oval cells), or type III (transitional hepatocytes with ductal and hepatocyte features) are viewed by some as "transit" cells that rapidly proliferate and eventually differentiate into hepatocytes or ducts (Hixson et al., 1992; Sirica, 1992, 1995; Sell, 1998). In relating these adult stem cell types to fetal progenitors in the rat, the 12- to 14-d progenitors described by several groups (Dabeva et al., 2000; Kubota and Reid, 2000; Yin et al., 2002) would seem to most closely resemble type 0 progenitors and the OC.10-positive, bipotent, fetal ductal cells isolated by Simper et al. (2001) from embryonic day (E) 16 fetal livers and the type II progenitors in adult liver. Gordon et al. (2000), Chise et al. (Tateno and Yoshizato, 1996; Tateno et al., 2000), and others have also described what have been called "small hepatocyte-like progenitor" (SHP) cells. SHPs are present in normal liver in low numbers but expand rapidly in retrorsine-treated rats. SHPs would seem to fit the characteristics ascribed by Sell (1998) to type III progenitors because they initially express both hepatocyte and bile ductal markers and seem to be restricted to differentiation along a hepatocyte lineage.

Because hepatocytes and bile ductal epithelial cells appear to have an immense capacity for proliferation and "self-renewal" (Fausto, 1997; Overturf et al., 1999), it has been suggested that they are in reality "unipotent" stem cells (Thorgeirsson, 1996), a concept that seems at odds with the basic properties assigned to true stem cells. For example, proliferation of hepatocytes and bile duct epithelial cells appears to involve symmetric cell division. In essence, the two daughter cells are identical to their parent. True stem cells, by contrast, not only undergo symmetric cell divisions but also divide asymmetrically. This results in self-renewal as well as the production of daughter cells that can undergo further differentiation (Hixson et al., 1992). How asymmetric division is accomplished is only partly understood at present. Recent investigations suggest that this process is mediated by inhibitors or promoters of differentiation that are retained during division by the parental stem cells or the differentiation-committed daughter cell, respectively (Chenn and McConnell, 1995; Knoblich et al., 1995; Mello et al., 1996). Herein lies a second distinction between "stem cells" and hepatocytes—the absence of differentiation and lineage progression, a hallmark of stem cells that is unnecessary for hepatocytes or mature bile duct epithelial cells. Under conditions in which the replicative capacity of hepatocytes has not been compromised, such as restitutive proliferation following partial hepatectomy, regeneration appears to be a purely replicative process mediated by mature hepatocytes or ductal cells (Fausto, 1990, 1997). In addition, true stem cells have a self-renewal capacity that renders them immortal relative to the normal life-span of the animal. Hepatocytes, by contrast, have a finite longevity and, over the course of a year, are completely replaced by "new" hepatocytes through the normal process(es) of renewal (Steiner et al., 1966). In this respect, hepatocytes are more akin to progenitor cells, a bi- or multipotent cell population distinguished from stem cells by their more limited capacity for self-renewal (Fausto, 1990).

When all of the current information on hepatic progenitor cells is viewed as a whole, the picture that emerges is a multitiered system of renewal designed to provide alternative pathways to liver regeneration that will ensure the retention or reestablishment of liver functionality even when mature hepatocytes and ductal cells have been severely compromised. Recent reports documenting the ability of hematopoietic stem cells (HSCs) to differentiate into hepatocytes (Theise, N.D. et al., 2000; Petersen, 2001) indicate that infiltrating HSCs can supplement hepatic stem/progenitor cell–mediated renewal to a currently unknown degree. In this regard, Van der Kooy and Weiss (2000) have made the intriguing suggestion that true hepatic stem cells may not even be present during the initial stages of liver formation but may appear later during organogenesis and thus may be distinct from fetal stem cells that create embryonic tissues and organs. Ductal progenitors corresponding to the type II progenitors of Sell (1998) are consistent with this view because they are not present in the liver primordium but derive from hepatoblasts that enter a ductal lineage following their contact with and subsequent invasion into the portal mesenchyme (Van Eyken et al., 1988; Shiojiri et al., 1991). The same may also be true of hematopoietic progenitors. The yolk sac hematopoietic progenitors that initially colonize the liver are thought to give rise primarily to fetal blood cells such as nucleated red blood cells. Adult HSCs, on the other hand, appear to derive from HSCs that migrate from the aorta–gonad–mesonephros region later in liver development (Medvinsky and Dzierzak, 1996). Adult HSCs may represent yet another stem cell population distinguished from fetal HSCs by their greater developmental flexibility that allows differentiation into hepatocytes, astroglia, and skeletal muscle cells, a plasticity not displayed by fetal or adult HSCs in the fetal microenvironment. As suggested by Van der Kooy and Weiss (2000) the multilineage flexibility permitted by the adult environment may enhance survival. Moreover, because the fetal liver is the major site of hematopoiesis during fetal development and retains sites of hematopoiesis in the adult, it is not surprising that adult HSCs, which are constantly migrating through the liver, would take up long-term residence and thus be available to respond when other front-line systems of renewal and repair have been compromised.

Because hepatic stem/progenitor cells are few and thus not usually apparent in the normal adult liver, development of methods to induce their activation, expansion, and differentiation in a reproducible fashion are essential for understanding the molecular events that determine their ultimate fate, particularly during the course of liver carcinogenesis and chronic disease. In the remainder of this review, I discuss a number of the most commonly used animal models for studying hepatic progenitor cell populations and strategies used for lineage analysis.

32.2. SELECTIVE ACTIVATION OF RODENT LIVER PROGENITOR CELLS

In most animal models, activation of progenitor cells requires treatment with agents that severely compromise the proliferative capacity of hepatocytes but have little or no effect on progenitor cells, presumably because progenitors lack the cytochrome p450s needed to generate mitoinhibitory metabolites. 2-Acetylaminofluorene (2-AAF), a liver carcinogen that is effective in blocking the regenerative capacity of hepatocytes at noncarcinogenic doses, has been used for this purpose in several oval cell induction protocols (Sell et al., 1981; Evarts et al., 1989; Petersen et al., 1997). When coupled with partial hepatectomy (PH) or treatment with CCl_4, 2-AAF induces a moderate oval cell expansion that peaks between 7 and 9 d after PH/CCl_4 (Fig. 1) (Evarts et al., 1996; Petersen et al., 1997). The reproducibility and short induction

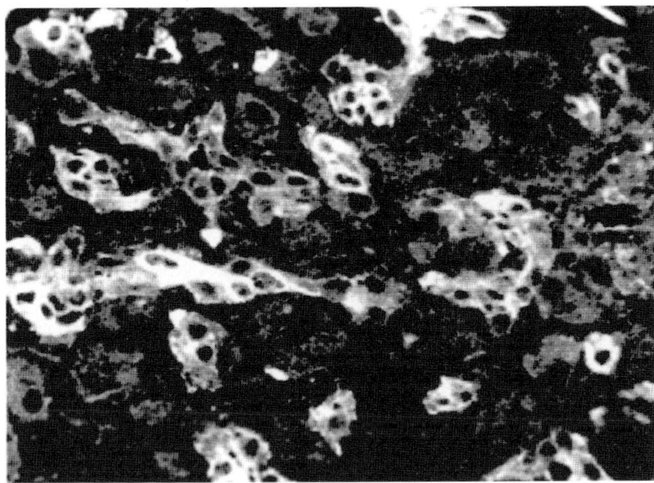

Fig. 1. Oval cells in liver of Fisher rat 7 d after treatment with 2-AAF and partial hepatectomy. Oval cells are brightly stained by indirect immunofluorescence with monoclonal antibody (MAb) OV6, an antibody that recognizes oval cells and bile duct epithelial cells.

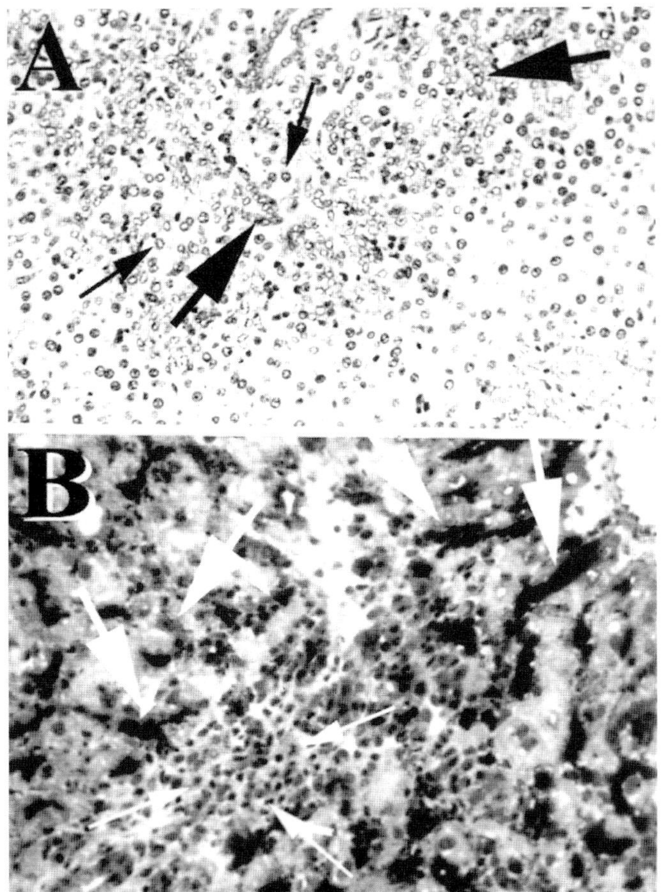

Fig. 2. Oval cells induced by 2-AAF/CD diet. **(A)** H&E-stained frozen section from liver of rat treated with 2-AAF/CD diet for 4 wk. Large arrows designate oval cell ducts and small arrows, individual oval cells. **(B)** Frozen section from same liver stained by a histochemical protocol for γ-glutamyl transpeptidase. Oval cell ducts (large arrows) are strongly positive whereas scattered oval cells (small arrows) are only weakly stained.

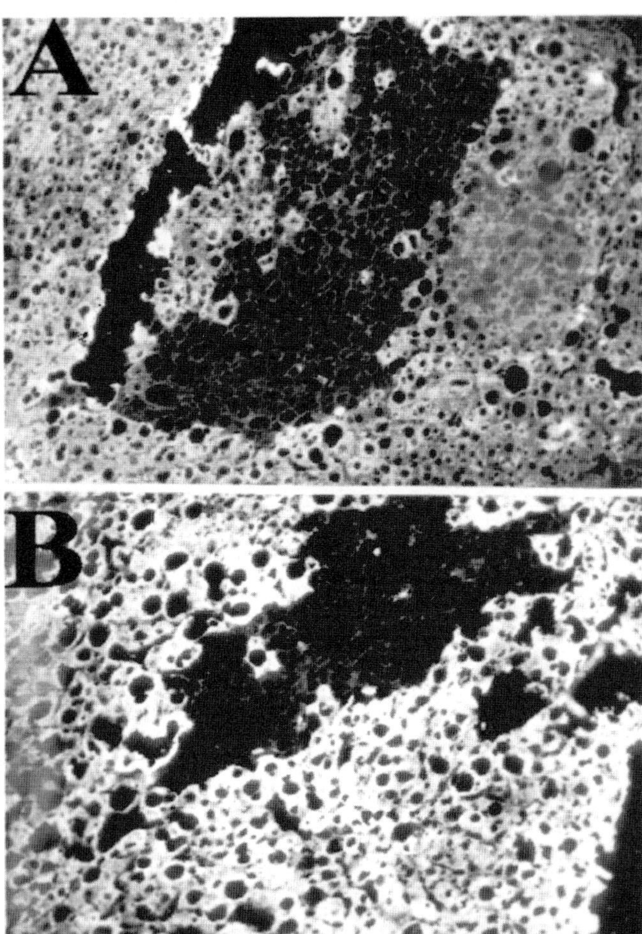

Fig. 3. Frozen sections containing colonies of hepatocytes formed in F_1 Long Evans (LE) X ACI progeny transplanted with ACI oval cells isolated from rats maintained on 2-AAF–CD diet for 3 wk. **(A, B)** Host rats were maintained on the CD diet 1 wk prior to and 12 wk following transplantation. Sections were stained by indirect immunofluorescence. Donor cell colonies appear as dark areas against a background of brightly stained host parenchymal cells. (Photograph provided by Ronald Faris, Rhode Island Hospital, Providence, RI.)

period have led to wide use of this protocol. A considerably larger oval cell expansion can be obtained by combining 2-AAF with a choline-deficient (CD) diet (Sell et al., 1981). With this protocol, as much as 80% of the liver is occupied by oval cells and atypical oval cell ducts within 2–4 weeks (Fig. 2). A major drawback to this system is the high mortality in treated animals (100% by 5 wk), a figure that does not improve significantly when animals are placed on a choline-sufficient diet (Sell et al., 1981). It could be that in the AAF/CD-treated liver, an unfavorable microenvironment shortens the life-span or hinders the differentiation of oval cells, the end result being inadequate liver function due to a failure to regenerate functional hepatocytes. Whatever the mechanism, the lethality appears to require the combined effects of 2-AAF and a CD diet. This conclusion is based on our own studies showing that oval cells recovered from AAF/CD-treated rats form GGT^+ colonies of hepatocytes when transplanted into rats maintained on a CD diet (Faris and Hixson, 1989) (Fig. 3).

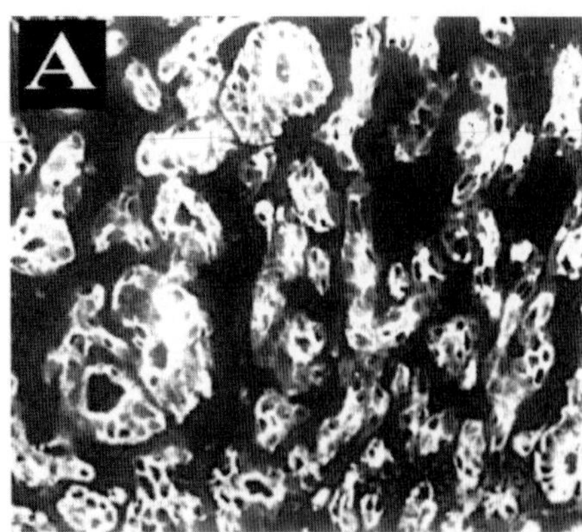

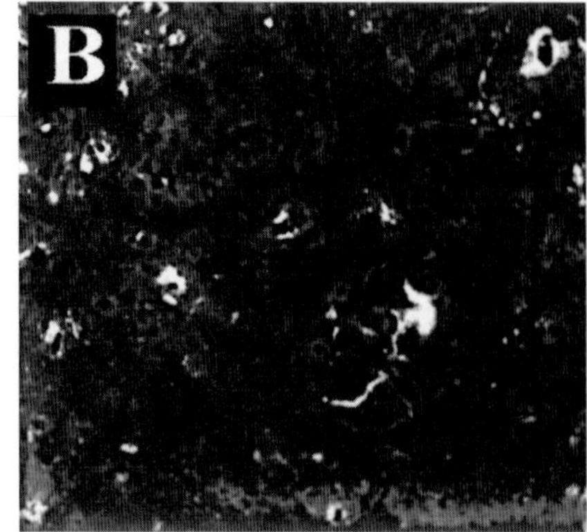

Fig. 4. Oval cells induced by CDE diet. Frozen sections prepared from the liver of rat maintained on the CDE diet for 4 mo were stained by indirect immunofluorescence with MAb OV6 **(A)** or MAb BD.1 **(B)** OV6-positive oval cell ducts were negative for BD.1, a ductal marker expressed on fetal and adult bile duct epithelial cells. (Photograph provided by Li Yang, Emery University, Atlanta, GA.)

Oval cells can also be induced by placing animals on a CD diet containing 0.05–0.1% ethionine (Fig. 4), a low-toxicity protocol with very low mortality (Shinozuka et al., 1978; Sirica and Cihla, 1984; Lenzi et al., 1992; Hixson et al., 1996). Interestingly, a CD diet with or without ethionine destroys the acinar pancreas. This causes the activation of ductal progenitor cells that, for unclear reasons, differentiate into hepatocytes instead of acinar cells (Fig. 5). Ethionine in a CD diet (CDE) appears to act indirectly to cause DNA damage (Rushmore et al., 1986) by elevating levels of reactive lipid peroxidation products, free radicals, and reactive oxygen species. Ethionine is also known to cause changes in gene expression by altering DNA methylation patterns (Shivapurkar et al., 1984), which, in turn, alter differentiation and/or growth regulatory pathways. In addition, there is evidence that a CDE diet may decrease growth inhibitors and increase growth stimulators, a shift that could facilitate oval cell expansion (Lombardi et al., 1985). As one might expect, combining ethionine with a CD diet greatly enhances formation of hepatocellulr carcinoma (HCC) and cholangiocarcinomas (Lombardi and Shinozuka, 1979; Sells et al., 1979), relative to CD diet (Lombardi, 1971; Hoffman, 1984; Counts et al., 1996), which by itself is also carcinogenic in Fischer rats (Chandar and Lombardi, 1988; Nakae et al., 1992). Studies from our and other laboratories have shown that many CDE-induced HCCs express both hepatocyte and oval cell/ductal antigens, a phenotype consistent with a derivation from oval cells arrested at an intermediate stage in hepatocytic differentiation (Hixson et al., 1996, 2000).

From a practical standpoint, protocols that avoid the use of liver carcinogens would be highly desirable, and several of these have been devised. Intraperitoneal injection of galactosamine induces extensive liver damage that results in marked oval cell proliferation beginning within 24 h post injection, peaking at 5 d (Lemire et al., 1991; Dabeva and Shafritz, 1993) and diminishing thereafter as oval cells undergo hepatocytic differentiation or apoptosis. In Long-Evans Cinnamon (LEC) rats, an inbred strain that carries a defect in the Wilson's disease gene, oval cell expansion occurs spontaneously in response to the acute hepatitis that occurs at 20–23 wk of age (Takahashi et al., 1988; Betto et al., 1996). Survivors of acute hepatitis (the end result of copper accumulation caused by the genetic defect) develop chronic hepatitis that ultimately leads to the generation of cholangiocarcinomas and HCCs at a high incidence (Yasui et al., 1997). Yasui et al. (1997) have demonstrated that LEC oval cells differentiate into hepatocytes when transplanted into a normal rat liver, suggesting that at least a portion of HCCs in this model system derive from oval cell progenitors. Ethanol has also been reported to activate oval cells, but the time course of activation differs significantly from CDE or other rapid induction protocols. After 4 wk of treatment with ethanol, oval cells are barely detectable but increase steadily with time thereafter, multiplying by more than 15-fold over the next 23 mo (Smith et al., 1996). By comparison, animals maintained on a CDE diet attain oval cell densities at 4 wk that are fourfold higher than those induced by ethanol after 2 yr.

Although the rat has been the preferred animal model for studying oval cell activation, several excellent mouse models are now available. As in the rat, a number of these take advantage of the oval cell expansion that occurs following treatment with chemical carcinogens, pertinent examples being ethionine in a CD diet (Akhurst et al., 2001), diethylnitrosamine (He et al., 1994), and *N*-nitrosodimethylamine in combination with *Heliobacter hepaticus* (Diwan et al., 1997). SV40 transgenic mice offer a model system not only for studying the role of oval cells in hepatocarcinogenesis but also for examining the factors that lead to their spontaneous activation around wk 10 after birth (Bennoun et al., 1993). Oval cell proliferation also occurs in mouse models of chronic hepatitis induced by infection with cytomegalovirus (Cassell et al., 1998) or treatment with allyl alcohol (Lee et al., 1996). Interestingly, the repair process following exposure to allyl alcohol differs significantly from the rat. For one thing, necrotic areas in the mouse are not restricted to the periportal zone. In addition, restoration of necrotic areas is primarily mediated by hepatocytes instead of oval cells, the major agents of repair in the rat.

Treatment with dipin in combination with partial hepatectomy is another model system used for mouse oval cell activation, expan-sion, and differentiation. Of interest is the close resemblance between the dipin/PH protocol in mice and the more recently developed retrorsine/PH regimen used in rats (Laconi et al., 1998; Gordon et al., 2000). Both dipin and retrorsine are potent alkylating agents capable of causing irreversible damage to liver cell DNA that severely inhibits the replicative capacity of differentiated hepatocytes (Engelhardt al., 1990; Factor et al.,

1994). This creates a dire situation that leads to the activation of proliferation-competent, bipotent progenitors capable of hepatocytic differentiation and restoration of the liver mass. Herein lies a major difference between the two model systems: the nature and origin of these bipotent progenitors. Results from both autoradiographic and morphological analyses indicate that regeneration in dipin/PH treated mice is mediated by oval cells in the canals of Hering that express A6 and A7, two antigens common to oval and biliary epithelial cells, and transitional hepatocytes. By contrast, oval cells appear to play a relatively minor role in the retrorsine/PH model in which reconstitution seems to be mediated almost entirely by a population of SHPs (Gordon et al., 2000). Significantly, these progenitors are also present in normal rat livers, suggesting that they may play a role in normal repair and renewal processes (Tateno et al., 2000). SHPs are distinguished in several respects from oval cells, bile duct epithelial cells, and other liver progenitors. At the time of emergence, SHPs already express the hepatocyte-specific proteins albumin and transferrin, the latter a marker not found on oval cells or other liver progenitors (Gordon et al., 2000b). In addition, early SHPs transiently express the ductal markers OC.2 and OC.5 (Gordon et al., 2000b; Hixson et al., 2000) and display high levels of the transferrin receptor that is not detectable by indirect immunoflourescence (IIF) on oval cells or any other liver epithelial cell types (unpublished observation by D.C. Hixson). SHPs are also distinguished from oval cells and oval cell progenitors by their expression of H.4, a hepatocyte-specific epitope that is never found on oval cells or fetal/adult bile duct epithelial cells (Gordon et al. 2000).

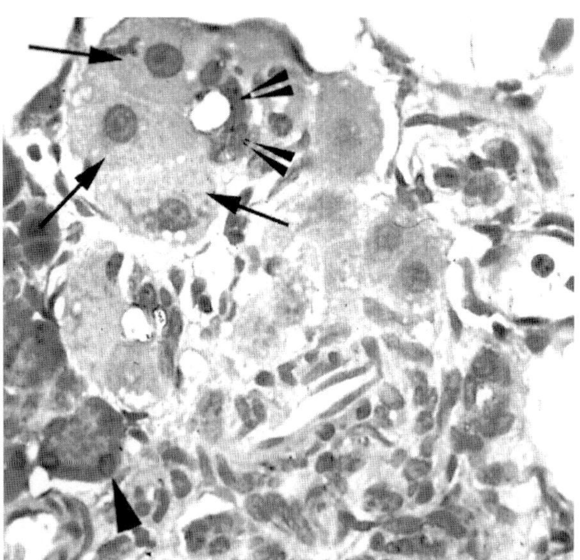

Fig. 5. Hepatocytes in pancreas of rat maintained on CDE diet for 6 mo. Frozen liver section stained by indirect immunoperoxidase with MAb against OC.2, a cell-surface/cytoplasmic antigen expressed on pancreatic ductal cells, oval cells, and bile duct epithelial cells. Split arrows show OC.2-positive ductal cells in a duct that contains three hepatocyes (arrows). The arrowhead designates a degenerating acinus.

32.3. HUMAN MODEL SYSTEMS FOR OVAL CELL ACTIVATION

It is impossible, for obvious reasons, to treat human subjects with agents that induce oval cell proliferation. Consequently, studies of human liver have to rely on biopsy specimens; excess tissue from tumor resections; tissue from liver transplants; or donated livers that, for logistical or immunological reasons, went unused. When taken together, results from a number of studies suggest that progenitor cells arise in the human liver under conditions similar to those that activate their rodent counterparts. As noted above, oval cell expansion is a consistent feature of liver tissue taken from patients infected with hepatitis C virus (HCV) or hepatitis B virus (HBV) (Hsia et al., 1992; Lowes et al., 1999; Dominguez-Malagon and Gaytan-Graham, 2001). Hepatic progenitor cells positive for C-kit and OV6 have been found in liver specimens from pediatric patients diagnosed with biliary atresia, $\alpha 1$ antitrypsin deficiency, or fulminant hepatic failure (Crosby et al., 1998; Baumann et al., 1999). As in rodents, oval cell activation also occurs in humans exposed to aflatoxin, a potent human carcinogen that in mice works synergistically with HBV to produce a more rapid and extensive oval cell proliferation and a shortened time of progression to HCC (Sell et al., 1991). The availability of human liver tissues containing proliferating hepatic progenitors raises the possibility of using strategies developed in rodent model systems for isolating these cells and characterizing their growth and differentiation under both in vivo and in vitro conditions.

32.4. TARGETED LIVER INJURY IN RAT

Hepatotoxins can be valuable tools to amplify rare pathways of differentiation. Their major drawback is a lack of specificity that results in varying degrees of collateral damage, which can often confound the interpretation of results. This is exemplified by 4,4'-methylene dianiline (DAPM), a polycyclic aromatic amine that produces a time- and dose-dependent ductal necrosis (Kanz et al., 1992). In rodents, DAPM treatment has been shown to inhibit carcinogenesis initiated in the liver by ethyl-hydroxy-ethyl-nitrosamine, 2-AAF, and methyl-dimethyl-amino-azobenzene (Fukushima et al., 1981), suggesting that DAPM was destroying initiated ductal progenitors. Support for this conclusion was provided by Peterson et al. (1997), who reported that treatment with DAPM prior to PH greatly attenuated oval cell expansion in response to AAF/PH. In subsequent studies, however, Chapman-Berry and Hixson (2002) did not see the marked inhibition of oval cell expansion under several different DAPM treatment regimens that produced the same time course and extent of biliary injury observed by Petersen et al. (1997). Taken together, these two studies suggest that there are unrecognized variables that influence the effects of DAPM, raising uncertainties about the usefulness of this biliary toxin for assessing the nature and origin of biliary progenitors.

Allyl alcohol, another hepatotoxin that induces progenitor cells is distinguished by its ability to target hepatocytes located in the periportal region (Yavorkovsky et al., 1995). Intraperitoneal injection of allyl alcohol initiates the proliferation of intraportal progenitors lacking both ductal cell and hepatocyte lineage markers. Following expansion into the region of periportal necrosis, these hepatic "stem cells" repair the damage by undergoing hepatocytic differentiation (Yavorkovsky et al., 1995). The restriction of stem cell–mediated regeneration to the periportal region is the key advantage offered by this system. Unfortunately, the marked lobular and intralobular variability in the extent of damage and stem cell activation is a major limitation that hampers analysis and complicates the interpretation of results.

In a series of reports, Sirica and colleagues have shown that under extreme conditions of liver injury or toxicity in which there is extensive damage to existing hepatocytes, progenitors other than oval cells can be activated. These investigators reported that CCl_4 treatment of bile duct–ligated rats produced ductlike structures composed of ductal cells and hepatocyte-like cells in various stages of differentiation, suggesting that under these severe conditions, bile duct epithelial cells (BDECs) had undergone hepatocytic differentiation (Sirica and Williams, 1992; Sirica, 1995). Severely hepatotoxic doses of furan also produced hepatocyte-containing ducts resembling the mixed ducts induced by bile duct ligation (BDL)/CCl_4 treatment (Sirica A. E., et al. 1994). Unfortunately, the high mortality associated with this protocol severely limits its application for lineage analysis at the molecular level. However, at lower nonlethal doses of furan, a very different outcome was observed. Under these conditions, ductal cells in the right and caudate lobes entered into an intestinal lineage leading to the appearance of well-developed intestinal glands (Elmore and Sirica, 1992). Although the mechanism for this dramatic change in lineage commitment remains unclear, it seems likely that furan treatment resulted in the activation of multipotent periductular stem cells. Under the influence of a microenvironment that was apparently unique to the right and caudate lobes, these multipotent progenitors entered an intestinal instead of a hepatocytic pathway of differentiation, a shift similar to that observed with pancreatic ductal cells following depletion/repletion of copper.

32.5. CARCINOGENESIS PROTOCOLS PRODUCING PROGENITOR CELL–DERIVED HCC

Recent studies indicate that progenitor cell activation is a constant feature of viral hepatocarcinogenesis. Patients with chronic HCV and HBV infections often exhibit extensive oval cell proliferation, usually in close proximity to regenerating nodules and HCC (Hsia et al., 1992; Lowes et al., 1999). Approximately 50% of the HCCs found in HBV-positive individuals express both oval cell and hepatocyte markers, a percentage similar to CDE diet–induced carcinomas in the rat. These observations suggest that HCCs resulting from viral carcinogenesis arise from oval cell progenitors. Interestingly, in the woodchuck experimental model systems for viral hepatocarcinogenesis, oval cells are nonpermissive for virus infection (Hsia et al., 1992; Hixson et al., 1996). This suggests that, similar to other induction protocols, activation of progenitors in the woodchuck model results from the ability of the virus to compromise directly or indirectly the viability/replicative capacity of hepatocytes. Indeed, based on their studies of the pathogenesis of HBV in human liver and the livers of mice transgenic for the HBV surface antigen, Dunsford et al. (1990) suggested that chronic liver injury and inflammation and the resulting compensatory mitogenesis play a central role in the carcinogenic process by causing mutations in proliferating hepatocytes or in chronically activated progenitors. A recent report by Lowes et al. (1999) indicates that similar mechanisms are also operative during hepatocarcinogenesis mediated by HCV.

In the rat, the CDE diet (described above) and the Solt–Farber protocol have been the most widely used treatments for studying the progenitor cell origin of HCC. In the Solt–Farber protocol, initiation is achieved by a single ip injection of diethylnitrosamine (DENA). The key element in this regimen is the delivery of a noncarcinogenic dose of 2-AAF to block hepatocyte proliferation following PH at 2 wk after DENA. This selectively promotes the growth of initiated cells that have become "resistant" to the mitoinhibitory activity of 2-AAF (Solt and Farber, 1976; Faris et al., 1991). Since oval cell proliferation occurs to a similar degree and with a similar time course as was described for oval cell induction with 2-AAF followed by PH/CCl_4, the initiation with DENA appears to have little or no effect on the proliferative response of progenitors. Dumble et al. (2002) have recently described a novel liver carcinogenesis model utilizing p53 knockout mice. They showed that oval cells from p53 knockout mice maintained on a CDE diet progressed to HCC following transplantation into nude mice, a finding that provided further support for oval cells as progenitors of HCC.

There are many additional hepatocarcinogens that induce oval cell expansion, but their discussion is beyond the scope of this review. However, it is interesting that much of what is known about the process of liver carcinogenesis (e.g., initiation, promotion, and progression) has been acquired using carcinogenesis protocols that induce minimal activation of progenitor cells and produce almost exclusively HCC derived from initiated hepatocytes (Scherer and Emmelot, 1975; Pitot et al., 1989; Sell and Dunsford, 1989; Farber, 1992). It is not surprising, therefore, that for many years, attention centered on hepatocytes, a bias that diminished the role of hepatic progenitor cells. In reality, however, both the progenitor cell and hepatocyte models of liver carcinogenesis may be closely linked if one considers the possibility that initiated oval cell progenitors undergo differentiation into hepatocytes prior to progression to HCC. This idea is suggested by recent studies demonstrating the ability of transplanted or endogenous oval cells to differentiate into hepatocytes (Golding et al., 1995; Dabeva et al., 1997). It also explains the presence of HCC displaying both hepatocyte and oval cell markers, a phenotype consistent with the blocked ontogeny model proposed by Sell and Pierce (1994).

If HCCs produced by treatment with hepatocarcinogens arise from progenitor cells, it follows necessarily that progenitor cells must become initiated at some point during exposure to the carcinogenic agent. Initiation requires DNA synthesis/cell proliferation and appropriate enzymes for activating the carcinogen. Farber and colleagues have argued that oval cells cannot be progenitors of HCC in the Solt–Farber protocol, because they do not fulfill either of these requirements (Farber, 1992; Gindi et al., 1994) at the time of DENA injection. However, at least two studies, one in rats and one in mice, showed a transient proliferation of oval cells following treatment with DENA (Smith et al., 1990; He et al., 1994). Rapid activation of progenitors also occurs following treatment with other liver carcinogens. Significantly, Bisgaard et al. (1996) and others have shown that oval cells undergo DNA synthesis within 24 h of a *noncarcinogenic* dose of 2-AAF. There are also a number of reports indicating that progenitor cells express cytochrome P450 enzymes (CYPs) needed for activating carcinogens (Weisburger, 1989). Small hepatocytes (an immature stemlike cell population found in retrorsine/PH-treated livers), transitional cells described by Golding et al. (1995), and bile duct epithelial cells all express CYPs capable of metabolizing 2-AAF and DENA, two of the most frequently used liver carcinogens (Degawa et al., 1995; Hagiwara et al., 1996; Cooper and Porter, 2000; Gordon et al., 2000a). In addition, Yang et al. (1995) and Ring et al. (1999) have detected expression of CYPs by human embryonic liver cells at very early stages of gestation, raising the possibility that immature stemlike cells such as oval cells may also have significant levels of CYPs. It is well known that oval cells are a heterogeneous cell population composed of undifferentiated periductal/periportal progenitors and bipotent oval cells in different stages of ductal differentiation (Hixson and Allison, 1985). The most primitive oval cells, designated by Sell as type 0 and type 1 progenitors, would seem to be the least likely to undergo initiation because they are sequestered in a protected niche free of contacts with surrounding cells and are most likely lacking the CYPs needed for carcinogen activation (Novikoff and Yam, 1998; Sell, 1998, 2001). By contrast, their hepatocyte or oval cell progeny, the type II and III progenitors (Sell, 1998), are more likely to express the appropriate CYPs. They will also differentiate and integrate into hepatic cords, a location that will expose them to the promoting agent and hasten their progression to HCC.

Consideration must also be given to the possibility that cells lacking essential CYPs may become initiated by indirect mechanisms. Previous studies by Novikoff and Yam (1998) have shown that oval cells can acquire a dual polarity by forming two apical domains, one with hepatocytes through a shared bile canaliculus and a second as part of the lumen of an oval cell duct. Since the canalicular domain is delineated by specialized junctions shared with hepatocytes, it is conceivable that oval cells lacking appropriate CYPs could be initiated by metabolites produced by their hepatocyte partner. A similar situation may exist between oval cells and bile duct epithelial cells at the interface between oval cells and intralobular ducts. It should also be considered that oval cells are capable of expressing either connexin 43 or connexin 32, the gap junction proteins found on BDEC and hepatocytes (Zhang and Thorgeirsson, 1994; Fallon et al., 1995). This raises the possibility that oval cells may be exposed to carcinogenic metabolites transferred through gap junctions.

32.6. ANIMAL MODELS FOR LINEAGE ANALYSIS

In the past, the fate of progenitor cells has been determined by inference using classic histochemical or immunological methods to identify oval cells and transitional hepatocyte/ductal cells with both oval cell and hepatocyte characteristics (Hixson and Fowler, 1997). One of the most widely used markers for lineage analysis has been α-fetoprotein (AFP), a fetal protein expressed at high levels by oval cells and many HCCs. Since AFP is a secreted protein, serum levels can be used as a minimally invasive means to measure appearance and expansion of AFP^+ oval cells and the growth of AFP^+ HCC. Labeling with tritiated thymidine or bromodeoxyuridine (BrdU) has also been used to provide evidence for hepatocytic differentiation of oval cells (Sell et al., 1981; Allison et al., 1982; Gerlyng et al., 1994; Evarts et al., 1996). In general, the labeled nucleotide is injected at time points during oval cell induction when there is minimal hepatocyte proliferation, thereby restricting labeling to oval cells. This timing is critical since the appearance of hepatocytes labeled with tritium or BrdU is the primary evidence in these studies for the differentiation of oval cells into hepatocytes. Significantly, transference of label from hepatocytes into oval cells or ductal cells has never been convincingly demonstrated, suggesting that the formation of CK19 ducts by hepatocytes in primary culture (Michalopoulos et al., 2001) may be a rare event that occurs with a relatively high frequency in vitro because of the inability to accurately reproduce conditions in vivo.

Several new strategies for following cell fates are based on the creation of subpopulations carrying endogenous or exogenous markers. Transduction in vivo with adenoviral or retroviral vectors has been used successfully by several investigators to transfer marker genes such as β-gal into hepatocytes, bile duct epithelial cells, and other nonparenchymal cell types. Yu et al. (2002) used an adenovirus encoding β-gal to transduce normal and injured rat livers and found that nonparenchymal cells were transduced more efficiently than hepatocytes over a range of adenoviral titers. A major advantage of adenovirus is the high efficiency of transduction, which can approach 100% in vivo and in vitro. However, this high infectivity is counterbalanced by the transient nature of expression (7–14 d) and the possibility of an immune response against viral structural proteins. The latter problem can be overcome with a baculoviral vector carrying a reporter gene driven by a mammalian promoter and viral genes controlled by insect promoters that are inactive in mammalian cells (Boyce and Bucher, 1996; Delaney and Isom, 1998). Although somewhat longer than adenoviral vectors, expression conferred by baculovirus is still transient in nature.

Retroviral vectors, on the other hand, produce long-term, stable expression following integration into the host genome. In vivo, however, there is often extinction or attenuation of the promoter driving transgene expression (Hafenrichter et al., 1994b), a problem that can be overcome greatly by using tissue-specific promoters such as the one for transthyretin, which Ponder and colleagues (Ponder et al., 1991; Ponder, 1996) showed was still driving the expression of activated *ras* in HCC 6 mo after transduction (Lin et al., 1995). To determine whether a retroviral vector could transduce the biliary tree, Cabrera et al. (1996) introduced a Gibbon ape leukemia virus (GALV) carrying the β-gal gene into the biliary tree of bile duct–ligated rats 24 h after ligation. Interestingly, their results showed that at 12 h after infection, most of the β-gal activity was located in CK19-negative peribiliary cells. Although the basis for this selectivity remains to be determined, it could be that Pit 1, a sodium-dependent phosphate transporter that acts as the GALV viral receptor (Macdonald et al., 2000), is differentially expressed on hepatic progenitors and hyperplastic bile duct epithelial cells. If so, this would suggest that Pit1 could be used to identify and isolate hepatic progenitors activated by BDL. In a report by Bralet et al. (Bralet et al., 1994, 1996), a retroviral β-gal vector was used to trace cell lineages in rats maintained on a CD diet containing 0.02% 2-AAF. When transduction was performed at the onset of oval cell proliferation after d 13–24 on the diet, clusters of β-gal-positive hepatocytes were readily detected but labeled oval cells were never observed, leading to the conclusion that the clusters of labeled hepatocytes were derived directly from proliferating hepatocytes and not oval cells. However, one of the limitations of retroviral transduction is the inability of the virus to pass through the basement membrane that surrounds oval cells and bile ducts. As noted by Bralet et al. (Bralet et al., 1994, 1996), there is also the possibility that the expression of the retroviral receptor is regulated by the microenvironment in vivo or by the stage of differentiation. Indeed, retroviral binding sites include receptors for growth factors/cytokines, amino acid transporters, and sodium-dependent phosphate transporters (Richardson and Bank, 1996).

32.7. LINEAGE ANALYSIS IN CHIMERIC LIVERS

As discussed above, transduction with retro- or adenoviral vectors can be used to generate what Cardiff has called *transgenic organs*, a term originally invented to describe mammary glands formed *in situ* by mammary epithelial cells expressing a transgene of interest (Cardiff and Aguilar-Cordova, 1988). In the case of the liver, most strategies produce a mosaic of transgene positive and negative cells and thus fall short of a completely transgenic organ (Ponder, 1996). Another approach to producing organ chimeras that has yielded valuable information regarding organogenesis and cell lineage relationships has been chimeric mice generated by implanting a fusion of two, eight-stage embryos into the uterus of a pseudopregnant female (Iannaccone, 1987). The resulting offspring have organs composed of cells from each of the two parental strains, thereby providing a means to identify morphological features that are clonally derived. Chimeras between mouse strains with H-2k and H-2b haplotypes, e.g., were used to show that individual crypts in the large and small intestine were composed entirely of cells of a single strain, suggesting that crypts were derived by clonal expansion of a single stem cell (Iannaccone, 1987). Villi, by contrast, often contained cells of both haplotypes, indicating that they were formed from progeny produced by two or more crypts. Khokha et al. (1994) used a similar approach to analyze liver organogenesis in chimeras generated from congenic rat strains differing only in their major histocompatibility complex (MHC). Their results suggested that the liver was fashioned with patches of cells from each strain that were fractal in nature. For the liver, this meant that the parenchyma was generated by repetitive application of a simple cell division program that required no bias in the spatial arrangement of daughter cells. The end result was a patchwork of cells from the two mouse strains arranged in patterns that bore no relationship to either the lobular or acinar architecture of the liver. This model of parenchymal organization is thus at odds with both streaming liver and stem cell–fed maturational models in which hepatocytes are thought to continually stream from the portal areas to the hepatic veins, a movement that would require biased siting of newly formed hepatocytes (Arber et al., 1988; Zajicek et al., 1988; Grisham, 1994). An even simpler mosaic liver model that avoids the complexities involved in the formation and implantation of fused embryos was recently reported by Shiojiri et al. (2001). In this model, mosaic livers form spontaneously as a result of random X-chromosome inactivation in female mice heterozygous for a wild type (WT) and an inactive OTC gene carrying the spf^{ash} mutation. Under specific immunofluorescence conditions, cells with WT OTC are intensely fluorescent whereas cells with mutant OTC remain dark. Using this system, Shiojiri et al. (2001) confirmed earlier work by Khokha et al. (1994) and showed that patches of negative or positive cells were connected with each other, forming cell aggregates with no definite orientations regarding to portal areas or central veins.

32.8. TRANSPLANTATION ANALYSIS WITH DONOR CELLS CARRYING ENDOGENOUS OR EXOGENOUS REPORTER GENES

Transplantation of donor cells that can be distinguished from host liver cells by endogenous markers or exogeneous reporter genes is another widely used method for generating chimeric livers suitable for lineage analysis and cell fate determinations. However, interpretation of events in chimeric livers generated by transplantation can be difficult as there are several confounding variables that can greatly influence the results. One issue is the stability of marker gene expression. If the marker is an endogenous gene unique to the donor cells, expression will be regulated in a normal manner. This is usually advantageous but can become a problem if expression of the marker gene is developmentally regulated, a characteristic that may preclude the analysis of early developmental stages. This is exemplified by dipeptidyl peptidase IV (DPPIV), a surface protein that appears late in fetal development (Petell et al., 1990), and is usually not expressed by progenitor cells in vitro. In vivo, expression levels can vary significantly depending on the origin of the promoter driving the marker gene, with tissue-specific promoters providing more stable expression than those of viral origin (CMV, retrovirus LTR).

Assessment of differentiation potential by transplantation also requires the development of protocols for isolating subpopulations of putative progenitors. Purity thus becomes an important consideration because the presence of contaminating cells in significant concentrations complicates interpretation by raising questions about the origin of the engrafted donor cells. In a number of studies, this issue has been resolved by showing that at low doses, the number of contaminating cells cannot account for all of the donor-derived colonies. Purified donor cells must also be able to stably integrate into hepatic cords or ducts, an ability that seems to be inherent in most liver cell types, both normal and malignant. The efficiency of engraftment, however, can vary significantly depending on the origin of the donor cells, the status of the host liver, and the mode of transplantation. Transplantation of donor cells into the liver via the spleen, e.g., increases overall viability relative to portal vein infusion but delivers only 50% of the injected cells to the liver (Rajvanshi et al., 1996). Analysis in the liver can be further complicated by lobular differences in the distribution of donor cells and variation in density related to the distance from the liver hilum.

In spite of these caveats, transplantation continues to be the method of choice for analyzing the differentiation capacity of hepatic progenitor cells. Details of a number of current transplantation models are reviewed in Chapter 36. All of these models, by necessity, share common features. In most models, the donor cells express a gene product that distinguishes them from host cells. Markers used successfully include MHC/alloantigens (Hunt et al., 1982), Y-chromosome (Theise, N. D. et al., 2000; Petersen, 2001), α1 antitrypsin (Hafenrichter et al., 1994a), and DPPIV (Thompson et al., 1991). Models relying on alloantigens, an approach pioneered by Hunt et al. (1982), utilize F1 progeny as hosts for donor cells from either of two inbred strains differing in their MHC haplotypes (ACI and Long-Evans or Wistar Furth and Fischer F344) (Hunt et al., 1982; Faris and Hixson, 1989). Donor cells are subsequently identified using alloantisera produced by immunizing ACI rats or F344 rats with Long-Evans or Wistar Furth spleen cells, respectively. In the past, we have used this model system to demonstrate that oval cells induced by 2-AAF and a CD diet formed GGT^+ colonies of hepatocytes following transplantation into partially hepatectomized F1 hosts (Faris and Hixson, 1989) (Fig. 3). One drawback to tracing donor cell fates using alloantisera is that it is the host cells, not the donor cells, that show positive staining with the alloantisera. Consequently, unstained donor cells have to be discerned against a positive background of host hepatocytes, making it difficult to detect small, single donor cells or even small donor cell clusters.

In a number of recent studies, the fate of donor cells from male rats or mice following transplantation into female hosts has been determined by using flourescent in situ hybridization to detect cells bearing a Y-chromosome (Petersen, 2001). In studies by Thiese, N. et al. (2000), this approach was used to demonstrate that circulating stem cells, most likely of bone marrow origin, can integrate into the liver and differentiate into hepatocytes and cholangiocytes, an observation that challenges current concepts of differentiation and commitment. However, a recent report suggesting that bone marrow cells can fuse spontaneously with other cell types and assume their identity (Terada et al., 2002) implies that the presence of the Y-chromosome in hepatocytes of female hosts could represent a fusion event rather than differentiation of bone marrow stem cells into hepatocytes.

Transplantation models with *Wt* donors and mutant hosts that lack or express an inactive form of a normal liver protein have been used extensively for lineage analysis. Transplantation of *Wt* hepatocytes or progenitors capable of hepatocyte differentiation has been shown to restore serum albumin levels in Nagase analbuminemic rats (Oren et al., 1999). Engrafted donor cells were readily detected in tissue sections by *in situ* hybridization with albumin cDNA probes or immunohistochemically with anti–rat albumin antibodies. One of the advantages of this model system is the ability to quantitate engraftment efficiency, the extent and rate of expansion of donor cells, and the duration of engraftment by measuring changes with time in the serum albumin levels. The DPPIV transplantation model developed by Thompson et al. (Thompson, 1991) offers similar advantages. In this case, the DPPIV-negative Fischer 344 rats used as hosts produce an enzymatically inactive form of DPPIV. This allows the localization of transplanted donor cells from *Wt* Fischer rats by a simple histochemical procedure for active DPPIV or by immunocytochemical staining methods with MAbs that recognize only the active form of DPPIV (Fig. 6). Although DPPIV is a type II transmembrane protein, it is susceptible to cleavage by extracellular proteases that release a soluble 200-kDa enzymatically active fragment into the serum (Shibuya-Saruta et al., 1996). Serum levels of DPPIV can thus be used to measure the same parameters noted above for albumin (Hanski et al., 1986). One limitation of the DPPIV model system is the late expression of the enzyme during fetal development (Petell et al., 1990) and its absence on most hepatic progenitors, a temporal pattern that limits its usefulness for analyzing early time points in liver development or progenitor cell differentiation. This is not an issue for albumin because it is one of the earliest development markers detected in fetal liver.

There are increasing numbers of excellent transplantation models that make use of transgenic mice carrying marker genes as a source of donor cells. These genetically marked cells are transplanted into hosts treated with chemicals that inhibit the proliferation of hepatocytes or into transgenic or mutant strains carrying genetic defects that severely compromise hepatocyte/ductal cell–mediated regeneration and repair. Mignon et al. (1998) transplanted donor cells from transgenic mice expressing human Bcl-2 into immunosuppressed host mice treated with a nonlethal dose of anti-fas/CD95 antibodies to induce apoptosis. Since the donor cells were protected from apoptosis by Bcl-2, they selectively expanded and gradually replaced as much as 16% of the host liver. Rhim et al. (1994, 1995) have used liver cells from β-gal mice to demonstrate the expansion of hepatocytes transplanted into the livers of transgenic mice expressing the urokinase-type plasminogen activator (uPA) under control of the albumin promoter. Overex-pression of uPA is cytotoxic to hepatocytes and greatly compromises their regenerative capacity, giving a large growth advantage to donor β-gal$^+$ hepatocytes or host hepatocytes that lose the uPA transgene. The end result

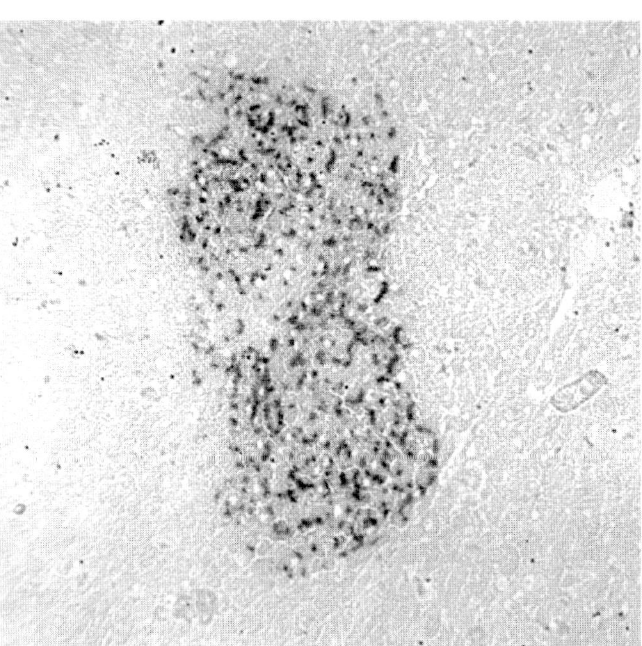

Fig. 6. Colony of DPPIV-positive hepatocytes in liver of DPPIV-deficient rat. Frozen liver sections prepared 2 mo after transplantation of E12–13 fetal liver cells depleted of TuAg1-positive hepatoblasts were stained histochemically for DPPIV. Shown is a colony of donor hepatoblasts were stained histochemically for DPPIV. Shown is a colony of donor hepatocytes strongly positive for DPPIV. (Photograph provided by Rhonda Simper, Rhode Island Hospital, Providence, RI.)

is complete clonal repopulation of the host liver by donor hepatocytes. Overturf et al. (1996, 1999) reported similar findings following transplantation of β-gal-positive mouse liver cells into host mice lacking the enzyme fumarylacetoacetate hydrolase (FAH). The lack of this gene produces symptoms similar to those observed for hereditary tyrosinemia, type 1, a condition that produces extensive liver damage that accelerates and promotes repopulation of the liver by *Wt* β-gal$^+$, FAH$^+$ donor cells. Another relatively new model system that is currently enjoying wide use is the retrorsine model of Laconi et al. (1998, 2001). In this system, DPPIV-negative host rats are pretreated with retrorsine, a DNA alkylating agent that promotes the selective expansion of donor liver cells by impairing the ability of host hepatocytes to proliferate following PH. After several weeks, donor cells repopulate as much as 80% of the liver. Similar effects can also be obtained by pretreatment with galactosamine (Gupta et al., 2000) or radiation, but the degree of repopulation by donor cells is considerably lower.

An unanswered question in all of these animal models is, Does the introduction of exogenous genes into donor cells or the abnormal host environment needed to promote donor cell expansion have any effect on the differentiation capabilities of donor cells? This may be particularly important for putative progenitor cells, which are likely to be more sensitive to changes in microenvironment than mature hepatocytes or ductal cells. Introduction of Bcl2, e.g., may enhance the ability of transplanted cells to survive genetic changes that would normally induce apoptosis, thereby creating a population of damaged cells that may be at higher risk for neoplasia. Wesley et al. (1999) have reported that DPPIV suppresses the malignant behavior of melanoma cells by reversing a block in differentiation and by restoring growth factor–dependent cell survival, activities mediated by serine proteases. Although previous investigations by Coburn et al. (1994) led to the conclusion that DPPIV proteolytic activity was not necessary for immune competence, more recent studies suggest that DPPIV enzymatic activity is essential for certain T-cell activation pathways and is involved in the inactivation of chemokines (Iwata and Morimoto, 1999).

32.9. FUTURE DIRECTIONS

In a lecture at the Experimental Biology 2001 meeting in Orlando, FL, Dr. David Stocum from the University of Indiana suggested that the ultimate goal for stem cell research should be to determine ways to induce regeneration by stem cells *in situ* rather than by transplantation. The ability to promote regeneration by endogenous progenitors eliminates many of the potential problems with transplantation therapy: rejection of the transplanted cells by the recipient, incomplete differentiation, limited availability of hepatic progenitors, poor engraftment, and limited expansion with prior treatment. Transplantation therapy, on the other hand, offers the possibility of ex vivo gene therapy, whose success will depend on the engraftment of transgene-bearing donor cells into a normal liver in sufficient numbers to produce a beneficial effect. Clearly, development of effective stem cell treatments will require an accurate inventory of genes that promote or inhibit stem cell expansion and/or commitment regardless of whether the therapeutic effects are mediated by endogenous or exogenous stem cells. The ability to commit HSCs to a hepatocyte lineage prior to transplantation, e.g., could enhance liver colonization sufficiently to eliminate the need for bone marrow ablation and donor cell engraftment. Similarly, insight into the mechanisms controlling proliferation during development could lead to new methods for inducing expansion following integration into hepatic cords or ducts. It is important, however, not to lose sight of the fact that an understanding of critical molecular events cannot by achieved without first gaining insight into the temporal changes that occur in the developmental potential of progenitor cells as they progress along a hepatocyte or ductal lineage. Identifying periods marked by shifts in developmental or proliferative potential will define time points and cell populations that should be targeted for detailed molecular analysis by proteomic or gene array methodologies. The realization of endogenous stem cell therapy will also require a better understanding of the activators of hepatic progenitors and the expansion and differentiation of their offspring.

Another key issue in the future will be the development of better methods for assessing the extent of differentiation of hepatic progenitors under in vitro or in vivo conditions. Do hepatocytes derived from progenitor cells express the full complement of hepatocyte genes or are they only partially differentiated, a status that could not be discerned by examining a small handful of hepatocyte or ductal markers that may or may not be appropriately expressed? Expression of CK19 and OV6, an epitope shared by CK19 and CK14, e.g., is not always an indicator of a ductal origin, a conclusion suggested by its expression on all of the neoplastic foci and carcinomas generated by the Solt–Farber protocol, most of which are of hepatocyte origin (Sell and Dunsford, 1989; Hixson et al., 1992). The same holds true for albumin and AFP, two early hepatocyte gene products that can also be expressed by immature ductal cells. Functional differentiation is another important consideration. Even if donor cells express a large number of hepatocyte-specific genes, they may fail to perform critical functions if proteins needed for important biosynthetic pathways are not expressed or are expressed inappropriately. It is also essential that donor cells respond to the microenvironmental cues that regulate cell growth and death, thus, creating a steady state that maintains the size and lobular structure of the liver. In this regard, there is evidence that hepatocyte colonies derived from fetal donor cells do not require a selective growth advantage such as that conferred by retrorsine/PH but continue to increase in size even when transplanted into a normal liver (Sandhu et al., 2001). This continued expansion is not observed with mature hepatocytes, suggesting a defect in the microenvironment or in the programming that prevents fetal progenitors from maintaining a G_0 status. Work by Curran et al. (1993) and Boylan and Gruppuso (1994) suggests a possible mechanism for this defect. These investigators have found that around gestation d 20, fetal hepatoblasts stop proliferating as they transition from a growth factor–independent to a growth factor–dependent status. Thus, donor fetal liver cells harvested at gestation d 14 have never undergone this fetal/neonatal transition, which may involve the acquisition of a key step in growth regulation. This possibility is supported by results reported by Simper et al. (2001) showing that gestation d 14 fetal hepatoblasts were composed of two cell types—TuAg1$^+$ hepatoblasts that integrated but failed to grow in the adult liver and TuAg1$^-$ fetal cells that formed numerous large DPPIV-positive colonies (Fig. 6).

A final issue that needs further study is the ability to distinguish between phenotypic modulation and true differentiation. As my laboratory and others have shown, true differentiation involves the loss or gain of tissue-associated genes in a well-ordered and invariant temporal process (Haruna et al., 1996; Hixson et al., 1996). This process has been demonstrated for T-cells; B-cells; keratinocytes; mammary epithelial cells; and, most recently, bile duct epithelial cells. Phenotypic modulation, on the other hand, does not appear to follow well-defined patterns. Wu et al. (1994) found that when a continuous line of mouse hepatocytes was maintained in serum-containing medium, hepatocyte-specific gene expression was rapidly extinguished. After only 24 h in serum-free medium, however, a well-differentiated hepatocyte phenotype was restored. These findings suggest that one way to distinguish between phenotypic modulation and true differentiation is to determine whether the appearance of tissue-specific markers occurs in a fashion that recapitulates fetal liver ontogeny. Differentiation in experimental systems must also be defined in the context of the microenvironment to which the cells of interest are exposed. This is exemplified by pancreatic ductal cells, which can switch to a hepatocyte lineage when the acinar pancreas has been destroyed by copper depletion, thereby exposing ductal cells to a vastly different microenvironment (Reddy et al., 1991).

All of these concepts are testable using currently available technologies and the animal models described in this review. Completeness of differentiation in vitro and in vivo could be tested using the DPPIV transplantation model. DPPIV-specific antibodies could be used to isolate DPPIV-positive hepatocytes and compare them by proteomic or gene microarray analysis to DPPIV-host hepatocytes isolated from the same liver. This would be particularly interesting for donor hepatocytes derived from fetal progenitors. The issue of phenotypic modulation vs a temporal process of differentiation could be analyzed using similar techniques.

REFERENCES

Akhurst, B., Croager, E. J., et al. (2001) A modified choline-deficient, ethionine-supplemented diet protocol effectively induces oval cells in mouse liver. *Hepatology* 34(3):519–522.

Allison, J. P., Hixson, D. C., et al. (1982) *Monoclonal Antibodies as Probes of Surface Antigenic Alterations During Experimental Carcinogenesis in the Rat.* Elsevier, Amsterdam.

Arber, N., Zajicek, G., et al. (1988) The streaming liver II. Hepatocyte life history. *Liver* 8:80–87.

Baumann, U., Crosby, H., et al. (1999) Expression of the stem cell factor receptor c-kit in normal and diseased pediatric liver: identification of a human hepatic progenitor cell? *Hepatology* 30(1):112–117.

Bennoun, M., Rissel, M., et al. (1993) Oval cell proliferation in early stages of hepatocarcinogenesis in Simian Virus 40 large T transgenic mice. *Am. J. Pathol.* 143:1326–1336.

Betto, H., Kaneda, K., et al. (1996) Development of intralobular bile ductules after spontaneous hepatitis in Long-Evans mutant rats. *Lab. Invest.* 75(1):43–53.

Bisgaard, H. Nagy, C., P., et al. (1996) Proliferation, apoptosis, and induction of hepatic transcription factors are characteristics of the early response of biliary epithelial (oval) cells to chemical carcinogens. *Hepatology* 23:62–70.

Boyce, F. M. and Bucher, N. L. (1996) Baculovirus-mediated gene transfer into mammalian cells. *Proc. Natl. Acad. Sci. USA* 93(6): 2348–2352.

Boylan, J. and Gruppuso, P. (1994) *In vitro* and *in vivo* regulation of hepatic mitogen-activated protein kinases in fetal rats. *Am. J. Physiol.* 267:G1078–G1086.

Bralet, M. P., Branchereau, S., et al. (1994) Cell lineage study in the liver using retroviral mediated gene transfer: evidence against the streaming of hepatocytes in normal liver. *Am. J. Pathol.* 144: 896-905.

Bralet, M.-P., Calise, D., et al. (1996) *In vivo* cell lineage analysis during chemical hepatocarcinogenesis using retroviral-mediated gene transfer. *Lab. Invest.* 74(5):871–881.

Cabrera, J. A., Wilson, J. M., et al. (1996) Targeted retroviral gene transfer into the rat biliary tract. *Somatic Cell. Mol. Genet.* 22(1):21–29.

Cardiff, R. and Aguilar-Cardova, E (1988) Proto-neoplasia revisited: the molecular biology of mouse mammary hyperplasia. *Anticancer Res.* 8:925—933.

Cassell, H. S., Price, P., et al. (1998) The association between murine cytomegalovirus induced hepatitis and the accumulation of oval cells. *Int. J. Exp. Pathol.* 79(6):433–441.

Chandar, N. and Lombardi, B. (1988) Liver cell proliferation and incidence of hepatocellular carcinomas in rats fed consecutively a choline-devoid and a choline-supplemented diet. *Carcinogenesis* 9:259–263.

Chenn, A. and McConnell, S. K. (1995) Cleavage orientation and the asymmetric inheritance of Notch1 immunoreactivity in mammalian neurogenesis. *Cell* 82(4):631–641.

Coburn, M. C., Hixson, D. C., et al. (1994) *In vitro* immune responsiveness of rats lacking active Dipeptidylpeptidase IV. *Cell. Immunol.* 158:269–280.

Cooper, M. T. and Porter, T. D. (2000) Mutagenicity of nitrosamines in methyltransferase-deficient strains of salmonella. *Mutat. Res.* 454: 45–52.

Counts, J. L., Sarmiento, J. I., et al. (1996) Cell proliferation and global methylation status changes in mouse liver after phenobarbital and/or choline-devoid, methionine-deficient diet administration. *Carcinogenesis* 17(6):1251–1257.

Crosby, H. A., Hubscher, S. G., Joplin, R. E., et al. (1998) Immunolocalization of OV-6, a putative progenitor cell marker in human fetal and disease pediatric liver. *Hepatology* 28:980–985.

Curran, T. R. J., Bahner, R. I., et al. (1993) Mitogen-independent DNA synthesis by fetal rat hepatocytes in primary culture. *Exp. Cell Res.* 209:53–57.

Dabeva, M. D. and Shafritz, D. A. (1993) Activation, proliferation, and differentiation of progenitor cells into hepatocytes in the D-galactosamine model of liver regeneration. *Am. J. Pathol.* 143:1606–1620.

Dabeva, M., D., Hwang, S.-G., et al. (1997) Differentiation of pancreatic epithelial progenitor cells into hepatocytes following transplantation into rat liver. *Proc. Natl. Acad. Sci. USA* 94:7356–7361.

Dabeva, M., D., Petkov, P. M., et al. (2000) Proliferation and differentiation of fetal liver epithelial progenitor cells after transplantation into adult rat liver. *Am. J. Pathol.* 156(6):2017–2031.

Degawa, M., Miura, S., et al. (1995) Altered expression of hepatic CYPIA enzymes in rat hepatocarcinogenesis. *Jpn. J. Cancer Res.* 86 (6):535–539.

Delaney, W. and Isom, H. (1998) Hepatitis B virus replication in human HepG2 cells mediated by hepatitis B virus recombinant baculovirus. *Hepatology* 28(4):1134–1146.

Diwan, B. A., Ward, J. M., et al. (1997) Promotion by Helicobacter hepaticus-induced hepatitis of hepatic tumors initiated by N-nitrosodimethylamine in male A/JCr mice. *Toxicol. Pathol.* 25(6): 597–605.

Dominguez-Malagon, H. and Gaytan-Graham, S. (2001) Hepatocellular carcinoma: an update. *Ultrastruct. Pathol.* 25(6):497–516.

Dumble, M. L., Croager, E. J., et al. (2002) Generation and characterization of p53 null transformed hepatic progenitor cells: oval cells give rise to hepatocellular carcinoma. *Carcinogenesis* 23(3):435–445.

Dunsford, H. A., Sell, S., et al. (1990) Hepatocarcinogenesis due to chronic liver cell injury in hepatitis B virus transgenic mice. *Cancer Res.* 50(11):3400–3407.

Elmore, L. W. and Sirica, A. E. (1992) Sequential appearance of intestinal mucosal cell types in the right and caudate liver lobes of furan treated rats. *Hepatology* 16:1220–1226.

Engelhardt, N. V., Factor, V. M., et al. (1990) Common antigens of mouse oval and biliary epithelial cells: expression on newly formed hepatocytes. *Differentiation* 45(1):29–37.

Evarts, R. P., Nagy, P., et al. (1989) In vivo differentiation of rat liver oval cells into hepatocytes. *Cancer Res.* 49(6):1541–1547.

Evarts, R. P., Hu, Z., et al. (1996) Precursor-product relationship between oval cells and hepatocytes: comparison between tritiated thymidine and bromodeoxyuridine as tracers. *Carcinogenesis* 17:2143–2151.

Factor, V. M., Radaeva, S. A., et al. (1994) Origin and fate of oval cells in Dipin-induced hepatocarinogenesis in the mouse. *Am. J. Pathol.* 145(2):409–422.

Fallon, M., Nathanson, M., et al. (1995) Altered expression and function of hepatocyte gap junctions after commmon bile duct ligation in the rat. *Am. J. Pathol.* 268:C1186–C1194.

Farber, E. (1992) On cells of origin of liver cell cancer. In: *The Role of Cell Types in Hepatocarcinogenesis.* (Sirica, A. E., ed.), CRC, Boca Raton, FL, pp. 1–28.

Faris, R. A. and Hixson, D. C. (1989) Selective proliferation of chemically altered rat liver epithelial cells following hepatic transplantation. *Transplantation* 48(1):87–92.

Faris, R. A., Monfils, B. A., et al. (1991) Antigenic relationship between oval cells and a subpopulation of hepatic foci, nodules, and carcinomas induced by the "resistant hepatocyte" model system. *Cancer Res.* 51(4):1308–1317.

Fausto, N. (1990) Hepatocyte differentiation and liver progenitor cells. *Curr. Opin. Cell Biol.* 2:1036–1042.

Fausto, N. (1997) Hepatocytes break the rules of senescence in serial transplantation studies: is there a limit to their replicative capacity? *Am. J. Pathol.* 151(5):1187–1189.

Fukushima, S., Hirose, M., et al. (1981) Inhibitory effect of 4,4'-diaminodiphenylmethane on liver, kidney and bladder carcinogenesis in rats ingesting N-ethyl-N-hydroxyethylnitrosamine or N-butyl-N-(4-hydroxybutyl) nitros-amine. *Carcinogenesis* 2(10): 1033–1037.

Gerlyng, P., Grotmol, T., et al. (1994) Flow cytometric investigation of a possible precursor-product relationship between oval cells and parenchymal cells in the rat liver. *Carcinogenesis* 15:53–59.

Gindi, T., Ghazarian, D. M. D., et al. (1994) An origin of presumptive preneoplastic foci and nodules from hepatocytes in chemical carcinogenesis in rat liver. *Cancer Lett.* 83:75–80.

Golding, M., Sarraf, C. E., et al. (1995) Oval cell differentiation into hepatocytes in the acetylaminofluorene-treated regenerating rat liver. *Hepatology* 22:1243–1253.

Gordon, G. J., Coleman, W., Grisham, J. W., et al. (2000a) Temporal analysis of hepatocyte differentiation by small hepatocyte-like progenitor cells during liver regeneration in retrorsine-exposed rats. *Am. J. Pathol.* 157(3):771–786.

Gordon, G. J., Coleman, W. B., Hixson, D. C., Grisham, J. W. (2000b) Liver regeneration in rats with retrorsine-induced hepatocellular injury proceeds through a novel cellular response. *Am. J. Pathol.* 156(2):607–619.

Grisham, J. (1994) Migration of hepatocytes along hepatic plates and stem cell-fed hepatocyte lineages. *Am. J. Pathol.* 144(5):849–854.

Grisham, J. W. (1980) Cell types in long-term propagable cultures of rat liver. *Ann. NY Acad. Sci.* 349:128–137.

Grisham, J. W. and Porta, E. A. (1964) Origin and fate of proliferated hepatic ductal cells in the rat: electron microscopic and autoradiographic studies. *Exp. Mol. Pathol.* 3:242–261.

Gupta, S., Rajvanshi, P., et al. (2000) Cell transplantation causes loss of gap junctions and activates GGT expression permanently in host liver. *Am. J. Physiol. Gastrointest. Liver Physiol.* 279(4):G815–G826.

Hafenrichter, D. G., Ponder, K. P., et al. (1994a) Liver-directed gene therapy: evaluation of liver specific promoter elements. *J. Surg. Res.* 56(6):510–517.

Hafenrichter, D. G., Wu, X., et al. (1994b) Quantitative evaluation of liver-specific promoters from retroviral vectors after in vivo transduction of hepatocytes. *Blood* 84(10):3394–3404.

Hagiwara, A., Matsuda, T., et al. (1996) Dose-related increased in quantitative values for altered hepatocytic foci and cytochrome P-450 levels in the livers of rats exposed to phenobarbital in a medium-term bioassay. *Cancer Lett.* 110:155–162.

Hanski, C., Zimmer, T., et al. (1986) Increased activity of dipeptidyl peptidase IV in serum of hepatoma-bearing rats coincides with the loss of the enzyme from the hepatoma plasma membrane. *Experientia* 42:826–828.

Haruna, Y., Saito, K., et al. (1996) Identification of bipotential progenitor cells in human liver development. *Hepatology* 23(3):476–481.

He, X. Y., Smith, G. J., et al. (1994) Short-term diethylnitrosamine-induced oval cell responses in three strains of mice. *Pathology* 26(2): 154–160.

Hixson, D. C. and Allison, J. P. (1985) Monoclonal antibodies recognizing oval cells induced in the liver of rats by N-2-fluorenylacetamide or ethionine in a choline-deficient diet. *Cancer Res.* 45(8): 3750–3760.

Hixson, D. C. and Fowler, L. C. (1997). Development and phenotypic heterogeneity of intrahepatic biliary epithelial cells. In; *Biliary and Pancreatic Ductal Epithelium.* (Sirica, A. E. and Longnecker, D. S. eds.), Marcel Dekker, NY, pp. 1–40.

Hixson, D. C., Faris, R. A., et al. (1992). Antigenic clues to liver development, renewal and carcinogenesis. In: *The Role of Cell Types in Hepatocarcinogenesis.* (Sirica, A. E., ed.), CRC, Boca Raton, FL, pp. 151–182.

Hixson, D. C., Affigne, S., et al. (1996) Delineation of antigenic pathways of ethionine induced liver cancer in the rat. *Pathobiology* 64:79–90.

Hixson, D. C., Brown, J., et al. (2000) Differentiation status of rat ductal cells and ethionine-induced hepatic carcinomas defined with surface-reactive monoclonal antibodies. *Exp. Mol. Pathol.* 68:152–169.

Hoffman, R. M. (1984) Altered methionine metabolism, DNA methylation and oncogene expression in carcinogenesis. A review and synthesis. *Biochim. Biophys. Acta* 738:49–87.

Hsia, C. C., Evarts, R. P., et al. (1992) Occurrence of oval-type cells in hepatitis B virus-associated human hepatocarcinogenesis. *Hepatology* 16(6):1327–1333.

Hunt, J. M., Buckley, M. T., et al. (1982) Liver cell membrane alloantigens as cellular markers in genotypic mosiac rat livers undergoing chemically-induced hepatocarcinogenesis. *Cancer Res.* 42:227–236.

Iannaccone, P. (1987) The study of mammalian organogenesis by mosaic pattern analysis. *Cell Differ.* 21:79–91.

Iwata, S. and Morimoto, C. (1999) CD26/Dipeptidyl peptidase IV in context: the different roles of a multifunctional ectoenzyme in malignant transformation. *J. Exp. Med.* 190:301–305.

Kanz, M., Kaphalia, L., et al. (1992) Methylene dianiline: acute toxicity and effects on biliary function. *Tox. App. Pharm.* 117:88–97.

Khokha, M. K., Landini, G., et al. (1994) Fractal geometry in rat chimeras demonstrates that a repetitive cell division program may generate liver parenchyma. *Devel. Biol.* 165:545–555.

Knoblich, J. A., Jan, L. Y., et al. (1995) Asymmetric segregation of Numb and Prospero during cell division. *Nature* 377(6550):624–627.

Kubota, H. and Reid, L. M. (2000) Clonogenic hepatoblasts, common precursors for hepatocytic and biliary lineages, are lacking classical major histocompatibility complex class 1 antigen. *Proc. Natl. Acad. Sci. USA* 97:12,132–12,137.

Laconi, E., Oren, R., et al. (1998) Long-term, near total liver replacement by transplantation of isolated hepatocytes in rats treated with retrorsine. *Am. J. Pathol.* 153:319–329.

Laconi, S., Pillai, S., et al. (2001) Massive liver replacement by transplanted hepatocytes in the absence of exogenous growth stimuli in rats treated with retrorsine. *Am. J.Pathol.* 158(2):771–777.

Lee, J. H., Ilic, Z., et al. (1996) Cell kinetics of repair after allyl alcohol-induced liver necrosis in mice. *Int. J. Exp. Pathol.* 77(2):63–72.

Lemire, J. M., Shiojiri, N., et al. (1991) Oval cell proliferation and the origin of small hepatocytes in liver injury induced by D-galactosamine. *Am. J. Pathol.* 139(3):535–552.

Lenzi, R., Liu, M. H., et al. (1992) Histogenesis of bile duct-like cells proliferating during ethionine hepatocarcinogenesis: evidence for a biliary epithelial nature of oval cells. *Lab. Invest.* 66(3):390–402.

Lin, Y.-Z., Brunt, E. M., et al. (1995) *Ras*-transduced diethylnitrosamine-treated hepatocytes develop into cancer of mixed phenotype *in vivo*. *Cancer Res.* 55:5242–5250.

Lombardi, B. (1971) Effects of choline deficiency on rat hepatocytes. *Fed. Proc.* 30(1):139–142.

Lombardi, B., Ove, P., et al. (1985) Endogenous hepatic growth-modulating factors and effects of a choline-devoid diet and of phenobarbital on hepatocarcinogenesis in the rat. *Nutr. Cancer* 7(3):145–154.

Lombardi, B. and Shinozuka, H. (1979) Enhancement of 2-acetylaminofluorene liver carcinogenesis in rats fed a choline-devoid diet. *Int. J. Cancer* 23:565–570.

Lowes, K., Brennan, B., et al. (1999) Oval cell numbers in human chronic liver diseases are directly related to disease severity. *Am. J. Pathol.* 154:537–541.

Macdonald, C., Walker, S., et al. (2000) Effect of changes in expression of the amphotropic retroviral receptor PiT-2 on transduction efficiency and viral titer: implications for gene therapy. *Hum. Gene Ther.* 11(4): 587-595.

Medvinsky, A. and Dzierzak, E. (1996) Definitive hematopoiesis is autonomously initiated by the AGM region. *Cell* 86(6):897–906.

Mello, C. C., Schubert, C., et al. (1996) The PIE-1 protein and germline specification in C. elegans embryos. *Nature* 382(6593):710–712.

Michalopoulos, G. K., Bowen, W. C., et al. (2001) Histological organization in hepatocyte organoid cultures. *Am. J. Pathol.* 159(5):1877–1887.

Mignon, A., Guidotti, J. E., et al. (1998) Selective repopulation of normal mouse liver by Fas/CD95-resistant hepatocytes. *Nat. Med.* 4: 1185–1188.

Nakae, D., Yoshiji, H., et al. (1992) High incidence of hepatocellular carcinomas induced by a choline deficient L-amino acid defined diet in rats. *Cancer Res.* 52:5042–5045.

Novikoff, P. M. and Yam, A. (1998) Stem cells and rat liver carcinogenesis: contributions of confocal and electron microsccopy. *J. Histochem. Ctochem.* 46:613–626.

Oren, R., Dabeva, M. D., et al. (1999) Restoration of serum albumin levels in nagase analbuminemic rats by hepatocyte transplantation. *Hepatology* 29:75–81.

Overturf, K., Al-Dhalimy, M., et al. (1996) Hepatocytes corrected by gene therapy are selected in vivo in a murine model of hereditary tyrosinaemia type I. *Nat. Genet.* 12(3):266–273.

Overturf, K., Al-Dhalimy, M., et al. (1999) The repopulation potential of hepatocyte populations differing in size and prior mitotic expansion. *Am. J. Pathol.* 155:2135–2143.

Petell, J. K., Quaroni, A., et al. (1990) Alteration in the regulation of plasma membrane glycoproteins of the hepatocyte during ontogeny. *Exp. Cell Res.* 187:299–308.

Petersen, B. E. (2001) Hepatic "stem" cells: coming full circle. *Blood Cells Mol. Dis.* 27(3):590–600.

Petersen, B. E., Zajac, V. F., et al. (1997) Bile ductular damage induced by methylene dianiline inhibits oval cell activation. *Am. J. Pathol.* 151 (4):905–909.

Pitot, H. C., Campbell, H. A., et al. (1989) Critical parameters in the quantitation of the stages of initiation, promotion, and progression in one model of hepatocarcinogenesis n the rat. *Toxicol. Pathol.* 17(4): 594–612.

Ponder, K. P. (1996) Analysis of liver development, regeneration, and carcinogenesis by genetic marking studies. *FASEB J.* 10(7):673–682.

Ponder, K. P., Dunbar, R. P., et al. (1991) Evaluation of relative promoter strength in primary hepatocytes using optimized lipofection. *Hum. Gene Ther.* 2(1):41–52.

Rajvanshi, P., Kerr, A., et al. (1996) Studies of liver repopulation using the dipeptidyl peptidase IV–deficient rat and other rodent recipients: cell size and stucture relationships regulate capacity for increased transplanted hepatocyte mass in the liver lobule. *Hepatology* 23(3): 482–496.

Reddy, J. K., Rao, M. S., et al. (1991) Pancreatic hepatocytes: an *in vivo* model for cell lineage in pancreas of adult rat. *Dig. Dis. Sci.* 36(4): 502–509.

Rhim, J. A., Sandgren, E. P., et al. (1994) Replacement of diseased mouse liver by hepatic cell transplantation. *Science* 263:1149–1152.

Rhim, J. A., Sandgren, E. P., et al. (1995) Complete reconstruction of mouse liver with xenogeneic hepatocytes. *Proc. Natl. Acad. Sci. USA* 92(May):4942–4946.

Richardson, C. and Bank, A. (1996) Developmental-stage-specific expression and regulation of an amphotropic retroviral receptor in hematopoietic cells. *Mol. Cell. Biol.* 16(8):4240–4247.

Ring, J. A., Ghabrial, H., et al. (1999) Fetal hepatic drug elimination. *Pharmacol. Ther.* 84:429–445.

Rushmore, T. H., Farber, E., et al. (1986) A choline-devoid diet, carcinogenic in the rat, induces DNA damage and repair. *Carcinogenesis* 7 (10):1677–1680.

Sandhu, J. S., Petkov, P. M., et al. (2001) Stem cell properties and repopulation of the rat liver by fetal liver epithelial progenitor cells. *Am. J. Pathol.* 159(4):1323–1334.

Sarraf, C., Lalani, E.-N., et al. (1994) Cell behavior in the Acetylaminofluorene-treated regenerating rat-liver. *Am. J. Pathol* 145(5):1114–1125.

Scherer, E. and Emmelot, P. (1975) Kinetics of induction and growth of precancerous liver-cell foci, and liver tumour formation by Diethylnitrosamine in the rat. *Eur. J. Cancer* 11:689–696.

Sell, S. (1998) Comparison of liver progenitor cells in human atypical ductular reactions with those seen in experimental models of liver injury. *Hepatology* 27:317–331.

Sell, S. (2001) Heterogeneity and plasticity of hepatocyte lineage cells. *Hepatology* 33(3):738–750.

Sell, S. and Dunsford, H. A. (1989) Evidence for the stem cell origin of hepatocellular carcinoma and cholangiocarcinoma. *Am. J. Pathol.* 134:1347–1363.

Sell, S. and Pierce, G. B. (1994) Maturation arrest of stem cell differentiation is a common pathway for the cellular origin of teratocarcinomas and epithelial cancers. *Lab. Invest.* 70(1):6–22.

Sell, S. and Salman, J. (1984) Light- and electromicroscopic autoradiographic analysis of proliferating cells during the early stages of chemical hepatocarcinogenesis in the rat induced by feeding N-2-fluorenylacetamide in a choline deficient diet. *Am. J. Pathol.* 114: 287–300.

Sell, S., Leffert, H. L., et al. (1981) Rapid development of large numbers of alpha-fetoprotein-containing "oval" cells in the liver of rats fed N-2-fluorenylacetamide in a choline-devoid diet. *Gann* 72(4):479–487.

Sell, S., Hunt, J. M., et al. (1991) Synergy between hepatitis B virus expression and chemical hepatocarcinogens in transgenic mice. *Cancer Res.* 51(4):1278–1285.

Sells, M. A., Katyal, S. L., et al. (1979) Induction of foci of altered, gamma-glutamyltranspeptidase-positive hepatocytes in carcinogen-treated rats fed a choline-deficient diet. *Br. J. Cancer* 40(2): 274–283.

Shibuya-Saruta, H., Kasahara, Y., et al. (1996) Human serum dipeptidyl peptidase IV (DPPIV) and its unique properties. *J. Clin. Lab. Anal.* 10(6):435–440.

Shinozuka, H., Lombardi, B., et al. (1978) Enhancement of DL-ethionine-induced liver carcinogeneis in rats fed a choline-devoid diet. *J. Natl. Cancer Inst.* 61:813–817.

Shiojiri, N., Inujima, S., et al. (2001) Cell lineage analysis during liver development using the spf(ash)-heterozygous mouse. *Lab. Invest.* 81(1):17–25.

Shiojiri, N., Lemire, J. M., et al. (1991) Cell lineages and oval cell progenitors in rat liver development. *Cancer Res.* 51(10):2611–2620.

Shivapurkar, N., Wilson, M. J., et al. (1984) Hypomethylation of DNA in ethionine-fed rats. *Carcinogenesis* 5:989–992.

Simper, R., McBride, A., et al. (2001) Phenotypic analysis of putative embryonic progenitors for rat hepatocyte and ductal cells. *FASEB J.* 15:A943.

Sirica, A., Cole, S., et al. (1994b) A unique rat model of bile ductular hyperplasia in which liver is almost totally replaced with well-differentiated bile ductules. *Am. J. Pathol.* 144:1257–1268.

Sirica, A. E., ed. (1992). *The Role of Cell Types in Hepatocarcinogenesis*, CRC, Boca Raton, FL,

Sirica, A. E. (1995) Ductular hepatocytes. *Histol. Histopathol.* 10: 433–456.

Sirica, A. E. and Cihla, H. P. (1984) Isolation and partial characterizations of oval and hyperplastic bile ductular cell-enriched populations from the livers of carcinogen and noncarcinogen-treated rats. *Cancer Res.* 44(8):3454–3466.

Sirica, A. E. and Williams, T. W. (1992) Appearance of ductular hepatocytes in rat liver after bile duct ligation and subsequent zone three necrosis by carbon tetrachloride. *Am. J. Pathol.* 40: 129–136.

Sirica, A. E., Gainey, T. W., et al. (1994) Ductular hepatocytes, evidence for a bile ductular cell origin in furan-treated rats. *Am. J. Pathol.* 145:375–383.

Smith, G. J., Kunz, H. W., et al. (1990) Histopathology and cell culture characteristics of liver cells from grc- and grc+ rats given diethylnitrosamine [published erratum appears in *Cell Biol. Toxicol.* 1990;6(4):423]. *Cell Biol. Toxicol.* 6(2):205–217.

Smith, P. G., Tee, L. B., et al. (1996) Appearance of oval cells in the liver of rats after long-term exposure to ethanol. *Hepatology* 23(1): 145–154.

Solt, D. and Farber, E. (1976) New principle for the analysis of chemical carcinogenesis. *Nature* 263:701–703.

Steiner, J., Perz, A., et al. (1966) Cell proliferation dynamics in the liver: a review of quantitative morphological techniques applied to the stud of physiological and pathological growth. *Exp. Mol. Pathol.* 5: 146–181.

Takahashi, H., Oyamada, M., et al. (1988) Elevation of serum alpha-fetoprotein and proliferation of oval cells in the livers of LEC rats. *Jpn. J. Cancer Res.* 79(7):821–827.

Tateno, C. and Yoshizato, K. (1996) Growth and differentiation in culture of clonogenic hepatocytes that express both phenotypes of hepatocytes and biliary epithelial cells. *Am. J. Pathol.* 149:1593–1605.

Tateno, C., Takai-Kajihara, K., et al. (2000) Heterogeneity of growth potential of adult rat hepatocytes in vitro. *Hepatology* 31:65–74.

Terada, N., Hamazaki, T., et al. (2002) Bone marrow cells adopt the phenotype of other cells by spontaneous cell fusion. *Nature* 416 (6880):542–545.

Theise, N. D., Nimmakayalu, M., et al. (2000) Liver from bone marrow in humans. *Hepatology* 32:11–16.

Theise, N., Badve, S., et al. (2000) Derivation of hepatocytes from bone marrow cells in mice after radiation-induced myeloablation. *Hepatology* 31(1):235–240.

Thompson, N. L., Hixson, D. C., et al. (1991) A Fischer rat substrain deficient in dipeptidyl peptidase IV activity makes normal steady state RNA levels and an altered protein. *Biochem. J.* 273:497–502.

Thorgeirsson, S. S. (1996) Hepatic stem cells in liver regeneration. *FASEB J.* 10:1249–1256.

Van der Kooy, D. and Weiss, S. (2000) Why stem cells? *Science* 287(5457):1439–1441.

Van Eyken, P., Sciot, R., et al. (1988) Intrahepatic bile duct developement in the rat: a cytokeratin-immunohistochemical study. *Lab. Invest.* 59(1):52–59.

Weisburger, E. K. (1989). Chemical carcinogenesis in experimental animals and humans. In: *The Patholbiology of Neoplasia*. (Sirica, A. E., ed.), Plenum Press, New York, pp. 39–52.

Wesley, U. V., Albino, A. P., et al. (1999) A role for dipeptidyl peptidase IV in suppressing the malignant phenotype of melanocytic cells. *J. Exp. Med.* 190(3):311–322.

Wilson, J. W. and Leduc, E. H. (1958) Role of cholangioles in restoration of the liver of the mouse after dietary injury. *J. Pathol. Bacteriol.* 76: 441–449.

Wu, J. C., Merlino, G., et al. (1994) Establishment and characterization of differentiated, nontransformed hepatocyte cell lines derived from mice transgenic for transforming growth factor α. *Proc. Natl. Acad. Sci. USA* 91:674–678.

Yang, H., Namkung, M., et al. (1995) Expression of functional cytochrome P4501A1 in human embryonic hepatic tissues during organogenesis. *Biochem. Pharmacol.* 49:717–726.

Yasui, O., Miura, N., et al. (1997) Isolation of oval cells from long-evans cinnamon rats and their transformation into hepatocytes *in vivo* in the rat liver. *Hepatology* 25:329–334.

Yavorkovsky, L., Lai, E., et al. (1995) Participation of small intraportal stem cells in the restitutive response of the liver to periportal necrosis induced by allyl alcohol. *Hepatology* 21(6):1702–1712.

Yin, L., Sun, M., et al. (2002) Derivation, characterization, and phenotypic variation of hepatic progenitor cell lines isolated from adult rats. *Hepatology* 35(2):315–324.

Yu, Q., Que, L. G., et al. (2002) Adenovirus-mediated gene transfer to nonparenchymal cells in normal and injured liver. *Am. J. Physiol. Gastrointest. Liver Physiol.* 282(3):G565–G572.

Zajicek, G., Ariel, L., et al. (1988) The streaming liver III. Littoral cells accompany the streaming hepatocyte for the biliary tree. *Liver* 8:213–218.

Zhang, M. and Thorgeirsson, S. (1994) Modulation of connexins during differentiation of oval cells into hepatocytes. *Exp. Cell Res.* 213:37–42.

33 Normal Liver Progenitor Cells in Culture

KATHERINE S. KOCH, PhD AND HYAM L. LEFFERT, MD

The stem cell nature of normal liver progenitor cells (LPCs) is addressed by studies of normal LPCs in culture. Several questions are addressed such as: What are the patterns of proliferation, lineage commitment, differentiated gene expression, plasticity, and responses to epigenetic and environmental signals? Early studies were interpreted to show that propagable LPCs were derived from dedifferentiated or retrodifferentiated mature liver cells. The recognition that oval cells seen in hepatocarcinogenesis models in rats had characteristics of LPCs suggested that these cells might actually be LPCs or descendants of LPCs. A comparison of more than 30 publications over three decades reporting explants; clonal lines; fresh isolates or strains of cells from noncarcinogen-exposed normal mouse, rat, pig, and human liver; or embryonic tissues indicates that small, immature LPCs, which have the plasticity to mature into ductal cells or hepatocytes, can be obtained from embryonic and fetal tissues, as well as adult liver. A wide variation in the methods of isolation, culture media, feeder layers, growth factors, and substrata used to study putative LPCs in vitro makes comparisons of results from different laboratories difficult. Although the liver is endodermally derived, putative coexpression of primitive hematopoietic and hepatocytic markers is consistent with LPCs in hepatic as well as in blood-forming tissues. In addition to bile duct and hepatocytic differentiation, LPCs have been reported to express markers of pancreatic and endothelial cells in vitro, and to differentiate into bile ducts, hepatocytes, pancreatic islet and acinar epithelial cells, intestinal epithelial cells, and cardiac myocytes after transplantation in vivo. Culture of LPCs on STO embryonic fibroblast feeder layers maintains primitive phenotypes, but requirements of feeder layers appear not to be absolute and are poorly understood. Emerging trends suggest HGF, Flt-3 ligand, SCF, EGF, and DMSO promote hepatocyte differentiation of LPCs, and that transforming growth factor-β, Na^+-butyrate, and culture on Matrigel promote biliary differentiation; however, exceptions have been reported. Critical studies on proliferation kinetics have not convincingly shown self-renewal and asymmetric cell division expected of tissue stem cells, but long-term doublings (up to 150 generations) without spontaneous transformation suggest considerable growth potential. The source of LPCs in normal liver remains unknown and controversial. LPCs may be derived from a liver tissue progenitor cell located in the duct or periductal tissue; from retrodifferentiation of more mature hepatocytes; from bone marrow–derived cells, which circulate through the liver; or from bone marrow remnants of intrahepatic embryonic development. Given the lack of well-defined markers for LPCs, incomplete knowledge of their growth characteristics and regulation signals, their apparent heterogeneity, their apparent plasticity, the possibilities of transdifferentiation or retrodifferentiation of other cells to LPCs, or fusion of LPCs with other cells, as well as their potential for tumorigenesis, much research needs to be conducted to understand what LPCs are and how to use them.

33.1. INTRODUCTION

Current views about mammalian stem cells are based on classic studies of murine hematopoiesis in radiation chimeras. These studies showed that adult stem cells are sparse; undifferentiated; and, following asymmetric cell division, capable of self-renewal and commitment to one or more lineages of cells that proliferate into differentiated tissue (Till and McCulloch, 1961; Siminovitch et al., 1963; Till et al., 1963; McCulloch, 2003). Nuclear transplantation and cloning studies have expanded these views (Gurdon et al., 1975; Campbell et al., 1996), and it now appears that, given the proper environment, stem cell–like behavior is intrinsic even to fully differentiated cells from many distinct animal tissues (Byrne et al., 2002; Hochedlinger and Jaenisch, 2002).

In support of the physiological prevalence and diversity of stem cells, investigations during the last three decades suggest that liver stem cells can be obtained not only from primordial foregut endoderm (Le Douarin, 1975), but also from blastocyst inner cell mass (ICM) (Jones et al., 2002); fetal, neonatal, and adult liver (reviewed in Sell, 2001); extrahepatic endodermally

From: *Stem Cells Handbook*
Edited by: S. Sell © Humana Press Inc., Totowa, NJ

derived pancreas (Rao et al., 1986; Dabeva et al., 1997; Wang et al., 2001) and mesodermally derived adult bone marrow (Petersen et al., 1999; Lagasse et al., 2000; Theise et al., 2000a, 2000b; Avital et al., 2001).

Although liver stem cell systems express hepatic and nonhepatic markers in vitro and in vivo, rigorous and consistent evidence of retention of a full complement of stem cell properties is lacking in most systems. Thus, we refer to extant systems of liver stem cells as *liver progenitor cells* (LPCs), and we pose the following question: Are the patterns of proliferation, lineage commitment, differentiated gene expression, plasticity, and responses to epigenetic and environmental signals in LPCs consistent with authentic stem cell behavior? Studies of normal LPCs in culture are one approach toward solving the authenticity problem, and they have begun to provide answers to questions regarding this problem.

Normal uncommitted and self-renewing LPCs might have been cultured inadvertently more than 30 yr ago in attempts to grow authentic hepatocytes in vitro. These attempts were intensive, because many investigators believed—correctly, in retrospect—that growth control studies with normal hepatocytes in vitro would provide key insights into the physiological control of liver regeneration in vivo (Leffert and Paul, 1972, 1973; Koch and Leffert, 1974, 1979, 1994; Leffert, 1974a, 1974b; Leffert and Koch, 1978; Kruijer et al., 1986; Lu et al., 1992).

In one laboratory, e.g., adult rat livers were dissociated with trypsin, and the surviving cells were plated into plastic tissue culture dishes using standard media. Clusters of rapidly proliferating epithelial cells were observed, but the cells did not look or function like hepatocytes. Instead, they were characterized by small diameters (<10 μm), relatively high nuclear:cytoplasmic ratios, "clear" cytoplasms, and the inability to express ornithine transcarbamylase (OTC), an intrahepatic marker of fully differentiated hepatocytes (unpublished observations). Propagable OTC$^-$ cell lines with similar features and limited hepatocytic functions were reported concurrently from adult rat (Coon, 1969; Gerschenson et al., 1970; Iype, 1971; Williams et al., 1971; Chessebeuf et al., 1974; Takaoka et al., 1975; Grisham et al., 1975) and human liver (Kaighn and Prince, 1971). Nevertheless, since differentiated primary hepatocytes displayed properties of limited proliferation (Leffert and Paul, 1972, 1973; Koch and Leffert, 1974; Leffert, 1974a, 1974b; Leffert et al., 1977, 1978b) and retrodifferentiation (Leffert et al., 1978b; Sirica et al., 1979), most investigators believed that propagable rat liver epithelial cell lines were derived from dedifferentiated or retrodifferentiated adult hepatocytes, not from LPCs (Bissell, 1976; Herring et al., 1983; Tsao et al., 1984).

This belief was altered by observations that suggested lineal relationships between hepatic "oval" cells and differentiated hepatocytes. Parallel in vitro cell culture and in vivo physiological studies of α_1-fetoprotein (AFP) production played significant roles in this paradigm shift. First, it was observed that AFP, an abundant fetal plasma protein expressed constitutively in the yolk sac and in most hepatomas (Abelev, 1968), is also expressed in a growth state–dependent manner in normal hepatocytes. Expression was repressed in quiescent hepatocytes, constitutive in normal proliferating fetal and neonatal hepatocytes, and reactivated in subpopulations of normal adult hepatocytes during retrodifferentiation after 70% hepatectomy or proliferative growth transitions in vitro (Leffert and Sell, 1974; Sell et al., 1975, 1976; Leffert et al., 1978a; Koch and Leffert, 1979, 1980). Second, elevated plasma AFP levels induced by hepatocarcinogens (Watabe, 1971; Becker and Sell, 1974) were traced immunohistochemically to hepatic AFP$^+$ oval cells *in situ* (7–15 μm in diameter), not to hepatocytes or to hepatic nodule cells (Sell, 1978; Sell et al., 1981b). Third, small periportal "oval" cells were massively induced in the livers of adult rats fed choline-deficient (CD) diets supplemented with the hepatoprocarcinogen *N*-2-acetylaminofluorene (AAF-CD diets). Like normal differentiated fetal hepatocytes in vitro (Watabe et al., 1976), the oval cells induced by such dietary regimens proliferated rapidly and coexpressed AFP and albumin (Sell et al., 1981a, 1981b). They also repopulated damaged liver by extensive migration through liver sinusoids, they displayed high nuclear : cytoplasmic ratios, and they appeared to differentiate into hepatocytic- and bile duct–like cells.

Taken together, these findings suggested that oval cells behaved like LPCs or descendants of LPCs. Accordingly, initial investigations in vitro, motivated by efforts to evaluate cellular lineages in the pathogenesis of hepatocellular carcinoma, shifted to attempts to culture nonparenchymal liver "oval" cells. Soon, authentic oval cells were isolated from the livers of AAF-CD-treated adult rats and grown in primary culture, where they expressed AFP and proliferated rapidly (Sell and Leffert, 1982). Oval-like cells were also cultured from normal adult rat livers under enzymatic harvesting conditions designed to destroy hepatocytes (Grisham, 1980; Koch and Leffert, 1980; Marceau et al., 1980; Herring et al., 1983). In fact, one investigator concluded that oval-like rat liver epithelial cells originated from "hepatic stem cells" (Grisham, 1980), a speculation unsupported by the available evidence at the time.

This review focuses on "normal" LPCs in culture. Sixteen different systems of clonal lines and cell strains, and freshly isolated dispersed cells, cell clusters, or tissue explants are summarized in Table 1. Except for "small" and "large" adult hepatocytes, and a genetically engineered system of embryonic stem (ES) cells, we define normal LPCs as wild-type nonparenchymal liver or extrahepatic cells derived from tissues that have *not* been exposed to known carcinogens or mutagens before or during cell isolation and cell plating. More information about liver oval cells or LPCs cultured from carcinogen- or mutagen-treated animals, or similar cells cultured from the livers of transgenic or knockout mice, is provided elsewhere (Sell and Leffert, 1982; Tsao et al., 1985; Sirica et al., 1990; Steinberg et al., 1994; Hixson et al., 1997; Isfort et al., 1997; Lazaro et al., 1998; Spagnoli et al., 1998; Dumble et al., 2002).

33.2. CELL CULTURE: SOURCES, SYSTEMS AND METHODS

In this section, we consider the animal and tissue sources of cultured LPCs, the 16 specific systems reported thus far, and the methods used to isolate, plate, passage, and store the various cells.

33.2.1. SOURCES

33.2.1.1. Animals Normal LPCs have been obtained from various animals including inbred mice and rats, abattoir pigs, and humans (*see* Table 1). To our knowledge, there are no reports to date of normal LPC cultures from fish, reptiles, amphibians, or birds.

33.2.1.2. Tissues Normal LPCs have been cultured from various tissues: blastocyst ICM (ES cells); dispersed cells from untreated fetal, neonatal, and adult liver, or from chemically treated adult liver; and suspensions of adult bone marrow. Embryonic foregut endoderm and fetal liver organ explants have also been investigated (*see* Table 1).

Table 1
Normal LPCs in Culture

CL, FI, S[a]	*Initial Phenotypic Markers*	*Species of Origin*	*Tissue sources*	*Reported plasticity pathways* *In vitro*	*In vivo*
*Gtar*I.114 (CL)[1]	AFP+, ALB+, Transferrin+	Mouse	ES cells (blastocyst ICM)	Hepatocytic	To be determined
FI (explant)[2]	AFP+, ALB+, TTR+	Mouse	E8–E8.5 ventral foregut endoderm	Hepatoblast	To be determined
HBC-3 (CL)[3]	AFP+, ALB+ CK14+	Mouse	E9.5 liver diverticulum	Hepatocytic Bile duct	To be determined
FI (explant)[4]	AFP+, ALB+, *c-kit*+	Mouse	E9.5 liver diverticulum	Hepatocytic Bile duct	To be determined
FI (dispersed cells)[5]	AFP+, ALB+	Rat	E12 fetal liver	Hepatocytic Bile duct	To be determined
FI (dispersed cells)[6]	AFP+, ALB+, ICAM-1+, OX18^{low}, RT1A^{l-}	Rat	E13 fetal liver	Hepatocytic Bile duct	To be determined
H-CFU-H (dispersed clones)[7]	ALB−, CD49f$^{+/LOW}$, CD45−, CK19−, *c-kit*−, *c-met*+, TER119−	Mouse	E13.5 fetal liver	Hepatocytic Bile duct	Hepatocytic Bile duct Pancreatic Intestinal
FNRL/NRLM (CL, S)[8]	AFP−, ALB−, Transferrin−, G-6-Pase−, H-4−, OV-6+	Rat (♂)	Neonatal liver	Hepatocytic	Hepatocytic (intrasplenic only)
WB-F344 (CL)[9]	AFP+, ALB+, DPPIV−, GGTweak+, G-6-Pase−	Rat (♂)	Adult liver (terminal ducts?)	Hepatocytic Bile duct	Hepatocytic Cardiac myocyte
LA (CL)[10]	AFP+, ALB−, *c-kit*−, *c-met*+, Cx26−, Cx32+, Cx43+, A6+	Mouse	Adult liver	Hepatocytic	To be determined
NPEC (dispersed clusters)[11]	AFP+, ALB+, CK7+, CK18+, *c-met*+, OV6+, Transferrin+	Pig	Adult liver	Hepatocytic Bile duct	To be determined
FI (dispersed cells)[12]	ALB+, CK8+ CK18+, Transferrin+	Rat	Parenchymal Adult Liver (small hepatocytes)	Hepatocytic Bile duct	To be determined
FI (dispersed cells)[13]	ALB+, P450 IIB1+	Rat	Adult hepatocytes (large hepatocytes)	Hepatocytic Bile duct	To be determined
13-1 (CL)[14]	ALB^{weak+}, CK14+, OV-6+	Rat	D-Galactosamine-treated adult liver	Hepatocytic	To be determined
FI (dispersed cells)[15]	CD34+, *c-kit*+, OV-6+	Human	Adult liver	Bile duct Endothelial (?)	To be determined
FI (dispersed cells)[16]	β_2-m−, Thy-1+, IL-3+	Rat	Adult bone marrow	Hepatocytic	Hepatocytic

[a]CL, clonal line; FI, fresh isolate; S, Strain(s). The LPC systems described are listed in the order of developmental time scales (embryonic → fetal → neonatal → adult). The references are as follows: [1]Jones et al. (2002); [2]Jung et al. (1999), Deutsch et al. (2001), Rossi et al. (2001); [3]Rogler (1997); [4]Monga et al. (2001), Shiojiri and Mizuno (1993); [5]Germain et al. (1988); [6]Kubota and Reid (2000); [7]Suzuki et al. (2002); [8]Herring et al. (1983), McMahon et al. (1986), Ott et al. (1999); [9]Tsao et al. (1984); [10]Richards et al. (1997); [11]Kano et al. (2000); [12]Tateno and Yoshizato (1996a, 1996b), Mitaka et al. (1999); [13]Block et al. (1996); [14]Dabeva and Shafritz (1993), Petkov et al. (2000); [15]Crosby et al. (1998, 2001); [16]Avital et al. (2001).

33.2.2. SYSTEMS

33.2.2.1. ES Cells The differentiated fate of cultured murine ES cells can be followed by trypsinizing undifferentiated ES cells from solid surfaces and replating them onto bacterial Petri dishes to induce formation of free-floating embryoid bodies (EBs) (Rosenthal et al., 1970). Similar experiments were performed with normal ES cells genetically tagged with I.114, a gene trap vector insertion element that contains a β-galactosidase (β-gal) reporter cassette downstream of a consensus splice acceptor site (Jones et al., 2002). I.114 integrated 3' to an endogenous promoter of the ankyrin repeat-containing gene, *Gtar*; in this configuration, it provided a reporter-based method for the specific detection of early hepatocyte differentiation in developing embroyonic day (E) E10.5 liver bud (Watt et al., 2001). AFP and albumin (ALB) mRNA expression were detected by reverse transcriptase polymerase chain reaction (RT-PCR) 3 and 8 d, respectively, after the in vitro induction of EB formation; transferrin was also detected by immunofluorescence at 4 d in β-gal$^+$ hepatocyte-like cells. Bile duct markers were not reported.

33.2.2.2. Embryonic Explants of Gut Endoderm Investigations with chicken (Le Douarin, 1975; Parlow et al., 1991) and mouse embryos (Crossley and Martin, 1995; Zhu et al., 1996) have shown that liver cells develop from unspecified ventral foregut endoderm. These observations were extended in vitro with anterior portions of mouse E8–E8.5 embryo explants. The endoderm was separated from cardiac mesoderm, transferred to microwell cultures, and examined for tissue-specific differentiation (Jung et al., 1999). Following treatment with purified polypeptide growth factors (*see* Subheading 33.4.2.), three hepatocyte-specific functions were detected by RT-PCR: ALB, AFP, and transthyretin (TTR). None of these functions were detected under similar conditions in neural tube, midsection, or head explants. These experiments suggest that ventral foregut endoderm is a source of "hepatoblasts."

33.2.2.3. Fetal Rodent Liver Clonal, explant, and dispersed cell cultures of LPCs have been investigated at different stages of fetal development. To varying extent, all of these systems have provided LPCs that generate hepatocytic or bile duct–like cells in vitro.

For example, surgically dissected explants of mouse E9.5 liver diverticulum were employed to make a clonal AFP$^+$ ALB$^+$ CK14$^+$ line, HBC-3 (Rogler, 1997). Confluent 3-wk-old HBC-3 cells cultured on feeder layers displayed ultrastructural and enzymatic markers of hepatocytes; on Matrigel, bile duct–like structures were formed.

Similar explants were also investigated en bloc in more complex systems (Shiojiri and Mizuno, 1993) using the marker PRAJA-I, a RING finger protein restricted to liver expression in putative AFP$^+$ ALB$^+$ *c-kit*$^+$ hepatic stem cells (Monga et al., 2001).

Dispersed rat E12 (Germain et al., 1988) or E13 (Kubota and Reid, 2000) AFP$^+$ ALB$^+$ liver cell systems were investigated either in mass cultures or following flow cytometry fluorescence-activated cell sorting (FACS) for cell membrane major histocompatibility complex (MHC) class I, ICAM-I, and integrin β_1 markers in putative clonal cultures, respectively. In the former system, parallel in vivo liver marker studies were performed *in situ*, suggesting that the in vitro findings were not artifacts of isolation and culture. In the latter system, which reported LPCs lacking MHC class I antigens, parallel *in situ* marker studies were not performed. Dispersed mouse E13.5 ALB$^-$ CK19$^-$ liver cells, separated by FACS for surface markers CD49f$^{+/LOW}$ CD45$^-$ *c-kit*$^-$ *c-met*$^+$ TER119$^-$, were also examined in clonal cultures, and the clones obtained (hepatic colony-forming unit in culture [H-CFU-C]) showed properties of self-renewal (Suzuki et al., 2002). After 5–21 d in vitro, only a small fraction of H-CFU-C expressed ALB and/or CK19, whereas most colonies remained ALB$^-$ CK19$^-$. Nevertheless, the in vitro–generated uncommitted or dedifferentiated clones were capable of differentiating into liver, pancreas, and intestinal epithelial cells following transplantation in vivo (*see* Subheading 33.3.5.).

33.2.2.4. Neonatal Rodent Liver Nonparenchymal liver epithelial cells, rederived from 10-d-old neonatal rats ("NRLM"; McMahon et al., 1986) according to standard procedures ("FNRL"; Herring et al., 1983), have been shown to express OV-6 (Ott et al., 1999), a well-known oval cell marker (Dunsford and Sell, 1989). FNRL cells also expressed restricted hepatocytic markers under special conditions in vitro (*see* Subheading 33.4.3.) and in vivo (*see* Subheading 33.3.5.).

33.2.2.5. Untreated Adult Rodent Liver

33.2.2.5.1. Nonparenchymal Epithelial Cells WB-F344, a clonal line of rat liver epithelial cells, and its oval cell–like derivatives have been studied systematically for more than 20 yr in vitro (Grisham, 1980; Lee et al., 1989; Grisham et al., 1993) and in vivo (Coleman et al., 1993, 1997; Malouf et al., 2001). Initially characterized by immunocytochemical and histochemical staining as AFP$^+$ ALB$^+$ GGT^{weak+} (Tsao et al., 1984), WB-F344 cells expressed hepatocytic and cardiac myocyte markers following transplantation (*see* Subheading 33.3.5.).

Another oval-like cell system, a clonal mouse LPC line designated LA (or MOCLA1), has also been reported (Richards et al., 1997). The LA line was developed as a control to study the growth regulatory behavior of a tetratricopeptide repeat–containing protein encoded by the liver tumor suppressor gene *Tg737* (Richards et al., 1996; Isfort et al., 1997). LA cells expressed A6, a mouse oval and erythroid lineage cell marker (Engelhardt et al., 1993), and, as determined by RT-PCR, AFP, Cx32 and Cx43, and *c-met* mRNAs.

In a third system, clusters of porcine AFP$^+$ ALB$^+$ OV-6$^+$ nonparenchymal epithelial cells (NPECs) have been harvested from 6- to 7-mo-old pigs and examined in primary cultures (Kano et al., 2000). As determined by immunocytochemistry and electron microscopy, these clusters underwent morphogenetic changes over 9 d in vitro suggestive of differentiation into hepatocytic and bile duct–like cells.

33.2.2.5.2. "Small" and "Large" Hepatocytes Growth control studies of hepatocytes in primary culture, using serum and nicotinamide-supplemented arginine-free media, have demonstrated that fetal hepatocytes are smaller and proliferate faster than adult hepatocytes, that the expression of hepatocyte-differentiated functions including functional bile canalicular transport is growth state dependent, that nonparenchymal cell survivors insinuate themselves underneath and beside parenchymal cells (unpublished observations), that hepatocyte "scattering" (defined as the movement of hepatocytes away from monolayer aggregates, along with morphological cellular changes from cuboidal to flattened structures with ruffled edges) occurs early in the proliferative log phase, and that adult rat hepatocytes display size heterogeneity and a capacity to retrodifferentiate (Leffert and Paul, 1972, 1973; Leffert, 1974a; Leffert and Sell, 1974; Sell et al., 1975; Leffert et al., 1977, 1978a, 1983; Leffert and Koch, 1979, 1980, 1985; Sirica et al., 1979; Sell and Leffert, 1982).

Recent work in three different laboratories has extended these findings and suggested that in primary culture "small" and "large-sized" adult hepatocytes might have properties of LPCs and a lineal relationship between them. Two systems of "small" hepatocytes have been investigated. In one, colonial growth was followed in "small" hepatocytes (Tateno and Yoshizato, 1996a, 1996b): Growth state dependence of adult phenotypic markers was observed, and some hepatocytes coexpressed mature hepatocyte (ALB) and biliary markers (CK7 and CK19). In the other system (Mitaka et al., 1999), in which mass cultures of "small" and "large" hepatocytes, as well as Kupffer, sinusoidal endothelial, and stellate cells, were examined. About 10% of $CK8^+$ cells formed colonies of "small" ALB^+CK18^+transferrin$^+$ hepatocytes. Within 14–21 d, nonparenchymal cells surrounded these colonies, and markers of immature hepatocytes (AFP, CK7, CK19), including glutathione S-transferase placental type (GST-P), and morphological changes in the shapes of the cells in the colonies were observed. The latter changes were associated with enhanced expression of ALB, Cx32, and tryptophan 2,3-dioxygenase messenger RNA, and with the formation of duct- or cystlike structures consisting of differentiated hepatocytes. Although performed in the presence of dimethylsulfoxide (DMSO), the transitions exhibited by "small" hepatocytes in both systems might reflect a physiological process, since in vivo studies of repair of adult rat liver injury following 70% hepatectomy and exposure to retrorsine, a pyrrolizidine alkaloid that blocks the differentiated hepatocyte cell cycle, also suggest a role for small LPC-like hepatocytes in the restitution of hepatocytes and bile duct epithelium during liver regeneration (Laconi et al., 1998; Gordon et al., 2000a, 2000b). Similar primary culture findings—mature hepatocyte → bile duct epithelial cell—were reported with "large" hepatocytes (Block et al., 1996) although no parallel in vivo studies were provided to support the physiological significance of these observations.

33.2.2.6. Chemically Pretreated Adult Rodent Liver One laboratory has investigated a clonal LPC line, 13-1, derived by standard procedures (*see* Subheading 33.2.3.) from the livers of chemically pretreated adult rats. A single dose of the noncarcinogenic chemical D-galactosamine was injected intraperitoneally to induce liver injury and periportal oval cell formation, thereby providing an enriched putative source of normal LPCs (Dabeva and Shafritz, 1993; Petkov et al., 2000). The ALB^+ $CK14^+$ $OV\text{-}6^+$ gene expression profile of 13-1 was consistent with an LPC that differentiates along a hepatocytic pathway (Petkov et al., 2000).

33.2.2.7. Adult Human Liver Based on prior in vivo observations (Haruna et al., 1996), putative *c-kit*$^+$ and $CD34^+$ LPCs have been identified in or near portal tracts by immunofluorescence confocal microscopy and isolated from enzymatically dispersed preparations before or after immunodepletion of HEA-125$^+$ bile duct epithelial cells (Crosby et al., 2001). The frequencies of such LPCs was difficult to estimate, but a lower limit of ~300 of either type per gram wet weight of tissue was reported. If it is assumed that each gram of tissue contains 1×10^8 parenchymal cells, there is ~1 LPC per 3×10^5 hepatocytes. Although the possibility of bile duct CK19 expression was not eliminated by immunodepletion at the time of plating, individual primary cultures of both types of normal cells gave rise by d 7 to colonies expressing CK19 (bile duct) or CD31 (endothelial) cell markers. Thus, these findings suggest but do not prove the presence of human LPCs with the capacity to differentiate into bile duct and endothelial cell lineages.

33.2.2.8. Adult Rodent Bone Marrow from Cholestatic Rats Transplantation studies have suggested that normal bone marrow cells from rats (Petersen et al., 1999), humans (Alison et al., 2000; Theise et al., 2000a, 2000b), and mice (Lagasse et al., 2000) can differentiate into hepatocytes in vivo. These observations were confirmed recently in rats made cholestatic following complete bile duct ligation. The bone marrow and livers of these animals were enriched for populations of $\beta_2\text{-m}^-/\text{Thy-1}^+$ cells; following isolation by a two-step magnetic bead cell-sorting procedure, these cells were reported to differentiate into mature hepatocyte-like cells in vitro in cocultures with primary hepatocytes in the presence of cholestatic serum (Avital et al., 2001). In vitro evidence of bone marrow–derived hepatocytic maturation of these cells consisted of albumin expression (measured by RT-PCR) and metabolism of ammonia into urea. The frequency of $\beta_2\text{m}^-/\text{Thy-1}^+$ cells in the adult rat liver, 1×10^{-4}, is higher than that reported for oval-like LPCs in human livers (Crosby et al., 2001). However, if these results are correct, the frequency of hepatocyte-like cell formation from cholestatic bone marrow LPCs is likely to be further reduced.

33.2.3. METHODS

33.2.3.1. Isolation Procedures Where applicable, a series of standard steps is usually employed that includes surgery; tissue perfusion and digestion; cell centrifugation and washing; cell fractionation; and determinations of final cell yields, concentrations, and viability. Surgical procedures for isolating explants of liver diverticulum require microdissection with a tungsten needle (Jung et al., 1999); in particular, care must be taken to avoid cross contamination of hepatic and gut endoderm with unwanted tissue containing cardiac mesoderm and septum transversum (Rogler, 1997). Owing to access problems, human liver biopsy tissue often requires lengthy storage, 48 h at 4°C in Dulbecco's Modified Eagle's medium (DMEM) before use (Crosby et al., 2001). To minimize contamination with microorganisms, aseptic conditions are needed; rat bone marrow cells are flushed from femurs under laparotomy (Avital et al., 2001).When possible, owing to hemoglobin and serum toxicity, and to the inhibition of digestion enzymes by serum components, blood must be removed from source tissues by perfusion with physiological salt solutions often supplemented with Ca^{2+} and/or Mg^{2+}. The inclusion or exclusion of such divalent cations depends on the degree of dispersion required and the digestion enzymes employed. For adult livers, anterograde or retrograde perfusions are variably used with solutions containing organic buffers such as HEPES, or with buffers supplemented with glucose, insulin, inosine, and hydrocortisone (Leffert et al., 1977, 1979). Blood is removed from fetal and neonatal livers by rapid exsanguination.

Uncut or delicately minced source tissues are dispersed with solutions of one or more enzymes, with or without divalent cation chelators. Enzyme purity, concentrations, and incubation times, as well as the speeds of magnetic stirring bars or shakers, can be critical. Conditions to pretest commercial preparations and optimize these variables should be established. Some of the many enzymes include pronase (Koch and Leffert, 1980), trypsin (Grisham, 1980); collagenase and Dispase I (Herring et al., 1983) plus hyaluronidase (Marceau et al., 1980; Germain et al., 1988); pancreatin, which is a mixture of trypsin, lipase, and amylase (Jung et al., 1999); and collagenase (Richards et al., 1997; Crosby et al., 2001). Collagenase has also been combined with thermolysin (a metalloendopeptidase), DNase I, and trypsin-EDTA followed by sieving with a sterile 30-μm filter (Kubota and Reid, 2000). Although many commercial preparations of various types of collagenase are available, representing various degrees of substrate specificity and purity, with widely variable effectiveness of dispersal, it remains to be determined whether the effects of this enzyme are due instead to one or more enzymatic impurities.

Centrifugation speeds employed by different investigators vary from 100 to 1000*g*. Unlike primary hepatocyte isolations, which require low centrifugal forces to minimize hepatocyte breakage and nonparenchymal cell contamination, centrifugal speed used to separate LPCs may not be critical. By contrast, washing steps, which vary by report, may well be critical in order to remove cytotoxic debris from lysed cells.

Several groups have reported fractionation steps for rat and mouse liver–derived LPCs using FACS (Kubota and Reid, 2000; Suzuki et al., 2002), and for rat bone marrow– and human liver–derived LPCs using immunomagnetic beads and Percoll gradients (Avital et al., 2001; Crosby et al., 2001). In each case, specific classes of putative LPCs were enriched significantly. However, except for bone marrow–derived LPCs, the practical utility of these procedures is unclear because LPCs have been isolated from rat and mouse livers without prior fractionation (Table 1).

33.2.3.2. Plating, Passage and Longterm Storage Conditions Initial plating conditions for culturing LPCs vary considerably among laboratories. Some of these conditions, including passage and storage protocols, are summarized here along with reported rationales.

Feeder layers of STO embryonic fibroblasts were originally used to support the growth of ES cells (Martin and Evans, 1975). STO cells have also been used in two different LPC systems: the outgrowth of putative HBC-3 LPCs from initial liver diverticulum required mitomycin C–treated growth-inhibited $STON^+$ (neo^R STO cells) (Rogler, 1997), and dispersed rat E13 liver cells were grown on growth-inhibited STO5, a STO cell subclone (Kubota and Reid, 2000). Rat hepatocytes were also employed as a top feeder layer for the bottom layer of rat bone marrow–derived LPCs (Avital et al., 2001). Problems with feeder layers are discussed in Subheading 33.5.2.

Different solid surfaces and macromolecular coatings have been employed. Mouse ES cells were cultured on 0.1% gelatin-coated flasks (Jones et al., 2002); dissected ventral endoderm explants and NPEC clusters of pig adult liver cells were cultured on Type I collagen–coated plastic microwells (Jung et al., 1999) and plastic dishes (Kano et al., 2000), respectively, or en bloc on nucleopore mixed-ester filters (Monga et al., 2001). Bile duct–like HBC-3 cells cultured on Matrigel were shifted to uncoated or coated glass cover slips prior to immunofluorescence studies; similar shifts were performed with rat E12 liver cells cultured on fibronectin-coated glass slides (Germain et al., 1988). Laminin- or Type IV collagen–coated plastic plates were used for fractionated mouse E13.5 liver cells, with similar results (Suzuki et al., 2002). In the most complex system reported thus far, dual-chamber Transwell plastic dishes were used for cocultures of adult rat hepatocytes and bone marrow–derived LPCs: rat tail Type I collagen was applied underneath the hepatocytes (top layer) and Matrigel was applied underneath the bottom layer of bone marrow–derived LPCs (Avital et al., 2001).

Rich nutrient media have generally been employed, such as supplemented minimal essential medium, DMEM, and Dulbecco and Vogt's modified Eagle's medium. Simple and complex supplements have been used. Most systems require 5–20% fetal bovine serum. Leukemia inhibitory factor (LIF) was used to maintain the growth of ES cells in an undifferentiated state, prior to induction of differentiation (Jones et al., 2002). HBC-3 was cultured with β-mercaptoethanol, a glutathione-stabilizing reagent (Rogler, 1997). The H-CFU-Cs in the mouse E13.5 liver system required epidermal growth factor (EGF) and hepatocyte growth factor (HGF) (Suzuki et al., 2002). This system also had a requirement of conditioned media for clonal expansion from a single cell: standard media supplemented with 50% CM from d 7 mass fetal liver cell cultures (Suzuki et al., 2000). WB-F344 rat liver cells were grown in Richter's improved minimal essential medium with zinc ± insulin (Tsao et al., 1984). Pig liver cells were grown in NAIR-1 medium (DMEM/Ham's F12 [1:1]) supplemented with insulin, glucagon, and hydrocortisone (Leffert et al., 1977, 1979), plus EGF, transferrin, triiodothyronine (T_3), sodium selenate, ascorbic acid, α-tocopherol acetate, and linoleic acid (Kano et al., 2000). Gibco's ITX solution (insulin, transferrin, selenium) was used for rat bone marrow–derived cocultures (Avital et al., 2001).

Most cultures were incubated at 37°C in humidified air incubators gassed with 5–10% CO_2. Routine passages of cell lines were performed at weekly frequencies; HBC-3 cells were passaged 1:5 every 3–5 d (Rogler, 1997). The times at which the cells were held at confluence were critical (*see* Subheading 33.4.5.). Most cell lines were frozen in DMSO-supplemented media by standard procedures and stored in liquid N_2.

33.3. PHENOTYPIC PROPERTIES

In this section, we review the morphological; ultrastructural; and, where known, karyotypic details of normal LPCs in culture. We also review gene expression profiles of microarray findings and markers defining primitive and differentiated LPC status, as well as in vitro plasticity and potential, and the effects of in vivo microenvironment.

33.3.1. MORPHOLOGY AND ULTRASTRUCTURE Nine groups have examined the microscopic properties of undifferentiated or differentiated LPCs by transmission electron microscopy. In the mouse ES system, 12-d-old EBs contained cells with the ultrastructural appearance of cells similar to embryonic hepatocytes; unlike embryonic hepatocytes, some of the induced cells were polyploid (Jones et al., 2002). After 3 wk without passage, mouse HBC-3 cells showed spontaneous multilayering of cells with predominant hepatocytic morphology: bile canaliculi, peroxisomes, and glycogen granules (Rogler, 1997). Cultures of nonparenchymal rat E12 liver cells showed two types of cells: on day 4, bile duct–like cells were observed with free polysomes, oval nuclei, and diffuse nucleoli; on d 6, hepatocyte-like cells were observed with abundant long cisternae, rough endoplasmic reticulae, round nuclei, large nucleoli, large mitochondria with long transverse cristae, Golgi apparatuses, and glycogen granules (Germain et al., 1988). NPEC porcine clusters showed similar hepatocyte-like structure (Kano et al., 2000). In 21-d-old cultures of mouse H-CFU-Cs, derived from E13.5 sorted, cloned, and resorted and recloned progeny, tight junctional complexes, well-developed ovoid mitochondria, and ductlike structures with cells containing high nuclear:cytoplasmic ratios and luminal microvilli were reported (Suzuki et al., 2002). FNRL cells showed numerous mitochondria, tight junctions, and occasional lipid droplets (Herring et al., 1983). Polygonal rat WB-F344 cells contained numerous intercellular desmosomes and nexus junctions; few lysosomes; abundant free ribosomes; sparse smooth endoplasmic reticulum and Golgi membranes; microbodies; and small, pleomorphic mitochondria; no glycogen particles were seen (Tsao et al., 1984). Small hepatocyte LPCs also displayed typical hepatocyte ultrastructure (Tateno and Yoshizato, 1996a). Rat bone marrow–derived LPCs were reportedly present in the liver, where they were clustered periportally; they displayed blastlike morphologies with simple ultrastructure, small diameters (5–8 μm), and high nuclear:cytoplasmic ratios (Avital et al., 2001).

33.3.2. KARYOTYPE Rat WB-F344 and FNRL cells were reported to be diploid ($2N = 42$) following weekly subcultures through passage 12 and, as determined by flow microfluorimetry (FMF), through passage 20 (Tsao et al., 1984), and through passage 25 (Herring et al., 1983), respectively. Mouse HBC-3 cells remained euploid ($2N = 40$) through passage 40 (Rogler, 1997). However, FMF cannot directly detect ploidy status, nor can it detect chromosome changes, such as the chromosome 1 trisomy that appeared in WB-F344 cells at passage 16 (Tsao et al., 1987), or cell fusion. Karyotypes for human ($2N = 46$) or pig ($2N = 38$) LPCs have not been reported to date.

33.3.3. GENE EXPRESSION PROFILES

33.3.3.1. Microrarray Studies Two reports have appeared: mouse HBC-3 cells (Plescia et al., 2001) and rat LPC line 13-1 (Petkov et al., 2000). HBC-3 cells were cultured for 7 d in the presence of DMSO in order to augment the induction of differentiation toward a hepatocyte lineage (*see* Subheading 33.4.3.). Initially, undifferentiated cells expressed muscle, neuron, myeloid, and lymphoid-specific genes; this pattern attenuated over time along with cell proliferation. Biphasic bursts of gene regulation occurred on d 1 and 7, and several groups of genes showed complex changes. Downregulated genes included the Wnt/β-catenin pathway, T-cell transcription factor family target genes, cellular receptors for fibronectin and laminin, and other extracellular matrix molecules, and, as expected from studies of adult rat liver regeneration (Lu et al., 1992) and in vitro hepatocyte growth control (Koch et al., 1994), the cell-cyle proteins cyclin B1 and cyclin D. Upregulation was observed for growth inhibitory genes encoding cyclin I and p18, and, as expected from in vitro studies of hepatocyte growth control (Leffert and Sell, 1974; Leffert et al., 1978b,a; Lad et al, 1982), differentiated functions such as apolipoprotein C-IV, phosphoenolpyruvate carboxykinase, alcohol dehydrogenase, and asialoglycoprotein receptor, and transcriptional regulators including Twist, Snail, HNF1a, and GATA-6. The physiological significance of these findings is unclear, because parallel studies without DMSO were not reported.

Studies with 13-1 cells suggest that 13 unknown genes are highly expressed in vitro. These conclusions were obtained using differentially expressed cDNA clones from fetal rat liver isolated using suppression subtractive hybridization. Four categories of genes were defined by the subtracted clones, as expressed by (1) hepatoblasts; (2) hematopoietic cells; (3) hepatoblasts, hematopoietic cells, and other tissues at varying levels; and (4) fetal liver (in abundance), activated LPCs, and other epithelial progenitor cell lines. Among the unknown genes expressed by 13-1, one DNA clone (1A10) was expressed exclusively in fetal liver, and five or eight clones were expressed in fetal liver, spleen, and bone marrow, or fetal liver, respectively, with various expression patterns in other organs and cell lines. One DNA clone, 43\22, a mesodermal gene of unknown function (GenBank accession no. NM_008590), was the most abundantly expressed DNA clone observed. Interpretation of these findings would have been aided by information regarding the growth state of the cultures studied.

33.3.3.2. Protein and Antigenic Markers Defining Primitive and Differentiated LPC Status Primitive liver- specific markers were expressed by most but not all LPC systems. Primitive markers usually disappeared as the expression of differentiated functions emerged. In general, these time intervals ranged between 4 and 21 d in vitro.

More than half of the systems expressed AFP^+ ALB^+, markers of embryonic hepatoblasts and oval cells, but several did not (Table 1). Other oval cell–associated markers, including *c-kit*$^+$, as well as $A6^+$ and $OV\text{-}6^+$ (mouse and rat cytoskeletal antigens, respectively, localized to oval cells and bile duct epithelial cells [Dunsford and Sell, 1989; Engelhardt et al., 1993]), were also observed (Dabeva and Shafritz, 1993; Richards et al., 1997; Ott et al., 1999; Kano et al., 2000; Crosby et al., 2001; Monga et al., 2001). In some investigations, *c-kit* expression was absent (Richards et al., 1997; Suzuki et al., 2002), suggesting a more primitive but not a dysfunctional state. For example, two systems of *c-kit*$^-$ phenotypes (Richards et al., 1997; Suzuki et al., 2002) differentiated along one or more hepatic lineage pathways, and one displayed bipotential capabilities (Suzuki et al., 2002). Three systems expressed no primitive liver-specific markers (adult rat bone marrow, human liver, and mouse E13.5 fetal liver), as summarized in Table 1 (Avital et al., 2001; Crosby et al., 2001; Suzuki et al., 2002), yet mouse H-CFU-H cells displayed pluripotentiality in vivo (*see* Subheading 33.3.5.). Clearly, the presence or absence of known markers examined to date is not necessarily predictive of progenitor cell capacity.

Most LPCs in culture are presumably known to be of endodermal origin, yet mesodermal markers have been localized inside some of them. For example, vimentin, localized previously in human fetal liver ductal plate and biliary epithelial cells, but not in hepatoblasts or hepatocytes in vivo (Haruna et al., 1996), was also observed in hepatic NPECs in vitro (Kano et al., 2000). Of particular interest and possible significance, a surface antigen associated with hematopoietic stem cells (HSCs) such as A6 (Engelhardt et al., 1993), was found on hepatic LPCs derived from adult mouse (Richards et al., 1997). Cultures of mouse E13.5 H-CFU-Hs also expressed CD34 and thy-1 (Suzuki et al., 2002). In vitro findings of HSC antigens in liver-derived LPCs are provocative, because they suggest that resident or recruited adult liver-derived bone marrow cells, or primitive liver-derived epithelial cells with bone marrow–like properties, may serve as LPCs.

The capacity for differentiation along one or more hepatic lineages was revealed by the expression of many common markers used to define hepatocytic and bile duct pathways (*see* Table 1). For example, hepatocytic pathways were frequently monitored by expression of AFP, ALB, and transferrin, whereas bile duct pathways were monitored by expression of CK19 (Block et al., 1996; Kano et al., 2000; Kubota and Reid, 2000; Crosby et al., 2001; Suzuki et al., 2002) and γ-glutamyl-transpeptidose (GGT) (Tsao et al., 1984). In one system, AFP^+ and ALB^+ expression in WB-F344 cells was reported to be higher in log than in stationary phase (Tsao et al., 1984). This anomalous behavior might have been an artifact of immunostaining commonly caused by antibody crossreactions with adherent or pinocytosed serum proteins. The coexpression and growth dependence of some of these markers were also examined in cultures of "small" adult hepatocyte LPCs; some cells displayed only hepatocytic markers (CK8 and CK18) whereas others expressed dual phenotypes (ALB, and the biliary markers CK7 and CK19), and still other cells were identified as biliary-like, since they ceased to express ALB and in turn became positive for CK19 or CK7 (Tateno and Yoshizato, 1996a; Mitaka et al., 1999). Adult bone marrow–derived rat LPCs did not express GGT (Avital et al., 2001).

Prominent but less commonly investigated markers have also been reported. For example, H-CFU-C cells were characterized extensively by RT-PCR (Suzuki et al., 2002), and a variety of hepatocyte (GST, glucose-6-phosphatase [G-6-Pase], dipeptidyl peptidase IV [DPPIV]) and bile duct markers were detected (thymosin β4, biliary glycoprotein, GGT, vinculin). In the NPEC LPC pig liver system, hepatocytic and bile duct–like pathways were suggested by α1-antitrypsin and GGT expression, respectively; weak GGT expression was reported in WB-F344 cells as well (Tsao et al., 1984). Hepatocytic pathways were also suggested in AFP^- "large" adult hepatocyte LPCs that expressed cytochrome P450IIB1 (Table 1), and in 7-d-old cultures of bone marrow–derived LPCs that expressed CK8, CK18, and metabolized urea (Avital et al., 2001). In addition, human LPCs differentiated along endothelial pathways, as suggested by $CD31^+$ expression (Crosby et al., 2001).

33.3.4. PLASTICITY AND POTENTIAL IN VITRO The demonstrated plasticity pathways of normal LPCs are summarized in Table 1. All of the systems displayed hepatocytic plasticity, except human LPCs, which displayed only bile duct and endothelial cell plasticity (Crosby et al., 2001). Regardless of the times of harvest from in vivo sources, half of the LPCs displayed both hepatocytic and bile duct plasticity. Included in this group are cloned mouse HBC-3 cells, mouse E9.5 explants; dispersed cells from rat E12 and E13, and mouse E13.5 fetal liver; adult NPEC clusters; small and large primary adult rat hepatocytes; adult bone marrow–derived cells; and cloned adult WB-F344 cells. Further studies with human systems, and with explant systems maintained for longer periods in vitro might reveal much broader plasticity.

Three LPC categories of varying potential were revealed. Two included cultures of unipotential (ES, FNRL, LA, 13-1, and rat bone marrow–derived LPCs) and bipotential capacity (ventral foregut endoderm and mouse E9.5 explant, HBC-3, dispersed rat E12 and E13 liver, H-CFU-H, WB-F344, NPEC, "small" and "large" adult primary hepatocytes, and adult human liver LPCs). The third was defined by adult bone marrow–derived LPCs, which appeared capable of transdifferentiation (Avital et al., 2001). Further work is needed to identify the culture conditions that best facilitate, stabilize, and simulate in vivo behavior. Efforts along these lines are discussed in Subheadings 33.3.3 and 33.4.

33.3.5. EFFECTS OF MICROENVIRONMENT IN VIVO Several groups have investigated the plasticity of cultured LPCs following transplantation in vivo. One demonstrated pluripotential effects following intrasplenic injections of E13.5 mouse H-CFU-H cells into 4-wk-old Balb/cA mice (Suzuki et al., 2002). The cells migrated into the livers and, within 38 d, differentiated into hepatocytes (in animals pretreated with CCl_4 to induce centrolobular hepatocyte regeneration) or cholangiocytes (in animals pretreated with 4,4'methylene dianiline [4,4'-diaminodiphenylmethane] to induce bile duct regeneration). Following administration into the common bile duct (to deliver cells retrograde into the pancreas) or by direct injection into the duodenal wall, the cells differentiated into pancreatic ductal and acinar epithelium, and intestinal epithelial cells, respectively. Thus, H-CFU-H cells were bipotential in vitro but pluripotential in vivo.

One group has studied clones of different liver-derived epithelial LPCs under various conditions in vivo. A derivative line of WB-F344 genetically engineered to express *Escherichia coli* β-gal and neo^R reporter genes differentiated into β-gal^+ hepatocytes between 1 and 17 wk post-intrahepatic or intrasplenic transplantation. Attempts to recover and culture transplanted β-gal^+ hepatocytes were inconclusive although propagable β-gal^+ colonies were isolated that contained neo^R sequences (Coleman et al., 1993; Grisham et al., 1993). To rule out possible effects of the integrated reporter genes, the genetically $DPPIV^+$ parental line (which did not express DPPIV in vitro) was injected into genetically deficient German Fischer $DPPIV^-$ adult rats. Within 1 to 2 mo post-transplantation, the injected cells integrated into hepatic plates and differentiated into hepatocyte-like cells that expressed ALB (and purportedly transferrin and α1-antitrypsin [the data of which were not shown]) and were morphologically and functionally indistinguishable from mature hepatocytes (Coleman et al., 1997). Thus, it would appear that cloned LPCs can survive and differentiate into mature hepatocytes in quiescent microenvironments of normal liver. However, WB-F344 cells have not yet been shown to differentiate into bile duct cells in vivo.

Two groups have reported transdifferentiation of LPCs. In one case, following intraventricular injections into the cardiac muscles of adult ♀ nude mice, β-gal+ ♂ WB-F344 cells were reported to form cardiac myocytes within 6 wk after transplantation (Malouf et al., 2001). The donor cells were identified *in situ* by three assays: by the presence of a rat Y-chromosome-specific DNA sequence in β-gal^+ myocytes (using PCR and fluorescence *in situ* hybridization); by immunohistochemical expression of cardiac troponin T; and by ultrastructural analysis, which confirmed a cardiac myocyte phenotype of the LPC-derived myocytes. In the other case, MHC $C3^-$ bone marrow LPCs isolated from cholestatic inbred ♂ Lewis rats and injected into the portal veins of MHC $C3^+$ D'Agouti ♀ rat livers transplanted into Lewis inbred ♂ MHC $C3^+$ recipients gave rise to mature hepatocytes that integrated in liver plates, as revealed by light-level microscopy (Avital et al., 2001). Results of the latter study are difficult to evaluate because the times of hepatocyte formation were not stated, and, in both cases, the possibilities of fusion artifacts were not eliminated (*see* Subheading 33.5.2.).

Paradoxical expression has been observed in vivo in one system: FNRL cells expressed G-6-Pase intrasplenically but not intrahepatically (Ott et al., 1999). The mechanisms of ectopic G-6-Pase expression are unknown.

33.4. GROWTH REGULATORY CONDITIONS IN VITRO

In this section, we review experimental evidence of the regulatory properties of feeder layers, growth factors, chemicals, and substrata on the growth patterns, function, and mortality of normal LPCs in culture. Little is known about genes involved in normal LPC growth control, but inferences can be made from preliminary work with mutant systems.

33.4.1. FEEDER LAYERS Two groups have employed growth-inhibited mouse embryonic STO, STON$^+$, or STO5 fibroblasts to culture mouse HBC-3 (Rogler, 1997) and dispersed E13 fetal rat liver cells (Kubota and Reid, 2000). STO cells were considered essential in the first system, particularly during the isolation process, and they sustained primitive expression and delayed or attenuated differentiation of hepatocytic or bile duct lineages until the LPCs were subcultured without the feeder layers. However, strict dependence on such feeder cells remains to be demonstrated, and the precise role of the feeder cells remains unclear. Neither cell-cell contact, nor cell-cell fusion, nor the production of positive or negatively acting conditioning factors were eliminated. The latter possibility was supported by a report of a role for embryonic fibroblast conditioned medium on the growth and survival of H-CFU-Hs (Suzuki et al., 2000), but no conditioning factors have been identified thus far. A third group employed adult rat hepatocytes from cholestatic animals as an essential top-layer feeder system for bone marrow–derived LPCs (*see* Subheading 33.2.2.8.). The role of hepatocyte top feeder layers and potential artifacts resulting from such configurations requires further investigation (*see* Subheading 33.5.2.).

33.4.2. GROWTH FACTORS Knowledge of growth factor effects can provide working hypotheses of signal transduction pathways involved in the mechanisms of commitment and proliferation of LPCs. For example, in growth control studies of primary fetal and adult rat hepatocytes in serum-free media, complex sets of growth factors—including type I (EGF, transforming growth factor-α [TGF-α], TGF-β, fibroblast growth factor-2 [FGF-2]) and type II polypeptides (insulin, insulin-like growth factor-1 [IGF-1], IGF-2, glucagon), lipophilic hormones (hydrocortisone), lipids, lipoproteins, nutrients (inosine, arginine, ornithine), and conditioning factors—interact in dose-dependent, time duration–dependent, and time interval of action–dependent ways to regulate the initiation of DNA synthesis and mitosis (Koch and Leffert, 1974, 1979, 1980; Leffert and Weinstein, 1976; Leffert and Koch, 1977, 1980, 1982a, 1982b; Koch et al., 1982; reviewed in Leffert et al., 1988, and Koch et al., 1990). Not surprisingly, many of these and other known polypeptide growth factors regulate proliferation and gene expression of normal LPCs in vitro. This kind of information has been reported in 11 systems, including cells from early embryos, midphase fetal liver, postnatal liver, and adult bone marrow. However, direct genetic evidence of molecular mechanisms and delineation of cellular responses is limited.

Two groups have described findings in early embryonic LPC systems: ES cells and explants of mouse E8–E8.5 ventral foregut endoderm. In the ES system, LIF removal was essential for EB formation (Jones et al., 2002), but molecular signals freed from the inhibitory effects of LIF remain to be elucidated. In the explant system, four different FGF receptor ligands and one or more bone morphogenetic proteins (BMPs) appear to be involved in "liver organogenesis" in vitro. In the proposed sequence of events, exogenously added FGF-1 and FGF-2, putatively supplied in vivo by cardiac mesoderm, together commit ventral foregut endoderm, which constitutively produces FGF-1 and FGF-4, toward a hepatic lineage (Jung et al., 1999). Apart from the apparent redundant requirement of FGF-1, the commitment events were explained partly by the induction of sonic hedgehog (*Shh*) and GATA-4 transcription factor gene expression by cardiac mesoderm FGF-2 and septum transversum mesenchyme BMPs; together, these growth factors concertedly diverted the default fate of ventral foregut endoderm from pancreas to liver (Deutsch et al., 2001; Rossi et al., 2001). These events were followed by morphogenetic outgrowth of the hepatic-specified endoderm that was stimulated by exogenously added FGF-8 (Jung et al., 1999). It is unclear if these events describe the "first" specifying signals, because unspecified endoderm constitutively expressed FGF-1-stabilized TTR, a hepatocyte marker, suggesting that still earlier signals preactivated the explants prior to culture.

In studies of LPCs derived from midphase fetal liver tissues, HGF, Flt-3 ligand, and stem cell factor (SCF) increased the frequencies of hepatocytic cells in explants of mouse E9.5 liver diverticulum; in the same system, TGF-β increased the frequency of bile duct–like cells, and dexamethasone increased the frequencies of both lineages (Monga et al., 2001). HGF and EGF regulated differentiation in both rat E13 liver (Kubota and Reid, 2000) and mouse E13.5 H-CFU-H cell systems (Suzuki et al., 2002). In the former system, EGF was required for hepatocyte-like differentiation but not for bile duct–like expression (it was unclear if dexamethasone was required in both instances); in the latter system, distinct for its lack of expression of any specific hepatic markers in its uninduced state, addition of EGF and HGF led to H-CFU-C formation 21 d later. In primary cultures of dispersed rat E12 liver (Germain et al., 1988), LPC formation was not induced by FBS, insulin, and dexamethasone, yet all of these factors were required for induction of different lineages by chemicals in the presence of dexamethasone (*see* Subheading 33.4.3.)

Observations with LPC systems derived from postnatal liver have been reported by three groups, but most have tended to be phenomenological in nature. Tsao et al. (1986) investigated WB-F344 cells in detail. They reported that EGF affected growth-state dependence of expression of several glycolytic enzymes, by increasing the activities of NADH-diaphorase, pyruvate kinase, glucose-6 phosphate dehydrogenase, GGT, and lactate dehydrogenase, and by decreasing the activity of alkaline phosphatase (Tsao et al., 1986). In addition, TGF-β reversibly inhibited the proliferation of early passage cells, whereas at higher passages the cells became progressively less sensitive to its inhibitory effects (Lin et al., 1987). Late-passage WB-F344 cells were also trisomic for chromosome 1, and the insensitivity to the inhibitory effects of TGF-β was explained by constitutive production of TGF-β and IGF-2 in serum-free improved minimal essential medium with zinc option (IMEMZO) medium supplemented with iron-free transferrin. The effects of EGF were attributed to its potentiation of the growth-stimulatory effects of transferrin (Tsao et al., 1987), via stimulation of phosphoinositide hydrolysis and EGF receptor protein synthesis and mRNA levels (Earp et al., 1988).

Another group has surveyed the effects of more than 12 families of growth factors on LA cells (Isfort et al., 1998). Mitogenic stimulation and inhibition were defined by enhanced (≥50%) or attenuated (≤70%) rates of [^{3}H]dT uptake in treated cultures compared with cultures in serum-free media. Using these assays, growth stimulation was observed following treatments with interleukins (interleukin-4 [IL-4], IL-9, and IL-13), chemokines (macrophage inflammatory protein-1α [MIP-1α] and MIP-1β, MCP-1/MCAF, GRO/MGSA, RANTES, IL-8), stem cell factors

(NGF), EGF (EGF, TGF-α, AR, BC, HB-EGF), FGF (FGF-1, FGF-2, and FGF-4; βECGF, keratinocyte growth factor), platelet-derived growth factor (PDGF) (PDGF A/A, A/B, and B/B), the insulin family (insulin, IGF-1 and IGF-2), angiotensin II and III, erythropoietin, β-estradiol, glucagon, HGF, hydrocortisone, progesterone, retanol acetate, T_3, and vasopressin. Growth inhibition was observed with stem cell factors (ciliary neurotrophic factor, pleotropin), the FGF family (FGF-5 and FGF-6), TGF-β (TGF-β1, TGF-β2, and TGF-β3), vascular endothelial growth factor (placental growth factor), granulocyte macrophage colony-stimulating factor, and platelet-derived endothelial cell growth factor. While some of these trends are consistent with findings in other systems, the data are difficult to evaluate because thymidine incorporation assays are subject to artifacts of transport and alterations in intracellular pool sizes, as well as to cell density dependence (Paul et al., 1972), parameters that were not investigated in these assays.

In studies reminiscent of retrodifferentiation and redifferentiation of adult hepatocytes (Leffert et al., 1978a; Sirica et al., 1979), a third group studied colonial growth of "large adult hepatocytes" (Block et al., 1996). These cells were stimulated by HGF/SF (scatter factor); EGF; and, as reported earlier (Brenner et al., 1989), TGF-α. As expected, the proliferating hepatocytes lost expression of a differentiated function like ALB, and a nontissue-specific function like P450IIB1 (Srivastava et al., 1989); when the cells were blocked from redifferentiating by virtue of the substratum employed (*see* Subheading 33.4.4.), they expressed markers of bile duct epithelium (CK19), produced TGF-α and FGF-1, and assumed a simple ultrastructural phenotype. During this dedifferentiating transition, the expression of transcription factors HNFI, HNF3, and HNF4 remained constant, while reciprocal changes occurred in the ratio of transcription factors C/EBP-α to C/EBP-β (which decreased) and the levels of transcription factors AP1 and nuclear factor-κB (which increased).

Finally, culture media supplemented with cholestatic serum, insulin, transferrin, and selenium stimulated the transdifferentiation of rat bone marrow cells into hepatocytes in cocultures with adult hepatocytes (Avital et al., 2001). This transition, which peaked 7 d after plating, was monitored by morphological changes and by urea production. Although no arginine was added to the serum-free medium used during the 8-h ureogenesis assay period, the bone marrow–derived cells had previously been cultured in media containing nondialyzed sera, a potential source of significant amounts of intracellular free arginine. Thus, intracellular arginine from such cells may have been exported and available as a substrate for urea-synthesizing macrophages in the cultures (Avital et al., 2001).

33.4.3. CHEMICALS The ubiquitous differentiation-inducing effects of DMSO and Na^+-butyrate have been studied extensively, with varying results from different laboratories. Four qualitative categories have emerged. In category I, *DMSO → hepatocytic, and Na^+-butyrate → bile duct–like lineages*. Two systems behaved precisely this way. In explants of mouse E9.5 liver diverticulum, DMSO augmented hepatocytic development, and Na^+-butyrate augmented ductular expression (Monga et al., 2001). In primary cultures of rat E12 liver cells plated in serum-, insulin-, and dexamethasone-supplemented media, DMSO and Na^+-butyrate stimulated HES_6^+ and BDS_7^+ expression, respectively (Germain et al., 1988; Blouin et al., 1995).

In category II, *DMSO → hepatocytic lineage*. This trend was observed in HBC-3 cells. In the presence of DMSO, the cells stopped growing, ALB expression was maintained, G-6-Pase expression appeared, and AFP and CK14 expression disappeared (Rogler, 1997).

In category III, *Na^+-butyrate → hepatocytic lineages*. Two systems exemplified this trend. Na^+-butyrate inhibited WB-F344 cell growth reversibly and facilitated dexamethasone induction of tyrosine aminotransferase activity in cells between passage numbers 7 and 12 (Coleman et al., 1994). In addition, Na^+-butyrate induced AFP and ALB mRNA expression in FNRL cells, as revealed by assays of run-on transcription, as well as *c-fos* and histone H3 expression (Ott et al., 1999).

In category IV, *DMSO → both lineages*. This was seen in cultures of "small hepatocytes," in which DMSO was required for LPC-like differentiation in cells expressing hepatocytic (ALB, CPS-I, CK8 and CK18, transferrin) *and* bile duct–like properties (Tateno and Yoshizato, 1996a; Mitaka et al., 1999). No reports have appeared thus far of systems in which *Na^+-butyrate → both lineages*.

The mechanisms by which these chemicals exerted their effects in these different culture systems are poorly understood. Roles for changes in patterns of methylation, acetylation, or other enzymatic and covalent modifications of DNA and chromatin might be postulated, but no available evidence yet supports such speculation.

33.4.4. SUBSTRATA The phenomenological effects of various substrata have been investigated in LPCs in culture. In the ES system, gelatin-coated plastic appeared to replace the requirement of feeder cells (Jones et al., 2002).

Matrigel, a commercially available basement membrane matrix preparation derived from Engelbreth-Holm-Swarm tumor cell culture fluids has been examined in two systems. On Matrigel-coated plastic, HBC-3 cells formed ductular structures that contained AFP^+, GGT^+, weak $CK14^+$, and $CK19^+$ cells (Rogler, 1997). In the "large hepatocyte" system, after population expansion and clonal growth, Matrigel stimulated proliferating hepatocytes to reexpress mature hepatocyte phenotypes, defined by ultrastructure and ALB^+ expression (Block et al., 1996).

Collagen and laminin have been investigated in H-CFU-Hs, and in "small" and "large" primary hepatocyte systems. In the former (Suzuki et al., 2002), the colonies were cultured on laminin- or type IV collagen–coated surfaces, but no significant differences were reported (and Matrigel was not needed). In the "small" hepatocyte system (Tateno and Yoshizato, 1996a, 1996b), nonparenchymal cells migrated into and around the hepatocyte clusters and, as suggested by additional observations (Mitaka et al., 1999), gradually deposited type I and type IV collagen, and laminin; this was followed by morphological changes in the "small" hepatocytes from relatively flat to cuboidal or rectangular shapes, expression of ALB, Cx32, and TO mRNAs, and the formation of duct- or cystlike structures consisting of hepatocytes. In the "large" hepatocyte system (Block et al., 1996), the cultures were also induced in the presence of HGF/SF and type I collagen gels to form acinar/ductular structures akin to bile ductules, transformations that reportedly affected the entire hepatocyte population even when DNA synthesis was inhibited.

Current interpretation of these results is difficult. On the one hand, purified materials were not always used. For example,

Matrigel is a complex mixture of collagens, laminins, proteoglycans, polypeptide growth factors, and matrix metalloproteinases (Mackay et al., 1993). In addition, the constitutive production of substrata macromolecules by the LPCs, such as the constitutive production of fibronectin by WB-F344 cells (Tsao et al., 1984), is difficult to monitor in real time and can be further complicated by the mixed populations in the cultures.

33.4.5. PROLIFERATION AND PROLIFERATIVE KINETICS The proliferative behavior of LPCs is one of the most important yet least studied aspects of the biology of these systems. Properties of "self-renewal" have been observed in mouse E13.5 H-CFU-C cells (Suzuki et al., 2002), but no quantitative evidence of asymmetric cell division was reported.

The proliferation and differentiation of all normal LPCs in culture appear to require anchorage dependence. The population doubling times were ~24 h in early passage WB-F344 cultures, but the doubling times fell ~50% and colony-forming efficiency increased with increasing passage number (Tsao et al., 1984). These changes may be related to age-dependent attenuation of G_2 checkpoint function as WB-F344 cells age (Kaufmann et al., 2001). After frequent passage, WB-F344 cells became spontaneously tumorigenic after 8–10 cycles of what was termed *selective growth* (3 wk at "confluence arrest" between weekly passages); this behavior was attributed to the slower growth of diploid or pseudo-diploid cells during prolonged confluence, following unstable periods of aneuploidy, as measured by FMF (Lee et al., 1989; Hooth et al., 1998). Alternatively, confluent cells might be more susceptible to accumulated genetic damage than growing cells (Rubin, 2001).

33.4.6. MORTALITY AND AGING Sustained proliferation and limited aging are key attributes of stem cells. These attributes were difficult to evaluate in complex LPC explant systems, the survival of which was limited to 7 d in vitro (Shiojiri and Mizuno, 1993); they have received little attention except for studies with WB-F344 and HBC-3 cells. Cells from both lines exceeded the Hayflick 75–80 population doubling limit: HBC-3 > 150 doublings (Rogler, 1997), and WB-F344 ≤ 90 doublings, with no evidence of spontaneous transformation as assayed by growth in soft agar or FMF analysis if the cells were transferred immediately on reaching confluence (Tsao et al., 1984; Lee et al., 1989). Telomeres and telomerase have been studied during aging in WB-F344 cultures (Golubovskaya et al., 1999). Telomere length declined with passage number; however, at passage numbers ≤13, 14–40 and >104, low, repressed, and reexpressed levels of telomerase activity were observed, respectively.

33.4.7. MORPHOGENESIS Little information is available on morphogenetic changes in normal LPC cultures. ES cells form EBs (*see* Subheading 33.2.2.1.). Mouse HBC-3 and H-CFU-Cs form hepatocyte-like plates and aggregates (Rogler, 1997; Suzuki et al., 2002); both of these systems, as well as "small" and "large" hepatocytes, form bile duct–like cords.

Mixed colonies of aggregated cells and ductlike structures (DLSs) have been reported for NPEC clusters (Kano et al., 2000). The process appeared to start with cell scattering, which was suggested to be a good predictor of DLS formation. Expression of hepatocyte markers was observed at the periphery of the colonies but was lost in DLSs that acquired bile epithelial markers. In cells surrounded by DLSs, mixed expression of both markers was found. These findings are difficult to interpret because the possibility of clonal generation of such structures was not distinguished from the possibility of generation from cell aggregates during isolation or postplating.

33.4.8. GENES Apart from the phenotypic characterization of gene expression by microarray profiling (*see* Subheading 33.3.3.1.) and checkpoint studies with WB-F344 cells (Kaufmann et al., 2001), which, as expected from prior studies of normal rat liver regeneration (Lu et al., 1992), suggested growth regulatory roles of cyclin, cyclin-dependent kinase, and cyclin-dependent kinase inhibitor genes, little is known about genes that regulate the growth and function of normal LPCs in culture. Direct evidence from studies of mutant mouse and oval cells has implicated growth-controlling roles for the type 1 tumor necrosis factor-α (TNF-α) receptor gene (Knight et al., 2000), *p53* (Dumble et al., 2002), *c-met* (Spagnoli et al., 1998), and the tumor suppressor *Tg737* gene (Richards et al., 1996, 1997; Isfort et al., 1997). Findings of genetic regulation of LPC proliferation by ubiquitous p53 and the HGF/SF receptor, *c-met*, are consistent with growth regulatory studies thus far (*see* Subheading 33.4.2.). However, the requirement of the type 1 TNF-α receptor for the emergence of proliferation-competent oval cells in mice is difficult to understand given the lack of a mitogenic response in mouse LA cells exposed to TNF-α (Isfort et al., 1997). No mechanism studies of the potential role of the *Tg737* gene were reported.

33.5. CONCEPTUAL AND EXPERIMENTAL ISSUES

In this final section, we discuss key issues associated with the origins, isolation and culture, proliferation patterns, phenotypic and genotypic diversity, plasticity, and potential clinical applications of cultured LPCs.

33.5.1. LPC ORIGINS All cultured LPCs are derived from solid tissues or semisolid suspensions such as bone marrow. The complex cellular composition of these sites changes throughout life with respect to time, microenvironmental conditions, vascularization, and lymphatic drainage. Given these dynamics, it is reasonable to ask from where precisely do experimentally isolated LPCs come? Are they native inhabitants of liver, and/or nonresident immigrants transiently moving into and out of the organ? Or, are they activated or recruited only from extrahepatic sites during extreme situations of environmental hepatocellular stress?

The available evidence is inconclusive. It is confounded by the cellular anatomy of the liver, the capacity of hepatocytes for retrodifferentiation, the putative transdifferentiation of bone marrow cells into hepatocytes (*see* Subheadings 33.5.2. and 33.5.6.), and the normal physiology of fetal liver hematopoiesis. For example, morphological and marker studies have suggested that oval cell–like LPCs are constitutively located in periportal spaces and terminal bile ducts (reviewed in Sell and Ilic, 1997, and Sell, 2001). Although it is likely that several ALB^+, AFP^+, and/or $OV\text{-}6^+$ LPC lines (FNRL, WB-F344, and 13-1) were derived from these sites, no direct evidence supports this conclusion. The ancestral origins of cells at both sites is also unclear, and it is possible that these cells were mesenchymal or hepatic remnants from early development. Uncertainty of microanatomical origin also applies to systems of dispersed nonparenchymal LPCs (Table 1), although in some cases hematopoietic cells were excluded intentionally during fractionation procedures (Kubota and Reid, 2000; Suzuki et al., 2002).

In retrodifferentiating primary adult hepatocyte cultures, mature differentiated functions are transiently lost and fetal functions are reexpressed as hepatocytes emerge from monolayer aggregates and proliferate; as scattered proliferating cells become quiescent or when they are induced to differentiate, the hepatocytes reexpress mature functions (Leffert et al., 1978a; Sirica et al., 1979; Block et al., 1996). If retrodifferentiation is interrupted by blocking redifferentiation, dedifferentiation occurs instead and the hepatocytes express biliary markers and less complex ultrastructure (Block et al., 1996). Thus, in some instances, owing to the fact that liver oval cell formation occurs in response to hepatotoxins that do not block hepatocyte proliferation, the possibility that some cultured LPC lines were derived from dedifferentiated hepatocytes expressing oval cell markers cannot be eliminated (*see* Subheading 33.5.4.).

The reports of transdifferentiation of bone marrow cells into hepatocytes (*see* Subheading 33.2.2.8.) suggest that a cultured LPC line such as LA, which displays the bone marrow stem cell marker A6$^+$, might be derived from intrahepatic bone marrow "contaminants" (Taniguchi et al., 1996). Such contaminants might be blood-borne but they might also be remnants of normal fetal hematopoiesis. These observations seem paradoxical: on the one hand, there is no developmental evidence that fetal bone marrow cells are direct LPC participants during liver organogenesis; on the other, in several experimental models transplanted adult bone marrow cells migrate into physiologically stressed livers of recipient animals and differentiate into hepatocytes. Thus, either transdifferentiation events are either incompatible within fetal liver microenvironments, or stressed mature livers facilitate them. In all cases, the cellular and humoral signaling mechanisms that control these events (or artifacts) are poorly understood and warrant further investigation.

33.5.2. POSSIBLE ARTIFACTS OF LPC ISOLATION AND CULTURE Several unresolved problems deserve attention in further studies of normal LPCs in culture. First, although *in situ* phenotypes of putative rat and human oval cell–like LPCs have been reported (Sell, 2001; Crosby et al., 2001), further work is needed to identify potential losses or alterations of physiological markers and functional attributes of LPCs during and after isolation procedures, including the possibilities of subsequent selection of variants and aging during the growth and subculture of such cells. In one system, e.g., parallel *in situ* marker studies were not performed prior to isolation of cells that gave rise to putative clonal cultures (Kubota and Reid, 2000), yet the fractionated cells contained LPCs lacking MHC class I antigens, raising the possibility of selective destruction of these surface markers during the isolation procedure.

Second, the presence of mixed cell types in a single system complicates interpretations of cellular and biochemical changes intrinsic to the LPCs. For example, it is difficult to eliminate or determine the short- or long-term effects of interactions between cell types in a coculture system such as HBC-3 with growth-inhibited feeder layers. In addition, the possibility of inadvertent contamination or selection of STO feeder layer cells that have escaped growth inhibition by mitomycin C treatment or γ-irradiation must be examined before and after purification of LPCs from feeder cells (Zhang et al., 2003).

Third, explant LPC systems are also examples of mixed populations in which it is difficult to identify precisely the mechanisms of growth factor actions that have been reported in these systems. Moreover, it is not yet clear that in vitro explants can be kept alive and intact long enough, with outgrowth prevented and three-dimensional structure maintained, in order to produce mature hepatocytes and bile duct epithelium.

Fourth, claims of colony-forming and clonogenic assays (Kubota and Reid, 2000) should be rigorously founded on time-lapse or real-time photomicroscopy as in the single cell mouse ED13.5 system (Suzuki et al., 2002). Such studies alone can eliminate the possibilities of *in situ* contamination, as a result of random or directed migration, aggregation, or diffusion of other plated cells into the developing colony, and alone can ensure that bipotentiality is the property of one cell type. The claims of colonial growth of "small" and "large" primary hepatocytes are subject to similar problems of interpretation (Block et al., 1996; Tateno and Yoshizato, 1996a, 1996b; Mitaka et al., 1999). Thus, "bipotentiality" in such systems may be owing not to one but to two different cell types initially present in the "clones."

Fifth, primary cultures of mouse cells are subject to crisis under conditions of repeated subculture (Rubin, 2001). The events surrounding crisis involve multiple genetic and abnormal chromosomal changes, and they often lead to transformation and malignant properties in the clones that survive it. The mouse LPC line LA was derived following crisis (Richards et al., 1997). Therefore, this cell line might be abnormal, and its biological properties might not simulate normal physiology. This might explain why the TNF-α type 1 receptor was necessary for oval cell proliferation in vivo, whereas LA cell mitogenesis was nonresponsive to exposure to TNF-α in vitro. By contrast, rat cells do not undergo crisis, and no reports of significant aneuploidy have been reported for early or late passage cultures of WB-F344 cells if they are subcultured at or before reaching confluence.

Finally, apart from the origin issues raised in Subheading 33.5.1., it is possible that the transdifferentiation of adult "liver-derived" WB-F344 and bone marrow–derived stem cells into cardiac myocytes (Malouf et al., 2001) and mature hepatocytes (Avital et al., 2001), respectively, is caused by artifacts of in vitro culture or in vivo engraftment (Subheading 33.5.6). Similar artifacts might apply to observations of plasticity in several LPC coculture systems. These doubts were raised by recent findings of spontaneous fusions between pluripotent embryonic feeder layer cells and neuronal (Ying et al., 2002) or bone marrow stem cells (Terada et al., 2002), which suggested that induced plasticity changes or transdetermination resulting from cell fusion might underlie many observations attributed to the intrinsic plasticities of LPCs. These problems are unlikely to have accounted for the in vitro plasticity of HBC-3 cells in cocultures, given the karyotypic findings suggestive of diploidy in this LPC system. However, until definitive karyotypic observations are made on cells cloned off feeder layers, this conclusion must remain tentative. The in vitro transdifferentiation claim of bone marrow → liver also seems problematic, since autologous hepatocytes (top feeder layer) were cocultured with a bottom layer of autologous bone marrow cells (Avital et al., 2001). Under these conditions, the possibility was not eliminated that feeder layer hepatocytes migrated or fell through the pores of the proprietary filter, situated in the top layer of the dual chamber system (the pore sizes were not stated in the report but are known to range between 0.1 and 12 μm), and settled onto the bottom surface layer, where the fallen cells would have masqueraded as transdifferentiated ureogenesis-competent hepatocytes.

33.5.3. PATTERNS OF LPC PROLIFERATION With the exception of WB-F344 cells, no growth curves or defined studies of proliferative kinetics of LPCs have been reported thus far. WB-F344 cells exhibited logarithmic growth curves with gradual rises to stationary phase, but the data were limited to cell counts alone. H-CFU-Cs proliferated to confluency, but kinetic patterns were not analyzed. Thus, there is no evidence of asymmetric cell division in normal LPC cultures, and other modes of kinetic behavior remain possible. Growth control studies in defined media, DNA synthesis initiation assays in quiescent systems (if LPCs can assume such states), quantification of cell-cycle interval times, and real-time microcinematography studies should together provide answers to this long-standing problem.

33.5.4. PHENOTYPIC DIVERSITY Normal LPCs in culture expressed many phenotypes after initial isolation and plating (Table 1). Some of these phenotypic patterns depended on the times of in vivo LPC isolation with respect to in vivo development; they ranged from the expected familiar patterns of markers of oval cells (Sell and Leffert, 1982), early embryonic layers, and fetal liver (Hixon and Allison, 1985; Hixson et al., 1997; Sell and Ilic, 1997; Ott et al., 1999), to the unexpected yet typical patterns of surface antigens associated with HSCs (Petersen et al., 1998; Crosby et al., 2001). However, despite such common patterns and time-dependent trends (*see* Subheading 33.3.3.2.), there are as yet no definitive biological or mechanistic explanations of this variability (Table 1).

Clearly, species and animal strain differences, in addition to artifacts of measurement and isolation (*see* Subheading 33.5.2.) might be involved. For example, different markers were displayed by rat bone marrow–derived LPCs (β_2-m$^-$ Thy-1$^+$) compared to their mouse counterparts (c-*kit*$^+$ Lin$^-$ Sca-1$^+$). Immunofluorescence or Western blot assays of cellular markers abundantly present in serum-supplemented culture media, such as ALB and AFP, are also vulnerable to crossreactions with adherent serum proteins. Furthermore, no parallel measurements of cellular immunofluorescence and studies of biosynthesis and secretion of such proteins have been reported. RT-PCR studies employed by some investigators would tend to eliminate immunofluorescence false positives but would not prove directly that such a protein (or proteins) was synthesized and secreted by LPCs. Since considerable variability exists among hepatoblast, oval cell, and HSC markers (including c-*kit* and Thy-1), it would appear that none of these explanations would account fully for phenotypic diversity. It seems more plausible, therefore, that, barring artifact, such phenotypic variation is intrinsic to LPCs.

If this conclusion is correct, is such variation predetermined genetically? Or, does it depend on the in vivo microenvironments in which the LPCs were situated before they were isolated? These kinds of questions were recently addressed by transplantation studies with normal differentiated adult hepatocytes in different models of periportal and centrolobular liver injury. The results suggested a major role for *in situ* position and microenvironment as determinants of phenotypic expression patterns (Gupta et al., 1999). These results have significant implications for studies of normal LPCs in culture, particularly for systems of propagable normal cell lines such as WB-F344, and its derivatives. One corollary is that phenotypes of normal LPCs in vitro may initially reflect in vivo states; unless in vitro conditions are provided to stabilize these states, such phenotypes may dissipate with time and cell division in vitro. Similar problems were encountered in early studies of primary fetal hepatocytes, in which maintenance of differentiated functions in vitro depended on low rates of hepatocyte proliferation and selection against the overgrowth of faster-growing nonhepatocytes (Leffert and Paul, 1972). Notably, these transplantation studies did not consider the outcomes of engrafted clonal populations, compared to the engraftment of mass populations. Thus, fluctuation analyses and other direct experiments will be needed to distinguish the roles of environment and prior differentiation states on phenotypic diversity of LPCs in vitro.

33.5.5. REGULATION OF COMMITMENT AND STABILIZATION OF DIFFERENTIATION Comparative work in vivo and in vitro is needed to understand when and how genetic and epigenetic changes required of lineage commitment and stabilization of differentiation occur in normal LPCs *in situ*, and whether or not these changes are perturbed during isolation, culture, and confluent states. These studies are needed not only to understand the physiological roles of LPCs in development and fulminant liver injury, but also the potential roles of normal LPCs in therapeutic stem cell cloning (Rideout et al., 2002) and tumorigenesis (Sell and Leffert, 1982). The availability of normal cultures of LPCs provides an opportunity to investigate directly the hypothesis that liver-derived LPCs are progenitors of hepatocellular carcinoma (*see* Fig. 1).

33.5.6. PLASTICITY OF TRANSPLANTED CULTURED LPCS Four LPC culture systems have been transplanted into rats or mice: three were derived from liver and were reported to differentiate along hepatocytic, biliary, pancreatic, intestinal, and cardiac lineages; one was derived from bone marrow and was reported to generate hepatocytes (*see* Subheading 33.3.5.). Further work is needed to better quantify the efficiencies of engraftment in these various systems, particularly for bone marrow cells (Avital et al., 2001), as well as the extrahepatic fates of intrasplenically injected donor cells. Although adult hepatocytes are capable of phagocytosis (Seglen, 1997), this process is unlikely to explain the in vivo development of the injected LPCs into various hepatic lineages; its possible role at other homing sites is unknown.

A more troublesome issue concerns the potential of donor cell fusion with recipient host cells (*see* Subheading 33.5.2.) or in vitro feeder layer cells, or the activation of endogenous tissue stem cells by insertion of the transplanted cells at or near such endogenous sites. The latter possibility was suggested recently in two different studies: first, although it had been reported that transplanted bone marrow cells generated new endothelium in models of transplant arteriosclerosis, it was shown subsequently that such endothelial cells were actually derived from extant endothelium (Hillebrands et al., 2002); second, it was shown directly that previously assumed muscle-derived HSCs were actually derived from bone marrow stem cell contaminants in muscle, and not from extant myogenic stem cells (McKinney-Freeman et al., 2002). Past and future transplantation studies with cultured LPCs must consider these artifacts.

Little is known about individual LPC potentials for lineage commitments. Do all cells have equal probabilities of producing differentiated cells of one kind or another? Or do prior culture conditions or tissue microenvironments select for individual cells with such properties? Similar questions apply to the possibility that the nuclei of LPCs are totipotential; plasticity of this kind has not been demonstrated and awaits anticipated reports of the cloning of an LPC nucleus. In each case, fluctuation analyses should help to answer these fundamental questions.

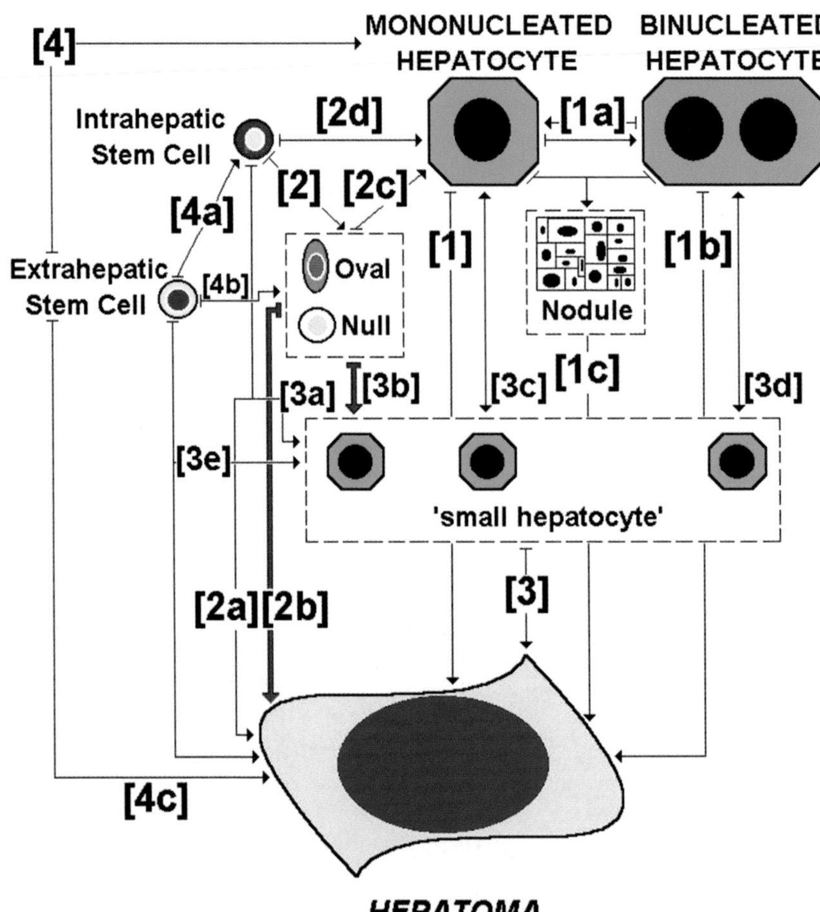

Fig. 1. Carcinogenic lineage model: a role for liver stem or liver progenitor cells. Hepatomas might develop from mono- (pathway [1]) or binucleated (pathways [1a] or [1b]) mature hepatocytes, or from preneoplastic nodules derived from hepatocytes (pathway [1c]), or from intrahepatic stem cells directly (pathway [2a]) or indirectly (pathway [2]) from "oval" or "null" cell intermediates (pathway [2b]) formed from such stem cells. Hepatomas might also develop from "large" mononucleated hepatocytes derived from oval or null cells (pathway [2c]) or directly from intrahepatic stem cells (pathway [2d]); or from "small hepatocytes" (pathway [3]) derived from intrahepatic stem cells (pathway [3a]), oval or null cells (pathway [3b]), mono- (pathway [3c]) or binucleated (pathway [3d]) hepatocytes, or extrahepatic stem cells (pathways [3e], [4] or [4a–4c]). The red arrows indicate likely oval cell ♑ hepatoma lineages ([2b] and [3b]).

ACKNOWLEDGMENTS

Work in the authors' laboratories was supported by grants from National Institutes of Health (CA71390) and National Institutes of Environmental Health Sciences (5-P42-ES10337-02). This paper is dedicated to one of our parents, Margaret Shroyer Koch.

REFERENCES

Abelev, G. I. (1968) Production of embryonal serum alpha-globulin by hepatomas. Review of experimental and clinical data. *Cancer Res.* 28: 1344–1350.

Alison, M. R., Poulsom, R., Jeffery, R., et al. (2000) Hepatocytes from non-hepatic adult stem cells. *Nature* 406:257.

Avital, I., Inderbitzin, D., Aoki, T., et al. (2001) Isolation, characterization, and transplantation of bone marrow-derived hepatocyte stem cells. *Biochem. Biophys. Res. Commun.* 288:156–164.

Becker, F. F., and Sell. S. (1974) Early elevation of alphafetoprotein in *N*-2-fluorenylacetamide carcinogenesis. *Cancer Res.* 34:2489–2494.

Bissell, D. M. (1976) Study of hepatocyte function in cell culture. *Prog. Liver Dis.* 5:69–82.

Block, G. D., Locker, J., Bowen, W. C., et al. (1996) Population expansion, clonal growth, and specific differentiation patterns in primary cultures of hepatocytes induced by HGF/SF, EGF and TGF alpha in a chemically defined (HGM) medium. *J. Cell Biol.* 132:1133–1149.

Blouin, M. J., Lamy, I., Loranger, A., et al. (1995) Specialization switch in differentiating embryonic rat liver progenitor cells in response to sodium butyrate. *Exp. Cell Res.* 217:22–30.

Brenner, D. A., Koch, K. S., and Leffert, H. L. (1989) Transforming growth factor alpha (TGF-α) initiates proto- oncogene c-*jun* expression and a mitogenic program in primary cultures of adult rat hepatocytes. *DNA J. Cell. Mol. Biol.* 8:279–285.

Byrne, J. A., Simonsson, S., and Gurdon, J. B. (2002) From intestine to muscle: nuclear reprogramming through defective cloned embryos. *Proc. Natl. Acad. Sci. USA* 99:6059–6063.

Campbell, K. H., McWhir, J., Ritchie, W. A., and Wilmut, I. (1996) Sheep cloned by nuclear transfer from a cultured cell line. *Nature* 380:64–66.

Chessebeuf, M., Olsson, A., Bournot, P., et al. (1974) Long term cell culture of rat liver epithelial cells retaining some hepatic functions. *Biochimie* 56:1365–1379.

Coleman, W. B., Wennerberg, A. E., Smith, G. J., and Grisham, J. W. (1993) Regulation of the differentiation of diploid and some aneuploid rat liver epithelial (stemlike) cells by the hepatic microenvironment. *Am. J. Pathol.* 142:1373–1382.

Coleman, W. B., Smith, G. J., and Grisham, J. W. (1994) Development of dexamethasone-inducible tyrosine aminotransferase activity in WB-F344 rat liver epithelial stemlike cells cultured in the presence of sodium butyrate. *J. Cell Physiol.* 161:463–469.

Coleman, W. B., McCullough, K. D., Esch, G. L., et al. (1997) Evaluation of the differentiation potential of WB-F344 rat liver epithelial stem-like cells in vivo: differentiation to hepatocytes after transplantation into dipeptidylpeptidase-IV-deficient rat liver. *Am. J. Pathol.* 151: 353–359.

Coon, H. (1969) Clonal culture of differentiated cells from mammals: rat liver cell culture. *Carnegie Inst. Washington Year Book* 67:419–421.

Crosby, H. A., Hubscher, S. G., Joplin, R. E., Kelly, D. A., and Strain, A.J. (1998) Immunolocalization of OV-6, a putative progenitor cell marker in human fetal and diseased pediatric liver. *Hepatology* 28: 980–985.

Crosby, H. A., Kelly, D. A., and Strain, A. J. (2001) Human hepatic stem-like cells isolated using c-kit or CD34 can differentiate into biliary epithelium. *Gastroenterology* 120:534–544.

Crossley, P. H. and Martin, G. R. (1995) The mouse Fgf8 gene encodes a family of polypeptides and is expressed in regions that direct out-growth and patterning in the developing embryo. *Development* 121: 439–451.

Dabeva, M. D. and Shafritz, D. A. (1993) Activation, proliferation, and differentiation of progenitor cells into hepatocytes in the D-galactosamine model of liver regeneration. *Am. J. Pathol.* 143:1606–1620.

Dabeva, M. D., Hwang, S-G., Vasa, S. R. G., et al. (1997) Differentiation of pancreatic epithelial progenitor cells into hepatocytes following transplantation into rat liver. *Proc. Natl. Acad. Sci. USA* 94: 7356–7361.

Deutsch, G., Jung, J., Zheng, M., Lora, J., and Zaret, K.S. (2001) A bipotential precursor population for pancreas and liver within the embryonic endoderm. *Development* 128:871–881.

Dumble, M. L., Croager, E. J., Yeoh, G. C., and Quail, E. A. (2002) Generation and characterization of p53 null transformed hepatic progenitor cells: oval cells give rise to hepatocellular carcinoma. *Carcinogenesis* 23:435–445.

Dunsford, H. A. and Sell, S. (1989) Production of monoclonal antibodies to preneoplastic liver cell populations induced by chemical carcinogens in rats and to transplantable Morris hepatomas. *Cancer Res.* 49: 4887–4893.

Earp, H. S., Hepler, J. R., Petch, L. A., et al. (1988) Epidermal growth factor (EGF) and hormones stimulate phosphoinositide hydrolysis and increase EGF receptor protein synthesis and mRNA levels in rat liver epithelial cells: evidence for protein kinase C-dependent and -independent pathways. *J. Biol. Chem.* 263:13,868–13,874.

Engelhardt, N. V., Factor, V. M., Medvinsky, A. L., Baranov, V. N., Lazareva, M. N., and Poltoranina, V. S. (1993) Common antigen of oval and biliary epithelial cells (A6) is a differentiation marker of epithelial and erythroid cell lineages in early development of the mouse. *Differentiation* 55:19–26.

Germain, L., Blouin, M. J., and Marceau, N. (1988) Biliary epithelial and hepatocytic cell lineage relationships in embryonic rat liver as determined by the differential expression of cytokeratins, alpha-fetoprotein, albumin, and cell surface–exposed components. *Cancer Res.* 48: 4909–4918.

Gerschenson, L. E., Andersson, M., Molson, J., and Okigaki, T. (1970) Tyrosine transaminase induction by dexamethasone in a new rat liver cell line. *Science* 170:859–861.

Golubovskaya, V. M., Filatov, L. V., Behe, C. I., et al. (1999) Telomere shortening, telomerase expression, and chromosome instability in rat hepatic epithelial stem-like cells. *Mol. Carcinog.* 24:209–217.

Gordon, G. J., Coleman, W. B., Hixson, D. C., and Grisham, J. W. (2000a) Liver regeneration in rats with retrorsine-induced hepatocellular injury proceeds through a novel cellular response. *Am. J. Pathol.* 156: 607–619.

Gordon, G. J., Coleman, W. B., and Grisham, J. W. (2000b) Temporal analysis of hepatocyte differentiation by small hepatocyte-like progenitor cells during liver regeneration in retrorsine-exposed rats. *Am. J. Pathol.* 157:771–786.

Grisham, J. W. (1980) Cell types in long-term propagable cultures of rat liver. *Ann. NY Acad. Sci.* 349:128–137.

Grisham, J. W., Thal, S. B., and Nagel, A. (1975) Cellular derivation of continuously cultured epithelial cells in normal rat liver. In *Gene Expression and Carcinogenesis in Cultured Liver* (Gerschenson, L. E. and Thompson, E. B., eds.), Academic, New York, pp. 1–23.

Grisham, J. W., Coleman, W. B., and Smith, G. J. (1993) Isolation, culture, and transplantation of rat hepatocytic precursor (stem-like) cells. *Proc. Soc. Exp. Biol. Med.* 204:270–279.

Gupta, S., Rajvanshi, P., Sokhi, R. P., Vaidya, S., Irani, A. N., and Gorla, G. R. (1999) Position-specific gene expression in the liver lobule is directed by the microenvironment and not by the previous cell differentiation state. *J. Biol. Chem.* 274:2157–2165.

Gurdon, J. B., Laskey, R. A., and Reeves, O. R. (1975) The developmental capacity of nuclei transplanted from keratinized skin cells of adult frogs. *J. Embryol. Exp. Morphol.* 34:93–112.

Haruna, Y., Saito, K., Spaulding, S., Nalesnik, M. A., and Gerber, M. A. (1996) Identification of bipotential progenitor cells in human liver development. *Hepatology* 23:476–481.

Herring, A. S., Raychaudhuri, R., Kelley, S. P., and Iype, P. T. (1983) Repeated establishment of diploid epithelial cell cultures from normal and partially hepatectomized rats. *In Vitro* 19:576–588.

Hillebrands, J. L., Klatter, F. A., van Dijk, W. D., and Rozing, J. (2002). Bone marrow does not contribute substantially to endothelial-cell replacement in transplant arteriosclerosis. *Nat. Med.* 8:194–195.

Hixson, D. C. and Allison, J. P. (1985) Monoclonal antibodies recognizing oval cells induced in the liver of rats by N-2-fluorenylacetamide or ethionine in a choline-deficient diet. *Cancer Res.* 45:3750–3760.

Hixson, D. C., Chapman, L., McBride, A., Faris, R., and Yang, L. (1997) Antigenic phenotypes common to rat oval cells, primary hepatocellular carcinomas and developing bile ducts. *Carcinogenesis* 18:1169–1175.

Hochedlinger, K. and Jaenisch, R. (2002). Monoclonal mice generated by nuclear transfer from mature B and T donor cells. *Nature* 415: 1035–1038.

Hooth, M. J., Coleman, W. B., Presnell, S. C., Borchert, K. M., Grisham, J. W., and Smith, G. J. (1998) Spontaneous neoplastic transformation of WB-F344 rat liver epithelial cells. *Am. J. Pathol.* 153:1913–1921.

Isfort, R. J., Cody, D. B., Doersen, C. J., et al. (1997) The tetratricopeptide repeat containing Tg737 gene is a liver neoplasia tumor suppressor gene. *Oncogene* 15:1797–1803.

Isfort, R. J., Cody, D. B., Richards, W. G., Yoder, B. K., Wilkinson, J. E., and Woychik, R. P. (1998) Characterization of growth factor responsiveness and alterations in growth factor homeostasis involved in the tumorigenic conversion of mouse oval cells. *Growth Factors* 15: 81–94.

Iype, P.T. (1971) Cultures from adult rat liver cells. I. Establishment of monolayer cell-cultures from normal liver. *J. Cell Physiol.* 78:281–288.

Jones, E. A., Tosh, D., Wilson, D. I., Lindsay, S., and Forrester, L. M. (2002) Hepatic differentiation of murine embryonic stem cells. *Exp. Cell Res.* 272:15–22.

Jung, J., Zheng, M., Goldfarb, M., and Zaret, K. (1999) Initiation of mammalian liver development from endoderm and fibroblast growth factors. *Science* 284:1999–2003.

Kaighn, M. E. and Prince, A. M. (1971) Production of albumin and other serum proteins by clonal cultures of normal human liver. *Proc. Natl. Acad. Sci. USA* 68:2396–2400.

Kano, J., Noguchi, M., Kodama, M., and Tokiwa, T. (2000) The *in vitro* differentiating capacity of nonparenchymal epithelial cells derived from adult porcine livers. *Am. J. Pathol.* 156:2033–2043.

Kaufmann, W. K., Behe, C. I., Golubovskaya, V. M., et al. (2001) Aberrant cell cycle checkpoint function in transformed hepatocytes and WB-F344 hepatic epithelial stem-like cells. *Carcinogenesis* 22: 1257–1269.

Knight, B., Yeoh, G.C.T., Husk, K.L., et al. (2000) Impaired preneoplastic changes and liver tumor formation in tumor necrosis factor receptor type 1 knockout mice. *J. Exp. Med.* 192:1809–1818.

Koch, K. and Leffert, H. L. (1974) Growth control of differentiated fetal rat hepatocytes in primary monolayer culture. VI. Studies with conditioned medium and its functional interactions with serum factors. *J. Cell Biol.* 62:780–791.

Koch, K. S. and Leffert, H. L. (1979) Increased sodium ion influx is necessary to initiate rat hepatocyte proliferation. *Cell* 18:153–163.

Koch, K. S. and Leffert, H. L. (1980) Growth control of differentiated adult rat hepatocytes in primary culture. *Ann. NY Acad. Sci.* 349:111–127.

Koch, K. S. and Leffert, H. L. (1994) Hepatic regeneration and gene expression: a tribute to Hidematsu Hirai. *J. Tumour Marker Oncol.* 9: 35–56.

Koch, K. S., Shapiro, P., Skelly, H., and Leffert, H. L. (1982) Rat hepatocyte proliferation is stimulated by insulin-like peptides in defined medium. *Biochem. Biophys. Res. Commun.* 109:1054–1060.

Koch, K. S., Lu, X. P., Brenner, D. A., Fey, G. H., Martinez-Conde, A., and Leffert, H. L. (1990) Mitogens and hepatocyte growth control *in vivo* and *in vitro*. *In Vitro Cell. Dev. Biol.* 26:1011–1023.

Koch, K. S., Lu, X. P., and Leffert, X. P. (1994) Primary rat hepatocytes express cyclin D1 messenger RNA during their growth cycle and during mitogenic transitions induced by transforming growth factor-alpha. *Biochem. Biophys. Res. Commun.* 204:91–97.

Kruijer, W., Skelly, H., Botteri, F., et al. (1986) Proto-oncogene expression in regenerating liver is simulated in cultures of primary adult rat hepatocytes. *J. Biol. Chem.* 261:7929–7933.

Kubota, H. and Reid, L. M. (2000) Clonogenic hepatoblasts, common precursors for hepatocytic and biliary lineages, are lacking classical major histocompatibility complex class I antigen. *Proc. Natl. Acad. Sci. USA* 97:12,132–12,137.

Laconi, E., Oren, R., Mukhopadhyay, D. K., et al. (1998) Long-term, near-total liver replacement by transplantation of isolated hepatocytes in rats treated with retrorsine. *Am. J. Pathol.* 153:319–329.

Lad, P. J., Shier, W. T., Skelly, H., de Hemptinne, B., and Leffert, H. L. (1982) Adult rat hepatocytes in primary culture. VI. Developmental changes in alcohol dehydrogenase activity and ethanol conversion during the growth cycle. *Alcohol. Clin. Exp. Res.* 6:64–71.

Lagasse, E., Connors, H., Al-Dhalimy, M., et al. (2000) Purified hematopoietic stem cells can differentiate into hepatocytes in vivo. *Nat. Med.* 6:1229–1234.

Lazaro, C. A., Rhim, J. A., Yamada, Y., and Fausto, N. (1998) Generation of hepatocytes from oval cell precursors in culture. *Cancer Res.* 58: 5514–5522.

Le Douarin, N. M. (1975) An experimental analyisis of liver development. *Med. Biol.* 53:427–455.

Lee, L. W., Tsao, M. S., Grisham, J. W., and Smith, G. J. (1989) Emergence of neoplastic transformants spontaneously or after exposure to N-methyl-N'-nitro-N-nitrosoguanidine in populations of rat liver epithelial cells cultured under selective and nonselective conditions. *Am. J. Pathol.* 135:63–71.

Leffert, H. L. (1974a) Growth control of differentiated fetal rat hepatocytes in primary monolayer culture. V. Occurrence in dialyzed fetal bovine serum of macromolecules having both positive and negative growth regulatory functions. *J. Cell Biol.* 62:767–779.

Leffert, H. L. (1974b) Growth control of differentiated fetal rat hepatocytes in primary monolayer culture. VII. Hormonal control of DNA synthesis and its possible significance to the problem of liver regeneration. *J. Cell Biol.* 62:792–801.

Leffert, H. L. and Koch, K. S. (1977) Control of animal cell proliferation. In: *Growth, Nutrition and Metabolism of Cells in Culture* (Rothblat, G. H. and Cristofalo, V. J., eds.), vol. 3, Academic, New York, pp. 225–294.

Leffert, H. L. and Koch, K. S. (1978) Proliferation of hepatocytes. In: *Hepatotrophic Factors,* vol. 55, CIBA Foundation Symposium, London, UK, pp. 61–83.

Leffert, H. L. and Koch, K. S. (1979) Regulation of growth of hepatocytes by sodium ions. *Prog. Liver Dis.* 6:123–134.

Leffert, H. L. and Koch, K. S. (1980) Ionic events at the membrane initiate rat liver regeneration. *Ann. NY Acad. Sci.* 339:201–215.

Leffert, H. L. and Koch, K. S. (1982a) Hepatocyte growth regulation by hormones in chemically-defined medium: a two-signal hypothesis. *Cold Spring Harbor Symp. Cell Prolif.* 9:597–613.

Leffert, H. L. and Koch, K. S. (1982b) Monovalent cations and the control of hepatocyte proliferation in chemically defined medium. In: *Ions, Cell Proliferation and Cancer* (Boynton, A., McKeenan, W. L. and Whitfield, J. F., eds.), Academic, New York, pp. 103–130.

Leffert, H. L. and Koch, K. S. (1985) Experimental issues in hepatocyte growth control studies in primary culture. In: *Growth and Differentiation of Cells in Defined Environment* (Murakami, H., Yamane, I., Barnes, D. W., Mather, J. P., Hayashi, I., and Sato, G. H., eds.), Springer-Verlag, Berlin, pp. 9–18.

Leffert, H. L. and Paul, D. (1972) Studies on primary cultures of differentiated fetal liver cells. *J. Cell Biol.* 52:559–568.

Leffert, H. L. and Paul, D. (1973) Serum dependent growth of primary cultured differentiated fetal rat hepatocytes in arginine-deficient medium. *J. Cell Physiol.* 81:113–124.

Leffert, H. L. and Sell, S. (1974) Alpha$_1$-fetoprotein biosynthesis during the growth cycle of differentiated fetal rat hepatocytes in primary monolayer culture. *J. Cell Biol.* 61:823–829.

Leffert, H. L. and Weinstein, D. B. (1976) Growth control of differentiated fetal rat hepatocytes in primary monolayer culture. IX. Specific inhibition of DNA synthesis initiation by very low density lipoprotein and possible significance to the problem of liver regeneration. *J. Cell Biol.* 70:20–32.

Leffert, H. L., Moran, T., Boorstein, R. T., and Koch, K. S. (1977) Procarcinogen activation and hormonal control of cell proliferation in differentiated primary adult rat liver cell cultures. *Nature* 267:58–61.

Leffert, H., Moran, T., Sell, S., et al. (1978a) Growth state dependent phenotypes of adult hepatocytes in primary monolayer culture. *Proc. Natl. Acad. Sci. USA* 75:1834–1838.

Leffert, H. L., Koch, K. S., Rubalcava, B., Sell, S., Moran, T., and Boorstein, R. (1978b) Hepatocyte growth control: *in vitro* approach to problems of liver regeneration and function. *Natl. Cancer Inst. Monogr.* 48:87–101.

Leffert, H. L., Koch, K. S., Moran, T., and Williams, M. (1979) Liver cells. In: *Methods in Enzymology*, vol. 58 (Jakoby, W. and Pastan, I., eds.), Academic, New York, pp. 536–544.

Leffert, H. L., Koch, K. S., Sell, S., Skelly, H., and Shier, W. T. (1983) Biochemistry and biology of N-acetyl-2-aminofluorene in primary cultures of adult rat hepatocytes. In: *Application of Biological Markers to Carcinogen Testing*, vol. 29 (Milman, H. A. and Sell, S., eds.), U.S. Environmental Protection Agency, Environmental Sciences Research, Washington, DC, pp. 119–133.

Leffert, H. L., Koch, K. S., Lad, P. J., Shapiro, P., Skelly, H., and de Hemptinne, B. (1988) Hepatocyte regeneration, replication and differentiation. In: *The Liver: Biology and Pathobiology* (Arias, I., Popper, H., Schacter, D., and Shafritz, D., eds.), 2nd ed., Raven, New York, pp. 833–850.

Lin, P., Liu, C., Tsao, M. S., and Grisham, J. W. (1987) Inhibition of proliferation of cultured rat liver epithelial cells at specific cell cycle stages by transforming growth factor-beta. *Biochem. Biophys. Res. Commun.* 143:26–30.

Lu, X. P., Koch, K. S., Lew, D. J., et al. (1992) Induction of cyclin mRNA and histone H1-kinase during liver regeneration. *J. Biol. Chem.* 267: 2841–2844.

Mackay, A. R., Gomez, D. E., Cottam, D. W., Rees, R. C., Nason, A. M., and Thorgeirsson, U. P. (1993) Identification of the 72-kDa (MMP-2) and 92-kDa (MMP-9) gelatinase/type IV collagenase in preparations of laminin and Matrigel. *Biotechniques* 15:1048–1051.

Malouf, N. N., Coleman, W. B., Grisham, J. W., et al. (2001) Adult-derived stem cells from the liver become myocytes in the heart in vivo. *Am. J. Pathol.* 158:1929–1935.

Marceau, N., Goyette, R., Deschenes, J., and Valet, J. P. (1980) Morphological differences between epithelial and fibroblast cells in rat liver cultures, and the roles of cell surface fibronectin and cytoskeletal element organization in cell shape. *Ann. NY Acad. Sci.* 349:138–152.

Martin, G. R. and Evans, M. J. (1975) Differentiation of clonal lines of teratocarcinoma cells: formation of embryoid bodies in vitro. *Proc. Natl. Acad. Sci. USA* 72:1441–1445.

McCulloch, E. A. (2003) Normal and leukemia hematopoietic stem cells and lineages. In: *Stem Cells Handbook* (Sell, S., ed.), Humana, Totowa, NJ, pp. 119–132.

McKinney-Freeman, S. L., Jackson, K. A., Camargo, F. D., Ferrari, G., Mavilio, F., and Goodell, M. A. (2002) Muscle-derived hematopoietic stem cells are hematopoietic in origin. *Proc. Natl. Acad. Sci. USA* 99: 1341–1346.

McMahon, J. B., Richards, W. L., del Campo, A. A., Song, M.-K. H., and Thorgeirsson, S. S. (1986) Differential effects of transforming growth factor-β on proliferation of normal and malignant rat liver epithelial cells in culture. *Cancer Res.* 46:4665–4671.

Mitaka, T., Sato, F., Mizuguchi, T., Yokono, T., and Mochizuki, Y. (1999) Reconstruction of hepatic organoid by rat small hepatocytes and hepatic nonparenchymal cells. *Hepatology* 29:111–125.

Monga, S. P., Tang, Y., Candotti, F., et al. (2001) Expansion of hepatic and hematopoietic stem cells utilizing mouse embryonic liver explants. *Cell Transplant.* 10:81–89.

Ott, M., Rajvanshi, P., Sokhi, R. P., et al. (1999) Differentiation-specific regulation of transgene expression in a diploid epithelial cell line derived from the normal F344 rat liver. *J. Pathol.* 187:365–373.

Parlow, M. H., Bolender, D. L., Koran-Moore, N. P., and Lough, J. (1991) Localization of bFGF-like proteins as punctate inclusions in the preseptation myocardium of the chicken embryo. *Dev. Biol.* 146: 139–147.

Paul, D., Leffert, H., Sato, G., and Holley, R. W. (1972) Stimulation of DNA and protein synthesis in fetal rat liver cells by serum from partially hepatectomized rats. *Proc. Natl. Acad. Sci. USA* 69: 374–377.

Petersen, B. E., Goff, J. P., Greenberger, J. S., and Michalopoulos, G. K. (1998) Hepatic oval cells express the hematopoietic stem cell marker Thy-1 in the rat. *Hepatology* 27:433–445.

Petersen, B. E., Bowen, W. C., Patrene, K. D., et al. (1999) Bone marrow as a potential source of hepatic oval cells. *Science* 264:1168–1170.

Petkov, P. M., Kim, K., Sandhu, J., Shafritz, D. A., and Dabeva, M. D. (2000) Identification of differentially expressed genes in epithelial stem/progenitor cells of fetal rat liver. *Genomics* 68:197–209.

Plescia, C., Rogler, C., and Rogler, L. (2001) Genomic expression analysis implicates Wnt signaling pathway and extracellular matrix alterations in hepatic specification and differentiation of murine hepatic stem cells. *Differentiation* 68:254–269.

Rao, M. S., Subbarao, V., and Reddy, J. K. (1986) Induction of hepatocytes in the pancreas of copper-depleted rats following copper repletion. *Cell Differ.* 18:109–117.

Richards, W. G., Yoder, B. K., Isfort, R. J., et al. (1996) Oval cell proliferation associated with the murine insertional mutation TgN737Rpw. *Am. J. Pathol.* 149:1919–1930.

Richards, W. G., Yoder, B. K., Isfort, R. J., et al. (1997) Isolation and characterization of liver epithelial cell lines from wild-type and mutant TgN737Rpw mice. *Am. J. Pathol.* 150:1189–1197.

Rideout, W. M. III, Hochedlinger, K., Kyba, M., Daley, G. Q., and Jaenisch, R. (2002) Correction of a genetic defect by nuclear transplantation and combined cell and gene therapy. *Cell* 109:17–27.

Rogler, L.E. (1997) Selective bipotential differentiation of mouse embryonic hepatoblasts in vitro. *Am. J. Pathol.* 150:591–602.

Rosenthal, M. D., Wishnow, R. M., and Sato, G. H. (1970) In vitro growth and differentiation of clonal populations of multipotential mouse cells derived from a transplantable testicular teratocarcinoma. *J. Natl. Cancer Inst.* 44:1001–1014.

Rossi, J. M., Dunn, N. R., Hogan, B. L., and Zaret, K. S. (2001) Distinct mesodermal signals, including BMPs from the septum transversum mesenchyme, are required in combination for hepatogenesis from the endoderm. *Genes Dev.* 15:1998–2009.

Rubin, H. (2001) Multistage carcinogenesis in cell culture. *Dev. Biol. (Basel)* 106:61–66.

Seglen, P. O. (1997) DNA ploidy and autophagic protein degradation as determinants of hepatocellular growth and survival. *Cell Biol. Toxicol.* 13:301–315.

Sell, S. (1978) Distribution of alphafetoprotein and albumin containing cells in the livers of Fischer rats fed *N*-2- fluorenyl-acetamide. *Cancer Res.* 38:3107–3113.

Sell, S. (2001) Heterogeneity and plasticity of hepatocyte lineage cells. *Hepatology* 33:738–750.

Sell, S. and Ilic, I. (1997) *Liver Stem Cells*, R. G. Landes and Chapman & Hall, New York.

Sell, S. and Leffert, H. L. (1982) An evaluation of cellular lineages in the pathogenesis of experimental hepatocellular carcinoma. *Hepatology* 2:77–86.

Sell, S., Leffert, H., Mueller-Eberhard, U., Kida, S., and Skelly, H. (1975) Relationship of the biosynthesis of alpha$_1$-fetoprotein, albumin, hemopexin and haptoglobin to the growth state of fetal rat hepatocyte cultures. *Ann. NY Acad. Sci.* 259:45–48.

Sell, S., Becker, F. F., Leffert, H. L., and Watabe, H. (1976) Expression of an oncodevelopmental gene product (alpha$_1$-fetoprotein) during fetal development and adult oncogenesis. *Cancer Res.* 36:4239–4249.

Sell, S., Osborn, K., and Leffert, H.L. (1981a) Autoradiography of oval cells appearing rapidly in the livers of rats fed N-2-fluorenylacetamide in a choline-devoid diet. *Carcinogenesis* 2:7–14.

Sell, S., Leffert, H. L., Shinozuka, H., Lombardi, B., and Gochman, N. (1981b) Rapid development of large numbers of AFP-containing "oval" cells in the liver of rats fed N-2-fluorenylacetamide in a choline-devoid diet. *GANN* 72:479–487.

Shiojiri, N. and Mizuno, T. (1993) Differentiation of functional hepatocytes and biliary epithelial cells from immature hepatocytes of the fetal mouse in vitro. *Anat. Embryol. (Berl.)* 187:221–229.

Siminovitch, L., McCulloch, E. A., and Till, J. E. (1963) The distribution of colony-forming cells among spleen colonies. *J. Cell. Comp. Physiol.* 62:327–336.

Sirica, A. E., Richards, W., Tsukada, Y., Sattler, C. A., and Pitot, H. C. (1979) Fetal phenotypic expression by adult rat hepatocytes on collagen gel/nylon meshes. *Proc. Natl. Acad. Sci. USA* 76:283–287.

Sirica, A. E., Mathis, G. A., Sano, N., and Elmore, L. W. (1990) Isolation, culture, and transplantation of intrahepatic biliary epithelial cells and oval cells. *Pathobiology* 58:44–64.

Spagnoli, F. M., Amicone, L., Tripodi, M., and Weiss, M. C. (1998) Identification of a bipotential precursor cell in hepatic cell lines derived from transgenic mice expressing cyto-met in the liver. *J. Cell Biol.* 143:1101–1112.

Srivastava, G., Bawden, M. J., Anderson, A., and May, B. K. (1989) Drug induction of P450IIB1/IIB2 and 5-aminolevulinate synthase mRNAs in rat tissues. *Biochim. Biophys. Acta* 100:192–195.

Steinberg, P., Steinbrecher, R., Radaeva, S., et al. (1994) Oval cell lines OC/CDE 6 and OC/CDE 22 give rise to cholangiocellular and undifferentiated carcinomas after transformation. *Lab. Invest.* 71:700–709.

Suzuki, A., Taniguchi, H., Zheng, Y.W., et al. (2000) Clonal colony formation of hepatic stem/progenitor cells enhanced by embryonic fibroblast conditioning medium. *Transplant. Proc.* 32:2328–2330.

Suzuki, A., Zheng, Y.-W., Kaneko, S., et al. (2002) Clonal identification and characterization of self-renewing pluripotent stem cells in the developing liver. *J. Cell Biol.* 156:173–184.

Takaoka, T., Yasumoto, S., and Katsuta, H. (1975) A simple method for the cultivation of rat liver cells. *J. Exp. Med.* 45:317–326.

Taniguchi, H., Toyoshima, T., Fukao, K., and Nakauchi, H. (1996) Presence of hematopoietic stem cells in the adult liver. *Nat. Med.* 2:198–203.

Tateno, C. and Yoshizato, K. (1996a) Growth and differentiation in culture of clonogenic hepatocytes that express both phenotypes of hepatocytes and biliary epithelial cells. *Am. J. Pathol.* 149:1593–1605.

Tateno, C. and Yoshizato, K. (1996b) Long-term cultivation of adult rat hepatocytes that undergo multiple cell divisions and express normal parenchymal phenotypes. *Am. J. Pathol.* 148:383–392.

Terada, N., Hamazaki, T., Oka, M., et al. (2002) Bone marrow cells adopt the phenotype of other cells by spontaneous cell fusion. *Nature* 416: 542–545.

Theise, N. D., Badve, S., Saxena, R., et al. (2000a) Derivation of hepatocytes from bone marrow cells in mice after radiation-induced myeloablation. *Hepatology* 31:235–240.

Theise, N. D., Nimmakayalu, M., Gardner, R., et al. (2000b) Liver from bone marrow in humans. *Hepatology* 32:11–16.

Till, J. E. and McCulloch, E. A. (1961) A direct measurement of the radiation sensitivity of normal mouse bone marrow cells. *Radiat. Res.* 14:1419–1430.

Till, J. E., McCulloch, E. A., and Siminovitch, L. (1963) A stochastic model of stem cell proliferation based on the growth of spleen colony-forming cells. *Proc. Natl. Acad. Sci. USA* 51:29–36.

Tsao, M. S., Grisham, J. W., and Nelson, K. G. (1985) Clonal analysis of tumorigenicity and paratumorigenic phenotypes in rat liver epithelial cells chemically transformed in vitro. *Cancer Res.* 45:5139–5144.

Tsao, M. S., Earp, H. S., and Grisham, J. W. (1986) The effects of epidermal growth factor and the state of confluence on enzymatic activities of cultured rat liver epithelial cells. *J. Cell Physiol.* 126:167–173.

Tsao, M.S., Smith, J.D., Nelson, K.G., and Grisham, J.W. (1984) A diploid epithelial cell line from normal adult rat liver with phenotypic properties of "oval" cells. *Exp. Cell Res.* 154:38–52.

Tsao, M.S., Sanders, G.H., and Grisham, J.W. (1987) Regulation of growth of cultured hepatic epithelial cells by transferrin. *Exp. Cell Res.* 171:52–62.

Wang, X., Al-Dhalimy, M., Lagasse, E., Finegold, M., and Grompe, M. (2001) Liver repopulation and correction of metabolic liver disease by transplanted adult mouse pancreatic cells. *Am. J. Pathol.* 158: 571–579.

Watabe, H. (1971) Early appearance of embryonic alpha-globulin in rat serum during carcinogenesis with 4-dimethylaminoaxobenzene. *Cancer Res.* 31:1192–1194.

Watabe, H., Leffert, H., and Sell, S. (1976) Developmental and maturational changes in $alpha_1$-fetoprotein and albumin production in cultured fetal rat hepatocytes. In: *Oncodevelopmental Gene Expression*, Fourth Meeting, International Study Group for Carcinoembryonic Proteins (Fishman, W. and Sell, S., eds.), Academic, New York, pp. 123–130.

Watt, A. J., Jones, E. A., Ure, J. M., Peddie, D., Wilson, D. I., and Forrester, L. M. (2001) A gene trap integration provides an early in situ marker for hepatic specification of the foregut endoderm. *Mech. Dev.* 100:205–215.

Williams, G. M., Weisburger, E. K., and Weisburger, J. H. (1971) Isolation and long-term cell culture of epithelial-like cells from rat liver. *Exp. Cell Res.* 69:106–112.

Ying, Q.-L., Nichols, J., Evans, E. P., and Smithe, A. G. (2002) Changing potency by spontaneous fusion. *Nature* 416:545–548.

Zhang, M., Sell, S., and Leffert, H. L. (2003) Hepatic progenitor cell lines from allyl alcohol-treated adult rats are derived from γ-irradiated mouse STO cells. *Stem Cells* 21:449–458.

Zhu, X., Sasse, J., McAllister, D., and Lough, J. (1996) Evidence that fibroblast growth factors 1 and 4 participate in regulation of cardiogenesis. *Dev. Dyn.* 207:429–438.

34 Permanent Lines of Stem Cells from the Liver

HÉLÈNE STRICK-MARCHAND, PhD AND MARY C. WEISS, PhD

The origins and properties of putative liver stem cell lines, sometimes designated LPC or liver progenitor cells, are discussed. Stem cells are considered to be clonogenic, self-renewing, and to have more than one pathway of differentiation. Cells lines that were among the first to be isolated, such as WB-F334, LE/2, LE/6, and OC/CDE are presented, as well as more recently isolated IPFLS, HBC-3, MMH, BMEL, and $p53^{-/-}$ lines. Each appears to have at least some properties of liver progenitor or oval cells. If obtained from a carcinogen-treated animal or transformed in vitro, most of these cell lines produce tumors on transplantation. Pancreatic oval cells from copper-treated rats also have LPC properties. A number of the lines discussed are bipotential and can form bile ducts or hepatocytes.

34.1. INTRODUCTION

This chapter deals with liver cell lines mainly from rodent species, and in particular those cases in which the cells have been shown to demonstrate properties of stem cells. Here we use the term *stem cell* to mean cells that are clonogenic, are self-renewing, and have the capacity to follow more than one pathway of differentiation.

Taking into account the embryonic and adult organism, the liver is thought to contain four cell types of endodermal origin. In the developing liver, there are three endodermal cell types: hepatoblasts, defined as cells that express α-fetoprotein (AFP) and albumin and are recognizable in the mouse embryo at embryonic day (E) 10, and their derivatives hepatocytes and bile duct cells (cholangiocytes), which begin to emerge at E14. In the adult liver, a cell type designated "the oval cell" is distinguished. Since oval cells can give rise to both hepatocytes and bile duct cells, their filiation within the endodermal compartment is presumed (*see* below for references). However, it has recently been found that hematopoietic stem cells (HSCs) may contribute to several highly differentiated tissues, including the liver (Petersen et al., 1999), Theise et al., 2000). Indeed, such a possibility of destiny switching (transdifferentiation) of cells derived from dissimilar embryonic germ layers has reinforced the importance of clarification of hepatic lineages.

It is not known how long hepatoblasts remain in the liver, nor whether they give rise to oval cells. Indeed, the hepatoblast population could disappear entirely during development as its progeny differentiate into hepatocytes or bile duct cells. Alternatively, hepatoblasts could remain in the liver, being recognized as oval cells in the adult. An alternative view of the origin of oval cells comes from their localization, in proximity to the canals of Hering, where hepatocytes and bile duct cells are contiguous. Consequently, it can be argued that oval cells arise from the neighboring bile duct cells, from nearby hepatocytes, or even from a transitional cell that is not yet committed.

Some of the work on liver-derived cell lines has been undertaken to clarify these questions of lineage relations among hepatic cells. One of the difficulties in identifying cell types of the liver resides in the fact that some of them possess few if any diagnostic markers (Table 1). This problem does not concern the hepatocyte for which markers are numerous, including liver-enriched transcription factors (LETFs), serum proteins, specialized enzymes of metabolism of sugars, lipids, and amino acids, and surface receptors. However, the situation is less simple for hepatoblasts, bile duct cells, and oval cells. In particular, few studies address directly the question of markers of hepatoblasts. In addition, oval cells are known to be variable in the markers that they express, such that they are often referred to as comprising a "heterogeneous cell compartment." Bile duct cells are identified by a combination of morphological features and the markers that are expressed, but most of the latter are shared with oval cells. It is clear from Table 1 that distinction between hepatoblasts and oval cells is difficult, and that both cell types share many markers in common with bile duct cells.

It should also be stressed that subtle differences in the properties of cell lines could be related to the species of origin. For example, it is well known that primary cultures of hepatocytes

From: *Stem Cells Handbook*
Edited by: S. Sell © Humana Press Inc., Totowa, NJ

Table 1
Markers for Identification of Liver Cells

Markers	*Hepatoblast*	*Hepatocyte*	*Bile duct cell*	*Oval cell*	*References*
Oval					
Thy-1	NF	–	–	+	Petersen et al. (1998)
Bile duct/oval					
GGT IV	+	–	+	+	Shiojiri et al. (1991), Petersen et al. (1998), Holic et al. (2000)
CD34	NF	–	+	+	Omori et al. (1997)
c-kit	NF	–	+	+	Fujio et al. (1996), Omori et al. (1997)
IB 4	–	–	+	+	Couvelard et al. (1998)
CX 43	NF	–	+	+	Zhang and Thorgeirsson (1994)
CK 7 and 19	–	–	+	+	Germain et al. (1988), Shiojiri et al. (1991), Petersen et al. (1998)
OC.2, OC.4, OC.5, OC.10*	+	–	+	+	Hixson et al. (2000)
OV6*	–	–	+	+	Yang et al. (1993), Hixson et al. (1997)
BD1*	–	–	+	–	Yang et al. (1993)
Bile duct/hepatocyte					
HNF6	+	+	+	NF	Landry et al. (1997), Rausa et al. (1997), Clotman et al. (2002)
HNF1β	+	+	+	+	Ott et al. (1991), Nagy et al. (1994), Barbacci et al. (1999), Coffinier et al. (1999, 2002)
CK 8 and 18	+	+	+	+	Germain et al. (1988), Shiojiri et al. (1991)
LETF					
HNF1α	+	+	–	+	Ott et al. (1991), Nagy et al. (1994)
HNF3 α, β, γ	+	+	–	+	Ang et al. (1993), Nagy et al. (1994), Rausa et al. (1999)
HNF4	+	+	–	+	Duncan et al. (1994), Nagy et al. (1994)
C/EBP α, β	NF	+	–	+	Kuo et al. (1990) Nagy et al. (1994)
Hepatocyte					
AFP	+	+	–	+	Shiojiri et al. (1991), Rogler (1997), Petersen et al. (1998)
Albumin	+	+	–	+	Germain et al. (1988), Shiojiri et al. (1991), Rogler (1997)
Apo AIV	NF	+	–	NF	Glickman and Sabesin (1994)
ADH	NF	+	–	NF	Weiner et al. (1994), Tietjen et al. (1994), Vonesch et al. (1994)
Apo B	NF	+	–	NF	Glickman and Sabesin (1994)
Aldolase B	NF	+	–	NF	Guikkouzo et al. (1981), Numazaki et al. (1984), Seifter and Englard (1994)

NF, not found in the literature; –, not expressed; +, expressed, *, rat-specific marker.

Identification of marker expression by hepatoblasts and oval cells is delicate because only individual cell analysis procedures (immunocytochemistry, immunohistochemistry, and in situ hybridization) are conclusive. Indeed, these two cell types are often minority and are surrounded by other cell types. If conclusive evidence of expression (or absence thereof) was not found, NF is used.

from the rat are more resistant to short-term culture than those of the mouse. Rat cells grow more vigorously and clone more easily than mouse cells, and mouse cells are more tractable than primate and human cells. In addition, diploid cell lines are more frequently isolated from rat than from mouse tissues. There may also be differences in the stringency of regulation of specific genes in cultured cells from different species, and even differences in the morphological aspects of cultures from rat, mouse and human. For all of these reasons, we cite the species used by different investigators.

Early attempts to isolate liver cell lines were almost exclusively biased to obtaining well-differentiated hepatocytes in culture. The lines that fulfilled this criterion were mainly derived from hepatomas (Darlington, 1987). Those from normal liver, whether fetal or adult, were frequently of poorly differentiated cells, some of them now recognized as being stem cells. The interest in liver stem cells was generated mainly from the cancer research community, because oval cells were thought by many to be the target cell of hepatic carcinogenesis (Fausto, 1994, and references therein).

A few investigators used immortalization directed by oncogenic viruses, including temperature-sensitive mutants, to obtain hepatocyte cell lines. A new generation of attempts to obtain more normal differentiated hepatocyte lines emerged with the technology of transgenic mice. Investigators used overexpression of growth factors or receptors relevant for the liver, or of immortalizing transgenes whose expression was directed by hepatocyte-specific promoters, to obtain immortalized hepatocytes.

In recent years, there has been a more concerted effort to make lines of undifferentiated (stem?) liver cells and to use manipulation of growth media and culture conditions to obtain differentiation. Some of these attempts have led to the description of bipotential and even pleuripotential lines. Although this chapter focuses on bipotential hepatic cell lines, work on hepatocyte lines that has contributed to development of the arsenal of techniques presently available to the liver research community are also considered.

34.2. RAT LIVER EPITHELIAL STEM-LIKE CELLS: WB-F344

The pioneering work of Coon (1968) established that epithelial cell colonies can be cloned from low-density cultures inoculated with collagenase-digested suspensions of rat liver. This work was then extended by J. W. Grisham and coworkers, who have contributed an exhaustive characterization of WB-F344 rat epithelial cells from adult liver. This section provides an overview of this work as a preamble to discussing other cell lines.

In 1980, Grisham suggested that long-term propagable hepatic cells were derived from a facultative stem cell residing in the terminal bile ductules. Parenchymal and nonparenchymal cell fractions of adult rat liver were plated and only the nonparenchymal fraction led to colonies of surviving cells; no clones survived from the hepatocyte fraction (Grisham, 1980). The WB-F344 cell line isolated from an adult rat is diploid and nontransformed, expressing AFP, albumin, aldolase A and C, pyruvate kinase L, alkaline phosphatase, lactate dehydrogenase, and weakly γ-glutamyl-transpeptidase (GGT) (Tsao et al., 1984). Although originally the WB-F344 cells were nontransformed, Hooth et al. (1998) showed that maintaining the cells at confluence in culture with infrequent passages led to selective enhancement of spontaneous transformation.

WB-F344 cells that had been transformed in vitro were injected subcutaneously and shown to give rise to hepatocellular carcinomas (HCCs), adenocarcinomas of biliary or intestinal type, and hepatoblastomas, thus showing the pleuripotency of the cells (Tsao and Grisham, 1987). In addition, WB-F344 cells tagged with β-galactosidase (β-gal) and a Neo selection cassette (BAG2-WB cells) were transplanted into the liver and shown to integrate into the liver plate, forming clusters of hepatocytes, which express albumin, transferrin, α1-antitrypsin, and tyrosine amino transferase (TAT) (Coleman et al., 1993; Grisham et al., 1993). Transplantation of WB cells into dipeptidyl peptidase IV (DPPIV) deficient rats showed that WB cells integrated into hepatic plates and differentiated into mature hepatocytes with functional bile canaliculi (Coleman et al., 1997). However, no transplanted cells were seen in bile ductules. BAG2-WB cells that were retrieved from the transplanted livers and replated in culture resembled the original WB cells rather than mature hepatocytes, implying that they dedifferentiate to their original phenotype when grown in culture (Grisham et al., 1993). It is still possible that a fraction of the BAG2-WB cells remained undifferentiated in the liver, and that it was this population that proliferated upon inoculation in culture.

In conclusion, in vivo transplanted WB cells differentiate as mature hepatocytes, and chemically transformed WB cells differentiate into multiple carcinomas: hepatocytic, bile ductular, hepatoblastic, and intestinal.

34.3. OVAL CELL LINES

34.3.1. OVAL CELL PROLIFERATION IN VIVO Oval cells are induced to proliferate in the adult liver when parenchymal damage has occurred and hepatocyte proliferation is inhibited. Numerous experimental protocols in rats induce oval cell proliferation, which eventually leads to hepatocarcinogenesis (Sell, 2001). Proliferation of oval cells is observed at the junctions between hepatocytes and bile ducts, an area known as the canal of Hering or the terminal bile ductule. Furthermore, the origin of oval cells was confirmed by the observation that bile ductular damage, induced by 4,4'-methylene dianiline exposure, inhibited oval cell proliferation (Petersen et al., 1997). Oval cells proliferate, invade the parenchyma, and differentiate as mature hepatocytes, fully replacing the damaged parenchyma. However, the embryonic origin of the oval cell compartment remains a matter of debate, because there are no markers unique to oval cells (Table 1).

Proliferation of oval cells in vivo is accompanied by the onset or upregulation of hepatocyte-specific gene expression in the terminal bile ductules. It has been shown, by *in situ* hybridization, that the LETFs HNF1α and HNF1β, HNF3γ, HNF4, and C/EBPβ, as well as the serum proteins AFP and albumin, display increased expression during oval cell proliferation (Nagy et al., 1994; Bisgaard et al., 1996). Bile duct ligation induces the proliferation of bile duct cells, but not of oval cells: indeed, no induction of the LETFs or of serum proteins is observed (Bisgaard et al., 1996). The overexpression of LETFs and serum proteins in oval cells is accompanied by the increased expression of the growth factors: hepatocyte growth factor (HGF), acidic fibroblast growth factor (aFGF), transforming growth factor-α (TGF-α), and interferon-γ receptor complex and its secondary response genes (Evarts et al., 1993; Alison, 1998; Bisgaard et al., 1999). However, the signals that induce the onset of oval cell proliferation are unknown.

34.3.2. BIPOTENTIAL OVAL CELL LINES DERIVED FROM ADULT RAT LIVERS

3.2.1. LE Cell Lines Adult rats were fed a choline-deficient (CD) diet containing ethionine, which induces oval cell proliferation, and two oval cell lines LE/2 and LE/6 were isolated following centrifugal elutriation (Braun et al., 1987). In basal culture conditions, these cells express AFP, aldolase A, lactate dehydrogenase, but not GGT. The early passage cell lines are not transformed, yet at later passages (over 50) the cells become transformed. As with the WB-F344 cells, LE cells grown at confluence with infrequent passages become transformed. The transformed cells constitutively express c-myc but not p53 mRNA (Braun et al., 1989).

LE cells were transformed by transfection with the ras oncogene and injected into nude mice, where they formed trabecular HCCs. Tumor cell lines derived from the nude mice expressed AFP and GGT, suggesting that LE cells are bipotential (Braun et al., 1987).

To determine whether the LE cells could differentiate in vitro along both the hepatic and biliary pathways, they were cocultured with mesenchymal feeder cells (NIH3T3) in three-dimensional collagen gels. The LE cells morphologically resembled mature hepatocytes and expressed albumins, CK8 and CK18 (Lazaro et al., 1998). LE/2 cells cultured in a collagen gel with HGF formed arborizing branching structures that resembled bile duct tubules (Lazaro et al., 1998). However, no expression of the bile duct–specific markers GGT or CK19 was detected, suggesting incomplete differentiation into bile duct epithelial cells.

34.3.2.2. OC/CDE Cell Lines The OC/CDE oval cell lines were established from rats fed a CD diet with ethionine after 6, 14, or 22 wk of treatment (Pack et al., 1993). In basal culture conditions, the cells express CKs 8 and 18 but not AFP, albumin, GGT, or CKs 7 and 19. OC/CDE cells can be induced to differentiate as immature hepatocytes and as bile duct epithelial cells by culture with dimethylsulfoxide (DMSO) or with sodium butyrate, which induces expression of albumin, GGT, glucose-6-phosphatase, and alkaline phosphatase. As for the previously described cell lines, the OC/CDE cells became transformed after maintenance for extended periods of time in culture at confluency with occasional transfers (Radaeva and Steinberg, 1995).

34. 3.2.3. LPC Cell Lines These cell lines were derived from rats treated with allyl alcohol, which induces periportal injury and oval cell proliferation. Eleven liver progenitor cell (LPC) lines were isolated, characterized, and shown to express the embryonic hepatocyte markers AFP, albumin, and CK14, as well as the HSC/oval cell markers c-kit, Thy-1, and CD34 (Yin et al., 2002). LPC cells were shown to be bipotential: culture with basic FGF (bFGF) induced expression of hepatocyte markers H4 and CYPIAII, whereas culture in Matrigel-induced formation of bile duct–like structures and expression of the bile duct cell marker BD1.

34.3. 3. BIPOTENTIAL OVAL CELL LINES DERIVED FROM ADULT MICE The TgN737 gene was identified by a transgene-induced insertional mutation in the mouse (Moyer et al., 1994). In humans, mutation in the gene causes liver abnormalities and a syndrome similar to autosomal recessive polycystic kidney disease (Yoder et al., 1995). The TgN737Rpw mouse develops biliary hyperplasia, associated with proliferation of cells that morphologically and immunologically resemble oval cells, which then give rise to ductular structures (Richards et al., 1996). This model is particularly interesting in that it induces oval cell proliferation without damaging the parenchyma.

Hepatic cell lines derived from TgN737Rpw mice express albumin, AFP, and c-Met (Richards et al., 1997). Immunoreactivity of the cell lines to the A6 antibody confirmed that they were oval cells. To test for bipotentiality, the cells were injected into interscapular fat pads of isogeneic mice and it was found that they developed ductular structures with columnar epithelial cells surrounding a central lumen.

Dumble et al. (2002) have isolated oval cell lines from $p53^{-/-}$ adult mice that had been subjected to a CD-ethionine-supplemented diet. Oval cells were purified by centrifugal elutriation of collagenase-perfused liver cell suspensions. Cells were first inoculated into serum-containing medium, but after 24 h they were switched to chemically defined medium without serum. Two of the five lines came from pure cultures of oval cells whereas three emerged from mixed hepatocyte–oval cell cultures. All five lines were characterized for hepatic functions and for tumorigenicity. The phenotypes of the lines were relatively homogeneous: the hepatocytic markers albumin, AFP, transferrin, aldolase B, and phenylalanine hydroxylase were expressed. In addition, the oval cell marker antigen A6 was expressed in at least some of the cells of each population. Three of the five lines formed tumors in nude mice.

Although each oval cell line is unique in its gene expression pattern, these cell lines share a number of common traits, including oval shaped nuclei, scant cytoplasm, expression of fetal hepatocyte genes, formation of bile ductular structures, and progressive transformation in culture.

34.3.4. ROLE OF HSCs IN OVAL CELL PROLIFERATION During embryogenesis, hematopoiesis originates in the extraembryonic yolk sac and subsequently in the fetal liver. The HSCs rapidly proliferate in the liver, then undergo maturation and differentiation leading to erythropoiesis, myelopoiesis, B and T lymphopoiesis, and production of megakaryocytes. Thereafter, the bone marrow replaces the fetal liver as the primary site of hematopoiesis. Oval cells and HSCs share several common markers—CD34, Thy1, c-kit, and flt-3 receptor—suggesting that HSCs could be a parental cell of the oval cell compartment.

In 1999, Petersen et al. showed, by several approaches using genetic markers, that bone marrow–derived cells could differentiate as mature hepatocytes in vivo. In 2000, Lagasse et al. refined these studies by isolating a specific population of HSCs by fluorescence-activated cell sorting (FACS) that were then shown to engraft and proliferate in the liver of fumarylacetoacetate hydrolase–deficient mice.

Taken together, these studies show the competence of HSCs to contribute to liver regeneration, yet the degree to which these cells participate in a "normal" context of liver regrowth is not known. Furthermore, there has not yet been proof of filiation between the HSCs and oval cells.

34.3.5. TRANSDIFFERENTIATION OF PANCREAS TO HEPATIC CELLS: OVAL CELL COMPARTMENT IN PANCREAS An intriguing observation is the presence of oval cells in the pancreas after acinar cell injury. A copper depletion–repletion diet induces severe acinar cell necrosis as well as ductular epithelial and oval cell proliferation (Rao and Reddy, 1995). The remaining acinar cells do not proliferate to reconstitute the loss; rather, oval cells proliferate and differentiate as mature hepatocytes in the pancreas. Repletion of copper to the diet eventually leads to the disappearance of oval cells in the pancreas (Dabeva et al., 1995). Along with pancreatic oval cell proliferation,

C/EBPα, C/EBPβ, C/EBPδ, HNF3β, stem cell factor (c-kit ligand), and albumin become expressed (Rao et al., 1996). A comparaison of oval cell proliferation in the liver and the pancreas showed that AFP and albumin expression, which are absent in the pancreas, increase during hepatic and pancreatic oval cell proliferation and decrease when oval cells cease to proliferate (Dabeva et al., 1995).

Pancreatic oval cell lines isolated from adult rats show different degrees of hepatic differentiation. None of the cells differentiate into acinar or islet cells. One line, when cultivated in collagen gels, formed ductlike structures with tubules and expressed albumin (Ide et al., 1993). Another cell line derived from the pancreas of a carcinogen-treated rat was injected subcutaneously and shown to differentiate as hepatocytes, expressing albumin, AFP, transferrin, TAT, and aldolase B, and formed bile duct–like structures (Chen et al., 1995). Finally, a pancreatic oval cell line isolated from a copper depletion–repletion-treated rat was transplanted into a DPPIV-deficient rat liver and shown to differentiate as mature hepatocytes with functional bile canaliculi (Dabeva et al., 1997).

In addition, the pancreatic acinar cell line AR42J, as well as organ cultures of pancreatic buds, were shown to transdifferentiate to hepatocytes by treatment with dexamethasone, this transdifferentiation being dependent on C/EBPβ (Shen et al., 2000). However, to date there is no evidence to suggest that hepatocytes can transdifferentiate to pancreatic cells.

The role of pancreatic oval cells, which do not differentiate as acinar cells and thus cannot regenerate the pancreas after acinar cell necrosis, remains unclear. The oval cells that differentiate as hepatocytes do not produce the same enzymes as acinar cells, and their function in vivo is unknown.

The properties of the oval cell population in the liver, and the cell lines that have been derived from them, provoke questions regarding their embryonic origin, the pathways that lead to differentiation into either hepatocytes or bile duct epithelial cells, their possible filiation to HSCs, their role in the pancreas, and the signals and events that lead to transformation.

34.4. AN IMMORTALIZED BIPOTENTIAL CELL LINE FROM PRIMATE EMBRYO

Bipotential hepatic cell lines from primates or humans had not been described due to the difficulty in obtaining immortalized cells. Recently, however, hepatic cell cultures were established from primate embryos perfused through the umbilical vein (Allain et al., 2002). These cultures were transduced with two retroviral vectors, the first containing simian virus 40 large T antigen for immortalization, and the second containing LacZ for tracing. Of the 1.8×10^7 cells plated, one clone survived after 1 yr in culture and is considered an immortal cell line: immortalized primate fetal liver stem (IPFLS) cells. Interestingly, the IPFLS cells express nuclear p53 (which is undetectable in normal cells) and high telomerase activity (which is presumed to be low in normal cells).

In vitro the cells expressed AFPs, albumin, and CKs 7, 8, 18, and 19. Some cells coexpressed albumin and CK19, thus implying that IPFLS cells are bipotential. The IPFLS cells were injected into nude mice and did not form tumors. To determine whether these cells could contribute to the liver in vivo, they were injected into nude mice through the portal vein, and liver biopsies were taken after 7 and 21 d. The cells were shown to have engrafted into the liver parenchyma—some were binucleated, others expressed albumin and AFP, but none were found in bile ducts or expressed CK19. Thus in vivo the IPFLS cells, as all the rodent cell lines previously described, differentiate into the hepatocyte lineage, yet are restricted from the bile duct cell lineage.

The isolation of IPFLS cells illustrates the difficulty of obtaining hepatic cell lines from primates and is a first indication of the problems that could lie ahead for isolating human bipotential hepatic cell lines.

34.5. HEPATOCYTE CELL LINES

Although strictly hepatocyte cell lines do not fall within the scope of this chapter, it is necessary to mention some of them briefly, for their study has led to techniques and approaches that would later be useful for the appreciation of bipotential cell lines. Indeed, many of the hepatocyte lines, especially the less well-differentiated ones and/or those whose differentiation is conditional, may be bipotential without this having been realized when the work was performed. Since the focus was on hepatocyte differentiation, the search for expression of markers of cholangiocytes was often not carried out.

34.5.1. CONDITIONAL DIFFERENTIATION OF HEPATOCYTE CELL LINES While the isolation of oval cell lines proved straightforward, obtaining well-differentiated lines that were certifiably from hepatoctyes proved to be more difficult. Several innovative strategies were attempted, based on the notion of conditional transformation. For example, in 1981, Chou and Schlegel-Haueter used a temperature-sensitive mutant of SV40 (ts SV40) to obtain rat hepatocyte lines. The reasoning was that at the temperature permissive for expression of viral functions the transformed cells should grow progressively, while at the nonpermissive temperature viral functions would be shut down, resulting in the inhibition of growth coupled with expression of differentiated functions. Indeed, this prediction proved to be accurate, including for albumin, AFP, and transferrin. A slightly different strategy was emplolyed by Stutenkemper et al. (1992), using a retrovirus expressing large T of SV40, they obtained hepatocyte lines from E18–20 mouse fetal liver cells and used serum withdrawal to induce the cells to undergo differentiation.

A particularly informative article concerning the differentiation of hepatocyte lines in culture is by DiPersio et al. (1991), who employed ts SV40 to derive a line of mouse hepatocytes. They showed by transient expression assays that transcription directed by the albumin enhancer was dependent on culture conditions. When cells were cultivated on a gel, albumin expression was augmented; the effect depended on the upstream albumin enhancer, presumably mediated by enhanced expression of HNF3, providing a first indication of a mechanism underlying the observation that substrate modification can influence gene expression.

More recently, Fiorino et al. (1998) used ts SV40 large T to obtain E14 mouse liver lines. Selecting the most epithelial clone for study, they went on to show that a shift to the nonpermissive temperature provoked a more differentiated hepatocytic phenotype, leading to morphological change and glucocorticoid-dependent regulation of a series of hepatocyte functions.

34.5.2. TRANGENIC MICE AS A SOURCE OF HEPATIC CELL LINES The success of conditional transformation/differentiation led investigators to extrapolate this strategy to mice expressing a cytokine that is mitogenic for the liver, or a transforming transgene whose transcription is directed by a hepatocyte-specific promoter. In these cases, it was reasoned that even if the immortalizing transgene tends to provoke transformation and consequently dedifferentiation, the transgene would be expressed only by the cells that initiate transcription from the promoter employed (Antoine et al., 1992).

Jhappan et al. (1990) employed the metallothionein 1 promoter to direct the expression of TGF-α; the resulting transgenic mice showed increased hepatocyte proliferation, liver hypertrophy and were prone to the development of liver tumors. Wu et al. (1994b) describe the derivation of two lines from adult mice, AML 12 and 14, both well differentiated to start with; one retained these characteristics, and the other evolved with continuous passage to a less well-differentiated phenotype. Both lines showed enhanced expression of hepatocyte functions when passaged in serum-free medium, and neither expressed the traits of transformed cells: growth in soft agar or the ability to form tumors.

In a further study, Wu et al. (1994a) derived hepatocyte cell lines from both TGF-α transgenic and wild-type (WT) adult mice using serum-free medium supplemented with growth factors and hormones. Cell lines from both types of mice expressed serum proteins and the hepatocyte connexins, but not the bile duct connexin 43. Interestingly, the generation time of the transgenic cells was shorter than that of the WT cells, and the two types of lines also differed in karyotype, with only the WT cells retaining a near-diploid rather than hypotetraploid chromosome number. Finally, cells from both WT and transgenic mice formed colonies in soft agar (with a very low efficiency), but only the transgenic cells were tumorigenic in vivo. The investigators speculate that growth in serum-free medium may contribute to the selection of transformed lines, by comparison with the nontransformed lines from the same transgenic mice (Wu et al., 1994b).

Among the hepatocyte lines isolated from transgenic mice expressing a transforming gene under control of a liver-specific promoter, examples can be found in Perraud et al. (1991), Antoine et al. (1992), Levrat et al. (1993) and Courjault-Gautier et al. (1997). In this work, the emphasis is on obtaining the most highly differentiated hepatocytes.

34.6. SPONTANEOUS IMMORTALIZATION OF EMBRYONIC RAT LIVER CELLS IN SERUM-FREE MEDIUM

Kubota and Reid (2000) have derived E15 rat embryonic liver cell lines on collagen- or laminin-coated plates in hormonally defined medium containing a large cocktail of growth factors and other additives plus 0.2% bovine serum albumin. After several weeks of growth, the cells were switched to culture on feeder layers. Three lines were derived from three different preparations of embryonic liver cells. Characterization of the three lines permitted them to conclude that the one that (1) gave rise to the highest frequency of colonies on feeder layers and (2) showed homogeneous intense expression of albumin and AFP produced only colonies of morphology designated "packed or aggregated" in opposition to "flattened" monolayers. They then went on to show that cells of this rhel4321 line are deficient in expression of classic major histocompatibility complex class I (MHC I) antigens. Going back to fresh liver suspensions, these investigators separated by FACS those cells that were negative for expression of MHC I antigens combined with enhanced granularity and were able to enrich in "clonogenic" normal cells. These gave rise to colonies of cells expressing albumin or CK19, used as diagnostic markers for hepatocytes and cholangiocytes, respectively. Hence, Kubota and Reid concluded that they had identified normal bipotential precursor cells from liver tissue.

34.7. A CELL LINE FROM LIVER DIVERTICULUM

A deliberate attempt to isolate a line of liver progenitor hepatoblast cells has been described by Rogler (1997). She dissected E9.5 mouse liver diverticuli and treated them as was formerly done to obtain embryonic stem cells: dissociation and growth on fibroblast feeder layers. Colonies of polygonal cells, judged to be hepatoblasts, were selected for study. Only one retained this morphology and was able to grow at low density. Cells of this HBC-3 line presented a euploid karyotype. Several methods were employed to elicit the differentiation of the cells: long-term (several weeks) culture on feeders, growth in the presence of sodium butyrate or DMSO, or growth in Matrigel. Aged cultures were positive for albumin, GGT, CK14, and AFP. In addition, electron microscopy revealed the presence of bile canalicular structures, of bodies resembling peroxisomes and rare glycogen granules. The chemical inducers provoked loss of AFP and GGT expression, the retention of albumin synthesis, and the activation of glucose-6-phosphatase expression. Finally, in Matrigel, ductular structures that showed GGT activity were observed. These results leave no doubt that the HBC-3 cells isolated by Rogler (1997) are indeed bipotential LPCs. In addition, in her articles, she attempts to define the culture conditions adequate to elicit the differentiation of progenitor cells.

Recently, Plescia et al. (2001) have used HBC-3 cells induced by DMSO to differentiate for a kinetic analysis using microarrays for nearly 9000 cDNAs, including both identified genes and expressed sequence tags. They observed changes in gene expression profiles during the first few hours of induction, and again after several days. Globally, after differentiation the gene expression profile of HBC-3 cells resembled much more closely adult liver than before. An attempt was made to identify genes more highly expressed in HBC-3 cells than in the liver and whose downregulation during differentiation would make them candidates for maintenance of the stem cell phenotype. Since DMSO treatment blocks cell proliferation, it is satisfying to observe that genes whose up- or downregulation is anticipated do indeed follow the predicted behavior, including cyclins and cyclin-dependent kinases. A careful analysis of transcription factor expression and of molecules implicated in signaling pathways indicated that downregulation of the wnt pathway may be critical for liver differentiation, just as had been found to be the case for epidermal and intestinal cell differentiation.

34.8. BIPOTENTIAL HEPATIC CELL LINES FROM TRANSGENIC MICE

Amicone et al. (1995, 1997) have established lines of transgenic mice that express a truncated constitutively active form of human c-Met, the receptor for HGF. The cDNA was placed within the transcription unit of the human α1-antitrypsin gene to ensure its expression in the liver. These transgenic mice did not form liver tumors but, instead, proved to have a discrete phenotype: their livers were more resistant than controls to induction of apoptosis by antibody binding of the fas receptor. Furthermore, primary cultures of liver cells from embryonic or newborn transgenic mice were able to give rise reproducibly to immortalized hepatic cell lines. When tested for the expression of hepatic functions, cells of the Met Murine Hepatocyte (MMH) lines were

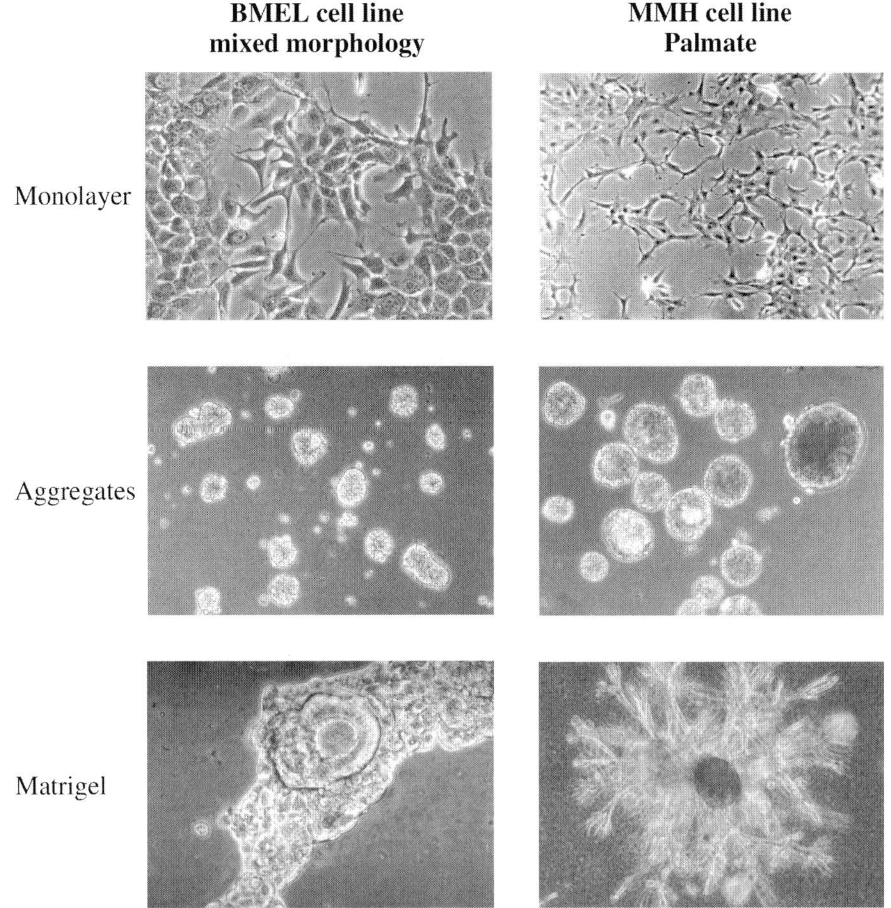

Fig .1. BMEL and MMH cell lines display different morphologies due to culture conditions. Monolayer cultures show the mixed morphology of BMEL cell lines containing palmate-like and epithelial cells, whereas MMH palmate cell lines are of homogeneous morphology. Cells cultured in bacteriological dishes form aggregates (5 d of culture). BMEL cells cultured in Matrigel form bile duct units, and MMH cells form branching bile duct tubules (10 d of culture). All magnifications are 100× except the upper left (200×).

found to express LETFs as well as some liver functions. In addition, they were nontransformed.

Unexpectedly, cloning of the MMH lines revealed that they harbored two cell types: the anticipated epithelial cells that form colonies with smooth borders; and a second cell type, designated palmate to describe its fibroblast-like form (Fig. 1). The epithelial cells expressed LETFs and were competent to express hepatic functions when induced by chemical inducers or substrate modification. Palmate cells, which, like the epithelial cells, express hepatic CKs 8, 18, and 19, were deficient in the expression of liver functions and of some of the LETFs such as HNF4α and HNF1α. In addition, they were shown to be bipotential precursor cells. When grown in Matrigel, palmate cells formed spheroids from which emerged branching tubules composed of a circular monolayer of cells interconnected by tight junctions and surrounding a lumen decorated by microvilli, closely resembling bile duct tubules (Fig. 1). In addition, palmate cells were shown by serial cloning to give rise to epithelial cells expressing the LETFs and competent to express liver functions when induced (Spagnoli et al., 1998).

Although palmate cells can give rise to epithelial cells, the reverse was not observed. A partial transition of epithelial to palmate morphology was obtained when epithelial cells were treated with low concentrations of TGF-β, accompanied by downregulation of HNF4α and HNF1α. However, the phenotype was entirely reversible. This is in contrast to the consequences of treatment with aFGF, whose differentiation-promoting activity had a heritable effect on the epithelial cells, inhibiting the effects of TGF-β (Spagnoli et al., 2000).

These bipotential hepatic precursor cells have already proved to be useful as feeder layers for hematopoietic cell differentiation (Aiuti et al., 1998), for dissecting the pathways of epithelial-mesenchymal transition (Gotzmann et al., 2002); and for definition of the factors that influence their differentiation, via modulation of expression of transcription factors (Spagnoli et al., 2000).

34.9. BIPOTENTIAL HEPATIC CELL LINES FROM MICE OF MULTIPLE GENOTYPES

The cell lines described above originated from transgenic mice. However, the role of the transgene in cell line emergence was unknown. Recently, we were able to isolate similar cell lines from nontransgenic mouse embryos (Strick-Marchand and Weiss, 2002). The technique is simple, easily reproducible with mice of any genetic background, and thus widely applicable. Embryonic livers are dissociated and plated, and one waits until proliferating colonies of epithelial cells appear 6–12 wk later; the method is called Plate and Wait. These cell lines are of two types, designated epithelial or mixed morphology, the former containing only epithelial cells and the latter displaying epithelial cells and palmate-like cells (Fig. 1). All of the cell lines analyzed express HNF1α, HNF3α,β, and HNF4α, GATA-4, CKs 8, 18, and 19 and AFP, yet only cell lines of epithelial morphology express ApoB and albumin. Although these cells express the hematopoietic stem and oval cell marker CD34, the lack of expression of c-kit and Thy-1 distinguishes them from oval cells (Table 1). Karyotype analysis of the cells showed two populations: near diploid and quasi-tetraploid. The cell lines do not grow in soft agar and are not transformed.

The undifferentiated phenotype of cell lines of mixed morphology prompted us to determine whether differentiated progeny could arise from these cells. Numerous culture conditions were tested: growth on gelatin, without serum, or in the presence of aFGF, dexamethasone (DEX), DEX + cyclic adenosine monophosphate, or, oncostatin M. The most efficient inducer of hepatic differentiation was found to be aggregation (Fig. 1). Reverse transcriptase polymerase chain reaction (RT-PCR) analysis of cells cultured as aggregates for 5 d revealed an impressive upregulation of AFP and aldolase B and induction of albumin, Apo B, Apo AIV, and alcohol dehydrogenase (ADH) expression. Concomitant with the induction of hepatocyte-specific genes, the bile duct/oval cell markers integrin B4 (IB4), CD34, and Cx43 were downregulated.

Differentiation of cell lines of mixed morphology along the bile duct lineage was obtained by culture in Matrigel. The cells formed doughnut-like structures with a central lumen surrounded by a columnar epithelium (Fig. 1), similar to primary cultures of bile duct units (Mennone et al., 1995; Cho et al., 2001). RT-PCR analysis of 10-d cultures showed an induction of HNF6, GGTIV, c-kit, and Thy-1, all markers of bile duct epithelial cells. Thus, the bipotential mouse embryonic liver (BMEL) cell lines isolated from WT mice are similar to stem cells: self-renewing, nontransformed, and bipotential.

Furthermore, we were able to reverse the differentiation of cells cultured in aggregates or in Matrigel by simply replating them in their basal culture conditions. In cultures of replated aggregates, cells cease to express albumin, Apo AIV, and ADH, and downregulate expression of aldolase B. Furthermore, the bile duct/oval cell markers IB4, CD34, c-kit, and connexin 43 (Cx43) return to their basal expression levels, confirming the return to an undifferentiated state. Cells from replated Matrigel cultures downregulate HNF6 and extinguish GGTIV expression (Strick-Marchand and Weiss, 2002).

The reversible differentiation of these cell lines demonstrates that a stem cell can be induced to differentiate or dedifferentiate at will, without commitment to a differentiated state. This idea has been proposed by Potten and Loeffler in their concept of the transit stem cell: a stem cell that differentiates to a transit state where it is capable of self-maintenance, or differentiation and maturation, or renewal/rejuvenation to an undifferentiated stem cell. Such transit stem cells have been described in the crypts of the small intestine (Potten and Loeffler, 1990) but had not yet been described in the liver. We propose that the BMEL cell lines originate from a transit stem cell population in the embryonic liver that is most probably the hepatoblast (Strick-Marchand and Weiss, 2003). Alternatively, E14 mouse embryonic liver cells are endowed with differentiation plasticity.

34.10. CONCLUSION

34.10.1. IDENTITY OF THE TARGET CELL FOR IMMORTALIZATION AND ITS DIFFFERENTIATION POTENTIAL In most cases, the identity of the target cell for immortalization cannot be affirmed. There are a few exceptions. In the cases of transgenic mice in which a mitogenic factor for the hepatocyte is overexpressed, or in which an immortalizing transgene is under control of a hepatocyte-specific promoter, the properties of the cells obtained as well as the rationale of the experimental system make it likely that a mature hepatocyte is the target cell. A second situation in which identification of the target cell can be assumed concerns cell lines obtained from animals that had been treated to obtain oval cell proliferation. Finally, the cell line obtained by Rogler from dissected liver buds is almost certainly derived from a hepatoblast (the only endodermal cell type present at the relevant stage). In essentially all other cases, the cell of origin of the line can only be argued on the basis of the properties of the cells. However, simply removing the cells from their in vivo environment leads to phenotypic changes, which makes it even more difficult to define the cell of origin. Furthermore, as mentioned above, there are few diagnostic markers for hepatoblasts, oval cells and bile duct cells. Among the papers cited above, a few give careful descriptions of the derivation of cell lines and the properties of the cultures from an early stage; *see* in particular Wu et al. (1994b) and Kubota and Reid (2000).

Are HSCs at the origin of any of the cell lines mentioned? The answer to this question is likely to be no, although careful investigation of possible contributions of such cells to the liver is only beginning to be undertaken (Alison et al., 2000). The possible participation of HSCs in pathological circumstances, such as oval cell proliferation, merits further investigation. Nevertheless, under most circumstances, the bulk of the liver cells most likely descend in linear fashion from the endodermal liver bud.

34.10.2. IS THE BIPOTENTIALITY OF HEPATIC CELLS LIMITED TO HEPATOBLASTS AND OVAL CELLS While the bipotentiality of oval cells and hepatoblasts is well established, to our knowledge, careful tests of the possible capacity of adult hepatocytes to form bile ducts or to express bile duct markers have not been carried out. While systematic searches for the expression of markers of hepatocyte differentiation are usual for the literature of the last 20 yr, analysis of bile duct cell markers is more recent. Hence, many cell lines that have not been recognized as bipotential may turn out to be so when carefully investigated.

Other cell types in the liver such as endothelial cells, Kupffer cells, and Ito cells (assumed to descend from mesenchymal cells) could participate in the maintenance of a stem cell population in vivo. The isolation of stem cell lines could perhaps be enhanced by the presence of these cell types in culture. Cocultures of different cell types have been shown to enhance the differentiation potential of some cells in vitro (Lazaro et al., 1998; Avital et al., 2001; Michalopoulos et al., 2001).

34.10.3. INFLUENCE OF SERUM-CONTAINING VS CHEMICALLY DEFINED MEDIUM Some investigators avoid medium-containing serum because this incompletely defined and variable supplement can inhibit the expression of differentiation of hepatic cells (Stutenkemper et al., 1992; Wu et al., 1994a). While most hepatic cell lines were initially isolated in medium-containing serum, a few have been derived in defined medium containing a varied cocktail of supplements and growth factors (e.g., *see* Wu et al., 1994a; Groen de et al., 1998; Kubota and Reid, 2000; Dumble et al., 2002). Does the condition in which the cell lines are isolated influence their differentiation potential? In most cases, the answer seems to be no. However, it is interesting in this regard to recall the conclusion of Wu et al. (1994a) that derivation of cell lines in serum-free medium may result in additional stress, the final effect of which is to select for cells having undergone transformation. An additional consideration concerns human cell lines that might ultimately be used in medical treatments; here, it is clear that serum should be avoided if possible.

34.11. CHALLENGES FOR FUTURE RESEARCH

34.11.1. IMPROVING MARKERS AVAILABLE FOR MEASURING THE DIFFERENTIATION STATE OF HEPATIC CELLS As can be seen from Table 1, there is a real problem of markers permitting the classification of cells, particulary among hepatoblasts, bile duct-cells, and oval cells. Judging from the careful analysis of Plescia et al. (2001), microarrays will come into increasing use for the definition of gene expression patterns of hepatic cell lines and their differentiated derivatives. Once the gene expression pattern determined for cell lines can be checked on developing, adult, and pathological liver, the spectrum of markers for classifying cells into one category or another should be greatly enlarged. In addition, while some progress has been made (Kubota and Reid, 2000; Suzuki et al., 2000, 2002), the attempt to define cell-surface markers that will permit FACS of cell populations of different growth and differentiation potential remains a constant challenge.

34.11.2. DEFINITION OF CONDITIONS FOR OBTAINING OPTIMAL DIFFERENTIATION OF HEPATIC CELLS The problem of defining culture conditions for optimal differentiation of hepatic cells is a fairly recent endeavor that can be considered to be in its infancy. Chemical inducers have been used, as well as growth factors, hormones, and culture substrates (e.g., *see* Rogler, 1997; Lazaro et al., 1998; Spagnoli et al., 1998; Auth et al., 2001; Yin et al., 2002; Strick-Marchand and Weiss, 2003). Rapid and striking advances can be anticipated here.

34.11.3. LINEAGE ANALYSIS WITHIN THE LIVER COMPARTMENT Lineage analysis implies genetic marking. In the adult, this can be achieved by retroviral marking, as was done by Bralet et al. (1994), to disprove the model of hepatocyte streaming. In the embryo, the task is more difficult, and the logical approach is to use a reporter transgene whose expression is directed by a promoter active from the moment of derivation of the cell type of interest, and in all descendants thereafter. An additional level of sophistication is added by the use of a reporter transgene carrying an internal duplication that results in an inactive protein; in this case active β-gal is synthesized only by the rare cell (and its progeny) in which an event of intragenic homologous recombination has deleted the duplication (Eloy-Trinquet et al., 1999). For lineage analysis, the crippled transgene is under control of a tissue-specific promoter; hence, the event of recombination is sufficiently rare that a maximum of one "clone" per embryo is obtained *within the territory where the promoter is active*. For analysis of the liver lineages, expression of the transgene could be directed by any promoter that is active in all endodermal derivatives.

Clarification of lineage relationships among liver cells, both *in situ* and in culture, will resolve many of the ongoing disagreements among researchers in the field. It is only after the lineage issue is resolved that satisfactory experiments can be designed to determine the target cell of permanent lines, the target cell for different regimens of hepatocarcinogenesis, and at what stages liver cells become irreversibly committed to undergo one and not an alternative differentiation.

REFERENCES

Aiuti, A., Cicchini, C., Bernardini, S., et al. (1998) Hematopoietic support and cytokine expression of murine-stable hepatocyte cell lines (MMH). *Hepatology* 28:1645–1654.

Alison, M. (1998) Liver stem cells: a two compartment system. *Curr. Opin. Cell Biol.* 10:710–715.

Alison, M. R., Poulsom, R., Jeffery, R., et al. (2000) Hepatocytes from non-hepatic adult stem cells. *Nature* 406:257.

Allain, J. E., Dagher, I., Mahieu-Caputo, D., et al. (2002) Immortalization of a primate bipotent epithelial liver stem cell. *Proc. Natl. Acad. Sci. USA* 99:3639–3644.

Amicone, L., Galimi, M. A., Spagnoli, F. M., Tommasini, C., De Luca, V. and Tripodi, M. (1995) Temporal and tissue-specific expression of the MET ORF driven by the complete transcriptional unit of human A1AT gene in transgenic mice. *Gene* 162:323–328.

Amicone, L., Spagnoli, F. M., Späth, G., et al. (1997) Transgenic expression in the liver of truncated Met blocks apoptosis and permits immortalization of hepatocytes. *EMBO J.* 16:495–503.

Ang, S. L., Wierda, A., Wong, D., et al. (1993) The formation and maintenance of the definitive endoderm lineage in the mouse: involvement of the HNF3/ Forkhead proteins. *Development* 119:1301–1315.

Antoine, B., Levrat, F., Vallet, V., et al. (1992) Gene expression in hepatocyte-like lines established by targeted carcinogenesis in transgenic mice. *Exp. Cell Res.* 200:175–185.

Auth, M.K., Joplin, R.E., Okamoto, M., et al. (2001) Morphogenesis of primary human biliary epithelial cells: induction in high-density culture or by coculture with autologous human hepatocytes. *Hepatology* 33:519–529.

Avital, I., Inderbitzin, D., Aoki, T., et al. (2001) Isolation, characterization, and transplantation of bone marrow-derived hepatocyte stem cells. *Biochem. Biophys. Res. Commun.* 288:156–164.

Barbacci, E., Reber, M., Ott, M. O., Breillat, C., Huetz, F., and Cereghini, S. (1999) Variant hepatocyte nuclear factor 1 is required for visceral endoderm specification. *Development* 126:4795–4805.

Bisgaard, H. C., Nagy, P., Santoni-Rugiu, E. and Thorgeirsson, S. S. (1996) Proliferation, apoptosis, and induction of hepatic transcription factors are characteristics of the early response of biliary epithelial (oval) cells to chemical carcinogens. *Hepatology* 23:62–70.

Bisgaard, H. C., Muller, S., Nagy, P., Rasmussen, L. J. and Thorgeirsson, S. S. (1999) Modulation of the gene network connected to interferon-gamma in liver regeneration from oval cells. *Am. J. Pathol.* 155:1075–1085.

Bralet, M.P., Branchereau, S., Brechot, C. and Ferry, N. (1994) Cell lineage study in the liver using retroviral mediated gene transfer: evidence against the streaming of hepatocytes in normal liver. *Am. J. Pathol.* 144:896–905.

Braun, L., Goyette, M., Yaswen, P., Thompson, N. L. and Fausto, N. (1987) Growth in culture and tumorigenicity after transfection with the ras oncogene of liver epithelial cells from carcinogen-treated rats. *Cancer Res.* 47:4116–4124.

Braun, L., Mikumo, R., and Fausto, N. (1989) Production of hepatocellular carcinoma by oval cells: cell cycle expression of *c-myc* and p53 at different stages of oval cell transformation. *Cancer Res.* 49: 1554–1561.

Chen, J. R., Tsao, M. S., and Duguid, W. P. (1995) Hepatocytic differentiation of cultured rat pancreatic ductal epithelial cells after in vivo implantation. *Am. J. Pathol.* 147:707–717.

Cho, W. K., Mennone, A., and Boyer, J. L. (2001) Isolation of functional polarized bile duct units from mouse liver. *Am. J. Physiol. Gastrointest. Liver Physiol.* 280:G241–G246.

Chou, J. Y. and Schlegel-Haueter, S. E. (1981) Study of liver differentiation in vitro. *J. Cell Biol.* 89:216–222.

Clotman, F., Lannoy, V. J., Reber, M., et al. (2002) The onecut transcription factor HNF6 is required for normal development of the biliary tract. *Development* 129:1819–1828.

Coffinier, C., Barra, J., Babinet, C., and Yaniv, M. (1999) Expression of the vHNF1/HNF1 homeoprotein gene during mouse organogenesis. *Mech. Dev.* 89:211–213.

Coffinier, C., Gresh, L., Fiette, L., et al. (2002) Bile system morphogenesis defects and liver dysfunction upon targeted deletion of HNF1β. *Development* 129:1829–1838.

Coleman, W. B., Wennerberg, A. E., Smith, G. J., and Grisham, J. W. (1993) Regulation of the differentiation of diploid and some aneuploid rat liver epithelial (stem-like) cells by the hepatic microenvironment. *Am. J. Pathol.* 142:1373–1382.

Coleman, W. B., McCullough, K. D., Esch, G. L., et al. (1997) Evaluation of the differentiation potential of WB-F344 rat liver epithelial stem-like cells in vivo: differentiation to hepatocytes after transplantation into dipeptidylpeptidase-IV-deficient rat liver. *Am. J. Pathol.* 151: 353–359.

Coon, H. G. (1968) Clonal culture of differentiated rat liver cells. *J. Cell Biol.* 39:29A.

Courjault-Gautier, F., Antoine, B., Bens, M., et al. (1997) Activity and inducibility of drug-metabolizing enzymes in immortalized hepatocyte-like cells (mhPKT) derived from a L-PK/Tag1 transgenic mouse. *Exp. Cell Res.* 234:362–372.

Couvelard, A., Bringuier, A. F., Dauge, M. C., et al. (1998) Expression of integrins during liver organogenesis in humans. *Hepatology* 27: 839–847.

Dabeva, M. D., Hurston, E., and Sharitz, D. A. (1995) Transcription factor and liver-specific mRNA expression in facultative epithelial progenitor cells of liver and pancreas. *Am. J. Pathol.* 147:1633–1648.

Dabeva, M. D., Hwang, S. G., Vasa, S. R., et al. (1997) Differentiation of pancreatic epithelial progenitor cells into hepatocytes following transplantation into rat liver. *Proc. Natl. Acad. Sci. USA* 94:7356–7361.

Darlington, G. J. (1987) Liver cell lines. In: *Molecular Genetics of Mammalian Cells*, vol. 151 (Gottesman, M., ed.), Academic, San Diego, CA, pp. 19–37.

de Groen , P. C., Vroman, B., Laakso, K., and LaRusso, N. F. (1998) Characterization and growth regulation of a rat intrahepatic bile duct epithelial cell line under hormonally defined, serum-free conditions. *In Vitro Cell. Dev. Biol.* 34:704–710.

DiPersio, C. M., Jackson, D. A., and Zaret, K. S. (1991) The extracellular matrix coordinately modulates liver transcription factors and hepatocyte morphology. *Mol. Cell Biol.* 11:4405–4414.

Dumble, M. L., Croager, E. J., Yeoh, G. C. T., and Quail, E. A. (2002) Generation and characterization of p53 null transformed hepatic progenitor cells: oval cells give rise to hepatocellular carcinoma. *Carcinogenesis* 23:435–445.

Duncan, S. A., Manova, K., Chen, W.S., et al. (1994) Expression of transcription factor HNF-4 in the extraembryonic endoderm, gut, and nephrogenic tissue of the developing mouse embryo: HNF-4 is a marker for primary endoderm in the implanting blastocyst. *Proc. Natl. Acad. Sci. USA* 91:7598–7602

Eloy-Trinquet, S., Mathis, L., and Nicolas, J. F. (1999) Retrospective tracing of the developmental lineage of the mouse myotome. *Curr. Top. Dev. Biol.* 47:33–79.

Evarts, R. P., Hu, Z., Fujio, K., Marsden, E. R., and Thorgeirsson, S. S. (1993) Activation of hepatic stem cell compartment in the rat: role of transforming growth factor α, hepatocyte growth factor, and acidic fibroblast growth factor in early proliferation. *Cell Growth Differ.* 4: 555–561.

Fausto, N. (1994) Liver stem cells. In: *The Liver Biology and Pathobiology* (Arias, I. M., Boyer, J. L., Jakoby, W. B., et al. eds.), Raven, New York, pp. 1501–1518.

Fiorino, A. S., Diehl, A. M., Lin, H. Z., Lemischka, I. R., and Reid, L. M. (1998) Maturation-dependent gene expression in a conditionally transformed liver progenitor cell line. *In Vitro Cell. Dev. Biol.* 34: 247–258.

Fujio, K., Hu, Z., Evarts, R. P., Marsden, E. R., Niu, C. H., and Thorgeirsson, S. S. (1996) Coexpression of stem cell factor and c-kit in embryonic and adult liver. *Exp. Cell Res.* 224:243–250.

Germain, L., Noël, M., Gourdeau, H., and Marceau, N. (1988) Promotion of growth and differentiation of rat ductular oval cells in primary culture. *Cancer Res.* 48:368–378.

Glickman, R. M. and Sabesin, S. M. (1994) Lipoprotein metabolism. In: *The Liver Biology and Pathobiology* (Arias, I. M., Boyer, J. L., Jakoby, W. B., Fausto, N., Schachter, D., and Shafritz, D. A., eds.), Raven, New York, pp. 391–415.

Gotzmann, J., Huber, H., Thallinger, C., et al. (2002) Hepatocytes convert to a fibroblastoid phenotype through the cooperation of TGF-beta1 and Ha-Ras: steps toward invasiveness. *J. Cell Sci.* 115:1189–1202.

Grisham, J. W. (1980) Cell types in long-term propagable cultures of rat liver. *Ann. NY Acad. Sci.* 349:128–137.

Grisham, J. W., Coleman, W. B., and Smith, G. J. (1993) Isolation, culture, and transplantation of rat hepatocytic precursor (stem-like) cells. *Proc. Soc. Exp. Biol. Med.* 204:270–279.

Guillouzo, A., Weber, A., Le Provost, E., Rissel, M., and Schapira, F. (1981) Cell types involved in the expression of foetal aldolases during rat azo-dye hepatocarcinogenesis. *J. Cell Sci.* 49:249–260.

Hixson, D. C., Chapman, L., McBride, A., Faris, R., and Yang, L. (1997) Antigenic phenotypes common to rat oval cells, primary hepatocellular carcinomas and developing bile ducts. *Carcinogenesis* 18:1169–1175.

Hixson, D. C., Brown, J., McBride, A. C., and Affigne, S. (2000) Differentiation status of rat ductal cells and ethionine-induced hepatic carcinomas defined with surface-reactive monoclonal antibodies. *Exp. Mol. Pathol.* 68:152–169.

Holic, N., Suzuki, T., Corlu, A., et al. (2000) Differential expression of the rat gamma-glutamyl transpeptidase gene promoters along with differentiation of hepatoblasts into biliary or hepatocytic lineage. *Am. J. Pathol.* 157:537–548.

Hooth, M. J., Coleman, W. B., Presnell, S. C., Borchert, K. M., Grisham, J. W., and Smith, G. J. (1998) Spontaneous neoplastic transformation of WB-F344 rat liver epithelial cells. *Am. J. Pathol.* 153:1913–1921.

Ide, H., Subbarao, V., Reddy, J. K., and Rao, M. S. (1993) Formation of ductular structures in vitro by rat pancreatic epithelial oval cells. *Exp. Cell Res.* 209:38–44.

Jhappan, C., Stahle, C., Harkins, R. N., Fausto, N., Smith, G. H., and Merlino, G. T. (1990) TGF alpha overexpression in transgenic mice induces liver neoplasia and abnormal development of the mammary gland and pancreas. *Cell* 61:1137–1146.

Kubota, H. and Reid, L.M. (2000) Clonogenic hepatoblasts, common precursors for hepatocytic and biliary lineages, are lacking classical major histocompatibility complex class I antigen. *Proc. Natl. Acad. Sci. USA* 97:12,132–12,137.

Kuo, C. F., Xanthopoulos, K. G., and Darnell, J. E. (1990) Fetal and adult localization of C/EBP: evidence for combinatorial action of transcription factors in cell-specific gene expression. *Development* 109: 473–481.

Lagasse, E., Connors, H., Al-Dhalimy, M., et al. (2000) Purified hematopoietic stem cells can differentiate into hepatocytes in vivo. *Nat. Med.* 6:1229–1234.

Landry, C., Clotman, F., Hioki, T., et al. (1997) HNF-6 is expressed in endoderm derivatives and nervous system of the mouse embryo and participates to the cross-regulatory network of liver-enriched transcription factors. *Dev. Biol.* 192:247–257.

Lazaro, C. A., Rhim, J. A., Yamada, Y., and Fausto, N. (1998) Generation of hepatocytes from oval cell precursors in culture. *Cancer Res.* 58: 5514–5522.

Levrat, F., Vallet, V., Berbar, T., Miquerol, L., Kahn, A., and Antoine, B. (1993) Influence of the content in transcription factors on the phenotype of mouse hepatocyte-like cell lines (mhAT). *Exp. Cell Res.* 209: 307–316.

Mennone, A., Alvaro, D., Cho, W., and Boyer, J. L. (1995) Isolation of small polarized bile duct units. *Proc. Natl. Acad. Sci. USA* 92: 6527–6531.

Michalopoulos, G. K., Bowen, W. C., Mulè, K., and Stolz, D. B. (2001) Histological organization in hepatocyte organoid cultures. *Am. J. Pathol.* 159:1877–1887.

Moyer, J. H., Lee-Tischler, M. J., Kwon, H. Y., et al. (1994) Candidate gene associated with a mutation causing recessive polycystic kidney disease in mice. *Science* 264:1329–1333.

Nagy, P., Bisgaard, H. C., and Thorgeirsson, S. S. (1994) Expression of hepatic transcription factors during liver development and oval cell differentiation. *J. Cell Biol.* 126:223–233.

Numazaki, M., Tsutsumi, K., Tsutsumi, R., and Ishikawa, K. (1984) Expression of aldolase isozyme mRNAs in fetal rat liver. *Eur. J. Biochem.* 142:165–170.

Omori, N., Omori, M., Evarts, R.P., et al. (1997) Partial cloning of rat CD34 cDNA and expression during stem cell-dependent liver regeneration in the adult rat. *Hepatology* 26:720–727.

Ott, M.O., Rey-Campos, J., Cereghini, S., and Yaniv, M. (1991) vHNF1 is expressed in epithelial cells of distinct embryonic origin during development and precedes HNF1 expression. *Mech. Dev.* 36:47–58.

Pack, R., Heck, R., Dienes, H. P., Oesch, F., and Steinberg, P. (1993) Isolation, biochemical characterization, long-term culture, and phenotype modulation of oval cells from carcinogen-fed rats. *Exp. Cell Res.* 204:198–209.

Perraud, F., Dalemans, W., Gendrault, J. L., et al. (1991) Characterization of trans-immortalized hepatic cell lines established from transgenic mice. *Exp. Cell Res.* 195:59–65.

Petersen, B. E., Zajac, V. F., and Michalopoulos, G. K. (1997) Bile ductular damage induced by methylene dianiline inhibits oval cell activation. *Am. J. Pathol.* 151:905–909.

Petersen, B. E., Goff, J. P., Greenberger, J. S., and Michalopoulos, G. K. (1998) Hepatic oval cells express the hematopoietic stem cell marker Thy-1 in the rat. *Hepatology* 27:433–445.

Petersen, B. E., Bowen, W. C., Patrene, K. D., et al. (1999) Bone marrow as a potential source of hepatic oval cells. *Science* 284:1168–1170.

Plescia, C., Rogler, C., and Rogler, L. (2001) Genomic expression analysis implicates Wnt signaling pathway and extracellular matrix alterations in hepatic specification and differentiation of murine hepatic stem cells. *Differentiation* 68:254–269.

Potten, C. S. and Loeffler, M. (1990) Stem cells: attributes, cycles, spirals, pitfalls and uncertainties. Lessons for and from the crypt. *Development* 110:1001–1020.

Radaeva, S. and Steinberg, P. (1995) Phenotype and differentiation patterns of the oval cell lines OC/CDE 6 and OC/CDE 22 derived from the livers of carcinogen-treated rats. *Cancer Res.* 55:1028–1038.

Rao, M. S. and Reddy, J. K. (1995) Hepatic transdifferentiation in the pancreas. *Semin. Cell Biol.* 6:151–156.

Rao, M. S., Yukawa, M., Omori, M., Thorgeirsson, S. S., and Reddy, J. K. (1996) Expression of transcription factors and stem cell factor precedes hepatocyte differentiation in rat pancreas. *Gene Express.* 6: 15–22.

Rausa, F., Samadani, U., Ye, H., et al. (1997) The cut-homeodomain transcriptional activator HNF-6 is coexpressed with its target gene HNF-3 beta in the developing murine liver and pancreas. *Dev. Biol.* 192:228–246.

Rausa, F. M., Galarneau, L., Belanger, L., and Costa, R. H. (1999) The nuclear receptor fetoprotein transcription factor is coexpressed with its target gene HNF-3beta in the developing murine liver, intestine and pancreas. *Mech. Dev.* 89:185–188.

Richards, W. G., Yoder, B. K., Isfort, R. J., et al. (1996) Oval cell proliferation associated with the murine insertional mutation TgN737Rpw. *Am. J. Pathol.* 149:1919–1930.

Richards, W. G., Yoder, B. K., Isfort, R. J., et al. (1997) Isolation and characterization of liver epithelial cell lines from wild-type and mutant TgN737Rpw mice. *Am. J. Pathol.* 150:1189–1197.

Rogler, L. E. (1997) Selective bipotential differentiation of mouse embryonic hepatoblasts in vitro. *Am. J. Pathol.* 150:591–602.

Seifter, S. and Englard, S. (1994) Energy metabolism. In: *The Liver Biology and Pathobiology* (Arias, I. M., Boyer, J. L., Jakoby, W. B., Fausto, N., Schachter, D., and Shafritz, D. A., eds.), Raven, New York, pp. 323–365.

Sell, S. (2001) Heterogeneity and plasticity of hepatocyte lineage cells. Hepatology 33:738–750.

Shen, C. N., Slack, J. M. W., and Tosh, D. (2000) Molecular basis of transdifferentiation of pancreas to liver. *Nat. Cell Biol.* 2:879–887.

Shiojiri, N., Lemire, J. M., and Fausto, N. (1991) Cell lineages and oval cell progenitors in rat liver development. *Cancer Res.* 51:2611–2620.

Spagnoli, F. M., Amicone, L., Tripodi, M., and Weiss, M. C. (1998) Identification of a bipotential precursor cell in hepatic cell lines derived from transgenic mice expressing Cyto-Met in the liver. *J. Cell Biol.* 143:1101–1112.

Spagnoli, F. M., Cicchini, C., Tripodi, M., and Weiss, M. C. (2000) Inhibition of MMH (Met murine hepatocyte) cell differentiation by TGF(beta) is abrogated by pre-treatment with the heritable differentiation effector FGF1. *J. Cell Sci.* 113:3639–3647.

Strick-Marchand, H. and Weiss, M. C. (2002) Inducible differentiation and morphogenesis of bipotential liver cell lines from wild-type mouse embryos. *Hepatology* 36:794–804.

Strick-Marchand, H. and Weiss, M. C. (2003) Embryonic liver cells and permanent lines as models for hepatocyte and bile duct cell differentiation. *Mech. Dev.* 120:89–98.

Stutenkemper, R., Geisse, S., Schwarz, H. J., et al. (1992) The hepato cyte-specific phenotype of murine liver cells correlates with high expression of connexin32 and connexin26 but very low expression of connexin43. *Exp. Cell Res.* 201:43–54.

Suzuki, A., Zheng, Y. W., Kondo, R., et al. (2000) Flow-cytometric separation and enrichement of hepatic progenitor cells in the developing mouse liver. *Hepatology* 32:1230–1239.

Suzuki, A., Zheng, Y. W., Kaneko, S., et al. (2002) Clonal identification and characterization of self-renewing pluripotent stem cells in the developing liver. *J. Cell Biol.* 156:173–184.

Theise, N. D., Nimmakayalu, M., Gardner, R., et al. (2000) Liver from bone marrow in humans. *Hepatology* 32:11–16.

Tietjen, T. G., Mjaatvedt, C. H., and Yang, V. W. (1994) Expression of the class I alcohol dehydrogenase gene in developing rat fetuses. *J. Histochem. Cytochem.* 42:745–753.

Tsao, M. S. and Grisham, J. W. (1987) Hepatocarcinomas, cholangiocarcinomas, and hepatoblastomas produced by chemically transformed cultured rat liver epithelial cells. *Am. J. Pathol.* 127: 168–181.

Tsao, M. S., Smith, J. D., Nelson, K. G., and Grisham, J. W. (1984) A diploid epithelial cell line from normal adult rat liver with phenotypic properties of "oval" cells. *Exp. Cell Res.* 154:38–52.

Vonesch, J. L., Nakshatri, H., Philippe, M., Chambon, P., and Dolle, P. (1994) Stage and tissue-specific expression of the alcohol dehydrogenase 1 (Adh-1) gene during mouse development. *Dev. Dyn.* 199: 199–213.

Weiner, F. R., Degli Esposti, S., and Zern, M. A. (1994) Ethanol and the Liver. In: *The Liver Biology and Pathobiology* (Arias, I. M., Boyer, J. L., Jakoby, W. B., Fausto, N., Schachter, D., and Shafritz, D. A., eds.), Raven, New York, pp. 1383–1413.

Wu, J. C., Merlino, G., Cveklova, K., Mosinger, B. Jr., and Fausto, N. (1994a) Autonomous growth in serum-free medium and production of hepatocellular carcinomas by differentiated hepatocyte lines that overexpress transforming growth factor alpha 1. *Cancer Res.* 54: 5964–5973.

Wu, J.C., Merlino, G., and Fausto, N. (1994b) Establishment and characterization of differentiated, nontransformed hepatocyte cell lines derived from mice transgenic for transforming growth factor a. *Proc. Natl. Acad. Sci. USA* 91:674–678.

Yang, L., Faris, R. A., and Hixson, D. C. (1993) Phenotypic heterogeneity within clonogenic ductal cell populations isolated from normal adult rat liver. *Proc. Soc.Exp. Biol. Med.* 204:280–288.

Yin, L., Sun, M., Ilic, Z., Leffert, H. L., and Sell, S. (2002) Derivation, characterization, and phenotypic variation of hepatic progenitor cell lines isolated from adult rats. *Hepatology* 35:315–324.

Yoder, B. K., Richards, W. G., Sweeney, W. E., Wilkinson, J. E., Avener, E. D., and Woychik, R. P. (1995) Insertional mutagenesis and molecular analysis of a new gene associated with polycystic kidney disease. *Proc. Assoc. Am. Physicians* 107:314–323.

Zhang, M. and Thorgeirsson, S. S. (1994) Modulation of connexins during differentiation of oval cells into hepatocytes. *Exp. Cell Res.* 213: 37–42.

35 Biology of Human Liver Stem Cells

ALASTAIR J. STRAIN, PhD, SARBJIT S. NIJJAR, PhD,
AND HEATHER A. CROSBY, PhD

Although until recently the existence of the liver stem cell was questioned, now the race is on to obtain and characterize the human liver stem cell for clinical therapeutic use for gene replacement, liver repopulation, drug development, and bioartificial liver support systems. Experimental models have identified cells at different stages in the hepatocyte lineage, including mature hepatocytes, bipotential ductular cells (so-called oval cells), as well as blood-derived periductular cells, as possible liver progenitor cells (LPCs). In the human liver, many cholestatic diseases, associated with loss of bile ducts and proliferation of "ductular proliferative cells" with an appearance similar to oval cells, are seen. The origin of these cells is not clear; possibilities considered are canals of Hering, mature bile duct cells, metaplasia of hepatocytes, and blood-borne stem cells that locate in the liver. Specific markers for these cells have not been identified, but OV-6, a marker identified by a monoclonal antibody to a cytokeratin in rat oval cells and bile ducts, has been shown to identify a presumptive human LPC. Other markers seen on human oval cells are CD34 and c-*kit*, markers shared by hematopoietic stem cells, and indeed, after bone marrow transplantation, donor cells can be found in the recipient liver, albeit rarely. Cells from injured or normal human liver expressing c-*kit* have been selected, cultured in vitro, and shown to express biliary phenotype; cells expressing CD34 have been shown to give rise to hepatic epithelial cell phenotype; and cells expressing endothelial phenotype have also been identified. In addition, a $CD34^-$, c-kit^- mesenchymal cell from the bone marrow (so-called multipotential adult progenitor cell) can differentiate into hepatocytes. Human fetal hepatoblasts may also serve as a source for LPCs. A number of different factors may control differentiation of liver precursor cells, including Jagged/Notch signaling. Much further work is required before the potential of the therapeutic use of liver stem cells can be realized.

35.1. INTRODUCTION

For years the very existence of the liver stem cell was questioned (Sell, 1990). However, the concept has now become firmly established in the literature, and the race is on to characterize liver stem cells and to exploit their potential for clincal therapeutic uses. If successful, such a cell could be of use clinically in treating patients, e.g., cell transplantation for gene replacement therapy or liver repopulation in efforts to accelerate liver repair in the critical time period following acute liver failure (Strom et al., 1999). Additionally, hepatic stem cells would potentially prove a valuable source from which to generate human hepatocytes for drug development (Guillouzo et al., 1997) or use in bioartificial liver support systems (Strain and Neuberger, 2002). Clinical initiatives such as primary hepatocyte transplantation and bioartificial liver systems have had only limited success due to the severe constraints imposed by the shortfall of quality primary human hepatocytes. In addition, drug development is highly dependent on direct toxicological and pharmacokinetic data from human cells, as opposed to rodent-based in vivo and in vitro studies relied so heavily on in the past.

35.1.1. LIVER REGENERATION The ability of the liver to regenerate fully is a unique feature of what is a structurally and functionally highly complex organ (Bucher and Farmer, 1998). Regeneration begins with growth activation of the primary hepatocyte population. Hepatocytes in the adult are fully differentiated and normally quiescent, and yet retain this remarkable capacity to enter the growth cycle on demand (Fausto, 2001). Subsequently, they return to a quiescent, fully functional state. The other cell types in the liver, including biliary epithelium, sinusoidal endothelium, stellate cells, and Kupffer cells, also proliferate, but with a lag period that in the rat is approx 24 h (Grisham, 1962), but is less well defined in humans. The process is precisely regulated resulting, under normal conditions, in full restoration of tissue mass and cellular architecture. Thus, not only has it served as a good model of liver cell growth control, but as a means to investigate the control of tissue-specific cellular morphogenesis—cell polarization and three-dimensional organisation (Bucher and Farmer, 1998; Fausto, 2001).

From: *Stem Cells Handbook*
Edited by: S. Sell © Humana Press Inc., Totowa, NJ

The primary hepatocyte has been included as a contributor to the liver stem cell pool by some investigators (Alison, 1998) although this does not fit with the normally accepted definition of a stem cell—a cell that can self-renew and is multipotential (Watt and Hogan, 2000). Hepatocytes are unipotential, and although during the pathogenesis of liver disease there is some evidence of metaplasia of hepatocytes giving rise to ductular epithelium (Sirica, 1995), this is not thought to be of major significance in normal cellular replacement mechanisms that occur in the liver. If in the experimental animal the liver is subjected to repeated rounds of partial hepatectomy, hepatocytes demonstrate the capacity for continual replacement of differentiated cells (Simpson and Finkh, 1963). This is not to say, however, that they form a continual cycling cell population as in other tissues in which there are recognizable stem cell compartments such as skin, bone marrow, or gastrointestinal tract. Following a two-thirds partial hepatectomy, each remaining hepatocyte need only divide at most twice to replenish liver mass. Under normal conditions, therefore, liver regeneration occurs from primary hepatocytes and the other existing liver cell types without the need to invoke a stem cell compartment (Fausto, 2001).

35.2. OVAL CELLS

Stem cells do, however, fulfill an important role when liver damage is severe and parenchymal hepatocytes are effectively eradicated or, for some reason, are prevented from growth initiation. This has been studied extensively in a wide variety of experimental animal models and occurs following treatment with numerous liver toxins or carcinogens (reviewed in Sell, 2001, and in Chapters 32 and 33). This work has culminated in the clear consensus view that, under certain specific circumstances, liver stem cells do indeed fulfill an important cellular replacement role in the liver. A major problem over the years, however, has been identifying the cellular origin and location of hepatic stem cells.

Farber (1956) was the first to describe the cellular response in the rat liver to toxins such as acetylaminofluorene (AAF) or diethylnitrosamine, and to coin the term *oval cell*. These cells are small, characterized by oval nuclei and scant cytoplasm, and are suggested to arise from a stem cell pool (Fausto et al., 1993; Thorgeirsson, 1996). Following activation and growth expansion, they migrate from the margins of the portal tracts into the parenchyma, where they differentiate into hepatocytes. They can also differentiate into BEC or abnormal ductular proliferative cells (Germain et al., 1988; Sirica et al., 1990; Lenzi et al., 1992; Dabeva and Shafritz, 1993; Golding et al., 1995) The indication, therefore, is that oval cells represent the 'activated ' progeny of the stem cell.

However, despite years of effort to resolve the problem, the origin of oval cells remains contentious, the overriding problem being that oval cells or their stem cell precursors cannot readily be identified in normal liver. Various investigators have proposed that they are derived from the cells lining the terminal bile ductules (the canals of Hering) and in close proximity to periportal hepatocytes (Paku et al., 2001; Sell, 2001). There is no doubt that the pattern following activation and appearance of oval cells in injured liver and their expansion and subsequent migration into the parenchyma are compatible with this proposal. Others suggest they arise not just from the canals of Hering but that all mature biliary cells (Thung, 1990; Alison et al., 1996) have the capability to dedifferentiate into oval cells and hence hepatocytes. The evidence for this, however, is less established. A third mechanism, that of metaplasia of hepatocytes, is also a possibility, and there is some documented evidence in rodents (Vandersteenhoven et al., 1990; Sirica 1995).

35.3. HEMATOPOIETIC STEM CELLS

With the now widely recognized plasticity of hematopoietic stem cells (HSCs), clearly indicating their ability to give rise to cell types of all three primordial germ cell layers (Anderson et al., 2001), the possibility that hepatic oval cells are actually derived from bone marrow–derived stem cells has introduced further speculation. In rodents, Petersen et al. (1999) first demonstrated that transplanted bone marrow stem cells traffic to the liver of animals lethally irradiated and whose livers had received treatments (e.g., AAF) designed to cause oval cell activation. By following the fate of sex mismatched cells (Y-chromosome analysis of donor cells transplanted into a female recipient), they found evidence of oval cells, hepatocytes, and some ductal cells, which were thought to be derived from transplanted bone marrow cells (Petersen et al., 1999). This work was soon confirmed by Theise et al. (2000a) also in bone marrow–transplanted mice but using a regimen that did not induce any apparent degree of liver damage.

Further confirmation of the principle quickly followed. Lagasse et al. (2000), using an experimental mouse model of hereditary tyrosinemia, showed that whole bone marrow and even purified HSCs could rescue animals from the fatal metabolic disease. Although in this model massive liver repopulation took place, the investigators indicated that the number of initiating cells may in fact be quite low, but that in the presence of a selective growth advantage, such as is the case here in animals with serious liver damage, cell expansion occurred albeit over a relatively long time scale (Lagasse et al., 2000).

In a recent elegant study, Mallet et al. (2002) have reported similar transplantation experiments using Bcl-2 transgenic mice with a hepatocyte-specific promoter that gives cells a protective effect against fas-mediated apoptosis. Bone marrow from the transgenic animals was transplanted into normal mice that subsequently were treated with anti-Fas antibodies. Livers of such animals were shown to have bone marrow–derived clusters of Bcl-2 positive hepatocytes. They conclude that transdifferentiation of bone marrow cells into mature hepatocytes is a very rare event, but that by genetically modifying bone marrow cells it is possible to selectively expand their hepatocyte progeny from this stem cell pool (Mallet et al., 2002). Thus, although their study indicated that the process was a very inefficient one under physiological conditions, it may be possible to exploit this technology and transplant genetically modified bone marrow cells with a selective growth advantage.

35.4. HUMAN LIVER STEM CELLS

35.4.1. HISTOLOGICAL OBSERVATIONS It is against this background that the question of whether there is an equivalent liver stem cell population in humans arises. In response to many cholestatic liver disorders in humans, often but not always associated with loss of the bile ducts, a classic histological cellular reaction occurs (Fig. 1). First reported more than 40 yr ago, and commonly called ductular proliferation (Masuko et al., 1964), these cells arise in the portal and periportal spaces, express markers of biliary phenotype, but form disorganized cords, aggregates, and individual groups of cells. Because they consistently fail to form lumenal polarized ducts, they do not serve to drain bile from

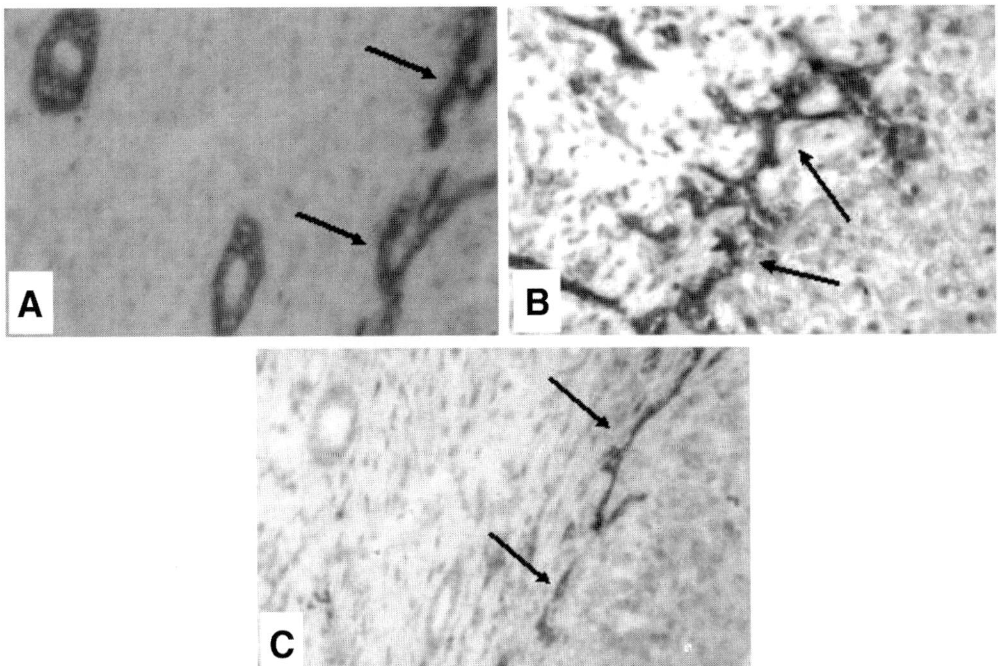

Fig. 1. Immunohistochemical analysis of diseased human liver tissue. Sections were immunostained using the biliary-specific marker HEA 125. Tissue was from end-stage livers removed at the time of orthotopic transplantation from patients with **(A)** α1-Anti-trypsin deficiency, **(B)** biliary atresia, and **(C)** primary biliary cirrhosis. Note the presence of aberrant ductular proliferative cells in each case (arrows) with differing morphological features.

the liver, with the consequence of cholestasis and the secondary damage that ensues, leading to fibrosis and cirrhosis. These cells can be distiguished from lumenal duct cells by their aberrant morphology and expression of markers not seen on intact ducts (*see* below).

A wide array of terms and nomenclature has been used to describe these cells (Sell, 2001). One of the most commonly used—*ductular proliferative cells* (Van Eyken et al., 1989; Thung, 1990)—can be somewhat misleading because it tends to be interpreted to suggest that all the cells in question are in a state of active proliferation. Although the cells clearly increase in number, attempts to correlate with proliferation markers such as Ki-67 indicate that at any one time only a small number are actively proliferating (Fabris et al., 2000). In most cases, even within histological or ostensibly well-recognized diagnostic categories of liver disease, often a wide range of morphologies is encountered. Other terms used to describe the cellular histological pattern seen include ductular reaction (De Vos and Desmet, 1992); *ductular hepatocytes*, (Vandersteenhoven et al., 1990; Gerber et al., 1995); *neocholangioles* or *atypical ductular cells* (Sell, 1998); and *type I, II*, or *III cells* (Popper, 1990).

This confusion in terminology has in turn led to a lack of clarity in determining the origin of such cells. In human hepatic pathology, speculation is equally strong in the debate on the possible origin or location of these cells: (1) cells residing in the canals of Hering (Theise et al., 1999), (2) proliferation of mature bile duct cells themselves (Slott et al., 1990; Alison et al., 1996), or (3) metaplasia of hepatocytes (Van Eyken et al., 1989; Fabris et al., 2000). More recently, of course, a fourth source of stem cells with human hepatic potential is the bone marrow–derived cell compartment as described in relation to the observations made in numerous experimental animal transplant models already described.

Oval-like stem cells have been described in human liver in numerous reports by either light or electron microscopic studies (Thung, 1990; De Vos and Desmet, 1992; Ruck et al., 1996). However, morphology alone is insufficient to determine the properties or phenotype of such cells. Therefore, immunohistochemistry has been utilized more recently in an effort to more precisely define the origin of the various disparate cell populations. Consequently, a number of different antigenic determinants including CK19 (Van Eyken et al., 1989; Hsia et al., 1992; Crosby et al., 1998a), NCAM (Roskams et al., 1990; Fabris et a, 2000; Van den Heuvel et al., 2001), PthrP (Roskams et al., 1993), and M2PK(Lowes et al., 1999) have been used to identify human oval-type cells. While often useful indicators of cell lineage potential—for example, oval cell expression of a specific isozyme of M2PK has been reported to correlate with the severity of disease (Lowes et al., 1999)—none of these determinants are, of course, stem cell specific.

Even in experimental animal studies, there are few definitive stem cell–specific markers. To date, one of the most widely used is the mouse monoclonal antibody OV-6. This was originally raised against nodular hepatocytes isolated from the livers of rats treated with diethylnitrosamine (Dunsford and Sell, 1989). It is one of the best rat liver oval cell markers but notably also reacts with normal bile duct epithelium in the rat (Dunsford and Sell, 1989) and therefore cannot be considered stem cell specific. OV-6 recognizes an epitope shared on cytokeratins (CKs) 19 and 14 (Bisgaard et al., 1993). In humans, it has also proven widely useful, recognizing oval-like cells in a number of adult and

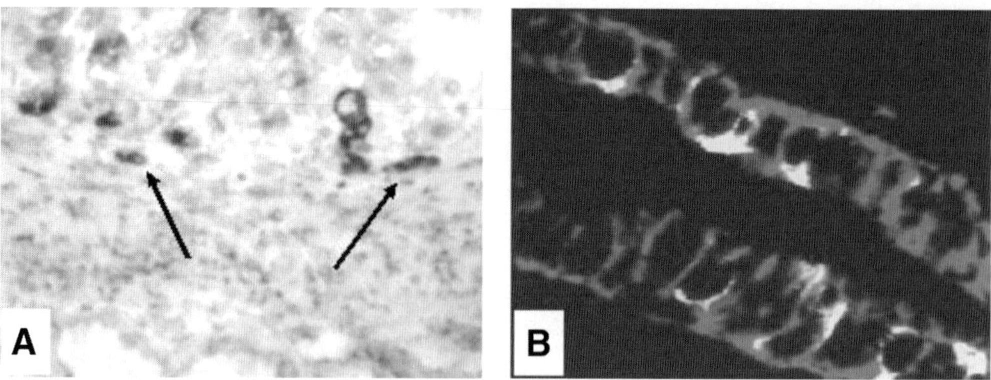

Fig. 2. Human primary biliary cirrhosis liver tissue sections immunostained with antibody OV-6. **(A)** Oval-like OV-6-positive cells (arrows) are found in the periportal margins adjacent to hepatocyte lobules. **(B)** Intact bile duct showing cells mature integrated biliary cells that coexpress OV-6 and the biliary marker HEA 125. (Reproduced with permission from Crosby et al. 1998b.)

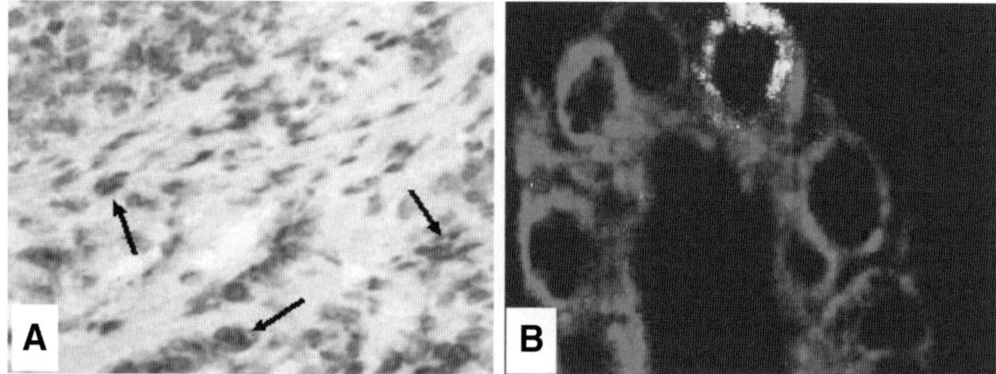

Fig. 3. Human diseased liver tissue sections immunostained with an antibody against the SCF receptor c-*kit*. **(A)** Individual oval-like c-*kit*-positive cells (arrows) can be seen scattered throughout the portal tracts of a liver section from a patient with biliary atresia. **(B)** Bile duct from a patient with acute liver failure showing a cell integrated into the duct that coexpresses c-*kit* and the mature biliary marker CK19 (white). (Reproduced with permission from Baumann et al. 1999.)

pediatric chronic viral and cholestatic liver diseases (Fig. 2A) (Van Eyken et al., 1988; Crosby et al., 1998a, 1998b; Roskams et al 1998a). The antigen recognized in human liver by OV-6 has not been determined, although the immunostaining pattern differs markedly from CK19 (Crosby et al., 1998a, 1998b). Additionally, there is some debate as to whether OV-6 recognizes normal bile duct epithelium in humans (Crosby et al., 1998a; Roskams et al., 1998a). Nevertheless, careful immunohistochemistry reveals that cells positive for OV-6 with either hepatocyte or biliary phenotype also have led investigators to conclude that these oval-like cells can and do differentiate in pathological human liver conditions. Indeed, cells that coexpress OV-6 and mature biliary markers can be found integrated into intact polarized lumenal ducts (Crosby et al., 1998a) (Fig. 2B).

35.4.2. HUMAN HSCs Other markers of particular relevance are c-*kit*, the stem cell factor (SCF) receptor, and CD34, which are bone marrow–derived stem cell markers (Andrews et al., 1989; Hassan and Zander, 1996). Both are upregulated on oval cells during the activation process following liver damage or carcinogen treatment regimes used in experimental rodent studies (Fujio et al., 1994; Omori et al., 1997). In human liver, although c-*kit* is expressed on, e.g., mast cells, the close proximity of c-*kit*$^+$ cells to surviving bile ducts in diseased liver (Baumann et al., 1999) (Fig. 3A) and even colocalization with cells expressing mature biliary epithelial markers integrated into bile ducts (Baumann et al., 1999) (Fig. 3B) led us to speculate that these cells may be acting as liver stem cells. Additionally, CD34-positive cells can also be found adjacent to bile ducts and in the periportal margins (Crosby et al., 2001). Together with the ability to separate these cells out and induce their differentiation in vitro into biliary phenotype (Crosby et al., 2001) (*see* below), this provided compelling evidence that this is a potentially important additional population of stem cells with hepatic potential.

That HSCs are found in human liver has been recognized for some time (Taniguchi et al., 1996; Crosbie et al., 1999). Clearly, this can form part of the normal trafficking of immune cells throughout tissues. What has now been convincingly shown, however, is that as initially determined in rodents, transdifferentiation of bone marrow cells into hepatic epithelium also occurs in humans, in both liver and bone marrow transplant recipients (Alison et al., 2000; Theise et al., 2000b; Korbling et al., 2002). One might speculate that (stem) cells that encounter a specific

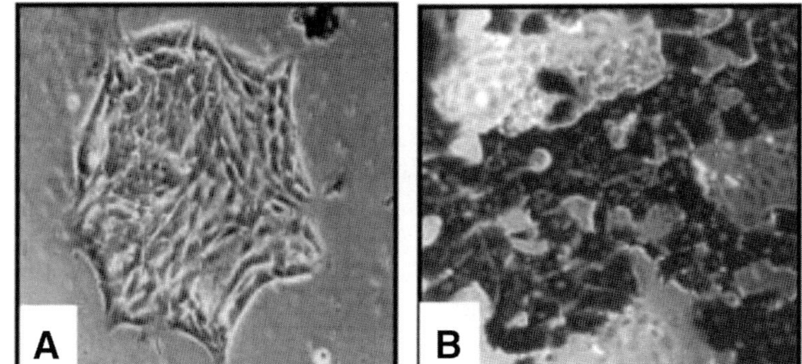

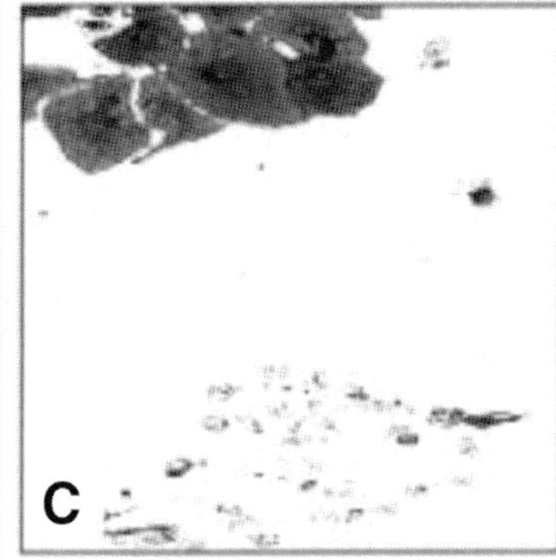

Fig.4. Cell colonies derived from individual c-*kit* immunoisolated liver-derived cells. **(A)** Phase-contrast micrograph of c-*kit*$^+$ cells following culture in media containing 10% Fetal calf serum and 10 ng/mL of hepatocyte growth factor (HGF) for 7 d showing a discrete colony formed comprising cells with two distinct morphologies; **(B)** colony fixed and immunostained with biliary marker CK19 showing one population of cells strongly positive and a second cell population negative; **(C)** similar culture immunostained with endothelial cell marker CD31.

microenvironment respond to the signals found there (growth factor, cytokine, or even matrix derived), by differentiation down a certain lineage pathway. There is currently considerable debate concerning the extent to which the this process occurs clinically or experimentally. Indeed, as indicated, many investigators now believe that the number of cells that do undergo the transition from bone marrow stem cells to hepatic, neural, or other tissues (Lagasse et al., 2000; Hillebrands et al., 2002; Mallet et al., 2002; Morshead et al., 2002) or the other direction (i.e., neural to blood) (Anderson et al., 2001) is actually quite small and repopulation is relatively slow. Therefore, physiologically, in vivo they may well be relatively rare events and there is doubt that the phenomena, although clearly demonstrable in vivo, play any significant role in the response of the human liver to tissue injury. Nevertheless, the principles of transdifferentiation are now established and subsequently may well prove to be of clinical importance.

Several major questions arise from these recent observations and are the subject of intensive investigation: Do all hepatic stem cells, and therefore oval cells, derive from HSCs? If so, have these cells migrated into the liver from the circulation, as is the case when leukocytes traffic in and out of tissues (Lalor and Adams, 1999)? Or, alternatively, are these HSCs with hepatic potential perhaps dormant in the liver? As yet there is no final answer, but it seems likely that in the adult, at least, cells capable of giving rise to hepatic epithelial cell phenotype derive from more than one source: resident liver stem cells that give rise to oval cells and their differentiated progeny; and, under some circumstances, cells derived from the bone marrow and present in the circulation.

35.5. HUMAN HEPATIC STEM CELLS IN VITRO

One of the approaches to resolve the question of the physiological relevance and clinical potential of hepatic stem cells is to develop in vitro strategies that can be used to investigate the molecular mechanisms and regulatory events leading to stem cell expansion and differentiation. The central aim of many laboratories now is to attempt to exploit the potential of these stem cells and use them to generate populations of differentiated hepatic epithelium, especially hepatocytes, which continue to be in great demand but in short supply.

35.5.1. c-*kit* and CD34-POSITIVE CELLS Cells expressing OV-6 cannot be isolated readily because this antibody recognizes an intracellular antigen (Crosby et al., 2001). Cells expressing c-*kit*, however, can be immunoselected using magnetic bead technology or fluorescence-activated cell sorting (FACS) and when maintained in culture, differentiation into cells expressing biliary phenotype in vitro can be observed (Crosby et al., 2001). Initially, single cells (approx 10% of those isolated) attach to culture plates and form small discrete colonies. If allowed to proliferate, they can give rise to colonies of cells with mixed morphologies (Fig. 4). These colonies when rigorously phenotyped demonstrate that discrete cell populations express either biliary (CK19 and CK18) or endothelial phenotype (*see* below) and are negative for hepatocyte, or mesenchymal markers. In parallel, remarkably, CD34$^+$ cells selected from a similar variety of normal and diseased human liver were also found to give rise to hepatic epithelial cell phenotype (Crosby et al., 2001).

Another striking finding from the in vitro model studies is the presence of CD31$^+$ cells (endothelial cell phenotype) in c-*kit*-derived cultures (Crosby et al., 2001), raising the possibility, as discussed above, that the cells targeted also have hemangioblast (Choi et al., 1998; Gehling et al., 2000) (endothelial precursor) potential. This is postulated to be an early progenitor cell in the bone marrow during embryogenesis in several species including mice, with the capability of giving rise to either endothelial lineage or cells of the hematopoietic system. Since CD34 is also expressed on endothelial cells, it is not surprising that CD31$^+$ cells can also be found in culture (Crosby et al., 2001). However, following depletion with CD31 and further positive selection with CD34, endothelial cells were still detected in subsequent cultures (not shown), adding weight to the hypothesis that cells with hemangioblast properties were being targeted. Using FACS analysis, the number of cells coexpressing c-*kit* and CD34 is quite small (<0.2% of liver-derived mononuclear cells), and, therefore, it is still unclear as to whether this represents a highly specialized population of cells or whether all cells have the potential to differentiate into different cell types, perhaps influenced by their microenvironment.

It is noteworthy that these c-*kit* and CD34 cells with hepatic stemlike properties can be isolated from liver of patients with an

Table 1
Comparison of Growth of Cells Isolated Using C-*kit* or HEA125

Tissue[a]	Age range (yr)	Isolation	Growth from c-kit(%)	Growth from HEA125 (%)
Adult PBC	39–60	7	100	100
Adult ALD	47–61	4	100	100
Pediatric normal	1.5–14	8	87	100
Adult normal	18–51	4	0	100
Pediatric EHBA	0.2–2	10	20	40

[a]PBC, primary biliary cirrhosis; ALD, alcoholic liver disease; EHBA, extra hepatic biliary artresua.

extensive range of disease pathology. This includes patients transplanted for acute liver failure, alcoholic liver disease, primary biliary cirrhosis, $\alpha 1$ antitrypsin deficiency, and biliary atresia (Table 1) and from normal human liver from a wide age range (2–65 yr). Furthermore, it is interesting to note that cells isolated from adult normal tissue grow less successfully. It is likely that this is because they have different growth requirements not met by the current tissue culture regimen. It is also possible that cells from younger donors have a greater inherent propensity to proliferate and/or differentiate.

The observation that cells with this differentiation potential are present not only in multiple types of diseased liver tissue with widely differing etiologies, but also in normal liver is significant because it suggests that it reflects a general property—the plasticity of cells. This may reflect the maturity of the organ, which continues to grow and develop in the younger donor age range, and that these cellular properties reemerge during disease pathogenesis. As with other stem cell approaches, means to allow clonal expansion of these cells without spontaneous differentiation occurring must be determined.

Recently, Avital et al. (2001) isolated a population of β_2-microglobulin$^-$, Thy1$^+$ cells from human (and rat) bone marrow with the potential to differentiate into mature hepatocytes. Interestingly, these cells appeared to be negative for both CD34 and c-*kit*. Since these investigators did not report any differentiation into cells of biliary phenotype, it is likely that the cells represent a population of hematopoietic-derived cells that differ from those that can be isolated from normal and diseased human liver. Although not described as such (Avital et al., 2001), CD44$^-$, class HLA$^-$, and β_2-microglobulin$^-$, are characteristics of classic mesenchymal stem cells (Bianco et al., 2001; Pittenger et al., 1999).

Recent extensive work with another cell population, distinct from the hematopoietic lineage but present in human bone marrow, with diverse stem cell potential, has become the focus of attention—mesenchymal (or stromal-derived) stem cells. These cells have been known to be progenitors of skeletal tissue components such as bone, cartilage, the hematopoietic-supporting stroma, and adipocytes (Bianco et al., 2001). Now, Verfaillie has extensively characterized a population of cells given the name multipotential adult progenitor cells (MAPCs) generated by depleting bone marrow of all glycophorin-expressing cells (erythrocytic lineage) and of all CD45 (leukocyte common antigen) and therefore hematopoietic lineage–expressing cells (Reyes et al., 2001). Again, these cells are also CD34$^-$ and, c-*kit*$^-$ and therefore differ from the primitive hematopoietic progenitors described earlier (Crosby et al., 2001).

These MAPCs have remarkable multipotential characteristics. When cultured under carefully defined conditions, they appear to differentiate into functional, mature hepatocytes (Schwartz et al., 2002). This differentiation response requires a complex set of conditions and growth factor and media additions, but principally the presence of HGF and Fibroblast growth factor-4 (FGF-4). Significantly, also, it was dependent on matrix-derived signals, as the investigators used Matrigel to support the clonal expansion and subsequent differentiation (Schwartz et al., 2002). Although the chief constituents of Matrigel incorporate the three main basement membrane components—laminin, heparan sulfate proteoglycan, and collagen IV (Bissell et al., 1987)—it is also known to contain significant levels of contaminating growth factor cytokines and other macromolecules. Thus, the differentiation signals clearly are complex.

Even more striking is the observation that the same MAPC population, when exposed to a different set of conditions, in particular in response to vascular endothelial growth factor, acts as endothelial cell progenitors (Reyes et al., 2002). Verfaillie therefore has additionally categorized them as angioblasts. Paradoxically, however, they are reported to be AC133+ (Reyes et al., 2002), a marker previously described as marking a subset of hematopoietic CD34-positive stem cells, which leaves some confusion as to their true lineage and origin.

35.5.2. HUMAN FETAL HEPATOBLASTS During development, the ventral foregut endoderm gives rise to the early hepatic epithelium, through differentiation signals from the cardiac mesoderm and from cells of the septum transversum (Zaret, 2000; Duncan and Watt, 2001). The emerging hepatoblasts are considered at least bipotential liver progenitor cells since they differentiate into mature hepatocytes or those in contact with the portal mesenchyme, into biliary epithelium (Roskams et al., 1998b). Hepatoblasts there represent a potentially highly valuable source of human liver stem cells.

Experimental animal studies have generated useful models on which the more difficult human hepatoblast work can be built. Working with mouse embryonic hepatoblasts, Rogler et al (1997) succeeded in establishing a bipotent cell line. Subsequently, two groups used positive selection to purify what they described as fetal rat or mouse progenitor cells. That these are also from the hepatoblast compartment is clear because they were selected using epithelial markers including antibodies against histocompatibility antigens and integrins and, conversely, negatively selected out all hematopoietic-expressing cells (Kubota and Reid, 2000; Suzuki et al., 2000).

Human fetal tissue is much more difficult to work with and, consequently less is known about the potential of human fetal liver cells. In a purely morphological immunohistochemical study, Gerber was the first to describe the characteristics of human fetal hepatoblasts and their lineage potential using markers including CK14, CK19, vimentin, and Hepar1 (Haruna et al., 1996). While we know that human fetal hepatoblasts can be isolated and grown in culture (Strain et al., 1987), it is only very recently that their true clonal and differentiation capacity has been recognized. Mahli et al. (2002) have reported that following transplantation into severe combined immunodeficiency disease mice, human fetal hepatoblasts have the ability to give rise to fully mature hepatocytes and in vitro can be expanded extensively. Theirs is an important study that confirms, at least experimentally, the consistent behavior in vivo of human fetal liver cells and opens the way for further assessment of their potential for clinical transplantation therapies.

An interesting question and one that has not been addressed, from the current human (or rodent-based) fetal liver work, is, does the hematopoietic cell compartment, which forms a bulk of the tissue at the gestational ages under investigation, harbor cells with the same plasticity as has been shown with their adult bone marrow–derived counterparts? Although this is an interesting intellectual challenge, the difficulty associated with regular tissue supply of human fetal liver precludes a rigorous systematic investigation of this question, which is more readily addressed with cells from bone marrow or umbilical cord blood.

35.6. MOLECULAR CONTROL OF HUMAN LIVER STEM CELL GROWTH AND DIFFERENTIATION

Several families of cytokines and growth factors are potentially important in regulation of liver stem cell responses. In a series of comprehensive studies from Thorgeirsson's laboratory, a number of receptor ligand families were reported to be upregulated during activation or progression of rat oval cell growth and differentiation. These include FGFs, leukaemia inhibitory factor, SCF, transforming growth factor-α (reviewed in Grisham and Thorgeirsson, 1997). Important clues also come from studies of hepatic development during embryogenesis (*see* Chapter 31) highlighting the involvement of, e.g., members of the FGF (Zaret, 2000) and bone morphogenetic protein families (Duncan and Watt, 2001).

Although much is known about the individual growth requirements of human hepatocytes and BEC (Strain et al., 1991; Blanc et al., 1992; Joplin et al., 1992; Matsumoto et al., 1994; Yokomuro et al., 2000), relatively few studies to determine the factors important in the control of growth and differentiation of human liver stem cells have been undertaken. As described, since cells isolated using the antibody c-*kit* lose expression of this protein shortly thereafter, SCF is not a significant player in the current in vitro strategy to expand these cells. Other factors of likely relevance include members of the Gp130 receptor family such as OSM, which is known to regulate maturation of fetal hepatoblasts into hepatocytes during development (Kamiya et al., 1999). Interestingly, it is the fetal hematopoietic cell population that produces OSM, underlining the importance of local cell and tissue paracrine regulatory mechanisms.

35.6.1. NOTCH PATHWAY One of the molecular signaling pathways of special interest is that of the Notch ligand/receptor system. This is an important pathway during embryogenesis (first described in *Drosophila*) (Artavanis-Tsakonas et al., 1999) that regulates cell fate determination during development. In eukaryotes their properties are still being defined, but it is increasingly clear that they occupy a central position in biological control processes in mammalian biology (Mumm and Kopan, 2000). Mutations in the Jagged 1 gene are associated with the condition Alagille syndrome (Oda et al., 1999), a congenital condition leading to severe developmental abnormalities of the heart, musculoskeletal system, liver, and eyes (Spinner et al., 2001). The liver manifestations of the disease are loss of bile ducts leading to cholestasis, liver fibrosis, and cirrhosis and thereby requiring liver transplantation to alleviate the life-threatening consequences of liver dysfunction.

There are four Notch receptors in humans and several ligands with widespread distribution. Both receptor and ligand family members are large molecular weight membrane-spanning proteins, indicating that biological activation is restricted to cells in immediate contact with one another. It is known that they act as key players in, e.g., the regulation of early mammalian hematopoiesis (Karunu et al., 2001a) and angiogenesis (Krebs et al., 2000). Our current hypothesis is that Jagged/Notch signaling may be important in hepatic cell responses seen during disease pathogenesis and in some of the early events in hepatic stem cell activation.

Jagged 1 is expressed ubiquitously in human liver in multiple cell types (Nijjar et al., 2001, 2002). However, we now know that the recently characterized ligand Delta-4, as well as the Notch receptors, are expressed with greater selectivity in human liver cell isolates (Fig. 5) (Nijjar et al., 2001, 2002). BEC and endothelial cells are capable of post–Notch receptor signaling through the transcription factors Deltex and Hes-1 and, interestingly, may also self-regulate bioactivity of Notches through expression of extracellular mediator proteins called Fringe proteins (Nijjar et al., 2002). Because the isolation technique used to separate endothelial cells (using CD31) cannot differentiate between vascular and sinusoidal endothelium, we cannot be sure which cells in particular are expressing Delta-4 (Fig. 5), although the immunohistochemistry with J1/CD31 colocalization indicated expression was on neo-vessels in the portal tracts (Nijjar et al., 2002). At the time of the study, antibodies against Delta-4 were unavailable.

In view of the cellular patterns of expression determined in vitro and the interesting coexpression patterns in tissue sections, we have postulated that Notch signaling plays a role in epithelial cell responses and in controlling hepatic neovascularization during adult liver disease pathogenesis (Nijjar et al., 2001, 2002). We have demonstrated colocalization of J1 on ductular proliferative cells and on endothelial cells in portal tract fibrous septa where neovessel formation occurs. Although the two pathogenetic cellular responses are distinct, these observations indicate that they may be linked mechanistically and that one of the controlling elements may be through the Jagged Notch pathway.

The expression of Notch receptors by hematopoietic bone marrow–derived stem cells has been well documented (Varnum-Finney et al., 2002; Karanu et al., 2001b) Furthermore, freshly isolated liver-derived stem cells do express Notch receptors (Fig. 6). If Notch-expressing cells migrate into the liver and come into contact with cells in the liver microenvironment that express Jagged 1 or Delta-4 (Nijjar et al 2002), then the consequence could be a switch to epithelial and/or endothelial lineage. Since the c-*kit*- and CD34-isolated cells clearly have the potential to differentiate to either lineage apparently 'spontaneously' in vitro (Crosby et al., 2001), it is reasonable to speculate that these cell responses under investigation are under Notch receptor control.

Further strength is added to the hypothesis with recently reported exciting knockout (KO) mouse technology. Although Jagged 1 homozygous KO mice die during embryogenesis, and heterozygous mice display eye defects, they have little other phenotype (McCright et al., 2002). However, doubly heterozygous mice created by mating the J1 KO with a mouse carrying a Notch 2 mutant allele display many of the features of Alagille syndrome including bile duct defects and cholestasis. This adds strength to the hypothesis that this pathway is important in mammalian biliary tree morphogenesis (McCright et al., 2002; Nijjar et al., 2002).

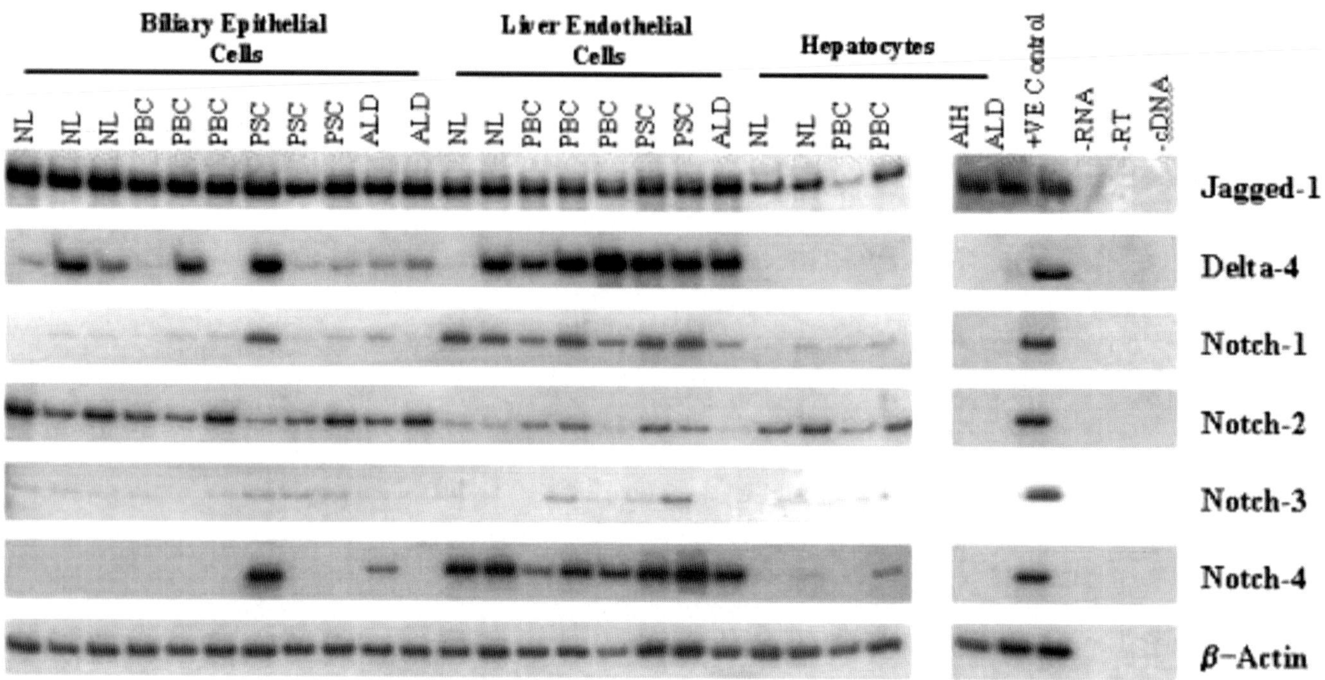

Fig. 5. Reverse transcriptase polymerase chain reaction (RT-PCR) analysis of Notch receptor and ligand mRNA expression in human liver cell isolates. cDNA was prepared from human liver cells isolated from normal liver (NL), primary biliary cirrhosis (PBC), primary sclerosing cholangitis (PSC), alcoholic liver disease (ALD), and autoimmune hepatitis (AIH). Specific primer sets were used against human Jagged 1, Delta-4, and Notch 1–4. (Reproduced with permission from Nijjar et al. 2001, 2002.)

While the relationship of HSCs (with hepatic potential) to oval cells remains to be clarified, it is apparent that adult tissue–derived stem cells have enormous potential as multipotent cells. Currently, relatively little is known about the specific factors that might regulate their expansion and/or differentiation, but the Jagged/Notch pathway is likely to play a central role. If the microenvironment of the liver (or any other organ/tissue of choice) in vivo can be defined and mimicked in vitro, this will facilitate the ability to induce the differentiation of stem cells into the appropriate phenotype.

35.7. CELL FUSION OR TRANSDIFFERENTIATION?

Recently, the transdifferentiation debate has been thrown open with the observations that chromosome transfer (and therefore Y-chromosome positivity in the nucleus of female cells) can arise as a result of fusion of cells rather than transdifferentiation, as has been assumed and frequently reported. This evidence comes from experiments in two independent laboratories and is based on observations made in vitro. Mouse embryonic stem cells were cocultured with either neurospheres (Terada et al., 2002) or bone marrow–derived cells (Ying et al., 2002), and using clearly distinct genetic markers, in both cases apparently spontaneous fusion led to the establishment of hybrid cell colonies. It must be emphasized that there is as yet no indication that this phenomenon occurs in vivo, and it is not yet certain how this would translate into the likelihood of such an event in vivo. However, the presence in, e.g., the livers of female bone marrow transplant recipients receiving male-derived donor bone marrow of hepatocytes with Y-chromosome signals could in theory arise as a result of cell fusion. This clearly requires careful scrutiny and indicates that genotyping of cells suggested to have transdifferentiated may be the only way to exclude cell fusion as an entity and indeed before any final conclusions can be drawn on the extent of cell fusion ves transdifferentiation phenomena.

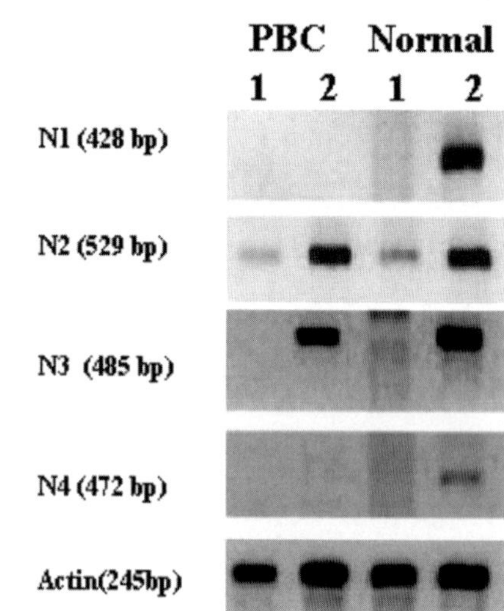

Fig. 6. RT-PCR analysis of Notch receptor expression in liver-derived c-*kit*+ cells immediately following isolation. cDNA was prepared from purified cells immediately following immunoisolation using c-*kit* antibodies. PBC, primary biliary cirrhosis.

35.8. FUTURE CLINICAL USE OF HUMAN LIVER STEM CELLS

At present, the general feeling is that with the now widely recognized plasticity of adult stem cells, the goal of using them as a therapeutic tool to treat liver disease is a little closer. Primary human hepatocytes have been used, with some partial measure of success (Strom et al., 1999), for single enzyme replacement therapy, as in Crigler–Najjar syndrome. These patients suffer from hyperbilirubinemia due to the congenital deficiency of the liver enzyme responsible for the conjugation and therefore excretion of bilirubin, UDP-glucuronyl transferase. In a single case report, a young woman was given a single bolus transplantation of 10^9 primary human hepatocytes, which led in subsequent weeks to a significant (>50%) reduction in serum bilirubin levels (Fox et al., 1998). Unfortunately this clinical benefit was relatively short-lived, although it did indicate that the transplanted hepatocytes were functional for at least some time.

The prospect, however, of directly transplanting stem cells themselves, in the hope that they will find a niche and differentiate into the desired cell phenotype in vivo, is a little more distant, since some means of controlling the cellular responses must be sought. Nevertheless, the future for cell/gene therapy is very much brighter than even only a few years ago.

NOTE ADDED IN PROOF

Evidence for fusion in vivo of donor bone marrow-derived hematopoietic stem cells to recipient hepatocytes has recently been reported in mice (Wang et al., 2003).

ACKNOWLEDGMENTS

Work in our laboratories is supported by the Wellcome Trust, The Biotechnology and Biological Sciences Research Council, and The United Birmingham Hospitals Research Endowment Fund.

REFERENCES

Alison, M. (1998) Hepatic stem cells. *J. Hepatol.* 29:676–682.

Alison, M., Golding, M., Sarraf, C., Edwards, R., and Lalani, E. (1996) Liver damage in the rat induces hepatocyte stem cells from biliary epithelial cells. *Gastroenterology* 110:1182–1190.

Alison, M., Poulsom, R., Jeffery, R., et al. (2000) Hepatocytes from non-hepatic adult stem cells. *Nature* 406:257.

Anderson, D. J., Gage, F. H., and Weissman, I. L. (2001) Can stem cells cross lineage boundaries? *Nat. Med.* 7:393–395.

Andrews, R. G., Singer, J. W., and Bernstein, I. D. (1989) Precursors of colony forming cells by expression of the CD33 and CD34 antigens and light scattering properties. *J. Exp. Med.* 169:1721–1731.

Artavanis-Tsakonas, S., Rand, M. D., and Lake, R. J. (1999) Notch signaling: cell fate control and signal integration in development. *Science* 284: 770–776.

Avital, I., Inderbitzin, D., Aoki, T., et al. (2001) Isolation, characterisation and transplantation of bone marrow derived hepatocyte stem cells. *Biochem. Biophys. Res. Commun.* 288:156–164.

Baumann, U., Crosby, H. A., Ramani, P., Kelly, D. A., and Strain, A. J. (1999) Expression of the stem cell factor receptor c-kit in normal and diseased pediatric liver: identification of a human hepatic progenitor cell? *Hepatology* 30:112–117.

Bianco, P., Riminucci, M., Gronthos, S., and Robey, P. G. (2001) Bone marrow stromal stem cells: nature, biology and potential applications. *Stem Cells* 19:180–192.

Bisgaard, H. C., Parmelee, D. C., Dunsford, H. A., Sechi, S., and Thorgeisson, S. S. (1993) Keratin 14 protein in cultured nonparenchymal rat hepatic epithelial cells: characterization of keratin 14 and keratin 19 as antigens for the commonly used mouse monoclonal antibody OV-6. *Mol. Carcinogen* 7:60–66.

Bissell, D. M., Arenson, D. M., Maher, J. J., and Roll, F. J. (1987) Support of cultured hepatocytes by a laminin-rich gel: evidence for a functionally significant subendothelial matrix in normal rat liver. *J. Clin. Invest.* 79:801–812.

Blanc, P., Etienne, H., Daujat, M., et al. (1992) Mitotic responsiveness of cultured adult human hepatocytes to epidermal growth factor, transforming growth factor alpha and human serum. *Gastroenterology* 102: 1340–1350.

Bucher, N. L. R. and Farmer, S. (1998) Liver regeneration after partial hepatectomy: genes and metabolism. In: *Liver Growth and Repair* (Strain, A. J. and Diehl, A. M., eds.), Chapman & Hall, London, UK, pp. 3–27.

Choi, K., Kennedy, M., Kazarov, A., Papadimitriou, J. C., and Keller, G. (1998) A common precursor for hematopoietic and endothelial cells. *Development* 125:725–732.

Crosbie, O. M., Reynolds, M., McEntee, G., Traynor, O., Heggarty, L. E., and O'Farrelly, C. (1999) In vitro evidence for the presence of hematopoietic stem cells in the adult human liver. *Hepatology* 29: 1193–1198.

Crosby, H. A., Hubscher, S., Joplin, R., Kelly, D. A., and Strain, A. J. (1998a) Immunolocalisation of OV-6, a putative stem cell marker in human fetal and diseased liver. *Hepatology* 28:980–985.

Crosby, H. A., Hubscher, S., Fabris, L., et al. (1998b) Immunolocalisation of putative human liver progenitor cells in liver of patients with end-stage primary biliary cirrhosis and sclerosing cholangitis using the monoclonal antibody OV-6. *Am. J. Pathol.* 152:771–779.

Crosby, H. A., Kelly, D. A., and Strain, A. J. (2001) Use of *C-kit* or CD34 to isolate putative stem cells from human liver which can differentiate into biliary epithelial cells *Gastroenterology* 120: 534–544.

Dabeva, M. A. and Shafritz, D. A. (1993) Activation, proliferation and differentiation of progenitor cells into hepatocytes in the D-galactosamine model of liver regeneration. *Am. J. Pathol.* 143:1606–1620.

De Vos, R. and Desmet, V. (1992) Ultrastructural characteristics of novel epithelial cell types identified in human pathologic liver specimens with chronic ductular reaction. *Am. J. Pathol.* 140:1441–1450.

Duncan, S. A. and Watt, A. J. (2001) BMPs on the road to hepatogenesis. *Genes Dev.* 15:1879–1884.

Dunsford, H. A. and Sell, S. (1989) Production of monoclonal antibodies to preneoplastic liver cell populations induced by chemical carcinogens in rats and to transplantable Morris hepatomas. *Cancer Res.* 49: 4887–4893.

Fabris, L., Strazzabosco, M., Crosby, H. A., et al. (2000) Characterization and isolation of ductular cells co-expressing neural cell adhesion molecule and Bcl-2 from primary cholangiopathies and ductal plate malformations. *Am. J. Pathol.* 156:1599–1612.

Farber, E. (1956) Similariies in the sequence of early histological changes induced in the livers by ethionine, 2-acetylaminofluorene and 3-methyl4-dimethylaminoazobenzene. *Cancer Res.* 16:142–148.

Fausto, N. (2001) Liver regeneration: from laboratory to clinic. *Liver Transplant.* 7:835–844.

Fausto, N., Lemire, J. M., and Shiojiri, N. (1993) Cell lineages in hepatic development and the identification of progenitor cells in normal and injured liver. *Proc. Soc. Exp. Biol. Med.* 204:237–241.

Fox, I. J., Chowdhury, J. R., Kaufman, S. S., et al. (1998) Treatment of the Crigler-Najjar Syndrome Type 1 with hepatocyte transplantation. *N. Engl. J. Med.* 338:1422–1426.

Fujio, K., Evarts, R. P., Hu, Z., Marsden, E. R., and Thorgeirsson, S. S. (1994) Expression of stem cell factor and it's receptor, c-kit, during liver regeneration from putative stem cells in adult rat. *Lab. Invest.* 70: 511–516.

Gehling, U. M., Ergun, S., Schumacher, U., et al. (2000) In vitro differentiation of endothelial cells from AC133-positive progenitor cells. *Blood* 95:3106–3112.

Gerber, M. A., Thung, S. N., Shen, S., Stromeyer, F. W., and Ishak, K. G. (1983) Phenotypic characterization of hepatic proliferation: antigenic expression by proliferating epithelial cells in fetal liver, massive hepatic necrosis, and nodular transformation of the liver. *Am. J. Pathol.* 110:70–74.

Gerber, M. A., Thung, S. N., Gerlach, J., et al. (1995) Electron microscopic studies on a hepatocyte culture model with woven multicompartment capillary systems. *Hepatology* 22:546.

Germain, L., Noel, M., Gourdeau, H., and Marceau, N. (1988) Promotion of growth and differentiation of rat ductular oval cells in primary culture. *Cancer Res.* 48:368–378.

Golding, M., Sarraf, C. E., Lalani, E., et al. (1995) Oval cell differentiation into hepatocytes in the acetylaminofluorene-treated regenerating rat liver. *Hepatology* 22:1243–1253.

Grisham, J. W. (1962) A morphologic study of DNA synthesis and cell proliferation in the regenerating liver. *Cancer Res.* 22:842–849.

Grisham, J. W. and Thorgeirsson, S. S. (1997) Liver stem cells. In: *Stem Cells* (Potten, C. S., ed.), Academic, London, UK, pp. 233–282.

Guillouzo, A., Fabrice, M., Langouet, S., Maheo, K., and Maryvonne, R. (1997) Use of hepatocyte cultures for the study of hepatotoxic compounds. *J. Hepatol.* 26(Suppl.):73–80.

Haruna, Y., Saito, K., Spaulding, S., Nalesnik, M. A., and Gerber, M. A. (1996) Identification of bipotential progenitor cells in human liver development. *Hepatology* 23:476–481.

Hassan, H. T. and Zander, A. (1996) Stem cell factor as a survival and growth factor in normal and malignant hematopoiesis. *Acta Haematol.* 95:257–262.

Hillebrands, J., Klatter, F. A., van Dijk, W., and Rozing, J. (2002) Bone marrow does not contribute to endothelial cell replacement in transplant atherosclerosis. *Nat. Med.* 8:194–195.

Hsia, C. C., Evarts, R. P., Nakatsukasa, H., Marsden, E. R., and Thorgeirsson, S. S. (1992) Occurrence of oval-type cells in hepatitis B virus-associated human hepatocarcinogenesis. *Hepatology* 16: 1327–1333.

Joplin, R., Hishida, T., Tsubouchi, H., et al. (1992) Human intra-hepatic biliary epithelial cells proliferate in vitro in response to human hepatocyte growth factor. *J. Clin. Invest.* 90:1284–1289.

Kamiya, A., Kinoshita, T., Ito, Y., et al. (1999) Fetal liver development requires a paracrine action of oncostatinM through the gp130 signal transducer. *EMBO J.* 1999; 18:2127–2136

Karanu, F. N., Murdoch, B., Gallacher, L., et al. (2001a) The notch ligand jagged-1 represents a novel growth factor of human hematopoietic stem cells. *J. Exp. Med.* 192:1365–1372.

Karanu, F. N., Murdoch, B., Miyabayashi, T., et al. (2001b) Human homologues of Delta1 and Delta 4 function as itogenic regulators of primitive human hematopoietic cells. *Blood* 97:1960–1967.

Korbling, M., Katz, R., Khanna, A., et al. (2002) Hepatocytes and epithelial cells of donor origin in recipients of peripheral blood stem cells. *N. Engl. J. Med.* 346:738–746.

Krebs, L. T., Xue, Y., Norton, C. R., et al. (2000) Notch signaling is essential for vascular morphogenesis in mice. *Genes Dev.* 14:1343–1352.

Kubota, H. and Reid, L. M. (2000) Clonogenic hepatoblasts, common precursors for hepatocytic and biliary lineages, are lacking classical major histocompatability complex class I antigen. *Proc. Natl. Acad. Sci.* 97:12,132–12,137.

Lagasse, E., Connors, H., Al-Dhalimy, M., et al. (2000) Purified hematopoietic stem cells can differentiate into hepatocytes in vivo. *Nat. Med.* 6:1229–1234.

Lalor, P. F. and Adams, D. H. (1999) Adhesion of lymphocytes to hepatic endothelium. *Mol. Pathol.* 52:214–219.

Lenzi, R., Liu, M. H., Tarsetti, F., et al. (1992) Histogenesis of bile duct-like cells proliferating during ethionine carcinogenesis: evidence for a biliary epithelial nature of oval cells. *Lab. Invest.* 66: 390–402.

Lowes, K. N., Brennan, B. A., Yoeh, G. C., and Olynyk, J. K. (1999) Oval cell numbers in human chronic liver diseases are directly related to disease severity. *Am. J. Pathol.* 154:537–541.

Malhi, H., Irani, A. N., Gagandeep, S., and Gupta, S. (2002) Isolation of human progenitor liver epithelial cells with extensive replication capacity and differentiation into mature hepatocytes. *J. Cell Sci.* 115:2679–2688.

Mallet, V. O., Mitchell, C., Mezey, E., et al. (2002) Bone marrow transplantation in mice leads to a minor population of hepatocytes that can be selectively amplified in vivo. *Hepatology* 35:799–804.

Masuko, K., Rubin, E., and Popper, H. (1964) Proliferation of bile ducts in cirrhosis. *Arch. Pathol.* 78:421–431.

Matsumoto, K., Fuji, H., Michalopoulos, G., Fung, J. J., and Demetris, A. J. (1994) Human biliary epithelial cells secrete and respond to Interleukin-6, HGF and EGF promote DNA synthesis in vitro. *Hepatology* 20:376–382.

McCright, B., Lozier, J., and Gridley, T. (2002) A mouse model of Alagille syndrome: Notch2 as a genetic modifier of Jag1 haploinsufficiency. *Development* 129:1075–1082.

Morshead, C. M., Benveniste, P., Iscove, N. N., Van der Koov, D. (2002) Hematopoietic competence is a rare property of neural stem cells that may depend on genetic and epigenetic alterations. *Nat. Med.* 8: 268–273.

Mumm, J. S. and Kopan, R. (2000) Notch signalling from the outside in. *Dev. Biol.* 228:151–165.

Nijjar, S. S., Crosby, H. A., Wallace, L., Hubscher, S. G., and Strain, A. J. (2001) Notch receptor expression in adult human liver: a possible role in normal bile duct formation and endothelial cell function. *Hepatology* 34:1184–1192.

Nijjar, S. S., Wallace, L., Crosby, H. A., Hubscher, S. G., and Strain, A. J. (2002) Altered notch ligand expression in human liver disease: further evidence for a role of the Notch signaling pathway in hepatic neovascularisation and biliary ductular defects. *Am. J. Pathol.* 160: 1695–1703.

Oda, T., Elkahloun, A. G., Pike, B. L., et al. (1999) Mutations in the human Jagged1 gene are responsible for Alagille syndrome. *Nat. Genet.* 16:235–242.

Omori, N., Omori, M., Evarts, R., et al. (1997) Partial cloning of rat CD34 cDNA and expression during stem cell-dependent liver regeneration in the rat. *Hepatology* 26:720–727.

Paku, S., Schnur, J., Nagy, P., and Thorgeirsson, S. S. (2001) Origin and structural evolution of the early proliferating oval cells in rat liver. *Am. J. Pathol.* 158:1313–1323.

Petersen, B. E., Bowen, W. C., Patrene, K. D., et al. (1999) Bone marrow as a potential source of hepatic oval cells. *Science* 284:1168–1170.

Pittenger, M. F., Mackay, A. M., Beck, S. C., et al. (1999) Multilineage potential of adult human mesenchymal stem cells. *Science* 284:143–147

Popper, H. (1990) The relation of mesenchymal cell products to heaptic epithelial systems. *Prog. Liver Dis.* 9:27–38.

Reyes, M., Lund, T., Lenvik, T., Aguiar, D., Koodie, L., and Verfaillie, C. M. (2001) Purification and ex vivo expansion of postnatal human bone marrow mesodermal progenitor cells. *Blood* 98:2615–2625.

Reyes, M., Dudek, A., Jahagirdar, B., Koodie, L., Marker, P. H., and Verfaillie, C. M. (2002) Origin of endothelial progenitors in human postnatal bone marrow. *J. Clin. Invest.* 109:337–346.

Rogler, L. E. (1997) Selective bipotentila differentiation of mouse embryonic hepatoblasts in vitro. *Am. J. Pathol.* 150:591–602

Roskams, T., van den Oord, J. J., De Vos, R., and Desmet, V. J. (1990) Neuroendocrine features of reactive bile ducts in cholestatic liver disease. *Am. J. Pathol.* 137:1019–1025.

Roskams, T., Campos, R. V, Drucker, D. J., and Desmet, V. (1993) Reactive human bile ducts express parathyroid hormone-related peptide. *Histopathology* 23:11–19.

Roskams, T., De Vos, R., Van Eyken, P., et al. (1998a) Hepatic OV-6 expression in human liver disease and rat experiments: evidence for hepatic progenitor cells in man. *J. Hepatol.* 29:455–463.

Roskams, T, Van Eyken, P., and Desmet, V. (1998b) Human liver growth and development. In: *Liver Growth and Repair* (Strain, A. J. and Diehl, A. M., eds.), Chapman and Hall, London, pp.541–557.

Ruck, P., Xiao, J., and Kaiserling, E. (1996) Small epithelial cells and the histogenesis of hepatoblastoma: electron microscopic, immunoelectron microscopic, and immunohistochemical findings. *Am. J. Pathol.* 148:321–329.

Schwartz, R. E., Reyes, M., Koodie, L., et al. (2002) Multipotent adult progenitor cells from bone marrow differentiate into functional hepatocyte-like cells. *J. Clin. Invest.* 109:1291–1302.

Sell, S. (1990) Is there a liver stem cell? *Cancer Res.* 50:3811–3815.

Sell, S. (1998) Comparison of liver progenitor cells in human atypical ductular reactions with those seen in models of liver injury. *Hepatology* 27:317–332.

Sell. S. (2001) Heterogeneity and plasticity of hepatocyte lineages. *Hepatology* 33:738–750.

Simpson, G. E. C. and Finkh, E. S. (1963) Pattern of regeneration of rat liver after repeated partial hepatectomies. *J. Pathol. Bacteriol.* 86: 361–370.

Sirica, A. E., Mathis, G. A., Sano, N., and Elmore, L. (1990) Isolation, culture and transplantation of intrahepatic biliary epithelial cells and oval cells. *Pathobiology* 58:44–64.

Sirica, A. E. (1995) Ductular hepatocytes. *Histol. Histopathol.* 10: 433–456.

Slott, P. A., Liu, M. J., and Tavaloni, N. (1990) Origin, pattern and mechanism of bile duct proliferation following biliary obstruction. *Gastroenterology* 99:466–477.

Spinner, N. B., Colliton, R. P., Crosnier, C., Krantz, I. D., Hadchouel, M., and Meunier-Rotival, M. (2001) Jagged 1 mutations in alagille syndrome. *Hum. Mutat.* 17:18–33.

Strain, A. J. and Neuberger, J. M. (2002) A bioartificial liver: state of the art. *Science* 295:1005–1009.

Strain, A. J., Hill, D. J., Swenne, I., and Milner, R. D. G. (1987) The regulation of DNA synthesis in human fetal hepatocytes by placental lactogen growth hormone and insulin-like growth factor I/somatomedin-C. *J. Cell. Physiol.* 132:33–40.

Strain, A. J., Ismail, T., Tsubouchi, H., et al. (1991) Native and recombinant human hepatocyte growth factors are highly potent promoters of DNA synthesis in both human and rat hepatocytes. *J. Clin. Invest.* 87: 1853–1857.

Strom, S. C., Chowdhury, J. R., and Fox, I. J. (1999) Hepatocyte transplantation for the treatment of human liver disease. *Semin. Liver Dis.* 19:39–48.

Suzuki, A., Zheng, Y., Kondo, R., et al. (2000) Flow cytometric separation and enrichment of hepatic progenitor cells in developing mouse liver. *Hepatology* 32:1230–1239.

Taniguchi, T., Toyoshima, T., Fukao, K., and Nakuchi, H. (1996) Presence of hematopoietic stem cells in the adult liver. *Nat. Med.* 2:198–203.

Terada, N., Hamazaki, T., Oka, M., et al. (2002) Bone marrow cells adopt the phenotype of other cells by spontaneous cell fusion. *Nature* 416: 542–545.

Theise, N. D., Saxena, R., Portmann, B. C., et al. (1999) The canals of Hering and hepatic stem cells in humans. *Hepatology* 30:1425–1433.

Theise, N., Badve, S., Saxena, R., et al. (2000a) Derivation of hepatocytes from bone marrow cells in mice after radiation induced myeloablation. *Hepatology* 31:235–240.

Theise, N. D., Nimmakayalu, M., Gardner, R., et al. (2000b) Liver from bone marrow in humans. *Hepatology* 32:11–16.

Thorgeirsson, S. S. (1996) Hepatic stem cells in liver regeneration. *FASEB J.* 10:1249–1256.

Thung, S. N. (1990) The development of proliferating ductular structures in liver disease. *Arch. Pathol. Lab. Med.* 114:407–411.

Van den Heuvel, M. C., Sloof, J. H., Visser, L., et al. (2001) Expression of anti-OV6 antiboby and anti-NCAM antibody along the biliary line of normal and diseased human livers. *Hepatology* 33:1387–1393.

Vandersteenhoven, A. M., Burchette, J., Michalopoulos, G. (1990) Characterization of ductulat hepatocytes in end-stage cirrhosis. *Arch. Pathol. Lab. Med.* 114:403– 406.

Van Eyken, P., Sciot, R., Paterson, A., Callea, F., Kew, M. C., and Desmet, V. J. (1988) Cytokeratin expression in hepatocellular carcinoma *Hum. Pathol.* 19:562–568.

Van Eyken, P., Sciot, R., Calea, F., and Desmet, V. J. (1989) A cytokeratin immonohistochemical study of focal nodular hyperplasia of the liver: further evidence that ductular metaplasia of hepatocytes contributes to ductular proliferation. *Liver* 9:372–377.

Varnum-Finney, B., Xu, L., Brashem-Stein, C., et al. (2002) Pluripotent cytokine dependent hematopoietic stem cells are immortalized by constitutive Notch1 signaling. *Nat. Med.* 6:1278–1281.

Wang, X., et al. (2003) Cell fusion is the principal source of bone marrow-derived hepatocytes. *Nature* 422:897–901.

Watt, F. M. and Hogan, B. L. (2000) Out of Eden: stem cells and their niches. *Science* 287:1427–1430.

Ying, Q., Nichols, J., Evans, E. P., and Smith, A. G. (2002) Changing potency by spontaneous fusion. *Nature* 416:545–548.

Yokomuro, S., Tsuji, H., Lunz, J. G., et al. (2000) Growth control of human biliary epithelial cells by interleukin-6, hepatocyte growth factor, transforming growth factor beta1, and activin. *Hepatology* 32: 26–35.

Zaret, K. S. (2000) Liver specification and early morphogenesis. *Mech. Dev.* 92:83–88.

36 Transplantation of Hepatic Stem Cells and Potential Applications for Cell Therapy

SANJEEV GUPTA, MBBS, MD, FACP, FRCP AND JAE-JIN CHO, DVM, PhD

Transplantation of liver stem/progenitor cells offers a new approach for treating genetic diseases or liver failure. Transplantation of mature liver cells or whole liver is limited by severe shortages of donor organs. Extensive studies indicate that optimal function of transplanted cells may require cell transplantation into the liver itself rather than into ectopic sites. Cells may be transplanted into the liver by intraportal or intrasplenic injection, which leads to the entry of transplanted cells in liver sinusoids, and eventually transplanted cells become incorporated in the liver parenchyma. During this process, transplanted cells disrupt sinusoidal endothelial cells to enter the liver plate and reestablish plasma membrane structures by joining with native liver cells. Some liver progenitor cells (LPCs) are less able to integrate and may remain within the vascular spaces. Approximately 1 to 2% of the normal liver can be repopulated in animals when a somewhat larger cell number is transplanted. However, donor hepatocytes can eventually replace virtually the entire liver of animals, if transplanted cells are spared from disease processes that threaten the survival of native hepatocytes. In fact, in FAH mice with chronic liver injury owing to extensive tyrosinemia, normal adult hepatocytes have been demonstrated to divide more than seven times each, as they repopulate the liver. Extensive repopulation will also occur in recipient animals treated with hepatotoxic chemicals, such as retrorsine in rats, which damages and inhibits proliferation of native hepatocytes. Putative LPCs in the form of oval cells produced following some types of liver injury may differentiate into hepatocytes after transplantation. Similarly, putative oval cell lines and a hepatoblast cell line derived from the embryonic mouse liver may differentiate into hepatocytes and additional cell types. Primary cells isolated from the fetal rat liver and the fetal human liver can differentiate into mature hepatocytes in animals. Finally, cells from even other organs, especially the bone marrow, can differentiate into hepatocytes after transplantation into animals. Liver cell transplantation has been used to correct a number of metabolic diseases in experimental animals and cell therapy shows enormous potential. Use of liver stem/ progenitor cells will thus be applicable for cell therapy in multiple human diseases, and further efforts in this area will certainly be worthwhile.

36.1. INTRODUCTION

In vivo analysis of hepatic stem/progenitor cells is required for defining the fate of specific cell populations. Moreover, if stem cells are to be utilized for cell therapy, it is essential to demonstrate mechanisms concerning cell engraftment, proliferation, and function, similar to studies with primary hepatocytes (Gupta et al., 1999b). The correction of diseases in specific animal models following stem cell transplantation will constitute an important step in developing cell therapy applications.

Liver-directed cell therapy is of much interest for several reasons (Malhi and Gupta, 2001). Numerous genetic diseases are amenable to cell therapy. These conditions include situations in which the liver itself is affected by disease, such as, copper toxicosis in Wilson's disease characterized by ATP7B gene mutations, which impair hepatic excretion of copper into bile. In other situations, the liver itself may not be affected by disease but hepatic gene expression is necessary for preventing disease in nonhepatic organs, such as disorders of bilirubin conjugation with bilirubin toxicity in the brain. Furthermore, expression of some genes in the liver can prevent or ameliorate disease in additional nonhepatic target organs, such as, familial hypercholesterolemia, in which extensive cardiovascular disease occurs in people due to mutations in the low-density lipoprotein receptor. In addition, extraordinarily large numbers of people carry the burden of acquired liver disease, such as, chronic viral hepatitis, which afflicts some 350 million people worldwide. In principle, these patients could benefit from cell therapy, especially if hepatitis virus infection and subsequent cytotoxicity could be prevented in transplanted cells, the so-called "disease-resistance cell" model.

Orthotopic liver transplantation (OLT), which indicates replacing the native liver with a donor liver, is quite effective in curing many genetic and acquired diseases and provides rationale for

From: *Stem Cells Handbook*
Edited by: S. Sell © Humana Press Inc., Totowa, NJ

undertaking cell therapy for many such conditions. However, OLT has been greatly hampered by organ shortages. For instance, only approx 5000 donor livers become available in the United States annually, whereas the number of people on waiting lists for OLT is three to four times greater. And these numbers do not include many people who would be suitable for OLT if patient selection criteria for OLT were broadened. Additional strategies, such as living-related liver transplantation, have not made a significant impact in decreasing waiting lists for OLT. Therefore, it is obviously imperative to develop alternative strategies, such as cell therapy. Although mature hepatocytes can be isolated from donor livers that either are not transplanted timely or are considered unsuitable for OLT, the supply of these organs is again limited. As reviewed elsewhere here, isolation of stem/progenitor cells from the liver itself, or from nonhepatic organs, and differentiation of embryonic stem (ES) cells or their derivatives into hepatocytes, constitute increasingly viable options for further consideration.

This discussion focuses on how issues in liver stem cell biology may be approached by intact animal systems. Although animal models were initially established for studying the fate of mature hepatocytes, the systems are applicable to stem cells, especially where cell differentiation into hepatocytes is concerned. On the other hand, while hepatocytes are the most abundant liver cell type (approx 60%), other cell types include biliary epithelial cells; hepatic stellate cells (found in the space of Disse, juxtaposed between hepatocytes and liver sinusoids); endothelial cells lining sinusoids; fibroblasts; pit cells; and littoral cells, including Kupffer cells. Excellent systems are available to analyze cell differentiation along hepatocyte lineages, although animal models to analyze differentiation along other hepatic lineages, such as bile ducts, are not well established (also *see* discussion in Chapter 32).

36.2. ANIMAL MODELS FOR DEMONSTRATING CELL ENGRAFTMENT, FUNCTION, AND PROLIFERATION

It should be noteworthy that liver cells display major histocompatibility complex (MHC) antigens, and mismatched cells are rejected with first-order kinetics in adult animals (Bumgardner et al., 2000; Reddy et al., 2001). On the other hand, during fetal development prior to thymic maturation, allogeneic cells (MHC incompatible from same species) or even xenogeneic cells (MHC incompatible from another species) are tolerated, as shown by studies in fetal sheep and rats (Ouyang et al., 2001; Porada et al., 2002). In other situations, this consideration implies that cells must be transplanted into either MHC-compatible or immunodeficient animals, especially for human cells, such as mice in the background of defective recombination activating genes (Rag) or severe combined immunodeficiency disease (SCID) due to T- and B-lymphocyte deficiency (Brown et al., 2000; Dandri et al., 2001; Mercer et al., 2001). Whether stem cells will be rejected less well is presently unknown, although MHC antigens are not as well expressed in stem/progenitor cells (Kubota and Reid, 2000).

36.2.1. ECTOPIC SITES FOR CELL TRANSPLANTATION Stem cells may potentially be transplanted into a variety of extrahepatic locations where liver cells survive. In early studies, the fate of transplanted hepatocytes could be examined in only ectopic locations because distinction between transplanted and native cells in the liver required genetic markers, which were not then available.

The spleen, interscapular or dorsal fat pad, and peritoneal cavity were especially supportive of transplanted liver cells (Gupta et al., 1996a). Survival of mature hepatocytes in the splenic pulp was shown 20 yr ago (Kusano and Mito, 1982). Further work established that unique features of the splenic architecture and extracellular matrix (ECM) components were likely contributors in this process (Darby et al., 1986). For example, the rat and human spleens display venous spaces and sinuses in the red pulp, which are absent in the mouse spleen. Consequently, significant numbers of transplanted hepatocytes become entrapped in these vascular spaces, and cells proliferate over several months in the rat spleen, but not in the mouse spleen. Nonetheless, transplantation of progenitor cells in the spleen has been associated with differentiation of cells into mature hepatocytes (Sigal et al., 1995a). Transplantation of cells in the interscapular fat pad has been effective in analyzing cell differentiation along hepatocytic as well as biliary lineages (Jirtle et al., 1980; Germain et al., 1985). Large numbers of cells may be transplanted into the peritoneal cavity (Demetriou et al., 1986; Selden et al., 1995). However, survival of hepatocytes in the peritoneal cavity requires microcarriers or additional liver cell types that presumably provide ECM components, growth factors, and other types of support. Similarly, hepatocyte transplantation under the kidney capsule has shown some efficacy, although cotransplantation of pancreatic islets, which secrete hepatotrophic hormones, such as insulin and glucagon, promotes hepatocyte survival (Ohashi et al., 2000).

Nonetheless, despite the simplicity of transplanting cells into sc sites, concerns have arisen about the potential of these sites in supporting hepatic differentiation. In one type of study, hepatic function was determined in mature mouse hepatocytes in the spleen, liver, peritoneal cavity, and dorsal fat pad (Gupta et al., 1994). These studies showed that liver gene expression was superior in the liver and spleen compared with that ineither peritoneal cavity or dorsal fat pad. In another type of study, progenitor rat liver epithelial cells containing a reporter transgene were transplanted into various ectopic sites (Ott et al., 1999a). These studies showed selective extinction of transgene expression in the peritoneal cavity or dorsal fat pad, whereas the transgene was expressed in the liver. Interestingly, transgene expression was restored when cells were removed from these nonpermissive sites. Further analysis elucidated that cytokine-mediated regulation of transgene expression was responsible for this difference. In addition, the abundance of both positive and negative transcriptional regulators in progenitor cells may determine whether specific genes will or will not be expressed (Aragona et al., 1996; Ott et al., 1999a). These findings are in agreement with varying differentiation potential in the WB344 rat epithelial liver cell line (*see* Chapter 32) following transplantation into the liver or ectopic sites (Hooth et al., 1998; Malouf et al., 2001). Similarly, pancreatic ductal cells have been found to exhibit differences in their differentiation potential into hepatocytes in the liver vs in nonhepatic sites (Dabeva et al., 1997). These studies imply that analysis of stem cell differentiation should take into consideration the site of cell transplantation, and that induction of hepatic differentiation should be optimally achieved in the liver itself.

36.2.2. CELL TRANSPLANTATION IN THE LIVER An advantage of injecting cells into the splenic pulp concerns immediate deposition of large numbers of cells in liver sinusoids (Gupta et al., 1999c). This makes it quite convenient for transplanting cells in small animals. Although cells can be deposited in liver sinusoids following injection into the portal vein or one

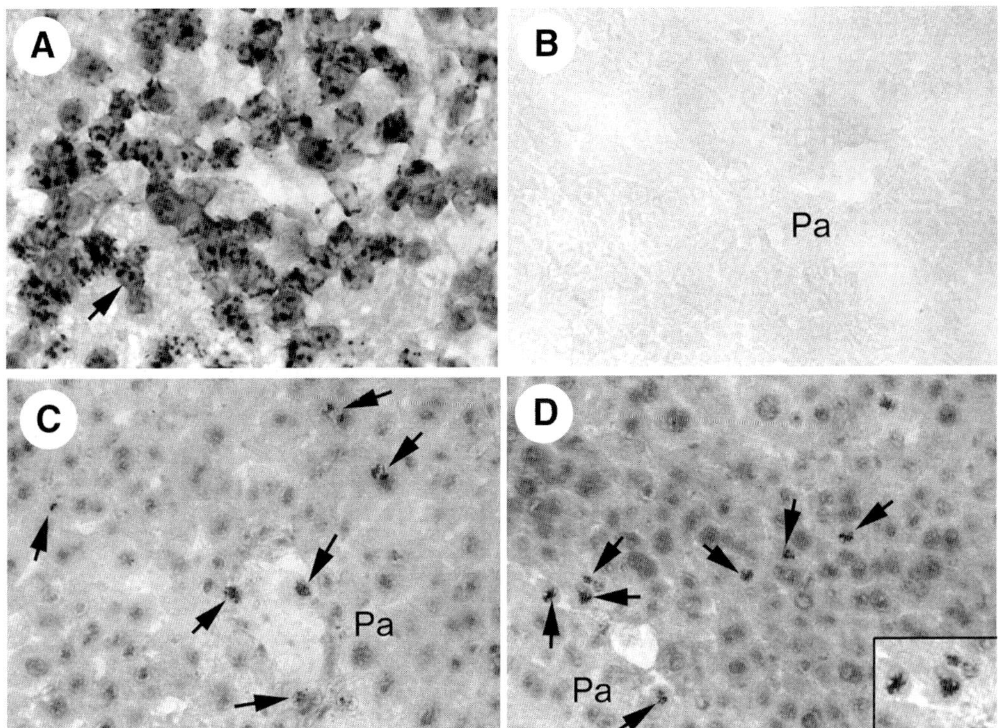

Fig. 1. Localization of transplanted human hepatocytes in liver of NOD-SCID mice with ISH. A pancentromeric probe binding to each of the 23 centromeres in diploid cells was digoxigenin labeled. Color visualization after ISH and antidigoxigenin binding utilized the diaminobenzidine peroxidase substrate. **(A)** Human liver showing multiple signals in nuclei (arrow). **(B)** Negative control mouse tissue showed no signal. **(C)** Two hours after intrasplenic transplantation, several human hepatocytes (arrows) are located within a portal vein and in periportal sinusoids. **(D)** Human hepatocytes (arrows) were present in the liver plates several days following transplantation. The inset in (D) shows ISH signals in transplanted cells at higher magnification. (A) (C) and (D) were lightly counterstained with toluidine blue to visualize cell nuclei. Pa, portal area. Original magnification: ×1000.

of its tributaries, intrasplenic injection of cells lends itself to superior hemostasis and far less mortality in small animals. Injection of cells into arterial circulations, including splenic artery, hepatic artery or pulmonary artery, does not produce significant transplanted cell survival.

Several animal models are now available to demonstrate survival of transplanted cells in the liver (*see* review by Gupta and Rogler, 1999). The essential approach concerns marking of transplanted cells with genetic reporters. The earliest models utilized inbred transgenic mice, in which unique foreign genes were expressed. In one system, mouse hepatocytes containing hepatitis B virus surface antigen (HBsAg) gene were transplanted into recipient congeneic mice (Gupta et al., 1990). In another system, transplanted cells contained either human α1 antitrypsin (hAAT) gene or *Escherichia coli* β-galactosidase (LacZ) gene (Ponder et al., 1991). Transplanted cells could then be identified in various organs by localizing transgene expression by molecular methods, including *in situ* hybridization (ISH), or by demonstrating transgene products by histochemistry in tissues or by various assays in the peripheral blood. The success of this approach for cell marking led to additional strategies, in which transplanted cells contained other reporters, such as green fluorescent protein, sex chromosome markers, or unique proteins deficient in recipient animals. In the case of human cells, the most appropriate strategies are likely to involve the use of either sex-chromosome markers (for localizing transplanted cells in people) or additional chromosomal probes for localizing cells in immunodeficient animals (Zahler et al., 2000; Malhi et al., 2002) (Fig. 1).

An alternative strategy for cell localization has concerned the use of inbred Fischer 344 rats exhibiting a spontaneous mutation in the dipeptidyl peptidase IV (DPPIV) gene (Thompson et al., 1991). The DPPIV– rat was initially utilized for analyzing the oncogenic potential of cells and has been extensively applied subsequently for numerous studies concerning transplanted cell biology. DPPIV is an ectopeptidase with expression in cells throughout the body, although the precise role of this enzyme is unclear, and DPPIV– animals lack specific phenotype. DPPIV is expressed particularly abundantly in the bile canalicular domain of hepatocytes and on the cell membrane of some bile duct cells, which provides convenient assays for demonstrating integration of transplanted cells in the liver parenchyma (Gupta et al., 1995). The power of this system will be extended further by the recent availability of DPPIV– mice, in which the DPPIV gene has been inactivated by homologous recombination. These mice are currently being crossed into immunodeficient backgrounds to generate universal recipients, and these efforts will provide especially useful tools to undertake studies of stem cells derived from human material.

Our present understanding of how transplanted cells engraft in the liver is largely obtained from studies in the DPPIV– rat (Gupta et al., 1999b). Subsequent to the deposition of cells in liver sinusoids, transplanted hepatocytes enter the liver plate with disruption of sinusoidal endothelial cells (Gupta et al., 1999c). It is noteworthy that transplanted cells are deposited predominantly in the periportal areas of the liver and require several days to become fully integrated in the liver parenchyma. However, the process of cell integration involves reconstitution of plasma membrane structures, including development of conjoint structures, such as bile canaliculi and gap junctions, with adjacent hepatocytes (Gupta et al., 1995, 1999c). Additionally, transplanted hepatocytes exhibit physiologically regulated gene expression patterns and become responsive to mitogenic stimuli present in the liver microenvironment, albeit after completing the integration process in the liver parenchyma (Gupta et al., 1999c, 2000). Studies with liver progenitor cells (LPCs) demonstrate similar mechanisms, although many LPCs might stay within vascular spaces, without entering the liver plate, and some LPCs might integrate in portal areas proximal to the limiting plate, with biliary gene expression (Sigal et al., 1995a, Dabeva et al., 2000).

With respect to cell proliferation, note that transplanted cells do not proliferate in the normal rat or mouse liver (Gupta et al., 1991; Ponder et al., 1991; Sokhi et al., 2000). Exceptions to this rule exist in the developing rat liver, where transplanted cells exhibit some proliferation, which ceases when the liver matures, and the aged F344 rat liver, where transplanted cells exhibit spontaneous proliferative activity, which is presumably related to the propensity of F344 rats to develop hepatic hyperplasia and biliary proliferation in old age (Sokhi et al., 2000).

After transplantation of 20 million cells in the rat or 2 million cells in the mouse liver, approx 0.5–1% of the normal adult liver can be repopulated, and transplanted cells survive the entire lifespan of these animals (Sokhi et al., 2000). On the other hand, transplanted cells begin to proliferate in the liver when exposed to mitogenic activity, as may be induced with chemicals, such as carbon tetrachloride, that selectively injure native hepatocytes and spare transplanted hepatocytes (Gupta et al., 1996b, 2000). This concept of selective hepatic injury was dramatically established in transgenic mice expressing the urokinase-type plasminogen activator under control of the albumin promoter (alb-uPA mouse) (Rhim et al., 1994). In this situation, normal hepatocytes, which were unaffected by the toxic alb-uPA transgene, showed extensive proliferation with virtually complete replacement of the liver. These mice have been useful for xenotransplantation, including human hepatocytes, with the establishment of permissive hepatitis B and C virus infections (Brown et al., 2000; Dandri et al., 2001; Mercer et al., 2001). Another mutant, the FAH mouse, which constitutes a model of hereditary tyrosinemia type 1, shows extensive liver disease, such that transplantation of normal hepatocytes leads to near-total liver repopulation (Overturf et al., 1996). In mice deficient in the ABCB4 gene (adenosine triphosphatase–binding cassette B4 gene/multidrug resistance [mdr] 2 gene/P-glycoprotein 2 gene), which impairs biliary phopholipid excretion with hepatobiliary injury, especially following addition of toxic bile salts to the diet, normal hepatocytes proliferate extensively (De Vree et al., 2000). Depending on the extent of liver injury in additional situations, either limited or more extensive transplanted cell proliferation can be induced. Examples of these approaches include expression of a prodrug activating herpes simplex virus thymidine kinase transgene, adenoviral expression of the cell-cycle regulator Mad1 gene, and induction of apoptosis in native hepatocytes by the Fas ligand (Mignon et al., 1998; Braun et al., 2000; Gagandeep et al., 2000). Obviously, models that can be applied to stem cells, in which genetic manipulation of transplanted cells is unnecessary, would be most desirable for avoiding confounding due to the regulation of transgene expression or unscheduled differentiation following activation of manipulation-induced additional events.

In the rat liver, use of DNA-damaging agents has been helpful for inducing liver repopulation with transplanted cells. Obviously, such liver injury should be insidious without precipitating acute liver failure. Use of the pyrollizidine alkaloid retrorsine, which alkylates DNA and inhibits proliferation in native hepatocytes, has been extraordinarily effective for liver repopulation in the DPPIV– rat (Laconi et al., 1998). It is noteworthy that retrorsine induces extensive polyploidy in the rat liver, which is also induced by two-thirds partial hepatectomy and the thyroid hormone triiodothyronine (T_3) (Sigal et al., 1999; Torres et al., 1999). Polyploid hepatocytes exhibit DNA damage with evidence for the activation of cell-aging events, increased apoptosis, and decreased proliferation capacity in vitro, as well as in vivo after transplantation into DPPIV– rats (Sigal et al., 1999; Gorla et al., 2001). Therefore, it should not be surprising that the combination of either partial hepatectomy or T_3 with retrorsine served synergistic roles in inducing hepatic polyploidy and liver repopulation with normal transplanted cells (Oren et al., 1999). This system offers yet another animal model to analyze the fate of transplanted stem cells. Use of retrorsine served synergistic roles in the Long-Evans Cinnamon (LEC) rat, in which liver disease occurs due to copper toxicosis following mutation in the atp7b gene and impairment in biliary excretion of copper (Irani et al., 2001). A characteristic feature of copper-induced liver disease concerns oxidative damage in the liver, which in the presence of retrorsine and partial hepatectomy-induced perturbations leads to extensive proliferation of transplanted hepatocytes and correction of the disease. Because of species-specific differences in xenobiotic metabolism, retrorsine is not very effective in promoting transplanted cell proliferation in mice. However, additional ways to induce liver injury in rodents for developing further models are under active investigation in several laboratories. In this respect, use of ionizing radiation to induce hepatic DNA damage has been as effective as retrorsine in repopulating the rat liver (Guha et al., 1999). Again, radiation of the liver becomes highly effective when combined with partial hepatectomy.

36.3. TRANSPLANTATION ANALYSIS OF REPRESENTATIVE TYPES OF LIVER STEM CELLS

Several types of stem cell populations can obviously be tested to determine their differentiation potential. While the issue of stem cell plasticity has engaged significant attention lately, it is worth commencing this discussion with the consideration of extensive proliferation capacity exhibited by hepatocytes themselves. The Hayflick limit should predict attenuation of proliferation capacity in mature hepatocytes after a certain number of cell divisions. However, when Grompe and colleagues serially transplanted normal hepatocytes in the FAH mouse, transplanted cells proliferated indefinitely, repopulating the liver of recipient animals down more than seven generations (Overturf et al., 1997). Even conservative estimates of the number of divisions each transplanted hepatocyte underwent during this period suggested >80 cell divisions, without exhaustion of the replication potential, which was considerably beyond the suggested Hayflick limit, and led to considering this behavior of hepatocytes as "stem cell-like." However, it was unclear whether these cell preparations could have included specific hepatocyte subpopulations with greater proliferation capacity. As already indicated, highly polyploid rat hepatocytes were found to have attenuation of proliferative capacity (Sigal et al., 1999; Gorla et al., 2001). However, analysis of diploid vs tetraploid mouse hepatocytes indicated that the proliferation capacity in these hepatocyte subsets was similar (Overturf et al., 1999; Weglarz et al., 2000). Transplanted hepatocytes were not observed to generate bile duct cells in any of these studies, presumably because the injury in mice affected only the hepatocyte compartment, without stimulating biliary proliferation.

Transplantation of hepatic "oval cells," which originally referred to poorly differentiated cholangiolar cells following carcinogenic induction (Farber, 1956), has been under investigation as a potentially useful progenitor cell type. Oval cells arise during liver regeneration induced by chemicals, such as carbon tetrachloride and D-galactosamine (GalN), as well as following acute liver failure in humans (Sell, 2001). As reviewed elsewhere, oval cells exhibit a variety of hepatocytic and biliary markers, including albumin, (-fetoprotein (AFP), glycogen, glucose-6-phosphatase, and hybrid isoenzymes. Various studies have shown that primary oval cells isolated from the normal rat liver, as well as the diseased liver of LEC rats with copper toxicosis, can differentiate into mature hepatocytes (Yasui et al., 1997; Hooth et al., 1998; Ott et al., 1999a; Malouf et al., 2001). Similarly, isolation of oval cell–like ductular cells from the pancreas of rats subjected to a regimen of copper depletion and repletion was associated with the development of mature hepatocytes in the DPPIV– rat liver, although the overall frequency of this event in the normal rat liver was relatively limited (Dabeva et al., 1997). Oval cells isolated from the mouse pancreas have also been shown to generate hepatocytes (Wang et al., 2001), indicating some degree of cellular plasticity, although liver and pancreas share a common embryological origin from the primitive foregut, with retention of multiple transcriptional regulators even subsequently.

Similarly, transplantation of an F344 rat liver–derived oval cell line, designated FNRL, led to the production of hepatocytes following transplantation into the liver, although far more transplanted cells appeared either to show incomplete lineage maturation into hepatocytes or not to differentiate into hepatocytes at all (Ott et al., 1999b). In this respect, it was noteworthy that differentiation of transplanted cells into hepatocytes was promoted by the liver microenvironment, whereas transplantation of cells into extrahepatic sites, such as spleen and peritoneal cavity, was less effective in inducing hepatic differentiation. Another highly studied epithelial nonparenchymal liver cell line from the F344 rat, designated WB344 cells, shows progenitor cell properties, including differentiation into hepatocytes, as well as into other cell types, depending on the microenvironmental context (Hooth et al., 1998; Malouf et al., 2001). Obviously, the precise signals regulating cell differentiation are unknown at present and this area requires further study.

Another area of active investigation concerns isolation of cells from the fetal liver. In contrast to the adult liver, where progenitor cells are encountered rarely, the fetal liver contains large numbers of hepatoblasts (Sigal et al., 1995b; Dabeva et al., 2000; Malhi et al., 2002). The fetal human liver arises at 4 wk of gestation and matures rapidly with bile formation by 12 wk and establishment of lobular architecture shortly thereafter. Studies of the rat liver showed that fetal cells are predominantly diploid, whereas with increasing age hepatocytes mature and acquire cytoplasmic complexity, ploidy, specialized gene expression profiles, and so on (Sigal et al., 1995b, 1999). Cell lines have occasionally been derived from the fetal liver, e.g., HBC-3 cells from the 9.5d embryonic mouse liver with the capacity to differentiate along biliary and hepatocyte lineages (Ott et al., 1999b). Similarly, hepatic progenitor cell populations have been enriched and characterized from the fetal rat liver (Sigal et al., 1995a; Dabeva et al., 2000). Cells isolated from the fetal rat liver differentiate into hepatocytes after transplantation into DPPIV– rats. In other studies, fetal hepatocytes completed differentiation programs after transplantation into animals, with acquisition of cytochrome P450 activity, which reflects advanced hepatic differentiation (Kato et al., 1996). Fetal rat hepatoblasts can be readily isolated from mid- to late-term gestation livers, and these cells generate mature hepatocytes and (sometimes) bile duct cells in DPPIV– rats, along with greater replication in the intact liver compared with mature hepatocytes (Sandhu et al., 2001).

These considerations reinforced our investigations of the potential of fetal human hepatocytes (Malhi et al., 2002). A specific advantage of using the fetal human liver as a source of clinically useful cells concerns lack of interference with OLT programs, e.g., by further encroaching on the already limited supplies of donor livers. Unlike the somewhat unresolved ethical debate concerning the derivation of human ES cells, fetal tissues are from cadaveric material designated for discarding as medical waste. Initial studies have shown that the fetal human liver cells can be isolated quite reproducibly. Epithelial cells isolated from fetal human liver in the second trimester showed expression of several oval cell markers, including AFP, γ-glutamyl transpeptidase, plasminogen activator inhibitor type-1, and various cytokeratins, capacity to proliferate extensively in cell culture; and excellent recovery following cryopreservation. Moreover, when transplanted in SCID mice, fetal cells engrafted in the liver parenchyma and produced mature hepatocytes. These findings suggest that yet another type of progenitor cell could be useful for cell therapy.

Finally, the issue of hematopoietic stem cells (HSCs) derived from the bone marrow or peripheral blood has generated considerable interest. Of course, the fetal liver is a major site of extramedullary hematopoiesis (Petersen et al., 1999; Alison et al., 2000; Lagasse et al., 2000; Theise et al., 2000a, 2000b; Korbling et al., 2002). By using the DPPIV– rat, Petersen et al. initially demonstrated that HSCs derived from the bone marrow differentiated into hepatocytes. This finding has been extended in the mouse, as well as humans, where several groups have now established the presence of liver cells originating from donor-derived HSCs (Alison et al., 2000; Lagasse et al., 2000; Theise et al., 2000a, 2000b; Korbling et al., 2002). Particularly convincing evidence of such properties of HSCs is found in studies in the FAH mouse (Lagasse et al., 2000). Here, transplantation of specific subsets of HSCs that are characterized by c-kit and sca-1 antigen expression with low expression of Thy-1 antigen leads to extensive liver repopulation with mature hepatocytes and correction of the tyrosinemic metabolic abnormalities. Taken together, these recent findings of cell transplantation in animals suggest that a variety of

Table 1
Selected Conditions Amenable to Liver-Directed Cell Therapy

Liver is primary target of disease	*Other organs are targets of disease*
Metabolic diseases	Metabolic deficiency states
Wilson disease	Congenital hyperbilirubinemia, e.g., Crigler-Najjar syndrome
α1 antitrypsin deficiency	Familial hypercholesterolemia
Erythropoietic protoporphyria	Hyperammonemia syndromes
Lipidoses, e.g., Gaucher disease, Niemann-Pick disease	Defects of carbohydrate metabolism
Tyrosinemia, type 1	Oxalosis
Chronic viral hepatitis, cirrhosis, and liver failure	Coagulation disorders
	Hemophilia
	Factor IX deficiency
Fulminant liver failure owing to viral hepatitis, drug toxicity, and so on.	Immune disorders
	Hereditary angioedema

Table 2
Animal Models Particularly Useful for Establishing the Value of Liver Cell Therapy

Nature of animal model	*Animal designation*	*Animal origin or manipulation*	*Disease reproduced*
Genetically determined	Nagase analbuminemic rat (NAR)	Spontaneous mutation	Analbuminemia
	FAH mouse	Gene knockout	Tyrosinemia, type 1
	Gunn rat	Spontaneous mutation	Crigler–Najjar syndrome
	LEC rat	Spontaneous mutation	Wilson disease
	P-glycoprotein (mdr) 2 mouse	Gene knockout	Bile phospholipid transport defect
	Watanabe heritable hyperlipidemic rabbit (WHHL)	Spontaneous mutation	Familial hypercholesterolemia
Iatrogenically produced	Various mice, rats, rabbits	Acetaminophen, carbon tetrachloride, D-galactosamine, other chemicals, subtotal partial hepatectomy	Acute liver failure
	Various rats	Repeated carbon tetrachloride	Liver cirrhosis

candidate stem/progenitor cells will be suitable for cell therapy applications in humans. Of course, progress in the area of ES cell differentiation into hepatobiliary cells is eagerly awaited.

36.4. ANIMAL MODELS AVAILABLE FOR ASSESSING THERAPEUTIC POTENTIAL OF STEM CELLS

Hepatocyte transplantation can potentially correct or ameliorate many disorders (Table 1). These conditions extend from genetic deficiency states to metabolic and acquired liver diseases. As discussed earlier, liver may be the primary target of injury. On the other hand, liver may be entirely normal but injury in distant organs may arise from either inactivation of a toxic protein, such as unconjugated bilirubin (brain injury), or production of toxin, such as oxalosis (renal injury). Yet, other disorders may be corrected by hepatocyte transplantation, such as coagulation disorders or hypercholesterolemia, in which manifestations arise from systemic perturbations.

To develop appropriate clinical strategies, it is necessary to demonstrate the magnitude of liver repopulation necessary for correcting specific diseases, as well as to understand the kinetics of such disease correction. Several animal models have been especially helpful in such analyses with hepatocyte transplantation (Table 2) (Gupta and Rogler, 1999). These models should be similarly effective in defining the potential of stem cell transplantation. If the goal is to correct genetically determined metabolic disorders, transplantation of normal hepatocytes will be required. Alternatively, one could modify autologous stem cells in vitro and transplant corrected cells (ex vivo gene therapy). Treatment of acquired disorders, such as chronic viral hepatitis, requires transplantation of hepatocytes resistant to hepatitis virus infection. By contrast, patients with acute liver failure could be treated with healthy cells from any source.

Detailed discussion of how specific diseases are reproduced in various animal models listed and others not listed in Table 2 is beyond the scope here, and the reader is referred to specialized texts. Changes in the host liver microenvironment, especially in the presence of circulating toxins, could affect initial cell engraftment and subsequent proliferation. In many instances, animals are inbred and normal donors exist in the same genetic background, which permits cell transplantation without rejection. Examples include the Gunn rat (Wistar-RHA rats are normal syngeneic donors), the LEC rat (LEA rats are normal syngeneic donors), and mice created by homologous recombination in established inbred lines. In other circumstances, animals are outbred and will reject cells from one another, such as NAR and WHHL rabbits, which necessitates use of either immunosuppression to prevent cell rejection or autologous cell transplants following introduction of reporter genes. In some animal models, such as NAR rats, Gunn rats, LEC rats, and WHHL rabbits, outcomes of cell therapy can be monitored by measuring changes in specific serum proteins (albumin, bilirubin, ceruloplasmin, and total cholesterol, respectively). In other situations, animals must be sacrificed to localize transplanted cells in the liver and to analyze histological improvement following cell therapy. Assessment of improved outcomes in manifestations of liver disease, such as encephalopathy, coagulopathy, and jaundice, requires appropriate physiological parameters or mortality. Nonetheless, the availability of excellent animal models indicates that exciting progress can be made in this area.

36.5. CONCLUSION

The availability of excellent animal models of human disease and insights into how transplanted cells engraft, function, and proliferate in the rodent liver offers unique opportunities to analyze stem cell biology. Advances in isolating stem/progenitor cells by reproducible and convenient ways further facilitate insights into this area. The ability to work with enriched or highly purified populations of human stem/progenitor cells will help translate this area into exciting clinical applications.

ACKNOWLEDGMENTS

We gratefully acknowledge the contributions of our colleagues in many of the studies cited herein. This work was supported in part by National Institutes of Health grants RO1 DK46952, P33DK41296, P30 CA13330, and MO1 RR12248.

REFERENCES

Alison, M. R., Poulson, R., Jeffrey, R., et al. (2000) Hepatocytes from non-hepatic adult stem cells. *Nature* 406:257.

Aragona, E., Burk, R. D., Ott, M., Shafritz, D. A., and Gupta, S. (1996) Cell-type specific mechanisms regulate hepatitis B virus transgene expression in liver and other organs. *J. Pathol.* 180:441–449.

Braun, K. M., Degen, J. L., and Sandgren, E. P. (2000) Hepatocyte transplantation in a model of toxin-induced liver disease: variable therapeutic effect during replacement of damaged parenchyma by donor cells. *Nat. Med.* 6:320–326.

Brown, J.J., Parashar, B., Moshage, H., et al. (2000) A long-term hepatitis B viremia model generated by transplanting nontumorigenic immortalized human hepatocytes in Rag-2-deficient mice. *Hepatology* 31: 173–181.

Bumgardner, G. L., Gao, D., Li, J., Baskin, J. H., Heininger, M., and Orosz, C. G. (2000) Rejection responses to allogeneic hepatocytes by reconstituted SCID mice, CD4, KO, and CD8 KO mice. *Transplantation* 70:1771–1180.

Dabeva, M. D., Hwang, S. G., Vasa, S. R., et al. (1997) Differentiation of pancreatic epithelial progenitor cells into hepatocytes following transplantation into rat liver. *Proc. Natl. Acad. Sci. USA* 94: 7356–7361.

Dabeva, M. D., Petkov, P. M., Sandhu, J., et al. (2000) Proliferation and differentiation of fetal liver epithelial progenitor cells after transplantation into adult rat liver. *Am. J. Pathol.* 156:2017–2031.

Dandri, M., Burda, M. R., Török, E., et al. (2001) Repopulation of mouse liver with human hepatocytes and in vivo infection with hepatitis B virus. *Hepatology* 33:981–988.

Darby, H., Gupta, S., Johnstone, R., Selden, A. C., and Hodgson, H. J. F. (1986) Observations on rat spleen reticulum during the development of syngeneic hepatocellular implants. *Br. J. Exp. Path.* 67: 329–339.

Demetriou, A. A., Levenson, S. M., Novikoff, P. M., et al. (1986) Survival, organization and function of microcarrier-attached hepatocytes transplanted in rats. *Proc. Natl. Acad. Sci. USA* 83:7475–7479.

De Vree, J. M., Ottenhoff, R., Bosma, P. J., Smith, A. J., Aten, J., Oude Elferink, R. P. (2000) Correction of liver disease by hepatocyte transplantation in a mouse model of progressive familial intrahepatic cholestasis. *Gastroenterology* 119:1720–1730.

Farber, E. (1956) Similarities in the sequence of early histological changes induced in the liver of the rat by ethionine, 2-acetylaminofluorene and 3'-methyl-4-dimethyl-aminoazobenzene. *Cancer Res.* 16:142–149.

Gagandeep, S., Sokhi, R., Slehria, S., et al. (2000) Hepatocyte transplantation improves survival in mice with liver toxicity induced by hepatic overexpression of Mad1 transcription factor. *Mol. Ther.* 1: 358–365.

Germain, L., Goyette, R., and Marceau, N. (1985) Differential cytokeratin and alpha-fetoprotein expression in morphologically distinct epithelial cells emerging at the early stage of rat hepatocarcinogenesis. *Cancer Res.* 45:673–681.

Gorla, G. R., Malhi, H., and Gupta, S. (2001) Polyploidy associated with oxidative DNA injury attenuates proliferative potential of cells. *J. Cell Sci.* 114:2943–2951.

Guha, C., Sharma, A., Gupta, S., et al. (1999) Amelioration of radiation-induced liver damage in partially hepatectomized rats by hepatocyte transplantation. *Cancer Res.* 59:5871–5874.

Gupta, S. and Rogler, C. E. (1999) Liver repopulation systems and study of pathophysiological mechanisms in animals. *Am. J. Physiol.* 277: G1097–G1102.

Gupta, S., Chowdhury, N. R., Jagtiani, R., et al. (1990) A novel system for transplantation of isolated hepatocytes utilizing HBsAg producing transgenic donor cells. *Transplantation* 50:472–475.

Gupta, S., Aragona, E., Vemuru, R. P., Bhargava, K., Burk, R. D., and Roy-Chowdhury, J. (1991) Permanent engraftment and function of hepatocytes delivered to the liver: implications for gene therapy and liver repopulation. *Hepatology* 14:144–149.

Gupta, S., Vemuru, R. P., Lee, C.-D., Yerneni, P., Aragona, E., and Burk, R. D. (1994) Hepatocytes exhibit superior transgene expression after transplantation into liver and spleen compared with peritoneal cavity or dorsal fat pad: Implications for hepatic gene therapy. *Hum. Gene Ther.* 5:959–967.

Gupta, S., Rajvanshi, P., and Lee, C.-D. (1995) Integration of transplanted hepatocytes in host liver plates demonstrated with dipeptidyl peptidase IV deficient rats. *Proc. Natl. Acad. Sci. USA* 92:5860–5864.

Gupta, S., Rajvanshi, P., Bhargava, K. K., and Kerr, A. (1996a) Hepatocyte transplantation: progress toward liver repopulation. *Prog. Liver Dis.* 14:199–222.

Gupta, S., Rajvanshi, P., Aragona, E., Yerneni, P. R., Lee, C.-D., and Burk, R. D. (1996b) Transplanted hepatocytes proliferate differently after CCl4 treatment and hepatocyte growth factor infusion. *Am. J. Physiol.* 276:G629–G638.

Gupta, S., Rajvanshi, P., Sokhi, R. P., et al. (1999a) Entry and integration of transplanted hepatocytes in liver plates occur by disruption of hepatic sinusoidal endothelium. *Hepatology* 29:509–519.

Gupta, S., Bhargava, K. K., and Novikoff, P. M. (1999b) Mechanisms of cell engraftment during liver repopulation with transplanted hepatocytes. *Semin. Liver Dis.* 19:15–26.

Gupta, S., Rajvanshi, P., Sokhi, R., Vaidya, S., Irani, A. N., and Gorla, G. R. (1999c) Position-specific gene expression in the liver lobule is directed by the microenvironment and not by the previous cell differentiation state. *J. Biol. Chem.* 274:2157–2165.

Gupta, S., Rajvanshi, P., Irani, A. N., Palestro, C. J., and Bhargava, K. K. (2000) Integration and proliferation of transplanted cells in hepatic parenchyma following D-galactosamine-induced acute injury in F344 rats. *J. Pathol.* 190:203–210.

Hooth, M. J., Coleman, W. B., Presnell, S. C., Borchert, K. M., Grisham, J. W., and Smith, G. J. (1998) Spontaneous neoplastic transformation of WB-F344 rat liver epithelial cells. *Am. J. Pathol.* 153:1913–1921.

Irani, A. N., Malhi, H., Slehria, S., et al. (2001) Correction of liver disease following transplantation of normal hepatocytes in LEC rats modeling Wilson's disease. *Mol. Ther.* 3:302–309.

Jirtle, R. L., Biles, C., and Michalopoulos, G. (1980) Morphologic and histochemical analysis of hepatocytes transplanted into syngeneic hosts. *Am. J. Pathol.* 101:115–126.

Kato, K., Kasai, S., Onodero, K., et al. (1996) Developmental expression of cytochrome P450S within intrasplenically transplanted fetal hepatocytes. *Cell Transplant.* 5:S27–S30.

Korbling, M., Katz, R. L., Khanna, A., et al. (2002) Hepatocytes and epithelial cells of donor origin in recipients of peripheral-blood stem cells. *N. Engl. J. Med.* 346:738–746.

Kubota, H. and Reid, L. M. (2000) Clonogenic hepatoblasts, common precursors for hepatocytic and biliary lineages, are lacking classical major histocompatibility complex class I antigen. *Proc. Natl. Acad. Sci. USA* 97:12,132–12,137.

Kusano, M. and Mito, M. (1982) Observations on the fine structure of long survived isolated hepatocytes inoculated into rat spleen. *Gastroenterology* 82:616–628.

Laconi, E., Oren, R., Mukhopadhyay, D. K., et al. (1998) Long-term, near-total liver replacement by transplantation of isolated hepatocytes in rats treated with retrorsine. *Am. J. Pathol.* 153:319–329.

Lagasse, E., Connors, H., Al Dhalimy, M., et al. (2000) Purified hematopoietic stem cells can differentiate into hepatocytes in vivo. *Nat. Med.* 6:1229–1234.

Malhi, H. and Gupta, S. (2001) Hepatocyte transplantation: new horizons and challenges. *J. Hepatobiliary Pancreat. Surg.* 8:40–50.

Malhi, H., Irani, A. N., Gagandeep, S., and Gupta, S. (2002) Isolation of human progenitor liver epithelial cells with extensive replication capacity and differentiation into mature hepatocytes. *J. Cell Sci.* 115: 2679–2688.

Malouf, N. N., Coleman, W. B., Grisham, J. W., et al. (2001) Adult-derived stem cells from the liver become myocytes in the heart in vivo. *Am. J. Pathol.* 158:1929–1935.

Mercer, D. F., Schiller, D. E., Elliott, J. F., et al. (2001) Hepatitis C virus replication in mice with chimeric human livers. *Nat. Med.* 7:927–933.

Mignon, A., Guidotti, J. E., Mitchell, C., et al. (1998) Selective repopulation of normal mouse liver by Fas/CD95-resistant hepatocytes. *Nat. Med.* 4:1185–1188.

Ohashi, K., Marion, P. L., Nakai, H., et al. (2000) Sustained survival of human hepatocytes in mice: a model for in vivo infection with human hepatitis B and hepatitis delta viruses. *Nat. Med.* 6:327–331.

Oren, R., Dabeva, M. D., Karnezis, A. N., et al. (1999) Role of thyroid hormone in stimulating liver repopulation in the rat by transplanted hepatocytes. *Hepatology* 30:903–913.

Ott, M., Rajvanshi, P., Sokhi, R., et al. (1999a) Differentiation-specific regulation of transgene expression in a diploid epithelial cell line derived from the normal F344 rat liver. *J. Pathol.* 187:365–373.

Ott, M., Ma, Q., Li, B., Gagandeep, S., Rogler, L. E., and Gupta, S. (1999b) Regulation of hepatitis B virus expression in progenitor and differentiated cell-types: evidence for negative transcriptional control in nonpermissive cells. *Gene Exp.* 8:175–186.

Ouyang, E. C., Wu, C. H., Walton, C., Promrat, K., and Wu, G. Y. (2001) Transplantation of human hepatocytes into tolerized genetically immunocompetent rats. *World J. Gastroenterol.* 3:324–330.

Overturf, K., Muhsen, A.-D., Tanguay, R., et al. (1996) Hepatocytes corrected by gene therapy are selected in vivo in a murine model of hereditary tyrosinemia type 1. *Nat. Genet.* 12:266–273.

Overturf, K., Al-Dhalimy, M., Ou, C.-N., Finegold, M., and Grompe, M. (1997) Serial transplantation reveals the stem-cell-like regenerative potential of adult mouse hepatocytes. *Am. J. Pathol.* 151: 1273–1280.

Overturf, K., Al-Dhalimy, M., Finegold, M., and Grompe, M. (1999) The repopulation potential of hepatocyte populations differing in size and prior mitotic expansion. *Am. J. Pathol.* 155:2135–2143.

Petersen, B. E., Bowen, W. C., Patrene, K. D., et al. (1999) Bone marrow as a potential source of hepatic oval cells. *Science* 284:1168–1170.

Ponder, K.P., Gupta, S., Leland, F., et al. (1991) Mouse hepatocytes migrate to liver parenchyma and function indefinitely after intrasplenic transplantation. *Proc. Natl. Acad. Sci. USA* 88:1217–1221.

Porada, C. D., Tran, N. D., Almeida-Porada, G., et al. (2002) Transduction of long-term-engrafting human hematopoietic stem cells by retroviral vectors. *Hum. Gene Ther.* 7:867–879.

Reddy, B., Gupta, S., Chuzhin, Y., et al. (2001) Differential effect of CD28/B7 blockade with CTLA4Ig on alloreactive CD4+, CD8+ and B cells in murine isolated hepatocyte transplantation. *Transplantation* 71:801–811.

Rhim, J. A., Sandgren, E. P., Degen, J. L., Palmiter, R. D., and Brinster, R. L. (1994) Replacement of diseased mouse liver by hepatic cell transplantation. *Science* 263:1149–1152.

Rogler, L. E. (1997) Selective bipotential differentiation of mouse embryonic hepatoblasts in vitro. *Am. J. Pathol.* 150:591–602.

Sandhu, J. S., Petkov, P. M., Dabeva, M. D., and Shafritz, D. A. (2001) Stem cell properties and repopulation of the rat liver by fetal liver epithelial progenitor cells. *Am. J. Pathol.* 159:1323–1334.

Selden, C., Calnan, D., Morgan, N., Wilcox, H., Carr, E., and Hodgson, H. J. F. (1995) Histidinemia in mice: a metabolic defect treated using a novel approach to hepatocellular transplantation. *Hepatology* 21: 1405–1412.

Sell, S. (2001) Heterogeneity and plasticity of hepatocyte lineage cells. *Hepatology* 33:738–750.

Sigal, S., Rajvanshi, P., Reid, L.M. and Gupta S. (1995a) Demonstration of differentiation in hepatocyte progenitor cells using dipeptidyl peptidase IV deficient mutant rats. *Cell. Mol. Biol. Res.* 41:39–47.

Sigal, S., Gupta, S., Gebhard, D. F. Jr, Holst, P., Neufeld, D., and Reid, L. M. (1995b) Evidence for a terminal differentiation process in the liver. *Differentiation* 59:35–42.

Sigal, S. H., Rajvanshi, P., Gorla, G. R., et al. (1999) Partial hepatectomy-induced polyploidy attenuates hepatocyte replication and activates cell aging events. *Am. J. Physiol.* 276:G1260–G1272.

Sokhi, R. P., Rajvanshi, P., and Gupta, S. (2000) Transplanted reporter cells help in defining onset of hepatocyte proliferation during the life of F344 rats. *Am. J. Physiol. Gastroint. Liver Physiol.* 279: G631–G640.

Theise, N. D., Badve, S., Saxena, R., et al. (2000a) Derivation of hepatocytes from bone marrow cells in mice after radiation-induced myeloablation. *Hepatology* 31:235–240.

Theise, N. D., Nimmakayalu, M., Gardner, R., et al. (2000b) Liver from bone marrow in humans. *Hepatology* 32:11–16.

Thompson, N.L., Hixson, D.C., Callanan, H., et al. (1991) A Fischer rat substrain deficient in dipeptidyl peptidase IV activity makes normal steady-state RNA levels and an altered protein: use as a liver-cell transplantation model. *Biochem. J.* 273:497–502.

Torres, S., Diaz, B. P., Cabrera, J. J., Diaz-Chico, J. C., Diaz-Chico, B. N., and Lopez-Guerra, A. (1999) Hormone regulation of rat hepatocyte proliferation and polyploidization. *Am. J. Physiol.* 276:G155–G163.

Wang, X., Al-Dhalimy, M., Lagasse, E., Finegold, M., and Grompe, M. (2001) Liver repopulation and correction of metabolic liver disease by transplanted adult mouse pancreatic cells. *Am. J. Pathol.* 158: 571–579.

Weglarz, T. C., Degen, J. L., and Sandgren, E. P. (2000) Hepatocyte transplantation into diseased mouse liver: kinetics of parenchymal repopulation and identification of the proliferative capacity of tetraploid and octaploid hepatocytes. *Am. J. Pathol.* 157:1963–1974.

Yasui, O., Miura, N., Terada, K., Kawarada, Y., Koyama, K., and Sugiyama, T. (1997) Isolation of oval cells from Long-Evans Cinnamon rats and their transformation into hepatocytes in vivo in the rat liver. *Hepatology* 25:329–334.

Zahler, M. H., Irani, A., Malhi, H., et al. (2000) The application of a lentiviral vector for gene transfer in fetal human hepatocytes. *J. Gene Med.* 2:186–193.

37 Plasticity of Adult-Derived Pancreatic Stem Cells

AMMON B. PECK, PhD AND VIJAYAKUMAR K. RAMIYA, PhD

The pancreas develops from fusion of dorsal and ventral evaginations from the primitive gut. Active notch signaling leads to expression of transcription factors for exocrine cells, and lack of notch to expression of factors for endocrine cells. The islets of Langerhans develop from a pool of undifferentiated precursor cells associated with the ductal epithelium that have the capacity to produce progeny that differentiate into each of the four islet-associated endocrine cells: glucagon producing α-cells, insulin-producing β-cells, somatostatin-producing δ-cells, and pancreatic polypeptide–producing γ-cells. Islets are formed by migration of the islet progenitor cells into the surrounding exocrine tissue associated with angiogenesis to provide a rich arteriolar blood supply. In the adult, the pancreatic ducts contain precursor cells, which are able to self-renew and differentiate into functional islets; exocrine cells; and, under certain conditions, hepatocytes. The isolation, culture, and transplantation of the progeny of these precursor cells for treatment of experimental diabetes are described and prospects for human use are discussed.

37.1. INTRODUCTION

Ductal structures of the adult pancreas contain multipotent stem cells that, under controlled in vitro conditions, are able to self-renew and differentiate into functional islets of Langerhans. In vitro–generated islets exhibit temporal changes in mRNA transcripts similar to islet-associated markers observed during fetal pancreatic organogenesis, as well as regulated insulin responses following glucose challenge. In experiments in which in vitro–generated mouse islets have been implanted into diabetic mice, the implanted islets reversed insulin-dependent diabetes without invoking the autoimmune response. The possibility of growing functional islets from adult stem cells provides new opportunities to produce large numbers of islets, even autologous islets, for use as implants.

37.2. IN VIVO DIFFERENTIATION OF PANCREATIC STEM CELLS TO ISLETS OF LANGERHANS

Islets of Langerhans are compact clusters of endocrine hormone–producing cells that, for most vertebrate species, are embedded within the acinar tissue of the pancreas (Fig. 1A). These clusters have evolved a cellular organization that optimizes rapid and highly regulated responses to elevated blood glucose levels. During embryogenesis, the pancreas forms from a fusion of the dorsal and ventral primordia. These two primitive glands, which appear as evaginations from the gut, develop independently, then merge during midgestation. The region of the gut epithelium fated to form the dorsal pancreatic bud is initially in direct contact with the notochord. Notochord-derived factors such as activin-β B and fibroblast growth factor-2 (FGF-2) have been implicated in the initial repression of sonic hedgehog (*Shh*) expression in the presumptive dorsal pancreatic endoderm (Wessels and Cohen, 1967; Kim et al., 1997). By contrast, the ventral gut epithelium does not contact notochord, and ventral exclusion of *Shh* and *Ihh* expression is achieved by a notochord-independent mechanism. Repression of the hedgehog genes induces homeobox transcription factor *Pdx1* that, along with transcription factor *Hblx9*, determines pancreas commitment (Harrison et al., 1994), which occurs in mice at embryonic day (E) 9.5.

Factors inducing morphogenesis and differentiation of the ventral pancreatic bud are less known. The mesenchyme surrounding the dorsal pancreatic bud can be distinguished from that surrounding the ventral bud by the selective expression of the LIM-homeodomain gene *Isl1* (Ahlgren et al., 1997). Pancreas differentiation begins in epithelial cells expressing *Hnf6*, *Hlxb9*, *Hnf3*β, and *Pdx1*. *Hnf6* is required to induce *Hnf3*β and *Ngn3*. *Hnf3*β is an important regulator of the *Pdx1* gene (Wu et al., 1997). After this early developmental period, a group of transcription factors appears, including *Isl1*, *Pax6*, *Nkx2.2*, β*2/NeuroD*, and *Ngn3*. *Isl1* expression is critical for both exocrine and endocrine development. While Pax6 expression is distributed in all endocrine cell types, *Pax4* is a key factor in the differentiation of β cells. Ngn3 induces β2/NeuroD, a transcription factor implicated in both

From: *Stem Cells Handbook*
Edited by: S. Sell © Humana Press Inc., Totowa, NJ

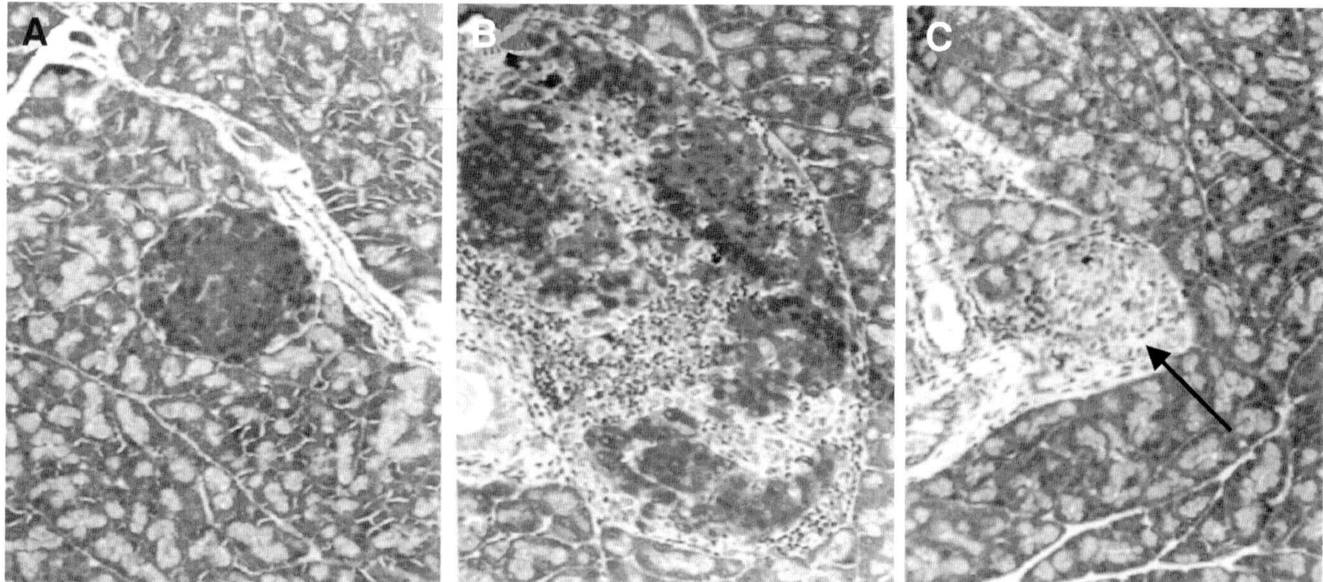

Fig. 1. Destruction of β-cell mass within the islets of Langerhans by invading autoimmune response. Photomicrographs show the developmental stages in the insulitis lesion of the pancreatic islets of Langerhans in nonobese diabetic (NOD)/Uf mice. **(A)** Normal islet of prediabetic 4-wk-old NOD mouse devoid of leukocytic infiltrates. **(B)** At 5 wk of age, the early stages of the autoimmune attack are evident, showing mostly periislet leukocyte infiltrates that progress to an advanced stage of autoimmune attack showing increased numbers of intraislet leukocyte infiltrates and β-cell destruction. **(C)** Once all β-cells have been destroyed, the leukocyte infiltrates disappear, leaving an islet containing only α-, γ-, and/or δ-cells. Each histological section was first stained with hematoxylin and eosin (H&E) dye, then counterstained with antiinsulin antibody and avidin-biotin.

insulin gene expression and islet cell differentiation (Naya et al., 1997). Exocrine cells are seen in the mouse at d E14.5, a stage at which endocrine cells are already found as cell clusters around pancreatic ducts. Apparently, notch signaling controls the choice between endocrine and exocrine fates during pancreatic cell differentiation. Lack of notch signaling results in *Ngn3* expression that promotes an endocrine fate. By contrast, active notch signaling leads to expression of *Hes1* and *p48-Ptf1*, two factors critical for the generation of exocrine cells (Jensen et al., 2000). These regulatory events are detailed in Fig. 2.

Development of the islets of Langerhans *per se* within the pancreas is initiated from a pool of undifferentiated precursor cells associated with the pancreatic ductal epithelium (Pictet and Rutter, 1972; Hellerstrom, 1984; Weir and Bonner-Weir, 1990; Peters et al., 2000). These precursor cells, in turn, appear to be derived from stem cells, and these stem cells possess the capacity to differentiate into each of the four islet-associated endocrine cell populations: glucagon-producing α-cells, insulin-producing β-cells, somatostatin-producing δ-cells, and pancreatic polypeptide-producing γ-cells (Teitelman et al., 1993). Whether all four differentiated cell populations arise from a single stem cell or from multiple stem cell populations is still questioned. During differentiation of endocrine tissue the progenitor cells are programmed to coexpress multiple hormones prior to final maturation into cells expressing a single hormone. Once islets have formed from the proliferating precursor cell populations, the islets migrate short distances into the surrounding exocrine tissue. Angiogenesis results in vascularization that leads to direct arteriolar blood flow to mature islets, which also has the effect of increasing further the number of β-cells, probably since blood glucose can stimulate β-cell mass expansion. Finally, neurogenesis leads to the innervation of the islets with sympathetic, parasympathetic, and peptidergic neurons (Bonner-Weir and Orci, 1982; Menger et al., 1994; Teitelman et al., 1998), however, the role of neural innervation in the function of the islets, if any, remains unknown.

In individuals genetically predisposed to type 1 (or autoimmune) diabetes, the organization and function of the islet clusters is disrupted due to the specific destruction of the βcell population by a progressive and relentless attack from the body's own immune system (Fig. 1B). Since pancreatic β-cells *per se* have a limited capacity to proliferate after post-fetal development of the islets of Langerhans, the β-cell mass is rapidly lost despite comprising more than 60% of the total islet mass. Once the autoimmune attack is completed, the inflammatory cells exit the islet, leaving islet remnants containing only the non-β-cell populations (Fig. 1C). It is interesting to speculate that an increased need for insulin in a type 1 diabetic patient undergoing an autoimmune response promotes β-cell proliferation by the hyperglycemia, thereby promoting the autoimmune response and ensuring the complete loss of β-cell function. The result is that most diabetic patients today require lifelong insulin therapy. While insulin therapy offers a means to achieve normoglycemia, only ecto-pancreas transplants or islet implants represent a real "cure" for the insulin-dependent diabetic patient. Unfortunately, availability of donor pancreata for transplantation or as a source of islets for implantation is acutely limited. Furthermore, because any transplanted tissue will be, in all likelihood, allogeneic (non-self), the transplanted recipient may have to take immunosuppressive drugs for the life of the transplant.

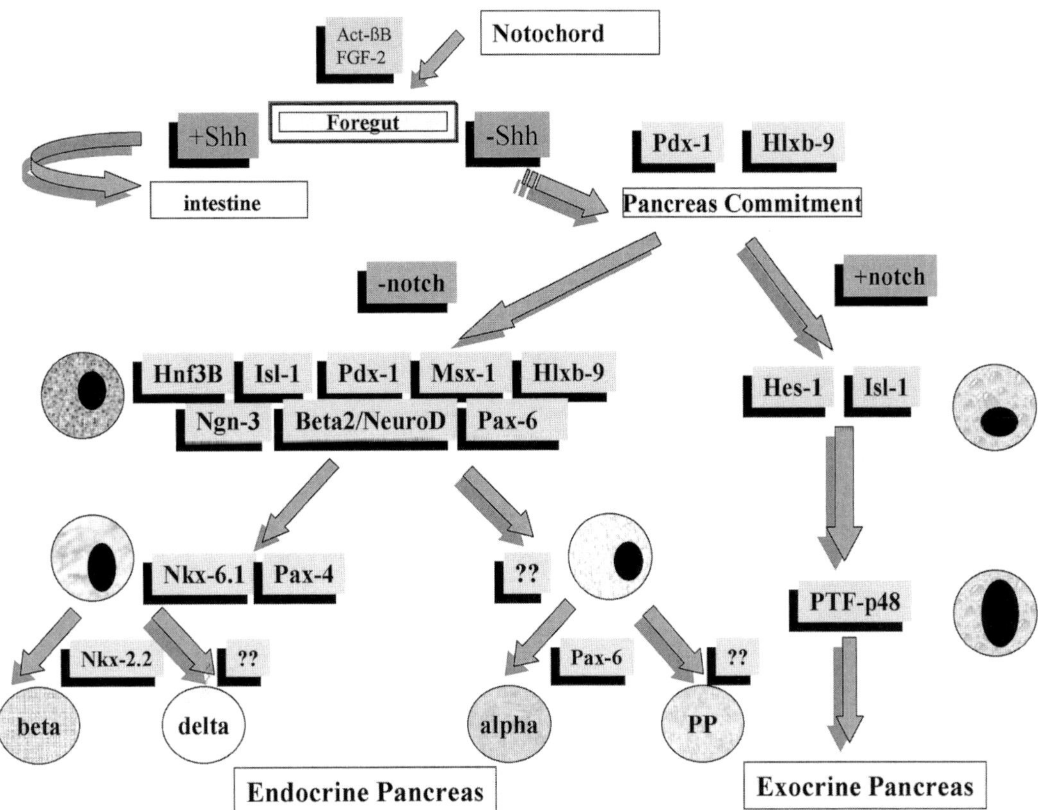

Fig. 2. Sequential expressions of transcription factors controlling pancreas organogenesis and potentially adult stem cell differentiation to endocrine pancreas. Pancreatic buds are formed by repression of *Shh* in the foregut destined to become pancreas. Entry into the pancreatic pathway is committed by expression of *Pdx-1* and *Hlxb9*. Active notch signaling promotes exocrine pancreas through activation of *Hes-1*, *Isl-1*, and *PTF-48*. Inactive notch signaling promotes endocrine pancreas. *Ngn-3* and *Pax-6* expression favor α-cell differentiation; *Pdx-1*, *Ngn-3*, *Pax-4*, and *Nkx-6.1* expression favors δ-cell differentiation, while *Nkx2.2* and *Pdx-1* expression favors β-cell differentiation.

37.3. PRESENCE OF PANCREATIC STEM CELLS

In recent years, there has been an increasing interest in the possible use of cell implants derived from stem cells to treat a variety of diseases (*see* Vogel et al., 2000). Stem cells are currently defined as cells that are capable of self-renewal and differentiation into a defined set of progeny. Pluripotent epidermal, hematopoietic, pancreatic islet, and mesenchymal stem cell populations have all been shown to possess the capacity for both self-renewal and differentiation along multiple cell lineages. In addition, these properties have been described as well for embryonic, fetal, and adult pancreatic stem cells. More recently, the functional plasticity of stem cells has become more apparent. For instance, neural stem cells may transdifferentiate into blood (Bjorson et al., 1999), or muscle cells (Vescovi, 2000) or different germ layer–derivative cells (Clarke et al., 2000). However, both plasticity and transdifferentiation of stem cells have to be proven under stringent conditions in vitro (Morshead et al., 2002) and the extent of occurrence of these phenomena in vivo. Nevertheless, use of epithelial skin grafts (Navsaria et al., 1995) and hematopoietic cell transplants (Siena et al., 2000) are already well-established clinical procedures, thus showing the utility of stem cell therapies.

Because the mature pancreatic β-cell is considered, for the most part, to be a differentiated end-stage cell, it is believed that the body has a limited capacity to regenerate new β-cells once the β-cell mass is destroyed, as in the case of type 1 diabetes (Lieter et al., 1987). However, in adults, islet cells have been shown to replicate and respond to stimuli known to initiate neonatal islet cell growth, such as glucose, growth hormone, several peptide growth factors, and especially hepatocyte growth factor (HGF) (Swenne, 1992; Hayek et al., 1995). These observations suggest that normal β-cell growth in the adult can accommodate functional demand. Furthermore, conditions such as obesity and pregnancy result in reversible increase in β-cell mass (Marynissen et al., 1983; Bonner-Weir et al., 1989; Brelje et al., 1993). There is also evidence for a slow turnover of adult cells through stem cell differentiation as demonstrated by a number of experimental models: cellophane wrapping and partial duct obstruction (Rosenberg and Vinik, 1992), alloxan treatment (Korcakova, 1971), streptozotocin (STZ) treatment (Cantenys et al., 1981), partial pancreatectomy (Bonner-Weir et al., 1993), steroid injection (Kern and Logothetopoulos, 1970), insulin antibody injection (Logothetapoulos and Bell, 1966), copper deficiency condition (Rao et al., 1989), overexpression of interferon-γ (Gu and Sarvetnick, 1993), soybean trypsin inhibitor treatment (Weaver et al., 1985), and specific growth factor treatments (Otonkoski et al., 1994). Furthermore, in normal conditions, nearly 15% of all β-cells that are smaller compared with mature islet β-cells were located in or along ductules in human adult pancreata in immunohistochemical analysis, indicating the presence of numerous sites with a potential for β-cell neogenesis (Bouwens and Pipeleers, 1998). A number of groups, including us, have reported isolation and differentiation of stem cells derived from adult pancreatic ductal structures (Peck and Cornelius, 1995; Cornelius et al., 1997; Bonner-Weir et al., 2000; Ramiya et al., 2000) that express endocrine hormones. These observations would suggest the existence of adult pancreatic stem cells.

All three major cellular compartments of the adult pancreas contain either stem cells or cell populations capable of transdifferentiating or dedifferentiating into cells that have the potential to become endocrine cells. For instance, purified rat acinar cells can transdifferentiate into ductlike cells expressing cytokeratin 20 (CK20), CK7, and fetal liver kinase (Flk-1) receptors. During this process, exocrine transcription factor (PTF-P48) remains detectable, pancreatic duodenal homeobox–containing transcription factor PDX-1 is induced, while the acinar phenotype is gradually lost (Rooman et al., 2000). These ductlike cells appear identical to precursor cells of β-cell neogenesis. Similar conversion of human exocrine tissue into CK19- and CK7-expressing ductal cells has also been reported (Gmyr et al., 2001). In a rat model of copper deficiency, endocrine α- and β-cells were shown to appear from exocrine and ductular compartments of the pancreas (Al-Abdullah et al., 2000). Hamster islets, with in vitro culturing, gave rise to ductal and acinar cells, eventually resulting in exocrine cells that were negative for PDX-1, Nkx6.1, Pax-6, and β2/NeuroD (Schmied et al., 2000). Such transdifferentiation of human islets into exocrine cells has also been documented (Schmied et al., 2001). Recently, both rat and human islets were shown to contain cells expressing nestin (a neural cell marker). These islet stem cells expressed an extended proliferative capacity when cultured in vitro and could be cloned repeatedly. After reaching confluent growth, they were able to differentiate into cells that expressed liver and exocrine pancreas markers, yet displayed ductal and endocrine phenotype with expression of CK19, insulin, glucagons, and PDX-1 (Zulewski et al., 2001). It is not clear whether these islet stem cells are the same as those reported in STZ-induced diabetic mice in which STZ-mediated elimination of existing insulin and PDX-positive cells induces the appearance of special precursor cells expressing somatostatin, PDX, and insulin (Fernandes et al., 1997).

37.4. IN VITRO ORGANOGENESIS OF ISLETS OF LANGERHANS FROM STEM CELLS

As early as 1995, Peck and coworkers (Cornelius et al., 1997; Ramiya et al., 2000) reported the successful generation of functional islets within long-term cultures of islet-producing stem cells (IPSCs) initiated from ductal epithelial cells of digested pancreatic tissue freshly explanted from either human organ donors or prediabetic NOD/Uf mice. The protocol that proved successful required several sequential steps that permitted successive growth, differentiation, and maturation of the cultured cell populations. Today, this process has been divided into four major steps, as summarized in Table 1. First, ductal epithelial cells, isolated from digested pancreas, are cultured in a growth-restrictive medium to enrich for an epithelial cell subpopulation that can form monolayers with a neuroendocrine cell phenotype. Second, islet progenitor cells (IPCs) are induced to bud from the epithelial-like monolayers, and these IPCs provide a favorable growth medium that permits proliferation into spheroid structures. Third, cells of the spheroid structures are stimulated with high concentrations of glucose, forcing the development of well-organized islet-like structures containing differentiated cells, the vast majority of which stain weakly for insulin plus glucagon. Fourth, the islet-like structures are implanted into an in vivo environment to promote final maturation of the β-cells expressing insulin.

Table 1
Protocol for Production of Functional Islets of Langerhans from Pancreatic Ductal Epithelial Stem Cells

Step	*Procedure*
1	Isolation of pancreatic ducts from partially digested pancreas and growth of stem cells in cultures of serum-free EHAA medium
2	Induction of IPCs from established stem cell cultures by addition of serum from diabetic individuals
3	Enhancement of cell proliferation and differentiation, especially β-cells, by addition of glucose
4	Implantation to an in vivo environment to complete β-cell maturation

The growth of islet cell clusters from the ductal epithelial monolayer representative of the four differentiation stages of IPSC growth into immature (preimplanted) islets is presented in Fig. 3 for mouse, pig and human tissues. The immature islets maintain their structural integrity anywhere from a few days to several weeks but can easily be dissociated into single-cell suspensions that, if cultured, start the process again. Thus, the clusters contain cells retaining the IPSC/IPC phenotype, including the capacity for self-renewal. Interestingly, the IPCs bud from the epithelial cells and proliferate into three-dimensional (3D) cell clusters (Fig. 4). We have maintained mouse IPSC cultures nearly 3 yr with constant expansion via repeated serial transfers. Each subculture exhibited the ability to produce increasing numbers of islet-like structures provided that culture expansion was not pushed too rapidly. Based on the number of islets produced within the secondary cultures, more than 10,000 pancreas equivalents were produced within the 3 yr of growth from five pancreata used for the starting tissue preparation.

During differentiation of ductal epithelial stem cells to islet-like clusters, the IPSC- and IPC-derived cells temporally express various endocrine hormones and islet cell–associated factors, listed in Table 2. These include insulin-I and -II, insulin receptor, HGF and its receptor c-MET, glucagon, somatostatin, GLUT-2 receptor, and insulin-like growth factors 1 and 2 (Bonner-Weir et al., 2000; Ramiya et al., 2000) In addition, genes related to development and differentiation were also detected, including those for REG-1, PDX-1, β-galactosidase, tyrosine hydroxylase, and β2/neuroD. Despite expression of these islet-associated genes, the cells within the islet structures remain mostly immature. The fact that islets generated in vitro fail to achieve full maturation prior to implantation into an in vivo environment is supported by the observations that (1) many of the cells continue to express during the culture phases more than a single endocrine hormone, and (2) insulin secretion in response to glucose challenge remains minimal. When nicotinamide, a reagent known to enhance proliferation and maturation of mouse β-cells, is added to differentiating cultures of mouse IPSCs, the number of islets increases (Fig. 5), and the islet cells exhibit increased insulin

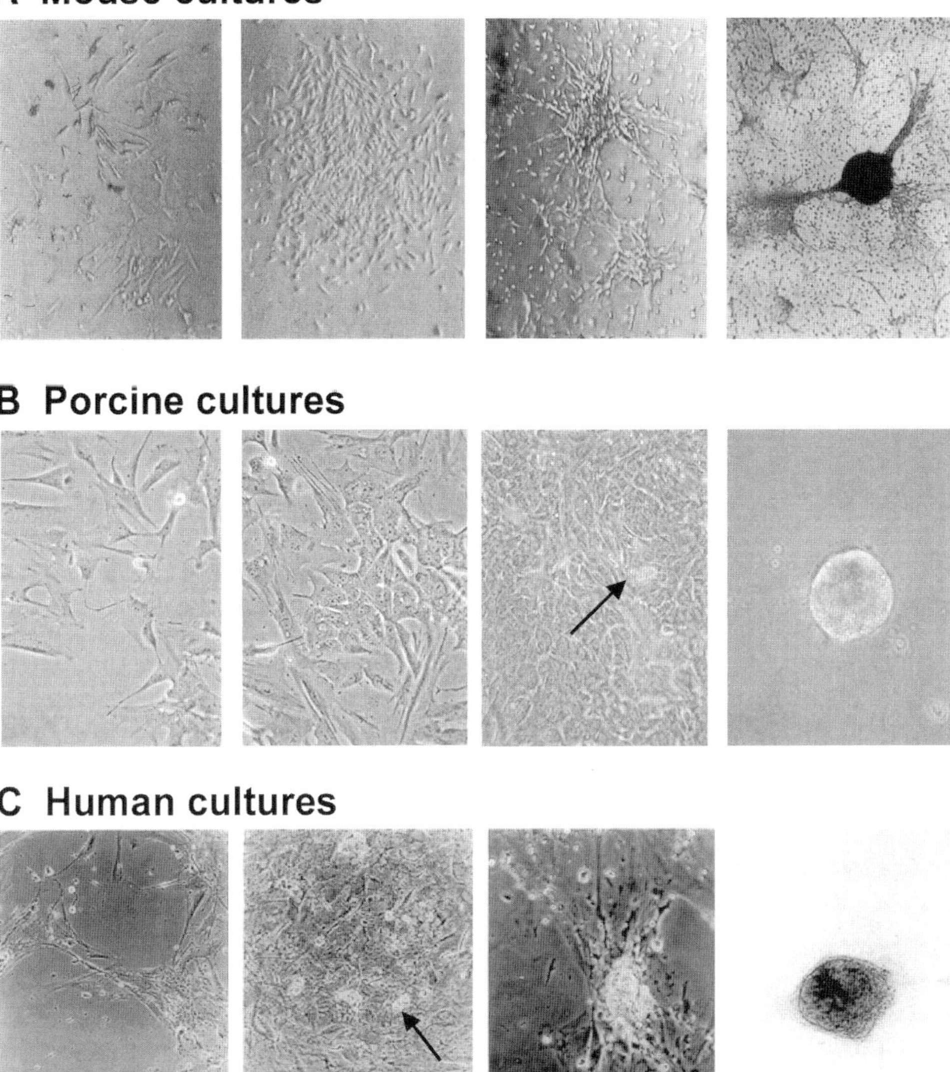

Fig. 3. Differentiation of human ductal epithelial stem cells into immature, functional islet-like structures. Single-cell suspensions of adult (**A**) mouse, (**B**) pig, and (**C**) human pancreatic ducts were cultured in glucose-reduced medium until primary cultures were established (left panels). After foci of epithelioid cells appeared (IPSCs), the cells were induced into production of small rounded IPCs (arrows), the progenitors of islet formation, that underwent rapid proliferation to form organized clusters of cells (center panels). The cell clusters expanded to form tightly grouped glucagon-, insulin-, and somatostatin-producing cells, but the β-cells remained immature (right panels).

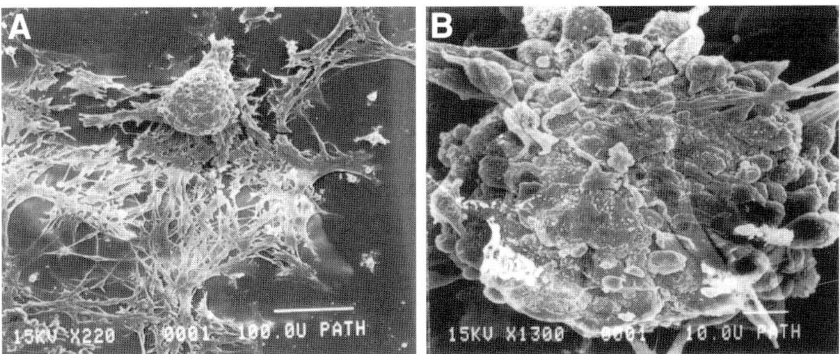

Fig. 4. Ultrastructure of stem cell–derived, in vitro–grown islets. (**A**) A scanning electron micrograph of an in vitro–grown islet cell cluster shows its development as a 3D structure above the adherent stromal epithelial monolayer. (**B**) Higher magnification reveals the complex structure.

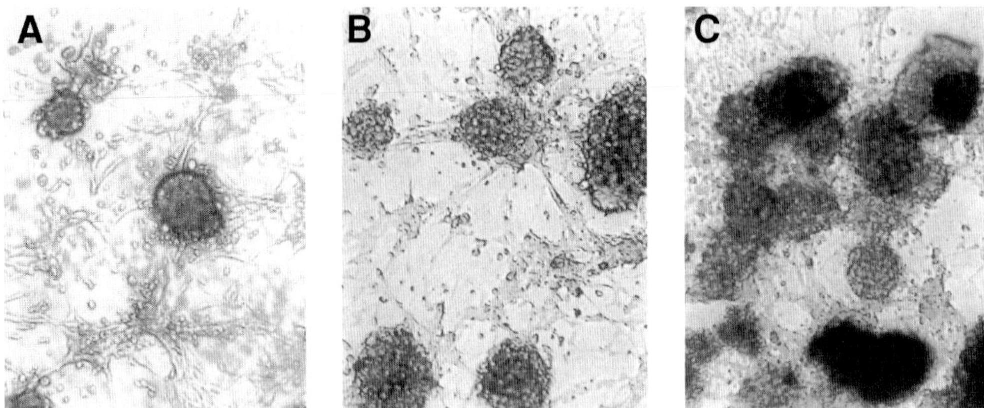

Fig. 5. Increased islet cell cluster formation in presence of nicotinamide. IPSC/IPC cultures were established. **(A)** Under normal conditions, differentiation led to 50–75 islets/mm^2. An increase in islet formation occurred in the presence of 10 ng/mL of EGF, 10 ng/mL of HGF, and nicotinamide at 5 **(B)** or 10 m*M* **(C)**. Reprinted with permission of Nature Publishing Group.

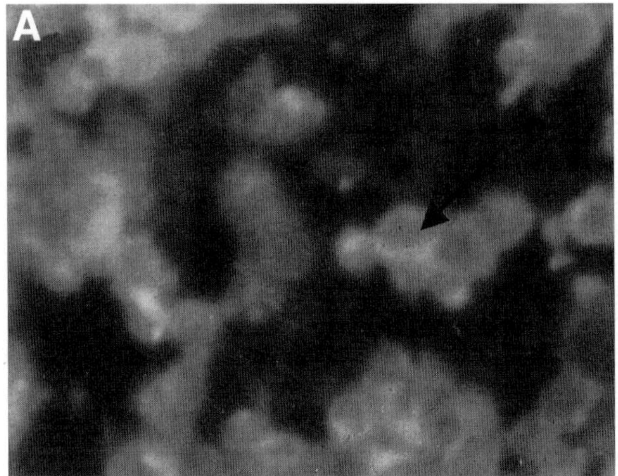

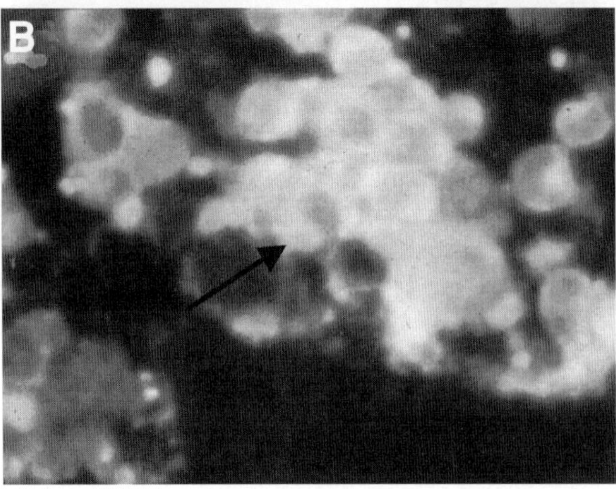

Fig. 6. Synthesis of insulin by cells present in in vitro–generated islets. Islets isolated from **(A)** untreated or **(B)** nicotinamide-treated in vitro cultures were collected, the clustered cells dispersed by gentle pipeting, and the dispersed cells cytocentrifuged onto glass slides. The slides were then stained with fluorescein isothiocyanate (FITC)-conjugated antiinsulin antibody. A marked increase in fluoresence staining occurred within cells cultured in the presence of nicotinamide.

synthesis (Fig. 6) and respond with increased insulin secretion following glucose challenge (Table 3). Whether other factors, such as islet neogenesis-associated protein (INGAP) (Rafaeloff et al., 1997), might prove efficacious in promoting enhanced proliferation and differentiation of β-cells in vitro would be of interest. To date, however, only combinations of epidermal growth factor (EGF) plus FGF appear to enhance growth, mostly proliferation and not differentiation.

37.5. FUNCTIONAL EFFICACY OF STEM CELL–DERIVED, IN VITRO–GENERATED ISLETS

One potential drawback has been the inability to identify a culture condition that permits full differentiation of the immature β-cells within the islet clusters to mature cells. This has required exposure to an in vivo environment, indicating that we still need to identify factors present in the in vivo environment permitting the last differentiation step. The functional capacity of in vitro–grown, stem cell–derived mouse islets has been investigated in a set of implantation experiments (Ramiya et al., 2000). In the first set of experiments, female diabetic NOD mice that had been maintained on daily insulin injections for >3 wk were each implanted with 300 IPSC-derived islets into the subcapsular region of the left kidney and then weaned from their insulin injections. In addition, other mice received implants to other sites, including the spleen and leg muscle. Within 1 wk, mice implanted with islets to the kidney or spleen exhibited stable blood glucose levels of 180–220 mg/dL and remained insulin-independent until euthanized for analysis of the implants. Diabetic mice that had islets implanted into the muscle or had not received any islet implants exhibited severe wasting syndrome when weaned from their insulin and had to be euthanized.

In a second experiment, female diabetic NOD mice maintained on insulin for 4 wk were implanted subcutaneously with 1000 in vitro–grown, stem cell–derived mouse islets with similar results, except that the blood glucose stabilized at more normal levels (100–150 mg/dL). Islets placed subcutaneously required about 2–4 wk to achieve a homeostatic state with the recipient. We have speculated (Ramiya et al., 2000) that this might be the time required for the islets to become fully vascularized and establish the necessary glucose-sensing machinery for rapid and regulated insulin responses.

Table 2
Gene Expression During Differentiation of IPSC- and IPC-Derived Islets of Langerhans in Long-Term Cultures

Endocrine pancreas development and differentiation markers	*Endocrine hormones and islet cell–associated factors*
Reg-1, IPF-1 (PDX-1), tyrosine hydroxylase, Isl-1, β-galactosidase, Pax-4, Pax-6, Nkx6.1, β2/neuroD-factor	Insulin-I, insulin-2, glucagon, somatostatin,insulin receptor, GLUT-2, HGF, c-MET

Table 3
Increased Synthesis and Release of Insulin Following Treatment of In Vitro–Generated Islets with Nicotinamide

Treatment	*Insulin Synthesis* [a]	*Increase (%)*	*Insulin Secretion*[a]	*Increase (%)*
No treatment	42 ± 2	—	8.7 ± 6	—
Nicotinamide (10 m*M*) [b]	146 ± 60	248	152 ± 57	1647
Arginine (10 m*M*) [c]	52 ± 2	23	62 ± 3	613
GLP-1 (1 m*M*) [c]	53 ± 2	26	65 ± 2	647

[a] Cells stimulated with 17.5 m*M* glucose; insulin measured by enzyme-linked immunosorbent assay (pgm/300 islets)
[b] IPC-derived cells were incubated for 5 d with or without nicotinamide.
[c] IPC-derived cells were incubated for 3 h with the secretagogue.

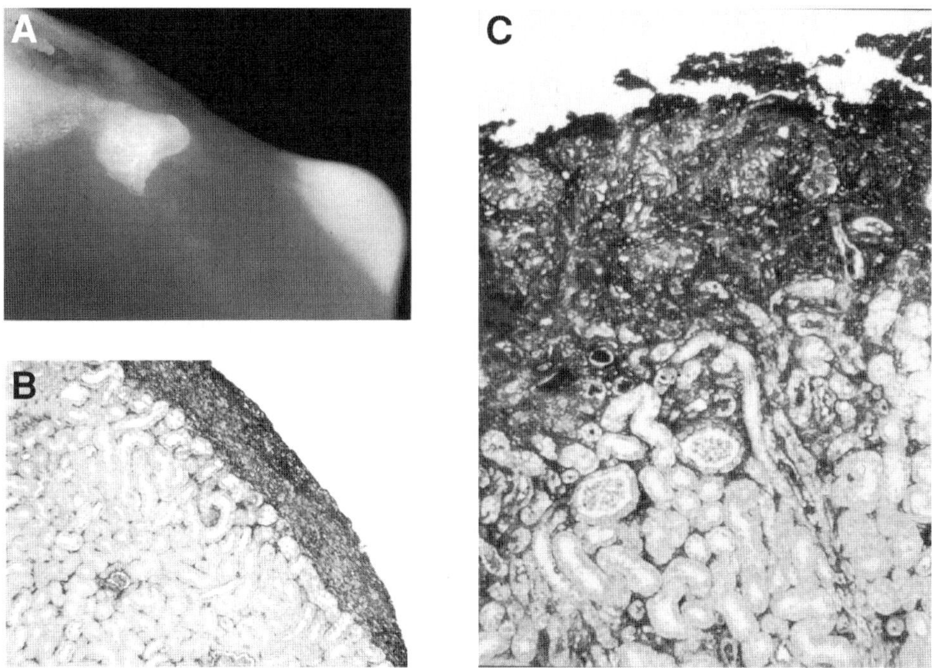

Fig. 7. Histological characteristic of intrarenal implant site. A diabetic mouse was implanted with 300 in vitro–grown islets in the subcapsular region of the kidney, and then taken off daily insulin injections. At 55 d postimplantation, the mouse was euthanized. The implanted kidney was explanted (**A**), fixed in formalin, embedded in paraffin, and sectioned (0.4 μ). Each section was stained with antiinsulin antibody and then counterstained with H&E dye. Although the islets lost their structure, forming a more continuous mass, there was intense staining with antiinsulin antibody (**B**). The implant site showed no signs of leukocytic infiltrates (**B,C**). Reprinted with permission of Nature Publishing Group.

One of the more interesting observations from the implant studies has been the fact that none of the implants of the in vitro–grown islets reactivated the autoimmune response, as evidenced by a lack of immune cells in histological examinations of implant sites at the time of euthanization (Fig. 7). We have speculated that this may be due to a loss of expression of the β-cell autoantigen(s) as the result of in vitro culturing, development of peripheral tolerance following the restimulation of the autoimmune response by the newly implanted islets, an insufficient time allowed for the reactivation of an autoimmune response, or implant site specificity. In addition, lack of reactivation of the autoimmune response in the implanted diabetic mice might be related to the derivation of the islets (i.e., ductal epithelial stem cells). Regardless, understanding why the autoimmune response is not reactivated in this system could prove valuable for islet transplantation, in general.

37.6. STEM CELL-DERIVED ISLETS AND INTERVENTION THERAPY

Over the past several years, we have presented evidence indicating that immature, yet functional islet cell clusters can be grown in vitro from stem cells isolated from the pancreatic ducts presumed to contain the epithelial stem cells from which islets of Langerhans are derived during embryogenesis. Using the NOD mouse as a model for human type 1 diabetes, we have shown the feasibility of obtaining such stem cells from prediabetic, postdiabetic, or normal adult mice; the potential for establishing stem cell cultures for the growth of immature functional islet-like structures; and the efficacy of the in vitro–generated islets to reverse insulin-dependent diabetes when implanted into the diabetic animals. In addition, these stem cell–derived in vitro–grown islet cell clusters exhibit a cellular organization similar to natural islets of Langerhans. These results may circumvent the need for artificially regulating insulin responses in engineered surrogate cell populations. Although our ability to control the growth and differentiation of islet stem cells potentially provides an abundant source for β-cell reconstitution in type 1 and specific forms of type 2 diabetes, obtaining pancreatic tissue for autologous implants poses additional and interesting challenges, especially for the physician.

Despite the high profile of type 1 diabetes and the implementation of new genetic screening programs for families and newborns to identify "high-risk" individuals, the incidence of type 1 diabetes is increasing worldwide. Type 1 diabetes is an especially insidious disease with clinical symptoms usually not being detected until after the patient's own immune system has destroyed >90% of the total insulin-producing β-cells of the endocrine pancreas (Eisenbarth, 1986; Harris, 1999). Although routine insulin injections can provide diabetic patients with their daily insulin requirements, blood glucose excursions are common, resulting in hyperglycemic episodes. Hyperglycemia represents the major health problem for the diabetic patient, especially long term. When inadequately controlled, chronic hyperglycemia can lead to microvascular complications (e.g., retinopathy and blindness, nephropathy and renal failure, neuropathy, foot ulcers, and amputation), and macrovascular complications (e.g., atherosclerotic cardiovascular, peripheral vascular, and cerebrovascular disease). Both the Diabetes Control and Complications Trial (DCCT) and UK Prospective Diabetes Study (UKPDS) demonstrated a strong relationship between good metabolic control and the rate/progression of complications (DCCT Research Group, 1993; UKPDS Group, 1998). Unfortunately, adequate control of hyperglycemic excursions cannot be attained by most patients and attempts at maintaining euglycemia through intensive insulin treatment lead to increased incidences of hypoglycemia.

A "cure" for type 1 diabetes relies on replacement of the β-cell mass. Currently, this is accomplished by either ectopancreas or islets of Langerhans transplantation (Kendall and Robertson, 1997; Silverstein and Rosenbloom, 2000). Pancreas organ transplantation often results in normalization of fasting and postprandial blood glucose levels, normalization of HbA1C, as well as secretion of insulin and C-peptide in response to glucose. However, this procedure requires long-term immunosuppression, thereby restricting it to patients who have end-stage renal disease listed to receive a kidney transplant, or to those with long-standing type 1 diabetes who have failed insulin therapy due to extremely poor compliance (Lanza and Chick, 1997). Islet implantation, although historically less efficacious, possesses several notable advantages over pancreatic organ transplantation: First, the islets can be isolated and delivered by percutaneous catheterization of the portal vein under local anesthesia, thus relieving the patient from undergoing general anesthesia. Second, genetically, islets can be manipulated in vitro to resist immune attack—either by reducing potential islet immunogens or by expressing immunomodulators. Third, the islets can be encapsulated in order to be protected, theoretically, from the immune system, but still release insulin (Scharp et al., 1994; Hering et al., 1996; Sutherland et al., 1996; Lanza and Chick, 1997; Drachenberg et al., 1999). Drawbacks to islet cell transplantation have included the inability to obtain sufficient viable numbers of islets and the need for immunosuppressive agents (Hering et al., 1996; Carel et al., 1997; Drachenberg et al., 1999; Hering and Ricordi, 1999; Ricordi and Hering, 1999). Recent success in islet implantation protocols has brought this intervention to the forefront; however, such success portends to an even greater shortage of implantable islets.

The possibility that stem cell–derived, in vitro–generated islets may soon be an alternative to cadaver-derived islets for treating diabetic patients is gaining credibility, now that others have started to reproduce this work. For example, Bonner-Weir et al. (2000) have now shown that adult human ductal tissue can be expanded and differentiated in-vitro to form islet clusters. The ability of this group to achieve differentiation of ductal cells to endocrine pancreas also appears to be dependent on the sequential stimulation of growth and differentiation through the use of specific media, growth factors (e.g., keratinocyte growth factor), and a Matrigel environment. Although insulin, glucagon, pancreatic polypeptide, and somatostatin were expressed and the islets were responsive to glucose challenge, CK19, pancytokeratin, and islet-promotor factor (IPF-1) expression remained. Insulin-positive cells were widely scattered within the islet structures, and IPF-1-positive cells without insulin staining were present, while some cells double-stained for insulin and non-β-cell hormones. Pattou and coworkers (Kerr-Conte et al., 1996; Gmyr et al., 2000) previously reported similar results. Considered together, these studies confirm the fact that in vitro–grown islets only achieve an immature β-cell phenotype, indicating that much more work is essential to understand fully the differentiation process.

Although the presence of stem cells associated with the pancreatic ducts has been known in both healthy and diabetic individuals, the ability to stimulate these stem cells in vitro into functional islets represents a major breakthrough that has unlimited potential for the therapeutic intervention of type 1 diabetes. However, results from a number of laboratories indicate that although we are able to initiate expansion and differentiation of such stem cells, we still lack the basic understanding to fully control the process. Identification of temporal gene expressions during the developmental stages of islets may provide the information necessary to sequentially stimulate stem cells to mature to end-stage islets using cocktails of appropriate growth factors. Alternatively, the immature islets may represent the ideal reagent for implantation since they appear to have the potential to respond to the in vivo environment, thereby establishing homeostasis with the recipient through vascularization, expansion of the β-cell mass, and differentiation to insulin-secreting cells. Understanding fully each of these stages will require further investigations.

A recurrent question is, How and when can diabetic patients expect to benefit from this new stem cell technology? Except for ethical issues, implantation of in vitro–grown allogeneic islets derived from cadaveric donors could be performed now. While the use of allogeneic islets is most expedient, autologous islets, if available, would no doubt be preferred by the patient over either allogeneic or xenogeneic islets. Thus, the ideal situation will be to isolate the recipient's own stem cells and grow autologous islets in order to provide autologous implants. With the advent of new technology, such as transesophageal endosonographically monitored fine-needle tissue sampling of the pancreas (Bhutani, 1999), it may someday be possible to grow autologous islets from overtly diabetic patients (Fig. 8). In addition, genetic modification of the isolated endocrine pancreas stem cells may allow the opportunity to build even healthier and stronger islets. Thus, although several major hurdles remain before stem cell therapy will be a routine procedure, we are rapidly nearing that goal.

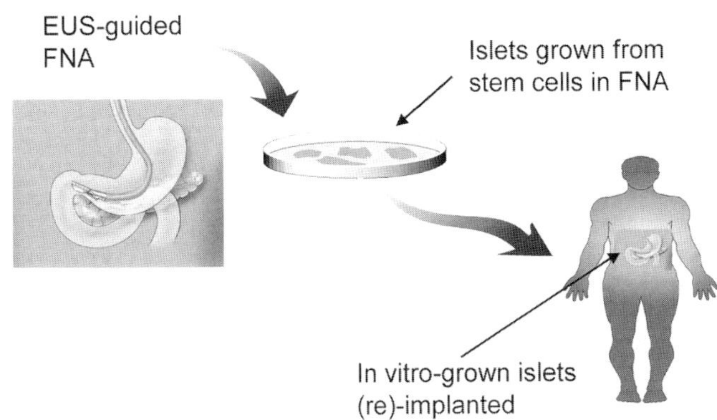

Fig. 8. Proposed therapy with autologous in vitro–grown, stem cell–derived islets. Using developing technologies such as ultrasound-guided endoscopy, pancreatic ducts can be biopsied directly from the patient through the sterile stomach environment. The biopsy material can be taken to the laboratory, where IPSC/IPC cultures are initiated. Islet clusters grown in the laboratory can then be returned to the patient at predetermined sites, possibly even the pancreas, using ultrasound-guided endoscopy. In establishing the in vitro IPSC/IPC cultures, the stem/progenitor cells can be genetically manipulated to function better or be resistant to immunological attack. IPSC/IPC cultures can be maintained in case the patient requires additional islets over time.

37.7. FUTURE DIRECTIONS

Stem cells present in the pancreatic ducts of both healthy and diabetic individuals capable of being stimulated in vitro into functioning, although immature, islets represent a major breakthrough that has unlimited potential for therapeutic intervention in type 1 diabetes. However, results from a number of laboratories indicate that while we are able to initiate expansion and differentiation of such stem cells, we still lack the basic understanding to fully control the process (Kerr-Conte et al., 1996; Cornelius et al., 1997; Fernandes et al., 1997; Al-Abdullah et al., 2000; Bonner-Weir et al., 2000; Gmyr et al., 2000, 2001; Ramiya et al., 2000; Rooman et al., 2000; Schmied et al., 2000; Zulewski et al., 2001). Identification of temporal gene expressions during the developmental stages of islets may provide the information necessary to sequentially stimulate stem cells to mature to end-stage islets. Alternatively, the immature, yet functional islets achievable in culture may represent the ideal reagent for implantation since the immature islets appear to have the potential to respond to the in vivo environment, expanding β-cell mass and establishing homeostasis with the recipient. These are areas that require further understanding. In light of a recent report (Yang et al., 2002) that liver oval cells can be pushed to transdifferentiate to endocrine cells, is it possible to push pancreatic stem cells to liver cells?

The second aspect of concern is the use of autologous or allogeneic islets. Although syngeneic islets generated in vitro may be masked from the autoimmune response, allogeneic islets are definitely rejected by the T-cell-mediated allogeneic response. This would mean that the recipient of allogeneic stem cell–derived implants would still require immunosuppression or the implant would have to be isolated via encapsulation. Thus, the ideal situation would be to isolate the recipient's own stem cells in order to provide autologous implants. This will require new approaches to access the pancreas by a minimally invasive procedure (e.g., transesophageal endoscopically monitored fine-needle sampling of the pancreas). This may allow the growth of autologous islets from severely diabetic patients and subsequent transplantation of these islets in the future. However, it is crucial to resolve most, if not all, important aspects of the in vitro differentiation process before approaching the possibility of autologous islet derivation. Thus, major hurdles remain before stem cell therapy will be a routine procedure, but researchers continue to be optimistic of its potential and believe it will soon occur.

REFERENCES

Ahlgren, U., Plaff, S., Jessel, T. M., et al. (1997) Independent requirement for ISL1 in the formation of the pancreatic mesenchyme and islet cells. *Nature* 385:257–260.

Al-Abdullah, I. H., Ayala, T., Panigrahi, D., et al. (2000) Neogenesis of pancreatic endocrine cells in copper-derived rat models. *Pancreas* 21: 63–68.

Bhutani, M. S. (1999) *Interventional Endoscopic Ultrasound*, Harwood, Amsterdam, The Netherlands.

Bjorson, C. R. R., Reitza, R. L., Reynolds, B. A., et al. (1999) Turning brain into blood: a hemopoietic fate adopted by adult neural stem cells in vivo. *Science* 283:534–537.

Bonner-Weir, S. and Orci, L. (1982) New perspectives on the microvasculature of the islets of Langerhans in the rat. *Diabetes* 41:93–97.

Bonner-Weir, S., Deery, D., Leahy, J. L., et al. (1989) Compensatory growth of pancreatic beta-cells in adults after short-term glucose infusion. *Diabetes* 38, 49–53.

Bonner-Weir, S., Baxter, L. A., Schuppin, G. T., et al. (1993) A second pathway for regeneration of adult exocrine and endocrine pancreas: a possible recapitulation of embryonic development. *Diabetes* 42: 1715–1720.

Bonner-Weir, S., Taneja, M., Weir, G., et al. (2000) In vitro cultivation of human islets from expanded ductal tissue. *Proc. Natl. Acad. Sci. USA* 97:7999–8004.

Bouwens, L. and Pipeleers, D. G. (1998) Extra-insular beta cells associated with ductules are frequent in adult human pancreas. *Diabetologia* 41:629–633.

Brelje, T. C., Scharp, D. W., Lacy, P. E., et al. (1993) Effect of homologous placenta lactogens, prolactins, and growth hormones on islet B-cell division and insulin secretion in rat, mouse, and human islets: implication for placental lactogen regulation of islet function during pregnancy. *Endocrinology* 132:879–887.

Cantenys, D., Portha, B., Dutrillaux, M. C., et al. (1981) Histogenesis of the endocrine pancreas in newborn rats after destruction by streptozotocin: an immunocytochemical study. *Virchows Arch. B Cell Pathol. Incl. Mol. Pathol.* 35:109–122.

Carel, J. C., Lotten, C., and Bourgneres, P. (1997) Prediction and prevention of type 1 diabetes: what can be expected from genetics. *Diabetes Metab.* 2:29–33.

Clarke, D. L., Johansson, C. B., Wilbertz, J., et al. (2000) Generalized potential of adult neural stem cells. *Science* 288:1660–1663.

Cornelius, J. G., Tchernev, V., Kao, K. J., et al. (1997) In vitro generation of islets in long term cultures of pluripotent stem cells from adult mouse pancreas. *Horm. Metab. Res.* 29:271–277.

Diabetes Control and Complications Trial Research Group. (1993) The effects of intensive treatment of diabetes on the development and progression of long-term complications in insulin-dependent diabetes mellitus. *N. Engl. J. Med.* 329:977–986.

Drachenberg, C. B., Klassen, D. K., Weir, M. R., et al. (1999) Islet cell damage associated with tacrolimus and cyclosporin: morphological features in pancreas allograft biopsies and clinical correlation. *Transplantation* 68:396–402.

Eisenbarth, G. S. (1986) Type 1 diabetes: a chronic autoimmune disease. *N. Engl. J. Med.* 314:1360–368.

Fernandes, A., King, L. C., Guz, Y., et al. (1997) Differentiation of new insulin-producing cells is induced by injury in adult pancreatic islets. *Endocrinology* 138:1750–1762.

Gmyr, V., Kerr-Conte, J., Belaich, S., et al. (2000) Adult human cytokeratin 19-positive cells reexpress insulin promoter factor 1 in vitro. *Diabetes* 49:1671–1680.

Gmyr, V., Kerr-Conte, J., Vandewalle, B., et al. (2001) Human pancreatic duct cells: large-scale isolation and expansion. *Cell Transplant.* 10: 109–121.

Gu, D. and Sarvetnick, N. (1993) Epithelial cell proliferation and islet neogenesis in IFN-g transgenic mice. *Development* 118:33–46.

Harris, M. I. (1999) Newly revised classification and diagnostic criteria for diabetes mellitus. In: *Current Review of Diabetes* (Taylor, S. T., ed.), Current Medicine, Philadelphia, PA, pp.1–9.

Harrison, K. A., Druey, K. M., Deguchi, Y., et al. (1994) A novel human homeobox gene distantly related to proboscipedia is expressed in lymphoid and pancreatic tissues. *J. Biol. Chem.* 269:19,968–19,975.

Hayek, A., Beattie, G. M., Cirulli, V., et al. (1995) Growth factor/matrix-induced proliferation of human adult beta cells. *Diabetes* 44: 1458–1460.

Hellerstrom, C. (1984) The life story of the pancreatic B cell. *Diabetologia* 28:393–400.

Hering, B. J. and Ricordi, C. (1999) Islet transplantation in type 1 diabetes: results, research priorities and reasons for optimism. *Graft* 2:12–27.

Hering, B. J., Brendel, M. D., Schultz, A. O., et al. (1996) *Newslett. No. 7. Int. Islet Transplant Registry* 6:1–20.

Jensen, J., Pedersen, P., Galante, P., et al. (2000) Control of endodermal endocrine development by Hes-1. *Nat. Genet.* 24:36–44.

Kendall, D. and Robertson, R. P. (1997) Pancreas and islet transplantation: challenges for the twenty first century. *Endocrinol. Metab. Clin. North Am.* 26:611–30.

Kern, H. and Logothetopoulos, J. (1970) Steroid diabetes in the guinea pig. Studies on islet-cell ultrastucture and regeneration. *Diabetes* 19: 145–154.

Kerr-Conte, J., Pattou, F., Lecomte-Houcke, M., et al. (1996) Ductal cyst formation in collagen-embedded adult human islet preparations. *Diabetes* 45:1108–1114.

Kim, S. K., Hebrok, M., and Melton, D. A. (1997) Notochord to endoderm signaling is required for pancreas development. *Development* 124: 4243–4252.

Korcakova, L. (1971) Mitotic division and its significance for regeneration of granulated B-cells in the islets of Langerhans in allozan-diabetic rats. *Folia Morphol. (Praha)* 19:24–30.

Lanza, P. P. and Chick, W. L. (1997) Transplantation of encapsulated cells and tissues. *Surgery* 121:1–9.

Lieter, E. H., Prochazka, M. and Coleman, D. L. (1987) The non-obese diabetic (NOD) mouse. *Am. J. Pathol.* 128:380–383.

Logothetopoulos, J. and Bell, E. G. (1966) Histological and autoradiographic studies of the islets of mice injected with insulin antibody. *Diabetes* 15:205–211.

Marynissen, G., Aerts, L., and Van Assche, F. A. (1983) The endocrine pancreas during pregnancy and lactation in the rat. *J. Dev. Physiol.* 5: 373–381.

Menger, M. D., Vajkoczy, P., Beger, C., and Messmer, K. (1994) Orientation of microvascular blood flow in pancreatic islet isografts. *J. Clin. Invest.* 93:2280–2285.

Morshead, C.M., Benveniste, P., Iscove, N.N., et al. (2002) Hematopoietic competence is a rare property of neural stem cells that may depend on genetic and epigenetic alterations. *Nat. Med.* 8:268–273.

Navsaria, H. A., Myers, S. R., Leigh, I. M., et al. (1995) Culturing skin in vitro for wound therapy. *Trends Biotech.* 13:91–100.

Naya, F. J., Huang, H. P., Qui, Y., et al. (1997) Diabetes, defective pancreatic morphogenesis, and abnormal enteroendocrine differentiation in BETA2/NeuroD-deficient mice. *Genes Dev.* 11:2323–2334.

Otonkoski, T., Beattie, G. M., Rubin, J. S., et al. (1994) Hepatocyte growth factor/scatter factor has insulinotropic activity in human fetal pancreatic cells. *Diabetes* 43:947–953.

Peck, A. B. and Cornelius, J. G. (1995) In vitro growth of mature pancreatic islets of Langerhans from single, pluripotent stem cells isolated from prediabetic adult pancreas. *Diabetes* 44:10A.

Peters, J., Jurgensen, A., and Kloppel, G. (2000) Ontogeny, differentiation and growth of the endocrine pancreas. *Virchows Arch.* 436:527–538.

Pictet, R. L. and Rutter, W. J. (1972) The endocrine pancreas. In: *Handbook of Physiology* (Steiner, D. and Frienkel, N., eds.), Williams & Wilkins, Baltimore, MD, pp. 25–66.

Rafaeloff, R., Pittenger, G., Barlow, S., et al. (1997) Cloning and sequencing of the pancreatic islet neogenesis associated protein (INGAP) gene and its expression in islet neogenesis in hamsters. *J. Clin. Invest.* 9:2100–2109.

Ramiya, V. K., Maraist, M., Arfors, K. E., et al. (2000) Reversal of insulin dependent diabetes using islets generated in vitro from pancreatic stem cells. *Nat. Med.* 6:278–282.

Rao, M. S., Dwivedi, R. S., Yeldandi, A. V., et al. (1989) Role of periductal and ductular epithelial cells of the adult rat pancreas in pancreatic hepatocyte lineage: a change in the differentiation commitment. *Am. J. Pathol.* 134:1069–1086.

Ricordi, C. and Hering, B. J. (1999) Pancreas and islet transplantation. In: *Current Review of Diabetes* (Taylor, S., ed.), Current Medicine, Philadelphia, PA, pp. 49–60.

Rooman, I., Heremans, Y., Heimberg, H., et al. (2000) Modulation of rat pancreatic acinoductal transdifferentiation and expression of PDX-1 in vitro. *Diabetologia* 43:907–914.

Rosenberg, L. and Vinik, A. I. (1992) Trophic stimulation of the ductular-islet cell axis: a new approach to the treatment of diabetes. *Adv. Exp. Med. Biol.* 321:95–104.

Scharp, D. W., Swanson, C. J., Olack, B. J., et al. (1994) Protection of encapsulated human islets implanted without immunosuppression in patients with type I diabetes or type II diabetes and in nondiabetic control subjects. *Diabetes* 43:1167–1170.

Schmied, B.M., Liu, G., Matsuzaki, H., et al. (2000) Differentiation of islet cells in long-term culture. *Pancreas* 20:337–347.

Schmied, B.M., Ulrich, A., Matsuzaki, H., et al. (2001) Transdifferentiation of human islet cells in a long-term culture. *Pancreas* 23:157–171.

Siena, S., Schiavo, R., Pedrazoli, P., et al. (2000) Therapeutic relevance of CD34 cell dose in blood cell transplantation for cancer therapy. *J. Clin. Oncol.* 18:1360–1377.

Silverstein, J. and Rosenbloom, A. (2000) New Developments in type 1 (insulin dependent) diabetes. *Clin. Pediatr.* 39:257–66.

Sutherland, D. E., Gores, P. F., Bernhard, H. J., et al. (1996) Islet transplantation: an update. *Diabetes Metab. Rev.* 12:137–150.

Swenne, I. (1992) Pancreatic beta-cell growth and diabetes mellitus. *Diabetologia* 35:193–201.

Teitelman, G., Alpert, S., Polak, J.M., et al. (1993) Precursor cells of mouse endocrine pancreas coexpress insulin, glucagon and the neuronal protein tyrosine hydroxylase and neuropeptide Y, but not pancreatic polypeptide. *Development* 118:1031–1039.

Teitelman, G., Alpert, S., and Hanahan, D. (1998) Proliferation, senescence, and neoplastic progression of β cells in hyperplasic pancreatic islets. *Cell* 52:97–105.

UK Prospective Diabetes Study (UKPDS) Group. (1998) Effect of intensive blood-glucose control with metformin on complications in overweight patients with type 2 diabetes (UKPDS-34). *Lancet* 352:854–65.

Vescovi, A. L. (2000) Skeletal myogenic potential of human and mouse neural stem cells. *Nat. Neurosci.* 3:986–991.

Vogel, G., Marshall, E., Barinaga, M, et al. (2000) Stem cell research and ethics. *Science* 287:1417–1442.

Weaver, C. V., Sorenson, R. L., and Kaung, H. C. (1985) Immunocytochemical localization of insulin-immunoreactive cells in the pancreatic ducts of rats treated with trypsin inhibitor. *Diabetologia* 28: 781–785.

Weir, G. C. and Bonner-Weir, S. (1990) Islets of Langerhans: the puzzle of intraislet interactions and their relevance to diabetes. *J. Clin. Invest.* 85:983–987.

Wessels, N. K. and Cohen, N. H. (1967) Early pancreas organogenesis: morphogenesis, tissue interactions and mass effects. *Dev. Biol.* 15: 237–270.

Wu, K., Gannon, M., Peshavaria, M., et al. (1997) Hepatocyte nuclear factor 3beta is involved in pancreatic beta-cell specific transcription of the Pdx-1 gene. *Mol. Cell Biol.* 17:6002–6013.

Yang, L.-J., Li, S.-W., Hatch, H., et al. (2002) In vitro trans-differentiation of adult hepatic stem cells into endocrine hormone-producing cells. *Proc. Natl. Acad. Sci. USA* 99:8078–8083.

Zulewski, H., Abraham, E. J., Gerlach, M. J., et al. (2001) Multipotential nestin-postitive stem cells isolated from adult pancreatic islets differentiate ex vivo into pancreatic endocrine, exocrine and hepatic phenotypes. *Diabetes* 50:521–533.

38 Islet Cells

LUC BOUWENS, PhD

A common progenitor cell for liver and pancreas gives rise to progeny, which become determined transit-amplifying cells in different mesenchymal microenvironments. They can become primitive ducts that bud to form hepatocytes in the liver, or exocrine or endocrine pancreas. The differentiation of pancreatic progenitor cells is determined by controlled expression of specific transcription factors. The extracellular factors that regulate this expression are beginning to be unraveled. Injury or surgical removal of part of the adult pancreas may be followed by proliferation of ductal cells, which may lead to partial regeneration of the lost tissues. Stem cells in the ducts of adult animals may be inferred from the reexpression of some embryonic markers during repair of injury. However, transdifferentiation of exocrine cells has also been described and is considered to be equally likely to be responsible for restoration of islet cells. The appearance of hepatocytes in the pancreas during repair of some injury models raises the possibility of a common hepatopancreatic stem cell in the adult pancreas that reflects the common precursor cell for liver and pancreas during embryonic life. However, transdifferentiation of exocrine cells to hepatocytes has also been described. Culture of ductal pancreatic stem cells or selection of embryonic stem cells may lead to effective treatment for diabetes, but much work needs to be done to achieve this goal.

38.1. HISTOLOGY OF PANCREAS

The mammalian pancreas is composed of six epithelial cell types that can be discriminated on the basis of phenotypic and functional characteristics: the two types of exocrine cells, acinar and ductal cells, and four types of endocrine cells, α, β, δ, and PP-cells. These cell types can be recognized by their major secretory products, which are the digestive proenzymes in the case of acinar cells (e.g., amylase, trypsin, chymotrypsin, lipase, ribonuclease), and the hormones glucagon, insulin, somatostatin, and pancreatic polypeptide for the four types of endocrine cells. Ductal cells form the single-layered lining of the ductules and ducts leading the exocrine secretory products to the duodenum. They can be identified by the expression of cytokeratin (CK) proteins that form their intermediate filaments, such as CK7, CK19, and CK20, depending on the species (Bouwens et al., 1994, 1995; Bouwens and Pipeleers, 1998). Ductal cells contain more intermediate filaments, probably because they are required for the mechanical strength of the ductal lining. In addition, ductal cells secrete bicarbonate and other ions, and water, making them recognizable by their expression of carbonic anhydrase enzyme and the CFTR chloride pump (Githens, 1988). Another marker of duct cells is the carbohydrate antigen CA-19.9 or syalyl-Lewis-a antigen (Bouwens and Pipeleers, 1998). More than 90% of the pancreatic volume is occupied by exocrine tissue. Exocrine tissue is arranged in acini interconnected by centroacinar cells to small ducts, or ductules, which branch from larger ducts. Centroacinar cells express markers similar to ductular cells. The endocrine tissue is organized in islets of Langerhans interspersed within the exocrine tissue (Fig. 1). The endocrine islets are all in close contact with both acini and ducts. Especially in human pancreas, there are also many endocrine cells that are not located within islets but occur as single cells (Bouwens and Pipeleers, 1998). Islet cells secrete their hormones directly into fenestrated islet capillaries to regulate glucose homeostasis. Both in terms of cell number and of pathophysiology, the insulin-producing β-cells represent the major endocrine cell type in the pancreas.

38.2. DIABETES

Diabetes is the most common disease of the pancreas. It afflicts 120 million people worldwide, and the World Health Organization estimates that the number will rise to 300 million by 2025. Type 1 diabetes, also known as juvenile diabetes, is an autoimmune disease characterized by the gradual destruction of the β-cell mass. Type 2 diabetes, also known as adult-onset diabetes, is the most common form of diabetes (≥90% of cases). It is characterized by insufficient insulin production and reduced sensitiv-

From: *Stem Cells Handbook*
Edited by: S. Sell © Humana Press Inc., Totowa, NJ

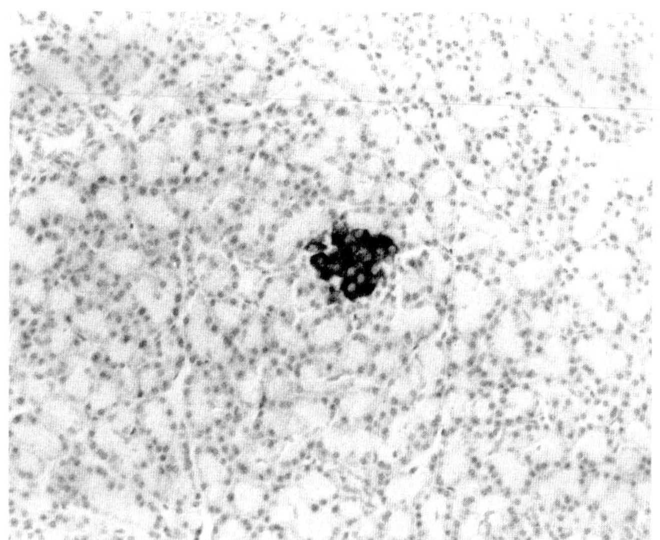

Fig. 1. Histological picture of normal pancreas, immunostained for insulin. A small islet of Langerhans is seen in the middle of a sea of exocrine acini. Original magnification: ×250.

ity to insulin by the responding tissues. All patients with type 1 must take insulin to live, and many patients with type 2 benefit from treatment with insulin. However, insulin treatment does not cure any type of diabetes nor prevent the possibility of its eventual and devastating effects: kidney failure, blindness, nerve damage, amputation, heart attack, and stroke.

Since the 1990s, clinical trials have been set up to treat type 1 diabetes by transfer of islet cells isolated from allogeneic organ donors in order to replace the endogenous insulin-producing cell mass of the patient. There have been some recent breakthroughs in this field (Keymeulen et al., 1998; Ryan et al., 2001; Bertuzzi et al., 2002) indicating β-cell transfer to represent a very promising therapy for diabetes. Two major obstacles still hamper a large-scale application of this form of therapy—the need for continuous immunosuppression and the serious shortage in donor organs. To solve these problems, several approaches are being investigated, including the genetic modification of pancreatic cells or cell lines, or the stimulation of in vivo or ex vivo growth of the β-cell mass. Because human β-cells have a very limited, if any, mitotic potential (Bouwens et al., 1997; Bouwens and Pipeleers, 1998) there is now much interest in the possibility of generating β-cells from stem cells to overcome the problem of organ shortage.

38.3. ONTOGENY OF PANCREAS

All the epithelial cell types of the pancreas, from both the exocrine and endocrine tissue, originate from a common pool of endodermal progenitor cells during embryogenesis (Slack, 1995; Percival and Slack, 1999). Recently, a common pool of progenitor cells was demonstrated for liver and ventral pancreas, which become determined by a gradient of morphogen (fibroblast growth factor [FGF]) derived from the cardiac mesoderm (Deutsch et al., 2001). Starting at embryonic d (E) 9 in rodents, or after about 35 d in humans, first the dorsal pancreatic bud and later the ventral bud are formed from the embryonic foregut and eventually fuse (Slack, 1995). The endodermal cells in these buds soon organize into ductal structures that become surrounded by abundant mesenchymal tissue. Immature acinar structures as well as endocrine cell groups bud off from the fetal ducts. Mesenchymal factors have been shown to regulate the epithelial morphogenesis. Mesenchymally derived FGF-10 (Bhushan et al., 2001), and epidermal growth factor (Cras-Meneur et al., 2001), promote the proliferation of undifferentiated progenitor cells. FGF signaling via the FGF-2b receptor promotes exocrine differentiation (Miralles et al., 1999). Stimulation of the proliferation of pancreatic progenitors in these experiments was accompanied by a reduced endocrine differentiation.

Recent studies have started uncovering the "master switch" genes that encode for transcription factors regulating cell differentiation in the embryonic pancreas (reviewed in Sander and German, 1997; Edlund, 1998). The first and major pancreatic gene is Pdx-1 (synonyms: Ipf-1, Stf-1), which belongs to the family of homeobox-containing genes and which transactivates the insulin gene (Ohlsson et al., 1993). In the early pancreatic buds, all epithelial cells express Pdx-1. Later in development, Pdx-1 expression is downregulated in most cells, but a high level of expression is retained in all β-cells and a small subset of δ-cells. In adult human pancreas, however, Pdx-1 is still expressed in ductal cells but shows a different pattern of activation and DNA binding compared with β-cells (Heimberg et al., 2000). Another major pancreatic gene is p48/Ptf-1, a helix-basic-helix loop type of gene that is initially expressed by all pancreatic epithelial cells but later becomes restricted to exocrine cells (Krapp et al., 1996). In the case of both Pdx-1 and p48, targeted ablation of the gene leads to pancreatic agenesis, which indicates the importance of both transcription factors for pancreatic cell differentiation (Jonsson et al., 1994; Krapp et al., 1998).

Lateral inhibition seems to be involved in the specification of cells toward either an endocrine postmitotic cell population or an exocrine/progenitor cell population. Similarly to the development of the neural system, this lateral specification involves expression of the Delta-Notch system. In a pool of undifferentiated progenitor cells, lateral inhibition is initiated by the expression of the Delta ligand by a differentiating cell. Ligation of Notch, the receptor for Delta, on a neighboring progenitor cell leads to activation of the HES genes that encode for basic helix-loop-helix (bHLH) transcription factors. HES proteins repress pro-endocrine genes and thus prevent endocrine cell differentiation but promote maintenance of a pool of undifferentiated progenitor cells. Ngn-3, another bHLH protein, is thought to represent the major proendocrine gene that is directly antagonized by the Delta-Notch system (Apelqvist et al., 1999; Gradwohl et al., 2000; Jensen et al., 2000; Schwitzgebel et al., 2000). Based on this model, it has been proposed that the direct progenitors of islet cells are Pdx-1+, p48–, Ngn-3+, Hes-1– cells (Jensen et al., 2000). Pdx-1+, p48+ Ngn-3–, Hes-1+ cells would represent exocrine or not-yet-committed progenitor cells. If these phenotypes were observed in cells located in the adult pancreas, one might speak of "undifferentiated" pancreatic stem cells, but this has not been the case yet. The proendocrine gene Ngn-3 is never expressed in mature endocrine (i.e., hormone-positive) cells, and its expression stops at about E17 in mouse pancreas. It apparently cannot serve as an adult islet stem cell marker. By contrast, in the intestinal crypts, the proliferating progenitor cells remain Ngn3 positive (Gradwohl, G., personal communication). Transcription factors as yet have not permitted researchers to prospectively identify putative stem cells in the adult pancreas.

38.4. ARE ADULT STEM CELLS LOCATED IN PANCREATIC DUCTS?

Where could the pancreatic stem cells be located in the adult organ? Most investigators have been looking for islet progenitor cells (IPCs) in the pancreatic ducts, based on several arguments:

1. The ducts are reminiscent of the early embryonic stage wherein pancreatic endodermal cells are organized in ducts or ductlike structures, at a time when there are not yet islets or acini.
2. Both in the fetal pancreas and in islet regeneration models, islet cells are seen to bud from the ductal lining into the periductal tissue, and in some cases even into the duct lumen (*see* below).
3. Cellular markers of fetal and adult ducts, such as CK19 and CK7 (human) or CK20 (rat), are transiently expressed in immature islet cells, both in the fetal period (Bouwens et al., 1994, 1996, 1997) and during tissue regeneration (Wang et al., 1995).

An additional argument may be that most human pancreatic tumors (carcinomas) have a ductal phenotype (Luttges and Kloppel, 2001). In general, duct cells are considered poorly differentiated, because they produce fewer secretory products than the other exocrine cells and endocrine cells. However, this view is misleading because duct cells are differentiated cells and express particular sets of genes to fulfill their secretory and transducing functions (Githens, 1988). A putative stem cell subset residing in the duct lining may theoretically be identifiable by the lack of functional characteristics (e.g., no expression of carbonic anhydrase?), a more loose embedding in the lining (and lower content of CK filaments?), or the retention of fetal characteristics (i.e., less differentiated phenotype). Such a discrete subpopulation of cells that could qualify as stem cells has not yet been observed in adult pancreatic ducts. There may be no direct evidence for the presence of stem cells in the pancreatic ducts of adults, but this may be because appropriate cellular markers for this purpose have not yet been discovered. On the other hand, some embryonic markers have been demonstrated to become reexpressed in duct cells during conditions of tissue injury and regeneration; this is dealt with in the next section.

38.5. MECHANISMS OF PANCREAS REGENERATION: CAN MATURE CELLS TRANSDIFFERENTIATE?

Regeneration of pancreatic tissue, and especially of islets, has been studied using several experimental rodent models. In general, it can be said that pancreas does not seem to possess as impressive a regenerative potential as, e.g., liver. Nevertheless, chemically or mechanically caused injury is followed by at least a partial regeneration in rodents. When massive acinar cell necrosis is induced by administration of agents such as ethionin or caerulein, acute pancreatitis is induced. Numerous acinar cells die but proliferation of surviving acinar cells results in the restoration of normal histology and function (reviewed in Varricchio et al., 1977; Webster et al., 1977; Willemer et al., 1992). Acinar and ductal exocrine cells are normally slowly dividing cells, but in conditions of tissue damage they can revert to independently proliferating cells.

Chemical ablation of islet β-cells by selectively cytotoxic drugs such as streptozotocin (STZ) or alloxan is followed only by a partial restoration of the β-cell mass, depending on the dose of the toxic agent. At high doses, the β-cell mass is nearly completely destroyed and does not regenerate, leading to overt diabetes. At lower doses, there may be some regeneration of β-cells, depending on the species used; for instance, mice are more resistant to these toxic agents than rats. In rats, significant (but yet incomplete) regeneration takes place only when the injury is caused immediately after birth, but not in older rats (Portha et al., 1989; Wang et al., 1994, 1996). It has been suggested that adult mouse islets contain progenitor cells that can replace β-cells destroyed by STZ (Fernandes et al., 1997). The IPCs that proliferate following islet injury were characterized as a population of hormone-negative cells expressing the Glut-2 glucose transporter normally found in β-cells, and as a second population of islet cells expressing somatostatin (Guz et al., 2001). In this model, both the glucagon producing α-cells and the somatostatin-producing δ-cells of the islets appeared to revert to an immature phenotype which has also been observed in fetal pancreas. It is not clear from these studies whether non-β islet cells can still transdifferentiate to β-cells, or whether islets retain a small population of endocrine progenitor cells that could be considered as immature or undifferentiated.

Another pancreas regeneration model consists of surgical removal of part of the gland. Subtotal (90%) pancreatectomy is followed by proliferation in the remaining ducts. These form new ductules, acini, and islets (Bonner-Weir et al., 1993). It was proposed that differentiated duct cells can act as facultative or functional stem cells (Sharma et al., 1999), i.e., mature functional cells that retain the facultative capacity to function as or be a source of multipotential stem cells. This hypothesis was initially proposed in the case of hepatocytes (Block et al., 1996; Thorgeirsson, 1996).

With the use of transgenesis, interesting models have been introduced such as mice that produce interferon-γ under control of the insulin promotor. In these transgenic mice, lymphocytic infiltration of the islets (i.e., where interferon is produced) leads to β-cell destruction. This is compensated, however, by a continuous regeneration of β-cells that appear in or alongside the lining of ducts, and even within the duct lumen (Gu and Sarvetnick, 1993). In these mice, "transitional cells" were observed that coexpressed amylase and insulin, markers of acinar and β-cells, respectively (Gu et al., 1994, 1997). In this model, there also seems to be damage to the exocrine tissue, and tissue remodeling is observed that may be very similar to what is observed in other models of major pancreatic tissue injury, such as partial pancreatectomy. This remodeling involves the exocrine tissue which undergoes profound changes, namely, the formation of "ductal complexes" (*see* below).

Another experimental model of pancreas regeneration consists of the complete obstruction, by ligation, of a part of the exocrine ductal system. A ligature is placed whereby the exocrine secretion products cannot be evacuated from part of the gland. In the unligated (or downstream) part of the gland, the tissue remains unaffected. In the ligated part there occurs an impressive tissue remodeling whereby the normally predominantly acinar tissue is converted into ductal complexes (also termed *pseudoductal*, or *tubular complexes*) (Figs. 2 and 3). This conversion is also commonly observed in pathological conditions and has been referred to as "acinoductal metaplasia." The switch from acinar to ductal organization of the exocrine cells is caused by three mechanisms: (1) transdifferentiation of acinar cells to ductlike cells, (2) proliferation of the latter and of the original duct cells, and (3) apoptotic

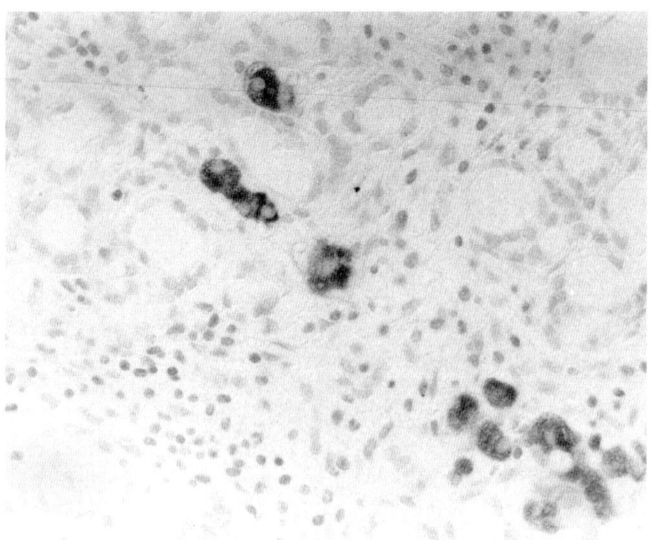

Fig. 2. Histological picture of regenerating pancreas, 7 d after duct ligation, immunostained for insulin. Several single, extraislet, insulin-positive cells are seen within a "metaplastic" ductal complex. Original magnification: ×500.

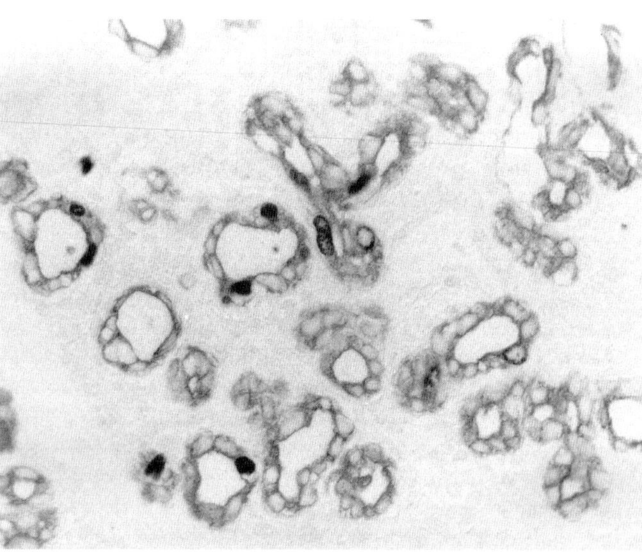

Fig. 3. Histological picture of regenerating pancreas, 7 d after duct ligation, immunostained for ductal cell marker CK20 and proliferation marker bromodeoxyuridine. A ductal complex is seen with several proliferating cells (labeled with a dark nucleus). It is from these metaplastic ductal structures that new islet cells are budding (*see* Fig. 2). Original magnification: ×500.

death of part of the acinar cells. This conversion is accomplished within 5 d postligation and remains for about 2 wk, after which the original histology becomes gradually restored. Although islets do not seem to be injured by this treatment, an important increase in the number of islet β-cells occurs between 4 and 7 d after ligation (Wang et al., 1995). Since β-cells only have a very limited mitotic activity, the oberved increase is due to neogenesis of β-cells from progenitor cells. New β-cells are seen in and around duct cells of the ductal complexes. Transitional cells coexpressing a ductal marker (CK20) and insulin were also observed and taken as evidence for a ductlike phenotype of the β-progenitor cells.

In the same model of ductligation, but at earlier time points, others have observed transitional cells coexpressing the acinar marker α-amylase and insulin (Bertelli and Bendayan, 1997). This would indicate a transition from acinar to β-cells. In this respect, it is important to note that purified acinar cells have been shown to transdifferentiate into ductlike cells in vitro, and in a time frame of 5 d, which is completely consistent with the in vivo observations of the duct-ligation model (Rooman et al., 2000). Thus, the observation of transitional cells that coexpress β-cell characteristics (insulin) and ductal characteristics (CK20) do not preclude the possibility that the β-cell precursors were originally acinar cells that first transdifferentiated to a ductal phenotype and then to β-cells.

Interestingly, the ductal cells of the metaplastic complexes in ligated pancreas start to express some proteins, such as the neuroendocrine marker PGP9.5 and the CCK-B receptor for gastrin, which are not expressed in the unligated part or in the normal adult pancreas. The same "fetal" proteins are also expressed by acinar cells that transdifferentiate in vitro. In vitro, furthermore, it was shown that the transdifferentiating acinar cells start to express the Pdx-1 (pancreas and duodenum homeobox) transcription factor (Rooman et al., 2000). Pdx-1 is involved as a master switch gene in embryonic pancreas differentiation and targeted ablation of the Pdx-1 gene results in pancreatic agenesis, but it is also a transactivator of the insulin gene (Ohlsson et al., 1993; Jonsson et al., 1994; Sander and German, 1997; Edlund, 1998). In adult pancreas it is expressed in all β-cells and in a subset of somatostatin cells, but in the embryonic pancreas all "protodifferentiated" epithelial cells express it. Thus, expression of Pdx-1, and of several other fetal markers, indicates that dedifferentiation of mature exocrine cells to fetal-like cells may be involved in the general process of transdifferentiation and tissue regeneration (Rooman et al., 2000).

One of the fetal proteins that is reexpressed during pancreas regeneration is the CCK-B receptor. Gastrin hormone and its CCK-B high-affinity receptor are normally expressed in the stomach but are first expressed in fetal pancreas (when there is still no gastrin production in the stomach) and also in regenerating pancreas after duct ligation (Wang et al., 1997; Rooman et al., 2001). Thus, during the transdifferentiation process occurring in injured tissue, the metaplastic exocrine cells become responsive to the endocrine and auto- or paracrine action of gastrin. It was also shown that gastrin is mitogenic for the CCK-B-positive transdifferentiated/metaplastic exocrine cells both in vivo and in vitro (Rooman et al., 2001). Furthermore, gastrin administration in duct-ligated rats stimulates β-cell neogenesis from the metaplastic ductal complexes (Rooman et al., 2002). In the unligated tissue, gastrin shows no effect at all. These observations indicate not only that gastrin is an important regulator of islet progenitor activity, but also that the progenitors are not normally responsive to this action. First they must alter their pattern of gene expression (i.e., dedifferentiate), and start to express gastrin receptor.

These data are consistent with the hypothesis that transdifferentiation may be operating during conditions of tissue remodeling and regeneration in the adult pancreas (Bouwens, 1998). According to this hypothesis, mature differentiated cells (possibly a subpopulation) can act as tissue progenitors. This means that tissue injury can induce their transdifferentiation or phenotypic switch or, in other words, that mature cells can become reprogrammed depending on the needs (e.g., from acinar to β-cell phenotype). In this context, it is interesting that the immortalized acinar cell line AR42-J can be induced to transdifferentiate to a β-cell phenotype by the addition of certain growth factors in the culture medium (Mashima et al., 1996a, 1996b; Zhou et al., 1999).

We can conclude this section by stating that a subpopulation of (ductal) cells that could qualify as stem cells has not been identified in pancreas regeneration models. At the least, the question of whether differentiated acinar and/or duct cells retain the potential to revert into pancreatic progenitor cells should be taken as seriously as the question of whether pancreas tissue contains a population of adult stem cells (Bouwens, 1998). This question is similar to the long-standing question of whether epithelial cancers in general arise either from dedifferentiated cells gaining proliferative potential or from stem cells with arrested differentiation (Sell and Pierce, 1994).

38.6. IS THERE A COMMON HEPATO-PANCREATIC STEM CELL IN PANCREAS?

Probably the most remarkable kind of differentiation plasticity in adult pancreas is the formation of pancreatic hepatocytes. When rats or hamsters are subjected to treatments that cause chronic injury to the exocrine pancreas, the development of foci of hepatocytes has been observed in the pancreas (Reddy et al., 1984; Rao et al., 1989; Makino et al., 1990). Transgenic overexpression of keratinocyte growth factor in the mouse pancreas also leads to the appearance of pancreatic hepatocytes (Krakowski et al., 1999). Transplantation of pancreatic cells was shown to lead to liver repopulation and restoration of metabolic liver defects in the recipients (Wang et al., 2001). Although initially it was thought that acinar exocrine cells were the precursor cells for pancreatic hepatocytes (Reddy et al., 1984), later studies proposed that duct cells or "periductal" cells, or "oval" cells might be the precursors (Rao et al., 1989; Makino et al., 1990). A population of "bipotential" stem cells was demonstrated in a region of mouse ventral foregut endoderm that can differentiate to either liver or pancreas, depending on a gradient of FGFs released from the cardiac mesoderm (Deutsch et al., 2001). Thus, a common hepatopancreatic stem cell may be retained in the embryonic pancreas, but it is not known whether such a stem cell still resides in postnatal pancreas.

It is difficult to establish conclusively the progenitor/precursor cell type of the hepatocytes from whole-animal experiments. Do the pancreatic hepatocytes emerge due to transdifferentiation of pancreatic exocrine, acinar, and/or duct cells, or to the alternative specification of an as yet undefined stem cell population? An in vitro model to investigate the origin of hepatocytes from adult pancreas should be developed. Recently, the glucocorticoid dexamethasone was shown to induce hepatocyte differentiation in mouse dorsal pancreatic buds in vitro and in the AR42-J pancreatic acinar cell line (Shen et al., 2000). When the AR42-J pancreatic cell line is cultured in the presence of dexamethasone and oncostatin M, the cells develop hepatocyte characteristics such as expression of albumin, α-fetoprotein, and several hepatocyte enzymes. The same cell line was previously shown to develop characteristics of insulin-producing islet cells when cultured in the presence of growth factors such as activin, hepatocyte growth factor (HGF), or glucagon-like peptide-1 (Mashima et al., 1996a, 1996b; Zhou et al., 1999). The AR42-J cell line is derived from an acinar exocrine tumor, and these observations thus seem to support the notion of acinar cells having multipotential transdifferentiation capacity.

38.7. ARE THERE CLONOGENIC ISLET STEM CELLS?

An important characteristic of stem cells is their clonogenic potential. Long-term cultures of epithelial cells have been derived from normal pancreas of adult rats, and these cells expressed ductal and acinar characteristics (Tsao and Duguid, 1987). Human duct tissue obtained from donor pancreas can be stimulated to proliferate and expand by the addition of growth factors such as HGF (Lefebvre et al., 1998). Furthermore, it was reported that human expanding duct cell cultures could generate a limited amount of new islet cells in the presence of extracellular matrix (Kerr-Conte et al., 1996; Bonner-Weir et al., 2000). Still, these in vitro studies could not establish whether ductal cell were undergoing dedifferentiation to become islet cell precursors or whether a subset of islet stem-like cells was present and activated in these cultures.

Starting from a whole enzymatic digest of pancreatic tissue obtained from prediabetic NOD mice, a recent study reported that long-term cultures (>3 yr) of epithelial cells could be obtained in which islet-like cell clusters could be formed (Ramiya et al., 2000). Islet differentiation was stimulated by low glucose, low serum, nicotinamide, and several growth factors. When these in vitro–generated islet cells were transplanted into diabetic mice, the diabetes was reversed. This study by Ramiya et al (2000) demonstrates that normal epithelial cells from adult pancreas can be controlled to grow and form differentiated islet cells in vitro. However, the exact nature of these epithelial progenitor cells still needs to be determined. Can these cells be prospectively identified, isolated, and purified? In what numbers are they present in the normal pancreas, if they are? Do they arise from dedifferentiation, or from true adult stem cells present in pancreatic tissue?

It has also been reported that pancreatic tissue contains nestin-expressing cells (Hunziker and Stein, 2000; Zulewski et al., 2001). In one study, these nestin-positive cells were shown to be clonogenic, and within the expanded cultures, endocrine and other cell markers were detected by reverse transcriptase polymerase chain reaction (Zulewski et al., 2001). It remains to be determined whether these nestin-positive cells were induced to show a low level of expression for the genes that were considered or whether they can really undergo complete differentiation to islet or other cell types. Nestin is generally considered a marker of neural stem cells (Lendahl et al., 1990), but one should be cautious with the interpretation that all nestin-positive cells would be stem cells. There are also other cell types that can express nestin, such as hepatic stellate cells (Niki et al., 1999). Stellate cells are mesenchymal cells that are also present in pancreas tissue, and they are clonogenic cells that can be easily expanded in vitro (Apte et al., 1998; Bachem et al., 1998).

38.8. EMBRYONIC STEM CELLS

Another potential source for cell therapy in diabetic patients is the establishment of pluripotent human embryonic stem (ES) cells. A recent success along this line has been the production of glucose-sensitive insulin-secreting cells from mouse ES cells, which

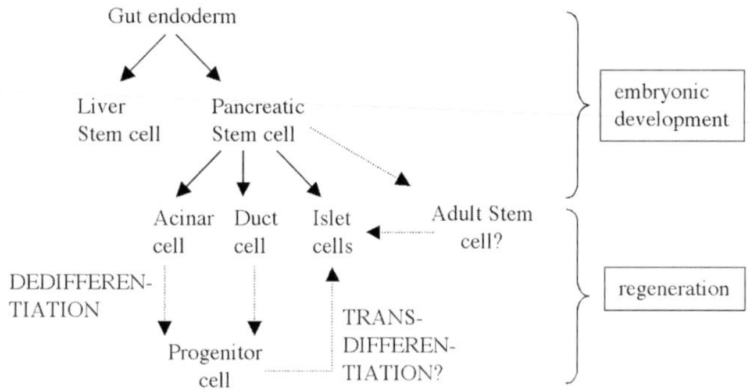

Fig. 4. Schematic representation of two hypotheses concerning islet cell neogenesis: transdifferentiation from dedifferentiated exocrine cells, and putative adult stem cells.

were able to restore normal glycemia in diabetic mice (Soria et al., 2000). In that study, a gene trap strategy was applied to isolate insulin-secreting cells from ES cells. More recently, culture conditions and cellular markers were found that allowed selective expansion of IPCs from undifferentiated mouse ES cells (Lumelsky et al., 2001). The different islet endocrine cell types differentiated and became associated with the typical islet organization which is also found in vivo, but they produced only small amounts of insulin, indicating an incomplete differentiation. β-Cell differentiation was also recently observed in human ES cell cultures, but their glucose responsiveness was not tested (Assady et al., 2001).

The available data thus indicate that ES-based cell therapy may be feasible and, in fact, that diabetes could be among the first applications of stem cell therapy. However, several important issues remain to be solved. The procedures to obtain, select, and stimulate maturation of functional β-cells need improvement to obtain pure populations of cells with the desired phenotype and degree of maturation (current protocols give mixed populations that are not suitable for transplantation). The exact nature of the signals that regulate ES cell differentiation, and those that regulate β-cell maturation, are not known. It seems that gene expression under in vitro conditions does not follow the strict and regulated schedule observed during in vivo development (which is regulated by intercellular interactions). It is not evident whether information gathered from studies with ES cells will be directly relevant to our understanding of embryonic developmental mechanisms or physiological repair mechanisms and the cells involved therein. For instance, clonal analysis of differentiating ES cells revealed a common progenitor to neural and insulin-producing cells (Lumelsky et al., 2001). This is surprising in view of the neuroectodermal vs endodermal origins of neurons and islet cells, respectively. These progenitors also expressed nestin, a marker that is expressed in neural stem cells (Lendahl et al., 1990) but that so far has not been reported in embryonic pancreas. Nevertheless, the results indicate that coaxing (human) ES cells into functioning β-cells may soon become possible and might find applications in replacement therapy for diabetes.

38.9. CONCLUSION

It is clear that islets of Langerhans can regenerate in the adult mammalian pancreas, most probably also in humans. This means that therapeutic applications are in reach, such as to regenerate or expand the islet mass in diabetic patients, or to provide adequate numbers of islet cells for transplantation. However, the mechanisms of regeneration are still poorly understood so that it is not yet feasible to control at will the expansion of islet cells in vivo or in vitro. It is still unclear whether IPCs remain present in the pancreas (or elsewhere in the body) under the form of classic tissue stem cells, or if only differentiated pancreatic cells can revert to a multipotential (and clonogenic) state (Fig. 4). These two types of stem cells, which could be termed *actual stem cells* and *facultative/functional stem cells*, respectively, are not mutually exclusive. So far there is much, mostly indirect, evidence for transdifferentiation ("facultative/functional stem cells"). However, there still is no way to identify a putative population of adult stem cells in the pancreas.

As far as ES cells are concerned, it is clear that these stem cells have the potential to differentiate, at least partly, to an islet phenotype in vitro. It remains to be elucidated which inductive conditions are necessary to direct this type of cell differentiation and to generate functional islet cells.

REFERENCES

Apelqvist, A., Li, H., Sommer, L., et al. (1999) Notch signalling controls pancreatic cell differentiation. *Nature* 400:877–881.

Apte, M. V., Haber, P. S., Applegate, T. L., et al. (1998) Periacinar stellate shaped cells in rat pancreas: identification, isolation, and culture. *Gut* 43:128–133.

Assady, S., Maor, G., Amit, M., Itskovitz-Eldor, J., Skorecki, K. L., Tzukerman, M. (2001) Insulin production by human embryonic stem cells. *Diabetes* 50:1691–1697.

Bachem, M. G., Schneider, E., Gross, H., et al. (1998) Identification, culture, and characterization of pancreatic stellate cells in rats and humans. *Gastroenterology* 115:421–432.

Bertelli, E. and Bendayan, M.(1997) Intermediate endocrine-acinar pancreatic cells in duct ligation conditions. *Am. J. Physiol.* 273: C1641–C16419.

Bertuzzi, F., Grohovaz, F., Maffi, P., et al. (2002) Succesful transplantation of human islets in recipients bearing a kidney graft. *Diabetologia* 45:77–84.

Bhushan, A., Itoh, N., Kato, S., et al. (2001) Fgf10 is essential for maintaining the proliferative capacity of epithelial progenitor cells during early pancreatic organogenesis. *Development* 128:5109–5117.

Block, G. D., Locker, J., Bowen, W. C., et al. (1996) Population expansion, clonal growth, and specific differentiation patterns in primary cultures of hepatocytes induced by HGF/SF, EGF and TGF alpha in a chemically defined (HGM) medium. *J. Cell Biol.* 132:1133–1149.

Bonner-Weir, S., Baxter, L. A., Schuppin, G. T., and Smith, F. E. (1993) A second pathway for regeneration of adult exocrine and endocrine pancreas. A possible recapitulation of embryonic development. *Diabetes* 42:1715–1720.

Bonner-Weir, S., Taneja, M., Weir, G. C., et al. (2000) In vitro cultivation of human islets from expanded ductal tissue. *Proc. Natl. Acad. Sci. USA* 97:7999–8004.

Bouwens, L. (1998) Transdifferentiation versus stem cell hypothesis for the regeneration of islet beta-cells in the pancreas. *Microsc. Res. Tech.* 43:332–336.

Bouwens, L. and De Blay, E. (1996) Islet morphogenesis and stem cell markers in rat pancreas. *J. Histochem. Cytochem.* 44:947–951.

Bouwens, L. and Pipeleers, D. G. (1998) Extra-insular beta-cells associated with ductules are frequent in adult human pancreas. *Diabetologia* 41:629–633.

Bouwens, L., Wang, R. N., De Blay, E., Pipeleers, D. G., and Kloppel, G. (1994) Cytokeratins as markers of ductal cell differentiation and islet neogenesis in the neonatal rat pancreas. *Diabetes* 43:1279–1283.

Bouwens, L., Braet, F., and Heimberg, H. (1995) Identification of rat pancreatic duct cells by their expression of cytokeratins 7, 19 and 20 in vivo and after isolation and culture. *J. Histochem. Cytochem.* 43: 245–253.

Bouwens, L., Lu, W. G., and De Krijger, R. (1997) Proliferation and differentiation in the human fetal endocrine pancreas. *Diabetologia* 40:398–404.

Cras-Meneur, C., Elghazi, L., Czernichow, P., and Scharfmann, R. (2001) Epidermal growth factor increases undifferentiated pancreatic embryonic cells in vitro: a balance between proliferation and differentiation. *Diabetes* 50(7):1571–1579.

Deutsch, G., Jung, J., Zheng, M., Lora, J., and Zaret, K. S. (2001) A bipotential precursor population for pancreas and liver within the embryonic endoderm. *Development* 128:871–881.

Edlund, H. (1998) Transcribing pancreas. *Diabetes* 47:1817–1823.

Githens, S. (1988) The pancreatic duct cell: proliferative capabilities, specific characteristics, metaplasia, isolation, and culture. *J. Pediatr. Gastroenterol. Nutr.* 7:486–506.

Gradwohl, G., Dierich, A., LeMeur, M., and Guillemot, F. (2000) Neurogenin3 is required for the development of the four endocrine cell lineages of the pancreas. *Proc. Natl. Acad. Sci. USA* 97:1607–1611.

Gu, D. and Sarvetnick, N. (1993) Epithelial cell proliferation and islet neogenesis in IFN-g transgenic mice. *Development* 118(1):33–46.

Gu, D., Lee, M. S., Krahl, T., and Sarvetnick, N. (1994) Transitional cells in the regenerating pancreas. *Development* 120:1873–1881.

Gu, D., Arnush, M., and Sarvetnick, N. (1997) Endocrine/exocrine intermediate cells in streptozotocin-treated Ins-IFN-gamma transgenic mice. *Pancreas* 15:246–250.

Guz, Y., Nasir, I., and Teitelman, G. (2001) Regeneration of pancreatic beta cells from intra-islet precursor cells in an experimental model of diabetes. *Endocrinology* 142(11):4956–4968.

Fernandes A, King LC, Guz Y, Stein R, Wright CV, Teitelman G. (1997) Differentiation of new insulin-producing cells is induced by injury in adult pancreatic islets. *Endocrinology* 138:1750–1762.

Heimberg, H., Bouwens, L., Heremans, Y., Van De Casteele, M., Lefebvre, V., and Pipeleers, D. (2000) Adult human pancreatic duct and islet cells exhibit similarities in expression and differences in phosphorylation and complex formation of the homeodomain protein Ipf-1. *Diabetes* 49:571–579.

Hunziker, E. and Stein, M. (2000) Nestin-expressing cells in the pancreatic islets of Langerhans. *Biochem. Biophys. Res. Commun.* 271:116–119.

Jensen, J., Heller, R. S., Funder-Nielsen, T. Et al. (2000) Independent development of pacreatic alpha- and beta-cells from neurogenin3-expressing precursors. A role for the Notch pathway in repression of premature differentiation. *Diabetes* 49:163–176.

Jonsson, J., Carlsson, L., Edlund, T., and Edlund, H. (1994) Insulin-promoter-factor 1 is required for pancreas development in mice. *Nature* 371:606–609.

Kerr-Conte, J., Pattou, F., Lecomte-Houcke, M., et al. (1996) Ductal cyst formation in collagen-embedded adult human islet preparations: a means to the reproduction of nesidioblastosis in vitro. *Diabetes* 45: 1108–1114.

Keymeulen, B., Ling, Z., Gorus, F. K., et al. (1998) Implantation of standardized beta-cell grafts in a liver segment of IDDM patients: graft and recipient characteristics in two cases of insulin-independence under maintenance immunosuppression for prior kidney graft. *Diabetologia* 41:451–459

Krakowski, M. L., Kritzik, M. R., Jones, E. M., et al. (1999) Pancreatic expression of keratinocyte growth factor leads to differentiation of islet hepatocytes and proliferation of duct cells. *Am. J. Pathol.* 154:683–691.

Krapp, A., Knofler, M., Frutiger, S., Hughes, G. J., Hagenbuchle, O., and Wellauer, P. K. (1996) The p48 DNA-binding subunit of transcription factor PTF1 is a new exocrine pancreas-specific basic helix-loop-helix protein. *EMBO J.* 15:4317–4329

Krapp, A., Knofler, M., Ledermann, B., et al. (1998) The bHLH protein PTF1-p48 is essential for the formation of the exocrine and the correct spatial organization of the endocrine pancreas. *Genes Dev.* 12(23): 3752–3763.

Lefebvre, V. H., Otonkoski, T., Ustinov, J., Huotari, M. A., Pipeleers, D. G., and Bouwens, L. (1998) Culture of adult human islet preparations with hepatocyte growth factor and 804G matrix is mitogenic for duct cells but not for beta-cells. *Diabetes* 47:134–137.

Lendahl, U., Zimmerman, L. B., and McKay, R. D. (1990) CNS stem cells express a new class of intermediate filament protein. *Cell* 60:585–595.

Lumelsky, N., Blondel, O., Laeng, P., Velasco, I., Ravin, R., and McKay, R. (2001) Differentiation of embryonic stem cells to insulin-secreting structures similar to pancreatic islets. *Science* 292:1389–1394.

Luttges, J. and Kloppel, G. (2001) Update on the pathology and genetics of exocrine pancreatic tumors with ductal phenotype: precursor lesions and new tumor entities. *Dig. Dis.* 19:15–23.

Makino, T., Usuda, N., Rao, S., Reddy, J. K., and Scarpelli, D. G. (1990) Transdifferentiation of ductular cells into hepatocytes in regenerating pancreas. *Lab. Invest.* 62:552–561.

Mashima, H., Ohnishi, H., Wakabayashi, K., et al. (1996a) Betacellulin and activin A coordinately convert amylase-secreting pancreatic AR42J cells into insulin-secreting cells. *J. Clin. Invest.* 97:1647–1654,

Mashima, H., Shibata, H., Mine, T., and Kojima, I. (1996b) Formation of insulin-producing cells from pancreatic acinar AR42J cells by hepatocyte growth factor. *Endocrinology* 137:3969–3976.

Miralles, F., Czernichow, P., Ozaki, K., Itoh, N., and Scharfmann, R. (1999) Signaling through fibroblast growth factor receptor 2b plays a key role in the development of the exocrine pancreas. *Proc. Natl. Acad. Sci. USA* 96:6267–6272.

Niki, T., Pekny, M., Hellemans, K., et al. (1999) Class VI intermediate filament protein nestin is induced during activation of rat hepatic stellate cells. *Hepatology* 29:520–527.

Ohlsson, H., Karlsson, K., and Edlund, T. (1993) IPF1, a homeodomain-containing transactivator of the insulin gene. *EMBO J.* 12:4251–4259.

Percival, A. C. and Slack, J. M. (1999) Analysis of pancreatic development using a cell lineage label. *Exp. Cell Res.* 247:123–132.

Portha, B., Blondel, O., Serradas, P., et al. (1989) The rat models of non-insulin dependent diabetes induced by neonatal streptozotocin. *Diabetes Metab.* 15:61–75.

Ramiya, V. K., Maraist, M., Arfors, K. E., Schatz, D. A., Peck, A. B., and Cornelius, J. G. (2000) Reversal of insulin-dependent diabetes using islets generated in vitro from pancreatic stem cells. *Nat. Med.* 6:278–282.

Rao, M. S., Dwivedi, R. S., Yeldandi, A. V., et al. (1989) Role of periductal and ductular epithelial cells of the adult rat pancreas in pancreatic hepatocyte lineage. A change in the differentiation commitment. *Am. J. Pathol.* 134:1069–1086.

Reddy, J. K., Rao, M. S., Qureshi, S. A., Reddy, M. K., Scarpelli, D. G., and Lalwani, N. D. (1984) Induction and origin of hepatocytes in rat pancreas. *J. Cell Biol.* 98:2082–2090.

Rooman, I., Heremans, Y., Heimberg, H., and Bouwens, L. (2000) Modulation of rat pancreatic acinoductal transdifferentiation and expression of PDX-1 in vitro. *Diabetologia* 43:907–914.

Rooman, I., Lardon, J., Flamez, D., Schuit, F., and Bouwens, L. (2001) Mitogenic effect of gastrin and expression of gastrin receptors in duct-like cells of rat pancreas. *Gastroenterology* 121:940–949.

Rooman, I., Lardon, J., and Bouwens, L. (2002) gastrin stimulates beta-cell neogenesis and increases islet mass from transdifferentiated but not from normal exocrine pancreas tissue. *Diabetes* 51:686–690.

Ryan, E. A., Lakey, J. R. T., Rajotte, R. V., et al. (2001) Clinical outcomes and insulin secretion after islet transplantation with the Edmonton protocol. *Diabetes* 50:710–719.

Sander, M. and German, M. S. (1997) The beta cell transcription factors and development of the pancreas. *J. Mol. Med.* 75:327–340.

Schwitzgebel, V. M., Scheel, D. W., Conners, J. R., et al.(2000) Expression of neurogenin3 reveals an islet cell precursor population in the pancreas. *Development* 127:3533–3542.

Sell, S. and Pierce, G. B. (1994) Maturation arrest of stem cell differentiation is a common pathway for the cellular origin of teratocarcinomas and epithelial cancers. *Lab. Invest.* 70:6–22.

Sharma, A., Zangen, D. H., Reitz, P., et al. (1999) The homeodomain protein IDX-1 increases after an early burst of proliferation during pancreatic regeneration. *Diabetes* 48:507–513.

Shen, C. N., Slack, J. M. W., and Tosh, D. (2000) Molecular basis of transdifferentiation of pancreas to liver. *Nat. Cell Biol.* 2:879–887

Slack, J. M. (1995) Developmental biology of the pancreas. *Development* 121:1569–1580.

Soria, B., Roche, E., Berna, G., Leon-Quinto, T., Reig, J. A., Martin, F. (2000) Insulin-secreting cells derived from embryonic stem cells normalize glycemia in streptozotocin-induced diabetic mice. *Diabetes* 49:157–162.

Thorgeirsson, S. S. (1996) Hepatic stem cells in liver regeneration. *FASEB J.* 10:1249–1256.

Tsao, M.-S. and Duguid, W. P. (1987) Establishment of propagable epithelial cell lines from normal adult rat pancreas. *Exp. Cell Res.* 168: 365–375.

Varricchio, F., Mabogunje, O., Kim, D., Fortner, J. G., and Fitzgerald, P. J. (1977) Pancreas acinar cell regeneration and histone (H1 and H10) modifications after partial pancreatectomy or after a protein-free ethionine regimen. *Cancer Res.* 37:3964–3969.

Wang, R. N., Bouwens, L., and Kloppel, G. (1994) Beta-cell proliferation in normal and streptozotocin-treated newborn rats: site, dynamics and capacity. *Diabetologia* 37:1088–1096.

Wang, R. N., Kloppel, G., and Bouwens, L. (1995) Duct- to islet-cell differentiation and islet growth in the pancreas of duct-ligated adult rats. *Diabetologia* 38:1405–1411.

Wang, R. N., Bouwens, L., and Kloppel, G. (1996) Beta-cell growth in adolescent and adult rats treated with streptozotocin during the neonatal period. *Diabetologia* 39:548–557.

Wang, R. N., Rehfeld, J. F., Nielsen, F. C., and Kloppel, G. (1997) Expression of gastrin and transforming growth factor-alpha during duct to islet cell differentiation in the pancreas of duct-ligated adult rats. *Diabetologia* 40:887–893.

Wang, X., Al-Dhalimy, M., Lagasse, E., Finegold, M., and Grompe, M. (2001) Liver repopulation and correction of metabolic liver disease by transplanted adult mouse pancreatic cells. *Am. J. Pathol.* 158: 571–579.

Webster, P. D., 3 rd, Black, O., Jr, Mainz, D. L., and Singh, M. (1977) Pancreatic acinar cell metabolism and function. *Gastroenterology* 73: 1434–1449 .

Willemer, S., Elsasser, H. P., and Adler, G. (1992) Hormone-induced pancreatitis. *Eur. Surg. Res.* 24 (Suppl. 1):29–39.

Zhou, J., Wang, X., Pineyro, M. A., and Egan, J. M. (1999) Glucagon-like peptide 1 and exendin-4 convert pancreatic AR42J cells into glucagon-and insulin-producing cells. *Diabetes* 48:2358–2366.

Zulewski, H., Abraham, E. J., Gerlach, M. J., et al. (2001) Multipotential nestin-positive stem cells isolated from adult pancreatic islets differentiate ex vivo into pancreatic endocrine, exocrine, and hepatic phenotypes. *Diabetes* 50:521–533.

39 Mammary Epithelial Stem Cells

GILBERT H. SMITH, PhD

A major interest in mammary stem cells relates to their potential role in the development of mammary cancer. The role of stem cells in development of premalignant and malignant lesions is exemplified by the study of the lesions, that develop in the mammary glands of mice infected with mouse mammary tumor virus. In addition, fragments or dispersed cells from the mammary glands of old or young mice possess equivalent ability to repopulate a cleared mammary fat pad through serial passages before reaching growth senescence. Limiting dilution experiments reveal lobule-limited and duct-limited progenitor cells in the cell mixtures, and both of these arise from a common mammary epithelial stem cell. Human mammary epithelial cells may be separated into myoepithelial (CALLA+) and luminal (MUC+). The luminal population is able to give rise to both populations, whereas the myoepithelial population only gives rise to myoepithelial cells. The luminal cell population contains pale- or light-staining cells that appear to have stem cell properties. By electron microscopy, five populations of mammary epithelial cells are identified: primitive small light cells (SLCs), undifferentiated large light cells (ULLCs), very differentiated large light cells, classic cytologically differentiated luminal cells, and the myoepithelial cell. The SLC represent only 3% of the total cells from puberty through postlactation, but these and ULLCs are absent from growth-senescent mammary cell transplants. Mammary epithelial cell progenitors have been tentatively identified by expression of stem cell antigen-1. Mammary cancer appears to arise from clonal expansion of a stem cell–derived cell population through progression from premalignant lesions. The reduction of breast cancer risk associated with early pregnancy appears to be related to an absence of a proliferative response of parous epithelium to environmental carcinogens, suggesting that parity produces a new cell population that is committed to the secretory fate, perhaps owing to expansion of a committed transit-amplifying population of cells. Little is known about the signaling that is responsible for the maintenance and control of the mammary stem cell population in the normal gland or in "immortalized" premalignant lesions, and this may be critical for preventing or controlling breast cancer.

39.1. INTRODUCTION

The mammary gland has often been considered a relatively dispensable organ except during the period of pregnancy and lactation. Despite this, a long history of scientific interest is associated with this organ because of its seminal role in infant nutrition and well-being and because it is often afflicted by cancer development. In fact, before the beginning of the twentieth century, there were already more than 10,000 scientific references to published articles relating to mammary biology (Lyons, 1958). It was an interest in cancer and cancer development in the breast that brought about the first series of experiments that led to our current concept of tissue-specific mammary epithelial stem cells and their important role in tissue renewal and in the development of breast cancer.

Interest in premalignant lesions of the breast led DeOme and his colleagues (DeOme, J. et al., 1959) to develop a biological system to recognize, characterize, and study hyperplastic nodules in the mammary glands of mouse mammary tumor virus (MMTV)–infected mice. In the quest for a means to demonstrate that these structures were precursors to frank mammary adenocarcinoma, these investigators developed a method for removing the endogenous mammary epithelium from a mammary fat pad. Subsequently, the "cleared" pad was used as a site of implantation where suspected premalignant lesions could be placed and their subsequent growth and development could be observed. Using this approach, they were able to show that both premalignant and normal mammary implants could grow and fill the empty fat pad within several weeks. During this growth period, the premalignant implants recapitulated their hyperplastic phenotype, whereas normal implants produced normal branching mammary ducts. Serial transplantation of normal and premalignant outgrowths demonstrated that while normal gland invariably showed growth senescence after several generations, hyperplastic outgrowths did not. It soon became apparent that any portion of the normal mammary

From: *Stem Cells Handbook*
Edited by: S. Sell © Humana Press Inc., Totowa, NJ

parenchyma could regenerate a complete mammary tree over several transplant generations suggesting the existence of cells capable of reproducing new mammary epithelium through several rounds of self-renewal. However, it was some time later before this property was recognized as representative of the presence of mammary epithelial stem cells (Williams and Daniel, 1983; Smith and Medina, 1988).

39.2. BIOLOGY

The discovery that all portions of the mouse mammary gland appeared competent to regenerate an entire new gland on transplantation triggered a series of articles relating to the reproductive lifetime of mammary cells (Daniel et al., 1968; Daniel and Young, 1971; Young et al., 1971). It was determined that no difference existed in the regenerative ability of mammary tissue taken from very old mice vs that taken from very young mice during serial transplantation. In addition, neither reproductive history nor developmental state had a significant impact on the reproductive longevity of mammary tissue implants. The ability of grafts from old donors to proliferate equivalently to those from young in young hosts suggested that the life-span of mammary cells is primarily affected by the number of mitotic divisions rather than by the passage of chronological or metabolic time. The authors in a series of experiments tested this, where mammary implants were serially transplanted. In one series, fragments were taken from the periphery of the outgrowth for subsequent transplantation, and in the other the fragments for transplant were removed from the center. The supposition was that the cells at the periphery had undergone more mitotic events than those in the center, and, therefore, peripheral tissue should show growth senescence more quickly than tissue near the center. This was shown to be the case (Daniel and Young 1971). The authors concluded that growth senescence in transplanted mammary epithelium was related primarily to the number of cell divisions. On the other hand, mouse mammary epithelial cells could be transformed to unlimited division potential either spontaneously, by MMTV infection, or by treatment with carcinogens (Daniel et al., 1968, 1975). At the time, this observation was taken to signify that "immortalization" (i.e., attainment of unlimited division potential) was an important early step in malignant transformation.

To what extent do these observations extend to the glands of other mammals? With respect to transplantation of mammary fragments to epithelium-free fat, extensive studies indicate that rat mammary epithelium shows clonogenic activity similar to that of the mouse. In fact, rat mammary implants grow extensively to complete glandular structures within the mouse mammary fat pads (Welsch et al., 1987). Human mammary fragments were maintained and could be stimulated to functional differentiation in mouse mammary fat pads but did not grow extensively (Sheffield and Welsch, 1988).

39.3. CULTURE

Dispersed mouse mammary epithelial cells have been shown to be able to recombine and grow to form a new gland within the epithelium-free mammary fat pad (Daniel, 1965; DeOme et al., 1978a, 1978b; Medina et al., 1986). In these experiments, both normal and transformed mammary outgrowths were developed, indicating that both normal and abnormal mammary cells could exist within any given apparently normal glandular population. Alternative interpretations included the acquisition of an abnormal developmental phenotype as a result of the in vitro culture conditions. More recently, irradiated feeder cells have been employed to propagate primary cultures of mouse mammary epithelium. Under these conditions, the cells were maintained for nine passages and produced normal mammary outgrowths on introduction into cleared mammary fat pads (Ehmann et al., 1987). The number of dispersed mammary cells required to produce a positive take (i.e., form a glandular structure within the fat pad) increased with increasing passage number. This observation applies to all mouse mammary epithelial cell lines that have been developed in vitro and maintained through serial passages. Eventually, with passage, as with the fragment implants, either no growth is attained or neoplastic development is achieved when the cells are placed in cleared fat pads (Kittrell et al., 1992). Some mouse mammary cell lines that were grown for various periods in culture demonstrated an extended reproductive life-span when reintroduced into cleared mammary fat pads and transplanted serially. The resulting outgrowths appeared in every way to be normal and did not exhibit hyperplastic or tumorigenic growth (Medina and Kittrell, 1993). It was concluded that the immortalization phenotype could be dissociated from the preneoplastic phenotype, and it was suggested that these mammary cell lines may represent an early stage, perhaps the earliest, in progression to mammary tumorigenesis.

During the last decade, a number of researchers have investigated the end point of the clonogenic capacity of dispersed rodent mammary epithelial cells in limiting dilution transplantation experiments (Smith, 1996; Kamiya et al., 1998, 1999; Kordon and Smith, 1998). Both in the mouse and in the rat, 1000–2000 mammary epithelial cells represent the smallest number required for the establishment of epithelial growth in a fat pad. Earlier it was shown that genes could be introduced into primary mammary epithelial cell cultures with retroviral vectors. Subsequently, the genetically modified epithelial cells were reintroduced into cleared mammary fat pads for evaluation, in vivo (Smith et al., 1991). Although stable transduction of gene expression could be achieved in a high percentage of mammary cells in culture, recovery of these retroviral-marked cells in regenerated glandular structures was only possible when virtually 100% of the implanted cells were stably modified. It was determined that this resulted from the fact that only a very small proportion of the primary epithelial cells inoculated was capable of contributing to tissue renewal in vivo. This was the first indication that only a subset of the mammary epithelial population possessed the capacity to regenerate mammary tissue on transplantation. From this followed the possibility that this cellular subset represented the mammary epithelial stem cell compartment.

For two entirely different purposes, dispersed rat and mouse mammary cells were tested for their ability to form epithelial structures in empty fat pads at limiting dilution. This author investigated the possibility that lobule and ductal-lineage-limited cells existed among the mouse mammary epithelial population based on the common observation that lobular development could be suppressed in transgenic mouse models when ductal branching morphogenesis was unaffected. The results of this study provided evidence for distinct lobule-limited and ductal-limited progenitors in the mouse mammary gland (Smith, 1996). In an effort to establish the total number of clonogenic cells in the rat mammary gland as a measure of radiogenic susceptibility to cancer induction, Kamiya (1998, 1999) conducted similar experiments. They found that, like the mouse, rat mammary glands possessed distinct lobule-committed and duct-committed progenitors. In the mouse, it was shown in clonal-dominant mammary populations that both of these progenitors arose from a common antecedent, i.e., a primary mammary epithelial stem cell (Kordon and Smith, 1998).

39.4. GROWTH AND DIFFERENTIATION IN VITRO

As described above, efforts to propagate mammary epithelial cells in continuous culture and subsequently demonstrate their ability to reconstitute the mammary gland in vivo have met with limited success. A different approach to understanding mammary epithelial cell lineage was applied by using cell-surface markers to distinguish basal (myoepithelial) from luminal (secretory) epithelial cells. Using fluorescence activated cell sorting (FACS), Stingl et al. (1998, 2000) separated human mammary epithelial cells into myoepithelial (CALLA-positive) and luminal (MUC-positive) populations and evaluated for their respective capacity to produce mixed colonies in cloning assays. They reported that individual epithelial cells bearing luminal markers alone or both luminal and myoepithelial surface markers could give rise to colonies with a mixed lineage phenotype. Cells bearing only the CALLA marker (basal/myoepithelial) were only able to produce like epithelial progeny. Using a similar approach, Pechoux et al. (1999) demonstrated that CALLA-positive (myoepithelial) and MUC-1-positive (luminal) mammary epithelial cells could be purified to essentially homogeneous populations and maintained as such under certain specific culture conditions in vitro. Expression of distinctive keratin gene patterns and other genetic markers also characterized these disparate cellular populations. It was further demonstrated that only the luminal epithelial cell population was able to produce both luminal and myoepithelial cell progeny in vitro, providing further evidence that the multipotent cellular subset in mammary epithelial cells resided among the luminal rather than the myoepithelial lineage. More recently, this same group has shown that unlike myoepithelial cells from normal glands, tumor-derived myoepithelial cells were unable to support three-dimensional growth when combined with normal luminal cells in vitro (Gudjonsson et al. 2002b). This deficiency was shown to be the inability of the tumor myoepithelial cells to express a specific laminin gene (LAM1) product.

Mouse mammary epithelial cells have been FACS separated according to their luminal or myoepithelial surface markers. Subsequent study of these different populations in vitro provided results that agree with those reported for human cells. The cells capable of giving rise to mixed colonies in cloning studies were only found among the cells bearing luminal epithelial cell markers (Smalley et al., 1999).

39.5. MAMMARY STEM CELL MARKERS

Several recent studies have demonstrated that the multipotent cells in mammary epithelium reside within the luminal cell population in human and mouse (Pechoux et al., 1999; Smalley et al., 1999). However, no specific molecular signature for mammary epithelial stem cells was revealed. Smith and Medina (1988) presented an earlier marker that held promise for identifying mammary stem cells in the ultrastructural description of mitotic cells in mammary epithelial explants. These investigators noticed that mouse mammary explants, like mammary epithelium *in situ*, contained pale or light-staining cells, and that only these cells entered mitosis when mammary explants were cultured.

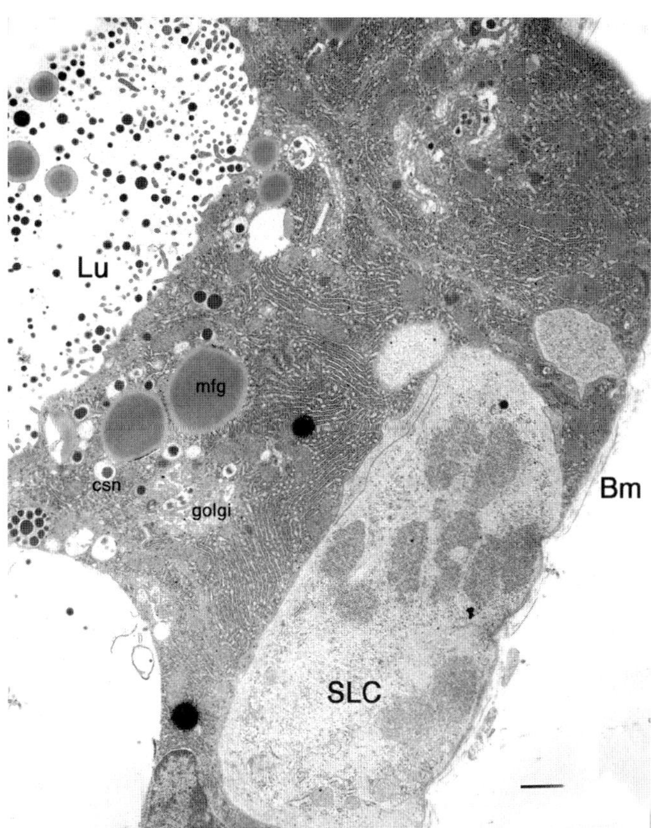

Fig. 1. Electron micrograph depicting a small light cell (SLC) in basal location near basement membrane (Bm) of secretory acinus in lactating mouse mammary gland. The SLC is in the process of dividing and contains condensed mitotic chromosomes seen as dense regions within the cell. No nuclear envelope is present. Bordering the dividing SLC are portions of two differentiated secretory cells (DSCs), whose cytoplasm contains abundant profiles of RER, an active Golgi apparatus (golgi), casein secretory granules (csn), and milk fat globules (mfg). The lumen (Lu) of the acinus is at left and contains numerous secretory elements. Bar = 1.0 μm.

Chepko and Smith (1997) analyzed light cells in the electron microscope utilizing their ultrastructural features to distinguish them from other mammary epithelial cells. The following basic features expected of stem cells were applied in the ultrastructural evaluation: division-competence (presence of mitotic chromosomes) and an undifferentiated cytology (Fig. 1). The pale-staining (stem) cells are therefore of distinctive morphology; their appearance in side-by-side pairs or in one-above-the-other pairs (relative to the basement membrane) was interpreted as the result of a recent symmetric mitosis. In addition to pairs, other informative images would be of juxtaposed cells that were morphologically intermediate between a primitive and differentiated morphology based on the number, type, and development of cytoplasmic organelles. Cells were evaluated for cytological differentiation with respect to their organelle content and distribution; for example, cells differentiated toward a secretory function might contain specific secretory products, such as milk protein granules or micelles, which have been ultrastructurally and immunologically defined (Hogan and Smith, 1982). In addition, the presence

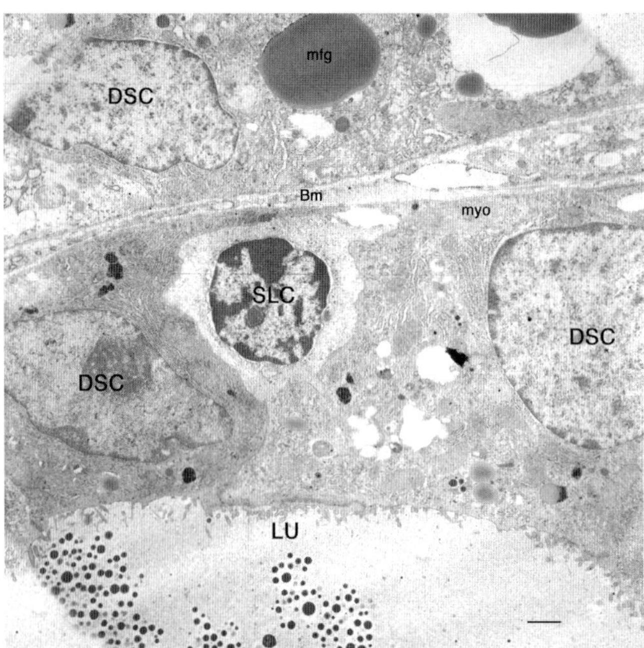

Fig. 2. In another secretory acinus from a lactating mouse mammary gland, a quiescent SLC rests on the basement membrane (Bm). Differentiated secretory mammary epithelial cells (DSCs) lie on either side in an adjacent acinus. Milk fat globules (mfg) and casein micelles are present within the DSCs and in the lumen (LU). A portion of a myoepithelial cell cytoplasm (myo) also appears near the SLC. Bar = 1.0 μm.

and number of intracellular lipid droplets, the extent and distribution of Golgi vesicles, and rough endoplasmic reticulum (RER) attest to the degree of functional secretory differentiation of a mammary epithelial cell. These features are characteristically well developed in the luminal cells of active lactating mammary gland (Fig. 2). Myoepithelial cells are flattened, elongated cells located at the basal surface of the epithelium, and their prominent cytoplasmic feature is the presence of many myofibrils and the absence of RER or lipid droplets.

In a retrospective analysis of light and electron micrographs, a careful and detailed scrutiny of mammary tissue was performed to determine the range of morphological features among the cell types that had previously been reported. The samples evaluated included mouse mammary explants; pregnant and lactating mouse mammary glands; and rat mammary gland from 17 stages of development beginning with nulliparous through pregnancy, lactation, and involution (Smith and Vonderhaar, 1981; Vonderhaar and Smith, 1982; Smith et al., 1984; Chepko and Smith, 1997). From this analysis, we were able to expand the number of cell types in the epithelium from two—secretory (or luminal) and myoepithelial cells—to five distinguishable structural phenotypes or morphotypes. Our observations strengthen the conclusion that only the undifferentiated (light) cells enter mitoses. The undifferentiated cells were found in two easily recognized forms: small (approx 8 μm) and large (15–20 μm). Mitotic chromosomes were never found within the differentiated cells, namely secretory and myoepithelial cells, suggesting that they were terminally differentiated and out of the cell cycle. Using all of these features, we were able to develop a more detailed description of the epithelial subtypes that comprise the mammary epithelium.

The characteristics used to develop a standardized description of five mammary epithelial cellular morphotypes were staining of nuclear and cytoplasmic matrix, cell size, cell shape, nuclear morphology, amount and size of cytoplasmic organelles, location within the epithelium, cell number, and grouping relative to each other and to other morphotypes. These characteristics were used to perform differential cell counts and morphometric analysis of the cell populations in rat mammary epithelium (Chepko and Smith, 1997). Figure 3 illustrates that each population can be used on both the light and electron levels to help form a search image for recognizing them *in situ*. The five morphotypes that we recognize in rodent mammary epithelium are a primitive SLC, an undifferentiated large light cell (ULLC), a very differentiated large light cell (DLLC), the classic cytologically differentiated luminal cell, and the myoepithelial cell. We described three sets of division-competent cells in rodent mammary epithelium and demonstrated that mammary epithelial stem cells and their downstream progenitors are morphologically much less differentiated than either the secretory or the myoepithelial cells. We counted a total of 3552 cells through 17 stages of rat mammary gland development and calculated the percentage of each morphotype. This analysis showed that the population density (number of cells/mm^2) of SLCs among mammary epithelium did not change from puberty through postlactation involution. The proportion of SLCs remained at 3%. This means that although the number of mammary epithelial cells increased by 27-fold during pregnancy in the mouse (Nicoll and Tucker, 1965; Kordon and Smith, 1998), the percentage of SLCs in the population does not change. Therefore, SLCs increase and decrease in absolute number at the same relative rate as the differentiating epithelial cells. Neither SLCs nor ULLCs cells were observed in an extensive study of growth-senescent mouse mammary transplants. Examination of growth-competent implants in the same host reveals easily detectable SLCs and ULLCs (Smith et al., 2002). These observations lend further support to the conclusion that SLCs and ULLCs may represent mammary epithelial stem cells *in situ*.

Gudjonnson et al. (2002a) predicted that if human mammary epithelium contained cells similar to the SLC and ULLC described in rodents, then these cells would be low or negative for the luminal surface marker, sialomucin (MUC-1), since they do not commonly contact the luminal surface. Coincidentally, such cells would be positive for epithelial-specific antigen (ESA) but negative for the basal myoepithelial cell marker, smooth muscle actin (SMA). Using this approach, they isolated two luminal epithelial cell populations. One, the major population, coexpressed MUC-1 and ESA. The other, a minor population, was found in a suprabasal location in vivo and expressed ESA but not MUC-1 or SMA. The latter cells formed elaborate branching structures, composed of both luminal and myoepithelial lineages, both in vitro and in vivo. These outgrowths resembled terminal duct lobular units both by morphology and by marker expression. These data provide strong evidence for the presence of mammary epithelial stem cells in the human breast with characteristics similar to those described above in rodent mammary gland.

At present there are no specific cellular markers that identify the "stemness" of any particular mammary epithelial cell. Several features known to define stem cells in other organs have been applied to the mammary gland. For example, the property of retaining DNA synthesis incorporated label over a long chase following labeling, i.e., long label-retaining cells (LRCs) in

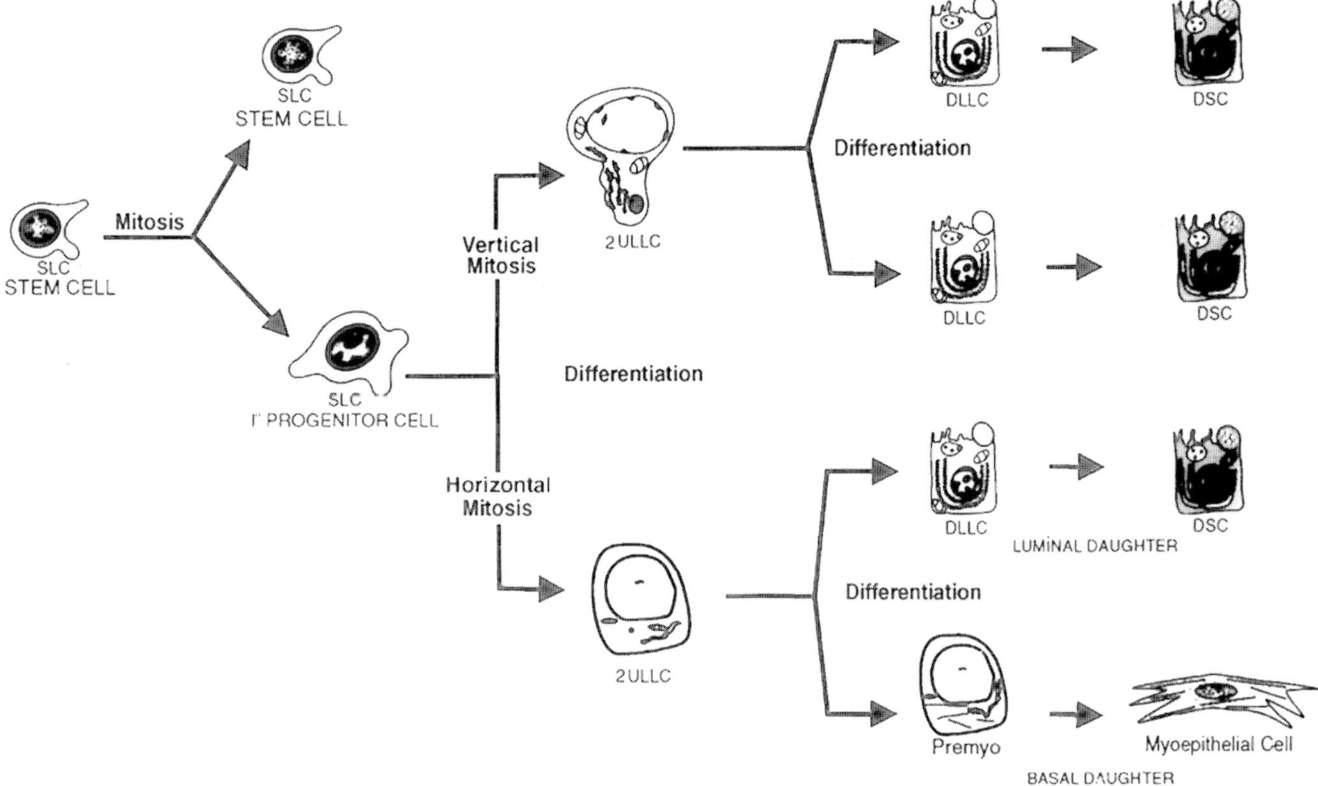

Fig. 3. This illustration portrays the various morphological forms, that make up the fully differentiated murine mammary epithelium. It is not drawn to scale; rather, it indicates our interpretation of the lineal relationships among the mammary stem cell (SLC) at the left, the lineage-limited progenitors (ULLC), and the fully differentiated secretory (DSC) and myoepithelial cells, as determined from an extensive electron microscopic study of differentiating murine mammary glands.

^{3}H-thymidine- or 5BrdU-pulsed mammary tissue. Mammary cells in the mouse with this property have been identified and were found scattered along the mammary ducts. Zeps et al. (1998) showed that estrogen receptor (ER) immune staining suggests that these cells are often ER positive. These investigators made no further characterization of these cells.

In a very recent attempt to further characterize LRCs in the mouse mammary gland, specific cellular markers were applied to mammary cells pulsed for 14 d in vivo with BrdU and chased for 9 wk (Welm et al., 2002). Two characteristics, efficient efflux of Hoechst dye (SP) and the presence of stem cell antigen-1 (Sca-1), known to associate with stem cells in other organ systems, were used to identify putative mammary epithelial stem cells. Isolation of mammary cells with these properties revealed the following. LRCs isolated from primary cultures were enriched for Sca-1 expression and SP dye–effluxing properties. *In situ*, LRCs represent 3–5% of the population after 9 wk, in good agreement with the number of SLCs (Chepko and Smith, 1997). In addition, the SP mammary cells, which showed increased clonogenic activity, in vivo, possessed a frequency and size distribution that was very similar to SLCs. The Sca-1-positive mammary cells showed a greater regenerative potential in cleared fat pads than similar numbers of Sca-1-negative cells. Taken together, this study represents a first important step in isolating mammary epithelial progenitors and may permit the identification of additional markers useful in determining the biological potential of mammary stem cells.

39.6. MAMMARY STEM CELLS IN CARCINOGENESIS

Contiguous portions of the human mammary gland possess the identical pattern of X chromosome inactivation. Thus, local portions of the gland are derived from a single antecedent (Tsai et al., 1996). In a further study of human mammary tissue, this same group (Deng et al., 1996) showed that mammary cancer *in situ* and the apparently normal tissue surrounding the lesion shared similar genetic alterations. This was interpreted to indicate that mammary lesions arise as a result of the clonal expansion of previously affected epithelium subsequent to further genetic change. The results imply that local genetically damaged mammary stem cells may give rise to premalignant lesions, which may progress to frank malignancy. Studies by several other laboratories (Lakhani et al., 1996, 1999; Rosenberg et al., 1997) have confirmed and extended these observations, supporting the concept of clonal progression in the development of breast cancer in humans. Therefore, it is conceivable that mammary hyperplasia and tumors develop locally from damaged clonogenic epithelial progenitors (stem cells).

Experimental evidence from MMTV-induced mouse mammary hyperplasia and tumorigenesis (reviewed in Callahan and Smith, 2000) provides strong support for the concept of clonal progression from normal through premalignant to malignant epithelium in the rodent mammary gland. In an effort to provide a proof of principle (i.e., mammary stem cells may contribute to mammary tumor development), mice exhibiting a mammary growth-senescent phenotype in transplant experiments were challenged with MMTV (Boulanger and Smith, 2001). Only one tumor was induced by MMTV in these mice, whereas more than half of their MMTV-infected wild-type female littermates developed mammary tumors. The result indicates that premature senescence in mammary epithelial stem cells may reduce the subsequent risk for mammary tumorigenesis in MMTV-challenged mice.

Previous experimentation with retrovirus-marked (MMTV) clonal-dominant mammary populations demonstrated that an entire functional mammary glandular outgrowth might comprise the progeny of a single antecedent (Kordon and Smith, 1998). These populations have been transplanted serially to study the properties of aging, self-renewing, mammary clonogens derived from the original progenitor. Premalignant, malignant, and metastatic clones arose from these transplants during passage. All of these bore a lineal relationship with the original antecedent because all of the original proviral insertions were represented in each of these lesions (Smith and Boulanger, 2002). Although this does not prove that mammary stem cells may directly give rise to cancerous lesions within the mammary gland, it does demonstrate that normal, premalignant, and malignant progeny are all within the "repertoire" of an individual mammary cell.

In mice, rats, and humans a single early pregnancy provides lifelong reduction in mammary cancer risk. In rats and mice, the protective effect of pregnancy can be mimicked through hormonal application in the absence of pregnancy. This refractoriness to chemical induction of mammary tumorigenesis has recently been linked to the absence of a proliferative response in the parous epithelium when confronted with the carcinogen as compared with the nulliparous gland (Sivaraman et al., 1998, 2001). Concomitant with the reduction in proliferative response is the appearance of stable activation of p53 in epithelial cell nuclei. This suggests that in response to the hormonal stimulation of pregnancy a new cellular population is created with an altered response to carcinogen exposure. Employing the Cre recombinase/loxp system to identify mammary cells *in situ*, which have differentiated during pregnancy and expressed Cre from the whey acidic protein (WAP) promoter, a new parity-induced mammary epithelial cell population was discovered (Wagner et al., 2002). Transplantation studies indicate that this parity-specific epithelium has the capacity for self-renewal and can contribute extensively to regeneration of mammary glands in cleared fat pads. This population does not appear in nulliparous females at any age but accumulates on successive pregnancies in parous females. The evidence shows that, *in situ*, these cells are committed to secretory cell fate and contribute extensively to the formation of secretory lobule development on successive pregnancies. Transplantation of dispersed cells indicates that this population is preferentially included in growth-competent mammary cell reassembly and has an individual capacity to undergo at least eight cell doublings. Studies are in progress to isolate and characterize these cells and to determine their contribution to the refractoriness of parous mammary tissue to cancer development.

39.7. FUTURE PROSPECTS

The vast array of genetic models and manipulations developed in the mouse has yet to be fully employed in the dissection of stem cell biology in the mammary gland or, for that matter, in a number of other organ systems. To date, most investigators have been satisfied with describing the effect of the deletion or dysregulation of their favorite genes on mammary growth, differentiation, and tumorigenesis or with providing evidence for the importance of their pet signaling pathway in mammary function. However, this attitude may be changing, especially with the increased awareness of multipotent cells in adult organs and the mounting evidence for the importance of somatic cell signaling on stem cell behavior in tissue-specific stem cell niches (Spradling et al., 2001). The application of conditional gene deletion or expression in stem cell populations in the epidermis provides an example of this approach (Arnold and Watt, 2001). Here, conditional activation of myc, even transiently, in epidermal stem cells commits them to the production of sebaceous epithelial progeny at the expense of hair follicle progeny. In the mammary gland, only indirect evidence supports the possible role of somatic cell control of stem cell behavior. In mice expressing constitutively active transforming growth factor-β1 (TGF-β1) from the WAP gene promoter, mammary implants show early growth senescence on transplantation into cleared mammary fat pads (Kordon et al., 1995; Boulanger and Smith, 2001). This suggests that expression of active TGF-β1 in adjacent hormonally responsive epithelial progeny may result in a reduction in the capacity for stem cell self-renewal. The premature loss of stem cell reproductive activity in this model also results in a decreased risk for mammary tumor induction by MMTV (Boulanger and Smith, 2001). Modulation of stem cell behavior holds exceptional promise for a new prophylactic approach for controlling mammary cancer risk. An important step toward the achievement of this control will be the characterization of the stem cell niche in the rodent mammary gland and ultimately in humans.

REFERENCES

Arnold, I. and Watt, F. M. (2001) c-Myc activation in transgenic mouse epidermis results in mobilization of stem cells and differentiation of their progeny. *Curr. Biol.* 11(8):558–568.

Boulanger, C. A. and Smith, G. H. (2001) Reducing mammary cancer risk through premature stem cell senescence. *Oncogene* 20:2264–2272.

Callahan, R. and Smith, G. H. (2000) MMTV-induced mammary tumorigenesis: gene discovery, progression to malignancy and cellular pathways. *Oncogene* 19(8):992–1001.

Chepko, G. and Smith, G. H. (1997) Three division-competent, structurally-distinct cell populations contribute to murine mammary epithelial renewal. *Tissue Cell* 29(2):239–253.

Daniel, C. W. and Young, L. J. (1971) Influence of cell division on an aging process: life span of mouse mammary epithelium during serial propagation in vivo. *Exp. Cell Res.* 65(1):27–32.

Daniel, C. W., DeOme, K. B. (1965) Growth of mouse mammary gland in vivo after monolayer culture. *Science* 149:634–636.

Daniel, C., DeOme, K., et al. (1968) The in vivo life span of normal and preneoplastic mouse mammary glands: a serial transplantation study. *Proc. Natl. Acad. Sci. USA* 61:53–60.

Daniel, C. W., Young, L. J., et al. (1971) The influence of mammogenic hormones on serially transplanted mouse mammary gland. *Exp. Gerontol.* 6(1):95–101.

Daniel, C. W., Aidells, B. D., et al. (1975) Unlimited division potential of precancerous mouse mammary cells after spontaneous or carcinogen-induced transformation. *Fed. Proc.* 34(1):64–67.

Deng, G., Lu, Y., et al. (1996) Loss of heterozygosity in normal tissue adjacent to breast carcinomas. *Science* 274(5295):2057–2059.

DeOme, K. B., et al. (1959) Development of mammary tumors from hyperplastic alveolar nodules transplanted into gland-free mammary fat pads of female C3H mice. *J. Natl. Cancer Inst.* 78:751–757.

DeOme, K. B., Miyamoto, M. J., et al. (1978a) Detection of inapparent nodule-transformed cells in the mammary gland tissues of virgin female BALB/cfC3H mice. *Cancer Res.* 38(7):2103–2111.

DeOme, K. B., Miyamoto, M. J., et al. (1978b) Effect of parity on recovery of inapparent nodule-transformed mammary gland cells in vivo. *Cancer Res.* 38(11 Pt. 2):4050–4053.

Ehmann, U. K., Guzman, R. C., et al. (1987) Cultured mouse mammary epithelial cells: normal phenotype after implantation. *J. Natl. Cancer Inst.* 78(4):751–757.

Gudjonsson, T., Ronnov-Jessen, L., et al. (2002b) Normal and tumor-derived myoepithelial cells differ in their ability to interact with luminal breast epithelial cells for polarity and basement membrane deposition. *J. Cell Sci.* 115(Pt. 1):39–50.

Gudjonnson, T., Villadsen, R., et al. (2002a) Isolation, immortalization, and characterization of a human breast epithelial cell line with stem cell properties. *Genes Dev.* 16:693–706.

Hogan, D. L. and Smith, G. H. (1982) Unconventional application of standard light and electron immunocytochemical analysis to aldehyde-fixed, araldite-embedded tissues. *J. Histochem. Cytochem.* 30(12): 1301–1306.

Kamiya, K., Gould, M. N., et al. (1998) Quantitative studies of ductal versus alveolar differentiation from rat mammary clonogens. *Proc. Soc. Exp. Biol. Med.* 219(3):217–225.

Kamiya, K., Higgins, P. D., et al. (1999) Kinetics of mammary clonogenic cells and rat mammary cancer induction by X-rays or fission neutrons. *J. Radiat. Res. (Tokyo)* 40(Suppl.):128–137.

Kittrell, F. S., Oborn, C. J., et al. (1992) Development of mammary preneoplasias in vivo from mouse mammary epithelial cell lines in vitro. *Cancer Res.* 52(7):1924–1932.

Kordon, E. C. and Smith, G. H. (1998) An entire functional mammary gland may comprise the progeny from a single cell. *Development* 125 (10):1921–1930.

Kordon, E. C., McKnight, R. A., et al. (1995) Ectopic TGF beta 1 expression in the secretory mammary epithelium induces early senescence of the epithelial stem cell population. *Dev. Biol.* 168(1):47–61.

Lakhani, S. R., Slack, D. N., et al. (1996) Detection of allelic imbalance indicates that a proportion of mammary hyperplasia of usual type are clonal, neoplastic proliferations. *Lab. Invest.* 74(1):129–135.

Lakhani, S. R., Chaggar, R., et al. (1999) Genetic alterations in "normal" luminal and myoepithelial cells of the breast. *J. Pathol.* 189(4):496–503.

Lyons, W. R. (1958) Hormonal synergism in mammary growth. *Proc. Roy. Soc. Lond.* 149:303–325.

Medina, D. and Kittrell, F. S. (1993) Immortalization phenotype dissociated from the preneoplastic phenotype in mouse mammary epithelial outgrowths in vivo. *Carcinogenesis* 14(1):25–28.

Medina, D., Oborn, C. J., et al. (1986) Properties of mouse mammary epithelial cell lines characterized by in vivo transplantation and in vitro immunocytochemical methods. *J. Natl. Cancer Inst.* 76(6):1143–1156.

Nicoll, C. S. and Tucker, H. A. (1965) Estimates of parenchymal, stromal, and lymph node deoxyribonucleic acid in mammary glands of C3H/Crgl-2 mice. *Life Sci.* 4(9):993–1001.

Pechoux, C., Gudjonsson, T., et al. (1999) Human mammary luminal epithelial cells contain progenitors to myoepithelial cells. *Dev. Biol.* 206(1):88–99.

Rosenberg, C. L., Larson, P. S., et al. (1997) Microsatellite alterations indicating monoclonality in atypical hyperplasias associated with breast cancer. *Hum. Pathol.* 28(2):214–219.

Sheffield, L. G. and Welsch, C. W. (1988) Transplantation of human breast epithelia to mammary-gland-free fat-pads of athymic nude mice: influence of mammotrophic hormones on growth of breast epithelia. *Int. J. Cancer* 41(5):713–719.

Sivaraman, L., Stephens, L. C., et al. (1998) Hormone-induced refractoriness to mammary carcinogenesis in Wistar-Furth rats. *Carcinogenesis* 19(9):1573–1581.

Sivaraman, L., Conneely, O. M., et al. (2001) p53 is a potential mediator of pregnancy and hormone-induced resistance to mammary carcinogenesis. *Proc. Natl. Acad. Sci. USA* 98(22):12,379–12,384.

Smalley, M. J., Titley, J., et al. (1999) Differentiation of separated mouse mammary luminal epithelial and myoepithelial cells cultured on EHS matrix analyzed by indirect immunofluorescence of cytoskeletal antigens. *J. Histochem. Cytochem.* 47(12):1513–1524.

Smith, G. H. (1996) Experimental mammary epithelial morphogenesis in an in vivo model: evidence for distinct cellular progenitors of the ductal and lobular phenotype. *Breast Cancer Res. Treat.* 39(1):21–31.

Smith, G. H. and Boulanger, C. A. (2002) Mammary stem cell repertoire: new insights in aging epithelial populations *Mech. Ageing Dev.* 123: 1505–1519.

Smith, G. H. and Medina, D. (1988) A morphologically distinct candidate for an epithelial stem cell in mouse mammary gland. *J. Cell Sci.* 90(Pt 1):173–183.

Smith, G. H. and Vonderhaar, B. K. (1981) Functional differentiation in mouse mammary gland epithelium is attained through DNA synthesis, inconsequent of mitosis. *Dev. Biol.* 88(1):167–179.

Smith, G. H., Vonderhaar, B. K., et al. (1984) Expression of pregnancy-specific genes in preneoplastic mouse mammary tissues from virgin mice. *Cancer Res.* 44(8):3426–3437.

Smith, G. H., Gallahan, D., et al. (1991) Long-term in vivo expression of genes introduced by retrovirus-mediated transfer into mammary epithelial cells. *J. Virol.* 65:6365–6370.

Smith, G. H., Strickland, P., and Daniel, C. W. (2002) Putative epithelial stem cell loss corresponds with mammary growth senescence. *Cell Tissue Res.* 310:313–320.

Spradling, A., Drummond-Barbosa, D., et al. (2001) Stem cells find their niche. *Nature* 414(6859):98–104.

Stingl, J., Eaves, C. J., et al. (1998) Phenotypic and functional characterization in vitro of a multipotent epithelial cell present in the normal adult human breast. *Differentiation* 63(4):201–213.

Stingl, J., Eaves, C. J., et al. (2001) Characterization of bipotent mammary epithelial progenitor cells in normal adult human breast tissue. *Breast Cancer Res. Treat.* 67:93-109.

Tsai, Y. C., Lu, Y., Nichols, P. W., Zlotnikov, G., Jones, P. A., and Smith, H. (1996) Contiguous patches of normal human epithelium derived from a single stem cell: implications for breast carcinogenesis. *Cancer Res.* 56:402–404.

Vonderhaar, B. K. and Smith, G. H. (1982) Dissociation of cytological and functional differential in virgin mouse mammary gland during inhibition of DNA synthesis. *J. Cell Sci.* 53:97–114.

Wagner, K.-U., Boulanger,C. A., Henry, M. D., Sagagias, M., Hennighausen, L., and Smith, G. H. (2002) An adjunct mammary epithelial cell population in parous females: its role in functional adaptation and tissue renewal. *Development* 129:1377–1386.

Welm, B. E., Tepera, S. B., Venezia, T., Graubert, T. A., Rosen, J. M., and Goodell, M. A. (2002) Sca-1^{pos} cells in the mouse mammary gland represent an enriched progenitor cell population. *Dev. Biol.* 245:42–56.

Welsch, C. W., O'Connor, D. H., et al. (1987) Normal but not carcinomatous primary rat mammary epithelium: readily transplanted to and maintained in the athymic nude mouse. *J. Natl. Cancer Inst.* 78(3): 557–565.

Williams, J. M. and Daniel, C. W. (1983) Mammary ductal elongation: differentiation of myoepithelium and basal lamina during branching morphogenesis. *Dev. Biol.* 97(2):274–290.

Young, L. J., Medina, D., et al. (1971) The influence of host and tissue age on life span and growth rate of serially transplanted mouse mammary gland. *Exp. Gerontol.* 6(1):49–56.

Zeps, N., Bentel, J. M., et al. (1998) Estrogen receptor-negative epithelial cells in mouse mammary gland development and growth. *Differentiation* 62(5):221–226.

40 Morphogenesis of Prostate Cancer

HELMUT BONKHOFF, PhD

The three epithelial cell types of the prostatic epithelium—secretory luminal cells, basal cells, and neuroendocrine cells—arise from a common pluripotent stem cell in the basal layer through transit-amplifying cells that display intermediate phenotypes. The cellular diversity of the prostatic epithelium is maintained through a network of hormonal control, growth factors, and interactions with the basement membrane. Severe differentiation and proliferation disorders occur during the malignant transformation of the prostatic epithelium. In high-grade prostatic intraepithelial neoplasia (HGPIN), basal cells lose their proliferative capacity while luminal cells acquire increased proliferative activity. This process is associated with an abnormal expression of oncogenes (erbB-2, erbB-3, and c-*met*), the apoptosis-suppressing Bcl-2, and the classic estrogen receptor α (ERα). Conversely, the ERβ which mediates chemopreventive effects of phytoestrogens is partially lost in HGPIN. Neoplastic progression to invasion is associated with loss of cell adhesion proteins and formation of new tumor-associated basement membranes, which provide a matrix for invasion. Common prostatic adenocarcinoma is composed of exocrine cell types expressing prostate-specific antigen and cytokeratins 8 and 18, as well as androgen receptors (Ars), making exocrine tumor cells androgen responsive even in androgen-insensitive stages of the disease. The only phenotype of common prostatic adenocarcinoma lacking the nuclear AR shows neuroendocrine differentiation. These endocrine tumor cells do not proliferate or undergo apoptosis, indicating that such tumor cells are androgen-insensitive and escape radiation therapy and other cytotoxic drugs. In addition, endocrine tumor cells secrete a number of endocrine growth factors that can maintain proliferative activity in exocrine tumor cells through a paracrine mechanism. After androgen deprivation therapy, prostate cancer cells acquire estrogen and progesterone receptors and may use the pertinent steroids to survive in an androgen-deprived milieu. This warrants clinical trials to test the efficacy of antiestrogens in the medical treatment of advanced prostate cancer.

40.1. INTRODUCTION

Prostate cancer is one of the most commonly diagnosed cancers in North America and Europe. Despite its clinical magnitude and the recent progress made in molecular biology, the pathogenesis of prostate cancer remains poorly understood. This reflects several factors including the complex composition of the prostate gland by different anatomical, cellular, and functional compartments (Figs. 1 and 2); the heterogeneous and multifocal nature of prostate cancer; the limited number of established cell lines for in vitro studies; and the lack of suitable animal models that faithfully recapitulate all stages of disease progression. This chapter focuses on current morphogenetic factors implicated in the development of prostate cancer and tumor progression. The concepts discussed herein refer to recent data obtained in human prostate tissue.

From: *Stem Cells Handbook*
Edited by: S. Sell © Humana Press Inc., Totowa, NJ

40.2. CELLULAR BIOLOGY OF PROSTATIC EPITHELIUM

The prostatic epithelium has a complex composition of three cell types differing in their hormonal regulation and marker expression (Fig. 2). The most prevalent phenotype consists of secretory luminal cells expressing prostate-specific antigen (PSA) and cytokeratins (CKs) 8 and 18 (Nagle et al., 1991; Xue et al., 1998). Basal cells, the second most important phenotype, maintain normal epithelial-stromal relation and express high-molecular-weight cytokeratins (Wernert et al., 1987; Bonkhoff, 1996). The third phenotype shows neuroendocrine differentiation. Neuroendocrine cells express chromogranins and secrete a number of neurosecretory products that may have growth-promoting properties (Di Sant'Agnese, 1992; Di Sant'Agnese and Cockett, 1996). Although these basic cell types clearly differ in their biological functions, they obviously share a common origin from pluripotent stem cells located in the basal cell layer (Bonkhoff, 1996a;

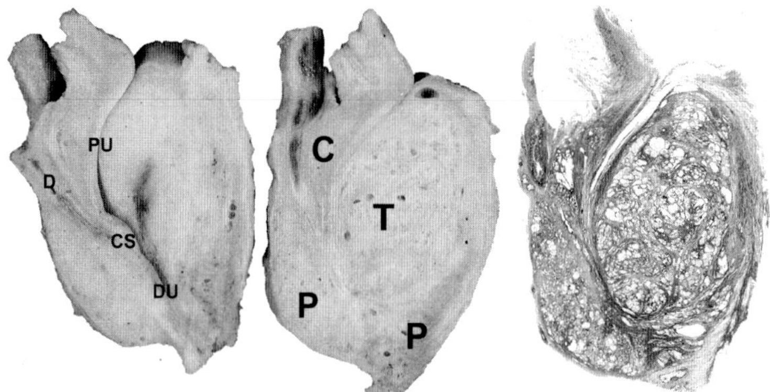

Fig 1. Zonal anatomy of prostate. The prostate of a 65-yr-old patient with benign prostatic hyperplasia (BPH) is shown. The prostate gland is composed of three anatomical zones. The central zone (C) located between ejaculatory ducts (D), and the proximal urethral segment (PU) accounts for about 25% of normal prostatic volume. Only about 10% of carcinomas arise in this zone. The transition zone (T) located around the proximal urethral segment accounts for about 5% of normal prostatic volume but steadily increases with age. Virtually all forms of BPH arise here. Transition zone cancer accounts for about 15–20% of prostatic adenocarcinoma and is commonly diagnosed in transurethral resection specimens from patients with BPH (incidental carcinoma). The peripheral zone (P) located around the distal urethral (DU) segment represents approx 70% of the normal gland. The majority (70–75%) of prostatic adenocarcinomas and high-grade prostatic intraepithelial neoplasias (HGPINs) arise in this zone. CS, colliculus seminalis.

Bonkhoff et al., 1996b, 1998b). This concept is based on the occurrence of intermediate differentiation among the three basic cell types making up the prostatic epithelium (Bonkhoff et al., 1994a; Xue et al., 1998), and some biological properties of basal cells (Bonkhoff, 1996a; Bonkhoff et al., 1996b, 1998b). Cell kinetic studies indicate that the proliferation compartment of the normal and hyperplastic epithelium is located in the basal cell layer (Bonkhoff et al., 1994b). Seventy percent of proliferating epithelial cells express basal cell–specific cytokeratins, while the remaining 30% of cycling cells are identified in secretory luminal cell types (Bonkhoff et al., 1994b). Chromogranin A (ChrA)–positive neuroendocrine cells lack proliferative activity and represent a terminal differentiated cell population within the prostatic epithelial cell system (Bonkhoff et al., 1991a, 1995b). It is therefore most unlikely that neuroendocrine cells present in the normal or dysplastic epithelium are precursors of prostate cancer cells with neuroendocrine features (Bonkhoff, 1998). In benign prostate tissue, apoptotic cell death is androgen regulated and mainly occurs in secretory luminal cells. Basal cells uniformly express the apoptosis-suppressing Bcl-2 oncoprotein, which obviously protects the proliferation compartment from programmed cell death (Bonkhoff et al.,1998a).

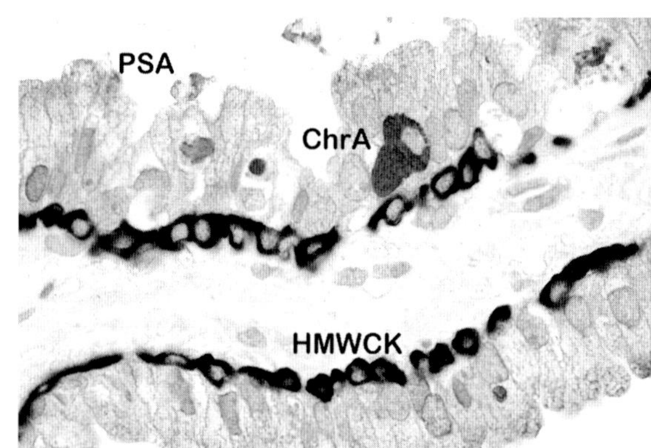

Fig. 2. Cellular composition of prostatic epithelium. The prostatic epithelium is composed of three basic cell types: secretory luminal cells, basal cells, and endocrine-paracrine cells. Simultaneous demonstration of PSA (secretory luminal cells), high-molecular-weight cytokeratins (HMWCK) (basal cells), and ChrA (endocrine cells).

The cellular diversity of the prostatic epithelium is maintained through a network of hormonal control, growth factors, and adhesive interactions with the underlying basement membrane. The differentiation compartment of the prostatic epithelium is made up of secretory luminal cells that are androgen dependent but have a limited proliferative capacity (Bonkhoff et al., 1994b). Luminal cells strongly express the nuclear androgen receptor (AR). The proliferation compartment (basal cells) is androgen independent but harbors androgen-responsive target cells. In fact, subsets of basal cells express the nuclear AR at high levels (Bonkhoff et al., 1993b). Basal cells also contain the 5α-reductase isoenzyme 2, which is crucial for the dihydrotestosterone-forming process (Bonkhoff et al., 1996c). It is likely that androgen-responsive basal cells are committed to differentiate toward luminal cells under appropriate androgen stimulation (Bonkhoff, 1996a; Bonkhoff et al., 1996b). This differentiation process is balanced by estrogens. Estrogen treatment leads to basal cell hyperplasia and atrophy of luminal cells by preventing basal cells to differentiate toward luminal cells (Bonkhoff, 1996a; Bonkhoff et al., 1996b). This process is mediated by the classic estrogen receptor α (ERα), which is expressed in stromal and basal cells but not in secretory luminal cell types (Bonkhoff et al., 1999b). On the other hand, the new ERβ is expressed extensively in luminal cells and at lower levels in basal cells (Fixemer et al., 2003). The ERβ binds phytoestrogens with high affinity and is a promising target for chemoprevention of BPH and prostate cancer (Chang and Prins, 1999; Steiner et al., 2001). It has been shown that ERβ knockout mice develop BPH with age, indicating that a functional ERβ protects the prostatic epithelium from hyperplastic changes (Krege et al., 1998).

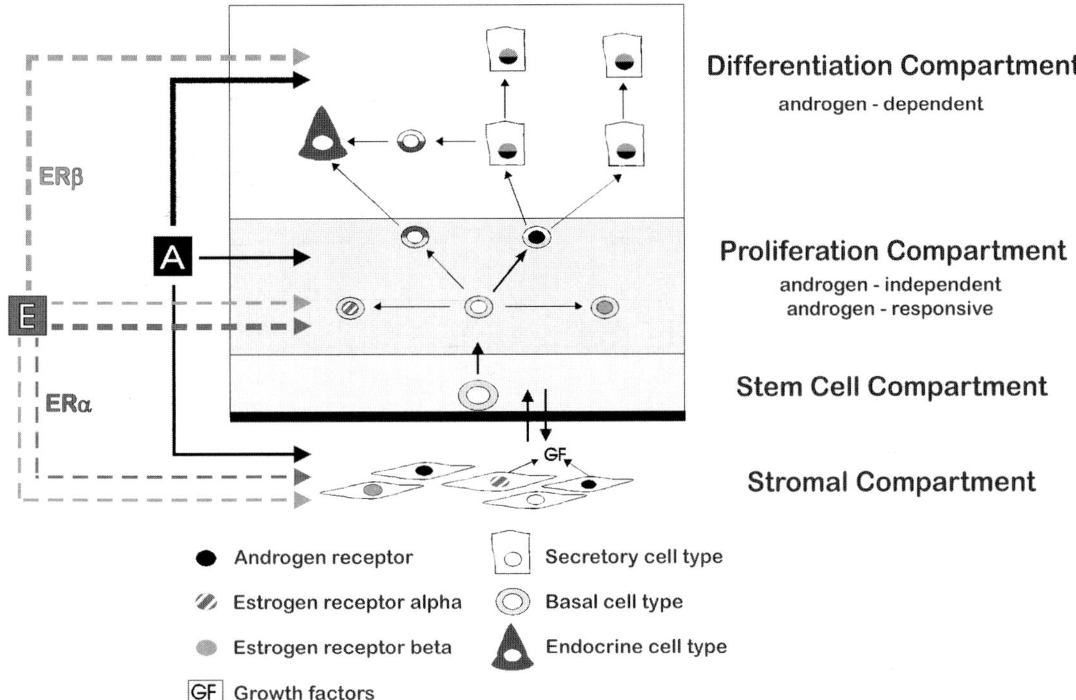

Fig. 3. Stem cell model for organization of the prostatic epithelium. Three functional compartments can be identified within the complex prostatic epithelial cell system. The differentiation compartment consists of secretory luminal cells that are androgen dependent but have a limited proliferative capacity. Luminal cells express high levels of the nuclear AR and the ERβ, which may exert antiproliferative effects on luminal cells. The basal cell layer is androgen independent and makes up the proliferation compartment. The proliferation function of basal cells is maintained by several growth factors (e.g., EGF), oncogenes (erbB-2, erbB-3, c-*met*), and tumor suppressor gene products (nm23-H1), while Bcl-2 protects basal cells from apoptotic cell death. The basal cell layer houses a small stem cell population that gives rise to all epithelial cell types through a process of intermediate differentiation. These differentiation processes depend on a delicate balance between estrogens and androgens. Estrogens prevent basal cell differentiation toward luminal cells, thus leading to basal cell hyperplasia and atrophy of luminal cells. When basal cells require androgens to give rise to androgen-dependent luminal cells, the turnover of the luminal epithelium depends on the presence of androgen-responsive target cells in the basal cell layer. The age-related decrease in circulating androgens may hypersensitize basal cells to the reduced levels of bioavailable androgens by upregulation of the nuclear AR in the basal cell layer. This may lead to glandular hyperplasia by accelerating the turnover of luminal cells from basal cells. Alternatively, other stroma-derived factors may be involved to control the differentiation process from basal to luminal cell types.

Besides estrogens and androgens, a number of nonsteroidal growth factors are involved in the regulation of benign glandular growth. Most growth factor receptors (e.g., epidermal growth factor receptor [EGF-R]), oncogenes (erbB-2, erbB-3, c-*met*, Bcl-2), and tumor suppressor genes (nm-23-H1) of the prostatic epithelium are expressed in basal cells (Bonkhoff et al., 1998b; Myers and Grizzle, 1996). The interplay among these factors may ultimately determine the growth fraction within the basal cell layer. On the other hand, differentiation processes within the prostatic epithelial cell system most likely depend on a hormonal balance between circulating androgens and estrogens (Bonkhoff, 1996a; Bonkhoff et al., 1996a, 1998b). Another important factor implicated in benign prostatic growth is the role of basal cells mediating adhesive interactions with epithelial basement membranes (Bonkhoff, 1998c). Prostatic epithelial cells require basement membrane components for their in vitro growth and differentiation (Fong et al., 1991). In human prostate tissue, basal cells exhibit polarized distribution of integrin receptors (α6β1, α2β1, α6β4) and hemidesmosome-associated proteins (BP180, BP220, HD1) (Bonkhoff et al., 1993a; Knox et al., 1994; Nagle et al., 1995). It seems likely that formation of stable hemidesmosomes and adhesive interactions with basement membrane contribute significantly to the integrity and biological functions of basal cells.

In summary, basal cells play a pivotal role in benign prostate growth. The basal cell layer houses pluripotent stem cells and maintains cell proliferation and normal epithelial-stromal relations (Fig. 3). Genetic and epigenetic factors interfering with the normal function of basal cells are therefore crucial for the development of prostate cancer.

40.3. DIFFERENTIATION AND PROLIFERATION DISORDERS IN EARLY PHASES OF PROSTATIC CANCEROGENESIS

HGPIN is the most likely precursor of prostate cancer (Bostwick, 1996; Montironi et al., 1996). This lesion usually arises in preexisting ducts and duct–acinar units of the peripheral zone and shares cytological features with intermediate and high-grade carcinoma but retains basal cell differentiation (Bostwick, 1996). Autopsy studies indicate that PIN precedes carcinoma by 10 yr and more. HGPIN is currently the most significant risk factor for prostate cancer. Its identification in prostatic biopsy specimens warrants further searches for concurrent cancer (Bostwick, 1996).

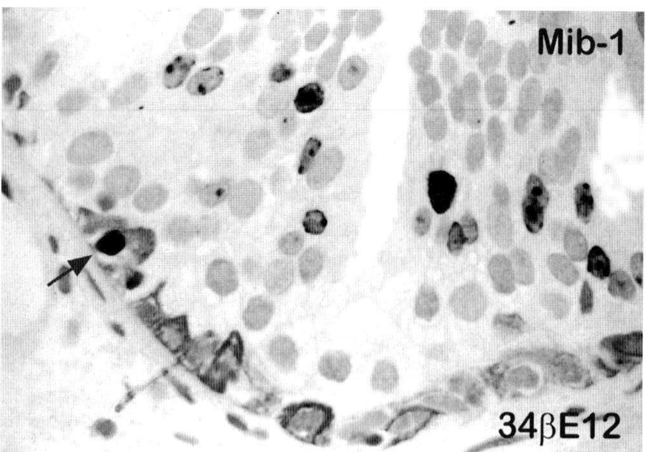

Fig. 4. Proliferation abnormalities detected in HGPIN. During the malignant transformation of the prostatic epithelium, the proliferation zone extends to secretory luminal cells (differentiation compartment). Less than 10% of cycling cells are identified in the basal cell layer (arrow), the proliferation compartment of the normal prostatic epithelium. Simultaneous demonstration of the proliferation marker Mib-1 and basal cell–specific cytokeratins (34βE12) in HGPIN.

Severe differentiation and proliferation disorders occur during the malignant transformation of the prostatic epithelium. In HGPIN, the basal cell layer (proliferation compartment) loses its proliferation function, whereas secretory luminal (dysplastic) cells acquire increased proliferative activity (Bonkhoff et al., 1994a). Less than 10% of cycling cells detected in HGPIN belong to the former proliferation compartment (Bonkhoff et al., 1994a) (Fig. 4). Extension of the proliferative zone to luminal cells in the differentiation compartment is a typical feature of well established premalignant lesions such as colorectal adenomas. The premalignant proliferation disorders encountered in HGPIN are associated with an aberrant expression of oncogenes (erbB-2, erbB-3, c-*met*) and tumor suppressor genes (nm23-H1) in the differentiation compartment of the transformed epithelium (Myers and Grizzle, 1996; Bonkhoff et al., 1998b). Restricted to basal cells in normal conditions, these biomarkers are implicated in the malignant transformation of the prostatic epithelium. In addition, severe regulatory disorders of the programmed cell death have been identified. At least 20% of HGPINs express high levels of the apoptosis-suppressing Bcl-2 oncoprotein in the differentiation compartment of the transformed epithelium and thus prevent dysplastic cells from the apoptotic cell death (Bonkhoff et al., 1998a). The resulting prolonged life-span of transformed cells, together with their high proliferation rate, provides an excellent environment in which genetic instability can occur. The most common genetic alterations in HGPIN include gain of chromosome 7, particularly 7q31; loss of 8p and gain of 8q; and loss of 10q, 16q, and 18q. The overall frequency of numeric chromosomal anomalies reported is remarkably similar in HGPIN and invasive cancer, suggesting that they have a similar pathogenesis (Qian et al., 1998; Foster et al., 2000).

Clinical studies suggest that HGPIN lesions are androgen dependent and generally regress after androgen deprivation (Bostwick, 1996). This observation obviously reflects the fact that most of these precursors express the nuclear AR and Bcl-2 as described in benign acini. Conversely, HGPIN with aberrant expression of Bcl-2 in the differentiation compartment tends to downregulate the AR, as documented by markedly reduced levels of detectable receptor proteins (Bonkhoff et al., 1998a). It is likely that such premalignant lesions escape the androgen-regulated programmed cell death and do not regress after androgen deprivation. Accordingly, Bcl-2 may be a promising biomarker to define the virulence of HGPIN.

The role of estrogens in the malignant transformation of the prostatic epithelium appears even more complex. Epidemiological and experimental data suggest that estrogens may exert cancerogenic and chemopreventive effects on the prostatic epithelium (Chang and Prins, 1999; Griffiths, 2000; Steiner et al., 2001). This apparent contradiction obviously reflects the presence of two distinct estrogen receptors (ERα and ERβ). In human prostate tissue, ERα gene expression is restricted to basal and stromal cells in normal conditions. In HGPIN, high steady-state levels of ERα mRNA are detected in the dysplastic epithelium (Bonkhoff et al., 1999b). At least 10% of HGPINs express the ERα at the protein level (Bonkhoff et al., 1999b). The estrogen-inducible PS2 has been identified in a significant number of benign and dysplastic prostate tissues from patients with locally advanced prostate cancer, but not in prostate tissue from patients without evidence of malignant disease (Bonkhoff et al., 1995a). These data suggest that the cancerogenic effects that estrogens may exert on the prostatic epithelium are mediated by the classic ERα. On the other hand, the ERβ binds phytoestrogens, which have antiproliferative and chemopreventive properties in animal models (Chang and Prins, 1999; Steiner et al., 2001). Expressed in luminal cell types at high levels in normal conditions, the ERβ is downregulated in HGPIN. At least 64% of these precursors reveal decreased or markedly decreased levels of the ERβ in the dysplastic epithelium (Fixemer et al., 2003). This indicates that the ERβ is a tumor suppressor that is partially lost during the malignant transformation of the prostatic epithelium.

In summary, virtually all phenotype and genotype data amassed in recent years suggest that HGPIN is the precursor of intermediate and high-grade cancer arising in the peripheral zone (Bostwick, 1996; Foster et al., 2000; Montironi et al., 1996). Conversely, the significance of atypical adenomatous hyperplasia (AAH) as a precursor of low-grade transition zone cancer is not well established (Montironi et al., 1996). Although the proliferative activity of AAH is increased compared with hyperplastic lesions, the proliferation zone and Bcl-2 expression are restricted to basal cells as described in benign prostate tissue (Bonkhoff et al., 1994b, 1998a). Thus, AAH does not reveal typical premalignant proliferation and differentiation abnormalities as found in HGPIN. Nevertheless, allelic imbalance may occur, indicating a genetic link between AAH and prostatic adenocarcinoma (Cheng et al., 1998). Much more work is needed to define the morphogenesis of low-grade transition zone cancer.

40.4. PATHOGENESIS OF STROMAL INVASION

Adhesive interactions in premalignant lesions do not differ significantly from those encountered in benign prostate tissue. Dramatic changes occur during early stromal invasion when the transformed epithelium loses basal cell differentiation (Bonkhoff, 1998c). This process is associated with the loss of a number of hemidesmosome-forming proteins and associated adhesive molecules, including collagen VII, β3 and γ2 subchains of laminin 5, and α6β4 integrins (Knox et al., 1994; Nagle et al., 1995). It is quite clear that benign acini and HGPIN require these adhesive elements to maintain basal cell differentiation and normal epithelial–stromal relations. Alternatively, the inability of transformed cells to express hemidesmosome-associated proteins obviously presents a key step in the neoplastic progression of HGPIN to early invasive cancer.

Another important event in early stromal invasion refers to the synthesis of tumor-associated basement membranes (Bonkhoff et al., 1991b, 1992). Invasive prostate cancer cells produce basement membrane–like matrices to invade the host tissue, and express associated integrins (α6β1, α2β1) that mediate attachment to this newly formed matrix (Bonkhoff et al., 1991b, 1992, 1993a) (Fig. 5). This particular tumor–host relation encountered in prostate cancer is maintained through the various stages of the disease, including high-grade, metastatic, and recurrent lesions (Bonkhoff et al., 1991b, 1992, 1993). Neoplastic basement membranes differ from their normal counterparts in their differential susceptibility to pepsin treatment and lack hemidesmosome-associated laminin 5, collagen VII, and type IV collagen α5 and α6 chains (Bonkhoff et al., 1993a; Knox et al., 1994; Nagle et al., 1995). Recent *in situ* hybridization analysis showed that these basement membranes are produced by tumor cells and not by the host tissue (Pföhler et al., 1998). High steady-state levels of laminin and type IV collagen mRNA are detected in metastatic lesions when compared with primary tumors (Pföhler et al., 1998). This indicates that the basement membrane–forming process increases with tumor progression. Their functional significance for the process of stromal invasion has also been demonstrated in vitro, showing that prostate cancer cell lines generally require reconstituted basement membrane (Matrigel) to be tumorigenic in athymic mice (Bonkhoff, 1998c).

In summary, *de novo*–synthesized basement membrane and adhesion via specific receptors significantly contribute to the ability of prostate cancer to penetrate the extracellular matrix (ECM) during stromal invasion and metastasis (Bonkhoff, 1998c).

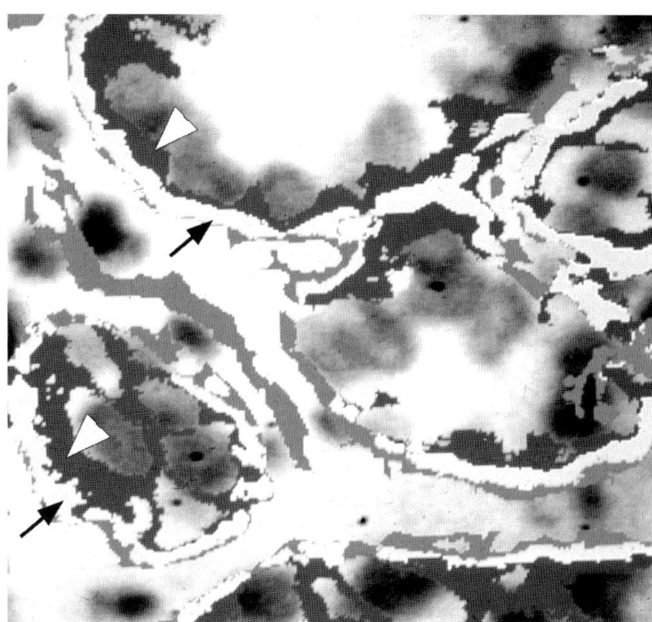

Fig. 5. Epithelial-stromal relation in prostate cancer. Invasive tumor cells are separated from the host tissue by pericellular and periacinar basement membranes expressing laminin and other basement membrane components. Pertinent receptors such as α6β1 integrins (arrowheads) mediate attachment to these newly formed matrices. Computer-assisted double staining reveals coordinate expression of the extracellular receptor domain and its corresponding ligand in basement membranes (arrows).

40.5. MORPHOGENETIC FACTORS IMPLICATED IN PROSTATE CANCER PROGRESSION AND HORMONE THERAPY FAILURE

Common prostatic adenocarcinoma is mainly composed of exocrine tumor cells that express PSA and CKs 8 and 18 and share phenotype similarity with secretory luminal cells of the normal prostatic epithelium (Nagle et al., 1991). These exocrine tumor cells generally express the nuclear AR and 5α reductase isoenzymes 1 and 2 in primary, metastatic, and recurrent lesions (Bonkhoff et al., 1993c, 1996c; Koivisto et al., 1998). This observation suggests that exocrine tumor cells are androgen responsive and maintain the dihydrotestosterone-forming process even in hormone-refractory stages of the disease. The continuous expression of the nuclear AR in androgen-insensitive tumors can be explained partially by AR gene amplification, which has been identified in at least 30% of recurrent lesions (Koivisto et al., 1998). The presence of the nuclear AR in prostate cancer tissue, however, does not imply androgen-dependent growth. Point mutations in the steroid-binding domain of the AR gene can seriously interfere with the normal function of the receptor protein (Koivisto et al., 1998; Culig et al., 2000). Mutant AR can bind estrogens and other steroids that maintain transcription of androgen-regulated genes even in the absence of androgens (Culig et al., 2000). AR gene mutations, however, are rather infrequent in prostate cancer. A significant number of hormone-refractory tumors have been reported to have apparently normal AR gene (Koivisto et al., 1998). Other factors are certainly involved in the multifactorial process of androgen insensitivity. Alternative pathways by which prostate cancer cells can escape androgen deprivation include their ability to acquire neuroendocrine differentiation (Bonkhoff, 1998d, 2001b) or to use estrogens for their own growth (Bonkhoff et al., 1999, 2000, 2001a).

40.5.1. NEUROENDOCRINE DIFFERENTIATION The second most prevalent phenotype encountered in prostate cancer shows neuroendocrine differentiation (Di Sant'Agnese and Cockett,1992; Di Sant'Agnese et al., 1996). Virtually all prostatic adenocarcinomas reveal at least focal neuroendocrine features as assessed by immunohistochemical markers such as ChrA. Tumors with extensive and multifocal neuroendocrine features (accounting for approx 10% of all prostatic malignancies) tend to be poorly differentiated, more aggressive, and resistant to hormonal therapy (Di Sant'Agnese, 1992; Di Sant'Agnese and Cockett, 1996). Several pathways have been described showing how neuroendocrine differentiation can affect tumor progression and hormone therapy failure (Bonkhoff, 1998d, 2001b). It has been shown that prostate cancer cells expressing ChrA consistently lack the nuclear AR in primary, metastatic, and recurrent lesions (Bonkhoff et al., 1993c) (Fig. 6). This clearly indicates that neuroendocrine phenotypes constitute an androgen-insensitive cell population in all stages of the disease. Neuroendocrine tumor cells most likely derive from exocrine phenotypes through a process of intermediate differentiation. This obviously reflects

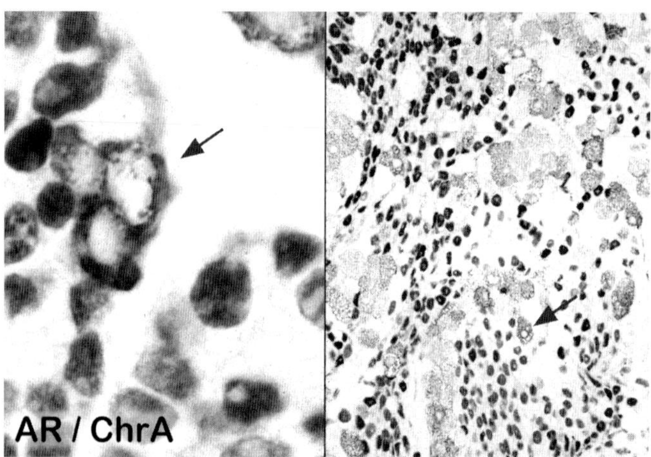

Fig. 6. AR status of endocrine and exocrine prostate cancer cells. AR expression is restricted to exocrine tumor cells. Neuroendocrine tumor cells identified by ChrA (arrows) consistently lack the nuclear AR in both primary (left) and recurrent (right) lesions.

the differentiation repertoire of prostatic stem cells. In fact, neuroendocrine foci frequently harbor amphicrine cell types expressing both endocrine (ChrA) and exocrine (PSA) markers (Bonkhoff et al., 1994a). Despite their androgen insensitivity, neuroendocrine tumor cells have no proliferative capacity. It has been shown that neuroendocrine differentiation predominantly occurs in the G0 phase of the cell cycle and is lost when tumor cells reenter the cell cycle (Bonkhoff et al., 1991a, 1995b). Although neuroendocrine tumor cells lack proliferative activity, they may exert growth-promoting stimuli on adjacent (exocrine) tumor cells. Endocrine tumor cells secrete a number of neurosecretory products with mitogenic properties in vitro, including serotonin, bombesin, calcitonin, and parathyroid hormone–related peptides (Di Sant'Agnese, 1992; Di Sant'Agnese and Cockett, 1996). It is likely that these neuroendocrine growth factors can maintain cell proliferation of adjacent (exocrine) tumor cells through a paracrine (androgen-independent) mechanism.

Recent studies also indicate that neuroendocrine (ChrA-positive) tumor cells escape the apoptotic cell death, as assessed by DNA fragmentation assays (Bonkhoff et al., 1999a; Fixemer et al., 2003). The absence of proliferative and apoptotic activity in neuroendocrine phenotypes may have some therapeutic implications, since radiation therapy and other cytotoxic drugs mainly affect cycling cells. Given their cell kinetic features, it will be very difficult to kill neuroendocrine tumor cells by endocrine and other cytotoxic treatments currently available. Recent clinical studies lend credence to this concept. Elevated serum levels of ChrA in patients with prostate cancer correlate with poor prognosis and are scarcely influenced by either endocrine therapy or chemotherapy (Berruti et al., 2000).

40.5.2. ROLE OF ESTROGENS IN ANDROGEN-INSENSITIVE PROSTATIC GROWTH Since the time of Higgins, estrogens have been widely used in the medical treatment of advanced prostate cancer to reduce the testicular output of androgens. The recent discovery of the classical ERα and estrogen-regulated proteins such as the progesterone receptor (PR) and the heat-shock protein HSP27 clearly shows that prostate cancer cells can use estrogens for their own growth (Bonkhoff et al., 1999b, 2000, 2001a). In apparent contrast with breast cancer, the presence of the ERα and the estrogen-inducible PR and HSP27 is a late event in prostate cancer progression. The most significant levels of these markers are detectable in recurrent and metastatic lesions (Bonkhoff et al., 1999b, 2000, 2001a) (Fig. 7). This indicates that metastatic and androgen-insensitive tumors are estrogen responsive and can use estrogens for their maintenance and growth to survive in an androgen-deprived milieu. It is noteworthy that the expression of the ERα and PR in the normal prostatic epithelium is restricted to the basal cell layer, which is androgen independent, proliferative active, and harbors prostatic stem cells (Bonkhoff et al., 1998b). The reappearence of the ERα and PR in advanced and androgen-insensitive tumors suggests that prostate cancer cells expressing these steroid receptors recapitulate some biological features of basal cells or prostatic stem cells.

In apparent contrast with the classic ERα, the ERβ variant is expressed extensively in primary and metastatic lesions without any clear correlation with the histological grade or the pathological stage (Fixemer et al., 2003). Nevertheless, the ERβ is partially lost in androgen-insensitive stages of the disease, which may reflect the androgen dependence of ERβ gene expression in prostate cancer (Fixemer et al., 2003). Although the precise role of the ERβ in prostate cancer remains to be established, most studies suggest that the ERβ exerts anti-proliferative effects by counteracting the stimulating effects of the ERα (Krege et al., 1998; Chang and Prins, 1999; Steiner et al., 2001). Irrespective of possible explanations, the progressive emergence of the ERα and PR during tumor progression provides a theoretical background for studying the efficiency of antiestrogens and antigestagens in the medical treatment of advanced prostate cancer. The current morphogenetic factors implicated in prostate cancer development and progression are summarized in Fig. 8.

40.6. CONCLUSION

Prostate cancer is a complex disease process involving phenotype, epigenetic, and genetic factors. In each methodological approach, knowledge of prostatic heterogeneity, histology, and

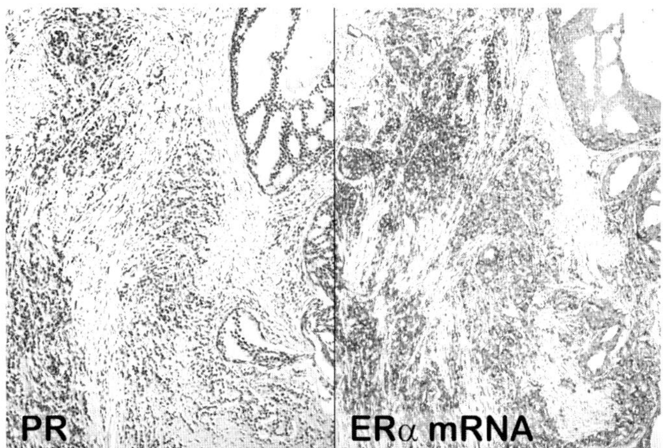

Fig. 7. ER and PR expression in recurrent prostatic adenocarcinoma. At least 30% of androgen-insensitive prostatic adenocarcinomas express the PR at significant levels (Bonkhoff et al., 2001a). The presence of the PR is associated with high steady-state levels of ERα mRNA. This indicates that such lesions have a functional ERα able to induce the PR.

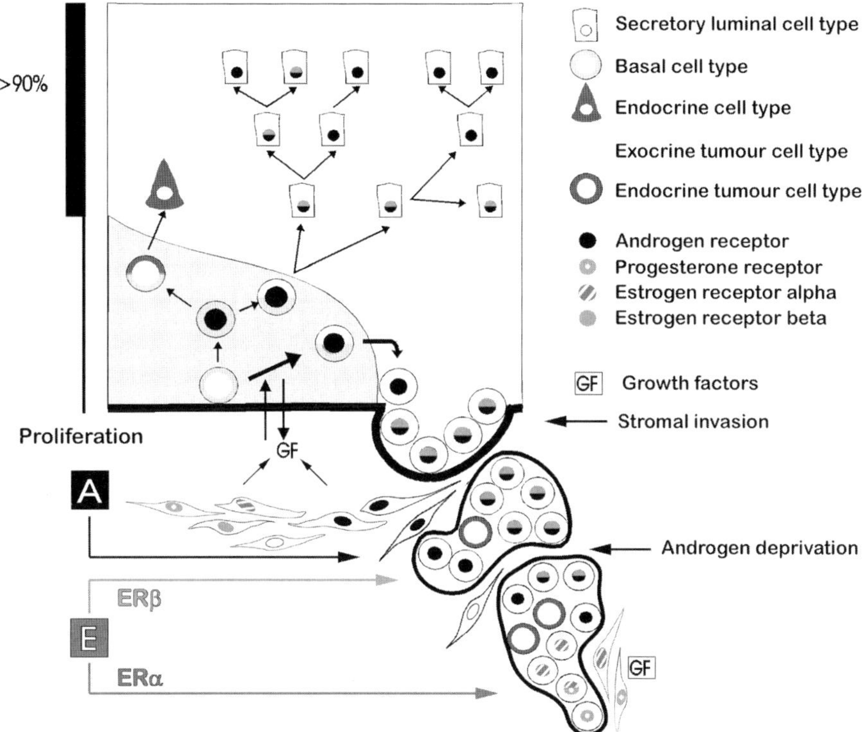

Fig. 8. Morphogenetic pathways implicated in prostate cancer development and tumor progression. Preinvasive phases of prostatic cancerogenesis are characterized by severe differentiation and proliferation disorders within the prostatic epithelial cell system (*see* text). Transformed precursor cells originating from the basal cell layer acquire exocrine features and produce an altered ECM. These newly formed (tumor-associated) basement membranes provide a supporting scaffold for penetration of the host tissue during stromal invasion and metastasis. Exocrine tumor cells (the most prevalent phenotype of prostatic adenocarcinoma) generally express the nuclear AR, 5α reductase isoenzymes 1 and 2, and the androgen-regulated ERβ. Thus, exocrine tumor cells remain androgen responsive even in hormone refractory stages of the disease. Point mutations in the steroid-binding domain of the AR gene, however, can seriously interfere with the normal function of the receptor protein. The progressive emergence of neuroendocrine tumor cells during tumor progression obviously reflects the differentiation potency of prostatic stem cells. Devoid of the nuclear AR, neuroendocrine tumor cells are androgen insensitive but produce neuroendocrine growth factors that can exert growth-promoting effects on adjacent exocrine tumor cells through a paracrine mechanism. The lack of proliferative and apoptotic activity in neuroendocrine tumor cells further contributes to their drug resistance. After androgen deprivation, prostate cancer cells acquire the ability to use estrogens and prosgestins for their own growth. The presence of the ERα and the estrogen-inducible PR in metastatic and recurrent lesions recapitulates some biological properties of basal cells and prostatic stem cells and provides a novel target for antiestrogens and antigestagens in the medical treatment of advanced prostate cancer.

morphogenesis is essential. This highlights the role of pathologists in contemporary prostate cancer research. The morphogenetic data reviewed here may provide a conceptual framework for studying the impact of biochemical and genetic factors on prostate cancer development and tumor progression.

ACKNOWLEDGMENT

This work was supported by the Deutsche Forschungsgemeinschaft.

REFERENCES

Berruti, A., Dogliotti, L., Mosca, A., et al. (2000) Circulating neuroendocrine markers in patients with prostate carcinoma. *Cancer* 88(11): 2590–2597.

Bonkhoff, H. (1996) Role of the basal cells in premalignant changes of the human prostate: a stem cell concept for the development of prostate cancer. *Eur. Urol.* 30:201–205.

Bonkhoff, H. (1998a) Analytical molecular pathology of epithelial-stromal interactions in the normal and neoplastic prostate. *Anal. Quant. Cytol. Histol.* 20(5):437-442.

Bonkhoff, H. (1998b) Neuroendocrine cells in benign and malignant prostate tissue: morphogenesis, proliferation, and androgen receptor status. *Prostate Suppl.* 8:18–22.

Bonkhoff, H. (2001) Neuroendocrine differentiation in human prostate cancer: morphogenesis, proliferation and androgen receptor status. *Ann. Oncol.* 12(Suppl. 2):141–144.

Bonkhoff, H. and Remberger, K. (1993) Widespread distribution of nuclear androgen receptors in the basal layer of the normal and hyperplastic human prostate. *Virchows Archiv. A Pathol. Anat.* 422: 35–38.

Bonkhoff, H. and Remberger, K. (1996) Differentiation pathways and histogenetic aspects of normal and abnormal prostatic growth: a stem cell model. *Prostate* 28(2):98–106.

Bonkhoff, H. and Remberger, K. (1998) Morphogenetic concepts of normal and abnormal growth of the human prostate. *Virchows Arch.* 433: 195–202.

Bonkhoff, H., Wernert, N., Dhom, G., and Remberger, K. (1991a) Basement membranes in fetal, adult normal, hyperplastic and neoplastic human prostate. *Virchows Archiv. A Pathol. Anat.* 418:375–381.

Bonkhoff, H., Wernert, N., Dhom, G., and Remberger, K. (1991b) Relation of endocrine-paracrine cells to cell proliferation in normal, hyperplastic and neoplastic human prostate. *Prostate* 18:91–98.

Bonkhoff, H., Wernert, N., Dhom, G., and Remberger, K. (1992) Distribution of basement membranes in primary and metastatic carcinomas of the prostate. *Hum. Pathol.* 23:934–939.

Bonkhoff, H., Stein, U., and Remberger, K. (1993a) Androgen receptor status in endocrine-paracrine cell types of the normal, hyperplastic, and neoplastic human prostate. *Virchows Arch. A Pathol. Anat. Histopathol.* 423(4):291–294.

Bonkhoff, H., Stein, U., and Remberger, K. (1993b) Differential expression of α-6 and α-2 very late antigen integrins in the normal, hyperplastic and neoplastic human prostate: simultaneous demonstration of cell surface receptors and their extracellular ligands. *Hum. Pathol.* 24: 243–248.

Bonkhoff, H., Stein, U., and Remberger, K. (1994a) Multidirectional differentiation in the normal, hyperplastic and neoplastic human prostate: simultaneous demonstration of cell specific epithelial markers. *Hum. Pathol.* 25:42–46.

Bonkhoff, H., Stein, U., and Remberger, K. (1994b) The proliferative function of basal cells in the normal and hyperplastic human prostate. *Prostate* 24(3):114–118.

Bonkhoff, H., Stein, U., Welter, C., and Remberger, K. (1995a) Differential expression of the pS2 protein in the human prostate and prostate cancer: association with premalignant changes and neuroendocrine differentiation. *Hum. Pathol.* 26(8):824–828.

Bonkhoff, H., Stein, U., and Remberger, K. (1995b) Endocrine-paracrine cell types in the prostate and prostatic adenocarcinoma are postmitotic cells. *Hum. Pathol.* 26:167–170.

Bonkhoff, H., Stein, U., Aumüller, G., and Remberger, K. (1996) Differential expression of 5 α-reductase isoenzymes in the human prostate and prostatic carcinoma. *Prostate* 29:261–267.

Bonkhoff, H., Fixemer, T., and Remberger, K. (1998) Relation between Bcl-2, cell proliferation and the androgen receptor status in prostate tissue and precursors of prostate cancer. *Prostate* 34(4):251–258.

Bonkhoff, H., Fixemer, T., Hunsicker, I., and Remberger, K. (1999a) Estrogen receptor expression in prostate cancer and premalignant prostatic lesions. *Am. J. Pathol.* 155(2):641–647.

Bonkhoff, H., Fixemer, T., Hunsicker, I., and Remberger, K. (1999b) Simultaneous detection of DNA fragmentation (apoptosis), cell proliferation (MIB-1), and phenotype markers in routinely processed tissue sections. *Virchows Arch.* 434(1):71–73.

Bonkhoff, H., Fixemer, T., Hunsicker, I., and Remberger, K. (2000) Estrogen receptor gene expression and its relation to the estrogen—inducible HSP27 heat shock protein in hormone refractory prostate cancer. *Prostate* 45(1):36–41.

Bonkhoff, H., Fixemer, T., Hunsicker, I., and Remberger, K. (2001) Progesterone receptor expression in human prostate cancer: correlation with tumor progression. *Prostate* 48:285–291.

Bostwick, D. G. (1996) Prospective origins of prostate carcinoma: prostatic intraepithelial neoplasia and atypical adenomatous hyperplasia. *Cancer* 78(2):330–336.

Chang, W. Y. and Prins, G. S. (1999) Estrogen receptor-beta: implications for the prostate gland. *Prostate* 40:115–124.

Cheng, L., Shan, A., Cheville, J. C., Qian, J., and Bostwick, D. G. (1998) Atypical adenomatous hyperplasia of the prostate: a premalignant lesion? *Cancer Res.* 58(3):389–391.

Culig, Z., Hobisch, A., Bartsch, G., and Klocker, H. (2000) Androgen receptor—an update of mechanisms of action in prostate cancer. *Urol. Res.* 28(4):211–219.

Di Sant'Agnese, P. A. (1992) Neuroendocrine differentiation in carcinoma of the prostate: diagnostic, prognostic and therapeutic implications. *Cancer* 70:254–268.

Di Sant'Agnese, P. A. and Cockett, A. T. (1996) Neuroendocrine differentiation in prostatic malignancy. *Cancer* 78(2):357–361.

Fixemer, T., Remberger, K., and Bonkhoff, H. (2002) Apoptosis resostamce of neuroendocrine phenotypes in prostatic adenocarcinoma. *Prostate* 53:118–123.

Fixemer, T., Bonkhoff, H., and Remberger, K. (2003) Differential expression of the estrogen receptor beta (ER() in human prostate tissue, premalignant changes, and in primary, metastatic and recurrent prostatic adenocarcinoma. *Prostate* 54:79–87.

Fong, C., Sherwood, E., Sutkowski, D., et al. (1991) Reconstituted basement membrane promotes morphological and functional differentiation of primary human prostatic epithelial cells. *Prostate* 19: 221–235.

Foster, C. S., Bostwick, D. G., Bonkhoff, H., et al. (2000) Cellular and molecular pathology of prostate cancer precursors. *Scand. J. Urol. Nephrol. Suppl.* 205:19–43.

Griffiths, K. (2000) Estrogens and prostatic disease: International Prostate Health Council Study Group. *Prostate* 45:87–100.

Knox, J. D., Cress, A. E., Clark, V., et al. (1994) Differential expression of extracellular matrix molecules and the alpha 6-integrins in the normal and neoplastic prostate. *Am. J. Pathol.* 145:167–174.

Koivisto, P., Kolmer, M., Visakorpi, T., and Kallioniemi, O. P. (1998) Androgen receptor gene and hormonal therapy failure of prostate cancer. *Am. J. Pathol.* 152:1–9.

Krege, J. H., Hodgin, J. B., Couse, J. F., et al. (1998) Generation and reproductive phenotypes of mice lacking estrogen receptor beta. *Proc. Natl. Acad. Sci. USA* 95:15,677–15,682.

Montironi, R., Bostwick, D. G., Bonkhoff, H., et al. (1996) Origins of prostate cancer. *Cancer* 78:362–365.

Myers, R. B. and Grizzle, W. E. (1996) Biomarker expression in prostatic intraepithelial neoplasia. *Eur. Urol.* 30:153–166.

Nagle, R. B., Brawer, M. K., Kittelson, J., and Clark, V. (1991) Phenotypic relationship of prostatic intraepithelial neoplasia to invasive prostatic carcinoma. *Am. J. Pathol.* 138:119–128.

Nagle, R. B., Hao, J., Knox, J. D., Dalkin, B. C., Clark, V., and Cress, A. E. (1995) Expression of hemidesmosomal and extracellular matrix proteins by normal and malignant human prostate tissue. *Am. J. Pathol.* 146:1498–1507.

Pföhler, C., Fixemer, T., Jung, V., Dooley, S., Remberger, K., and Bonkhoff, H. (1998) In situ analysis of genes coding collagen IV a1 chain, laminin b1 chain, and S-laminin in prostate tissue and prostate cancer: increased basement membrane gene expression in high grade and metastatic lesions. *Prostate* 36(3): 143–150.

Qian, J., Jenkins, R. B., Bostwick, D. G. (1998) Determination of gene and chromosome dosage in prostatic intraepithelial neoplasia and carcinoma. *Anal. Quant. Cytol. Histol.* 20(5):373–380.

Steiner, M. S., Raghow, S., and Neubauer, B. L. (2001) Selective estrogen receptor modulators for the chemoprevention of prostate cancer. *Urology*57(4 Suppl. 1):68–72.

Wernert, N., Seitz, G., and Achtstätter, T. (1987) Immunohistochemical investigations of different cytokeratins and vimentin in the prostate from fetal period up to adulthood and in prostate carcinoma. *Pathol. Res. Pract.* 182:617–626.

Xue, Y., Smedts, F., Debruyne, F. M., de la Rosette, J. J., and Schalken, J. A. (1998) Identification of intermediate cell types by keratin expression in the developing human prostate. *Prostate* 34(4):292–301.

41 Stem Cells in Lung Morphogenesis, Regeneration, and Carcinogenesis

ANK A. W. TEN HAVE-OPBROEK, MD, PhD, SCOTT H. RANDELL, PhD, AND BARRY R. STRIPP, PhD

Two levels of epithelial progenitors are involved in lung morphogenesis: multipotent undifferentiated cells (lung primordial cells) and pluripotent regiospecific (bronchial, bronchiolar, and alveolar) stem cells. The trachea and bronchi are lined by pseudostratified columnar epithelium (ciliated, mucous and basal cells). The bronchioles are lined by simple columnar epithelium (ciliated and Clara cells). Pulmonary neuroendocrine cells (PNECs) are also present. The alveolar ducts and sacs and the alveolar zone of the respiratory bronchioles are lined by cuboid alveolar type II cells and squamous alveolar type I cells. These regions all originate from evagination of ventral foregut endoderm containing lung primordial cells, into the surrounding visceral mesoderm with budding and branching, with their specific determination established as early as the pseudoglandular period of lung development. Tracheobronchial glands arise from specialized outpockets of basally situated surface cells. Bronchial stem cells are identified as label-retaining cells after pulsing with ^{3}H thymidine or BrdU and can be found after injury in tracheal gland ducts as well as systematically distributed along the surface of the trachea and bronchi. In vitro and in vivo studies suggest that basal cells and columnar cells retain plasticity to regenerate a complete mucociliary epithelium. Clara cells represent the principal progenitor pool of the bronchiolar epithelium and are an example of a transit-amplifying cell population. True stem cells may proliferate after injury and depletion of the Clara cell population and are located in association with PNECs and also at the junction between the bronchioli and alveolar ducts. The stem cell for the alveolar epithelium is the early embryonic type II cell, which can give rise to progeny that differentiate to type I or type II cells. The preferential distribution of type II cell clusters in the adult lung supports the existence of type II stem cell niches. According to the current theory, the regiospecific stem cells are the most relevant targets for transformation and thus the source of lung cancer. The mixed phenotype of many lung carcinomas and studies of bronchial carcinogenesis suggest origin from a common un- or retrodifferentiated stem cell. Prospective treatments for common lung diseases, transplantation, gene transfer, and tumor therapy would all be advanced by a greater understanding of lung stem cells.

41.1. INTRODUCTION

The lung epithelium is a key environmental interface whose normal function is essential to host well-being. It is susceptible to phenotypic modulation during the course of common maladies, is involved in genetic diseases such as cystic fibrosis, and is the source of the world's most prevalent lethal cancer. Elucidation of epithelial stem cells and their transit-amplifying progeny is a prerequisite toward improving our understanding of the pathogenesis, detection and treatment of significant diseases. However, compared to organs such as skin and gut, progenitor cells in the lung are poorly understood. This is likely due to regional complexity of lung cell populations, a low cell turnover in the steady state, difficulties associated with progenitor cell isolation, and inadequate culture models. The bulk of research in this area of lung biology has involved analysis of progenitor cells and their contribution to repair following injury, without discrimination between stem and transit-amplifying populations. However, recent evidence indicating that epithelial cells with stem cell–like properties exist has prompted renewed enthusiasm for their further characterization.

This chapter summarizes current knowledge about epithelial progenitors in the lung. We first review the cellular composition and developmental origin of the lung epithelium. Subsequently, we examine the process of regiospecific (i.e., bronchial, bronchiolar, and alveolar) differentiation, focusing on data regarding stem cell niches, stem and transit-amplifying populations, and phenotypic plasticity in the developing and adult lung. We examine recent advances in the field of cell isolation, culture systems, and animal models for testing growth capacity and differentiation potential. As will be shown, two levels of epithelial progenitors are probably involved in lung morphogenesis: (1) multipotent undifferentiated cells (called lung primordial cells) that design the

From: *Stem Cells Handbook*
Edited by: S. Sell © Humana Press Inc., Totowa, NJ

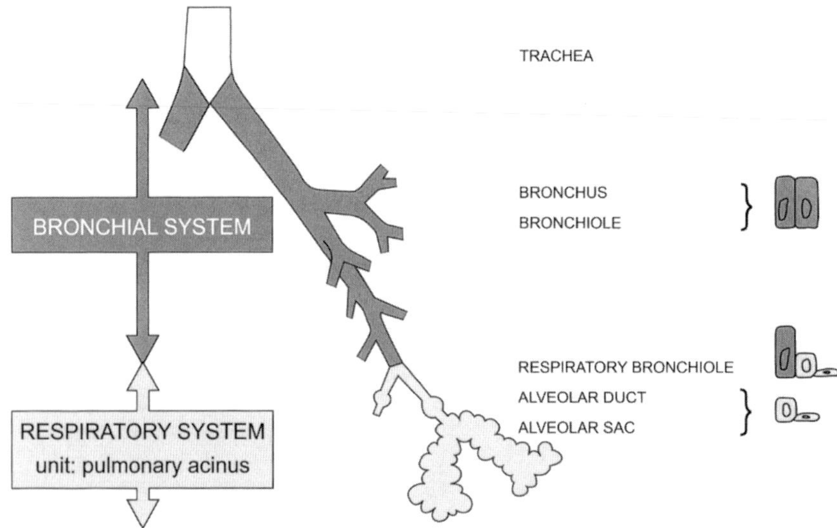

Fig. 1. Structural plan of the lower respiratory tract in mammals. The respiratory tract in the lung is divided into the air-conducting bronchial system and the gas-exchanging respiratory system (unit: pulmonary acinus). The bronchial system consists of the trachea, bronchi, and bronchioles (pseudostratified or simple columnar epithelium). The pulmonary acinus consists of respiratory bronchioles (columnar, cuboid, and squamous epithelium), and alveolar ducts and sacs (cuboid and squamous epithelium).

primitive structural plan of the respiratory tract, and (2) pluripotent lung-specific (i.e., bronchial, bronchiolar, and alveolar) stem and transit-amplifying populations necessary for regiospecific growth, differentiation, maintenance, and repair. We review the potential role of multi- and pluripotent progenitors in the development of bronchogenic carcinomas based on studies in humans and animal models and conclude with a discussion of relevant diseases for stem cell–mediated therapies and needs for future research.

It is beyond the scope of this chapter to review regulatory mechanisms involved in lung morphogenesis, regeneration, and carcinogenesis. However, where applicable, the reader is directed to recent reviews.

41.2. CELLULAR COMPOSITION AND ORIGIN OF THE LUNG EPITHELIUM

41.2.1. CELL TYPES AT DIFFERENT AIRWAY LEVELS The respiratory tract in the mammalian lung (Fig. 1) is subdivided into the air-conducting bronchial system (components: trachea, bronchi, and bronchioles) and the gas-exchanging respiratory system with its units, the pulmonary acini (components: respiratory bronchioles, alveolar ducts, and alveolar sacs). Light and electron microscopy have identified notable differences in cell composition in different airway generations between humans and common laboratory species. However, most of the relevant studies have been performed in experimental animals, and it is widely assumed that cell lineage is fundamentally the same in airways containing similar gland structures and cell types. The trachea and bronchi are lined by a pseudostratified columnar epithelium consisting primarily of ciliated, secretory (mucous or serous), and basal cells (Figs. 2A and 3). Tracheobronchial submucosal glands evaginate from the surface epithelium. Distally, glands disappear and the pseudostratified columnar epithelium becomes simple columnar or low columnar in the smallest bronchioles and comprises ciliated cells and nonciliated secretory Clara cells (Fig. 4A,B). Solitary pulmonary neuroendocrine cells (PNECs) and clusters of neuroendocrine cells known as neuroepithelial bodies (NEBs) are present throughout the bronchi and bronchioles (*see* Fig. 9). The alveolar ducts and sacs are lined by cuboid alveolar type II cells and squamous alveolar type I cells (Figs. 2C,D; 4E,F).The most proximal component of the pulmonary acinus, the respiratory bronchiole, is lined by a composite type of epithelium consisting of columnar ciliated and Clara cells, cuboid (immature) type II cells, and squamous type I cells (Figs. 2B; 4C,D). Studies in humans (ten Have-Opbroek et al., 1991; Plopper and ten Have-Opbroek, 1994), other primates (Tyler and Plopper, 1985; Plopper et al., 1989), and rodents (ten Have-Opbroek, 1986) have demonstrated that these bronchiolar and alveolar cell populations occupy distinctly different zones of the respiratory bronchiole's wall (pulmonary artery zone vs remaining wall). The transition between these bronchiolar and alveolar zones is abrupt (Figs. 2B and 4D).

41.2.2. DEVELOPMENTAL ORIGIN OF LUNG EPITHELIUM Lung development in mammals initiates with the formation of a protrusion from the ventral foregut endoderm into the surrounding visceral mesoderm (Balfour, 1881; Spooner and Wessels, 1970). It has been thought for more than a century (Balfour, 1881; Dubreuil et al., 1936; Loosli and Potter, 1959; Boyden, 1972; Hislop and Reid, 1974) that the lung anlage gives rise to the bronchial tree with its adult number of generations of branches (pseudoglandular period of lung development). After its completion, the pulmonary acini would arise at the distal ends of the tree. The respiratory bronchioles develop first (canalicular period). Alveolar type II cells appear at about the same time as alveolar type I cells, i.e., in the period just before birth when primitive alveoli (i.e., smooth-walled ducts and sacs) are formed, called the terminal-sac period. Alveolar outpocketings develop in the postnatal or alveolar period (reviewed by ten Have-Opbroek, 1981). Determination of the airways present in each period using light and electron microscopy and immunocytochemistry for SP-A has shown that primitive pulmonary acini lined by alveolar type II cells first appear in the pseudoglandular period. Such type II cells can only originate from the undifferentiated epithelium present at that time (ten Have-Opbroek, 1979, 1991).

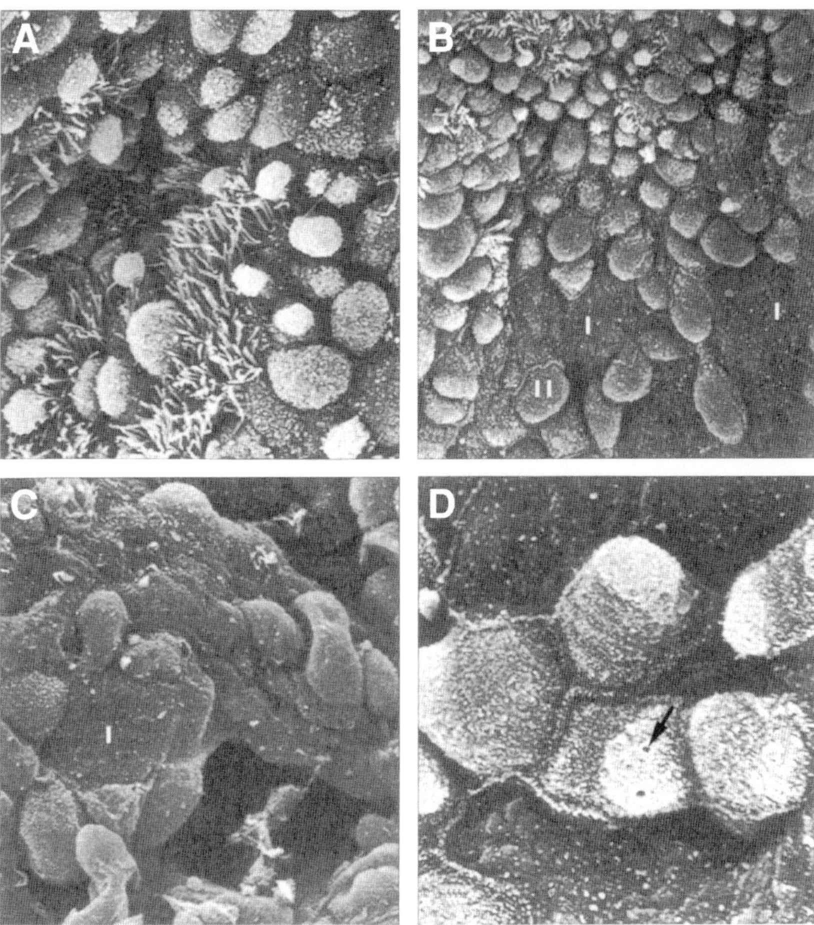

Fig. 2. Surface view of human lung epithelium at different airway levels (scanning electron microscopy). **(A)** Bronchus. The bronchial epithelium consists of columnar ciliated cells and nonciliated secretory cells (×3200). **(B)** Respiratory bronchiole. The bronchiolar zone with columnar ciliated and secretory cells (top left) joins tightly to the alveolar zone with type I and II cells (I, II) (×2300). **(C)** Alveolar wall showing squamous type I cells and globular type II cells (×4000). **(D)** Cluster of type II cells stuffed with microvilli. Pores (arrow) indicate where multilamellar bodies (MLBs) have been secreted into the alveolar lumen (×6000).

According to the present view (ten Have-Opbroek, 1981), the critical cell type in mammalian lung development is the undifferentiated progenitor cell that appears during the conversion of foregut endoderm into the epithelium of the lung anlage. We use the term *lung primordial cells* to indicate these *multipotent* undifferentiated progenitors of the adult lung epithelium. The lung primordium with its two lung buds (Fig. 5) appears in the pseudoglandular period of lung development, i.e., in the mouse embryo by d 9.5 and in the human embryo between the third and fourth weeks after conception. Each lung bud develops into a branching tubular system called the primordial system of the lung. This primordial system is lined by pseudostratified or simple columnar epithelium consisting of multipotent primordial cells. Because of regiospecific differences in the microenvironment, the proximal part of the primordial system differentiates into the air-conducting bronchial system (columnar bronchial and bronchiolar epithelium), and the distal part into the gas-exchanging respiratory system (cuboid, and later also squamous, alveolar epithelium). As indicated by the timing of type II cell differentiation, the latter process starts in the pseudoglandular period, i.e., on d 14 after conception in the mouse and between 11 and 12 wk after conception in the human.

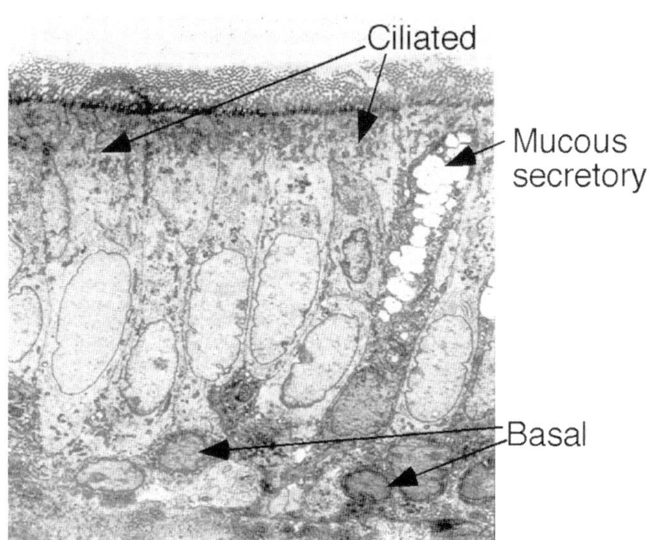

Fig. 3. Transmission electron micrograph of human bronchial epithelium illustrating predominant cell types: ciliated, mucous, and basal cells (× 3000). (Kindly provided by Dr. Robert R. Mercer).

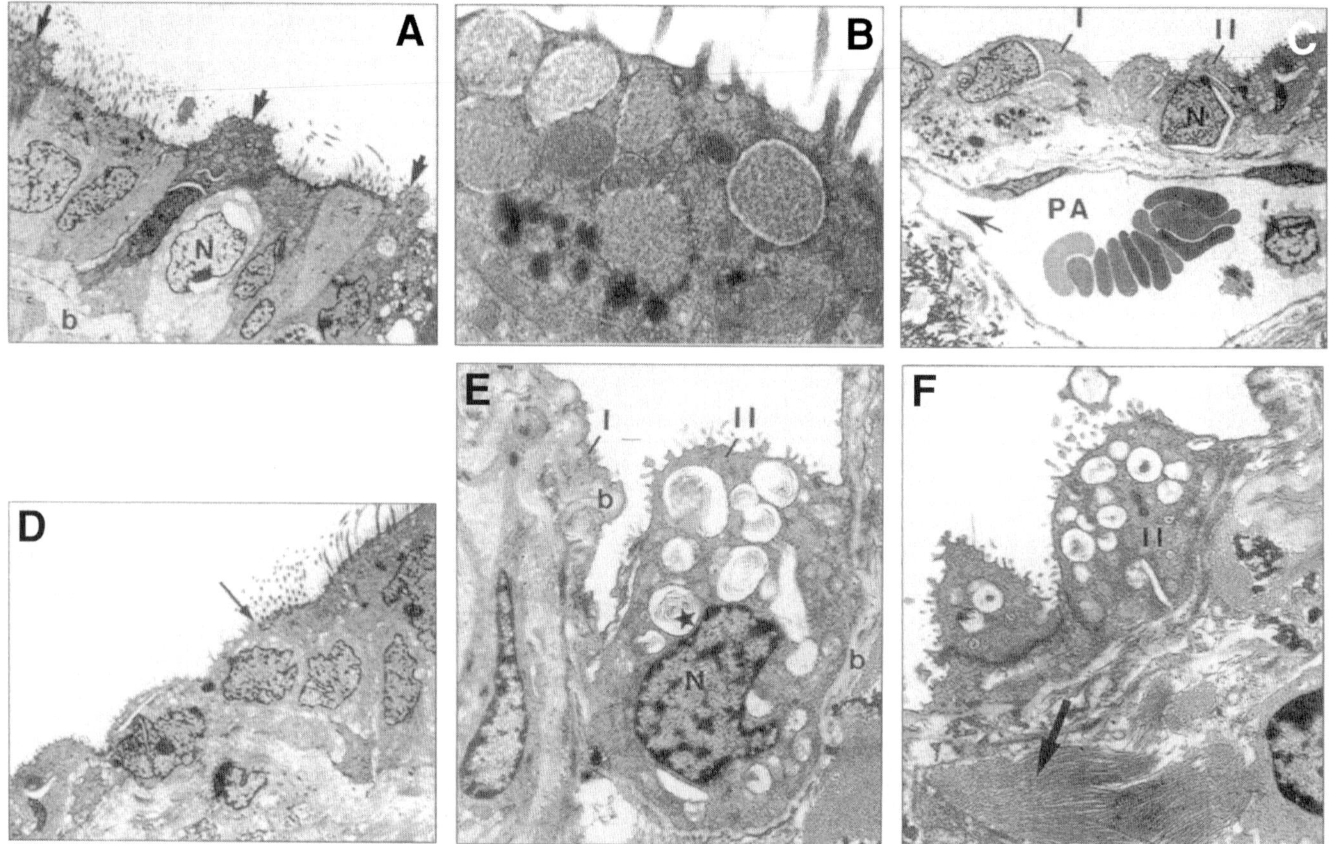

Fig. 4. Transmission electron micrographs of human bronchiolar and alveolar epithelium. **(A)** Terminal bronchiole lined by columnar ciliated and Clara (arrows) cells. N, nucleus; b, basal lamina (× 1400). **(B)** Detail of mucous granules in human Clara cells (×6000). **(C)** The alveolar zone of the respiratory bronchiole is lined by immature type I and II cells (I, II). PA, pulmonary artery branch flowing into a capillary (arrow) (×1500). **(D)** The transition (arrow) between the alveolar (left) and bronchiolar epithelial zones is abrupt (×1600). **(E,F)** Alveolar wall. The mature alveolar epithelium shows type II cells with MLB (E, star) and type I cells. Note the bundles of collagen fibers (F, arrow) in the human alveolar wall. (E) ×3500; (F)×2000.

As further explained in Subheading 41.3. there is increasing evidence that at least *three pluripotent regiospecific stem cell types* (i.e., bronchial, bronchiolar, and alveolar) that differentiate locally from the pulmonary primordial system are involved in the establishment of the adult lung epithelium. The differentiation of these stem cell types seems to be initiated at proximal airway locations and to proceed distally (Hogan et al., 1997; Shannon et al., 1998; Weaver et al., 2000; Wert et al., 1993). The detection of mRNA expression for SP-C in transgenic mouse lungs before d 14 after conception (Wert et al., 1993) suggests that the differentiation of primordial epithelium into alveolar epithelium is a gradual process and that a first step in the molecular determination may start (locally) at an earlier time point. Very likely, the differentiation, proliferation, and spatial organization of uncommitted epithelial progenitor cells are tightly regulated through reciprocal paracrine interactions with adjacent mesenchymal cells (Shannon et al., 1998; Lebeche et al., 1999; Warburton et al., 2000; Weaver et al., 2000). For other regulatory mechanisms, the reader is directed to recent reviews (Perl and Whitsett, 1999; Cardoso, 2001; Cardoso and Williams, 2001; Warburton et al., 2001, 2002).

41.2.3. DEFINITIONS FOR THE PROGENITOR CELLS OF THE NORMAL LUNG EPITHELIUM In this chapter, as before (ten Have-Opbroek, 1981), the term *lung primordial cells* refers to the multipotent undifferentiated epithelial progenitor cells that appear during the conversion of foregut endoderm into the epithelium of the lung primordium. We use the terms *bronchial, bronchiolar, and alveolar stem cells* to indicate the pluripotent lung-specific cell types that arise from the lung primordial epithelium and give rise to the bronchial, bronchiolar, and alveolar epithelia of the lung. Their lineages include transit-amplifying populations that perform most of the proliferation in the steady state and in response to injury.

41.2.4. DEFINITIONS FOR THE PROGENITOR CELLS OF LUNG CARCINOMA As defined previously (ten Have-Opbroek et al., 1993, 1994), we use the term *primordial-like cells of origin* to indicate the multipotent undifferentiated epithelial progenitor cells that appear during the conversion of normal adult lung epithelium to lung carcinoma. The term *tumor stem cell* refers to the pluripotent lung cell types that arise from the primordial-like cells of origin, occupy the dividing layers of the preneoplastic and neoplastic lesions, and generate the neoplasm.

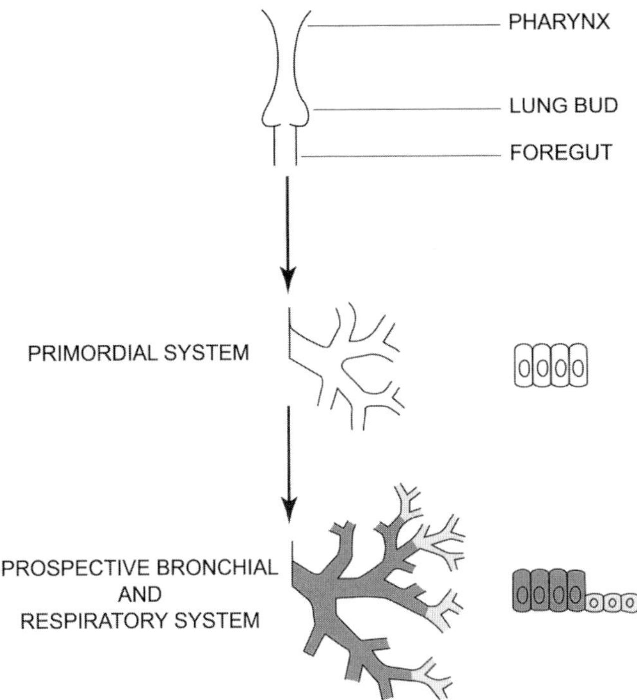

Fig. 5. Novel pathway of lung development proposed by ten Have-Opbroek (1981). Each of the two lung buds arising from the primitive foregut develops into a branching tubular system. This constitutes the development of the primordial system of the right and left lungs. The primordial system is lined by multipotent undifferentiated lung primordial cells (pseudostratified or simple columnar epithelium). Because of regiospecific differences in the microenvironment, the epithelium of the proximal part of the primordial system differentiates into the epithelium of the air-conducting portion (columnar bronchial and bronchiolar epithelium), and that of the distal part into the epithelium of the respiratory portion (cuboid, and later also squamous, alveolar epithelium). There is always a sharp demarcation between the air-conducting system and the respiratory system in developing (and adult) lungs.

41.3. REGIOSPECIFIC STEM CELLS IN LUNG MORPHOGENESIS AND REGENERATION

41.3.1. PLURIPOTENT BRONCHIAL STEM CELLS IN PROXIMAL CONDUCTING AIRWAYS (TRACHEA AND BRONCHUS)

41.3.1.1. Proximal Airway Development and Lineage Specification Embryologically, differentiated tracheobronchial epithelial cells arise from the multipotent columnar primordial cells that line the primordial system of the lung (*see* Subheading 41.2.2.). Such primitive cells may already express markers of multiple differentiated adult lung cells (Wuenschell et al., 1996). PNECs, once thought to be of ectodermal origin, but now considered endoderm derivatives, are the first recognizable differentiated cell type and are scattered within the epithelium (Boers et al., 1996). Ciliated and mucous goblet cells appear after PNECs (Jeffery et al., 1992). Typical basal cells appear last and thus are not the ontogenic precursors of secretory and ciliated cells (Plopper et al., 1986).

During development, tracheobronchial glands derive from specialized outpockets of basally situated surface cells (Bucher and Reid, 1961; Tos, 1968). Specific gland progenitor cells have not been characterized, and there are very few data regarding ongoing patterns of cell renewal in the adult gland acinar and duct system. It is possible that an interaction of multiple epithelial progenitors (Engelhardt et al., 1995) in combination with mesenchymal signals inducing transcription factors such as LEF-1 (Duan et al., 1999) is necessary to form glands. However, one report suggests clonal derivation of tracheal glands (Borthwick et al., 1999) and glandlike structures form in culture in the absence of mesenchyme. Cells in gland duct necks may serve as a reservoir of progenitors for the surface epithelium (Borthwick et al., 2001).

41.3.1.2. Lineage Specification in the Adult Trachea and Bronchus In normal humans, the tracheobronchial surface epithelium from just below the vocal cords down to 1-mm-diameter bronchioles is of the pseudostratified columnar type and consists primarily of basal, secretory, and ciliated cells (Mercer et al., 1994) (*see* Fig. 3). Although metabolic-labeling studies identifying dividing cells in this portion of the respiratory tract have been performed in animals for more than 35 yr (Blenkinsopp, 1967), cell lineage remains poorly understood. Cell turnover in the normal epithelium is extremely slow, with an approx 1% labeling index following a 1- or 2-h pulse of ^{3}H-thymidine or BrdU (Blenkinsopp, 1967; Donnelly et al., 1982; Breuer et al., 1990). However, there is vigorous proliferation and migration following injury (Erjefalt and Persson, 1997), and progressive phenotypic remodeling subsequent to chronic stimulation or injury (hyperplasia → metaplasia → dysplasia → cancer). Slow steady-state turnover and marked plasticity following injury have complicated the identification of stem cells among the proliferative cell compartment.

Of the three major cell types comprising the adult pseudostratified surface epithelium, basal and secretory cells are known to divide whereas ciliated cells are generally considered terminally differentiated. However, the latter point is controversial, and it has been proposed that injured ciliated cells, having lost their cilia, may reenter the dividing pool (Lawson et al., 2002). In repair of injury, secretory cells are often highly proliferative, and there is evidence that ciliated cells are derived from secretory cells (McDowell et al., 1985). Progenitor–progeny relationships during steady-state renewal of the pseudostratified epithelium in the adult are uncertain, but most investigations support a progenitorial role for basal cells (Blenkinsopp, 1967; Donnelly et al., 1982; Breuer et al., 1990; Boers et al. 1998), perhaps via an intermediate phenotype (Donnelly et al., 1982). The discrepancy between late developmental appearance of basal cells noted above, and their presumed progenitorial role in adult cell renewal suggests that repair of injury does not necessarily recapitulate ontogeny. Perhaps different paths are followed for the initial establishment of the respiratory epithelial phenotype vs its maintenance in the adult.

In tissues such as the epidermis and intestinal epithelium, it is widely appreciated that stem cells occupy distinct niches spatially proximal to the pool of dividing and differentiated progeny. The property of stem cells to cycle infrequently leads to the development of so-called label-retaining stem cells (LRCs). Following tracheal injury in mice, LRCs were found in tracheal gland ducts and systematically distributed along the surface of the lower, glandless trachea, suggesting the presence of stem cell niches (Fig. 6) (Borthwick et al., 2001). Characterization of the niches remains a goal of future research.

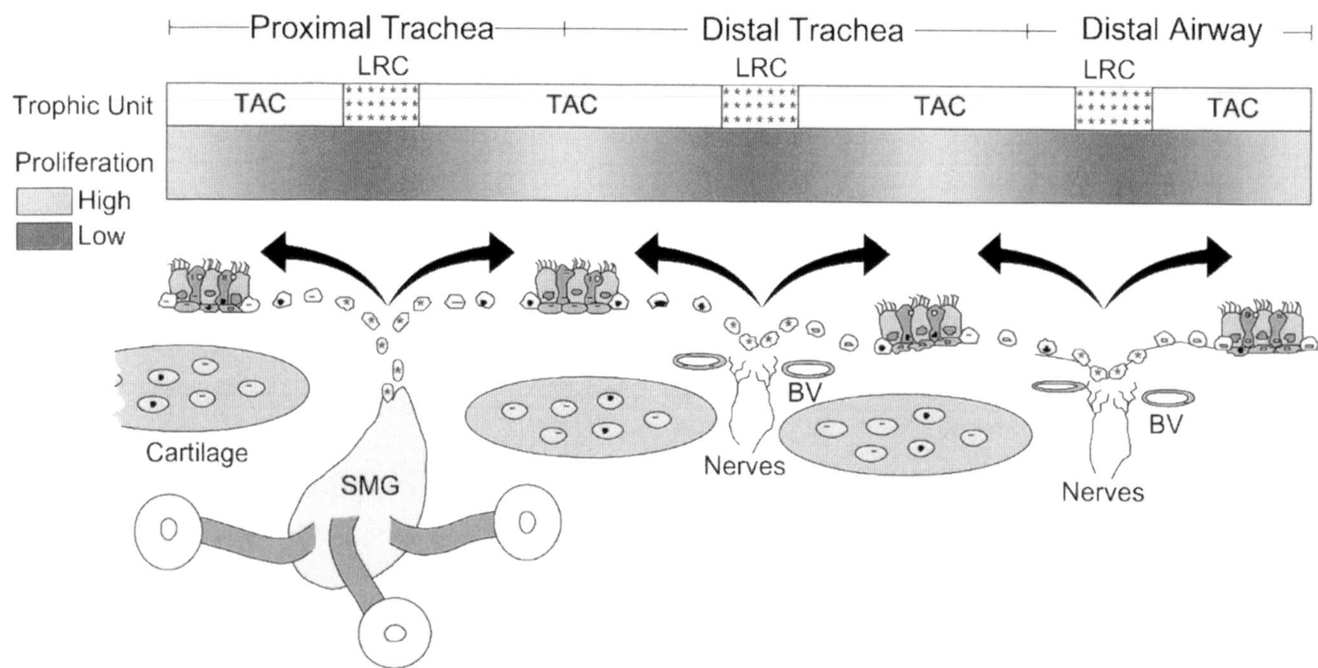

Fig. 6. Putative stem cell niches and progenitor-progeny relationships in murine airway epithelium. In the gland-containing upper trachea, LRCs that give rise to the transiently amplifying compartment (TAC) localize to submucosal gland (SMG) ducts. LRCs in the glandless lower trachea are systematically arrayed near the cartilage–intercartilage junction where blood vessels (BV) and nerves approach the epithelium. (Modified from Engelhardt, 2001, with permission.)

41.3.1.3. In Vivo Localization and Characterization of Bronchial Stem Cells Differentiated characteristics of bronchial cells in the steady-state airway have been exploited for the study of stem cell populations in the bronchus and trachea. As noted by Basbaum and Jany (1990), the tracheobronchial epithelium is strikingly remodeled in disease. Such plasticity can be accomplished in one of two ways. Several cell types proliferate during remodeling and daughter cells may follow different differentiation pathways from their parents. This is a type of plasticity, but within the specific tissue compartment. More recently, plasticity has grown to include differentiation across barriers of embryonic derivation previously thought impenetrable, such as from bone marrow to hepatocyte (Blau et al., 2001). A study by Krause et al. (2001) demonstrated integration of bone marrow–derived cells into the bronchus in lethally irradiated mice. However, the role of bone marrow–derived progenitors in normal turnover or repair in the trachea or bronchi in humans remains to be determined.

To study plasticity within the tracheobronchial epithelium, cells were separated into subpopulations for testing of growth and differentiation potential. Tracheal graft studies of basal cell fractions obtained by centrifugal elutriation (Inayama et al., 1988) and single-cell cloning (Inayama et al., 1989) strongly suggested that basal cells were capable of regenerating a complete pseudostratified mucociliary epithelium. Sorting based on cell-surface glycoconjugates showed that subsets of cells within both the basal and secretory cell compartment also regenerated a complete mucociliary epithelium (Randell et al., 1991; Liu et al., 1994). Others have concluded that only secretory cells (Johnson and Hubbs, 1990) or only basal cells (Ford and Terzaghi-Howe, 1992) were pluripotent stem cells. The reasons for the discrepancies are not clear, but the preponderance of the evidence is that cells within both the basal and secretory cell compartments of the pseudostratified portion of the airways divide and can serve as an actively proliferating and pluripotent stem cell pool in epithelial renewal and repair. Using retroviral vectors, Engelhardt et al. (1995) tagged human bronchial cells proliferating in vitro and studied their behavior in denuded rat tracheas placed in Nu/Nu mice. A restricted pattern of colony types suggested that basal cells and intermediate cells were the predominant progenitors and that goblet cells did not undergo extensive self-renewal. This study relied, however, on an initial cell culture step for retroviral labeling. Additional evidence for enrichment of progenitors within a specific morphological category is that the in vitro colony-forming efficiency of basal–cell enriched fractions of rat tracheal epithelial cells was, on average, fivefold greater than that of non–basal cells (Randell et al., 1991). However, cell culture may not have been truly indicative of the in vivo potential of the cells. Despite their potential flaws, these studies suggest a spectrum of progenitorial capacities among proliferation-competent airway epithelial cells and substantial proliferative reserve in diverse airway epithelial cell types. However, the assays currently used to measure proliferation capacity and differentiation potential may not discriminate between cells with modest and profound progenitorial ability and do not rule out the existence of bona fide stem cells that may be recruited in vivo under extreme conditions of epithelial damage.

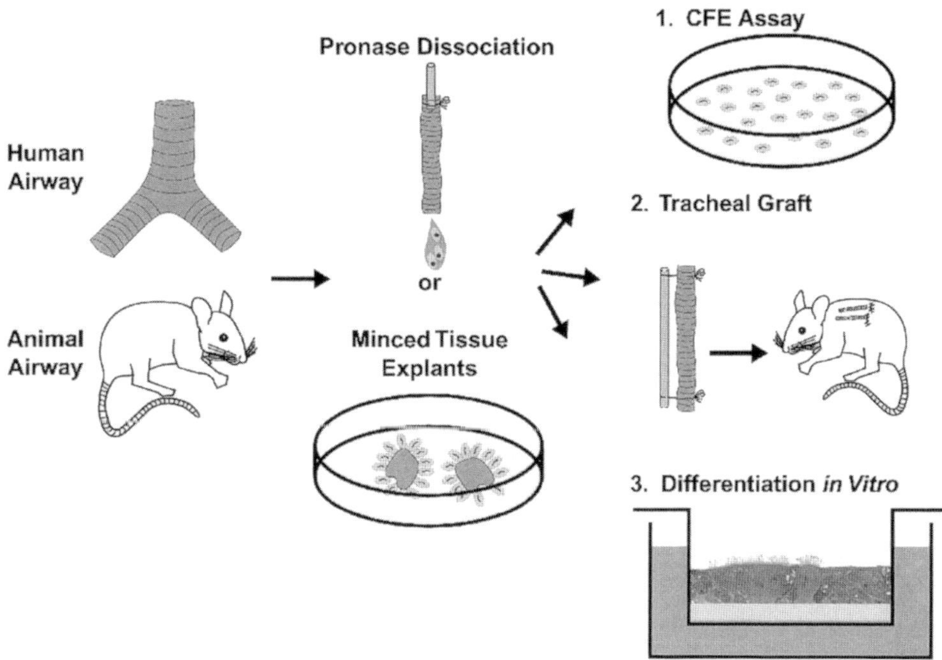

Fig. 7. Methods for defining the growth capacity and differentiation potential of tracheobronchial epithelial cells. Cells may be harvested from human or animal airways by protease dissociation or as outgrowths from tissue fragments. Cells can then be plated on plastic for assessment of colony-forming efficiency (CFE) or for monolayer growth to an undifferentiated phenotype. When the same cells are inoculated into denuded, devitalized tracheas and transplanted heterotopically in vivo or when grown in vitro on a porous support, the cells differentiate to a mucociliary epithelium.

41.3.1.4. Culture Systems for Tracheobronchial Epithelial Cells

41.3.1.4.1. Primary Cultures: Harvesting and Characterization Including Identifying Markers There is a long history of epithelial cell culture from normal human (Lechner et al., 1981) and animal (Heckman et al., 1978) airways, either as outgrowths from explanted tissue fragments or from protease-dissociated cell suspensions (reviewed in Gruenert et al., 1995) (*see* Fig. 7). Cells harvested by either method are usually mixed populations of epithelial, fixed mesenchymal, and migratory cells (leukocytes), and verification of epithelial derivation is usually performed by using the epithelial marker, cytokeratin, and/or by morphological characteristics during passage in defined medium. Although primary cells on plastic can be maintained for up to 30 passages (Lechner et al., 1981), transit-amplifying populations with stem cell properties such as the ability to form large colonies or identifying markers including specific receptors, telomerase, p63, or dye efflux have not yet been identified.

41.3.1.4.2. Cell Lines vs Primary Cultures There exist very large collections of cell lines originating from human lung cancers (Gazdar and Minna, 1996; Oie et al., 1996), or following chemical or physical mutagenesis of normal cells (Stoner et al., 1980), or by introduction of oncogenes (Yoakum et al. 1985; Masui et al., 1986a; Gruenert et al., 1988). Some of these cell lines were likely derived from the tracheobronchial portion of the airways or from glands and have proven invaluable for studies of cancer and lung cell biology, but they have generally not provided keen insights into normal progenitor–progeny relationships.

41.3.1.4.3. Growth and Differentiation In Vitro: Conditions for Differentiation and Markers of Differentiation Human tracheobronchial epithelial cells may be cultured as monolayers in conventional plastic dishes with or without feeder layers, generally using supplemented, serum-free, or low-serum medium (Lechner et al., 1981) due to the powerful terminal differentiating effects of serum on these cells, presumably because of transforming growth factor-β (TGF-β) (Masui et al., 1986b). On plastic, the original differentiated cells revert to a poorly differentiated cell type. In marked contrast, if the cells are grown as three-dimensional spheroids (Jorissen et al., 1989), or within collagen gels (Benali et al., 1993), or on specialized porous supports (Whitcutt et al., 1988), or in heterotopic tracheal grafts in vivo (Terzaghi et al., 1978) the cells undergo mucociliary differentiation characteristic of the epithelium in vivo (Fig. 8). This profound transition involves induction of tight junctional structures, differentiation of mucous and ciliated cells, and acquisition of characteristic epithelial physiological properties. Numerous genes and proteins are induced, including mucins (e.g., Bernacki et al., 1999) and proteins inherent to the apparatus of motile cilia (e.g., *see* Zhang et al., 2002). Retinoic acid is required to suppress or reverse squamous differentiation (Koo et al., 1999). In the near future, gene array and proteomics technology will likely reveal numerous additional differentiation-related genes and proteins and may elucidate the complex network of transcriptional factors underlying the dramatic phenotypic changes.

41.3.2. PLURIPOTENT BRONCHIOLAR STEM CELLS IN DISTAL CONDUCTING AIRWAYS

41.3.2.1. Distal Airway Development and Lineage Specification During early lung development, primitive airway epithelial cells that line the primordial system (*see* Subheading 41.2.3.) promiscuously express genes that are later restricted to distinct subsets of maturing epithelial cell types of the adult lung (Wert et al., 1993; Khoor et al., 1996; Wuenschell et al., 1996). It is unclear at what stage during development microenvironments (or niches) are established for the sequestration and maintenance of stable stem cell populations through adulthood. Two microenvironments have been identified in the adult bronchiolar epithelium that maintain cells with characteristics of stem cells, the NEB and the bronchoalveolar duct junction (BADJ) (Reynolds et al., 2000b; Hong et al., 2001; Giangreco et al., in press). More important, PNECs are among the first to differentiate in the developing airway epithelium (Sorokin et al., 1982), raising the possibility that their early appearance may function to sequester stem cells at this stage in airway maturation.

41.3.2.2. Lineage Specification in the Adult Bronchiole Lung progenitor cells have been identified based on their contribution to epithelial renewal following pollutant-induced injury to terminally differentiated cell populations (Evans et al., 1976, 1978; Lum et al., 1978). These studies have been performed principally within lungs of rodent species and led to the conclusion that nonciliated bronchiolar (Clara) cells serve as the progenitor population for renewal of the airway epithelium following injury to ciliated cells. In response to airway injury, residual Clara cells undergo a process of morphological dedifferentiation, involving loss of secretory granules and smooth endoplasmic reticulum, prior to entry into the cell cycle (Evans et al., 1978). Even though this process has not been as well studied in the human lung, Clara cells do represent the principal progenitor cell pool of the distal bronchiolar epithelium of humans (Boers et al., 1999). However, methods used for identification of progenitor cells do not allow for the distinction between stem and transit-amplifying populations. Moreover, the bulk of the progenitor pool that is activated to enter the cell cycle either in the steady state or following injury targeted to terminally differentiated cells is represented by the transit-amplifying population. The location, identity, and even existence of cells of the bronchiolar epithelium with properties of stem cells have only recently been investigated. Approaches that have been used for the identification and localization of airway stem cell populations are discussed in depth next.

41.3.2.3. In Vivo Localization and Characterization of Bronchiolar Stem Cells Differentiated characteristics of Clara cells in the steady-state airway have been exploited for the study of bronchiolar stem cell populations. Properties of Clara cells that identify them as transit-amplifying cells include their differentiated phenotype, the ease with which they can be activated to proliferate, and their ubiquitous distribution within bronchioles of the mammalian lung (Plopper et al., 1980, 1983). Analysis of stem cell populations within regenerative epithelia has required the development of strategies for stem cell activation and methods to distinguish their contribution toward epithelial renewal from that of the more abundant transit-amplifying cell populations.

41.3.2.3.1. Chemically Induced Depletion of Transit-Amplifying Cells Differentiated Clara cells of the adult rodent lung are particularly active in the metabolism of lipophilic compounds through phase I oxidation reactions catalyzed by cytochrome P450 (CYP) isoenzymes. Among epithelial cells of the lung, members of the CYP superfamily of genes are expressed either preferentially or exclusively within differentiated Clara cells in a species-specific manner. One such CYP isoenzyme in mice, CYP2F2, metabolizes the xenobiotic compound naphthalene to a toxic epoxide (Buckpitt et al., 1992). Conjugation of bioactivated naphthalene metabolites to intracellular macromolecules results in dose-dependent Clara cell toxicity leading to epithelial cell necrosis and a dramatic decline in the abundance of Clara cell–specific gene products within the lung (Plopper et al., 1992; Stripp et al., 1995; Van Winkle et al., 1995). Bronchiolar repair that follows severe Clara cell depletion involves proliferation of naphthalene-resistant epithelial cells that are located at discrete sites within conducting airways (Stripp et al., 1995). Regenerative foci localize to clusters of pulmonary neuroendocrine cells, termed *NEB's* (Fig. 9), that tend to localize to airway branch points of all but terminal bronchiolar airways (Reynolds et al., 2000b; Giangreco et al., 2002). Regeneration of injured terminal bronchiolar epithelium occurs at sites defined by the junction between bronchiolar and alveolar ducts, the BADJ. Unlike NEB-associated regenerative domains within more-proximal bronchioles, BADJ-associated epithelial renewal occurs independently of NEB involvement (Giangreco et al., 2002). However, a property of subpopulations of cells maintained within both NEB and BADJ microenvironments is a low frequency of proliferation as measured by the long-term retention of labeled DNA precursors incorporated during S-phase traversal (Hong et al., 2001; Giangreco et al., 2002). As such, chemically induced progenitor cell ablation has revealed that stem cells of the conducting airway epithelium exhibit properties of pollutant resistance, multipotent differentiation potential, and long-term retention of DNA precursors. A limitation of these studies is the inability to identify precisely the molecular phenotype of cells that exhibit some or all of these characteristics of stem cells, and the inability to demonstrate definitively a precursor-progeny relationship among cells contributing to the early proliferative response to naphthalene exposure. This caveat is exemplified by the finding that both Clara cell secretory protein (CCSP) and calcitonin gene–related peptide (CGRP)-expressing populations of NEB-associated regenerative cells show proliferative potential after naphthalene-induced Clara cell ablation in vivo and each population includes a labeled DNA precursor-retaining subset (Reynolds et al., 2000a; Hong et al., 2001). A model describing cellular interactions and differentiation potential within the NEB microenvironment, in which PNECs may serve as either the stem cell or a population of cells contributing to the microenvironment necessary for maintenance of a variant, pollutant-resistant population of CCSP-expressing (CE) cells, is proposed in Fig. 10.

41.3.2.3.2. Transgenic and Knockout Mouse Models to Investigate Bronchiolar Stem Cells Transcriptional regulatory elements have been defined for a number of genes expressed within the pulmonary epithelium that allow cell type–specific expression of heterologous genes in lungs of transgenic mice (reviewed by Ho, 1994). Most relevant to the analysis of bronchiolar progenitor and stem cell populations are elements associated with the CCSP gene (DeMayo et al., 1991; Hackett and Gitlin, 1992; Stripp et al., 1992; 1994; Sandmoller et al., 1994; Ray et al., 1995). Transgenic mice have been generated in which the SV40 large T antigen has been placed under the regulatory control of the CCSP promoter,

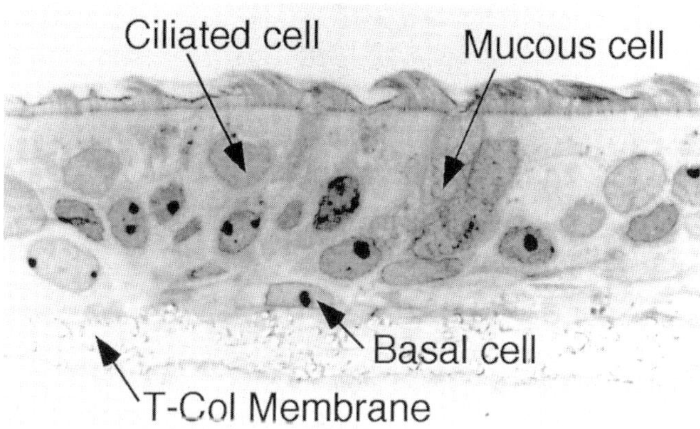

Fig. 8. Light micrograph of human bronchial epithelial cells cultured at air-liquid interface on porous support showing differentiation to mucociliary phenotype resembling in vivo epithelium (×1500).

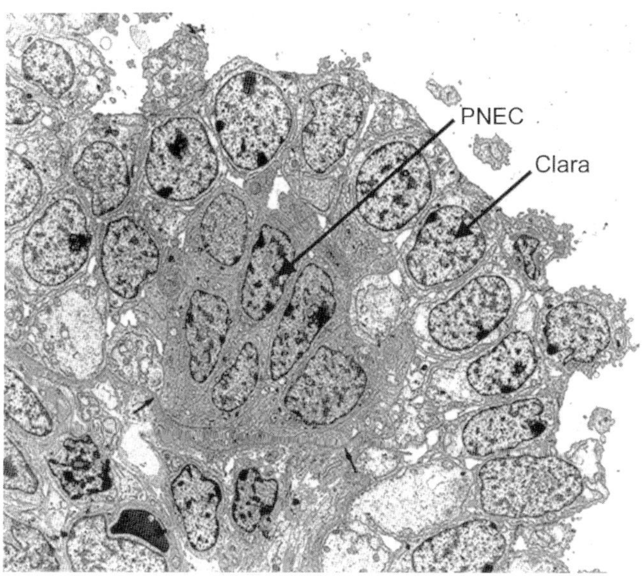

Fig. 9. Epithelial cell types present within NEB of human lung. An innervated cluster of PNECs is located in close association with the basement membrane. Overlying nonciliated Clara cells form a cap limiting access of the PNEC population to the airway lumen (×3000). (Contributed by Dr. M. Stahlman, from Stahlman and Gray, 1984.)

leading to the generation of lung tumors originating from CE airway cells (DeMayo et al., 1991; Sandmoller et al., 1994). In other studies, transgenic mice expressing herpes simplex virus thymidine kinase (HSVtk), a pro-toxin gene, under the control of the mouse CCSP promoter have been used to define contributions made by CE cells to epithelial renewal (Reynolds et al., 2000b; Hong et al., 2001). Complete ablation of CE cells could be achieved through exposure of CCSP-HSVtk transgenic mice to gancyclovir, allowing assessment of the potential for CGRP-expressing populations to proliferate independently of CE cells and effect repair of the injured epithelium. Under these conditions, CGRP-expressing cells proliferated but failed to restore depleted CE cell populations (Hong et al., 2001). Based on these studies, it can be concluded that CCSP-expressing cells of the NEB microenvironment serve as stem cells for bronchiolar renewal.

At present, reciprocal experiments have not been performed to define functional roles for PNECs in stem cell maintenance and function in adulthood. Mice have been established that lack functional copies of the proneural gene achete-scute homolog 1 (*ash1*), a consequence of which is not only aberrant development of the nervous system but also the absence of PNEC differentiation during lung development (Borges et al., 1997). More important, even though mash1–/– mice die of respiratory failure at birth, development and differentiation of the conducting airway epithelium appears to be unaffected by the lack of PNECs. However, the inability to perform functional assays and the lack of molecular markers unique to bronchiolar stem cells has precluded careful evaluation of stem cell maintenance in airways of mash1–/– mice.

41.3.2.4. Culture Systems for Bronchiolar Epithelial Cells

41.3.2.4.1. Primary Cultures The lack of detailed information regarding the molecular phenotype of bronchiolar stem cells represents a barrier to the development of culture methods for their in vitro maintenance. However, as shown next, in vitro culture methods have been established for the short-term maintenance of bronchiolar progenitor cell populations including both Clara cells and PNECs.

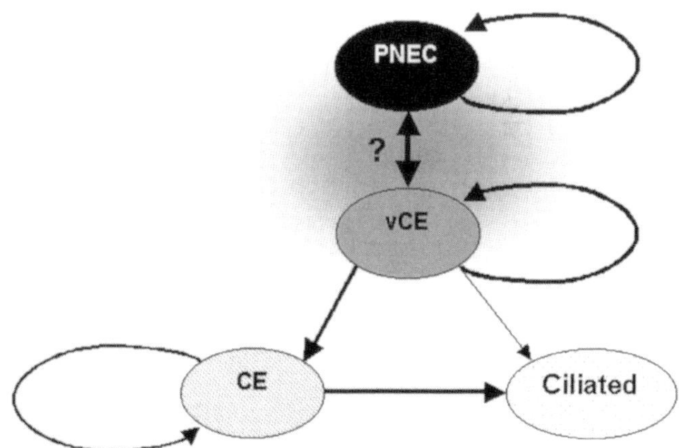

Fig. 10. Hypothetical model of cell lineage relationships within bronchiolar epithelium. Clara cells represent an abundant population of cells that serve as progenitor cells for renewal of the bronchiolar epithelium following injury to ciliated cells (Evans et al., 1978). Naphthalene-induced depletion of CE cells is associated with proliferation of latent PNECs located within neuroepithelial bodies (gray shaded zone) (Reynolds et al., 2000b). Proliferating CE cells that are associated with NEBs may be regenerated either through differentiation of PNECs or through proliferation of variant (naphthalene-resistant) CE cells (vCE) located within the NEB microenvironment. Studies using transgenic mice allowing conditional ablation of all CE populations implicate vCE cells as the regenerative population contributing to renewal following naphthalene exposure and suggest that PNECs and/or other unique properties of the NEB microenvironment contribute to their maintenance in the steady-state airway (Hong et al., 2001).

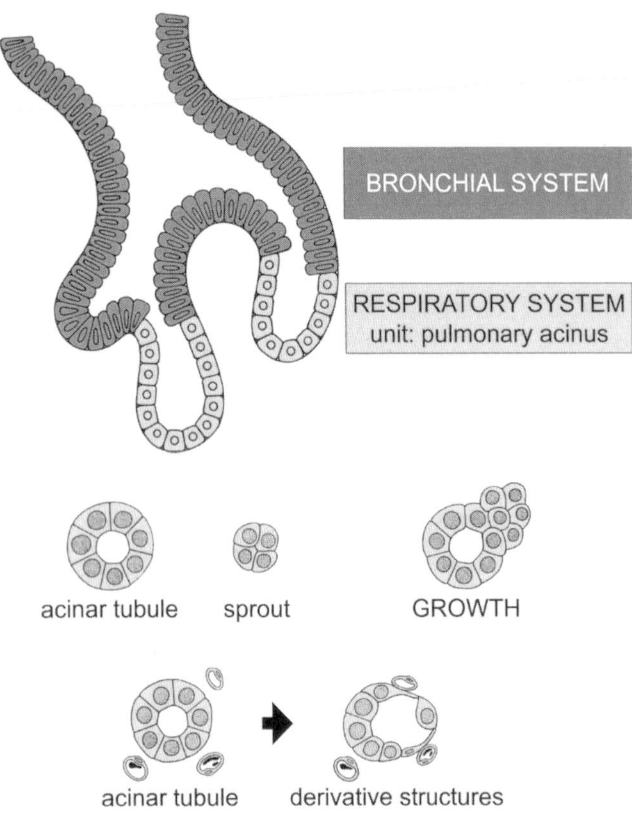

Fig. 11. Development of pulmonary acinus in mammals. Pluripotent type II stem cells, which first appear in the pseudoglandular period of lung development, play a key role in the development and maintenance of the respiratory system (ten Have-Opbroek, 1981). The basic structure in pulmonary acinus formation before and after birth is the acinar tubule or sprout (lining: type II cells). Growth of the pulmonary acinus takes place by budding (proliferation) of type II cells. The cuboid type II cells also give rise to squamous type I cells, a process that takes place in association with developing endothelium.

41.3.2.4.2. Methods for Isolation and Propagation of Clara Cells Clara cells have been isolated from rabbit, mouse, and rat lungs at a purity that varies from 70 to 85%. Methods typically involve limited proteolysis of the airway epithelium through intratracheal instillation of a protease-containing solution, followed by centrifugal elutriation for the purification of a Clara cell–enriched fraction of epithelial cells (Haugen and Aune, 1986; Hook et al., 1987; Walker et al., 1989; Belinsky et al., 1995; Nord et al., 1998). Corti et al. (1996) developed an alternative approach involving the instillation of low-melting-temperature agarose prior to introduction of dispase to limit the proteolysis of alveolar epithelia, yielding preparations of Clara cells with similar purity to those isolated through centrifugal elutriation methods. However, once isolated, Clara cells rapidly lose their differentiated characteristics during culture (Haugen and Aune, 1986; Nord et al., 1998). Isolation and maintenance of Clara cells in their differentiated state requires the use of reconstituted in vivo models such as denuded rat tracheas that are seeded with isolated Clara cells and maintained subcutaneously in the flank of nude mice (Hook et al., 1987). Alternative approaches that have been employed to characterize bronchiolar epithelial cells in vitro have involved the development of methods for airway microdissection and culture (Plopper et al., 1991). Microdissected mouse airways maintain differentiated functions for prolonged periods in culture in addition to retaining the competence to effect epithelial renewal following injury (Plopper et al., 1991; Van Winkle et al., 1996).

41.3.2.4.3. Methods for Isolation and Propagation of PNECs Methods for the isolation of PNECs are very similar to those just described for Clara cells. However, their source and selective enrichment procedures are critical due to the low abundance within the normal lung (Speirs and Cutz, 1993). PNECs have most commonly been isolated from either neonatal rabbit or hamster lung with subsequent purification of isolated PNECs from crude enzymatic digests through either gradient centrifugation or use of monoclonal antibodies coupled to magnetic beads (Cutz et al., 1985; Speirs et al., 1992). Culture conditions have been established that maintain many differentiated characteristics of PNECs such as the formation of dense core granules and production of various neuroendocrine cell–specific markers such as neuron-specific enolase, calcitonin, and serotonin (Cutz et al., 1985; Linnoila et al., 1993).

41.3.3. PLURIPOTENT ALVEOLAR TYPE II STEM CELLS IN THE PULMONARY ACINI

41.3.3.1. Development and Lineage Specification of Pulmonary acinus As reviewed in Subheading 41.2.2., type II cells differentiate from lung primordial cells at an early stage of lung development (pseudoglandular period). Because these early type II cells are the first and only type of alveolar epithelial cell to occur in the early prenatal lung, they must also generate the alveolar type I cell, and thus the entire pulmonary acinus lining. This conclusion for the mouse (ten Have-Opbroek, 1979) also holds true for the rat (Otto-Verberne and ten Have-Opbroek, 1987), the rhesus monkey (ten Have-Opbroek and Plopper, 1992), and the human (Otto-Verberne et al., 1988, 1990; Tebar et al., 2000b). It is supported by autoradiographic studies of alveolar epithelial renewal in late prenatal and postnatal rat (Kauffman et al., 1974; Adamson and Bowden, 1975; Evans et al., 1975) and adult mouse (Adamson and Bowden, 1974) lungs.

These findings support the view (Fig. 11) that pluripotent type II stem cells play a key role in the development and maintenance of the respiratory system in connection with its function (i.e., respiration). First, these cells generate the basic components of the primitive pulmonary acinus (i.e., the acinar tubule or sprout). Growth of the acinus takes place by proliferation ("budding") of type II cells. Second, these cells may give rise to type II cell lineages or become terminally differentiated type I cells. The transformation from cuboid type II cells to flatter type I cells takes place locally, in association with developing endothelium. As a result, the acinar tubules transform into wider derivative structures (smooth-walled ducts and sacs) lined by cuboid type II cells and more or less squamous type I cells. After birth, the smooth-walled ducts and sacs acquire alveolar outpocketings according to the same principle—type II cell budding. The type I cells in the lining of the pulmonary acini form part of the blood-air barrier of the lung, whereas the type II cells are responsible for further alveolar epithelial growth and regeneration. Alveolar epithelial differentiation is possibly mediated by wild-type p53 (Tebar et al., 2001b) and the Wnt signaling pathway (Tebar et al., 2001a). For other regulatory mechanisms, *see*, e.g., the review by Cardoso (2001).

41.3.3.2. Lineage Specification in the Adult Acinus Type II cells display a preferential distribution in normal adult lungs. In the alveolar septa, they usually occur grouped in twos or threes, often in a region where several adjacent septa meet (Sorokin, 1966). Other sites of type II concentrations include the borders of the connective-tissue septa of the lung and the pleura. These observations support the existence of type II stem cell niches. In agreement with this view, we found type II cell patterns as found in embryonic lungs within the septa of lungs from patients with pulmonary fibrosis (A. A. W. ten Have-Opbroek, personal communication).

In studies on alveolar regeneration in adult rabbits (Ebe, 1969), regenerating alveoli were found to originate from respiratory bronchioles. This is not surprising because these airways contain a specific zone with immature type II cells, which may function as a stem cell reservoir for type I cell renewal and alveolar regeneration (*see* Subheading 41.2.1.). This new insight into the architecture and cellular composition of respiratory bronchioles is relevant for studies of, e.g., small airway diseases and carcinogenesis.

Together, the developmental (*see* Subheading 41.3.3.1.) and adult data support the concept that type II cells are an actively proliferating population of alveolar stem cells that most likely include a stem cell pool necessary for development and regeneration of the acinus.

41.3.3.3. In Vivo Localization and Characterization of Type II Stem Cells

41.3.3.3.1. Adult and Developing Lungs Differentiated characteristics of type II cells in the steady-state acinus have been exploited for the study of type II stem cell populations. One of the main functions of type II cells in the adult lung is the synthesis and secretion of pulmonary surfactant, a lipid-protein complex that prevents alveolar collapse at end expiration after birth (Goerke, 1974; Batenburg, 1992). The role of the surfactant-associated proteins therein is the subject of several reviews (*see*, e.g., Weaver, 1991). Distinctive features of type II cells in the adult lung include the approximately cuboid cell shape, the large and roundish nucleus, cytoplasmic staining for the surfactant proteins SP-A and SP-C, and the presence of MLBs and/or their precursory forms (ten Have-Opbroek, 1975; ten Have-Opbroek et al., 1991). MLBs are involved in surfactant storage and secretion (*see* Figs. 2D and 4E,F). Based on these criteria, pluripotent type II stem cells can be detected in the developing lung as early as the pseudoglandular period. At that time, the pulmonary acinus consists of a few generations of acinar tubules lined by type II cells (Figs. 12A,B). Such early type II cells already stain for SP-A (Fig. 12C) and SP-C (Wert et al., 1993) and contain precursory forms of MLB in their cytoplasm (ten Have-Opbroek, 1979; ten Have-Opbroek et al., 1988, 1990b; ten Have-Opbroek and Plopper, 1992).

In studies in humans and other larger mammals, the term *Clara cell* is frequently used to indicate cuboid cells present in centriacinar airways. As a result, there is much confusion between Clara cells and type II cells. According to Clara's (1937) description, Clara cells have a tall columnar shape, an oblong nucleus, and discrete small granules in the apical cytoplasm. Moreover, they occur within the ciliated/nonciliated columnar epithelium (*see* Fig. 4A). In this sense, the term *Clara cell* is already used in other species such as rodents. Finally (Singh and Katyal, 1984; ten Have-Opbroek and De Vries, 1993), these cells stain for Clara cell–specific protein (CCSP). Together, these criteria allow for a clear distinction from type II cells.

41.3.3.3.2. Animal Models to Investigate Type II Stem Cells Several groups have made use of the rat (Brandsma et al., 1994a) or mouse (Cilley et al., 1997) model of pulmonary hyperplasia and diaphragmatic hernia to study the role of type II cells in normal and abnormal lung development. Hyperplasia was induced in fetuses by administration of the herbicide 2,4-dichlorophenyl-*p*-nitrophenylether (Nitrofen) to pregnant Sprague-Dawley rats (100 mg on d 10 of gestation) or to CD-1 mice (25 mg on d 8 of gestation). The rat studies suggest that lung hypoplasia is caused by a retardation in the differentiation of type II cells into type I cells and thus in the development of the air spaces. Factors mediating this differentiation process include thyroid hormone (Brandsma et al., 1994b) and vascular endothelial growth factor (Zeng et al., 1998). Very likely, Nitrofen decreases the binding of triiodothyronine to its receptor (Brandsma et al., 1994b).

41.3.3.4. Culture Systems for Type II Cells

41.3.3.4.1. Primary Cultures: Harvesting and Characterization Including Identifying Marker and Cell Lines Type II cells have been isolated from fetal rat (Batenburg et al., 1988) and mouse (Oomen et al., 1990) lungs using density gradient centrifugation and differential adherence in monolayer culture. The purity was 91 and 79% for cells isolated from 19- and 21-d fetal rat lungs, respectively, and at least 75% for those from fetal mouse lungs. Alcorn et al. (1997) isolated type II cells by collagenase digestion of human fetal lung tissue that had been maintained in organ culture in the presence of dibutyryl cyclic adenosine monophosphate for 5 d. Other studies have made use of organoid (Zimmermann, 1987) and organ (Chinoy et al., 2000; Schwarz et al., 2000; Bragg et al., 2001; Keijzer et al., 2001) culture systems. In the organoid culture, lungs from mouse fetuses were first dissociated enzymatically into single cells and then grown at high density at the medium/air interface. In the organ cultures, whole fetal lungs were placed on membranes, whereas one study (Schwarz et al., 2000) performed such cultures in an allograft model. In addition, studies have used existing type II cell lines (*see*, e.g., Stearns et al., (2001).

Primary markers used to identify type II cells in the cultures included presence of type II cell characteristics as shown by light or phase contrast microscopy, presence of MLBs or their precursory forms as shown by histochemical osmiophilic body staining or transmission electron microscopy, positive immunostaining for SP-A or SP-C proteins and cytokeratins, SP-A or SP-C mRNA expression, positive staining for *Maclura pomifera* lectin, determination of the phospholipid composition (major class: phosphatidylcholine), and/or a high degree of disaturation in phosphatidylcholine as shown by [*Me*-^{3}H]choline incorporation.

41.3.3.4.2. Type II Cell Cultures and Viability The type II cell isolations from fetal lungs (Batenburg et al., 1988; Oomen et al., 1990) provided a high yield of the desired cells and easy access to viable type II cells that had been in culture for a short period (1.5–2 d). The isolations by Alcorn et al. (1997), which started from an organ culture for 5 d, facilitated the maintenance of morphological and biochemical properties of type II cells for up to 2 wk. Type II cells present in organoid fetal mouse lung cultures

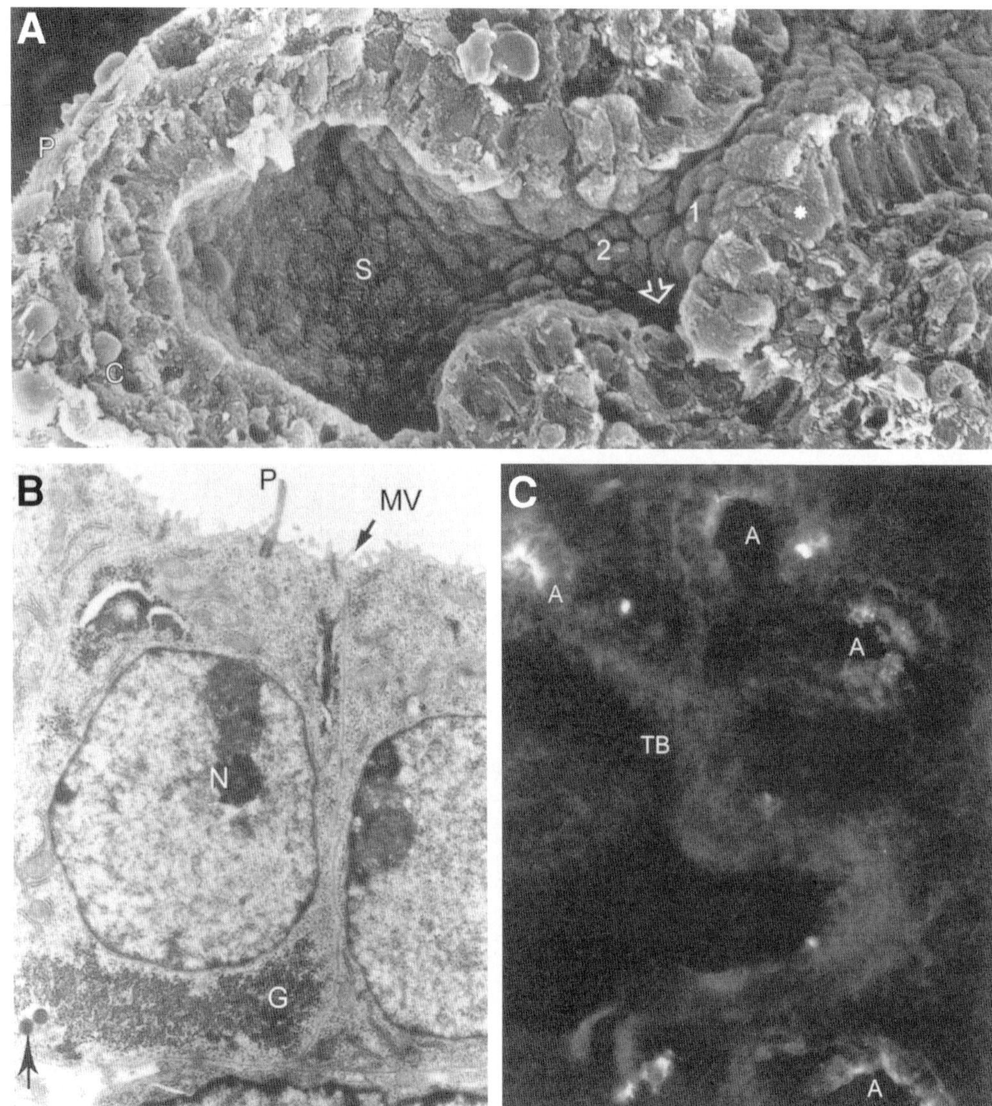

Fig. 12. Cellular composition of pulmonary acinus in early embryonic lung. **(A)** Mouse embryo on d 15 after conception (scanning electron microscopy). The pulmonary acinus consists of a short first-order (1) and two second-order (2) acinar tubules One of which has been cut off, arrow) with a saccular ending (S). The cuboid epithelium of the acinar tubule is sharply demarcated from the columnar epithelium of the connecting bronchiole. The most proximal type II cell is indicated (asterisk) (×1400). **(B)** Detail of early embryonic type II cells (transmission electron microscopy), showing cuboid cell shape, large and roundish nucleus, and presence of precursory forms of MLB (i.e., dense bodies; arrow) in their cytoplasm. G, glycogen; MV, microvilli; P, primary cilium (×6200). (C) Rhesus monkey embryo age 62 d (immunofluorescence). The early type II cells stain for SP-A, whereas the bronchiolar epithelium is unstained. A, acinar tubule; TB, terminal bronchiole (×330).

were viable up to 3 to 4 wk, depending on the stage used for cultivation (Zimmermann, 1987). The cultures of whole fetal lungs were kept for approx 4–7 d.

41.3.3.4.3. Type II Cell Growth and Differentiation In Vitro The cultures of type II cell isolations from fetal lungs consisted mainly of type II cells arranged in epithelial sheets (Batenburg et al., 1988; Oomen et al., 1990; Alcorn et al., 1997). The sheets from fetal d-19 rat lungs were relatively large, but they were progressively smaller for fetal d 20 and 21 and neonatal d 1. In the organoid cultures (Zimmermann, 1987), epithelial cell reaggregation was followed by formation of histotypic structures with an alveolar-like lumen. Alveolar type II cell differentiation manifested itself by SP-A and MLB production, a basal lamina was formed, and the mesenchyme developed into a mature connective tissue. Morphogenesis and differentiation depended on the stage of development from which the lung cells were derived (Zimmermann, 1989; Hundertmark et al., 1999). This conclusion is in line with our findings (A. A. W. ten Have-Opbroek, personal communication) (*see* Fig. 13).

The morphogenesis of the pulmonary acinus and differentiation of type II cells can be influenced by in vitro treatment with specific factors such as dexamethasone (Chinoy et al., 2000) in combination with thyroid hormone (Hundertmark et al., 1999), TGF-β (Bragg et al., 2001), transcription factor GATA-6 (Keijzer et al., 2001), and cytoskeleton-disturbing drugs such as colchicine and cytochalasin B (Zimmermann, 1989).

41.4. PROGENITOR CELLS FOR BRONCHOGENIC CARCINOMA IN HUMANS AND ANIMAL MODELS

Bronchogenic carcinoma is the most important type of human lung carcinoma. Tobacco smoking is a high risk factor for the disease. The developmental pathway of the two major varieties of bronchogenic carcinomas, non–small cell lung cancer (NSCLC) and small cell lung cancer (SCLC), has been investigated in humans and animal models. It is hypothesized (McDowell, 1987) that regiospecific stem cells in renewing lung tissues, thought to retain many aspects of those during embryonic lung development, are the targets for initiation during multistage lung carcinogenesis. In the adult bronchial epithelium, the cells which divide are mucous cells and basal cells and PNECs. Mucous cells (McDowell, 1987) and basal cells (Nasiell et al., 1987) are therefore considered to be the major stem cells for NSCLC in humans. The role of PNECs in the development of SCLC in humans and other species is still poorly understood (McDowell, 1987; Sunday et al., 1995; Van Lommel et al., 1999; Linnoila et al., 2000a). At this time, it is unclear which cells in the tracheobronchial epithelium are most susceptible to oncogenic transformation in vivo. Recent studies in a canine model identified basal cells as specific targets. According to the same principle, carcinomas of the small peripheral airways arise from the dividing cells in the distalmost airways (i.e., Clara cells) (Schüller, 1987; Rehm et al., 1989, 1991; Linnoila et al., 2000b) and alveolar type II cells (Schüller, 1987; Rehm et al., 1988; Linnoila et al., 1992; Oomen et al., 1992; ten Have-Opbroek et al., 1996, 1997).

The varied types of lung carcinomas and multiple pathways of differentiation, even within a single tumor, suggest that diverse lung carcinomas in humans may derive from a common undifferentiated progenitor cell. Support for this idea comes from studies of bronchial carcinogenesis performed in a canine model (sc bronchial autografts exposed to 3-methylcholanthrene) (ten Have-Opbroek et al., 1993, 1994). Very likely, chemical carcinogenesis in the adult bronchial epithelium leads to a maturation arrest (Sell and Pierce, 1994) followed by local retrodifferentiation (Fig. 14). This results in the appearance of undifferentiated primordial-like cells. Such multipotent cells are the cells of origin for bronchogenic carcinomas. They may give rise to specific tumor stem cell types. These are in principle of either the alveolar, bronchial, or primordial kind; the choice may depend on multiple (growth, genetic, and environmental) factors. However, it is possible that retrodifferentiation sometimes affects existing bronchial epithelial cells only incompletely or occurs simultaneously with novel differentiation in those cells. If that is the case, the primordial-like cells of origin and their tumor stem cell lineages may display either predominantly bronchial or mixed (primordial, bronchial, and/or alveolar) cell properties. Examples of carcinomas that may originate from such newly appearing alveolar tumor stem cells (Fig. 14) are carcinomas of the bronchioloalveolar, papillary, acinar, adenoid-cystic, and squamous types (ten Have-Opbroek et al., 1990a, 1993, 1994, 1996). The type II tumor stem cells involved probably arise from normal bronchial basal cells (Ten Have-Opbroek et al., 2000). Such unusual differentiation can also be induced in fetal rat tracheal epithelium in vitro (Shannon, 1994). The fact that second to seventh generation nude mouse transplants have patterns of type II cells similar to those found in the original tumors suggests that alterations in the genetic makeup induced by MCA may get fixed in the DNA and can be inherited (Ten Have-Opbroek et al., 1990a, 1993). There is some evidence that the oncofetal mechanism of differentiation described (retrodifferentiation followed by novel differentiation) may apply to human bronchial carcinogenesis (ten Have-Opbroek et al., 1997). For information about the molecular pathogenesis of lung carcinomas, the reader is directed to recent reviews, such as Zoechbauer-Mueller et al. (2002).

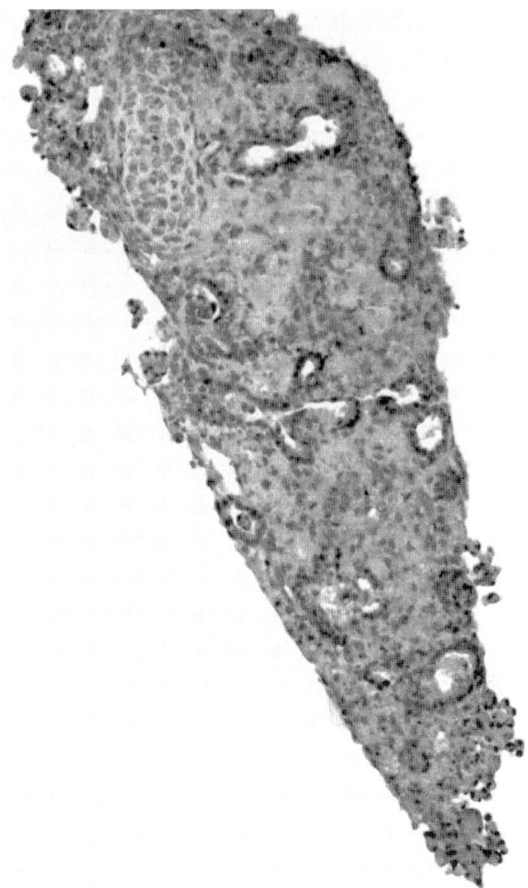

Fig. 13. Light micrograph of organoid culture of mixed fetal and adult mouse lung cells after 7 d in culture showing histotypical morphology of early fetal lung and cartilage formation. Hematoxylin and eosin staining (×70).

41.5. FUTURE PROSPECTS

According to the present view, two theoretical levels of epithelial progenitor cells are involved in lung morphogenesis. Multipotent lung primordial cells create the layout of the respiratory tract and three main types of pluripotent lung-specific (i.e., bronchial, bronchiolar, and alveolar) stem cells serve as the source of transit-amplifying cell populations for regiospecific differentiation, growth, maintenance, and repair. Although there have been incremental advances in our understanding of regiospecific stem cells, their precise locations and cellular phenotype are not yet known with certainty. According to the current theory, the existing or newly differentiated (*see* Subheading 41.4.) stem cells are the most relevant targets for transformation and are thus the source of lung cancer, but this has not yet been proven.

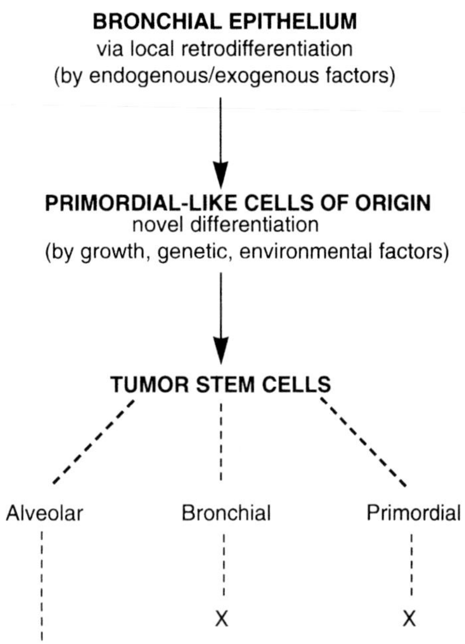

Fig. 14. Oncofetal concept of bronchogenic carcinoma development based on studies of NSCLC in canine model (sc bronchial autografts treated with 3-methylcholanthrene) (ten Have-Opbroek et al., 1993, 1994). Local retrodifferentiation in normal bronchial epithelium from carcinogens may result in the appearance of multipotent primordial-like cells of origin for bronchogenic carcinomas. These primordial-like cells of origin give rise to specific (alveolar, bronchial, or primordial) tumor stem cell types from growth, genetic, and/or environmental factors. Incomplete retrodifferentiation or simultaneous novel differentiation may result in carcinomas with mixed cell properties. Alveolar carcinomas that may derive from such newly differentiated type II tumor stem cells are shown. X, data about carcinomas derived from newly differentiated bronchial or primordial tumor stem cells are not yet available.

A greater understanding of the developmental pathway and the location and regulation of stem cells in the adult lung is important for several reasons. Heroic modern medical treatment enables improved survival of premature humans but also incurs a risk of chronic lung injury. Improved knowledge of stem cells is prerequisite for future interventions to speed lung maturation or to support regeneration of lung tissue damaged due to premature birth. Similarly, phenotypic changes in the epithelium, such as mucous cell hyperplasia, are hallmarks of common lung diseases including asthma, chronic bronchitis, and cystic fibrosis. A better understanding of cell lineages and the regulation of cell phenotype may lead to potentially beneficial therapies. Lung cancer is the world's most prevalent lethal cancer. Stem cells probably are at greatest risk of malignant transformation. Thus, identification of lung stem cells and their modification during multistage carcinogenesis may form the basis for improved detection and treatment of this prevalent disease. There is a pressing need to identify airway epithelial stem cells if gene therapy, e.g., for the prototypical monogenic disorder cystic fibrosis, is to succeed. Theoretically, integrating vectors offer the promise of permanent cure, but only if stem cells are identified and targeted. Lung transplantation is a viable therapy for end-stage lung disease, but the prevalence of the bronchiolitis obliterans syndrome, the manifestation of chronic rejection, decreases the odds of long-term survival to the worst of all solid organs. The development of bronchiolitis obliterans syndrome may represent alloimmune recognition and destruction of airway epithelial stem cells. Thus, a better understanding of cell dynamics in transplanted lungs may improve monitoring and treatment, and thus outcomes, after transplantation. In pulmonary emphysema, it is currently considered that lung tissue is permanently destroyed. However, novel stem cell–based therapies may offer the prospect of regenerating even the complex architecture of the lung. Thus, there are compelling reasons to apply current advances in stem cell biology toward the future identification of stem cell lineages of the lung.

REFERENCES

Adamson, L. Y. R. and Bowden, D. H. (1974) The type 2 cell as progenitor of alveolar epithelial regeneration: a cytodynamic study in mice after exposure to oxygen. *Lab. Invest.* 30:35–42.

Adamson, L. Y. R. and Bowden, D. H. (1975) Derivation of type 1 epithelium from type 2 cells in the developing rat lung. *Lab. Invest.* 32: 736–745.

Alcorn, J. L., Smith, M. E., Smith, J. F., Margraf, L. R., and Mendelsohn, C. R. (1997) Primary cell culture of human type II pneumonocytes: maintenance of a differentiated phenotype and transfection with recombinant adenoviruses. *Am. J. Respir. Cell Mol. Biol.* 17:672–682.

Balfour, F. M. (1881) A treatise on comparative embryology: the alimentary canal and its appendages. In: *The Chordata*, vol. 2, MacMillan, London, UK, pp. 628–630.

Basbaum, C. and Jany, B. (1990) Plasticity in the airway epithelium. *Am. J. Physiol.* 259:L38–L46.

Batenburg, J. J. (1992) Surfactant phospholipids: synthesis and storage. *Am. J. Physiol. (Lung Cell. Mol. Physiol. 6)* 262:L367–L385.

Batenburg, J. J., Otto-Verberne, C. J. M., ten Have-Opbroek, A. A. W., and Klazinga,W. (1988) Isolation of alveolar type II cells from fetal rat lung by differential adherence in monolayer culture. *Biochim. Biophys. Acta* 960:441–453.

Belinsky, S. A., Lechner, J. F., and Johnson, N.F. (1995) An improved method for the isolation of type II and Clara cells from mice. *In Vitro Cell. Dev. Biol. Anim.* 31:361–366.

Benali, R., Tournier, J. M., Chevillard, M., et al. (1993) Tubule formation by human surface respiratory epithelial cells cultured in a three-dimensional collagen lattice. *Am. J. Physiol.* 264:L183–L192.

Bernacki, S. H., Nelson, A. L., Abdullah, L., et al. (1999) Mucin gene expression during differentiation of human airway epithelia *in vitro*: muc4 and muc5b are strongly induced. *Am. J. Respir. Cell Mol. Biol.* 20:595–604.

Blau, H. M., Brazelton, T. R., and Weimann, J. M. (2001) The evolving concept of a stem cell: entity or function? *Cell* 105:829–841.

Blenkinsopp, W. K. (1967) Proliferation of respiratory tract epithelium in the rat. *Exp.Cell Res.* 46:144-154.

Boers, J. E., Ambergen, A. W., and Thunnissen, F. B. (1998) Number and proliferation of basal and parabasal cells in normal human airway epithelium. *Am. J. Respir. Crit. Care Med.* 157:2000–2006.

Boers, J. E., den Brok, J. L., Koudstaal, J., Arends, J. W., and Thunnissen, F. B. (1996) Number and proliferation of neuroendocrine cells in normal human airway epithelium. *Am. J. Respir. Crit. Care Med.* 154: 758–763.

Boers, J. E., Ambergen, A. W., and Thunnissen, F. B. (1999) Number and proliferation of clara cells in normal human airway epithelium. *Am. J. Respir. Crit. Care Med.* 159(5 Pt. 1):1585–1591.

Borges, M., Linnoila, R. I., Van de Velde, H. J., et al. (1997) An achaete-scute homologue essential for neuroendocrine differentiation in the lung. *Nature* 386:852–855.

Borthwick, D. W., West, J. D., Keighren, M. A., Flockhart, J. H., Innes, B. A., and Dorin, J. R. (1999) Murine submucosal glands are clonally derived and show a cystic fibrosis gene-dependent distribution pattern. *Am. J. Respir. Cell Mol. Biol.* 20:1181–1189.

Borthwick, D. W., Shahbazian, M., Todd, K. Q., Dorin, J. R., and Randell, S. H. (2001) Evidence for stem-cell niches in the tracheal epithelium. *Am. J. Respir. Cell Mol. Biol.* 24:662–670.

Boyden, E. A. (1972) Development of the human lung. In: *Brennemann's Practise of Pediatrics*, vol. 4, Harper & Row, Hagerstown, MD, pp. 1–12.

Bragg, A. D., Moses, H. L., and Serra, R. (2001) Signaling to the epithelium is not sufficient to mediate all of the effects of transforming growth factor β and bone morphogenetic protein 4 on murine embryonic lung development. *Mech. Dev.* 109:13–26.

Brandsma, A. E., ten Have-Opbroek, A. A. W., Vulto, I. M., Molenaar, J. C., and Tibboel, D. (1994a) Alveolar epithelial composition and architecture of the late fetal pulmonary acinus: an immunocytochemical and morphometric study in a rat model of pulmonary hypoplasia and congenital diaphragmatic hernia. *Exp. Lung Res.* 20:491–515.

Brandsma, A. E., Tibboel, D., Vulto, I. M., De Vijlder, J. J. M., ten Have-Opbroek, A. A. W., and Wiersinga, W. M. (1994b) Inhibition of T_3-receptor binding by nitrofen. *Biochim. Biophys. Acta* 1201:266–270.

Breuer, R., Zajicek, G., Christensen, T. G., Lucey, E. C., and Snider, G. L. (1990) Cell kinetics of normal adult hamster bronchial epithelium in the steady state. *Am. J. Respir. Cell Mol. Biol.* 2:51–58.

Bucher, U. and Reid, L. (1961) Development of the mucus secreting elements in human lung. *Thorax* 16:219–225.

Buckpitt, A., Buonarati, M., Avey, L. B., Chang, A. M., Morin, D., and Plopper, C. G. (1992) Relationship of cytochrome P450 activity to Clara cell cytotoxicity. II. Comparison of stereoselectivity of naphthalene epoxidation in lung and nasal mucosa of mouse, hamster, rat and rhesus monkey. *J. Pharmacol. Exp. Ther.* 261(1):364–372.

Cardoso, W. V. (2001) Molecular regulation of lung development. *Annu. Rev. Physiol.* 63:471–494.

Cardoso, W. V. and Williams, M. C. (2001) Basic mechanisms of lung development: Eighth Woods Hole Conference on lung cell biology 2000. *Am. J. Respir. Cell Mol. Biol.* 25 (2):137–140.

Chinoy, M., Zgleszewski, S. E., Cilley, R. E., and Krummel, T.M. (2000) Dexamethasone enhances ras-recision gene expression in cultured murine fetal lungs: role in development. *Am. J. Physiol. Lung Cell. Mol. Physiol.* 279:L312–L318.

Cilley, R. E., Zgleszewski, S. E., Krummel, T. M., and Chinoy, M. R. (1997) Nitrofen dose dependent gestational day-specific murine lung hypoplasia and left-sided diaphragmatic hernia. *Am. J. Physiol. Lung Cell. Mol. Physiol.* 272:L362–L371.

Clara, M. (1937) Zur Histobiologie des Bronchalepithels. *Z. Mikrosk. Anat. Forsch.* 41:321–347.

Corti, M., Brody, A. R., and Harrison, J. H. (1996) Isolation and primary culture of murine alveolar type II cells. *Am. J. Respir. Cell Mol. Biol.* 14(4):309–315.

Cutz, E., Yeger, H., Wong, V., Bienkowski, E., and Chan, W. (1985) In vitro characteristics of pulmonary neuroendocrine cells isolated from rabbit fetal lung. I. Effects of culture media and nerve growth factor. *Lab. Invest.* 53(6):672–683.

DeMayo, F. J., Finegold, M. J., Hansen, T. N., Stanley, L. A., Smith, B., and Bullock, D. W. (1991) Expression of SV40 T antigen under control of rabbit uteroglobin promoter in transgenic mice. *Am. J. Physiol.* 261(2 Pt. 1):L70–L76.

Donnelly, G. M., Haack, D. G., and Heird, C. S. (1982) Tracheal epithelium: cell kinetics and differentiation in normal rat tissue. *Cell Tissue Kinet.* 15:119–130.

Duan, D., Yue, Y., Zhou, W., et al. (1999) Submucosal gland development in the airway is controlled by lymphoid enhancer binding factor 1 (LEF1). *Development* 126:4441–4453.

Dubreuil, G., Lacoste, A., and Raymond, R. (1936) Observations sur le développement du poumon humain. *Bull. Histol. Appliq. Physiol. Pathol.* 13:235–245.

Ebe, T. (1969) Light and electron microscope studies on experimental pneumonitis induced by blasticidin-S, with special reference to alveolar regeneration. *Arch. Histol. Jpn* 30:149–182.

Engelhardt, J. F. (2001) Stem cell niches in the mouse airway. *Am. J. Respir. Cell Mol. Biol.* 24(6):649–652.

Engelhardt, J. F., Schlossberg, H., Yankaskas, J. R., and Dudus, L. (1995) Progenitor cells of the adult human airway involved in submucosal gland development. *Development* 121:2031–2046.

Erjefalt, J. S. and Persson, C. G. A. (1997) Airway epithelial repair: breathtakingly quick and multipotentially pathogenic. *Thorax* 52:1010–1012.

Evans, M. J., Cabral, L. J., Stephens, R. J., and Freeman, G. (1975) Transformation of alveolar type 2 cells to type 1 cells following exposure to NO_2. *Exp. Mol. Pathol.* 22:142–150.

Evans, M. J., Johnson, L. V., Stephens, R. J., and Freeman, G. (1976) Renewal of the terminal bronchiolar epithelium in the rat following exposure to NO2 or O3. *Lab. Invest.* 35(3):246-257.

Evans, M. J., Cabral-Anderson, L. J., and Freeman, G. (1978) Role of the Clara cell in renewal of the bronchiolar epithelium. *Lab. Invest.* 38(6): 648–653.

Ford, J. R. and Terzaghi-Howe, M. (1992) Characteristics of magnetically separated rat tracheal epithelial cell populations. *Am. J. Physiol.* 263:L568–L574.

Gazdar, A. F. and Minna, J. D. (1996) NCI series of cell lines: an historical perspective. *J. Cell. Biochem. Suppl.* 24:1–11.

Giangreco, A. S., Reynolds, D., and Stripp, B. R. (2002) Terminal bronchioles harbor a unique airway stem cell population that localizes to the broncho-alveolar duct junction. *Am. J. Pathol.* 161:173–182.

Goerke, J. (1974) Lung surfactant. *Biochim. Biophys. Acta* 344:241–261.

Gruenert, D. C., Basbaum, C. B., Welsh, M. J., Li, M., Finkbeiner, W. E., and Nadel, J. A. (1988) Characterization of human tracheal epithelial cells transformed by an origin-defective simian virus 40. *Proc. Natl. Acad. Sci. USA* 85:5951–5955.

Gruenert, D. C., Finkbeiner, W. E., and Widdicombe, J. H. (1995) Culture and transformation of human airway epithelial cells. *Am. J. Physiol.* 268:L347–L360.

Hackett, B. P. and Gitlin, J.D. (1992) Cell-specific expression of a Clara cell secretory protein-human growth hormone gene in the bronchiolar epithelium of transgenic mice. *Proc. Natl. Acad. Sci. USA* 89(19): 9079–9083.

Haugen, A. and Aune, T. (1986) Culture of rabbit pulmonary Clara cells. *Proc. Soc. Exp. Biol. Med.* 182(2):277–281.

Heckman, C. A., Marchok, A. C., and Nettesheim, P. (1978) Respiratory tract epithelium in primary culture: concurrent growth and differentiation during establishment. *J. Cell Sci.* 32:269–291.

Hislop, A. and Reid, L. (1974) Development of the acinus in the human lung. *Thorax* 29:90–94.

Ho, Y. S. (1994) Transgenic models for the study of lung biology and disease. *Am. J. Physiol.* 266(4 Pt. 1):L319–L353.

Hogan, B. L., Grindley, J., Bellusci, S., Dunn, N. R., Emoto, H., and Itoh, N. (1997) Branching morphogenesis of the lung: new models for a classical problem. *Cold Spring Harbor Symp. Quant. Biol.* 62:249–256.

Hong, K. U., Reynolds, S. D., Giangreco A., Hurley, C. M., and Stripp, B. R. (2001) Clara cell secretory protein-expressing cells of the airway neuroepithelial body microenvironment include a label-retaining subset and are critical for epithelial renewal after progenitor cell depletion. *Am. J. Respir. Cell Mol. Biol.* 24(6):671–681.

Hook, G. E., Brody, A. R., Cameron, G. S., Jetten, A. M., Gilmore, L. B., and Nettesheim, P. (1987) Repopulation of denuded tracheas by Clara cells isolated from the lungs of rabbits. *Exp. Lung Res.* 12:311–329.

Hundertmark, S., Ragosch, V., Zimmermann, B., Halis, G., Arabin, B., and Weitzel, H. K. (1999) Effect of dexamethasone, triiodothyronine, and dimethyl-isopropylthyronine on lung maturation of the fetal rat lung. *J. Perinat. Med.* 27:309–315.

Inayama, Y., Hook, G. E. R., Brody, A. R. et al. (1988) The differentiation potential of tracheal basal cells. *Lab. Invest.* 58:706–717.

Inayama, Y., Hook, G. E. R., Brody, A. R., et al. (1989) *In vitro* and *in vivo* growth and differentiation of clones of tracheal basal cells. *Am. J. Pathol.* 134:539–549.

Jeffery, P. K., Gaillard, D., and Moret, S. (1992) Human airway secretory cells during development and in mature airway epithelium. *Eur. Respir. J.* 5:93–104.

Johnson, N. F. and Hubbs, A. F. (1990) Epithelial progenitor cells in the rat trachea. *Am. J. Respir. Cell Mol. Biol.* 3:579–585.

Jorissen, M., Van der Schueren, B., van den Berghe, H., and Cassiman, J. J. (1989) The preservation and regeneration of cilia on human nasal epithelial cells cultured *in vitro. Arch. Otorhinolaryngol.* 246:308–314.

Kauffman, S. L., Burri, P. H., and Weibel, E. R. (1974) The postnatal growth of the rat lung. II. Autoradiography. *Anat. Rec.* 180:63–76.

Keijzer, R., Van Tuyl, M., Meijers, C., et al. (2001) The transcription factor GATA6 is essential for branching morphogenesis and epithelial cell differentiation during fetal pulmonary development. *Development* 128:503–511.

Khoor, A., Gray, M. E., Singh, G., and Stahlman, M. T. (1996) Ontogeny of Clara cell-specific protein and its mRNA: their association with neuroepithelial bodies in human fetal lung and in bronchopulmonary dysplasia. *J. Histochem. Cytochem.* 44(12):1429–1438.

Koo, J. S., Yoon, J. H., Gray, T., Norford, D., Jetten, A. M., and Nettesheim, P. (1999) Restoration of the mucous phenotype by retinoic acid in retinoid-deficient human bronchial cell cultures: changes in mucin gene expression. *Am. J. Respir. Cell Mol. Biol.* 20: 43–52.

Krause, D. S., Theise, N. D., Collector, M. I., Henegariu, O., Hwang, S., Gardner, R., Neutzel, S., and Sharkis, S. J. (2001) Multi-organ, multi-lineage engraftment by a single bone marrow-derived stem cell. *Cell* 105:369–377.

Lawson, G. W., Van Winkle, L. S., Toskala, E., Senior, R. M., Parks, W. C., and Plopper, C. G. (2002) Mouse strain modulates the role of the ciliated cell in acute tracheobronchial airway injury-distal airways. *Am. J. Pathol.* 160:315–327.

Lebeche, D., Malpel, S., and Cardoso, W. V. (1999) Fibroblast growth factor interactions in the developing lung. *Mech. Dev.* 86(1-2):125–136.

Lechner, J. F., Haugen, A., Autrup, H., McClendon, I. A., Trump, B. F., and Harris, C. C. (1981) Clonal growth of epithelial cells from normal adult human bronchus. *Cancer Res.* 41(6):2294–2304.

Lechner, J. F., Fugaro, J. M., Wong, Y., Pass, H. I., Harris, C. C., and Belinsky, S. A. (2001) Perspective: cell differentiation theory may advance early detection of and therapy for lung cancer. *Radiat. Res.* 155:235–238.

Linnoila, R.I., Mulshine, J.L., Steinberg, S.M., and Gazdar, A.F. (1992) Expression of surfactant-associated protein in non-small cell lung cancer: a discriminant between biologic subsets. *J. Natl. Cancer Inst. Monogr.* 13:61–66.

Linnoila, R. I., Gazdar, A. F., Funa, K., and Becker, K. L. (1993) Long-term selective culture of hamster pulmonary endocrine cells. *Anat. Rec.* 236(1):231–240.

Linnoila, R. I., Sahu, A., Miki, M., Ball, D. W., and DeMayo, F. J. (2000a) Morphometric analysis of CC10-hASH1 transgenic mouse lung: a model for bronchiolization of alveoli and neuroendocrine carcinoma. *Exp. Lung Res.* 26:595–615.

Linnoila, R. I., Szabo, E., Demayo, F., Witschi, H., Sabourin, C., and Malkinson, A. (2000b) The role of CC10 in pulmonary carcinogenesis: from a marker to tumor suppression. *Ann. NY Acad. Sci.* 923: 249–267.

Liu, J. Y., Nettesheim, P., and Randell, S. H. (1994) Growth and differentiation of tracheal epithelial progenitor cells. *Am. J. Physiol.* 266: L296–L307.

Loosli, C. G. and Potter, E. L. (1959) Pre- and postnatal development of the respiratory portion of the human lung. *Am. Rev. Respir. Dis.* 80: 5–23.

Lum, H., Schwartz , L. W., Dungworth, D. L., and Tyler, W. S. (1978) A comparative study of cell renewal after exposure to ozone or oxygen: Response of terminal bronchiolar epithelium in the rat. *Am. Rev. Respir. Dis.* 118(2):335–345.

Masui, T., Lechner, J. F., Yoakum, G. H., Willey, J. C., and Harris, C. C. (1986a) Growth and differentiation of normal and transformed human bronchial epithelial cells. *J. Cell Physiol. Suppl.* 4:73–81.

Masui, T., Wakefield, L. M., Lechner, J. F., LaVeck, M. A., Sporn, M. B., and Harris, C. C. (1986b) Type beta transforming growth factor is the primary differentiation-inducing serum factor for normal human bronchial epithelial cells. *Proc. Natl. Acad. Sci. USA* 83:2438–2442.

McDowell, E. M. (1987) Bronchogenic carcinomas. In: *Lung Carcinomas, Part IV, Neoplasms of the Human Lung* (McDowell, E. M., ed.), Churchill Livingstone, Edinburgh, pp. 255–285.

McDowell, E. M., Newkirk, C., and Coleman, B. (1985) Development of hamster tracheal epithelium: II. Cell proliferation in the fetus. *Anat. Rec.* 213:448–456.

Mercer, R. R., Russell, M. L., Roggli, V. L., and Crapo, J. D. (1994) Cell number and distribution in human and rat airways. *Am. J. Respir. Cell Mol. Biol.* 10:613–624.

Nasiell, M., Auer, G., and Kato, H. (1987) Cytologic studies in man and animals on development of bronchogenic carcinoma. In: *Lung Carcinomas, Part III, Progression to Neoplasia* (McDowell, E. M., ed.), Churchill Livingstone, Edinburgh, pp. 207–242.

Nord, M., Lag, M., Cassel, T. N., et al. (1998) Regulation of CCSP (PCB-BP/uteroglobin) expression in primary cultures of lung cells: involvement of C/EBP. *DNA Cell Biol.* 17(5):481–492.

Oie, H. K., Russell, E. K., Carney, D. N., and Gazdar, A. F. (1996) Cell culture methods for the establishment of the NCI series of lung cancer cell lines. *J. Cell. Biochem. Suppl.* 24:24–31.

Oomen, L. C. J. M., ten Have-Opbroek, A. A. W., Hageman, P. C., et al. (1990) Fetal mouse alveolar type II cells in culture express several type II cell characteristics found in vivo, together with Major Histocompatibility antigens. *Am. J. Respir. Cell Mol. Biol.* 3:325–339.

Oomen, L. C. J. M., Calafat, J., ten Have-Opbroek, A. A. W., Egberts, J., and Demant, P. (1992) Derivation of tumorigenic and non-tumorigenic mouse alveolar type II cell lines from fetal type II cells after a combined *in vivo/in vitro* carcinogen treatment. *Int. J. Cancer* 52: 290–297.

Otto-Verberne, C. J. M. and Ten Have-Opbroek, A. A. W. (1987) Development of the pulmonary acinus in fetal rat lung: a study based on an antiserum recognizing surfactant-associated proteins. *Anat. Embryol.* 175:365–373.

Otto-Verberne, C. J. M., ten Have-Opbroek, A. A. W., Balkema, J. J., and Franken, C. (1988) Detection of the type II cell or its precursor before week 20 of human gestation, using antibodies against surfactant-associated proteins. *Anat. Embryol.* 178:29–39.

Otto-Verberne, C. J. M., ten Have-Opbroek, A. A. W., and De Vries, E. C. P. (1990) Expression of the major surfactant-associated protein, SP-A, in type II cells of human lung before 20 weeks of gestation. *Eur. J. Cell Biol.* 53:13–19.

Perl, A. K. T. and Whitsett, J. A. (1999) Molecular mechanisms controlling lung morphogenesis. *Clin. Genet.* 56:14–27.

Plopper, C. G. and ten Have-Opbroek, A. A. W. (1994) Anatomical and histological classification of the bronchioles. In: *Diseases of the Bronchioles* (Epler, G. R., ed.), Raven, New York, pp. 15–25.

Plopper, C. G., Hill, L. H., and Mariassy, A. T. (1980) Ultrastructure of the nonciliated bronchiolar epithelial (Clara) cell of mammalian lung. III. A study of man with comparison of 15 mammalian species. *Exp. Lung Res.* 1(2):171–180.

Plopper, C. G., Mariassy, A. T., Wilson, D. W., Alley, J. L., Nishio, S. J., and Nettesheim, P. (1983) Comparison of nonciliated tracheal epithelial cells in six mammalian species: ultrastructure and population densities. *Exp. Lung Res.* 5(4):281–294.

Plopper, C. G., Alley, J. L., and Weir, A.J. (1986) Differentiation of tracheal epithelium during fetal lung maturation in the Rhesus monkey Macaca mulatta. *Am. J. Anat.* 175:59–71.

Plopper, C. G., Heidsiek, J. G., Weir, A. J., St. George, J. A., and Hyde, D. M. (1989) Tracheobronchial epithelium in the adult Rhesus monkey: a quantitative histochemical and ultrastructural study. *Am. J. Anat.* 184:31–40.

Plopper, C. G., Chang, A. M., Pang, A., and Buckpitt, A.R. (1991) Use of microdissected airways to define metabolism and cytotoxicity in murine bronchiolar epithelium. *Exp. Lung Res.* 17(2):197–212.

Plopper, C. G., Macklin, J., Nishio, S. J., Hyde, D. M., and Buckpitt, A. R. (1992) Relationship of cytochrome P-450 activity to Clara cell cytotoxicity. III. Morphometric comparison of changes in the epithelial populations of terminal bronchioles and lobar bronchi in mice, hamsters, and rats after parenteral administration of naphthalene. *Lab. Invest.* 67(5):553–565.

Randell, S. H., Comment, C. E., Ramaekers, F. C. S., and Nettesheim, P. (1991) Properties of rat tracheal epithelial cells separated based on expression of cell surface β-galactosyl end groups. *Am. J. Respir. Cell Mol. Biol.* 4:544–554.

Ray, M. K., Magdaleno, S. W., Finegold, M. J., and DeMayo, F. J. (1995) Cis-acting elements involved in the regulation of mouse Clara cell-specific 10-kDa protein gene: in vitro and in vivo analysis. *J. Biol. Chem.* 270(6):2689–2694.

Rehm, S., Ward, J. M., ten Have-Opbroek, A. A. W., et al. (1988) Mouse papillary lung tumors transplacentally induced by N-nitrosoethylurea: evidence for alveolar type II cell origin by comparative light microscopic, ultrastructural, and immunohistochemical studies. *Cancer Res.* 48:148–160.

Rehm, S., Takahashi, M., Ward, J. M., Singh, G., Katyal, S. L., and Henneman, J. R. (1989) Immunohistochemical demonstration of Clara cell antigen in lung tumors of bronchiolar origin induced by N-nitrosodiethylamine in Syrian golden hamsters. *Am. J. Pathol.* 134:79–87.

Rehm, S., Lijinski, W., Singh, G., and Katyal, S. L. (1991) Mouse bronchiolar cell carcinogenesis. *Am. J. Pathol.* 139:413–422.

Reynolds, S. D., Hong, K. U., Giangreco, A., et al. (2000a) Conditional clara cell ablation reveals a self-renewing progenitor function of pulmonary neuroendocrine cells. *Am. J. Physiol. Lung Cell. Mol. Physiol.* 278(6):L1256–L1263.

Reynolds, S. D., Giangreco, A., Power, J. H., and Stripp, B. R. (2000b) Neuroepithelial bodies of pulmonary airways serve as a reservoir of progenitor cells capable of epithelial regeneration. *Am. J. Pathol.* 156(1):269–278.

Sandmoller, A., Halter, R., Gomez-La-Hoz, E., et al. (1994) The uteroglobin promoter targets expression of the SV40 T antigen to a variety of secretory epithelial cells in transgenic mice. *Oncogene* 9(10):2805–2815.

Schüller, H. M. (1987) Experimental carcinogenesis in the peripheral lung. In: *Lung Carcinomas, Part III, Progression to Neoplasia* (McDowell, E. M., ed.), Churchill Livingstone, Edinburgh, pp. 243–254.

Schwarz, M. A., Zhang, F., Lane, J. E., et al. (2000) Angiogenesis and morphogenesis of murine fetal distal lung in an allograft model. *Am. J. Physiol. Lung Cell. Mol. Physiol.* 278:L1000–L1007.

Sell, S. and Pierce, G. B. (1994) Maturation arrest of stem cell differentiation is a common pathway for the cellular origin of teratocarcinomas and epithelial cancers. *Lab. Invest.* 70:6–22.

Shannon, J. M. (1994) Induction of alveolar type II cell differentiation in fetal tracheal epithelium by grafted distal lung mesenchyme. *Dev. Biol.* 166(2):600–614.

Shannon, J. M., Nielsen, L. D., Gebb, S. A., and Randell, S. H. (1998) Mesenchyme specifies epithelial differentiation in reciprocal recombinants of embryonic lung and trachea. *Dev. Dyn.* 212(4):482–494.

Singh, G. and Katyal, S. L. (1984) An immunologic study of the secretory products of rat Clara cells. *J. Histochem. Cytochem.* 32:49–54.

Sorokin, S. P. (1966) A morphologic and cytochemical study on the great alveolar cell. *J. Histochem. Cytochem.* 14:884–897.

Sorokin, S. P., Hoyt, R. F. Jr., and Grant, M. M. (1982) Development of neuroepithelial bodies in fetal rabbit lungs. I. Appearance and functional maturation as demonstrated by high-resolution light microscopy and formaldehyde-induced fluorescence. *Exp. Lung Res.* 3: 237–259.

Speirs, V. and Cutz, E. (1993) An overview of culture and isolation methods suitable for in vitro studies on pulmonary neuroendocrine cells. *Anat. Rec.* 236(1):35–40.

Speirs, V., Wang, Y. V., Yeger, H., and Cutz, E. (1992) Isolation and culture of neuroendocrine cells from fetal rabbit lung using immunomagnetic techniques. *Am. J. Respir. Cell Mol. Biol.* 6(1):63–67.

Spooner, B. S. and Wessels, N. K. (1970) Mammalian lung development: interactions in primordium formation and bronchial morphogenesis. *J. Exp. Zool.* 175:445–454.

Stahlman, M. T. and Gray, M. E. (1984) Ontogeny of neuroendocrine cells in human fetal lung. I. An electron microscopic study. *Lab. Invest.* 51(4):449–463.

Stearns, R. C., Paulauskis, J. D., and Godleski, J. J. (2001) Endocytosis of ultrafine particles by A549 cells. *Am. J. Respir. Cell Mol. Biol.* 24: 108–115.

Stoner, G. D., Katoh, Y., Foidart, J. M., Myers, G. A., and Harris, C. C. (1980) Identification and culture of human bronchial epithelial cells. *Meth. Cell Biol.* 21A:15–35.

Stripp, B. R., Sawaya, P. L., Luse, D. S., et al. (1992) cis-acting elements that confer lung epithelial cell expression of the CC10 gene. *J. Biol. Chem.* 267(21):14,703–14,712.

Stripp, B. R., Huffman, J. A., and Bohinski, R. J. (1994) Structure and regulation of the murine Clara cell secretory protein gene. *Genomics* 20(1):27–35.

Stripp, B. R., Maxson, K., Mera, R., and Singh, G. (1995) Plasticity of airway cell proliferation and gene expression after acute naphthalene injury. *Am. J. Physiol.* 269(6 Pt. 1):L791–L799.

Sunday, M., Willet, C. G., Graham, S. A., et al. (1995) Histochemical characterization of non-neuroendocrine tumors and neuroendocrine cell hyperplasia induced in hamster lung by 4-(methylnitrosamino)-1-(3-pyridyl)-1-butanone with or without hyperoxia. *Am. J. Pathol.* 147: 740–752.

Tebar M., Destrée, O., De Vree, W. J. A., and ten Have-Opbroek, A. A. W. (2001a) Expression of Tcf/Lef and sFRP and localization of β-catenin in the developing mouse lung. *Mech. Dev.* 109:437–440.

Tebar, M., Boex, J. J. M., and ten Have-Opbroek, A. A. W. (2001b) Functional overexpression of wild-type p53 correlates with alveolar cell differentiation in the developing human lung. *Anat. Rec.* 263: 25–34.

ten Have-Opbroek, A. A. W. (1975) Immunological study of lung development in the mouse embryo. I. Appearance of a lung-specific antigen, localized in the great alveolar cell. *Dev. Biol.* 46:390–403.

ten Have-Opbroek, A. A. W. (1979) Immunological study of lung development in the mouse embryo. II. First appearance of the great alveolar cell, as shown by immunofluorescence microscopy. *Dev. Biol.* 69:408–423.

ten Have-Opbroek, A. A. W. (1981) The development of the lung in mammals: an analysis of concepts and findings. *Am. J. Anat.* 162: 201–219.

ten Have-Opbroek, A. A. W. (1986) The structural composition of the pulmonary acinus in the mouse: a scanning electron microscopical and developmental-biological analysis. *Anat. Embryol.* 174:49–57.

ten Have-Opbroek, A. A. W. (1991) Lung development in the mouse embryo. *Exp. Lung Res.* 17:111–130.

ten Have-Opbroek, A. A. W., and De Vries, E. C. P. (1993) Clara cell differentiation in the mouse: ultrastructural morphology and cytochemistry for surfactant protein A and Clara cell 10 kD protein. *Microsc. Res. Technolol.* 26:400–411.

ten Have-Opbroek, A. A. W. and Plopper, C. G. (1992) Morphogenetic and functional activity of type II cells in early fetal Rhesus monkey lungs: a comparison between primates and rodents. *Anat. Rec.* 234: 93–104.

ten Have-Opbroek, A. A. W., Dubbeldam, J. A., and Otto-Verberne, C. J. M. (1988) Ultrastructural features of type II alveolar epithelial cells in early embryonic mouse lung. *Anat. Rec.* 221:846–853.

ten Have-Opbroek, A. A. W., Hammond, W. G., and Benfield, J. R. (1990a) Bronchiolo-alveolar regions in adenocarcinoma arising from canine segmental bronchus. *Cancer Lett.* 55:177–182.

ten Have-Opbroek, A. A. W., Otto-Verberne, C. J. M., and Dubbeldam, J. A. (1990b) Ultrastructural characteristics of inclusion bodies of type II cells in late embryonic mouse lung. *Anat. Embryol.* 181:317–323.

ten Have-Opbroek, A. A. W., Otto-Verberne, C. J. M., Dubbeldam, J. A., and Dijkman, J. H. (1991) The proximal border of the human respiratory unit, as shown by scanning and transmission electron microscopy and light microscopical cytochemistry. *Anat. Rec.* 229:339–354.

ten Have-Opbroek, A. A. W., Hammond, W. G., Benfield, J. R., Teplitz, R. L., and Dijkman, J. H. (1993) Expression of alveolar type II cell markers in acinar adenocarcinomas and adenoid-cystic carcinomas arising from segmental bronchi: a study in a heterotopic bronchogenic carcinoma model in dogs. *Am. J. Pathol.* 142:1251–1264.

ten Have-Opbroek, A.A.W., Benfield, J.R., Hammond, W.G., Teplitz, R.L., and Dijkman, J.H. (1994) In favour of an oncofoetal concept of bronchogenic carcinoma development. *Histol. Histopathol.* 9:375–384.

ten Have-Opbroek, A.A.W., Benfield, J.R., Hammond, W.G., and Dijkman, J.H. (1996) Alveolar stem cells in canine bronchial carcinogenesis. *Cancer Lett.* 101:211–217.

ten Have-Opbroek, A.A.W., Benfield, J.R., Van Krieken, J.H.J.H., and Dijkman, J.H. (1997) The alveolar type II cell is a pluripotential stem cell in the genesis of human adenocarcinomas and squamous cell carcinomas. *Histol. Histopathol.* 12:319–336.

ten Have-Opbroek, A.A.W., Shi, X.-B., and Gumerlock, P.H. (2000) 3-Methylcholanthrene triggers the differentiation of alveolar tumor cells from canine bronchial basal cells and an altered p53 gene promotes their clonal expansion. *Carcinogenesis* 21:1477–1484.

Terzaghi, M., Nettesheim, P., and Williams, M. L. (1978) Repopulation of denuded tracheal grafts with normal, preneoplastic, and neoplastic epithelial cell populations. *Cancer Res.* 38:4546–4553.

Tos, M. (1968) Development of the mucous glands in the human main bronchus. *Anat. Anz.* 123:376–389.

Tyler, N. K. and Plopper, C. G. (1985) Morphology of the distal conducting airways in Rhesus monkey lungs. *Anat. Rec.* 211:295–303.

Van Lommel, A., Bolle, T., Fannes, W., and Lauweryns, J. M. (1999) The pulmonary neuroendocrine system: the past decade. *Arch. Histol. Cytol.* 62:1–16.

Van Winkle, L. S., Buckpitt, A. R., Nishio, S. J., Isaac, J. M., and Plopper, C. G. (1995) Cellular response in naphthalene-induced Clara cell injury and bronchiolar epithelial repair in mice. *Am. J.Physiol.* 269(6 Pt. 1):L800–L818.

Van Winkle, L. S., Buckpitt, A. R., and Plopper, C. G. (1996) Maintenance of differentiated murine Clara cells in microdissected airway cultures. *Am. J. Respir. Cell Mol. Biol.* 14(6):586–598.

Walker, S. R., Hale, S., Malkinson, A. M., and Mason, R. J. (1989) Properties of isolated nonciliated bronchiolar cells from mouse lung. *Exp. Lung Res.* 15(4):553–573.

Warburton, D., Schwarz, M., Tefft, D., Flores-Delgado, G., Anderson, K. D., and Cardoso, W. V. (2000) The molecular basis of lung morphogenesis. *Mech. Dev.* 92:55–81.

Warburton, D., Tefft, D., Mailleux, A., et al. (2001) Do lung remodeling, repair, and regeneration recapitulate respiratory ontogeny? *Am. J. Respir. Crit. Care Med.* 164(10 Pt. 2):S59–S62.

Warburton, D., Wuenschell, C., Flores-Delgado, G., and Anderson, K. (2002) Commitment and differentiation of lung cell lineages. *Biochem. Cell Biol.* 76:971–995.

Weaver, M., Dunn, N. R., and Hogan, B. L. (2000) Bmp4 and Fgf10 play opposing roles during lung bud morphogenesis. *Development* 127(12):2695–2704.

Weaver, T. E. (1991) Surfactant proteins and SP-D. *Am. J. Respir. Cell Mol. Biol.* 5:4–5.

Wert, S. E., Glasser, S. W., Korfhagen, T. R., and Whitsett, J. A. (1993) Transcriptional elements from the human SP-C gene direct expression in the primordial respiratory epithelium of transgenic mice. *Dev. Biol.* 156(2):426–443.

Whitcutt, M. J., Adler, K., and Wu, R. (1988) A biphasic chamber system for maintaining polarity of differentiation of cultured respiratory tract epithelial cells. *In Vitro Cell. Dev. Biol.* 24:420–428.

Wuenschell, C. W., Sunday, M. E., Singh, G., Minoo, P., Slavkin, H. C., and Warburton, D. (1996) Embryonic mouse lung epithelial progenitor cells co-express immunohistochemical markers of diverse mature cell lineages. *J. Histochem. Cytochem.* 44(2):113–123.

Yoakum, G. H., Lechner, J. F., Gabrielson, E. W., et al. (1985) Transformation of human bronchial epithelial cells transfected by Harvey *ras* oncogene. *Science* 227:1174–1179.

Zeng, X., Wert, S. E., Federici, R., Peters, K. G., and Whitsett, J. A. (1998) VEGF enhances pulmonary vasculogenesis and disrupts lung morphogenesis in vivo. *Dev. Dyn.* 211:215–227.

Zhang, Y. J., O'Neal, W. K., Randell, S. H., et al. (2002) Identification of dynein heavy chain 7 as an inner arm component of human cilia that is synthesized but not assembled in a case of primary ciliary dyskinesia. *J. Biol. Chem.* 277:17,906–17,915.

Zimmermann, B. (1987) Lung organoid culture. *Differentiation* 36: 86–109.

Zimmermann, B. (1989) Secretion of lamellar bodies in type II pneumocytes in organoid culture: effects of colchicine and cytochalasin B. *Exp. Lung Res.* 15:31–47.

Zoechbauer-Mueller, S., Gazdar, A. F., and Minna, J. D. (2002) Molecular pathogenesis of lung cancer. *Annu. Rev. Physiol.* 64: 681–708.

42 Noninvasive Imaging in Stem Cell Therapies

Current State and Future Perspectives

Juri Gelovani, MD, PhD

Determination of the fate of putative stem cells after transplantation requires development of novel cell-labeling and tracing techniques that permit noninvasive whole-body monitoring. Previous methods used in vitro radiolabeling of cells followed by imaging the transplanted radiolabeled cells in vivo, or labeling of cells with supermagnetic agents followed by magnetic resonance imaging with near microscopic resolution. The ex vivo labeling methods provide excellent short-term results but are not suitable for long-term repetitive imaging because of loss of label owing to radiolabel decay or biological clearance of supermagnetic label. This chapter describes approaches for labeling cells with bioluminescent, fluorescent, and positron emission tomography (PET)-reporter genes for imaging adoptively transplanted cells in vivo. Genetic labeling of cells with different reporter genes allows for long-term, repetitive in vivo imaging using different imaging platforms. These imaging platforms include whole-body fluorescence imaging of green fluorescent protein (GFP) reporter gene expression (in rodents), bioluminescence imaging of the firefly luciferase (Luc) reporter gene expression using luciferin as reporter probe (in rodents), PET imaging of HSV1-tk or human TK2 reporter gene using different radiolabeled nucleoside analogs as reporter probes (in rodents and humans), and PET imaging of human D2 receptor with radiolabeled fluoroethylspiperone as reporter probe (in rodents and humans). The most developed approaches utilize herpes virus thymidine kinase (HSV1-tk) as a reporter gene, which produces a gene product (an enzyme) that can be identified by phosphorylation of a radiolabeled reporter probe, which is trapped within the gene-labeled cell and can be visualized by PET scanning. This process can be repeated within a short time because of the short half-life of the radiolabeled substrate. This approach has been further developed using a fusion gene between the thymidine kinase gene and a GFP or luciferase genes for multiplatform imaging. Linking the reporter gene to specific promoters allows imaging of different cell types because only the cells activating the promoter will express the reporter gene. Such systems have been used successfully to trace cells expressing p53, T-cells activated to express nuclear factor of activated T-cells, and cells activated to express transforming growth factor-β. Cells with weak promoters may be detected using a two-step amplification system in which promoter strength is increased by linking the weak promoter to a transcription transactivator. Transplanted stem cells may be followed using a dual-reporter system, which uses a constitutive promoter to image the localization and viability of transplanted cells and an inducible promoter, which is activated when the stem cell commits to a certain differentiation lineage. PET imaging has been used successfully for in vivo monitoring of transplanted T-cells and EMV-specific cytotoxic lympocytes in mice, and the first successful PET imaging of HSV1-tk gene expression in human gene therapy trials has been reported. Noninvasive whole-body imaging of various reporter genes is now being validated in different preclinical models and is expected to play an increasing role in monitoring the fate of transplanted stem cells in human stem cell therapy.

42.1. INTRODUCTION

Significant progress in stem cell research over the past decade is now gradually translating into a number of new and promising methods of therapy for various diseases, including cancer. It is quite obvious that continuous monitoring of the efficacy of many emerging stem cell–based therapeutic approaches cannot be performed using invasive methods (e.g., multiple biopsies), except for blood sampling. This dictates the need to develop techniques for noninvasive monitoring of the fate of stem cells after their administration to patients.

The aim of current molecular-genetic and cellular-imaging research is to develop novel approaches for repetitive in vivo imaging of the fate of stem cells after their administration into an organism. However, for the successful development of such imaging approaches, several components must be developed first, including vectors for transduction of progenitor cells, in vitro models for validation of tracer accumulation in transduced cells,

From: *Stem Cells Handbook*
Edited by: S. Sell © Humana Press Inc., Totowa, NJ

in vivo imaging of reporter gene expression in transduced tumor cells (as a model for validation of the reporter gene imageability in vivo), in vivo imaging of adoptively transferred stem cells, in vivo models of disease to study the therapeutic approaches using progenitor cells, and synthesis and validation of novel radiotracers for reporter gene imaging. In this chapter, several of these components are described to provide an overview of the current status of molecular/cellular imaging in application to monitoring the fate of stem cells after their in vivo transfer.

42.2. METHODS FOR IMAGING ADOPTIVELY TRANSFERRED CELLS

Until recently, immunostaining of the whole-body slices of small animals was the most straightforward, reliable, and traditional approach used for assessment of the localization and targeting of adoptively transferred cells (e.g., T-cells) (Reinhardt et al., 2001). However, the invasive nature of classic pathology techniques precludes the possibility of a repetitive monitoring of cellular trafficking in the same subject. The results obtained from multiple biopsies or from animals sacrificed at different time points, may be flawed by individual variations and require a large number of samples to achieve statistical significance of the results. Another major shortcoming of invasive methods is that they cannot be applied in a clinical setting, especially in biopsy-restricted organs (i.e., brain, heart). Therefore, a method for noninvasive and repetitive evaluation of stem cell trafficking, homing, targeting, differentiation, and persistence would significantly aid the development of many stem cell–based adoptive therapy approaches and facilitate their wider implementation into clinical practice.

42.2.1. EX VIVO RADIOLABELING AND NUCLEAR IMAGING Marking cells for monitoring adoptive therapies with radiolabeled tracers was first tested in the early 1970s. [^{3}H]- and [^{14}C]-labeled uridine was used to assess lymphocyte trafficking in a graft-versus-host disease (GVHD) model (Atkins and Ford, 1975). Other earlier studies used [^{3}H]-glycerol to label cells (Constantin et al., 1997). A similar method, using [^{3}H]-labeled glucose, was used in patients to compare the kinetics of lymphocytes in healthy and human immunodeficiency virus–infected individuals (Hellerstein et al., 1999). However, the use of β-emitting isotopes is limited to studies of cellular kinetics in the blood using sequential blood sampling.

Noninvasive imaging of lymphocyte trafficking dates back to the early 1970s, when the first experiments were performed using extracorporeal labeling of lymphocytes with ^{51}Cr or ^{99m}Tc (Papierniak et al., 1976; Gobuty et al., 1977). Different radioisotopes (e.g., ^{111}In, ^{67}Co, ^{64}Cu) were used for imaging of various ex vivo–labeled immune cells (Papierniak et al., 1976; Gobuty et al., 1977; Rannie et al., 1977; Korf et al., 1998; Adonai et al., 2002). [^{111}In], in particular, found a wide clinical application in oncology as an imaging agent for monitoring the effects of vaccination with tumor-infiltrating lymphocytes (Kasi et al., 1995; Dillman et al., 1997).

Imaging techniques that utilize ex vivo labeling of the adoptively transferred cells have several limitations. A major limitation of this approach is the eventual radiolabel decay, which limits the monitoring of the fate of labeled cells to a period of time corresponding to only a few half-lives of the radionuclide. Another limitation is a relatively low level of radioactivity per cell when labeling with passively equilibrating radiotracers, such as [^{111}In]oxime or [^{64}Cu]PTSM. Furthermore, both [^{64}Cu]PTSM and [^{18}F]FDG gradually efflux out of the labeled cells (Adonai et al., 2002), and a progressive loss of radiolabel occurs during cell division. Exposure of cells to higher doses of radioactivity during labeling is also limited by radiotoxicity.

Significantly higher levels of radioactivity per cell can be obtained with tracers such as [^{18}F]FDG that utilize facilitated transport, enzyme-amplified accumulation, and metabolic entrapment within a labeled cell. More stable labeling of cells can be achieved with radiolabeled thymidine analogs, such as [^{125}I]IUdR, which incorporate into DNA of proliferating cells. Successful noninvasive imaging with [^{131}I]IUdR and [^{124}I]IUdR of tumor proliferative activity has been demonstrated with single-photon emission computed tomography (SPECT) and positron emission tomography (PET), respectively (Tjuvajev et al., 1994; Blasberg et al., 2000) and could be applied to imaging of ex vivo–radiolabeled stem cells. However, such an approach has never been tested.

42.2.2. LABELING WITH SUPERMAGNETIC AGENTS AND MAGNETIC RESONANCE IMAGING Recent developments in magnetic resonance imaging (MRI) have enabled in vivo imaging at near microscopic resolution (Jacobs and Cherry, 2001). To visualize and track stem cells by MRI, it is necessary to tag the cells magnetically. Conventional cell-labeling techniques rely on surface attachment of magnetic beads ranging in size from several hundred nanometers to micrometers (Safarik and Safarikova, 1999). Although these methods are efficient for in vitro cell separation, cell-surface labeling is generally not suitable for in vivo use because of the rapid reticuloendothelial recognition and clearance of labeled cells. Alternatively, various cells have been labeled with small monocrystalline nanoparticles ranging from 10 to 40 nm using fluid-phase or receptor-mediated endocytosis (Weissleder et al., 1997). Unfortunately, the labeling efficiency of this procedure is generally low, particularly in differentiated and nondividing cells (Hawrylak et al., 1993; Schoepf et al., 1998; Bulte et al., 1999; Dodd et al., 1999; Weissleder et al., 2000).

TAT protein-derived peptide sequences have recently been used as an efficient way of internalizing a number of marker proteins into cells (Fawell et al., 1994; Nagahara et al., 1998; Schwarze et al., 1999). Lewin et al., (2000) demonstrated that hematopoietic CD34^{+} and neural progenitor cells (C17.2) can be tagged efficiently with a novel triple-label (magnetic, fluorescent, isotope) supermagnetic nanoparticle called CLIO-Tat. Furthermore, it was demonstrated that the labeled cells retain their capability for differentiation, can be visualized by high-resolution MRI, and can be retrieved from excised tissues and bone marrow using magnetic-sorting techniques. However, the potential for long-term imaging (months or years) using in vitro–labeled progenitor cells is impossible due to eventual biological clearance of the magnetic label.

42.2.3. GENETIC LABELING OF CELLS FOR ADOPTIVE THERAPIES Stable genetic labeling of adoptively transferred cells, such as lymphocytes, with various reporter genes has been used to circumvent the temporal limitations of in vitro radiolabeling or magnetic labeling of cells. A long-term circulation of infused Epstein-Barr virus (EBV)–specific cytotoxic donor-derived T-cells in patients treated for post–bone marrow transplantation (BMT) EBV-LPD was demonstrated by retrovirally transducing T-cells with neomycin resistance gene (Rooney et al., 1995) or LNGFR and HSV1-tk genes (Bonini et al., 1997; Verzeletti et al., 1998), which were detectable in the peripheral blood samples from the patients by polymerase chain reaction (PCR) or fluorescence-activated cell sorting (FACS) analysis.

The effectiveness of stem cell–mediated gene therapy largely depends on efficient gene delivery into long-term repopulating progenitors and on targeted transgene expression in an appropriate progeny of transduced progenitor cells. Different vector types have been used to achieve stable transduction of stem cells. To date, retroviral vectors have been most frequently used to stably transduce different cell types, including stem cells. However, the retroviral-mediated gene transfer has its own limitations. One, probably the most limiting factor, is the requirement of cellular proliferation for retroviral integration into the cell genome. Another limiting factor for retroviral-mediated transduction of many progenitor cells types, such as fetal liver stem cells, is the lack of amphotropic retroviral receptors on their surface (Richardson and Bank, 1996).

By contrast, self-inactivating (SIN) lentiviral vectors are capable of transducing mitotically inactive cells, including stem cells, and can accommodate various nonviral promoters to control therapeutic transgene expression in transduced cells. For example, in a recent report by Cui et al. (2002), two SIN lentiviral vectors were evaluated. The expression of green fluorescent protein (GFP) reporter gene in these vectors was controlled solely by the promoter of either a housekeeping gene EF-1alpha (EF.GFP) or the human HLA-DRalpha gene (DR.GFP), which is selectively expressed in antigen-presenting cells (APCs). It was demonstrated that both vectors efficiently transduced human pluripotent $CD34^+$ hematopoietic progenitor cells capable of engrafting in nonobese diabetic/severe combined immunodeficiency disease (NOD/SCID) mice. When the EF.GFP-encoding vector was used, constitutively high levels of GFP expression were observed in all of the human hematopoietic stem cell progeny (HSC) in NOD/SCID mice and in subsequent in vitro differentiation assays, indicating that engrafting human HPCs have been effectively transduced with GFP gene. By contrast, DR.GFP vector–mediated transgene expression was observed specifically in human $HLA\text{-}DR^+$ cells and highly in differentiated dendritic cells (DCs), which are critical in regulating immunity. Furthermore, human DCs derived from transduced and engrafted human cells potently stimulated allogeneic T-cell proliferation. This study demonstrated successful targeting of transgene expression to APCs/DCs after stable gene transduction of pluripotent HSCs and the ability to selectively detect the differentiation fate of the transduced hematopoietic progenitor cells (HPCs) (Cui et al., 2002). Nevertheless, reporter gene labeling of stem cells and measurement of their concentration in blood alone does not allow for the assessment of their spatial distribution in the body.

42.3. REPORTER GENE IMAGING APPROACHES

42.3.1. BIOLUMINESCENCE REPORTER GENE IMAGING Genetic labeling of lymphocytes with the firefly luciferase (Luc) reporter gene and noninvasive bioluminescence imaging (BLI) has been reported (Hardy et al., 2001). Although the BLI of Luc reporter gene–expressing cytotoxic T-lumphocyte (CTL) allows for the assessment of their distribution throughout the organism, it is semiquantitative at best, because the signal intensity largely depends on thickness and variable optical characteristics of different tissues. Because of this shortcoming, the BLI of Luc reporter gene expression has a limited applicability in the clinic. Nevertheless, BLI is a very convenient imaging platform for research applications, especially for imaging in mice (Contag and Bachmann, 2002).

42.3.2. PET REPORTER GENE IMAGING Tjuvajev et al. (1995) were the first to describe the PET reporter gene imaging paradigm. This paradigm involves administering a radiolabeled probe that is selectively bound or metabolized (e.g., phosphorylated) and trapped by interaction with the gene product (e.g., an enzyme) in the reporter gene–transduced cell. In this manner, the bound or metabolized (e.g., phosphorylated) probe accumulates selectively in transduced tissue, and the level of probe accumulation is proportional to the level of gene product being expressed. It may be useful to consider the reporter imaging paradigm as an enzymatic radiotracer assay that reflects reporter gene expression. Recent reviews by Gambhir et al. (2000) and Blasberg and Gelovani (Tjuvajev) (2002) provide a good overview of the founding principles of noninvasive reporter gene imaging.

42.3.2.1. Herpesviral Thymidine Kinase Herpes simplex virus type 1 thymidine kinase gene (*HSV1-tk*) is a good example of a "reporter gene" for several reasons. *HSV1-tk* has been extensively studied; it is essentially nontoxic in humans and is currently being used in clinical gene therapy protocols as a "susceptibility" gene for treatment of cancer in combination with ganciclovir administration (Banerjee, 1999). A gene product of *HSV1-tk* is an enzyme, HSV1 thymidine kinase (HSV1-TK). Several research groups have shown that *HSV1-tk* can be used as a reporter gene as well as a therapeutic gene (Tjuvajev et al., 1998, 1999a, 1999b, 2001; Bennett et al, 2001; Floeth et al., 2001; Jacobs, A., et al., 1999, 2001a, 2001b; Hackman et al., 2002). This represents an ideal situation in which the therapeutic and reporter genes are the same. In *HSV1-tk* gene therapy protocols, identifying the location and magnitude of HSV1-TK enzyme activity by noninvasive imaging would provide a highly desirable measure of gene expression following successful gene transduction. More important, the acquired sensitivity of *HSV1-tk*-transduced cells to ganciclovir provides an additional margin of safety, because the transduced cells can be easily eliminated by treatment with ganciclovir.

A general paradigm for reporter gene imaging using PET is illustrated in Fig. 1. In this paradigm, *HSV1-tk*, with specific upstream promoter/enhancer elements, is transfected into target cells by a vector. Note that imaging transgene expression is independent of the vector used to transfect/transduce target tissue; namely, any of several currently available vectors can be used (e.g., retrovirus, adenovirus, adenoassociated virus, lentivirus, liposomes). Inside transfected cells, the *HSV1-tk* gene is transcribed to *HSV1-tk* mRNA and then translated on the ribosomes to a protein (enzyme), HSV1-TK. After administration of a complementary radiolabeled reporter probe (FIAU or FHBG) and its transport into transduced cells, the probe is phosphorylated by HSV1-TK (gene product). The phosphorylated reporter probe does not readily cross the cell membrane and is "trapped" within the cell. Thus, the magnitude of reporter probe accumulation in transduced cells reflects the level of HSV1-TK enzyme activity and level of *HSV1-tk* gene expression. Enzymatic amplification of the signal (e.g., level of radioactivity) also facilitates imaging the location and magnitude of probe accumulation and reporter gene expression. It may be useful to consider this reporter-imaging paradigm as an in vivo enzymatic radiotracer assay that reflects reporter gene expression. Viewed from this perspective, *HSV1-tk* reporter gene imaging is similar to imaging hexokinase activity with fluorodeoxyglucose.

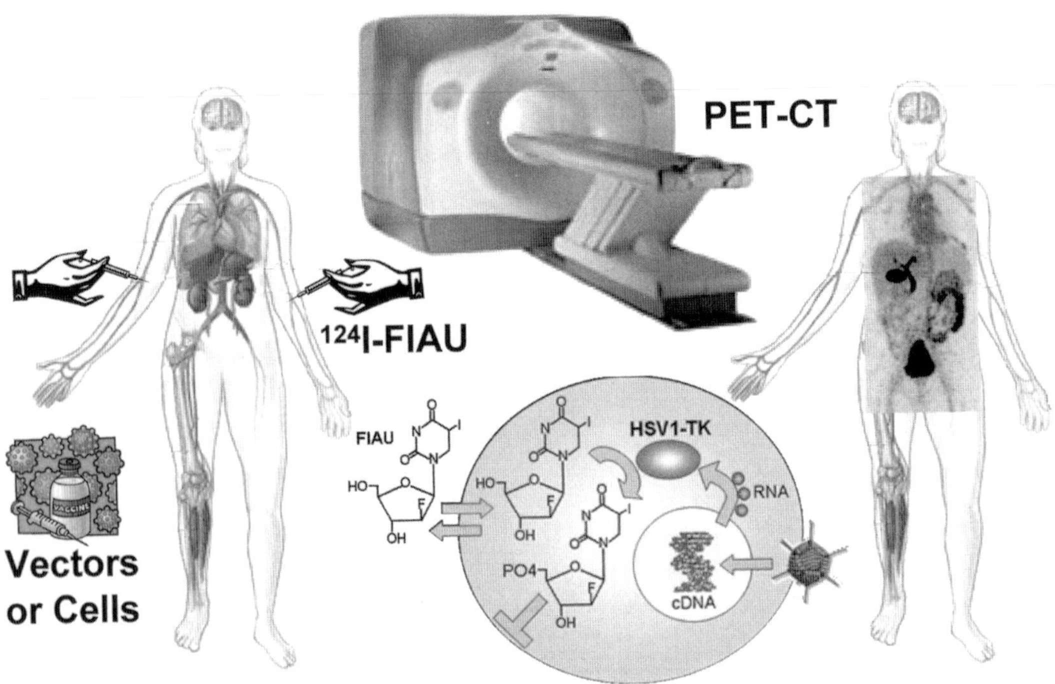

Fig. 1. Paradigm for imaging *HSV1-tk* reporter gene expression with FIAU reporter probe. The *HSV1-tk* gene complex is transfected into stem cells by a vector in vitro or is coexpressed along with another "therapeutic" gene after in vivo administration of a gene delivery vector (left). Inside the transfected cell, the *HSV1-tk* gene is transcribed to HSV1-tk mRNA and then translated on the ribosomes to a protein (enzyme), HSV1-TK. After administration of a radiolabeled FIAU and its transport into the cell, the FIAU is phosphorylated by HSV1-TK enzyme (gene product). The phosphorylated radiolabeled FIAU-MP does not readily cross the cell membrane and is "trapped" within the cell. Thus, the magnitude of probe accumulation in the cell (level of radioactivity) reflects the level of HSV1-TK enzyme activity and level of *HSV1-tk* gene expression. The images of radiolabeled FIAU accumulation can be visualized three-dimensionally with PET and coregistered with an anatomical image obtained with CT or MRI. The newer PET-CT tomographs represent a merger of CT and PET devices into one, and the images can be obtained during the same imaging study without moving the patient.

Currently, *HSV1-tk* is the most widely used reporter gene for experimental nuclear (PET) imaging, and the two most widely used radiolabeled reporter probes are [^{124}I]FIAU and [^{18}F]FHBG. The ability to image "late," 24 or more hours after administration of [^{124}I]FIAU, provides an advantage in the ability to achieve high image specificity due to physiological washout of background radioactivity and retention of the radioactivity in *HSV1-tk*-transduced tissue. In addition, the HSV1-tk and eGFP fusion reporter gene (TKGFP) was developed and allows for optical microscopic and whole-body fluorescence imaging, as well as PET imaging (Jacobs et al., 1999). Recent reports demonstrated the advantages of TKGFP reporter gene over the wild-type (WT) *HSV1-tk* when applied to multimodality imaging of transcription of p53-dependent endogenous genes (Doubrovin et al., 2001) and T-cell activation (Ponomarev et al., 2001b). Another advantage of the dual reporter system is the ability to compare the images of reporter gene expression obtained with PET, gamma camera, or autoradiography with corresponding *in situ* GFP fluorescence images. Comparison of GFP fluorescence and autoradiographic images, coupled with histology of corresponding tissue sections, provides for spatial and quantitative assessments of reporter gene expression at the microscopic as well as macroscopic level.

For PET imaging of *HSV1-tk* reporter gene expression, the FIAU is usually labeled with ^{124}I (or with ^{123}I or ^{131}I for gamma scintigraphy and SPECT) and "late" imaging is usually performed. The feasibility of "early" FIAU imaging has recently been demonstrated (Gelovani Tjuvajev et al., 2002). Good-quality images can be obtained within the first 10 min to 2 h following administration of radiolabeled FIAU. Thus, radiolabeling of FIAU with ^{18}F or ^{11}C could be used as an alternative tracer to [^{18}F]FHBG; it would provide additional positron-emitting FIAU probes with higher *HSV1-tk* sensitivity and dynamic range than [^{18}F]FHBG. In addition, the relatively short half-life of these radionuclides would facilitate repeated or sequential PET imaging of *HSV1-tk* expression in the same subject.

Already, the first successful reporter gene imaging study in humans was performed in Germany and reported by Jacobs et al. (2001b), as the result of technology transfer from the research group at MSKCC (New York, NY). In this study, a liposome-encapsulated plasmid-mediated direct intratumoral *HSV1-tk* gene delivery was performed in human patients with glioblastoma followed by administration of [^{124}I]FIAU and PET imaging of *HSV1-tk* gene expression. In another report, Yaghoubi et al. (2001) presented data on human pharmacokinetics of [^{18}F]FHBG and PET imaging in healthy volunteers. Therefore, the possibility of clinical PET imaging of *HSV1-tk* reporter gene expression for monitoring the fate of HSV1-tk-transduced stem cells in the nearby future is very realistic.

42.3.2.2. Nonimmunogenic Reporter Genes for Long-Term Repetitive Imaging in Humans and Mice An important requirement for long-term genetic labeling of progenitor cells is the nonimmunogenicity of a reporter gene product to avoid the possibility of rejection of transduced progenitor cell–derived tissues. From this perspective, the *HSV1-tk*–FIAU reporter gene imaging system is not quite suitable for imaging the adoptively transferred cells.

In a number of publications, it was demonstrated that human D2R receptor could be used as a reporter gene and that it could be imaged with [^{18}F]FESP and PET. The D2R-FESP imaging approach is sensitive and could be used for monitoring reporter gene expression in conjunction with various reporter systems (Liang et al., 2001). Recently, a mutant D2R80A reporter was developed in which the ligand binding is uncoupled from signal transduction. Consequently, D2R80A is a nonimmunogenic gene that is suitable for long-term expression in progenitor cells for repetitive imaging. Also note that [^{18}F]FESP has been used for many years for imaging dopaminergic status in patients with parkinsonism and depression (Satyamurthy et al., 1990).

Another nonimmunogenic reporter gene was developed only recently and is derived from the human mitochondrial thymidine kinase type two (ΔhTK2) in which the mitochondrial localization signal located in the N-terminal was deleted. It was found serendipitously that the ΔhTK2 enzyme does very effectively phosphorylate the radiolabeled FIAU and its 5-ethyl derivative of FIAU (FEAU), which are currently used for imaging *HSV1-tk* reporter gene expression (Gelovani Tjuvajev, et al., unpublished observations). Because hΔTK2 is nonimmunogenic in humans, it could be readily translated into the clinical setting for PET imaging of stem cells with [^{124}I]FIAU, [^{18}F]FEAU, or [^{11}C]FEAU.

In a recent report by Ponomarev et al. (2002), a fusion gene between the hΔTK2 cDNA and GFP was described (hΔTK2GFP). The hΔTK2GFP cDNA (fusion gene) was cloned into a retrovirus and used to transduce U87 human glioma cells. A mixed population of transduced U87/hΔTK2GFP cells was obtained by FACS in which a highly intensive green fluorescence could be visualized with fluorescence microscopy (Fig. 2). The transduced cells exhibited high rates of [^{14}C]FIAU and [^{14}C]FEAU accumulation compared with the WT cells, which did not show any dynamics of accumulation of either tracer. A bicistronic MoMLV-based retroviral vector in which the hΔTK2 and GFP cDNAs are linked by the polyviral IRES sequence (hΔTK2iresGFP) was developed as well, in which the resulting hΔTK2iresGFP mRNA encodes both proteins; the hΔTK2 and GFP proteins are translated as separate proteins (one by cap-mediated translation, and another by IRES-mediated translation). The transduced U87/hΔTK2iresGFP cells had intensity of green fluorescence and radiotracer (FIAU and FEAU) accumulation similar to that in U87/hΔTK2GFP cells. Then, the U87/hΔTK2GFP and WT U87 xenografts were established in the opposite shoulders of SCID mice. When the xenografts reached palpable size (approx 0.5 cm in diameter), [^{124}I]FIAU (150 μCi/animal) was injected intravenously and micro-PET imaging was performed at 2 and 24 h. In vivo PET imaging at 2 and 24 h showed highly specific localization of radioactivity to U87/hΔTK2GFP tumors, with only the background levels of radioactivity accumulation in the WT tumor xenografts (Fig. 3). Tissue sampling at 24 h post–tracer injection confirmed the results of noninvasive imaging. The U87/hΔTK2GFP xenograft/plasma and xenograft/muscle radiotracer accumulation ratios were 87 ± 17 and 27 ± 11, respectively; the U87 xenograft/plasma and xenograft/muscle ratios were 14 ± 2 and 4 ± 0.3, respectively.

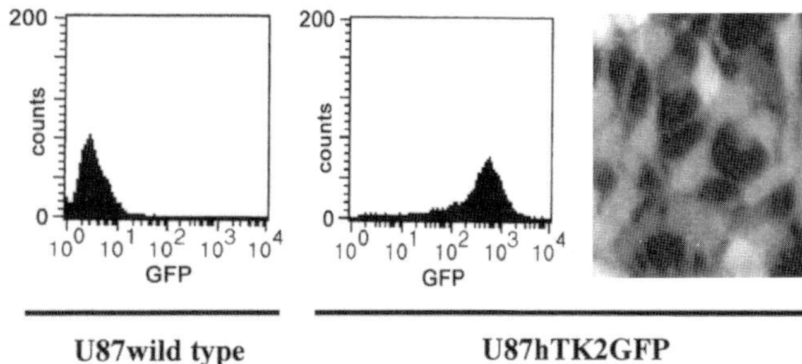

Fig. 2. FACS and fluorescence microscopy of WT and hTK2GFP transduced U87 human glioma cells. The WT U87 cells exhibit only background levels of fluorescence, whereas the U87hTK2GFP cells exhibit highly intensive green fluorescence that is pancellular (no mitochondrial localization).

Based on these results, PET imaging of ΔhTK2 reporter gene expression in transduced stem cells after their administration in humans should be feasible and could be readily applied in a clinical setting. The murine variants of the ΔTK2 reporter gene (mΔTK2, mΔTK2GFP, and mΔTK2iresGFP) have also been developed and are undergoing validation for imaging experimental stem cell therapies in mice.

It is noteworthy that both human and murine ΔTK2 enzymes cannot phosphorylate the radiolabeled acycloguanine analogs, including tritiated ganciclovir, acyclovir, penciclovir, and [^{18}F]FHBG. The inability to phosphorylate radiolabeled acycloguanosine analogs is an important feature of both hΔTK2 and mΔTK2 with respect to multi-gene-imaging applications. Further in this chapter the utilization of ΔTK2 reporter gene in different dual reporter gene–imaging approaches is discussed in greater detail.

Another nonimmungenic reporter gene is human type two somatostatin receptor (hSSTr), which can be imaged using different radiolabeled peptide ligands including [^{99m}Tc]P829 or [^{188}Re]P829, [^{99m}Tc]P2045, and [^{111}In]octreotide. These studies demonstrated that the hSSTr reporter gene–transduced solid tumors could be easily visualized and differentiated from normal (nontransduced) tissues using these radiolabeled peptide probes. The efficacy of the hSSTr reporter gene imaging was demonstrated following adenoviral-mediated in vivo gene delivery in a metastatic ovarian cancer model in mice using [^{111}In]octreotide (Rogers et al., 1999) and [^{99m}Tc]P2045 (Chaudhuri et al., 2001).

42.3.2.3. PET Reporter Imaging System for Central Nervous System Applications The radiolabeled nucleoside analogs are not suitable tracers for imaging reporter gene expression in central nervous system (CNS), because they do not cross the normal blood–brain barrier (BBB). To address this issue, several other novel reporter gene–reporter probe combinations are being developed in which the reporter probe readily crosses the BBB and cell membranes.

One of the novel reporter systems is based on the bacterial xanthine phosphoribosyl transferase gene (XPRT from *Escherichia coli*) and radiolabeled xanthine. Xanthine is a purine base

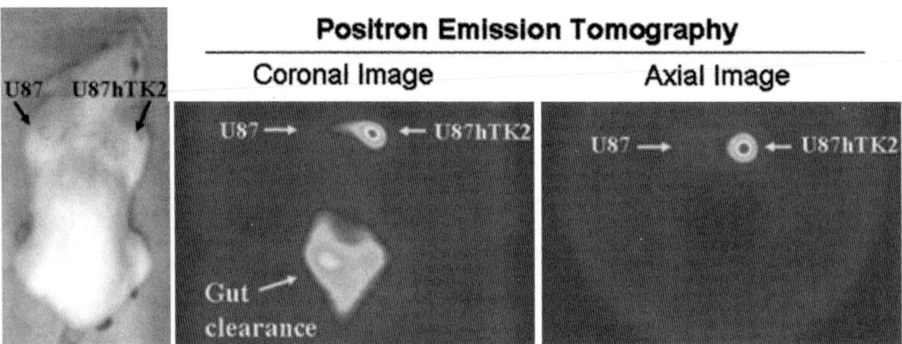

Fig. 3. In vivo PET imaging of ΔhTK2 expression in transduced and WT U87 xenografts in mice. The two tumor xenografts are located in the shoulder regions (left, arrows). The coronal and axial images, obtained at 2 h after [^{124}I]FIAU administration through both tumors, demonstrate highly specific accumulation of [^{124}I]FIAU in U87hTK2GFP tumor, whereas no accumulation of the radiolabeled probe is observed in the WT U87 xenograft.

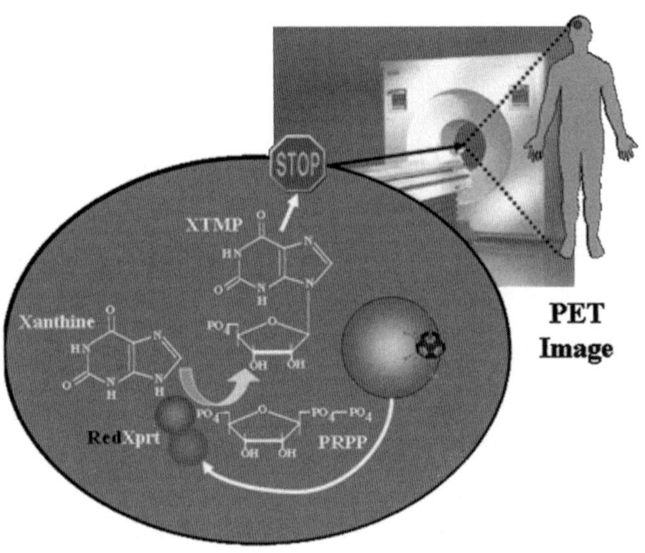

Fig. 4. Paradigm of RedXPRT-xanthine reporter gene imaging system for CNS applications. The RedXPRT fusion reporter gene is transfected into stem cells by a vector in vitro or is coexpressed along with another "therapeutic" gene after in vivo administration of a gene delivery vector. Inside the transfected cell, the RedXPRT gene is transcribed to RedXPRT mRNA and then translated on the ribosomes to a protein (red fluorescent enzyme), RedXPRT. After administration of a radiolabeled xanthine and its transport into the cell, the xanthine is ribophosphorylated by RedXPRT enzyme. The radiolabeled xanthosine-MP does not readily cross the cell membrane and is "trapped" within the cell. Thus, the magnitude of probe accumulation in the cell (level of radioactivity) reflects the levels of RedXPRT enzyme activity and RedXPRT gene expression. The images of radiolabeled [^{11}C]xanthine accumulation can be visualized three dimensionally with PET and coregistered with an anatomical image obtained with CT or MRI.

and xanthine derivatives (e.g., coffein, theophylin) readily cross the BBB (Spector, 1987; Habgood et al., 1998; Franke et al., 1999). The XPRT enzyme ribophosphorylates xanthine to xanthosine monophosphate, which is selectively trapped in XPRT-transduced cells. To facilitate the multimodality visualization of the reporter gene product, the fusion between XPRT and red fluorescent protein type 2 from a *Discosoma* corral (XPRTRed2) was developed (Fig. 4). Whole-body optical imaging and [^{14}C]-xanthine autoradiographic studies were performed in mice bearing intracerebral xenografts produced from either the RG2/TKGFP glioma cells or RG2/XPRTRed2 cells. Based on the results of previous studies, small intracerebral RG2 gliomas exhibit very little BBB disruption (if not at all) and could be used as a model for assessment of radiotracers for their permeability through the BBB (Miyagawa et al., 1998). At 40 min post–iv injection, radiolabeled [^{14}C]xanthine is accumulated in intracerebral RG2/XPRTRed tumors but not in intracerebral RG2/TKGFP tumors (which served as negative control for irrelevant gene expression; Fig. 5). These results indicate that the XPRT-xanthine reporter system could be applied to imaging genetically labeled stem cells localized in the CNS, even with no BBB disruption (Ponomarev et al., 2001b).

42.4. PET IMAGING OF REPORTER GENE EXPRESSION DRIVEN BY ENDOGENOUS MOLECULAR-GENETIC PROCESSES

Imaging regulation of endogenous gene expression in living animals (and potentially in human subjects) using noninvasive imaging techniques will provide a better understanding of normal and cancer-related biological processes. Modern molecular techniques provide the opportunity to design specific reporter gene constructs, in which the reporter gene is placed under the control of different promoter/enhancer elements. For example, they can be "always turned on" by constitutive promoters (such as LTR, RSV, CMV, PGK, EF1, etc.). Alternatively, the promoter/enhancer elements can be constructed to be "inducible" and "sensitive" to activation and regulation by specific endogenous transcription factors and promoters (factors that bind to and activate specific enhancer elements in the promoter region of the endogenous genes and reporter vector construct leading to the initiation of reporter gene transcription). Reporter gene expression can be made tissue specific, in which enhancer elements are activated by promoter/transcription factors that are selectively expressed in specific tissue, such as the PSMA promoter that is selectively active in prostate cancer cells (Zhang et al., 2002), the albumin promoter expressed selectively in liver (Sun et al., 2001), or the corcinoembryonic antigen (CEA) promoter selectively expressed in colorectal cancer (Qiao et al., 2002). The most simple approach for imaging endogenous gene expression regulation is based on a so-called *cis*-reporter system. In the *cis*-reporter system, the expression of a reporter gene is directly regulated by a target-specific enhancer/promoter element positioned upstream of the reporter gene; a reporter gene is expressed whenever the endogenous genes are transcriptionally activated via similar enhancer/promoter elements. Using this approach, several reporter systems have been constructed to noninvasively image specific endogenous molecular processes, including the regulation of endogenous gene expression (Doubrovin et al., 2001), the activity of specific signal transduction pathways (Ponomarev et al., 2001a), as well as specific protein–protein interactions (Luker et al., 2002), and posttranscriptional regulation of protein expression (Mayer-Kuckuk et al., 2002). In the context of stem cell therapies, an important question that can be answered by noninvasive imaging using tissue-specific (or differentiation step-specific) promoters to drive the expression of a reporter gene is, What are the types of progeny tissues (e.g., vascular, neuronal, glial, cardiomyocitic, cartilage)? Several examples of imaging endogenous gene expression mediated by the activities of different signal transduction pathways are described next.

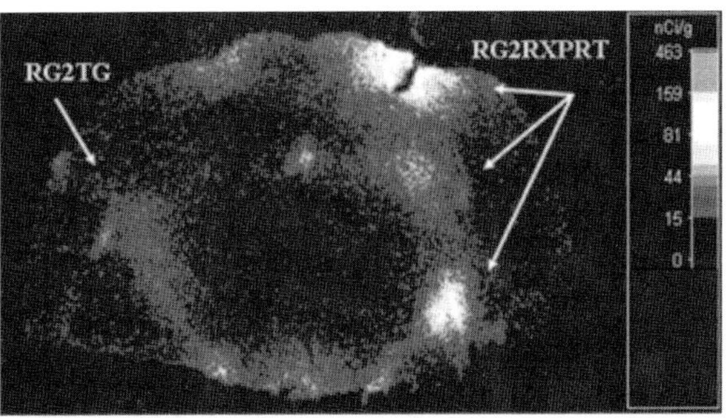

Fig. 5. Quantitative autoradiography of RedXPRT expression with [^{14}C]xanthine of an axial section through mouse brain. There are two small intracerebral tumors in this section: RG2/TKGFP on the left side (with green fluorescence) with no accumulation of [^{14}C]xanthine, and RG2/XPRTRed on the right side (with red fluorescence) that accumulates [^{14}C]xanthine. The autoradiograms were obtained at 40 min after [^{14}C]xanthine administration to model the clinical situation in which ^{11}C labeled xanthine with a 20-min decay half-life will be used.

42.4.1. IMAGING p53-MEDIATED SIGNALING A recent study by Doubrovin et al. (2001) was the first to demonstrate that noninvasive imaging of endogenous gene expression in rats is feasible using a *cis*-reporter system. Specifically, it was demonstrated that p53-dependent gene expression can be imaged in vivo with PET and by *in situ* fluorescence. A retroviral vector (*Cis*-p53/*TKeGFP*) was generated by placing the herpes simplex virus type 1 thymidine kinase (*tk*) and enhanced green fluorescent protein (*egfp*) fusion gene (*TKGFP*) under control of a p53-specific response element. DNA damage–induced upregulation of p53 transcriptional activity was demonstrated and correlated with the expression of p53-dependent downstream genes (including p21). These findings were observed in U87 (p53+/+) cells and xenografts, but not in SaOS (p53–/–) cells (Fig. 6). The PET images corresponded with upregulation of genes in the p53 signal transduction pathway (p53-dependent genes) in response to DNA damage induced by BCNU chemotherapy. Thus, PET imaging of p53 transcriptional activity in tumors using the *Cis*-p53TKGFP reporter system could be used to assess the effects of new drugs or other novel therapeutic paradigms that are mediated through p53-dependent pathways. A similar approach could be used to study p53-dependent apoptosis in stem cells (or their progeny) as the result of various physiological and local tissue environment or various therapeutic interventions (e.g., radiation, chemotherapy).

42.4.2. IMAGING NUCLEAR FACTOR OF ACTIVATED T-CELLS–MEDIATED SIGNALING T-cell activation is an essential component of the immune response in many normal and disease states. The objective of a recent study by Ponomarev et al. (2001a) was to detect and monitor T-cell receptor (TCR)–dependent activation of T-cells in vivo using noninvasive PET imaging. A retroviral vector (*Cis*-NFAT/TKeGFP) was generated by placing the fusion gene *TKGFP* under control of the nuclear factor of activated T-cells (NFAT) response element. A human T-cell leukemia Jurkat cell line that expresses a functional TCR was transduced with the *Cis*-NFAT/TKGFP reporter vector and used as a model of T-cell activation in these studies. Known activators of T-cells (anti-CD3 and anti-CD28 antibody) produced significantly higher levels of *TKGFP* reporter gene expression (increased GFP fluorescence, increased levels of HSV1-tk mRNA, and increased [^{14}C]FIAU accumulation in vitro) in *Cis*-NFAT/TKGFP+ Jurkat cells, in comparison with nontreated or nontransduced cells. In mice with sc *Cis* NFAT/TKGFP+ Jurkat cell infiltrates, similar results were observed using micro-PET imaging and in vivo fluorescence imaging. A strong correlation between *TKGFP* expression and upregulation of T-cell activation markers (CD69 and interleukin-2 production) was demonstrated both in vitro and in vivo (Fig. 7). These results demonstrated that activation of the NFAT signal transduction pathway occurs after TCR stimulation, and that PET imaging of T-lymphocyte activation following TCR engagement is feasible using the described *TKGFP*-based *Cis*-reporter system. This imaging paradigm could be used to assess the efficacy of novel antitumor vaccines and adoptive immunotherapy.

A similar approach could be applied to imaging stem cell–derived T-cell differentiation and activation post-BMT. Various therapeutic interventions (e.g., cytokine therapy, gene therapy) aimed at directing and stimulating the differentiation of stem cells in vivo toward the T-cell lineage posttransplantation and the activation capacity of the reconstituted T-cells could be monitored by noninvasive PET imaging.

42.4.3. TRANSFORMING GROWTH FACTOR-β (TGFβ) The cytokine transforming growth factor-β (TGF-β) plays a dual role in tissue differentiation and tumorigenesis. It stimulates the proliferation of mesenchymal cells while inhibiting the growth of most normal epithelial cells. A reporter system for noninvasive imaging of TGF-β receptor signaling activity during tumor development and growth has been recently reported by Serganova et al. (2002a). A Cis-TGF-β/TKGFP reporter vector was developed containing both AML1 and Smad consensus binding sites (Igα-Smad-AML-TKGFP reporter gene) as the enhancer elements to regulate the expression of TKGFP reporter gene. MDA-MB-231 cells were stably transduced with the Igα-Smad-AML-TKGFP reporter system and exposed to various concentrations of human TGF-β1 for 48 h; the level of TKGFP expression was assayed by fluorescence microscopy, quantified by FACS and a radiotracer assay ([^{14}C]FIAU and [^{3}H]dThd). Transcriptional activation of

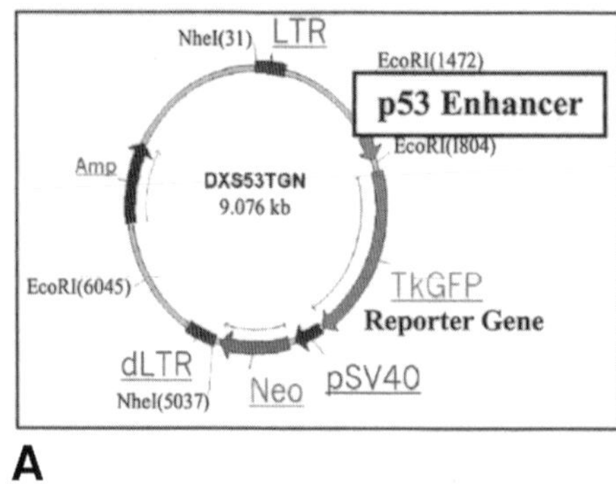

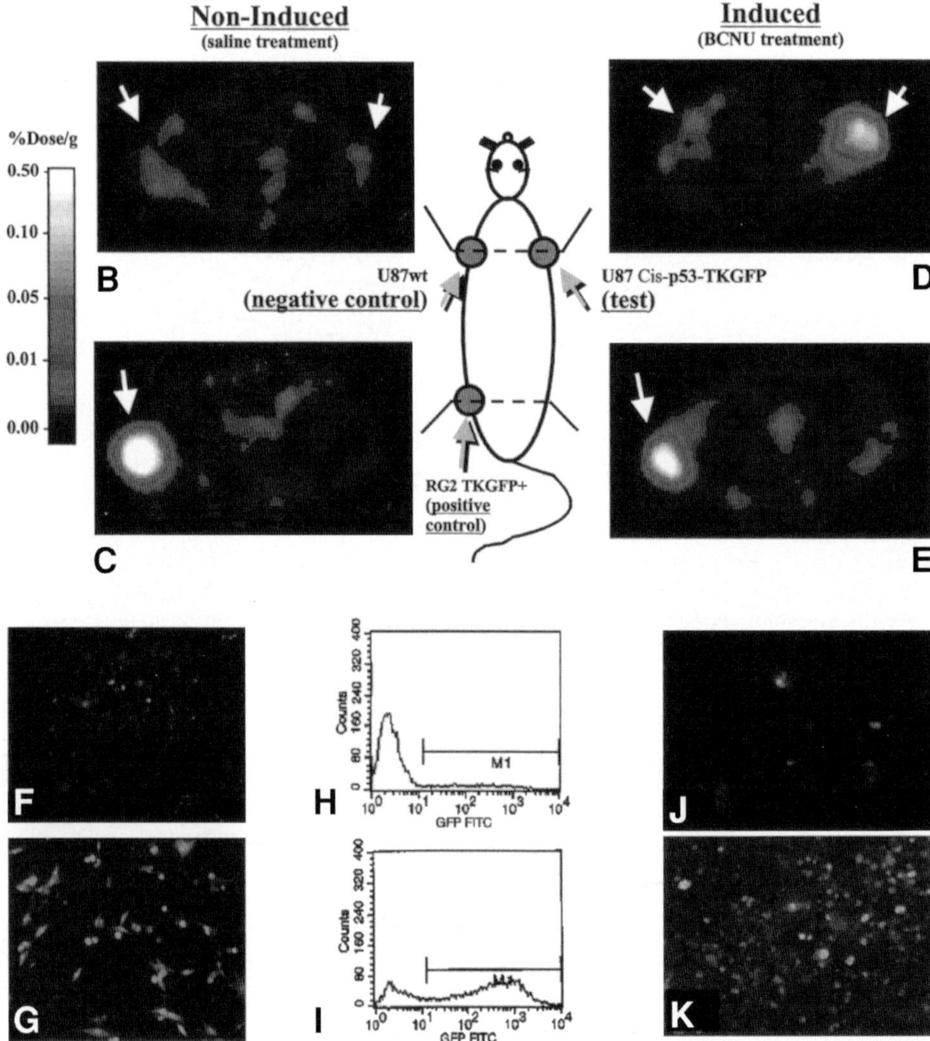

Fig. 6. PET imaging of endogenous p53 activation. **(A)** Vector structure. Transaxial PET images through the shoulder **(B,D)** and pelvis **(C,E)** of two rats are shown (top); the images are color coded to the same radioactivity scale (%dose/g). An untreated animal is shown on the left, and a BCNU-treated animal is shown on the right. Both animals have three sc tumor xenografts: U87p53TKGFP (test) in the right shoulder, U87 WT (negative control) in the left shoulder, and RG2TKGFP (positive control) in the left thigh. The nontreated animal on the left shows localization of radioactivity only in the positive control tumor (RG2TKGFP); the test (U87p53TKGFP) and negative control (U87wt) tumors are at background levels. The BCNU-treated animal on the right shows significant radioactivity localization in the test tumor (right shoulder) and in the positive control (left thigh), but no radioactivity above background in the negative control (left shoulder). Fluorescence microscopy and FACS analysis of a transduced U87p53/TKGFP cell population in the noninduced (control) state **(F,J)** (B), and 24 h after a 2-h treatment with BCNU 40 μg/mL **(G,K)** are presented. Assessment of the *cis*-p53/TKGFP reporter system in U87p53/TKGFP sc tumor tissue is shown (right). Fluorescence microscopy images of U87p53/TKGFP sc tumor samples obtained from nontreated rats **(H)** and rats treated with 40 mg/kg of BCNU intraperitoneally **(I)** are also shown. FITC, fluorescein isothiocyanate. (Adapted from Doubrovin et al., 2001.)

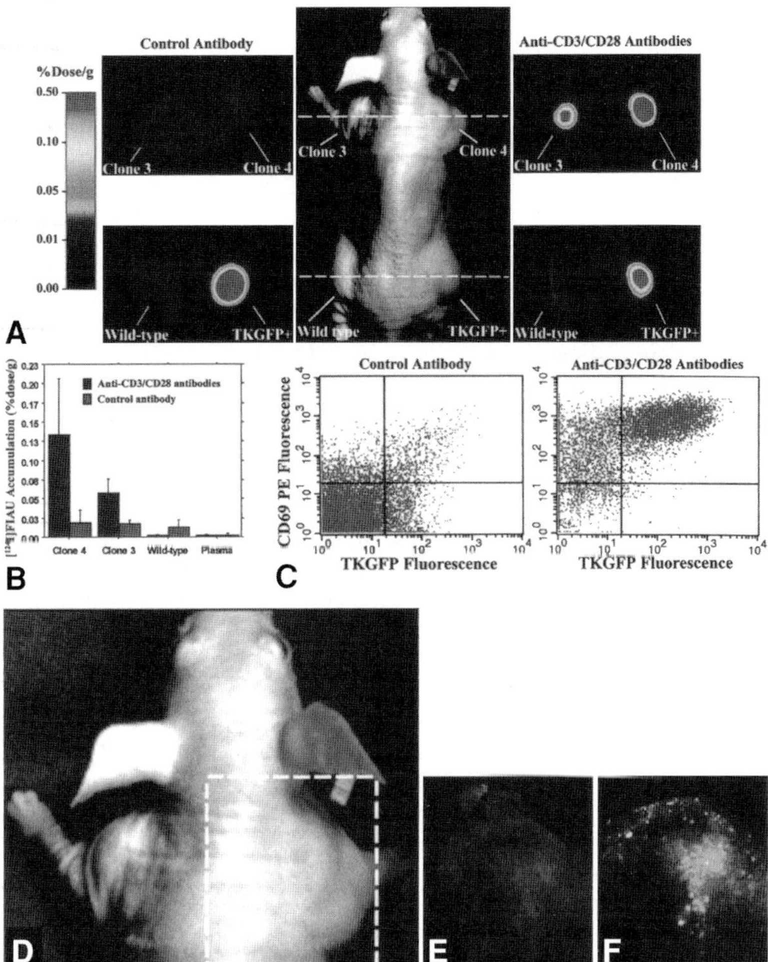

Fig. 7. Imaging NFAT-TKGFP reporter system activity with [^{124}I]FIAU and PET plus assessments in tissue samples. **(A)** Photographic images of a typical mouse bearing different sc infiltrates (middle), transaxial PET images (GE Advance tomograph) of TKGFP expression in a mouse treated with control antibody (left), and anti-CD3/CD28 antibodies (right) were obtained at the levels indicated by the dashed lines. **(B)** [^{124}I]FIAU accumulation (%dose/g) in tissue samples of the Jurkat/dcmNFATtgn clone 3 and 4 infiltrates, WT Jurkat infiltrates, and blood plasma, obtained after PET imaging is shown. **(C)** FACS profiles of TKGFP and CD69 expression in a tissue sample from the same Jurkat/dcmNFATtgn clone 4 infiltrate that was imaged with PET are presented. **(D)** Images of TKGFP fluorescence in an sc Jurkat/dcmNFATtgn (clone 4) infiltrate were obtained in the same animal; first before **(E)** and then after **(F)** treatment with anti-CD3/CD28 antibodies. (Adapted from Ponomarev et al., 2001.)

the TGF-β signaling pathway was observed after 16 h of exposure to minimal concentrations of TGF-β1 (2 ng/mL) and nutrient depletion. TGF-β1-treated cells exhibited high levels of TKGFP reporter protein as measured by FACS, and significantly higher levels of FIAU accumulation were observed compared to baseline, nonstimulated cells. Successful imaging of TGF-β receptor–mediated transcriptional activity was demonstrated in vivo with ^{124}FIAU and PET in Igα-Smad-AML-TKGFP-transduced tumor xenografts in mice after systemic administration of TGF-β.

The ability to image the activity of the TGF-β signaling pathway could be used to assess stem cell therapies directed at restoration of connective tissue, cartilage, and bone remodeling, or in other stem cell–derived progeny tissues in which the TGF-β signaling pathway may play an important role in homing or tissue differentiation.

42.4.4. IMAGING WEAK PROMOTERS Several attempts have been made to image reporter gene expression driven by weak promoters (e.g., CEA, prostate-specific antigen promoter [PSE]), but the results were usually hampered by the poor transcriptional activity of such promoters. In the original study, Qiao et al. (2002) validated methods to enhance the transcriptional activity of the CEA promoter using a *trans*- or so-called two-step transcriptional amplification (TSTA) system. To increase promoter strength while maintaining tissue specificity, a recombinant adenovirus was constructed that contained a TSTA system with a tumor-specific CEA promoter driving a transcription transactivator, which then activates a minimal promoter to drive expression of the *HSV1-tk* suicide/reporter gene. This ADV/CEA-binary-tk system resulted in equal or greater cell killing of transduced cells by ganciclovir in a CEA-specific manner, compared with ganciclovir killing of cells transduced with a CEA-independent vector containing a constitutive viral promoter driving HSV-tk expression (ADV/RSV-tk). It was demonstrated in vitro that the expression of the *HSV1-tk* gene mediated by the CEA-specific TSTA system was about 250-fold higher than that from the *cis*-CEA promoter alone. To test the imaging capacity of the ADV/CEA-binary-*tk* system for monitoring CEA-specific transgene expression in tumor-bearing animals, [^{131}I]FIAU and gamma camera imaging were used to assess adenovirus-mediated HSV-tk gene expression in vivo. After injection of the ADV/binary-tk virus into CEA-expressing liver tumors, [^{131}I]FIAU accumulation was observed only in the area of CEA-positive tumors. More important, there was significantly less *HSV1-tk* expression in adjacent liver tissue following ADV/binary-tk injection than after the constitutively expressing ADV/RSV-tk virus. A dose escalation therapeutic trial in these animals demonstrated significantly reduced hepatoxicity following ganciclovir administration with the ADV/binary-tk virus compared with the ADV/RSV-tk virus, with equal antitumor efficacy.

In another study, Iyer et al. (2001) validated methods to enhance the transcriptional activity of the androgen-responsive PSE using a similar TSTA approach to amplify expression of firefly luciferase, and mutant herpes simplex virus type 1 thymidine kinase (HSV1-sr39tk) in a prostate cancer cell line (LNCaP). An approx 50-fold (luciferase) and a 12-fold (HSV1-sr39tk) enhancement of transgene expression was demonstrated by using this TSTA system while retaining tissue selectivity. A cooled charge-coupled device (CCD) optical imaging system was used to visualize the signal generated from the luciferin–luciferase reporter system in living mice implanted with LNCaP cells that were transfected ex vivo. Further improvements of the androgen-responsive TSTA system for reporter gene expression were made using a "chimeric" TSTA system that uses duplicated variants of the prostate-specific antigen gene enhancer to express GAL4 derivatives fused to one, two, or four VP16 activation domains. The resulting activators were transduced into to prostate carcinoma cells with reporter templates bearing one, two, or five GAL4-binding sites upstream of firefly luciferase. The activity of the luciferase reporter gene in transfected cell extracts and in live nude mice was monitored using an *in situ* luciferase assay and cooled CCD imaging, respectively. It was found that luciferase expression in prostate cancer cells in vitro can be varied over an 800-fold range, and that a single plasmid bearing the optimized chimeric GAL4-VP16 enhancer expressed the firefly luciferase reporter at 20-fold higher levels than the CMV enhancer. A very encouraging result was the demonstration that the TSTA system was androgen concentration sensitive, suggesting a continuous rather than binary reporter response. However, as observed with the CEA-TSTA reporter system above, in vivo imaging comparison of the TSTA and *Cis*-reporter systems showed substantially less dramatic differences than that obtained by the in vitro analyses. Only about a fivefold difference in response to androgen stimulation was observed between PSE TSTA and *Cis*-reporter systems in the in vivo firefly luciferase optical imaging experiments.

42.5. DUAL-REPORTER SYSTEMS FOR IMAGING ENDOGENOUS CELLULAR SIGNALING ACTIVITY OR DIFFERENTIATION

Significant efforts are being made to develop novel approaches for imaging progenitor cell differentiation after their administration in vivo. One approach is based on the coexpression of a "beacon" reporter gene from a strong constitutive promoter to visualize the localization of transduced stem cells and their progeny tissues; another reporter gene (different from the "beacon") could be linked to a promoter element that is selectively activated at a certain stage of progenitor cell differentiation or to a promoter that is active only in a specific tissue type on terminal differentiation and serve as a "sensor" gene (e.g., for neurons, the neuron-specific T alpha 1 tubulin promoter; for glial cells, the GFAP promoter; for endothelial cells, Tau2 promoter; for T-cells, RAG or TCR promoters).

Several dual-reporter systems are currently being developed at MSKCC (New York, NY), in which one reporter gene (e.g., hΔTK2) will be expressed constitutively as a "beacon" gene and the HSV1-sr39tk will be expressed as a "sensor" gene in a stem cell differentiation–specific inducible manner. It should be possible to image the localization of transduced progenitor cells with PET using [^{11}C]FEAU because it is a good substrate for hΔTK2 "beacon" enzyme (and by the HSV1-sr39TK "sensor" enzyme) and has a fast radiolabel decay (^{11}C $t_{1/2}$ = 20 min). Four hours after PET imaging with [^{11}C]FEAU, the differentiation status could be selectively imaged using [^{18}F]FHBG via the expression of a tissue-specific promoter-regulated HSV1-sr39tk "sensor" gene, because the h_ΔTK2 "beacon" enzyme does not phosphorylate [^{18}F]FHBG or other acycloguanozine analogs. Alternatively, the mutant D2R80A (Liang et al., 2001) could be used as a "sensor" gene and its expression imaged with [^{18}F]FESP and PET.

The following is an example of such an imaging strategy. The dual-reporter Cis-HRE/TKGFP-cmv/RedXPRT vector was developed for imaging hypoxia-induced transcriptional upregulation of HIF-1α-responsive genes in vitro and in vivo (Serganova et al., 2002b). In this vector, the TKGFP "sensor" reporter gene is controlled by the eight tandem repeats of hypoxia response elements (8xHRE) from the vascular endothelial growth factor (VEGF) gene promoter; the RedXPRT fusion reporter gene serves as a "beacon" and is driven by the CMV promoter. Multicellular spheroids derived from C6-XPRTRed/HRE-TKGFP cells demonstrated a size-dependent induction of hypoxia-induced HIF-1α-mediated signaling. C6 spheroids >200 μ*M* developed a hypoxic core and could be visualized by fluorescence confocal microscopy. This observation is explained by a decrease in oxygen diffusion and development of a hypoxic core with the increasing spheroid size (Fig. 8).

C6 rat glioma cells were transduced with the reporter vector and the functionality of the dual-reporter system was verified by treating the cells with $CoCl_2$. Single cell–derived clones of transduced C6 cells with low background and high inducibility were obtained by FACS and expanded in vitro. Hypoxia was induced by exposing cells to various concentrations of $CoCl_2$ (1–500 μ*M*). TKGFP expression was assessed by fluorescence microscopy, FACS, and radiotracer assay (^{14}C-FIAU and ^{3}H-TdR). HIF-1α protein levels were assessed by Western blot. TKGFP and VEGF mRNAs were measured by quantitative reverse transcriptase (RT)-PCR. VEGF production was measured by enzyme-linked immunosorbent assay. A plateau of the HIF-1α signaling pathway activation was observed after 16 h of exposure to 100 μ*M* (IC_{50}) $CoCl_2$, as manifested by a high level of TKGFP fluorescence (10^2–10^3 rU) and a high FIAU/TdR accumulation ratio (0.34 ± 0.005, compared with 0.08 ± 0.001 in nontreated cells). The increase in HIF-1α corresponded with upregulation of VEGF and TKGFP gene expression at the mRNA and protein levels.

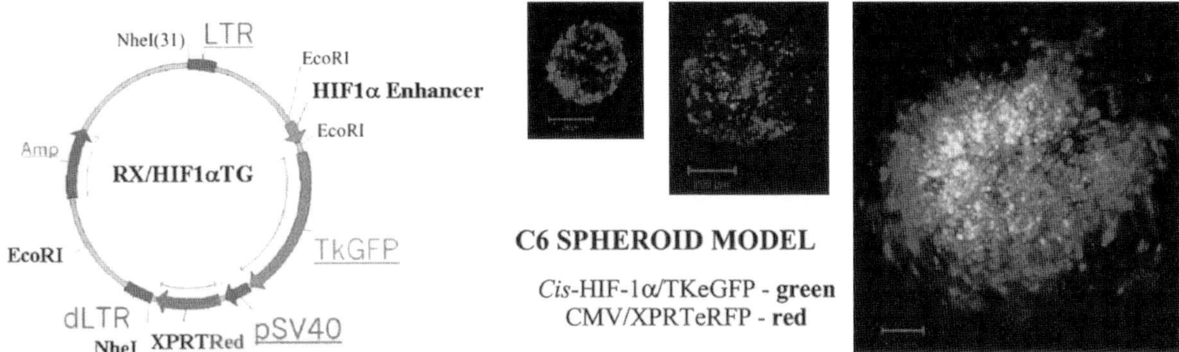

Fig. 8. Optical imaging of hypoxia. The HIF-1-α reporter vector is shown on the left. There are two elements in this reporter construct: one is a "sensor" element that consists of an artificial HIF-1-α enhancer/promoter controlling the expression of TKeGFP gene (a fusion between the *HSV1-tk* and the enhanced GFP, *eGFP*, genes) and the other is a "beacon" element that consists of the SV40 promoter that constitutively expresses the Red1/XPRT gene (a fusion between the Red1 fluorescent protein and *E. coli* xanthine phosphoribosyl transferase genes). Spheroids produced from transduced C6 cells are shown; red fluorescence is seen in a confocal image of each spheroid demonstrating constitutive expression of the Red1/XPRT fusion gene (beacon element). Green fluorescence is seen in the center of the larger spheroid indicating hypoxia-inducible expression of the TKeGFP gene (sensor element).

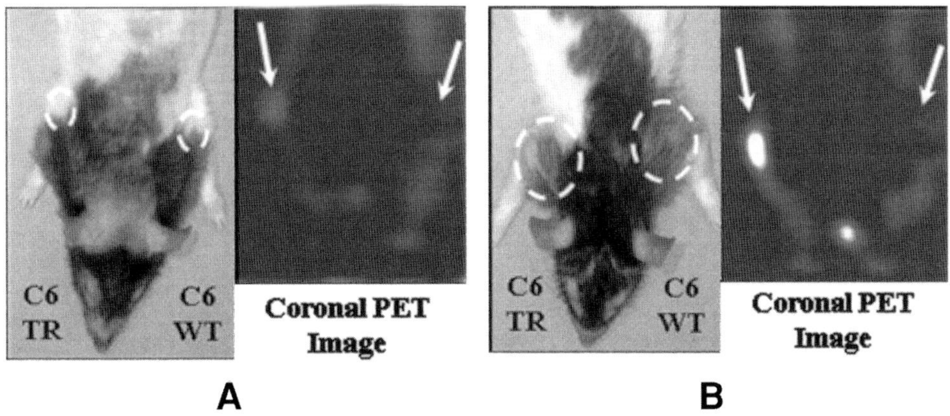

Fig. 9. PET MicroPET images of hypoxia-induced HIF-1α-mediated signal transduction activity. Studies were performed with small and large C6-XPRTRed/HRE-TKGPF sc tumors (right shoulder) and WT C6 tumors (left shoulder). **(A)** Large sc tumors; **(B)** small sc tumors. Images were acquired at 4 h after administration of [^{124}I]FIAU.

Rats with sc transduced C6-XPRTRed/HRE-TKGPF tumors were imaged with [^{124}I]FIAU and PET. PET imaging of XPRTRed "beacon" gene expression with [^{11}C]xanthine has not been done yet, because the radiotracer is still under development. However, PET images with [^{124}I]FIAU demonstrated the presence of hypoxic areas in large sc C6-XPRTRed/HRE-TKGPF tumors. Small C6-XPRTRed/HRE-TKGPF tumors had lower levels of HSV1-tkGFP "sensor" gene activity (Fig. 9). Tumor tissue sampling and fluorescence macroscopic analysis confirmed the in vivo PET imaging results. Namely, the large C6-XPRTRed/HRE-TKGPF sc tumor had indeed a central hypoxic core that showed green fluorescence (hypoxia-induced TKGFP expression). This hypoxic core was surrounded by a rim of dilated and pathologically enlarged vasculature (suggesting high levels of VEGF production). Fluorescence macroscopic analysis of different zones within this tumor confirmed these observations—the tumor core consisted of mainly green fluorescing (hypoxic) cells (TKGFP "sensor" gene expression); the transition zone was also clearly observed; and the periphery showed only red fluorescence (XPRTRed2 "beacon" gene expression) with no green fluorescing cells (indicating normoxic cells).

These results are very relevant to imaging stem cells and their progeny tissues, by demonstrating that dual–reporter gene imaging systems are feasible and could be applied to imaging the location and differentiation state of the adoptively transferred progenitor cells (using tissue-specific or differentiation step–specific promoters to drive "sensor" gene expression).

42.6. GENETIC LABELING OF MURINE STEM/PROGENITOR CELLS WITH DIFFERENT REPORTER SYSTEMS

Several retroviral and lentiviral vectors have been developed and used for transduction of stem/progenitor cells with different reporter genes. Recently, three retroviral vectors have been generated and

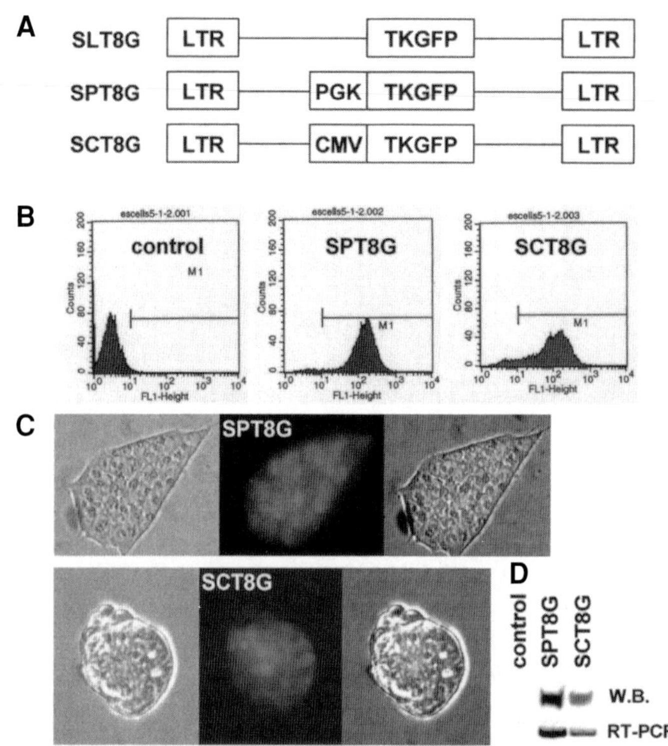

Fig. 10. Transduction of murine stem cells with different retroviral vectors. (**A**) Schematics of retroviral vectors; (**B**) FACS profiles of two functional vector-transduced stem cells and control stem cells; (**C**) dark-field, fluorescence, and merged images of transduced stem cells; (**D**) Western blots and RT-PCR of KTGFP expression in transduced embryonic stem (ES) cells.

assessed in which the constitutive expression of TKGFP reporter gene is driven by either the retroviral LTR (SLT8G), or the PGK promoter (SPT8G), or the CMV promoter (SCT8G) (Fig. 10). The vector-encoding plasmids were transfected into the Phoenix amphotropic retroviral vector producer cells, which produce retroviral particles at very high titers (Kinsella and Nolan, 1996; Grignani et al., 1998). After transduction, the murine stem/progenitor cells were sorted for high expression of TKGFP using FACS (based on GFP fluorescence at 488 nm above 100 rU). It is important to note that there was a significant difference in the efficacy of different promoters with respect to the stability of TKGFP expression in transduced stem/progenitor cells over time. TKGFP expression under control of retroviral LTR resulted in almost complete TKGFP downregu-lation at 1 wk posttransduction. Almost no change in the level of TKGFP expression was observed with PGK promoter. However, in stem/progenitor cells transduced with TKGFP gene under control of CMV promoter, 20–30% of cells exhibited downregu-lation of TKGFP expression by 1 wk after transduction (Fig. 10). Even when the cells were incubated in the medium without the leukemia inhibitory factor (an inhibitor of differentiation) and allowed to form the embryonal-like bodies, the expression of TKGFP under control of PGK promoter did not fade. Therefore, the PGK promoter is more suitable for imaging studies in stem cells (Gelovani Tjuvajev, unpublished data).

This in vitro study demonstrates the complex nature of the approach that is being proposed in the current application. The choice of the right promoter to drive the expression of a reporter gene is important in terms of long-term monitoring and imaging the fate of infused progenitor cells.

42.7. MULTIMODALITY REPORTER GENE IMAGING

Multimodality imaging approaches allow for different imaging technologies to be combined in the same study. For example, planar BLI of Luc reporter gene expression could be used for the initial screening and monitoring of the fate of adoptively transferred progenitor cells in murine models. An important advantage of BLI is that it does not have a nonspecific signal background in the images, because the photons are produced only where the luciferin–luciferase reaction takes place (in Luc-expressing cells). Once critical time points or conditions are determined, one could follow up with a study involving PET imaging of TKGFP expression. Therefore, another retroviral vector, L89gCluc, for expression of both the Luc and TKGFP genes was developed (Fig. 11). As a model for localization of progenitor cell–derived tissue, a transduced tumor xenograft model in mice was used to demonstrate efficacy of such an approach. RG2 rat glioma cells were transduced with the LT8gCluc vector (RG2/LT9gCluc cells) and established as sc xenografts in mice. When the tumors grew to approx 0.5 cm in diameter, sequential BLI and PET imaging studies were performed in the same animals. The sc RG2/LT8gCluc tumors could be easily visualized with two-dimensional BLI and whole-body fluorescence imaging of TKGFP expression. Thereafter, Micro-PET images were acquired in three-dimensionally 2 h postadministration of [^{124}I]FIAU and demonstrated localization

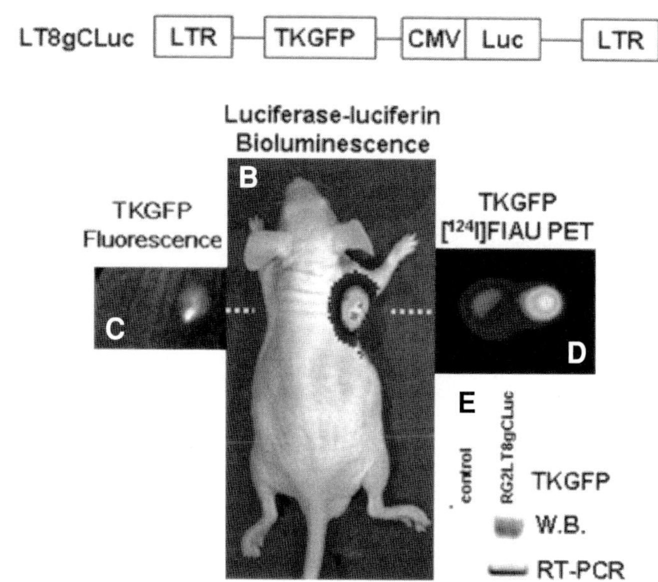

Fig. 11. Bioluminescence and PET multimodality imaging. (**A**) Schematics of LT8gCLuc retroviral vector for coexpression of TKGFP and Luc reporter genes; (**B**) bioluminescence image of Luc expression in sc RG2/LT8gCLuc tumor (**C**) and axial PET image of TKGFP expression in same animal (**D**); (**E**) RT-PCR and Western blot analysis of TKGFP expression in transduced tumor tissue.

of transduced RG2/LT8gCluc cells similar to that obtained using BLI or fluorescence imaging (Fig. 11).

In another study, T-cell trafficking and tumor targeting were imaged following the transfection of T-cells from B6 mice (unpublished data). The transduced T-cells were readministered intravenously into the allogeneic host B6/C3F1 animals (2 or 4 × 10^6 T-cells/animal) along with the BMT. Monitoring with BLI was performed at different times after the BMT. BLI demonstrated different patterns of the infused T-cell trafficking and homing that are consistent with various degrees of GVHD. Especially dramatic symptoms of GVHD with predominant localization in the gut and lungs were observed in animal no. 3 (Fig. 12, bottom row).

This is an ongoing study and is in a preliminary screening phase (using BLI), and PET imaging had not yet been performed. Nevertheless, there is no doubt that PET imaging will be successful in this model, based on the recent RG2/LT9gCluc tumor imaging results (described earlier) and previous studies on imaging T-cell trafficking and activation with PET (Doubrovina et al., 2002; Koehne et al., 2002). Such a multimodality approach could be easily extended to imaging stem cell fate by inserting suitable promoters (e.g., PGK, EF1) in vector constructs to drive the expression of different reporter genes.

42.8. PET IMAGING OF ADOPTIVE T-CELL TRANSFER

The recent work at MSKCC on imaging trafficking and targeting of genetically HSV1-tk-labeled T-cells after in vitro loading with [^{131}I]FIAU for gamma camera imaging (early trafficking and targeting) (Koehne et al., 2002) and repetitive administration of [^{124}I]FIAU in vivo for PET imaging (late trafficking and targeting) (Doubrovina et al., 2002) demonstrated the feasibility of implementing this combined approach, which could be easily applied to imaging stem cells in vivo.

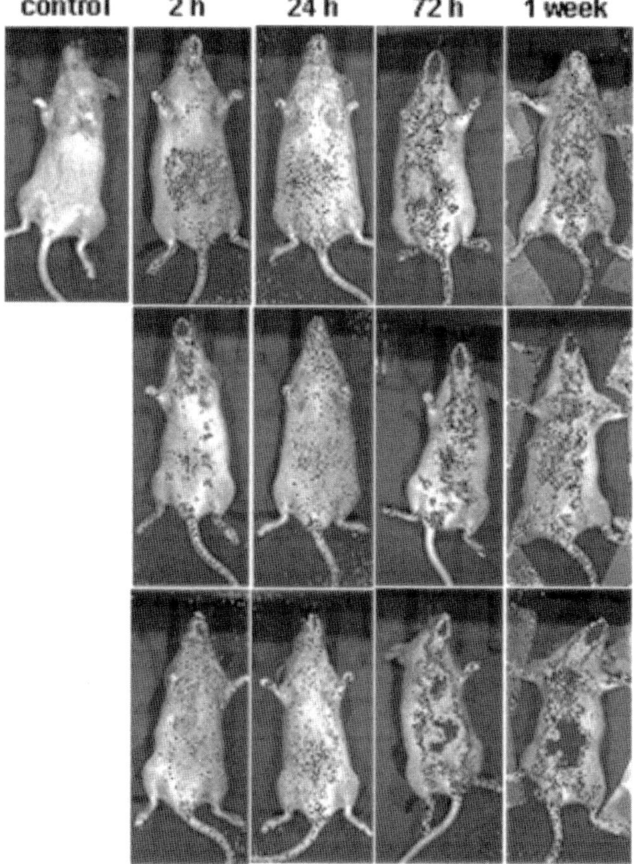

Fig. 12. BLI of adoptively transferred bone marrow cells along with reporter-transduced T-cells. Three animals are presented in rows. The third animal developed marked GVHD (last row) with preferential targeting of T-cells to the gut.

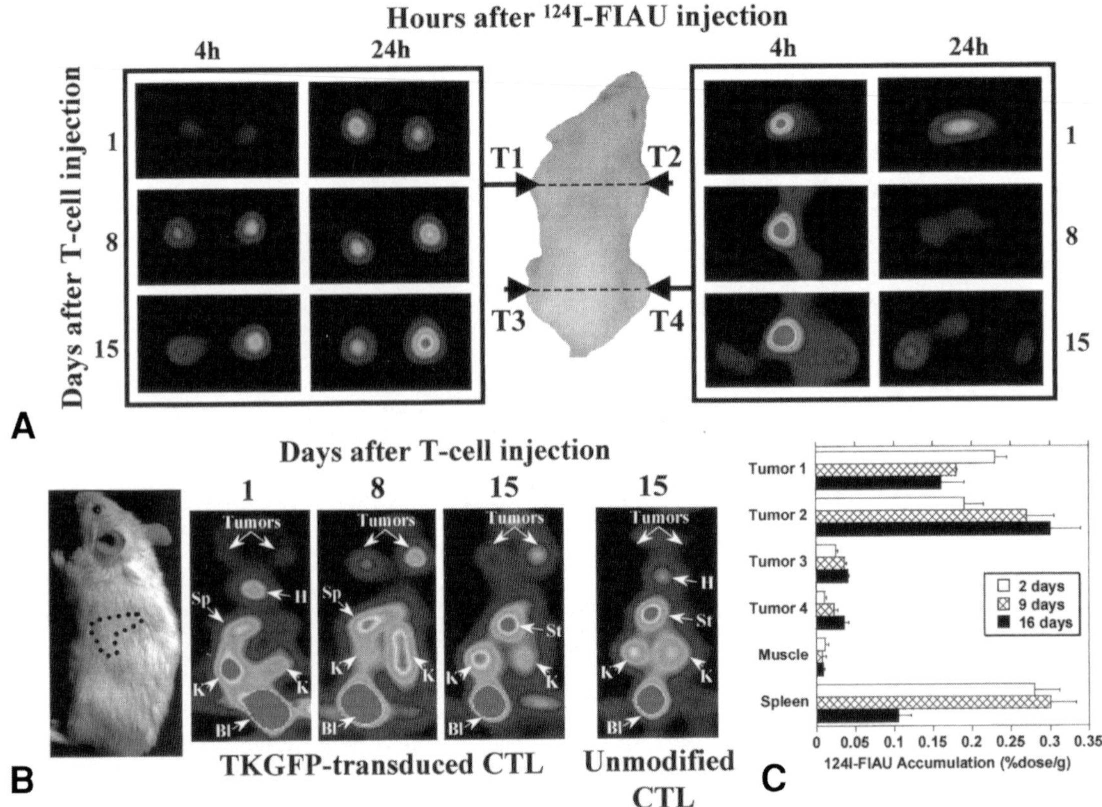

Fig. 13. MicroPET imaging of T-cell migration and targeting. **(A)** Axial images through shoulders (left) and thighs (right) of mice bearing autologous BLCL (T1), HLA-A0201-matched BLCL (T2), HLA-mismatched BLCL (T3), and HLA-A0201 ALL (T4) tumors in left and right shoulders and left and right thighs, respectively. All images are from a single representative animal. **(B)** Oblique projection of summed coronal images at a 45° angle to visualize spleen, 4 h after [^{124}I]FIAU injection. All images are from a single representative animal. K, kidneys; H, heart; St, stomach; Sp, spleen; Bl, bladder. **(C)** Radioactivity (%dose/g) in different tumors and tissue at three time points (2, 9, and 16 d) after EBV-specific TKGFP-CTL administration. Radioactivity was measured 24 h after [^{124}I]FIAU injection. (Adapted from Doubrovina et al., 2002.)

In a recent study by Dourovina et al. (2002), the feasibility of in vivo monitoring of the migration, targeting, and persistence of EBV-specific CTLs over an extended period of time was demonstrated using stable transduction of these cells with TKGFP reporter gene combined with repetitive administration of [^{124}I]FIAU and PET imaging. The feasibility of noninvasive visualization of the human leukocyte antigen (HLA)–restricted EBV-specific anti-EBV CTL targeting to the tumors was assessed. Noninvasive imaging results showed that the anti-EBV-specific TKGFP-CTLs preferentially accumulated in tumors presenting EBV antigens in the context of the dominant restricting HLA allele, A0201. By contrast, the tumors generated following inoculation of EBV BLCL from HLA-mismatched donors showed neither [^{124}I]FIAU accumulation nor TKGFP fluorescence. Similarly, transduced T-cells did not accumulate in control tumors developed from HLA A0201-matched leukemia cells, since they did not express EBV antigens. This study also demonstrated that the exposure of antigen-specific TKGFP-CTLs to [^{124}I]FIAU did not significantly affect their specific cytotoxicity in comparison to the unmodified CTL population.

The initially high FIAU accumulation in the autologous tumor, as compared to the allogeneic HLA A0201 homozygous BLCL, most likely reflects the broader spectrum of EBV antigens presented on these cells in the context of both HLA A0201 and the other HLA alleles expressed on these cells, but not on the HLA A0201 homozygous EBV BLCL. However, the significant level of radioactivity accumulated in the HLA A0201 homozygous allogeneic tumor and the extent of infiltration of TKGFP-CTLs in these tumors demonstrate the striking level of in vivo activity attributable to the immunodominant HLA A0201–restricted EBV-specific T-cell population. These conclusions are also supported by the pronounced response of the autologous BLCL tumor (T1) and the allogeneic HLA A0201 homozygous EBV tumor to the adoptive T-cell therapy, as reflected by the degree of tumor regression observed in this study. A gradual decrease in [^{124}I]FIAU signal in the autologous BLCL tumor, observed with subsequent [^{124}I]FIAU administrations, could reflect a decline in the number of infiltrating TKGFP-CTLs resulting from the apoptosis of T-cells following cytotoxic interaction with autologous EBV-positive targets. Continuous increase of [^{124}I]FIAU accumulation in HLA-A0201 homozygous EBV BLCL tumors after repeated injections may reflect either the differences in the recruitment of the immunodominant HLA A0201-restricted CTL, or the expansion of donor-derived transfected T-cells with other than EBV specificity, which respond to alloantigens not expressed by the CTL donor (Fig. 13A).

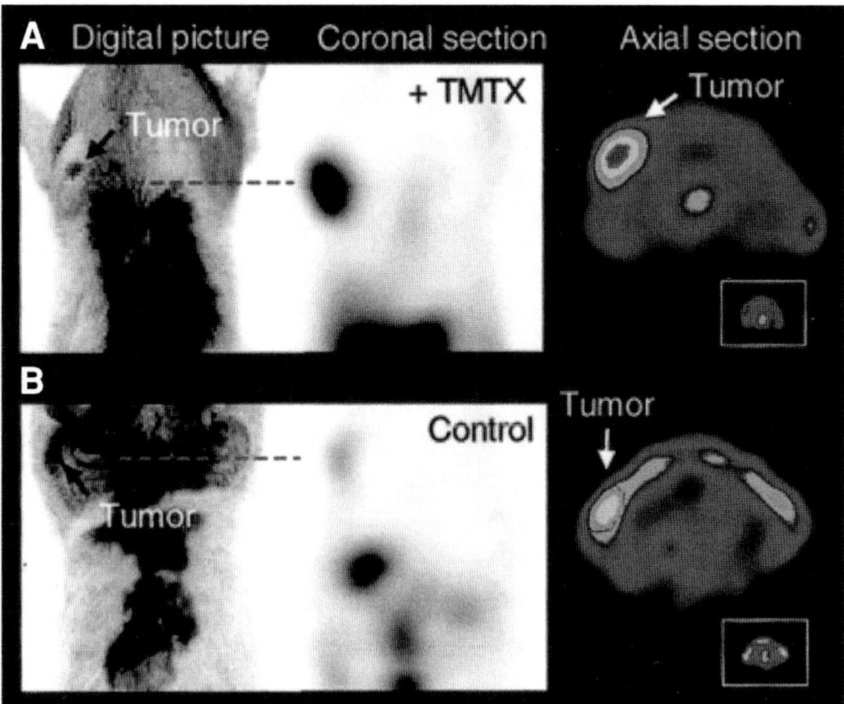

Fig. 14. PET imaging of DHFR-TK expression with [^{124}I]FIAU. Trimethotrexate treatment results in increased [^{124}I]FIAU accumulation in transduced tumor xenografts in vivo. Shown are digital pictures as well as axial and transaxial tumor PET scans obtained from **(A)** an antifolate-treated and **(B)** a water-treated (control) RNU rat. The inserted scans show the heart as control. (Adapted from Mayer-Kuckuk, 2002.)

Transduced TKGFP-CTL dynamics in the spleen and the rest of the body was assessed using images obtained at 4 h post-[^{124}I]FIAU injection, because the 24-h images reflect not only the results of washout of nonaccumulated radioactivity from the body (lower background), but also the redistribution of initially radiolabeled TKGFP-CTL. The initial homing of injected TKGFP-CTL to the spleen in 28 h after iv administration up to d 9 was followed by a gradual clearance and migration to the specific tumor targets by d 15 after TKGFP-CTL administration (Fig. 13B,C). These results are in agreement with the previously reported data on a limited time of circulation and rapid tumor-directed targeting by CTLs (Costa, et al. 2001; Adonai, et al. 2002), although the latter reports did not extend longer than 9 d after CTL injection (Costa et al., 2001).

The successful noninvasive imaging of the fate of genetically labeled cytotoxic T-cells using the radio-reporter gene approach demonstrated in the study by Doubrovina et al. (2002) warrants the feasibility of monitoring stem cell kinetics and proliferation in vivo.

42.9. PET IMAGING OF MUTANT DHFR/HSV1-tk FUSION GENE EXPRESSION IN VIVO

Dihydrofolate reductase (DHFR) (EC 1.5.1.3.) catalyzes the formation of tetrahydrofolate (THF) from dihydrofolate (DHF). The most significant consequence of DHFR inhibition by methotrexate (MTX) or trimethotrexate (TMTX) is a decrease in thymidylate biosynthesis by means of depletion of the N^5,N^{10}-methylene-THF pool resulting in DNA synthesis inhibition and cell death. MTX is used to treat acute lymphoblastic leukemia, lymphoma, gastrointestinal cancers, and breast cancer. However, antifolate resistance remains an obstacle to successful cancer treatment, and the significantly higher doses of MTX required to achieve tumor response also cause myelotoxicity. To address the issue of myelotoxicity, a double double-mutant (Phe22-Ser31) dmDHFR gene was developed at MSKCC and used to confer 10–12 times higher bone marrow resistance to MTX as compared to the mock-transfected NIH3T3 cells (Sauerbrey et al., 1999). The dmDHFR gene can be expressed in fusion with the *HSV1-tk* gene in a similar way as we produced a fusion gene between the WT DHFR and *HSV1tk*. A retroviral vector for transduction of various stem/progenitor cells bearing dmDHFR-hTK2 fusion gene under control of PGK promoter was also generated at MSKCC (SPdmDHFRhTK2 vector) and the CD34$^+$ hematopoietic progenitor cell transduced for imaging studies.

In a recent study, Mayer-Kuckuk et al. (2002) demonstrated the feasibility of in vivo imaging of WT human DHFR and *HSV1-tk* fusion gene (DHFRhsvTK) expression in sc xenografts using [^{124}I]FIAU and PET. Based on these results, imaging MTX-resistant dmDHFRhTK2-transduced CD34$^+$ progenitor cells and their progeny should also be feasible with radiolabeled FIAU or FEAU and PET. In clinical trials involving adoptive therapy with dmDHFR/HSV1-tk fusion gene–transduced chemoresistant bone marrow progenitor cells, noninvasive PET imaging could potentially be used for monitoring the persistence/viability of the engrafted cells during high-dose chemotherapy and complement blood-sampling analyses (Fig. 14).

42.10. POSSIBLE CLINICAL SCENARIOS FOR USE OF IMAGING IN DIFFERENT STEM CELL THERAPY APPROACHES

A simplistic scenario for stem cell therapy for an organ regeneration after injury or infarct would involve the following:

1. Generation of "tracer cells" by stably transducing the stem cells with a reporter gene.
2. Administration of unmodified stem cells mixed with the "tracer cells" to a patient either systemically or by a locoregional injection (e.g., intracerebral).
3. Monitoring of the localization, migration, homing, and persistence of stem cells by imaging "tracer cells" using reporter gene–specific radiolabeled probe and PET in the early period post–stem cell administration.
4. Repetitive PET imaging over several months or years to assess the viability (persistence) of progeny tissues coupled with functional (physiological) imaging and other tests.

A more-sophisticated scenario for using imaging in stem cell therapy that would include additional treatment to control and direct the process of stem cell differentiation into a desired tissue type would involve the following:

1. Generation of "tracer cells" by stably transducing the stem cells with a dual-reporter system—one bearing a constitutively expressed "beacon" reporter gene, and a "sensor" reporter gene expression that will be controlled by a tissue type–specific promoter (e.g., neuronal specific).
2. Administration of unmodified stem cells mixed with the "tracer cells" to a patient either systemically or by a locoregional injection (e.g., intracerebral).
3. Monitoring of the localization, migration, homing, and persistence of stem cells by imaging "tracer cells" using a "beacon" reporter gene–specific short-lived radiolabeled probe (e.g., labeled with ^{11}C) and 4 h later, followed by imaging of the differentiation status of the administered stem cells and progeny tissues (e.g., cortical neurons) using a "sensor" reporter gene–specific radiolabeled probe (e.g., labeled with ^{18}F).
4. Repetitive PET imaging over several months or years to assess the viability (persistence) and differentiation type of progeny tissues coupled with functional (physiological) imaging and other tests.

42.11. CONCLUSION

In perspective, noninvasive whole-body imaging will play an increasingly larger role in the development and clinical implementation of various stem cell therapeutic approaches by providing the means for noninvasive monitoring of the fate of the stem cells administered to patients over a long period of observation. Noninvasive imaging will be invaluable especially in monitoring the efficacy of stem cell therapies in biopsy-restricted critical organs such as brain and heart, although the advantages of using noninvasive imaging over multiple biopsies in other organs are obvious as well. Imaging could also help to address several questions related to stem cell migration and homing, their viability, subsequent differentiation, and the persistence of their progeny-derived tissues. Finally, imaging should aid in the development of new therapeutic strategies that would control and direct stem cell engraftment and differentiation processes in patients.

REFERENCES

Adonai, N., Nguyen, K. N., Walsh, J., et al. (2002) Ex vivo cell labeling with 64Cu-pyruvaldehyde-bis(N4-methylthiosemicarbazone) for imaging cell trafficking in mice with positron-emission tomography. *Proc. Natl. Acad. Sci. USA* 99:3030–3035.

Atkins, R. C. and Ford, W. L. (1975) Early cellular events in a systemic graft-vs.-host reaction. I. The migration of responding and nonresponding donor lymphocytes. *J. Exp. Med.* 141:664–680.

Banerjee, D. (1999) Technology evaluation: gene therapy (mesothelioma), NCI. *Curr. Opin. Mol. Ther.* 1:517–520.

Bennett, J. J., Tjuvajev, J., Johnson, P., et al. (2001) Positron emission tomography imaging for herpes virus infection: implications for oncolytic viral treatments of cancer. *Nat. Med.* 7:859–863.

Blasberg, R. and Gelovani (Tjuvajev), J. (2002) Molecular-genetic imaging: a nuclear medicine perspective. *Mol. Imaging* 1(3):1–22.

Blasberg, R. G., Roelcke, U., Weinreich, R., et al. (2000) Imaging brain tumor proliferative activity with [124I]iododeoxyuridine. *Cancer Res.* 60:624–635.

Bonini, C., Ferrari, G., Verzeletti, S., et al. (1997) HSV-TK gene transfer into donor lymphocytes for control of allogeneic graft-versus-leukemia. *Science* 276:1719–1724.

Bulte, J. W., Zhang, S., van Gelderen, P., et al. (1999) Neurotransplantation of magnetically labeled oligodendrocyte progenitors: magnetic resonance tracking of cell migration and myelination. *Proc. Natl. Acad. Sci. USA* 96:15,256–15,261.

Chaudhuri, T. R., Rogers, B. E., Buchsbaum, D. J., Mountz, J. M., and Zinn, K. R. (2001) A noninvasive reporter system to image adenoviral-mediated gene transfer to ovarian cancer xenografts. *Gynecol. Oncol.* 83:432–438.

Constantin, G., Laudanna, C., and Butcher, E. C. (1997) Novel method for following lymphocyte traffic in mice using [3H]glycerol labeling. *J. Immunol. Meth.* 203:35–44.

Contag, C. H. and Bachmann, M. H. (2002) Advances in in vivo bioluminescence imaging of gene expression. *Annu. Rev. Biomed. Eng.* 4: 235–260.

Costa, G.L., Sandora, M.R., Nakajima, A., et al. (2001) Adoptive immunotherapy of experimental autoimmune encephalomyelitis via T cell delivery of the IL-12 p40 subunit. *J. Immunol.* 167:2379–2387.

Cui, Y., Golob, J., Kelleher, E., Ye, Z., Pardoll, D., and Cheng, L. (2002) Targeting transgene expression to antigen-presenting cells derived from lentivirus-transduced engrafting human hematopoietic stem/progenitor cells. *Blood* 99:399–408.

Dillman, R. O., Hurwitz, S. R., Schiltz, P. M., et al. (1997) Tumor localization by tumor infiltrating lymphocytes labeled with indium-111 in patients with metastatic renal cell carcinoma, melanoma, and colorectal cancer. *Cancer Biother. Radiopharm.* 12:65–71.

Dodd, S. J., Williams, M., Suhan, J. P., Williams, D. S., Koretsky, A. P., and Ho, C. (1999) Detection of single mammalian cells by high-resolution magnetic resonance imaging. *Biophys. J.* 76:103–109.

Doubrovin, M., Ponomarev, V., Beresten, T., et al. (2001) Imaging transcriptional regulation of p53-dependent genes with positron emission tomography in vivo. *Proc. Natl. Acad. Sci. USA* 98:9300–9305.

Fawell, S., Seery, J., Daikh, Y., et al. (1994) Tat-mediated delivery of heterologous proteins into cells. *Proc. Natl. Acad. Sci. USA* 91: 664–668.

Floeth, F. W., Shand, N., Bojar, H., et al. (2001) Local inflammation and devascularization—in vivo mechanisms of the "bystander effect" in VPC-mediated HSV-Tk/GCV gene therapy for human malignant glioma. *Cancer Gene Ther.* 8:843–851.

Franke, H., Galla, H. J., and Beuckmann, C. T. (1999) An improved low-permeability in vitro-model of the blood-brain barrier: transport studies on retinoids, sucrose, haloperidol, caffeine and mannitol. *Brain Res.* 818:65–71.

Gambhir, S. S., Herschman, H. R., Cherry, S. R., et al. (2000) Imaging transgene expression with radionuclide imaging technologies. *Neoplasia* 2:118–138.

Gelovani Tjuvajev, J., Doubrovin, M., Akhurst, T., et al. (2002) Comparison of radiolabeled nucleoside probes FIAU, FHBG and FHPG for PET imaging of HSV1-tk gene expression. *J. Nucl. Med.* 43:1072–1083.

Gobuty, A. H., Robinson, R. G., and Barth, R. F. (1977) Organ distribution of 99mTc- and 51Cr-labeled autologous peripheral blood lymphocytes in rabbits. *J. Nucl. Med.* 18:141–146.

Grignani, F., Kinsella, T., Mencarelli, A., et al. (1998) High-efficiency gene transfer and selection of human hematopoietic progenitor cells with a hybrid EBV/retroviral vector expressing the green fluorescence protein. *Cancer Res.* 58:14–19.

Habgood, M. D., Knott, G. W., Dziegielewska, K. M., and Saunders, N. R. (1998) Permeability of the developing and mature blood-brain barriers to theophylline in rats. *Clin. Exp. Pharmacol. Physiol.* 25:361–368.

Hackman, T., Dubrovin, M., Balatoni, J., et al. (2002) Monitoring *E. coli* cytosine deaminase gene expression by noninvasive imaging of the HSV-1-tk marker gene: potential clinical applications in cancer gene therapy. *Mol. Imaging* 1:36–42.

Hardy, J., Edinger, M., Bachmann, M. H., Negrin, R. S., Fathman, C. G., and Contag, C. H. (2001) Bioluminescence imaging of lymphocyte trafficking in vivo. *Exp. Hematol.* 29:1353–1360.

Hawrylak, N., Ghosh, P., Broadus, J., Schlueter, C., Greenough, W. T., and Lauterbur, P. C. (1993) Nuclear magnetic resonance (NMR) imaging of iron oxide-labeled neural transplants. *Exp. Neurol.* 121:181–192.

Hellerstein, M., Hanley, M. B., Cesar, D., et al. (1999) Directly measured kinetics of circulating T lymphocytes in normal and HIV-1-infected humans. *Nat. Med.* 5:83–89.

Iyer, M., Wu, L., Carey, M., Wang, Y., Smallwood, A., and Gambhir, S. S. (2001) Two-step transcriptional amplification as a method for imaging reporter gene expression using weak promoters. *Proc. Natl. Acad. Sci. USA* 98:14,595–14,600.

Jacobs, A., Dubrovin, M., Hewett, J., et al. (1999) Functional coexpression of HSV1 thymidine kinase and green fluorescent protein: implications for noninvasive imaging of transgene expression. *Neoplasia* 1:154–161.

Jacobs, A., Tjuvajev, J. G., Dubrovin, M., et al. (2001a) Positron emission tomography-based imaging of transgene expression mediated by replication-conditional, oncolytic herpes simplex virus type 1 mutant vectors in vivo. *Cancer Res.* 61:2983–2995.

Jacobs, A., Voges, J., Reszka, R., et al. (2001b) Positron-emission tomography of vector-mediated gene expression in gene therapy for gliomas. *Lancet* 358:727–729.

Jacobs, R. E. and Cherry, S. R. (2001) Complementary emerging techniques: high-resolution PET and MRI. *Curr. Opin. Neurobiol.* 11: 621–629.

Kasi, L. P., Lamki, L. M., Saranti, S., Podoloff, D. A., and Freedman, R. S. (1995) Indium-111 labeled leukocytes in evaluation of active specific immunotherapy responses. *Int. J. Gynecol. Cancer* 5:226–232.

Kinsella, T. M. and Nolan, G. P. (1996) Episomal vectors rapidly and stably produce high-titer recombinant retrovirus. *Human Gene Ther.* 7:1405–1413.

Koehne, G., Doubrovin, M., Doubrovina, E., et al. (2003) Serial in vivo imaging of targeted migration of human HSV-TK-transduced antigen-specific lymphocytes. *Nat. Biotechnol.* 21:405–413.

Korf, J., Veenma-van der Duin, L., Brinkman-Medema, R., Niemarkt, A., and de Leij, L. F. (1998) Divalent cobalt as a label to study lymphocyte distribution using PET and SPECT. *J. Nucl. Med.* 39:836–841.

Lewin, M., Carlesso, N., Tung, C. H., et al. (2000) Tat peptide-derivatized magnetic nanoparticles allow in vivo tracking and recovery of progenitor cells. *Nat. Biotechnol.* 18:410–414.

Liang, Q., Satyamurthy, N., Barrio, J. R., et al. (2001) Noninvasive, quantitative imaging in living animals of a mutant dopamine D2 receptor reporter gene in which ligand binding is uncoupled from signal transduction. *Gene Ther.* 8:1490–1498.

Luker, G. D., Sharma, V., Pica, C. M., et al. (2002) Noninvasive imaging of protein-protein interactions in living animals. *Proc. Natl. Acad. Sci. USA* 99:6961–6966.

Mayer-Kuckuk, P., Banerjee, D., Malhotra, S., et al. (2002) Cells exposed to antifolates show increased cellular levels of proteins fused to dihydrofolate reductase: a method to modulate gene expression. *Proc. Natl. Acad. Sci. USA* 99:3400–3405.

Miyagawa, T., Oku, T., Uehara, H., et al. (1998) "Facilitated" amino acid transport is upregulated in brain tumors. *J. Cereb. Blood Flow Metab.* 18:500–509.

Nagahara, H., Vocero-Akbani, A. M., Snyder, E. L., et al. (1998) Transduction of full-length TAT fusion proteins into mammalian cells: TAT-p27Kip1 induces cell migration. *Nat. Med.* 4:1449–1452.

Papierniak, C. K., Bourey, R. E., Kretschmer, R. R., Gotoff, S. P., and Colombetti, L. G. (1976) Technetium-99m labeling of human monocytes for chemotactic studies. *J. Nucl. Med.* 17:988–992.

Ponomarev, V., Doubrovin, M., Lyddane, C., et al. (2001a) Imaging TCR-dependent NFAT-mediated T-cell activation with positron emission tomography in vivo. *Neoplasia* 3:480–488.

Ponomarev, V., Serganova, I., Ageyeva, L., Beresten, T., Doubrovin, M., and Tjuvajev, J. (2001b) A new reporter gene for multi-modality imaging: xanthine phosphoribosyl transferase and red fluorescent protein fusion. *J. Nucl. Med.* 42(Suppl.):70P.

Ponomarev, V., Doubrovin, M., Serganova, I., et al. (2002) A novel non-immunogenic reporter gene for non-invasive imaging in humans: human thymidine kinase type two. *J. Nucl. Med.* 43(Suppl.):70P.

Qiao, J., Doubrovin, M., Sauter, B. V., et al. (2002) Tumor-specific transcriptional targeting of suicide gene therapy. *Gene Ther.* 9:168–175.

Rannie, G. H., Thakur, M. L., and Ford, W. L. (1977) An experimental comparison of radioactive labels with potential application to lymphocyte migration studies in patients. *Clin. Exp. Immunol.* 29: 509–514.

Reinhardt, R. L., Khoruts, A., Merica, R., Zell, T., and Jenkins, M. K. (2001) Visualizing the generation of memory CD4 T cells in the whole body. *Nature* 410:101 105.

Richardson, C. and Bank, A. (1996) Developmental-stage-specific expression and regulation of an amphotropic retroviral receptor in hematopoietic cells. *Mol. Cell Biol.* 16:4240–4247.

Rogers, B. E., McLean, S. F., Kirkman, R. L., et al. (1999) In vivo localization of [(111)In]-DTPA-D-Phe1-octreotide to human ovarian tumor xenografts induced to express the somatostatin receptor subtype 2 using an adenoviral vector. *Clin. Cancer Res.* 5:383–393.

Rooney, C. M., Smith, C. A., Ng, C. Y., et al. (1995) Use of gene-modified virus-specific T lymphocytes to control Epstein-Barr-virus-related lymphoproliferation. *Lancet* 345:9–13.

Safarik, I. and Safarikova, M. (1999) Use of magnetic techniques for the isolation of cells. *J. Chromatogr.* 722:33–53.

Satyamurthy, N., Barrio, J. R., Bida, G. T., Huang, S. C., Mazziotta, J. C., and Phelps, M. E. (1990) 3-(2'-[18F]fluoroethyl)spiperone, a potent dopamine antagonist: synthesis, structural analysis and in-vivo utilization in humans. *Int. J. Radiol. Appl. Instrum. [A]* 41:113–129.

Sauerbrey, A., McPherson, J. P., Zhao, S. C., Banerjee, D., and Bertino J. R. (1999) Expression of a novel double-mutant dihydrofolate reductase-cytidine deaminase fusion gene confers resistance to both methotrexate and cytosine arabinoside. *Hum. Gene Ther.* 10: 2495–2504.

Schoepf, U., Marecos, E. M., Melder, R. J., Jain, R. K., and Weissleder, R. (1998) Intracellular magnetic labeling of lymphocytes for in vivo trafficking studies. *BioTechniques* 24:642–651.

Serganova, I., Ponomarev, V., Doubrovin, M., et al. (2002a) Imaging TGFβ signal transduction pathway activity with PET. *J. Nucl. Med.* 43(Suppl.):69P.

Serganova, I., Beresten, T., Ageyeva, L., et al. (2002b) Imaging hypoxia-induced HIF-1α signaling by PET. *J. Nucl. Med.* 43(Suppl.):69P.

Schwarze, S. R., Ho, A., Vocero-Akbani, A., and Dowdy, S. F. (1999) In vivo protein transduction: delivery of a biologically active protein into the mouse. *Science* 285:1569–1572.

Spector, R. (1987) Hypoxanthine transport through the blood-brain barrier. *Neurochem. Res.* 12:791–796.

Sun, X., Annala, A. J., Yaghoubi, S. S., et al. (2001) Quantitative imaging of gene induction in living animals. *Gene Ther.* 8:1572–1579.

Tjuvajev, J., Macapinlac, H., Scott, A., et al. (1994) Imaging of the tumor proliferative activity with [^{131}I]-iododeoxyuridine. *J. Nucl. Med.* 35:1407–1417.

Tjuvajev, J., Stockhammer, G., Desai, R., et al. (1995) Imaging gene transfer and expression in vivo. *Cancer Res.* 55:6121–6135.

Tjuvajev, J., Finn, R., Watanabe, K., et al. (1996) Noninvasive imaging of herpes virus thymidine kinase gene transfer and expression: a potential method for monitoring clinical gene therapy. *Cancer Res.* 56:4087–4095.

Tjuvajev, J. G., Avril, N., Oku, T., et al. (1998) Imaging herpes virus thymidine kinase gene transfer and expression by positron emission tomography. *Cancer Res.* 58:4333–4341.

Tjuvajev, J., Joshi, A., Callegari, J., et al. (1999a) A general approach to the non-invasive imaging of transgenes using cis-linked herpes simplex virus thymidine kinase. *Neoplasia* 1:315–320.

Tjuvajev, J.G., Chen, S.-S., Joshi, A., et al. (1999b) Imaging adenoviral mediated herpes virus thymidine kinase gene transfer and expression. *Cancer Res.* 59:5186–5193.

Tjuvajev, J., Blasberg, R., Luo, X., Zheng, L. M., King, I., and Bermudes, D. (2001) Salmonella-based tumor-targeted cancer therapy: tumor amplified protein expression therapy (TAPET) for diagnostic imaging. *J. Control Release* 74:313–315.

Verzeletti, S., Bonini, C., Marktel, S., et al. (1998) Herpes simplex virus thymidine kinase gene transfer for controlled graft-versus-host disease and graft-versus-leukemia: clinical follow-up and improved new vectors. *Hum. Gene Ther.* 9:2243–2251.

Weissleder, R., Cheng, H. C., Bogdanova, A., and Bogdanov, A. (1997) Magnetically labeled cells can be detected by MR imaging. *J. Magn. Reson. Imaging* 7:258–263.

Weissleder, R., Moore, A., Mahmood, U., et al. (2000) In vivo magnetic resonance imaging of transgene expression. *Nat. Med.* 6: 351–354.

Yaghoubi, S., Barrio, J. R., Dahlbom, M., et al. (2001) Human pharmacokinetic and dosimetry studies of [(18)F]FHBG: a reporter probe for imaging herpes simplex virus type-1 thymidine kinase reporter gene expression. *J. Nucl. Med.* 42:1225–1234.

Zhang, L., Adams, J. Y., Billick, E., et al. (2002) Molecular engineering of a two-step transcription amplification (TSTA) system for transgene delivery in prostate cancer. *Mol. Ther.* 5:223–332.

43 What Is the Future for Stem Cell Research?

Whether Entity or Function?

REGIS DOYONNAS, PhD AND HELEN M. BLAU, PhD

The goal of the intense research on stem cells is for human application. Recently, knowledge of stem cells has progressed rapidly and experimental therapies are already in clinical trials. However, for more far reaching application and successful therapy much more remains to be learned about stem cells. There are many more questions than answers. What is a stem cell? What different kinds are there? Can they be obtained and manipulated? What are the lineages that derive from stem cells, and how plastic are cells in a lineage? Are there circulating stem cells in adults? Do they participate in repair of injury? What type of stem cell is most appropriate for a given clinical application? New technologies need to be developed to apply to stem cells for effective gene delivery. What role, if any, does fusion of stem cells play in tissue regeneration? Can the differentiation potential of stem cells at different stages of determination be used to select cells for specific clinical applications? What are the signals that recruit, activate, and induce differentiation in stem cells? What are the signaling pathways for activation and differentiation of stem cells, and can they be manipulated to advantage? How can stem cells be used to understand carcinogenesis and developmental abnormalities? Are embryonic stem cells, which have the potential to produce progeny that can differentiate into any adult tissues the best cells for therapeutic use? Or in some instances, are adult stem cells resident in adult tissues or circulating adult bone marrow derived cells a better choice? How can immune rejection of transplanted cells be avoided or prevented? Will therapeutic cloning, whereby transfer of somatic nuclei to provide an embryonic cell line that matches the patient, become clinically applicable? Can embryonic germ cells be used to greater advantage that embryonic stem cells? Will adult stem cells, if they can be isolated and cultured, be a better choice for selected use, for example, in replacing cells in a specific damaged organ. Are there adult multipotent stem cells or can adult tissue-determined stem cells transdifferentiate to another tissue cell type? If so, how can this be controlled? Does fusion play a role in functional stem cell plasticity and can a way to use fusion to direct tissue repair or replacement be devised? Do human somatic cells have the capacity for dedifferentiation, such as found during regeneration of tissues in amphibians? Can this phenomenon be applied to mammalian tissues and eventually used clinically? How can genetic modification of human embryonic or adult stem cells be improved and applied? The potential of stem cell therapy has great promise and is only limited by our incomplete knowledge of what stem cells are and how they function.

43.1. INTRODUCTION

The use of human stem cells for clinical applications is one of the major scientific goals of the 21st century. This handbook includes an exciting inventory of the state-of-the-art, including what we know and don't know about diverse stem cells. Remarkably, the field has progressed so rapidly that many experimental therapies are now well into clinical trials. Thus, the hope that stem cells could some day constitute therapies for human diseases seems more likely to become a reality than ever previously deemed possible. Nonetheless, although cures for a wide variety of diseases due to the specific replacement of damaged tissues by stem cells are thought by some to be on the horizon, numerous questions remain to be answered before stem cells become an integral part of the armamentarium of approaches to the clinical practice of regenerative medicine.

First, we need a better understanding of the *biology of stem cells* and their inherent plasticity. What is a stem cell? Is it an entity or a function? How many different types exist? Are they interrelated? Can they all be prospectively isolated and their fate predicted given certain signals or environments? Where are they located and what potential for differentiation does each type of stem cell offer? Currently, the lineage of stem cells, or *hierarchy*, is quite unclear as diverse populations with different cell surface markers are being studied either after prolonged culture for many months or without any time in culture at all. Do mammalian cells, as in newts ever dedifferentiate spontaneously to regenerate tis-

From: *Stem Cells Handbook*
Edited by: S. Sell © Humana Press Inc., Totowa, NJ

sues? Is there a universal shared multipotent precursor that gives rise to tissue-specific intermediates and eventually well-defined tissue-specific stem cells in adults? If so, perhaps these cells are capable of circulating throughout the entire body in order to repair damaged tissues. The crucial issue is knowing their biological characteristics and how best one can exploit stem cells in order to attract them to particular sites where they are needed and to enhance their differentiation at will. A better understanding of stem cell biology may not only lead to new approaches to cell replacement or cell rescue, but also to new approaches to preventing cancers, which may constitute stem cells that have gone awry and continue to proliferate irrespective of environmental cues.

A better knowledge of stem cell biology will aid us in deciding which *type of stem cell* is most appropriate for a given medical application. Many questions remain to be answered. Are embryonic stem cells essential for certain therapeutic applications? Can adult stem cells become sufficiently effective to regenerate tissues, and, if so, what factors might stimulate their proliferation, targeting of tissues, and genetic reprogramming? Stem cells have been shown to find their way to the tissue requiring regeneration in some cases, whereas in others they are differentiated in tissue culture and then implanted. Very likely, different cells and different means of delivery of those cells will be suitable for different therapeutic purposes and investigation of all types in parallel will be the best way to find out what is optimal on a case by case basis.

Concurrent with increasing our understanding of stem cell biology is a need to develop *new technologies*. For example, stem cells may be useful in correcting genetic diseases, as vehicles for gene therapy, but methods for effective gene delivery are lagging. Retroviral vectors are advantageous when dividing cells are the target, although the recent clinical trial that showed an unexpected selective integration of these vectors resulting in cancer has led to new concerns. Adenoviral vectors that held great promise in early clinical trials have resulted in severe immune responses, although "gutted" adenoviruses may overcome such problems. Adeno-associated viruses are still difficult to produce in quantity and are also somewhat immunogenic and limited to delivery of relatively short DNA sequences. On the other hand, lentiviral vectors deserve further study to render them absolutely safe, as they are readily delivered to all species and to non-dividing cells. Perhaps the answer will lie in the use of DNA that can be delivered orally and targeted to specific integration sites. A major goal of stem cell therapy would be not only cell delivery but also genetic alteration that could allow repair of heritable defects. Will it be possible to use stem cell genetic engineering technology not only to deliver missing genes in those individuals with genetic disease, but also as a regulatable localized drug delivery system? Will we need to use therapeutic cloning continuously to supply sufficient cells that are immunologically matched in order to deliver therapeutic genes or can we build a stem cell reserve with sufficient immunocompatibility diversity? Perhaps a stem cell bank with extensive immunodiversity can be generated that will preclude the need for ongoing "therapeutic cloning," i.e. the derivation of immunologically matched cells from fertilized eggs.

Finally, stem cells isolated from both embryonic and adult tissues have been shown to possess extensive potential for self-renewal and differentiation and can be used as a source of non-mutated genetic material that can be introduced into a deficient tissue. Two mechanisms have been observed by which cells aid in *tissue regeneration*: *de novo* formation of tissue specific cells and fusion to existing cells. To what extent do diverse stem cells contribute to the regeneration of adult tissues by producing cells *de novo* which are reprogrammed in response to environmental cues? To what extent do the cells become reprogrammed following fusion with resident tissue-specific cells? Both would constitute mechanisms for enlisting the inherent plasticity of stem cells and inducing them to reprogram gene expression in order to repair tissue damage.

43.2. TOWARD A BETTER UNDERSTANDING OF STEM CELL BIOLOGY

43.2.1. STEM CELL DEFINITION In order to progress, stem cell research has met with diverse problems that require ongoing investigation. The term "stem cell" has often been loosely used in the scientific literature, which complicates our understanding of what a true stem cell is, e.g., are the criteria functional or are stem cells specific entities that can be prospectively isolated and their differentiative behavior reliably controlled (Blau et al., 2001). To be designated as a stem cell, a cell must be able to proliferate, to replace itself, and also give rise to at least one differentiated cell type (Anderson et al., 2001). This final requirement entails an asymmetric division or proliferation, which should allow the regeneration of at least one cell type throughout the lifetime of a defined tissue. However, this concept must be extended to include the establishment of subcategories that distinguish among stem cells that (i) may only differentiate into a few closely related cell types within a tissue, or unipotent stem cells, (ii) may only differentiate into a range of phenotypes within a single tissue, or multipotent stem cells, (iii) may differentiate into a greater range of phenotypes extending across several tissues, or pluripotent stem cells, and, finally, (iv) have the ability to differentiate into the range of cells present in all tissues, or totipotent stem cells (Lovell-Badge, 2001). Although the last category is limited to the blastomeres of the early morula, all of these subcategories of stem cells possess therapeutic potential. For example, unipotent keratinocyte stem cells have been used for almost three decades for the treatment of burn injuries (Rheinwald and Green, 1975). Clearly, such unipotent stem cells will be found in a large variety of tissues and will be used more and more often in regenerative medicine for damaged tissues. The current challenge is to enhance our knowledge of specific unipotent or multipotent stem cells in order to allow their efficient isolation from patients.

43.2.2. STEM CELL PROLIFERATION AND DIFFERENTIATION Critical to using stem cells is a better understanding of how to facilitate their proliferation and self-renewal as stem cells and how to direct their differentiation in appropriate locations with great precision. The prototypes of stem cell renewal in mature mammals are cells from tissues such as blood and skin that undergo continuous turnover as part of their normal function. It is unclear whether cells with similar proliferative potential will be found in other tissues that are less dedicated to continuous cell production (Partridge, 2003), although unexpected sources such as one located in the central nervous system continue to be discovered (van Praag et al., 2002). Nonetheless in certain cases, pluripotent stem cells may need to be forced to contribute to the regeneration of tissues that are not capable of self-renewal. The extent to which such cells are still present in the adult is not known. However, remarkable examples continue to emerge, such as the recent discovery of the ability of adult differentiated oligodendrocytes to generate all three neuronal cell types (oligodendrocytes, astrocytes, and neurons) in vitro (Kondo and Raff, 2000), a finding that suggests that dedifferentiation of at least some cell types can occur. On the other hand, cells may need to be isolated from unrelated tissue sources in order to fully regenerate damaged organs. An elucidation of the signals that recruit stem cells, activate their self-renewal capabilities, and induce them toward a pathway of differentiation is of vital importance to future therapeutic applications of stem cells. Embryonic stem cells (ES cells) from the inner cell mass of mouse and human blastocysts represent pluripotent cells that are highly advantageous from several perspectives. Targeted alteration of gene expression in these cells will shed light on the molecular mechanisms that allow them either to proliferate without differentiation or to commit to a specific differentiation program. To date, three major areas of investigation have proved to be particularly useful. One is an analysis of the affect of secreted factors such as cytokines on ES cell differentiation (Dani et al., 1998; Niwa et al., 1998). The second is an analysis of transcription factors such as Oct-4 and nanog that maintain the pluripotency of ES cells (Nichols et al., 1998, Chambers et al., 2003, Mitsui et al., 2003). Third is an elucidation of the signaling pathways that must be tightly regulated within stem cells in order for them to remain in a self-renewing state (Wilson et al., 2001, Kielman et al., 2002). The most common approach to direct ES cell differentiation in vitro is to alter the culture media. Spontaneous differentiation occurs routinely in ES cell cultures and produces a wide variety of cell types. By adding specific growth factors and changing the environmental conditions, specific genes of mouse and human ES cells can be either activated or inactivated. Such changes initiate a series of molecular events that induce the cells to differentiate along particular pathways.

43.2.3 FUNCTIONAL GENOMICS, DISEASE PROGRESSION AND DRUG SCREENING It is highly likely that as techniques become available for creating targeted genetic modifications in human stem cells, these cells will serve as a powerful tool for understanding human functional genomics. This will aid our understanding of how diseases such as cancer occur and will also enable drug screening on cellular models of human diseases rather than on animal surrogates. If human ES cells can be directed to form specific cell types, these ES-derived cells may be more likely to mimic the in vivo response of a specific tissue to a drug. This should offer safer and potentially cheaper models for drug screening.

Human ES cells can also be used to study early events in human development. Detailed analyses of human ES cells in vitro may contribute to the identification of genetic, molecular, and cellular events that lead to congenital birth defects, placenta abnormalities, and spontaneous abortion. Better knowledge of these events will facilitate the elaboration of methods to prevent them (Rathjen et al., 1998).

43.3. STEM CELL SOURCES AND PREPARATION BEFORE THERAPEUTIC USE

43.3.1. EMBRYONIC STEM CELLS What is and what will be the best source of stem cells for future therapeutic use? Stem cells have been detected in many adult and embryonic tissues. With very few exceptions, adult stem cells are generally thought to give rise to only a limited range of differentiated cell types. This limit is imposed by powerful molecular constraints on gene expression that are critical to the preservation of tissue integrity during normal tissue maintenance and repair. By contrast, embryonic stem cells (ES cells) are derived from a very early embryonic cell population that is not yet committed to form a particular tissue of the body. Thus, ES cells are by nature thought to be more versatile than most of their adult counterparts. Mouse ES cells have now been used to treat several model diseases such as Parkinson's disease (Kim et al., 2002), myocardial infarction (Min et al., 2002), spinal injury (McDonald et al., 1999), and severe genetic immune disorder (Rideout et al., 2002). However, significant advances are still required before these approaches can be applied in the clinic. First, we need to be able to generate a sufficient number of the desired cell types in a homogeneous population and cautiously remove all undifferentiated cells in order to decrease the risk of teratoma formation after implantation (Thomson et al., 1998). Second, more progress needs to be made in order to identify what cell type or what intermediate precursor should be isolated and delivered to correct a specific pathology. Protocols will need to be elaborated in order to maintain and expand these cells in vitro before eventual therapeutic transplantation. Finally, the host immune mechanisms must be controlled to avoid immune rejection of the implanted cells. A number of solutions to this problem have been considered such as the elaboration of large banks of stem cells to create a wide array of histocompatibility backgrounds or by the manipulation of the host T cell activity (Waldmann, 2001). Others have suggested the use of a combined transplant, which will replace the patient's hematopoietic and lymphoid systems with ES-derived cells, followed by engraftment of the target cell type (Kaufman et al., 2001). Recently, a novel solution that has attracted a great deal of interest is "therapeutic cloning." This powerful technique combines cloning by somatic cell nuclear transfer with the creation of a clonally derived ES cell line that is the perfect match of a patient's own cells (Rideout et al., 2002). Patients' genetic defects can be corrected in the ES cells using homologous recombination and specific cell types can be derived from these ES cells before implantation into the patient. Although conceptually simple, multiple difficulties, such as the wide variety of developmental defects in cloned animals, still need to be understood and overcome before this technology can be used in humans. However, the fact that this technique has been successfully achieved in mice is encouraging.

43.3.2. EMBRYONIC GERM CELLS Most attention has been given to human embryonic stem cells that are derived from the inner cell mass of blastocysts (Thomson et al., 1995), but other embryonic cells possess stem cell capabilities and could potentially prove to have better regenerative capabilities than pure ES cells. Embryonic germ (EG) cells, for example, have been isolated from primordial germ cells located in the gonadal ridge and mesenchyma of 5–9-wk-old human fetal tissue (Shamblott et al., 1998). Like the ES cells, EG cells are capable of long-term self-renewal, while retaining a normal karyotype. When culture conditions are adjusted to permit differentiation, both ES and EG cells can spontaneously differentiate into derivatives of all three primary germ layers—endoderm, mesoderm, and ectoderm (Amit et al., 2000, Itskovitz-Eldor et al., 2000, Shamblott et al., 2001). However, although EG cells have about five times less capacity for proliferation in vitro, they do not generate teratomas in vivo, whereas human ES cells do (Shamblott et al., 1998, Thomson et al., 1998). This advantage could be crucial in their potential use in transplant therapy. At this stage, any therapy based on the use of human embryonic cells would require their direct differentiation into specific cell types prior to transplantation. Because it is the undifferentiated ES cells, rather than their differentiated progeny, that have been shown to induce teratomas, tumor formation might be avoided by either removing the undifferentiated ES cells or by using other cells such as EG cells that do not have the potential to be tumorigenic.

43.3.3 ADULT STEM CELLS Although tremendous progress is being made in embryonic stem cell research, efforts are now underway to take advantage of the newfound capabilities of adult stem cells. After years of work on tissue repair mechanisms, there is new evidence that stem cells are present in far more adult tissues and organs than once thought and that these cells are capable of generating more kinds of cells than previously imagined. Sources of adult stem cells have been found in the bone marrow, blood stream, cornea and retina, dental pulp, blood vessels, skeletal muscle, liver, skin, gastrointestinal tract, pancreas, spinal cord, and brain. Unlike ES cells, there is no evidence thus far that isolated adult stem cells are capable of pluripotency. Adult stem cells are also thought to have less potential for self-renewal, in part because they lack a high level of telomerase [for review *see* Verfaillie et al. (2002)]. The hematopoietic stem cells (HSC) have been considered as the prototype of adult stem cells and are perhaps the best characterized. Even a single HSC transplanted into a lethally irradiated mouse has the potential to rescue the animal from death reconstituting all of the blood cell lineages of a mouse (Osawa et al., 1996). In complementary experiments, random chromosomal integration sites of retroviral vectors as unique clonal markers in hematopoietic stem cells demonstrated that a single HSC can give rise to multiple daughter cells with the ability to differentiate into multiple lineages in both mice (Dick et al., 1985, Szilvassy et al., 1989) and humans (Nolta et al., 1996).

Although many technical challenges have now been overcome in adult stem cell research, the rare occurrence of these cells among other differentiated cells has remained a problem as they are not easily isolated making further characterization difficult. The proliferation of adult stem cells ex vivo has also proved difficult. New ways to identify adult stem cells in vivo and to propagate them in vitro have been developed (Reya et al., 2003, Willert et al., 2003), but further efforts are needed to make these cells amenable to in-depth analysis and experimentation. New sources of adult stem cells with potential pluripotent capabilities also show promise, for example, in tissues such as skin (Toma et al., 2001) and fat (Zuk et al., 2001), where relatively non-invasive techniques can be used to harvest these cells. In order to refine and improve stem cell culture systems, as with ES cells, a major challenge is developing means for allowing adult stem cells to proliferate in vitro without differentiating (Odorico et al., 2001, Jiang et al., 2002). Another important avenue of investigation relevant to both types of cells is whether their genetic imprinting status plays a role in their maintenance, directing their differentiation, potentially limiting their plasticity and their suitability for certain types of therapeutic applications [for review *see* Surani (2001) and Reik et al. (2001)]. Do changes in gene imprinting occur when cells proliferate and are expanded in number in vitro? If so, what are the effects of gene imprinting or other changes, on the differentiation capabilities of these cells in vitro and in vivo and on the behavior of any differentiated cell types derived from cultured cells? What are the intracellular signaling pathways that are involved in cell maintenance and that correlate with differentiation along a cell lineage? Can these potential limitations be altered by pretreatment with demethylating or acetylation agents? Are there other ways to make the chromatin more amenable to changes in gene expression, more capable of plasticity? Answers to these questions may increase the utility of adult stem cells by increasing the frequency with which they can take part in tissue regeneration.

Finally, the lineage or derivation of adult stem cells will be important to discern. Are the bone-marrow-derived cells hematopoietic stem cells (HSCs), or other circulating cells? What are the intermediate stages along the cell differentiation pathway these cells take in vitro and in vivo? Typically, such intermediate cells are designated as precursors or progenitors, they are partly differentiated, are usually "committed" to differentiating along a particular cellular developmental pathway, but are still able to give rise to more than one specialized cell type. Can these precursors or progenitors be isolated, maintained in culture, and expanded? Would such cells be more useful for therapeutic transplantation than the adult tissue-specific stem cells to which they give rise? In vivo, such cells are found in many organs and are thought to maintain the integrity of the tissue by replacing dying cells. It is often difficult, if not impossible, to distinguish adult unipotent stem cells from progenitor cells, and some may claim that progenitor or precursor cells in those tissues do not exist. Thus, endothelial progenitor cells, satellite cells in the skeletal muscle and epithelial precursors in the skin and the digestive system are often considered "the tissue-specific stem cell." As mentioned above, a stem cell should be capable of self-renewal for the lifetime of the organism, while progenitors are by definition not able to self-renew and thus are exhausted if not replaced by the progeny of a stem cell. Although this concept is fundamental to the currently accepted definition of a stem cell, it is difficult to prove that this criterion is met in vivo.

43.4. STEM CELLS, HIERARCHY, PLASTICITY, AND REGENERATION POTENTIAL

43.4.1. STEM CELL HIERARCHY Unlike embryonic cells, which are defined by their origin, the origin of adult stem cells in any mature tissue remains unknown. It has been proposed that adult stem cells are somehow set aside during fetal development and restrained from differentiating. These cells are generally thought to be uni- or multipotent and by definition only have the potential to differentiate into a restricted range of cell types within a tissue. However, if these cells are "late embryonic" cells, then they might have retained their potential for pluripotency. Are such "universal cells" still present in the adult body? If so, it will be of great interest to determine how they are maintained in this undifferentiated state. If these cells exist, are they present in each tissue or do they circulate in the blood, extravasate from the blood, and repopulate various adult tissues? And if so, what are the signals responsible for the recruitment and the migration of these cells?

The developmental process has always been associated with the tacit assumption that any daughter cell does not acquire a greater range of differentiative capacities than its mother. This dogma stipulates that the hierarchy of cells is unidirectional and that one given cell can only be derived from a higher, or at least equivalent, level in the hierarchy. If pluripotent adult stem cells exist somewhere in the body, then it should be possible to obtain this population from any tissue, irrespective of which of the three germ layers from which it derived.

However, this dogma and associated hypotheses have recently been challenged by the discovery that cells from one adult tissue can generate cells of an unrelated tissue (*see* below). Are these findings due to pluripotent cells that are able to give rise to other tissues or are the specific adult stem cells of one tissue sufficiently plastic to change their fate and become the cells of another tissue? Finally, are these cells really stem cells?

43.4.2. ADULT STEM CELL PLASTICITY: TRANSDIFFERENTIATION? Adult stem cells have long been thought to be able only to differentiate into cell types associated with the tissue from which they were isolated. In the past few years, this supposition has been called into question by increasing evidence that adult stem cells exhibit unexpected plasticity. For example, cells isolated from bone marrow, generally considered as a hematopoietic tissue (mesodermal), have been reported to generate skeletal muscle, heart, liver, brain, and various epithelial tissues. This plasticity extends to a range of mesodermal tissues such as skeletal muscle (Ferrari et al., 1998, Gussoni et al., 1999, LaBarge and Blau, 2002; Brazelton et al., 2003), cardiac muscle (Kocher et al., 2001, Orlic et al., 2001), bones (Pereira, 1998, Reyes et al., 2001), cartilage (Pereira, 1995), and even heart and blood vessels (Jiang et al., 2002, Reyes et al., 2002). Moreover, cells of different germ layers, or lineages, such as neural tissues (ectodermal) (Brazelton et al., 2000, Mezey et al., 2000, Weimann et al., 2003), hepatic tissues (endodermal) (Alison et al., 2000, Lagasse et al., 2000) or epithelial tissues (ectodermal and endodermal) (Krause et al., 2001) can also be found to which bone marrow has contributed. These findings either prove the existence of pluripotent stem cells within the bone marrow or suggest that multi- or unipotent stem cells within the bone marrow are capable of plasticity. Plasticity has previously been ascribed to bone-marrow-derived cells, brain-derived cells (Bjornson et al., 1999, Clarke et al., 2000), and muscle-derived cells (Seale et al., 2000, Jiang et al., 2002). In order to study plasticity within or across germ layers, the source of the adult stem cells must be highly purified and, if possible, must be analyzed at the single cell level, i.e., a single cell must be shown to have the capacity to give rise to diverse progeny. In addition, to be able to claim that adult stem cells or progenitor cells purified from a specific tissue demonstrate plasticity, it is also important to show that these cells have the ability to give rise to the differentiated cell types that are characteristic of the tissue of origin, as well as giving rise to cell types that normally occur in a different tissue. Even more important, these differentiated cell types must be shown to be functional. To date, only some of these criteria have been met in the avalanche of reports that have been generated by the "stem cell plasticity field" during the past few years. Indeed, as yet none of the reports meets the rigorous criteria for plasticity presented above, based on transplantation experiments in vivo. Thus, additional research is clearly required.

Until a single cell and its progeny are shown to be capable of multipotency with respect to gene expression and function, the possibility remains that the donor bone-marrow-cell population hosts a variety of dedicated tissue-specific stem cells, such as muscle stem cells, neural stem cells and hepatic progenitors, that are able to migrate to the recipient tissue and give rise to their normal progeny. For example, given that the bone marrow is known to contain various types of stem cells, it is possible that the reported conversion of bone marrow stem cells into muscle, endothelial, and brain cells after bone marrow transplantation (Ferrari et al., 1998, Brazelton et al., 2000, Mezey et al., 2000, Jackson et al., 2001, Orlic et al., 2001) could originate from mesenchymal stem cells and angiogenic precursors. Recent evidence suggests that plasticity is exhibited in vitro. For example, progenitors of astrocytes and oligodendrocytes can, under special tissue culture conditions, reacquire the ability to develop into neurons, a property normally restricted to the neural stem cell precursors of these progenitors (Kondo and Raff, 2000). In another set of experiments, the overexpression of the *MSX1* gene within differentiated skeletal myotubes in vitro can cause multinucleated myotubes to dissociate into individual mononucleated cells that are able to proliferate and then differentiate into a variety of cell types (Odelberg et al., 2000). Evidence for such plasticity in vivo using single cells is a critically important step in this field.

To show plasticity in adults requires that the donor cells not only integrate into a mature host tissue or turn into a different cell type, but that they also become mature and fully functional cells. Many experiments designed to assess the plasticity of adult stem cells have involved injury and/or strong selective pressure due to a lethal genetic defect (Bjornson et al., 1999, Lagasse et al., 2000, Orlic et al., 2001). In these cases, the cells that are capable of plasticity must be able to restore a loss of function. This has been readily possible to show for embryonic cells, but is still under investigation for adult stem cells. The low incidence with which adult stem cells change phenotype has been the subject of much discussion, and could constitute a limitation to their therapeutic application. As a result, a search for the factors that could improve recruitment and differentiation is extremely active and important. To date, it remains to be shown whether the signals that induce stem cell plasticity circulate through the blood stream in order to recruit stem cells and alert the body to a malfunction. Alternatively, local signals could be present in the microenvironment of the injury site that are able to attract circulating cells that cruise the whole body.

Because the experimental paradigms are extreme and the frequencies are generally low, it is not yet possible to determine whether and to what extent plasticity normally occurs in vivo. Also, it is not yet known how this phenomenon can be harnessed to generate tissues and whether it could serve a useful therapeutic function. The models and techniques will need to be more reproducible and reliable in order to ascertain the extent to which adult stem cells are actually plastic. Nonetheless, the exciting possibility remains that if and when such methods are established, it should be possible to use easily accessible stem cells such as bone marrow, skin, fat, or umbilical cord blood–derived stem cells to "repair" tissues with low or nonexistent regenerative capabilities.

43.4.3. FUSION OR NOT FUSION? The discovery that embryonic stem cells can fuse in vitro either with the precursors of neuronal cells or with adult bone marrow cells has initiated a new debate around stem cell plasticity (Terada et al., 2002, Ying et al., 2002). Although these experiments were carried out in tissue culture with strong selective pressure, they spawned the new idea that fusion could be possible in vivo between deficient cells from one tissue and genetically non-mutated cells from another tissue. This raised the possibility that in a stressed situation, some cells might fuse and their nuclei be "reprogrammed" for a different function in adulthood (Blau, 2002). As a result, numerous earlier reports claiming that stem cells can change their morphology and activate new genes characteristic of a different cell type *de novo* are now being questioned.

However, fusion does not really change the notion that there is a stem cell with plasticity. It simply alters the mechanism whereby this may occur. A cell, for example, from bone marrow, that fuses with a defective cell in a tissue may be reprogrammed. Thus, the cell assumes a new genetic fate, albeit due to intracellular signals not extracellular signals as in the case of *de novo* adoption of a tissue-specific fate. Both can constitute repair mechanisms: the *de novo* generation of a cell type is a form of regeneration of lost cells, whereas fusion to a dying cell is rescue of a cell that would otherwise be lost. As such, the cell that fuses may well prevent cells from destruction by providing healthy genetic material. Highly specialized or complex cells of tissues such as liver and brain might more easily be "rescued" than induced to form *de novo* in the adult. This could constitute a previously unrecognized mechanism whereby nomadic cells could physiologically gain access to injured cells and repair them. Cell fusion has long been known to cause reprogramming of cells. In non-dividing "heterokaryons," obtained by the fusion of mature cells into skeletal-muscle cells, muscle genes have been previously shown to be activated in primary human keratinocytes, fibroblasts, and hepatocytes that had no need for those genes (Blau et al., 1983, Chiu and Blau, 1984, Blau et al., 1985) [for review *see* Blau and Baltimore (1991)]. These findings showed that the phenotype of a differentiated cell is not terminal but is plastic and can be altered. Genetic reprogramming occurs due to the balance of regulators in a cell at any given time, is dynamic, and can change the differentiation state of a cell. The recent experimental evidence that heterokaryons can exist in vivo is exciting. Understanding the underlying mechanism that triggers a cell to fuse with another cell, and knowledge of which cells are capable of this fusion, will be of fundamental importance. Moreover, an increase in the fusion efficiency may lead to the rescue of genetic diseases characterized by protein deficiencies. Once the signals that allow this process to occur are known, cells such as bone-marrow-derived cells, from either a genetically normal heterologous source or from a genetically engineered autologous source, could be used to deliver genes to specific targeted cells. Thus, fusion, instead of being an artifact that happens in vitro under selective pressure, could become a process of repair in adulthood that helps maintain life.

43.4.4. DEDIFFERENTIATION AND REGENERATION POTENTIALITIES Stem cells are by definition able to regenerate at least one tissue in response to degeneration or injury. In mammals, they are the only cells that are ultimately responsible for all cell replacement within a tissue during the lifetime of an animal. However, for many years, it has been known that some species such as amphibians possess the amazing ability to regenerate lost body parts. This phenomenon remains poorly understood, but it is now fully accepted that this occurs via dedifferentiation. This process implies that somatic dedifferentiation happens at the site of injury followed by new differentiation and regeneration. This can be linked to the activation of resident cells rather than recruitment of new stem cells. Replacement of cells within an organ must be regulated by some mechanism to selectively remove the current "damaged" resident cells and encourage the repopulation by healthy precursors or stem cells. A major challenge for stem cell research over the next decade is to identify the genes and hence the proteins or regulatory factors that control the regeneration process. The more we know about dedifferentiation and regeneration, the more likely the possibility that we may be able to regenerate tissues just by triggering the right signals that would either attract stem cells or increase the plasticity of the somatic and precursor cells present near the site of injury.

43.5. GENETIC MODIFICATION OF STEM CELLS

43.5.1. GENE THERAPY TECHNOLOGY Gene therapy uses genetic engineering to alter or supplement the function of an abnormal gene by either providing a non-abnormal copy of that gene or by directly repairing such a gene. Clinical efforts to apply genetic engineering technology to the treatment of human diseases initially focused on genetic diseases and disorders in which only one gene is abnormal, such as cystic fibrosis, Gaucher's disease, severe combined immune deficiency (SCID), Fanconi's anemia, Fabry's disease, and leukocyte adherence deficiency. Now efforts are also underway that are directed toward more complex, chronic diseases, such as heart disease, blood flow deregulation, arthritis, and Alzheimer's disease. These new gene therapies consist of providing genes that add new functions or regulate the activity of other genes. In other cases, it may be desirable to program the cells to produce therapeutic agents in response to certain signals and turn on and off the therapeutic transgene when needed. These introduced genes can direct differentiation to a specific cell type or rescue a cell type that was deficient in a specific protein.

Two major technologies have been used for delivering therapeutic transgenes into human recipients. The first is to "directly" infuse the gene into a patient using genetically engineered viruses as delivery vectors. This method is fairly imprecise, as cells are randomly targeted and uptake is limited to the specific types of human cells that the viral vector can infect. The second strategy consists of removing specific cells from the patient, gene-tically modifying the cells ex vivo, and reimplanting the genetically modified cells into the original patient. This technology offers several advantages over direct gene transfer. In particular, the genetic modification of the cells takes place ex vivo and can be tightly controlled for the production of the therapeutic agent. Novel approaches for gene transduction are required and are pivotal for developing strategies that will increase the therapeutic potential of stem cells. In order to achieve clinical success, new knowledge must be gained in several domains, such as the design of new viral and nonviral vectors for introducing transgenes into cells, the ability to direct where in the genome the transgene is introduced, the ability to target the diseased tissues using the genetically modified stem cells, and the ability to control the immunological response to the implanted cells. If these goals can be attained in a relatively pure population, stem cells are a very desirable target for gene therapy. First, the correction of a genetic defect at a stem cell level will allow the expansion and transplantation of a large number of either a specific cell type or cells at a less differentiated stage (*see* above stem cell expansion and differentiation) to correct this genetic disease. Second, since stem cells are able to self-renew, introduction of genetic modifications to stem cells will ensure the longevity of the therapeutic effect.

43.5.2. GENE THERAPY AND EMBRYONIC STEM CELLS It may be possible to use human ES cells as the basis for the development of new methods for genetic engineering. Using homologous recombination, genes can be removed, exchanged, or introduced into ES cells. Adult stem cells are very often quiescent and are not very receptive to the introduction of therapeutic transgenes into their chromosomal DNA using retroviral vectors, which only infect dividing cells. New viral vectors that can infect nondividing cells, such as lentivirus or adeno-associated viruses, have been developed but are not completely satisfactory because of their low transduction efficiency (Park et al., 2000). Moreover, the manipulation of adult stem cells ex vivo can change some of their most important properties such as plasticity, self-renewal, and engraftment. Thus, since embryonic stem cells might overcome these technological problems, they may become more useful than adult stem cells when genetic engineering is required.

Human embryonic stem cells would constitute a very powerful tool for avoiding immune reaction problems. The potential immunological rejection of human ES-derived cells could be avoided by genetically engineering the ES cells to express the MHC antigens of the transplant recipient, by creating an extensive bank of human ES cells with different MHC genes, or by creating a universal ES cell that would be compatible with all patients.

However, because ES cells can give rise to teratomas, it may be preferable to use a differentiated derivative of genetically modified embryonic stem cells. Since ES cells can be expanded, large populations of these differentiated cells can be produced, saved, and given to the patient when needed. Even if they are not used directly as cell carriers for gene therapy, embryonic stem cells may still constitute a useful test system for evaluating the efficiency of new viral and nonviral vectors and also for testing the level of therapeutic agent produced after ES cell differentiation in culture.

43.5.3. GENE THERAPY ON ADULT STEM CELLS

Most of the clinical trials developed for ex vivo gene therapy to date have been carried out using hematopoietic stem cells. These cells have been used in several clinical trials during the last couple of years to treat diseases such as cancer and AIDS. Hematopoietic stem cells have been a delivery cell of choice for several reasons. They are generally easy to isolate when mobilized in circulating blood or in bone marrow in adults or in the umbilical cord blood of newborn infants. In addition, they are easy to manipulate ex vivo and can be returned to the patients by simple injection. Hematopoietic stem cells can also migrate and home to a number of strategic locations in the body such as bone marrow, liver, spleen, and lymph nodes, where delivery of therapeutic agents can be crucial.

Although hematopoietic stem cells have been the predominant cells used in ex vivo gene therapy, other types of stem cells are being studied as gene-delivery candidates. They include muscle-forming cells (myoblasts) (Barr and Leiden, 1991, Dhawan et al., 1991, Blau and Springer, 1995b, Springer et al., 1998, Mohajeri et al., 1999), bone-forming cells (osteoblasts) (Laurencin et al., 2001), and neural stem cells (Aboody et al., 2000). These cells have the advantageous property that they become an integral part of the body after engraftment. They are then able to produce the therapeutic agents not only at the precise location of their integration but also to distribute such agents indirectly through the blood and nerve systems in order to treat a range of diseases (Blau and Springer, 1995a, Ozawa et al., 2000a, 2000b). Stem-cell-based drug delivery is being investigated and the use of tissue-specific or tumor-restrictive gene promoters for the delivery of therapeutic agents to sites of tumor metastasis may induce cancer regression.

Although more clinical trials using gene therapy are planned, further research is needed to improve the control of crucial stem cell mechanisms such as the modulation of their proliferation, differentiation, and tissue targeting. In addition, genetic engineering can be used to force stem cell plasticity by genetic and epigenetic alteration. More progress needs to be made in order to control the gene-silencing mechanism and ensure that the therapeutic transgene introduced into the cells is not turned off over a period of time. Attention to dosage of the gene product delivered will be crucial and will vary with age, genetic background, and disease status. Non-invasive imaging methods will be critical in conjunction with regulatable gene therapy vectors in order to achieve adequate treatments and avoid iatrogenic disease.

43.6. CONCLUSION

Although stem cells are highly unlikely to contribute to human fantasies of immortality and eternal youth, tremendous progress has been made in the past few years in the potential use of these cells as therapeutic agents, which may lead to prolonged life with less suffering and higher quality. Certainly the strides toward the cure of various diseases have been remarkable. Stem cells may be the key to replacing cells lost in many devastating diseases such as Parkinson's disease, diabetes, chronic heart disease, muscular dystrophy, end-stage kidney disease, and liver failure, and normal cells lost due to cancer. For many of these life-shortening diseases there are no effective treatments as yet and the goal is to find a way to replace those natural processes that have been lost. Despite recent advances in tissue and organ transplantation, there is already a shortage of donor organs that makes it unlikely that the growing demand for organ replacement will be fully met through organ donor strategies alone. Neurological diseases such as spinal cord injury, multiple sclerosis, Parkinson's disease, and Alzheimer's disease are among those diseases for which the concept of replacing damaged or dysfunctional cells is a practical goal. The carefully directed differentiation of adult and embryonic stem cells into highly specialized cells for a physiological function such as pancreatic islets should allow the treatment of chronic diseases such as diabetes. Current challenges include the control of the differentiation process of stem cells such as ES cells into specialized cell populations and of their development and proliferation once they have been implanted into patients. In order to safely use stem cells or their differentiated progeny, methods of purification and methods of cell-death control will need to be developed. Another important aspect of stem-cell-based therapies will be the necessity of preventing the rejection of the donated cells by the immune system.

In summary, much basic research lies ahead before application of a stem cell therapy to patients in a rigorous therapeutic manner is realized. However, mankind will surely benefit enormously by conducting research in this important area.

REFERENCES

Aboody, K. S., Brown, A., Rainov, N. G., et al. (2000) Neural stem cells display extensive tropism for pathology in adult brain: evidence from intracranial gliomas. *Proc. Natl. Acad. Sci. USA* 97:12,846–12,851.

Alison, M. R., Poulsom, R., Jeffery, R., et al. (2000) Hepatocytes from non-hepatic adult stem cells. *Nature* 406:257.

Amit, M., Carpenter, M. K., Inokuma, M. S., et al. (2000) Clonally derived human embryonic stem cell lines maintain pluripotency and proliferative potential for prolonged periods of culture. *Dev. Biol.* 227:271–278.

Anderson, D. J., Gage, F. H., and Weissman, I. L. (2001) Can stem cells cross lineage boundaries? *Nat. Med.* 7:393–395.

Barr, E., and Leiden, J. M. (1991) Systemic delivery of recombinant proteins by genetically modified myoblasts. *Science* 254:1507–1509.

Bjornson, C. R., Rietze, R. L., Reynolds, B. A., Magli, M. C., and Vescovi, A. L. (1999) Turning brain into blood: a hematopoietic fate adopted by adult neural stem cells in vivo. *Science* 283:534–537.

Blau, H. M. (2002) A twist of fate. *Nature* 419:437.
Blau, H. M., and Baltimore, D. (1991) Differentiation requires continuous regulation. *J. Cell Biol.* 112:781–783.
Blau, H. M., Brazelton, T. R., and Weimann, J. M. (2001) The evolving concept of a stem cell: entity or function? *Cell* 105:829–841.
Blau, H. M., Chiu, C. P., and Webster, C. (1983) Cytoplasmic activation of human nuclear genes in stable heterocaryons. *Cell* 32:1171–1180.
Blau, H. M., Pavlath, G. K., Hardeman, E. C., et al. (1985) Plasticity of the differentiated state. *Science* 230:758–766.
Blau, H. M., and Springer, M. L. (1995a) Gene therapy—a novel form of drug delivery. *N. Engl. J. Med.* 333:1204–1207.
Blau, H. M., and Springer, M. L. (1995b) Muscle-mediated gene therapy. *N. Engl. J. Med.* 333:1554–1556.
Brazelton, T. R., Rossi, F. M., Keshet, G. I., and Blau, H. M. (2000) From marrow to brain: expression of neuronal phenotypes in adult mice. *Science* 290:1775–1779.
Brazelton, T. R., Nystrom, M., and Blau, H. M. (2003) Significant differences among skeletal muscles in the incorporation of bone marrow-derived cells. *Dev. Biol.*, in press.
Chambers, I., Colby, D., Robertson, M., et al. (2003) Functional expression cloning of nanog, a pluripotency sustaining factor in embryonic stem cells. *Cell* 113:643–655.
Chiu, C. P., and Blau, H. M. (1984) Reprogramming cell differentiation in the absence of DNA synthesis. *Cell* 37:879–887.
Clarke, D. L., Johansson, C. B., Wilbertz, J., et al. (2000) Generalized potential of adult neural stem cells. *Science* 288:1660–1663.
Dani, C., Chambers, I., Johnstone, S., et al. (1998) Paracrine induction of stem cell renewal by LIF-deficient cells: a new ES cell regulatory pathway. *Dev. Biol.* 203:149–162.
Dhawan, J., Pan, L. C., Pavlath, G. K., Travis, M. A., Lanctot, A. M., and Blau, H. M. (1991) Systemic delivery of human growth hormone by injection of genetically engineered myoblasts. *Science* 254:1509–1512.
Dick, J. E., Magli, M. C., Huszar, D., Phillips, R. A., and Bernstein, A. (1985) Introduction of a selectable gene into primitive stem cells capable of long-term reconstitution of the hemopoietic system of W/Wv mice. *Cell* 42:71–79.
Ferrari, G., Cusella-De Angelis, G., Coletta, M., et al. (1998) Muscle regeneration by bone marrow-derived myogenic progenitors. *Science* 279:1528–1530.
Gussoni, E., Soneoka, Y., Strickland, C. D., et al. (1999) Dystrophin expression in the mdx mouse restored by stem cell transplantation. *Nature* 401:390–394.
Itskovitz-Eldor, J., Schuldiner, M., Karsenti, D., et al. (2000) Differentiation of human embryonic stem cells into embryoid bodies compromising the three embryonic germ layers. *Mol. Med.* 6:88–95.
Jackson, K. A., Majka, S. M., Wang, H., et al. (2001) Regeneration of ischemic cardiac muscle and vascular endothelium by adult stem cells. *J. Clin. Invest.* 107:1395–1402.
Jiang, Y., Jahagirdar, B. N., Reinhardt, R. L., et al. (2002) Pluripotency of mesenchymal stem cells derived from adult marrow. *Nature* 418: 41–49.
Kaufman, D. S., Hanson, E. T., Lewis, R. L., Auerbach, R., and Thomson, J. A. (2001) Hematopoietic colony-forming cells derived from human embryonic stem cells. *Proc. Natl. Acad. Sci. USA* 98:10,716–10,721.
Kielman, M. F., Rindapaa, M., Gaspar, C., et al. (2002) Apc modulates embryonic stem-cell differentiation by controlling the dosage of beta-catenin signaling. *Nat. Genet.* 32:594–605.
Kim, J. H., Auerbach, J. M., Rodriguez-Gomez, J. A., et al. (2002) Dopamine neurons derived from embryonic stem cells function in an animal model of Parkinson's disease. *Nature* 418:50–56.
Kocher, A. A., Schuster, M. D., Szabolcs, M. J., et al. (2001) Neovascularization of ischemic myocardium by human bone-marrow-derived angioblasts prevents cardiomyocyte apoptosis, reduces remodeling and improves cardiac function. *Nat. Med.* 7:430–436.
Kondo, T., and Raff, M. (2000) Oligodendrocyte precursor cells reprogrammed to become multipotential CNS stem cells. *Science* 289: 1754–1757.
Krause, D. S., Theise, N. D., Collector, M. I., et al. (2001) Multi-organ, multi-lineage engraftment by a single bone marrow-derived stem cell. *Cell* 105:369–377.
LaBarge, M. A. and Blau, H. M. (2002) Biological progression from adult bone marrow to mononucleate muscle stem cell to multinucleate muscle fiber in response to injury. *Cell* 111:589–601.
Lagasse, E., Connors, H., Al-Dhalimy, M., et al. (2000) Purified hematopoietic stem cells can differentiate into hepatocytes in vivo. *Nat. Med.* 6:1229–1234.
Laurencin, C. T., Attawia, M. A., Lu, L. Q., et al. (2001) Poly(lactide-co-glycolide)/hydroxyapatite delivery of BMP-2-producing cells: a regional gene therapy approach to bone regeneration. *Biomaterials* 22:1271–1277.
Lovell-Badge, R. (2001) The future for stem cell research. *Nature* 414: 88–91.
McDonald, J. W., Liu, X. Z., Qu, Y., et al. (1999) Transplanted embryonic stem cells survive, differentiate and promote recovery in injured rat spinal cord. *Nat. Med.* 5:1410–1412.
Mezey, E., Chandross, K. J., Harta, G., Maki, R. A., and McKercher, S. R. (2000) Turning blood into brain: cells bearing neuronal antigens generated in vivo from bone marrow. *Science* 290:1779–1782.
Min, J. Y., Yang, Y., Converso, K. L., et al. (2002) Transplantation of embryonic stem cells improves cardiac function in postinfarcted rats. *J. Appl. Physiol.* 92:288–296.
Mitsui, K., Tokuzawa, Y., Itoh, H., et al. (2003) The homeoprotein nanog is required for maintenance of pluripotency in mouse epiblast and ES cells. *Cell* 113:631–642.
Mohajeri, M. H., Figlewicz, D. A., and Bohn, M. C. (1999) Intramuscular grafts of myoblasts genetically modified to secrete glial cell line-derived neurotrophic factor prevent motoneuron loss and disease progression in a mouse model of familial amyotrophic lateral sclerosis. *Hum. Gene Ther.* 10:1853–1866.
Nichols, J., Zevnik, B., Anastassiadis, K., et al. (1998) Formation of pluripotent stem cells in the mammalian embryo depends on the POU transcription factor Oct4. *Cell* 95:379–391.
Niwa, H., Burdon, T., Chambers, I., and Smith, A. (1998) Self-renewal of pluripotent embryonic stem cells is mediated via activation of STAT3. *Genes Dev.* 12:2048–2060.
Nolta, J. A., Dao, M. A., Wells, S., Smogorzewska, E. M., and Kohn, D. B. (1996) Transduction of pluripotent human hematopoietic stem cells demonstrated by clonal analysis after engraftment in immune-deficient mice. *Proc. Natl. Acad. Sci. USA* 93:2414–2419.
Odelberg, S. J., Kollhoff, A., and Keating, M. T. (2000) Dedifferentiation of mammalian myotubes induced by msx1. *Cell* 103:1099–1109.
Odorico, J. S., Kaufman, D. S., and Thomson, J. A. (2001) Multilineage differentiation from human embryonic stem cell lines. *Stem Cells* 19: 193–204.
Orlic, D., Kajstura, J., Chimenti, S., et al. (2001) Bone marrow cells regenerate infarcted myocardium. *Nature* 410:701–705.
Osawa, M., Hanada, K., Hamada, H., and Nakauchi, H. (1996) Long-term lymphohematopoietic reconstitution by a single CD34-low/negative hematopoietic stem cell. *Science* 273:242–245.
Ozawa, C. R., Springer, M. L., and Blau, H. M. (2000a) Ex vivo gene therapy using myoblasts and regulatable retroviral vectors. In *Gene Therapy: Therapeutic Mechanisms and Strategies*, (N. S. Templeton, and D. D. Lasic, eds.) Marcel Dekker, New York, NY, pp. 61–80.
Ozawa, C. R., Springer, M. L., and Blau, H. M. (2000b) A novel means of drug delivery: myoblast-mediated gene therapy and regulatable retroviral vectors. *Annu. Rev. Pharmacol. Toxicol.* 40:295–317.
Park, F., Ohashi, K., Chiu, W., Naldini, L., and Kay, M. A. (2000) Efficient lentiviral transduction of liver requires cell cycling in vivo. *Nat. Genet.* 24:49–52.
Partridge, T. A. (2003) Stem cell route to neuromuscular therapies. *Muscle Nerve* 27:133–141.
Pereira, R. F., Halford, K. W., O'Hara, M. D., et al. (1995) Cultured adherent cells from marrow can serve as long-lasting precursor cells for bone, cartilage, and lung in irradiated mice. *Proc. Natl. Acad. Sci. USA* 92:4857–4861.
Pereira, R. F., O'Hara, M. D., Laptev, A. V., et al. (1998) Marrow stromal cells as a source of progenitor cells for nonhematopoietic tissues in transgenic mice with a phenotype of osteogenesis imperfecta. *Proc. Natl. Acad. Sci. USA* 95:1142–1147.

Rathjen, P. D., Lake, J., Whyatt, L. M., Bettess, M. D., and Rathjen, J. (1998) Properties and uses of embryonic stem cells: prospects for application to human biology and gene therapy. *Reprod. Fertil. Dev.* 10:31–47.

Reik, W., Dean, W., and Walter, J. (2001) Epigenetic reprogramming in mammalian development. *Science* 293:1089–1093.

Reya, T., Duncan, A. W., Ailles, L., et al. (2003) A role for Wnt signalling in self-renewal of haematopoietic stem cells. *Nature* 423:409–414.

Reyes, M., Dudek, A., Jahagirdar, B., Koodie, L., Marker, P. H., and Verfaillie, C. M. (2002) Origin of endothelial progenitors in human postnatal bone marrow. *J. Clin. Invest.* 109:337–346.

Reyes, M., Lund, T., Lenvik, T., Aguiar, D., Koodie, L., and Verfaillie, C. M. (2001) Purification and ex vivo expansion of postnatal human marrow mesodermal progenitor cells. *Blood* 98:2615–2625.

Rheinwald, J. G., and Green, H. (1975) Serial cultivation of strains of human epidermal keratinocytes: the formation of keratinizing colonies from single cells. *Cell* 6:331–343.

Rideout, W. M., 3rd, Hochedlinger, K., Kyba, M., Daley, G. Q., and Jaenisch, R. (2002) Correction of a genetic defect by nuclear transplantation and combined cell and gene therapy. *Cell* 109:17–27.

Seale, P., Sabourin, L. A., Girgis-Gabardo, A., Mansouri, A., Gruss, P., and Rudnicki, M. A. (2000) Pax7 is required for the specification of myogenic satellite cells. *Cell* 102:777–786.

Shamblott, M. J., Axelman, J., Littlefield, J. W., et al. (2001) Human embryonic germ cell derivatives express a broad range of developmentally distinct markers and proliferate extensively in vitro. *Proc. Natl. Acad. Sci. USA* 98:113–118.

Shamblott, M. J., Axelman, J., Wang, S., et al. (1998) Derivation of pluripotent stem cells from cultured human primordial germ cells. *Proc. Natl. Acad. Sci. USA* 95:13,726–13,731.

Springer, M. L., Chen, A. S., Kraft, P. E., Bednarski, M., and Blau, H. M. (1998) VEGF gene delivery to muscle: Potential role for vasculogenesis in adults. *Mol. Cell* 2:549–558.

Surani, M. A. (2001) Reprogramming of genome function through epigenetic inheritance. *Nature* 414:122–128.

Szilvassy, S. J., Fraser, C. C., Eaves, C. J., Lansdorp, P. M., Eaves, A. C., and Humphries, R. K. (1989) Retrovirus-mediated gene transfer to purified hemopoietic stem cells with long-term lympho-myelopoietic repopulating ability. *Proc. Natl. Acad. Sci. USA* 86:8798–8802.

Terada, N., Hamazaki, T., Oka, M., et al. (2002) Bone marrow cells adopt the phenotype of other cells by spontaneous cell fusion. *Nature* 416: 542–545.

Thomson, J. A., Itskovitz-Eldor, J., Shapiro, S. S., et al. (1998) Embryonic stem cell lines derived from human blastocysts. *Science* 282: 1145–1147.

Thomson, J. A., Kalishman, J., Golos, T. G., et al. (1995) Isolation of a primate embryonic stem cell line. *Proc. Natl. Acad. Sci. USA* 92: 7844–7848.

Toma, J. G., Akhavan, M., Fernandes, K. J., et al. (2001) Isolation of multipotent adult stem cells from the dermis of mammalian skin. *Nat. Cell Biol.* 3:778–784.

van Praag, H., Schinder, A. F., Christie, B. R., Toni, N., Palmer, T. D., and Gage, F. H. (2002) Functional neurogenesis in the adult hippocampus. *Nature* 415:1030–1034.

Verfaillie, C. M., Pera, M. F., and Lansdorp, P. M. (2002) Stem cells: hype and reality. *Hematology (Am. Soc. Hematol. Educ. Program)*:369–391.

Waldmann, H. (2001) Therapeutic approaches for transplantation. *Curr. Opin. Immunol.* 13:606–610.

Weimann, J. M., Charlton, C. A., Brazelton, T. R., Hackman, R. C., and Blau, H. M. (2003) Contribution of transplanted bone marrow cells to Purkinje neurons in human adult brains. *Proc. Natl. Acad. Sci. USA* 100:2088–2093.

Willert, K., Brown, J. D., Danenberg, E., et al. (2003) Wnt proteins are lipid-modified and can act as stem cell growth factors. *Nature* 423: 448—452.

Wilson, S. I., Rydstrom, A., Trimborn, T., et al. (2001) The status of Wnt signalling regulates neural and epidermal fates in the chick embryo. *Nature* 411:325–330.

Ying, Q. L., Nichols, J., Evans, E. P., and Smith, A. G. (2002) Changing potency by spontaneous fusion. *Nature* 416:545–548.

Zuk, P. A., Zhu, M., Mizuno, H., et al. (2001) Multilineage cells from human adipose tissue: implications for cell-based therapies. *Tissue Eng.* 7:211–228.

Index

D

E

F

G

H

W

Y